THE PICTURE OF THE TAOIST GENII PRINTED ON THE COVER of this book is part of a painted temple scroll, recent but traditional, given to Mr Brian Harland in Szechuan province (1946). Concerning these four divinities, of respectable rank in the Taoist bureaucracy, the following particulars have been handed down. The title of the first of the four signifies 'Heavenly Prince', that of the other three 'Mysterious Commander'.

At the top, on the left, is Liu *Thien Chün*, Comptroller-General of Crops and Weather. Before his deification (so it was said) he was a rain-making magician and weather forecaster named Liu Chün, born in the Chin dynasty about +340. Among his attributes may be seen the sun and moon, and a measuring-rod or carpenter's square. The two great luminaries imply the making of the calendar, so important for a primarily agricultural society, the efforts, ever renewed, to reconcile celestial periodicities. The carpenter's square is no ordinary tool, but the gnomon for measuring the lengths of the sun's solstitial shadows. The Comptroller-General also carries a bell because in ancient and medieval times there was thought to be a close connection between calendrical calculations and the arithmetical acoustics of bells and pitch-pipes.

At the top, on the right, is Wên *Yuan Shuai*, Intendant of the Spiritual Officials of the Sacred Mountain, Thai Shan. He was taken to be an incarnation of one of the Hour-Presidents (*Chia Shen*), i.e. tutelary deities of the twelve cyclical characters (see p. 262). During his earthly pilgrimage his name was Huan Tzu-Yü and he was a scholar and astronomer in the Later Han (b. +142). He is seen holding an armillary ring.

Below, on the left, is Kou *Yuan Shuai*, Assistant Secretary of State in the Ministry of Thunder. He is therefore a late emanation of a very ancient god, Lei Kung. Before he became deified he was Hsin Hsing, a poor woodcutter, but no doubt an incarnation of the spirit of the constellation Kou-Chhen (the Angular Arranger), part of the group of stars which we know as Ursa Minor. He is equipped with hammer and chisel.

Below, on the right, is Pi *Yuan Shuai*, Commander of the Lightning, with his flashing sword, a deity with distinct alchemical and cosmological interests. According to tradition, in his earthly life he was a countryman whose name was Thien Hua. Together with the colleague on his right, he controlled the Spirits of the Five Directions.

Such is the legendary folklore of common men canonised by popular acclamation. An interesting scroll, of no great artistic merit, destined to decorate a temple wall, to be looked upon by humble people, it symbolises something which this book has to say. Chinese art and literature have been so profuse, Chinese mythological imagery so fertile, that the West has often missed other aspects, perhaps more important, of Chinese civilisation. Here the graduated scale of Liu Chün, at first sight unexpected in this setting, reminds us of the ever-present theme of quantitative measurement in Chinese culture; there were rain-gauges already in the Sung (+12th century) and sliding calipers in the Han (+1st). The armillary ring of Huan Tzu-Yü bears witness that Naburiannu and Hipparchus, al-Naqqās and Tycho, had worthy counterparts in China. The tools of Hsin Hsing symbolise that great empirical tradition which informed the work of Chinese artisans and technicians all through the ages.

SCIENCE AND CIVILISATION
IN CHINA

Every blade of grass and every tree possesses its own pattern-principle, and should be examined.   Chu Hsi & Lü Tsu-Chhien, +1175
(*Chin Ssu Lu*, III, 12, tr. Chhen Jung-Chieh (11), p. 93)

I am in a position to assure you [said Plutarch] that Hegesander of Delphi nowhere mentions the citron, for I read through the whole of his 'Memorials' with the express purpose of finding out.   Athenaeus of Naucratis, +228,
Deipnosophistae, III, 25–9

In a treatise on Chinese music, by the late Mr Tradescant Lay, that gentleman has remarked: 'It has been asserted that the Chinese have no science; but of a surety, if we advance in the free and scholar-like spirit of antiquarian research, we shall be obliged to set our feet upon the head of this assertion at every step of our progress.'   W. H. Medhurst, in the preface to his
translation of the *Shu Ching*, 1846, p. vi.

I know that the stationary state which it is claimed that Chinese science and civilisation show, has often been attributed to their script; but this opinion, which is getting less and less convincing every day, derives from a time when people judged the Chinese as well as their sciences and their writing just by hearsay. The ideographic and pictographic script seems on the contrary to have been marvellously appropriate for the study of natural history ...
J. P. Abel Rémusat, in his 'Discours sur
l'Etat des Sciences Naturelles chez
les Peuples de l'Asie Orientale'
in *Mélanges Posthumes* (13), p. 211.

Just as the Western countries have certainly never heard of the teaching of the Sages of the Central Country, so we too have never heard tell of the books of their ancient Saints and Scholars which they are circulating in China. But now both enlighten and benefit each other. In this way the whole world is becoming a single family, and minds are mutually comprehending each other, so that there will no longer be misunderstandings between the people of the far-flung East and West.
From the preface of Fêng Ying-Ching
to Matteo Ricci's fourth world map,
+1603, tr. d'Elia (15), p. 134.

Thus more and more the science and learning of the Sages of East and West will fuse into one.
From Giulio Aleni's *Hsi Hsüeh Fan*,
tr. d'Elia (9).

中國科學技術史

李約瑟 著

冀朝鼎

SCIENCE AND CIVILISATION IN CHINA

BY

JOSEPH NEEDHAM, F.R.S., F.B.A.

SOMETIME MASTER OF GONVILLE AND CAIUS COLLEGE, CAMBRIDGE
FOREIGN MEMBER OF ACADEMIA SINICA

With the collaboration of

LU GWEI-DJEN, PH.D.

FELLOW EMERITUS OF ROBINSON COLLEGE, CAMBRIDGE

and a special contribution by

HUANG HSING-TSUNG, D. PHIL.

PROGRAM DIRECTOR
NATIONAL SCIENCE FOUNDATION
WASHINGTON, D.C.

VOLUME 6

BIOLOGY AND BIOLOGICAL TECHNOLOGY

PART I: BOTANY

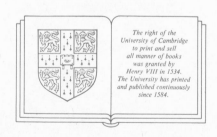

The right of the University of Cambridge to print and sell all manner of books was granted by Henry VIII in 1534. The University has printed and published continuously since 1584.

CAMBRIDGE UNIVERSITY PRESS

CAMBRIDGE

LONDON   NEW YORK   NEW ROCHELLE

MELBOURNE   SYDNEY

Published by the Press Syndicate of the University of Cambridge
The Pitt Building, Trumpington Street, Cambridge, CB2 1RP
32 East 57th Street, New York, NY 10022, USA
10 Stamford Road, Oakleigh, Melbourne, 3166, Australia

First published 1986

Printed in Great Britain at the
University Press, Cambridge

Library of Congress catalogue card number: 54–4723

British Library Cataloguing in Publication Data
Needham, Joseph
Science and Civilisation in China.
Vol 6: Biology and biological technology
Pt. 1: Botany
1. Science—China—History   2. Technology
—China—History
I. Title   II. Lu Gwei-Djen, Huang Hsing-Tsung
509 .51   Q127,C5

ISBN 0 521 08731 7

A.O.

To the memory of

SHIH SHÊNG-HAN
Professor of Botany and Mycology
National North-West Agricultural College
Wukung, Shensi

in gratitude for much inspiration and blithe discourse,
and recalling *Homomycetes* that rainy day at Chiating,

and of

WU SU-HSÜAN
Director of the Department of Cytology
Institute of Botany, Academia Sinica, Peking

in gratitude for so sympathetic a welcome, at the United
Universities (Lien Ho Ta Hsüeh) in war-time Kunming,
and recalling the *yang mei* at the Anning Hot Springs,

this volume is dedicated.

CONTENTS

The following subsections, by Georges Métailié, are not yet ready for publication

---

(f) Treatises on traditional botany, and the development of classification

(g) The development of plant description and illustration

(h) Chinese knowledge of the life of plants

    (1) Plant physiology

    (2) Sexuality in plants

    (3) Parasitism and the epiphytic condition

    (4) Phycology (the algae)

    (5) Plant pathology

    (6) Mycology (the fungi)

(i) Horticulture and its techniques

    (1) Treatises on horticulture

    (2) Gardens and botanic gardens

    (3) Fruit production and preservation

    (4) Grafting and vegetative propagation

(j) The influence of Chinese flora and botany on modern plant science

    (1) Ancient and medieval plant transmissions

    (2) The influence of Chinese gardens on Europe

(k) Conclusions

LIST OF ILLUSTRATIONS

LIST OF TABLES

LIST OF ABBREVIATIONS

The following abbreviations are used in the text. For abbreviations used for journals and similar publications in the bibliographies, see pp. 555 ff.

B	Bretschneider, E. (*1*), *Botanicon Sinicum* (successive volumes indicated as B I, B II, B III).
CC	Chia Tsu-Chang & Chia Tsu-Shan (*1*), *Chung-Kuo Chia Wu Thu Chien* (Illustrated Dictionary of Chinese Flora), 1958.
CFP	(Hui phu, Hua phu, Kuo phu, etc.) Wang Hsiang-Chin, *Chhun Fang Phu* (The Assembly of Perfumes), a botanical thesaurus, +1630.
CHS	Pan Ku (and Pan Chao), *Chhien Han Shu* (History of the Former Han Dynasty), c. +100.
CKI	Hsieh Li-Hêng (*2*) (ed.), *Chung-Kuo I-Hsüeh Ta Tzhu Tien* (Great Chinese Medical Encyclopaedia). 1st ed. (2 vols.), Shanghai, 1921; new ed. (4 vols.), 1954.
CMYS	Chia Ssu-Hsieh, *Chhi Min Yao Shu* (Important Arts for the People's Welfare), between +533 and +544.
CSHK	Yen Kho-Chün (ed.), *Chhüan Shang-ku San-Tai Chhin Han San-Kuo Liu Chhao Wên* (complete collection of prose literature [including fragments] from remote antiquity through the Chhin and Han Dynasties, the Three Kingdoms, and the Six Dynasties), 1836.
CT & W	Clapham, A. R., Tutin, T. G. & Warburg, E. F. (*1*), *Flora of the British Isles*, Cambridge, 1962.
HTNC/SW	*Huang Ti Nei Ching, Su Wên* (The Yellow Emperor's Manual of Corporeal (Medicine): Questions (and Answers) about Living Matter).
HTS	Ouyang Hsiu & Sung Chhi, *Hsin Thang Shu* (New History of the Thang Dynasty), +1061.
ICK	Taki Mototane: *I Chi Khao* (Iseki-Kō) (Comprehensive Annotated Bibliography of Chinese Medical Literature [Lost or Still Existing]), finished c. 1825, pr. 1831; repr. Tokyo 1933, Shanghai 1936.
IWLC	Ouyang Hsün, *I Wên Lei Chü* (Art and Literature Collected and Classified), an encyclopaedia, c. +640.
K	Karlgren, B. (*1*), *Grammata Serica* (dictionary giving the ancient forms and phonetic values of Chinese characters).
KHTT	Chang Yü-Shu (ed.), *Khang-Hsi Tzu Tien* (Imperial Dictionary of the Khang-Hsi reign-period), +1716.

KSP	Ku Chieh-Kang & Lo Ken-Tsê (eds.), *Ku Shih Pien* (Discussions on Ancient History and Philosophy); a collective work.
KYCC	Chhüthan Hsi-Ta, *Khai-Yuan Chan Ching* (The Khai-Yuan reign-period Treatise on Astrology and Astronomy), +729.
MCPT	Shen Kua, *Mêng Chhi Pi Than* (Dream Pool Essays), +1086.
NCCS	Hsü Kuang-Chhi, *Nung Chêng Chhüan Shu* (Complete Treatise on Agriculture), +1639.
PTCCC	Thao Hung-Ching, *Pên Tshao Ching Chi Chu* (Collected Commentaries on the Classical Pharmacopoeia [of the Heavenly Husbandman]), +492.
PTKM	Li Shih-Chen, *Pên Tshao Kang Mu* (The Great Pharmacopoeia), +1596.
R	Read, Bernard E., *et al.*, Indexes, Translation and Précis of Certain Chapters of the *Pên Tshao Kang Mu* of Li Shih-Chen. If the reference is to a plant, see Read (1); if to a mammal, see Read (2); if to a bird, see Read (3); if to a reptile, see Read (4 or 5); if to a mollusc, see Read (5); if to a fish, see Read (6); if to an insect, see Read (7).
RP	Read & Pak (1), Index, translation and précis of the mineralogical chapters in the *Pên Tshao Kang Mu*.
SKCS/CMML	Chi Yün (ed.) *Ssu Khu Chhüan Shu Chien Ming Mu Lu*, (Brief Explanatory Catalogue of the *Complete Library of the Four Categories*), +1782; the shorter bibliography of the imperial MS collection ordered by the Chhien-Lung emperor in +1772, giving particulars only of those books actually copied into the corpus.
SKCS/TMTY	Chi Yün (ed.), *Ssu Khu Chhüan Shu Tsung Mu Thi Yao* (Analytical Catalogue of the *Complete Library of the Four Categories*), +1782; the great bibliographical catalogue of the imperial MS collection ordered by the Chhien-Lung emperor in +1772.
SSIW	Toktaga (Tho-Tho) *et al.*; Huang Yü-Chi *et al.* & Hsü Sung *et al.*, *Sung Shih I Wên Chih, Pu, Fu Phien* (A Conflation of the Bibliography and Appended Supplementary Bibliographies of the History of the Sung Dynasty). Com. Press, Shanghai, 1957.
SWCY	Kao Chhêng, *Shih Wu Chi Yuan* (Records of the Origins of Affairs and Things), *c.* +1085.
TCTC	Ssuma Kuang, *Tzu Chih Thung Chien* (Comprehensive Mirror [of History] for Aid in Government), +1084.
TH	Wieger, L. (1), *Textes Historiques*.
TPYL	Li Fang (ed.), *Thai-Phing Yü Lan* (the Thai-Phing reign-period [Sung] Imperial Encyclopaedia), +983.
TSCC	Chhen Mêng-Lei *et al.* (ed.), *Thu Shu Chi Chhêng*, (the Imperial Encyclopaedia of +1726). Index by Giles, L. (2). References to

ACKNOWLEDGEMENTS

LIST OF THOSE WHO HAVE KINDLY READ THROUGH SECTIONS IN DRAFT

The following list, which applies only to Vol. 6, pt 1, brings up to date those printed in Vol. 1, pp. 15 ff., Vol. 2, p. xxiii, Vol. 3, pp. xxxix ff., Vol. 4, pt 1, p. xxi, Vol. 4, pt 2, p. xli, Vol. 4, pt 3, pp. xliii ff., Vol. 5, pt 2, p. xvi, Vol. 5, pt 3, p. xviii, and Vol. 5, pt 4, p. xxix.

Mr E. F. Allen	Ipswich
Prof. Derk Bodde	Philadelphia
Dr Chiang Jun-Hsiang (Y. C. Kong)	Hongkong (CUHK)
Prof. John Corner	Cambridge
Dr David Coombe	Cambridge
Prof. Frank N. Egerton	Pittsburgh, now Parkside, Wisconsin
Dr D. M. Henderson, FRSE	Edinburgh
Dr L. Andrew Lauener	Edinburgh
Prof. Li Hui-Lin	Philadelphia
Prof. A. G. Morton	Edinburgh
Dr R. M. S. Perrin	Cambridge
Dr Max Walters	Cambridge

AUTHOR'S NOTE

Thirty years have now elapsed since the first volume of *Science and Civilisation in China* was passing through the press; and so at last we can turn from all the inorganic sciences and technologies to the realms of the biological. There were, it is true, some adumbrations of the level of living things when we were talking about Chinese organic philosophy in Vol. 2, but now, with Vol. 6, we shall plunge into the study of what the Chinese of ancient and medieval times thought, did, and knew, about the phenomena of life.

The present volume contains most of Section 28, on the plant sciences. We cannot say all, because there will still be more to come in a following volume, the work of our collaborator Dr Georges Metailié of the Centre Nationale de la Recherche Scientifique and the Musée d'Histoire Naturelle at Paris.[a] It would no doubt have been preferable to bring it all out together in one volume, but the necessities of collaboration and the interlocking of commitments have made it impossible. This volume must also contain another first—we have had to give up the numbering of the figures in one continuous order. The highest number we attained was Fig. 1632 at the end of Vol. 5, pt. 5. But now that volumes are being produced by various collaborators as well as ourselves, one cannot foresee how many numbers should be allowed for particular Sections, and it is therefore simpler to start each Volume with Fig. 1

It has been a particular pleasure, however, to be able to include in the present volume a contribution on naturally-occurring plant insecticides, and on the ever-memorable Chinese invention of biological pest control, by my old friend, Dr Huang Hsing-Tsung[1]. He and I were immediate colleagues in the Sino-British Science Cooperation Office (Chung-Ying Kho-Hsüeh Ho-Tso Kuan[2]) during the greater part of the Second World War,[a] and afterwards he spent a lifetime in the field of biological and chemical pest control.

After the botany has been all completed, we plan to proceed on our pertinacious way with zoology and with biochemical technology (including nutritional science and the fermentation industries). Next comes agriculture, a subject of incalculable importance for the Chinese past—or rather I should say it has already come, since Section 41 (forming Vol. 6, pt. 2) prepared by another collaborator, Miss Francesca Bray, has already achieved publication. Associated with it will be some material in due course on animal husbandry and pisciculture, as well as on the agricultural arts and industries (Section 42); a subject for which Mr Christian Daniels has now joined our group.

---

[a] This can be seen from the division of the Table of Contents.
[a] His account of those years has recently been published; Huang Hsing-Tsung (1).

[1] 黃興宗　　[2] 中英科學合作舘

We then come to the whole vast subject of medicine and pharmacy. We expect to deal first with the 'institutes of medicine' (as they used to be called), anatomy, physiology and embryology, in Sect. 43. We shall go on in the following Section to present the development of the classical theories of Chinese clinical medicine, including diagnosis and prognosis, pathology and epidemiology. We must then treat of the many specialties of medicine, such as paediatrics, obstetrics, gynae-cology, physiotherapy and medical gymnastics.[a] The role of China in the birth of all immunology will be carefully reviewed,[b] and the characteristic techniques and theories of acupuncture and moxibustion historically portrayed.[c] Then will follow surgery (or external medicine, as the Chinese thought of it), dermatology, ophthal-mology, and psychotherapy.[d] Quite a lot will have to be said about hygiene and preventive medicine in ancient and medieval China,[e] as also on medical education and administration, including the government dispensaries and the Medical Colleges, much older than any in the West, as well as the Imperial Medical Service.[f]

With Section 45 on pharmaceutics this part of our grand design will attain its completion. I have often said in the past that preparing the material for a new chapter is like watching the picture come up on a large negative in the developing bath of a dark-room. Once the innumerable confusions, misunderstandings, mis-interpretations and erroneous ideas have been cleared away, the pattern gradually appears, reaching in the end as much clarity as we are capable of giving it at this particular point of time.

But there is also another analogy which may be used. In order to arrange the material in the clearest way, it is desirable to have some sort of trellis-work on which to hang all the information; and this often means that one has to know how to ask the essential questions. For nearly twenty years now I have been privileged to lecture annually to the Cambridge Pharmacology Part II Tripos class on the history of Chinese pharmaceutics. This opportunity has shown that the right question to ask is this; did the Chinese have and use the most important active principles known classically in the Western world since the time of Dioscorides, or not? As it turns out, the answer is almost never no; but there are several modified forms of the affirmative answer. It can be just yes, but in some cases one has to say 'yes, and much earlier', in other cases 'yes, but from a different plant source', or in others yet again 'yes, but a different active principle'. Only occasionally does one have to say 'yes, but later', and of course there remain many interesting cases of pharmacological active principles never known to the traditional West at all. Finally, there was never anything in China like the Galenic insistence on the use of

---

[a] Something on this last will be found in Vol. 5, pt 5 of the series, on physiological alchemy.

[b] It will be based on our careful study of smallpox inoculation and its Chinese origins in Needham (85).

[c] Our already published monograph, *Celestial Lancets*, by Lu Gwei-Djen & Needham (5), cf. (6), will be the basis for this.

[d] For this last our collaborator is Dr Hans Ågren of Uppsala.

[e] It will be an enlargement of Needham & Lu Gwei-Djen (1).

[f] This again will be a revised and enlarged form of Lu Gwei-Djen & Needham (2).

plant drugs alone; there the pharmacopoeias and the pharmaceutical natural histories always from the beginning allowed room for other drugs, both mineral and animal in origin.

Mention of the pharmaceutical natural histories reminds us that there is a great deal about them in the present volume, as was inevitable on account of the mass of botanical knowledge they contained. Later on herein we shall give good reasons for refusing to call them 'herbals'; and incidentally we shall show that the oldest official pharmacopoeia was the *Hsin Hsiu Pên Tshao* of +659, and not *Pharmacopoeia Londiniensis* of nearly a thousand years later.

But before embarking further on a brief description of what the present volume contains, it may be desirable to mention a few of the more recent developments in the *Science and Civilisation in China* enterprise. Francesca Bray's remarkable study of the history of agriculture in China (Vol. 6, pt. 2) has already been mentioned, but at the present time there is another volume passing through the press, namely Vol. 5, pt. 1, which contains the epic history of paper and printing in Chinese civilisation, the work of Professor T. H. Tsien (Chhien Tshun-Hsün[1]) of Chicago. This will fill the gap in Vol. 5 which has existed for some time past. We are happy that so distinguished a scholar was willing to join our group, and to fit in with our design. While speaking of this volume it may be well to say something of the remaining parts of Vol. 5. The account of military technology for Vol. 5, pt. 6 only needs pulling together, for most of the sub-sections of Section 30 have now been completed, with the collaboration of the late Professor Lo Jung-Pang[2], Professor Ho Ping-Yü[3], Dr Robin Yates and Dr Krzszytof Gawlikowski. This includes archery and the cross-bow, in which we have had the help of Mr Edward McEwen; and the fabulous story of gunpowder, where we have had much assistance from Mr Howard Blackmore, until his recent retirement Deputy Master of the Armouries at the Tower of London. Then in Toronto Professor Ursula Franklin and Dr John Berthrong are making good progress with the non-ferrous metallurgy in Section 36, while the same may be said of Dr Donald Wagner of Copenhagen, whose contribution on ferrous metallurgy, the important history of iron and steel, will embody a revision and updating of the monograph which Dr Wang Ling[4] and I did together nearly thirty years ago.[a] I mention here only a few of our far-flung company of collaborators, because their work is nearing completion, but this abates nothing of our gratitude towards all those who are labouring away in the earlier stages of their adopted Sections. And some there are, I am sorry to say, who have been looking for an outlet which we have not yet been able to give them—I am thinking particularly of the late Professor Lo Jung-Pang, who collaborated directly with me many years ago on the history of the salt industry in China and the epic of deep borehole drilling; this Section 37, with all its rich illustration, has been waiting for

---

[a] Needham (32).

[1] 錢存訓　　[2] 羅榮邦　　[3] 何丙郁　　[4] 王鈴

space and is still waiting. But without doubt it will find its place in one of the parts of Vol. 5 eventually.

In previous volumes we have often given a few words of help to the prospective reader, intended as a sort of waywiser to guide him through those pages not always possible to lighten by some memorable illustration. This was not meant to be a substitute for the contents table, but rather as some useful tips of 'inside information' to indicate where the really important paragraphs could be found, and to distinguish them from the mass of supporting detail secondary in significance, though often fascinating in itself. I confess that in this Volume on the botanical sciences I feel incapable of doing this, for I believe that every lover of plants and their ways, opening at random any page of the various sub-sections, will be at once enthralled, and unable to put the book down until he or she has finished the topic in question—provided always that there is a willingness to bear with strange names and strange book-titles, landmarks of a civilisation quite other than that of Europe. Instead of a waywiser, then, I propose to go very rapidly through the contents of the volume, explaining briefly what each part of it covers.

First, in the Introduction, we take a lively canter through the history of botany as it has been understood in the Western world, aiming thus to put the history of botany in China into the right perspective. We reach the same conclusions as so often before, namely that in China there was a slow and rather steady growth in knowledge about plants, with no Dark Ages at all. Sometimes one can even represent this in graphical form. Indeed the dead period in Europe, between the $+4$th and the $+14$ centuries, when the number of plants that could be described sank to an absolute minimum, was filled in China with botanical monographs, devoted to particular families, genera and species, the like of which remained totally unknown in Europe until post-Linnaean times. On the way we emphasise that from the beginning of Chinese woodcut illustrations of plant forms, this art preceded that of the German fathers of botany in the $+16$th century by more than five hundred years. Indeed, there may even be a direct connection, because we know that the imperial prince Chu Hsiao[1], whose *Chiu Huang Pên Tshao*[2] was first published in $+1406$, as the result of work in his botanical garden and nutritional laboratory, was well acquainted with the Jewish community settled in the city where he dwelt and had his estates, Khaifêng.[3] Like all such communities, it had its physicians and its travelling merchants, who may well have been in touch with their confrères in the Far West; so when Conrad of Megenberg in $+1475$ made his first illustrations of plants, the idea might have been in fact transmitted. The point was identifiability; how to make a woodcut illustrate a plant so well that other gatherers or herbalists, or men we might be willing to call botanists, could find it reliably in the wild, and be sure of it for their own (probably pharmaceutical, or at least nutritional) purposes.

---

[1] 朱橚   [2] 救荒本草   [3] 開封

Next we try to make clear what this volume is *not* trying to do. It is not an exhaustive and systematic account of the plants of China; it is not in any sense an East Asian flora. Nor is it a survey of the origins and transmissions of cultivated plants. It does not attempt to replace the phyto-geographies and oecologies which already exist. What it is designed to do is to follow the development of botany and the plant sciences in China, from the earliest proto-scientific stages onwards, until they fuse with the oecumenical botany of modern science. How far did they really get before this absorption took place? As will be seen, we find that indigenous Chinese botany reached a Magnolian or Tournefortian level, rather than a Linnaean one; just as we found in earlier volumes that the physical sciences there reached a Vincean rather than a Galilaean stage. Even so, Linnaeus was not the Galileo of botany, still less its Newton, and maybe we are still looking for this biological messiah, for even Darwin was not he.

Geo-botany has two divisions, plant geography on the one hand and oecology on the other, the former dealing with floristic quality, the latter with the types of vegetation found in the various habitats distinguished by soil, moisture and climate. The Chinese flora is much richer than that of Europe, as also is that of North America, doubtless because the ice ages of the Pleistocene bore down more heavily on the western part of the Eurasian land-mass. The total number of plant species known in the world is some 225,000, with an average of eighteen species to the genus; and of these China has some 30,000, so that the Sino-Japanese area is by far the richest of the whole north temperate zone. We argue that geo-botany is to be found in China *in statu nascendi*, for there is a great deal of oecological observation in the Warring States philosophers, and especially the *Kuan Tzu*[1] book. It was Chinese farmers and economists too who laid the foundations of pedology or the science of soils, for many different kinds of soils are described in the Yü Kung[2] chapter of the *Shu Ching*[3], which can hardly be later than the early −5th century, as also in the *Kuan Tzu*, which may be dated in the −4th. By comparison we show how the Roman agriculturalists practically gave up the attempt to classify soil-types. This is why we make so bold as to say that along with oecology and plant geography, pedology too was born in China. As a concluding part of this sub-section we consider the case of the *chü*[4] (*Citrus reticulata*) and the *chih*[5] or thorny limebush (*Poncirus trifoliata*). The universal ancient saying was that north of the River, the *chü* turns into the *chih*; one of those *loci communes* which haunted us for many years, and which we have been very glad to clear up. In early centuries it was thought to be a true metamorphosis (like the darnel story), but not much after +1150, when people realised that it was just a matter of species distribution. The use of the limebush as a root-stock for the grafting of the orange may well have contributed to the legend, however.

After this we take up the whole range of botanical linguistics—first plant terminology, the phytographic language, and then plant nomenclature, the taxo-

¹ 管子   ² 禹貢   ³ 書經   ⁴ 橘   ⁵ 枳

nomic language, distinguishing between the common and the learned names. Botanists have to have highly technical ways of talking about plants and their parts. One can see the beginnings of this in the text of Theophrastus, and we find that his contemporaries, in the China of the −3rd century, had embarked on a very similar process of coining the terms of the art; this clearly appears from dictionaries such as the *Erh Ya*[1]. Among the points from where we start is the study of the 'botanical radicals', for it must always be remembered that Chinese was an ideographic language, and therefore contains within itself many of the most ancient plant drawings, some of which fell by the wayside while others took their place in the developed language and script, even though progressive stylisation, simplification and systematisation might somewhat disguise them.

One of the greatest differences between China and Europe was that the latter had a dead language in its past history from which scientifically defined names could be formed, set off permanently, as it were, from the common names of countryfolk and farmers. 'The idea often prevalent (we remark in one place) that traditional Chinese botanical nomenclature was in some sense "unscientific" is closely connected with the prejudice in the European, and now in the modern, mind, that nothing can be scientifically identifiable unless it bears a Latin name.' But this fission of the learned from the popular plant appellations took place remarkably late in European history, not before +1500. In China, where there was no background of another tongue which could be drawn upon, the distinction had been present from a very early time, and was never lost. Again, it proved interesting to tabulate all the various characteristics and properties of plants which Westerners used in establishing the taxonomic vocabulary, and it was easily possible to show that just the same range of special features was utilised in Chinese. I always remember receiving in my rooms at Caius College in 1967 some of the members of the International Commission of Botanical Nomenclature, who had been meeting in Cambridge, and appreciating the shock which they experienced upon realising that if the rules of botanical nomenclature were rigorously applied, there would be a vast influx of Chinese names, generic, specific and personal, into the world literature of botany.

Eventually we go so far as to suggest that a suitably improved ideographic name for a plant would be much better than the quasi-numerical ones which have been proposed for computerisation. I always remember how puzzled I was when as a biochemical embryologist I was informed that such and such a number of the eggs of *Pila globosa* had been taken for an experiment, and it took me quite some time to ascertain that this was neither a moth nor a mammoth but in fact a terrestrial mollusc. The specific names beloved of traditional botanists such as *rehderi* or *japonica* would have been no more informative, but if the ideographic principle were employed, it would be possible to see at a glance that the thing (*Tilia leptocarya*) was a tree, in fact one of the lindens, and its specific name might well be

---

[1] 爾雅

derived from the shape of its nuts—all this in two or three straightforward characters. And once computers get used to scanning patterns, as some of them are already doing, then there would be no particular advantage in strings of numbers and letters.

Next comes the spacious story of the Chinese botanical literature, until now almost unknown to Westerners and others who lack acquaintance with the script in which it was written. First we talk about the lexicographic and encyclopaedic texts, because there is in them a vast wealth of botanical information, hitherto very little drawn upon by historians of the plant sciences. Then follow the imperial florilegia (unique to the Chinese tradition), the classified compendia, and the dictionaries of origins, together with those based on script, sound or phrase. If these could all be made to yield up the knowledge about plants which they contain, then in spite of a certain amount of mutual copying, and allowing for a modicum of legendary lore, great benefit would accrue to the oecumenical history of mankind's understanding of the plant world.

After this we try to deal as thoroughly as possible with the tradition of the Pên Tshao[1], i.e. the pharmaceutical natural histories, many dozens of which were produced between the Warring States period and the nineteenth century, by which time modern science was well under way. We call them the Pandects of Natural History, because they always aimed at completeness in covering the kingdoms of minerals, plants and animals; and their essentially biological nature, often free from too much application to human use, is clearly betrayed by the characteristic phrase *yu ming wu yung*[2], 'this thing has a name but no utility'. We deal with them by order of date and period, from the *Shen Nung Pên Tshao Ching*[3] of the −2nd or −1st century, not the oldest text known but the oldest surviving one, down to the *Pên Tshao Kang Mu Shih I*[4], which was not finished till after the nineteenth century had started. This was explicitly an addendum to the great work of Li Shih-Chen[5], published in +1596, which is rightly considered the apogee of the whole tradition. In every case we study the botanical content of the Pên Tshao books, giving examples of the illustrations which they contain. For the most part we avoid the term 'pharmacopoeias', reserving it fairly strictly for those works only which were commissioned by imperial decree, but we make exceptions for Shen Nung and Li Shih-Chen, exceptions which have been hallowed by so much usage.

As time went on, the Pên Tshao books began to diverge into several specialities, among which only two or three may be mentioned. Some specialised in nutritional science; others concentrated on those plants only which it was safe for the country people to eat in time of famine. There thus developed what we call the esculentist movement, from about +1400 onwards. Writers gradually dropped the Pên Tshao title, as they diverged more and more from plants which had pharmacological properties, and it is interesting that this kind of work necessarily involved labo-

---

[1] 本草　　　[2] 有名無用　　　[3] 神農本草經　　[4] 本草綱目拾遺
[5] 李時珍

ratory experiments, for in some cases poisonous substances had to be got rid of by previous extraction, or measures had to be taken to get rid of dangerous crystals and raphides in the plant tissue.

In due course we go on to the study of that fascinating literature, the botanical monographs. It was perhaps the bamboos that first attracted attention, and in +460 Tai Khai-Chih[1] in his *Chu Phu*[2] gave a poetical description of a large number of genera and species. But the type-specimen of them, we have always felt, was the book on the citrous fruits and trees that Han Yen-Chih[3] wrote in +1178, the *Chü Lu*[4]. Many other groups were monographed equally carefully, for example, the tree-peonies by Ouyang Hsiu[5], in his *Lo-Yang Mu-Tan Chi*[6] of +1034, and the chrysanthemums by Chou Shih-Hou[7] in his *Lo-Yang Hua Mu Chi*[8] of +1082. The orchids, the Rosaceae, and many others, followed or preceded these, and all this was during the millennium before Linnaeus came upon the scene.

Eventually we take up two genres of botanical literature for which, perhaps, there were no exact counterparts in the Western world. First there was the exploration of the border-lands, the charting of the unfamiliar plants, fruits, herbs, shrubs and trees which the Chinese came to know as they spread outwards to occupy the Chinese came to know as they spread outwards to occupy the whole of the East Asian continental *oikoumene*. Here the type-specimen (out of many examples) was surely the *Nan Fang Tshao Mu Chuang*[9] which Hsi Han[10] wrote probably in +304. Secondly there was the elucidation of the ancient, i.e. the investigation of the different kinds of plants which had been referred to in the classics, and the real nature of which had been obscured by changes in linguistic usage. Here the best example might be the study of the plants and trees, the birds and animals, and the insects and fishes referred to in Mao Hêng's[11] edition of the Book of Odes (the *Shih Ching*[12]), the *Mao Shih Tshao Mu Niao Shou Chhung Yü Su*[13] indited by Lu Chi[14] in +245, some four hundred years later. The early date of both these types of writing will be noted, but both interests continued right down to the end of the Chhing dynasty, indeed into quite modern times.

Finally we discuss two particularly interesting discoveries which Chinese people made in ancient times, first the fact that certain plants contained powerfully insecticidal substances, and secondly that it was possible to use certain insects to control other insects and so to protect crop plants important for mankind. Texts of Han and pre-Han date have many references to naturally occurring insecticides, such as artemisia, and their use must have been of considerable importance for hygiene and public health. As for biological plant protection, as it is now called, this was an outstanding first for Chinese science and technology; by the third century if not before, citrus farmers in the south were accustomed, at the right time of year, to go to the market-places and buy bags containing a particular kind of ant.

---

[1] 戴凱之   [2] 竹譜   [3] 韓彥直   [4] 橘錄   [5] 歐陽修
[6] 洛陽牡丹記   [7] 周師厚   [8] 洛陽花木記   [9] 南方草木狀   [10] 嵇含
[11] 毛亨   [12] 詩經   [13] 毛詩草木鳥獸蟲魚疏   [14] 陸璣

When these were hung upon the orange-trees the ants preyed upon all the insect pests, the spiders, etc. which were otherwise capable of damaging and spoiling the crop completely. In modern times practices of this kind have spread throughout the world, and in China today many such techniques are in use, but even there few people realise that the first discovery of biological plant protection was Chinese.

With this we conclude the first portion of our story, and indeed the present volume itself, but we may well say, in the words of the Asian storytellers 'if you want to know how it all ended, come back here at the same time next Saturday'. For it is Georges Metailié who will be carrying on the tale, and what he will be discussing can be seen from the contents table. In the meantime, we cannot forbear from singling out three of our friends whose names are in the Table of Acknowledgements, and to whom we feel particularly indebted for the large amount of work and time which they put in to help us. First, we would like to pay our tribute to Dr R. M. S. Perrin, who guided us through the intricacies of the science of soils, and provided the first draft of pp. 56–75. Secondly, it was Dr L. Andrew Lauener of the Edinburgh Botanic Garden who modernised all our generic and specific names in accordance with the rules of botanical nomenclature. Thirdly, Dr Frank N. Egerton, formerly of the Pittsburgh Botanic Garden and now of Parkside, Wisconsin, who went through the entire text with a fine-tooth comb, as it were, and offered us hundreds of ameliorations and corrections.

Needless to say, we have greatly benefited over the years from conversations with botanists in many countries. We cannot hope to name them all, but in China we should like to mention the late Professor Shih Shêng-Han[1], Dr Ching Li-Pin[2], Dr Hsia Wei-Ying[3], Professor Thang Phei-Sung[4] and Dr Phei Chien[5]; in Hongkong Dr Robert Whyte and the historian Professor Lo Hsiang-Lin[6]. In England we have had great advantage from the conversation and writings of Dr W. T. Stearn, as also from the late Professor Harold Godwin, founder of palynology, or pollen analysis, with all its archaeological implications. Many others have helped us too, such as the late Dr J. S. L. Gilmour, some time Director of the Cambridge Botanic Garden, and Mr E. F. Allen of the Royal National Rose Society. Others from the United States encouraged us in correspondence and in person, for instance Dr Egbert H. Walker of Takoma, Maryland, and Professor Jerry Stannard of Rutgers University at Newark, New Jersey. On the continent of Europe we shall never forget the assistance of Dr André Haudricourt of Paris, the doyen of ethnographic botany, pharmacy and agriculture; and it was through Professor E. Hintzsche of Bern that we came to know of some of the oldest Chinese botanical books that arrived in Europe.

Meanwhile we take pleasure in recording out indebtedness to our panel of advisers on languages, who kindly correct our inevitable mistakes. For Sanskrit Professor Shackleton Bailey still acts, for Syriac we have Dr Sebastian Brock, for Arabic we can always rely on Professor Douglas M. Dunlop, and for Korean we are

---

[1] 石聲漢   [2] 經利彬   [3] 夏緯瑛   [4] 湯佩松   [5] 裴鑑
[6] 羅香林

helped by Professor Gari Ledyard of New York. Finally, Dr Charles Sheldon corrects our Japanese, and Prof. E. J. Wiesenberg keeps us on the rails where Hebrew is concerned.

Time was when the number of people in our active group was so small that in each of our forewords we could name every one. That is no longer the case. But it will be indispensable to pay homage to those who have actually contributed to the preparation of the present volume. Our venture as a whole is covering such a long period, so many years, that age has perforce overtaken some of our most valued assistants, yet others are now stepping into the breach. The present volume was begun more than a dozen years ago, and has been kept under revision ever since. When the time came to select all the bibliographical reference cards, this was done by our secretary-amanuensis Mr Stefan Cooke, with the guidance of my personal assistant and collaborator Mr Gregory Blue. Page proof is now checked by Mrs Patricia Corbett. My sister-in-law Miss Muriel Moyle having retired, the indexes are being prepared by Mrs Christine Outhwaite, wife of another of the Fellows of Caius, the College in which so much of this book was written; and no survey of our activities would be complete without meed of honour to Mrs Diana Brodie, who now acts as Secretary of the East Asian History of Science Trust (U.K.) as well as my own personal secretary-typist.

Mention of our three supporting foundations or Trusts leads me to report that the British one now has as Chairman Lord Roll of Ipsden, with Mr Peter Burbidge, the tutelary administrative genius of our whole series of volumes, as Executive Vice-Chairman. The East Asian History of Science Inc. of New York has as its Chairman the distinguished business man Mr John Diebold, world-renowned as an expert in management organisation and computerisation. The East Asian History of Science Foundation of Hongkong is chaired by the distinguished surgeon Dr Philip Mao[1]. And it was from Hongkong that there came the greatest individual benefaction we have so far had, presented by the Croucher Foundation,[a] with the personal intermediation of its chairman, Lord Todd of Trumpington. It has always been our custom to acknowledge with most grateful thanks benefactions and subventions,[b] among which the continued support of the National Science Foundation of Washington, D. C. is outstanding. Both these are intended for the completion of the *Science and Civilisation in China* project, and for the endowment of the Library and Research Centre. Here in this category we cannot refrain from thanking the Wellcome Foundation of London, which generously supported one of us during the earlier phases of the writing of this book; as also the Coca Cola Company of Atlanta, Georgia, which has given for some years useful contributions to our project. Finally, in 1981 we were invited to Japan by the National Institute for Research Advancement, in which country we met many

---

[a] In former days I had the pleasure of the personal acquaintance of the late Mr Noel Croucher himself, and his munificent benefaction to us is now commemorated by a permanent plaque in one of the rooms of the Library, recording his gift with an inscription in both Chinese and English.

[b] See, for example, Vol. 5, pt 5, pp. xxxii ff.

[1] 毛文奇

people of the greatest distinction, and after due deliberation NIRA contributed a most useful subvention for the preparation of our Vol. 7, on the social and economic background of science, technology and medicine in Chinese culture.

In days to come, our Library will be built in the precincts of Robinson College in Cambridge, that new foundation which bears the name of Mr David Robinson of Newmarket. The construction of the permanent home of the Library is now imminent, and plans for all the fund-raising required are now well advanced. Indeed, the start of the building may be synchronous with the appearance of this volume itself. Our first Librarian was Mrs Philippa Hawking Hufton, and our second Dr Michael Salt; now the collection is in the charge of Miss Li Chia-Wên[1] (Carmen Lee), a graduate of the London University School of Library, Archive and Information Studies. Here may be the place to mention with gratitude the large numbers of important recent Chinese books, maps, and facsimile reproductions, otherwise unobtainable, which scholars and publishing houses in that country have bestowed upon us. At the same time we should like to record our profound thanks to Academia Sinica, which has financed several of our visits in China, enabling us not only to study the tropical botany of Hainan and the southern provinces, but also to visit many of the most important botanical gardens and institutes in the whole country.

In conclusion Dr Huang Hsing-Tsung wishes to thank the following scholars for much useful information and references: Dr Chin Shin-Foon (Kuangchow, China), Dr Paul de Bach (Riverside, California), Mr Hu Tao-Ching (Shanghai, China), Mr C. A. Lanciani (Gainesville, Florida), Prof. Li Hui-Lin (Philadelphia, Pennsylvania), Dr Ma Tai-Loi (Chicago, Illinois), Mr Allan Smith and Ms Diane Secoy (Regina, Saskatchewan), Mr Bruce Smith (Toronto, Ontario), and Dr Yang Phei and Dr Phu Chê-Lung (Kuangchow, China). We also wish to express our thanks to the National Science Foundation for allowing him to spend part of his official time working on the project, and to Mrs Dolores Taylor for typing his manuscript.

Long ago we quoted the words which Simon Stevin placed at the opening of his book on decimals and decimal fractions, *Le Disme* (+1585): 'To all astronomers, surveyors, measurers of tapestry, barrels and other things, to all mintmasters and merchants, good luck!'.[a] These words have never ceased to ring in our minds, and now we can apply them to a different tribe of men, all botanists, gardeners, phyto-geographers, explorers, horticulturists, oecologists, and all those who love and care for the world of plants—may they have much pleasure in reading our book, and seeing how men of a far land and culture, from ancient days onwards, totally unknown to our ancestors, people of whom they themselves never heard, treated the world of living plants, and wrote about it in their books which now lie open to us too.

1 August 1983

[a] Vol. 3, p. 167.

[1] 李嘉雯

38 BOTANY

(a) INTRODUCTION

'Botany', wrote the famous missionary S. Wells Williams in 1841, 'in the scientific sense of the word, is wholly unknown to the Chinese.'[a] Such a statement could only have been made by one of a generation totally ignorant of the history and pre-history of science as it had unrolled in the development of his own culture, and prone to suppose, no doubt, that almost by definition nothing could be 'scientific' unless it was clothed in the Latin tongue.[b] Today we have different criteria. That there was a turning-point, when modern plant science forged ahead of the botanical learning of all cultures in which modern science had not spontaneously come to birth, must of course be agreed, but that time came later than most Westerners may be inclined at first sight to think.[c] In this Section we propose to show that for more than two millennia before that moment the development of the botanical sciences in China had constituted a veritable epic, almost wholly uninfluenced by the botany of other peoples, and flowering from an assemblage of plant forms distinctly different from those which the Greeks, Persians or Indians studied. Justice has never yet been done to it in a world literature, in spite of the devoted labours of a handful of men (not all of whom were fully liberated from the prejudice of necessary European superiority); and botanists the world over have for the most part no idea of the vastness, complexity and high achievements of the Chinese botanical literature,[d] an earnest account of which will form one of the most indispensable parts of what follows.

What would one say of the history of botany if one were wanting to describe it in summary terms to a friend from a quite different world of activity such as trade, engineering or letters? Botany, that is to say, as it grew up in the civilisation of the West. First, though something is known of plant study in the cultures of ancient Egypt and Babylonia,[e] the main emphasis falls on the Greeks,[f] for even at the

[a] In his contribution (1841) to Bridgman's *Chinese Chrestomathy*, (1), p. 436. Williams (1812 to 1884) later spent many years in the American diplomatic service in China, and ended as Professor of Chinese at Yale. In his influential *Middle Kingdom* (1) he did much to expound China to the West, and wrote at length on Chinese science and technology, but in the absence of a sophisticated history of science in general, this could not be evaluated. Bridgman (1801–1861) was the first American missionary in China, and like the younger man a notable scholar in his day.

[b] Whewell's *History of the Inductive Sciences* (1), the first modern work on the subject (Sarton (12), pp. 49 ff.), had been published only a very few years before. And without the subsequent labours of classical scholars and medievalists it would hardly have helped Williams much.

[c] We may even be able to illustrate this in part graphically; cf. Fig. 1. below.

[d] Only a few days before beginning the drafting of this Section we were privileged to receive in our rooms at Caius five distinguished members of the International Commission on Botanical Nomenclature. As botanists and bibliophiles they were delighted to view the Chinese botanical treatises which we unfolded before them, but evidently appalled at the task which would present itself to their successors if the Code of Priority rules were ever to be taken with oecumenical seriousness.

[e] On Egypt see Woenig (1). On Babylonia see Thompson (2).

[f] Out of a large literature we may mention the works of Capelle (1); Lenz (2); Thomson (1); Langkavel (1).

I

time of the pre-Socratic philosophers there were men who investigated and spoke
about plants, whether the timber-trees which the woodmen felled for houses or
shipbuilding, or the medicinal plants which the 'root-diggers' or 'root-cutters'
(*rhizotomoi*, ῥιζοτόμοι) sought and collected.[a] The names and ideas of some of these
writers have been preserved in the works of Theophrastus of Eresus (−371 to
−287)[b], the pupil of Aristotle,[c] and his successor as head of the Peripatetic school,
a man without doubt among the greatest botanists of any age and any civilisation.
To his pair of long discursive books, still intensely stimulating today though
available to us only in somewhat incomplete form, there is no parallel in the
Chinese literature, just as Chinese culture produced no one comparable in the
encyclopaedic analysis of natural phenomena to Aristotle himself. Nevertheless
that does not mean that the Chinese naturalists of Warring States, Chhin and Han
times were not talking at length about very similar things, for this can be demonstrated by a study of the technical terms which their lexicographers have fully
preserved for us.[d] Thus the foundations of the science were laid by Greeks and
Chinese at the two ends of the Old World at about the same time. Though the true
manner of plant nutrition, and the true meaning of the flower, remained unknown
to all of them, they made a beginning of vegetative organography, the distinctions
between the many different sorts of parts of plants, they initiated anthology and
carpology, the study of flowers and fructifications, and above all they began the
elaboration of that basic essential of botany, the accurate phytographic language.
It is in these early times also that we find the first beginnings of the natural
classification of plants. Theophrastus is interesting, too, for his tentative steps in
plant physiology, which included a refinement of earlier attempts to divide them
into two great groups in accordance with primitive element theory.

The idea of the 'dark ages' is not so far off the mark where botany is concerned,
for Theophrastus had no real successors until the Renaissance. The only work of
importance in the intervening sixteen centuries was the treatise on *materia medica*
by Pedanius Dioscorides of Anazarba (*fl.* +40 to +55), a botanist and military

[a] Their 'superstitions' were not as superstitious, perhaps, as they may have seemed to some of our forefathers.
Their directions to pick a particular plant at a particular time of day must be seen in the light of modern
knowledge of the great diurnal variations in chemical composition. Saveliev (1), for example, has shown that the
carotene content of fodder plants varies by a factor of nearly ten between the maximum at sunrise and the
minimum at midnight. Cf. our discussion of circadian rhythms in man Lu Gwei-Djen & Needham (5), pp. 149 ff.

[b] The standard account of him is doubtless Regenbogen (1); we shall recur very shortly to the best studies of
his scientific achievements (p. 9). On the school, cf. Brink (1). See also of course Meyer (1), vol. 1, pp. 146 ff.;
Sarton (1), vol. 1, pp. 143 ff.

[c] Aristotle's own statements about plants, quite numerous in his authentic writings, have been quoted and
discussed by Meyer (1), vol. 1, pp. 81 ff. But he, or some other close pupil, may have written a book specifically on
plants, since there is evidence that the *De Plantis* of Nicholas, the friend of Herod, king of Judaea, mentioned
immediately below, has Aristotelian connections. For example, some of the Arabic translations attribute it
distinctly to the great philosopher, making Nicholas the commentator; and there are in the text cross-references to
other works of Aristotle, e.g. the *Meteorologica*. Perhaps Nicholas abridged a work which had come down from the
Peripatetic Academy. Cf. Egerton (1), pp. 20–1, and Senn (2).

[d] See pp. 126 ff. below.

physician under Claudius and Nero.[a] Some other books have of course come down to us, such as the *De Plantis* of the Syrian diplomat Nicholas of Damascus (b. −64),[b] or the toxicological *Theriaca* and *Alexipharmaca* of Nicander of Colophon (*fl.* −275), a priest of Apollo,[c] or the *De Herbarum Virtutibus* of the obscure +5th-century writer Apuleius Platonicus,[d] but still a thousand years were to pass before anyone in the West produced botanical work comparable with that of the −4th-century Greek. This last text of Apuleius has generally been regarded as the first of the 'herbals'.[e] Together with that of Dioscorides, it circulated in innumerable manuscript forms, with illustrations varying to the extreme of crudeness, throughout the European middle ages. Only in a few recensions, such as the outstanding Juliana Anicia Codex of Vienna, written and illustrated for an imperial princess in +512, are the drawings of botanical merit,[f] and the best of these, it seems, were derived directly from an illustrated Greek botany long since lost, the work of Krateuas.[g] Pseudo-Apuleius knew only some 130 plants, a flora four or five times more restricted than that with which Theophrastus had been familiar,[h] and even the encyclopaedist Isidore of Seville (*c.* +560 to +636) could muster no more than about half of those in the Peripatetic's list.[i] By the time that we come to the +12th century, the time of Odo of Meung (d. +1161) pseud. Macer Floridus,[j] the number of plants described has dropped as low as 77. It is striking that during this long period of decline or quiescence among the Franks and Byzantines, China knew a continual growth in botanical knowledge. One can find no backsliding in the number of plant species recorded as the centuries pass. The time of Apuleius saw in China the beginnings of a great literature of monographic type devoted to particular groups or even individual genera, a movement which was to continue through the time of Isidore and reach

[a] General accounts and bibliography in Meyer (1), vol. 2, pp. 96 ff.; Sarton (1), vol. 1, pp. 258 ff. We have used the translation edited by Gunther (3).

[b] Sarton (1), vol. 1, p. 226; Meyer (1), vol. 1, pp. 324 ff.

[c] Sarton (1), vol. 1, p. 158; Meyer (1), vol. 1, pp. 227 ff.; Greene (1), p. 144. Fr. Gow & Scholfield (1).

[d] Arber (3), 1st ed., p. 11; Sarton (1), vol. 1, p. 296; Meyer (1), vol. 2, pp. 316 ff.

[e] We find some difficulty in the definition of this term, and later on discuss it (pp. 225 ff.). The Monte Cassino Codex of Pseudo-Apuleius, as he is often called, together with the first printed edition of +1481, has been edited in facsimile by Hunger (1).

[f] See Singer (14, 15, 17); Arber (3), 1st ed. p. 8; and the facsimile edition of Karabacek (1). Its sources are analysed in Singer (15), pp. 60 ff.

[g] Krateuas the Rhizotomist was royal physician to Mithridates, King of Pontus, a ruler whom we have already encountered in Vol. 4, pt 2, p. 366 in connection with the first Western water-mills. His life must have spanned a similar period to that of his master, −120 to −63 (cf. Sarton (1), vol. 1, p. 213). And we shall meet both again in the pharmaceutical Section, for with Attalus of Pergamon, Mithridates was one of the pair called by Meyer 'die gekrönten Giftmischer' (1), vol. 1, p. 284. See also pp. 228, 230 below.

[h] Regenbogen (1). For Dioscorides Stadler (1) counted 537 kinds of plants.

[i] See Sarton (1), vol. 1, p. 471; Meyer (1), vol. 2, p. 389.

[j] See Arber (3), 1st ed., p. 40, 2nd ed. p. 44; Rohde (1), p. 42; Singer (15), p. 73; Sarton (1), vol. 1, p. 765. The mid +12th-century date (accepted by Singer) is the latest possible; it depends on an identification with Odo of Morimont. Sarton accepts a *floruit* of about +1095, and Thorndike, who has the best discussion, (1), vol. 1, pp. 612 ff., shows that the book must have been circulating by +1112. Arber held to the +10th. century, but this is surely too early.

its height before Odo. Contemporary with Apuleius too were the beginnings of books devoted to the exotic plants which the Chinese found as they pushed towards the south, and other scholarly works on the identification of the plants which had been mentioned in the classics of the early − 1st millennium. The almost complete absence of any 'dark age' in China's scientific history is a phenomenon now quite familiar to us, for we have had striking examples of it in earlier volumes, notably with geography and magnetical science (Sections 22 and 26*i*). Botany shows it clearly.[a]

But to whatever depths European plant lore had sunk it had a Renaissance, and this happened rather suddenly towards the end of the + 15th century, stimulated in part, no doubt, by the recovery of classical learning after the fall of Byzantium, and also in part by the possibilities of mass diffusion which the newly arrived (Chinese) invention of printing offered.[b] The first botanical woodcuts intended for specific recognition rather than ornament were probably those contained in the *Pûch der Natur* of Konrad von Megenberg,[c] printed in + 1475. After that things happened rather fast. The 'German Herbarius', written by an Anonymus who travelled in the Middle East and was assisted by a physician named Johann von Cube, came out in + 1485; it contained naturalistic and clearly identifiable plant illustrations in bold outline, sometimes quite beautiful, but always showing the general habit of the plant and the rough character of leaf and flower rather than those details of the inflorescence which later botanists would want.[d] During the next few decades many other 'herbals' of this kind were produced, mostly un-illustrated or with cruder woodcuts,[e] but after + 1530 a new era dawned with the work of the German 'fathers of (modern) botany', Otto Brunfels (+ 1464 to + 1534), Leonhard Fuchs (+ 1501 to + 1566), Jerome Bock (Hieronymus Tragus, + 1498 to + 1554)[f] and Valerius Cordus (+ 1515 to

[a] Here we should not overlook, however, the contribution of Albertus Magnus (Albert of Bollstadt, + 1193 to + 1280, Bp. of Regensburg) whose *De Vegetabilibus*, written about + 1265 was printed first in + 1517. More important that any medieval register of plants, it was the only theoretical botany of the Latin West, and thus a link between Theophrastus and Cesalpino. On Albert as a biologist see Meyer (1, 2); Fischer (1), pp. 34 ff., 159 ff.; Balss (2, 3); Needham (2), pp. 67 ff.: further literature in Egerton (1), pp. 77, 177–8.

Botany in Islamic culture is also a different story, but we can do no more than touch on it here. The eminent philosopher Ibn Bājja (*c.* + 1080 to + 1138) wrote much on plants (cf. Palacios, 1), but Ibn Mufarraj al-Nabātī + 1165 to + 1240) surpassed him. The latter's pupil, Ibn al-Baythār (d. + 1248), was the greatest botanist and pharmacist of all the Arabs; his *Kitāb al-Jāmi' fī al-Mufrada* (Book of Simple Drugs) ran to 1400 entries, some 300 of which were new. Cf. Mieli (1), p. 212; Hitti (1), pp. 575–6.

[b] Cf. p. 278 below.

This was an old text. Konrad's life-work had been between *c.* + 1309 and + 1374 (Sarton (1), vol. 3, p. 817) and his writing had been but a paraphrase of the still older *De Rerum Natura* of Thomas of Cantimpré (*c.* + 1186 to + 1210, see Sarton (1), vol. 2, p. 592). A few years before the printed *Pûch der Natur* (i.e. in + 1470) the *De Proprietatibus Rerum* of Bartholomaeus Anglicus, not to be confused with Bartholomew de Glanville (cf. Stannard (2); Arber (3), 1st ed. pp. 10–1, 37; 2nd ed., pp. 13–4, 41) appeared in print. Senior to Konrad, he had written his book just before + 1250, but he was interested mainly in words, and there is no evidence that he ever actually looked at any plants himself. Still, he knew 154 of them (more than twice the number that Odo could muster), though he arranged them just in alphabetical order. The woodcuts are rather stylised, like repeating fresco patterns. To gain a measure of the extraordinary backwardness of Europe at this time, one has only to recall that + 1249 was the date of the *editio princeps* of the voluminous and beautifully illustrated *Chêng Lei Pên Tshao* (cf. p. 291 below).

[d] Arber (3) gives the best account.　　　[e] Cf. Nissen (1).　　　[f] Cf. Hoppe (1).

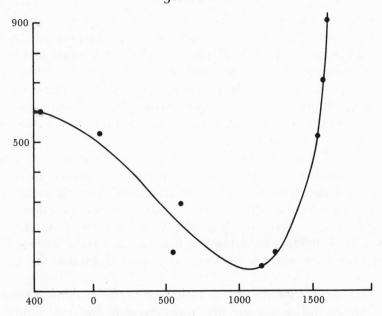

Fig. 1. Graph of the number of plant forms known in the Western world from Theophrastus to the German Fathers of Botany.

+1544).[a] Plant description, illustration, and to a lesser extent classification, took great strides forward in this period. Other countries quickly reinforced the effort the Germans had begun. In Italy Pierandrea Mattioli (+1501 to +1577) wrote a famous commentary on Dioscorides which clarified the phytogeographic situation as between northern and southern Europe, and Fabio Colonna (+1567 to +1650) introduced copper-plate engraving to illustrate his advanced plant descriptions.[b] Meanwhile the philosophical quality of Theophrastus reappeared in the writings of Andrea Cesalpino (+1519 to +1603)[c] and Joachim Jung of Lübeck (+1587 to +1657).[d] Then for all modern botany the stage was set (and the metaphor, it will be seen, is a most appropriate one) by Kaspar Bauhin (+1560 to +1624) who produced in the year before his death the *Pinax Theatri Botanici*,[e] the first complete and methodical concordance of the names of plants, amounting in number to no less than 6000 species.[f]

[a] See Greene (1) for their work in detail, together with Arber (3). As Greene points out, the four were very different; Brunfels and Fuchs were fathers of perfected illustration, but Bock and Cordus developed the new idea that it should be possible to describe plants comprehensibly without any illustrations at all. On Brunfels the best work is that of Sprague (1); on Fuchs Sprague & Nelmes (1); and on Cordus Sprague & Sprague (1).

[b] Details in Arber (3).

[c] See Sachs (1); Reed (1); Arber (3); Miall (1), p. 36;

[d] Sachs (1), pp. 58 ff. and Greene (1), p. 81. There are special studies by Meyer (4, 5) and Schuster (1). There was a great difference between Cesalpino and Jung, however, for the former was very Aristotelian, the latter a geometrical spirit and an atomist who aimed at a 'Botanica Democritea'.

[e] Arber (3), 1st ed., pp. 94 ff.; Reed (1), p. 73; Savage (1).

[f] It should not be thought that the great advances of this period were confined to Western Europe. There was, for example, the fine work of Szymon Syreniusz, published at Cracow in Poland in +1613.

As we are now in the modern period, only a few words more will suffice to sketch the subsequent developments. With the +17th century, the botanical afflatus shifted to some extent to England, where Robert Morison and John Ray laid further foundations of taxonomy, Nehemiah Grew initiated 'phytotomy' or plant morphology and microscopical anatomy, and Stephen Hales established plant physiology.[a] From now onwards botanists sought assiduously for a (or, as some of them would have said, idealistically, the) natural classification of plant genera into the more inclusive wider groups of families and orders. Here the greatest figure was J. P. de Tournefort, whose *Institutiones Rei Herbariae* appeared in +1700, but another systematist too is especially important for us, Pierre Magnol. To him is due, it seems, the very word 'family', and still today, in those which do not end with the suffix -aceae we may recognise the time-honoured groups which Magnol accepted, e.g. the Palmae, Gramineae, Cruciferae, Leguminosae, Umbelliferae, Labiatae and Compositae.[b] His *Prodromus Historiae Generalis Plantarum in quo Familiae Plantarum per Tabulas Disponuntur* was published in +1689.

Just at this time came another great turning-point, the definitive discovery of sexuality in plants, which for the first time revealed the true nature of the flower. The terms 'male' and 'female' had, it is true, been applied to plants for many centuries, in China as well as in the West, and the idea had been 'in the air' for a number of decades previously, but R. J. Camerarius was the man who demonstrated the function of pollen in fertilisation for the first time,[c] and reported it in his letter *De Sexu Plantarum* of +1694. Here we shall not elaborate on a story which requires telling in more detail later on,[d] because of its Chinese echoes, but one can fairly say that botany was never the same again. During the last decades of the +17th century and the first of the +18th there was a great expansion in European knowledge of plants from the far parts of the earth, and this helped to make the time ripe for the greatest botanical synthesiser in history, the Swede Linnaeus (Carl von Linné, +1707 to +1778).[e] Aided by great gifts of mental tenacity, visual memory and incredible industry, he imposed order upon the vegetable world to such good purpose that in the present century the 1st edition of his *Species Plantarum* (+1753, von Linné, 13) and the 5th. of his *Genera Plantarum* (+1754, von Linné, 6) have become the internationally accepted starting-points for all modern botanical nomenclature. 'Botanicorum facile princeps' he was called, yet his greatest taxonomic achievement was not that for which he won

---

[a] Sachs (1), pp. 66 ff. 230 ff., 476 ff.; Raven (2, 3); Miall (1).
[b] See especially Stearn (1), p. xc. Cf. Fig. 1.
[c] Sachs (1), pp. 382 ff.; Reed (1), pp. 95 ff.
[d] Sect. 38 (h), 2.
[e] The best introduction to Linnaeology, as it may almost be called, is surely the admirable treatise which Stearn (3) prefixed to the facsimile edition of von Linné (13) issued by the Ray Society. Cf. Heller & Stearn (1). Most of the histories of botany are very inadequate on Linnaeus. Biographies by Fries (1) abbreviated in translation by Johnson (1); also Hagberg (1) and Gourlie (1). More recent is Goerke (1) in German. In Chinese there is Wang Chen-Ju (1).

most renown in his own lifetime; it was simply his determination to restrict all plant names to two words only, those of genus and species, a binomial system which restored to science the simplicity of unlettered countryfolk and has never since been departed from.[a] In his more dazzling accomplishment he built upon the work of Camerarius and invented a system of classification which employed the arrangements of gynoecium and androecium, pistils and stamens, almost exclusively as critical characters.[b] Linnaeus knew of course that this sexual system was fundamentally artificial since it cut across some of the most obvious lines of natural division,[c] but it had its uses as a pigeon-holing or finding device; and he himself, after all, published at one point a *Fragmenta Methodi Naturalis* ($+$1738, von Linné, 8) which he laid on one side as too difficult to pursue, though often returning to the problem, as in his mature lectures of $+$1764 and $+$1771 (von Linné, 6, 14). In many countries the sexual system reigned for some decades after the death of Linnaeus, but was given up everywhere early in the nineteenth century.[d] In others it never gained popularity at all, as in France, where the search for a natural classification went on continuously, making great advances in the work of the famous de Jussieu family, for five generations Botanists-Royal at the Trianon and other gardens as well as the Museum at Paris.[e] Alongside theirs must always be mentioned the work of a great tropical botanist Michel Adanson ($+$1727 to 1806), whom we shall meet again in various contexts.[f]

Here, at the time of French Revolution, with Citizen Adanson seeing it through in age and poverty, we may finish this brief introductory sketch. The nineteenth century ushers in the understanding of the living cell with the fundamental work of Schleiden and Schwann,[g] and many new sciences—cytology, histology,[h] plant biochemistry, bacteriology and plant pathology—come into being. We are in our own world. But to know how our Chinese colleagues came into it too, it was necessary to recall for comparison the stages passed through by botany in the West.

Naturally the foregoing paragraphs are no substitute for regular histories of botany, and anyone who wants to go further than these pages themselves will

[a] This began in certain minor works (von Linné, 9, 10, 11) as an indexer's paper-saving device, and Linnaeus himself hardly realised the importance of it. Cf. Stearn (8).

[b] The quickest way to understand this is to look at Anon (76), the Lichfield translation of the *Systema Vegetabilium* ($+$1774, von Linné, 1a), in conjunction with the explanations in Stearn (3) pp. 24 ff. The famous broadside plate of Ehret which illustrates it was many times reproduced, as in Lee (1), after p. 355.

[c] On the other hand, in other cases it was not opposed to natural divisions. 'It is significant', say Heller & Stearn (1), p. 93, 'that many genera which Linnaeus associated in his artificial (sexual) system have remained associated in later classifications based on different principles and more detailed knowledge. The correspondence is much too great to be a matter of chance.' This of course arises simply from the fact that the flower structure is indeed among the most important of all the features which can be used in determining affinities.

[d] Here Bernal unaccountably nodded, (1), p. 463.

[e] See Stearn (1), pp. lxxxvi ff.; Sachs (1), pp. 115 ff.; Reed (1), pp. 101 ff.; Fée (3). As primary sources, A. L. de Jussieu (1, 2).

[f] See Stearn (1), pp. xcii ff.; Stafleu (1). Biography by Chevalier (1).

[g] This is treated of at some length by Sachs (1), pp. 256 ff., 311 ff. On Schwann see Florkin (1, 2). An interesting early evaluation is that of Darapsky (1).

[h] G. M. Smith (1) has written well on the development of this.

need some help of this kind in the study of what the Chinese did.[a] The first that came my way was that of Reed (1), and it is perhaps of all now available the best balanced, but as it devotes much space to the relatively recent branches such as those just mentioned, its treatment of antiquity and the middle ages could be but summary. Nevertheless, it remains the only history of botany in a Western language which attempts to tell of some of the Chinese contributions.[b] Penetrating further, one finds that histories of botany have tended to complement one another neatly.[c] One may set side by side the account of medieval European botany by Fischer (1),[d] the well-known book on the Renaissance herbals by Arber (3),[e] and the history of later plant illustration by Blunt & Stearn (1).[f] Similarly the four-volume treatise by Ernst Meyer (1),[g] still indispensable though published a century and a decade ago, takes the story down from the *rhizotomoi* with painful care through the dark ages only to the Renaissance 'fathers of botany', just where the other famous history, that of Sachs (1) begins. Von Sachs, however, had strong prejudices which did not please everyone; he favoured histology, morphology and physiology, disliking taxonomy and oecology, so that his accounts of the two

[a] It is a strange circumstance that, so far as we are aware, no Chinese scholar has attempted a book on the history of botany in China, though many other sciences have had monographs devoted to their history in that language. The rich, if very condensed, bibliographical survey of Sun Chia-Shan (1) might be regarded as the nearest approach. Cf. also Wu Chêng-I (1). Perhaps the Chinese botanists have been too busy describing their marvellous flora. It is true that there are books in Chinese dealing with the history of botany in the West, notably that of Hu Hsien-Su (2), based on Harvey-Gibson (1), but these do not fill the bill.

[b] This was because Howard Reed, whom I knew in his last years in Berkeley, was a friend of Michael J. Hagerty (cf. p. 368c below), who had been for a large part of his life official Chinese translator for the U.S. Department of Agriculture. Him I am also glad to have known. Hagerty's story was a striking one. He had originally been a bookbinder, and then one day had been given some Chinese books to bind. Like St Paul on the road to Damascus, he was, as it were, thunderstruck by the fascination of the ideographic language, and somehow succeeded in getting help to learn it. Then his interest in botany and his agricultural background led naturally to his life-work in the Department. Both Reed and Hagerty showed me great help and kindness.

The new history of the plant sciences by A. G. Morton (1) also treats well of Chinese botany, but he read our work in typescript, as well as giving us valuable advice. Derivative in a sense too is the recent book of Chuang Chao-Hsiang, Huan Pei-Shêng & Chiang Jun-Hsiang (1, 1), for the last-named worked in our library and had welcome access to our text.

[c] Certain histories of botany we have not been able to use, e.g. Möbius (1), who deals mainly with modern botany by topics and families; and of others we have been able to trace only a single copy, e.g. Schultes (1), at Geneva. This interesting work, by the author of *Flora Austriaca*, appeared in 1817, and devotes most of its space to the eighteenth century. A unique position is occupied by the well-known history of Singer (1) for it treats of botany within the framework of biology as a whole. One might call it an indispensable background for everything that is contained in the present volume.

Concerning the history of botany in India there are valuable works by Burkill (4) and Chowdhury, Ghosh & Sen (1).

[d] The fundamental work of Singer (14) on the herbal in antiquity we have already quoted.

[e] The second edition of this was considerably enlarged and re-written. Again I should like to acknowledge personal indebtedness to Agnes Arber, for many years a legendary *yin shih*[1] in Cambridge. She was always pleased to help and advise us. After her death I inherited valuable books of hers on the history of botany. Her work may now be supplemented by the monographs of Schmid (1) and Nissen (1).

[f] Much larger, though not perhaps so readable, is the valuable compilation of Nissen (2).

[g] This must be the most learned history of botany ever written, for Meyer studied all the original texts in Latin, Greek and Byzantine Greek, and did a bit of Sanskrit for the beginning of his third volume, most of which was devoted to botany in the Arabic tongue. He was also well qualified as a philologist. The history of botany by Jessen (1) may be regarded as Meyer evaporated down into one volume, but nevertheless contains some original material not found elsewhere.

[1] 隱士

greatest figures of all, Theophrastus and Linnaeus, are not recommendable. This deficiency can readily be supplied for the latter by reading Stearn (3) and the excellent account of 18th-century botanical gardens and literature given by Stearn (1). For the former we have a golden book, that of Greene (1), written in a characteristic and distinguished English style reminiscent of Sherrington's, and dealing also with the German fathers. But this too must be supplemented by the monograph of Senn (1).[a] Besides all these there are of course many works which have attempted to identify the plants mentioned in ancient books.[b] Lastly, one must remember that some of the greatest botanists—de Tournefort (1); Adanson (1); Sprengel (1)—prefixed to their works historical studies of substantial size;[c] so that botany has been truly (to coin a word in the botanical manner) autoscopic. For so continuously growing a science this was most natural.[d]

In the present Section we have to face the same problem that arose once before, in the study of mineralogy (Sect. 25), a purely descriptive science, at least in its earlier phases.[e] Since the limits of our survey preclude absolutely any exhaustive or systematic account of the subject-matter of such sciences, we can only choose (as we did with ores and minerals) some particularly interesting examples of plant families, genera and species for discussion. These will present themselves naturally enough as we go on. But one great difference between ancient and medieval Chinese mineralogy and botany is that the latter has a far more abundant, indeed a voluminous, literature, still very little known in the West. A second difference is that the Chinese (as would be expected from the comparative history of the two sciences elsewhere) made much more progress in the scientific classification of plants than of minerals. Here, at the beginning of our biological volume, it is doubtless the right place to remark that the name which all modern science bears

[a] Because Greene dealt only with the *Historia Plantarum* while there is much of interest in the shorter *De Causis Plantarum*, and this is brought out by Senn. For the former we have used the edition of Hort (1), for the latter that of Dengler (1).

[b] A. L. A. Fée (1, 2, 4) was industrious at this, but he worked early in the nineteenth century, and his conclusions have now to be accepted with reserve. Presently (p. 463) we shall see how extremely Chinese this interest was.

The book of Lenz (2) differs much from other historical botanical writings, being arranged more like an encyclopaedia. The first part has entries for a variety of subjects such as timber woods, grafting technique, dyes, fruit preservation, etc. each with a bunch of quotations in translation from authors of Greek and Latin books. The second and larger part extends the system to individual plant species arranged in families under the heads of monocotyledons, dicotyledons and cryptogams.

[c] See the interesting introduction in Greene (1) on this.

[d] The reader may remember a remark on p. 181 of Vol. 3 that the spherical astronomy of the ancient Greeks and the ancient Chinese was still in use in modern positional astronomy. This was why it was possible to write, as R. Wolf (1) did, an excellent manual of general astronomy in which both the spherical geometry and the observations of the ancients could find their place in a story continuous with the post-Renaissance discoveries, which depended on new sciences such as mathematical optics and electricity, unknown to the ancients. Such a presentation, I wrote, would be difficult in biology and impossible in medicine. It is true that for all the physiological sciences there is a great gulf fixed, but where the taxonomic part of botany is concerned, and the gradual extension of the boundaries of the world of known plants, there continuity does still reign, and what a Wolf did for the cosmic sciences was very much like what Adanson had already done for systematic plant science. The same must be true, on a minor scale, for zoology. At a later stage we shall try to present this continuity in graphical form (Fig. 66a).

[e] Vol. 3, p. 636.

in Chinese, *kho hsüeh*[1], means essentially 'classification knowledge', and that the word *kho* (K 8n) has a etymology clearly botanical. The left-hand component (Rad. 115), which means growing grain, derives from a pictograph of a grami-naceous plant which shows three roots, the culm, a couple of leaves and a droop-ing ear or nodding inflorescence.[a] The right-hand one (Rad. 68) is an ancient drawing of uncertain meaning which always signified a bushel or ladle, (K 116). Bronze, but not oracle-bone, forms of the character are known, so it dates back at least as early as the beginning of the −1st millennium. About +120 the *Shuo Wên* defines it as *chhêng*[2], i.e. measure, dimension, quantity, capacity, model, pattern, to weigh, to examine, to regulate, to fix, to arrange, to classify.[b] Since the word *kho* occurs in the *I Ching* and *Mêng Tzu* in the sense of the hollow of a tree or any hollow cavity, one wonders whether its semantic significance may not have been connected first of all with someone 'pigeon-holing' things which clearly differed from one another. The term *kho hsüeh* for natural science in general never occurs, so far as we know, in the Sung period, or the Jesuit times of Ming and early Chhing, but seems to have come into use during the nineteenth century.[c] Classi-cally *kho* was a term primarily connected with the bureaucratic examinations system (*kho chü*[3]) and its rules and regulations (*kho tsê*[4])[d]. From this idea of selection and classification after close examination it was easy to draw a phrase for the natural sciences in general,[e] though those who did so would probably not have had the botanical background consciously in mind. Of course 'classification knowledge' itself had existed in Chinese science from the beginning. The first star catalogues, probably pre-Hipparchan, open its story.[f] Then, as we shall shortly see, it is exemplified by the long line of pandects of natural history which start with the −2nd-century *Shen Nung Pên Tshao Ching*[5] (Pharmacopoeia of the Heavenly Husbandman).[g] It helped to lay the basis of our knowledge of chemical affinity in the theories of polarities (*i*[6]) and categories (*lei*[7]) found in treatises such as the +5th-century *Tshan Thung Chhi Wu Hsiang Lei Pi Yao*[8] (Arcane Essentials of the Categories and Reactivities of the Five (Fundamental Substances) in the 'Kinship of the Three').[h] If systematic classifications of parhelic

---

[a] The botanical radicals will be discussed in detail later on (cf. p. 117).
[b] Ch. 7A (p. 146. 1).
[c] If Suzuki Shiuji (1) is right, it originated as an expression in Japanese.
[d] Sun Jen I-Tu (1), nos. 1130, 1310, from the *Liu Pu Chhêng Yü Chu Chieh*[9] (Terminology of the Six Boards, with Explanatory Notes). Cf. entry no. 400 on the Liu Kho[10] or Six Sections into which the Imperial Censorate was divided (Mayers (7), no. 188), the body which scrutinised the work of the Six Ministries throughout the empire.
[e] All the more so, perhaps, as in the field of medicine, the Three, Six or Nine Kho, or specialities, had been a customary phrase for centuries (cf. Lu & Needham, 2).
[f] Vol. 3, pp. 263 ff.
[g] See p. 235 below.
[h] See Ho Ping-Yü & Needham (2), or better Vol. 5, pt 4, pp. 305 ff.

[1] 科學   [2] 程   [3] 科擧   [4] 科則   [5] 神農本草經
[6] 義   [7] 類   [8] 參同契五相類秘要   [9] 六部成語註解
[10] 六科

phenomena in the heavens,[a] and of the diseases of mankind on earth,[b] were worked out a full millennium before Scheiner and Sydenham, it was but an expression of the firm hold which the Chinese had on this basic form of scientific activity.

In this connection it may be as well to say clearly before going further what this Section will not be trying to do. Obviously it will not be a study of the Chinese and East Asian flora; that can only be done by professional botanists, and they are hard at work on it, as the vast bibliography of Merrill & Walker (1) shows. It will not be a systematic survey of the origin of cultivated plants in East Asia and elsewhere in the tradition established by Auguste de Candolle (1) and his successors such as Vavilov (1, 2), nor will it always track down the oldest mentions of particular plant species in Chinese literature. It will not give a perfect account of the transmissions of cultivated plants to and from the Chinese culture-area such as Laufer (1) attempted in his unique 'Sino-Iranica', or Schafer (13) in his fascinating work on Thang exotica. All these topics will be involved incidentally. But the proper task of this Section is to trace the development of botany in China from its proto-scientific stages to the status of true science, and to try to answer the question how far in this direction did it get before the period of unification of modern world science.

The first thing to do will be to take a look at the position of China in the world from the point of view of phyto-geography, for the floristic background of Chinese botanists was very different from that of Europeans, and in many ways richer. So also before modern times, as we have said, was the literature which they produced.[c] Let us then follow the pattern used in the mathematical and astronomical Sections[d] and survey this great body of writing; first the lexicographic texts, then the so-called pharmacopoeias, the pandects (as we call them) of natural history, then certain special branches of these, notably the works which concerned themselves with vegetable foods which could be used in emergencies by the people. After this we must turn to the books in which the Chinese reported on the strange plants which they met with in their conquests and travels in the southern regions, and another branch of study which sought to identify the plants named in the ancient classics. The monographic literature forms perhaps the most extraordinary difference from Europe, for in the days when Western botany was struggling in the depths with Isidore of Seville, Thomas of Cantimpré and Konrad of Megenberg, Chinese scholars were devoting elaborate treatises and

---

[a] As in the +7th-century *Chin Shu* (History of the Chin Dynasty) astronomical chapters; see Ho Ping-Yü & Needham (1).

[b] Notably in the *Chu Ping Yuan Hou Lun*[1] (Treatise on Diseases and their Aetiology), finished by Chhao Yuan-Fang[2] in +610. On this see Sect. 44 below.

[c] One of the greatest hindrances in these studies is that no one has made a collection of the biographies of the great figures of natural history in China analogous to the *Chhou Jen Chuan* (cf. Vol. 3, p. 3) or the *Chê Chiang Lu*, works which dealt with the lives of the mathematicians, astronomers and engineers.

[d] Vol. 3, pp. 18 ff., 194 ff.

[1] 諸病源候論   [2] 巢元方

elegant tractates to particular genera of cultivated plants, both useful and ornamental, recording the results of rich hybridisations and naming varieties literally
by the hundred. The woodcut illustrations of this period, too, were eminently
superior. Later, in times corresponding to those of the Renaissance herbalists and
the early European travellers, these interests diverged to produce large encyclopaedias of agricultural[a] and horticultural knowledge.[b] Lastly, we propose to
end this story with an account of the influence which Chinese flora and Chinese
botany had upon modern plant science. Here it is interesting that the Jesuit
influence in China was nothing like so strong in botany as it was in the physical,
mathematical and cosmological sciences,[c] indeed it was almost nil, and what the
Jesuits transmitted was mainly from China to the West.

In the course of this study of the science, scientific thought, and technology of
ancient and medieval China, my collaborators and I have many times re-lived a
similar experience. Approaching a new subject—non-ferrous metallurgy, astronomical instruments or physiological theory, we open first the dossier that has
built itself up through the years, and then embark upon the last weeks or months
of research which has to precede the writing of each Section. When I describe this
process to enquirers I have often been tempted to use the analogy of the development of a photographic negative, remembering how I used to stand in dark-rooms
watching the image come gradually into view as the developer did its work upon
the film or plate. At last one sees the whole picture clearly, or as clearly as one can
ever expect to see it, and then there is no difficulty in the writing of the Section.[d]
With all the tenseness of eclipse observers we awaited this clarification in the
matter of systematic botany. Sciences were variously 'strong' in different parts of
the medieval world. Dynamics, for example, as Dugas said, is a field where one
can speak neither of 'the Greek miracle' nor of 'the night of the Middle Ages', for
Greek mechanics was fallacious, and medieval European (though not Chinese)
progress rather striking.[e] Conversely, medieval Europe was lamentably 'weak' in
the study of magnetism, and all the fundamental work was done in China.[f] Now
we shall see that Chinese botany was definitely a 'strong' science in pre-
Renaissance times. In the physical sciences it was appropriate to coin a *mot* and
say that indigenous Chinese achievement attained a Vincean, not a Galilean,
level.[g] This pin-points more clearly the evident fact that it was given to Europe,
but not to China or India, to develop modern science.[h] From what follows it will
emerge, perhaps, that Chinese botany attained a Magnolian or Tournefortian,

---

[a] A preliminary account of these has already been given in Vol. 4, pt 2, pp. 165 ff.

[b] Of vegetative propagation, insecticides, plant protection, grafting and the like, much will be found below.
So too some gleams of plant physiology and sexuality.

[c] Here see Vol. 3, pp. 103 ff., 145, 155 ff., 437 ff., 583 ff.; Vol. 4, pt 2, pp. 211 ff.; Or Needham (35).

[d] Sometimes, of course, since life is short and the art long, new regions show up in sharp focus after the
Section is written, or—worse still—after it has been printed and published. This was the case with the steam-
engine's ancestry, the full analysis of which was only implicit in Vol. 4, pt 2; but that I could repair in my
Newcomen Centenary Lecture (Needham, 48).

[e] See Vol. 4, pt 1, p. 57.

[f] See Vol. 4, pt 1, pp. 330 ff.

[g] See Vol. 3, p. 160.

[h] See Vol. 3, pp. 154 ff.

not a Linnaean, level. In other words, as I had long before suspected, the line between medieval and modern science, sharp enough no doubt in mathematics and physics, turns out to be not at all so sharp in biology, and the decisive overtaking-point comes later.[a] In +1600 the hour struck, so to say, with Galileo and his contemporaries, but for botany, at any rate, it was much nearer +1700. Also Linnaeus was not the Galileo of biology, certainly not its Newton—perhaps neither has yet appeared.[b] The binomial nomenclature, though so incredibly useful, was not a work of intellectual genius, and the sexual system, illuminating though for a time it seemed, was really a loopline or siding which led nowhere. Everyone returned to the search for the best 'natural' system, a search in which the Chinese, too, had always more or less consciously been engaged.

Just as there are certain objectives not to be expected of this Section, so also there are certain branches of botanical science which are not to be anticipated in an account of Chinese botany. The first essential is to have some historical perspective, and not to expect post-Renaissance developments in traditional Chinese sciences. For example, no decisive advance in the understanding of plant nutrition and assimilation could be contemplated before the rise of pneumatic chemistry in the +18th century,[c] and the source of the nitrogen of plant proteins was not cleared up until the time of Liebig well into the nineteenth.[d] At this time, too, the cell theory of Schleiden and Schwann came into its own; before that there could be no real comprehension of the structures that Hooke and Grew and Malpighi had first seen with their primitive microscopes, in a word no histology.[e] In parallel, without the dissecting microscope, plant morphology and embryology could not begin;[f] without the immense advances of the last hundred years there could be no knowledge of flower colours, no analysis of plant hormones.[g] All the more perspicaceous, therefore, were those gleams of phytological natural philosophy which we sometimes find in old Chinese writings.[h] As for theories of evolution, the case is strangely different, for here Chinese anticipations were rather striking, but this we shall beg leave to postpone till the Section on zoology, (Vol. 6, pt. 3 below), since the changes which plants and animals were thought to have undergone are somewhat inextricable. Here also we may add that the plants discussed in the present Section will be mainly those of particular scientific interest, with some of the economic plants, wild flora and ornamentals. Drug plants, in

[a] I.e. the point at which one can say, for any particular branch of science, that the level of sophistication of its European form overtook and surpassed that of its Chinese form. We shall have more to say about this in Vol. 7; meanwhile the reader is referred to a preliminary treatment of the whole historical pattern in Needham (59).

[b] As J. H. Woodger used to say: 'It will be time enough to talk about the Newton of biology when it has found its Galileo.' Darwin measures up better, perhaps, vast though the benefits which have flowed from von Linné's systematic mind and giant industry, and endearing though his character.

[c] Cf. Sachs (1), pp. 491 ff.; Reed (1), pp. 106 ff., 197 ff.

[d] Cf. Sachs (1), pp. 521 ff.; Reed (1), pp. 215 ff., 241 ff.

[e] Cf. Sachs (1), pp. 229 ff., 311 ff.; Reed (1), pp. 87 ff., 154 ff.

[f] Cf. Sachs (1), pp. 155 ff., 182 ff. Reed (1), pp. 135 ff.

[g] Cf. Haas & Hill (1); Thimann (1); Went & Thimann (1); Boysen-Jensen (1), to introduce the current literature.

[h] See pp. 137 ff. below.

the main, will be deferred to Section 45 on pharmaceutics, cereal and vegetable crop plants to Section 41 on agriculture, trees to forestry in the same Section, and dye plants to Section 31 on textile technology.

We come now to the question of the modern literature, and what helps one can use in the study of the history of botany in China.[a] Two points stand out: first, the regrettable lack of any work on the subject by a modern Chinese historian of science—naturally, in the absence of such a secondary source, our task has been much more difficult than it would otherwise have been;[b] and secondly the domination of a towering figure among the pioneer sinologists, Emil Vasilievitch Bretschneider (1833–1901), a botanist and geographer of great merit, who was physician at the Russian Legation and Ecclesiastical Mission at Peking from 1866 to 1883. No one can do anything on the history of Chinese botany without a copy of his *Botanicon Sinicum* (1) on one's desk, but the fact remains that as his work was done three quarters of a century ago it has to be taken with various reserves; some of his identifications are certainly wrong, many plants which he could not identify have been resolved since, and many of his Latin binomials are out of date. Nevertheless, his three volumes remain indispensable today.[c] They constituted, as it were, the climax of a movement which had been going on since the +17th century, the translation of Chinese plant-names into Western and then into modern Latin binomial nomenclature. The earlier phases of this, in which the

[a] On botanical bibliography see the remarkable essay of Stearn (7).

[b] There are, of course, flora, iconographies, bibliographies, anthologies of quotations from the Chinese literature about particular families or genera, even important articles on the history of botany, and all these will be mentioned in due course. But if there were a book on the history of botany as a science in ancient and mediaeval China, we should, I feel certain, have come across it.

[c] Bretschneider's first volume was introductory. He begins with a historical account of the Chinese literature on materia medica and botany, throwing his net much wider than the *Pên Tshao* pandects so as to include some of the dictionaries, books on exotica, and later horticultural encyclopaedias; then in a separate section he reviews some of the agricultural treatises, often giving lists of the plants which can be identified in them. There follows a brief section on geographical compendia and local gazetteers to show how they contain much botanical information, and an interesting study of the 'early acquaintance of the Chinese with Indian and West Asian plants'. After this the historical bibliography is resumed for the literatures in Japanese, Korean, Manchu, Mongolian and Tibetan, but these pages could hardly be more than sketchy in the state of knowledge at the time. Next comes a valuable account of the progress in the scientific determination of the plants described in Chinese books, with all the difficulties which it entailed for the pioneers. The rest of the volume is occupied by a bibliography of Chinese botanical books, 1148 in number, and drawn from many sources, full of mistakes no doubt, and awkward because of its antique romanisations, but a superb effort for its time and not yet superseded. Bretschneider certainly never had leisure to study the monographic literature which he here listed, for otherwise his opinion of the Chinese contributions would have been higher, but this he left for others. An appendix on the celebrated mountains of China finally brings up the rear.

The second volume contains two parts, first an examination of the plants mentioned in the *Erh Ya* dictionary (cf. pp. 183 ff. below), entry by entry up to 334; secondly the same systematic treatment of those in the classics (the *Shih Ching*, the *Shu Ching*, the *Li Chi*, the *Chou Li*, etc.) from 335 to 571. This material was annotated by the German missionary Ernst Faber (1839 to 1899), who gave some useful tables but was perhaps unduly influenced by the Japanese identifications in the first edition of Matsumura Jinzō (1). It remains an invaluable research tool.

The third volume is devoted entirely to the plants in the oldest of the materia medica, the −2nd-century *Shen Nung Pên Tshao Ching*, and to those in the late +5th-century *Ming I Pieh Lu* (cf. pp. 248 below). The entries, some of which are lengthy, number 358. An appendix on historical geography, dealing with 430 ancient place-names, completes the work.

All three volumes are enriched to the maximum with Chinese characters. As they are rarely found today outside the great libraries it seemed worth while to describe them with some care.

initiative was taken mainly by Western travellers and missionaries, we shall briefly recount later on in connection with the influence of Chinese plants and books (Sect. 38 (*j*), the later phases, in the present century, when modern Chinese botanists have been extremely active, belong rather to modern botany itself, and we shall be satisfied with bibliographical references.[a] Besides old Bretschneider, of course, the scholar who bends his steps into these fascinating fields without the advantage of professional botanical training will need other guides, for the fundamentals of plant science,[b] the taxonomy of the flora of his own region as a background,[c] the taxonomy of floras related to that of the Chinese region,[d] and last but not least, some valuable encyclopædias of economic botany.[e] Here we can only name what we ourselves have used.

If Bretschneider's *magnum opus* was not at all centered on medicinal plants, later works of encyclopaedic quality tended to become so, partly no doubt for pressing practical reasons, but probably also because of the well-deserved renown of the late + 16th-century *Pên Tshao Kang Mu* (see p. 308 below), which it was convenient to go through systematically. This was done by Read (1), (properly speaking,

[a] See for instance those given immediately below. The greatest bibliography of East Asian botany in modern times is that of Merrill & Walker (1) already referred to, which comes down as recently as 1960.

[b] I was brought up on Strasburger's treatise (1), so often revised and reprinted, and certainly I have found no other large work suitable for beginners which bears so good a balance as between taxonomy, morphology, histology and plant physiology. We supplement it, however, with the admirable treatise of Lawrence (1) which describes systematically the characteristics of the majority of the families of the pteridophytes and phanerogams. This also opens with an excellent account of botanical taxonomy and its history. Both these works are organised in accordance with the Engler-Diels evolutionary system rather than the older English Bentham-Hooker system (which dates from the pre-evolutionary period), but for our purposes this matters little; the really important units are the families, genera and species, which remain the same in all the great systems, since these concern themselves rather with the higher, more theoretical, units of classes, sub-classes, orders, etc.

For the phytographic language we rely upon the introduction of Lawrence (2) with its clear illustrations. The book of Stearn (5) on botanical Latin, eagerly awaited, appeared too late to help us in the initial stages. But on all questions of nomenclature and terminology the Chinese scholar, as well as the non-botanical Westerner, can get much help from small handbooks such as those of Bailey (1) Johnson & Smith (1) and Jaeger (1). Conversely the Western botanist will find the standard Chinese equivalents of the technical terminology in a manual by Ting Kuang-Chhi & Hou Khuan-Chao (1) as well as at the back of Chhen Jung (1) and in Chêng Tso-Hsin (1); besides which there are of course the usual standard glossaries issued by Academia Sinica, as for all other sciences. As long ago as 1841 Bridgman & Williams (1) made a beginning with a glossary of Chinese phytographic terminology.

[c] The names of Bentham & Hooker (1); Clapham, Tutin & Warburg (1); and Tansley (1) will suffice. Sowerby's famous compendium (Boswell, Brown, Fitch & Sowerby, 1) has been available to us in the Cory Library of the Cambridge Botanic Garden. Add such small handbooks at that of Step (1) for sylva.

[d] Obviously, for Malaya, Ridley (1); for Vietnam, Pham-Hoang Ho & Nguyen-Van-Du'o'ng (1). Any comparative literature that one may happen to have at one's disposal will come in useful, as in our case the book of Collett (1) on Himalayan flora, or Armstrong & Thornber (1) on Western American wild flowers. Moldenke & Moldenke (1) is a particularly well-arranged discussion of the flora of Palestine. The book of Merrill (6) on the plant life of the Pacific world, though semi-popular, is stimulating.

[e] This is a particularly important branch of literature, since really great scholars have worked in it. Burkill (1), ostensibly on the economic plants of Malaya, embraces in its two volumes almost the whole of tropical botany, with abundant references to Chinese plants. It is not illustrated, but a pictorial companion to it is afforded by the two books on edible and medicinal plants compiled by Watanabe Kiyohiko (1, 2) and issued by the Japanese army when in Singapore. Their 'survival manuals' on Malayan animals and plants, probably also by him (Anon, 59, 60) are useful addenda. The merit of Burkill for our purpose is shared in somewhat lesser degree by the three-volume treatise of Brown (1) on Philippine useful plants. Then there is the nine-volume encyclopaedia of Watt (1) on Indian plants, fortunately available also in a handier one-volume abridgment. One may add mention of J. Smith's dictionary of the popular names of useful plants (1), and the vocabularies of Steinmetz (1, 2).

Read & Liu Ju-Chhiang), giving Latin binomial equivalents for every one of Li Shih-Chen's entries, with very full synonymy both in Latin and Chinese, abundant pharmacological bibliography, and notice of active principles reported. Their work has often been criticised, and must not be used without circumspection, but there is as yet nothing to put in its place, and it will probably not soon be superseded. Its total lack of illustrations has been made up for in recent years by the appearance of a number of works by Chinese botanists, notably two 4-volume drug-plant iconographies, one by Phei Chien & Chou Thai-Yen (*1*), the other by a collective group (Anon. *57*). To these may be added a well-illustrated one-volume study of the drug-plants used by Chinese countryfolk, but not properly dealt with in the old pandects and pharmacopoeias (Anon. *58*). These splendid aids have arisen from a long-standing movement for the accurate botanical description of the rich Chinese *materia medica* begun before the first world war, the earlier contributions to which have naturally now only limited uses.[a] Recent works by Europeans may fall into the purely pharmaceutical class,[b] but one or two have devoted up-to-date botanical knowledge to the interpretation of the pandects, especially that of Roi (1), which we have found very useful.[c]

Thirty years ago E. H. Walker (1) bewailed the absence of any comprehensive Flora of China, but it seems he could not have known of the illustrated dictionary of Chia Tsu-Chang & Chia Tsu-Shan (*1*), published just before the outbreak of the second world war, and several times revised and reprinted in recent years. This was for long the best and most complete hand flora of China that we came across, and we have made constant use of it.[d] Walker knew and tentatively approved a similar work, a good deal older (1918) by Khung Chhing-Lai *et al.* (*1*), also illustrated though not so systematically.[e] Khung and his collaborators, however, allowed their work to be largely influenced by Japanese identifications, and therein lay a snag. During the nineteenth century Japanese botany was much more strongly influenced by modern science than was Chinese botany;[f] and at

[a] See, e.g., Chao Yü-Huang (*1, 2*) (1); Nakao & Kimura (*1*). We tread here upon the borderline between botany and pharmacognosy, and in which category one classifies a book depends, I suppose, on how far it describes fresh, or complete herbarium, material.

[b] E.g. Chamfrault & Ung Kang-Sam (1), vol. 3.

[c] Unfortunately it is not provided with Chinese characters, but these may be obtained from another source (see bibliography).

[d] Larger in format, with 250 fine folio drawings, but more restricted in scope, was the *Icones Plantarum Sinicarum* of Hu Hsien-Su & Chhen Huan-Jung (*1*). The two-volume *Silva* (*2*) is also important. A full-dress Flora of the Chinese Republic has now at last been set on foot; it is edited by the veteran botanist Chhien Chhung-Shu with Chhen Huan-Jung (*1*) and twenty-one volumes out of the eightly planned have now appeared. Meanwhile a parallel project has been hopefully initiated at Harvard University (see Hu Hsiu-Ying, 9), and the monograph on the first of the 237 families (Malvaceae) has been published (Hu Hsiu-Ying, 1).

[e] Chia & Chia (*1*) is arranged by families, genera and species in taxonomic order on the Engler system under the Latin binomials (with 2602 entries); it has full indexes. Khung *et al.* (*1*) is arranged according to the stroke order of the Chinese plant names in the usual ideographic lexical way, but it includes in its entries Japanese, German and English names as well as the Latin.

[f] Presently, Sect. 38 (*f*), we shall describe important works down to as late as 1850 which were purely in the Chinese tradition and almost wholly uninfluenced by the modern science which had arisen in the West.

least three major works were produced which we cannot class as traditional (though to the superficial observer they may seem so because of their format).[a] It was therefore natural enough that foreigners working in China, impressed by this more modern approach, assumed that identifications established by the Japanese for Japan could be applied without much scrutiny to the Chinese flora, a proceeding which soon led to considerable difficulties. So it was then often pointed out by the judicious[b] that Japanese identifications can only be accepted for Chinese plants with great caution. While it is true that for roughly corresponding climatic and edaphic areas there is considerable similarity in the flora of the island and the mainland, complete reliance on this is most unsafe. Often the genera may be the same but the species quite different, perhaps vicarious, as the phytogeographers say, and what is even worse, the same Chinese plant name may in Japan be applied by long custom to a plant of a totally different alliance. Consequently, Japanese botanical publications can only be taken as valid for their home area,

[a] The first was the enormous and rare *Honzō Zufu*[1] (Illustrated Manual of Medicinal Plants) of Iwasaki Tsunemasa[2], which began to appear in 1828 and was not completely finished till 1856 (see Rudolph, 6). Shortly afterwards, from 1832 on, came the *Sōmoku Zusetsu*[3] (Iconography of Japanese Plants) of Iinuma Yokusai[4]. By the time one reaches the *Yūyō Shokubutsu Zusetzu* (Illustrations and Descriptions of Useful Plants) by Tanaka Yoshio & Ono Motoyoshi (1) there is not much left of traditional botany but the Japanese-style printing and binding. Immediately afterwards, with the dictionary of Matsumura Jinzō (1), already mentioned, we are entirely in the contemporary world. Thus the productions of modern botany in Japan blend back into the traditional period with perhaps greater continuity than was achieved on the more sorely troubled mainland. Another important flora, which includes many non-Japanese East Asian plants, is that of Murakoshi Michio (1). Murakoshi (2) is arranged rather like Chia & Chia (1) but with numerous drawings in colour on one page, with text facing. More recent works of a similar sort are those of Nakai Takenoshi *et al*. (1) and Makino Tomitaro (1, 2).

If one asks what was the earliest Japanese work that could be regarded as a flora, the answer would be the *Yamato Honzō*[5] (Medicinal Natural History of Japan) by Kai bara Ekiken[6], which was printed in +1708 and again in +1715. We have come across this famous scholar already, in another connection, in Vol. 5, pt 5. Then the *Ka-i*[7] (Classified Selection of Flowering Plants) was a further start towards a flora; composed by Shimada Mitsufusa[8] and Ono Ranzan[9], it was finished by +1759. The latter was the eminent commentator of the *Pên Tshao Kang Mu* (cf. Vol. 5, pt 2, p. 160). The *Ka-i* was translated into French in 1875 by Savatier (1), and together with the works of Iwasaki and Iinuma formed the basis for the *Enumeratio Plantarum in Japonia sponte crescentium* ... (1879) of Franchet & Savatier (1). Meanwhile, another European scholar, de Rosny (4), was translating excerpts from similar early books, notably the *Ka-i*, the *Wakan Sanzai Zue* encyclopaedia, and the *Sōmoku Kinyōshū*[10] (Collection of Trees with Ornamental Foliage) by Mizuno Chūkyō[11], published in 1829. We cannot enlarge further here on the pre-modern Japanese literature, but in the libraries of European botanical gardens one is liable to find copies of such useful works as the *Sōmoku Seifu*[12] (Treatise on the Natures of Plants) by Kiyohara Shigeomi[13], published in 1823, and the *Nihon Sambutsushi*[14] (Record of the (Plant) Products of Japan) by Itō Keisuke[15], published in 1872.

As far back as the +17th century, Japanese scholars realised that the plants in their country were often different from the related species in China, but they simply chose for illustration the nearest equivalent. This was done, for example, by Nakamura Tekisai[16] in his *Kimmōzui*[17] (Illustrated Compendium for the Relief of Ignorance), an encyclopaedia for young people issued in +1666; see Kimura Yōjiro (1); Bartlett & Shohara (1), pp. 101 ff. This last-named work is indispensable for the study of the history of botany in Japan.

[b] E.g. Faber, in Bretschneider (1), vol. 3, p. 403. Bretschneider himself saw the danger; cf. vol. 1, pp. 99, 124. For a striking statement by Matsumura Jinzō himself see Chao Yü-Huang (1, 2), p. 6.

[1] 本草圖譜	[2] 岩崎常正	[3] 草木圖說	[4] 飯沼慾齋	[5] 大和本草
[6] 貝原益軒	[7] 華彙	[8] 島田充房	[9] 小野蘭山	[10] 草木錦葉集
[11] 水野忠敬	[12] 草木性譜	[13] 清原重巨	[14] 日本產物志	[15] 伊藤圭介
[16] 中村暢齋	[17] 訓蒙圖彙			

Japan itself;[a] a fact all the more regrettable because modern Japanese botanists have put colour photography and colour printing to great use in their publications,[b] the like of which we do not yet have for China.[c] It may be germane to notice here a circumstance very well known to botanists, though not perhaps to other scientific men and humanists, that many species bearing the name *japonica*, or a Japanese vernacular name such as *mume*, or even the term *indica*, are really native to China, the misleading appellations arising from the fact that Western botanists first knew them from those other countries, hence the subsequent fixation of their names by the international rules of priority.

All these floras have been demoted to second rank by the new *Iconographia Cormophytorum Sinicorum* in course of preparation by the Botanical Institutes of Academia Sinica; of this, five volumes have already appeared, Anon. (*109*). There has also been in recent years a great proliferation of drug-plant floras, some of which cover compendiously all regions, e.g. Anon. (*110, 190*); others particular areas, such as the North in Anon. (*178*) or the North-east in Anon. (*181*); and since 1970 there have been at least a dozen devoted separately to each province.[d] Meanwhile writers in Western languages have not been idle, and there are important works by Li Hui-Lin (8) on ornamental flowering plants grown in gardens, and by Unschuld (1) on the history and coverage of the pandects of natural history.[e]

The only comprehensive enumeration of all the plant species of China ever attempted in a Western language, so far as we know, is contained in the three-volume work of Forbes & Hemsley (1) with its supplements by Smith (1) and Dunn (1). F. B. Forbes was an American merchant in Shanghai in the seventies who delighted in collecting the plants of Chekiang, Chiangsu, Anhui and Chiangsi, and persuaded Hemsley of Kew to prepare a list of all known Chinese plants. Even with the subsequent addenda, it is now of course a good half-century out of date; and only very partially supplemented by Koidzumi Genichi's studies of Chinese plants in occidental herbarium material (1). If one turns over the pages of this classical work one is struck by two things: the absence of any illustrations, and (more astonishingly) the lack of a single Chinese plant name, even in romanised form, let alone with its proper characters.[f] The former deficiency may well pass, since we know that according to high doctrine the phyto-

---

[a] This is of course less true of Japanese works about Chinese plants written in China, as for example the medical botany of the North by Sato Junpei (*1*) which we have used a good deal. Sato's book is noteworthy in that it reproduces many woodcuts from the *Ta Kuan Pên Tshao* edition printed in +1211 and conserved in Japan (see p. 282 below).

[b] E.g. for medicinal plants Kimura Koichi (*2*), Kimura & Kimura (*1*); for herbaceous plants Kitamura, Murata, Hōri & Hirasuke (*1*); for trees and shrubs Kitamura & Okamoto (*1*); for alpine flora Takeda Hisayoshi (*1, 1*), Miyoshi & Makino (*1*), Kano Reizō (*1*); for orchids Nagano Yoshio (*1*).

[c] But Yü Tê-Chün (*1*) is a beginning.

[d] Anon. (*176, 177, 179, 180, 182–186, 188, 189*). Inner Mongolia is covered in Anon. (*187*).

[e] Cf. pp. 220 ff. below.

[f] Coming from the Chinese world one gets the same kind of shock as on opening Pritzel's famous botanical bibliography (*2*), *Thesaurus Literaturae Botanicae Omnium Gentium* ... (1871) and finding that the Chinese were apparently not among the *omnium gentium*. Truly, they might as well have been on the moon.

graphic language designedly obviates necessity for icons, but the latter seems nowadays rather extraordinary; it must have sprung partly from an *idée fixe* that 'vernacular' Chinese was not the language of a learned people,[a] and partly from the practice established by Linnaeus of rejecting all plant names taken from languages which he considered 'barbarous' as opposed to the 'civility' of Greek and Latin.[b] No doubt the exclusion of the Chinese written names of Chinese plants made the work much easier for botanists both amateur and professional not linguistically inclined,[c] but more sagacious practitioners have subsequently condemned it as even scientifically dangerous. Thus Burkill (3) pursued the names of the yams of the *Dioscorea* genus through many a South-east Asian language as well as Chinese. 'A high percentage of such names', wrote Merrill (3),

are really safe guides to the identification of genera, species and even varieties. Very many of them are even more fixed, as designating certain definite units, than are many of our Latin binomials. They have been used for many centuries to indicate definite species, and will be used for many centuries to come, regardless of the vagaries of the binomial system. They have not been changed because of priority rules, or because of varying conceptions of what constitutes a genus or a species, or because of the personal idiosyncracies of this or that botanist, but have persisted generation after generation as definite designations for definite plant forms; some may be of very local application, others are applied to the same species over a very wide geographical range.[d]

Nevertheless modern botany has persistently continued to neglect what Merrill calls 'the great wealth of philological material' in Asian, and for our purpose especially Chinese, plant names.[e] Whether it be monographs of particular provinces or floristic regions[f], or monographs and revisions of particular families or genera,[g] the Chinese nomenclature in characters, and even in romanised forms, has in nine cases out of ten been ignored.[h] Sometimes the least pretentious contributions to the

[a] Thus the work of the entire Jesuit mission, and the devoted labours of generations of sinologists, had all been in vain. Today it is almost inconceivable that the illustrious traditions of Chinese scholarship could have been so blandly ignored.

[b] To this point we shall return in due course (p. 144).

[c] In the last century it was not customary for treaty-port merchants to learn Chinese, so Hemsley could hardly have obtained sinological help from Forbes.

[d] Minor stylistic alterations ours.

[e] Take as a typical example the contributions of Finet & Gagnepain (1) to the flora of East Asia at the beginning of the present century, in spite of the excellent drawings they contained.

[f] E.g. on Hongkong and Kuangtung: Bentham (1); Hance (2); Jarrett (1); Dunn & Tutcher (1); Hu Hsiu-Ying (10, 11); on Yunnan and Kweichow: Leveillé (1, 2); on Fukien: Metcalf (2); on North China and Manchuria: Garven (1); Noda Mitsuzō (1); on Thaiwan: Li Hui-Lin (2), the trees, complemented by Liu Thang-Jui (1) and Kanehira Ryōzō (1); on Indochina: Lecomte, Gagnepain & Humbert (1); on Kansu, Tibet, Sinkiang and Mongolia: Maximowicz (1, 2); Ostenfeld & Paulsen (1); on Hupei: Pampanini (1).

[g] For families e.g. Fang Wên-Phei (1) on Aceraceae; Irmscher (1) on Begoniaceae; Hu Hsiu-Ying (3) on Compositae; Kudo Yushun (1) on Labiatae; Lecomte (1) and Liu Ho (1) on Lauraceae; Wilson (1) on Liliaceae; Johnstone (1) on Magnoliaceae; Steward (1) on Polygoneae; Hiroe Minosuke (1) on Umbelliferae; Wu, Huang & Phêng (1) on Polypodiaceae. For genera e.g. Hu Hsiu-Ying (2) on *Philadelphus*; Hu Hsiu-Ying (7) on *Clethra*; Hao Chin-Shen (1) on *Salix*; Gardener (2) on *Cornus*.

[h] Needless to say, the same applies to nearly all the works which have been devoted to the results of the plant explorers (on whom see Sect. 38 (*j*) below). Naturally it does not apply to the publications of East Asian botanists brought out in East Asia, as e.g. Mori (1) on the plants of Korea; Wu Khuan-Chao & Chhien Chhung-Shu (1) on Chinese eucalypts; Hu Hsien-Su & Chhin Jen-Chhang (1) and Fu Shu-Hsia (1) on Chinese Pteridophyta;

literature are the most meritorious from this point of view.[a] Sometimes two volumes complement one another—Li Shun-Chhing (1), a Chinese writing in English in China, gave no links with Chinese botany in his book on forest trees, but this was repaired by Chhen Jung (1) two years later in a Chinese textbook.[b] So powerful in the thirties was Western prestige and the Linnaean mesmerism that one of China's most distinguished botanists, Fang Wên-Phei (1), could monograph the Chinese maples in 1939 and give not a single Chinese name or character; though of course in other works such as his celebrated Flora of O-mei Shan (1, 2), all those relevant were included and discussed. Liu Ho (1), who again had omitted them a few years before in his study of the Chinese Lauraceae, clung only to a preface in the mother tongue. But now the tide has turned. Great credit is due to workers such as A. N. Steward, for many years professor of botany at the Nanking Agricultural College, who in 1958 furnished his beautifully printed manual of vascular plants of the lower Yangtze valley (2) with an abundance of Chinese plant names in their proper characters. During the past dozen years, too, Chinese botanists working in the Western world are taking care to include data on the Chinese botanical nomenclature in their monographic work; see e.g. Hu Hsiu-Ying (1, 4, 5, 8).[c] This improved state of affairs owes much to both E. D. Merrill and E. H. Walker, who gave the right example in their works of more than thirty years ago.[d]

By way of ending to this introduction I should like to revert once more to the work of Emil Bretschneider. We are now more than a decade beyond the centenary date of his little book (6) *On the Study and Value of Chinese Botanical Works, with Notes on the History of Plants and Geographical Botany from Chinese Sources* printed at Fuchow in 1871. It is safe to say that since that time no Western scholar has written better on the history of Chinese botany, none half so well. Bretschneider first reviewed the classification system of Li Shih-Chen (cf. p. 317 below), touched upon the botanical radicals (p. 117) and technical phytographic terms (p. 126), described some of the more important Chinese treatises (cf. p. 278), and then discussed in turn cereals, other crops (with mention of the recognition of certain dioecious plants), fruits, yams, flowering ornamentals, Chang Chhien[e] and the introductions from Western Asia, with other similar topics. Then, after a sketch of the progress of Western knowledge of Chinese plants, he turned to the Palmae as an

---

Liu Shen-O et al. (1) on the Convolvulaceae, Gentianaceae, Caprifoliaceae, Chenopodiaceae and Polygonaceae of North China; Anon. (61) on Leguminosae; Hu Hsien-Su (1) and Tshui Yü-Wên (1) on economic plants; Hou Khuan-Chao & Hsü Hsiang-Hao (1) on Hainan plants; Hou Khuan-Chao et al. (1, 2) on Kuangtung plants; Chia Liang-Chih & Kêng I-Li (1) on South China's economic grasses; Sun Tai-Yang & Liu Fang-Hsün (1) on weeds of arable land. We need not be exhaustive.

[a] E.g. Wickes' unassuming *Flowers of Pei-ta-ho* (1), intended for foreign holiday-makers at Peking's sea resort. Cf. Porterfield (1) on Shanghai plants; Chou Han-Fan (1) on Hopei trees; and Thrower (1) on Hongkong plants.

[b] The validity of all his Chinese names is another question; cf. Hu Hsiu-Ying (4), p. 9 on his *Paulownia* spp. Cf. p. 88a below.

[c] Not always, however, is the linkage with traditional Chinese botany all that it might be, even when local names are given, as in Liu I-Jan's book (1) on the angiosperms of Hopei.

[d] See e.g. Merrill's enumeration of Hainan plants (7), the island where Boym had worked so long before, and Walker's monograph on the ornamental trees of Lingnan University park in Kuangtung (2).

[e] See Vol. 1, pp. 173 ff.

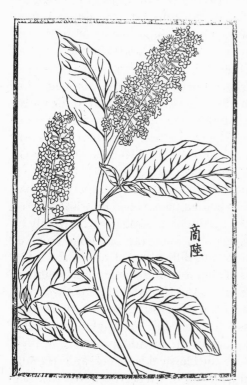 

Fig. 2a. *Phytolacca acinosa* (*shang lu*), a drawing by a Chinese artist made for Bretschneider (6).   Fig. 2b. *Citrus sarcodactylus* (*fo shou kan*), a drawing by a Chinese artist made for Bretschneider (6).

example of what could be done, and gave an extended account of all the species of this family information on which he could extract from the Chinese literature. This pioneer work concludes with a bibliography and eight Chinese botanical woodcuts especially prepared for Bretschneider by a Peking artist (Figs. 2a, b) following those in the *Chih Wu Ming Shih Thu Khao* (cf. Sect. 38 *f*). Fifty years ago it was honoured with a partial translation into Chinese by my friend the mycologist Shih Shêng-Han, who added a preface commenting on some of Bretschneider's criticisms of traditional Chinese science.[a]

It is true that Bretschneider (like other early sinological botanists) sometimes expressed a poor opinion of the Chinese literature, but the embryonic state of sinology at the time accounts for this quite as much as the ingrained sense of European superiority which here could base itself only on a single century of Linnaean systematism. One can see this well from the *cri de coeur* with which Bretschneider describes his linguistic and philological struggles. The seeming lack

[a] Some of the sinological and botanical mistakes were also corrected in footnotes by Shih Shêng-Han and Hu Hsien-Su.

of punctuation[a] and the alleged ambiguity[b] of classical Chinese dismayed him; the lack of all indexes (except the contents tables, *mu lu*[1]) and indexed bibliographies[c] exasperated him. Without the convenient chronological tables of today, he stumbled among the innumerable reign-period titles (*nien hao*[2])[d] and lost himself in the confusions of toponymic changes in different dynasties,[e] as well as the maze of obscure names of ancient foreign countries.[f] 'I need not observe', he said, 'that you often seek in vain, and that the demand for some explanation from the native scholars is equally fruitless.'[g] Here the fact simply was (as we have seen in other instances already)[h] that Bretschneider never knew the best Chinese scholars and botanists of his time—China is a vast country, and the good Russian doctor never attained an intimacy with Chinese scholars such as that of Ricci and Hsü Kuang-Chhi, or Johann Schreck and Wang Chêng.[i] Perhaps the age of imperialism had dimmed the freshness of good Father Ricci's golden days. Then he was bothered (until he solved the puzzle) by Li Shih-Chen's habit of quoting everybody only by their given names, as if we should always speak of Leonhardus instead of Fuchs or Caspar instead of Bauhin.[j] Then there were the difficulties of travel, for a foreigner, to collect plants in the remote parts of China, and the near impossibility of determining species from the dried material in the shops of the apothecaries.[k] Truly the pioneer had appalling hindrances to face. And some there may have been of which he was not himself conscious—first, as Shih Shêng-Han points out, the disintegration of Chinese society in the age of the invading mercantile West which must have hampered his contacts with the classical learning, and secondly the altogether inadequate perspective of the history of science in the West which must have engendered an inflated conception of European superiority.[l] But Bretschneider was always eager to learn, and he changed his point of view strikingly as he went on. In 1870 he wrote:[m]

---

[a] (6), p. 4.

[b] (6), p. 6 (1) vol. 1, p. 19.

[c] (6), pp. 19, 20, (1) vol. 1, p. 66. He made his own alphabetical indexes of Chinese plant names and synonyms, but what afterwards happened to them we do not know. Today sinology rejoices, of course, in a wealth of invaluable index works.

[d] (6), p. 19. On the reign-periods see Vol. 1, p. 77. The tables of Perny (1) appeared in 1872, but that Kweichow missionary was a better field naturalist than an accurate scholar.

[e] (6), p. 19, (1) vol. 1, p. 67 ff.

[f] (6), p. 20, (1) vol. 1, p. 69.

[g] Speaking of Su Sung's *Pên Tshao Thu Ching* (see p. 281 below), he remarks, 'It is almost in vain that you ask your native teacher about such works.' (6), p. 19.

[h] Cf. Vol. 4, pt 1, p. 309, a case which prompted the remark that it would be paralleled by the efforts of a Chinese scholar to find out what Englishmen knew of nuclear physics by interrogating fishwives and traditional morris dancers.

[i] Cf. Vol. 3, pp. 52, 106, 437 ff., Vol. 4, pt 2, pp. 170 ff.

[j] (6), p. 19, (1) vol. 1, p. 67.

[k] (6), pp. 2, 20, (1), vol. 1, p. 106. Cf. Roi (1), pp. vii, viii.

[l] Here again he learnt as he proceeded. He came to recognise that the medieval Chinese woodblock illustrations were eminently scientific, and 'date from a time when engravings on wood were altogether unknown in Europe' (1), vol. 1, p. 50.

[m] (6), p. 6.

[1] 目錄   [2] 年號

It is true that the Chinese possess very little talent for observation and zeal for truth, the principal conditions for the naturalist. The Chinese style is inaccurate and often ambiguous. In addition to this the Chinese have an inclination to the marvellous, and their opinions are often very puerile. None of the Chinese treatises can be compared with the admirable works of the ancient Romans and Greeks, Plinius [of all people!], Dioscorides (both in the first century), etc. Nevertheless the Chinese works on natural science are very interesting, not only for sinologues, but also for our European naturalists.[a]

Eleven years later he had changed his mind. In the parallel passage of *Botanicon Sinicum* he left the first three sentences unaltered (after all, this was the small change of Old China Hand conversation) and then continued:[b]

But not withstanding these deficiencies, met also in all the other branches of Chinese literature, their works on botany, if critically studied and rightly understood and appreciated, will be found to be replete with interest, and to present much valuable information, especially in elucidating the history of cultivated species. These treatises have no less claim to be translated into European languages, and to be commented upon, than the works of Theophrastus, Dioscorides and Plinius.

(b) THE SETTING; CHINA'S PLANT GEOGRAPHY

Before entering into the question of what the ancient and traditional Chinese botanists did, it seems obvious that we should take a look at the natural background of their enquiries. This is all the more necessary because the flora of China is very different from that to which we are accustomed in Northern Europe, and incomparably richer.[c] Whoever wishes to follow the brief exposition which we shall now present would do well to turn back to Section 4 and re-read the account there given of Chinese topography and human geography in general,[d] then to look up Section 21 and remind himself of what was said of the particularities of the Chinese climate as a whole.[e] This will obviate unnecessary repetition here.

Plant geography, which deals with the flora, i.e. the floristic quality,[f] of different regions of the earth's surface, is of course only half of the wider subject of geo-botany, the other half being constituted by plant oecology, which deals with their types of vegetation. Vegetation is a structural and quantitative concept, concerned with the different sorts of plant associations which tend to recur in

---

[a] Even then he was moving towards a better appreciation. In a footnote on the following page he said: 'It seems that the Chinese have a predilection for investigating the origin of natural objects' and then described the *Ko Chih Ching Yuan* (see Vol. 1, pp. 48, 54) and the *Mao Shih Ming Wu Thu Shuo* (see p. 467 below).

[b] (1), vol. 1, p. 66. After complaining of the misprints in his earlier work, he wrote (p. 18): 'I would therefore feel quite disposed to disavow this my first scientific essay, all the more since at the time I wrote it I had not yet sufficiently mastered the subject, and many of my former statements require modification.'

[c] The reader may naturally wish to read books on the subject of phyto-geography far more informative than what we have space for in these pages; he may therefore be referred to the treatises of Good (1); Wulff (1) and Cain (1). These are general surveys; references of a more special character will be given below.

[d] Vol. 1, pp. 55 ff. Among key references may be added Cressey (1, 3); Shen Tsung-Han (1); Li Ssu-Kuang (1); Roxby (2, 5) and Roxby & O'Driscoll (1), the last-named copious but now very dated.

[e] Vol. 3, pp. 462 ff. Among key references may be added Sion (1); Chu Kho-Chen (3, 4, 5);

[f] The actual species of plants in occupation, and their family relationships.

similar climatic, edaphic and historical conditions, e.g. deciduous woodland or littoral scrub. These associations are essentially physiological, having to do with the relations of the plants to the environment[a] and to one another. Such dominant growth patterns, however, are not always made up of the same plants; on the contrary, their several components may consist of quite different species in different parts of the globe. Thus the nature of the flora, and the vegetation, in any place are two quite distinct questions, and phyto-geography and oecology two distinct branches of science.

In the West they have been differentiating gradually during the past century and a half.[b] Then plant geography itself acquired a double motif, first the straightforward description of the floras characteristic of the widely different parts of the world, later, more fascinatingly, the inferences about the history and evolution of plant life which it has seemed possible to draw therefrom. Hence the enthusiasm of Darwin, who wrote in 1845 of 'that grand subject, that almost keystone of the laws of creation, geographical distribution'.[c] The first originator of this kind of botany was C. L. Willdenow, who in +1792 called it *Geschichte der Pflanzen*, a more prophetic title perhaps than he himself realised.[d] In his pioneer work he discussed seed dissemination, plant associations and peculiarities of flora, drawing attention to the remarkable affinities often shown by plants of widely separated regions. In this connection he remarked upon some extremely interesting resemblances between the trees and shrubs of Northern Asia and North America, a fundamental discovery to which we shall presently return.[e] From the former region he noted a maple *Acer cappadocium* (= *chhing phi tsu*[1])[f], the beech *Fagus sylvatica* ($\simeq$ *chü*[2][g], = *mi hsin shu*[3]), the alder *Alnus glutinosa* ($\simeq$ *chhi mu*[4]) and the elder *Sambucus nigra* ($\simeq$ *chieh ku mu*[5]);[h] from the latter the closely related species *Acer saccharinum, Fagus latifolia, Alnus serrulata* and *Sambucus canadensis*. He also noticed shrub similarities between Australia and the Cape of Good Hope, and floristic parallels between the Bahamas and the neighbouring continent. Willdenow's ideas were much extended by the celebrated traveller Alexander von

---

[a] The reader may remember our account of geo-botanical and bio-geochemical prospecting in relation to the history of science in China, given at the end of Vol. 3 (pp. 657 ff.) in Section 25.

[b] For good historical accounts see Wulff (1); Reed (1), pp. 126 ff.

[c] Letter to J. D. Hooker, 10 Feb. 1845, in *Life and Letters*, vol. 1, p. 336.

[d] Actually there had been a long previous history of such thinking, as can be followed in the monograph of von Hofsten (1).

[e] P. 44 below. In a sense he had been preceded by Linnaeus himself, who in a dissertation of +1750 defended by J. P. Halen, pointed out similarities between North American and Siberian floras, 11 species being supposed to be common to the two regions.

[f] A modern name.

[g] So CC 1635, but in its ancient and proper meaning, as in the *Erh Ya*, the name denotes *Biota* (*Thuja*) *orientalis*. See B 11 225, 505; Lu Wên-Yü (1).

[h] We shall henceforward use the sign $\simeq$ to indicate probable or approximate equivalence. There may be e.g. some disagreement of authorities, sinological or botanical; residual uncertainty of identification; or doubts due to historical change in Chinese usage, fluctuation of local Chinese nomenclature, or unresolved Latin synonymy. We shall also use the sign $\simeq$ to indicate incomplete equivalence. There may be assurance only for the familial placing, or only for the generic name.

[1] 青皮槭   [2] 椈   [3] 米心樹   [4] 橙木   [5] 接骨木

Humboldt,[a] whose *Essai sur la Géographie des Plantes* (5) appeared in 1807. Then it was Augustin P. de Candolle who in 1820 first introduced the idea of endemics, i.e. plant species or other taxonomic units confined completely to one floristic region and never found outside it.[b] The first classification of floras was attempted by J. F. Schouw in 1823; he spoke of North Eurasia as the 'Kingdom of Umbelliferae and Cruciferae', of Northern North America as that of *Aster* and *Solidago* (Compositae), and of China and Japan as 'the kingdom of Camellias and Celastraceae' (the spindle-tree family).[c] Schouw's book was strictly geographical, not historical or evolutionary, but a few decades later, just before the appearance of the *Origin of Species*, Unger (1) produced the first book on what would now be called palaeobotany (1852).[d] The first real synthesis of the whole subject, the *Geographie Botanique Raisonnée* (2) of Alphonse de Candolle, also appeared at this pregnant moment (1855). Thus in the fifties (the time of Lyell, Forbes, Hooker and Darwin), the full import of the extant distribution of plants was entirely clear, and as soon as the doctrine of the immutability of species was swept away, research and speculation could have free flow. By the end of the century much more sophisticated world-regional floristic classifications were advanced, as in the works of Delpino (1) and Engler (1).

(1) FLORISTIC REGIONS

If we look at a map of the world showing the distribution of the basic type of vegetation in the different regions (Fig. 3), we shall see that there is nothing very special about the Chinese culture-area. The vast North Eurasian belt of coniferous forest extends to about the northern frontier of the Manchurian provinces, but everywhere south of that, from Peking and the Shantung peninsula to the borders of Indo-China and beyond, the land was originally covered with deciduous forests and woodlands.[e] There was thus a close analogy with the primitive state of Europe at the other end of the Old World.[f] The northwest corner of the country, however, comprising part of Shansi and all Shensi and Kansu (the loess

---

[a] Biography by Kellner (2), and, more fully, Beck (1).

[b] Note the difference in this usage from current medical terminology. There the original meaning was also no doubt 'peculiar to' a particular locality, but now the word means rather 'normally prevalent in', and unfortunately not only 'peculiar to'.

[c] This contains the whole *Euonymus* genus, and an interesting plant, the thunder-god vine *Tripterygium wilfordi* (= *lei kung thêng*[1]), which contains a powerful insecticide long known and used in China. Cf. Walker (1), p. 359; Feinstein & Jacobson (1), p. 460; Anon. (*109*), vol. 2, p. 686.1. I encountered this during the second world war; Needham (*25*); Needham & Needham (1), p. 224.

[d] His other writings (2, 3, 4) were also epoch-making. Of course the study of fossil plants as such went back to the beginning of the eighteenth century with Scheuchzer (+1709), and later Brongniart, von Sternberg and von Schlotheim, all in the eighteen twenties. Cf. von Zittel (1), pp. 368 ff. The term 'geo-botany' seems to have been first used by Grisebach in 1866.

[e] Cf. Chang Kuang-Chih (1), pp. 45, 94, (2), pp. 12–3; Ho Ping-Ti (*1*), pp. 65 ff.

[f] And most of the ancient Chinese, like the ancient inhabitants of Europe, had to begin their agriculture by the clearing of the forests.

[1] 雷公藤

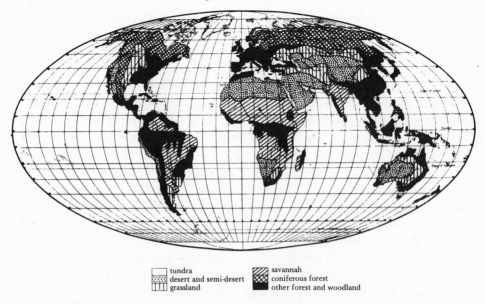

tundra
desert and semi-desert
grassland

savannah
coniferous forest
other forest and woodland

Fig. 3. Map of the world showing the distribution of vegetation; Good (1), pl. 2.

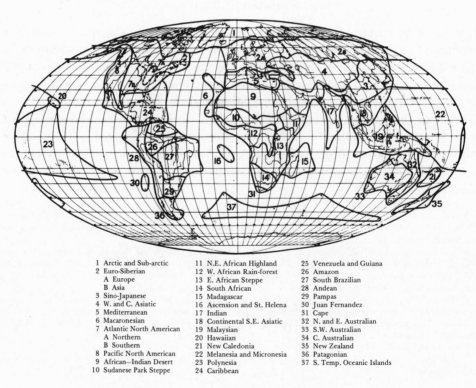

1 Arctic and Sub-arctic
2 Euro-Siberian
  A Europe
  B Asia
3 Sino-Japanese
4 W. and C. Asiatic
5 Mediterranean
6 Macaronesian
7 Atlantic North American
  A Northern
  B Southern
8 Pacific North American
9 African–Indian Desert
10 Sudanese Park Steppe

11 N.E. African Highland
12 W. African Rain-forest
13 E. African Steppe
14 South African
15 Madagascar
16 Ascension and St. Helena
17 Indian
18 Continental S.E. Asiatic
19 Malaysian
20 Hawaiian
21 New Caledonia
22 Melanesia and Micronesia
23 Polynesia
24 Caribbean

25 Venezuela and Guiana
26 Amazon
27 South Brazilian
28 Andean
29 Pampas
30 Juan Fernandez
31 Cape
32 N. and E. Australian
33 S.W. Australian
34 C. Australian
35 New Zealand
36 Patagonian
37 S. Temp. Oceanic Islands

Fig. 4. Map of the world showing the floristic regions; Good (1), pl. 4.

areas) was grass- or scrub-land,[a] in fact the eastern end of a belt which stretched across north of the Gobi desert, over Sinkiang lapping around the Thien Shan mountains, and as far west as the north shore of the Black Sea and the highlands of Asia Minor. In this there lies, no doubt, great historic significance, for the south Shensi area was the well-known cradle of Chhin and Han civilisation, while the east-west stretch of this grassland zone was threaded by the Old Silk Road and later by the many overland routes which linked China and Europe in the days of the Pax Mongolica.[b] The third type of vegetation zone in or near China was the desert and semi-desert with its xerophytic life, and this anciently covered the whole of Tibet as well as the Gobi of today, separated only by the Persian grasslands from the lower-latitude deserts of Mesopotamia, Arabia and Saharan North Africa. The only coniferous forests in China were belts at suitable altitudes on the great western mountains of the Tibetan massif,[c] and there were two fundamental types of zone not present at all in China, the tundra of the circum-polar regions and the savannah of the tropics.

When we turn from this to a world-map showing the floristic regions (Fig. 4) the contrast is striking. For one can see at once that China consists broadly of two regions with flora so unlike that existing anywhere else as to justify classification apart—thus there was oecological similarity but profound phyto-geographical difference. Following the exposition of Good (1), the constructor of this map,[d] we find that China participates in three of the thirty-seven world floristic regions, notably ③ known as the Sino-Japanese Region, including Korea, and ⑱ known as the Continental South-East Asiatic Region. Along the 50th parallel of latitude, about the northern border of Manchuria, the former of these regions joins Region ㉒Ⓑ the Asiatic portion of the Euro-Siberian Region. To the west it is separated from the third Chinese domain, Region ④, the Western and Central Asiatic Region (which runs as far west as the north shore of the Black Sea and the interior of Asia Minor), by a line passing south-west from the Amur River[e] across the Yellow River and following approximately the course of the inner Great Wall (cf. Fig. 711 in Vol 4, pt. 3) between the Ordos Desert to the north and Shensi province to the south within the great bend, then making for the similar great bend of the Brahmaputra River[f] and continuing westwards in such a way as to leave a finger of Region ③ along the Himalayas between Region ④ and the Indian Region ⑰. The Indian Region in its turn has a small area of contact with

[a] Pollen analyses of ancient levels in this area show 95.4% of all species herbaceous, and less than 5% arboreal; references in Ho Ping-Ti (1), pp. 26, 28, (5), pp. 25 ff., 49. Milpa (slash-and-burn) cultivation was therefore not needed, only the burning of the scrub over the high-fertility earth.

[b] On these topics the reader is referred to Sections 5, 6 and 7 in Vol. 1.

[c] One of the most obvious features of plant distribution in mountain regions is that successively higher zones of altitude give suitable conditions for the growth of plants characteristic of more northerly sea-level latitudes. See on this the diagram in Good (1), p. 23.

[d] See especially pp. 30 ff.

[e] Very nearly along the borders between Inner Mongolia and the Chinese provinces of Chhilin, Liaoning, Hopei and Shansi, from the point where the Amur crosses the 50th parallel.

[f] I.e. crossing Kansu, skirting Szechuan to the north, and traversing Sikang at right angles to the great river gorges.

the south Chinese (or Continental South-East Asiatic) Region ⑱. This includes, of course, all Burma, Indo-China (Siam, Cambodia, Laos and Vietnam), as well as Malaya down to a little below the 10th parallel of latitude, and the island of Thaiwan (Formosa). The line followed by the border between ③ and ⑱ in China is essentially that of the Nan Ling range (cf. Vol. 1, pp. 57, 64) which joins the mountains of Fukien in the east to the upland plateaus of Kuangsi, Kweichow and Yunnan in the west.[a] Thus the seaward-looking east- and south-facing amphitheatres of Chekiang, Fukien, Kuangtung and Kuangsi, with the island of Hainan, all sub-tropical or tropical in climate (natural provinces 12, 15, 16, 17 and 18; cf. Sect. 4a in Vol. 1) are distinctly separate from the rest of China, and this is borne out by their affiliation to a different floristic region. Immediately we have the key to a whole genre of Chinese botanical writings, those on the 'strange plants of the south', of which the type-specimen is certainly the *Nan Fang Tshao Mu Chuang*, a literary landmark due for closer examination in due course (p. 447). Greatly were the Chinese botanists stimulated when they travelled, as they did in very early centuries, from Region ③ to Region ⑱, with the expansion of Sinic civilisation among the tribal peoples of the southern lands.[b] Two more Regions only remain for mention, Region ⑲ the Malaysian, incorporating all Indonesia, Bali, Borneo, New Guinea, the Celebes and the Philippines, and therefore bordering Chinese ⑱ to the southeast—and Region ㉒ the Melanesian-Micronesian, this somewhat of a formality, since it joins with Chinese ⑱ only in the Pacific ocean eastwards of Thaiwan. For our present purpose we may neglect all the other regions shown on the map, though of course in strict science no phyto-geographer studying one of them can afford to neglect any of the others on the face of the earth.

The next thing to do will be to have a look at the nature of the plant life in these great floristic regions, and this we shall most most conveniently do by viewing successively three taxonomic levels, first the families, then the genera and then the species. Before this, however, it should be noted that the floristic regions with which we are concerned have been divided into a number of sub-regions, according to the scheme in Table 1. In studying these taxonomically there will be little to say about orders, for these are categories too large to be of much practical importance. The natural families, however, are the largest groups in which general resemblance reveals close relationship between all the members, this again implying community of evolutionary origin and similarity of history. They are something like the (less numerous) phyla and sub-phyla of the animal kingdom. A century ago Bentham & Hooker recognised some 200 families, but forty years

---

[a] Many travellers of former centuries noticed this as a great divide, and I too was impressed by it when I came down through the passes on the Hankow-Canton railway to Kukong (Shao-kuan) in Kuangtung during the second world war.

[b] A considerable amount of information on the history of this movement from Chhin times onwards has been brought together by Wiens (3), in a work warped by certain prejudices; these may be corrected by books such as those of Winnington (1); Fitzgerald (11). For the story of the Chinese beyond China's frontiers in South Asia Purcell (1) is reliable and definitive.

Table 1. *Division of floristic regions*

SUB-REGIONS	
3 Sino-Japanese Region	a) Manchuria and South-east Siberia b) North Japan and South Sakhalin c) Korea and Southern Japan d) North China e) Central China f) Sino-Himalayan-Tibetan Mountains
2B East Euro-Siberian Region	a) Western Siberia b) Altai and Trans-Baikalia c) North-eastern Siberia d) Kamchatka
4 Western and Central Asiatic Region	a) Armenian and Persian Highlands b) South Russia and Trans-Caspia c) Turkestan and Mongolia (including much of Kansu) d) Tibetan Plateau (including Chhing-hai and part of Sikang)
18 Centinental South-East Asiatic Region	a) Eastern Assam and Upper Burma b) Lower Burma c) South China and Hainan Island d) Taiwan and the Liu-Chhiu Islands e) Siam and Indo-China (Cambodia, Laos, Vietnam)
17 Indian Region	a) Ceylon b) Malabar Coast and Southern India c) Deccan d) Ganges Plain e) Flanks of the Himalayas
19 Malaysian Region	a) Malay Peninsula b) Java, Sumatra and the Sunda Islands c) Borneo d) Philippines e) Celebes and Moluccas f) New Guinea and Aru
22 Melanesian-Micronesian Region	No sub-regions.

ago this had risen to 411 (Hutchinson), and Good accepts as many as 435. As for their content, it varies enormously, ranging from the Compositae with 1000 genera and 20,000 species[a] to a number of monotypic families containing only a single genus.[b] Some are cosmopolitan or sub-cosmopolitan, some tropical, others temperate, while many exhibit an intriguing discontinuity of distribution. So,

---

[a] An average figure for family size is some 600 species.

[b] And some of these only a single species, e.g. the Cercidiphyllaceae, quite unlike any other extant angiosperm. *Cercidiphyllum japonicum* (= *tzu ching yeh*[1] = *lien hsiang shu*[2]) is Chinese, a deciduous dioecious flowering tree which gives good lumber. CC 1441.

[1] 紫荊葉   [2] 連香樹

for example, certain families occur only in America and Eastern Eurasia, the Magnoliaceae, the Hydrangeaceae,[a] the Paeoniaceae,[b] the Schisandraceae,[c] and others.[d] This phenomenon relates to Willddenow's discovery already mentioned (p. 24), and the meaning of it will later appear (p. 44). Then other families are strictly endemic, i.e. confined to particular parts of the world; of these continental Asia with Japan and Thaiwan has 13 (mostly monotypic), Asia and Malaysia have 9 (one, the Daphniphyllaceae,[e] with as many as 30 genera), and Malaysia alone 5. A few have extremely anomalous distributions, such as the boxwood family, Buxaceae,[f] of which there are several representatives in China.

A few words now about some families with special Chinese connections. Primulaceae is a cosmopolitan family, and *Primula* its largest genus, with hundreds of species, widely ranging, but the vast majority of its forms are confined to the great Sino-Himalayan mountain region. Here is already one case of the immense contribution made by Chinese plants to European gardens during the past century. Elsewhere in China monotypic genera of this family have been discovered and described in recent times.[g] Another one, the Proteaceae, is typically of the southern hemisphere, with abundant species in South America, South Africa and Australia, but it reaches up into China with one large genus, *Helicia*, and a lesser one *Grevillea*[h]. The Berberidaceae are quite different; *Berberis* is characteristic of the Andean mountain-chain on the one hand, and of the Sino-Himalayan complex on the other, more than half its total species being found in the Chinese zone. But *Mahonia* links China rather with North America in a much more restricted range (Fig. 5)[i]. Finally the Bambuseae (though not formally recognised as a separate family, a highly distinctive sub-family, the woody Gramineae) bring us to one of China's most typical floristic components.[j] Again in illustration of the Willdenowian find, there are no native species in Europe at all, and of the 500 or so that comprise the family at least 90% are Asian or

[a] Segregated from the Saxifragaceae.

[b] Segregated from Ranunculaceae.

[c] Segregated from Magnoliaceae. It contains the interesting and officinal woody vines *Schisandra chinensis* (= (*pei*) *wu wei tzu*[1], very anciently used) CC 1359; and *Kadsura japonica* (= *nan wu wei tzu*[2]) CC 1341.

[d] A useful table is found in Good (1), pp. 62 ff.

[e] Segregated from the Euphorbiaceae. A good Chinese example is *Daphniphyllum macropodum* CC 859 (= *chiao jang mu*[3]), so called because the old leaves do not fall off until the new ones are fully formed in spring.

[f] Also segregated from the Euphorbiaceae. *Pachysandra* genus is again common to America and the Sino-Japanese region.

[g] *Bryocarpum* and *Pomatosace* in the Sino-Himalayan zone, and *Stimpsonia* in Central China and the Liu-Chhiu islands.

[h] *H. lancifolia* (or, if we pay homage to Loureiro, *cochinchinensis*) CC 1574 ≏ *shan lung yeh*[4] = *hung yeh shu*[5]). *G. robusta* (= *yin hua shu*[6]).

[i] *Mahonia bealii* or *japonica* with its spiny leaves and yellow-purple flowers, is an interesting shrub, CC 1378 (≏ *shih ta kung lao*[7]). The name commemorates its numerous valued uses, and indeed it is a powerful agricultural and horticultural insecticide long used by farmers and gardeners.

[j] On this see Vol. 1, p. 86, Vol. 4, pt 2, pp. 61 ff., Vol. 4, pt 3, pp. 90, 134, 391, 393 ff. 595ff.

[1] 北五味子　　[2] 南五味子　　[3] 交讓木　　[4] 山龍眼　　[5] 紅葉樹
[6] 銀華樹　　[7] 十大功勞

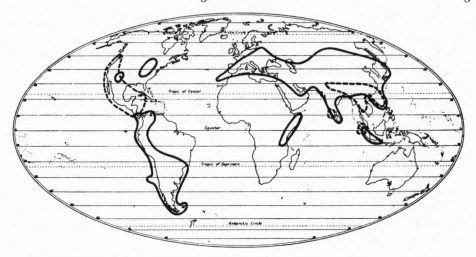

Fig. 5. Map showing the distribution of *Berberis* (continuous line) and *Mahonia* (broken line); Good (1), fig. 17

American, the latter far less numerous than the former. Besides this, a few are African or Australian.

The genus may, and sometimes has been, called a natural category.[a] Many families are still not free from the suspicion of being unnatural, heterogeneous, but the smaller taxon, the genus, arises from characteristics so clear that it can reliably be regarded as a true natural group. Of genera now recognised there are about 12,500, with an average size of *c.* 18 species each. Again the range is considerable; there are 14 genera with more than 1000 species each,[b] and 470 with more than 100.[c] Apart from cosmopolitan, pan-tropical and pan-temperate genera there is group known as 'Asiatic wide genera', about 375 in number, which are restricted to the Regions ③ ⑰ ⑱ and ⑲ above-mentioned. Some radiated from India. The Dipterocarpaceae provide good examples of this,[d] and there are genera such as the taro *Colocasia* (Araceae),[e] the mango *Mangifera* (Anacardiaceae),[f] and some rattans[g] which also show it. Others radiated from China and Japan. The interest-

[a] Good (1), p. 7. Even so, it is difficult in some cases to decide upon the genus to which a given species ought to belong.
[b] The vetch *Astragalus* (Leguminosae) and the ragwort *Senecio* (Compositae) lead the way, followed by *Solanum, Rhododendron, Euphorbia*, etc.
[c] Lemée (1).
[d] For instance the sal-tree *Shorea robusta*, famed for its timber (cf. Watt (1), p. 990) and attributed to India in its Chinese name ≃ *po lo shu*[1], Brahmin or Brahma tree, CC 698. Cf. Vol. 1, p. 128, Vol. 3, p. 202.
[e] *C. antiquorum*, R 710, CC 1926, = *yü*[2], a name certainly ancient.
[f] *M. indica*, Watt (1), p. 764, CC 836, = *hsien*[3], and many phonetic transliterations also. This is the family that includes, unexpectedly, *Rhus* and *Pistacia*.
[g] E.g. *Daemonorhops*, especially *D. draco*, of the Palmae, which produces one of the sorts of dragon's-blood resin, only in China we call it = *chhi-lin hsüeh thêng*[4] i.e. unicorn's blood rattan. See Lu Khuei-Shêng (1), p. 79; Burkill (1), vol. 1, pp. 747 ff.; Brown (1), vol. 1, pp. 299 ff.

[1] 婆羅樹   [2] 芋   [3] 樣   [4] 麒麟血藤

ing genus *Dichroa* (Saxifragaceae)[a] is a case of this kind, as also *Daphniphyllum* already cited, and *Broussonetia* the paper-mulberry.[b]

Let us now examine our three main Chinese regions at the generic level. Neighbouring Siberian ②B affects us little; its climate, with an almost continuously frozen subsoil, is hard for plants; its endemics are few and unimportant, and the lay of the mountains has always prevented easy infiltration into China on the south. *Rheum*, however, grows in these northern parts as well as on the borders of Tibet and in Northern China, historically a most important plant because of the medicinal rhubarb trade with the ancient Roman world.[c] The leguminous bush or pea-tree *Caragana*[d] was the subject of a classical monograph by Komarov (1). It ranges across this border, but with constantly more xerophytic species as the fertile land of China is quitted. *C. sinica* there is replaced by *C. rosea*, *leveillei* and *opulens* in Mongolia, by the desert species *C. polourensis* and *turfanensis* in Sinkiang, and so forth.

Our main region ③ consists of three main parts, the Sino-Himalayan-Tibetan mountains, the rest of China north of the Nan Ling range, and the insular area of Japan. For the whole region there are more than 300 endemics, and if non-alpine China alone be considered, more than 100, many of which are of much interest.[e] In view of the fact that the whole region has been the native land of so many highly-valued garden plants now grown throughout the world, these figures may seem rather lower than might be expected, but in fact most of the characteristic genera have rather wider ranges. China, however, was the focus of civilisation in this part of the world, and two millennia of Chinese horticulture paved the way for the dispersal of these ornamentals, while in relatively recent times explorers were tempted to collect in the high mountains using Chinese settlements as their bases. The region has considerable linkage with its southern neighbours, but an interesting barrier can be discerned between Korea and Japan, the former having 60 genera not in the latter, and conversely 260 genera insular but not peninsular.

Region ④ which includes Kansu and the eastern Tibetan provinces, has about 150 endemic genera. Mostly a high and dry area, its flora is limited and specialised, rich naturally in halophytes and xerophytes. Chenopodiaceae and Umbelliferae are numerous, and some genera like *Exochorda*[f] extend eastwards into China. The Continental South-east Asiatic Region ⑱ though including more of China, and luxuriant in vegetation, is not floristically outstanding, and

---

[a] Esp. *Dichroa febrifuga*, R 353, CC 2541, = *chhang shan*[1], an important anti-malarial in the Chinese pharmacopoeia (cf. Sect. 45).

[b] *B. papyrifera*, the paper-mulberry, an important textile plant very anciently used in China (cf. Sect. 31). See B II 503, III 333, CC 1591 (= *chhu*[2] and other, probably later, names). On Broussonet's life see Granel (1).

[c] See Vol. 1, p. 183. On *R. officinale* = *ta huang*[3], see B III 130, CC 1551.

[d] In China, *C. chamlagu* = *chin chi erh*[4], CC 964; Chhen Jung (1), p. 548.

[e] Without delaying to describe them now, since they may arise in other contexts later, we may mention *Akebia* (Lardizabalaceae), *Liriope* (Liliaceae), *Litchi* and *Xanthoceras* (Sapindaceae), *Paulownia* and *Rehmannia* (Scrophulariaceae), *Poncirus* (Rutaceae), *Reevesia* (Sterculiaceae), *Stachyurus* (Stachyuraceae). Some are very strange relicts such as *Euptelea* (Eupteleaceae).

[f] E.g. *E. serratifolia* = *chhih yeh pai chüan mei*[5], CC 2534, the 'white cuckoo plum'.

[1] 常山   [2] 楮   [3] 大黃   [4] 錦鷄兒   [5] 齒葉白鵑梅

may be considered intermediate between China proper and Malaysia ⑲. There are about 250 endemics, mostly small and localised. Region ⑲ itself, on the other hand is extremely rich in genera, with about 500 endemics, and celebrated of course for the various divisions within it such as Wallace's Line and Weber's Line which mark the separation between the Asiatic and Australasian flora and fauna.[a]

It now only remains to view the three regions at the specific level. There is, perhaps, general agreement that the species is not a very satisfactory taxonomic unit. The term has come down to us from pre-evolutionary times (cf. 'special' creation), but since this is the level at which hybridisation can occur there is no real practical criterion of what constitutes a species.[b] While the differences between genera are easily demonstrable,[c] botanists are by no means always agreed about the significance of the minor similarities and differences within the genus, so that there may be a number of opinions on what constitutes specific rank. We are all no doubt 'lumpers' or 'splitters', with a tendency either to reduce the slightly different plant forms to the level of varieties carrying 'horti-cultural' labels, or to elevate them to the rank of species dignified by Latin names.[d] Linnaeus himself was often uncertain what to do. In the *Species Plantarum* of +1753 he had only one species of *Magnolia*, with five varieties, but by +1762 he listed four species, three of them being former 'varieties'. Albrecht von Haller had warned him in +1746 of the danger of excessive 'lumping', and perhaps he saw this change as a justification of his viewpoint.[e] Many other Linnaean varieties are species today.

The total number of plant species in the world is now assessed as of the order of 225,000, the average number of species in each genus being about 18. China's flora contains some 30,000 species, with a relative density of 0.005.[f] Tropical Asia is a little richer in species than tropical America, but tropical Africa much poorer than either—why remains unknown. The Sino-Japanese Region ③ is one of the most interesting of all as regards species, for its flora is by far the richest of the whole northern temperate zone. The number of species of trees in China, indeed, is greater than that of all the other parts of this zone put together,

[a] Cf. Good (1), p. 141.

[b] Of course it is true that in most cases (apart from orchids and a few other exceptions) inter-specific hybrids are apparently infertile, and this might be regarded as a convenient practical criterion.

[c] But even here botanists differ much in their tendency to mass or divide, as in the well-known case of the fruit-trees *Pyrus, Malus, Cydonia*, which some unite in *Pyrus*. Cf. Bailey (1), pp. 64 ff.

[d] This is of particular importance for the history of botany in China, where horticultural hybridisation was going on through many centuries, and where hundreds of variant forms of the same genus were actually named long before Europeans made such distinctions. See p. 398 below.

[e] See Stearn (3), p. 160, and on the whole attitude of Linnaeus to varieties, pp. 156 ff. Linnaeus, of course, knew nothing of the modern genetical background to systematics. He early dismissed as having no taxonomic value modifications of a non-heritable nature. But by +1760 he became convinced that new species could arise by hybridisation. Eventually he thought that the Omnipotent had blended the Classes or Orders to form the genera, that Nature subsequently had blended the genera to form the species, and that Chance or Man had blended the species to form varieties.

[f] This figure is obtained by dividing the total area in question by the number of species, so that it represents the number of species per square mile. Generally the smaller the area the greater the species density. The highest densities occur in certain limited areas of the southern warm-temperate regions and sub-tropics. See further in Good (1), pp. 154 ff., who gives values ranging from 23 to 0.0003 for different parts of the world.

and endemism is high.[a] Hence, once again, the great contributions of the region to the ornamental plants of Europe. If one takes the case of *Rhododendron* (*tu chüan shu*[1])[b], for example, no less than 700 species, or more than two-thirds of the total known, are native to the mountain ranges where India, Burma, Tibet and China meet—the Sino-Himalayan Node of Ward (17). So also Region ③ has made great additions to the world's important economic plants; the Chinese were fine domesticators and their culture continuous longer than any other still extant, hence many crops originated here. Among species of particular interest may be mentioned the famous 'mandrake' of the East, the araliaceous tonic plant *Panax ginseng*[c], the caprifoliaceous shrub *Weigela (Diervilla) florida*, the *Camellias* and *Theas*, *Lilium tigrinum*, the mulberry *Morus alba* used from high antiquity for silkworm culture, and the persimmon *Diospyros kaki*.

The Western and Central Asiatic Region ④ it will be remembered, covers the northwest corner of China and Manchuria including the Kansu panhandle and Chhinghai. However, this region's interesting species are abundant only at the western end, which does not concern us. Apart from halophytes like the leafless saxaul tree *Haloxylon ammodendron* and another Chenopod *Salsola arbuscula* one may mention the tamaricaceous *Myricaria prostrata* and *dahurica*.[d] In passing from these relatively arid scenes to the Continental Southeast Asiatic Region ⑱, one comes over the passes from poverty into richness. Here the vegetation is exuberant and the number of endemic species hardly less than in the Malaysian Region ⑲. This was probably the home of some of man's most valuable economic plants, rice *Oryza sativa*, tea *Camellia sinensis*, and, especially towards the hillier western part, all the fruits of the genus *Citrus*. Other interesting species centering in this region are the leguminous *Bauhinia purpurea*.[e] and *japoncia*, *Cassia nodosa*, *tora* and their relatives; then the camphor tree *Cinnamomum Camphora* among the laurels, besides the adventive Borneo camphor tree *Dryobalanops aromatica* with *Dipterocarpus pilosus*[f] and *turbinatus*, all belonging to the family named from this last genus. The Guttiferae

---

[a] There is a convenient (if perhaps now rather outdated) catalogue of the trees and shrubs of China by Chung Hsin-Hsüan (1).

[b] A genus very anciently known to Chinese botanists. The *Pên Ching* (p. below) knew two species, *R. sinense* = *molle* = *yang chih chu*[2], containing andromedotoxin and giving the 'staggers' lethal for sheep, hence the name (R 203, CC 523); and (*pace* Roi, 1) *R. hymenanthes* = *pentamerum* ≃ *shih nan*[3] (R 202, CC 522). I too have witnessed the glory of rhododendrons in Chinese forests, recalling especially a journey with Dr Lin Jung of Amoy University through the Fukienese mountains between Chhangting and Yung-an in 1944. *R. mariae* = *ling nan tu chüan*[4], and *R. indicum* = *simsii* = *tu chüan*[5], var. *ignescens* = *ying shan hung*[6] (CC 530) were unforgettable. Cf. Needham & Needham (1), p. 214.

[c] See Sect. 45 on pharmacology. One of the earliest notices of it in the West was the Danzig thesis of J. P. Breyn (1), first published in +1700 and several times afterwards reprinted.

[d] Chhen Jung (*1*), p. 852 (= *shui po chih*[7]). This specific name invites the explanation (since it has been used by botanists not infrequently) that 'Dahuria' is the Nonni R. valley in eastern Manchuria running down beside Chhi-chhi-har to cross the trans-Manchurian Chinese Eastern Railway at right angles. The name comes from that of a Manchu tribe (see Gibert (1), 1. 825). I visited Dahuria in 1952.

[e] Cf. Hu Hsiu-Ying (10).

[f] This is called *chieh pu lo hsiang*[8], a transliteration of (Skr.) *karpūra*, i.e. camphor, perfume. Chhen Jung (*1*), p. 832. *Dryobalanops*, *lung nao shu*[9] (CC 697).

---

[1] 杜鵑屬   [2] 羊躑躅   [3] 石南   [4] 嶺南杜鵑   [5] 杜鵑
[6] 映山紅   [7] 水柏枝   [8] 羯布羅香   [9] 龍腦樹

give *Garcinia hanburyi* and *cochinchinensis*, the Rubiaceae *Gardenia jasminoides*.[a] *Lagerstroemia indica* (= *chinensis*) is lythraceous, like the henna (*Lawsonia inermis*) with which the girls of Canton used to dye their hair and paint their finger-nails.[b] Some musaceous plants (the bananas) belong here, like *Musa sapientum*[c] and *coccinea*[d]. Finally one might mention the combretaceous vine *Quisqualis indica* (= *sinensis*) prized for its anti-helminthic principle helpful in paediatrics.[e]

Region ⑲ is not our business exactly, but it shows, after all, that maximum intensity of floristic abundance, long embodied in the expression 'the spice islands', of which the richness of the South China area is the foretaste. Originating here are the jackfruit *Artocarpus incisa*[f] allied to the mulberry, rattans like *Daemonorhops draco*, taro species like *Colocasia esculenta*, palms proper like *Metroxylon rumphii*, and certain plants the history of which has given them a deeply Chinese flavour such as *Zingiber officinale*. *Dryobalanops* has just been mentioned, but one could not forget the true pepper *Piper nigrum*, which entered China in the Thang period to replace, at least partially, the indigenous fagara pepper of the ancients.[g] Nor can one forget ornamental species such as *Hibiscus Rosa-sinensis*, beloved of Chinese gardeners, which bears the highly evocative name of the *fu sang*[7] tree,[h] and the plant with the largest flower in the world, parasitic *Rafflesia arnoldi*, called by them just *ta hua tshao*[8], in a master-piece of under-statement. Here in ⑲ endemism is at its height; with approximately 27,000 species, as many as 70% may be confined completely to the region, largely no doubt because of its multi-insular character.

Thus far we have been looking at our China with, as it were, the coarse adjustment, considering the three world floristic regions which cover her territory in relation to those which adjoin them on every side. We must now quickly turn on the fine adjustment, and see how Chinese and Sinicolous botanists have divided the sub-continent itself. This may be followed from the map in Fig. 6. We shall then consider one or two studies of particular families or genera in the light of these divisions, and end by a solution of the Willdenow phenomenon. Last but not least we shall see what premonitions of phyto-geography there were in ancient and medieval Chinese writings.

[a] This again I have seen in full bloom in Chinese virgin forests; it was on the journey with Dr Lin Jung just referred to. We found masses of *Gardenia angustifolia* ≃ *jasminoides* (= *hsia yeh chih tzu*[1]), CC 221, growing wild. Cf. Needham & Needham (1), p. 213.

[b] Cf. Vol. 1, p. 180. The name *chih chia hua*[2] witnesses it. See Stuart (1), p. 232, CC 639, R 248.

[c] CC 1771, = *kan chiao*[3]. Some separate the Chinese form as *Musa cavendishii*.

[d] CC 1773, = *mei jen chiao*.[4]

[e] See Stuart (1), p. 368, R 245, CC 623. This 'Rangoon creeper' gets its Chinese name *shih chün tzu*[5] from a physician of early Sung or pre-Sung times Kuo Shih-Chün[6], who was famed for his use of it. Cf. Burkill (1), vol. 1, p. 1859.

[f] See CC 1588, 1589 and Bretschneider (6), p. 6.

[g] *P. nigrum* CC 712. The fagara is *Xanthoxylum piperitum* of the Rutaceae, CC 923. The former is *hu chiao*[9] the latter *chhin chiao*[10], doubtless after the State of Chhin (cf. Vol. 1, pp. 96 ff.). An excellent discussion of the subject will be found in Schafer (13), pp. 149 ff.

[h] See Vol. 4, pt 3, pp. 540 ff.; Vol. 3, pp. 436, 567, Fig. 242; Vol. 4, pt 1, p. 1141; Li Hui-Lin (1). Horticulturally, Li Hui-Lin (8), p. 137; CC 741.

[1] 狹葉梔子	[2] 指甲花	[3] 甘焦	[4] 美人焦	[5] 使君子
[6] 郭使君	[7] 扶桑	[8] 大花草	[9] 胡椒	[10] 秦椒

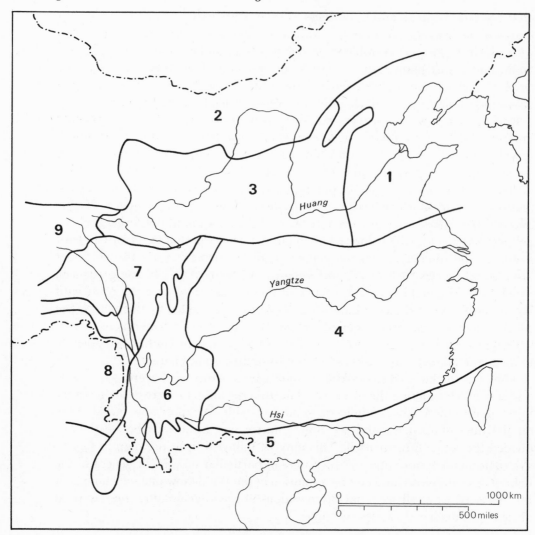

1 Forest region of north-east China and Korea
2 Gobi desert region
3 Loess steppe region of north China
4 Overlap region of mid-China and mid-Japan
5 Tropical region of south China and Indo-China
6 High plateau and mountain region of Yunnan and western Szechuan
7 East Tibetan grasslands
8 Mountain region of north-east Burma and west Yunnan
9 Tibetan high-altitude desert region

Fig. 6. General map of the floristic regions of China; after Handel-Mazzetti (7)

At the beginning of this century Diels (1) discussed the question but without proposing any precise regions. During the following thirty years several attempts were made, delimiting four,[a] five,[b] or six[c] regions, but the best and most widely approved is that put forward in 1927 by H. Handel-Mazzetti, an Austrian botanist very familiar with South-west China and the Sino-Himalayan Node.[d] This recognises eight regions, or, if the High Tibetan Wasteland is included, nine. They may be listed as follows:

(1) *North-east Sino-Korean Mixed Woodland Region.*
This has as its northern border a line approximately the same as that which separates in Fig. 4 World Regions ④ and ③. Apart from a western tongue which takes in Wu-thai Shan, its boundary runs down the mountain escarpment of the Thai-hang Shan between Hopei and Shansi, crosses the Yellow River and turns sharply east near Ju-ning to follow the Huai R. to the coast. Climate very severe but not extremely continental, temp. variation range -40 to $+24$,[e] precipitation 449 to 672.[f] Many southern plants will grow. Endemics include *Euphorbia lucorum, Deutzia grandiflora, Gleditsia heterophylla, Saussurea odontocalyx*, and many more. If Dahuria is included, the list increases. Three Bambuseae as far as North Korea. Some affinities with Northern Japan.

(2) *Southern Gobi Steppe Region.*
This is all in World Region ④ and takes in Sinkiang, Inner and Outer Mongolia, and the Tsaidam Basin, as well as the Ordos Desert within the great bend of the Yellow R., but not Chhinghai. It is less cold in winter than Manchuria, but extremely dry with a precipitation of only 46. Endemism is very slight, but *Ranunculus cuneifolius, Tilia mongholica* and *Convolvulus tragacanthoides* may be mentioned. The vegetation highly xerophilic.[g]

(3) *North Chinese Loess Steppe Region.*
Bounded on the south by the Chhin-ling Shan[h] and its eastern continuations as far as Nanyang, and on the north by the line of the Great Wall marking the north edge of Shensi province.[i] To the east the hills of Shansi and the Fên R. valley are all included; to the west the northern border passes north of Lanchow and includes most of Chhinghai. Here hot extremely dry summers, yet precipitation 338 to 472. *Populus euphratica* very characteristic of the watered valleys and oases. Endemism again slight, but there is *Macleaya microcarpa* and *Nothoscordum nerinifolium.*

(4) *Middle Sino-Japanese Laurel Region.*
The largest of the eight, stretching from O-mei Shan in Szechuan to the sea, and from Ju-ning and Nanyang in Honan south to the line of the Nan Ling range.[j] There is consider-

---

[a] Chhin Jen-Chhang (1) for ferns, and Chhien Chhung-Shu & Fang Wên-Phei (1) for maples.

[b] Engler (2) for conifers in the *Natürliche Pflanzenfamilien*; Hu Hsien-Su (1, 2, 3) for forest trees; Liu Shen-O (2), but for North China alone and geographically obvious.

[c] Hu Hsien-Su & Chaney (1) for the Miocene flora, based on Shantung.

[d] (1) is the main peper, (2) an English abridgment, and (7) contains the map as well as a number of floristic photographs of his region 8. Botanical results of his expeditions will be found in (9).

[e] Degrees centigrade.

[f] Annual in millimetres. Reference may be made to the map Fig. 861 in Vol. 4, pt 3.

[g] Just as I was deeply impressed by the divide of the Nan Ling range, so also in 1943 I found the desert character of northwestern Kansu, China's 'Wild West', extremely inspiring. Cf. Needham & Needham (1), p. 131.

[h] Cf. Vol. 4, pt 3, pp. 22 ff.

[i] Cf. Vol. 4, pt 3 pp. 46 ff.

[j] Handel-Mazzetti actually draws his boundary between regions 4 and 5 rather south of the Nan Ling (see Fig. 6) indeed along the Tropic of Cancer, but this is surely too low, even though the flora of the northernmost parts of Kuangtung may belong to region 4.

able temperature variation, but though the summers are liable to be very hot (up to 38) the winter minima do not fall below −6. Precipitation very variable (880 to 2072). Here the flora is rich, though not always strongly endemic. Sharp distinction has to be made regarding altitude. In the sub-tropical zone, up to 1500 ft, sclerophyllous trees, *Cunninghamia* and pines are more common than Lauraceae, but these predominate strikingly in the warm temperate zone which goes up to 6000 ft with Cupuliferae at the higher levels. A temperate zone follows, up to 9000 ft, with *Liquidambar formosana* and *Pinus massoniana*, and this is succeeded by a cold temperate zone separated at 12,000 ft from the high alpine zone, this being only found on the borders of region 6. *Davidia involucrata* and *Mahonia fortunei* inhabit the sub-tropical zone uniquely, *Artemisia anomala* the warm temperate one, while *Abies* spp. are rich in the cold temperate zone.

(5) *Tropical Chinese Region.*

Includes Thaiwan and the whole of the coastal lands through Fukien, Kuangtung and Kuangsi, the boundary from (4) running along the Nan Ling range, and therefore identical with that between World Regions ③ and ⑱. This 5th region is of course continuous as well as contiguous with Vietnam. Lowest winter temperature +13, summers quite tropical, precipitation 1270 to 2170. Endemism quite pronounced, with *Quercus jenseniana*, *Schima crenata*, etc.

(6) *Yunnan and West Szechuan Temperate and Warm-Temperate High Mountain Steppe and Woodland Region.*[a]

This is essentially the eastward-facing slopes of the Tibetan massif, hence it extends northwards in a gradually narrowing strip from the neighbourhood of Mandalay to the uppermost waters of the Chialing Chiang in southern Kansu. Besides this, a corridor leads north-westwards up the gorge of the Yangtze to Batang and the Tang-la Mountains. Relatively dry (precipitation 883 to 1040) and not extremely cold in winter (min. −6). Like region 4, this has naturally to be divided according to altitude levels. The subtropical zone here reaches to 5400 ft on plateaus and to 8000 ft between high mountain ranges. Here this is very dry, and also very rich endemically, with species such as *Caesalpinia morsei*, *Erythrina stricta*, etc. and a number of monotypic genera such as *Delavaya* and *Corallodiscus*. The warm temperate zone then takes over until 8700 ft on plateaus, full of sclerophyllous trees and shrubs with oaks and oak-allies of the laurel type. Endemics here include *Neocheiropteris* and *Xystrolobus*. Above this there is the temperate zone until 10,200 ft, the richest of all; it consists either of xerophilous pine and oak woods with heath-meadows and highland fens, or mesophilous evergreen forests joined by bush-meadows and tall perennials. In the cold temperate zone, thereto succeeding, up to the tree limit at 12,600 ft the vegetation is mainly coniferous, but there are forests of gnarled rhododendrons, tall herbaceous perennials, leaf 'mould-mats' and 'lush-meadows'. Here the affinities are strongest with the Himalayas; the westward tongue of World Region ③ in Fig. 4 will be remembered. Indeed, this region 6, together with the mountains of region 8, forms the Sino-Himalayan Node. Lastly comes the high alpine zone, with little snow and not very cold winters (−17 to +17), bearing, somewhat surprisingly, an enormous wealth of species, many endemic, such as *Phlomis rotata* and *Saussurea gossipiphora*.

(7) *East Tibetan Grassland Region.*

This includes Chhinghai south of the upper waters of the Yellow River, and all Sikang except the Batang Tang-la enclave, pointing down to Chung-tien in the sharp bend of the Yangtze. Natural meadows, scrub, coniferous forests in the valleys, tree limit formed by

---

[a] Further descriptions and many photographs of the floristic landscape in Handel-Mazzetti (4, 5, 6).

the red larch *Larix potanini*. Endemism rich, including species such as *Rheum alexandrae*, *Arenaria kansuensis*, *Meconopsis quintuplinervia* and *Hypericum przewalskii*.

(8) *Upper Burma West Yunnan Monsoon Region*.[a]

This forms the westward-looking amphitheatre of the mountains of the Sino-Himalayan Node, and overlaps with Upper Assam, Manipur, Kohima, etc. It consists mostly of warm temperate rain woodland (precipitation 1480). Here too one must distinguish altitudinal strips. Sub-tropical up to 6600 ft with Indian orchids and wild *Thuja* forests. Warm temperate up to 8400 ft containing *Taiwania* and *Pseudotsuga* (disjunct from Formosa). Temperate mixed rain forest up to 10,200 ft, the most characteristic, with many species, *Strobilanthes* prominent, and abundant mosses and epiphytes. Cold temperate zone with *Abies* woods up to 12,500 ft, the tree line, but many bamboos ascending well beyond it; at this level the snowfall is heavy in winter. Lastly, the high alpine zone, with many plants from Sikkim which go no further east.

To these eight regions Handel-Mazzetti added in some presentations a ninth, the High Tibetan Wasteland lying eastwards of regions 3 and 7. Others also have thought in terms of rather more regions than eight. Li Shun-Chhing (2) added an East Central Hardwood Region as 4B, to comprise the forests of Chekiang and Fukien. The more recent analysis of Walker (1) gives ten (Fig. 7) but only because he includes the whole of Tibet. W 1 (if we may so abbreviate) is very like HM 1 but includes a good deal more of Shansi and runs as far west as Sian.[b] There is no difference in their regions 2[c] and 4,[d] but W 3, the loess steppe region, is much more restricted than HM 3. For tropical and sub-tropical China there is no great difference save that the northern boundary of W 5 runs clearly along the Nan Ling range and includes all the eastern coastline up to a point somewhat north of Wênchow in Chekiang. W 6, Southwest tropical Yunnan,[e] is only the southern

[a] Photographs for this region in Handel-Mazzetti (7). His book (8) covers both regions 6 and 8.

[b] The characteristic plants of this region, says Walker, are those familiar to Europeans, such as oak, birch, beech, ash, walnut, elm, willow, with many conifers. But he adds a word about the maples, significant in view of the parallelism with America (see pp. 24, 44), noting that in Manchuria and northeastern Asia there occurs 'that gorgeous phenomenon of autumn leaf coloration, so familiar in the northern United States and Canada, but nowhere else in the world found to such a degree.' Great deforestation has occurred in this region (cf. Vol. 4, pt 3, pp. 224, 239) and Lowdermilk & Wickes (3).

[c] Walker draws attention to the low mountain ranges which accompany the Yellow River to its north all round the great bend and protect it from the Gobi (cf. Vol. 1, Table 4); these are the Alashan (Ho-lan Shan) and Yin Shan (Lang Shan) ranges. High enough to intercept in summer the remnants of the monsoon winds from the southeast, they gain enough moisture to support forests of spruce, pine and poplar; this in the midst of far-flung desert both north and south of the River. Cf. Vol. 4, pt 3, pp. 232 ff. This region is the home of the famous drug-plant *Ephedra* (see p. 239 and Sect. 45). Desiccation seems to be spreading here (cf. Vol. 1, p. 184), and since poplars, elms and willows withstand it best, these are the trees most characteristic of the towns and oases of China's Northwestern lands, together with the jujube or Chinese date *Zizyphus Jujuba* (= *tsao*[1], *ta tsao*[2]), and the smaller 'sand-jujube', *Z. spinosus* (= *chi*[3], *shan tsao*[4]), (Fig. 8).

[d] Here the trees of more northerly parts are replaced by trees less familiar to Europeans, such as *Liquidambar*, *Paulownia*, *Catalpa*, *Dalbergia*, *Ailanthus*, *Ginkgo* and the bamboos. We shall speak of most of these in other connections as we go on. South of the Yangtze more southern components enter in, *Cunninghamia* the southern fir, *Cedrela sinensis* a northward-extending member of the mahogany family, *Phoebe nammu* the famous lauraceous timber-tree, *Cinnamomum camphora*, etc. *Ficus retusa*, the Chinese banyan (*jung shu*[5]), reaches up into this region from the tropical, and often used to house field-god shrines (Fig. 9).

[e] Walker points out that of all the Chinese political provinces Yunnan is the richest in plant species, having some 6500. This is not surprising in view of the diversity of climatic and edaphic conditions, the proximity of many other floras and the long uninterrupted geological history. No wonder then that Chinese botanists were attracted to the region, as we shall see from such works as the *Tien Nan Pên Tshao* (p. 300 below).

[1] 棗　　　[2] 大棗　　　[3] 棘　　　[4] 山棗　　　[5] 榕樹

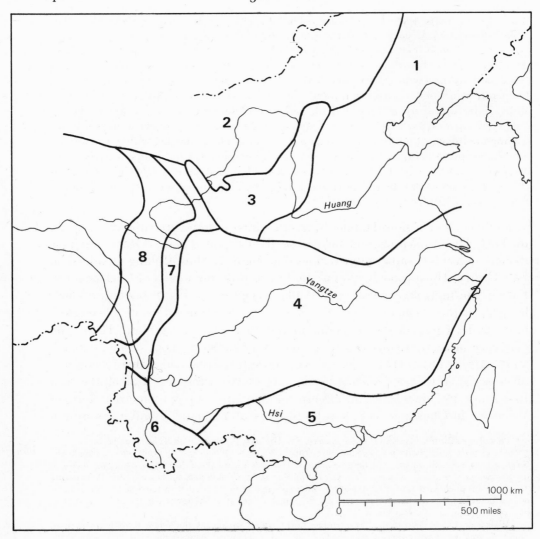

1 North-eastern China
2 Gobi desert-region
3 Loess region
4 Mid-China
5 South China
6 South-western Yunnan
7 West China highlands
8 Tibetan grasslands

Fig. 7. The principal floral regions of China; after Walker (1)

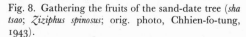

Fig. 8. Gathering the fruits of the sand-date tree (*sha tsao*; *Ziziphus spinosus*; orig. photo, Chhien-fo-tung, 1943).

Fig. 9. A Chinese banyan tree (*jung shu*; *Ficus retusa*) sheltering a field-god shrine (*thu ti miao*) on the path to Peiphei, north of Chungking (orig. photo. 1944)

half of HM 6; the northern half, which Walker calls the Highlands of West China, is represented by W 7 which runs up so far as to meet with region 2 northwest of Lanchow, in other words it includes the range of the Nan Shan or Chhi-lien Shan[a] those snow-covered peaks which march with the Old Silk Road on the south-west as the Gobi marches with it on the north-east. The East Tibetan Grasslands (W 8, HM 7) are much the same in both schemes, but Walker neglected the Burmese borderlands of HM 8. Finally W 9, the North Tibetan Plain, agrees with HM 9, and Walker adds a 10th region, that of Western or Outer Tibet, not considered by Handel-Mazzetti. It is certainly not for us to express any opinions on the relative merits of these phyto-geographical systems, which will not doubt be modified in the light of further research, but it seemed essential that the reader should be aware of them for the full understanding of all that will follow in this Section.

As more becomes known, the regions will probably become smaller and more clearly defined. On the basis of a revision of the Araliaceae family (4), Li Hui-Lin (3) could distinguish fourteen regions. This is the family which contains the famous ginseng, *Panax ginseng* (*jen shen*[1]).[b] A relative, *Acanthopanax spinosus* (*wu chia* (*phi*)[2]) was also included in the − 2nd-century *Pên Ching* (cf. p. 235).[c] This genus is endemic in East Asia, and from *Panax* nine further genera with derived names are

[a] Cf. Vol. 1, pp. 56, 59, 67, 173, 181. The divergence here between Walker and Handel-Mazzetti is perhaps not so great, for the latter (2) spoke of certain valleys in the Nan Shan, quite free of loess, forming 'exclaves' of the grasslands. It would be most reasonable to expect that the flora of the West Chinese Highlands should extend northwards in suitable altitude levels and shelter conditions along this great curving extension of the escarpment of the Tibetan massif. After experiencing the poverty of the Alashan Gobi, the great traveller von Przywalski was entranced by the floristic richness of the Nan Shan when he penetrated into them crossing the Old Silk Road at right angles near Yungtêng, (1), pp. 279, 283.

[b] Already mentioned, p. 34. Jean Breyn wrote on it as early as + 1700. Occurs in *Erh Ya* as well as *Shen Nung Pên Tshao Ching*; see R 237, B II 226, III 3, CC 594.

[c] This is what gives its name to the famous herbal wine of Canton. See R 234, B III 344, CC 588, Chhen Jung (*1*), p. 924.

[1] 人參   [2] 五加皮

known.[a] Among their species there is *Tetrapanax* (= *Fatsia*) *papyriferus (thung tho mu*[1], the pith-paper plant).[b] Besides these there are the lianous *Hedera* (ivy) species, and a number of other genera.[c] Li's recension is set out in Table 2.

Table 2. *Li Hui-Lin's floristic regions*

Li's region	correspondence W regions	correspondence HM regions	no. of spp.	no. of endemic spp.
1 South China Maritime Region	5	5	25	7
2 Tonkin Gulf Region (Kuangsi)	5	5	23	5
3 Mid Mekong Region (Yunnan)	6	6	25	5
4 Sino-Himalayan Region	6	6	56	32
5 Southwest China Plateau Region (Kweichow)	4	4	40	12
6 Upper Yangtze Region (Szechuan, Gorges)	4	4	18	3
7 Middle Lake Region	4	4	7	0
8 East China Maritime Region (Chekiang)	5	4	12	0
9 North China Plain Region	1	4	9	0
10 North China Loess Highland Region	3	3	10	1
11 Northeast Sino-Korean Region	1	1	6	1
12 Mongolian Desert-Grassland Region	2	1	6	1
13 Sinkiang Basin Region	2	2	0?	0?
14 Tibetan Highland Region	8	7	0?	0?

From this it can be seen that Li felt the necessity of splitting up several of the Walker and Handel-Mazzetti regions. He divided (cf. Fig. 10) the tropical zone into two (his 1 and 2) and made a great segmentation of the vast central China zone (HM 4) into five pieces (his 5, 6, 7, 8 and 9). The other areas changed less, Li's 11 corresponds to W 1 and HM 1, his 10, the loess region, is more or less identical with W 3 though much less extensive than HM 3, his 12 and 13 correspond to W 2, HM 2, while his southwestern regions 3 and 4 are much the same as W 6, HM 6. Li evidently saw no reason, however, for the long mountain-slope strip intervening in Walker and Handel-Mazzetti between the Tibetan grassland plateaus and the main central area of China, including its contents no doubt in his regions 6 and 10. As for the Araliaceae, his analysis clearly showed that they

[a] *Pentapanax, Heteropanax, Kalopanax, Nothopanax, Merrilliopanax, Macropanax, Dendropanax, Diplopanax, Tetrapanax.* Fifteen at least of the species of these genera have Chinese names (see Chhen Jung (*1*), pp. 922 ff.), though not all are old ones.

[b] Sometimes called the rice-paper plant; see R 238, B III 184, CC 597. It was much discussed in Li Kao's[2] +13th-century *Yung Yao Fa Hsiang*[3] (cf. p. 287).

[c] *Aralia* (hence the family name), *Brassaiopsis, Schefflera, Trevesia, Gilibertia,* and *Tupidanthus.* Seventeen of the species of these genera are named in Chinese (Chhen Jung, *loc. cit.*), some no doubt only in modern times.

[1] 通脫木   [2] 李杲   [3] 用藥法象

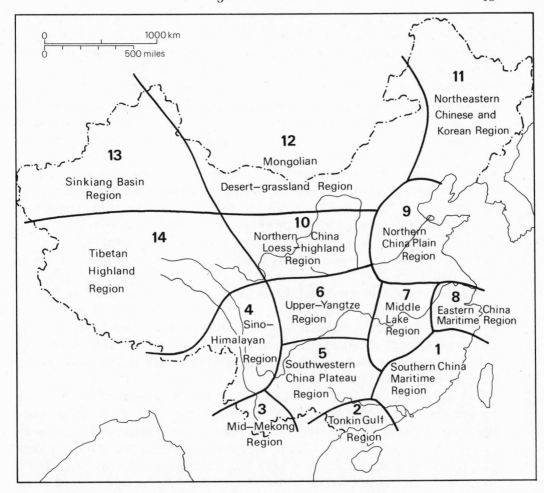

Fig. 10. Phyto-geographic regions of China; Li Hui-Lin (3).

are primarily plants of the south,[a] with endemism reaching its height in the Sino-Himalayan mountain region, though still very high in Fukien and Kuangtung, and also in the upland plateau of Kweichow. It also to some extent showed that the larger the number of species in any given region, the larger number of endemic ones there are likely to be. This subject may be closed with the remark that a nearly related genus *Oplopanax* (= *Echinopanax*) occurs in America, striking a note thus that we have heard before, and must now sound clearly and definitively.

A learned study somewhat similar to that of Li Hui-Lin was made by Stearn (6) on *Epimedium* and *Vancouveria*. The name of the former derives from the *epimēdion* (ἐπιμήδιον) of Dioscorides and Pliny, but the barrenwort *E. alpinum*,

---

[a] Yet *Panax ginseng*, the most famous among them, grows endemically in the northeast Sino-Korean region.

which lives in Yugoslavia and on the south coasts of the Black Sea, was probably not the same as the herb so named by the ancient botanists. *Epimedium* species are perennial woodland herbs of the Berberidaceae, spreading vegetatively with interlacing rhizomes, and very various of flower colour. They are particularly abundant in China and Japan, where two species became officinal, *E. sagittatum* (= *macranthum*) under the name of *yin yang huo*[1] (quite the opposite of barrenwort)[a] from the time of the *Pên Ching* onwards, and *E. grandiflorum* under such names as *chhien liang chin*[2] and *fang chang tshao*[3]. The records reveal that *Epimedium* grows in four of the Handel-Mazzetti provinces, the Northeast Sino-Korean woodlands (region HM 1), the North China loess steppe region (HM 3), the middle Sino-Japanese laurel province wherever warm temperate forests occur (HM 4), and in the Yunnan and West Szechuan temperate and warm temperate high mountain forests (HM 6). It does not penetrate either the desert Gobi areas, nor the grass-lands of Eastern Tibet, nor any of the semi-tropical and tropical regions of the south, whatever the altitude. Thus it is essentially a temperate genus, avoiding the cold and dry as well as the low, hot and moist, its preference being for the dappled shade of light woodland. Counting the few species of Southern Europe and Western Asia *Epimedium* has 23 in the Old World as a whole, but there is a closely related genus, *Vancouveria*, which has 3 species living abundantly all along the west coast of North America, where they often grow in the shade of sequoias. Why should the representation in Europe be so weak, when Asia and America alike have so many of these plants?

The answer lies of course in the incidence of Pleistocene glaciation.[b] After the pioneer observations of Willdenow (cf. p. 24 above) the strange similarity between the floras of North America and Eastern Asia greatly impressed the American botanist Asa Gray, who from 1846 to 1878 discussed it in a long series of papers (1–3).[c] He noticed that many genera such as *Liquidambar*,[d] *Sassafras*, *Aralia*, *Magnolia*, *Liriodendron*,[e] *Taxodium*,[f] *Sequoia*[g] and others, represented in America by various species but quite absent in Europe, had had in the Tertiary period an extensive distribution, and that many of these genera were also found today in Northeastern Asia. He therefore suggested that the great glaciations of rather less

---

[a] The name implies an aphrodisiac for ungulates, but the folk-name in either Europe or China might be ironical. The plant is now considered in Chinese medicine to have a valuable anti-hypertension principle (cf. Section 45) and on the last occasion when my friend Dr Chi Chhao-Ting (the calligrapher of our title-page) visited me in Cambridge, he told me that he was taking it. See R 521, CC 1374, 1375, B III 17.

[b] An excellent discussion will be found in Good (1), pp. 323 ff. See also the important papers of Li Hui-Lin (9, 10).

[c] See the account in Wulff (1), p. 21.

[d] *L. Styraciflua* in America, *formosana* in China.

[e] *L. Tulipifera* in America, *chinense* in China.

[f] *T. distichum* and others in America, the closely related genus *Glyptostrobus pendulus* and others in China. Cf. Walker (1), p. 345.

[g] Long afterwards the discovery of *Metasequoia glyptostroboides* in China brought a brilliant confirmation of Gray's hypothesis. See on this Miki Shigeru (1).

[1] 淫羊藿   [2] 千兩金   [3] 放杖草

than a million years ago had wiped out all such species in Europe, while permitting them to survive in America and China. This view is today a botanical commonplace. As Good puts it:[a]

The great ice-cap of the Pleistocene was not symmetrical about the present North Pole but had its centre in what is now the southern part of Greenland. As a result the ice reached particularly low latitudes in eastern North America and in Europe, but covered only a small part of Asia, in fact it may have made itself felt there little more than does the smaller ice-cap of today... There is good reason for believing that prior to the Pleistocene a single great flora, characterised by the prevalence of woody types, was found throughout the northern temperate regions or at least at the lower latitudes... It is therefore justifiable to suggest that the flora of Eastern Asia was comparatively little affected by the Pleistocene Ice Ages, and hence that the present Sino-Japanese flora is in fact a relatively little-changed descendant of it, giving a picture of the kind of vegetation which, before the glaciation, encircled the whole northern hemisphere.

Abundant proof of this has been provided by palaeo-botanists such as the Reids.[b] By the study of fossil seeds they found that the 'Pliocene epoch had witnessed the existence and extinction in West Europe of a flora closely allied to the living floras of the Far East of Asia and of North America'. Even *Incarvillea* was in Europe during the Oligocene of the Tertiary era. A forest-belt extended over all Europe and China, then when the Quaternary (Pleistocene) ice-cap covered all Scandinavia, the British Isles and North Germany, most of the Tertiary vegetation perished. The Pyrenees barred migration into Spain but the Balkans became an area of refuge. The greatest area of refuge, however, was West China with its north-south mountain ranges and feebly glaciated slopes, which plants could climb up and down as the ice-ages fluctuated.[c] Hence *Epimedium* lives on as a relict north of the Mediterranean.[d] Hence too the significance of all those genera which the phyto-geographers call 'north temperate discontinuous'. The *Magnolia* species are a typical case (Fig. 11), and we have just named a number of others. *Liquidambar formosana* (= *fêng*[3]) of the Hamamelidaceae came into prominence in Chinese botany during the Thang; it is allied to *L. altingiana* (= *Altingia excelsa*) which in the East Indies gives the storax (*rasa-māla*, 'rose malloes') of commerce.[e] Again, *Menispermum* has but two species, one in America and one in

---

[a] (1), p. 194. The paragraph may be illustrated by his figs. 73, 78, 79, 80. The four successive waves of glaciation are known by the names of Günz, Mindel, Riss and Würm successively; there were interglacial periods between them, and the second was the most intense.

[b] Reid (1); Reid & Reid (1).

[c] On the ice-age in West China see Richardson (3).

[d] Another example, *Meconopsis* among the poppies has 45 Sino-Himalayan species and others in America, but only one, relict, in Europe. A quite similar case is afforded by the monotypic family Clethraceae, segregated from Ericaceae and Theaceae, and recently monographed by Hu Hsiu-Ying (7). Species were first found in America, but there are many Chinese ones, e.g. *Clethra barbinervis* = *canescens* = *shan liu*[1] ≃ *ling fa*[2] (Jap.); cf. Chhen Jung (1), p. 941. One species is native to Madeira (in what Good calls Macaronesia), which might be its refuge, though Hu thinks it may have been adventive from the East Indies in historical time because of the spice trade.

[e] Vol. 1, p. 203. R 463, CC 1182. On *L. altingiana* see R 462 and Burkill (1), vol. 1, pp. 116 ff.

[1] 山柳   [2] 令法   [3] 楓

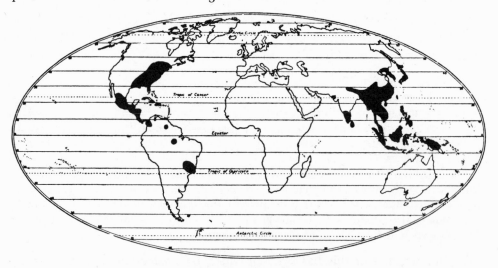

Fig. 11. Map showing the world distribution of the Magnolieae; Good (1), fig. 23.

Asia, *dauricum* (= *pien fu ko*[1], the bat creeper, and officinal as *han fang chi*[2]).[a] Lastly, *Liriodendron* the tulip tree, has one species in America and one in China, the latter *chinense* (= *o chang chhiu*[3], later *hua pai ho mu*[4]).[b]

If the Chinese in ancient and medieval times knew nothing of the traces of those prehistoric glaciations which other parts of the world had undergone, their writings show abundant evidence of careful observation of the distribution of plants, and many thoughts about it. Thus they made clear distinctions between habitats, and defined the types of plants associated with them, taking account of climate, altitude, aridity—and, as no other ancient people did, the range of edaphic factors concerned. Technical terms for different kinds of soil arose at an early time, and these were correlated, as we shall see, with wild vegetation as well as with the appropriate plants grown as crops. So much was this the case that pedologists generally recognise China to have been the first home of their science. Of course, no ancient people were in a position to make inter-continental comparisons between floristic regions, but in so far as the ancient Chinese writers specified the plants they were talking about at what corresponded to generic and

[a] There is some liability to confusion among these wild vines. On this identification see Anon. (*109*), vol. 1, p. 782, fig. 1564; CC 1366; R 515; Chhen Jung (*1*), p. 281). *Fang chi* was adopted as the term for the Menispermaceae, and the eponymous species is *Sinomenium acutum* (Anon. (*109*), vol. 1, p. 782, fig. 1563; CC 1367; Chhen Jung (*1*), p. 279). But the species more important pharmacologically is *fên fang chi*[5] or *thu fang chi*[6], i.e. *Stephania tetrandra* (Anon. (109), vol. 1, p. 784, fig. 1568; (*110*), p. 149, fig. 84). The alkaloid tetrandrine is a strong skeletal muscle relaxant, blocking the neuro-muscular junctions.

[b] CC 1343; Chhen Jung (*1*), p. 300. Magnoliaceae. Its showy flowers are figured by Good (1) in pl. 18, opp. p. 288, where the shape of the leaves explains at once why it was called in Chinese the goose web-foot *chhiu*. He gives a map of its distribution in fig. 74.

[1] 蝙蝠葛   [2] 漢防己   [3] 鵝掌楸   [4] 華百合木   [5] 粉防己
[6] 土防己

even specific level, and this they often did, they were taking at least the first steps towards phyto-geography, as it presented itself within their own oikoumene. This sub-section was intended as a setting of the stage for Chinese botany, but suddenly we find that the dramatis personae are already upon it.

(2) GEO-BOTANY *in statu nascendi*

For a biologist looking into the *Khung Tzu Chia Yü*[1] (Table Talk of Confucius)[a] it is rather astonishing to find at the very outset[b] the following words:

> In the second year (of Confucius' governorship of Chung-tu[2]) Duke Ting made (him) Minister of Public Works (Ssu Khung[3]). He then (for the first time)[c] distinguished the natures of the five types of soil (*pieh wu thu chih hsing*[4]);[d] so that the (living) things (i.e. crops and plantations) were sown and grown in the most appropriate places—everything indeed was rightly situated.[e]

Confucius is not generally thought of as a soil scientist, and indeed we have already encountered texts in which he expressly disclaimed any knowledge of agriculture,[f] but this intriguing passage is part of the Han traditions about him. Good government in ancient China, the economic high command of incipient feudal-bureaucratic rule, depended (to use a characteristic Chinese sorites)[g] upon the right use of land, the right use of land involved a knowledge of the natures of crop plants and trees, and knowledge of this kind implied an understanding of the diversity of soils on which one or another would most effectively grow. Therefore it was quite natural that Wang Su,[5] about +240, should have accepted into his compilation[h] the tradition that Confucius, greatest of human sages, was the first to attain this understanding. That Khung Chung-Ni[6] did become governor of Chung-tu in −501 is historically acceptable, but his nomination as Minister of Works of the State of Lu is not so,[i] nor is it at all likely that the first beginnings of plant geography and soil science were really due to him.[j]

The *Khung Tzu Chia Yü* text may be compared with two others not far distant in date. The *Chou Li*[7] (Record of Institutions of the Chou Dynasty), compiled from

---

[a] Here we retain our translation of this title previously adopted, but it must be understood in the sense of 'about Confucius'. Sinological rigour would prefer something like 'Sayings and Discourses by and about Confucius handed down in the Confucian School', but a mild *jeu d'esprit* taken from an English literary genre need not come amiss.

[b] Ch. 1, p. 1*b*.   [c] Such is the nuance implicit in the wording.

[d] On the traditional interpretations of the phrase *wu thu* see Morohashi's dictionary, vol. 1, p. 502, no. 374.

[e] Tr. auct. adjuv. Kramers (1), pp. 202, 253.   [f] See Vol. 2, p. 9.

[g] See Vol. 4, pt 1, p. 205, and Sect. 49.

[h] The prototype of the book was first put together in the −2nd or −1st century, in part probably from earlier sources, but then extensively remodelled by Wang Su early in the +3rd century to uphold the view of Confucius as a human sage rather than a supernatural being.

[i] Cf. Dubs (9), p. 276. Kramers thinks he may have been Vice-Minister.

[j] As we shall see in what follows, some of the relevant texts may be even older than the time of Confucius.

[1] 孔子家語   [2] 中都   [3] 司空   [4] 別五土之性   [5] 王肅
[6] 孔仲尼   [7] 周禮

older materials most probably in the −2nd century, has an entry for the Ta Ssu Thu,[1] Director-General of the Masses;[a] one of whose duties it was to 'distinguish the five types of land, and the plants, animals and people that grow on them' (*pien wu ti chih wu sêng*[2]). Details need not now concern us, but one of the five is characterised by a term *fên*[3] which later became very important in soil classifications.[b] Similarly the *Po Wu Chih*[4] (Record of the Investigation of Things) written by Chang Hua[5], one of Wang Su's younger contemporaries, about +280, has a passage[c] concerning the five types of soil couched in wording reminiscent of that quoted above. Again it uses the term *fên*[3]; for the rest, it is highly schematic in accordance with the five colours—'yellow' and 'white' soils are good for cereal grains, 'black' *fên* gives good yields of wheat and millet, dark brownish 'red' soils favour beans and yams, while low-level land (*hsia chhüan*[6])[d] is right for rice. If the soil is suitable, says Chang Hua in conclusion, there will be profit a hundredfold.

(i) *Oecology and phyto-geography in the* Kuan Tzu *book*

We can be present at the birth, as it were, of oecology, phyto-geography and soil science, in East Asian culture, by examining a number of texts, but it will be convenient to begin with the *Kuan Tzu*[7] (Book of Master Kuan), one of the most interesting of all the ancient works of natural philosophy and economics which have come down to us. Though it bears the name of Kuan Chung[8], the −7th-century minister of the State of Chhi, it stems from many different dates, and not much of the contents could now safely be attributed to the time of its putative author. Nevertheless, the traditions of his country are probably to be found in it, since there is reason for thinking that much of the present text may have been compiled in the Chi-Hsia Academy of Chhi[e] in the −4th century. We have met with this book on a number of occasions already,[f] and found it worth while in one connection to give a whole chapter of the work (ch. 39, Shui Ti[9]) in translation.[g] Now again we must consider a chapter in detail, but as it bristles with technical terms an analysis and tabulation of the content will be more convenient than an integral translation. About its date there is some difference of opinion, as we shall see, but it cannot be later than the −2nd century and may be as old as the −5th; difference also there is correspondingly about the region of China to which it applies. We shall come back to this.

---

[a] Ch. 3, pp. 10*b* ff. (ch. 9), tr. Biot (1), vol. 1, pp. 192 ff. The words quoted are on p. 11*b* (tr. p. 194). On the official himself, cf. Vol. 3, p. 534.

[b] Cf. pp. 85, 90 below.   [c] Ch. 1, p. 7*b*.

[d] No colour given in this case, but it would correspond to caerulean or glaucous; cf. p. 99 below in connection with the altar of the God of the Soil. The name must refer to the nearness of the low-lying land to the water-table.

[e] See Vol. 1, pp. 95 ff.

[f] Vol. 1, p. 150; Vol. 2, pp. 36, 60, 69; Vol. 3, p. 64, 535, 674; Vol. 4, pt 1, p. 30, 240.

[g] Vol. 2, pp. 42 ff.

[1] 大司徒   [2] 辨五地之物生   [3] 墳   [4] 博物誌
[5] 張華   [6] 下泉   [7] 管子   [8] 管仲   [9] 水地

Ch. 58 of the *Kuan Tzu* is entitled Ti Yuan[1]. The meaning of this gnomic phrase is not at first sight obvious, but the word *yuan* can mean the number of similar things in a group, and as the chapter discusses the numbers of plants which grow on the land, the best interpretation would probably be 'On the Variety of what Earth Produces'. This follows Sung Hsiang-Fêng,[2] but another commentator, Yin Chih-Chang,[3] took the words to mean the natural aspects of land, whether high or low, and of water sources, whether deep or shallow.[a] In fact the chapter discusses the different types of plant-clothed terrain, listing the quality of the soil and the accessibility of water, grouping plant species in scales of aridity and altitude, and considering their edaphic relations. Like some of the chapters in the *Mo Tzu* book, the text here had become very corrupt, but it has been restored by the brilliant work of a botanist, Hsia Wei-Ying (*2*), building on that of a succession of previous literary commentators.[b]

This part of *Kuan Tzu* must be one of the oldest writings on geo-botany in any civilisation, bearing every evidence of compilation following actual surveys of territory, farmland, neighbouring wilds, hill and mountain. It divides into five distinct parts. The first classifies the land of a gently sloping plain surrounding a great river into five categories depending on the depth of the natural water-table beneath them, and describes the plants and trees typical of such oecological situations; it is thus a gradation of aridity, but at the same time specific soil types are mentioned in each case, so that the generalisation is not entirely freed from the particular. The second part gives the names of fifteen sorts of hill land classified according to altitude and depth of water-table, but says nothing of their characteristic plants. The third separates mountain country into five grades of altitude and gives a full list of trees and plants which one may expect to find at the different heights. Then in the fourth part comes a very interesting oecological gradient, plants being arranged in an order ranging from lake water to dry ground, in other words an environmental humidity scale or 'hydrological sequence'. Lastly a long section divides the soils of the Nine Provinces[c] into three classes of productivity and each of these classes into six sub-classes, all named in a peculiar way, with details of the agricultural yields to be expected and a wide selection of the trees and plants which do best on them. Let us take a closer look at some of these statements.

First comes the categorisation of potentially arable farmland (lit. irrigable land, *tu thien*[4]). Each of the five different soil-types is considered in relation to the depth of the water-table (*shui chhüan*[5]), this being measured in ells or fathoms of

---

[a] Others have made guesses quite wide of the mark. Thus Than Po-Fu *et al.* (*1*), taking a passing glance, wrote 'Land Officials'; they did not attempt a translation. Sung was writing *c.* 1800; Yin was a Thang scholar of Empress Wu's time.

[b] Valuable contributions have also been made in the papers of Yu Yü (*1, 2*).

[c] I.e. those of the classical provinces in the Yü Kung chapter of the *Shu Ching*, the pedology of which we shall shortly discuss (pp. 82 ff. below).

[1] 地員   [2] 宋翔鳳   [3] 尹知章   [4] 瀆田   [5] 水泉

7 ft (*shih*[1]), and the sequence proceeds from the 'driest' to the 'wettest' type (Fig. 13). What now follows is an analysis almost amounting to a translation.[a] Type 1 is the *hsi-thu*[2,3], not further described, but considered by Hsia to be fertile loess of silty texture;[b] it is said to lie 5 *shih* (35 ft) above the water-table. All cereal crops grow well on this land. Characteristic plants are the thatch-grass, *tu jung*[4], *Miscanthus sinensis*[c], and the *yuan lun*[5], some herbaceous species not now identifiable. Characteristic shrubs and trees are the *ching thiao*[6], *Vitex chinensis* (= *negundo* var. *incisa*)[d], related to verbena, and the small wild date-tree (better, jujube-tree) *chi*[7], *Zizyphus jujuba* var. *spinosa*. This soil (when dry) gives out a ringing sound corresponding to the musical note *chio*[e], its water is dark-coloured and the people who live in those parts are robust.[f] These last three properties are repeated, *mutatis mutandis*, for each of the five edaphic and oecological regions, but we shall disregard them here as irrelevant to our present interests.

Type 2 is the *chhih lu*[8], with ground water at 28 ft, the only one of the five for which the text itself has preserved a soil characterisation—reddish, crumbly, hard and fertile (*li, chhiang, fei*[9]). On this all cereals succeed, but it is especially good for hemp, which gives a white fibre that weaves into fine cloth. Characteristic plants are *pai mao*[10] (floss-grass, *Imperata cylindrica* = *arundinacea*)[g] and the reed *chui*[11] or *wei*[12] (*Phragmites communis*); a characteristic tree the *chhih thang*[13], wild pear, *Pyrus betulifolia*.[h] Type 3 is *huang thang*[14], about 21 ft above the water-table, identified from other texts as yellowish brittle soil both salty and alkaline, occurring on land liable to flooding. This is good only for the millets (*chi*[15], *Panicum miliaceum*,[i] and *liang*[16], *Setaria italica*,[j] both of the glutinous variety).[k] The floss-grass grows there abundantly, and three trees are typical, the *hsiang chhun*[17], the toona or Chinese acajou *Toona sinensis* (Meliaceae)[l], the *chhiu*[18], *Catalpa bungei*[m], and the wild mulberry, *sang*[19], *Morus alba* or *mongolica*.[n]

[a] The technical terms given below are mainly those identified by Hsia Wei-Ying (2) and therefore do not necessarily appear either in the original text or in the text as emended by him.

[b] Our own conclusions about the identification of these soils will be found below, see p. 102.

[c] CC 2027; Burkill (1), p. 1479. Coville, in Safford (1), pp. 325, 399, renames this *Xiphagrostis japonica*.

[d] Chhen Jung (1), p. 1090.

[e] See Vol. 4, pt 1, pp. 140, 157 ff. A typical piece of symbolic correlation (cf. Vol. 2, pp. 261 ff.).

[f] The Hippocratic theme again (cf. Vol. 2, pp. 44, 45).

[g] Cf. Burkill (1), p. 1228, and Sect. 40 below.   [h] Chhen Jung (1), p. 413.

[i] Brown corn millet, panicled; R 751, CC. 2040.

[j] Spiked millet; R 758, CC. 2056. Some texts read *shu*[20] synonymously.

[k] Here the text includes some rather obscure phrases indicating that land of this sort, though often flooded, has to be irrigated with difficulty in dry seasons or years; something may be accomplished by the aid of movable dams (cf. Vol. 4, pt 3, p. 348) but the establishment of villages with their walls is not to be recommended.

[l] Sometimes called the Chinese mahogany, giving a valuable wood. The massive hanging inflorescences, delicately perfumed, can be appreciated each summer in the Botanic Gardens at Geneva, where there is a fine specimen. Cf. Chhen Jung (1), p. 602; Burkill (1), p. 499.

[m] Chhen Jung (1), p. 1112. Bignoniaceae.

[n] Chhen Jung (1), pp. 229, 230; *mongolica* is preferred here by Wang Ta (1), p. 226.

[1] 施	[2] 悉徒	[3] 息土	[4] 杜榮	[5] 蚖蕃
[6] 荊條	[7] 棘	[8] 赤壚	[9] 歷彊肥	[10] 白茅
[11] 藋	[12] 葦	[13] 赤棠	[14] 黃唐	[15] 稷
[16] 梁	[17] 香椿	[18] 楸	[19] 桑	[20] 秫

The fourth soil type is called *chhih chih*[1], undoubtedly argillaceous in quality and said by Hsia to be rather saline; its ground water is 14 ft below the surface. This land is good for wheat and for the soya bean (*ta shu*[2]) *Glycine maxima = soja, hispida*[a]; its characteristic plants are *Phragmites* as above, and the nutgrass *Cyperus rotundus* (*fu*[3])[b]; its trees are the willows *Salix purpurea* (*hung phi liu*[4]) and *cheilophila* (*khuang liu*[5]).[c] The fifth and last type is the *hei chih*[7], a dark 'black' sticky saline clay soil, with water only 7 ft down.[d] It grows both rice and wheat well. Typical plants are the *phing*[8], a composite, either *Anaphalis margaritacea* or some species of *Artemisia*,[e] and the *thi*[9], probably the dock *Rumex crispus*.[f] The most characteristic tree is the *pai thang*[10], perhaps *Crataegus sanguinea*[g] or, as Hsia thought, a white variety of *Pyrus betulifolia*. Such is the description of the five oecological zones which the farmers of ancient China could turn into cultivated land, a description surely unique for its period in any civilisation because of its clear appreciation of the importance and the distinguishability of a variety of soils and subsoils, always in relation to the their natural plant cover. We must shortly return to it, but first it is necessary to complete the description of the contents of the *Kuan Tzu* chapter.

The second part goes on to name 15 further types of land and soil of the nature of hills, low and high, measured in each case as before by the number of fathoms (*shih*) above ground water. The sequence is continuous with the former, for the lowest of these hill lands lies at 6 *shih* (42 ft) and the highest at 20 (140 ft). The lowest bears the same name (*fên yen*[11,12]) as that of one of the five divisions of land types in the *Chou Li*,[h] indicating foothills of eroded alluvial earth; but from the 13 *shih* level upwards the word *shan*[13] is added to the names, in order to show, say the commentators, that more rocky terrain, exposed and craggy, is intended.[i] The name of the highest, however, *kao ling thu shan*[14], implies perhaps that cultivation is still possible. No plant names accompany this orographical exercise now, but most probably (judging by the form of the whole chapter) typical vegetation was defined in the original text, and this has dropped out.

The third part, however, is of great oecological interest, for we find a definition of five mountain zones of different heights each with its characteristic plants and trees. The sequence, which descends from higher to lower, is assembled in Table 3 (cf. Fig. 12). The name of the first zone (from 6000 to 9000 ft as assessed by Hsia Wei-Ying), the 'hanging springs', suggests a region of sparse vegetation

---

[a] CC 989   [b] R 724.
[c] Chhen Jung (*1*), pp. 129, 130. The text has *chhi*[6], willows in general.
[d] It will be seen that the ground water depths have varied between 7 and 35 ft. This is almost uncannily accurate, for current estimates of the variation in North China give from 9 to 30 ft (Kovda (*1*), p. 20).
[e] Cf. Lu Wên-Yu (*1*).   [f] Wang Ta (*1*), p. 226.   [g] Chhen Jung (*1*), p. 443.
[h] Ch. 3, p. 11*a*; Biot (*1*), vol. 1, pp. 192, 193.
[i] The penultimate four have riders attached in the form of rather obscure phrases conveying that because of the bedrock ground water cannot be reached at all. In one case copper ore seems to be concerned.

[1] 斥埴   [2] 大菽   [3] 蕡   [4] 紅皮柳   [5] 筐柳
[6] 杞   [7] 黑埴   [8] 苹   [9] 蓨   [10] 白棠
[11] 埴延   [12] 埴衍   [13] 山   [14] 高陵土山

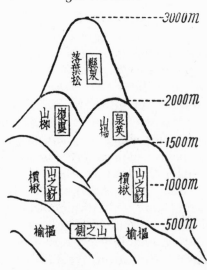

Fig. 12. Oecological altitude zones described in the *Kuan Tzu* book; from Hsia Wei-Ying (*2*), p. 36.

above the deciduous tree level and constantly penetrated with mist and rain, implying considerable run-off in streams and waterfalls, with marshy parts in the flat upper valleys. The name of the fourth distinctly refers to 'forested mountains', and that of the fifth might be translated 'foothills'. Paradoxically at first sight, the ground water depth now decreases as one ascends, being only 2 ft in the highest zone, because of the thin soil cover of its rock, but as much as 21 ft in the zones of accumulation in the valleys of the approaches[a] The identification of all the plants is not absolutely clear, but it is certain enough as a whole to make good sense of the ancient writer's classification. Taking the highest zone first, there is no doubt about the tree given, *Larix gmelini*[b], and indeed this is the region of the larch forests. *Ju mao* is less sure; it might mean two graminaceous plants[c], but the greater philological probility would point to *Rubia cordifolia*, 'Indian' madder or earthblood,[d] which does or did come into commerce from the Himalayas.[e] *Tshu* is unidentifiable; the *Erh Ya*[f] only says that shoes or grass-sandals are made from it, and no one in later times was quite sure what it was.[g] *Shan liu* in the second highest zone is presumably *Salix*, willows growing in the wet valleys[h], probably *grisea* (= *walichiana*), perhaps also *caprea* and *purpurea*.[i] *Nü wan* (in the ancient text *yü*

---

[a] See p. 55 and Fig. 13.
[b] Chhen Jung (*1*), p. 26; B II 507 has *leptolepis*, in any case the *lo yeh sung*[1].
[c] Cf. p. 50 above for *mao*.　　　　　[d] B II 22; Steward (*1*), p. 371; Stuart (*1*), p. 381; CC 228.
[e] Burkill (*1*), p. 1917; Watt (*1*), p. 926.　　[f] See p. 186.
[g] B II 86. Lu Wên-Yu (*1*) makes it a turmeric *Curcuma longa* ≃ *domestica*; cf. B II 408; Safford (*1*), p. 252; Burkill (*1*), pp. 704 ff., but this can hardly be right for such an altitude.
[h] Chhen Jung (*1*), pp. 127, 130.
[i] But the pepperbush *Clethra* is also called *shan liu*, *canescens* with the related species *delavayi, fargesii, monostachys* and *barbinervis* (Chhen Jung (*1*), p. 941; CC 538). This genus certainly belongs to mountain regions, as is clear from the monograph which Hu Hsiu-Ying (*7*) has devoted to it.

[1] 落葉松

Table 3. *The five altitude zones and their characteristic plants* (Kuan Tzu)

TECHNICAL NAME	ASSESSED HEIGHT (ft)	CHARACTERISTIC PLANTS	TREES	Depth to water, *chhih* (ft)
1 *Shan chih shang; Hsien chhüan*[1]	6000–9000	*ju mao*[6] *tshu*[7]	*man*[8]	2
2 *Shan chih shang; Fu lou*[2]	5000–6500	*nü wan*[9] *yu*[10]	*shan liu*[11]	3
3 *Shan chih shang; Chhüan ying*[3]	4500–6000	*shan chhi*[12a] *pai chhang*[13, 14]	*shan yang*[15]	5
4 *Shan chih chhai*[4]	1500–4500	*lien*[16] *chhiang mei*[17]	*chia chhiu*[18] (*chia* [19])	14
5 *Shan chih tshê*[5]	150–1500	*fu*[20] *lou hao*[21]	*chhu yü*[22] (or *shu yü*)	21

chhang[24]) is *Aster fastigiatus*[b], and *yu* may be the verbenaceous glorybower shrub *Clerodendrun foetidum* (= *bungei*)[c] which from its disagreeable smell has been called 'stinking peony' (*chhou mu tan*[25]). Then at the third level, the lowest of the high mountain zones proper but much overlapping with the second, comes the poplar *Populus tremula*[d], with the medicinal plant *tang kuei*[21] (*shan chhi*) *Angelica polymorpha*[e] and the sweet flag *Acorus Calamus* (*pai chhang*), presumably growing in marshy hanging valleys[f], as its associated herbaceous species. Now comes the thick forested zone (*chhai*), where *Catalpa bungei*[g] is the outstanding tree, accompanied by the composite *Sigesbeckia pubescens*[h] (*lien*) and some liliaceous plant (*chhiang mei*) not easy to pin down now, for *Liriope, Asparagus* or *Ophiopogon* might any one or all have been intended.[i] Finally, in the last of the zones, again clearly deciduous forest, an elm, *Hemiptelea davidi* (*chhu*) is chosen as the characteristic tree[j], while

[a] Anciently *chhin*.
[b] B III 103; Steward (*1*), p. 396.
[c] B II 85; Steward (*1*), p. 328; Burkill (*1*), pp. 581 ff.; cf. CC 367.
[d] CC 1695; Chhen Jung (*1*), p. 114 thinks var. *davidiana*.
[e] R 210; B II 5 and 89; Stuart (*1*), pp. 41, 133; CC 552; Sato (*1*), pp. 40 ff. The pharmacognosy of *Angelica* spp. seems still to be a little confused.
[f] B II 376; Steward (*1*), p. 499.
[g] CC 259; Chhen Jung (*1*), p. 1112.
[h] CC 136. One cannot mention this genus without recalling the background of its nomenclature. In +1737 Joh. Georg Sigesbeck, the director of the botanic gardens at St Petersburg, made a violent attack on the sexual system of Linnaeus, the lewdness and licentiousness of which, in his view, ruined the study of botany as an occupation for polite youth or young ladies. Linnaeus did not reply, but, says Stearn, Siegesbeck 'is remembered today only through the unpleasant small-flowered weed which Linnaeus named *Sigesbeckia*' (*3*), pp. 24 ff. Steward (*1*), p. 400, prefers *orientalis* as the specific name.
[i] B II 108; cf. CC 1864; Steward (*1*), pp. 516 ff. See the discussion on p. 256 below.
[j] Chhen Jung (*1*), p. 225.

[1] 山之上縣泉	[2] 山之上複巋	[3] 山之上泉英	[4] 山之嶽	[5] 山之側
[6] 茹茅	[7] 蓲	[8] 構	[9] 女菀	[10] 猶
[11] 山柳	[12] 山蕲	[13] 白昌	[14] 白菖	[15] 山楊
[16] 菼	[17] 薔蘼	[18] 檟楸	[19] 棍	[20] 菖
[21] 蔞蒿	[22] 樞楡	[23] 當歸	[24] 魚腸	[25] 臭牡丹

Table 4. *The Twelve Orders or Precedences*

ANCIENT NAME	PLANT	COMMON NAME	REFS.	FAMILY
1 *yeh*[1]	*Nelumbo nucifera* or *Euryale ferox*	Indian lotus foxnut, chickenhead	CC 1449 CC 1448	Nympheaceae Nympheaceae
2 *yu*[2]	*Trapa natans* or *Zizania aquatica*	water-calthrop wild water-rice	CC 2457, 8 CC 2067; B II 455	Hydrocaryaceae Gram-
3 *huan*[3]	*Scirpus lacustris*	lamp-wick rush	CC 1982; B II 455	Cyper-
4 *phu*[4]	*Typha latifolia*	cat-tail rush	CC 2113	Typh-
5 *wei*[5 a]	*Phragmites communis*	reed	B II 210, 455; Steward (1), 455;	Gram-
6 *kuan*[6]	*Metaplexis stauntoni* = *japonica*	milkweed vine	B II 93, 468; R 165; Steward (1), p. 441	Asclepiad-
7 *lou*[7]	*Artemisia vulgaris* or perhaps *campestris* = *mongolica*	wormwood, southernwood	B II 430; CC 16; Steward (1), p. 408	Comp-
8 *phing*[8]	*Kochia scoparia*	summer cypress, earthskin	CC 1516; B II 9, 36	Chenopod-
9 *hsiao*[9]	*Artemisia* sp.	wormwood, southernwood	B II 196; 435; Steward (1), p. 408	Comp-
10 *pi*[10] if *pi li*[11] if *pai chhi*[12] or *shan chhi*[13] if *shan ma*[14 d]	*Ficus pumila*	small climbing fig	CC 1599; Steward (1), p. 87	Mor-
	Angelica polymorpha	angelica	B II 5, 49, 80	Umb-
	Some urticaceous plant or *Cannabis sativa*	wild hemp	B II 168, 388	Urt- or Cann-
11 *thui*[15] or *lei*[15]	*Leonurus sibiricus*	motherwort	B II 244; CC 334 R 743; CC 2016;	Labiat-
12 *mao*[16]	*Imperata arundinacea* = *cylindrica*	floss-grass	Steward (1), p. 478	Gram-

[a] Mention may be made here of the interesting paper of Mizukami Shizuo (1) on the 'worship of reeds' in ancient China, particularly by the people of the southern and south-eastern cultural strain (cf. Vol. 1, p. 89). For the folk of the rivers and marshes, reeds supplied clothing, shelter and food. The reeds (*wei*) were used to 'envelop' (*wei*[17]) the houses, the thatched roofs of which were perhaps originally signified pictographically by *thu*[18] (alternatively, the drawing of a ploughshare, K 82a, x), a word later acquiring the sense of 'a bitter plant'. *Thu*[19], glutinous rice, is related to this, very naturally, since flooded fields would first have arisen in marshy country. Mizukami sees in *thiao*[20] and *chia*[21] ancient names of holy reeds used by shamans in high antiquity.

[b] This by no means exhausts the possibilities of what the ancient author meant here. Several trees have similar names, as we know from Chhen Jung (1). For example there is the *shan ma liu*[22], a kind of walnut, *Pterocarya paliurus* (p. 140); the *shan huang ma*[23], a kind of elm, *Trema orientalis* (p. 224); the *shan ma tzu*[24], one of the hackberry group, another ulmaceous tree, *Celtis koraiensis* (p. 219); and the *shan ma kan*[25], a euphorb, *Alchornea davidi* (p. 617). One may feel justified in dismissing the second, which seems to belong only to the Thaiwan flora, but about the others one cannot feel so sure; they may well have been unknown to the botanists of Chou, Chhin and Han times, but the memory of the folk is very tenacious, and local names surviving in particular provinces may give us the answer sometimes to an ancient problem, especially where, as here, Bretschneider himself had to give up in despair.

[1] 葉 [2] 薯 [3] 莞 [4] 蒲 [5] 葦 [6] 雚 [7] 蔞 [8] 荓
[9] 蕭 [10] 萉 [11] 薜荔 [12] 白蘄 [13] 山蘄 [14] 山蘵 [15] 蘱 [16] 茅
[17] 圍 [18] 荼 [19] 稌 [20] 蓨 [21] 葮 [22] 山蘵柳 [23] 山黃蘵 [24] 山蘵子
[25] 山蘵桿

round about grow the *fu* (*Convolvulus arvensis* or *Calystegia sepium*)[a] and various species of *Artemisia*, the adjectival *lou* in *lou hao* probably indicating *campestris* ($\simeq$ *mongolica*).[b] Such then is the delineation of the altitude zones in this ancient text. If Handel-Mazzetti and other modern botanists have studied plant life in the Sino-Himalayan Node up to 15,000 ft or even to nearly double the altitudes envisaged in the *Kuan Tzu* book (cf. p. 48 above), it must nevertheless be admitted that so far as the middle ranges within easier reach of the populated parts of China are concerned, such as the Chhin-ling Shan, the old Taoist naturalists had gone ahead of them by more than two thousand years.[c]

In the next paragraph the oecological theme continues. Suppose we consider a gently sloping piece of land extending from the bottom of a 6 ft deep lake to a relatively dry prairie or savannah a few miles away. Then it will be possible to define a whole series of favourite habitats occupied normally by particular groups of plants; in fact a 'topo-sequence' or 'catena'.[d] The *Kuan Tzu* writer specifies twelve of these habitats, which he calls the Shih-erh Chhui[1], the twelve orders or precedences.[e] It is indeed what we should think of now as an oecological gradient of humidity. He says:

As regards the Tao of plant (growth) and soil (conditions), every place has its own peculiarity where one or another crop will do well; whether lower or higher, every place has plants which are characteristic of it. For example, plant *x* prefers a place lower than plant *y*, and plant *y* prefers a place lower than plant *z* . . .[f]

So among all the sorts of plants there are twelve (oecological) areas in an order of precedence, and every one has a particular place (or area) to which it is confined (lit. to which it reverts or returns).

Again, among the Nine Provinces there are ninety different (sorts of) plants growing on their soils. Every type of soil has its regular characteristics, and every plant can be graded in an order (of luxuriance).[g]

This may be illustrated by Table 4 and Fig. 13. The plants represented by the letters in the quotation can be assembled as follows, passing from the wettest to the Yü Kung chapter of the *Shu Ching*, which we must shortly consider in detail) habitat 10) are easily identifiable. There can be no doubt that the writer meant them to be representative, and had in mind that there were many other plants which would preferentially occupy ('revert to') the several habitats. One is usually prepared to admit that the Greeks had a greater penchant for systematisation than the Chinese, as among the biological Peripatetics or the Alexandrian

---

[a] B II 442; cf. CC 397 ff.; Steward, p. 318. The species is not clear.

[b] B II 430; cf. CC 16.

[c] For a good account of the modern gradient analysis of vegetation, see R. H. Whittaker (1).

[d] Cf. p. 64 below.

[e] The usual pronunciation is *shuai*, with a meaning which would give 'the twelve down-slidings', but this is less fitting here.

[f] Actually here the writer details the first few plants of the series.

[g] Ch. 58, pp. 3*b*, 4*a*, tr. auct.

[1] 十二衰

Fig. 13. Oecological humidity zones described in the *Kuan Tzu* book; from Hsia Wei-Ying (*2*), p. 45.

mechanicians, but it would be hard to beat the laconic clarity of this so early oecology, based as it obviously was on extensive and thorough observations in the field.

It now only remains to discuss the last two sentences of the quotation just given. They refer to what takes up the remainder of the chapter (the fifth part of the whole), namely a division of the soils of the nine provinces (presumably those of the Yü Kung chapter of the *Shu Ching*, which we muct shortly consider in detail) into four grades, and each of these again into classes, making eighteen in all. The first three classes are described much more fully than the others, with a detailed account of the herbs, crop-plants, trees and fruit-trees which grow on them;[a] furthermore they are given a productivity rating of 100 %. Accordingly each of the following 15 soil classes is given a lesser rating, down to the last two, which yield only 30 %.[b] Each of these contains 'the Five Sorts of …', and while some of the soils are easily recognisable (e.g. *wu hsi*[1], no. 1; *wu jang*[2], no. 5; *wu lu*[3], no. 8; *wu sha*[4], no. 11; and *wu chih*[5], no. 15) most of them (e.g. *wu wei*[6], no. 3, or *wu hu*[7], no. 16) are much less so.[c] Since there is considerable overlap between the plant species mentioned an elaborate analysis would be necessary, and we shall omit it here.[d]

(ii) *China and the science of soils*

In all the foregoing, particular interest attaches to what the old Chinese texts say about the soil. To make the best sense of them and to extract their full signifi-

[a] Ninety herbs and trees are listed, with thirty-six cereals. A few animals are also brought in.

[b] This is an early instance of the approach to decimal fractions (cf. Vol. 3, pp. 21, 35, 64, 81 ff.; Wang Ling (*3*), etc.). The expression is *pu ju san thu i shih fen chih chhi*[8], 'it is not as good as the Three (Standard) Soils by the seventh part in ten'.

[c] Some rather precise identifications are given in the paper of Wang Ta (*1*). On the face of it the first five mean 'loessial', 'silty loessial', 'clayey', 'sandy' and another kind of 'clayey' respectively, but the last two characters mean 'position' or 'seat', and some kind of 'poor soil' not now easily identifiable. They must, however, have been technical terms.

[d] A full discussion of the soil science aspects of this text will be found in Wan Kuo-Ting (*2*).

[1] 五息   [2] 五壤   [3] 五壚   [4] 五沙   [5] 五埴
[6] 五位   [7] 五穀   [8] 不如三土以十分之七

cance, it is necessary to know something about modern soil science, pedology as it is often called, so that one can have a shot at identifying what they were talking about. This tempting procedure, however, brings one into a rather specialised branch of scientific knowledge with an intriguing, if unfamiliar, terminology.[a] It needs an introduction, bearing specially in mind the Chinese scene, but more than a few paragraphs would exceed the space at our disposal, and the reader must have recourse to the standard works on the subject.[b]

Pedology (from πέδον, *pedon*, the ground) is really the study of all aspects of the soil from the standpoint and by the methods of pure science, but the term is sometimes used to refer only to the morphology, classification and genesis of soils. It is this aspect with which we are here primarily concerned. The origins of applied pedology and empirical classification must be sought, as we shall see,[c] in China of the −1st millennium; but unlike rocks, plants and animals, soils were hardly studied as objects of intrinsic interest in modern science until the middle of the nineteenth century with the pioneer work of Dokuchaiev and his school in Russia.[d]

These investigators emphasised the importance of studying the whole soil 'profile', or section seen in a pit, which generally consists of a succession of layers or 'soil horizons' differing in physical, chemical and biological characteristics.[e] Some of these, such as colour, texture,[f] calcium carbonate content or root development, are easily observed (but not always easily described or measured) in the field; others, such as the nature of the exchangeable cations[g], clay minerals[h] or soil micro-organisms, are revealed only, and often imperfectly, by laboratory analysis. The material, generally rock or rock-debris, but occasionally organic accu-

---

[a] Since the Russians were great pioneers in soil science, many of the technical terms have come into use internationally straight from their language, as we shall see. The Chinese language, too, has provided more than one.

[b] The most obvious in English are the books of Robinson (1) and Joffe (1). More modern approaches will be found in those of Duchaufour (1) and Gerasimov & Glazovskaia (1). More biological are Eyre (1); Daubenmire (1). The work of Good (1) already mentioned, also considers soils.

[c] See pp. 87, 98 below.

[d] See the review by Margulis (1).

[e] They are often given letter symbols, such as A, B, or C, with suffixes to indicate the dominant process which appears to have take place. On this see especially Duchaufour (1).

[f] Texture is dependent on, though not necessarily defined by, the relative abundance of particle sizes. In international usage, gravel or stones are particles larger than 2.0 mm. in diameter, sand ranges from 2.0 to 0.02 mm., silt from 0.02 (20μ) to 2μ, and clay from 2μ downwards. Pedologists and agriculturists often refer to soils containing a predominance of particles in the appropriate ranges as 'sands', 'silts' or 'clays'. 'Loams' are those soils in which there is a fair amount of admixture of particle sizes. The use of these terms is sometimes confusing to the layman. Thus 'silt' may mean a range of particle sizes, a soil of a particular composition, or alternatively a riverine deposit or alluvium.

[g] The colloidal clay and humus fractions of soils, as defined below, are predominantly negatively charged because of their crystalline or molecular structures respectively. These charges are neutralised, in varying proportions, according to soil type, by calcium, magnesium, sodium, potassium, hydrogen, aluminium, and smaller amounts of many other, ions, which are sorbed on the surface of the colloidal particles but which can exchange with each other and with those in solution. Such ions are termed exchangeable. Those soils, especially clays and highly organic soils, which can hold large numbers of exchangeable cations are said to have high cation exchange capacities.

[h] Aluminosilicate minerals especially characteristic of the clay fraction of a soil, i.e. the particles effectively less than 2μ in diameter.

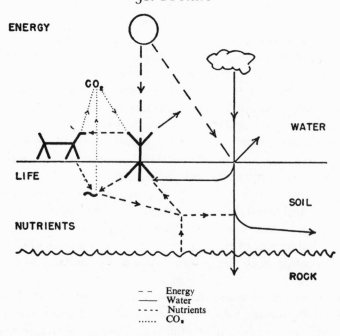

ENERGY

CO₂

WATER

LIFE

SOIL

NUTRIENTS

ROCK

- - Energy
—— Water
- - - - Nutrients
...... CO₂

Fig. 14. The soil and its environment (omitting the nitrogen and oxygen cycles); from Perrin (1).

mulation as in peat soils, which has given rise to the soil by the action of the soil-forming processes is termed 'parent material'; this may approximate closely to the relatively unweathered mass lying below, or it may have been wholly altered to its full depth and is thus no longer available for inspection in its original form.

The agriculturalist regards the biologically most active surface layer (sometimes consisting of more than one former horizon mixed by cultivation) as topsoil, and those below, relatively 'raw' or biologically inert, but which strongly influence root penetration, drainage, etc., as subsoil. The engineer, on the other hand, considers soil to be the whole local mantle of unconsolidated deposits including the topsoil and the subsoil often to a considerable depth.[a]

The soil profile at any site results from the local effects on (a) the parent material of (b) the climate, or succession of climates,[b] which has acted over a period of time depending on the age and history of the land surface; of (c) the topography and (d) drainage conditions at the site, and of (e) the activity of the associated living organisms including man. The lettered phrases specify what are

---

[a] Cf. Section 28 on hydraulic engineering, in Vol. 4, pt 3.

[b] The early Russian pedologists emphasised the significance of the climatic factor in the term 'zonal', used for soils which had essentially formed under the influence of, and had come to equilibrium with, the present-day zonal (i.e. more or less latitudinal) climate. Cf. here the Hellenistic *climata* in Vol. 3, pp. 527 ff. 'Intra-zonal' soils were those occurring in a particular zone but owing their formation to the dominance of some local factor such as a calcareous parent material, the presence of soluble salts, or a high ground-water level. 'Azonal' soils were those still young, consisting of little-altered parent material. In recent years there has been much more appreciation of the role of past climates.

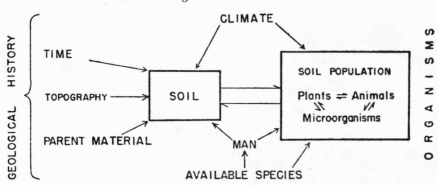

Fig. 15. The soil-forming factors; from Perrin (1).

known as the soil-forming factors. Their interplay is shown in simple terms in Figs. 14 and 15. The contribution of each factor will emerge in the ensuing short discussion of soil-forming processes, but in view of their overall importance it will be useful first to refer briefly to the present-day topography and climate of China.

The main features of the former have already been considered in various contexts,[a] but they are shown in a new way in Fig. 16. Two points are especially worthy of attention. First, the relatively low proportion of land lying at less than 1500 ft (500 m.) above sea-level, notably the Manchurian and East China plains, and the coastal strip extending from Nanking to the borders of Vietnam. Secondly, the remoteness from the ocean of the high mountains and tectonic basins of West and North-west China.

On the meteorological side also something has already been said in an earlier volume,[b] but some refreshment of the memory will not come amiss when thinking about soils. In East and Central China we can recognise four main temperature belts: (a) a cold belt in North and North-east China with a mean annual temperature range from 0–6°C.; (b) a temperate belt from approximately the latitude of Peking to that of Nanking, in which the mean annual temperature ranges from about 7° in the north to 16°C. in the south; (c) a sub-tropical belt southwards from the Yangtze with mean annual temperatures between 15° and 20°C.; and lastly (d) a tropical belt in the extreme south (including Hainan and Thaiwan islands) with mean annual temperatures between 21° and 25°C. A special feature of the Chinese climate is the frequent intrusion of cold air masses from the north (often bearing much fine dust), so that at a given latitude it is frequently cooler than in other countries.

Next we must consider the rainfall and evapo-transpiration.[c] In passing from north to south in Eastern China, the mean annual rainfall increases in the same sense as the temperature from about 15–26 in. (380–660 mm.) in the north up to

---

[a] Cf. Section 4 in Vol. 1, Sect. 22 in Vol. 3, and Sect. 28 in Vol. 4, pt 3.
[b] Especially Section 21 in Vol. 3.
[c] I.e. the combined loss of water from soils by evaporation and plant transpiration.

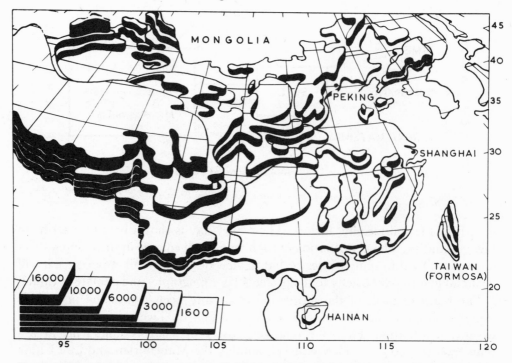

Fig. 16. Diagrammatic map of China (after Lo Kai-Fu) to show the continental character of the amphitheatres facing the Pacific Ocean to the east, and backed by the Tibetan massif to the west. The issue routes of the Yellow River and the Yangtze can be seen, with the basin of Szechuan west of the gorges. Latitudes and longitudes give the scale, with altitudes in feet.

20–80 in. (500–2000 mm.) in the subtropical belt between Fuchow and Canton, falling again, however, in the extreme south; cf. Fig. 861 in Vol. 4, pt 3. This parallel increase of moisture and temperature has important consequences in weathering and soil-formation.[a] At any given latitude there is also a general decrease in precipitation from east to west as one moves away from oceanic influences, falling to extremely low values in the deserts of the far west. An important feature of the rainfall generally is its monsoonal and irregular character, a basic cause of erosion and flooding throughout China's recorded history.[b]

In considering patterns of vegetation and soils,[c] mean annual rainfall is not the best index of moistness of climate since it is necessary to allow for evapotranspiration, which depends on thermal energy from the sun, wind-speed and relative humidity. It has been estimated in different ways.[d] Figs. 17a, b, based on the work of Chu Khang-Khung & Yang Jen-Chhang,[e] shows the broad distribution of zones of differing effective moistness. They should be compared with Figs. 6, 7,

[a] Cf. below, p. 64.
[b] See further on this Vol. 4, pt 3, pp. 217 ff.
[c] For studies of plant-soil associations in China, see e.g. Hou Hsüeh-Yü, Chhen Chhang-Tu & Wang Hsien-Phu (1, 1); Gordeev & Jernakov (1)
[d] Cf. Thornthwaite (1, 2); Penman (1); Chang Jen-Hu (1)
[e] In Kovda (1), p. 45.

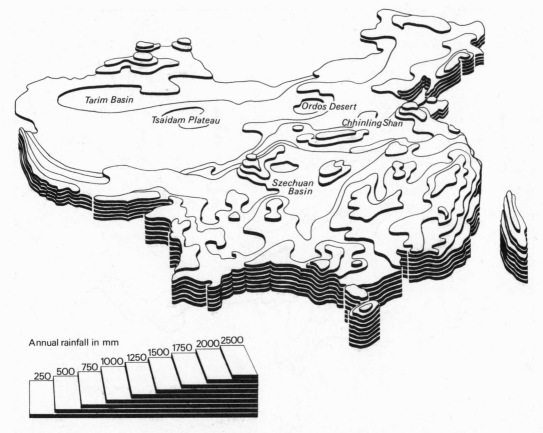

Fig. 17a. Rainfall distribution and zones of differing effective moistures in China (after Chu Khang-Khung &
Yang Jen-Chhang (1), in Kovda (1), p. 45).

and 10. Although much modified by the local interplay of other factors, the
broadly zonal pattern of soils is easily seen.

We are now in a position to examine the chief processes of soil-formation with
particular reference to China. In a much simplified way these may be considered
under five (slightly overlapping) headings: (i) erosion and deposition; (ii) movement of water and gases; (iii) weathering, translocation and precipitation; (iv)
structure development; and (v) biological processes.

First come erosion and deposition. At any point in the landscape, and depending on the nature of the surface soil, steepness of slope, intensity and duration of
rainfall, wind-speed, and vegetative cover, there are two opposing processes—loss
of material by the action of wind, water or gravity, and gain by deposition of, for
example, aeolian dusts[a] or blown sand, alluvium, 'colluvium' (hill-wash), or

[a] At the beginning of the present work a good deal was said about the 'loess highlands' of Shansi, Shensi and
Kansu (Vol. 1, pp. 68 ff.). The classical view was that the bedrocks of these provinces were blanketed by a thick
layer of calcareous wind-blown dust from Central Asia and the Gobi Desert. In recent years the theory of wind-
carriage and deposition has come in for much criticism; cf. Popov (1) a collective work; Berg (1), and the
discussion in Kovda (1), pp. 429, 433, 445. The latter regards the aeolian process as secondary to hydro-
accumulation. Consideration of the available evidence suggests to us that both processes occurred.

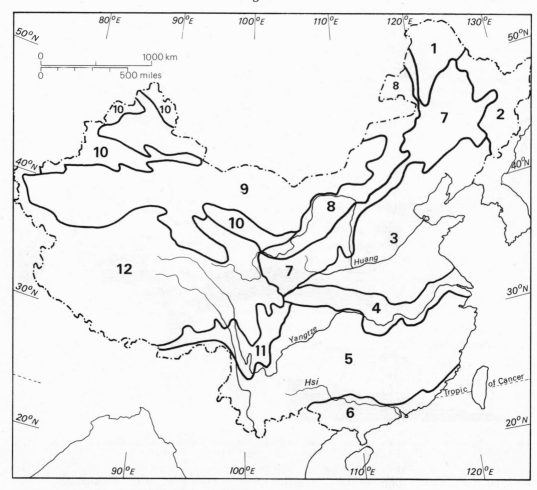

Fig. 17*b*. Schematic map of China's phyto-geographical areas, after Kovda (1), p. 57.
  1 Coniferous montane forests of the north; podzol and brown podzol soils.
  2 Coniferous and deciduous forests of the north-east; brown podzol and brown forest soils.
  3 Deciduous forests; brown and cinnamon forest soils.
  4 Mixed forests; yellow podzol and brown forest soils.
  5 Evergreen broad-leaved forests; yellow podzolic and purple soils and rendzinas. On the Yunnan uplands, red
    podzolic soils, red soils and terra rossa soils.
  6 Tropical monsoon rain forests; lateritic red and yellow soils.
  7 Forest-steppes, steppes and prairies; chernozem-meadow soils and loess carbonate soils.
  8 Steppes; various varieties of chestnut and salted soils.
  9 Semi-deserts and deserts; sierozems and desert soils, often salted.
 10 Montane areas of the north-west.
 11 Montane areas of eastern Tibet.
 12 Tibetan plateau.

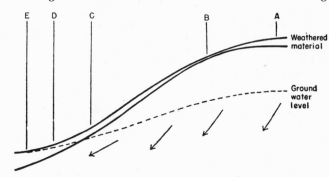

A erosion not very marked; water
  percolates deeply.
B maximum erosion.
C soils derived from colluvium
  (hill-wash).
D colluvium soils poorly drained.
E wet soils, anaerobic and peaty.

Fig. 18. Effects of topography on soil formation; from Perrin (1).

volcanic ash. When erosion dominates, soil profiles are feebly developed; with long-continued deposition, either a very deep almost continuous profile, or else a series of buried soils,[a] may be formed. These processes can occur on a regional scale; the thin rocky montane soils and the catastrophically eroded tracts of China afford many examples of the first, while the second is well displayed in the loess region (cf. Fig. 6 in Vol. 1) and the great alluvial plains of the east. On any individual hillside, however, even when the slope is very gentle, no soil is static, and a pattern of denudation and deposition is found. Fig. 18 shows an idealised version of this, but on an actual hillside it may be much complicated by topographic variation in the nature of the parent material. In areas of great amplitude, climate may also vary considerably in different parts of the landscape, thereby affecting both vegetation and soil-pattern. Furthermore, regional uplift or depression may take place during the soil-forming period—the continued rise of the loess area throughout the quaternary would be a case in point.

According to the relative magnitudes of erosion and deposition, the parent material of a soil may be either (a) the underlying rock, continually exposed by erosion, (b) a mantle, very variable in depth, of rock-debris weathered essentially *in situ*, (c) material, usually already weathered to some extent, which has been transported from elsewhere, and (d), some combination of (b) and (c). An outstanding example of (c) would be the process just mentioned, the addition of wind-borne calcareous dust to the surface of provinces adjoining desert areas in North-west China. This was considered by Thorp (1) to be of great local importance in soil-formation.

Next a few words on (ii) the movement of water and gases. The effective moistness of the climate acting on the soil and the associated vegetation depends on the annual rainfall, its intensity distribution, and the surplus water, which infiltrates into and percolates through the soil[b] after run-off and evapo-transpiration have

---

[a] Soils overlain by strata of later accumulation, 'fossil soils' or 'palaeosols'; cf. Gerasimov & Glazovskaia (1), pp. 168, 197 ff.

[b] The differential washing out of soil constituents (especially anions and cations) by water is called leaching. The term 'lessivage' is used by French pedologists, e.g. Duchaufour (1), specifically for the translocation of fine clay particles in a downward direction. Together with 'sols lessivés', soils in which this appears to have been an important process, the term has achieved some currency in English pedological literature.

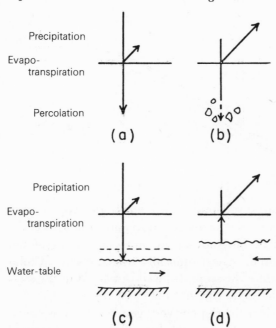

(a) moist climate; precipitation exceeding evapo-transpiration, soluble ions removed in drainage-water.
(b) drier climate; sparingly soluble products carried down and repreapated.
(c) ground-water accumulating and producing anaerobic conditions.
(d) high water-table; concentrating soluble products.

Fig. 19. Patterns of movement of soil water; from Perrin (1).

taken place. The former depends on the nature of the surface soil and the topography, and the overall effects of the latter in China have just now been seen (Fig. 17b). One may visualise various moisture régimes in a simplified way by the aid of Fig. 19. Comparison with Fig. 13 shows that different régimes typically occur in different parts of the landscape in a hydrological sequence. This pattern is superimposed on that of erosion and deposition mentioned above, and the resulting topographically controlled sequence of soils is known as a 'catena'. Since gases do not diffuse readily through pores which are filled with water, the soil atmosphere differs from that outside by an amount which depends on the moisture régime. In particular, poor drainage conditions produce ratios of carbon dioxide to oxygen many times that in the outside air.

We can now approach the more complicated subject (iii) of weathering, translocation and precipitation. Rocks 'weather' primarily by processes both physical and chemical, but biological processes also take part. However, disintegration by root pressure, surface plucking by lichens, micro-biological oxidations and the like, may be regarded as special aspects of physical or chemical weathering. Purely physical processes include disintegration by temperature changes or by the freezing of water in pores, and attrition by transport of rock-debris under the action of wind, water and glacier ice. No new minerals are formed but specific surfaces are increased, facilitating the more important chemical weathering.[a] The latter takes place in the presence of water, and includes solution, hydrolysis and

[a] Exfoliation is now considered a mixed process.

oxidation. In very cold mountain or dry desert conditions chemical weathering is feeble, and then there is little profile development.

The most obvious solution process is that of calcite[a] or dolomite[b] (in limestones and marble) by carbonic and other acids. The formation of 'karst' topography, so characteristic of Kuangsi (cf. p. 62 and Fig. 4 in Vol. 1), is an extreme case, but in all calcareous soils[c] there are important reversible equilibria between solid calcium carbonate, calcium bicarbonate in solution, exchangeable calcium ions and carbon dioxide. According to the moisture régime, which may vary seasonally, solution or deposition of calcite dominates. Under leaching conditions (*a* in Fig. 19) calcite, if present initially, and exchangeable calcium and other ions are progressively lost. On a limestone, chalk, or other calcareous parent material, the partly decalcified residue forms first a shallow calcareous slightly alkaline soil called 'rendzina'[d] (examples of this are found in Kuangsi, Kweichow and Yunnan)[e], which in time develops into a deeper acid soil from which all calcite has been removed (e.g. the 'terra rossa' or red limestone soils found locally in Shantung and Yunnan).[f] On the other hand, partial leaching (*b* in Fig. 19) produces horizons of calcite accumulation, or actual isolated concretions; these are found in some 'chernozems'[g] (Northern Manchuria, and Inner and Outer Mongolia), in the 'chestnut' soils,[h] usually loessial (as in Shensi, Shansi and Kansu)[i], and in the 'sierozems'[j] and desert soils of China's drier areas (notably the Gobi, North Shensi and the Kansu panhandle). Such soils are often referred to collectively as 'pedocals'.[k] The great tracts of alluvium of the East China plain

---

[a] One of the main forms of crystalline calcium carbonate.

[b] The double carbonate of calcium and magnesium $(Ca, Mg) CO_3$.

[c] I.e. soils containing free calcium carbonate. In the classification of Kovda (see below, p. 79), such soils are referred to as 'carbonate soils'.

[d] This esoteric term, of Polish folk origin, denotes the type of soil which characteristically appears on a bedrock of chalk.

[e] Cf. Thorp (1), pp. 213, 215.

[f] The term is derived from Italian. Such soils, formed on limestone, are widespread in the Mediterranean area. They often occur in 'dolina' (closed hollow) and 'polje' (blind valley) formations, brought about by subsidence after total solution of surface limestone; here the terms are most appropriately Hrvatski since remarkable examples can be seen in Croatia (Yugoslavia).

[g] A Russian term meaning 'black earth'. The chernozems are characteristic of grassy steppe lands. The Graminaceae utilise much more calcium relatively than trees, so by annual decay the soil remains comparatively rich in lime. In China they are generally less calcareous, and frequently under some ground-water influence, having affinity with meadow soils.

[h] The 'kashtanovy' of the Russians, again named from the colour. Cf. Kovda (1), pp. 446 ff., 462 ff.

[i] In these regions the bizarre shapes of the concretions have led to the term 'loess puppets'.

[j] A Russian term meaning 'grey soil'. Cf. Kovda (1), pp. 481 ff.; Daubenmire (1), p. 67.

[k] Marbut in the twenties (cf. Robinson (1), p. 354; Eyre (1), p. 41) distinguished the incompletely leached 'pedocals' containing calcium carbonate from the fully leached 'pedalfers' containing no carbonate but with free iron and aluminium oxides. Pedalfer is a less satisfactory term than pedocal, and is now little used.

Very broadly speaking, his two categories correspond with North and South China respectively. With certain exceptions, everywhere south of the Chhin-ling Shan, the Fu-niu Shan and the Huai River, acid and lime-depleted soils are found. This can be well seen in the extremely simplified map given by Tregear (1), fig. 16, and in more detail by the aid of the legend to Fig. 21.

As might be expected there are borderline cases. The chernozems, and the cinnamon soils (cf. p. 68), are examples where within a particular soil group there are found some well leached and some less well leached varieties. Thus we have both 'leached' and 'carbonate' cinnamon soils.

C. F. Marbut (1863 to 1935), founder of the American soil survey, died on a scientific mission in Manchuria.

(Shantung, Anhui, Honan) frequently contain in lower horizons concretions of calcite looking like the roots of ginger plants, hence their name, *sha chiang*[1], but sometimes attaining large size. The concretions in these 'sajong' soils have probably precipitated from rising calcareous ground-water under a régime such as *d* in Fig. 19.

Other important solution-reprecipitation equilibria are those of gypsum ($CaSO_4 . 2H_2O$) and the chlorides and sulphates of sodium and other ions which are normally readily leached but which accumulate where streams enter closed desert basins (e.g. Inner Mongolia and Sinkiang), or where ground-water lies close to the surface in a dry climate (D in Fig. 18). In the latter case upward capillary movement causes concentration, and in extreme cases precipitation, of salts. These effects form the saline or 'solonchak'[a] soils, found both inland in desert basins, and in relatively dry areas on alluvium, and along the coasts (Chhinghai, Southern Gobi, Sinkiang oases; Shensi, Shansi, Honan; Eastern Chiangsu, Eastern Hopei, Shantung).[b] In coastal areas the process is often greatly accentuated by occasional inundations of saline ground-water or salt brought in by typhoon winds.[c] The progressive removal of sodium salts by rainfall or irrigation and draining can lead to some replacement of exchangeable sodium on colloid surfaces by hydrogen ions, and a sharp rise in pH. This is associated with the development of the characteristic columnar structures in sub-surface horizons, often with surface crusts, found in these alkaline 'solonetz' soils.[d] They are found extensively in North-east Manchuria (Kirin),[e] outside the ancient Chinese *oikoumene*.

So much for solution. What about hydrolysis? The hydrolysis of primary[f] silicates is probably initiated by exchange of hydroxonium ions (H_3O^+) for metallic cations at the surface of the silicate lattice, which then becomes unstable, liberating into solution silicon, aluminium, and varying amounts of iron, sodium, potassium, magnesium and calcium, according to its initial composition. Most of the aluminium and part of the silicon recombine to form the layer-lattice clay minerals[g] which are generally crystalline but of colloidal dimensions. Provided that adequate moisture is present, these changes become more pronounced at higher temperatures and with the passage of time.

When leaching is not too strong (e.g. case B in Fig. 18), or inhibited by poor drainage (cases C or D), silicon and soluble cations are conserved, and silicon-rich clay minerals with high cation exchange capacities, such as hydrous mica con-

---

[a] A term adopted from the Russian.

[b] See e.g. I. Sun (*1*); Lêng Fu-Thien & Chao Shou-Jen (*1*).

[c] We have already encountered salinisation as a great perennial problem in the irrigation systems of Iraq (Vol. 4, pt 3, p. 366). On all aspects of saline soils and saline water-supplies, see the collective work edited by Boyko (*1*).

[d] Term modified from the Russian. On surface crusting see Chiang Chien-Min & Tshang Tung-Chhing (*1*).

[e] Cf. Kovda (*1*), pp. 102, 121 ff., 164, 176, 324.

[f] I.e. those found in igneous rocks formed by the solidification of magma at depth (plutonic types), or at the surface (volcanic). Quartz, feldspars, micas, amphibolers, pyroxenes and olivine are the most important.

[g] The distribution of clay minerals in the soils of China is briefly described by Hsiung & Jackson (*1*).

[1] 沙薑

taining potassium, or montmorillonite containing magnesium, are formed. Kaolinite, which is poorer in silicon, contains no cation other than aluminium, and has a much lower exchange capacity, is characteristic of strong leaching conditions (A in Fig. 18) where silicon and cations are easily lost.[a] In the very rapid weathering and leaching of humid sub-tropical and tropical areas under free drainage conditions, especially on old land surfaces where weathering has operated over a long time, the loss of silica and bases is intensified, and aluminium hydroxide (gibbsite) and oxyhydroxides, more or less mixed with kaolinite and with iron oxides liberated by oxidation, become important weathering products. These advanced weathering and base-depletion processes are found in the yellow and red soils[b] and laterites[c] of South China (Fukien, Southern Chekiang, Chiangsi, Hunan, Kuangsi, Kuangtung, Hainan and Yunnan),[d] but not to such

[a] Because of its importance for the pottery and porcelain industry the Chinese loan-word 'kaolin' is much more familiar internationally than, for example, sajong. The mineral kaolinite is the most important constituent of the white china-clay or kaolin deposits found in our own country around the granite masses of Dartmoor, Bodmin Moor and the St. Austell district. It was formed by deep-seated hydro-thermal alteration of feldspar in the granite during the closing stages of the igneous activity. The process is analogous to weathering in that it involves a loss of silicon and bases such as potassium, calcium and sodium. The alumino-silicates of the North-eastern Chiangsi deposits were almost certainly formed in the same way.

The *locus classicus* for the original term *kao ling*[1] (lit. high mountain-ridge) will be found in Lan Phu's[2] *Ching Tê Chen Thao Lu*[3] (Ching-tê-chen Record of the Potteries of China), written c. +1795, where he says (ch. 4, pp. 2b ff., tr. St Julien (7), pp. 250 ff.; Sayer (1), pp. 29 ff.) that it was the name of a mountain east of Ching-tê-chen (in Chiangsi), the capital of the late medieval Chinese ceramics industry. Four families settled in the hills and produced both the kaolin (also called *pai o*[4] and, more learnedly, *pai shih chih*[5], RP 57d) and the more fusible felspathic clay (*pai tun tzu*[6], hence 'petuntse' and other early European terms) needed for the porcelain manufacture. On this see Section 35 in Vol. 5, as also Honey (2), pp. 13 ff.; Savage (1), pp. 26 ff.; Rosenthal (1), pp. 25, 235. A picture of the mountains of Kao Ling[1], with the water-triphammers at work (cf. Vol. 4, pt 2, pp. 390 ff. and Fig. 617) pounding the clays, is given in Thang Ying's[7] *Thao Chih Thu Shuo*[8] (Illustrations of the Pottery Industry, with Explanations) of +1743, incorporated in *Ching Tê Chen Thao Lu*, ch. 1, pp. 7b, 8a, b (the accompanying text paraphrased by St Julien (7), pp. 116 ff., cf. pl. 1; and tr. Sayer (1), p. 4).

[b] Sometimes termed 'krasnozems', a term of obvious Russian derivation.

[c] The term 'laterite' was originally used by F. Buchanan (1) in 1807 for a relatively soft highly ferruginous deposit in Malabar, lying a few feet below the surface, which hardened irreversibly when cut and exposed to the air. The name derives from Lat. *later*, a brick, from the local use to which the material was put. The definition was afterwards extended to include vesicular or concretionary ironstones, sometimes gravelly, sometimes forming massive surface crusts which appeared to be of similar origin but to have hardened naturally. The conditions for laterite formation seem to have been very prolonged tropical weathering with advanced hydrolysis and oxidation (see pp. 66, 68), together with movement and concentration of iron favoured by low relief with high ground-water levels.

The word 'lateritic' has also been used loosely to refer to soils which are simply red from oxidative weathering of iron-bearing minerals in a hot climate, but without any appreciable movement and concentration of iron; or to contemporary soils formed by the re-weathering of old laterite sheets.

What the Chiangsi farmers say is: 'A slab of copper on clear days and a pool of slime when it rains.'

[d] These yellow, orange and red soils are deeply characteristic of the southern Chinese provinces. One finds them as one drives over the high rolling hills of Yunnan, or on a journey in Hunan—Chairman Mao's home country southwest of Chhang-sha—or across the moorlands between Nanchhang and Ching-tê-chen south of the Poyang Lake in Northern Chiangsi. In their upland stretches these soils are not at all fertile. This is primarily because of the prolonged period of intense leaching and weathering to which they have been subjected. From Vol. 3, pp. 463 ff. it will be remembered that in prehistoric and ancient times the Chinese climate was much hotter and wetter than now. On this a recent paper by China's most eminent phenologist, Chu Kho-Chen (9), may be consulted. Secondly, widespread erosion has been caused by unchecked clearance. Good crops can, however, be raised in the zones of accumulation in the valleys where the rice-field soils are only weakly leached because of restricted drainage. Such land responds well to fertilisers; cf. Pei Tê-An (1); Kovda (1), pp. 664 ff.; K. Buchanan (1); Anon. (168).

On the chemical composition of the sub-tropical red soils see Li Chhing-Khuei & Chang Hsiao-Nien (1).

[1] 高嶺   [2] 藍浦   [3] 景德鎮陶錄   [4] 白堊   [5] 白石脂
[6] 白不子   [7] 唐英   [8] 陶治圖說

an extreme degree as in some more southerly parts of South-east Asia. The relative seasonality of climate and a long history of erosion in South China are probable reasons.

The susceptibility to hydrolysis of individual minerals varies widely.[a] The potentiality of a parent material in soil-formation thus depends on its mineralogical composition; a basic igneous rock, such as basalt[b], will weather fairly rapidly to clay minerals and iron oxides, liberating Mg^{2+} and Ca^{2+}, while at the other extreme, non-calcareous blown sands or quartzites hardly weather chemically at all. The particle-size distribution and the chemical constitution of the soil thus reflect the proportions of susceptible and resistant minerals in the parent material as well as the intensity and duration of weathering. A high proportion of coarse resistant minerals leads to a low clay content with a high permeability, and thus more ready leaching in a humid climate. This in turn favours more rapid depletion of exchangeable cations such as Na^+, K^+, Mg^{2+} and Ca^{2+}, their replacement by H_3O^+ and Al^{3+} and a fall in pH, associated with vegetation of tolerant species. Such oligotrophic conditions inhibit the decomposition of organic residues.[c]

Lastly comes oxidation. Oxidation mainly affects divalent iron and the less abundant manganese in ferro-magnesian silicates[d], where it contributes to the breakdown of mineral structures in hydrolysis; and also iron and sulphur in sulphides such as pyrite (FeS_2). It is in part micro-biological. Most ferric iron is rapidly reprecipitated as yellow or brown oxy-hydroxides or hydrous oxides such as goethite or 'limonite'.[e] These, together with varying amounts of humus, are the main pigments in most freely drained soils, the intensity of colouring in the subsoil giving a rough indication of the degree of oxidative weathering, as for example in yellow and brown forest soils, cinnamon soils[f] and chestnut soils.[g] In hot climates, especially monsoonal, the red anhydrous oxide haematite becomes an important constituent—as in the red earths and laterites of South China already mentioned. In some regions, however, red or purple colours are not due to contemporary or recent weathering but to the character of the parent rocks. The outstanding example of this is the earth which provided the fertile fundament of the glorious kingdom of Shu, in other words the 'Red Basin' of Szechuan.[c] Pedologically this was always a misnomer, inviting confusion with the quite different and much less fertile red and lateritic soils of the south-east, so one should speak of the purple-

---

[a] For example, olivine, augite, and lime-rich plagioclase feldspar, are unstable; hornblende, biotite and sodium-rich plagioclase are intermediate; potash feldspars and muscovite are resistant; while quartz is virtually unweatherable, except in very extreme tropical humidity.

[b] Basalt consists mainly of augite and lime-rich plagioclase.

[c] As noted below.

[d] Such as olivine, augite, hornblende and biotite.

[e] Ferric iron in the form of oxides is very insoluble, and in many soils there is no evidence for its overall movement, but in some, local re-solution or actual translocation can be extremely important processes.

[f] The Russian-derived word 'korichnevyi' is often used by Chinese pedologists writing in English, e.g. Ma Yung-Chih (1).

[g] Cf. Vol. 1, pp. 61, 63, 72, Vol. 4, pt 3, p. 24. More in Cressey (1), pp. 310 ff.; and Tregear (1), pp. 232 ff.; Wang Chun-Hêng (1), pp. 170 ff.

red or purple-brown basin of Szechuan, its soils[a] derived from the environing and underlying shales and sandstones—often indeed seen there in revealing outcrops.[b]

In horizons where anaerobic conditions are caused by excess moisture (C and D in Fig. 18), the normal weathering direction is partially reversed and iron is reduced to ferrous forms, either simple salts or organic complexes, and characteristic grey, greenish or bluish colours develop. The phenomenon is termed 'gleying'.[c] Once in solution, ferrous iron may remain more or less *in situ* or it may be translocated in a direction depending on the moisture régime. Change of conditions, usually seasonal, may bring about re-oxidation, and iron is then precipitated locally as rusty streaks, mottles or concretions.[d] Gleying and reprecipitation, singly or together, are the essential features of the meadow soils[e] which occur in conditions of restricted drainage all over China, particularly along spring-lines, in valley bottoms and on alluvium generally. They vary considerably in detailed morphology,[f] often showing close similarities to the zonal soils in better-drained sites in the same region; thus one finds meadow cinnamon soils, meadow chernozems, etc. As would be expected, meadow soils are often associated with, or merge into, peaty or saline soils, the meadow-bog and meadow solonchak soils respectively. Iron may re-precipitate in many different forms, for example (as we saw) in association with calcite in sajong concretions, or else as a continuous horizon in many old irrigated or rice-field soils. Some of the lateritic soils (Fukien, Kuangtung) are extreme examples of long-continued concentration of this kind.

'Podzolisation'[g] is the name given to the process by which iron or humus or both together are translocated from an upper, or 'eluviated' horizon to a lower, or 'illuvial', one.[h] Although it is favoured by strong leaching and acid conditions, it appears to be often brought about by chelating agents, probably polyphenols,

---

[a] These soils are the 'violet brunizems' of Kovda (1), pp. 519 ff., though no son of Szechuan or honorary Szechuanese would, I am sure, recognise the land of Shu under this guise, due perhaps to the studious rush-wick of Kovda's translator. There is a great deal about them in the monograph of Richardson (2). Though often almost without calcium, they may harbour lime concretions at low horizons, and they have high fertility, being rich in phosphorus, potassium and some trace elements notably lacking in the red soils of South China. The purple-red soils of Szechuan are comparable with the red soils of Devon and the English Midlands, the colour of which is due to iron precipitation in the weathering conditions of the Permian and Trias when the parent rocks were deposited. Some of their other properties also derive from this history.

[b] For instance those which shelter the cave-temples at Ta-tsu, and elsewhere.

[c] A term for this process, adopted from the Russian.

[d] Sometimes, too, as a continuous horizon or 'iron-pan'.

[e] In Western usage now generally called 'gley' soils.

[f] Individual types are given names to indicate more precisely their main morphological features; for example carbonate light meadow soils are those containing calcium carbonate and of fairly light colour due to low humus content (cf. pp. 71, 72).

[g] Another term adopted from the Russian.

[h] Perhaps the method of chromatographic analysis, which has so greatly revolutionised modern biochemistry by making it possible to separate and identify organic chemical compounds, whether pigments or not, in extremely small quantities, adsorbing them differentially on columns of solid substances in powder or granular form, or upon strips of filter paper, and then eluting (eluviating) them with different solvents (cf. Stein & Moore (1); Lederer & Lederer (1), etc.), was one of the greatest indirect contributions of soil science. In view of the outstanding position of Russian scientists in the development of modern pedology it may be no coincidence that the central figure in chromatographic history was also a Russian, Michael Simeonovitch Tswett (1872 to 1920). We know too little of his life (cf. Dhéré, 1) to be able to say exactly what influences acted upon him, but in a series of classical papers (1–3) and a remarkable book (4) of 1910, he described the separation of the two

present in the litter of the associated vegetation. These appear to reduce the iron as well as forming chelates with it, so no clear distinction can be drawn between the solution of iron in podzolisation and in gleying, the two processes merging into each other in many soils. Well-developed podzols always occur under coniferous forest, in China now mainly confined to the northern part of Manchuria, but the term 'podzolic' is often used rather loosely to indicate feeble or incipient podzolisation as defined above, the translocation of fine clay in leached soils to form 'clay-pans'[a], or simply strong leaching with an associated low pH. All these processes are associated with a humid climate and free drainage, being therefore common in the moister regions of China.[b]

Structure development as a soil-forming process (iv) can be dealt with much more succinctly. Soils differ from weathered or transported rock-debris, or mixtures of such, in the rearrangement of the particles of sand, silt and clay into aggregates known as 'peds', characteristic of the conditions of formation. These soil structures, often descriptively named 'platy', 'prismatic', 'blocky' or 'crumb' are formed by cyclical processes of wetting and drying, or freezing and thawing; by aggregation with humus or iron oxides; or by the activity of roots, worms or other living things. As noted below, the maintenance of stable surface structures is essential in the prevention of erosion and the continuance of any permanent system of agriculture.

Thus we are confronted with the last group of processes (v) those of biological, including human, origin. Soil-formation effectively commences when inert rock-debris is colonised by living organisms; from this point the soil, together with its associated flora and fauna, develops in a particular direction depending on the local factors of parent meterial, climate, topography, and availability of plant and animal species. With the passage of time the system tends to an approximate

---

chlorophylls and the resolution of the four xanthophylls. A long latent period followed, but after the differentiation of the carotins by Kuhn & Lederer (1) in 1931, the method became one of the basic tools of biochemistry, with immense consequences for medicine and all the biological sciences.

The early history of chromatography has been the subject of some debate (cf. Lederer & Lederer (1), p. xix; Zechmeister (1–3); Williams & Weil (1) and others). But although as a result of this, Tswett, while retaining his pioneer importance, can no longer be regarded as its sole *fundator et primus abbas*, others who were active in the nineteenth century have also to be seen against a background of soil science. Paper chromatography doubtless originated in the practices of the medieval dyeing industry and even has roots in Roman times (cf. Pliny, *Nat. Hist.* XXXIV, 112, noted by Yagoda, 1), but it was greatly furthered by F. F. Runge from 1850 (Weil & Williams), and Schönbein and Goppelsroeder from 1861 onwards (Farradane, 1). Goppelsroeder referred to Justus von Liebig's work on soil water and the retention of salts in specific horizons, and Tswett referred to Goppelsroeder's writings. Most probably he also knew of the pioneer experiments of D. T. Day in America (1897) and Engler & Albrecht (1901) on the differential adsorption of petroleum fractions on mineral powder columns (cf. Weil & Williams), experiments which recalled those of H. S. Thompson and J. T. Way with salts and soils in England (1850). Thus all the lines seem to lead back to the phenomena of precipitation, adsorption and elution of substances and liquids travelling in the earth's crust, i.e. to soil science and its cognate field petroleum geology.

At the technical level Tswett may be praised as the first to develop a chromatogram with pure solvent. But his real originality lay rather in his application of the method to plant pigments, i.e. its introduction into biochemistry, whence have flowed in subsequent decades its most brilliant triumphs useful to man.

[a] English contributes at any rate one colourful technical term here—'fox-benches'. Such layers are uncemented when wet but almost stone-like when dry. The translocation process is the 'lessivage' of French usage noted above (p. 63).

[b] But they may not necessarily always be obvious at first sight, since, for example, an eluviated horizon may be totally removed by erosion.

equilibrium state in which the zonal soil and climax vegetation are closely related to the environment and to each other. However, influences such as change of climate, rise of ground-water, or human modifications of the vegetation, frequently deflect development along different paths; many, and perhaps all, soils, other than very immature ones, are polygenetic. This is especially true of countries like China with age-long agricultural histories.

In climates sufficiently warm and moist for the development of temperate broad-leaved deciduous forest (e.g. South Manchuria, Shansi uplands, Hopei, Shantung, Honan, South Shensi, Szechuan, Hupei, Anhui, Chiangsu)[a], the dead organic material was added mainly at the surface. Such material contains many almost unmodified compounds, e.g. carbohydrates, lignins, fats, waxes and proteins, as well as inorganic ions taken up selectively, and thus cycled, by the vegetation.[b] In eutrophic conditions, where the debris as a whole is relatively rich in nitrogen, when nutrients such as calcium and phosphorus are adequate, and if acidity, aeration, moisture and temperature are equable, these residues are fairly rapidly converted by the bacterial, fungal and invertebrate soil population[c] into carbon dioxide, simple salts and slowly decomposable colloidal mild humus or 'mull'[d]. This is intimately mixed with the mineral fabric, in particular by earthworms, and, in tropical soils, by termites.

The amount of humus depends on the balance between production and decomposition of organic matter, and thus on the vegetation and the climate. The colour of virgin brown forest soils associated with temperate woodland is due to a fairly high content of mull mixed with hydrated iron oxides, but the former is often sharply reduced by cultivation. In drier or warmer forests, lower levels of humus and lighter colours prevail (e.g. the cinnamon[e] soils of Shantung, Shansi, the escarpment between Hopei and Shansi, South Shensi, South Kansu, Hupei, Anhui). In moist sub-tropical or tropical forests, on the other hand, levels may be higher, though here the humus does not so effectively mask the colours of the oxides, especially after cultivation (e.g. the yellow and red laterite soils of South China). Mull is also the characteristic humus form of the grassland and steppe soils of the drier regions, where, in contrast to forests, it is added mainly below the surface. Here there is a regular decrease in intensity of colour in passing from the dark chernozems[f] through the 'kheilutu'[g] and chestnut[h] soils to the grey-brown

[a] Areas III and IV in Fig. 21 (Table).

[b] Here the reader will recall the contents of our sub-section on geo-botanical and bio-geochemical prospecting in Vol. 3, pp. 675 ff. [c] Pedologically, the role of the soil protozoa is a minor one.

[d] A term adopted from the Danish. It was introduced by Müller (1) in 1887. An accessible simplified account of his work will be found in Burges (1).

[e] Cf. p. 68 above. [f] Cf. p. 65 above.

[g] Here is a third example of a term which is entering international soil science nomenclature from Chinese. The deep yellow earth called *hei lu thu*[1] is found in Shensi above the loess from which it derives (cf. Rozanov (1, 2); Kovda & Kondorskaia (1); Kovda (1), pp. 446 ff., 449). Since these soils do not seem to have any parallel in other countries, their nearest relations being perhaps the chestnut chernozems of the Caucasus, a special name has proved necessary. The kheilutu soils are essentially pedocals, formed originally under steppe, but subjected to cultivation with the addition of earth fertilisers for immensely long periods. The topsoil is thus not very rich in humus, but is abnormally deep, and coloured a rather uniform dark yellowish grey. For another example of a Chinese-derived term see Kovda (1), p. 348. [h] Cf. p. 65 above.

[1] 黑壚土

and grey soils (sierozems) of the semi-desert regions. This decrease in coloration is of course due also to less intense liberation of iron oxides by weathering.

In oligotrophic conditions such as those associated with strongly leached soils and tolerant species, for example podzols under coniferous forest, organic matter accumulates as a surface mat termed raw humus or 'mor'[a], but this form is not common in China owing to the lack of surviving forest of this kind except in Manchuria. Peat is an extreme form of organic accumulation caused by anaerobic conditions associated with excess moisture, especially at low temperatures. It is destroyed by drainage and cultivation. It is also uncommon in China, being confined to montane soils and lowland swamps, where it is sometimes a consequence of prolonged irrigation with lack of adequate drainage. Thick peat deposits are called 'bog soils'; soils with peaty surfaces but with profiles otherwise essentially of mineral origin are 'meadow-bog soils'.[b]

So far no mention has been made of the human factor. Yet this has been of vital significance in the development of the present-day soil pattern of China, which, perhaps more than any other country, displays the whole history of man's ingenuity and man's folly in the modification of the soil.[c] The initial impact of man on an existing system of soil and vegetation is the alteration of the latter in clearance by grazing animals or by burning,[d] the tending of useful plant species,[e] and ultimately weeded and fertilised cultivation. Even without this last, alteration of the vegetation may profoundly affect the soil, for example in moist areas, where the replacement of deep-rooted trees by grass and crops can induce poorer drainage conditions and alter the distribution of humus.

In general the effect of tillage is to produce uniform top-soils lower in humus, lighter in colour, and with soil structures less stable than those in the original surface horizons. In moist or seasonally moist régimes, especially where, as in much of China, the rainfall is irregular and often intense, surface soil aggregates, in the absence of a permanent canopy, are easily disrupted by the mechanical effects of falling raindrops. Rainfall then tends to run off rather than penetrate, and soil erosion is initiated, with concomitant flooding and deposition of masses of alluvium downstream, often accompanied by seasonal desiccation in the eroded area. Such processes have occurred on a spectacular scale throughout China's history,[f] and over great tracts of erstwhile forest the original upper horizons have been completely stripped away.

[a] A term adopted from the Danish, and again due to Müller's work on the woodlands of his country.
[b] Cf. p. 69 above. The meadow soils have been classified by Kovda into 'rich', 'dark', 'normal' and 'light' types, reflecting the decreasing amounts of humus in the topsoil associated with decreasing intensity or frequency of saturation with water.
[c] Ingenuity in the heroic story of river-control and irrigation which has already been told in Sect. 28 in Vol. 4, pt 3, and in the multifarious agricultural techniques discussed in Sect. 41. Folly in the deforestation universally permitted through the ages (cf. Vol. 4, pt 3, pp. 239 ff.).
[d] We have already referred to the existence of *milpa* cultivation (often called 'slash-and-burn' or 'burn-and-sow') in ancient China. Cf. e.g. Vol. 1, p. 89, Vol. 2, p. 255. The convenient word *milpa* is Amerindian.
[e] See below, pp. 82–3.
[f] See Figs. 871, 872, and Vol. 4, pt 3, pp. 253. For a general perspective see Jacks & Whyte (1). On China in particular see the works of Lowdermilk and his collaborators cited in Vol. 4, pt 3.

Wind erosion is more characteristic of dry, or seasonally dry, areas where over-grazing or cultivation has produced bare surfaces, particularly where surface structures are poor and there is a predominance of particle sizes in the susceptible range for blowing. Sands and loess are obvious important examples.

In many cultivated soils of China, however, the influence of man has been the opposite of erosion; the practice of adding turf, earth and mud fertiliser,[a] marl[b] and sand or silt[c] to improve texture or chemical fertility over very long periods of time has locally produced great thicknesses of man-made topsoil. We have already seen an instance of this process in the kheilutu soils, but it has gone on all over the Chinese culture-area. At least as early as the Thang and Liu Chhao periods[d], and probably already in the Han,[e] the technique of contour terracing enabled crops to be grown, and the erosion and moisture régimes more or less controlled, on slopes and even steep hillsides. In some regions the whole landscape is taken up with terraces (e.g. Szechuan, Shensi, Kansu, Yunnan; cf. the air-view in Fig. 20).[f] Long-continued flooding with irrigation water has often induced secondary phenomena not present in the original soils, such as gleying and translocations of iron with re-precipitation as concretions, or the ferruginous 'pans' very typical of many old irrigated soils.[g] Permanent raising of the water-table by irrigation has locally given rise to peaty soils, particularly along the lower Yangtze valley, and where evaporation is high, to salinity in limited areas all over the East China plain.[h] On the other hand, centuries of river embankment and drainage by ditches and canals have maintained huge areas of alluvium and other meadow soils in cultivation.

A final facet of man's activity is the control of chemical fertility. A major contribution could be written on this aspect of China's history. Here it is only possible to note very briefly a few important topics. Apart from the pedocals and the soils on calcareous parent materials, lime, for the control of acidity and as a source of nutrient calcium, is required in varying amounts on nearly all cultivated

[a] Often from the beds of rivers and canals; cf. King (3), p. 153; Gourou (1a), p. 68; Wagner (1), pp. 223 ff.

[b] The use of various kinds of limestone (shih hui[1]) came into prominence particularly during the Sung period, from the +10th century onwards; cf. Yang Min (1), p. 79.

[c] Cf. Vol. 4, pt 3, p. 227. The lime content of the loessial soils persists when they are re-deposited in alluvial form on the North China plain after passing down that colossal belt-conveyor the Yellow River. Silt was therefore in much demand in medieval times as fertiliser. See also Vol. 1, p. 68, and further in Cressey (1), pp. 158 ff.; Kovda (1), pp. 92 ff., 102; Tregear (1), pp. 215 ff.

[d] See Vol. 4, pt 3, p. 247.

[e] See Shih Shêng-Han (2), his annotated translation of Fan Shêng-Chih Shu[2] (the Book of Agriculture by the author whose name it bears) written in the late −1st century (pp. 17, 19, 62). A rice-field may be visualised as a shallow artificial pool with a flat bottom and surrounded by permanent mud-earth ridges (shêng[3]). The small 'retaining-walls' of the terraces (strengthened with stone where that was available) would thus have arisen naturally as cultivation spread up the valleys of the head-waters.

[f] Cf. Vol. 1, pp. 72, 89.

[g] The 'rice-paddy podzols' of Thorp (1) were so named only because of their superficial resemblance to podzols of more normal kinds in possessing a horizon of iron accumulation.

[h] It is doubtful, however, whether any such cases have approached the almost catastrophic salinisation phenomena of the Tigris valley in Mesopotamia. On this see Jacobsen & Adams (1), who distinguish three periods of abandonment of cultivation due to this cause, with concomitant historical consequences.

[1] 石灰　　　[2] 氾勝之書　　[3] 塍

Fig. 20. Agricultural hillside terracing. Buchanan, Fitzgerald & Ronan (1), p. 36; cf. Kaplan, Sobin & Andors
(1), p. 20; Hook (1), p. 44.

soils. Chinese farmers in the past have shown considerable ingenuity in the use of
calcareous sands, oolite, marl,[a] molluscan shells,[b] and other locally available
liming materials. Many agricultural soils are therefore far less acid than they
would be if they had remained under virgin forest. These base-rich conditions
tend to increase the rate of micro-biological breakdown of organic matter. In

[a] See immediately above, p. 73.
[b] Cf. King (3), p. 155; Wagner (1), p. 233; Amano Motonusuke (4), p. 309. This form of lime began to be
prominent as a fertiliser from Sung times onwards (Yang Min, 1).

some cases, for example the brown forest and cinnamon soils of Shantung, long-continued addition of calcareous materials may be an explanation alternative to the addition of windborne calcareous dust proposed by Thorp (1), for the persistence of calcium carbonate in the profile.[a]

Although losses have been checked or mitigated by the age-old use of green manures, especially leguminous,[b] earth fertilisers and organic wastes,[c] nitrogen-deficiency has become country-wide with intensive cropping, over-cultivation and marginal erosion. With present population levels, it will be rectified only with the advance of the chemical industry. Much the same applies to phosphorus, which is widely deficient, particularly in acid soils with abundant free iron oxides that readily fix phosphorus in forms unavailable to crops (e.g. brown forest soils, yellow and red laterite soils).[d] Other deficiencies are also known.[e]

The widespread lowering of humus levels and impoverishment in nutrients has probably never applied to the intensively used market-garden soils adjacent to large towns. Here the use of great quantities of organic waste, and notably human manure or 'night-soil',[f] has maintained very high levels of humus and plant nutrient substances, representing in effect the continual transfer of fertility from rural areas.[g]

In order to know how 'the good earth' lies, geo-chemistry must join with geography, and the results of soil surveys must be plotted on maps. Before the present century, nothing was done about soils in China on modern lines, but in 1930 Shaw (1), working for the Chinese Geological Survey, published a simplified map of part of the eastern provinces.[h] During the following decade field studies were intensified under the leadership of Thorp, who produced in due course a

---

[a] Cf. Kovda (1), pp. 488, 508, 518.

[b] The systematic ploughing in of weeds can be documented from several ancient texts, notably the *Yüeh Ling* and the *Chou Li* (see further in Sect. 41), but the conscious and purposive use of legumes grown as a manure-crop appears not later than the *Chhi Min Yao Shu*[1] (Important Arts for the People's Welfare) written by Chia Ssu-Hsieh[2] about +540. See the annotated text and translation by Shih Shêng-Han (1), pp. 11, 17, 44 ff.; and cf. his (2), pp. 5, 7, on what is said in the book of Fan Shêng-Chih five and a half centuries earlier. Chia Ssu-Hsieh also recommends the use of the old mud walls of byres—good nitrate sources, as we would recognise. On clovers and vetches in contemporary use see King (3), pp. 163, 241, 258, 354; Wagner (1), pp. 226 ff.

[c] E.g. chaff, from Sung times onwards (King (3), p. 260).

[d] In the ancient agricultural books there is much on the use of animal bones as fertilisers (calcium and phosphorus). See Shih Shêng-Han (1), pp. 11, 12, (2), pp. 11 ff., 58. Silkworm excrement from the sericultural industry is also emphasised.

[e] In general, strongly weathered and leached soils of the humid sub-tropics and tropics are characterised by multiple nutrient deficiencies, so that potassium, sulphur, magnesium, and trace elements such as copper, zinc, cobalt and molybdenum may be simultaneously inadequate. Piecemeal correction is often useless, and can even be detrimental.

[f] Cf. the classical book of King (3) and a special study of much interest by J. C. Scott (1).

[g] Compare for example the international transfer that took place in the pre-war agriculture of the U.K. when stock fed on foreign-derived concentrates, such as oil-cake, and maize, produced great quantities of farmyard manure. In complete contrast to China, however, the entire nutrient content of human excrement was lost to the sea except in some local sewage-farms.

[h] This became widely known because it was later reproduced in the geography of Cressey (1), fig. 45 and pp. 86 ff. Cf. Shaw (2).

[1] 齊民要術   [2] 賈思勰

much more complete map, omitting only some of the remoter western areas.[a] In his accompanying treatise (Thorp, 1) he used a classification of soils based on that current in America at the time. This remained the standard work for two decades and is still a valuable source.[b] After that things hung fire until the fifties.[c] More intensive soil surveys were then started, and a number of important soil maps brought out in Peking,[d] culminating in the great map of 1957 which has been described by Kovda & Kondorskaia (1).[e] This was followed two years later by Kovda's *Soils and the Natural Environment of China* (1), the most comprehensive modern work on the subject in any Western language.[f] Two other generalised maps are fairly readily accessible in the West; that of Gerasimov (1)[g], and another, in colour,[h] in the *National Atlas* edited by Chang Chhi-Yün (1). At the present time pedology and its cartography are actively cultivated throughout China, especially in the institutes of Academia Sinica,[i] and the knowledge of the Chinese soil pattern is beginning to contribute to an understanding of the soils of the Old World as a whole.[j]

In preparing a map for the present work it was found that that of Gerasimov, though suitably simplified, was on too small a scale for easy interpretation and reproduction, while that of Chang Chhi-Yün was based too closely on the now rather outmoded classification of Thorp. We therefore decided to simplify the map and classification of Kovda (1). Such a procedure has obvious drawbacks, but we believe that the essential features of his original map are preserved. It is, however, most important to remember that any small-scale map can depict only the broadest aspects of the soil pattern; catenary relationships in particular, and local variations, must be added mentally by the user. For example, in a region shown as consisting predominantly of cinnamon soils and brown forest soils, it can

---

[a] Scale 1 : 7,500,000.   [b] Cf. Thorp (2, 3); Thang (1); Wolff (1).

[c] But in 1947 the Soils Division of the Chinese Geological Survey produced a map on a scale of 1 : 6,000,000; this we have not seen.

[d] In 1956 there was a 1 : 20,000,000 soils map by Ma Yün-Chih; a 1 : 16,000,000 one by Hou Hsüeh-Yu, Chang Tu-Chen & Wang Hsien-Pu relating soil types to vegetation; and a similar one at 1 : 4,000,000 by Hou Hsüeh-Yu & Ma Yün-Chih. Further details of maps and literature will be found in Kovda & Kondorskaia (1); Yang Chih-Hou & Kovda (1).

[e] By Ma Yung-Chih (Director of Academia Sinica's Soil Science Institute), Sung Ta-Chhêng, Li Chhang-Khuei, Hsiung I, Hou Kuang-Chhiung, Hou Hsüeh-Yu, Li Lien-Chieh, Wên Chen-Wang & Wang An-Chiu; apparently with the collaboration of V. A. Kovda & N. I. Kondorskaia.

[f] We are much indebted to Mr Buttress of the Cambridge Agricultural School Library for locating the English translation for us. After the publication of his fine book, my friend Viktor Abramovitch succeeded me as Director of the Natural Sciences Division of UNESCO, but he has now returned to the pedological field.

In the same year came Ma Yung-Chih's (1) discussion of taxonomic units of soils appropriate to more detailed mapping than that considered here.

[g] On a scale of 1 : 20,000,000.   [h] On a scale of 1 : 13,000,000.

[i] Of modern Chinese works on soil science (apart from papers in *Acta Pedologica Sinica*), only the brief introduction of Li Lien-Chieh *et al.* (1) has been available to us. The chapter in Shen Tsung-Han (1) is, however, useful. Cf. also Wang Yü-Hsin (1).

[j] The soils of all Asia, with a map, have been considered in a paper by Lobova & Kovda (1). See also the discussion in Kovda (1), pp. 675 ff. of Ma Yung-Chih's 'law of horizontal zonality' of soils in the Eurasian continental land-mass. Soils and agriculture in the Hongkong territories have been reported on by Grant (1). The soils of China are also shown at a scale of 1:5,000,000 in the FAO–UNESCO soil map of the world (sheet VIII). This uses a classification differing from that of Kovda.

be assumed without hesitation that the latter will occupy the moister sites, those at higher altitudes and on north-facing slopes.[a] Where one particular soil type is shown as dominant, the more leached varieties will occupy interfluves, while the less leached members, together with their associated gley or meadow soils will be found in valley bottoms. Likewise important variations from the median type must be anticipated according to local differences in parent material or in the past history of cultivation, addition of lime or manure, erosion, terracing or irrigation. Bearing these points in mind, we think that the shortened list of soil groups in the caption of Fig. 21 will be adequate for our purposes.

(iii) *Pedology in the* Kuan Tzu *and the* Shu Ching *(Yü Kung)*

We are now in a better position to compare the soil types of the *Kuan Tzu* book with what is known about the Chinese soils of today.[b] Unfortunately, the ancient descriptions are insufficiently precise to allow of sure identification, and indeed two divergent opinions have been proposed. Hsia Wei-Ying (*2*) took the expression *tu thien*[1] to refer to the *ssu tu*[2] or four great rivers (the Yangtze, the Huai, the Chi and the Yellow River); and his idea therefore was that the five soil types were representative of those found up and down the great plain of eastern China,[c] covering almost everywhere north of the Nan Ling mountains with the exception of Kuanchung.[d] Hence he dated the text in the Warring States period (−5th or −4th century), believing that it was written by the citizen of some eastern State who had not been able to study the soils of Chhin. Wang Ta (*1*), on the other hand, following Lo Kên-Tsê but basing himself on more recent technical studies of the soils of the Kuanchung area, believed that the description applied to the irrigated (or irrigable) land (*tu thien*[1]) of that region only, sufficient diversity of soil types being found therein.[e] Consequently he was inclined to date the text rather in the Chhin and Han periods (−3rd or −2nd century), when Kuanchung, with its many estates of patrician families and high civil servants, was great in importance; thus the writer would have been a Chhin or Han official rather than a Chhi or Yen philosopher.[f] There is no easy way of deciding between these two views, but it is interesting to see what possibilities the ancient text contains.

[a] Cf. Vol. 2, p. 274. The distinction between north- and south-facing mountain-sides goes very deep in the ancient Chinese *Weltanschauung* of Yin and Yang. Soil differences may well have been noted.

[b] The operation can only be approximative, for it must be remembered that some two thousand years of natural development (and human cultivation) have passed since the *Kuan Tzu* was written.

[c] Natural provinces nos. 7, 11, 13 (see Vol. 1, p. 62).

[d] I.e. the Wei River valley, modern Shensi and E. Kansu, home of the Chhin State and afterwards the metropolitan domain of the Han. See particularly Vol. 1, pp. 58, 70 on this.

[e] Wang Ta had taken part in soil surveys himself, of all areas concerned, with a collaborator Sang Jun-Shêng. See particularly his fig. 3.

[f] Various other arguments have entered into the debate, such as the presence of Yin-Yang doctrine or symbolic correlations in fives (cf. Sect. 13 in Vol. 2) but inconclusively.

[1] 瀆田   [2] 四瀆

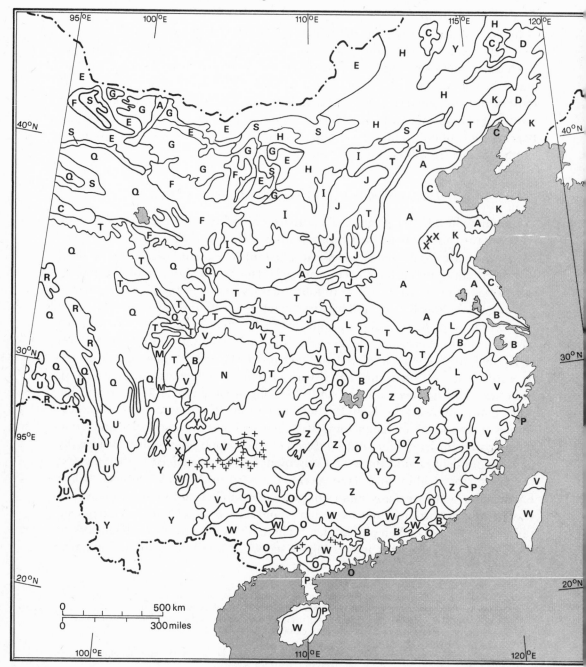

Fig. 21. Soil map of China (prepared by Dr R. M. S. Perrin). After Kovda & Kondorskaia (1).
Cf. Chang Chi-Yün (1), vol. 5, A7, 8.

MAP SYMBOL	SOIL GROUPS AND CHARACTERISTICS	SOIL NUMBERS IN KOVDA SYSTEM	GEO-BOTANICAL REGION	REGION NUMBER IN KOVDA SYSTEM
A	Light meadow soils, calcareous or with calcite concretions 'Sajong' soils), more or less halomorphic	5, 6	Alluvial plains	I
B	Bog (peat) and meadow-bog (peaty gley), often old irrigated, soils, occasionally acid	7, 8	Alluvial plains	I
C	Solonchaks, locally gleyed or boggy	10, 11, 12	Alluvial plains, littoral or inland	II
D	Calcareous dark meadow soils, locally halomorphic	4, 27	Alluvial plains and temperate steppe	I and IV
E	Grey-brown soils, with or without gypsum, locally stony	16, 17	Temperate desert and desert steppe	III
F	Sierozems and dry grey-brown soils	18, 19	Temperate desert and desert steppe	III
G	Blown and partly stabilised sands	13	Temperate desert and desert steppe	III
H	Young or mature chestnut soils, light or dark, locally with loose sands	20, 22, 23	Temperate steppe and prairie	IV
I	'Khei-lutu' soils on loess, dark or poorly developed or eroded	24, 25	Temperate steppe and prairie	IV
J	Cinnamon soils, calcareous or leached, locally irrigated variants	30	Temperate and warm temperate forest	V
K	Cinnamon and brown forest soils, the latter often podzolized	31, 32	Temperate and warm temperate forest	V
L	Yellow-brown and brown forest soils	29	Temperate and warm temperate forest	V
M	Yellow soils	33	Subtropical and tropical montane forest	VI
N	Purple brunizems, some yellow and red soils, old irrigated variants	34	Subtropical and tropical montane forest	VI
O	Red and podzolized red soils	35, 36	Subtropical and tropical montane forest	VI
P	Tropical red soils with or without laterite	37, 38	Subtropical and tropical montane forest	VI
Q	Montane meadow, meadow steppe and tundra soils, montane podzols	41, 42, 43	High montane areas	VII
R	High montane steppe soils	40	High montane areas	VII
S	Montane chestnut soils and brown steppe soils or cinnamon soils	44, 45	Temperate, montane steppe	VIII
T	Montane brown forest and cinnamon soils, locally immature stony soils	51, 52, 53	Warm temperate montane forest	X
U	Montane brown soils	54	Subtropical and tropical montane forest	XI
V	Montane yellow and yellow-brown soils	55, 56	Subtropical and tropical montane forest	XI
W	Montane tropical yellow and red soils	57	Subtropical and tropical montane forest	XI
Y	Montane red, yellow and purple soils, more or less eroded	58, 60	Subtropical and tropical montane forest	XI
Z	Montane brown, yellow and podzolized red soils	59	Subtropical and tropical montane forest	XI
+	Humus-carbonate (rendzina) soils on limestones	61	} Local occurrences on limestones	
X	Red limestone soils (terra rossa)	62		

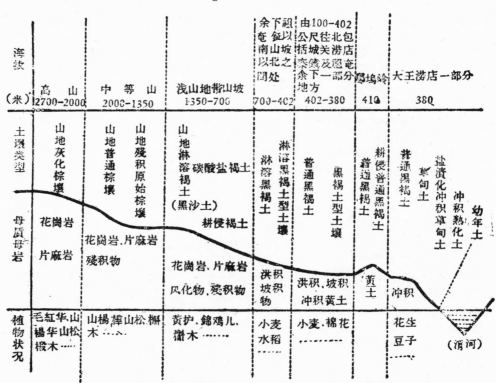

圖3　鄠縣地形土壤母質與植被關係示意圖(抄自鄠縣:土普)

Fig. 22. Diagrammatical cross-section of the Wei Valley; Wang Ta (*1*).

Since we regard the interpretation of Wang Ta as the more plausible geograph-ically (though not necessarily in dating) we shall give what follows from it first, listing afterwards the soil identifications implicit in the view of Hsia Wei-Ying. To aid our thesis we add a drawing (Fig. 22) giving a diagrammatical cross-section of the Wei Valley; this may be considered representative of the more important agricultural lands of Kuanchung. The first type then, *hsi thu*[1], may be translated well-drained loessial carbonate soil, including the irrigated grey cinnamon and kheilutu soils. This is the highest position of the five above ground-water. The second, *chhih lu*[2], is specified in the text as 'reddish, crumbly, hard and fertile'. Probably we have to do here with outcrops of red loess and ruddy-coloured loam, which in general underlie the normal yellowish loess soils of the area, as described in Popov (1).[a] The third type, *huang thang*[3], yellowish, brittle, somewhat salty and

---

[a] Although this soil is listed in our text as being 21 ft above the water-table it is to be noted that commonly associated plants are *Imperata cylindrica* and *Phragmites communis*, both normally indicative of moist soil conditions. A possible explanation might be that surface wetness, unconnected with the water-table, resulted from pro-longed irrigation and deterioration of drainage due to compaction of the loamy subsoil under cultivation.

[1] 息土   [2] 赤壚   [3] 黄唐

alkaline, on land, says Hsia, liable to flooding, would probably be colluvial footslopes, low alluvial terraces, or the normally better-drained parts of the flood plain. In the first case at least, periodic flooding would not necessarily be due to rise in river level but to seepage from irrigation works at higher levels. Here would be Kovda's carbonate (calcareous) light meadow soils, partially salted. Next comes the *chhih chih*[1], argillaceous and, as the adjective implies,[a] salty; most probably we could regard this as poorly drained solonchak soil on alluvium. Lastly the fifth type is *hei chih*[2], said to be dark, black, sticky, argillaceous and saline. This land, lowest above the ground-water, could be considered very poorly drained swampy and solonchak soil, some of it perhaps peaty in character. Wên Huan-Jan & Lin Ching-Liang (*1*) have made a special study of ancient Chou and Han records of the procedures used in ameliorating saline soils both in the Wei Valley and the North China plain. The construction of irrigation canals, the improvement of drainage, the planting of rice and the use of river silt fertiliser, all were brought into use.

In view of the multiplicity of the soils of China, it must be even braver guess-work to fix upon those which the ancient writer had in mind if, as Hsia supposes, he was referring to the whole of the north and east. *Hsi thu*, however, could then very reasonably be the loess-derived silt pedocals in the better-drained sites on the alluvial plain,[b] together with carbonate cinnamon soils on a basis of loess at the higher locations. For the reddish *chhih lu*, if the Kuanchung outcrops are excluded, one might think of the local regions of terra rossa[c] in Western Shantung; an identification all the more attractive because the purple-red soils of Szechuan would have been outside the likely area of discourse at the time, and in any case not in harmony with the simple water-table data. Both the next two classes would again be alluvial. *Huang thang* would agree with the soil of transitional zones between low-lying calcareous light-textured meadow soils and the carbonate cinnamon soils of the uplands. *Chhih chih* could be the more saline calcareous light-textured meadow soils, as also the sajong soils with their calcareous concretions. Finally, *hei chih* could be either meadow bog, old irrigated ricefield soils, or the coastal chloride solonchak meadow and bog soils.

This then is about as far as we can get with the identification of the soils in the

---

[a] This word *chhih* (K 792) is extremely ancient as a technical term for saline soils. Significantly, perhaps, one of its other meanings is to spread or enlarge extensively. Just so may the ancient farmers have started with the best, expanding cultivation into less good land as time went by and population grew.

[b] We have already met with this technical term for a type of soil, at a much earlier point, in Vol. 4, pt 3, p. 253, when discussing some special earth used by the successful semi-legendary engineers and culture-heroes of high antiquity for building dams and embankments (cf. Granet (1), pp. 266, 485, and Maspero (8), p. 49). *Hsi* was taken to mean 'alive', 'breathing' or 'swelling', but it also means 'to stop' or 'come to rest', and silt does just that when it is deposited; moreover, when built into earth dams it stop the flow of water very effectively. Something about the choice of the best soil for earth barrages is involved here, a chapter of great interest for the most ancient history of civil engineering. Light is now being thrown on it by modern experimentation; cf. Huang Wên-Hsi & Chiang Phêng-Nien (1).

[c] Cf. p. 65 above.

[1] 斥埴   [2] 黑埴

Kuan Tzu book, a text which retains all its fascination as one of the earliest in the world on geo-botany. But we can go still further back. Behind it there stands another, even older, which combines soil science with plant geography, the Yü Kung[1] (Tribute of Yü) chapter (ch. 6) of the *Shu Ching*[2] (Historical Classic), and to this we must now turn.

Whoever has followed our writing through as far as this will remember the account of the Yü Kung at the beginning of the Section on cartography, where it was called the oldest Chinese geographical document which has come down to us.[a] Later on there was much to say about the legendary hydraulic engineer and hero-emperor Yü the Great in the Section on civil engineering.[b] In the Yü Kung, then, we find mention of the mountains and communicating waterways within the traditional nine provinces of the Shang kingdom,[c] together with their revenues, predominating types of soil, vegetation, tribute commodities and characteristic products.[d] Formerly we gave as example a translation of the whole text concerning the first two provinces, but now for our present purpose we propose to set forth its words about the soils and products of all the provinces,[e] omitting altogether the purely geographical detail.[f] It is, as we shall see, the epitome of a kind of Domesday Book.[g]

Nearly forty years have now passed since I first discussed the soil science of the Yü Kung chapter with Professor Têng Chih-I[3] at Li-yuan-pao in wartime Kuangtung, and with Dr H. L. Richardson, then pedological adviser in Szechuan to the National Agricultural Research Bureau. For a long time the Yü Kung was connected primarily in my mind with the history of Chinese agriculture, but when one looks at it more closely one sees that settled agriculture is only a part, and not perhaps the greater part, of the wealth of the Nine Provinces, and the tribute which they yielded to the High King of the Shang, as also later to the first Chou emperors. Hence we can no longer postpone this text until Section 41 on agriculture. It seems to be a common mistake of laymen to suppose that a plant must be either 'cultivated' or 'wild'; but this is not so, and there is need for at least one other category, that of 'tended'. All the plants useful to man were wild in the first place, but where the coconut, the mulberry, the rubber-tree or the tea-plant grew in nature abundantly, man for centuries tended them by removing adventi-

---

[a] Vol. 3, pp. 500 ff. The Yü Kung has been the perennial paradigm of Chinese geographers. Numerous commentaries on it were assembled about +1160 in a book still useful today, the *Yü Kung Shuo Tuan*[4] by Fu Yin[5]. In our own time the Chinese journal of historical geography was named after it.

[b] Vol. 4, pt 3, pp. 247 ff.

[c] Cf. Vol. 1, pp. 83 ff.

[d] For the economic plants of the *Shih Ching* (Book of Odes), that ancient document of parallel, perhaps rather older, date, see the interesting paper of Kêng Hsüan (1).

[e] Certain differences will be noticed between the two translations, but the significance of these will be apparent from our present comments.

[f] Including the semi-legendary material on hydraulic engineering achievements.

[g] Only at least a millennium and a half earlier than England's. It would be fair to say that the compilation of the Yü Kung was symptomatic of the early tendencies of Chinese feudalism towards bureaucratic forms; cf. Sect. 48, and meanwhile Needham (53).

[1] 禹貢   [2] 書經   [3] 鄧植儀   [4] 禹貢說斷   [5] 傅寅

tious neighbours, so that eventually he came into the possession, as it were, of 'plantations' which he himself had not planted. True plantations ensued. Excellent examples of this principle are to be found in Africa.[a] The shea butter nut tree *Butyrospermum parkii* (Sapotaceae) is a very useful oilseed producer which grows where the oil palm does not, and each tree 'in the wild' is in fact protected and inherited.[b] Again the locust bean, the leguminous *Parkia filicoidea*, though a typical free-growing tree of park savannah country, is individually owned, its proprietor tending it and profiting from a remarkable variety of useful applications.[c] In the following translation of the Yü Kung we shall be able to pick out many examples of just such a situation in China in the middle or beginning of the −1st millennium.[d] This dating we continue to adopt advisedly, placing the text not later than the first part of the −5th century.[e]

It runs as follows:[f]

Yü disposed the lands (in order). Going along the mountains, he put the forests to use, felling the trees. He determined the high mountains and the great rivers.

---

[a] We are much indebted to Dr David Coombe of Christ's College, Cambridge, for advising us on this question.

[b] Dalziel (1), p. 350; Burkill (1), p. 385. The fat goes into margarine.

[c] Dalziel (1), p. 218; Burkill (1), p. 1668. The bark is used for tanning, the wood is easily worked, and the roots make a sponge. The pods give a fish poison, their lining is cut into strips for binding tape, the pulp is dried, powdered and eaten, and the seeds themselves are fermented into a cheese-like product.

[d] Indeed it still persists in some of the outlying parts of the Chinese culture-area. Visiting Hsishuangbana, the Thai autonomous region in Yunnan in 1958, Alley (9) found that the tea industry largely depended upon picking from wild tea bushes growing in profusion on the hillsides. Some were planted, but most came up through the undergrowth spontaneously. Yet the production is quite large.

[e] The question of its time of origin has been greatly debated. No one now accepts the traditional ascription to the −3rd millennium, but some Chinese philologists bring it down as late as the −3rd century, regarding it as an archaistic product of the late Warring States period; and some have even made it Han. Ku Chieh-Kang, who initiated the modern discussions (*KSP*, vol. 1, p. 206), considered that the Yü Kung could not be earlier than Confucius' time, the beginning of the −5th century, and Western sinologists (e.g. Creel, 4) have generally agreed with him. Whenever the material was first collected, it was during a period of unity, so the Warring States time itself can hardly qualify. During the centuries historical geographers tended to enlarge unduly the areas of the provinces, thus leading others to propose late dates on that gratuitous ground (cf. Herrmann (10), p. 10) but this may now be disregarded. The mention in the text of iron, however, and perhaps steel, is one feature which seems to us to preclude any frankly pre-Confucian date (cf. Needham, 32), unless it was a later interpolation. On the whole, recent Chinese scholarship agrees with the view that the Yü Kung chapter was written during the first half of the −5th century, about the time of the *Lun Yü*, after the five opening chapters of the *Shan Hai Ching*, and well before *Mo Tzu* and *Mêng Tzu*; see Tshao Wan-Ju (*1*); Chhü Wan-Li (1); Tshen Chung-Mien (2), pp. 72 ff. But the text is undeniably archaic, and one might say that it embodies what people in Confucius' own time, making use of perennial oral tradition, and wood or bamboo documents long since lost, knew of Shang and early Chou geographical, pedological and botanical stock-taking (c. −1100 to −700). Elaborate arguments in favour of their transmission of a genuine early Chou official document (perhaps −10th century) have recently been advanced by Hsin Shu-Chhih (1), in consultation with Ku Chieh-Kang himself and leading Japanese scholars. Moreover, the entries for the nine provinces which we shall give in what follows are interspersed with details of travel; while Karlgren (12) assumed that these reported a peregrination of Yü the Great himself, Legge (1) and Medhurst (1) had taken them to refer to the routes of the tribute-bearers making their way to the court. We think that here the older view was the better. Now Chavannes (5), p. 458, noted long ago that the routes all converged upon the southern part of Shansi, precisely the region where the Chou capital was fixed between −842 and −770, after its removal from Sian and before its establishment at Loyang. And perhaps it was hardly coincidence that the city of An-i, in the south-western corner of Shansi, was traditionally the capital of Yü the Great. The whole problem is certainly not settled, but there is much to support the historical interpretation which we adopt.

[f] Tr. auct., adjuv. Medhurst (1); Legge (1); Karlgren (12).

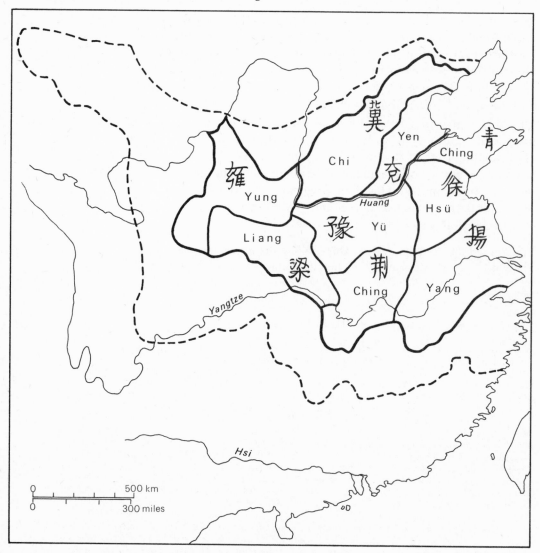

Fig. 23. Map of the Nine Provinces of the Yü Kung chapter of the *Shu Ching*; simplified from Herrmann (1), 1st ed., pp. 10–11. The line of dashes indicates the extent of the Shang Kingdom according to the commentators (now thought mistaken) of later Confucian tradition.

[1]^a Chi-chou¹ ...^b The soil of this province is of the *pai jang*² type. It yields revenues (*fu*³) of the upper first grade^c, with some admixture of lower grades. Its fields (*thien*⁴) lie on intermediate land (neither uplands nor lowlands).^d ... The Niao-I barbarians bring clothes made of skins and furs.

[2] Yen-chou⁵...^e The (wild) mulberry groves had many silkworms (*sang thu chi tshan*⁶), so that the people came down from the hills and dwelt in the land below. The soil of this province is of the *hei fên*⁷ type, its (characteristic) plant is the milk-vetch (*yao*⁸)^f, its (characteristic) tree the catalpa (*thiao*⁹).^g Its fields are on lower intermediate land. It yielded revenues of the lower third grade, and only after thirteen years of improvement work could they be raised to a level more like those of the other provinces.^h Its tribute is lacquer (*chhi*¹⁰)ⁱ and silk (*ssu*¹¹)^j, together with ornamentally coloured woven fabrics packed in baskets.^k

^a The numbering of the Nine Provinces is inserted by us.

^b Essentially all of modern Hopei north and west of the line of the Grand Canal and the Wei R. (cf. Vol. 4, pt 3, p. 269), i.e. the then course of the Yellow R.; together with nearly all of modern Shansi. Reference may be made to Herrmann (1) for a good conjectural map of the boundaries of the Nine Provinces (Fig. 23).

^c Nine grades are distinguished, by appropriate combinations of the characters *shang*¹², *chung*¹³ and *hsia*¹⁴.

^d Nine levels are distinguished, by the same combinations of the same characters. Our interpretation here departs from that which has been usual among commentators and translators, and we shall say a word about it at the conclusion.

^e This land, truly an 'isle-land' (cf. Schafer (4), p. 342), lay between the old course of the Yellow R. just mentioned, and the course of the Chi R. which it later captured or appropriated (cf. Fig. 859 in Vol. 4, pt 3). It was thus NW. Shantung and a little of E. Hopei.

^f Emending the *yu*^{15,16} of the text to *yao*, as in *yao chhê*¹⁷ and *chhiao yao*¹⁸ the later names of this vetch, *Astragalus sinicus* (CC 959; Lu Wên-Yü (1), p. 79). We have just noted that vetches were traditionally ploughed in as green manure (p. 75), but we must wait till Sect. 41 to consider how far back in history this practice went. Here is the first implementation of our view that the writer intended to refer to particular plants throughout his text, but that their names were blunted by centuries of literary, non-botanical, erudition into closely similar words denoting 'flourishing', 'abundance', etc. in general. We shall say something more about this also at the conclusion. The present emendation draws strength from a tradition about the wording recorded in *Chin Wên Shang Shu*, p. 9b, which indicates that the words concerned are nouns rather than adjectives.

^g This is only the most likely of several possibilities. The *Erh Ya* (cf. p. 187 below) says that *thiao* is the same as *thao*¹⁹, *fei*²⁰ and *chia*²¹, all being synonyms of *chhiu*²², which is certainly *Catalpa bungei* (CC 259), the tree pitied by Chuang Tzu (ch. 4, tr. Legge (5), vol. 1, p. 219), and often elsewhere here referred to (pp. 53, 128). But *thao*¹⁹ was also, according to the *Erh Ya*, a synonym of *yu*²³, the largest of the citrus fruits, *Citrus maxima* (CC 901), the pomelo or shaddock, which I used to enjoy in Szechuan. The latitude makes it less probable here, but one must remember (cf. Vol. 3, p. 463) that the climate of North China was formerly much hotter and wetter than now, so it cannot be quite excluded. And there is a third possibility. *Thiao* was also the name of a group of trees; as we shall see when dealing with nomenclature, it meant the deliquescent as opposed to the excrescent habit. So, says the *Erh Ya* (ch. 14, p. 11b), 'the mulberry and the willow belong to the *thiao* group (*sang liu chhou thiao*²⁴)'. This would fit in with the fact that silk was a product of the province.

^h The text is particularly obscure at this point, but we follow the interpretation of Chin Jen-Shan (see Tsêng Yün-Chhien (1), p. 54). Cf. *YKST*, ch. 1, p. 30a, b.

ⁱ From *Rhus vernicifera* (or *verniciflua*) of course (CC 844; Steward (1), p. 220). On the history of the lacquer industry see Sect. 42 below.

^j This fits in with the first sentence of our translation of the description of this province. One envisages tended, not planted, mulberry trees.

^k Karlgren's 'patterned woven stuffs', (12), p. 14, will not do at all here, for we dare not put the cardinal invention of the drawloom so early (see Sect. 31 below), though it may well be rather earlier than has previously been supposed (see Vol. 1, *sub voce*, Vol. 4, pt 2, p. 69). The implication is surely embroidery, or the use of various weft and warp colours, as in tartan patterns. We therefore prefer a phrase like that of Legge. Cf. *YKST*, ch. 1, p. 36a.

¹ 冀州	² 白壤	³ 賦	⁴ 田	⁵ 兗州
⁶ 桑土既蠶	⁷ 黑墳	⁸ 搖	⁹ 條	¹⁰ 漆
¹¹ 絲	¹² 上	¹³ 中	¹⁴ 下	¹⁵ 繇
¹⁶ 蘇	¹⁷ 搖車	¹⁸ 翹搖	¹⁹ 楢	²⁰ 檓
²¹ 椵	²² 楸	²³ 柚		²⁴ 桑柳醜條

[3] Chhing-chou[1]...[a] The soil of this province is of the *pai fên*[2] type, but along the sea coasts there are wide tracts of salt land (*kuang chhih*[3]). Its fields are on lower uplands, and its revenues are of the upper second grade. Its tribute articles are salt,[b] fine white vine-cloth (*chhih*[4])[c], and sea produce of various kinds; together with silk, hemp (*hsi*[5])[d], lead, pine-wood[e] and curious stones[f] from the valleys of Mount Tai.[g] The Lai-I barbarians, who live a pastoral life, bring baskets full of silk spun by worms from the mountain-mulberry tree (*yen (sang)*[6]).[h]

[4] Hsü-chou[7]...[i] The soil of this province is of the *chhih chih* and *fên*[8] types, its (characteristic) tree is a kind of *oak* (*chan*[9])[j], its (characteristic) plant a kind of spear-grass (*pao*[10]).[k] Its fields are on middle uplands, and it yields revenues of the middle second grade. Tribute comes as follows: earth of the five colours,[l] variegated pheasant (feathers)

[a] The north-eastern half of modern Shantung.

[b] On sea coast salterns and the refining process see Sect. 37 below.

[c] This was (and is) woven from the fibres of a leguminous climbing vine, *ko*[11], *Pueraria thunbergiana* (= *Pachyrhizus thunbergi*), CC 1038, R 406; Burkill (1), p. 1838. The most information will be found in B II 390. This was the 'dolichos cloth' of the older sinologists, so named from a botanical appellation now obsolete, *Dolichos hirsutus*. Besides the fine cloth called *chhih*, there was also anciently a coarse variety called *chhi*[12]. The vine is sometimes termed '(Chinese) arrowroot' (though omitted under this head by Willis (1) in his great dictionary), and indeed a starch (*ko fên*[13]) is extracted from the roots even now. A vigorous specimen of the plant may be seen in the Botanic Garden at Geneva.

[d] *Cannabis sativa*, at least co-ancient with silk in Chinese textile technology; cf. Lu Wên-Yü (*1*), p. 49.

[e] Very likely for the resin, and for the smoke obtained on combustion, from which the characteristic Chinese ink was made (see Vol. 3, p. 609, and Sect. 32 below). *Pinus sinensis* (= *tabulaeformis*), CC 2132; Lu Wên-Yü (*1*), p. 39.

[f] Cf. Vol. 3, p. 645. The love of *kuai shih*[14], curious stones and rocks (the very words used in this ancient text) became a permanent feature of Chinese aesthetics, and as we shall see, had a strong effect upon Europe when rock-gardens began there (Sect. 38 (*j*), 2 below). See also Schafer (11).

[g] I.e. Thai Shan, chief of China's sacred mountains.

[h] *Morus mongolica*, CC 1612, a good example of the 'tended', or half-wild state. On the 'Shantung silk' of our own times, see Sect. 31. Mountain-mulberry silk was considered to be especially good for the strings of musical instruments because of its high tensile strength.

[i] The land between the Chi and the Huai Rivers, i.e. the south-western half of Shantung, the northern third of Chiangsu and Anhui, and an eastern part of Honan.

[j] Emending the *chien*[15] of the text to *chan*, in accordance with *Ku Wên Shang Shu*, ch. 1, p. 19a, as noted in *Khang-Hsi Tzu Tien* (p. 575 or 713). *Chan* may be taken as equivalent to *chhien*[16], a board on which an inscription is cut, but if the word refers to the wood itself it is read *chan* and considered equivalent to *phu*[17] (ref. *Shuo Wên*, ch. 1B, (p. 24), ch. 6A, (p. 124)). Phu is most probably *Quercus dentata* (= *obovata*), CC 1645, R 613; Chhen Jung (*1*), p. 196, though Lu Wên-Yü prefers *Q. acutissima* (= *glandulifera*), (*1*), p. 67.

[k] Emending the *pao*[18] of the text to *pao*, in accordance with *YKST*, ch. 2, p. 2b and *Ku Wên Shang Shu*, ch. 1, p. 19a, as in *pao tzu tshao*[19], a graminaceous plant now called *Themeda gigantea* or *triandra* (CC 2063, B II 460) formerly *Anthistiria ciliata* or *caudata*. The *Shuo Wên* says that it was plaited into shoes and mats in Nanyang. Cf. Needham & Needham (1), p. 249.

[l] On this a story hangs, not unconnected with soil science. One of the most important cults of the ancient Chinese cosmic religion was the sacrifice at the altar of the God of the Soil (*shê*[20] or *chung thu*[21]); on which see Granet (4), p. 70, (5), p. 91, and especially Chavannes (5), pp. 437 ff., 450 ff. The flat altar tumulus was 50 ft. square, with sides of green earth on the east, red earth on the south, white earth on the west, and black earth on the north, while the top or centre was of yellow earth (*Chhun Chhiu Ta Chuan*, p. 1a, b). It was at this altar that the investitures of feudal lords (or in after times, imperial princes) were carried out. In his commentary on the present passage Khung An-Kuo[22], about −85, explains how this was done. 'The emperor's altar mound of the God of the Soil', he says, 'was made of the earths of the five colours. When a lord was enfeoffed with territory (in one or other of the four directions) he was presented with a sod cut from the altar on the side, and of the colour,

[1] 青州	[2] 白墳	[3] 廣斥	[4] 絺	[5] 枲
[6] 檿桑	[7] 徐州	[8] 赤埴墳	[9] 斬	[10] 苞
[11] 葛	[12] 綌	[13] 葛粉	[14] 怪石	[15] 漸
[16] 槧	[17] 樸	[18] 包	[19] 苞子草	[20] 社
[21] 冢土	[22] 孔安國			

Table 5. *The soil science of the Yü Kung chapter of the Shu Ching (Historical Classic)*

	Province	Geographical area in modern terms	Soil type	Revenue grade[a]	'Fields' grade (altitude)	Commentators' explanations	Identification
1	Chi 冀	NW Hopei and Shansi	*pai jang*	1	5	Clodless, porous, soft, silty, sometimes saline (1/17a)[b]	Carbonate cinnamon soils on loess. Some carbonate light meadow soils, partly saline.
2	Yen 兗	E Hopei and NW Shantung	*hei fên*	9	6	Fertile soil of gentle hill slopes, with just the right amount of moisture (1/29b)	Carbonate light meadow soils, partly saline.
3	Chhing 青	NE Shantung	*pai fên* (*kuang*) *chhih*	4	3	Also hill slopes, salinised near coast (1/38a) Saline-alkaline shore land	Cinnamon and brown forest soils, a few montane. Solonchak soils along littoral. Terra rossa locally.
4	Hsü 徐	SW Shantung, N Chiangsu & Anhui, and E Honan	*chhih chih* & *fên*	5	2	Sticky argillaceous red soil (2/2a)	Carbonate, and leached, light meadow soils, some saline. Some sajong soils.
5	Yang 揚	Chiangsu, Anhui, NE Chekiang and perhaps NW Chekiang	*thu ni*	7	9	Muddy, damp and low (2/12b)	Meadow-bog soils and old irrigated ricefield soils, on non-calcareous alluvium.
6	Ching 荊	W Anhui, N Chiangsi, N Hunan and S Honan	*thu ni*	3	8[c]	Muddy, damp and low (2/23a)	Meadow-bog soils and old irrigated ricefield soils, on non-calcareous alluvium.
7	Yü 豫	All Honan south of the Yellow R. and north of the Huai R.	*jang fên lu* (&?)	2	4	Porous, mixed-colour above, loessial silty; hard black earth below (2/31a)	If territorial, *jang* = carbonate, and leached, cinnamon soils; *fên lu* = light meadow soils with sajongs. If vertical, sajong-containing soils alone.
8	Liang 梁	S Shensi and S Kansu, NW Hupei and the northern edge of Szechuan	*chhing li*	8	7	Dark 'blue-green-black' soil fertile but not very porous (2/34b)	Montane humose dark forest soils along Han R. valley.
9	Yung 雍	N Shensi and E Kansu	*huang jang*	6	1	Original loessial porous soil (2/46b)	Grey, carbonate, and montane cinnamon soils, and kheilutu soil, all on loess. Carbonate light meadow soils, sometimes saline.

a Omitting the 'admixtures'.
b References to the discussions in *YKST*.
c Lin Chih-Chhi remarked, about +1160, 'The soil of this province, though *thu ni* like of Yang-chou, lies on lands of a rather more elevated level, so it is ranked as one grade higher' (*YKST*, ch. 2, p. 23a).

from the valleys of Mount Yü, *ku thung*[7] trees from the southern slopes of Mount Yi,[a] chime-stones from the frothy rocks along the banks of the Ssu River,[b] oyster pearls brought by the Huai-I barbarians,[c] and fish. Silk fabrics, black, white, and a sarcenet woven with black warp and white weft (*hsien*[8]), come packed in baskets.[d]

[5] Yang-chou[9] . . .[e] Here the bamboos, both small and large (*hsiao tang*[10])[f] abound. The (characteristic) plant is a kind of thistle (*yao*[11])[g], the (characteristic) trees are all upward-branching (*chhiao*[12]).[h] The soil of this province is of the *thu ni*[13] variety. Its fields are on the lower lowlands, and it yields revenues of the upper third grade, though some parts do better than this. Its tribute consists of the three metals[i], precious stones called *yao*[14] and *kun*[15], fine and coarse bamboos[j], ivory, (rhinoceros) hide,[k] (coloured) feathers, hair (*mao*[16])[l]

appropriate to the location of his fief, after which he bore it home to form (the nucleus of) his own altar to the god of the soil. Some of the yellow earth (from the centre) was scattered over the sod, which was then wrapped in floss-grass (*pai mao*[1], *Imperata cylindrica*). The grass symbolised purity (loyalty), and the yellow earth the Ruler's imperium over all the Four Quarters.' Wang Su[2] says the same, about +230, in shortened form (*Shang Shu Wang shih Chu*, ch. 1, p. 11*b*). Since the authenticity of the similar passage in the *I Chou Shu* (ch. 48), purporting to be early Chou, is suspect, Khung's description may be the oldest we have (apart from the bare reference in the Yü Kung itself), for the detailed explanation in the *Shih Chi* (ch. 60, p. 10*a, b*) is one of the additions of Chhu Shao-Sun[3], *c.* +10, and not written by Ssuma Chhien. Other accounts, such as that in the *Tu Tuan*[4] (Imperial Decisions and Definitions) by Tshai Yung[5], *c.* +190, are all later. As Chavannes points out, the etymology of the character *fêng*[6], enfeoffment, shows a piece of land with a plant growing on it, alongside a length measure and a hand (the radical which now means an inch, cf. *Shuo Wên*, ch. 13B (p. 287.2); K 1197 i, j). Thus just as in the medieval Western world, enffeofment was *per herbam et terram*.

[a] Some variety, not now easily identifiable, of *Paulownia fortunei* (= *tomentosa*), CC 274. Chhen Jung (*1*), pp. 1105 ff. gives details of many varieties, but none with this particular name. This is the 'dryandra tree' of the older sinologists, so named from an obsolete botanical appellation of an entirely different tree, *Aleurites cordata*, CC 854, euphorbiaceous not bignoniaceous. In *YKST*, ch. 2, p. 3*b* they say the *ku thung* wood was good for lutes.

[b] See Vol. 4, pt 1, p. 148 and passim.

[c] See Sect. 29 (i), 4, in Vol. 4, pt 3, pp. 668 ff.

[d] Thin silk tissue. Couvreur, in his Dictionary, was the only translator who understood this phrase aright.

[e] Approximately the region between the Huai and the Chhien-thang Rivers, comprising most of modern Chiangsu and Anhui, with a piece of northeastern Chiangsi, i.e. the Lower Yangtze valley and its delta.

[f] *Hsiao* means dwarf bamboos, perhaps of several species, among which we might identify *Arundinaria hindsii* (CC 2069; cf. Hagerty (*2*), pp. 416 ff.) and *Bambusa tuldoides* (CC 2077; cf. Hagerty (*2*), p. 419). *Tang* means large bamboos, with long internodes, such as *Phyllostachys bambusoides* (CC 2080; cf. Hagerty (*2*), p. 412), which grows more than 60 ft tall.

[g] Emending the *yao*[17] of the text to *yao*, as in *khu yao*[18], typical of Chiang su, i.e. *Cirsium nipponicum* (= *Cnicus chinensis*), CC 68, R 28. This emendation was proposed in +1099 by Lu Tien[23] in his *Erh Ya Hsin I*[24] (Fresh Interpretations of the *Literary Expositor*), ch. 12 (p. 373), cf. ch. 15 (pp. 454, 455).

[h] The *Erh Ya* says (ch. 14, pp. 10*b*, 11*b*) that the *huai* and the *chi* belong to the group of excrescent trees (*huai chi chhou chhiao*[19]) and so does the *chhiu*.[20] Thus we have in order *Sophora japonica* (CC 1045), *Zizyphus spinosus* (CC 776) and *Catalpa bungei* (CC 259).

[i] Gold, silver and copper.

[j] The words used are the same as those at the opening of the entry.

[k] Cf. the point made just above (p. 85) that central China was in ancient times much hotter and wetter than now. Hide was for armour (see Sect. 30).

[l] This may be what it seems, and commentators followed by Tsêng Yün-Chhien (*1*), p. 60, talk about 'flags' (*mao*[21]) made somehow of oxtails. But this *mao* is also the name of a plant, one of those mentioned in the *Erh Ya*, the winter peach, identified with *Prunus persica* var. *hiemalis* (R 448, CC 1104). Even the 'oxtails' could refer to a plant, the oxtail southern-wood, a composite related to *Artemisia* and used for incense and dyeing in sacrifices (*niu wei hao*[22])—*Anaphalis yedoensis* (CC 7; Lu Wên-Yü (*1*), p. 47; B II 435).

[1] 孤桐	[2] 纖	[3] 白茅	[4] 王肅	[5] 褚少孫
[6] 獨斷	[7] 蔡邕	[8] 封	[9] 揚州	[10] 篠簜
[11] 芺	[12] 喬	[13] 塗泥	[14] 瑤	[15] 琨
[16] 毛	[17] 夭	[18] 苦芺	[19] 槐棘醜喬	[20] 楸
[21] 旄	[22] 牛尾蒿	[23] 陸佃	[24] 爾雅新義	

and timber.[a] The Tao-I barbarians bring garments made of grass-cloth.[b] Woven textiles (*chih pei*[1]) come packed in baskets,[c] and bundles containing oranges and pommeloes (*chü yu*[2])[d] are sent up when requested.

[6] Ching-chou[9] ... [e] The soil of this province is also of the *thu ni*[10] variety. Its fields are on middling lowlands and it yields revenues of the lower first grade. As tribute it provides feathers, hair,[f] ivory, (rhinoceros) hide, gold, silver and copper. Item, toona mahogany (*chhun*[11])[g], mulberry wood for bows (*kan*[12])[h], juniper (*kua*[13])[i], and arbor vitae (*po*[14])[j]. Item,

[a] Commentators, as in *YKST*, ch. 2, p. 13*b*, again followed by Tsêng, say that the timber was *kêng, tsu, yü-chang*[3]. *Kêng* is *Hemiptelea davidi*, one of the elms (CC 1618), and *tzu* is usually *Catalpa bungei* again, or sometimes *Celtis sinensis*, another elm (CC 1616), the more particular name of which is *chia*[4]. *Yü-chang* is the ancient term, thrice mentioned in the *Shan Hai Ching*, for the *chang*[8] i.e. the camphor-laurel, *Cinnamomum Camphora*, on which see B II 518 and Chhen Jung (*1*), p. 332. From its name one would have expected it to have been contributed by Yü-chou rather than Yang-chou, but no doubt it grew well in both provinces.

[b] Some commentators (*YKST*, ch. 2, p. 14*a*) say that this was the 'vine-cloth' just explained (p. 86), but more probably it was the stuff woven from the fibres of the urticaceous plant *Boehmeria nivea*, called *chu ma*[5],[6] or ramie. This gives a cloth agreeably cool to wear in summer. Parts of ch. 78 of the *Shou Shih Thung Khao* dealing with the industry were translated long ago by Julien & Champion (*1*), pp. 162 ff. The plant may be seen growing at the Chelsea Physic Garden in London.

[c] This is a proper crux. What was this textile? A prominent later name for cotton was *chi-pei*[7], but all the evidence shows that cotton was not grown in South China till the +4th or +5th century, and not in North China till the +13th. *Chi-pei* is a transliteration of Skr. *karpāsa*, and the home of cotton was undoubtedly India. As the orthography varies considerably, Medhurst felt justified in interpreting this phrase as meaning cotton, but is it possible to believe that the 'island barbarians' were already engaging in the Nanhai trade as early as Confucius' time? Legge and Karlgren, conscious of the difficulty, suggested 'stuffs woven in cowrie patterns', but the existence of the drawloom is at least as dubious for this period. On the complex history of cotton see Sect. 31, and meanwhile Wittfogel & Fêng Chia-Shêng (*1*), pp. 155 ff.

The possibility of southern products of the 'kapok' or 'tree-cotton' type, whether woven or as floss for padding, must also be borne in mind. Kapok is the floss from the fruits of *Ceiba pentandra* of the Bombacaceae, a tree (Burkill (*1*), vol. 1, pp. 501 ff.). The commentators (*YKST*, ch. 2, pp. 14*b*, 15*a*) are here particularly uncertain, and their other suggestions include *Boehmeria* grass-cloth and real cowrie-shells. The latter were certainly used as currency (see Yang Lien-Shêng (*3*), pp. 12 ff.). It is interesting that Chhen Tsu-Kuei (*1*) opens his anthology of quotations concerning cotton and tree-cotton in Chinese texts with this Yü Kung passage.

[d] *Chü* is a general term for all oranges; *yu* is *Citrus maxima* (CC 901).

[e] Western Anhui, northern Chiangsi and northern Hunan around the great lakes, and Honan south of the Huai R., plus eastern Hupei.

[f] See p. 88 just above.

[g] Equivalent to *chhun*[15] and here interpreted as the fragrant *chhun*, *Toona sinensis*, one of the Meliaceae, which gives such valuable wood. See CC 885; R 334; Burkill (*1*), p. 499; Steward (*1*), p. 207; Watt (*1*), p. 290; Collett (*1*), p. 83; Chhen Jung (*1*), p. 602. A fine specimen grows in the Geneva Botanic Garden. There is also a malodorous *chhun*, more properly called *chhu*[16] (anciently *shu*), one of the Simaroubaceae or quassia family, the 'tree of heaven', *Ailanthus glandulosa* (= *altissima*), CC 892, but its wood is not nearly so useful and we think it was not intended here. A tall one is in the Master's Garden at Gonville and Caius College, Cambridge.

[h] There is no doubt that *kan* was a term for bow-wood in general, and the *Chou Li* tells us exactly what it comprised (ch. 12, p. 24*b* (ch. 44), tr. Biot (*1*), vol. 2, p. 582). There were seven sorts, of which *chê*[17] was the best, and *yen sang*[18] the third best. The latter we have just met with, *Morus mongolica* (CC 1612); the former is another moraceous tree, *Cudrania tricuspidata* or *triloba* (CC 1594; Lu Wên-Yü (*1*), p. 118). On the bowyers mem. Vol. 4, pt 2, p. 16.

[i] Also read *kuai*, identical with *kuei*[19], i.e. *Juniperus chinensis* (B II 506; CC 2143). The habitat fits (Lu Wên-Yü (*1*), p. 38).

[j] In North China always *Biota (Thuja) orientalis* (CC 2146; B II 505; Lu Wên-Yü (*1*), p. 16) and this must be meant here. The name is sometimes applied in the south, however, to *Cupressus funebris* (CC 2141).

[1] 織貝	[2] 橘柚	[3] 梗梓豫章	[4] 檟	[5] 苧麻
[6] 紵麻	[7] 吉貝	[8] 樟	[9] 荆州	[10] 塗泥
[11] 杶	[12] 榦, 斡, 幹	[13] 栝	[14] 柏, 栢	[15] 椿
[16] 樗	[17] 柘	[18] 檿桑	[19] 檜	

grindstones and whetstones (*li chih*[1])[a] flint arrow-heads (*nu*[2])[b], and cinnabar. Three districts furnish the *chhün*[3] and *lu*[4] bamboos[c], and the wood of the arrowthorn shrub (*hu*[5]).[d] Most famous are the bundles[e] that come packed with sweet haws (*chhiu*[6])[f] and filtering sedge (*ching-mao*[7])[g]. There are also baskets of black and red silk fabrics, with silken cords for threading coarse pearls. From the Chiu-Chiang (Nine Rivers) districts great turtles are sent up when caught.[h]

[7] Yü-chou[15] ...[i] The soil of this province is of the *jang*[16] kind, but underneath (or, lower) there is *fên lu*[17] earth.[j] Its fields lie on the higher sort of intermediate land (neither uplands nor lowlands), and its revenues are of the middle first grade though (certain districts) yield less or more. Tribute from this province is paid in lacquer,[k] hempen cloth,[l] fine white vine-cloth, and ramie grass-cloth.[m] Black and white sarcenets[n] and quantities of floss silk (*khuang*[18])[o] come packed in baskets. Abrasive (sand and) stone (laminae) for working (jade and) chime-stones are also provided when requisitioned.[p]

[8] Liang-chou[19] ...[q] The soil of this province is what is called *chhing li*[20]. Its fields lie on the higher sort of lowlands; and it yields revenues of the middle third grade, though some

[a]　Mem. Vol. 4, pt 1, p. 117, pt 2, p. 55.

[b]　See on this the full discussion in Chang Hung-Chao (*1*), pp. 407 ff., and for a wider relevance Needham (56), p. 34. In historical times flint arrow-heads or flint for making them was found in at least two places in China, one in Szechuan (*Hua Yang Kuo Chih*, ch. 3; cf. Vol. 3, p. 517), and another in Chiangsi, the present province (*Yün Lin Shih Phu*, tr. Schafer (11), p. 78).

[c]　These are coupled also in the oldest extant monograph on bamboos, the *Chu Phu*[8] of Tai Khai-Chih[9] (tr. Hagerty (2), p. 413), on which see p. 378 below, but it is not very easy to identify them. *Chhün* is said to be a 'black' or dark bamboo used for arrows, so probably *Phyllostachys nigra* (= *nidularia*), CC 2082. *Lu* was also used for arrows. *Pace* Tai's weighty authority, *chhün-lu* could have been a binome.

[d]　*Vitex Negundo* var. *incisa* (CC 379, 380, *cannabifolia* and *trifolia*; see B II 521, 543). From the time of the *Kuo Yü* onwards, this particular wood was always associated with the flint arrow-heads (cf. the passages collected by Chang Hung-Chao (*1*), pp. 407 ff.); i.e. from at least as early as Chhin and Han.

[e]　Some of the commentators consider that the term *pao*[14] here implies that this province also sent up oranges and pomeloes (cf. Yeh Ching-Yuan (2), p. 13). See p. 363 below.

[f]　Emending the *kuei*[10] of the text to *chhiu*, in accordance with the equation in the *Shuo Wên Thung Hsün Ting Shêng*, ch. 6, p. 30b. Tsêng Yün-Chhien (*1*), p. 64, in substance concurs. The *chhiu* is the *chi mei*[11], now called *shan cha*[12], i.e. *Crataegus* (≃ *Mespilus*) *pinnatifida* (CC 1065; R 422. 423; Chhen Jung (*1*), p. 441). The fruits were made into preserves.

[g]　This was a kind of sedge used for filtering the wine used in State sacrifices, most probably *Cyperus iria* (CC 1962; Steward (*1*), p. 493). The characteristic triangular stems are mentioned in the *Kuan Tzu* book (ch. 83, pp. 9b, 10a) in an amusing story (Ching-Mao Mou[13]) with an economic moral (tr. Than Po-Fu *et al.* (*1*), p. 190).

[h]　Probably *Emys reevesii* (R 199).

[i]　All modern Honan south of the Yellow River.

[j]　Previous translators have always taken this as referring to the lower lying land; we see a possible reference to the subsoil.

[k]　As in the case of Yen-chou.

[l]　As in the case of Chhing-chou; see the explanations in its footnotes.

[m]　As in the case of Yang-chou; see footnotes.

[n]　The meaning of this has been explained under Hsü-chou.

[o]　Short-staple silk from broken cocoons was used from high antiquity for the making of padded clothes. See Sect. 31.

[p]　This is the traditionally accepted interpretation (cf. Tsêng Yün-Chhien (*1*), p. 66; I do not know why Karlgren departed from it. It makes excellent sense, as can be seen from Vol. 3, p. 666.

[q]　South Shensi and South Kansu, with Western Hupei north of the Yangtze R., and perhaps the northern edge of Szechuan but certainly not the 'Red Basin' itself. Liang-chou was centered on the Han River valley, and its boundary from the last province, Yung-chou, seems to have run along the top of the Chhin-ling Shan.

[1] 礪砥	[2] 砮	[3] 箘	[4] 簵	[5] 楛
[6] 杭	[7] 菁茅	[8] 竹譜	[9] 戴凱之	[10] 匭
[11] 繫梅	[12] 山楂	[13] 菁茅謀	[14] 包	[15] 豫州
[16] 壤	[17] 墳壚	[18] 纊	[19] 梁州	[20] 青黎

districts do as well as the upper third or lower second grade, and others fall to the lower third grade. By way of tribute it sends up jade (*chhiu*[1])[a], iron, silver, inlaid metal work (*lou*[2])[b], flint arrow-heads,[c] and chime-stones; together with the furs of bears, both black, and brown and white, foxes and wild cats.[d]

[9] Yung-chou[3]...[e] The soil of this province is *huang jang*[4]. Its fields are on high uplands and it yields revenues of the lower second grade. It contributes various kinds of jade (*chhiu lin*[5])[f] and the precious stone called *lang-kan*[6] [g]... Furs (*chih phi*[7]) are brought by the Khun-lun, the Hsi-chih and the Chü-sou peoples, all Western barbarians (Hsi Jung[8].)[h]

It seems obvious that this text has been truncated at some time or other, so that certain parts have dropped out. The fullest entries are nos. 2, 4, and 5; they give the characteristic plants of the region, the nature of the soil, the altitude of the farmland, the richness of the revenues, the articles of tribute regularly or periodically transmitted to the capital, and whatever was brought in by the peripheral barbarians. Nos. 3, 6 and 7 leave out the characteristic plants, but contain much of botanical interest because of the plant products contributed. Nos. 1, 8 and 9 speak of nothing but the soil, altitude, revenues, and mineral and animal tribute products. Nevertheless we believe that originally all the descriptions were as full as the three fullest ones which we still possess.

Comparison of the translation here given with versions previously available will make manifest some interesting differences about which a word must be said. The older sinologists accepted many words such as 'flourishing', 'bushy' and 'luxuriant', at their face value, not realising that these were cloaking (as we think)

[a] Presumably jadeite, for true nephrite seems not to occur in China proper; but see the discussion in Vol. 3, p. 665.

[b] This again is a crux. Most commentators and all translators say steel, but one cannot feel confident that steel-making from wrought iron blooms was understood at this early date (cf. Needham, 32), or indeed that the cementation process was ever practised in China at all. Steel-making may have taken an entirely different course. By the Han time, of course, the word *lou* meant steel as well as metal inlaying (as we know from *Shuo Wên*, ch. 14A, (p. 294.1)), but we suspect that in the time of the Yü Kung writer the second meaning was much more prominent. Archaeological evidence of great skill in the inlaying of bronze, iron and the precious metals during the Chou period is now coming to light (see Chêng Tê-Khun (9), vol. 3, pp. 69, 88).

[c] See the entry under Ching-chou.

[d] I.e. *Ursus torquatus* (R 361), *Ursus arctos* (R 361a), *Vulpes japonicus* (R 374) and *Felis catus* (R 372) in order. For further details see also Tu Ya-Chhüan, Tu Chiu-Thien *et al.* (*1*) *sub voce*.

[e] North Shensi up to the line of the later Great Wall across the Ordos Desert, and Eastern Kansu as far as the Yellow River. The province did not, as oftern thought, include the Kansu corridor.

[f] Possibly this was true nephrite, obtained by trade from the neighbourhood of Khotan.

[g] The nature of this has been widely discussed, as by Chang Hung-Chao (*1*), pp. 23 ff., who concluded that the red *lang-kan* of antiquity was ruby spinel imported from Badakhshan, but that the blue *lang-kan* of later times was coral or turquoise. The most recent view (Schafer (11), p. 95) is that most of it was always coral. But could red Mediterranean coral really have reached China in the time of the Yü Kung?

[h] The expression here used, *chih phi*, is the same as that in the case of Liang-chou, lit. 'woven skins' or perhaps 'textiles and skins'. Karlgren suggests felt in both cases, an ingenious conjecture not supported by the classical paper on the history of felt-making (Laufer, 24), but not incompatible with the evidence presented therein. It was contact with the Huns, Laufer thought, which familiarised the Chinese with felt, and he found no literary references to it earlier than about −300. Furs may therefore be a safer translation. On felt in China cf. Olschki (7). The standard term for felt is *chan*[9]. Another possibility in both places would be 'tanned leather', but neither furs nor woollen cloth nor tree-cotton cloth can be excluded.

[1] 璆   [2] 鏤   [3] 雍州   [4] 黃壤   [5] 球琳
[6] 琅玕   [7] 織皮   [8] 西戎   [9] 氈,毯,毡,旃

ancient but specific botanical names. Only after becoming really familiar with texts such as the *Erh Ya*[a] and the *Hsia Hsiao Chêng*[b] can one hope to identify them, for it must be remembered that in ancient times the 'plants' radical and the 'tree' radical[c] were frequently omitted—indeed the names date back to a time before these fixative and defining radicals were customarily put in. Thus it was natural that most of the literary commentators in subsequent centuries, erudite indeed but not botanically minded, missed the precision which the venerable catalogue had once possessed. As for the classification and terminology of predominating soil types, the old sinologists shuffled a pack of most peculiar cards—'miry' and 'briny', 'fat', 'mellow' 'loamy' and even 'mouldy'. Perhaps in the absence of adequate soil surveys there was little else they could do, but now at any rate we have the means of identifying what the writer of the Yü Kung chapter was talking about with some reasonable plausibility. Before doing this, however, there is one more element of new interpretation which needs explaining.

Everyone who reads the traditional translations is puzzled by the fact that they contain two separate quality ratings expressed in terms of nine combinations of *shang*, *chung* and *hsia*, one for the revenues and one for the 'fields'.[d] If the former measures productivity value, as it must, what is the latter doing? There is no case where the ratings coincide. For example, the province (Chi-chou, no. 1) most productive in revenue (*shang shang*, first grade) is only rated intermediate (*chung chung*, fifth grade) for its 'fields'. Or again, Yang-chou (no. 5) has 'fields' the lowest of the low (*hsia hsia*, ninth grade) yet its revenue is the highest of the low (*hsia shang*, seventh grade). Yet in this very case, other sources of high authority only slightly less ancient, such as the *Chou Li*, state that the land is excellent for rice-growing.[e] Thus there is a sharp contradiction. Similarly the Yü Kung rates the fields of Yü-chou (prov. no. 7) only as upper second grade (the fourth) while the *Chou Li* says that they are excellent for growing all the five grains (wheat, rice, the two millets, and beans).[f] Considerations such as these have therefore led us to propose that the second classification was really one of altitude, distinguishing uplands from lowlands. And one has only to look at the Nine Provinces on the map to see that this corresponds in fact very reasonably with the topography of the country.

The *Chou Li* passages are so interesting in themselves that one at least demands to be given in full. It concerns the official called Thu Hsün[1] (Geographer-Royal), already introduced to the reader at an earlier stage.[g] We shall now see that he was

---

[a] Cf. p. 192 below, where the 'Literary Expositor' will be discussed more fully.

[b] Cf. Vol. 3, p. 194 for an account of the 'Lesser Annuary of the Hsia Dynasty'.

[c] Cf. pp. 117 ff. below, on the 'botanical' radicals in the written characters.

[d] It is of course possible to try to resolve the discrepancies by introducing the factor of population density, but as no evidence exists on the basis of which figures could be presented, the argument must necessarily be rather unconvincing. Still, it was traditional, cf. *YKST*, ch. 1, p. 17*b*.

[e] Cf. 4, p. 34b.   [f] Cf. 4, pp. 25*a* ff.   [g] Vol. 3, p. 534.

[1] 土訓

very much of an economic and botanical geographer. His duties are explained as follows:[a]

He takes charge of the maps (of the regions and districts), and explains them (to the emperor) so that edicts can be issued appropriate to the agriculture in different places.

> [Comm.] (Chêng Khang-Chhêng): The maps record the topography (*hsing shih*[1]) of the Nine Provinces, and what (types of vegetation do) best among mountains or in river valleys. He tells the emperor so that allocation of undertakings can be made.[b] So for instance in Ching-chou and Yang-chou the land is good for planting rice, while in Yu-chou[2] and Ping-chou[3] the land is good for planting hemp.[c]

He explains the evils (*thê*[4]) of particular places in order to distinguish between their produce, and he shows how (plant products) originate and grow so that edicts can be issued appropriate to taxation demands in different places.

> [Comm.] (Chêng Khang-Chhêng): In considering the products of places where the land is bad (*thê*), as if in some way inhibited by disease, one must distinguish first what the land has and what it lacks,[d] and then (find out) what does originate and grow there (well) and at what season. On these two points report must be made to the emperor (to guide him in) taxation. What the land has not got, and will not produce well, that (the emperor will) not (want) to tax (or to plant).
>
> (Chêng Ssu-Nung) 'Bad lands' also means country which produces evil things that harm men, such as worms, parasites, poisonous snakes and the like.[e]

When the emperor is making a tour of inspection, the Thu Hsün (Geographer-Royal) rides close to the imperial vehicle.

Elsewhere in the *Chou Li*, as usual, we are told that the Thu Hsün has a staff of sixteen assistants.[f]

He has two graduates of the second class, four graduates of the third class, two secretaries and eight clerks.

> [Comm.] (Chêng Ssu-Nung): Their duties are to acquaint the emperor with the strange products which grow in regions far away.
>
> (Chêng Khang-Chhêng): They have to be able to explain and discuss the various good and bad properties of different soils and localities (*thu ti shan o chih shih*[5]).

In yet another place, when describing the Chih Fang Shih[6] (Directors of Regions) and their duties, the *Chou Li* launches out into an elaborate enumeration of the

---

[a] *Chou Li*, ch. 4, p. 34*b* (ch. 16); tr. auct. adjuv. Biot (1), vol. 1, pp. 368 ff.

[b] The 'economic high command' of ancient bureaucratic feudalism; cf. Sect. 48.

[c] These were two provinces separated off from the Chi-chou of the Yü Kung. The former was more or less equivalent to modern Liaoning and north Hopei; the latter to modern Shansi. The distinction between south and north is obvious here.

[d] How interested the writer would have been in our modern knowledge of deficiencies of phosphorus, nitrogen or trace elements.

[e] Cf. Sects. 39 and 44. I am reminded of a visit I paid in 1964 to the National Schistosomiasis Research Institute on the shores of Thai-hu Lake near Wuhsi. An epic struggle is now being conducted in China against this age-old scourge.

[f] Ch. 3, p. 7*a* (ch. 8), tr. auct. adjuv. Biot (1), vol. 1, p. 185.

[1] 形勢   [2] 幽州   [3] 并州   [4] 惡   [5] 土地善惡之勢
[6] 職方氏

Nine Provinces,[a] listing in each case the guardian mountain, the principal river, the largest lake, the most important irrigation reservoir, the commodities produced and the best farm crops.[b] It is here that we find contradictions with the usual Yü Kung interpretation all along the line; Ching-chou is splendid for rice—but its 'fields' are only eighth grade; Yung-chou can only manage the two millets—but its 'fields' are of the very first grade. Something must be wrong. But the other view makes sense, for Ching-chou was the region around the great lakes in the Yangtze valley while Yung-chou was the uplands of Kansu and Shensi.

The trouble arises, of course, from the ambiguity of the words *shang*, *hsia*, in Chinese. They can mean not only above and below, higher and lower in space, but also better or worse in quality, and before and after, earlier or later in time, to say nothing of verbic usages such as to go, or send, up or down, to exalt or to depreciate. In his study of the agronomic chapters of the *Lü Shih Chhun Chhiu* (Master Lü's Spring and Autumn Annals, −249) Hsia Wei-Ying draws attention to examples where just this primary meaning is applied to farm land; the upland fields (*shang thien*[1]) are high-lying and dry, the lowland ones (*hsia thien*[2]) are low-lying and damp.[c] A similar understanding emerges from an important passage in the *Nung Sang Chi Yao*[3] (Fundamentals of Agriculture and Sericulture) of +1273, which reads as follows:[d]

If you have *thu ni*[4] soil present, and the fields lie upon middle or lower (lowlands), rice can be planted. There is no need to restrict the planting of rice to Yang-chou and Ching-chou. In the same manner, if white or yellow *jang*[5] soils are present, and the fields lie upon upper or middle (uplands or midlands)[e], the two millets, beans and kaoliang can all be sown. There is no need to restrict the sowing of the two millets, beans and kaoliang to Yung-chou and Chi-chou.

Here again is a clear comparison between the south and the north. The passage is of much historical importance because in the Sung and Yuan periods people were breaking away from the age-old associations of particular crops with particular regions, and finding that within specific limiting factors of soil and climate, crops might do very well quite outside their conventional areas.[f] If *shang* and *hsia* were here interpreted as some kind of soil fertility rating instead of altitude (and hence

---

[a] Here Yu-chou and Ping-chou are included, but Hsü-chou and Liang-chou omitted, the number thus remaining the same.

[b] Ch. 4, pp. 24b ff. (ch. 33), tr. Biot (1), vol. 2, pp. 264 ff. Mem. Vol. 3, p. 534.

[c] (3), p. 67, cf. p. 38. The passage occurs in ch. 159. R. Wilhelm (3), p. 457, did not get it right. Of course other meanings also occur in the same book, cf. Hsia Wei-Ying, *loc. cit.* p. 10. Hsin Shu-Chhih (1), pp. 22, 23, sees altitude in the Yü Kung.

[d] Ch. 2, pp. 11b, 12a, tr. auct.

[e] The reader will understand, and excuse, the momentary adoption of a convenient word which normally has quite a different sense in English.

[f] Cf. Sect. 41.

[1] 上田   [2] 下田   [3] 農桑輯要   [4] 塗泥   [5] 壤

climate), the passage would be almost meaningless.[a] However, we recognise that the problem is still open, and our interpretation is given with all due reservations.[b]

Now we come to the identification of the soil types mentioned in the Yü Kung text, which in certain cases, as we have noted, associates them with particular herbs and trees. One must emphasise that any attempt of this kind, unavoidably seductive though it is, must necessarily contain a large element of speculation, for one is trying to identify soils which were described by observers nearly three thousand years ago; the point of greatest substance, valid whatever one may say about details, is that some time between the −8th and the −4th centuries serious study of soil types was made by Chinese naturalists, who worked out a schematic geography of them for a large part of their vast country. Efforts at identification started with a pioneer paper by Shih Ya-Fêng (1), and its most thorough treatment so far is perhaps that by Wan Kuo-Ting (2).[c] To facilitate understanding, the chief data are recapitulated and summarised in Table 5, which should be examined in connection with the map in Fig. 21. In principle, as noted earlier, our soil classification conforms with that of the school of Kovda.[d] An obvious point is that identifications have to refer to areas likely to have been under cultivation at the time when the Yü Kung was written; areas then probably forested, especially those at the higher altitudes, have deliberately been omitted from our interpretations. After we had come to our conclusions we gained access to the further paper of Chang Han-Chieh (1), and were glad to find that he had reached approximately the same ones.[e]

It may be convenient to begin with Yung-chou (province no. 9) where, if the metaphor be allowed, we are certain of our ground; it covered of course the high uplands of the loess plateau, together with the Wei and Ching river valleys. One can deduce therefore that 'yellow *jang*' (*huang jang*[4]) was a generic name for all the types of cinnamon and kheilutu soils, i.e. the loessial pedocals; including however as well all those modified forms which we described in the Wei valley

---

[a] Similar passages are easy to find in the agricultural treatises, for example *Nung Shu* (+1313), ch. 1, p. 7a. Knowledge of soils and climates was 'what the sages called differentiating between the profitability of lands (*tzhu shêng jen so wei fên ti chih li chê yeh*[1]).' Altitude is again prominent in Chêng Chhêng's[3] famous +3rd-century commentary on *Hsiao Ching*[2] (Filial Piety Classic), ch. 6, p. 5b. Millets for uplands, he says, rice and wheat for lowlands, and jujube and chestnut trees for hilly land; thus the farmer's filial son will plant. Cf. Legge (1), p. 472.

[b] If further research could demonstrate that the 'fields' scale of the Yü Kung really was a soil fertility rating, the ancient Chinese would at least gain the credit of having made a clear distinction between soil fertility as such and economic productivity, whatever the social factors which intervened between the two.

[c] In consultation with the soil scientists Chhen Ên-Fêng & Yü Thien-Jen. Dr Wan, alas no more with us, was among the most important of modern historians of agriculture and the biological sciences in China, and I am glad indeed that in 1958 Dr Lu Gwei-Djen and I had the pleasure of attending a specially arranged symposium in Nanking with him and his collaborators.

[d] And therefore indirectly with that of the soil scientists of Academia Sinica, for Kovda himself spent a number of years in China working in collaboration with them.

[e] The same is true of a much earlier paper by Wang Kuang-Wei (1). But as he was working in the thirties before even Thorp and his Chinese collaborators had reported, his insight was naturally weaker.

[1] 此聖人所謂分地之利者也   [2] 鄭佀   [3] 孝經   [4] 黃壤

(Kuanchung) according to the interpretation of the *Kuan Tzu* book by Wang Ta
(cf. p. 80 above). Following our principle we do not include those montane brown
forest and similar soils which doubtless existed in newly-cleared areas especially in
the western parts of the province. Province no. 1, Chi-chou, was also largely a
plateau, the uplands of modern Shansi,[a] including, however, a strip along the foot
of the Thai-hang Shan[b] escarpment and a corridor to the sea east of modern
Peking. The whole area was loessial (cf. Fig. 6 in Vol. 1), whether parental, as on
the uplands, or alluvial, as on the North China plain to the north and west of the
very northerly course which the Yellow River took at that time. Consequently we
must think of the cinnamon soils again, adding some carbonate light-meadow
soils, here and there saline, these from silt derived, and by their light colour
accounting for the description 'pale *jang*' (*pai jang*[1]).[c] We shall meet once more
with the term *jang* in Yü-chou presently, but it is already clear that it meant
basically what we should now call the pedocal soils, mostly developed from the
loess, either directly *in situ*, or after fluvial carriage and alluvial deposition.

Yen-chou (province no. 2) sets a more difficult problem. Comprising all the
land between the Yellow River and the Chi R., it was an 'isle-land' (as the
geographical term implies) in the strictest sense. The parent material of its soils
must have been loessial silt (alluvium), so that the area is characterised today by
carbonate light-textured meadow soils, partly saline; but this does not explain
why the Yü Kung text should speak of 'dark' or 'black' *fên* (*hei fên*[2]). However,
from the accompanying remarks (cf. p. 85) one can deduce that the area was rather
thickly forested in ancient times, so that dark humose topsoil was almost certainly
present, perhaps enduring for a considerable while after cultivation had begun.[a]
We share this interpretation with Shih Ya-Fêng and Chang Han-Chieh, a view
which implies that the clearing of the Hopei forests had only recently begun when
our text was written. Along the coasts there were doubtless then as now solonchaks of various sorts. In the next province (no. 3) Chhing-chou, all of it contained in Shantung, *fên* is most prominent again, though here the ancient survey
makes a clear distinction between the pale-coloured cinnamon soils and associated brown forest soils (*pai fên*[3]), all the paler no doubt because already at
that time longer cultivated; and the solonchak soils of the littoral, which it
characterises as of 'broad acres' (*kuang chhih*[4]).[b] The 'brownsoils' of Shantung
were already classical in the days of Shaw and Thorp,[c] so there can be little doubt
about the identification. Province no. 4, Hsü-chou, again presents difficulties.
First, one does not know whether *chhih chih fên*[5] was intended as a single technical
term, or whether one should read it as 'red *chih*' and '*fên*'; we adopt the latter

---

[a] Cf. Moyer (1).   [b] Cf. Vol. 1, p. 58.
    [c] The light colour persists along the foot of the hills today, says Chang Han-Chieh.
    [d] The proximity of areas of uncleared forest may well have encouraged the long-continued use of earth
fertiliser from the remaining woodlands, and thus the maintenance of humus levels.
    [e] I am reminded of the *près salés* of my youth in Northern France, where the famous mutton is still raised.
    [f] Cf. Thorp & Chao (1).

[1] 白壤   [2] 黑墳   [3] 白墳   [4] 廣斥   [5] 赤埴墳

possibility.[a] Nowadays in this region one finds carbonate, and leached, light meadow soils, some saline; and similar sajong-containing soils are present especially in the west of the area. Undoubtedly *chih* has the undertone of 'sticky'. Both Shih Ya-Fêng and Wan Kuo-Ting took the whole phrase to mean glutinous red argillaceous earth overlying limestone, and indeed it is true that in the northwest of the ancient province there are areas of terra rossa; but these are not very extensive, and unless one were to assume that they were the only parts the Yü Kung's farmers knew, the whole explanation could hardly here be found. Fig. 6 (in Vol. 1) shows that calcareous alluvium underlies only part of the western portion, so the bedrock in the east must be different, and Chang Han-Chieh asserts that it is red; in effect, a glance at Li Ssu-Kuang's treatise shows[b] that there are red shales and ochreous shaly limestones underlying Northern Chiangsu. Thus one may suppose that in ancient times the cultivators knew both kinds of red clayey soils developed from limestones, as also the humose recent forest soils not long cleared of vegetation—red *chih*, and *fên*.

With Yang-chou (no. 5) and Ching-chou (no. 6), we swim again into calmer waters, those in fact of the irrigated ricefield soils derived from the non-calcareous alluvium of the Yangtze valley. It is evident that the 'daub mud' (*thu ni*[2]) soils of these provinces were all of meadow and meadow-bog character, leached and lime-depleted. Marginally, by the delta, solonchaks would have already appeared, and towards the south the infertile red laterite soils may have been known, but the text says nothing of either of these.

Both the remaining provinces, on the other hand, keep unsolved problems. The seventh, Yü-chou, is said to have *jang*[3] 'above' and *fên lu*[4] 'below'. But what do these appellations mean? Most probably the distinction was between the uplands of the Fu-niu Shan to the west of the province and the lowlands, the flat watershed of the Huai R. to the east; but one would not quite like to exclude the possibility that 'above' might mean topsoil, and 'below', its lower horizons. Since the greater part of the province was based on loess (parental to the west, alluvial to the east) the term *jang* here would most appropriately fit in with the carbonate, and leached, cinnamon soils which are still there; while *fên lu* would mean the light meadow soils, together with those containing sajong concretions, which are very prominent in the eastern half. Chang Han-Chieh understands *lu* to be hard dark soil based on loess (cf. the kheilutu types), and he too thinks that by the present double term the meadow soils with their underlying sajong concretions were meant. On the other hand, if in such ancient times anyone was looking at a soil profile in a pit,[c] then the sajong pattern alone might qualify, since the

---

[a] Here we follow the commentator Lin Chih-Chhi[1], *c.* +1160 (*YKST*, ch. 2, p. 2*b*). He takes both the *chih* and the *fên* to have been red, and says in so many words that the *fên* was very fertile while the *chih* had been flooded and washed out by the Huai R. In his time it was less moist and fertility was returning.

[b] (1), p. 424.

[c] Here one remembers the great excavations that were made for royal and aristocratic tombs.

[1] 林之奇   [2] 塗泥   [3] 壤   [4] 墳壚

horizons are so markedly differentiated. Lastly the text may mean that there was *jang* on the uplands, and *fên and lu* in the lowlands, in which case recently cleared cinnamon and humose brown forest soils might come into consideration, as is always possible.

Queerest of all the terms is that characterising the soil of the penultimate province (no. 8), Liang-chou, namely *chhing li*[1]. What to make of the adjectives 'caerulean' or 'glaucous', perhaps just 'dark' if not 'blue-black', and 'light', 'fine-grained', 'loose', 'friable', possibly 'powdery'? How could this be reconciled with the statements of some commentators[a] that it was not very porous? Besides, where exactly was Liang-chou? Shih, Wan and Chang all agree in assuming that the 'Red Basin' with its purple-red and purple-brown brunizems was included in the province, but we prefer the more conservative view that it occupied really only the Han R. valley, separating modern Shensi from modern Szechuan.[b] We therefore think that the term *chhing li* referred to the montane humose dark forest soils on each side of the Han valley, and suggest that its curious colour components could be explained by the special character of the bedrock of these mountains, as also by the special features of the natural vegetation cover (cf. p. 90 above and Figs. 6, 7, 10).[c] At the same time we should not like to be too pragmatical about the geography of the ancient province, and if it could be shown by philologists and pedologists that *chhing li* would make even better sense as applied to the purple-brown forest soils of Szechuan with their original humose surfaces, we would accept a revision; indeed if the evidence were good enough it might even constitute a serious argument for the inclusion of the Szechuan basin in Liang-chou. So far, however, that has not been done.

This brings us to the end of what can be said about the most ancient soil geography in any civilisation. On the whole it can be seen that it is possible today to make quite good sense of the classification of soils established by the field wardens and revenue officials of the Chou period, and transmitted to posterity in the text of the Yü Kung.

Of course, as we have said, the assignment of precise modern meanings to the ancient terms can hardly be expected after the lapse of so long a time. One can only conclude that *jang*[2], for example, bears the general sense of loessial soils and their derivative alluvial silts, while *fên*[3] has that of humus-rich recent forest soil. *Lu*[4] means dark hard compact soil, carrying the implication of claypans and sajong horizons, *chih*[5] applies to all sticky soils containing much clay, while *chhih*[6] is unquestionably saline soil of the solonchak type. But the terminological rep-

---

[a] There was no consensus of opinion among them as to the meaning of *chhing li*, cf. *YKST*, ch. 2, pp. 34*b*, 35*a*.

[b] I.e. with the Chhin-ling Shan to the north, and the Mi-tshang Shan, the Ta-pa Shan and the Wu-tang Shan to the south.

[c] Kovda's soil map gives a particular pattern for the Han R. valley, but it contains no special component soil not found anywhere else.

[1] 青黎   [2] 壤   [3] 墳   [4] 壚   [5] 埴
[6] 斥

ertoire of ancient Chinese soil science is far from exhausted by these words. In the preceding discussion, including two-character terms, we have noted about fifteen altogether, yet Wan Kuo-Ting (2) has collected nearly forty more from some twenty Han and pre-Han texts. Thus we find *hsing*[1] for the hard red lateritic clay soils, *chiao*[2] for gravelly soil, and *fên*[3] for dusty friable light earth. The texts which yield such terms and their explanations are multifarious. Among them is the *Chou Li* (Record of the Institutions of the Chou Dynasty), compiled from earlier material probably in the −2nd century, in its entry for the *Tshao Jen*[4] (Agricultural Planning and Improvement Officers), who are concerned with fertilisers for different soils.[a] 'They use the techniques of transforming the earth, and put substances on the land (to fertilise it).[b] They scrutinise the fields to see what their best use would be, and then (direct the) planting accordingly.'[c] In fact, most of the entry is concerned with a curious and interesting technique of treating seeds before sowing with extracts of animal bones, discussion of which we must postpone till Sect. 41, but nine two-character names of soils are given, only four out of the eighteen overlapping with those already mentioned. Recognisable, however, are the red lateritic soils under *hsing kang*[7], the Szechuanese purple-red earths under *chhih thi*[8], solonchaks under *hsien hsi*[9], and peaty swamp-land under *ho tsê*[10]. Another source for soil terms is unexpectedly the *Chiu Chang Suan Shu*[11] (Nine Chapters on the Mathematical Art)[d] where it deals with calculations about public works; and even some of the Han apocrypha, such as the *Hsiao Wei Yuan Shen Chhi*[12] (Apocryphal Treatise on the *Filial Piety Classic*; Documents adducing the Evidence of Spirits), which embroiders on the proceedings of good farmers' sons,[e] add more.

Two comments remain before we can end all that we have space to say about ancient Chinese soil science. First let us recur to that strange scene not long ago evoked (p. 86 (*i*) above) in which a feudatory baron is presented by the emperor with a sod of black earth from the north face of the imperial altar-tumulus in token of the lordship he will exercise over his northern fief. At an earlier stage some obvious suggestions were proposed to account for the classical association of five colours with the four spatial directions and the centre,[f] but it has occurred to more than one mind that their arrangement may have had a pedological origin.[g]

---

[a] Ch. 3, p. 7*a* (ch. 8) and ch. 4, pp. 32*b*, 33*a*, (ch. 16), tr. Biot (1), vol. 1, pp. 184, 365.

[b] *Chang thu hua chih fa, i wu ti.*[5] The methods used are said by the commentator Cheng Hsüan (*c.* +180) to go back to those taught by Fan Shêng-Chih about −50.

[c] *Hsiang chhi i erh wei chih chung.*[6]

[d] Compiled in the +1st century but containing much older material (cf. Vol. 3, pp. 24 ff.). We shall refer to part of what it says immediately below.

[e] Ch. 2, p. 13*a* (in *YHSF*, ch. 58, p. 26*a*).   [f] Vol. 2, pp. 261, 262, 263.

[g] See Wan Kuo-Ting (2), p. 110. Kovda (1), p. 77 also refers to the association of colours with the cardinal points, giving an interpretation similar to the above, but astonishingly he attributes it to the *Yü Kung* itself. He must have misunderstood his informants.

[1] 垶   [2] 墽   [3] 坋   [4] 草人   [5] 掌土化之法以物地
[6] 相其宜而爲之種   [7] 骍剛   [8] 赤緹   [9] 鹹湯
[10] 渴澤   [11] 九章算術   [12] 孝緯援神契

The historical heartland of China was the yellow loess plateau and the alluvial plains, to the north were the black or dark forest soils and chernozems, to the west the 'white' (pale grey) sierozems of the desert and its edges, to the south the krasnozems or red soils and laterites.[a] Only the blue-green east presents a problem, but there in the great river valleys grew the brilliant green of the young rice-fields, and it is also true that many of the eastern meadow soils or gleys are greenish-grey when wet; to say nothing of the extensive marshes in the flat eastern lands. The idea contributes something, at least, to the history of the symbolic correlations.[b] Secondly there is the view that the whole traditional division of the ancient empire into the nine provinces was based on the nature and produce of soils rather than on political boundaries or ordinary topography. In the +12th century this was held by Lü Tsu-Chhien[1]. In his commentary on the Yü Kung he wrote:[c]

So in differentiating the grades of the Nine Provinces and their various soils such as the *fên* and the *jang*, he (Yü) followed the productivity of the earth, and so differentiated (the boundaries) ... In distinguishing the boundaries of the regions and their soils he did not take account of the absolute area of the lands in question, and he did not fix the limits according to the obstacles presented by the great mountains and rivers; he considered only the relative richness or poverty of the fields of the people, and balanced the matter according to that.... This is why three provinces were relatively small.... while four other provinces were wide open and expanded.

So much then may suffice for soil science *in statu nascendi*. Together with oecology and plant geography, it does really seem to have been born in China.[d] Theophrastus has never had any warmer admirer than Greene, but even he could find relatively little to draw from him on these subjects.[e] Of course Theophrastus (*c.* −300) distinguished between habitats according to altitude and according to humidity, enumerating trees of mountains and plains,[f] or herbs of marshes, streams, lake shores, coasts and the sea.[g] But on the different types of soils he made no contribution, unless in some writing now long lost.[h] Perhaps it was natural that among the practically minded agricultural Romans interest should have focussed more clearly on the nature of the soil in which the plants grew, and indeed this is the case with all of them, but it is striking to find that they developed no technical terminology comparable with that of the Chinese. Cato (*c.* −160) simply recommends 'heavy rich treeless soil' for grain, 'heavy warm soil' for

---

[a] Moreover, Szechuan with its brunizems was doubtless considered south rather than west at the time when the colour series was being incorporated in the system of symbolic correlations.

[b] Cf. Sect. 13*d*.

[c] *YKST*, ch. 4, pp. 39*b* ff., tr. auct.

[d] This is also the conclusion of a recent valuable paper by Wang Hsün-Ling (*1*).

[e] (*1*); pp. 125 ff., 130 ff.

[f] *Hist. Plantarum*, I, vi and vii, III, iii and iv, IV, i.

[g] *Hist. Plantarum*, IV, vii to xiv. Corals for him are under-water tree species.

[h] There is nothing to the point in the *De Causis Plantarum*; cf. Senn (*1*).

[1] 呂祖謙

olives, 'chalky open soil' for figs, and so on;[a] none of the later writers improved upon this primitive intuitive classification. Varro (c. −36) repeats the oecology of Theophrastus in brief, and goes somewhat further in listing the kinds of material which different soils may contain (rock, marble, rubble, sand, loam, clay, red ochre, dust, chalk, ash and 'carbuncle', i.e. 'charred' plant roots, possibly humus?)[b], and noting that colour (such as whitish or reddish) may matter considerably. But he too is content with many vague adjectives, such as loamy (*sabulosa*), poor (*macra*), rich (*pinguis*) and thin (*tenuus*).

Virgil, in the *Georgics*, written about −30, has further passages of interest. Not all soils can bear all fruits; there are great differences between the preferred ranges of herbs and trees.

> nec vero terrae ferre omnes omnia possunt.
> fluminibus salices crassisque paludibus alni
> nascuntur, steriles saxosis montibus orni;
> litora myrtetis laetissima; denique apertos
> Bacchus amat colles, Aquilonem et frigora taxi.[c]

And the same principle applies to far-off lands as well.

> divisae arboribus patriae. sola India nigrum
> fert hebenum, solis est turea virga Sabaeis ...
> quid nemora Aethiopum molli canentia lana,
> velleraque ut foliis depectant tenuia Seres?[d]

Here Virgil is in the tradition of Theophrastus, who was well acquainted with plants that grew outside his own culture-area. The Greek had known, for example, the mangrove vegetation of the Persian Gulf (e.g. *Avicennia marina*), he had spoken of the nyctitropic movements of *Tamarindus indica* growing on Bahrein Island, he had described the banyan (*Ficus bengalensis*) and the banana (*Musa*

---

[a] *De Agri Cultura*, v. VI, VII, XXXIV and XXXVII. Cato was always giving warnings about a special kind of soil called *cariosa* which had a kind of crust through which light rain after drought does not sink in—don't plough it at all, he said.

[b] *Rerum Rusticarum*, I, vi, vii, ix, xxiii, xxiv and xxv. Tr. Hooper & Ash (1), as for Cato.

[c] II, ll. 109 ff. Not every land can nourish every tree
> Rivers are fringed with willows; alders grow
> In thick morasses; rocky hills give birth
> To barren mountain-ashes; myrtle-groves
> Grow strongest by the shore; the grape-vine loves
> An open eminence, and yews prefer
> North winds and cold.   tr. Royds (1), p. 96, mod. auct.

[d] II, ll. 114 ff. Trees have allotted climes, to each its land.
> Black ebony knows India alone.
> Only Sabaeans grow the incense-spray ...
> Need I describe to thee the Ethiop groves
> Of woolly raiment soft and white, or how
> The Chinese comb a silky fleece from leaves?
>   tr. Royds (1), p. 97, mod. auct.

Virgil speaks of *Diospyros Ebenum* (Watt (1), p. 498) and of the frankincense or olibanum produced by the balsamiferous trees of Arabia and Africa such as *Boswellia carteri* (Watt (1), p. 173). The Abyssinian reference is considered to be to cotton, and of course Virgil did not know of the silkworm (cf. Vol. 1, pp. 157, 185, 233, and Sect. 31 below).

sapientum), the Median *Citrus medica* and the *Nerium indicum* of Baluchistan. Of course, Alexander's men had passed that way.[a] But Virgil is even more interesting on soils. 'Now give we place', he says, 'to the genius of soils, the strength of each, its colour, and its native power for bearing.'[b] He has no better terminology, it is true, than 'churlish ground' or 'lean clay', and tends to particular examples such as the Tarentine meadows good for pasture, or the salt-free Capuan soil; but he notes the value of black crumbly earth from which old woods have been cleared, and decries the chalky rendzinas tunnelled with burrows.[c] To detect a 'villainous cold' soil (*sceleratum frigus*) is difficult, often it is only given away by the presence of pitch-pines, baleful yews and black ivy.[d] Next Virgil describes three experimental tests for soil quality. 'Richness' of soil is to be measured by working a lump in the hands to see whether it grows sticky like pitch, and so viscous and cohesive that it does not crumble when thrown on the ground. Saltiness is measured by filtering pure spring water through a basketful of the earth and tasting the filtrate. Suitability for crops or pasture is thus assessed—a pit is dug deep in the ground, and the excavated earth then put back again. If it will not quite fill the hole it is considered 'light' (*rarum*) and therefore good for pasture and vines; if there is earth to spare when the pit is filled, it is 'heavy' (*spissus*) and likely to be good, when strongly ploughed, for the growing of crops.[e]

This strangely recalls a Chinese term already encountered, *hsi* (*thu*)[1], on which Wan Kuo-Ting has written an interesting discussion.[f] As we know from the *Huai Nan Tzu* book[g] and other sources, this term has an opposite, *hao* (*thu*)[2]. While the former implies 'breathing' and also 'coming to rest', in contrast with the latter it also means 'much' as opposed to 'little', 'full' as opposed to 'lacking', 'to increase' rather than 'to diminish', etc. Is it possible, therefore, that in ancient China just as in ancient Europe men were noting whether excavated earth would or would not fill up the pits it had been taken out of, in other words, that measurements of bulk density and compactibility were being made? What the *Chiu Chang Suan Shu* says makes this quite probable.[h] In its civil engineering chapter it considers the digging out and piling up of earth. When one digs out 10,000 ft of earth, one gets 12,500 ft if it is *jang*[3] soil (i.e. *hsi thu*, says the +3rd-century commentator, Liu Hui), but only 7500 ft if it is *chien*[4] (firm) soil (i.e. like *chu*[5], or tamped earth,[i] he says). The rule is: if the digging is as four, the *jang* obtained is five, the *chien* obtained is three, and the trench (*chhiu*[6]) remains four. The text goes on to give

---

[a] All this is excellently described in the monograph of Bretzl (1).

[b] II, ll. 177 ff. tr. Fairclough (1), p. 129; Royds (1), p. 99.

[c] Fairclough tr., p. 131; Royds tr. p. 101.

[d] II, ll. 256 ff., tr. Fairclough (1), p. 135; Royds (1), p. 103. Presumably this would have been highly leached podzolic land from former coniferous forest.

[e] II, ll. 226 ff., tr. Fairclough (1), p. 133; Royds (1), p. 102. This is still done as a matter of common observation and incidental demonstration, often in the process of studying soil profiles. Pedologists say that it serves well to confirm manual texture assessments.

[f] (2), p. 105.   [g] Ch. 4, p. 6a.   [h] Ch. 5, pp. 1a ff.

[i] On this see Vol. 4, pt 3, p. 38 ff.

[1] 息土   [2] 耗土, 耗土   [3] 壤   [4] 堅   [5] 築   [6] 墟

examples of various combinations. Thus it is clear that men who made calculations about the building of dykes and city walls in China took account of just those compactibility properties mentioned by Virgil, and if this is so one can hardly doubt that the agricultural botanists there drew their parallel conclusions. Further insight into the meaning of the term *hsi* is thus rather rewarding. *Hsi thu* would have been the soil that proved too much for the pit, and *hao thu* that which when restored was insufficient to fill it.

Columella (*c.* +65)[a] repeats the three tests of Virgil, the rheological one for clay, the filtration process for salt, and the pit test for compactibility.[b] He makes the usual distinctions regarding altitude (champaign, hilly and mountainous), and looks especially in soils for those which show a combination of opposite qualities, or what the Greeks call *syzygiai enantiotētōn* (συζυγίαι 'εναντιοτήτων)—a curious echo of Yin-Yang doctrine. But there is no advance in technical terminology; indeed on the whole Columella decides to abandon soil science. In each of the altitude levels, he says,[c]

there fall six species of soil—fat or lean, loose or compact, moist or dry.[d] And these qualities, in combination and alternation with one another, produce a very great variety of soils. To enumerate them is not the mark of a skilled farmer, for it is not the business of any art to roam about over the species, which are countless, but to proceed through the classes, for these can readily be connected in the imagination and brought within the compass of words.

At once a programme and an epitaph, these words throw into high relief the Chinese achievement.[e]

(3) THE CASE OF THE *chü* AND THE *chih*

'When the orange tree (*chü*[1]) crosses the Huai River it turns into the thorny limebush (*chih*[2]).'[f] One of those gnomic utterances, constantly repeated with numerous variations, and characteristic of classical Chinese literature,[g] this par-

---

[a] See the special paper on his pedology by Olson (1).

[b] *Rei Rusticae*, II, ii, tr. Ash (1), pp. 109 ff., 118 ff. He thinks the latter may fail in certain cases, as in the black earth of Campania called *pulla* (cf. also I, pref. 24). Was this perhaps a chernozem? Pliny, a decade later, doubted (strange to say) both the clay test, which he said would indicate that potter's clay was a good soil, and the pit test, which he believed never showed excess volume. See *Nat. Hist.* XVII, iii, 27.

[c] Ash tr. p. 109.

[d] *Pinguis vel macri, soluti vel spissi, umidi vel sicci.*

[e] We cannot follow the further course of soil science in all cultures here, but it is interesting that the agriculturists of Muslim Spain in the +11th and +12th centuries got well away from the limitations of Hellenistic thinking, and distinguished some thirty types of soil. The most famous of these writers of books on agriculture (*Kitāb al-Filāḥa*) was Ibn al'Awwām al-Ishbīlī (cf. Mieli (1), p. 205, Nasr (1), p. 112), but also of Seville was Abū al-Khayr al-Shajjar al-Ishbīlī. The books of Ibn Baṣṣāl and Ibn Wāfid are important too. For an interesting treatment of this subject see Bolens (1).

[f] *Po Wu Chih*, ch. 4, p. 2a (just 'the River').

[g] For another typical example see Vol. 5, pt 4, pp. 168, 207 below, based on Tshao Thien-Chhin, Ho Ping-Yü & Needham (1). Cf. Butler, Glidewell & Needham (1).

[1] 橘   [2] 枳

ticular statement has intrigued me for twenty-five years. Its elucidation in the context of soils, habitats and climates, seems to belong here.[a]

Citrous fruits are native to China, and opportunity may later offer for a discussion of their various species in some detail. Suffice it to say now that while *chü*[1] was without doubt a wide generic name applicable to all oranges in general, most authorities agree that the orange *par excellence* to which it tended to refer was the sweet loose-skinned tangerine, *Citrus reticulata = nobilis*, perhaps var. *deliciosa*.[b] The identity of the *chih* is also clear, this was the sour trifoliate orange or thorny limebush, *Poncirus trifoliata*, also called *kou chü*[1] (or *chü chü*[1]).[c] As we have just seen (pp. 6116) oranges are mentioned as tax or tribute in the Chou Domesday Book (middle of the −1st millennium), while in the not much later *Shan Hai Ching*, *chü* orange trees occur six times (in ch. 5) and *chih* limebushes once (in ch. 3).[d] The *Shuo Wên* says that the *chü* orange is a product of Chiang-nan (the country south of the Yangtze), while the *chih* limebush is a tree that looks like the orange.[e] The Han tradition was that citrous fruits had not been cultivated in the days of the legendary emperors Yao and Shun,[f] but they grew them most assiduously themselves, and the *Shih Chi*, in its chapter on the merchants and industrialists, says that 'people who own a thousand orange trees in Shu, Han or Chiang-ling … may live just as well as a marquis enfeoffed with a thousand households'.[g]

The central expression of the idea occurs in two parallel texts, both from the −2nd century, and we must give them side by side because they differ in an important way. The *Chou Li* text has already been translated[h] since it forms part of the introduction of the Khao Kung Chi (Artificers' Record) section, where comparison is made between the natures of raw materials and the skill and ingenuity of artisans. The *Huai Nan Tzu* passage is perhaps less well known.

Chou Li [i]	*Huai Nan Tzu* [j]
With good material and good workmen it still may happen that the product is not good; in this case the season has not been suitable, or the *chhi* of the earth has not been obtained.	Now people who transplant trees (know that) if the proper Yin-Yang balance of their nature (and environment) is lost, there is not one that will not winter and decay.

[a] Schafer (16), in another context, has touched upon these beginnings of oecology (p. 119).

[b] CC 905; Watt (1), pp. 320 ff.; B II 486, III 281; Anon (57), vol. 2, pp. 206, 209; Kimura & Kimura (1), pl. 25, fig. 3 and p. 50; R 347, 348; Richardson (2), p. 58; Chhen Jung (1), pp. 579, 582. Cf. Yeh Ching-Yuan (1, 2).

[c] CC 918; B II 488, III 334; Anon (57), vol. 2, p. 263; Kimura & Kimura (1), pl. 25, fig. 2 and p. 49; R 349; Chhen Jung (1), p. 564. Formerly *Citrus t.*

[d] Cf. Yeh Ching-Yuan (2), pp. 14, 51.

[e] Ch. 6A (pp. 114.2, 117.1).

[f] See the remark of Tshui Shih[2] in his *Chêng Lun*[3], quoted in *TSCC, Tshao mu tien*, ch. 229, *tsa lu*, p. 2a (though not in *YHSF*, ch. 71, pp. 67a ff.), *c.* +155.

[g] Ch. 129, p. 15a; tr. Watson (1), vol. 2, p. 493; Swann (1), p. 432. Hence the term *mu nu*[4], 'wooden slaves'; see *TSCC, loc. cit. chi shih*, p. 2a. Yang Fu[5], in his *I Wu Chih*[6] (p. 3), a Han book, says that in Chiao-chih there was a regular official, the Chü Kuan[7], in charge of production and taxation, the 'Orangery Superintendent'.

[h] Vol. 4, pt 2, p. 12.   [i] Ch. 11, p. 3a, tr. auct. cf. Biot (1), vol. 2, p. 460.

[j] Ch. 1, p. 6b; tr. auct. adjuv. Balfour (1), p. 81; Morgan (1), p. 10.

[1] 枸橘   [2] 崔寔   [3] 政論   [4] 木奴   [5] 楊孚
[6] 異物志   [7] 橘官

Thus, for example, if the sweet-fruited orange crosses the Huai River to the north, it becomes a thorny lime-bush (*chü yü Huai erh pei, wei chih*[1]).

And the crested mynah (bird) (*chhü yü*[3])[a] never crosses the Chi River (to the north). And the raccoon dog (*ho*[5])[b] dies if it crosses the Wên River.

This is quite natural because of the *chhi* of the earth.

So for example if the sweet-fruited orange is planted north of the River, it will metamorphose into the thorny limebush (*ku chü, shu chih chiang pei, tsê hua erh wei chih*[2]).

And the crested mynah (bird) (*chhü yü*[4]) never comes across the Chi River (to the north). And the raccoon dog (*chou*[6]) dies if it swims across the Wên River.

For neither the forms nor the natures of these creatures can change, nor may their surroundings and indigenous haunts be altered.

At once the difference declares itself. The *Chou Li* statement is relatively vague; its writer may have meant simply to point to differences of normal geographical distribution, implying the sanctions of Nature if things transgressed their usual ranges; at most he wanted to say that the orange or tangerine would never ripen properly in the north, giving only sour fruits like those produced by the thorny limebush. It would there be no better than a *chih*. But the Taoist writer of the *Huai Nan Tzu* was more precise, apparently believing in a true transmutation of species.[c] He was followed naturally by the writer in the *Lieh Tzu* book, who told the story a little differently, saying that the people of Chhi thought so highly of the pommelo (*Citrus maxima*) that they took some of these trees (*yu*[7]) home to plant in the north, but they all turned into *chih* trees, sour and thorny. After giving the same zoological examples, he goes on:[d]

Although the (various species) differ in form and *chhi*, their natures are complementary (lit. balanced), so that they cannot take each other's places (lit. mutually exchange with one another). Their life is a whole (with their environment); their portion (*fên*[8]) is sufficient for them.[e] How should I know of any (absolute) scale or measure of largeness and smallness, or length and shortness, or similarity and difference?

The story of the tangerine and the limebush provided the theme for one of the most amusing tales of the proto-feudal period. It is contained in the *Yen Tzu Chhun Chhiu*[9] (Master Yen's Spring and Autumn Annals), a collection of stories about

---

[a] *Aethiopsar cristatellus*, R 296.   [b] *Nyctereutes procyonoides*, R 375.

[c] This was nothing surprising for ancient Chinese biological thinkers, who had no theory of special creation to contend with. On the famous 'evolution' passage in *Chuang Tzu*, ch. 18, see Vol. 2, p. 78 above, or Needham & Leslie (1), and p. 138 below.

[d] Ch. 5, p. 8a, tr. auct. adjuv. R. Wilhelm (4), p. 51. Cf. Yeh Ching-Yuan (1), p. 134. The passage comes at the end of a discussion on relativity closely paralleling that in *Chuang Tzu*, ch. 1, see Vol. 2, p. 81. Cit. *TSCC*, *Tshao mu tien*, ch. 229, *chi shih*, p. 1a.

[e] Cf. the parallel with *moira* in Greek thought, Vol. 2, p. 107 and elsewhere *sub voce*.

[1] 橘踰淮而北爲枳   [2] 故橘樹之江北則化而爲枳   [3] 鸜鵒
[4] 鴝鵒   [5] 貉   [6] 貐   [7] 櫾   [8] 分
[9] 晏子春秋

the −6th-century statesman and sceptical naturalist Yen Ying[1], compiled in Late Chou, Chhin or Han times.[a]

(Yen) Ying (of Chhi) was nominated as ambassador to Chhu. (The Prince of) Chhu heard of it and said to his counsellors: 'Yen Ying is Chhi's most brilliant talker. Now he is coming here I should dearly like to take him down a peg. What shall we do?' The counsellors said: 'We suggest that we bind a man and bring him before you.' The Prince said: 'What, may I ask, is the good of that?' They replied: 'He shall be a native of Chhi.' The Prince said: 'What will he be supposed to have done, then?' To which they replied: 'He shall be a robber.'

When (Yen) Ying came to Chhu the Prince gave him a feast, and when they were all warmed up two officials led in a man in bonds and reported to the Prince. The Prince asked what he had done, and it was answered that he was a robber from Chhi. The Prince looked at (Yen) Ying and said: 'Chhi people are considered good, do they then rob?' Getting up, (Yen) Ying replied: 'I, Ying, have heard that when oranges (chü) grow south of the Huai (River) they bear sweet fruit, but when they are planted north of it they become sour and thorny (chih).[b] The leaves look alike, but fruitlessly, for the taste of the fruits is quite different. How can this be? It is because the water and the soil are not the same. So also folk who grow up in Chhi never rob, but of course when they come down here to Chhu they rob; is this not because the water and soil of Chhu make men good at robbing?'

The Prince laughed and said: 'Sages never fail to be brilliantly intelligent. Who am I to find fault with them?'

By the Sung period, when Lu Tien[2] wrote his *Phi Ya*[3] (New Edifications on the *Literary Expositor*) in +1096, the metamorphosis was taken as a commonplace, and he used the word *pien*[4] to describe it.[c] In the following century, however, another lexicographer, Lo Yuan[5], emphasised the factor of natural geographical distribution—'If you cross the River to the north there aren't any of them, and this is why it is said "south of the River orange-trees (chü) are planted, but north of the River they become sour and thorny (chih)".' This was in his *Erh Ya I*[6] (Wings for the *Literary Expositor*)[d] about +1170. And now Sung scepticism began to come into play. In the *Chü Lu*[7] (Orange Record) of Han Yen-Chih[8], first printed eight years later, after talking about how the *li-chih* variety of orange got its name, the president of botanical monographers goes on:[e]

There is a saying that when the *chü* orange-tree is brought across the River Huai, it becomes the thorny limebush (chih). But can creatures within the vegetable kingdom be transformed like this? I doubt it. Surely it looks like just a confusion of names. And there are many instances of this sort of thing.

---

[a] Ch. 6, p. 4a, b, tr. auct.; adjuv. Forke (20). Cit. *TSCC, Tshao mu tien*, ch. 229, *chi shih*, p. 1a, b. The botanical interest of the passage was first noticed by Shirai Mitsutarō (1), p. 349. The reader has encountered Yen Ying before, cf. Vol. 2, p. 365, Vol. 3, p. 401.
[b] The wording here is the same as that in the *Chou Li*.
[c] Cit. in *TSCC, Tshao mu tien*, ch. 227, *hui khao*, p. 2b. On the words *pien* and *hua* see Vol. 2, p. 74.
[d] Cit. in *TSCC, Tshao mu tien*, ch. 227, *hui khao*, p. 3a. On the *Erh Ya* see p. 187 below.
[e] Ch. 2, p. 2a, b, tr. auct., adjuv. Hagerty (1).

[1] 晏嬰   [2] 陸佃   [3] 埤雅   [4] 變   [5] 羅願
[6] 爾雅翼   [7] 橘錄   [8] 韓彥直

After that the idea of metamorphosis had little credence.[a] Naturally Li Shih-Chen denied it, saying that south of the river both species grew but north of the river only the *chih*, 'so in fact these two trees are different species, and no transformation or metamorphosis is involved'.[b]

A third hypothesis about the famous phrase was suggested by Hagerty when he made his translation of the citrus chapter in the *Nung Chêng Chhüan Shu* (11). It might well mean, he thought, that the *chü* orange was commonly grafted on to the *chih*, and that when in the north the frosts of winter killed the orange part, the root-stock of the *chih* alone survived and later produced sour fruits. Thus the *chü* would 'revert' to the *chih*, and this word he used in his translation of the above passage. Grafting of orange trees is undoubtedly an ancient art in China. Let us read further in the *Chü Lu*. Han Yen-Chih says:[c]

Beginning to plant (Shih Tsai[3]).

To start with, the seeds of the red orange (*chu luan*[4])[d] should be washed very clean and planted in rich soil. After a growth of one year the seedlings are called *kan tan*[5]. Their roots now have a vigorous growth, and in the following year they are transplanted further apart. After another year's growth the trunk of the (little) tree will have grown to the thickness of a child's fist, and in the spring months specimens of good varieties of *kan*[5] and *chü*[6] oranges are grafted on to this stock. (Cuttings from) branches one year old which have faced towards the sun are used for attaching (*thieh*[7]) as scions. (To make a graft) take a very fine saw and cut off the stock a foot or more above the ground. Next split the bark and join the cutting to the stock, taking care not to shake its roots. Now place a handful of earth about the place where the graft has been made, in order to keep out water. This earth should be wrapped round with *jo*[8] leaves,[e] and the whole bound up with hempen cloth. (It must not be bound) too tightly nor too loosely, (and the graft must not be made) either too high or too low; then it will await the *chhi* of the earth, and respond to it.[f] The method for grafting these trees was (already) given in the book called *Ssu Shih Tsuan Yao*[9] (Important Rules for the Four Seasons)[g]. Old gardeners can do this. When those who have the best skill use their cutting instruments, the *chhi* and the substance (*chih*[10]) will accommodate to their (mutual) strangeness (*sui i*[11]), and none (of the grafted plants) will fail to live. If done after the (proper) time (i.e. the spring) has passed, the graft will not take, and the blossom and fruit will revert to that of the *chu luan*[4]. Wherever human power blends with that of Nature (*tsao hua*[12])[h], it is like this.

---

[a] Except in literature. In the +13th century Hsieh Fang-Tê[1] used the image of the profoundly transforming effect of soil and climate to illustrate the power of the great and sagely ruler in transforming all creatures (*Shih Chuan Chu Su*[2], ch. 3, p. 2*a*).

[b] *PTKM*, ch. 36 (p. 82), tr. auct.

[c] Ch. 3, p. 1*a*, *b*, tr. Hagerty (1), mod auct. Cit. *TSCC*, *Tshao mu tien*, ch. 226, *hui khao*, pp. 3*b*, 4*a*.

[d] Some variety of *Citrus Aurantium*; cf. Chhen Jung (1), p. 573; CC 896; Watt (1), pp. 320 ff. Perhaps *decumana*.

[e] Undoubtedly *Typha latifolia*; see Lu Wên-Yü (1), p. 46.

[f] Hagerty did not understand the inwardness of these words, but we can appreciate it in the light of Vol. 4, pt 1, p. 189.

[g] By Han Ê[13] of the Thang period.

[h] On this expression see Vol. 2, pp. 564, 581, Vol. 3, p. 599.

[1] 謝枋得   [2] 詩傳注疏   [3] 始栽   [4] 朱欒   [5] 柑淡
[6] 橘   [7] 貼   [8] 蒻   [9] 四時纂要   [10] 質
[11] 隨異   [12] 造化   [13] 韓諤

All that remains is to show that the thorny limebush (*chih*[1]) was sometimes used as the rootstock. To do this one has only to turn to one of the framstead encyclopaedias such as the *Pien Min Thu Tsuan*[2], first printed in +1502, where we find that Kuang Fan[3] says:[a]

The golden orange (*chin chü*[4])[b] is grafted on to the thorny limebush (*chih*[1]) in the third month, and then in the eighth transplanted into good rich soil with the addition of liquid fertiliser. This gives beautiful (fruit).

Hagerty's suggestion is therefore quite plausible.[c]

Let no one reproach the Taoists for believing in the metamorphosis of one plant into another. Nearly everyone in the West would have agreed with them. Still today there are many farmers in various parts of the occidental world who believe that seeds of wheat and barley are liable to spring up into chess or cheatgrass (*Bromus arvensis* ≃ *secalinus*),[d] or else to turn into darnel (*Lolium temulentum*) the 'tares' of Holy Writ.[e] What happens is that chess and darnel are able to flourish in the low wet parts of grain fields where the cultivated cereals cannot, so that nothing was more natural for ancient husbandmen than to see in these patches a metamorphosis analogous to those that they knew so well in the life-histories of the insects on the trees and the amphibia in the ponds.[f] Theophrastus was a little uncertain about the truth of this; he denied absolutely the rumour that wheat and barley could sometimes change into each other,[g] and he generally referred to the wheat-barley/darnel[h] and the flax/darnel[i] transformations as being what 'people say ...', but occasionally he accepted in passing, as it were, at least the former metamorphosis.[j] In other passages he implicitly cast doubt on it. Seed wheat, he says, from Pontus, Egypt and Sicily is free from darnel, but the Sicilian is infested by a different weed, the purple cow-wheat (scrophulariaceous *Melampyrum arvense*)[k], harmless, however, and not the cause of poisoning and headache like *Lolium*.[l] This suggests that Theophrastus visualised the grain as being mixed with the seeds of the other plants. So also elsewhere he records the autumnal germination and winter growth of *Lolium*, up and blowing before even the wheat was sown.[m] But few Europeans shared the scepticism of Theophrastus even two thousand years later, when Scaliger his commentator took him to task for it,

---

[a] Ch. 4, p. 3a, tr. auct. Quoted in *NCCS*, ch. 30, p. 12a, tr. Hagerty (11).
[b] Cf. Hagerty (1), p. 24.
[c] On grafting in China and the West see further, Sect. 38 (*i*), 4 below.
[d] Bentham & Hooker (1), p. 533.
[e] Bentham & Hooker (1), p. 530; Moldenke & Moldenke (1), p. 134.
[f] Inter-specific metamorphosis of animals will be discussed in detail later on in the zoological Section (Vol. 6, pt 2 below). The Chinese had no *a priori* objection to them, and were willing to look into every case on its own merits.
[g] *Hist. Plant.* II, ii, 9, 10.   [h] VIII, viii, 3; VIII, vii, 1   [i] VIII, vii, 1
[j] II, iv, 1; VIII, viii, 3   [k] Bentham & Hooker (1), p. 342.
[l] VIII, iv, 6. *Lolium* is one of the only two or three known poisonous grasses, and this may well be due to a fungus which normally grows on it (refs. in Moldenke & Moldenke (1), *loc. cit.*).
[m] VIII, vii, 1

[1] 枳   [2] 便民圖纂   [3] 鄺璠   [4] 金橘

saying that he himself had witnessed the metamorphosis of wheat into barley.[a] Scaliger's contemporary Li Shih-Chen thus had the better of him.

The denial of the wheat/barley transformation in Bk. II occurs in the midst of a discussion on the effects of soils and climate which is distinctly akin to the case of the sweet orange and the thorny limebush. Theophrastus thought that these effects were more important than the factors of cultivation and tendance, saying that many trees and plants after transplantation to a different environment become unfruitful or even refuse to grow at all.[b] In Egypt a pomegranate tree of the acid kind would produce sweet fruit whether grown from seeds or cuttings, and at a certain place in Cilicia all the pomegranate fruits lacked stones. So also the persion[c] would not fruit outside Egypt, the sorb produced nothing in the warmth of the south, and date-palms grew fruitlessly in Greece. 'Things that change like this do so spontaneously; it is due to a change of position and not to any particular method of cultivation.'

Greene, who thought carefully about the old idea of plant metamorphoses,[d] noticed the obvious *non sequitur* in the analogy between tadpoles and darnel plants; the former change is ontogenetic, occurring naturally in each individual life-history,[e] the latter inter-specific, an assumed deviation of development into a quite different adult form. He pointed out that in some families there are veritable metamorphoses of foliage, the mature tree being almost unrecognisable in comparison with its seedlings.[f] Most of these cases neither Greeks nor Chinese could have known, but Theophrastus did discuss the common ivy *Hedera Helix* where the bushy flowering branches are so different in habit and foliage from the climbing stem that the ancients gave them distinct names, the bushy cissus being a product of transformation at the top of the climbing helix.[g] He inclined to the view that the change was ontogenetic, not inter-specific. How right he was!

Room must naturally be left for mutation and the spontaneous occurrence of new varieties perpetuated by selection, vegetative propagation, etc. The tradition about the *chü* and the *chih* was often cited to throw light on peculiar changes. For example, the *pien kan*[1] or transmogrified orange, mentioned by Tuan Kung-Lu[2] in his *Pei Hu Lu*[3] about +873, a book on the southern provinces and Vietnam.[h]

---

[a] J. C. Scaliger (+1484 to +1558), the commentary appeared posthumously in +1584.

[b] *Hist. Plant.* II, ii, 7–11.

[c] This is *Mimusops schimperi* (Sapotaceae) from Ethiopia, cultivated since ancient Egyptian times (cf. Burkill (1), vol. 2, p. 1475).

[d] (1), p. 137; Egerton (1), pp. 48 ff.

[e] Of course there was no adequate understanding of animal metamorphosis before the time of Swammerdam in the +17th century. And by the same token the idea of the fixity of species hardly became a doctrine before the time of John Ray—helped perhaps by what puritan theologians might have dug out of the bible; on this subject see Zirkle (1).

[f] Notably in the Leguminosae (Mimosoideae) and the Myrtaceae (*Eucalyptus* spp), Australia being particularly rich in these.

[g] III, xviii, 6–10.

[h] 'Records of (the Country where) the Doors (open to) the North (to catch the Sun) (i.e. Jih-Nan and Lin-I)', p. 6a, b, tr. auct.

---

[1] 變柑   [2] 段公路   [3] 北戶錄

The 'transmogrified orange' [he says] is grown in Hsinchow as round as a gourd and as large as a pint measure; it is thin-skinned like the Tung-thing oranges.[a] When asked, the people replied that it had been brought from a distance less than 100 *li* away, but its shape and taste had both changed completely, hence the name. This is like the story of the *chü* and the *chih*—all due to differences of water and soil.[b]

This must have been either a mutated variety or the effect of different soil conditions, possibly the presence of trace elements, a factor which the medieval Chinese would certainly have classified under the head of the *chhi* of the earth. The literature contains some strange stories of this kind. About +1230 Chang Shih-Nan[1], in his *Yu Huan Chi Wên*[2], a book of miscellaneous reminiscences,[c] told how his father had once planted in the garden of their home a kernel of the 'Chinese olive' (*kan lan*[3]).[d] As the young plant feared the frost, a kind of greenhouse was built for it, but when it came to maturity it produced drupes exactly like those of the soapberry or soapnut tree (*mu huan tzu*[4])[e] which were just as good as the regular soapbeans[f] for washing clothes, and could be made into Buddhist prayer-beads just like normal soapnuts. Chang ended with a rhapsodical passage on the power of environment to change the nature of things, giving the usual examples, but one feels there can be little doubt that his father planted a soapnut seed by mistake.[g]

Given the love of symmetry in Chinese cosmism one would expect to find somewhere an east-west parallel to the north-south displacement of the orange and the thorny limebush. And indeed the *Huai Nan Tzu* book is quoted in late writings as follows:[h]

Orange trees and pommelo trees have their natural habitat (*yu hsiang*[5]); but the orange tree withers when transplanted to the north, just as the pomegranate (*shih liu*[6])[i] is grieved (*yü*[7])[j] when transplanted to the east.

[a] Cf. Hagerty (1), p. 15.
[b] This expression, already met with on p. 106 above, is still in colloquial use and includes climatic (as well as edaphic) factors in general.
[c] 'Things Seen and Heard on my Official Travels', ch. 9, pp. 8*b*, 9*a*, tr. auct.
[d] *Canarium album* (Burseraceae), R 337; Chhen Jung (*1*), p. 595, produced in Fukien, Szechuan and some other provinces.
[e] *Sapindus Mukorossi* (Sapindaceae), R 304; Chhen Jung (*1*), p. 682; Burkill (1), vol. 2, p. 1958; Collett (1), p. 97; Steward (1), p. 232.
[f] *Gleditsia sinensis*; see Needham & Lu Gwei-Djen (1); Lu & Needham (3).
[g] This seems almost certain when one finds that the *Chhün Fang Phu* says (*Kuo phu*, ch. 2, pp. 45*b*, 46*a*) that the *kan lan* tree looks very like the *mu huan tzu* tree.
[h] *Chhün Fang Phu* (*Hui phu*), ch. *shou*, p. 1*b*; also in *Kuang Chhün Fang Phu*, cf. Yeh Ching-Yüan (*2*), p. 16. A much older source is the +12th-century *Hsü Po Wu Chih* which quotes the second two-thirds of the sentence, without attribution (ch. 10, p. 4*b*). Tr. auct.
[i] The full name, *an shih liu*[8], is considered to point to come connection with Arsacid Parthia, An-Hsi[9], but Laufer (1), p. 284, was hard put to it to find the Iranian term to which the last part corresponded. No doubt all the Chinese names for the pomegranate are transcriptions of Persian or Indian words.
[j] This word presents difficulties, for apart from its most usual meaning given above, it can also mean 'flourishing luxuriantly', and also both 'fragrant' and 'malodorous'. All one can be sure of is that some kind of change following upon transplantation was intended.

| [1] 張世南 | [2] 游宦紀聞 | [3] 橄欖 | [4] 木槵子 | [5] 有鄉 |
| [6] 石榴 | [7] 欝 | [8] 安石榴 | [9] 安息 | |

Only the first third of this sentence is found in *Huai Nan Tzu* now,[a] but there is no reason for doubting that the tradition is an ancient one. It is generally agreed that *Punica granatum* was a native of Persia, and became familiar to the Chinese early in the +3rd century,[b] but there is no positive evidence for the ancient tradition that it was one of the plant introductions effected by Chang Chhien in the −2nd century.[c] The authenticity of this *Huai Nan Tzu* fragment would depend on this very thing, and it is intriguing to recall that the great explorer returned to China in −126, just as the philosophers of Liu An were writing. The coincidence, some might feel, strengthens the credibility both of the fragment and the plant introduction.

Another interesting east-west parallel is found in what the *Nan Fang Tshao Mu Chuang* (Prospect of the Plants and Trees of the Southern Regions) says about jasmine (+304). Hsi Han wrote:[d]

Both the *yeh-hsi-ming*[5] flower[e] and the *mo-li*[6] flower[f] were transplanted from the western countries by foreigners (*hu jen*[7]) and grow in Kuangtung (Nan-hai[8]). Southern people love their perfume and compete with one another in cultivating them.

Lu Chia[9] in his *Nan Yüeh Hsing Chi*[10] (Records of Travels to Southern Yüeh) said: 'In the land of Yüeh-Nan the five cereals are (rather) tasteless, and the hundred flowers lack perfume, yet these two flowers have a particularly beautiful scent. They were brought there originally from foreign countries in the West (*hu kuo*[11]), but they did not change (their characteristics) in accordance with the soil and water (of their new habitat). How different is this case from that of the *chü* orange which when taken to the north becomes a *chih*! Women and girls (in the south) thread the flowers together through their centres, and use them as a decoration for their hair.'

The *mo-li* flower resembles the white variety of *chhiang mei*[12],[g] and its fragrance exceeds that of the *yeh-hsi-ming*.[5]

---

[a] Ch. 17, p. 9a.

[b] The literature was first analysed by Laufer (1), pp. 278 ff. Representative mentions occur in the *Wu Tu Fu* (Ode on the Capital of the State of Wu, i.e. Nanking) by Tso Ssu about +270; and in the +3rd-century work *Chin Kuei Yao Lüeh*[1] (Systematic Treasury of Medicine) by Chang Chung-Ching[2], assuming no interpolation.

[c] Cf. Vol. 1, pp. 173 ff. Laufer's chief reason for rejecting it was that alfalfa and the grape-vine are alone mentioned in the *Shih Chi*, but it may have more basis than he thought. Its first statement occurs in a letter from Lu Chi[3] to his brother Lu Yün[4] (both d. +303), preserved in *Chhi Min Yao Shu*, ch. 41, whence *TPYL*, ch. 970, p. 4b and *CSHK* (Chin sect.), ch. 97, p. 11a; if that in the *Po Wu Chih* by Chang Hua (+290, begun c. +270) is not still earlier. Present texts of this book do not contain the passage, but it is attested by *TPYL*, ch. 970, p. 4b. The tradition was always accepted in China, cf. *CFP* (*Kuo phu*), ch. 3, p. 18b.

[d] Ch. 1, p. 1b, tr. auct. adjuv. Laufer (1); Li Hui-Lin (12). Cit. *CFP* (*Hua phu*), ch. 2, p. 46a. The passage was first noted by Bretschneider (12).

[e] *Jasminum officinale* ≃ *grandiflorum* (CC 455). Some think *g.* a variety of *o.* (e.g. Steward (1), p. 311) but others regard them as two distinct species (Li Hui-Lin (8), pp. 126 ff.). The former opinion seems preferable.

[f] *Jasminum Sambac* (CC 457), from Arabic *zanbaq*.

[g] With the orthography given, this could be either the umbellifer *Cnidium* (= *Selinum*) *monnieri* (R 230; Steward (1), p. 289) or the liliaceous *Asparagus lucidus* (R 676; B II 108; Steward (1), p. 520). There was some fancied resemblance in the whiteness or white texture of the petals. See p. 148 below.

[1] 金匱要略	[2] 張仲景	[3] 陸機	[4] 陸雲	[5] 耶悉茗
[6] 末利	[7] 胡人	[8] 南海	[9] 陸賈	[10] 南越行記
[11] 胡國	[12] 薔蘼			

And on another page Hsi Han, speaking of the henna plant (*chih chia hua*[1])[a], says that like the two jasmines it was an introduction from abroad, in this case by foreigners (*hu jen*[3]) who brought it from Ta-Chhin[2] (the Hellenistic East Mediterranean region).[b] While the relevance of the above passage to our present theme is obvious (though Lu Chia might have reflected that a north-south transplantation must naturally be more difficult climatically than one within the subtropical latitudes), it has caused a good deal of discussion;[c] chiefly because the date seemed too early for the evident derivation of *yeh-hsi-ming*[1] from Arabic *yāsmīn*, and *mo-li*[2] from Sanskrit *mallikā*. All the more was this so if the *Nan Yüeh Hsing Chi* was really to be attributed to Lu Chia, the great envoy of Han Wên Ti to the semi-independent viceroyalty of Chao Tho in Kuangtung (−196 and −179). Recourse was therefore had to theories of interpolation,[d] but as *yāsmīn* also occurs in Pahlavi,[e] and ancient contacts with India are not at all incredible,[f] the text of Hsi Han may be allowed without too much reserve. So often nowadays it happens that the growth of knowledge moderates the excessive scepticism of former scholars, both Western and Chinese.

While thinking of the jasmines, we may add mention of an interesting study by Schafer (15), who describes the rise of an alternative name for *yeh-hsi-ming* (*Jasminum officinale*) from the +10th century, *su hsing*[3], 'the pure white penetrating fragrance'; this seems to have originated from Indian legends entering through Canton, and was associated with a palace beauty of the Southern Han dynasty. A much later introduction, from Madeira by way of obvious Iberian channels, was *J. odoratissimum* (*huang hsing*[4]).[g] But it should not be thought that no species were native to China, for the winter jasmine, another yellow-flowered sort, *J. nudiflorum*, belongs to the northern provinces.[h] Since the flowers appear before the leaves this is called *ying chhun hua*[5], the 'welcomer of spring'. Other indigenous species are mentioned by Li Hui-Lin[i], and indeed it now seems that China was one of the main centres of development of the genus, though most of the species

---

[a] On henna the old discussion of Mayers *et al.* (1) is still interesting. Cf. Hirth (1), pp. 268 ff.

[b] So we usually say, following the evidence given in Vol. 1, pp. 174, 186 *et passim*, but one must not overlook the fact that for ancient Chinese writers it often included much of the Persian culture-area—witness Chang Hua talking in the *Po Wu Chih c.* +280 about Chang Chhien's explorations in the West. He says that he crossed the Western Ocean (the Caspian Sea) and so came to Ta-Chhin (ch. 1, p. 5*a*). Thus the area comprised was roughly from Teheran to Ankara and from Tiflis to Damascus, the western limits being quite undefined. Persian plants could therefore certainly come from Ta-Chhin. Of course there is no contemporary −2nd-century evidence that Chang Chhien himself went there.

[c] See Laufer (1), pp. 329 ff.

[d] The question of the authenticity of the text of the *Nan Fang Tshao Mu Chuang* will be found discussed in detail in the sub-section on exotic and historical botany (pp. 447 ff. below).

[e] The name may therefore be as old as the −4th century. Hirth (1), p. 271 is therefore to be upheld against Laufer.

[f] Cf. Vol. 1, pp. 178 ff. and Vol. 4, pt 3, pp. 441 ff.

[g] CC 456.

[h] CC 458. This is very common in Cambridge gardens, and is flowering above the snow within the walls of Caius College as I write.

[i] (1), pp. 129 ff.

[1] 指甲花   [2] 大秦   [3] 素馨   [4] 黄馨   [5] 迎春花

remained wild in the mountainous west, and have only recently been brought into cultivation.[a] Finally, a plant of a different genus, *Nyctanthes Arbor-tristis*, also travelled from India to China by the +4th century—this is the night-blooming jasmine, called *su nai hua*[1] (by loan from the crab apple, cf. p. 423) or *hung mo-li*[2], pink sambac.[b]

All through later times Chinese scholars interested themselves in plant distribution, delighting to note the ranges and habitats of particular plants. Thus in the middle of the +12th century Chhen Shan[3] marvelled at the fact that the flowering plants of central and southern China were so different from those of the north. Most probably he had been brought up as a child in Khaifêng or north of the River, and then with his family migrated south to Chekiang or Chiangsi when in +1126 the capital fell to the Chin Tartars.[c] Among the flowers of the south which he discussed[d] in his *Mên Shih Hsin Hua*[4] were the *mo li hua*[5] (*Jasmimum Sambac*, just mentioned), the *han hsiao hua*[6] (half smiling flower), i.e. *Michelia figo*, one of the magnolias;[e] the *chhü na*[7], equivalent to the *chia chu thao*[8], in fact the oleander *Nerium odorum* (Apocyanaceae)[f]; and the *shê-thi-chu*[9] or *Jāti* flower, a Buddhist name for the pot marigold or golden-bowl, *Calendule officinalis* (= *arvensis*).[g]

The same applied to fruits. About +1230 Chang Shih-Nan wrote in his *Yu Huan Chi Wên*:[h]

The lichis of San-shan[10] (i.e. the Fuchow district in Fukien) are most beautiful to look at when they become red. In the 4th mouth, though still small and sour, they give the place the name of Fiery Hills (Huo-shan[11]). In the 5th month, when they begin to taste right they are called *chung-kuan*[12] and the last stage (of ripeness) is called *chhang-shu*[13]. The best of this *chung-kuan* produce is not inferior to that from Phu-chung.... Among the fruits of San-shan there are also the *huang tan tzu*[14], the *chin tou tzu*[15], the *phu-thi kuo*[16] and the *yang thao*[17], and these are to be found absolutely nowhere else. *Huang tan tzu* is the size of a small orange (*chü*[18]) but fawn-coloured, slightly acid but at the same time sweet. The Pên Tshao books put this under citrous fruits, but is it not really something special with a name of its own?

He forgot to say anything about the other specialities of this part of the country, but we know pretty well what they were. On the *huang tan tzu* he was really perceptive, for it is in fact an unusual and little-known member of the Rutaceae,

---

[a] See Kobuski (1). [b] Cf. Burkill (1), vol. 2, p. 1564.
[c] Cf. Vol. 4, pt 2, pp. 497 ff.; Needham, Wang & Price (1), pp. 122 ff.
[d] The passage is given in *TSCC, Tshao mu tien*, ch. 14, *tsa lu*, p. 3a.
[e] CC 1356; see also *CFP* (*Hua phu*), ch. 3, p. 17a.
[f] CC 428; see also *Hua Ching*, ch. 3, p. 16a.
[g] R 18. The usual name is *chin chan tshao*[17].
[h] Ch. 5, p. 7b, tr. auct.

[1] 素奈花	[2] 紅茉莉	[3] 陳善	[4] 捫蝨新話	[5] 茉莉花
[6] 含笑花	[7] 渠那	[8] 夾竹桃	[9] 闍提著	[10] 三山
[11] 火山	[12] 中冠	[13] 常熟	[14] 黃澹子	[15] 金斗子
[16] 菩提果	[17] 羊桃	[18] 橘	[19] 金盞草	

Clausena Wampi (= *lansium*),[a] the 'wampee' of Old China Hands in Fuchow a century ago, advisable for taking when one has eaten too many lichis, and also good for jam to remind one of home. Its common name is still *huang-phi kuo*[1], the yellow-skinned fruit, and the word *tan*, signifying pale, can be written in various ways. The *chin tou tzu*[2], as more usually written, is another citrous fruit, *Fortunella hindsii*, one of the cumquats,[b] so delicious in syrup. *Phu-thi kuo* is almost certainly *Ficus religiosa*, fruit of the Bodhi tree, which can be eaten dried or made into preserves.[c] Perhaps the most interesting is the 'sheep peach' (*yang thao*[3]), known rather absurdly to Westerners as the 'false, or Southern, Chinese goose-berry' contrasting with the 'true, or Northern' one, but neither having any resemblance to the common gooseberry except a green pulp. One night the Fellows of Caius found on their dessert table some strange egg-shaped objects with a thin brown rather hairy skin containing a bright green pleasant-tasting flesh within which were black seeds and a yellowish core. This was the fruit of *Actinidia chinensis* (= *rufa*),[d] best eaten out of an egg-cup, yielded by this climbing shrub native to Shensi in the north, and known as *chhang chhu*[4] or (*mao yeh*) *mi hou thao*[5] (monkey peaches), but also called 'sheep peaches'. They were first mentioned in Chinese botanical literature in the *Khai-Pao Pên Tshao* of +970, which recommended them for the 'gravel' and other healing purposes. What Chang Shih-Nan was talking about, however, was the other *yang thao*, better written *yang thao*[6] (*yang*, perhaps southern, *midi*, peaches),[e] i.e. fruit of *Averrhoa Carambola*, the Malayan *belimbing manis*, shaped externally like the ovoid stone roller harrows used by Chinese farmers,[f] and hence called *wu lien tzu*[9] or *wu lêng tzu*[10], the fruit with five ridges.[g] Although the *Pên Tshao Kang Mu* (+1596) was the first of the pharmaceutical natural histories to refer to this, saying that it promotes salivation and may be antipyretic, it had been known to the botanists for many centuries. Hsi Han gave a good description of it in the *Nan Fang Tshao Mu Chuang*[h] and noted that people in Fukien and the Kuang provinces enjoyed it preserved in honey.

Chang Shih-Nan's localisation of the fruit-bearing trees he so much admired

---

[a] See Yeh Ching-Yuan (*2*), pp. 48 ff.; Stuart (*1*), p. 117; Burkill (*1*), vol. 1, p. 577.

[b] Cf. Chhen Jung (*1*), p. 567; CC 914.

[c] CC 1601; cf. Burkill (*1*), pp. 1000 ff., 1013.

[d] R 269; CC 721; B II 198, 493; Stuart (*1*), p. 14. Chia & Chia make a distinction between the Chinese names, taking *mi hou thao* to be *A. arguta* and *yang thao A. chinensis*. Actinidiaceae have been segregated from the Dilleniaceae.

[e] This is one of those cases where the orthography is confused, not only 'sheep' and 'south' being interchanged, but 'willow' (*yang*[7]) also used indiscriminately for both plants. The *Kuei Hai Yü Hêng Chih* of +1175 agrees with Chang's parlance, as also *CFP* (*Kuo phu*), ch. 2, p. 4*b*, but judging by the *Erh Ya* and its commentaries, this name applied anciently to *Actinidia*, also then called 'devils' peaches' (*kuei thao*[8]). Cf. *WCC/TK*, ch. 31 (p. 676); Sun Yün-Yü (*1*), p. 23; Wu Tê-Lin (*1*), p. 35.

[f] *Nung Shu*, ch. 12, p. 14*a*. Cf. Sect. 41 in Vol. 6, pt 3.

[g] R 366; CC 933; Stuart (*1*), p. 59; Burkill (*1*), vol. 1, pp. 269 ff., 271. It belongs to the Oxalidaceae, a family segregated from the Geraniaceae.

[h] Ch. 3, p. 3*a*; for illustration see Anon. (*56*), pl. 53. Li Hui-Lin (*12*), p. 127.

---

[1] 黃皮果   [2] 金豆子   [3] 羊桃   [4] 萇楚   [5] 毛葉獼猴桃
[6] 陽桃   [7] 楊   [8] 鬼桃   [9] 五斂子   [10] 五稜子

was perhaps excessive, but it is a well known fact that (especially where the finer points of flavour and pigmentation are considered) specific varieties may be extremely restricted in range. Wang Chên gave an example of this in +1313, when speaking of a citrus fruit known as the *ju kan*[1] or 'cream orange'.[a]

The name *kan* implies sweetness (*kan*[2]) and is a general name for the sweet kind of orange. The stem and leaf of the *kan* is the same as that of the *chü*, but has no thorns, being different in this respect. Planting and tending are the same as for the *chü*. (*Kan* trees) grow much in the vicinity of Chiang-han (in Hupei) and Thang-hsien and Têng-hsien (in Honan). The *kan* oranges found in Ni-shan[3] (a place in Chekiang)[b] are called *ju kan*[3] (cream oranges). The area where this variety is grown comprises the land of only one village, yet the fruit produced there is double the size (of *kan* grown in other localities).

We must not multiply examples of the interest taken by Chinese botanical scholars in plant geography, but one more instance may be permitted, especially as it is quite an exercise to identify the angiosperms concerned. It comes from a book written by Chhen Chi-Ju[4] in +1620 or thereabouts, and called *Yen Chhi Yu Shih*[5] (Peaceful Occupations of a Mountain Hermitage), i.e. botanical and horticultural pursuits. Here he lists the flowers most characteristic of the different provinces and regions of his time.[c] For Fukien he chooses an easy one, *hung mo li*[6], undoubtedly a pink or pale red variety of *Jasmimum Sambac*.[d] For Szechuan, however, he chooses the 'purple embroidered ball' (*tzu hsiu chhiu*[7]), a name which at first sight would imply one of the viburnums,[e] particularly at home in west and south-west China; but this would be quite on the wrong scent, so to speak, for in fact the name was a derivative term for a purple variety of tree-peony, 'the acme of the flowers of Thien-phêng'.[f] Thien-phêng mên[8] was a place near a pass through the hills in the neighbourhood of modern Phêng-hsien in Szechuan, and in the +11th century was considered second only to Loyang as the centre of culture of these beautiful ornamentals.[g] And by a secondary transfer, *tzu hsiu chhiu* could also mean a special purple chrysanthemum,[h] though probably not here. The land of Yen[9] (northern Hopei) is represented in Chhen Chi-Ju's list by a yellow variety of pomegranate (*huang shih liu*[10]),[i] and Loyang of course by a yellow variety of the herbaceous peony (*Paeonia lactiflora*), *huang shao yao*.[11] Chhangchow (probably in Honan) he typifies by its scented crab apple (*hai thang*[12]), some

---

[a] *Nung Shu*, ch. 9, p. 12a, tr. auct., adjuv. Hagerty (11), who took it from *NCCS*, ch. 30, p. 11a.

[b] We shall remember this when describing Han Yen-Chih's *Chü Lu*, type-specimen of the Sung botanical monographs (p. 368 below).

[c] The passage appears in *TSCC*, *Tshao mu tien*, ch. 14, *tsa lu*, p. 3b.

[d] Cf. *CFP* (*Hua phu*), ch. 2, p. 44a, b.

[e] *V. fragrans*, for example (CC 2473) or *V. macrocephalum*. Cf. *CFP* (*Hua phu*), ch. 1, p. 46a. See also Liu Tzu-Ming (1), p. 101; Li Hui-Lin (8), pp. 131 ff.

[f] Cf. *CFP* (*Hua phu*), ch. 2, pp. 7a, 22a.   [g] Li Hui-Lin (8), pp. 28 ff.

[h] Cf. *CFP* (*Hua phu*), ch. 3, p. 49a.

[i] Cf. *CFP* (*Kuo phu*), ch. 3, pp. 18b, 19a; Li Hui-Lin (8), pp. 192 ff.

---

[1] 乳柑   [2] 甘   [3] 泥山   [4] 陳繼儒   [5] 巖棲幽事
[6] 紅茉莉   [7] 紫繡球   [8] 天彭門   [9] 燕   [10] 黃石榴
[11] 黃芍藥   [12] 海棠

variety of *Malus spectabilis*;[a] and two other sorts of *Malus* are confined, he implies, to Thien-thai in Chekiang, a yellow kind and a white, the latter possibly *M. floribunda* or *halliana*.[b] Chekiang also boasts the best *kuei hua*[1], *Osmanthus fragrans*[c], white, purple yellowish or jade-green; besides the white *mei kuei*[2] rose.[d]

This sub-section may suitably end with a quotation from Wang Hsiang-Chin's thesaurus of botany, written about +1628. In his *Chhün Fang Phu* he wrote:[e]

As for flowers, herbaceous plants, vegetables and fruit-trees, there are differences in the soil and in the land on which they can flourish. Those that grow in the north are more tolerant to cold weather, while those that inhabit the south rejoice in a warm or hot climate. Then the (way of) planting, and the (method of) irrigation (as to time and quantity), which produce the best results are far from identical; periods of flowering and setting seed are also different. Alpine vegetation and the plants of the plains are as unlike each other as day and night.

When plants of northern habitat are moved to the south they generally flourish, but when plants from southern regions are transplanted to the north they readily change—as for example (the famous case of) *chü* orange-trees growing south of the Huai River but turning into *chih* limebushes (with sour fruits) when moved to the north. (Contrariwise) *ching*[3] plants[f] flourish in the north, but when they are planted in the south they produce no more (large) roots.

So also the lungan (*lung-yen*[4]) tree[g] and the lichi (*li-chih*[5]) tree[h] are prolific in Fukien and Kuangtung, while the hazel-nut tree (*chen*[6])[i], the jujube-date tree (*tsao*[7])[j], and (all sorts of) gourds and melons (*kua lou*[8]) are plenteous in Hopei and Shantung. Plants cannot disobey the appropriate seasons (for their regular development). How could man force plants (to do the impossible)?

Expert horticulturists must pay attention to Master Liu (Liu Tzu[9])[k], who said: 'A thing must follow its natural endowment in order to fulfil its nature and express its properties; then it will grow well and live long.' This is the way to cultivate plants.

[a] *CFP* (*Hua phu*), ch. 1, p. 4*a*, and Li Hui-Lin (8), pp. 121 ff.

[b] *CFP* (*Hua phu*), ch. 1, p. 2*a*, and Li Hui-Lin (8), pp. 121 ff.

[c] Li Hui-Lin (8), p. 151. Long shall we remember the dishes of *kuei hua* beside our beds in the guesthouse of the Taoist temple of Chin Tzhu in Shansi.

[d] *Rosa rugosa* (CC 1147) or perhaps *Rosa banksiae* (CC 1133); cf. Li Hui-Lin (1), pp. 95 ff.

[e] *Hui phu*, ch. *shou*, p. 2*b*.

[f] As we find elsewhere (p. 90) this word in combination with others may denote plants or parts of plants in Liliaceae and Cyperaceae, but here Wang Hsiang-Chin surely had in mind the vegetable *wu ching*[10], one of the cultivated turnip varieties, not too easy to define botanically. Its common name is *ta thou tshai*[11], perhaps best called the rape-turnip; *Brassica Rapa-depressa* (R 477), *B. Rapa* (CC 1290), *B. Napus* (Bentham & Hooker (1), p. 37), presumably derived from *B. campestris*. No doubt the hot climate decreases the propensity for storage.

[g] *Euphoria* (or *Nephelium*) *Longana*, which gives delicious fruit (R 302; Chhen Jung (1), p. 683). One of the Sapindaceae, along with the soapnut *Sapindus Mukorossi* just mentioned (p. 110 above) and the lichi.

[h] *Litchi chinensis* (R 300; Chhen Jung (1), p. 685).

[i] *Corylus*, probably *heterophylla*, but several other species flourish in China, e.g. *chinensis*, *sieboldiana*, *tibetica*, etc. See CC 1673; R 618; Chhen Jung (1), pp. 174 ff.

[j] *Zizyphus Jujuba* (Chhen Jung (1), p. 749).

[k] This must be a reference to Liu Tsung-Yuan[12], whose *Chung Shu Kuo Tho-Tho Chuan*[13] (Biographical Essay on Camel-back Kuo the Horticulturist), written about +800, contains a sentence worded in almost the same terms. See, for example, *Chhün Fang Phu*, *tshê* 6, *Mu phu*, p. 1*b*.

[1] 桂花   [2] 玫瑰   [3] 菁   [4] 龍眼   [5] 荔枝
[6] 榛   [7] 棗   [8] 瓜蓏   [9] 柳子   [10] 蕪菁
[11] 大頭菜   [12] 柳宗元   [13] 種樹郭橐馳傳

(c) BOTANICAL LINGUISTICS

(1) PLANT TERMINOLOGY

How did the Chinese talk about plants, and how did they give them their names? These are the questions which this sub-section is designed to answer. We have to think here of two sorts of designations, first those for the particular parts and patterns of plants, then those for particular species of plants, i.e. words and phrases used to differentiate them from other species. One can thus distinguish (and we propose to do so) between terminology and nomenclature.[a] The second of these is of course closely bound up with the development of systems of classification, but these deserve a sub-section to themselves at a somewhat later stage (Sect. 38 *f* below).

All this, it must be remembered, evolved within the ideographic universe—a world so different from the alphabetic that the reader may like to turn back at this point and re-read the brief account of the Chinese language and script provided in Vol. 1.[b] Most of the terms and names are composite or 'molecular'[c] so that perhaps the 'atoms' of the script should claim our attention before the 'molecules'. Since the Chinese written language is made up of ideographs based very largely on ancient pictographs one would rather expect to find a substantial 'botanical' component in the radicals and phonetics[d] out of which the characters were formed, and so indeed one does. Let us first look then at this intrinsic botanical portion of the pictographic fund before going on to review the terms and character-complexes used in describing plants. We shall then be able to consider more easily the genesis and variety of the specific plant names, together with certain questions raised thereby, such as the incidence of 'cross-referencing' or the validity of the classical view that ancient Chinese was a rigorously monosyllabic language. Lastly we may return to the structure of the ideographs themselves and take up a piquant point—the possible advantages of the ideographic system for all botanical and zoological nomenclature.

(i) *Botanical radicals*

Any young scientist beginning the study of Chinese will immediately be struck with the fact that many of the radicals derive, more or less obviously, from pictures of plants and animals.[e] If he should be moved, as I was long ago, to tabulate them, he will obtain a result such as that which appears in Table 6. No less than fifteen of the present 214 radicals are in fact based on plant forms, as

---

[a] I.e. the phytographic and the taxonomic languages. Métailié (1) has given us an admirable monograph on the usages in both, employed by contemporary botanists writing in Chinese.

[b] Pp. 27 ff., in Section 2.

[c] Cf. Vol. 1, p. 31.

[d] Vol. 1, p. 30.

[e] For discussion of the 'zoological' radicals see Vol. 6, pt 3 below in Sect. 39.

Table 6. *Present botanical radicals*

no.	no. of strokes	Radical no.	present form	pronunciation	meaning and comments	botanical position	ancient graph K or SW	K no.[a]	no. of derivs. now in common use (Soothill, 1)	SW ref.	no. of derivs. in SW
1	3	45	屮	chhê	sprout; see Table 7.			(1052a)	2	1/15.1	6
2	4	65	攴	chih	branch; orig. to pluck the culms of bamboos			(865)	1	3/65.1	1
3	4	75	木	mu	wood, trees			1212	398	6/114.2	432
4	5	97	瓜	kua	melon, gourd	= Cucurbitac.		(41)	7	7/149.2	8
5	5	100	生	shêng[b]	birth, to be born			812	5	6/127.2	5
6	5	115	禾	ho[b]	growing grain, cereal plants	= Gramin.		8	89	7/144.1	91
7	6	118	竹	chu	bamboo	= Gramin. Bambus.		1019	164	4/95.1	148
8	6	119	米	mi	rice and other grains	= Gramin. (*Oryza* and *Panicum*)		598	59	7/147.1	41
9	6	140	艸 abbr.++	++tshao[c]	herbaceous plants, undershrubs and shrubs			(1052c)	365	1/15.1	457
10	7	151	豆	tou[d]	bean	= Legumin.		118	12	5/102.2	5
11	9	179	韭	chiu	leek	= Liliac. *Allium* spp.		(1065)	2	7/149.2	5
12	10	192	鬯	chhang[e]	sacrificial aromatic wine made from black millet and orchidaceous or other plants			719	2	5/106.2	4
13	11	199	麥	mai	wheat	= Gramin. in part		932	10	5/112.1	12
14	11	200	麻	ma	hemp	= Morac. *Cannabis sativa*		(17)	2	7/149.1	3
15	12	202	黍	shu	glutinous millet	= Gramin. *Panicum miliaceum*, var. *glut.*		93	8	7/146.2	7

a If the number in Karlgren (1) is placed in brackets, it indicates that no oracle-bone forms are known.

b Combined from *chi* and *hua* in Table 7, nos. 9 and 10.

c The more common synograph 草 *hua* is not a radical form. See discussion in text.

d This graph is a picture of an eating vessel for meat and beans. It replaced the word *shu*, no. 13 in Table 7.

e This graph is a picture of a wine cup or bowl, not a plant in spite of its flower-like look.

may be seen by comparing the modern fully stylised ideographs in column 4 with the bone or bronze forms of the −2nd and −1st millennia in column 9 to the right. The radical number (col. 3) is of course the position of the word in the current standard order of radicals, and the K number (col. 10) gives access to the ancient graph variants in the repertory of Karlgren (1). Then follows the reference to the *Shuo Wên Chieh Tzu*[1] dictionary of Hsü Shen[2], finished in +121,[a] together with two statistical counts of derivatives, one drawn from this most ancient of dictionaries, the other from a widely used modern pocket dictionary, that of Soothill (1). By derivative we mean characters formed by the addition of phonetics to the particular character in question,[b] whence it can immediately be seen how varied were the fortunes of those graphs upon which the lexicographers conferred the titular dignity of radical; some pictures generating many hundreds of words, others less than a dozen. It is not to be thought that the number of derivatives here registered are all those that the language contains, for a great many more will be found in the *Khang-Hsi Tzu Tien*[3] of +1716 and the occidental works based on it such as that of Couvreur (2), but it is safe to say that if all these were taken into account the differences would remain about the same as those shown here because elaboration and character-invention throughout the centuries increased proportionally. The fact that the derivative words in common use today approximately equal in number all those which the *Shuo Wên* mustered in the +2nd century is but a coincidence.

Let us now take a closer look at the symbolism of the primitive pictographs.[c] The simplest is the trunk, stem or stalk with branches, symbolic also of the petioles with their leaves, or the petiolules of a trifoliate leaflet. As we may see from Table 7, this drawing came in four forms, the basic graph being repeated up to four times and the pronunciation varying. The single drawing, which signified a seedling or sprout, had no great future, but the double one (no. 9, rad. 140) gave rise to an enormous number of derivatives, dominating indeed the whole world of plants which were not recognisably trees. So much so was this the case that it attained the degree of stardom indicated by universal abbreviation in all its compounds. But now how to account for the fact that its expressive primitive radical form was replaced by a homophonic synograph *tshao*[7] as early as the +2nd century, Hsü Shen's own time? As he himself tells us,[d] this word meant the

---

[a] A word on its later history. The work of the erudite Thang scholar Li Yang-Ping[4] *c.* +763 was not an unmixed blessing for it, but a great restoration was made by Hsü Hsüan[5] and his younger brother Hsü Chiai[6] in the first decades of the Sung dynasty (+960 to +990). From them it came almost unchanged into the hands of the scholars of the nineteenth century.

[b] Cf. Vol. 1, p. 30.

[c] There is of course a wealth of writing on ancient Chinese graphical etymology. It may suffice to refer the reader here to the books of Karlgren (4, 5), the papers of Hopkins (3, 4) and his translation (36) of the *Liu Shu Ku*[8] written by Tai Thung[9] about +1275; and discussions by Schindler (6, 7) and others. Attention was drawn to the vegetable and animal pictographs by von Takács (1).

[d] Ch. 1, p. 27.1

[1] 說文解字   [2] 許慎   [3] 康熙字典   [4] 李陽冰   [5] 徐鉉
[6] 徐鍇   [7] 草   [8] 六書故   [9] 戴侗

fruit of the *li* [1] or *hsiang* [2] oak,[a] i.e. an acorn. His early Sung editor, Hsü Hsüan, added that it was interchangeable with the related word *tsao* [3] [4] normally meaning black, or a black dye; and the reason for this connection undoubtedly was that in ancient China the tannin of acorns was used for dyeing.[b] Why *tshao* [1] so universally replaced *tshao* [5] we do not know, nor is it clear whether its phonetic *tsao* [6], meaning early morning or dawn, was itself derived, as Hopkins (25), (departing from Hsü Shen) maintained, from a drawing of an acorn and its cupule, only diverted to horary use by homophonic borrowing. The other view, which made it a drawing of the sun coming up over the world, has the disadvantage that it does not explain the curious connection with the dyers. Lastly, there are the three-plant and four-plant forms to consider. The three-plant one, pronounced (and now written) *hui* [7], is interchangeable with *tshao*, and presumably always was, meaning herbaceous plants in general; it lived on into modern times because of its stylistic value according to the 'principle of elegant variation' and we meet with it in the title of one of the parts of Wang Hsiang-Chin's *Chhün Fang Phu*. It never had the luck to be a radical, and the *Shuo Wên* regarded it simply as a derivative from the two-plant tshao.[c] As for the four-plant form, which will appear in Table 7, the *Shuo Wên* recognised it as a radical[d] but one with hardly any derivatives, and read it as *mang* [8]. It simply meant masses of vegetation in general, a vague denomination for which there were all too many competitors, as we shall see.

tshao tsao hui

From stems, branches and foliage three other standard radicals derive, *shêng* [9], to be born, *chiu* [10], the leek and leeks, and *mai* [11], wheat. Since the first of these shows the stem and two branches or petioles of a seedling rising out of horizontal ground, the men of the Shang skilled in wort-cunning deserve the honour of having endowed the language with its basic word for generation as such. The second has one of the most striking of the ancient pictographs, vividly recalling the sheaf of parallel-veined basal or cauline, lorate or equitant monocot leaves characteristic of Liliaceae. The third depicts the graminaceous culm and inflorescence with the characteristically drooping leaf blades, all supported by the footstep[e] of the harvester. Perhaps to this group we should add no. 2 (rad. 65), *chih* [12], the word that

---

[a] *Quercus acutissima* ≒ *serrata* (CC 1639; Chhen Jung (*1*), p. 200).

[b] The pyrogallol tannins absorb oxygen in alkaline solution and darken; all tannins give blue- or green-black dyes with ferric salts. Cf. St Julien & Champion (*1*), pp. 95 ff.; Haas & Hill (*1*), pp. 192 ff. According to Burkill (*1*), vol. 2, p. 1852, acorns particularly rich in tannin are given by *Quercus mongolica* = *robur* (CC 1647; Chhen Jung (*1*), p. 197).

[c] Ch. 1, p. 25.2   [d] *Ibid.* p. 27.2

[e] Cf. K 961, 1258*c*. As Hu Hsi-Wên (*1*) points out, the word *lai* [13], now 'to come', is the same graphic picture of a cereal plant, and with this meaning was originally written with rad. 60, i.e. a step with the left foot (cf. Vol. 2, p. 551). For Hsü Shen the idea was that the wheat had 'come' from heaven, i.e. that it was derived from heaven-sent wild wheat (*Shuo Wên*, ch. 5, p. 111.2).

[1] 櫟	[2] 橡	[3] 皂	[4] 皁	[5] 艸
[6] 早	[7] 卉	[8] 冈	[9] 生	[10] 韭
[11] 麥	[12] 支	[13] 來		

afterwards came to mean all sorts of branches, but which started as a verb, "to pluck the culms of bamboos', doubtless with a knife, which one can see in the hand at right angles to the culm.

Logically we should think next of any pictographs which show only roots and stem, and one indeed there is, *chu*[1], that appropriated from the beginning to the sub-family Bambuseae (no. 7, rad. 118). Most of the rest show other things besides the roots and stem—type-specimen of course is no. 3 (rad. 75) dendritic in every sense, one of the simplest graphs that every child first learns, even if later he can hardly see the wood for the trees. No. 8 is very similarly constructed, unless the horizontal line is the ear and the dashes the rice-grains nestling in it. Next comes no. 6, *ho*[2], which includes the roots, stem and leaves of a grass or cereal plant with its head or inflorescence nodding to the left. As Table 7 will show us, this is a conflation of two ancient radical characters which had the head nodding either to the left or to the right, and each signified something a little different. Then the entire plants are seen again in the word for hemp, *ma*[3] (no. 14), but now however drying under the roof of an outhouse. The outhouse or trellis appears too in the cucurbitaceous radical (no. 4) in which for the first time we see a picture of a fine round fruit. The climbing tendril-bearing stem is well seen here. Lastly the millet radical, *shu*[4], combines the roots, culms and nodding cereal heads with the ancient graph for water.[a]

So far we have considered only the 'botanical' members of the present 214 radicals. But one must remember that in earlier times the number of these was not so far reduced, and as was noted in Sect. 2, the *Shuo Wên* recognised about 540 of them.[b] Landmarks in the progressive reduction were the *Liu Shu Pên I*[5] (Basic Principles of the Six Graphs) produced by Chao Wei-Chhien[6] about +1380, which recognised 360 radicals; and the two late Ming dictionaries which established and stabilised the system of 214. These were the work first of Mei Ying-Tso[7] in his *Tzu Hui*[8] (The Characters Classified) of +1615, and then of Chang Tzu-Lieh[9] with his *Chêng Tzu Thung*[10] (Complete Character Orthography) in +1627. This piece of philological history has great interest of its own, but does not directly concern us here.[c] What is obvious is that the old lists must contain a

---

[a] It seems odd at first sight that millet rather than the characteristically wet ricefield crop should have had water built in to its pictograph, but perhaps this argues for the necessity of irrigation of millet agriculture in ancient North China.

[b] Vol. 1, p. 31. Counts vary slightly, 541 or 548 according to the criterion.

[c] The story is a good deal more complicated than could be told in Sect. 2. The 540–odd radicals of the *Shuo Wên* were retained, though rearranged, by Ku Yeh-Wang[11] in his *Yü Phien*[12] (Jade Page Dictionary) of +543, as also half a millennium later by the conservative Ssuma Kuang[13] in his *Lei Phien*[14] (Classified Dictionary) with 544 radicals (+1067). 'Radicalism' in reduction, however, went further in the Thang and Sung than it did later, for in the Thang (*c.* +770) Chang Shen[15] accepted only 160 for his *Wu Ching Wên Tzu*[16] (Characters of the Five Classics); and in the Sung, Li Tshung-Chou[17] in his *Tzu Thung*[18] (Complete Character Dictionary) slashed away

---

[1] 竹	[2] 禾	[3] 麻	[4] 黍	[5] 六書本義
[6] 趙撝謙	[7] 梅膺祚	[8] 字彙	[9] 張自烈	[10] 正字通
[11] 顧野王	[12] 玉篇	[13] 司馬光	[14] 類篇	[15] 張參
[16] 五經文字	[17] 李從周	[18] 字通		

Table 7. *Additional ancient botanical radicals*

no.	no. of strokes	present form	pronunciation	meaning and botanical comments	ancient graph K or SW	K no.[a]	SW ref.	no. of derivs. in SW
1	2	弓	*han*	bud in the earliest stage of formation		—	7/142.2	4
2	3	才	*tshai*	talent, genius; (orig.) a plant or tree seedling		943	6/126.2	0
3	3	乇	*chio*	drooping leaves of a plant 'like a drooping inflorescence (*chui sui*[1])'		(780)	6/127.2	0
4	4	屮	*pei, pho*	lush vegetation		—	6/127.2	5 (incl. *nan*[2])
5	4	之	*chih*	genitive particle, also to go or come out; (orig.) a plant which has passed through the 屮 stage (Table 6, no. 1) and begins to have a larger stem with branches. K thinks this character a variant of *chih*[3], which is almost certainly the drawing of a foot, hence its meaning, to stop. Of course, to stop, and stay in one place, is to behave like a tree. Cf. no. 13 in Table 6.		962	6/127.1	1
6	4 [but 3]	丰 丯	*chieh, chiai* / *fêng*	mixed vegetation / is a derivative of no. 5 in Table 6. It also means luxuriant growth of herbaceous plants. A similar graph is now the abbreviation of *fêng*[4], abundance.		— / 1196	4/93.1 / 6/127.2	1 / —]
7	4	朮	*pin*	to peel off the integument of the male hemp plant *hsi*[5] (*Cannabis sativa*) for textile fibre; 屮 plus 八. Now absorbed in no. 3 of Table 6.		—	7/149.1	1
8	5	出	*chhu*	to go forth; (orig.) going forward, sprouts coming out of the ground, 'having got the benefit of nutrition (*i tzu*[6])', more advanced than *shêng*, no. 5 in Table 6.		496	6/127.1	4
9	5	末	*ki, chi*	bent tip of a plant which can grow no more		—	6/128.1	2 (or 5 if no. 27 below be included)

No.	(b)	Char.	Romanization	Meaning	No.	Reference	Count
		蒼		in it. The reference is to the ancient idea that the sun rose each morning out of the Fu-Sang[9] tree on an island in the Eastern Ocean (cf. Granet (1), vol. 2, p. 435 et sub voce). We have already seen this island in Vol. 3, Fig. 242. Cf. Vol. 4, pt 3, p. 541.			
12	6	朿	tzhu	thorn	868	7/143.1	2 (both *Zizyphus* spp., Rhamnac.)
13	6	朮	shu	beans (Leguminosae). Replaced by no. 10 in Table 6.	(1031)	7/148.1	1
14	8	瓜	pha	general term for *pa*[10] or *pha*[10], the corolla of a flower, perhaps the whole perianth anciently. Did the semantic connection with fibres (see no. 7 above) arise from the apparatus of the gynoecium and androecium? Now absorbed in no. 15.	—	7/149.1 1/ 22.1	2
15	8	林	lin	a dense grove on level ground	655	6/126.1	9
16	9		thiao	adj. drooping of plant fruits and seeds		7/143.1	2
17	10		li	plants properly or comfortably spaced out. Hence mod. derivs.: history, calendar, etc.	858	7/146.2	1
18	10		shui	the 'leaves' of flowers, petals of the corolla	—	6/128.1	0
19	10		cho, chho	a dense clumped grove. But the pictograph suggests a composite or umbelliferous capitulum, umbel or corymb.	—	3/ 58.2	3
20	10		huo	drooping flowers and fruit of trees		7/142.2	2
21	11	桼	chhi	lacquer; tree juice, dripping down like water, that can be used for varnishing things'. Specific for *Rhus verniciflua*, Anacardiaceae.	(401)	6/128.2	2
22	12	茻	mang	vegetation		1/ 27.2	3
23	12	舜	shun	a malvaceous plant, *Hibiscus syriacus* (B II 542; CC 742; Lu Wên-Yü (1), p. 52, no. 60; Li Hui-Lin (1), p. 140). The 'rose of Sharon', native to China, in spite of its Linnaean name.	(469)	5/113.1	1
24	12	華	hua, huang	the flower as a whole	(44)	6/128.1	1
25	13	蕚	khua	the flower (cf. later *o*[11], calyx, and *jui*[12], bud or stamens)	—	6/128.1	1

Table 7. (*continued*)

no.	no. of strokes	present form	pronunciation	meaning and botanical comments	ancient graph K or SW	K no.[a]	SW ref.	no. of derivs. in SW
26	14	蓐	*ju*	perennial. Plants above, and apparently below, separated by seasonal thunder and lightning (the simple form of *chen*[13]). Later meant a sucker arising from a tree's root.		(1223d)	1/ 27.2 14/311.1	1
27	15	稽	*ki, chi*	the modern meanings, to stop, detain, investigate, were already current in Hsü Shen's time. The graph combines no. 9 above with *yu*[14], fault, blame, shown by a hand holding something, perhaps a stick. Was the semantic significance that of 'nipping the growth of some misdemeanour in the bud' or rather 'halting its growth'?		(5520)	6/128.1	5 (incl. three names of woody plants acc. to Hsü Shen's friend Chia Khuei[15] cf. Vol. 3, *passim*)
28	16	馨	*hsiang*	perfume. *Kan*[16], sweet, surmounted by *shu*, millet, no. 15 in Table 6. Already in Hsu Shen's time it was written in its modern form *hsiang*[17], but animal perfumes, if fragrant, were not excluded, though the etymology is botanical. Mod. rad. no. 186, with six derivatives in Soothill (1).		(717)	7/147.1	1

[a] If the number in Karlgren (1) is placed in brackets, it indicates that no oracle-bone forms are known.

[1] 垂穗 [2] 南 [3] 止 [4] 豐 [5] 枲 [6] 益遊 [7] 禾 [8] 枲甘
[9] 扶桑 [10] 葩 [11] 尊 [12] 糵 [13] 尤 [14] 尤 [15] 賈逵 [16] 甘
[17] 香

group of 'submerged' botanical ideographs, and indeed one may pick out from the *Shuo Wên* as many as 28 more (see Table 7). We need not describe them as systematically as those in Table 6, for they follow the same principles, and the reader may study them for himself. Some, however, are especially worth looking at, for instance the series of ontogenetic terms, of course very imperfectly defined (nos. 1, 2, 5, 8, 9); and the variety of morphological ones, which hardly appeared in Table 6, e.g. foliation (no. 3), defences (no. 12), anthology (nos. 14, 18, 20, 24, 25) and carpology (no. 16). A term for duration (no. 26) and one depending upon biochemical properties (no. 28) also have interest. Besides these there are certain items of nomenclature in the taxonomic sense (nos. 11, 21, 23) which have crept in because the characters chanced to be given radical rank; the vast majority of such terms were of course "molecular" derivatives from the 'atomic' graphs in the two Tables. Lastly there are four graphs which never came to mean anything more than lush vegetation in general. Let us make another tabulation summarising what we have found (Table 8).

We have now seen quite enough forms and patterns based on plant life in the old pictographs to justify the appellation of 'tree-writing' given to the Chinese script by a whimsical scientific colleague of ours long ago. What strikingly appears from the count in Table 8 is that no less than 17 of the radicals we have analysed are essentially taxonomic in quality.[a] Thus a principle of classification was enshrined from the start in the Chinese pictographic and ideographic script. This was never of course systematically or scientifically planned, it just grew naturally, like a plant itself, hence the overlaps which one can see in the Tables, the meanings of questionable use, the vagueness of definition, and so on. But now and then a particular graph, such as the rooted bamboos of Table 6, no. 7, served a particular sub-family, here the Bambuseae, fairly and squarely; and no

all but 89. Scholars in the Liao and J/Chin dynasties were almost as revolutionary, for in +997 the monk Hsing-Chün[1] retained only 240 radicals in his *Lung-Khan Shou Chien*[2] (Handbook of the Dragon Niche), and he was followed by Han Tao-Chao[3] in +1208, the compiler of the *Wu Yin Lei Chü Ssu Shêng Phien Hai*[4] (Ocean of Characters arranged according to the Five Rhymes and the Four Tones); cf. Ogawa Tamaki (1). These reductions were not generally accepted, however, and Tai Thung's *Liu Shu Ku* (History of the Six Types of Graphs), just mentioned (p. 119), kept 479; this at the end of the southern Sung, c. +1275. Then, at the end of the Yuan, came Chao Wei-Chhien. About +1590 Chao's 360 radicals were cautiously reduced to 314 in a work by Tu Yü[5], the *Lei Tsuan Ku Wên Tzu Khao*[6] (Study of the Classification of the Ancient Literary Characters). However, the 214-level target had very nearly been attained by the Liao and J/Chin lexicographers, and a ranging shot on the other side of it, 200, was scored by Hsü Hsiao's[7] *Ho Ping Tzu Hsüeh Chi Phien*[8] (Collected Papers on Unified Graphology), a +15th-century work.

As an 'honorary' Chinese who has been using dictionaries for forty years and more, I should like to record my impression that the 214 reduction was too great (though some reductions were not made that ought to have been made), and that 234 or 240 would perhaps have been better, or at least more convenient. The Liao monk and the J/Chin scholar would thus bear the palm, but both their works are so rare now that we have not been able to see them.

Mei Ying-Tso seems to have been unduly neglected. He it was, apparently, who first arranged the radicals and characters in the order of the number of their strokes.

[a] Two signifying very general groups, three at natural family level, three at approximately sub-family level, and nine denoting genera or species.

[1] 行均　　　　[2] 龍龕手鑑　　[3] 韓道昭　　[4] 五音類聚四聲篇海
[5] 都俞　　　　[6] 類纂古文字考　　　　[7] 徐孝　　　　[8] 合幷字學集篇

Table 8. *Count of the meanings of radical graphs*

Categories	modern radicals (214)	Shuo Wên radicals (548)	totals
Collective terms (vegetation in general)	—	4	4
Ontogenetic stages	2	5	7
Duration	—	1	1
Morphological parts	1	8	9
Nomenclatorial terms:			
general groups	2	—	2
approx. Family rank	2	1	3
approx. Sub-Family rank	1	1	2
Sub-Family rank	1	—	1
approx. Generic and Specific rank	5	4	9
Agro-horticultural arrangement	—	2	2
Biochemical property	—	1	1
Miscellaneous	1	1	2
totals	15	28	43

one could quarrel with a symbol for the whole of the Leguminosae like no. 13 in Table 7. Very little ingenuity would be (or would have been) required to multiply these radical graphs by ten in order to cover the 435 odd natural families of angiosperms recognised in modern works such as that of Good (1) or Hutchinson (1), providing each with an unambiguous cipher capable of heading further ideographic components denoting sub-family, genus and species. But our intention here is only to adumbrate this point, for we shall return to it at the conclusion of the present sub-section (p. 178).

(ii) *The phytographic language*

Having now scrutinised the participation of plant forms in the linguistic medium of Chinese discourse, we can take up the development of the way of describing plants in it, i.e. the phytographic language. It will be simplest, perhaps, to look at these technical terms at both ends of their evolution. We can handle some of them, rounded, as it were, after two or three millennia of use, as they appear in a convenient glossary of botanical words compiled by Bridgman & Williams in 1841;[a] and for the other end we can extract a sheaf of interesting definitions from the −3rd-century *Erh Ya* dictionary.[b] This latter exercise is of much importance, for it will show that although Chinese literature has not preserved for us any

---

[a] Cf. Chao Yuan-Jen (4), pp. 387, 388.
[b] On this see pp. 187 ff. below.

connected and discursive treatise paralleling in size that of Theophrastus, Chinese scholars and gardeners of his time must have been talking in a very similar manner, otherwise the technical terms would not have existed. In a way the European sciences benefited strangely from the death of the Greek and Roman cultures, since these languages were able to furnish a vast stock of terms similar to those of common speech but easily set apart from them and consecrated for precise technical usage—capable also (as practising scientists so well know) of infinitely varying combinations. Monolithic Chinese had no such advantage, but it could, and at various times with great spurts it did, invent new characters *ad hoc* by the permutation of radical and phonetic components. As in other sciences, the technical terms of phytography did not begin to proliferate until the Renaissance, and even then the German fathers of botany were slow in developing them. Cordus, says Greene,[a] was the first to excel in this, because he believed that one ought to abandon reliance upon illustrations. Some of his words are common coin today (umbel, corymb, bract, involucre, calyx), others (such as fulcrum for an adventitious root) did not succeed. He still made no distinction between petal and leaf, recalling the long-continued use of *hua yeh*[1] for petal in Chinese, and it was not till Fabio Colonna (+1592) that petalon was introduced. Then in the +18th century it was one of the great achievements of Linnaeus to stabilise the vocabulary of plant description, fixing the meanings of words such as corolla, calyx, pollen, pistil, style, and introducing many new usages.[b]

The common origin of both terminology and classificatory nomenclature is doubtless that distinction Theophrastus made between trees, shrubs, undershrubs (or half-shrubs or suffrutescent plants)[c] and herbs (or herbaceous plants).[d] The first two of these are fully reflected in Chinese, for *dendron* (δένδρον) *shu*[2] and for *thamnos* (θάμνος) *kuan mu*[3];[e] so also the last, *poa* (πόα) as *tshao*[4] or *hui*[5]. The undershrub, *phryganon* (φρύγανον) was quite a convenient category, its name being derived from the idea of a faggot, so *hsin*[6] would have served, but was not so applied. The division of the vegetable kingdom as a whole according to texture and duration of stem and root was a chapter in botany, it has been said, that Theophrastus wrote for all time,[f] but it is clear that his contemporary naturalists at the other end of the Old World saw things in just the same way. Where they

---

[a] (1), pp. 275 ff. Cf. pp. 172, 206, 223. Valerius Cordus was the young genius who lived only between +1515 and +1544. Otto Brunfels (+1464 to +1534) had made no advance, wishing only to reproduce what the ancients had said, Leonhard Fuchs (+1501 to +1566) sought to be popular, but made some good distinctions, and Hieronymus Tragus (+1498 to +1554) though a great describer, introduced few neologisms. On Fuchs' glossary see Choate (1); Sachs (1), pp. 20–1.

[b] On this see the special account of Stearn (4).

[c] Stems woody at the base, where they survive from year to year, and with herbaceous distal portions.

[d] Theophrastus added an extra category for 'tree-potherbs', herbaceous plants like the cabbage which have a one-stemmed and arboreal mode of growth, calling them *dendrolachana* (δενδρολάχανα). We have not found this in China. On the whole subject see Greene (1), pp. 68 ff, 107, 110; Strömberg (1).

[e] *Erh Ya*, ch. 14, p. 10*b*.

[f] Greene (1), p. 67.

[1] 華葉　　　[2] 樹　　　[3] 灌木　　　[4] 草　　　[5] 卉
[6] 薪

made less distinction, perhaps, was between the cultivated plant, *phyton* (φυτόν), and the wild weed, *botanē* (βοτάνη), just using the term *yeh*[1] for those things that grew up spontaneously. This was natural enough, for as we noticed already (p. 82 above) there were so many valuable trees and herbs in the Chinese plant cover which needed 'tending' rather than cultivation.[a] However, the term *pai*[2] did come to be used specifically for wild valueless weeds.[b]

Now for a few paragraphs on the Chinese technical vocabulary. Passing over the words of lushness and abundance as such, we shall speak first of some which indicated general habit, and then mention root, stem, leaves, flowers, fruits and seeds in due order. Words of Chhin and Han time concerning habit are easy to find in the *Erh Ya*, for example *wan*[3], prostrate (procumbent or repent), applied e.g. to the lianous stems (*shao*[4]) of cucurbits (*tieh*[5], *cho*[6]).[c] But the most remarkable case is the distinction between deliquescence and excurrence in trees. It is evident enough, once one thinks of it, that in some trees the stem loses itself 'dendritically' by repeated branching (such as the oak), while in others the main axis rises continuously and unbranched, yet giving off branching branches on all sides (the pine or fir).[d] Theophrastus described this dichotomy clearly in one of his most modern moments, anticipating nineteenth-century dendrology,[e] but ancient China provides something very similar. The *Erh Ya* says:[f]

When the branches of a tree bend downwards like the feathers of a bird, it is called *chhiu*[7] (or *chiu*[7]); when they ramify upwards, it is called *chhiao*[8]. The *chhiu*[9] is a *chhiao*[8] tree.
    KP:[g] By nature the *Catalpa* (*chhiu*[9]) rises high into the air.
(Similarly) the *Sophora* and *Zizyphus* belong to the class of *chhiao* (*Huai chi chhou chhiao*[10]),
    KP: Their branches all ramify aloft like birds' wings.
But the *Morus* and *Salix*, mulberry and willow, belong to the (opposite) class of *thiao* (*Sang liu chhou thiao*[11]).
    KP; Spreading down and drooping is called *thiao*[11].[h]

There may have been some thought here of the difference between erect and pendulous habit, but the contrast of *chhiao* as against *chhiu* and *thiao* seems to point unmistakably to the deliquescent deciduous tree as against the excurrent conifer. The root (*kên*[12], *pên*[13]) of a plant was sometimes called *shu thou*[14], a term strangely reminiscent of Cordus' caput or crown, the point of origin of the leaves

---

[a] There were, however, a number of cases, notably in the genus *Allium*, where quite different single and double names were given to the same plant according to whether it was wild or cultivated.

[b] Anciently it was, and indeed properly still is, the name for 'barnyard millet', *Panicum* (≃ *Echinochloa*) *Crus-galli* (cf. Bretschneider (6), p. 9; CC 2034; Burkill (1), vol. 1, p. 889; Forbes & Hemsley (1), vol. 3, pp. 328–9).

[c] Ch. 13, p. 3*a*.　　　[d] Jackson (1); Lawrence (2).　　　[e] Greene (1), p. 129.

[f] Ch. 14, p. 11*b*. B II 320–28 is slightly off the rails here.

[g] Commentary by Kuo Pho.

[h] On the following page the *Erh Ya* reinforces the definition of *chhiao*[6], and probably thinking of Palmae says that *hsi*[15] is the term for trees that have no branches at all.

[1] 野	[2] 稗	[3] 蔓	[4] 紹	[5] 䟡
[6] 㐌	[7] 朻	[8] 喬	[9] 楸	[10] 槐棘醜喬
[11] 桑柳醜條	[12] 根	[13] 本	[14] 樹頭	[15] 檄

in 'stemless' plants like the carrot.[a] Liliaceous bulbs and corms were called in Chhin and Han times *kai*[1] and *chhiang*[2], tubers *chhiu ching*[3], this probably a later term. For stems and branches many words were available. The trunk (*mei*[7]) of a tree divides into branches (*chih*[8]) and twigs (*shao*[9]); the stem (*kan*[10]) of a herbaceous plant bears small branches (*miao*[11]) with leaf petioles (*kêng*[12]), in the axils (*a*[13]) of which new buds (*yuan*[14]) may form. Culms of grasses were called *ching*[15], of cereal grains *kao*[16], and of bamboos *ku*[17], but particularly in this last case the terminology became rather complicated, and may be found as much in the books of artists and painters as in those of the botanists themselves (cf. p. 377). Nodes and joints at any rate are familiar as *chieh*[18], a word that found a wider use in the division and measurement of time.[b] Tree bark is *phi*[19], if very rugose *tso*[20] and *hsi*[21], its pith *thung*[22]; it may generate thorns (*tzhu*[23,24,25,26]),[c] spines (*li*[27]) or tendrils (*chi*[28]). The stipule, or sheath of a sessile parallel-veined graminaceous leaf is called *chia*[29][d], if large as in bamboos, *tho*[30]. It might be tedious to continue, but if anyone had the impression that Chinese scholars, whether of the time of Theophrastus or of Leonhard Fuchs, found themselves tongue-tied in the presence of plant forms, he would be sadly mistaken. Of course the precision which came with modern science is not to be expected, and in a moment we shall see how the ancient Chinese terminology was hypertrophied in some ways while still deficient in others, but that was a feature only to be expected in a natural growth. For the multifarious forms of leaves (*yeh*[31]) with their midribs (*yeh kên*[32]) there were few designations to compare with that battery of adjectives developed in modern botany; either simple descriptive words were used, or the less well known plants were characterised in terms of better known ones.[e] For example the *Erh Ya* tells us that the *lo*[33] is the same as the *kuan-chung*[34], its leaves are roundish and pointed (*yuan jui*[35]) and its stem is hairy and dark (*mao hei*[36]).[f] There is no doubt

---

[a] Greene (1), p. 281. As if a plant were like a man planted upside down, the roots his hair, the branches his limbs. In his monograph on the miniature gardens of East Asia (Sect. 38 (*i*), 2 below), R. A. Stein (2), pp. 85 ff. has a good deal to say about the theme of inversion, which he brings into relation with several aspects of ancient Taoism. For example, tree-spirits were represented by sorceresses with disheveled hair, and the longevity exercises included hanging upside down (cf. *Hou Han Shu*, ch. 112B, p. 18a, tr. Lu Gwei-Djen & Needham (3), p. 106, also in Vol. 5, pt 5). Yang Ying-Hsiang[4] commenting on Chhiu Chün's[5] *Yu Hsüeh Ku Shih Hsün Yuan* (*Hsiang Chieh*)[6], (Studies in the Historical Elements of Basic Culture) of *c.* +1480, wrote: 'Plants and trees live head downwards, animals go horizontally, man alone stands upright; this is why men have consciousness and plants do not, while animals have it only partially.' We are indebted to Mr Georges Métailié for remembering this.

[b] Cf. Vol. 3, p. 404.

[c] This is a good example of a case where natural orthographic variation could have been pressed into the service of more precise scientific description.

[d] Not to be confused with *chia*[37], the leguminous pod.

[e] Cf. p. 143.   [f] Ch. 13, p. 4b.

1 荄		2 薑		3 球莖		4 楊應象		5 邱濬	
6 幼學故事尋源詳解			7 枚		8 枝		9 梢		
10 幹		11 杪		12 梗		13 椏		14 蓮	
15 莖		16 稾		17 箇		18 節		19 皮	
20 樷		21 皵		22 蓪		23 束		24 茦	
25 刺		26 莿		27 芀		28 其		29 英	
30 籜		31 葉		32 葉根		33 濼		34 貫衆	
35 圓銳		36 毛黑		37 筴					

that this was a fern, almost certainly *Cyrtomium fortunei*.[a] Its leaves, though lanceolate, are in fact rounded in a deltoid manner at the base.

Not unexpectedly, perhaps, the richest Chinese word-store was reserved for the sexual parts of the plant, so prominent and so useful. The commonplace words *hua*[1,2] for flower, *kuo*[3,4] for fruit, and *shih*[5] for hard fruits and seeds, belong to 'basic Chinese', soon learnt, but the matter was much more complicated than this. The *Erh Ya*[b] reserved *hua*[2] for the flowers of trees, adding *jung*[6] for the flowers of herbaceous plants.[c] Plants bearing fruits and seeds without (obvious) flowers that bloomed were called *hsiu*[7], a word which came to be used for grain in the ear gracefully bending. Plants with blooming flowers that produced no seed, the lexicographer went on, were called *ying*[8], no doubt a reference to infertile male flowers, but the word came to be used for flowers before the fruit had set. Then we find a good distinction between *sui*[9] and *thai*[10], the spike or raceme versus the capitulum or flower-head. Talking of a polygonaceous plant, a dock *Rumex*, perhaps the sheep-sorrel *acetosella* (*thui*[11] = *niu thui*[12]) the *Erh Ya*[d] says it has a *sui*[8] of violet or blue flowers, while on the other hand[e] the *kou*[13] (= *yao*[14]), some kind of *Cnicus*, has a *thai*[10], as is the wont of thistles (Compositae). If later *sui*[9] got loosely applied to catkins and aments, other longish inflorescences, as well, there was a special term for these, seen for instance in the expression *liu hsü*[15]. Peduncles, pedicels and scapes had also their word, *ti*[16], a form which intrigued Hopkins (11) in one of his pictographic reconnaissances, where following Chêng Chhiao[17] in the + 12th century, and modern scholars such as Wu Ta-Chhêng, he recognised

萊 萊 ⼊

ti[18]   *ti*[18]   *pu*[20]
bone   bronze   bone

that the character for 'emperor' or 'sovereign ruler' (*ti*[18]) was nothing but the picture of a flower. The new sense had been acquired simply by homophonic borrowing.[f] Even more surprisingly, a similar history lay behind the most familiar of negative words, *pu*[20], originally *fu*[20], loaned as an abstraction from a drawing of a flower on its stalk.[g] Besides this, *tai*[21] (or *ti*[21]) was available to mean a pedicel or scape.

For the organography of the flower itself numerous words existed. Buds of flowers were *tshou*[22,h] *jui*[23,i] *lei*[24], *pei*[25], *wei*[26], *yü*[27] (these two last very ancient)—an all

---

[a] Anon (*58*), no. 5, pl. 3; Phei Chien & Chou Thai-Yen (*1*), vol. 2, no. 53; Roi (1), p. 44; B II 110. Formerly identified as *Woodwardia radicans* or *Aspidium falcatum*. Cf. below, pp. 150, 157.
[b] Ch. 13, p. 8*a*, *b*.   [c] The word *huang*[19] could be used for either.
[d] Ch. 13, p. 5*a*.   [e] Ch. 13, p. 3*a*.   [f] Cf. K 877.
[g] Cf. K 999, and Vol. 2, p. 220.   [h] Cf. Vol. 4, pt 1, p. 171.
[i] Interchangeable, unfortunately, with *jui*[15]. Modern scientific precision would have rigidly separated them.

[1] 花	[2] 華	[3] 果	[4] 菓	[5] 實
[6] 榮	[7] 秀	[8] 英	[9] 穗	[10] 臺
[11] 蕡	[12] 牛蕡	[13] 鉤	[14] 芺	[15] 柳絮
[16] 蒂	[17] 鄭樵	[18] 帝	[19] 葟	[20] 不
[21] 蔕	[22] 簇	[23] 蕤	[24] 蕾	[25] 蓓
[26] 芛	[27] 蓫			

too numerous stock. Here we begin to see the unnecessary proliferation of synonyms which could have been, but never were, devoted to specific structures in a
scientific terminology of modern type.[a] The corolla was named, as we have seen,
pa[1], and its petals *hua yeh*[2], flower-'leaves' (cf. p. 127 above), or *hua pan*[3], flower-
'segments'.[b] The calyx was called *o*[5,6] (cf. Table 7), but sepals seem not to
have been distinguished in traditional writing, unless they were sometimes called
o pan. The ovary was *phing*[7] ('the vase'—a Sung usage), as also *tzu fang*[8] or
hua fang[9], phrases probably of fairly recent origin. Since traditional Chinese
botany was pre-Camerarian, 'stamens', *jui*[10,11], included both stamens and pistils;
later on they were qualified by the prefixes male and female. Special kinds
of flowers had special names, for instance the cylindrical spadix of Typhaceae
(bulrushes,[c] reed-mace,[d] or cat-tails). The *Erh Ya* says, and Kuo Pho enlarges on
it,[e] that the spike of the *kuan*[12f] and the *phu*[13g] is called *li*[14h] In later times, since no
petals were visible, it was known as *phu o*[15], and its abundant golden pollen (*phu
huang*[16]) collected, mixed with honey and sold as a sweetmeat.[i] Normally pollen
was called (*hua*) *yao*.[21]

Fruit and seed terms were also numerous.[j] What *kuo* was to trees, *lo*[22] was to
herbaceous plants in general.[k] The distinction was almost that we make between
drupe and berry. A rosaceous drupe had its skin (*phi*[23]), flesh (*jou*[24]), stone (*ho*[25])
and kernel (*jen*[26]); a cucurbitaceous gourd (*kua*[27], *hu*[28]) had a coat (*phao*[29]) divided
into segments (*chhi*[30]) and containing pulp (*jang*[31]) as well as the seeds. Acorns sat
in cupules (*chhiu*[32]). Most fruits were simply called by the name of the tree, but
some, like those of the mulberry, had special names (*jen*[33]), or the leguminous

---

[a] This tends simply to translate Latin adjectives, e.g. *tun-chuang-ti*[17] for peltate, or *chung-hsing-ti*[18] for campanulate.

[b] Or *ying*[4] (*HCCC*, ch. 551, p. 5*a*).

[c] Not of course rushes at all, if we restrict the name to the Juncaceae; cf. Bentham & Hooker (1), pp. 418, 469.

[d] Here is a case of full parallelism in nomenclature, for one of the Chinese names was *phu chhui*[19].

[e] Ch. 13, p. 4*a*.

[f] Now the sedge *Scirpus lacustris* (CC 1982), but anciently probably *Typha angustata* (CC 2112). Cf. B II 98, 455.

[g] *Typha latifolia* (CC 2113). Cf. II 98, 375, 455; III 196; = *hsiang phu*.[20]

[h] This word also had employment in *Allium* nomenclature, cf. B II 4.

[i] Stuart (1), p. 447. At least from the +11th century, for Su Sung (cf. Vol. 4, pt 2, p. 446) refers to it in the *Pên Tshao Thu Ching*.

[j] It may be worth pointing out here that even today there is no truly systematic and scientific terminology for fruit types. This is partly due to the extreme complexity and variety of the post-fertilisation changes in the floral parts of plants.

[k] A variety of statements making this definition, from Han and Thang onwards, will be found in *TSCC*, *Tshao mu tien*, ch. 15, *hui khao* 1, p. 1*a*, *hui khao* 2, p. 5*b*; ch. 16, *tsa lu*, p. 2*a*.

[1] 葩	[2] 花葉	[3] 花瓣	[4] 英	[5] 蕚
[6] 蕚	[7] 瓶	[8] 子房	[9] 花房	[10] 蘂
[11] 蕊	[12] 莞	[13] 蒲	[14] 蒚	[15] 蒲蕚
[16] 蒲黃	[17] 盾狀的	[18] 鐘形的	[19] 蒲槌	[20] 香蒲
[21] 花藥	[22] 苽	[23] 皮	[24] 肉	[25] 核
[26] 仁	[27] 瓜	[28] 瓠	[29] 匏	[30] 棲
[31] 瓤	[32] 梂	[33] 葚		

pod we have already noted (*chia*[1]). The *Erh Ya* says[a] that the tufted head or capitulum at the top of the stalk of thistles and other Compositae (*ong thai*[2]) bore seeds called *fou*[3], which suggests that the sunflower seeds so popular for cracking in tea-houses in China today had ancient predecessors now disused.[b] Naturally terminology concreted round the useful cereal grains where *li*[4] was classifier,[c] *pi*[5] the unripe seeds, *kho*[6] and *fu*[7] the chaff or outer envelope, and *kho chhiang*[8] the awn or beard.[d] The value of some other terms is less obvious, but they originated spontaneously—*niu*[9] for the seeds of the leguminous vine *Rhynchosia volubilis*[e], *thi*[10] for those of the sedge *Cyperus*, probably *rotundus*[f], and *i*[11] for the disseminules of elms such as *Ulmus macrocarpa*.[g] Hemp seeds even had two names, *fên*[12] when unripe and *tzu*[13] when fully ripe.[h] Thus one can see the tendency in an ancient unsophisticated vocabulary to multiply specific names for parts of specific plants useful to man in one way or another, rather than to apply them, as developed science was later to do, to the differentiation of structure and function irrespective of the specific plants in which they might apply (e.g. drupe, achene, pappus). One can sense also the growth of an excessive number of synonyms by the confluence of dialectal and local usages—a factor very prominent in Chinese plant nomenclature, as we shall soon see.

And then there is the question of the minutiae of the character orthography. Inquisitive readers will have noticed that cucurbitaceous *hu* and *phao* just now were written with a *yü*[14] component, while *fou* and *o* (the seeds of some thistles, and the term for the calyx) were written with *yü*[15]. One of the difficulties of beginners in Chinese is that they never feel sure when minute differences in character structure are significant and when not; that there are in fact non-significant variations[i] is of course an imperfection (a redundancy) in the coding, but inevitable in an information system of such age and natural growth. If this present difference were a significant one it could have been used as a valuable terminological mark (for example, *yü*[15] to denote connate sepals, in synsepalous, gamosepalous, flowers), but in fact it is not, and the two forms of *yü* are equivalent. The basic form is *chhiao*[16] or *khao*[16] (K 1041), to sob (?), represented in Chou inscriptions by a curious squiggle of unknown meaning, possibly the stick on which the old man is leaning,

---

[a] Ch. 13, pp. 7b, 8a.
[b] For *Helianthus annuus* etc. were introductions from the New World. Cf. Anderson (1), ch. 11.
[c] Vol. 1, p. 39.
[d] For a close parallel in the West, see Varro's discussion of grain terms, *Rerum Rust.* 1, xlviii.; Hooper & Ash tr., p. 281.
[e] R 408; CC 1040, 1041; Burkill (1), vol. 2, p. 1906.
[f] CC 1964.   [g] B II 263.
[h] This was surely because the seeds are richest in oil at the half-ripe stage (Watt (1), p. 250).
[i] 'Variations springing from the nature of bursh-writing' *pi fa*[17].

[1] 筴	[2] 蓊臺	[3] 荂	[4] 粒	[5] 秕
[6] 穀	[7] 稃	[8] 穀鎗	[9] 莥	[10] 媞
[11] 荑	[12] 黂	[13] 芓	[14] 丂	[15] 亐
[16] 丂	[17] 筆法			

breathing heavily, in *khao*[1], to investigate,[a] a word extremely familiar by now to those who have looked into our earlier volumes. Many of its derivatives have to do with breathing out, such as the ! of Chou and Han poetry, *hsi!*[2],[b] and technical skill, *chhiao*[3], shown forth by the master-craftsman with his tool.[c] *Yü*[1,2] are both considered interchangeable with *yü*[4], to speak, chant, go, the bone form of which (K 97) remains highly enigmatic.[d] Very reasonably, when surmounted by 'large', it becomes *khua*[5,6], meaning to boast or talk big (K 43). But when 'mouths' are placed above it, it is referred to quite a different root, being considered a deformation of a plant pictograph we have not yet encountered, *yi*[7] (K 788). This yields a word of much importance in the philosophy of biology, *ni*[8], to go against (the order of Nature), and adjectival for whatever does so.[e] And then it follows that *o*[10,11], to startle, beat a drum or make a noise (K 788), can also be written *o*[12,13],[f] and to come across something unexpectedly (K 788) can be *o*[14] or *o*[15,16]. The upshot of all this is that the two forms of the *yü* component in the botanical terms with which we began have no significant difference in meaning. If the spirit of modern science had descended on the Chinese culture-area instead of on that of Renaissance Europe, such slight variations in orthography could perhaps have been accurately defined and given a significance, but this never happened. And when the tide of modern botany flowed through China also, scientific men preferred to invent new expressions,[g] compiling equivalents (as noted already) to the Latin,[h] rather than making the fullest possible use of the ideographs of past ages, associated as these doubtless were with medieval vagueness and ancient misapprehension. Yet the fact that the Chinese script sprang essentially from pictures meant that it lent itself (and will as long as it lasts lend itself) to subtle elaboration of this kind.[i]

khao *yi* *ni*
K 1041f K 788a K 788d

The most remarkable example of pin-point terminology of the parts of a plant

[a] Cf. *lao*, old, K 1055 and Vol. 2, p. 226.

[b] K 1241*d*. [c] Cf. Vol. 4, pt 2, p. 9.

[d] This has become interchangeable with *yü*[9], the grammatical particle meaning at, in, to, from, by, than.

[e] See especially Vol. 2, p. 571. The graph shows fairly obviously someone coming up along a path to a boundary hedge and breaking through it.

[f] This last graph is assimilated to a quite different group, based on *thun*[18], to accumulate (K 427), and stylising what may be yet another plant pictograph.

[g] As an example of a 'new' term take synsepalous—*ho-o-ti*[17]. One evident reason for their preference was that such polysyllabic combinations were easier for auditory communication than refinements of the written characters, which required the assignment of single sounds.

[h] Generic and specific names derived from individual personal names have been simply translaterated phonetically.

[i] Chemistry, of course, required and obtained a great many new and specially designed characters. See Vol. 5, pt 3, pp. 255 ff.

1 考 2 兮 3 巧 4 于 5 夸
6 夸 7 屰 8 逆 9 於 10 咢
11 咢 12 㗁 13 噩 14 𨗄 15 遻
16 遌 17 合萼的 18 屯

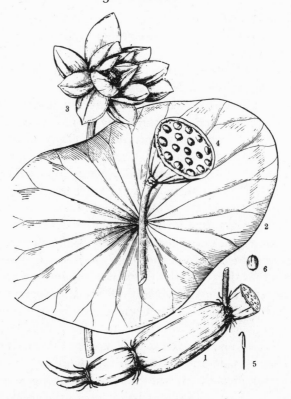

Fig. 24. Modern drawing of the lotus *Nelumbo nucifera*, from Anon. (57), p. 104.

is certainly found in what the *Erh Ya* says[a] about the 'Indian' or 'Buddhist' lotus, *Nelumbo nucifera*,[b] indigenous to China, and still a most prominent feature of Chinese garden art, (Fig. 24).[c] Everything had its name.

The *ho*[1] is the *fu-chhü*.[2]

    KP:[d] Another name for it is *fu-jung*.[3] Chiangtung people (still) call it *ho*.[1]

Its stem is called *chia*[4], its leaves *hsia*[5], and its roots *mi*.[6]

    KP: The *mi* is the white root (*pai jo*[7]) that lies in the mud (below).

Its (unopened) flower buds (*han*[8]) are called *tan*[9].

    KP: See the *Book of Odes*.[e]

Its fruit is called *lien*[10].

    [a] Ch. 13, p. 4*a*, *b*; B II 99–101.

    [b] The Latin synonymy is particularly bad, e.g. *Nelumbium nuciferum*, *speciosum* and *Nelumbo*; *Nymphaea Nelumbo*, etc. We follow Lawrence (1), p. 490 and CC 1449.

    [c] Cf. Anon. (57), vol. 2, nos. 30, 117.

    [d] Commentary by Kuo Pho.

    [e] Legge (8), vol. 1, p. 214.

[1] 荷   [2] 芙渠   [3] 芙蓉   [4] 茄   [5] 蕸
[6] 蔤   [7] 白蒻   [8] 菡   [9] 萏   [10] 蓮

KP: *Lien* means the *fang*[1] or receptacle.[a]

Its rhizome (*kên*[2]) is called *ou*[3]. What are found in the receptacle are the seeds (*ti*[4]).[b]

KP: Seeds (*tzu*[5]), in fact, inside.

What is inside the seeds is called the *i*[6] (i.e. the embryo).

KP: The bitter heart of the seeds.

Here then is a prime example of the minute attention which the naturalists of ancient China could give to a plant. All its parts were in this case useful; the rhizomes with their large longitudinal air-tubes can be eaten agreeably raw or cooked, and they furnish an excellent starch with special properties. The large seeds, which can be preserved and cooked in many ways, are a famous delicacy, and many of the other parts of the plant (petals, stamens, embryo) are medicinal.[c] The silky fibres of the stems and roots have symbolic significance, and the flower as a whole, so pure and perishable, became the very vexilla of Buddhist iconography. The scientific drawback was, of course, that the rich array of names belonged only to this particular plant, and could not easily become generalised as technical terms more widely useful. But in the first sentences one can see Kuo Pho already struggling against the divergent tendency of dialectal synonyms.

Li Hui-Lin[d] and the Moldenkes[e] seem to assume that the Chinese-Indian lotus was not known in ancient Egypt, and contrast it with the divers species of water-lily or 'Egyptian lotus' *Nymphaea stellata, alba, caerulea*, etc. But Burkill was right[f] in believing that the former reached Egypt at an early date (probably the time of the Persian conquest, *c.* −708), for Theophrastus has a long description of it under the name of the 'Egyptian bean', recording its numerous practical uses.[g] It is really instructive to compare his description with the laconic entries in the *Erh Ya*, for while it is long and discursive it uses no full-fledged technical terms. The air-tubes of the rhizomes are compared with a honeycomb, and the 'head' with a wasps' nest, in the cells of which are set the 'beans' or seeds. The peltate leaves are said to be like Thessalian hats, and most extraordinary perhaps of all, Theophrastus refers to the intensely bitter coiled-up embryo (*pilos*, πῖλος) just as the Chinese writer does. As he tells us that the plant will not ripen in Syria and Cilicia, these were probably way-stations on its introduction to Egypt from Persia. Immediately following there is an even longer description of the Nile water-lily *Nymphaea*. To his own account Li Hui-Lin adds details of a debate which arose some years ago about the Nymphaeaceae depicted in Mayan and

[a] The conical structure often called a fruit is really the enlarged receptacle; the 'seeds' are really the fruits or indehiscent nutlets.

[b] In another place (p. 7a) the writer of the *Erh Ya* recorded another word, *i*[7], as a synonym of *ti*[4]. Cf. B II 191.

[c] Cf. Burkill (1), vol. 2, pp. 1538 ff. The great economic value of the plant in East Asia was emphasised long ago by Palibin (1).

[d] (8), pp. 64 ff.   [e] (1), pp. 154 ff.   [f] (1), vol. 2, pp. 1538, 1565.

[g] IV, viii, 7, 8; Hort tr., vol. 1, pp. 351, 352.

[1] 房   [2] 根   [3] 藕   [4] 的   [5] 子
[6] 薏   [7] 薂

other Amerindian frescoes and reliefs.[a] Certain authors have taken these representations as strong evidence of pre-Columbian Asian-American culture-contacts,[b] but since there are indigenous American species of both genera (*Nelumbo lutea* and *Nymphaea ampla*)[c] botanists have tended to be sceptical of arguments based on this evidence. Meanwhile in China the *ho* or *lien-hua* has always had its enthusiastic admirers, among whom may be mentioned the pioneer of Neo-Confucianism Chou Tun-I (+1017 to +1073) who wrote a famous lapidary essay on the plant,[d] and Yang Chung-Pao (*1*) who devoted a monograph to it in 1808. Many poets wrote about it, taking it as a symbol of graceful beauty, spontaneity and strength; associating it, too, with the Chinese Helen, Hsi Shih, washing silk gauze in the Jo-yeh river, and with the famous apparition of the goddess of the River Lo, who might have been the spirit of Tshao Chih's lost love. Cf. Vol. 4, pt 3, p. 649. So Hsü Wei[1], for example, wrote verses such as these:[e]

> On the fifth of the fifth month, it's sultry, baking hot.
> Fanning oneself continually one can hardly bear the heat.
> O for a light boat to carry me away to Jo-yeh's stream
> Along ten *li* of cool breezes midst lotus flowers and leaves ...
>
> When the leaf is but five inches wide, the lotus bud is coy,
> Both float on the waves, bobbing with each dipping oar.
> But once the sweet winds of early summer have breathed upon them
> Bold and rude they shoot up, to nestle against the waist of some girl.

Talking of plant embryos reminds us of quite a different category of technical terms, those concerned with ontogenetic stages. A number of examples, denoting buds, shoots, etc. have already cropped up, but here again there was a tendency to use special appellations for stages of particular plants. For example, *sun*[2] is defined as the *mêng*[3] of bamboos, especially the edible ones,[f] and *tai*[4] as the *mêng*[3] of the 'arrow' (*chien*[5]) sorts of bamboo.[g] More interesting is the fact that the ancient Chinese naturalists recognised the existence of the temporary food-leaves or cotyledons in very young plants, a circumstance little known. A famous passage in Theophrastus[h] discusses the appearance of radicle, hypocotyl and cotyledons during germination, stating very clearly that some plants such as the bean and lupine have two cotyledons while others such as wheat and barley have

---

[a] (8), pp. 69 ff.

[b] Heine-Geldern & Ekholm (1); cf. Vol. 4, pt 3, pp. 540, 545.

[c] The botanical criticism is in Merrill (8) and Rands (1). Also of economic importance.

[d] The text will be found in *TSCC*, *Tshao mu tien*, ch. 94, p. 9*b*, and a partial translation in Li Hui-Lin (8), p. 66. On the philosopher himself, see Vol. 2, pp. 457, 460, 468.

[e] Tr. Chang Hsin-Tshang (3), mod. auct. Hsü Wei (+1521 to +1593) was a painter, calligrapher and playwright as well as a poet.

[f] B II 42 from *Erh Ya*.

[g] B II 174, 374 from *Erh Ya*.

[h] *Hist. Plant.* VIII, ii, 1–3; Hort tr. pp. 149 ff.

---

¹ 徐渭   ² 筍   ³ 萌   ⁴ 箁   ⁵ 箭

only one;[a] though perhaps he hardly realised the exhaustive character of the distinction between Dicots and Monocots.[b] But (in contemporary phrase) the Russians were on the moon too, for the *Erh Ya* says:[c] '*Kua*[1] (plants) are also called *mi-shê*[2] (deer's tongue)[d] (plants)'; and Kuo Pho comments: 'Now when *mi-shê*[2] plants (i.e. Dicots) germinate in the spring, the leaves are glossiform (hence the name)'. This is immediately followed by a definition of which the sense was long lost: '*Chhien*[3] (plants) are also called *chü-chhiu*[4] (plants).'[e] Since the first of these words means to pull out, the second means a gutter, and the third a long strip of dried meat, one can venture the obvious conclusion that this was the com-plementary definition of the Monocot coleoptile. Thus there can be little doubt that in ancient China people were looking at germinating seeds and talking about them, just as the Greeks were. Indeed, one can even find a pictograph of cotyledons in the oracle-bone script, the origin of the words *chüeh*[5], radicle, *ti*[6], root, foundation, and *shih*[7], clan, family.[f] So, at least, Hopkins (10), following Wang Yün, in a discussion worth reading.[g] And to this day, in some parts of China, cotyledons are called *shih yeh*[8], and the plumule following *mêng ya*.[9]

shih
bone

Another pictograph of a germinating seed leads us off in quite different direc-tions. This is *yao*[12] (see inset), defined by the *Shuo Wên*[h] as 'something minute; it portrays the form of seeds just beginning to sprout (*hsiao yeh, hsiang tzu chhu shêng chih hsing*[13])'. No oracle-bone example is given by Karlgren,[i] but the seal form seems to show two seeds. From a very early time this pictograph was assimilated to another which looked extremely like it, *mi*[14] (or *ssu*[14], see inset), probably intended originally as a drawing of a hank of silk; the *Shuo Wên* simply

yao seal	*mi* seal	*hsüan* seal	*yu* bone	*chi* bone	*chi* seal

---

[a] For the former the ancient Chinese word was *tou*[10], for the latter *mi*[11] (cf. Table 6 above). It might not be going too far to say that the forty orders of dicots with their broad and reticulately veined leaves would have been grouped by Chinese scholars under the former head, while all the twelve orders of monocots with their narrow parallel-veined leaves would have been placed under the latter.

[b] See Greene (1), pp. 95 ff. Malpighi's beautiful drawings of germinating seeds might almost have been done as illustrations of Theophrastus; they are reproduced in Singer (1), pp. 48 ff., who also gives an alternative translation. Cf. Percival (1), pp. 7 ff.

[c] Ch. 13, p. 7*b*. B II 205 did not understand it.

[d] The deer in question is *Cervus davidianus* or *Alces machlis* (R 365).

[e] Kuo Pho did not understand this, let alone B II 206.

[f] K 302, 590 and 867 respectively.

[g] *Shuo Wên*, ch. 12, pp. 265. 2, 266.1 gives a different, and much less convincing, explanation.

[h] Ch. 4B, p. 83.2.

[i] K 1115*a*. See, however, Hopkins (27).

[1] 菇   [2] 麋舌   [3] 搴   [4] 柜朐   [5] 乎
[6] 氐   [7] 氏   [8] 氏葉   [9] 萌芽   [10] 豆
[11] 米   [12] 幺   [13] 小也象子初生之形   [14] 糸

says[a] 'fine silk threads (*hsi ssu yeh*[1])'.[b] Fine distinctions may be attempted to differentiate the presence of the spike above or below the two circles,[c] as also the distance between them, but the writers of the Shang, Chou and Han were probably not consistent in their usage; indeed the though-connection between a delicate hypocotyl and a single fibre of new-spun silk was so close and natural that they felt no need to be, and the graphs became interchangeable. From *yao*[1] (rad. no. 52) the derivatives were important but relatively few, while *mi*[3] (rad. no. 120) gave off hundreds of words having to do with threads, continuity, silk and other textile fibres. The former have considerable scientific interest, not so much the obvious combination of 'sprout' with 'strength', *yu*[2], young and tender, often applied to plants, or *yu*[3], obscure and dark (bone form inset),[d] but particularly *chi*[4], the very word for the minute 'germs' of living things that we encountered already in such philosophical books as the *Chuang Tzu* and the *Kuan Tzu*[e] (bone and seal forms inset).[f] From this again, by the addition of the wood radical, came *chi*[6], the word for the loom,[g] and hence, by extension, for every kind of machine.[h] Here Hsü Shen had a truly lapidary definition which took him only two words, *chu fa*[7]—'controlled energy-application (or energy-output)'—*chu fa wei chih chi*.[8] Could any modern have done it more succinctly? The thought-connection must undoubtedly have been that a great deal of effect could be produced by a relatively small object, such as a crossbow-trigger,[i] a rudder,[j] or the control lever or sluice-gate bar of a ponderous water-mill. Similarly, 'tall oaks from little acorns grow'.[k] And from a pictograph of germinating seeds, much could germinate.

If in the foregoing pages many examples of technical terms have been taken from the Chhin and Han periods, it should by no means be supposed that botanical terminology in China underwent no development during the centuries. For example, in Thang and Sung writings one can find many approximations to the style of the Graeco-Latin adjectives of Europe. Speaking of the *ti chin*[9], a vine

[a] Ch. 13A, p. 271.1.

[b] This character is not in Karlgren, but the oracle-bone form of *hsi*[5] (K 876a, b), connection, filiation, hence 'dependent department', has a hand holding two hanks of silk.

[c] Some characters have both, e.g. *hsien*[6], the string of a lute or bow, *hsüan*[6], the hypotenuse (cf. Vol. 3, pp. 22, 95 ff.); *Shuo Wên*, ch. 12B, p. 270.2, see inset.

[d] K 1115*c, d, e.*

[e] See Vol. 2, pp. 43, 78 ff., 421, 469.

[f] *Shuo Wên*, ch. 4B, p. 84.1; K 547*a, b.* The other component is probably some kind of weapon (cf. K 1231*e*), here only a phonetic.

[g] *Shuo Wên*, ch. 6A, p. 123.1; K 547*c.* Note the association with the thin and delicate fibres of silk.

[h] As we took care to observe in Vol. 4, pt 2, pp. 9, 69 (cf. Needham, 34). Many years later, when I had occasion to live for some time in the noble city of Senglea in Malta, I used to pass Triq il-Macina (Machine Street) leading down to the quay. But here 'machine as such' meant the crane, not the loom. Hardly anything could symbolise better, I thought, the maritime-mercantile ethos of Europe, in contrast with agrarian-industrial China.

[i] Cf. Vol. 5, Sect. 30.

[j] Cf. Vol. 4, pt 3, p. 641, where the argument appears explicitly in an early text.

[k] *TTC*, ch. 64.

[1] 細絲也   [2] 幼   [3] 幽   [4] 幾   [5] 系
[6] 弦   [7] 機   [8] 主發謂之機   [9] 地錦

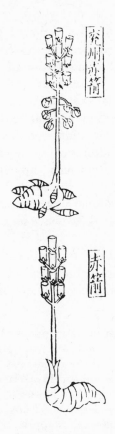

Fig. 25a. The orchid *Gastrodia elata (chhih chien)* depicted in the *Chêng Lei Pên Tshao*.

Fig. 25b. A modern drawing of the same plant (Chia & Chia (1), p. 999).

of medicinal value, *Parthenocissus* (= *Quinaria*) *tricuspidata*,[a] the *Chêng Lei Pên Tshao* (+1108) says that its leaves are like the web-feet of ducks (*ya chang*[1]), i.e. palmate.[b] Elsewhere the same work describes an orchid, *chhih chien*[2], *Gastrodia elata*,[c] quoting many previous writers.[d] Thao Hung-Ching (*c.* +500) said that its stem was reddish and stood up straight like the shaft of an arrow (*chien kan*[3], i.e. as we might say, sagittiform),[e] while its rhizome (*kên*[4]) swelled out into a tuber like a human foot (*jen tsu*[5], i.e. as we might say, pediform). Furthermore, like the *yü*[6] (the taro, *Colocasia antiquorum*),[f] this tuber had a dozen 'offspring', smaller tubers growing round about like bodyguards (*shih-erh tzu wei wei*[7], i.e. as we might say, phylacteric). Ko Hung, the great alchemist, *c.* +330, had already noticed

[a] CC 765; R 282. The name also applies to *Euphorbia hemifusa* (R 325).
[b] Ch. 7, p. 10*b*.   [c] CC 1733; R 636.
[d] Ch. 6, p. 52*a*. The picture in *CLPT* is worth comparing with a modern drawing (Figs. 25*a, b*).
[e] Cf. Hagerty (2), p. 411.   [f] CC 1926; R 710.

[1] 鴨掌   [2] 赤箭   [3] 箭䇯   [4] 根   [5] 人足
[6] 芋   [7] 十二子為衞

these and called them 'wandering offspring', *yu tzu*[1]; the connections between them, he said, were not essential, and only the *chhi* related them. Comparing this orchid with the dodder, *Cuscuta japonica* (= *sinensis*), he said that both had underground root-stocks, rhizomes or stolons (*fu thu chih kên*[2]),[a] without which they could not rise up or climb. Su Sung, in the *Pên Tshao Thu Ching, c.* +1070, however, preferred the ancient view that the dodder has no normal roots, and thought that Ko Hung must have been talking of some other plant not often seen. In any case, terminology continued to expand. The metaphorical or descriptive terms exemplified above are like Theophrastus' 'Thessalian hats', appellations well on the way to becoming technical terms but not yet quite arrived there. Where the boundary was passed was when a specific character was appropriated to the uses of phytography and not used for anything else. As we have seen, there is no lack of such characters in the Chinese language.

When Bretschneider was cataloguing, about 1865, the tribulations of the Western botanist who tries to make a serious study of Chinese botanical works (cf. p. 20 above), one of the things which distressed him was their systematic habit of 'cross-referencing description'. He wrote:[b]

On the whole it can be said of the *Pên Tshao (Kang Mu)*[c] that the descriptions of plants therein are very unsatisfactory [generally meagre and unsatisfactory].[d] We find statements of the native country (or province), of the form, the colour of the blossoms, the time of blooming, etc. (but) these accounts are insufficient, because the Chinese, in describing the parts [and organs] of plants, have not a botanical terminology. The blossoms, leaves, fruits, etc. are described by comparing them with the blossoms, leaves and fruits of other plants, which are often unknown to the reader [frequently unknown to European readers].

Besides these mentioned, there are also statements given about the utility of the plants for economical and industrial (as well as medical) purposes. The descriptions consist for the most part of successive quotations of authors, whereby the same statements are several times repeated. Finally Li Shih-Chen gives his own opinion, and generally it is the most reasonable of all.

Essentially this was a complaint about the inadequacy of the Chinese adjectival fund. By now it will have become very clear that technical terminology was far from lacking in traditional Chinese botany, but of course it could not compare with the wealth of adjectives specially coined in the European +18th century (e.g. lorate, runcinate). Bretschneider felt that the great disadvantage of traditional Chinese botany was that you had to know everything before you could

---

[a] There is doubtless some connection here with the old association between the dodder (*thu-ssu*[3] or *thu ssu tzu*[4]), a parasite of branches, and the *fu-ling*[5] (*Polyporus cocos*, R 838), a fungal parasite of roots. See Vol. 4, pt 1, p. 31 above. At the same time *fu-thu*[2] is an independent expression, and we have met with it in a quite unlikely place already (Vol. 4, pt 2, p. 251, Fig. 500), where it was the name for the axle-blocks under carts and chariots, *fu-thu*[6]. See *Khao Kung Chi Thu*, ch. 1 (p. 33). Presumably the botanical use preceded the technical—'a clamp of wood as long as a root-stock'.

[b] (6), p. 6. [c] Additions inserted by us.

[d] Changes of significance introduced by Bretschneider in his re-writing, (1), p. 65.

[1] 游子　　　　[2] 伏菟之根　　　[3] 菟絲　　　　[4] 菟絲子　　　　[5] 伏苓

[6] 伏兔

communicate anything; a reader had to be familiar with every plant before he could understand the descriptions of a writer. This criticism was not quite fair for several reasons. In his re-writing of 1882 Bretschneider made the very significant change to 'frequently unknown to European readers', realising then that the system had always been much easier for the Chinese themselves. And one has only to pick up a handbook such as that of Bailey (1) to find the same principle deeply embedded in the Linnaean binomials—e.g. *abietinus, aceroides, achilleaefolius; cannabinus, clethroides, cupressiformis; fraxinifolius.* Anyone professing not to know what pines, maples, hemp, cypresses and ashes look like will naturally be foxed by these simple cross-references.

If one turns to Greene as usual, one will find a lucid account of the two forms of description which he calls the natural versus the artificial, or the comparative versus the positive.[a] One might almost say that the spirit of modern science in the botanical field notably signalised itself by a turning away from the age-old principle of familiar acquaintance with certain specific types as standards of comparison, and the age-old ability to form a mental image of an unknown plant by the stated points of divergence from a given type known to both describer and reader; an adoption, on the contrary, of a vocabulary of very special terms, largely geometrical in origin, which could deal with any conceivable plant, banishing (at least theoretically) all need for illustrations and for previous knowledge of types. The Chinese were simply being Theophrastean. As Greene shows, he used for leaves four main reference-plants, the laurel (bay) or myrtle (lanceolate), the olive (oblong), the pear (suborbicular) and the box or ivy (ovate).[b] But he quite often used others, saying, for example, that the *Tamarindus indica* was 'many-leaved, after the manner of the rose-bush', i.e. with leaflets in a compound leaf; or that a bi-pinnate-leaved tree from Egypt, *Mimosa polyacantha*, had leaves 'like those of a fern'. All through the centuries down to the time of Linnaeus, Western botanists used no other method of description, only augmenting the number of indicative types. Hence it was very proper of Bretschneider to add, in his re-written version,[c] that this, the comparison system, 'was also the mode of describing plants adopted by the celebrated Dioscorides (first century of our era) and followed by our botanists down to the time of Linnaeus. Comp. e.g. Plukenet's *Amaltheum Botan.* 1705'.[d]

All that now remains on this subject is to give just one instance of the cross-referencing description, so widespread through all Chinese botanical literature. It may as well come from a date close to that of Theophrastus himself, exemplifying

---

[a] (1), pp. 79, 80, 101 ff. 104 ff.

[b] Only the adult leaves of ivy are ovate. The juvenile leaves, and those on non-flowering branches, are palmately lobed. This was known to Theophrastus, who distinguished the two forms as *helix* (ἕλιξ) and *kittos* (κιττος). What he did not know was that the change from one leaf form to the other depends on the level of the plant hormone giberellin in the tissues.

[c] (1), p. 65.

[d] He also added, very rightly, that the conscientious, if tedious, successive quotations of earlier authors all saying much the same thing, was also a feature of medieval Arabic and Western pharmaceutical and botanical writers.

the laconic character of the early Chinese writers. Two labiate plants are in question. The *Erh Ya* says:[a]

The *chia*[1] is the same as the *thui*[2].

KP:[b] This plant is now called *chhung-wei*[3]. Its leaves are similar to those of the *jen*[4]. It has a squarish stem and white flowers which grow out (in whorls) at the nodes (*chieh chien*[5]) (i.e. it is verticillate). According to the *Kuang Ya*[6] it is also called *i mu* (*tshao*)[7].[c]

Here the plant being described is *Leonurus sibiricus*,[d] and the *jen*[4] with which its leaves are compared is *Perilla frutescens*.[e] The *Kuang Ya* (Enlargement of the Erh Ya Dictionary) was written a couple of generations before Kuo Pho's time by a scholar of the Wei State in the Three Kingdoms period, Chang I.[8] Chinese botanists all through the ages after this, down to modern times, continued to use the comparative method of description alongside the positive phytographic terminology.[f]

(2) PLANT NOMENCLATURE

(i) *Names common and learned; binomes and multinomes*

The moment has now come at last to turn from terminology to nomenclature, how in Chinese one distinguished each wort from his fellow.[g] There can be no doubt that the single-character plant names (many of which have been encountered in the preceding pages) are the oldest of all, and they must have been stylised pictographs or phonetic inventions,[h] but their origin goes too far back into the Shang and early Chou periods for us to be able now to analyse at all clearly how they came to be. Then it was natural that as botanical knowledge grew the armoury of single sounds, even with tonal differences, could not suffice, so that by the Warring States period the full resources of the permutations and

---

[a] Ch. 13, p. 2*a*; B II 25.   [b] Commentary by Kuo Pho.

[c] The last sentence gives the pharmaceutical name, by which it is now best known.

[d] Siberian motherwort; R 126; CC 334; Steward (2), p. 336; Anon. (*57*), vol. 2, pp. 356 ff.; Anon. (*58*), p. 128 and pl. 72, no. 150; Stuart (1), p. 235.

[e] = *ocymoides*; R 135; CC 343; Steward (2), p. 343; Anon. (*57*), vol. 2, pp. 421 ff.; Anon. (*58*), p. 130 and pl. 73, no. 152; Stuart (1), p. 313.

[f] See, for example, the descriptions translated below—Chu Hsiao on *Adenophora stricta* (p. 338) and others on *Barringtonia* etc. (p. 428). Many more are in the monographic literature (pp. 355 ff.).

[g] The literature on this subject is not too abundant in any language. Apart from Bretschneider and other classics already cited, one may mention the interesting study of Chao Yuan-Jen (4), aimed mainly at a linguistic analysis of common-usage phraseology but well worth study here. The work of Kimura Koichi (3) we saw in Japan, but have not been able to use.

[h] One corpus of legend at least attributes the invention of the names of plants and animals to the culture-hero Sui Jen shih[9,10]. This tradition we find in the Han apocryphal book *Chhun Chhiu (Wei) Ming Li Hsü*[11], p. 4*a* (*YHSF*, ch. 57, p. 68*a*), often quoted afterwards, as in *Lu Shih (Chhien chi)*, ch. 5, p. 5*a*. Commentators say that it was not permissible to change the names thus given. Other legends give more prominence to Shen Nung (cf. p. 196).

[1] 萑   [2] 萑   [3] 茺蔚   [4] 荏   [5] 節閒

[6] 廣雅   [7] 益母草   [8] 張揖   [9] 遂人氏   [10] 燧人氏

[11] 春秋（緯）命歷序

combinations of two characters in binomes were being laid under contribution. Many of the single-character names persisted,[a] and still do, but already by the −4th century the *Erh Ya* scholars[b] found it necessary to explain them, using the formula of which we have seen several examples above: 'The *x* is the same as the *y*', or more often 'The *x* is the same as the *y-z*'. By the +3rd century Kuo Pho's commentary spoke almost exclusively in terms of binomes. Omitting from the estimate the items which he said he did not understand, as well as other aberrant entries, a count of chs. 13 and 14 (the herbs and trees) shows that out of 251 items no less than 209 or 84 % were single names requiring explanation.

It is important to notice that the explanations given are often in terms of more than one equivalent name, especially when Kuo Pho or one of the other early commentators are speaking. In late Chou, Chhin and Han times, dialectal variation and differences of provincial usage led, as already mentioned (p. 135) to a rich growth of synonymy, to some extent among the single names but much more so among the compounds, doublets or binomes. Then gradually by some consensus of the scholars, one of these was chosen as the primary learned name and handed down as such in the Pên Tshao and other botanical literature. In Kuo Pho's *Erh Ya* commentary the process is seen clearly at work, for some of the names he characterises as *su*[2] or popular. Thus for instance in the remarks about the habit of cucurbitaceous stems (p. 128 above), *tieh*[3] was the learned word and *cho*[4] the vulgar one.[c] Or in another place, after the entry: The *hung (tshao)*[5] is the same as the *lung ku*[6] . . .'; he writes, 'Common people call *hung tshao*[5] by the name of *lung ku*[7], it is just the countryfied speech that changed it.'[d] To elucidate the exact relation between such judgments as these and the ultimate stabilisation of nomenclature would require special research, but it is certain that every plant the Chinese knew did come to have a primary name in the literature, generally of two characters, with a number of admitted synonyms always, as it were, in attendance upon it.[e] Exactly what factors brought this choice about are lost in the mists of time.

Why was the bifurcation of names into learned and vulgar significant? The issue was not whether one should have an elegant, esoteric or 'establishment' language for natural science, but rather whether one should develop a clearly defined nomenclature for technical purposes, or let the natural evolution of spoken and written words take its wayward course. For the ancient Chinese this obviously meant inventing characters.[f] And it must have required a scientific, or

---

[a] For a striking example see an admonitory ode by Ma Jung[1] preserved in the *Hou Han Shu*, ch. 90A, pp. 4*b* ff. He enumerated a wealth of plants and animals which should be found in an ideal imperial park, as an allegory of the proper balance between civil and military activities which he wished to recommend. On him cf. Vol. 3, pp. 71, 334. Commentators could identify by means of *Erh Ya*.

[b] See p. 187 below.   [c] Ch. 13, p. 3*a*.

[d] P. 4*b*. *Polygonum orientale* (R 577; CC 1534) was the plant in question. Still today *hung tshao*[8].

[e] Perny appreciated and stated this in 1872, (1), vol. 2, *Nat. Hist.* sect., p. ii.

[f] As it certainly did two thousand years later, when the names for the elements and compounds of modern chemistry had to be coined and devised; cf. Vol. 5, pt 3, pp. 252, 259.

[1] 馬融　　[2] 俗　　[3] 眣　　[4] 吺　　[5] 紅（草）
[6] 蘢古　　[7] 蘢鼓　　[8] 荭草

at least a proto-scientific, tradition, in which men were interested in debating exactly what it was that someone else was talking about. This was a self-consciousness concerning appellations, closely analogous indeed to the 'rectification of names' (*chêng ming*[1]) which had been such a cardinal point in Confucian philosophy from −500 onwards.[a] There it had assumed a primarily political significance, being concerned with 'calling a spade a spade', however powerful the interests were which wanted it called a shovel; and it is not easy to find European equivalents for this grand subversive doctrine. But the same principle surely operated just as much among the botanists and pharmacists and naturalists of ancient China.[b]

The idea often prevalent that traditional Chinese botanical nomenclature was in some sense 'unscientific' is closely connected with the prejudice in the European, and now in the modern, mind, that nothing can be scientifically identifiable unless it bears a Latin name. But the fission of the learned from the popular plant nomenclature occurred remarkably late in Europe. According to Greene,[c] the distinction between Latin names and European vernacular names is hardly found before the time of Otto Brunfels (+1464 to +1534); and it is easy enough to see why. In medieval times Latin was a living language, the colloquial and international medium of all educated discourse,[d] but with the Reformation and the rise of nationalism and capitalism this state of affairs passed away, leaving only isolated remnants like island mountain-tops sticking up out of the sea—medical theses down to the end of the eighteenth century—and botanical and zoological nomenclature.[e] The time of Linnaeus gave this its great charisma, but one can immediately see that the usual European estimate of Chinese biological nomenclature needs revising; the idea of its 'unscientific' nature is really an illusion due to the fact that China did not have two separate cultural languages, the Latin and the vernacular.[f] In fact, as we can see in the *Erh Ya* commentaries a thousand years before Brunfels was born, processes were already at work which would distinguish scholarly from popular usages, and settle the primary and

---

[a] See Vol. 2, pp. 9–10, 29. The basic references are *Lun Yü*, XII, xi, xvii; XIII, iii (Legge (2), pp. 120, 122, 127–8). Each translator has his favourite phrase, e.g. 'to correct language' (Waley (5), p. 171), 'to render all designations accurately' (Ware (7), p. 82); 'préciser le sens des mots' (Leslie (9), p. 163). A whole chapter of the *Hsün Tzu* book ch. 22) is devoted to the rectification of names, tr. Dubs (8), pp. 282 ff. 'The use of a name', says Hsün Chhing, 'is to know the reality when one hears the name.'

[b] All of them must have known of the Confucian tenet, and this might be one striking example of Confucianism helping rather than hindering the growth of the sciences, though it was Taoism which usually played that role.

[c] (1), p. 191.

[d] The *kuan-hua*[2] of Europe (cf. Vol. 1, p. 33), but far more so. The 'King's English' of imperial China, the mode of speech of officials and educated men, was really 'English', and not a foreign language.

[e] One should add perhaps the liturgy and seminaries of the Latin Church, at least down to the time of the Second Vatican Council.

[f] A corollary of this was a greater blending of the two nomenclatures in China, a blurring, as it were, of their distinctness.

[1] 正名   [2] 官話

secondary names of plant species in all subsequent ages of Chinese literature.[a]

The *su* or 'ordinary' speech is still alive and kicking of course, in its contemporary form, and this colloquial is what Chao Yuan-Jen(4) has analysed. Of about 200 compound phrases which he lists, some 40 %, he finds, are noun-noun combinations (e.g. *wa sung*[1], tile-pine)[b] and a further 13 % the same, of the 'close apposition' type (e.g. *sung shu*[2], pine-tree); then 27 % adjective-noun (such as *hsiang tshai*[3] for coriander)[c], and lastly 20 % verb-noun or noun-verb 'exocentric constructions' (so *fang-fêng*[4] or 'wardwind', see p. 154 below; and *hua-shêng*[5] for the peanut).[d] In many cases the meaningless nominative suffixes *tzu*[6] and *erh*[7] are added according to custom. It is obviously much more difficult for the outsider to distinguish common names of this kind from the learned Chinese names, than Latin names from English ones. A parallel examination of the Chinese standard or learned names would be extremely desirable, but we know of no-one who has undertaken it, and it cannot be done here.[e]

If it is not possible now to ascertain how they were gradually chosen, it is also not possible now to be sure always (or even often) what they originally meant. Anciently, as we know from the *Erh Ya*, the binomes of plant nomenclature were written without the 'grass' radical (*tshao-thou*[8]) or the 'wood' radical (*mu-tzu-phang*[9]), but after the Han and Liu Chhao periods these became almost universal, giving pairs of *tshao-thou* words, and (to a rather lesser extent) pairs of *mu-tzu-phang* words.[f] We have already met with many of these, for instance *fu-jung*[10] when we were discussing the 'Buddhist' lotus, and only just now (p. 142) *chhung-wei*[11] for the labiate *Leonurus sibiricus*. They will be found everywhere throughout this Section. For these binomes it is only too easy to invent fanciful etymologies, and traditional explanations which may or may not be more convincing are widely found in Pên Tshao entries. Thus *chhung-wei*[11], if an adjectivenoun combination, could be imagined to mean the 'replete' or 'satisfactory' *wei*, or a plant 'impersonating' or 'substituting for' the *wei*; but Li Shih-Chen, no doubt representing long tradition, took it as doubly adjectival and referring to the seeds—the plant is so called, he says, because the seeds are numerous and abundant (*chhung shêng*[12]), lying close

---

[a] By the end of the +16th century Li Shih-Chen, in his *Pên Tshao Kang Mu*, shows a great concern not only for distinguishing basic standard names from synonyms or common names but for attributing priorities to previous writers (cf. pp. 312, 395 below); thus strangely in concert with the contemporary German fathers of botany, whose very names he never knew, as also with Linnaeus later.

[b] Not a pine at all, of course, but the crassulaceous roof-growing succulent *Sedum erubescens* var. *japonicum* (CC 1233, R 469), or *iwarenge*.

[c] *Coriandrum sativum* (R 217, CC 561).

[d] The various names for *Ginkgo biloba* would be relevant here, cf. Moule (17); Wedemeyer (1). See p. 39*d*.

[e] It would certainly give a very different result from Chao's census, with many more adjective-noun (epithetic) combinations.

[f] These are colloquial expressions in common use meaning 'with the grass (or plant) radical at the top of the character' and 'with the wood radical at the left of the character' respectively.

[1] 瓦松	[2] 松樹	[3] 香菜	[4] 防風	[5] 花生
[6] 子	[7] 兒	[8] 草頭	[9] 木字旁	[10] 芙蓉
[11] 茺蔚	[12] 充盛			

together and well-packed (*mi wei*[1]).[a] Or again some may have been tempted to see a geographical adjective in the name of the banana *Musa sapientum* (M. *Basjoo*)[b] *pa-chiao*[2] (from Pa[3], Szechuan), but Li Shih-Chen explains that the name comes from *pa*[3] as meaning 'dry', and *chiao*[4] as meaning 'scorched', for the leaves wither and go brown without falling off the tree.[c]

Perhaps the closest parallel to the Chinese binomes can be found in the European ancient and medieval binomes. These were really double generic names, not binomials of the Linnaean generic and specific sort, and they linger on now as rather despised, certainly demoted, species names. The Greek language had a genius for aggregation into compound single words, a strongly agglutinative tendency, so that a name like *Viola nigra* had been in Greek simply *Melanion* (*melan-ion*).[d] But even Theophrastus used a binome now and then, such as *Calamos-euosmos* (the sweet flag, now *Acorus Calamus*), or *Syce-Idaia*, the 'fig' of Mt Ida (in fact a service-tree, *Amelanchier rotundifolia*).[e] Afterwards, when Latin was a living language, the dogtooth violet was called *dens-canis*, and now we know it as *Erythronium Dens-canis*. Anyone can think of dozens of these cases—*Capsella Bursa-pastoris* for shepherd's-purse, *Taraxacum Dens-leonis* for the familiar dandelion (*dent-de-lion*), *Auricularia Auriculae-Judae* for the 'Jew's ear' tree-fungus, so well known as a constituent of Chinese cuisine. For seventeen centuries, says Greene, Latin botany admitted two-worded generic names as freely as the simpler kind; but when Latin began to become a dead language they were gradually eliminated in favour of the single names.[f] Linnaeus pronounced the decree of expulsion in Art. 242 of his *Philosophia Botanica* (12) in +1751, but the process had started a long time before, as may be seen from the practice, if not the expressed theory, of Otto Brunfels in +1530. Since Chinese never did become a dead language, this simplifying operation did not occur in Chinese culture. The effect it had in Europe was to make room for the 'fine adjustment' of coining names for individual species, as happened so explosively when modern science developed, but in China also there were the beginnings of specific appellations, as we shall later see (p. 313).

Those who study the biological binomes of Chinese are liable to find themselves involved in a classical controversy of philology and linguistics, the argument about whether this ancient language was always strictly monosyllabic.[g] Could it be that some at least of the binomes are the remains of polysyllabic words coming down from remote antiquity? For Chao Yuan-Jen (4) a binome is an unanalysed disyllable, whether it was once analysed or not. For G. A. Kennedy (2), in one of the wittiest articles he ever wrote, many of them never had been or could be

---

[a] *PTKM*, ch. 15, (p. 20).   [b] R 652; CC 1770.
[c] *PTKM*, ch. 15, (p. 61). The explanation seems to be at least as old as +1096, for Li Shih-Chen quotes Lu Tien's *Phi Ya* in connection with it. On the banana in general in Chinese literature there is a paper by Reynolds & Fang Lien-Chê (1).
[d] Greene (1), p. 184.   [e] *Loc. cit.* p. 123.   [f] *Loc. cit.* pp. 124, 185.
[g] We dispense with elaborate references here, citing only Kennedy (1) and Dubs (27). Cf. Vol. 1, pp. 27, 40.

[1] 密尉   [2] 芭蕉   [3] 巴   [4] 焦

analysed because derived from polysyllabic (disyllabic) names. For his type-specimen he took the term *hu-tieh*[1] (or *hu-thieh*[1]), the common general name for butterfly, entitling his monograph 'The Butterfly Case'. As everyone finds who looks through the dictionary entries under Rad. no. 142, *chhung*[2], 'insect' (as I myself did long ago), there are a very large number of words which always tend to appear in couples or doubles with others, making binomes of *chhung-tzu-phang*[3] words.[a, b] We shall see many of these in Section 39. They are, of course, quite equivalent to the pairs of *tshao-thou* words and *mu-tzu-phang* words which we are studying here. The questions raised by Kennedy were whether the in-dividual components alone could carry the full meaning of the binome, and whether in fact they were ever so used by themselves. First he considered the view of Karlgren that these binomes were synonym compounds, formed of two nouns of identical meaning when the confusion caused by the inadequacy of single sounds and homophones necessitated greater designative clarity, but he found it very strange that there should have been two of everything, coming out like the animals from the Ark of Noah. Taking a modern Chinese encyclopaedia and making a census of the words in the *chhung*[2] section, he found that no more than half (186) were defined alone, and the rest (187) only as pair-components with others. He therefore asserted[c] that neither *hu* or *tieh* were ever used alone, nor ever defined independently in Chinese dictionaries, so that the Western lexicographers had been guilty of introducing a 'fragmentation', breaking into two the 'halves of the same word-egg'. He also gave a number of other examples. But these conclu-sions had to be much modified as the result of remonstrances from Chao Yuan-Jen and L. C. Goodrich, who could point to a number of ancient and medieval texts where *tieh* demonstrably occurs alone with the full meaning of butterfly. Kennedy therefore finally stated two propositions—if it could be shown that *tieh* came first and *hu-tieh* later, then *hu* would be descriptive or epithetical, perhaps 'whiskered' or 'bearded'; but if it could be shown that the binome came first and the single usage of *tieh* later, then the latter must be considered an abbreviation of an originally disyllabic word. Beyond this he could not go, the second part of his monograph never appeared, and we still have no definitive answer to the question.

In matters like this there are always several alternatives. There is the possibility of double-noun collectives, like *sung-po*[6] for conifers, in this case 'butterflies and moths'; or of a hidden sexual distinction, like *fêng-huang*[7] for the male and female phoenix; and 'whiskered' was surely not the only epithetic meaning that could be conjectured, for 'powdery', 'long-lived' or even 'foreign' might also enter their claims. For our part, we remain very sceptical of any polysyllabic theories, and

---

[a] Again a colloquial expression meaning 'with the insect radical on the left of the character' that go together.
[b] For example *ming-ling*[4], a kind of bollworm, or *wu-kung*[5], a kind of centipede.
[c] P. 16.

[1] 蝴蝶   [2] 蟲字旁   [3] 虫   [4] 螟蛉   [5] 蜈蚣
[6] 松柏   [7] 鳳凰

fear that it is now too late to hope to reconstruct with any certainty the first
phases of Chinese biological nomenclature. The character store is like a vast box
of counters which can either stand alone or go together in a wide variety of
combinations; and where plant names are concerned the exact orthography and
the exact combination make a great deal of difference. For example, the *chhiang
mei*[1] of the *Erh Ya* was almost certainly *Asparagus lucidus*[a], and the proper way of
writing the name of the umbellifer *Cnidium* (= *Selinum*) *monnieri*[b] is almost certainly *chhiang mei*[2]. But *chhiang wei*[3] is something else again. This *chhiang* by itself
means the water-plant *Polygonum Hydropiper*, our smart-weed or water-pepper
(another old double generic name),[c] and has a preferable pronunciation of its
own (*sê* or *shih*). This *wei* by itself may mean either of two things, the vetch *Vicia
gigantea*,[d] or the fern *Osmunda regalis* (= *japonica*).[e] The two words together make
the well-known binome for the cultivated ornamental *Rosa multiflora*, ancestor of
all the 'rambler roses' of the West.[f] Whatever else Chinese botanical nomenclature may be accused of, it can certainly not be said to ignore subtle distinctions.[g]

Bored with binomes, the reader may be refreshed by trinomes and tetranomes.
Three-character names have made their appearance already, e.g. the henna
plant, *Lawsonia inermis*, as *chih chia tshao* (or *hua*)[4], the 'finger-nail plant', from its
cosmetic dyeing properties, known and used in Canton since the beginning of the
+2nd century.[h] Another instance would be a liliaceous rather hyacinth-like plant
much cultivated in Chinese town gardens and called 'millennial green', *wan nien
chhing*[5], because it springs up green so regularly and stays green so long; this is
Rohdea japonica.[i] And we may add, having spoken of the banana, *pa-chiao*, a three-
character derivative from this name covering a new genus, the *shan pa-chiao*[6], or
fan-banana *Ravenala madagascariensis* the leaves of which can be used as fans.[j]
Four-character names are rather rare, but one in particular insists on mention
because of the scientific philosophy which it enshrines.[k] This is the *wang-pu-liu-
hsing*[7], a name applied to the soapwort *Saponaria officinalis* (Caryophyllaceae)
because of its content of saponin, taken advantage of by medieval people.[l] The
words mean 'Even-the-king-can't-stop-it', and Li Shih-Chen explains this in so
many words: '(The juice of) this thing', he says, 'has a tendency to spread, and
even if the emperor himself were to command it to stop, he could not make it

[a] B II 108; CC 1830.   [b] R 230.
[c] R 573.   [d] R 414.   [e] CC 2183.
[f] R 456; cf. Li Hui-Lin (1), pp. 93 ff.
[g] We rarely have occasion in this work to mention the tonal pronunciations of the characters, but here it
would be very reasonable to ask whether these *chhiang* words were not distinguished linguistically by having
different tones. The answer is that all of them belong to the second tone. Or at least they do now.
[h] Cf. Vol. 1, p. 180, Vol. 4, pt 3, p. 498e, and p. 35 above. CC 639; R 248.
[i] CC 1874; Steward (2), p. 516.
[j] CC 1774; Burkill (1), vol. 2, p. 1886.
[k] Cf. Vol. 2, p. 131.
[l] Cf. pp. 160 ff. below, and also Needham & Lu Gwei-Djen (1).

[1] 蘠蘼   [2] 墻蘼   [3] 薔薇   [4] 指甲草（花）   [5] 萬年青
[6] 扇芭蕉   [7] 王不留行

stand still, hence the name.'[a] Now we shall speculate no more about the pre-
history of Chinese plant names, but simply show in tabular form how every
principle used in the devising of such appellations in the West can be paralleled
by what happened in China.[b]

(ii) *The taxonomic language*

If one was naming a plant oneself, what ideas would be likely to arise in the mind?
Obvious features and methods can be enumerated easily enough:

I   Shape and form
II   Size
III   Colour, (flower, leaf, stem, root)
IV   Aroma
V   Taste
VI   Special characteristics, e.g. milky sap
VII   Habitat, (oecological, epiphytic)
VIII   Geographical origin
IX   Duration and seasonal property
X   Climatic property
XI   Sex, (real, assumed)
XII   Usefulness, technological,
                        pharmaceutical,   striking property
                                            curative
                                            functional
                                            best locality of origin
XIII   Patronymic, (real, legendary)
XIV   Foreign origin, (barbarian, Iranian, sea-foreign, overseas)
                        transliterations

But this is a very abstract trellis. To clothe it with living green we must give a
few examples of each way of naming. These have been assembled in Table 9,
together with the following brief paragraphs of commentary.

The Table is divided into as many sections as those just listed, varying according to the motivation. The names explained are not all primary stabilised learned
names (these are distinguished by an asterisk), but often secondary or semi-
popular names, or else sometimes ancient ones dating from Chou, Chhin or Han
times and subsequently fallen out of use (these will generally be marked *EY*). The
explanation is given under Comments on the right hand side of the Table.[c]

---

[a] *PTKM* ch. 16, (p. 109); R 551; CC 1474; B III 113; coloured picture in Anon. (57), vol. 2, pl. 7. The plant
was in the oldest of the pharmaceutical natural histories, the *Shen Nung Pên Tshao Ching* (cf. p. 235 below) ch. 1
(p. 33).

[b] A beginning in this was made long ago by Bretschneider (6), p. 5; (1), vol. 1, pp. 63 ff.; appreciatively but
very unsystematically.

[c] How interested Li Shih-Chen would have been in the work now going on with *Xanthium* (p. 150) to elucidate
the phenomena of circadian rhythms—the biological clock by which plants recognise day and night. The cocklebur is yielding the essential chemical constituents of this, sensitive blue proteins named phytochromes. It was Dr
David Coombe who kindly drew our attention to this. See also especially Lu Gwei-Djen & Needham (5),
pp. 137 ff., 149 ff.

Table 9. *Origins of Chinese plant names.*

	Chinese name	Meaning	Identification; family	Common Eng. name	Comments	R	CC	Other refs.
I. SHAPE AND FORM								
shih i	石衣	stone's clothes	*Ceramium rubrum* or *Spirogyra lineata* Algae	pondweed	see pp. 383 ff. below	857	≏2338	*EY*
shih fa	石髮	stone's hair	*Ceramium rubrum* or *Spirogyra lineata* Algae			857	≏2411	*EY*
kou ku	狗骨	dog's bone	*Ephedra sinica* Gnetac-	sand cherry sea grape	the roots, from their form	783	2115	*PTKM* 15/(65), quoting Chang I's *Kuang Ya* (+230).
kou chi	狗脊	dog's backbone	*Woodwardia japonica* Filicales	—	from the form of the roots	≏800d	2253	*CLPT* (+1468 ed) 8/34a ff. (facsim., +1249 ed), 8/30a; *PTKM* 12/(11).
niu chhun	牛脣	ox-lips	*Alisma Plantago-aquatica* Alismatac-	water-plantain	from the leaves	780	2092 2093	*PTKM* 19/(89); *EY*; Lu Wên-Yü(1), no. 69.
niu shê	牛舌	ox-tongue	*Plantago major* Plantaginac-	lamb's tongue (*P. lanceolata*)	from the leaves	90	233	Hulme (1), pt. 3 p. 33.
ma wei	馬尾	horse-tail (for *Equisetum* see XII)	*Phytolacca esculenta* Phytolaccac-	pokeweed or pokeroot	from the tapering inflorescences	≏555	1494	*EY*; Anon. (58), no. 34.
têng lung tshao	燈籠草	lantern-basket herb	*Physalis Alkekengi* Solanac-	Chinese lantern	from the inflated calyx	116	307	*EY*; *PTKM* 16/(97) *CLPT* (facsim.)
suan chiang*	酸漿		*Physalis Alkekengi* Solanac-	Chinese lantern	sources of the shoots and fruits	116	307	8/37a; Kimura (2) vol. 2, pl. 223; Arber (3), figs. 90, 91
hsi erh*	枲耳	hemp ear	*Xanthium Strumarium* Comp-	cocklebur	from the fruit-head burrs looking like the dangling part of women's ear-rings, but the leaves like those of hemp	50	150	*EY*; *Kuang Ya*; *PTKM* 15/(50).
mu hu tieh*	木蝴蝶	tree-butterfly	*Oroxylum indicum* Bignoniac-	—	from the winged disseminules	—	—	Anon. (57), vol. 2, p. 53, no. 16;

the long stamens, echoing Pers. *gul-i abreshun* whence the Linnaean specific name.

				Latin name & Family	Common name	Notes			Reference
II.	**SIZE**								
	p'ing*	蘋	—	*Marsilea quadrifolia* Filicales	pepperwort	defined as the large sort of *p'ing* 萍	807	2182	*EY*
	t'ien tzu ts'ao	田字草	the herb like the character *t'ien*	*Marsilea quadrifolia* Filicales	pepperwort		807	2182	
	ssu yeh ts'ao	四葉草	four-leafed herb	*Marsilea quadrifolia* Filicales	pepperwort		807		
	kuei	虆	—	*Polygonum orientale* Polygonac-	prince's feather	defined as the large sort of *hung* (*tshao*) 葒草	577	1534	*EY*
III.	**COLOUR**								
	ma huang*	麻黃	hemp yellow	*Ephedra sinica* Gnetac-	{ sand cherry { sea grape	yellow flowers and stalk nodes	783	2115	B III 97; Kimura, (2), vol. I, pl. I
	tzu ching*	紫荊	purple thorn	*Cercis chinensis* Leg-	{ Judas tree { red bud	pink or purple flowers	380	969	Phei & Chou (*1*), no. 81
	pai fên	白粉	white meal-tree	*Ulmus pumila* = *campestris* Ulmac-	white elm	the inner bark gives a white mucilaginous meal; edible and used for making incense	606	1626	*EY*; Chhen Jung (*1*) pp. 209 ff.; Anon, (*58*), no. 17.
	pai t'ou ong*	白頭翁	white-headed old man	*Anemone cernua* Ranunculac-	pasque-flower	white hairs of stem, leaves and buds	528	1390	Anon. (*58*), no. 41
	lan*	藍	the blue	*Polygonum tinctorium* Polygonac-	—	gives an indigo dye, the most ancient source in China	579	1537	B II 392; *PTKM* 16/(126)
IV.	**AROMA**								
	ting hsiang*	丁香	nail perfume	*Eugenia aromatica* = *caryophyllata* = *Jambosa caryophylla* Myrtac-	clove (from *clavus*)	the dried unexpanded flower-buds used as spice	244	619	Watt (*1*), p. 526; Burkill (*1*), vol. I, p. 961.
	chhou su	臭蘇	stinking Perilla	*Mosla dianthera* Lab-	—	aromatic annual	130	339	Stewart (*1*), p. 339

Table 9. (*continued*)

	Chinese name	Meaning	Identification; family	Common Eng. name	Comments	R	CC	Other refs.
V.	**TASTE**							
kan tshao*	甘草	sweet herb	*Glycyrrhiza uralensis* = *glandulifera* Leg-	sweet root, liquorice		≃391	991	*PTKM* 12/(82); Chao Yü-Huang (1) vol. 1, pt. 1, pl. 2, fig. 6.
ta khu	大苦	great bitter	*Clematis recta* Ranunculac-	virgin's bower	bitter seeds	533	—	Stewart (1), p. 95
hsi hsin*	細辛	delicate acrid	*Asarum sieboldi* Aristolochiac-	wild ginger	medicinal rhizome	587	1564	Stewart (1), p. 95
VI.	**SPECIAL CHARACTERISTICS**							
chien chung	堅中	solid centre	unident., poss. *Arundinaria marmorea* Bam-	—	solid-centered bamboo, given as an explanation of *lin* 鄰	—	2071	*EY*; B II 171.
chhieh i*	竊衣	clothes stealer	*Torilis scabra* = *japonica* Umb-	≃ hedge-parsley	seeds as burrs	—	557	*EY*; B II 91
ku chêng tshao	鼓箏草	drums-in-the-wind	*Zoysia japonica* Gram-	—	from noise made by the leaves and stems	—	2068	*EY* (Kuo Pho); Stewart (1), p. 463
tsê chhi*	澤漆	marsh lacquer	*Euphorbia helioscopia* Euphorbiac-	sun-spurge	milky sap	324	861	*PTKM* 17/(15); Stewart (1), p. 216
ju chiang tshao	乳漿草	milk-juice herb	*Euphorbia Esula* Euphorbiac-	leafy spurge	milky sap	—	—	Stewart (1) ibid.
yang ju*	羊乳	goat's milk	*Codonopsis lanceolata* Campanulac-	—	milky juice in tuberous roots	—	160	Stewart (1), p. 385; Wu Chhi-Chün (1), p. 475.
hsiang jih khuei* han hsiu tshao*	向日葵 含羞草	sun-facing mallow bashful plant	*Helianthus annuus* Comp- *Mimosa pudica* Leg-	sunflower sensitive plant	heliotropism irritability of leaf pulvini	— —	95 1022	Khung et al. (1), p. 436.
chi hsing tzu	急性子	quick-natured one	*Impatiens Balsamina* Balsaminac-	touch-me-not	elastic bursting of seed capsules	296	777	
VII.	**HABITAT**							

					form and habitat			
hai tai	海帶	sea belt	*Laminaria religiosa* = *japonica* Laminariac-	kelp		863	2376	
yü mu	寓木	tree lodger	*Taxillus yadoriki* Loranthac-	'mistletoe'	mulberry epiphyte	588	1567	*EY*; B II 262; *PTKM* 37/(11).
*mu erh**	木耳	tree ear	*Auricularia-judae* Fungi	Jew's ear	tree fungus	827a	2309	
VIII. GEOGRAPHICAL ORIGIN								
*pa tou**	巴豆	bean of Szechuan	*Croton Tiglium* Euphorbiac-	croton	vesicant and intensely purgative; fish poison, arrow-poison.	322	857	*PTKM* 35/(63); Burkill (1), vol. I, pp. 688 ff.
chhu hêng	楚蘅	*hêng* of Chhu	*Pollia japonica* Commelinac-	a spiderwort	(Ar.) *habb al-Khiṭai* all. to *Tradescantia*	700	1906	*Kuang Ya*
IX. DURATION AND SEASONAL PROPERTY								
*pan hsia**	半夏	midsummer	*Pinellia tuberifera* = *ternata* Arac-	—	—	711	1929	*S.NPT*; *PTKM* 17/(53)
*tung chhing**	冬青	ripener green-in-winter	*Ilex pedunculosa* Aquifoliac-	a holly	evergreen, non-deciduous	310	832	
chhun tshao	春草	spring herb	*Cynanchum atratum* Asclepiadac-	—	—	160	415	
X. CLIMATIC PROPERTY								
kho tung	顆凍	can-stand-the-frost	{ *Tussilago Farfara* Comp- or *Petasites japonicus* or perhaps also: *Ligularia* (= *Farfugium*) *tussilaginea* (= *kaempferi*), *tho wu* 菟吾	coltsfoot a butterbur		49 49	— 122	*EY*; B II 160 CT&W, p. 829
*khuan tung hua**	款冬花	entertainer of winter		a butterbur			118	Stuart (1), pp. 172, 446.
XI. SEX								
*jen**	葚	—	*Morus alba* Morac-	Chinese or silk-worm mulberry	its ♀ pistillate flowers and fruit	605	1610	*EY*; *Erh Ya Chu Su* 9/4b
*chih**	梔	—	*Morus alba* Morac-	Chinese or silk-worm mulberry	its ♂ staminate flowers (this word came to be commonly used for *Gardenia florida* but that should probably be 巵)	605 82	1610 222	B II 302, 499

Table 9. (*continued*)

Chinese name		Meaning	Identification; family		Common Eng. name	Comments	R	CC	Other refs.
i sang	楱桑	—	*Morus alba*	Morac-	Chinese or silkworm mulberry	the female mulberry tree	605	1610	*EY*
hsi ma*	枲麻	—	*Cannabis sativa*	Morac-	hemp	plants with ♂ staminate flowers	598	1592	*EY* ⎱ B II 104, 140, 388
chü ma*	苴麻	—	*Cannabis sativa*	Morac-	hemp	plants with ♀ pistillate flowers	598	1592	*EY* ⎰

XII. USEFULNESS

Chinese name		Meaning	Identification; family		Common Eng. name	Comments	R	CC	Other refs.
mu tsei*	木賊	thief of wood	*Equisetum hiemale*	Equisetac-	horsetail	carpenters and wood-carvers used it for smoothing and polishing surfaces because of the silica in its nodes	797*a*	2174	*PTKM* 15/(68)
yang chih chu*	羊躑躅	{ paralyser of goats, 'goat staggerer', aphrodisiac for goats	*Rhododendron sinense* = *molle*	Ericac-	an azalea	toxic for sheep and goats, contains active principles	203	523	*SNPT*; Stuart (1), p. 375
yin yang huo*	淫羊藿		*Epimedium grandiflorum*	Berberidac-	a barrenwort	contains active principles	521	1374	*SNPT*; Stuart (1), p. 4
fang fêng*	防風	'ward-wind'	*Siler divaricatum*	Umb-	—	a defence against *fêng* illnesses (convulsive and paralytic affections, fainting, dizziness and cold in the extremities)	233	583	*SNPT*; Stuart (1), p. 407
i mu tshao	益母草	herb beneficial for mothers	*Leonurus sibiricus*	Lab-	Siberian motherwort or lion's tail	valued in women's diseases	126	334	*SNPT*; Stuart (1), p. 235
chieh ming (tzu)*	決明子	enlightener	*Cassia Tora*	Leg-	foetid cassia	seeds valued for eye diseases	379	968	*SNPT*; Stuart (1), p. 96
chhuan lien	川連	Szechuan lien	*Coptis chinensis*	Ranunculac-	a golden-thread	roots and bulbs, the best from	534*b*	1413	*SNPT*; Anon. (79) nos. 250

XIII. PATRONYMIC

					found ... Chhichow in Hupei			(1) p. 52; Braun & Lye (1), p. 2.
shih chün tzu*	使君子	personal name	Quisqualis indica = sinensis Combretac-	Rangoon creeper	from the name of Kuo 郭 Shih-Chün, an early Sung or pre-Sung physician	245	623	KPPT; PTKM, 18/(11); Stuart (1), p. 368
hsü chhang chhing*	徐長卿	personal name	Cynanchum paniculatum Asclepiadac-	—	from the name of Hsü Chhang-Chhing, a physician of Chou, Chhin or C/Han	166	423	S.NPT; PTKM 13/(73). B III 43
liu chi nu (tshao)*	劉寄奴草	personal name	Solidago Virga-aurea Comp-	golden rod	from the tzu name of the first Liu Sung emperor, Liu Yü. See text.	46	137	TPT; PTKM, 15/(26)
tu chung*	杜仲	personal name	Eucommia ulmoides Eucommiac-	hardy rubber tree	from the name of Tu Chung, a semi-legendary Taoist, hence the alternative names ssu chung 思仲 and ssu hsien 思仙. See text.	461	2463	S.NPT; B III 317; PTKM 35A/(9), tr. Hagerty (14).
tshao yü yü liang	草禹餘糧	food left behind by Yü	Heterosmilax japonica Lil-	China root	from the name of the legendary culture-hero Ta Yü (see text)	680	1845	PTKM 18B/(46)
yao huang	姚黃	Yao yellow, a variety of	Paeonia Moutan = suffruticosa Ranunculac-	tree-peony	from the name of a family of gardeners, Yao (see text)	537	1423	Li Hui-Lin (1), p. 25

Table 9. (continued)

Chinese name		Meaning	Identification; family		Common Eng. name	Comments	R	CC	Other refs.
XIV. FOREIGN ORIGIN									
jung shu	戎菽	barbarian pulse[a]	Pisum sativum	Leg-	common pea	of West Asian origin	402	1033	Laufer (1), p. 305; PTKM 22/(41)
hu ma*	胡麻	Iranian hemp	Sesamum indicum	Pedaliac-	sesame	of West Asian origin	97	257	Laufer (1), p. 288
hu thao*	胡桃	Iranian peach	Juglans regia	Juglandac-	walnut	of West Asian origin	619	1682	PTKM 30/(97); Laufer (1), p. 254
fan hung hua*	番紅花	sea-foreign red flower	Crocus sativus	Iridac-	saffron	of West Asian origin	654	1776	PTKM 15/(33); Laufer (1), p. 309; Burkill (1), vol. 1 p. 683
hai hung tou	海紅豆	overseas red bean	Erythrina Crista-galli	Leg-	Indian coral tree arbol madre, madre de cacao	tropical and sub-tropical SE Asia (genus first mentioned in KPPT)	≃384	983	PTKM 35/(68); Burkill (1). vol. 1 pp. 945 ff. On the prefix hai see TSCC, Tshao mu tien, ch. 10, p. 1b, quoting Hua Mu Chi and Chung Shu Shu.

ABBREVIATIONS:
B Bretschneider (1)
CC Chia & Chia (1)
CLPT Chêng Lei Pên Tshao
CT & W Clapham, Tutin & Warburg (1)
EY Erh Ya
KPPT Khai-Pao Pên Tshao
MIPL Ming I Pieh Lu
PTKM Pên Tshao Kang Mu (all refs. to mod. ed. here)
R Read (1)
SNPT Shen Nung Pên Tshao Ching
TPT Thang Pên Tshao

NOTE: The plant names explained in each section of the table are not necessarily the developed primary Chinese learned names (when they are, they are marked with an asterisk); some are secondary or semi-popular ones (cf. p. 149).

[a] Alternatively, hui hu tou 回鶻豆 Muslim pigeon beans, or yin-tu su 印度粟 Indian millet (Kao Jun-Shêng (1), cit. in Li Chhang-Nien (2), p. 44).

On shape and form the examples need little comment, but one ought to realise how apposite the names are. Bretschneider took care to emphasise this, mentioning the name *chi thou*[1] (chicken's head) for the water-plant *Euryale ferox* (R 541) and adding: 'Anybody who has seen the fruit of this plant will agree that the Chinese name is very significative.' *Lilium tigrinum* (R 682), he said, bears the Chinese name *pai ho*[2] (a hundred together), owing to the numerous scales which form the bulb; and indeed this name has become attached in modern Chinese botany to the family Liliaceae. One further example may be permitted, taken straight from the *Erh Ya*, because of its particular technological interest; it turns on the serrate character of the leaves of a fern.[a]

The *mien ma*[3] (continuous-thread horse) is the same as the *yang chhih*[4] (sheep tooth).

KP: This plant has delicate fine hairy leaves arranged in regular form. The leaves are shaped like the teeth of sheep. Nowadays Chiangtung people call it 'wild goose beak' (*yen chhih*[5]). It is used in the reeling off of silk (*sao chê i chhü chien hsü*[6]).

There is no doubt that the plant is *Aspidium Filix-mas*[b] with its deeply pinnatifid leaves broad at the base and narrowing to a tip. The tip could be thought of either as the root of a mammalian tooth or as the end of an avian beak, and both resulting names are graphic. But what is most interesting is to find that this natural object, a polypodiaceous leaf, was used in some way by those who drew off the raw silk from the cocoons. Perhaps they just used it as a pin to pick up the thread. But since it has a whole row of pinnae, they may well have found it convenient for making the rack on the ramping-arm of one of the most important of all the machines described in our earlier volumes, the *sao chhê*[7,8] or silk-winder, which by two simultaneous operations wound the fibres of fresh silk off from the cocoons and deposited them evenly on the receiving drum or barrel.[c] From a textual description we knew that this rather complicated machine was in use as early as the +11th century, and hints were given from time to time, largely on account of the known antiquity of the industry, that it might have to be envisaged a thousand years earlier than that; but here we get a glimpse of early usage which implies that its ancestral forms are to be looked for in the late, if not the early, Chou time (mid − 1st millennium). Presumably either the cut-off pinnae on the rachis, or perhaps more delicately the cut-off pinnules on the rachilla, were moved back and forth (automatically?) as the guide-racks for the fibres of the silk to pass through.

On size it need only be said that ancient Chinese botany had a habit of giving separate one-character names to species or varieties which were very similar in appearance but differed in size. We illustrate this by a couple of examples.

---

[a] Ch. 13, p. 7*b*, tr. auct.

[b] = *Dryopteris F-m.* (CT&W, p. 29), = *Nephrodium F-m.* (Khung *et al.* (*1*), p. 405; CC 2230).

[c] Vol. 4, pt 2, Fig. 409; see also pp. 2, 107–8, 116, 301, 382 and 404.

[1] 鷄頭　　　　　[2] 百合　　　　　[3] 騾馬　　　　[4] 羊齒　　　　　[5] 鴈齒

[6] 繰者以取繭緒　　　　　[7] 繰車　　　　[8] 繰車

The system was not the same as that used in the West, where names such as 'greater celandine', or *Megalostylis*, immediately betray their meaning. Nor did it ever become very important in Chinese nomenclature. On colour, as the Table shows, the most striking feature could be either the flowers, the stem, root or other parts, the vestiture, or the dye which could be extracted from the plant. We may here remark in passing that the 'indigo' shown, *Polygonum tinctorium*, was the ancient and indigenous blue dye-plant of China, the father and mother of all those millions of good blue garments that those who have lived in China know so well.[a] The 'true' indigo, *Indigofera tinctoria* (CC 997) a tropical or sub-tropical plant, passed from India to Persia about the +6th century, became known to China during the Thang period as *chhing tai*[1] (blue kohl) or *mu lan*[2], and was afterwards cultivated in the southern provinces.[b] Our English woad *Isatis tinctoria* (CC 1284), called *sung lan*[3], is also found in China, but was probably a much later introduction, arriving only in time for its mention by Li Shih-Chen at the end of the +16th century.

Leaving aside the obvious possibilities of smell and taste, a number of other special characteristics have been illustrated. A solid centre where one would expect hollowness, burrs that stick to one's clothes, plants that make a sound in the wind, plants with milky latex or thick juice in the roots, heliotropic plants and plants with striking muscle-like movements.[c] Then come the oecological specialists, whether forest or mountain dwellers or denizens of the sea, or epiphytic, parasites on other plants. Geographical origin, of which we give two examples, is really but a department of this; and we meet again with the same principle among the drug-plants, some of which got alternative names in accordance with the best locality of production (Section XII of the Table, the last two). To names based on duration, season and climatic property there is nothing to add. Sex is in a class by itself, and should be looked at in conjunction with our special sub-section on the subject (38 (*h*), 2, below); noteworthy, however, is the great age of the very correct distinctions which the Chinese made in the two dioecious plant species given in the Table. As in other civilisations 'male' and 'female' were applied to some plants not dioecious, but these erroneous assumed attributions will be reserved for that discussion. Again, the special names did not betray the sexual significance at sight, as a word like *Gynocardia* would do.

Coming to the category of usefulness, we inserted first the official name of

---

[a] Cf. Bretschneider (1), vol. 2, pp. 211 ff.

[b] Cf. Laufer (1), p. 370; Schafer (13), p. 212.

[c] Chinese ways of name-making continued as long as the traditional botany itself, as we may see from the example of the peanut *Arachis hypogaea* native to the Americas, which spread out all over the Old World in the +16th century (CC 955). In Chinese it is *lo-hua-shêng*[4], 'flowers that fall down and grow', à designation very similar to the Latin specific name, alluding to the curving downward of the fruit and the ripening of the pod in the ground. Later on we shall say something more about plant introductions to China from America (38 (*j*) 1). On *Arachis* see Burkill (1), vol. 1, pp. 205 ff. *Yen tshao*[5], the 'smoke-herb', is certainly a good deal more expressive than *Nicotiana Tabacum*, CC 304.

¹ 青黛   ² 木藍   ³ 菘藍   ⁴ 落花生   ⁵ 煙草

Equisetum in order to indicate that by no means all the names of this kind stem from pharmacological properties. Indeed the 'wood-thief', used for smoothing and polishing woodwork and carvings, is exactly paralleled by one of its designations in Europe, *Zihnkraut*, because of its value in polishing pewter.[a] Apart from this one can distinguish between a striking toxic effect exhibited very likely in animals, a well-known curative action, and a physiological function. Many Chinese plant names belong to this category. More surprising perhaps is the fact that Chinese names were also derived from the names of people, in a way analogous to that which gave us *Fuchsia* or *Sigesbeckia*. Thereby hangs a tale.

In Section XIII of the Table there is a selection ranging from the most historical of personalities to the most legendary. Kuo Shih-Chün[1], a physician of early Sung or pre-Sung date, gave his name not only to the valuable anthelminthic *Quisqualis indica* but by extension in modern times to the whole family Combretaceae. He may be dated by the fact that the drug appears first in the *Khai-Pao Pên Tshao*[2] (Pharmacopoeia of the Khai-Pao reign-period) composed by the physician Liu Han[3] and the Taoist Ma Chih[4] in +970. Kuo was by speciality a pediatrician, and in this drug he discovered a really safe and efficient vermifuge.[b] Hsü Chhang-Chhing[5], who gave his name to *Pycnostelma chinense*, must have belonged to a much earlier time, the late Chou, Chhin or early Han, because his drug-plant is one of those contained in the oldest of the pharmaceutical natural histories, the *Shen Nung Pên Tshao Ching*. From somewhat later dates the following item, named not after a physician but after an emperor of the Liu Sung dynasty. The story is worth telling by way of light relief as we go.[c]

Li Yen-Shou in his *Nan Shih* (History of the Southern Dynasties)[d] says that the emperor Kao Tsu of the (Liu) Sung dynasty, Liu Yü[6], had as his *tzu* name when young Liu Chi-Nu.[7] When cutting rushes in Hsinchow once, he met a big snake several tens of feet long, shot at it and wounded it. Next day in the forest he heard the sound of pestles and mortars pounding, so he followed and found a number of young men dressed all in blue-green clothes preparing drugs in a grove of hazelnut trees. He asked what was the matter, to which they replied that their master had been shot by one Liu Chi-Nu, and now they were compounding medicines to apply (to his wound). (Liu) Yü said, 'Why didn't the Spirit kill (the attacker) then?' but they answered 'Chi-Nu has the imperial charisma, and may not be killed.' As (Liu) Yü sighed with astonishment, the boys all disappeared, so he collected the medicine and went home ... Afterwards whenever anyone was wounded it was applied with great efficacy, hence the plant came to be known as Liu Chi-Nu's Herb.

---

[a] In Hegi (1), 2nd ed., vol. 1, p. 73, we find: 'Der Name Zihnkraut rührt davon her, dass die Schachtelhalme wegen ihres hohen Gehaltes an Kieselsäure zum Putzen von Geschirr, besonders von Zinnkannen, Zinntellern und Weberschiffchen gebraucht werden ... Der Winterschachtelhalm (*Equisetum hiemale*) speziell wird stellenweise von Tischlern beim Polieren von Möbeln und Parkettböden verwendet ...' It was Dr David Coombe who told us of this.

[b] Stuart (1), p. 368.

[c] *PTKM*, ch. 15, p. 35a, tr. auct.

[d] The story is given in ch. 1, pp. 1b, 2a in very slightly different form. Li Shih-Chen was perhaps quoting from memory, in any case we here conflate.

[1] 郭使君　　[2] 開寶本草　　[3] 劉翰　　[4] 馬志　　[5] 徐長卿
[6] 劉裕　　　[7] 劉寄奴

Assuming that Liu Yü was about twenty at the time, this story would date from the neighbourhood of +375. The next two examples are also quasi-legendary. Tu Chung[1] was a Taoist who became an immortal by habitually eating parts of the tree *Eucommia ulmoides*; he must have lived (if a real person) in Han or pre-Han times because the name is already in the *Shen Nung Pên Tshao Ching*. Though Tu Chung's name is now applied to the Eucommiaceae as such, we can hardly suppose that he was impressed by the great botanical interest of this unigeneric family of a single species; it was probably that he liked the spun-out longevity symbolism of the delicate silvery silky fibres which can be drawn to the length of an inch or more without breaking when the bark or stem is fractured.[a] He must have found it very indigestible, though the leaves are said to be still eaten.

The 'food left behind by Yü the Great' has reference, of course, to the culture-hero of legendary antiquity who 'controlled the waters', and about whom we have had much to say already.[b] That he passed by the doors of his home without ever going in while engaged on his sub-continental tasks became a household word for the selflessness of dedicated bureaucrats, and his picnics while 'on the job' left as remains not only the roots of this plant *Heterosmilax japonica* (= *Smilax pseudo-China*)[c] (Figs. 26a, b) but also the better-known *yü-yü-liang*, a mineral, nodular concretions of brown haematite.[d] It is closely related to other climbers, *S. China*[e] the famous Radix Chinae or 'China-root', and also to the American sarsaparilla, *S. officinalis*.[f] All these liliaceous roots owe what activity they have to the saponins and sapotoxins which they contain, but in spite of their great medieval and Renaissance reputations, both in East and West, they are not considered today as anything more than mild nauseants.[g] The use of the 'China-root' had gone back to the time of Thao Hung-Ching, but in the +16th, after the discovery of America, it became extremely fashionable for the treatment of syphilis, probably because it was a drug-plant already long in repute for diseases of the urino-genital system. Garcia da Orta had a good deal to say about it in +1563, devoting to it the 47th of his 'Colloquies';[h] as did his Chinese contemporaries. Da Orta refers to a striking cure with it accomplished in the year +1535, and only just a decade later, in +1546, the great anatomist Andreas Vesalius published at Basel a small but celebrated work upon it: *Epistola rationem modumque propinandi Racinis Chynae decocti ... pertractans*. At least six editions of this followed before the century was out (Fig. 27).[i] Thus we find a strange and unexpected link between the Western

---

[a] Stuart (1), p. 166.

[b] Cf. Vol. 1, pp. 87 ff., Vol. 3, pp. 500, 570, Vol. 4, pt 3, pp. 247 ff. *et passim*.

[c] R 680; CC 1845; Stuart (1), p. 410 and *TSCC*, Tshao mu tien, ch. 176, hui khao, p. 2a, b.

[d] RP 79.

[e] R 689; CC 1880; Ainslie (1), vol. 1, pp. 70, 592; Stuart (1), p. 409; Burkill (1), vol. 1, p. 2037; Otsuka Yasuo (1).

[f] Cf. Laufer (1), p. 556.   [g] Sollmann (1), p. 526.   [h] Markham ed., pp. 378 ff.

[i] E.g. *Radicis Chynae Usus* (Lyons, +1547). See further Cushing (1), pp. 154 ff., 160; Sarton (9), pp. 129, 212. The most complete studies of this curious subject are in the paper of Schmitz & Tan Tek-Tiong (1) and the Marburg dissertation of the latter author.

[1] 杜仲

Fig. 26a. Modern drawing of *Smilax China*; Anon. (*109*).   Fig. 26b. Modern drawing of *Smilax pseudo-China* = *Heterosmilax japonica*; Anon. (*109*).

scientific men of the Renaissance and the imperial hydraulic engineer of nascent Chinese culture.

One other item under patronymics calls for remark—the 'Yao family yellow'. This was the name for a variety, not a species, one of the innumerable terms for the cultivars of *Paeonia suffruticosa*, or tree-peony, produced in China's 'tulipomania'. We can read all about these varieties in the *Chhün Fang Phu*[a] which lists them in a micro-historical order. The *Yao huang*[1] was the culmination of a series, each named for a family either of gardeners or patrons, that ran through the Thang, Wu Tai and Sung periods. The flowers with single corolla were the earliest (e.g. *Su chia hung*[2], *Ho chia hung*[3] and *Lin chia hung*[4], all reds), then came as prizewinning doubles successively the *Tso hua*[5], the *Wei hua*[6], the *Niu huang*[7][b] and the *Yao huang*[1]. Mr Yao was a horticulturist, a man of the people, but the *Wei* flower was named for Wei Jen-Fu[8], prime minister of early Liao, *fl. c.* +950. This

---

[a] Hua Phu, ch. 2, pp. 1a, 2b, 6a, etc.

[b] This is a catch, for any unsuspecting translator meeting these words might easily take them to mean cow bezoar (R 337), gallstone concretions much favoured in medieval medicine. Cf. Ainslie (1), vol. 1, p. 35; Wootton (1), vol. 1, p. 111, vol. 2, pp. 15 ff.; Berendes (1), vol. 1, p. 12; Laufer (1), pp. 525 ff.; Schafer (13), pp. 191 ff. So is *wan-chüan-shu*[9], which has nothing to do with the scrolls and books of a great library, but designates a peach-coloured tree-peony flower variety the petals of which were rolled up like scrolls.

[1] 姚黃   [2] 蘇家紅   [3] 賀家紅   [4] 林家紅   [5] 左花
[6] 魏花   [7] 牛黃   [8] 魏仁浦   [9] 萬卷書

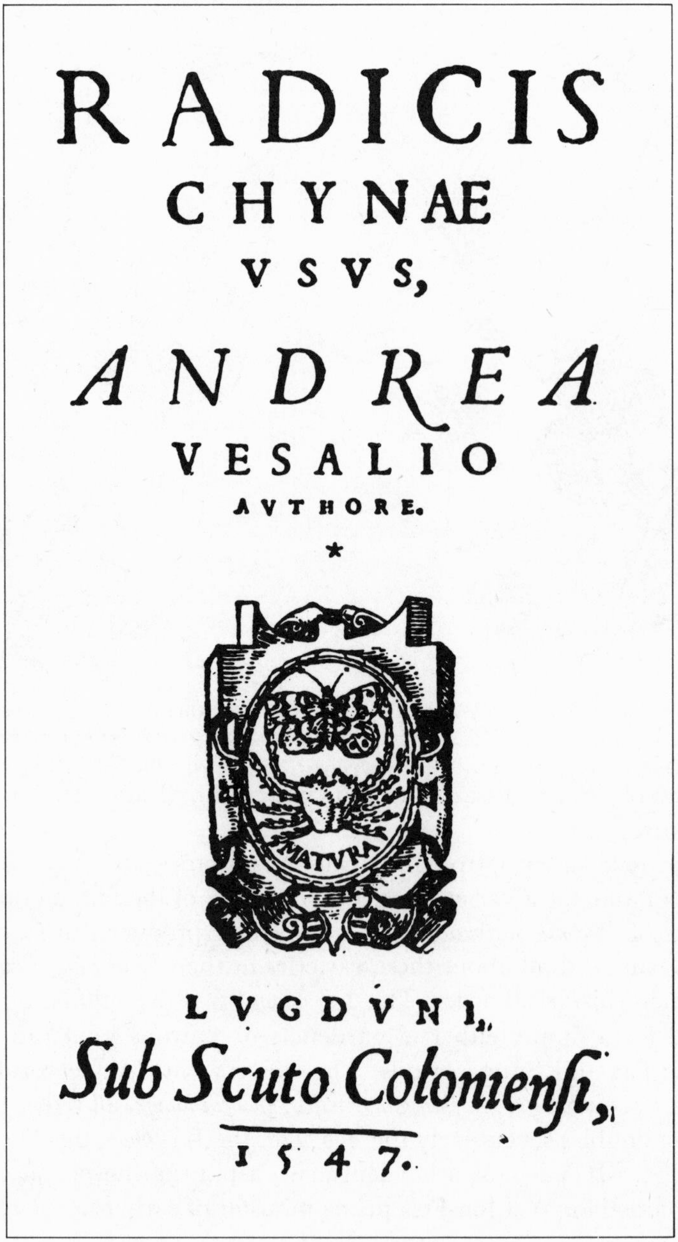

RADICIS
CHYNAE
VSVS,

ANDREA
VESALIO
AVTHORE.

*

LVGDVNI,
Sub Scuto Coloniensi,
1547.

Fig. 27. Title-page of the work of Andreas Vesalius on the 'China-root' (+1547).

was a double, flesh-pink in flower colour, not more than 4 ft high but with flowers five or six inches across and composed of more than 700 petals. It was also called *pao-lou-thai*[1], 'precious tower-platform'. If not well acquainted with the advanced state of horticultural selection in China a thousand years ago one is rather surprised to find the equivalents of varieties named like 'Mrs Banks' at that early time.

Lastly we come to the Chinese plant names of foreign origin. All the examples given in the Table are of the type which contain as prefix a word meaning foreign, but besides these there are of course many which simply transliterate foreign names. The identification of these is one of the favourite occupations of polymathic linguists and comparative philologists, the efforts of whom, when they attain correct solutions, can indeed make great contributions to our knowledge of culture contacts in history.[a] This is not the place, and indeed this work as a whole is not the right home, for any exhaustive survey of this subject, but brief reference to some of the transliterations cannot be omitted. At a somewhat earlier stage (p. 112) we stumbled upon the names for the jasmines, *yeh-hsi-ming*[2] for *Jasmimum officinale* $\simeq$ *grandiflorum*, and *mo-li-hua*[3] for *J. Sambac*; and satisfied ourselves that they were derived in Han times from Pahlavi and Sanskrit. Now we need mention only two other cases, the saffron pigment and the viniferous grape. The former is seen in the Table, but it bears two important subsidiary names, *tsa-fu-lan*[4] and *sa-fa-lang*[5][b] which clearly transliterate the Arabic *za'farān*. This is the dye, condiment, perfume and medicine prepared from the deep orange stigmas (and in part styles) of *Crocus sativus*. Though the plant was not described in the *pên tshao* literature till Li Shih-Chen's time, there had been knowledge of saffron in China as early as the +3rd century; between then and the end of the Thang it came sometimes as tribute.[c] Later, after the beginning of the Yuan, it was imported into China from Muslim countries and found habitual use among the people, though not grown there until the middle of the Ming. The names above would be of Sung or Yuan date; during the earlier period saffron was called *yü chin*[6], though not clearly distinguished from other yellow dye-plants,[d] because generally known only in the prepared powder form.

The grape-vine has afforded doubtless the greatest crux among early foreign loan-words in Chinese. Its introduction from the Bactrian regions by the great traveller and envoy Chang Chhien[7] in −126 is as well attested as any Chinese historical event.[e] Whatever other plants he may have brought along with him the

---

[a] Here the outstanding scholar was Laufer (1), still of mighty merit, though often criticised by Pelliot and others in detail. One cannot always accept Laufer's conclusions, but the material which he collected, when digested in one's own way and combined with new information, frequently leads to convincing results.

[b] Common orthographical deviations corrected here.

[c] We know that whole plants were sent to China from Kapiśa in +647.

[d] On the history of saffron in China see Laufer (1), pp. 309 ff.; Schafer (13), pp. 124 ff.

[e] Cf. Vol. 1, pp. 173 ff.

[1] 寶樓臺　　　[2] 耶悉茗　　　[3] 茉莉花　　　[4] 咱夫藍　　　[5] 撒法郎
[6] 鬱金　　　[7] 張騫

contemporary sources are clear that he introduced the vine *Vitis vinifera*[a] and the alfalfa grass *Medicago sativa*[b], the one for gladdening the hearts of men, the other for the fodder of brave horses. This place is as good as anywhere else for telling the story, so let us listen to Ssuma Chhien's words. The *Shih Chi* says:[c]

In the neighbourhood of Ferghana (Ta-Yuan[1]) wine was made from grapes (*phu-thao*[2]). Rich people stored ten thousand and more piculs[d] of it for several decades without spoiling.[e] The people (of Ta-Yuan) liked to drink wine, and their horses[f] liked to eat alfalfa grass (*mu-su*[3]). The Chinese envoy imported the seeds (of the grape-vine and alfalfa) on his way back (to China). The emperor thereupon caused alfalfa and grape-vines to be planted on rich tracts of ground. After some time he acquired large numbers of the 'Heavenly Horses' (*Thien ma*[4]), and so when various ambassadors from foreign countries arrived, by the side of imperial summer palaces and other retreats one might look out over wide prospects where the land was covered with vineyards and alfalfa fields.

A somewhat different but confirmatory account of the wine-growing of Ferghana and such places occurs in the *Ku Chin Chu*[g] about +300, and this may well be in the words of Chang Chhien himself since it is quoted from a small work called *Chang Chhien Chhu Kuan Chih*[5] (Record of Chang Chhien's Expeditions beyond the Passes) which still existed as late as the Sui period.[h] Now many have tried to reconstruct the name behind the sound *phu-thao*[6],[i] as we may learn from the recent and judicious survey of Chmielewski (1). First Tomaschek in 1877 and Kingsmill in 1879 suggested Greek *botrus* (βότρυς), a bunch of grapes, but no later philologists have found this acceptable. It cannot have anything to do with the Bodhi *phu-thi*[8] equation, for that is not found much before +400, the time of Kumārajīva. The theory of Yang Chih-Chiu (1) that it might be connected with a Phu-Thao Kuo[9] country somewhere near Bactria, mentioned incidentally in the *Chhien Han Shu*[j], runs into so many difficulties, that the best guess remains the Old Iranian word *budāwa* proposed in the long and classical discussion of Laufer.[k] Reconstructing the archaic pronunciation of the Chinese name, however, as **b'wo-d'ôg*[l], Chmielewski prefers a slightly different form *bādag(a)*

[a] R 288; CC 769.   [b] R 397; CC 1018.
[c] Ch. 123, p. 15a, b, tr. auct. adjuv. Hirth (2). The first translation was that of Brosset as long ago as 1828, but it is hardly usable now. Cf. *CHS*, ch. 96A, pp. 17b, 18b.
[d] C. 120 lb. wt each.
[e] See the discussion in Vol. 5, pt. 4, pp. 151 ff.
[f] These were the 'Heavenly Horses' just later referred to, a superior breed which the Chinese were anxious to get Cf. Sect. 39 below.
[g] Ch. 6, p. 2b (p. 23). The grape-vine also appears in the *Shen Nung Pên Tshao Ching*, which perhaps helps to date at least its compilation.
[h] *Sui Shu*, ch. 33, p. 23a.
[i] Here we write it in the form which later became generally used. The *Shih Chi* form appears again in e.g. *CHS*, ch. 96, p. 7a, referring to the vineyards of Cherchen (Chhieh-Mo[7]) a tiny city-State on the Nan Shan Pei Lu road (cf. Vol. 4, pt 3, p. 10) between Tunhuang and Khotan.
[j] Ch. 96A.   [k] (1), pp. 220 ff. Supported by Bailey (4).
[l] Based on K 102n' and K 1047d.

[1] 大宛   [2] 蒲陶   [3] 苜蓿   [4] 天馬   [5] 張騫出關志
[6] 葡萄   [7] 且末   [8] 菩提   [9] 樸桃國

meaning wine, and this is as far as we are likely to get. As for *mu-su*, there has been little improvement upon the suggestion in the accompanying chapter of Laufer[a] that it represents some word approximately *muk-suk or *buk-suk in Old Eastern Iranian, essentially a lost language. Thus the 'Medic weed' (*mēdikē, μηδική*) of Strabo was also a Median grass for the Chinese, though they adopted the Median word for it.

Many other similar transliterated loan-words could be adduced, if exhaustiveness were permitted in our plan.[b] A point worth making here, however, is that just as the Chinese received into their language the syllables of other tongues for naming plants that were not originally their own, so also the neighbouring civilisations of the Chinese culture-area which used Chinese as their literary language (like the *lingua franca* of Latin in our middle ages) systematically made up Chinese ideographic names in the Chinese style for plants which were not native to China but which grew in their own countries.[c] This applies particularly to Vietnam, Korea and Japan,[d] and can be followed in floras such as that of Phạm-Hoàng-Hộ and Nguyên-Văn-Du'óng (1) for the first of these regions.[e] There was thus an important radiation outward of traditional Chinese botany, checked only in the middle of the +15th century (+1446) by the introduction of the *han-gŭl*[1] or *ŏnmun*[2] (common speech) syllabary 'alphabet' in Korea,[f] and by the romanisation imposed on the Vietnamese after the French occupation from 1862 onwards.[g]

The foregoing pages have been slightly in the nature of a digression, and we must return to the principles of indigenous Chinese plant nomenclature. Our thesis is that there was an exact parallelism between East and West in the choice and construction of plant names. Let us look at Table 10 in which a representative variety of Greek and Latin plant names has been assembled.[h] A glance will show that every one of the categories into which the Chinese phytonyms were divided has its counterpart among the Western ones.[i] Whatever aura of advanced science may hang about the accepted Latin generic names of today is really illusory, for there is an exact analogy between the principles on which they were framed and those that gave the Chinese their nomenclature. If the impetus of modern science, with its urge to extend indefinitely the number of species de-

[a] (1), pp. 208 ff. Chmielewski (2) suggests Skr. *mrgaśāka*, animal feed, pronounced in a Kashmiri dialect.
[b] The process is continuing in China to this day, as is shown by Métailié (2).
[c] We owe a reminder of this to our friend Mr André Haudricourt.
[d] A warning has already been given (p. 17 above) that Chinese plant names were applied by the Japanese quite often to related species or even genera, when the Chinese ones did not occur in their islands.
[e] Most unfortunately this contains no characters.
[f] Cf. Vol. 4, pt 2, p. 516, and references there.
[g] Cf Lê Thanh-Khôi (1).
[h] Many of the Latin names later coined are, of course, based on Greek. The Theophrastean Greek colloquial ones are written without capital letters.
[i] We have hesitated whether or not to give the significances of each Latin name, no longer perhaps obvious in these days of the decline of classical education, when the Mancunian and the Malay are equal strangers to Marcus Tullius; but there are many botanical dictionaries and glossaries (such as that of Johnson & Smith, 1) which can be consulted for the derivations.

[1] 한글   [2] 諺文

Table 10. *Greek and Latin plant names, showing their origins*

I.   SHAPE AND FORM
     G.   batrachion   oxyacantha
     L.   Digitalis   Dracocephalum   Glottiphyllum   Hepatica
               Ptelea (from the disseminules)   Rhyncostigma
II. SIZE
     L.   Gigantochloa   Megalostylis   Microstylis
III. COLOUR
     G.   leucoion   melampyron
     L.   Galanthus   Glaucidium   Nigella   Porphyrodesme
               Rhododendron   Xanthoxylum
IV. AROMA
     G.   euosmos   myrrhis
     L.   Foetidia   Kakosmanthus   Lavandula (bath perfume)
               Oenanthe   Osmanthus   Sterculia
V.   TASTE
     L.   Capsicum   Oxalis   Piper   Saccharum
VI. SPECIAL CHARACTERISTICS
     G.   lithospermon   myriophyllon
     L.   Convolvulus (= Calystegia)   Impatiens (explosive dehiscence)
               Mimosa (sensitive leaves)   Mulgedium (milky juice)
               Nepenthes (insectivorous pitcher-plant)   Sanguinaria (red sap)
VII. HABITAT
     G.   anemone
     L.   Arenaria   Convallaria   Hylocereus (tree-candle, epiphytic)
               Littorella   Oreocereus (mountain-candle)
               Origanum (mountain-beauty)   Potamogeton
VIII. GEOGRAPHICAL REGION
     L.   Afrodaphne   Afrostyrax   Kalaharia (desert)   Mohavia
               (desert)   Taiwania
IX. DURATION AND SEASONAL PROPERTY
     L.   Hemerocallis   Primula   Sempervivum
X.   CLIMATIC PROPERTY
     L.   Arctagrostis
XI. SEX
     L.   Gynandropsis   Gynocardia   Thelymitra
XII. USEFULNESS
     technological
     L.   Hierochloe (holy grass, for church rushbearings)
     pharmacological
       property
     L.   Saponaria   Toxicodendron (= Rhus)
     curative
       function
     L.   Herniaria   Pyrethrum (anti-pyretic)   Tussilago
               Aristolochia (good for childbirth)   Exacum (driving out
               poisons)   Pancratium (panacea)
     best origin
     L.   Itatiaia   Parnassia
XIII. PATRONYMIC
     legendary
     G.   hyakinthos   narkissos
     L.   Satyrium

Table 10. (*continued*)

real					
L.	Aubrietia	Begonia	Clivia	Dahlia	Eschscholtzia
	Fuchsia	Garrya	Gesneria	Houstonia	Incarvillea
	Jussieua	Kaempferia	Lobelia	Magnolia	
	Montbretia	Stauntonia			
XIV. FOREIGN ORIGIN[a]					
G.	mēdikē-botanē (alfalfa)				
L.	Alkekengi	Azedarach	Basjoo	Batatas	Durio
	Icaco	Julibrissin	Litchi	Manihot	Medicago
	Moutan	Sassafras	Tabacum		
	(from foreign countries)				
	Luxemburgia	Malaisia	Sinowilsonia		

[a] The names of foreign origin are not all generic names now, but all or nearly all of them were so at one time.

scribed and known, had borne spontaneously upon Chinese botany, the need must have been felt for additional terms to denote specific rank, and doubtless some figure like Linnaeus would have arisen to insist on a limitation in the number of characters used. As it was, traditional Chinese botany simply developed a wealth of what were for the most part generic names. Yet qualifying adjectives analogous to specific epithets did develop to some degree, as we shall see in the sub-section on classification. In Table 10 no statistical comparison between the categories has been attempted, but we found when preparing it that the easiest one to fill up was the patronymic, and such names are indeed much more common in Latin binomials than they are in the Chinese nomenclature. Partly this may have sprung from a certain vanity or triviality of which botanists have sometimes been accused by other scientific men, but it might be fair to mention also a concern for the historical commemoration of discoverers, as also a declining familiarity with the Western classical languages.[a] Here the question of loan-words, which we thought we had finished with, intrudes again, for Western or 'modern' botany has been loth to accept into its nomenclature the 'native' names of non-European plants.

More than a hundred years ago Bretschneider wrote:[b]

Our botanists, who collect plants in foreign countries, do not trouble themselves generally about the indigenous names of the plants and their practical application, and they take no notice of the cultivated plants. Most of the systematic explorers endeavour only to discover new species or to create new genera in order to introduce their names into the science, or to call the newly discovered plants after the name of a friend. . . . In my opinion it would be more practical, in designating newly discovered plants, to preserve, if possible, the indigenous names, as has been done for instance with *Magnolia Yulan* and *Paeonia Moutan*, instead of giving them the names of savants or other persons which are often

[a] Perhaps it was easier to coin *Smithia* than a name like *Sericostachys*.
[b] (6), p. 21

dissonant or difficult to pronounce. Can anything more ridiculous be imagined than such names of plants as for instance *Turczaninowia, Heineckiana, Müllera, Schultzia, Lehmannia,* etc.[+]

> [+]The celebrated naturalist Agassiz is right in complaining (*vid.* the description of his travels on the Amazon River): 'Il est pitoyable d'avoir depouillé ces arbres' (palms) 'des noms harmonieux qu' ils doivent aux Indiens, pour les enregistrer dans les annales de la science sous les noms obscurcs des princes que la flatterie seule pouver vouloir sauver de l'oubli. L'inaja est devenu *Maximiliana,* le jara un *Leopoldina,* le pupunha un *Guilielma,* etc. . . .'

And in his revision a dozen years later he added further examples of conservation such as the lichi and the lungan.[a] It has to be admitted that Linnaeus was the evil genius of this Europocentrism.[b] In +1737 he laid down that he would admit no generic name unless it came from Greek or Latin, or looked as if it did, or commemorated a king or someone who had advanced the study of botany.[c] He was only prepared to accept 'barbarous' words as adjectival nouns forming a specific name, e.g. *Azedarach, Kali, Mays* (maize). Historically Linnaeus' transfer of old classical names to plants or groups unknown in antiquity was very unfortunate, and he also employed Latin names without any regard for their original use, e.g. *Cactus, Ceanothus.* In spite of all protests, these principles persisted, and have been built into the structure of the current internationally accepted codes of botanical nomenclature.[d] One such protest, however, may be quoted, that of the great Adanson in +1763, all the more colourful because written in his characteristic phonetic style:[e]

> *Employer les noms de païs.* A l'égard des noms de païs, que quelques Botanistes modernes apelent Barbares, il faut en doner ici l'explication; ils entendent, par ce terme, tous les noms Etrangers, Indiens, Afrikens, Amerikens, et même ceux de quelques nations Européenes. Mais si ces Auteurs Dogmatikes eussent voyajé, ils eussent reconu que dans ces divers païs on traite pareillement de Barbares nos noms Européens; ils sont tels, relativement à leur façon de prononcer, come les leurs le sont à la nôtre. Jujons donc autrement de l'acceptation d'un terme aussi impropre, et convenons que tous ces noms mis dans la balance équivalent les uns aux autres, et qu'ils doivent être adoptés toutes les fois qu'ils ne sont ni trop longs, ni trop rudes ou trop dificiles à prononser. C'est sur ce principe que nous rétablissons aux Genres, découverts par les Voiajeurs, leurs noms de païs, tels que celui de *Sialita* à la plante que M. Linnaeus a apelé *Dillenia,* celui d'*Upata* à la plante qu'il a nomé *Avicennia,* celui de *Panoe* à son *Vateria,* et beaucoup d'autres. Ces Auteurs, qui ont bien mérité de la Botanike, ne perdront rien à ces rèformes, on poura doner leurs noms à des Plantes qui n'en ont aucun; et à cet égard, on me permettra une réflexion, c'est que ces noms devienent si comuns et si triviaux, qu'on risque fort d'avilir la Botanike si l'on ne restreint cet honeur aux coryfés de cete science.

---

[a] (1), vol. 1, p. 110.   [b] See Stearn (3), p. 40.   [c] *Critica Botanica,* no. 229.
[d] True, the guiding afflatus of the Linnaean system was its practical functionality. It might have been difficult to decide, in the case of a plant growing in six different countries, which of the six different names should be adopted as official. But there was not even the will to consider this problem. Cf. pp. xxvii, 17–8, 312 on the rules of priority, and the undue specific emphasis on Japan.
[e] (1), vol. 1, p. clxxiii; cf. Stearn (1), p. xciv.

Thus the reception of transliterated foreign names in traditional Chinese botany was a much more international and oecumenical practice than the supercilious Europocentrism of eighteenth-century Western botany.

(iii) *Derivate complexes and coding redundancies*

There remain now only three things to be done before closing this disquisition on Chinese plant nomenclature: first to speak of the derivate complexes, the *mei*[1] that is not truly *mei*[1] and the *ma*[2] that are not exactly *ma*[2], then to glance at an example of the serious redundancies or confusions that the Chinese system occasionally fell into, and lastly to make a point already hinted at but strange to those who have not so far thought of it, namely the implicit superiority of the ideographic script over any other classificatory nomenclature for the species of the natural world.

Among European plant names we are very familiar with the fact that vernacular nomenclature offends sophisticated taxonomy. 'Nomenclature and taxonomy', wrote Greene[a], 'are almost inseparably connected. The name itself is but the expression of a taxonomic idea. Excepting those rare instances in which an individual historic tree has received a proper name[b], every plant name that ever was, in any language, is the name of a group. Naming is classifying.' Thus the cranesbill of ordinary speech (*su*[3], as it were) covers two different genera, of Geraniaceae, thus:

cranesbill	*Erodium cicutarium*
dove's-foot cranesbill	*Geranium molle*
meadow cranesbill	*Geranium pratense*[c]
shining cranesbill	*Geranium lucidum*[c]
musk cranesbill	*Erodium moschatum*[c]
sea cranesbill	*Erodium maritimum*[c]

But it is easy to find another instance where five vernacular names cover as many distinct Umbelliferae:

parsley	*Petroselinum crispum*
fool's parsley	*Aethusa Cynapium*
milk parsley	*Peucedanum palustre*
hedge parsley	*Torilis japonica*
cow parsley	*Anthriscus sylvestris*

This is very like what happened as the Chinese nomenclature developed. Probably it was largely the product of 'cross-referencing description'[d], so that

---

[a] (1), p. 122.

[b] The bo-tree of Bodh Gāyā or Anurādhapura, or (to descend from the sublime to the ridiculous) King Charles' oak.

[c] Note how here in each case the specific name reflects the vernacular.

[d] Cf. p. 117 above.

---

[1] 梅   [2] 麻   [3] 俗

hedge parsley was really hedge 'parsley', '*caroides*', as it were,[a] something that looked quite like common parsley but preferred the shade and shelter of hedges. If the wort-kenners had spoken Latin, they could have called it *Caroides septaceus*. Only the minute analysis of the various parts of the plant which eighteenth-and nineteenth-century botany would carry out could decide, at least provisionally, on the genus and species at the new level of precision.

So it was with the Chinese names. The *mei hua*[1] *par excellence* was the rosaceous *Prunus mume* (often improperly called the Japanese apricot, though its specific name comes indeed from the Japanese pronunciation of the character).[b] But as I write (in February) the *la mei*[2] (waxy *mei*) is blooming magnificently in the Cambridge Botanic Gardens; this is *Chimonanthus fragrans* (Calycanthaceae)[c]. People usually think that this wintersweet got its name because of the showy yellow beeswax-like and nearly translucent petals, but actually the first character is probably a corruption of the other form *la mei*[3], this *la* being the last month of the Chinese year, the time of its blooming and the month of the ancient winter solstice sacrifice (hence the 'meat' radical in the word). The flowers have a beautiful cool perfume, and are sometimes threaded on fine wire for women to put in their hair at the Chinese New Year. It is evident here that the word *mei* was allusive in character, as much as to say '*mumeoides*',[d] and indeed Li Shih-Chen states very clearly that *Chimonanthus* does not belong to the *mei* group (*mei lei*[4]) but spreads a similar fragrance at the same time of year.[e]

A much more complicated case is that of the 'hemp' or *ma*[5] nexus, for more than twenty plants and trees have this character in their names,[f] though belonging in terms of modern botany to more than a dozen families. The resemblances were perfectly real, whether in fibres fit for textiles, in oil extractable from the seeds, or in the shapes of leaves, the polygonal character of the stem cross-section, the position of the seeds in the capsules, etc. What appears to be the noun, therefore, *ma*[5], is really best thought of as adjectival *cannabinus*, and indeed as we shall see, one of the species concerned does actually bear this name today as half of its Linnaean binomial. To order the following material one could take several ways, alphabetical or tabulative, but it may be interesting to adopt a different method and look at the matter historically, dividing it thus into five phases.

With *ma*[5] as hemp, *Cannabis sativa* (Moraceae),[g] we are on firm, as well as very

---

[a] From *Carum*, the old generic name of *Petroselinum*.

[b] Cf. Li Hui-Lin (1), pp. 48 ff.

[c] R 504; CC 1338 (= *Meratia praecox*, = *Calycanthus praecox*). Li Hui-Lin (1), p. 166. A good text on it is *CFP*, Hua phu, ch. 1. p, 42a.

[d] Like *Narcissus cyclamineus* and many other such names.

[e] *PTKM* ch. 36 (p. 124). Li Shih-Chen himself favoured the waxiness explanation.

[f] In traditional Chinese botany, however, only half a dozen of these are included in the *ma* group or *lei* (*ta ma*, *hu ma*, *chhu ma*, *mêng* or *chhing ma*, *ya ma* and *huang ma*—see immediately below).

[g] R 598; CC 1592; B II 388. *PTKM*, ch. 22, pp. 1a ff.

[1] 梅花   [2] 蠟梅   [3] 臘梅   [4] 梅類   [5] 麻

ancient ground, for just as flax was so characteristic as the textile plant of the Westerners, all those inheritors of Ancient Egyptian culture, so hemp was the typical fibre-provider of China;[a] apart of course from the animal product which gave their silken clothes to the Shang and Chou aristocracy and the bureaucrats of Chhin and Han onwards.[b] It is mentioned in the *Shih Ching* and the *Shu Ching* during the early centuries of the − 1st millennium[c], and it occurs also in the first of the pharmacopoeias, the *Shen Nung Pên Tshao Ching* (c. − 1st cent.).[d] As other *ma* names developed, the primary *ma* acquired various double names,[e] such as *ta ma*[1]. 'the great hemp', *han ma*[2], Chinese (as opposed to foreign) hemp,[f] *huo ma*[3],[g] and *hsien ma*[4].[h] It has continued in cultivation in China for at least three thousand years.[i] Wild hemp, the *pi*[5] of the *Erh Ya*, was naturally called *shan ma*[6].

Still keeping to our first historical phase, which means more or less to the end of the Early Han period, we meet with another plant the identification of which is as difficult as that of *ma*[7] itself is straightforward. This is the *chü-shêng*[8,9,10] which appears in the *Pên Ching*, perhaps in quite ancient times also called *hu ma*[11] in the sense of 'admirable hemp',[j] and thus later inviting confusion with *hu ma*[11] in the sense of 'foreign hemp'. Chinese pharmaceutical and other botanists discussed the identity of *chü-shêng* (which probably means 'of large and abundant seeds') at length throughout the ages,[k] and modern botanists (partly guessing from occasional specimens of seeds in apothecaries' shops) have tentatively assigned it to the

[a] It is generally believed to be a plant of temperate Asian and North Indian origin, but the earliest domestication and cultivation may have been Chinese rather than Indian (cf. de Candolle (1), p. 148). It spread both east and west rather than north, so that the Teutonic peoples had it from the − 5th century, while (as we noted in Vol. 5, pt 2, p. 152–4) its use as a drug-plant among the Scyths and Sarmatians was described by Herodotus, c. − 420. Pharmacological usage has preserved the older name, *C. indica* (cf. Burkill (1), vol. 1. pp. 437 ff.; Watt (1), pp. 249 ff.). The hallucinogenic resin (taken as hashish, marijuana, etc.), heavy doses of which may induce comatose sleep (cf. Nahas, 1), was not used as an anaesthetic by physicians in ancient China, in spite of many opinions to the contrary (see Sect. 45 in vol. 6). Pollen analysis shows that in Eastern England hemp was cultivated from about + 400 onwards, with the Anglo-Saxon settlement, presumably as a fibre-plant; see Godwin (1, 2).
[b] Cf. Vol. 4, pt 2, p. 33.
[c] On its history and utilisation see the interesting papers of Li Hui-Lin (6, 7).
[d] We shall shorten this, as the Chinese do, to *Pên Ching*.
[e] We say nothing here of the words for the hemp sexes, which are treated of elsewhere (pp. 154, 158 above and Sect. 38 (*h*), 2 below).
[f] *Erh Ya I* (+1170)   [g] *Jih Yung Pên Tshao* (+14th cent.)   [h] A modern term.
[i] I have very pleasant memories of the hemp harvest as we saw it in the late summer of 1964 on our various ways through the countryside between Hangchow, Haining and Shao-hsing.
[j] But it could also have meant 'macrobiotic hemp', a plant which would confer longevity, perhaps immortality, *hu* being used in the sense of *hu kou*[12], an 'auncient man'. This is likely because one repeatedly encounters *chü-shêng* in alchemical connections, cf. Vol. 5, pt 2, p. 131, Vol. 5. pt 3, pp. 72, 93, 97.
[k] Many differentiating distinctions were made. Thao Hung-Ching said that the plant with a square stem was *chü shêng* while that with an eight-ribbed stem was *hu ma* (see immediately below). Su Kung in the Thang period said that the plant with eight ridges or angles (*lêng*[13]) on its capsules (*chio*[14]) was *chü shêng* while that with quadrangular capsules was *hu ma*. Kao Chhêng in the Sung (+ 1085) could not make out what the ancient *hu ma* was (*Shih Wu Chi Yuan*, ch. 10, p. 29*b*), because he thought the *Pên Ching* was much earlier than Chang Chhien's time, instead of being more or less contemporary.

[1] 大麻   [2] 漢麻   [3] 火麻   [4] 線麻   [5] 薜
[6] 山麻   [7] 麻   [8] 巨勝   [9] 鉅勝   [10] 苣勝
[11] 胡麻   [12] 胡耈   [13] 棱   [14] 角

Compositae, suggesting *Lactuca sibirica*[a] or some *Ixeris*,[b] but the problem may well be insoluble.[c]

The *Pên Ching* was compiled towards the end of this historical period, though most of its content is much older, and thus we have to reckon in it four other plants of the *cannabinus* quality. Two of these had *ma* in their learned names, and two had it only in their semi-popular names. The former were (a) *shêng ma*[1], 'rising-up hemp', i. e. *Cimicifuga foetida*[d] (Ranunculaceae), also called *chou ma*[2] from the name of the region, the ancient State of Chou, where it flourished; and (b) *ma huang*[3], 'hempy yellow', the celebrated *Ephedra sinica* of the Gnetaceae (cf. p. 151 and p. 239).[e] The two latter were (a) *yeh thien ma*[4], a subsidiary name for the Labiate *Leonurus sibiricus* (cf. p. 154), also called *chu ma*[5] because pigs like to eat it,[f] and (b) the *thien ma*[6] or 'heavenly, spontaneous, hemp' itself, a subsidiary name for the orchid *Gastrodia elata*[g] (cf. p. 139). There can be little doubt that most of these were so named because of the polygonal cross-section of the stems.

We come now to the last of the really ancient plants, *chu*[7,8] or *chhu*[7,8], usually called *chu ma*[9] or *chhu ma*[10], ramie or China-grass, from which a coarse textile is made. This, as we have already seen (p. 89 (b)), is the urticaceous *Boehmeria nivea*,[h] which, though it did not find its way into the Pên Tshao literature until the *Ming I Pieh Lu* about +500, had actually been known and used by retters and weavers for a full thousand years before; this we know from the references to it in the Yü Kung chapter of the *Shu Ching*.[i] One may see it growing well in the Chelsea Botanic Garden. It was probably not called *ma* until the Chhin and Han periods.

The second historical period we may let run from the beginning of the Later Han to the end of the Liu Chhao period. During this time three more plants joined the ranks of the *ma* group. By far the most important of these was *hu ma*[11], 'foreign hemp',[j] *Sesamum indicum*,[k] probably native originally to Africa but spreading through the Old World from cultivation in India. This member of the Pedaliaceae had certainly reached China by +500 when Thao Hung-Ching described it in the book just mentioned, saying that it had come from Ferghana,[l]

---

[a] B III 216; Stuart (1), p. 269; Laufer (1), p. 292.

[b] Khung *et al.* (1), pp. 303, 692, 1422.

[c] An elaborate discussion of it was given by Laufer (1), pp. 288 ff., but he was in captious mood, and wrote rather unfairly to the medieval Chinese botanists. His great *bête noire*, of course, was the assertion common in Chinese literature that Chang Chhien introduced many other plants besides the grapevine and alfalfa, but his criticism of this seems today rather excessive.

[d] R 529; CC 1402.　　[e] R 783; CC 2115.　　[f] R 126; CC 334.

[g] R 636; CC 1733.　　[h] R 592; CC 1576; B II 391, 458.

[i] Under Yang-chou, it will be remembered, the Tao-I barbarians brought garments of 'grass-cloth', *hui fu*[12], but under Yü-chou, the word *chu*[2] actually appears.

[j] Or 'Iranian, Persian' hemp.

[k] R 97; CC 257; Stewart (1), p. 358; Stuart (1), p. 404.

Pace Laufer, no Chinese reading this would have failed to get an allusion to Chang Chhien or some companion or slightly later envoy of the same kind.

[1] 升麻	[2] 周麻	[3] 麻黃	[4] 野天麻	[5] 豬麻
[6] 天麻	[7] 苧	[8] 紵	[9] 紵麻	[10] 苧麻
[11] 胡麻	[12] 卉服			

but we do not yet know the exact date of its arrival.[a] Chinese scholars were agreed, however, that its coming stimulated the formation of the distinctive names for hemp itself that we have just been examining.[b] Since its importance was rather as an oilseed producer than a fibre-plant, *hu ma*[1] acquired the synonyms of *yu ma*[1c] and *chih ma*[2].[d] Because of its square stalks Wu Phu[3] in his *Wu Shih Pên Tshao*[4] of *c.* +225 (further evidence as to dating) named it *fang ching*[5], and an additional name recorded by Thao Hung-Ching was *kou shih*[6], 'dog-lice', because of the shape of the seeds.

The second *ma* plant added in this period was malvaceous and fibre-yielding, *chhing ma*[10] or *mêng ma*[11],[e] i.e. *Abutilon theophrasti*[f], commonly known as Chinese jute or Tientsin jute, even American jute or Indian mallow. Its cultivation must be quite old in China because of a reference in the *Li Chi* (+1st century) to the making of ritual mourning girdles from it. *Chhing ma*[12,13,14] could be written in a number of ways, and *pai ma*[15] could be substituted for *mêng ma*. This plant is not to

---

[a] We remain doubtful, in spite of the possibility mentioned just above, that the name *hu ma* was really used anciently in any sense other than designating sesame. The oldest example appears to be that in a *Huai Nan Tzu* passage preserved in *TPYL*, ch. 989, p. 5*b*, which says that *hu ma* grows well near the Fên River (in Shansi); if genuine (*c.* −120) this is, in our view, just not too early for the bringing of it back by Chang Chhien—a story asserted also in a *Pên Tshao Ching* not further specified (but, *pace* Laufer, why not the *Pên Ching* itself?), as quoted in *TPYL*, ch. 841, p. 6*a*, *b*. However, modern editions of the *Huai Nan Tzu* book read, at the passage in question (ch. 4. p. 7*a*; *Chi Chieh*, p. 10*a*), *ma* and not *hu ma*, so *TPYL* may have misquoted. On the other hand they may be wrong. Another interesting place in *Huai Nan Tzu* speaks of Hu people seeing hemp and not realising that its fibres could be woven into cloth (ch. 11, p. 1*b*); this agrees with the botanical evidence that hemp spread westwards into Europe from Asia (cf. Burkill (1), vol. 1, p. 437). Perhaps these Hu were Greeks. The conviction that sesame was brought back to China by Chang Chhien was widespread in later Chinese literature; cf., e.g. *Chhi Min Yao Shu*, ch. 13, p. 11*b*; *Su Shen Liang Fang*, ch. 1, (p. 19).

Astronomical literature also contains what is probably a very early reference to sesame (noted by Kao Jun-Shêng (1), *cit.* in Li Chhang-Nien (2), p. 41). Outside the north polar region (Tzu wei kung) there is a constellation of eight stars, (S)W of Hua kai and N(E) of Wu chhê, called Pa ku (the Eight Grains, S 267). The grains concerned (and doubtless, for many ancient star-clerks, astrologically controlled) vary slightly as listed in the different recensions of the *Hsing Ching* (Star Manual), on which classical text see Vol. 3, pp. 197 ff., 248, 268, etc. Some of these (e.g. *TPYL*, ch. 837, p. 5*b*, and *KHTT*, p. 786) list *wu ma*[7] (black variety of sesame) as well as ordinary *ma* (hemp); those which do not (e.g. *KYCC*, ch. 69, p. 13*b*, and *Thien Wên Ta Chhêng*, ch. 18, p. 90*b*, and the source of Schlegel (5), p. 378) give instead of *wu ma*, *ma tzu*[8], probably signifying the edible sesame seeds. Now unlike most other texts the *Hsing Ching* can be dated by the internal evidence of its star-position measurements; the most recent researches have shown that its traditional date of *c.* −350 cannot be right, and that it must come from within thirty years on either side of −70; cf. Yabuuchi Kiyoshi (10, 2, 20, 28); Maeyama, Y. (1, 2). This is a good fit for the life-work of the astronomer Hsienyü Wang-Jen (cf. Vol. 3, pp. 216, 354), and what is here more important, quite acceptable for a mention of sesame if its introduction belonged to the time of Chang Chhien.

As for *chü shêng*, the +3rd or +4th-century *Lieh Hsien Chuan* (quoted in *TPYL*, ch. 989, p. 5*b*; cf. Kaltenmark (2), pp. 65 ff.) says that when Lao Tzu passed beyond the Shifting Sands he took the Guardian of the Gate, Kuanling Yin-Hsi[9], along with him; there they ate *chü shêng* seeds and were never seen again. Undoubtedly sesame was at first, quite naturally, though wrongly, identified with some indigenous plant called *chü shêng*; and its seeds were for a time regarded as a Taoist substitute for ordinary cereals, capable of conferring longevity or immortality.

[b] *MCPT*, ch. 26, p. 6*a*; *Shih Wu Chi Yuan*, ch. 10, p. 29*b*; both soon after +1080.

[c] *Shih Liao Pên Tshao* (+670).   [d] *Pên Tshao Yen I* (+1116).   [e] 'Fritillary hemp.'

[f] R 274; CC 734; B II 389; Burkill (1), vol. 1, p. 10. Alternative names have been *Abutilon avicennae* and *Sida Abutilon = tiliaefolia*.

[1] 油麻	[2] 脂麻	[3] 吳普	[4] 吳氏本草	[5] 方莖
[6] 狗虱	[7] 烏麻	[8] 麻子	[9] 關令尹喜	[10] 茼麻
[11] 茼麻	[12] 檾	[13] 蕢	[14] 青麻	[15] 白麻

be confused with another of the same alliance, *Hibiscus cannabinus*[a], which is Bimlipatam jute or Deccan hemp; grown nowadays in China to some extent, but of recent introduction and therefore called *yang ma*[1], a term with all the flavour of stove-pipe hats and fire-wheel ships. Evidently the fibre-crop character of *Abutilon* was quite sufficient to bring it within the *ma* group. Lastly some confusion was introduced in the +3rd century (possibly owing to adoption of local usages in the San Kuo period) when the name *shêng ma*[2], 'rising-up hemp', with its subsidiary *chou ma*[3] was given by the *Kuang Ya* dictionary to a plant quite different from *Cimicifuga*, namely the saxifrage *Astilbe chinensis*[b]. It never became an important drug, and was probably grown more as an ornamental.

If we may take the Thang dynasty as the third of our periods, there is only one new *ma* to record, but it was an important one—*Ricinus communis*, the castor-oil plant.[c] Mention of this euphorb occurs first in the *Thang Pên Tshao*[4], compiled *c.* +660 under the chairmanship of Su Ching[5], but just how long before that it had been known in China is uncertain. Its name was *pi ma*[6],[d] or *pi ma*[7], these names coming clearly from the shape of the seeds, which resembled the lice of the water-buffalo (*niu shih*[8], *niu pei*[9],[10]).[e] This useful plant is generally considered to have been African in origin, spreading from Egypt to India and Iran by the beginning of the present era, and it was probably from the Persian culture-area that it travelled to China because Su Ching mentioned the Hu[11] people as exporting it.

Our fourth period shall be the Sung, and all three of the next lot of *ma* plants received their first mention in the *Pên Tshao Thu Ching*[12] of +1062, the illustrated pharmaceutical natural history produced under the chairmanship of our old friend[f] Su Sung[13]. No doubt the most important of the three was familiar flax (cf. p. 108 above), *Linum usitatissimum* (Linaceae)[g], which received the scientific name of *ya ma*[14],[15];[h] though also known sometimes as Shansi *hu ma*[16],[i] or (appropriately

---

[a] CC 2508; Burkill (1), vol. 1, p. 1164.

[b] R 465; CC 1189.   [c] R 331; CC 877.

[d] 'Rumen hemp.' We vividly remember seeing whole fields of this growing, when walking in 1964 near the great Taoist temple of Chin-tzhu south of Thaiyuan in Shansi.

[e] Laufer (1), p. 403, points out the similarity between these names and *Ricinus* in Latin, which means a tick parasitic on domestic animals; but declines to give the Chinese credit for sufficient originality to think of a parallel name for themselves. If indeed the simile was an adaptive translation of a Latin name the case would be a highly unusual one; some transliteration like *li-hsi-nu* would have been far more likely. The buffalo louse is *Haematopinus tuberculatus* (R 44).

[f] Cf. Vol. 3, pp. 193 ff. and *passim*, Vol. 4, pt 2, pp. 446 ff. and *passim*. Also Needham, Wang & Price (1).

[g] R 365; CC 931.

[h] 'Disagreeable hemp', from the evil smell of the oil, and its inedibility (cf. Li Chhang-Nien (2), pp. 282 ff.). 'Duck' is a meaningless orthographic variant.

[i] 'Shansi foreign hemp.'

[1] 洋麻	[2] 升麻	[3] 周麻	[4] 唐本草	[5] 蘇敬
[6] 䕡麻	[7] 蓖麻	[8] 牛虱	[9] 牛蝨	[10] 牛蜱
[11] 胡	[12] 本草圖經	[13] 蘇頌	[14] 亞麻	[15] 鴉麻
[16] 山西胡麻				

from the shape of the seeds) as *pi-shih hu ma*[1], 'bed-bug hemp'.[a] In China it was always grown more for the oil, linseed oil, than for the fibre. A second plant also newly described by Su Sung and his collaborators was *than ma*[2] or *Urtica thunbergiana*[b] of obvious family. There is no reason to think that this was a foreign importation, rather it was a Chiangsu plant which had not previously been taken notice of. Though apparently a fish poison, it found little medicinal use except in dermatology, and probably owed its name to the fibrous character of its stems. The same might be said of the third new *ma*, *Corchoropsis tomentosa* (= *crenata*),[c] a tiliaceous plant allied to jute, which received the name of *thien ma*[3].[d] Though rather an insignificant herb, it has given its name in modern Chinese botany to the whole family of limes and lindens.

Fifthly comes the Ming, the Chhing and our own times. Counting Li Shih-Chen as an honorary Renaissance Man,[e] it has to be recorded that jute is first mentioned by him (+1596), *Corchorus capsularis*[f], one of a pan-tropic genus which provides a large part (90%) of the Indian commercial crop. Presumably it was introduced thence through the Ming trading contacts, and later on it was cultivated considerably in the lower Yangtze valley, under the name *huang ma*[4]. With this one might conclude the story of the *ma* group, but there are a few addenda. At some time since Li's work another species of *Corchorus* has been identified as native to China, the weedy and fibre-poor *acutangulus* (= *aestuans*),[g] with the name *tuan huang ma*[5] 'linden jute'. Then there come a number of trees which we have already encountered (p. 129 above)[h], mostly identified by fairly modern forest botanists in West China—*shan ma liu*[6], a kind of walnut (*Pterocarya paliurus*), *shan huang ma*[7], a kind of elm (*Trema orientalis*), *shan ma tzu*[8], another ulmaceous tree (*Celtis koraiensis*), and finally a euphorb *shan ma kan*[9] (*Alchornea davidi*). The Chinese names of these, however, cannot simply be written off as modern coinages, for some of them, at any rate, may have been used by the local people for many generations, perhaps many centuries, and one has to look for intrinsic evidence of dating. Thus the first, third and fourth of these might quite conceivably go back to the Han, but the second would not be likely to have antedated the coming of jute in the Ming.[i] At all events this is the sort of way in which one has to consider them.

---

[a] Laufer (1), pp. 289ff., 293 ff. was quite off the rails in asserting that Chinese botanical writers confounded *Linum* with *Sesame*. The former may well have been called *hu ma* in common parlance in some parts of China, but the scholars had their own name, or qualified *hu ma* as above. As for the bed-bug of China, about which we shall have more to say (Sects. 39, 44), it was the usual *Cimex lectularius* (R 43*a*). Meanwhile reference may be made to the interesting paper of Hoeppli & Chhiang I-Hung (1).
[b] R 595; CC 1587.   [c] CC 751; B II 388, Steward (2), p. 248.   [d] 'Field hemp.'
[e] And therefore suitable to head this period. Cf. Vol. 4, pt 1, p. 190, and pp. 308 ff. below.
[f] R 281; CC 750; B II 388; Burkill (1), vol. 1, p. 658; Watt (1), p. 406.
[g] Steward (2), p. 248; Burkill (1), vol. 1, p. 658.   [h] Therefore no references here.
[i] Except in so far as *huang ma* was an occasional synonym in earlier periods for true hemp itself.

[1] 壁虱胡麻   [2] 蕁麻   [3] 田麻   [4] 黃麻   [5] 椴黃麻
[6] 山麻柳   [7] 山黃麻   [8] 山麻子   [9] 山麻桿

The foregoing exercise has been rather long, but it may be particularly worth while if it helps us to pinpoint the exact level attained by traditional Chinese botany untouched by modern science.[a] One may indeed say that it became as sophisticated as it could possibly get without the tools that modern science provided and used—plant anatomy and comparative morphology, the investigation of minute parts by the aid of the lens and microscope, an understanding of the physiological meaning of flower, pollen and seed. This is illustrated, surely, by the case of the *ma* group within the *ku* or 'cereal' class,[b] naming of the components of which assuredly took place with 'cross-referencing' in mind. Running through the meanings of the Chinese names in the preceding paragraphs, we can translate them all into Latin binomials and see how they look. Taking *Cannabis sativa* as the fundamental type, we shall then have:

Cannabinoides exsurgens	for	*Cimicifuga* (and *Astilbe*)
Cannabinoides luteus		*Ephedra*
Cannabinoides porcallector		*Leonurus*
Cannabinoides coelestis		*Gastrodia*
Cannabinoides textilis		*Boehmeria*
Cannabinoides persicus		
or *oleaginus*		*Sesame*
Cannabinoides fritillarius		*Abutilon*
Cannabinoides barbaricus-oceanicus		*Hibiscus*
Cannabinoides ricinus		*Ricinus*
Cannabinoides foetidus		*Linum*
Cannabinoides fibrosus		*Urtica*
Cannabinoides campestris		*Corchoropsis*
Cannabinoides flavus		*Corchorus*

This was perfectly good classification in itself, serviceable but not, of course, 'natural' in the technical botanical sense. Nor was the sexual system, based entirely on the number and arrangement of flower parts, which Linnaeus tried to introduce. If everyone agrees nowadays that a natural system must be striven for, preferably such as to help to make sense when considered in the light of evolutionary ideas, it is because we have a picture in our minds of each plant as an organismic pattern far more complex than it was before the development of modern morphology, even though there may still not be perfect agreement about the weight to be laid on this or that structure or feature or on combinations of them. Perhaps the most striking comment that can be made concerning the Chinese knowledge of the hemp plant is that while they preceded everyone else in the appreciation of its dioeceous character, Chinese botany did not, down to the end of its time of independence, range hemp in the same family as the mulberry (Moraceae). But how many of us realise the lateness of this appreciation in

---

[a] We have had the chance before to be able to make some particularly precise comparisons between China and the West; cf. Vol. 4, pt 2, p. 124 on continuous screw and worm shapes in relation to tangent-plane helicoid structures.

[b] Cf. Sect. 38 (*f*) below.

Europe? De Tournefort (+1700, +1719) has them as far apart as Li Shih-Chen, the former (*Cannabis*) as genus 5 of section 6 of class 15 (herbs and suffruticose plants with apetalous or staminate flowers), the latter (*Morus*) as genus 4 of section 4 of class 19 (trees and fruit-trees with amentaceous flowers).[a] By +1763, however, Adanson is placing them together in his chestnut family, Castaneae (no. 47).[b] So one again acquires the impression that traditional Chinese botany attained a Magnolian or Tournefortian level, not an Adansonian one.[c] Only in the mid-eighteenth century did modern science begin in the plant world, not the seventeenth.

In a vast system of such long natural growth as the Chinese plant nomenclature, it was inevitable that there should arise what might nowadays be called coding redundancies—confused and over-lapping synonymic appellations. This was only too true of Greek and Byzantine names also,[d] and even today, with the Linnaean binomial system, the herculean efforts of permanent international commissions are required to reduce the wealth of names to standard usage. Anyone who might think that the Chinese were not conscious of this difficulty should look at one of the introductory portions of Li Shih-Chen's *Pên Tshao Kang Mu*,[e] where he lists the synonyms in decreasing order of redundancy from five to one. Of the worst case there is only a single example, five alternative meanings having been attached to a given three-character name (we shall describe it immediately); of the others a good many more. One could tabulate roughly as follows:

no. of alternative meanings	no. of cases
5	1
4	6
3	32
2	191[f]
1	76[f]

No doubt these difficulties arose partly because of local and provincial dialectal differences which led to variations in the writing, but partly also because of similar pharmacological actions and certain anatomical similarities. Thus *tu-yao-tshao*[1], properly *tu huo*[2], *Angelica grosseserrata* (Umbelliferae, R 208) had five other meanings (none of them primary):

[a] (1), vol. 1, pp. 535, 589, with the corresponding plates in vol. 2.

[b] (1), vol. 2, p. 377.

[c] Where the taxonomy of natural families is concerned, it is better to avoid the term Linnaean. To speak in this way is not to imply that the Chinese felt the need for a formalised system in the manner of Tournefort, but they perceived very clearly relationships between plant genera, even though these were often 'submerged' within their oecological and physiological classifications (see Sect. 38 (*g*) below). The parallel is with Chinese physical science and technology reaching a Vincian but not a Galilean level (Vol. 3, p. 160 and Needham (64), p. 405).

[d] See the useful book of Langkavel (1), just a century old and now reprinted.

[e] Hsü Li pref., ch. 2, p. 2*a* (p. 74).

[f] These figures include also data for mineral and animal names.

---

[1] 獨搖草   [2] 獨活

chhiang huo [1]	*Angelica sylvestris* (Umbelliferae)	R 211
kuei chiu [2]	*Diphylleia sinensis* (Berberidaceae)	R 520
kuei tu yu [3]	*Macroclinidium verticillatum* (Compositae)	R 41
thien ma [4] (*chhih chien* [5])	*Gastrodia elata* (Orchidaceae)	R 636
wei hsien [6]	*Senecio nikoense* (Compositae)	R 43

And so it goes on. Li Shih-Chen was busy unravelling this kind of thing a long time before international commissions were thought of, and much credit is due to him for it.[a]

(iv) *Taxonomy and ideography*

This sub-section may now conclude with a few reflections on the value (even the conceivable possible future value) of the ideographic system of writing as a vehicle of systematic nomenclature.[b] It might well be said that a language the dictionaries of which have to be based on taxonomic principles is a language remarkably congruent with the nature of biology.[c] I dare say that many working biologists besides myself have felt over the years a great sense of dissatisfaction with the Latin binomials, even when constructed from meaningful Greek and Latin roots, let alone those which derive from fortuitous personal names, whether of distinguished taxonomists or others. And this for the reason that there is nothing whatever to indicate what kingdom, class or order the thing belongs to; there is no conveyance of essential information. The names give no clue. I remember once receiving an interesting offprint on a subject in chemical embryology which stated that so many hundred eggs of *Pila globosa* had been used; as a biochemist I was put to no little inconvenience before finding that in fact this creature is a terrestrial mollusc. On another occasion, some twenty years ago, I attended a lecture about fibre-balls formed by *Poseidonia australis*. My classical education assured me that this must have something to do with the sea, but whether it was a shark, a mollusc or a protozoon I knew not; in fact it actually turned out to be one of the Potamogetonaceae, a monocotyledenous angiosperm allied to *Zostera* and growing richly underwater round the coasts of Australia. Its indestructible fibres, used for packing glass, were certainly of biochemical interest, but if the name had had a *tshao thou* (the 'grass' radical), I should have been clearer in the mind beforehand. Names exist, indeed, which do duty in both the vegetable and animal kingdoms, e.g. *Liparis*, which is both an orchid[d] and a butterfly.[e] Such a state of affairs would

---

[a] It should not be thought that Li Shih-Chen was the first to be concerned about synonymic redundancies; discussions of the kind go back at least to Thao Hung-Ching. Cf. for example, *CLPT*, ch. 6 (p. 166.2).

[b] The last page of Chao Yuan-Jen (5) seems to give some countenance to the propositions which will follow here.

[c] This point was forcibly made in conversation recently by Dr G. Herdan of Bristol.

[d] See Willis (1); the genus has 100 tropical species and one rare British one.

[e] Sherborn (1), vol. 4, p. 3607. The case of *Liparis* was commended to us by M. André Haudricourt.

[1] 羌活   [2] 鬼臼   [3] 鬼督郵   [4] 天麻   [5] 赤箭
[6] 薇蘅

have interested and perhaps surprised Li Shih-Chen, whose language had determinative radicals able to cope with it.[a]

A point which perhaps has not been made before in this connection is the
wealth of ideographic signs which the botanists (partly inspired by earlier alchemists and chemists)[b] proposed and used in the eighteenth and early nineteenth
centuries. A selection of these has been assembled in the accompanying Table 11.[c]
As a schoolboy, Linnaeus copied in his *Örtabok* of +1725 the alchemical signs
from the *Pharmacopoeia Leovardensis*,[d] and in later life applied them to botanical
uses, adding a good many more. By 1839 Lindley had to use several close-printed
pages to explain all the ideographs which Willdenow, de Candolle, Trattinick,
Loudon,[e] and others had introduced. If these find much less use today it does not
mean that they were never useful, nor that they might not be useful again. But
they were terminological, not nomenclatural. So far as we know, no Western
botanist ever thought of suggesting that it might be convenient to denote all plant
species by ideographs such as were used by those 'quaint Chinamen', whose
nursery gardens were providing such valuable spoils, and whose forests held such
allure of species in thousands yet unknown to Western science; ideographs which
would show at the merest glance to what division of the world of living things the
plant or animal belonged. It naturally never occurred to Western botanists, not
merely because European usages were implicitly those of the world of modern
science, but also because, not being sinologists, they knew nothing of the radical
system, nor of the possibility of its improvement beyond what the Chinese themselves had ever attempted.

During the past dozen years a number of biologists have concerned themselves
with the elaboration of coding systems which would act as permanent supplements to the binomial species names, and which might permit of use with computing machines, or at least punched-card storage and recovery apparatus.
Something of this kind began as long ago as 1907 when dalla Torre & Harms (1)
in their *Genera Siphonogamarum* gave numbers to the families and genera of vascular
plants, but failed to plan for expansion. Still, number arrangements were used by
many herbaria, and in some floras such as that of Deam (1).[f] More recently the
idea has been fairly thoroughly worked out by Gould (1) for plants, and by
Mullins & Nickerson (1) for insects. Thus in Gould's system the marsh marigold

---

[a] Cf. Vol. 1, p. 30.
[b] On these see Partington (6); Stearn (2, 5); Renkema (1) and Crosland (1).
[c] Nearly all the sciences and technologies in West as well as East have generated ideographic signs and
symbols. We have already noted the early appearance of systematic symbolism in cartography (Vol. 3, p. 552
and Fig. 233 on p. 555), and an extension of this (as Dr Raphael Loewe reminded us) was the ideograph series
used for military formations, flags of different kinds indicating headquarters or units of regiments, brigades or
divisions, with wheels to represent motorisation, and so on.
[d] See the frontispiece to the facsimile section of Stearn (3).
[e] One or two of Loudun's ideographs have an uncanny resemblance to Chinese. See Lindley (1), 1832 ed.
pp. 422 ff., 1839 ed. pp. 496 ff. The first edition is in many ways the more interesting, for in the latter chemistry
drove out much of taxonomy, morphology and plant geography.
[f] There is a recent treatise on the principles of such numerical classifications by Clifford & Stephenson (1).

Table 11. *Alchemical and botanical ideographs of the West*

Sun	☉ Au		annual	(Lin) monocarpous, bearing fruit but once ① (Can) (Can)
Moon	☽ Ag			⊙ monocarpous, at long intervals, like agave (Can)
Saturn	♄ ♄ Pb		woody	(Lin) ♄ undershrub (Can) ♄ shrub (Can)
		♄	shrub	(Lin) polycarpous, like fruit-trees (Can) caulocarpous (Can)
Jupiter	♃ Sn	♃	perennial	(Lin, Wil, Can, Tra) suffruticose (Can) rhizocarpous (Can)
Mars	♂ Fe		biennial	(Lin) male (general) ② (Can)
Venus	♀ Cu			female (general)
Mercury	☿ Hg			hermaphrodite
	△ fire			
	▽ water		aquatic	

♄ tree or shrub in general (Lin, Wil)
♄ small tree (Can)
♄ tree more than 25 ft high (Can)
⚈ deciduous tree (Lou)
⚇ evergreen tree (Lou)
♈ arborescent monocot, e.g. palm (Tra)
♎ tuber-propagated (Tra)
♏ runner-propagated (Tra)
⚢ propagated by falling buds, 'viviparous', e.g. *Lilium tigrinum* (Tra)
♐ with scape-borne flowers, e.g. *Hieraceum Pilosella* (Tra)
♓ with flowers and leaves on separate stems, e.g. *Curcuma Zedoaria* (Tra)
⚌ deciduous creeper (Lou)
⚍ evergreen plant (Lou)
⚎ aquatic plant (Lou)
⚹ fusiform-rooted (Lou)
♄ parasitic (Lou)

Linnaeus' herbarium signs for S. C. Bjelte's Siberian material (+1744)
⚺ Western edge of Asia
⚻ Eastern edge, Siberia (or ◦)
⚼ Kamchatka (G. W. Steller's specimens, +1746)
◦ Central Asia
⚆ East

* A good description at place cited in ref., *or*, a genus of which Linnaeus himself had seen the living material.

† An imperfectly known species, *or*, a genus Linnaeus knew only from herbarium material.

! Seen by the author.

× Hybrid.

Caltha palustris is represented by the following formula:

$$PA/1-21: 53-2369-4-x$$

where P is the plant kingdom, A the phylum, 1 the class, 21 the order; followed by 53 for Ranunculaceae, 2369 for the *Caltha* genus, 4 for the *palustris* species, and x, some number, for variety, if and when necessary. Similarly Mullins & Nickerson gave the example:

$$14 \cdot 13 \cdot 2 \cdot 9/1 \cdot 2/2/3 \cdot 1 \cdot 1$$

the numbers between the full stops representing in sequence phylum Arthropoda—class Hexapoda—sub-class Pterygota—order Odonata and sub-order Anisoptera—family Libellulidae, sub-family Libellulinae and tribe Libellulini—genus *Libellula*—and species *depressa*. Any innocent creature might perhaps be depressed at being reduced to such a series of impersonal numbers, but even organic molecules are going the same way, as witness the recent work of Lederberg (1, 2) which with the aid of topology has coded all possible structures, whether dendritic alkanes or the much more difficult aromatic ring-systems. Thus the highly complex 5-ring system of morphinane can be expressed as:

$$(8H\ N3, \$,,, 3, 0, 3,, 1)$$

It may well be that sets of numbers and letters like these could give to the initiated in centuries to come that immediate comprehension which I missed so much with *Pila* and *Poseidonia*, but whatever may be the preference of the scanners of computers I confess to a hankering for the concrete pattern, congruous in its way with the pattern of the living organism itself. This is exactly what the ideographic system could provide.[a] Of course it would have to be streamlined and subtilised for the international requirements of today and tomorrow, with all due safeguarding of determinative radicals and other components by precise semantic definition. Then the following patterns

椴,窄核 椴

would say to me quite clearly (however I pronounced them)[b] *Tilia leptocarya* (Tiliaceae), the species of linden-tree with the thin nuts, and they would say it even better than the Latin does because, for all I knew beforehand, *Tilia* might be a moth or a mammoth.[c] Here it could be only a tree, or a member of a pre-

---

[a] Here we must not attempt to do more than adumbrate the possibilities.

[b] They are in fact pronounced *tuan, chai-ho Tuan.* And the Chinese name of the tree is in fact *tshao-phi-tuan*[1], the coarse-bark linden (see Chhen Jung (*1*), p. 785, overlooked by Steward (*2*), p. 247). But for the present argument this is irrelevant.

[c] Here in one case the arthropod radical *chhung*[2] would spring to the eye, in the other the mammalian one, *chhüan*[3,4]. Especially by bringing into service the 'submerged' or forgotten radicals of the *Shuo Wên* (cf. p. 121 above), it would be easily possible to use, apply, or coin, as many characters as necessary for all phyla and sub-phyla of the animal kingdom. Genus and species would then need only two further characters. And the affiliation of the organism would be self-evident at a glance.

[1] 糙皮椴   [2] 虫   [3] 犬   [4] 犭

dominantly arborescent family, and its specific name could follow it rather than preceding it if the standard binomial sequence were retained. Of course, for all the namer knew, I might be totally ignorant of the meanings of *leptos* (λεπτός) and *caryon* (κάρυον), but I could find these out[a], which is more than I could manage if the name had been *Tilia rehderi*, as it only too easily might have been. We only need as many characters as there are natural families; and ingenuity could always manufacture more if need be. Perhaps in future centuries the biological sciences will move forward so fast that the scanners of computing machines will acquire a taste for patterns rather than numbers;[b] if so the ideographic script may yet conceivably perform a service for all humanity.

(*d*) THE LITERATURE

We can now go no further forward without having a look at the literature on natural history subjects which mightily flourished in China as the centuries went on. It has often been necessary to provide this kind of survey before—for mathematics[c] and for astronomy[d] in Vol. 3, for mechanical engineering[e] in Vol, 4, pt 2 and for civil engineering[f] in Vol. 4, pt 3—but it is doubtful if we have hitherto been faced with a literature of such vastness.[g] The task is all the heavier because in traditional China there was little differentiation of botanical from zoological literature, he sciences not having clearly emerged as in the post-Renaissance West, but this means that for the most part what we say here will cover the Section on zoology as well. How to divide the literary genres has presented a problem, but the following divisions seem to be the best: (*a*) lexicographical and encyclopaedic texts, (*b*) the compendia of pharmaceutical natural history, (*c*) works on wild plants edible in emergencies, (*d*) books on exotic and historical botany, (*e*) the numerous botanical monographs dealing with restricted groups of plants, a single genus or even the varieties of a single species, (*f*) treatises on horticulture and traditional botany. Then the sub-section may reasonably end with an account of the influence of Chinese flora and botany upon modern plant science, including the epic movement of taxonomic identification.

(1) LEXICOGRAPHIC AND ENCYCLOPAEDIC TEXTS

Why, one might ask, should this be thought the place for a disquisition, even though brief, on the role of dictionaries and encyclopaedias in the development of

[a] Even then I might think, since *caryon* means a nucleus as well as a nut, that the creature was some kind of protozoon.

[b] Soon after the time when this was written, Sir Gerald Clausen informed us (*c.* 1967) that 100 point assemblies within a small square could readily be scanned by computer, and these techniques must now have advanced much further.

[c] Sect. 19, in Vol. 3, pp. 18 ff.

[d] Sect. 20, in Vol. 3, pp. 186 ff.

[e] Sect. 27, in Vol. 4, pt 2, pp. 166 ff.

[f] Sect. 28, in Vol. 4, pt 3, pp. 323 ff.

[g] Especially if one includes the literatures on medical and agricultural subjects, with which we are not directly concerned here. On these see Sects. 44 and 41. below.

Chinese biological knowledge? The answer arises from the difficulty we encountered on the first page of the Section on mineralogy,[a] where we came, for the first
time in this book, to a science which was, in traditional China, essentially descriptive and classificatory. We are obviously precluded by our plan from dealing
with minerals, plants and animals individually; one can only choose a few examples for discussion, as indeed throughout the present Section we do, and apart
from that must necessarily confine the account to principles, special topics, ideas,
techniques, and the like. Already technical terms and taxonomic names have
been extensively under review. Since sciences of this kind are so overwhelmingly
composed of terms and names, and depend so much on the accurate correspondence of the terms and names with the actual things, the work of the lexicographer takes on a peculiar importance, much greater, for instance, than that
which it has in astronomy or mechanical technology. From what has already
been said (pp. 127, 137) it has been obvious that the compilers and commentators
of the *Erh Ya*, however deeply learned as philologists, were no mean naturalists
either; and all through Chinese history the lexicographers had their part to play
in keeping the descriptive biological sciences on the rails.[b] The fact that they were
imbued, like all Chinese scholars, with a 'strongly conservative spirit which
sought to perpetuate the values of antiquity', and that their encyclopaedias were
in turn the mainstays of the candidates in the imperial examination system, had
the incidental result of imparting great stability to the biological terminology and
nomenclature, and therefore of ensuring the age-long perpetuation of a gradually
expanding body of botanical knowledge. Here the contrast with Europe is striking. The Greek, the medieval Latin, the Norse and old Slavonic and Catalan
names of plants and animals have all gone long ago out of use,[c] but a Chinese
plant or animal name can still be employed in the same sense as it was in the − 1st
millennium.[d]

Anyone who sets out to talk about everything in the world will do one of two
things, he will arrange the material according to subjects, or he will arrange it
according to words and sounds. He will therefore produce either an encyclopaedia or a dictionary (see Fig. 28). The key ideas here are information on the one
hand, and mere meaning on the other. The encyclopaedist gives information
about facts, and while he cannot dispense with definitions, he will add descriptions as clear as he can make them, together with interpretations and sometimes

[a] Vol. 3, p. 636.
[b] There is no reason to think that they were not helped from time to time by the botanical and agricultural
experts, as of course happened also in the West.
[c] Except when embodied in a currently accepted scientific or vernacular name.
[d] The quotation in the above paragraph comes from Bauer (3), who has given us the most recent and
complete account of Chinese lexicography. It is part of a symposium containing much parallel material for other
civilisations by other authors, all of which is instructive for comparison. Sinologists seem to have written rather
less on Chinese lexicography than might have been expected, but there are certain treatments which though
now old are still worth reading, e.g. Mayers (2). The best-known general history of encyclopaedias (Collison, 1)
covers mainly the + 18th century and later. It made a gallant attempt to do justice to the Chinese contribution,
but here it is insufficiently complete, accurate and well informed to be of much value to us.

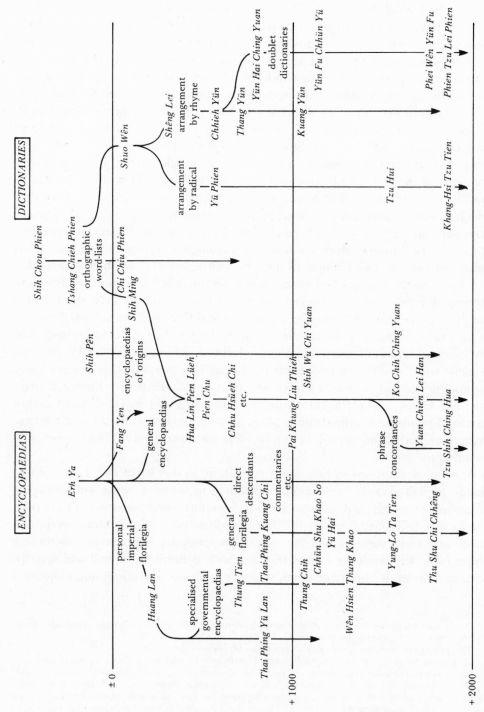

Fig. 28. Chart to show the filiation and descent of Chinese encyclopaedias and dictionaries.

evaluations, illustrated perhaps by long and abundant quotations. The writer of the dictionary, lexicon or glossary, however, is interested only in definition, and perhaps also in etymology (apart from bilingual or multi-lingual correspondences). An encyclopaedia, as Bauer says, provides complete pictures but makes the rapid location of information on particular points more difficult; a dictionary is good for the rapid location of subjects because of the large number of small entries, but it displaces material from the natural order and gives no complete picture of the things it mentions. Allied to encyclopaedias, on the other hand, are those large collections of quotations, or florilegia, of which most civilisations can provide examples.

Naturally all this takes on a somewhat different look within the ideographic universe. Although sharp lines between the genres were never drawn in classical Chinese literature, the two contrasted plans are clearly expressed by the terms *tzhu tien*[1] for encyclopaedia and *tzu tien*[2] for dictionary, familiar indeed to beginners in sinology yet not always fully understood by them. The great Chhing scholar, Juan Yuan[3], prefacing his edition of Hsing Ping's *Erh Ya Chu Su* (+1000)[a], said:

The *Erh Ya*[4] (the oldest encyclopaedia) contains characters which occur in the classical texts, yet some words do not agree with them; this is due to the mistakes of copyists since ancient times. But it also has some words the meanings of which do not agree with those in the *Shuo Wên*[5] (the oldest dictionary). This is because the *Shuo Wên* was concerned with deriving the meaning of characters from the forms of their ancient graphs, and has to do with the origins of words and their meanings. The object of the *Erh Ya* on the other hand was mainly to explain the names, terms and metaphorical phrases used by (the scholars of old in) the classical texts, and only a small number of entries are concerned with the original etymological meaning of the words.[b]

For example, 'red' was not merely red for the *Erh Ya*, because it might be part of the generic name of a particular plant if in permanent combination with one or two other words. Next, the assembly of masses of quotations took a particularly prominent place from an early time in China, certainly from the +3rd century onwards. 'All Chinese encyclopaedias', says Bauer (3) 'are anthologies, upon which were grafted greatly varying forms of dictionary arrangement.' This tendency he connects with a certain difficulty of formulating abstract concepts in Chinese, because of the basic paucity of morphemes in the language, a point to which we drew attention at the beginning of this work (Vol. 1, p. 36).[c] In this very 'isolating' or non-agglutinative tongue, moreover, it was not possible to coin abstractions such as 'odd-*ment*', empti-*ness*' or 'vacu-*ity*', and quite difficult to

---

[a] See immediately below, p. 192.
[b] Tr. auct.   [c] Cf. Potter (1).

[1] 辭典   [2] 字典   [3] 阮元   [4] 爾雅   [5] 說文

invent phrases to express them.[a] As we shall see later on,[b] there can be little doubt that this was to save Chinese philosophical thought from many *Schein-probleme* and false enigmas, yet abstract thought-patterns were certainly necessary, and it was found that the best way to attain them was to weave together a complex of historical, literary and scientific allusions by means of quotation.[c] Certain code words or phrases could then evoke a specific concept with a whole group of theoretical echoes.[d] Lastly, it may be said that the encyclopaedic genre, with the material arranged according to categories, greatly predominated over the strictly lexicographic in Chinese literature. At first sight it might be thought that this was owing to the use of the ideographic script instead of an alphabet, but that will hardly do because several methods were developed of ordering the characters in a single continuum. Indeed the disadvantage mentioned above, of displacement from natural order, was much less so in Chinese than in an alphabetic language, because of the radical system, which naturally linked together the words that signified various different kinds of plants and animals. In any case, the predominance of encyclopaedias in China constitutes a great help for the historian of any branch of natural history.

(i) The oldest encyclopaedias

Let us now take a brief bird's-eye view of the way in which Chinese encyclopaedias and dictionaries proliferated, and then, having fixed the date of the *Erh Ya*, trace out the times and qualities of its numerous progeny. All through the centuries this oldest of the encyclopaedias was enlarged by abundant commentaries, and that succession might be said to form the main line of descent from which many other kinds of 'category books' (*lei shu*[4]) branched off.[e] One line, starting in the +3rd century and reaching its climax in the +10th was that of the 'imperial florilegia', collections of quotations for the emperor's daily reading and reference, carefully sorted into categories. This generated in its turn specialised encyclopaedias concerned with government, starting in the +8th century, and others covering other fields, such as philosophy, only much later. General flori-

---

[a] For example, 'material-*ism*' eventually came out as *wei wu chu i*[1], 'matter only (lit. material things) as the guiding idea', but this happened rather late; earlier the reference was to people, as *wei wu phai*[2], 'the school of those who take ...', or *wei wu lun*[3], 'the discourses of those who take ...'.

[b] Sect. 49 in Vol. 7.

[c] This practice incidentally proved enormously valuable because it led to the preservation of literally millions of fragments of books which would otherwise have been completely lost. By the Ming and Chhing such rescue work became part of the avowed aims of encyclopaedists.

[d] The best among our own lexical works also have a richness of quotations. Sir James Murray's *New English Dictionary* gives not only the strict lexical meanings by direct definitions, but also the often almost equally important contextual senses, and this can only be done by the copious quotations, one for each century for each main use of a word. Cf. Potter (1), pp. 170 ff.

[e] The first appearance of this term as a class of books in its own right occurs in the dynastic bibliography of the *Hsin Thang Shu*, compiled in the middle of the +11th century.

[1] 唯物主義　　[2] 唯物派　　[3] 唯物論　　[4] 類書

legia followed from the +10th century onwards, reaching a climax early in the +18th. General encyclopaedias with lesser wealth of quotations were typical of the Sui and Thang, beginning in the late +6th century, but have continually been produced down to the present time. Another line of descent was formed by books which dealt with the origins of all sorts of things, inventions and customs; this genre had started almost as early as the *Erh Ya*, in the −2nd century, but was certainly influenced and stimulated in the +11th by the general encyclopaedia tradition. So much for the encyclopaedias. The dictionary pattern is in a way less complex. In the direct line of descent from the *Shuo Wên* came all those works which arranged the characters in accordance with their radicals, but from the +3rd century a quite different line diverged, those which arranged them in accordance with their sound, either by rhyme (*yün*[1]) or by initial syllable (*tzu mu*[2]). Since little identification or description follows the definition, the old dictionaries are only of minor help to the historian of science. 'Name of a plant' is not what he will be looking for. But the encyclopaedias are a very different matter.

The cardinal importance of the *Erh Ya*[3] for the history of natural history in ancient China has already emerged from the discussions on pp. 126 ff. and 183 ff. above, and the fixation of its date is thus particularly important. We call it by custom the 'Literary Expositor', but a better translation might be 'The Semantic Approximator', for from Chang Yen[4] in the +3rd century[a] onwards commentators explained that *erh* signified 'nearing' and *ya* the 'right meaning'. The grand design of its compilers was to explain in the more ordinary language of their time the meanings of words and names which colloquial development, dialectal divergence, or technical specialisation, had rendered unfamiliar. We place its composition between the late −4th and early −2nd centuries, i.e. in Chou, Chhin and Han (thus between the life-spans of Aristotle and Cato), but the considerations which justify this are rather complicated and need a short explanation.[b]

The traditional author was Chi Tan[5], alias Tan, Duke of Chi, or Chou Kung[6] (the Duke of Chou *par excellence*), a historical figure who died *c.* −1034. But it is interesting that this attribution was already doubted in the late −1st century, for the *Hsi Ching Tsa Chi*[7] (Miscellaneous Records of the Western Capital), probably compiled by Wu Chün[8] about +545, records[c] a conversation of some Han scholar with Kuo Wei[9] (*c.* −60 to +22), who pointed out that since a sage minister named Chang Chung[10], of the time of the Chou High King Hsüan[11] (r. −827 to −782), was mentioned in it, it could not have been composed by

---

[a] *Chhien Han Shu*, ch. 30, p. 12b.
[b] A longer one will be found in Chang Hsin-Chhêng (*1*), vol. 1, pp. 532 ff., here, as always, exhaustive and judicial.
[c] Ch. 3, p. 2b.

[1] 韻　　　　[2] 字母　　　　[3] 爾雅　　　　[4] 張晏　　　　[5] 姬旦
[6] 周公　　　[7] 西京雜記　　[8] 吳均　　　　[9] 郭威　　　　[10] 張仲
[11] 宣

Chou Kung.[a] The enquirer then went and asked the opinion of Yang Hsiung[1],[b] who said more cautiously that though there were later interpolations most of the text must be quite ancient because Confucius himself had taught it to Prince Ai. These two points of view, the more and the less radical, continued in vogue throughout the following two millennia of Chinese textual criticism.[c]

What exactly was the connection with Confucius? The key passage occurs in the *Ta Tai Li Chi*[2] (Record of Rites, compiled by Tai the Elder)[d] where Ai Kung[3] (Prince Ai of Lu, r. −493 to −467) is asking whether a ruler ought to study 'grammar and rhetoric'. As part of his reply Khung Chung-Ni says:

The *Erh Ya* explains the language of the men of old (lit. looks to what is old); and it suffices to distinguish the meanings and nuances of current speech.[e]

This is acceptable in itself, but the *Ta Tai Li Chi* was, as we know, a Han work, not put together till between +80 and +105, so it cannot tell us anything very certain about the times of Confucius and Ai Kung, though it must represent the traditions of the Former Han period.[f] An intimate association with didactic Confucianism was however perennially recognised. Chêng Hsüan[4] (+127 to +200), the great commentator of the *Chou Li*[5], understood clearly that the *Erh Ya* had not been the work of one hand (*fei i chia chih chu*[6]), but believed that it had emanated from the disciples of Confucius,[g] and this idea was not far wrong. The question was, what disciples? Kuo Pho[7] himself (Fig. 29), the greatest commentator of the *Erh Ya*, c. +310, said that its text was of 'middle antiquity' (*chung ku*[8]) but had been finished by Han scholars. A persistent but vague tradition connected it with Shusun Thung[19] (*f.* −201), the liturgiologist of the first Han emperor,[h] and something like what we have now must have existed about this time if it is true that Khung Fu[20], one of the descendants of the sage's house, who died in −208,

---

[a] We had occasion to mention at a much earlier stage this striking example of sceptical philology in the Han
[b] Naturalist, astronomer, mutationist, lexicographer; cf. Vol. 3, *sub voce, passim*.
period (Vol. 2, p. 391). It is interesting that the Chou Kung attribution does not occur in the *Chhien Han Shu* bibliography but became generally orthodox in the Later Han and afterwards.
[c] In the +11th century, for instance, Ouyang Hsiu[9] was prepared to deny that the *Erh Ya* contained any evidence of pre-Chhin authorship. In the late +18th, Tshui Shu[10], the sceptical philologist, agreed with this view. On the other hand, Yen Chih-Thui[11], c. +590 (in *Yen shih Chia Hsün*, ch. 6, p. 15a) criticised the *Erh Ya* in the same way as the *Shen Nung Pên Tshao Ching* (cf. p. 235 below), saying of the obvious interpolations—*fei pên wên yeh*[12], 'not in the original text'. In this reserved attitude he was followed by many later scholars, from Hsing Ping[13] (c. +990) to Shao Chin-Han[14] (c. +1780).
[d] Ch. 74, tr. auct. adjuv. Wilhelm (6), p. 89.
[e] It is true that the parallel prosodic phrase is not a book name, but to interpret the words *erh ya* as not a book name seems even more forced.
[f] The *Ta Tai Li Chi* text was often cited, as by Chang I[15], the compiler of the *Kuang Ya*[16], in his preface, c. +230.
[g] So also Kao Chhêng[17] in *Shih Wu Chi Yuan*[18] (Origins or Affairs and Things), c. +1085, ch. 17, p. 8a.
[h] Cf. Vol. 1, p. 103.

[1] 揚雄	[2] 大戴禮記	[3] 哀公	[4] 鄭玄	[5] 周禮
[6] 非一家之著	[7] 郭璞	[8] 中古	[9] 歐陽修	[10] 崔述
[11] 顏之推	[12] 非本文也	[13] 邢昺	[14] 邵晉涵	[15] 張揖
[16] 廣雅	[17] 高承	[18] 事物紀原	[19] 叔孫通	[20] 孔鮒

Fig. 29. A traditional depiction of Kuo Pho (+276 to +324), the great commentator of the *Erh Ya*; from *Lieh Hsien Chhüan Chuan*, ch. 4, p. 29a. A demon servitor behind him holds a banner inscribed with the words *Shui fu hsien po* (the Immortal Lord of the Bureau of the Wsters) because he edited the *Shan Hai Ching*, on which see Vol. 3, p. 504.

was the writer of the abridgement called *Hsiao Erh Ya*[1], which certainly circulated in Han times.[a] Another circumstance which pulls the *Erh Ya* back nearly into the −4th century is its presence (or that of something very like it) in the haul of 'bamboo books' made in +281 when Pu Chun[2] opened the tomb of An Li Wang[3], a former ruler of the State of Wei (r. −276 to −245).[b] There is also a statement to the effect that in the Chhin and Han there were Professors (Po-Shih[4]) for the *Erh Ya* and the *Mêng Tzu*; and though it lacks the authority of the *Chhien Han Shu*, occurring only in the *San Fu Chüeh Lu*[5] (A Considered Account of the Three Metropolitan Cities) of Chao Chhi[6], he was himself a man of Han times (d. +201 aet. 90+) and therefore a fairly reliable witness.

[a] It is not now preserved, for the extant book of the same name is thought to be a 'forgery' by Sung Hsien[7], c. +1060. The *CHS* bibliography has *Hsiao Ya*.

[b] See *Chin Shu*, ch. 51, pp. 15b, 16a and Vol. 3, p. 507 above.

[1] 小爾雅 [2] 不準 [3] 安釐王 [4] 博士 [5] 三輔決錄
[6] 趙岐 [7] 宋咸

Attempts to date the different parts of the book separately began in due course, Lu Tê-Ming[1] suggesting, about +600, that only the Shih Ku section (ch. 1) was Chou Kung's, while all the rest was later. Subsequent scholars elaborated this.[a] The Chhien-Lung bibliographers, for example, regarded the work as one of many periods ranging from the time of Chuang Chou[2] (late −4th century)[b] to that of Yang Hsiung (late −1st century). The most complete analysis of this kind has been made by Naitō Torajirō[4], who regards the basis of the work as originating from the early Warring States period, with the Chi-Hsia Academy (−325 onwards)[c] having a considerable hand in it, the text being enlarged and stabilised in the Chhin and Former Han. He connects the Shih Ku section (ch. 1) with the first generations of the Confucian School (−450 to −400), places the Shih Chhin (family relationships) to Shih Thien (astronomy and meteorology) sections (chs. 4 to 8) in the time of Hsün Chhing[5] (−300 to −230) with additions as late as −90, allocates the Shih Ti to Shih Shui (geographical) sections (chs. 9 to 12) to the late Warring States, Chhin and beginning of Han (−300 to −200), puts the natural history chapters (Shih Tshao to Shih Shou, chs. 13 to 18) between −300 and −160, and finally ascribes the last section, Shih Chhu, on domestic animals (ch. 19) to the time of Han Wen Ti or Ching Ti, i.e. −180 to −140.[d] This solution fits one thing which many scholars have remarked upon,[e] namely the close association of the natural history chapters with the *Shih Ching* (Book of Odes) which reached its definitive form in the redaction of Mao Hêng[6] (*fl.* −220 to −150).[f] The plant and animal names in these ancient folksongs were of course explained by the *Erh Ya*.

But this at once evokes an earlier event. Confucius himself in the *Analects* recommended the study of the *Odes* partly because they acquainted people with natural history.[g]

The Master said: 'You young men, why is it that none of you study the *Odes*? They stimulate one's mind ... they may be used at home in the service of one's father and

---

[a] E.g. Tshao Sui-Chung[3] c. +1120.

[b] See abundant references on him in Vol. 2 above.

[c] See Vol. 1, pp. 95 ff.

[d] Lü Ssu-Mien[7] again has pointed out that the zoological sections mention some animals from Central Asia and some from the far North or Korea, so that these passages could hardly be earlier than the Warring States period.

[e] E.g. Lü Nan-Kung[8] c. +1070, Yeh Mêng-Tê[9] and Tshao Sui-Chung[10] c. +1120, followed by the Chhien-Lung bibliographers (*Ssu Khu Chhüan Shu Tsung Mu Thi Yao*, ch. 40). Khang Yu-Wei, who of course believed that Liu Hsin forged the *Erh Ya*, pointed out a close connection between its Shih Yo (music) section (ch. 7) and the *Chou Li*.

[f] The great philologist Yao Chi-Hêng[11], c. +1695, remarked also, adopting the view of Chêng Chhiao in the Sung (cf. p. 202 below), that some at least of the *Erh Ya* must be later than the *Li Sao* of c. −295, since it explains some of the phrases contained therein (*Ku Chin Wei Shu Khao*, p. 60).

[g] *Lun Yü*, XVII, ix, 1–7. Tr. auct. adjuv. Legge (2), p. 187; Waley (5), p. 212.

[1] 陸德明	[2] 莊周	[3] 曹粹中	[4] 內藤虎次郎	[5] 荀卿
[6] 毛亨	[7] 呂思勉	[8] 呂南公	[9] 葉夢得	[10] 曹粹中
[11] 姚際恆				

abroad in the service of one's prince; and they widen one's acquaintance with the names of birds, beasts, plants and trees.'[a]

Now since this is in the *Lun Yü* it has much higher authority than the *Ta Tai Li Chi*, and obliges us indeed to believe that in the closing years of the −6th century canons of botanical and zoological nomenclature were being actively discussed by the learned. At the same time it goes some way to validate the *Ta Tai Li Chi*'s tradition, suggesting quite strongly that even Naitō Torajirō may have been somewhat too cautious—perhaps there really was some kind of lexical *Erh Ya* text in −500 which included definitions of natural history names and brief descriptions of their owners.

To sum up, therefore, the best dating for this first of all Chinese encyclopaedias, containing so much botanical and zoological terminology and nomenclature, is between the −4th and the −2nd centuries; provided that we do not exclude the possible existence of some nucleus of the work as early as the latter half of the −6th century, and the continued addition of material to it as late as the end of the −1st.

Although the only ancient *Erh Ya* commentary which has come down to us is that of Kuo Pho (*c.* +310), there were quite a number of others in early times.[b] Fan Kuang[1] did one, a second bore the name of Liu Hsin[2],[c] and a third was made by a court pundit called Li Hsün[3], all in the Han. The great grammarian Sun Yen[4] produced another in the San Kuo period (late +3rd century), and a court chamberlain of the Liang, Shen Hsüan[5], collected together all the commentaries early in the +6th. A special work on the pronunciation of the rarer characters had been written by Chiang Tshui[6] under the Chin or Liu Sung. Most of these books were still extant in the Thang but all had disappeared completely by the Sung. Noteworthy for the history of botanical illustration is the fact that Kuo Pho prepared an *Erh Ya Thu Tsan*[7] (Pictures with Explanations for the *Literary Expositor*), but (unless there was a change of title) the *Explanations* had gone by the Sui and the *Pictures* were lost by the Thang. Though all these works have long been swallowed up by time, some of their knowledge was doubtless handed down in other texts, and it is good at least to know the names of men who were certainly

---

[a] Waley believed that the 'names' here referred to were those correct in the ritual speech of the princely courts rather than local dialect peasant names; if so, we see perhaps one of the origins of the learned as opposed to the common nomenclature (cf. p. 143 above).

[b] See *Sui Shu*, ch. 32, p. 27b.

[c] Astronomer and bibliographer, see Vol. 3 *sub voce, passim*. A Shê-jen[8] is mentioned in the commentary, and Bretschneider (1), vol. 2, p. 21, took this to refer to Liu Hsin, but mistakenly. Shih Shêng-Han (4), p. 14, noticed that the *Chhi Min Yao Shu* (Important Arts for the People's Welfare, *c.* +540) quotes an *Erh Ya* commentary by Chien-wei Shê-jen[9], 'Mr. Secretary from Chien-wei' (a place in Szechuan). This explains both the rather obscure wording in the *Sui Shu* bibliography and the mysterious secretary in Kuo Pho's commentary. Whatever his real name, he was probably a man of the +1st century, and he is said to have been the first to punctuate the text.

[1] 樊光　　　[2] 劉歆　　　[3] 李巡　　　[4] 孫炎　　　[5] 沈璇
[6] 江灌　　　[7] 爾雅圖讚　[8] 舍人　　　[9] 犍爲舍人

more than literary scholars, for no-one unskilled in natural history could have ventured to tackle the *Erh Ya*.

Let us now take the *Erh Ya*'s derivative literature as the main line of descent of the encyclopaedia in China, and see what branched off from it at different times. The *Hsiao Erh Ya* we have already mentioned, but far more important was the *Kuang Ya*[1] (Enlargement of the *Literary Expositor*) produced by Chang I[2] in +230.[a] A glance at its biological chapters suffices to show that most of the entries are original and different, with little overlap, so that the 334 plants and trees of the classic were probably almost doubled by Chang I. There is rich material here for the history of botany, but hardly yet touched, so far as we know.[b] After Kuo Pho's work in the following century the biological nomenclature of the *Erh Ya* rested in authoritative repose for seven hundred years, but in the early Sung time there was a great revival of interest in it. Towards +1000 Hsing Ping[3] produced his *Erh Ya Chu Su*[4] (Explanations of the Commentaries on the *Literary Expositor*), pouring into it many descriptions and fresh quotations drawn from the *pên tshao* writings (cf. p. 220 below) down to the Wu Tai period, as well as from ordinary literature. A century later Lu Tien[5] continued the work, in two books, the *Phi Ya*[6] (New Edifications on the *Literary Expositor*) datable at +1096, and *Erh Ya Hsin I*[7] (New Interpretations of the *Literary Expositor*), +1099.[c] The movement ended with Lo Yuan's[8] *Erh Ya I*[9] (Wings for the *Literary Expositor*) of +1174, until modern scholarship returned to the subject. Works of this kind might be considered indispensable for any serious investigations of Chinese botany and zoology in detail, but really such studies hardly as yet exist.[d]

It was natural that the *Erh Ya* should have been among the books chosen when the classics were for the first time printed. From Section 32 it will be remembered that by the Later Thang[e] emperor's edict of +932, issued in the premiership of Fêng Tao[14], a commission was set up to carry out the task.[f] Under the adminis-

---

[a] A corrected text with much commentary was published by Wang Nien-Sun[10] in +1796 (*Kuang Ya Su Chêng*[11]).

[b] Nothing has appeared by modern writers at all equivalent to the work of Bretschneider (1) and Hsia Wei-Ying (1) on the *Erh Ya*.

[c] All these books follow the order of entries in the *Erh Ya*, but that of Hsing Ping is much more copious than those of his successors. Here belongs, perhaps, the *Shu Hsü Chih Nan*[12] (Literary South-Pointer, or Compass) of Jen Kuang[13], much less systematic, indeed rather muddled and arbitrary, yet containing a wealth of synonymic definitions of plant and animal names which have never received modern study.

[d] Certain late commentators, moreover, were rather good naturalists, notably Hao I-Hsing[15] (+1757 to 1825), whose *Erh Ya I Su*[16] (1808 to 1822) has been discussed by Chang Yung-Yen (1). Hao showed a scientific spirit quite unusual for his time.

[e] This State held all the Yellow and Wei river valleys, together with the southern part of Shansi.

[f] See *Tshê Fu Yuan Kuei*, ch. 608, p. 29b; and *Yü Hai*, ch. 43, p. 10b, which mentions the *Erh Ya* specifically. Although the work was seen at the time as a cheap substitute for engraving the texts on stone steles, it was in fact a truly epoch-making one.

[1] 廣雅	[2] 張揖	[3] 邢昺	[4] 爾雅注疏	[5] 陸佃
[6] 埤雅	[7] 爾雅新義	[8] 羅願	[9] 爾雅翼	[10] 王念孫
[11] 廣雅疏證	[12] 書叙指南	[13] 任廣	[14] 馮道	[15] 郝懿行
[16] 爾雅義疏				

Fig. 30. A page from the botanical section (Shih Tshao) of the edition of the *Erh Ya* printed in +953.

Fig. 31. The colophon of Li Ê to this edition.

tration of the Rector of the Imperial University (Kuo Tzu Chien[1])[a] Thien Min[2], and the scholarly superintendence of one of the professors, Ma Kao[3],[b] the eminent calligrapher Li Ê[4] was asked to write the script for the block-carvers, and all was finished by +953.[c] Now the only example of this truly first of editions that has survived is a copy of the *Erh Ya* itself, preserved in Japan, re-discovered by a Chinese ambassador there, and printed at the embassy in 1884.[d] Here in Figs 30, 31 we reproduce an opening of this beautiful book at the section on herbaceous plants, and beside it a portion of the last page carrying Li Ê's colophon.

[a] Cf. p. 274 below.

[b] Author of the *Chung Hua Ku Chin Chu*; cf. Vol. 4, pt 1, p. 274.

[c] The story is told in Carter (1), rev. Goodrich, pp. 69 ff., cf. Mao Chhun-Hsiang (1), pp. 20 ff. This was not the only printing of the classics at this time. From Sect. 32 it will be remembered that Wu Chao-I[5], the cultural leader in the adjoining southerly State of Shu, which had its capital at Chhêngtu, also produced an edition, starting probably about +944 and finishing probably in the same year (+953); cf. *TCTC*, ch. 291 (p. 9495). We do not know whether the *Erh Ya* was among the books which Wu Chao-I printed, nor have any simulacra of them survived. One need hardly stress that all this was five hundred years before Gutenberg and Caxton.

[d] We have been privileged to use constantly a copy of this Ku I Tshung Shu[6] reprint which Dr Lu Gwei-Djen bought in Hongkong in 1954.

[1] 國子監   [2] 田敏   [3] 馬縞   [4] 李鶚   [5] 毋昭裔
[6] 古逸叢書

According to expert opinion, based partly on stylistic evidence,[a] the Sung print from which this copy was taken had most probably been traced from the +10th-century original as a very exact reproduction, with only the few tabu characters changed.[b] We can thus take it to be essentially a replica of one of the volumes of the set that Wang Ming-Chhing[1] had in his home about +1140 and which he remembered in old age in his *Hui Chhen Lu*[2].[c]

It may be worth while to remind ourselves here that in previous contexts we have come across important works which show the penetration of the expository genre into specialised fields. Towards the close of the Thang period, in +806, Mei Piao[3] finished his *Shih Yao Erh Ya*[4], a title which we could translate as 'The Literary Expositor of Minerals and Drugs', or, more freely but equally truly, 'A Synonymic Dictionary of Chemical Physic'.[d] This was a book of cardinal importance in the history of alchemy and early chemistry in China, but we need only recall now the detailed treatment which we gave it at an earlier stage.[e] Presently (Sect. 44) we shall come across another title of the same kind, *Chhuan Ya Nei Pien*[5], the book which some thousand years later described the medical skills of the traditional rural medical practitioners.[f]

(ii) *Handy primers*

Our next plotted course is to take us through the genres of encyclopaedias which might be considered as deriving from the *Erh Ya*, but before embarking on this there must be a short excursus on certain lexical books almost as ancient as that. Singularly little attention has been given by sinologists to the small book called *Chi Chiu (Phien)*[6],[g] a curious work undoubtedly of Former Han date, composed by a court scholar, Shih Yu[7], between −48 and −33. It consists of 32 sections (*chang*[8]), all regarded as forming a single chapter. One hardly knows how to render its title, unless 'Handy Primer', for it is really a series of classified orthographic word lists, connected together with a thread of continuous text containing brief explanations. It was intended for the young (and perhaps also the not-so-young) to study the meanings of characters and how they should properly be written, it was suitable for memorising and recitation,[h] it formed a basis for verbal

---

[a] See Pelliot (51), pp. 316 ff.; Wang Kuo-Wei (5), pp. 143 ff.

[b] It was impermissible to print characters forming part of the name of the reigning emperor without omitting a stroke or so.

[c] Yü Hua sect. (p. 310).　　[d] *TT* 894.　　[e] Vol. 5, pt 3, pp. 151 ff.

[f] Meanwhile see Needham (64), pp. 265, 352, 391.

[g] The last word in this and a number of shortly following titles is bracketed because it is not really part of the title, though even in early times it was sometimes appended. The *CHS* bibliography says: '*Chi Chiu* i phien', i.e. 'Handy Primer, in one chapter'. The same applies to the rest. Cf. Shen Yuan (1).

[h] An obvious parallel is the later list of family names (*Pai Chia Hsing*) composed early in the Sung and repeated by many generations of boys at school. The *San Tzu Ching* (cf. Vol. 2, p. 21) was of course also memorised, but differed because of its considerable doctrinal content.

¹ 王明清　　　² 揮塵錄　　　³ 梅彪　　　⁴ 石藥爾雅　　　⁵ 串雅內編
⁶ 急就篇　　　⁷ 史游　　　⁸ 章

elaborations by the teacher, and could have served as a handy reference manual for scribes and copyists. The book has thus preserved a goodly number of technical terms and names of plants, animals, diseases, tools, materials and objects, which are important for the history of science, medicine and technology. By way of example one may take the following from *chang* 24; just following a list of diseases:

By moxa, acupuncture and the compounding of drugs we may drive out the malign (*chhi, pneumata*, that cause illness). (Of drugs and drug-plants there are:) *huang chhin, fu ling, yü, chhai hu, mu mêng, kan tshao, wan, li lu, wu hui, fu tzu, chiao, yuan hua, pan hsia, tsao chia, ai*[1]....[a]

These fifteen are quite easy to identify—they run as follows: *Scutellaria macrantha, Pachyma cocos*, the mineral arsenolite (arsenic oxide), *Bupleurum falcatum, Glycyrrhiza glabra, Aster tataricus, Veratrum nigrum*, two species of *Aconitum, Zanthoxylon* sp., probably *piperitum*, used before East Indian pepper came in, the strongly poisonous *Daphne genkwa, Pinellia ternata*, the soap-bean tree *Gleditsia sinensis*, and one or other species of *Artemisia*, used for moxa tinder.[b] One thus gets some idea of the lists which the book contains, and their value for those who wished to write correctly. In later centuries abundant explanations were given in the commentaries of Yen Shih-Ku[2] *c.* +620 and of Wang Ying-Lin[3] *c.* +1280.[c]

The penning of the *Chi Chiu (Phien)* by the famous calligrapher Wang Hsi-Chih[4] in the +4th century doubtless helped to conserve its text, which was thus copied by generation after generation of literati eager to perfect their hand. But though a unique survival, it was only one of a number of similar books which had circulated widely during the Chhin and Han periods.[d] The term *hsiao hsüeh*[5] (lesser learning) was now applied to fundamental knowledge of this kind.[e] As part of the Chhin programme of standardisation[f] and education for administration the great minister Li Ssu[6] produced about −220 an orthographic primer called *Tshang Chieh (Phien)*[7]. By the time of the *Chhi Lüeh*[8] bibliography (−6),[g] certainly by the time of that in the *Chhien Han Shu* (*c.* +100), this work had absorbed two

[a] Tr. and id. auct.

[b] As soon as one begins to check these identifications, one notices a parallelism with the *Shen Nung Pên Tshao Ching*, the earliest Chinese pharmaceutical natural history book, which contains them all. The dating of this we shall discuss presently, but such a correspondence adds weight to the belief that it must be at least −1st century. Cf. pp. 235 ff. below.

[c] Contemporary Chinese scholarship admires the *Chi Chiu Phien* because of its high factual content as contrasted with the much more moralising tendencies of similar later works such as the *San Tzu Ching*. (cf. Vol. 2, p. 21, Vol. 4, pt 3, p. 295). See Shen Yuan (*1*).

[d] A good account of the whole background may be read in Bodde (1), pp. 157 ff., where there is a translation of the basic relevant text, *Chhien Han Shu*, ch. 30, pp. 13b ff.

[e] In the Chou period this term had meant the Six Arts (*liu i*[9]) of ceremonial observance, music, archery, chariot-driving, writing and calculating; but now it was confined to the fifth of these.

[f] Cf. Vol. 2, p. 210, Vol. 4, pt 2, p. 250.

[g] Cf. van der Loon (1); Gardner (3), p. 33.

[1] 黃芩,伏苓,礜,柴胡,牡蒙,甘草,菀,黎蘆,烏喙,附子,椒,芫華,半夏,皁莢,艾
[2] 顏師古   [3] 王應麟   [4] 王義之   [5] 小學   [6] 李斯
[7] 蒼頡篇   [8] 七略   [9] 六藝

Fig. 32. Tshang Chieh the Nomenclator, with his four eyes, talking with some un-named person, probably Shen Nung, who holds a long-stalked plant; a scene from the I-nan reliefs (Tsêng Chao-Yü *et al.* (*1*), pl. 52).

Fig. 33. Tshang Chieh holding up a plant to Shen Nung, who is tasting another (perhaps of the same species). In the middle is Lao Tzu, then Confucius, and finally a disciple (*ti tzu*) holding a book of bamboo slips. Two other people of obscure name and story stand on the right, omitted here. A scene from one of the Szechuanese reliefs on stone boxes and coffins (Wên Yu (*1*), pl. 43.

others, the *Yuan Ma (Phien)*[1] (Explanation of Difficult Words) composed by the Grand Master of Equipages, Chao Kao[2], and the *Po Hsüeh (Phien)*[3] (Extensive Knowledge of Words) by the Chronologer- or Astronomer-Royal (Thai Shih Ling[4]) Humu Ching[5], both officials of the Chhin dynasty (*c.* −215). Since these are lost,[a] we do not know whether they all contained natural history names, or whether one specialised in one field and one in another. The title of Li Ssu's book needs a word of explanation, but this is not difficult, for Tshang Chieh[6] was a legendary person, none other than the reputed inventor of writing, one of the two 'recorders' or scribes of the mythical emperor Huang Ti, and soon to be, if not already, deified as its tutelary spirit (*tzu shen*[7]). Hence the interest of Fig. 32, taken from the I-nan tomb reliefs of *c.* +193, which shows Tshang Chieh, with his traditional four eyes, sitting under a tree and actually discussing botanical names, for an unidentified interlocutor sits before him and holds up a plant with a long stem.[b] And this demonstrative interest in botany, however symbolic, can be no coincidence, for in Fig. 33 we see Tshang Chieh again in a Han relief from Szechuan, holding up a plant and deliberating on its characteristics with Shen Nung himself (Fig. 34).

These Chhin and Han 'primers' were closely connected, in ways we cannot go into here, with the changes in the written script of Chinese which led to its standardisation and permanent stabilisation.[c] They were all apparently modelled on the *Shih Chou (Phien)*[8], a really ancient orthographic glossary which we shall meet with again (p. 232 below) in a different connection—'Chou the Chronologer-Royal, his book, in 15 chapters'—dating from between −827 and −782. It was doubtless known and used by Confucius but did not survive the Han.[d] However, during that period and after, the writings of the three Chhin experts were supplemented by others. About −140 Ssuma Hsiang-Ju[9][e] drew up a word list containing more characters but no duplicates and called *Fan Chiang (Phien)*[10] (Most Important Phrases);[f] then came Shih Yu's *Chi Chih (Phien)* of *c.* −40,

---

[a] A few bamboo strips, with about 40 characters in all, found on the *limes* in Kansu, constitute the only extant remnants of the once so popular *Tshang Chieh (Phien)*; cf. Wang Kuo-Wei (6). More recently Loewe (4), vol. 2, pp. 418 ff., has reported other fragments of the *Chi Chiu (Phien)* as well, and even some tablets the inscriptions on which were evidently exercises in copying.

Besides these, a reconstitution from all remaining literary fragments was undertaken by Ma Kuo-Han, and will be found in *YHSF*, ch. 59, pp. 18a ff. Whatever remains of the work of Chao Kao and Humu Ching is included in it. There are a certain number of words connected with plants and animals; as also a considerable number of dictionary definitions, but these seem to be due to the editorial activities of Chang I about +230 and then Kuo Pho about +300.

[b] Tsêng Chao-Yü, Chiang Pao-kêng & Li Chung-I (1), pl. 52.

[c] Bodde (1), *loc. cit.* explains these. He also notes certain doubts about the name of Shih Chou; the two words can also be interpreted as 'Historical Readings' rather than as a title and a personal name. But Hsü Shen, in his preface to the *Shuo Wên* (+121) certainly understood it in the latter sense.

[d] A partial reconstruction, however, is in *YHSF*, ch. 59, pp. 3a ff.

[e] Known best as a writer of poetry and prose, he was also an important builder of roads, cf. Vol. 4, pt 3, pp. 25, 36.

[f] Some earlier person, Chiang Fu-Shih[11], seems to be alluded to in this title, but we find *KHTT*, p. 222 obscure in its reference *sub voce chiang*. There is a fragmentary reconstruction in *YHSF*, ch. 60, pp. 3a ff.

---

[1] 爰應篇   [2] 趙高   [3] 博學篇   [4] 太史令   [5] 胡母敬
[6] 蒼頡   [7] 字神   [8] 史籀篇   [9] 司馬相如   [10] 凡將篇
[11] 將甫始

Fig. 34. Shen Nung (on the left) and Tshang Chieh; two standing figures in carved ivory (date uncertain, Wellcome Historical Medical Museum). Shen Nung wears his unmistakable garment of leaves, and behind Tshang Chieh can be seen gourds, traditional containers of drugs.

followed by a *Yuan Shang* (*Phien*)[1] (Ancient Traditional Terms) written by the Chief Engineer (Chiang Tso Ta Chiang[2]) Li Chhang[3] between −32 and −7. One cannot but remark here the prominence of men in the compilation of these word and name list who were obviously technical men, but this would have been natural enough in the period of crystallisation of the technical language.

In +5 the philologists of the empire, being gathered together at the capital, more than 100 of them, as part of a national cultural and scientific congress,[a] were asked to compile lists of the most useful characters, and these were pruned, conflated and edited by Yang Hsiung[4], forming the *Hsün Tsuan* (*Phien*)[5] (Instruction on Selected Words).[b] Later in the +1st century an addition to it was made by the great historian Pan Ku.[6] Meanwhile many of the sounds of the rarer words and names in the *Tshang Chieh* (*Phien*) had been forgotten, so about −60 a

---

[a] Fuller details about this will be found on p. 232 below.
[b] Fragmentary reconstruction in *YHSF*, ch. 60, pp. 6*a* ff. No definitions.

[1] 元尚篇   [2] 將作大匠   [3] 李長   [4] 揚雄   [5] 訓纂篇
[6] 班固

provincial expert, Chang Chhang[1], was brought to court and entrusted with the handing down of these. This knowledge he transmitted to one of his grandsons, Tu Lin[2], who before +47 wrote two books on the subject;[a] these lasted down to the Liang but were lost by the Sui.

In looking back over this branch of literature, if such it could be called, one has to reflect on the importance of correct character writing in an ideographic language aiming at any efficiency of information transfer. If someone recites a string of plant names such as 'ash, bedstraw, coltsfoot, daisy, elecampane, etc.' in an alphabetical language, the spelling by a writer will not diverge greatly because the script closely follows the sound, but in Chinese this is not entirely so. A given sound, as we saw at the beginning,[b] could be attached to a large number of written characters; Chinese was always rich in homophones. Conversely, a particular character, though not capable of bearing *any* sound, might appear as speech within a considerable range of morphemes, e.g. *yuan*[3] above[c] might have been *huan*, *nuan* or even *yün*.[d] These areas of phonetic fluctuation made the learning of a sophisticated terminology and nomenclature for the observational sciences quite difficult, especially in the earlier stages of its development, so that the orthographic word lists at the beginning of classical Chinese culture had a really important part to play. Before Li Ssu there undoubtedly existed a wide and confusing variety of word forms, the same word being written in several different ways; he and his colleagues simplified, standardised and disseminated. We have already had one instance of what they were up against, in the early failure to give a *tshao-thou* (the 'grass' or 'herb' radical) to plant names the characters of which clearly needed it (pp. 118 ff. above). Moreover, for the continued understanding of ancient texts such as the *Shih Ching*[e], and the handing down of the fund of ancient botanical and zoological knowledge, it was essential that the ancient names should be fixed and expounded. This was what the *hsiao hsüeh* experts, strongly encouraged by successive governments, succeeded in doing. Chinese natural history owes a great debt to these ancient systematisers.

Before returning to the main line of development, mention must be made of two Han books of lexicography which had no exact successors, the *Fang Yen*[4] and the *Shih Ming*[5]. The *Fang Yen*, finished by Yang Hsiung[6] in −15, was a dictionary of dialect expressions intended as a contribution to the unification of the language,[f] and indeed to some considerable extent in the *Erh Ya* tradition, as one can see from p. 143 above. Needless to say, the *Fang Yen* can provide priceless

---

[a] Cf., on him, Vol. 4, pt 2, p. 265, in a very different context. The fragmentary reconstruction in *YHSF*, ch. 60, pp. 9a ff. comes from materials which seem not to have been edited in the +3rd century, so as a number of definitions and explanations are included they must be Tu Lin's own, thus showing how the dictionary form gradually developed out of the orthographic word-list.

[b] Cf. Vol. 1, pp. 34, 36, 40.   [c] In *Yuan Ma Phien*, p. 197 above.

[d] Cf. Vol. 1, p. 33.   [e] Cf. here p. 463 below.

The best text is that of the *Fang Yen Su Chêng*[6], prepared by Tai Chen[7] in +1777.

[1] 張敞   [2] 杜林   [3] 爰   [4] 方言   [5] 釋名

[6] 揚雄   [7] 方言疏證   [8] 戴震

insights into Han words and things,[a] but for some reason Yang Hsiung had almost nothing to say about plant names, though a whole chapter is devoted to those of animals.[b] The *Shih Ming* (Explanation of Names; or, Interpretation of Concepts) was a strictly subject-classified encyclopaedia, rather later in date, begun by Liu Chen[1] about +100 and finished by Liu Hsi[2], whose name alone it usually bears, *c.* +180, perhaps +196. It was important in that it set the style for later classification in its 27 sections,[c] but unfortunately these did not include any natural history, the nearest department being the diseases of man, in a text admittedly of inestimable value for the history of pathology in Chinese culture though not so far adequately used by historians of medicine.[d] From our present point of view, therefore, the *Shih Ming* was a decline from the rich biological interest of the *Erh Ya*.

Still, they do both raise the question of the structuralisation of meaning.[e] Can we divide all knowledge into compartments in order to move from ideas and things to linguistic forms, instead of doing what most lexicographic works do, i.e. moving from forms to ideas and things? The *Onomasticon* of Pollux of Naucratis (not now extant) was a +2nd-century work of curiously Chinese flavour, for it was structured in just their way; starting with the gods, it proceeded to man, man's body, kinship terms, science, art, hunting, food, trade, law, administration, and utensils. So also in the +17th century the great Czech educationalist, Jan Amos Komenský (Comenius) had the idea of a *Theatrum Universitatis Rerum* which would keep the minds of children clear from the beginning, and pass from things to words, not vice versa.[f] And in a way the same line of thought has given rise to P. M. Roget's *Thesaurus of English Words and Phrases* (1852), which lies beside so many desks in the world, and has become so familiar that no-one pauses to notice the similarity between his classification system and those of the old Chinese lexicographers.

(iii) *Imperial florilegia*

After the Han, the encyclopaedia genre began to fork in several directions. The first was the tradition of imperial florilegia. Quotations of earlier texts had been

---

[a] See for example Vol. 4, pt 2, p. 267 above.

[b] Possibly the reason for this was philological, since Yang Hsiung devoted at least as much attention to words which normally bore the function of verbs and adjectives as he did to nouns.

[c] I.e. heaven, earth, mountains, waters, hills, roads, countries, man's body, its actions, his ages, relationships, speech, sustenance, colours and textiles, ornaments, clothing, buildings, camps and tombs, books and records, classics and arts, tools and utensils, musical instruments, war weapons, vehicles, ships and boats, diseases, mourning. The best text is the *Shih Ming Su Chêng Pu*[3]; cf. Wang Hsien-Chhien (3).

[d] Exapt Yü Yün-Hsiu (1). We have had occasion to use the *Shih Ming* often with advantage; cf. Vol. 4, pt 2, pp. 86, 96, Vol. 4, pt 3, pp. 600, 623, 639, 680.

[e] Cf. Potter (1), pp. 173 ff.

[f] Cf. Needham (63). The *lei shu* of this name was begun in +1612 and almost entirely destroyed in the sack of Leszno in +1656, but enough has remained to show its nature. For other aspects of Comenius and China cf. Chêng Tsung-Hai (1); Pokora (12).

[1] 劉珍   [2] 劉熙   [3] 釋名疏證補

made in the *Erh Ya* commentaries, but they were sparingly used and very brief;
now, however, in the +3rd century whole collections of them began to be put
together under appropriate heads for the convenience of the emperor and his
ministers. The first of these was the *Huang Lan*[1] (Imperial Speculum) commissioned by Tshao Phei[2] (Wên Ti of the Wei State)[a] and edited by Miu Shih-
Têng[3] about +220. Soon after +400 this was enlarged by an old friend of ours,[b]
the astronomer Ho Chhêng-Thien[4]. Later in the +5th century another work of
the same kind was prepared under the Northern Wei dynasty by Tsu Hsiao-
Chêng[5], the *Hsiu Wên Tien Yü Lan*[6] (Imperial Speculum of the Hall of the
Cultivation of Literature). These and others of similar times failed to survive, but
we can gain a good idea of what they must have been like from the magnificent
encyclopaedia produced by Li Fang[7] in +983, so often quoted in these pages, the
Thai-Phing Yü Lan[8] (Imperial Speculum of the Thai-Phing reign-period, i.e. the
Emperor's Daily Readings). Rounded off to 1000 chapters, natural history brings
up its rear, forming nearly 12 % of the whole, with a subject distribution as
follows:

	CHAPTERS		CHAPTERS
herbaceous plants	7	insects and worms	8
drug plants	10	reptiles and fishes	15
cereals	6	birds	15
vegetables	5	mammals	25
aromatic plants	3		
bamboos	2		
trees	10		
fruit trees	12		
	55		63

Since well over 2000 books were excerpted for this florilegium, and since some
70 % of them no longer exist as such today, the value of the collection for all
aspects of the history of science and culture is inestimable. The *Thai-Phing Yü Lan*,
however, stands rather alone as a monument in Chinese literature because of its
many-sidedness, and nothing quite like it was ever done again. Already by the
time of its compilation the florilegium genre was being polarised, as it were,

---

[a] The name of this ruler is always recurring in contexts of interest for the history of science and proto-science;
cf. Vol. 3, p. 659 (asbestos) and Vol. 4, pt 1, p. 327 (astronomical chess). He re-established the Imperial
University and organised a botanical and zoological garden (cf. Sect. 38 (*i*), 2). The *San Kuo Chih* bibliography
(Yao Chen-Tsung (*2*), p. 3263) records Tshao Phei himself as editor, and indeed he may have been the
chairman of an editorial board as well as doing some of the work himself, for he was clearly a man of real
intellectual curiosity and learning, not without scepticism.

[b] See Vol. 3, *passim*.

[1] 皇覽　　　[2] 曹丕　　　[3] 繆卜　　　[4] 何承天　　　[5] 祖孝徵
[6] 修文殿御覽　　[7] 李昉　　　[8] 太平御覽

and attracted irrevocably into the field of government administration.[a]

Liu Chih[1] was the man who first used the encyclopaedia of quotations as an aid for the bureaucracy, in his *Chêng Tien*[2] (Government Institutes) of +732. Most of this was incorporated into the larger *Thung Tien*[3] (Comprehensive Institutes) finished by Tu Yu[4] in +812.[b] From these works natural history was entirely excluded (perhaps in conformity with one of the less happy ancient Confucian attitudes)[c], as also from the later important production *Tshê Fu Yuan Kuei*[5] (Lessons from the Archives; the True Scapulimancy), edited by Wang Chhin-Jo[6] and Yang I[7] in +1013. Their book was a collection of material on the lives of emperors and ministers with a specifically moral purpose, illustrating the 'praise and blame' conception of history. One of the writers of this group, however, had a more original approach; this was Chêng Chhiao[8] whose *Thung Chih*[9] (Historical Collections) was finished about +1150 in the comparative peace of the south, after the capital had been yielded to the Chin Tartars and the government transferred beyond the Yangtze. The first part of this noble work was a continuous history of China down to the end of the Sui, and the last a great collection of biographies, while in between we find 20 'summary monographs' (*lüeh*[12]) modelled on those in the usual dynastic histories, but distinctly different from them. Nine indeed were on the usual subjects,[d] but the other eleven were new and unique.[e] Among them figured one on natural history, *khun chhung tshao mu*[13], animals, plants and trees, divided into two chapters[f] and eight categories (*lei*[14]): herbaceous plants, vegetables, beans and cereals (*tshao, shu, tao liang*[15]) then trees, fruit-trees, insects and fishes, birds, and mammals (*mu, kuo, chhung yü, chhin, shou*[16]). The entries are reminiscent of the *Erh Ya*, equipped with a copious synonymy but having only brief descriptions of shape and colour.[g]

[a] See the valuable monograph of Balazs (9).
[b] Here there is a close connection with the trend in Chinese historiography away from purely dynastic histories to narratives of continuous history covering long periods of time. Cf. Sect. 49 in Vol. 7, and meanwhile Needham (55, 56).
[c] See Vol. 2, p. 9. The same applies to the greatest work of the kind, Ma Tuan-Lin's[10] *Wên Hsien Thung Khao*[11] (Comprehensive Study of (the History of) Civilisation), published in +1319.
[d] I.e. astronomy, geography, ceremonies, music, bureaucracy, execution of justice, economic life, bibliography, and natural disasters.
[e] I.e. onomastics (clan-names), script, linguistics, towns and cities, posthumous titles, vestments and equipages, administrative memorialisation, textual criticism, geography and cartography, epigraphy and archaeology, together with natural history.
[f] Chs. 75 and 76 of the whole work, 51 and 52 of the *Thung Chih Lüeh*.
[g] Besides this admission of natural history, the subject was also included in the bibliography (*i wên lüeh*[17]), ch. 69 of the whole work, pp. 9*a* ff. Here were listed 39 books on *pên tshao*, pharmaceutical botany and zoology in general, 6 on the proper pronunciations of the taxonomic names, 6 books of illustrations, 26 books on the use of drugs (*yung yao*[18]) and 5 on pharmacognosy and the collection and preparation of *materia medica* (*tshai yao*[19]). In the +18th century all the 'Thung' books were given continuations bringing them down to the Ming and Chhing periods. Thus the *Hsü Thung Chih*,[20] complete to the end of the Ming, appeared about +1770, edited by Hsi

[1] 劉秩	[2] 政典	[3] 通典	[4] 杜佑	[5] 册府元龜
[6] 王欽若	[7] 楊億	[8] 鄭樵	[9] 通志	[10] 馬端臨
[11] 文獻通考	[12] 略	[13] 昆蟲草木	[14] 類	[15] 草, 蔬, 稻, 粱
[16] 木, 果, 蟲, 魚, 禽, 獸		[17] 藝文略	[18] 用藥	[19] 採藥
[20] 續通志				

Before this time it had been quite unheard-of to include anything on natural history in the great dynastic histories—and it must be admitted that the experiment was never repeated. There can be little doubt, however, as to whose ideas Chêng Chhiao was following. It was none other than Liu Chih-Chi[3], whose *Shih Thung*[4] (Summa Historiae) of +710 has been well described as the first treatise on historiography in Chinese or in any other civilisation.[a] Liu's ideas on the monographs or memoirs (*chih*[5]) in the dynastic histories;[b] what was suitable for them and what was not, were original and interesting. He was in favour of abbreviating, or omitting altogether, three of the traditional monographs, those on astronomy, bibliography and portents;[c] hence his views have a special interest for the history of scientific thought. For him, astronomy was a domain of ceaseless recurrences and eternal unchangingness, therefore discussions of it should not appear in books of history, though there was no harm in recording eclipses, comets and other unusual events which occurred during particular reigns or dynasties.[d] History consisted of non-repetitive changes while the stars in their courses were the same yesterday, today and for ever. Here Liu Chih-Chi overlooked two things; the heavens do in fact undergo saecular changes, and human knowledge of them was changing very fast by comparison,[e] so it was fortunate indeed that the majority of Chinese historians continued to insert memoirs on astronomy into their dynastic histories. But by contrast Liu recommended the inclusion of monographs on natural history, not indeed with any suspicion of biological evolution in mind, but rather because knowledge of plants and animals continually broadened as the surrounding tribal peoples and the rulers of countries far away sent tribute missions with strange presents to the Chinese imperial throne.[f] In other words, he would have banished basic and calendrical astronomy to special treatises, and confined natural history to those of *pên tshao* type,

Huang.[1] The natural history *lüeh* were greatly enlarged and are full of descriptions, forming a treasure which has not yet at all been drawn upon (chs. 174 to 180 inclusive). The bibliography, however (ch. 161, p. 4315.3), was increased very little. Later, about +1786, there appeared the *Chhing Chhao Thung Chih*[2] also edited by Hsi Huang[1]. Here the additions (chs. 125 and 126) were less significant, and the bibliography (ch. 102, p. 7334.2) has little new.

[a] See the fascinating study of him by Pulleyblank (7).

[b] Cf. Vol. 1, p. 74.

[c] These last were generally inserted under the rubric of *Wu hsing*[6], the Five Elements, since portents and natural disasters were thought of as disturbances in the normal interplay of the elements. Cf. Vol. 2, p. 247 ff.

[d] *Shih Thung*, ch. 8, p. 3*b*.

[e] Between −370 and +1742 no less than 100 'Calendars' or sets of astronomical tables were produced, embodying constants of ever greater accuracy, and dealing with the determination of solstices, day, month, and year lengths, the motions of sun and moon, planetary revolution periods, and the like. Each therefore represented an improvement on its predecessors, and every new emperor wanted to issue a new and better one. There was hardly one mathematician or astronomer of any eminence in ancient and medieval China who was not called upon at some time or other to help in the reconstruction of the calendar. In Vol. 3, pp. 390 ff. we minimised too much the importance of this great succession of ephemerides, but Yabuuchi Kiyoshi (6, 9, *1, 7, 9, 14, 15*) has gone far to repair this deficiency.

[f] *Shih Thung*, ch. 8, p. 10*b*.

[1] 劉知幾　　[2] 史通　　[3] 志　　[4] 秥璜　　[5] 清朝通志
[6] 五行

admitting only the occasional celestial happening on the one hand, and the appearance of unusual plants and animals on the other. Here he certainly struck a blow for exotic botany and zoology,[a] but Chêng Chhiao was the only historian who followed him to the extent of including a monograph on natural history in his work.

The reader may remember that we came across a contrast of a somewhat similar kind at an earlier stage, in the Section on seismology (Sect. 24) in Vol. 3. About +1290 the distinguished scholar Chou Mi[1] (+1232 to c. +1308) had difficulty in understanding how the seismographs of Chang Hêng and his successors could possibly have worked.[b] He could see that one could observe, measure and predict the regular motions of the heavens, but not perhaps realising the imperceptible passage of a shock wave in the earth he could not see how an instrument, even one set like a trap, could successfully measure the strength and direction of a perfectly unpredictable collision of chhi at a great distance away. There was a mental contrast here also between phenomena which show great natural regularity and those which require statistical methods for their treatment, perhaps even between the concepts of necessity and chance as leading principles in Nature.[c] So in an earlier century Liu Chih-Chi had wanted to remove systematic astronomy from the dynastic histories and to insert exotic natural history instead, for the unpredictable and unrepeatable seemed to him history while the regular and familiar was science.

As for the other suggestions of Liu Chih-Chi, the bibliographies seemed to him very overdone.[d] It was not necessary, he thought, to know the names of thousands of unimportant books, especially when they had long been lost.[e] The omens and portents, too, occupied unconscionable space.[f] Although Liu did not entirely disbelieve in them, he minimised their importance and objected to the elaborate theories of the Han about them, urging like another Hsün Chhing[g] that human actions were always far more important than presages. 'When one discusses the rise and fall of States', he wrote, 'one assuredly ought to take human actions as the essential; if one insists in bringing fate into one's discourse then reason is outraged.'[h]

Enough has now been said about the imperial florilegia and the collections

---

[a] Cf. pp. 443 ff. below.   [b] Vol. 3, p. 634.

[c] These were thoughts formulated, no doubt, in more sophisticated ages. Yet here we touch upon the differentiation which occurred historically between two functions in all ancient astronomy, the calendrical and the astrological. This has been clearly perceived in the monograph of Sivin (3), who points out that the growing ability to predict celestial events constantly tended to move them from the realm of the ominous-historical to that of the rhythmic-intelligible. Since in China omens were allusions to the ruler's moral and administrative imperfections, one social factor favouring the development of astronomical science can readily be discerned.

[d] Shih Thung, ch. 8, p. 4b.

[e] How fortunate for the history of Chinese culture it has been that no other historiographer agreed with Liu Chih-Chi on this.

[f] Shih Thung, ch. 8, p. 5b.   [g] See Vol. 2, p. 366.

[h] Shih Thung, ch. 43 (Tsa shuo 7), p. 7a, tr. Pulleyblank (7), mod. auct.

[1] 周密

which emanated from this style of compilation. They were intended, at least in theory, for the emperor's eye, but in due course many other eyes demanded consideration, those namely of the candidates in the civil service examinations. One might date the beginnings of this line of descent from the second Sung reign, when in +977 Li Fang[1] and an editorial committee were commissioned to make a great collection of anecdotes, stories, mirabilia and memorabilia. Again almost the only thing of its kind in Chinese literature, the *Thai-Phing Kuang Chi*[2] (Copious Records ...) appeared in the following year, but there is no section on natural history in it, nor was this treated of in the later *Chhün Shu Khao So*[3] (Critical Guide through the Multitude of Books) prepared by Chang Ju-Yü[4] about +1200. Matters of biological interest do appear, however, in the *Yü Hai*[5] (Ocean of Jade) encyclopaedia of quotations put together by the Sung dynasty's most learned scholar, Wang Ying-Lin[6], by +1267, but not printed until between +1337 and +1340, or even +1351, in the Yuan period. Here in ch. 197 we find plant phenomena prominent among auspicious signs periodically appearing (*hsiang jui*[7]), especially the occurrence of rich-yielding grain varieties (*chia ho*[8]), and also other plants such as orange-trees, lotuses, willows, mushrooms, etc. Here are to be found records of selection procedures practised from ancient times onwards (cf. Sect. 39, 41 (*d*) 5).[a] Another valuable work of Wang Ying-Lin, though very small in comparison,[b] was the *Hsiao Hsüeh Kan Chu*[9] (Useful Treasury of Elementary Knowledge), done before +1270 but not printed till +1299; here everything goes by numerical categories, and the customary ways of enumerating plants and animals in groups are duly given.[c]

When the Yung-Lo emperor came to the throne in +1403, he commissioned a work which turned out to be more of a collectaneum than a florilegium, since entire books and whole tractates were written straight into it, not excluding Taoist and Buddhist treatises. The main object was to preserve texts that might otherwise be lost. A band of 147 scholars under Hsieh Chin[10] presented some result in the following year, to be entitled *Wên Hsien Ta Chhêng*,[11] but it was considered insufficient, and a greatly increased effort was ordered, under the co-editorship of Yao Kuang-Hsiao[12].[d] By the assiduity of no less than 2169 scholars the work was finished and presented towards the end of +1407 under the title *Yung-Lo Ta Tien*[13]; it comprised 22,877 chapters in 11,095 volumes (*pên*[14]), and being too large to print, was simply conserved in the Imperial Library. A single further copy was made between +1562 and +1567. The original and about a tenth of

---

[a] Chs. 198, 199 on auspicious animals may also have zoological interest.
[b] It has often been drawn upon in earlier volumes, see their indexes, *sub voce*.
[c] Ch. 10, pp. 25*a* ff.
[d] A remarkable man, scholar, imperial counsellor, royal tutor, eminent painter, keen opponent of Neo-Confucianism, and eventually a Buddhist about with the name in religion of Tao-Yen[15].

[1] 李昉   [2] 太平廣記   [3] 羣書考索   [4] 章如愚   [5] 玉海
[6] 王應麟   [7] 祥瑞   [8] 嘉禾   [9] 小學紺珠   [10] 解縉
[11] 文獻大成   [12] 姚廣孝   [13] 永樂大典   [14] 本   [15] 道衍

the copy were burnt in the disturbances at the end of the Ming, much of what remained was scattered during the Chhing, and the last thousand volumes destroyed at the time of the Boxer Rising in 1900 when the Imperial Archives (Huang Shih Chhêng[1]) were shelled by European artillery from the Legation Quarter. However, 385 books had been copied into the *Ssu Khu Chhüan Shu* collection by +1782, and there still remain just over 370 *pên*[5] scattered in libraries all over the world.[a] How much natural history lost by the decimation of this great collection is not easy to say, certainly it would have been difficult to use, for we know that the material was classified according to the 76 standard 'rhyme'-sounds accepted at the time (cf. p. 218 below), book-titles, chapter-headings and even section-headings or key-words being placed in the succession of these index-characters.[b]

When one comes to the +18th century one finds a work of compilation which put all previous florilegia completely in the shade, outdoing indeed anything contemporary in other civilisations. This was the *Chhin-Ting Ku Chin Thu Shu Chi Chhêng*[4] (Imperial Encyclopaedia and Florilegium; lit. Synthetic Collection of Pictures and Writings Old and New, compiled by Imperial Command), enormous in size (6109 subjects in 10,000 chapters) and rather strange in its history.[c] Chhen Mêng-Lei[5] was a scholar of great erudition whose life was dogged by the misfortune of having been unwillingly involved in a rebellion in his home province of Fukien when in his twenties. However, after serving an exile in Manchuria, he won the goodwill of the emperor in +1698 and became the secretary of his third son Prince Yin-Chih[6], in which position he began to organise the copying of portions of a vast number of books, especially those that were rare and risked being lost. By +1716 the enterprise had been adopted by the State, with all aid and facilities, and it was essentially complete, even perhaps in part already printed, by the death of the Khang-Hsi emperor in +1722. Unfortunately there then occurred a vicious internecine strife about the succession; Yin-Chih backed the losing party and died ten years later disgraced and in prison,[d] while Chhen Mêng-Lei was exiled to the North again and never returned home.

---

[a] Assemblies of the extant books have recently been printed both in Peking and Thaipei, the latter being the larger as it includes some MSS only available in Thaiwan.

[b] Corr. by this paragraph Vol. 1, p. 145. We do not repeat the references there given, but add Yang Lien-Shêng (4); Bauer (3). Cf. Vol. 3, Figs. 65, 81. The contents-table of the *Yung-Lo Ta Tien* is printed in the Lien Yün I Tshung-Shu[2] edited by Chang Mu[3] (1848).

[c] Cf. Vol. 1, pp. 47, 48. We constantly quote from it in the present book. For longer accounts of it cf. Mayers (2); O. Franke (9); Têng & Biggerstaff (1), pp. 126 ff.; Hummel (2), pp. 93 ff., 142 ff., 922 ff. Indexes by L. Giles (2); Takizawa Toshizuke (1).

[d] This prince deserves the notice of historians of science. In +1702 he acted as chief collaborator of the Jesuit Antoine Thomas (An To[7]) in the measurement of the degree of terrestrial latitude on a meridian line near Peking, which led to the fixing of 200 *li* to the degree. Thomas said of him that he was 'a very clever observer familiar with apparatus, and a quick and accurate computer'. Besides Chhen Mêng-Lei, he also had the collaboration of the distinguished mathematician, astronomer and hydraulic engineer Ho Kuo-Tsung[8], then a young man.

[1] 皇史宬   [2] 連筠簃叢書   [3] 張穆   [4] 欽定古今圖書集成
[5] 陳夢雷   [6] 胤祉   [7] 安多   [8] 何國宗

The new Yung-Chêng emperor placed Chiang Thing-Hsi[1] in charge of the project with instructions to revise it, but very little change was made before the work was finally presented in +1726 and printing completed by +1728.

In this mammoth florilegium natural history came into its own, with three relevant main sections, accounting for nearly 20% of the entire volume of subjects treated.

Section (*tien*[2]) no.	Department		Subjects (sub-sections, *pu*[3])	% of whole	Chapters (*chüan*[4])	% of whole
19	animals	(*chhin chhung*[5])	317	5.2	192	1.9
20	plants	(*tshao mu*[6])	700	11.5	320	3.2
27	foodstuffs	(*shih huo*[7])	83	1.35	360	3.6
			1100	18.05	872	8.7
	total		6109		10,000	

As we shall later see (p. 399) many valuable botanical monographs and other texts are preserved in the *Thu Shu Chi Chhêng* and cannot now be found anywhere else, but it is always necessary to remember that the editors freely abbreviated and omitted without giving warning of when they were doing so, hence it is always desirable to check their versions against the originals or others if this is at all possible. The natural history sections are among those well illustrated—we may see one example of the drawings in Fig. 35. Chhen Mêng-Lei and his assembly of copyists should never be forgotten by historians of science.

(iv) *Classified compendia*

Littleton in his dictionary of +1677 attributes the word *florilegus* to Ovid, and says of it: 'That gathers flowers, or out of flowers, as bees do.' Flowers of literature are collected into 'anthologies', as is right enough, but an encyclopaedia need not have any quotations at all, or if it does, they can be much more brief than those in a florilegium, descending in fact to the minimum length required for a concordance. As we have already seen (p. 185), there were reasons why Chinese lexical works could never dispense with the amassing of quotations, but the group of them we must now mention might be said to diverge on the other side, as it were, of the straight line of descent of the successors of the *Erh Ya*, constituting private (i.e. non-or semi-governmental) encyclopaedias of knowledge in quotations very brief and strictly classified. For these the Thang and the Chhing were the great periods of activity, though the movement had begun a good deal earlier, probably with Hsü Mien's[8] *Hua Lin Pien Lüeh*[9] (An Arrangement of the Whole Company of Flowers) produced in the Liang period about +530. Of this nothing

---

[1] 蔣廷錫 [2] 典 [3] 部 [4] 卷 [5] 禽蟲
[6] 草木 [7] 食貨 [8] 徐勉 [9] 華林遍略

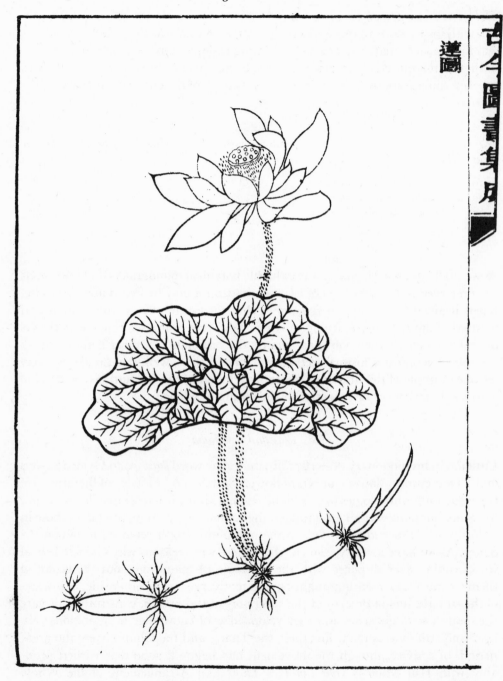

Fig. 35. Drawing of the lotus *Nelumbo nucifera* from *Thu Shu Chi Chhêng*, Tshao mu tien, ch. 93, p. 1*b*.
Cf. Fig. 24 above.

is left now, but parts of the extant *Pien Chu*[1] (Strung Pearls of Literature) finished by Tu Kung-Chan[2] about +605 are believed to be genuine.[a]

After that four famous encyclopaedias came out almost in a rush, doubtless because of the greatly increased emphasis placed by the Thang dynasty upon the imperial examinations as the channel of intake for the civil service.[b] No longer now was this obligatory only for the lower echelons—everyone had to pass through it. There resulted an insatiable thirst for general knowledge. Since the examinations were essentially literary, the knowledge of the classics was the first consideration, but there was a vast wealth of later literature, highly esteemed, as well, and whether a man was well educated or not would appear by his capacity for 'capping' a 'tag' from the mouth of his superior official. This, of course, was not the milieu from which a scientific revolution could arise, but it did signify a very great increase in the spread of polite learning; only matched perhaps by the later great increase which took place just after the Thang when the new technique of printing disseminated knowledge even more widely among the people for the benefit of aspiring sons. And so it came about that hardly a dozen years after the beginning of the dynasty, in +630, Yü Shih-Nan[3], later the President of the Imperial Academy, produced the first of the four, the *Pei Thang Shu Chhao*[4] (Excerpts from the Books in the Northern Hall). Astronomy was included in this but not natural history. Within a decade a second encyclopaedia had appeared, the *I Wên Lei Chü*[5] of Ouyang Hsün[6], an eminent scholar and Censorate official noted also as a calligrapher. Again natural history was omitted, but it figured quite prominently in the third of the series, the *Chhu Hsüeh Chi*[7] (Entry into Learning), compiled in +727 by Hsü Chien[8], a learned friend of the great historiographer Liu Chih-Chi just mentioned (p. 203 above). Conceivably it was this association which led to three of his 30 chapters being on plants and animals; an illustration of one of the entries in these is given on p. 210. First however, we must note the fourth of the encyclopaedias, written by the famous poet Pai Chü-I[9] either during the year +802 or between +840 and +845. Its very title throws a light on the need which these works met, *Liu Thieh Shih Lei Chi*[10] (The 'Six Slips' Collection of Classified Quotations); for the reference was to the six slips of paper on which the examination candidates had to complete whole sentences or passages chosen by the examiner, the text of the work being covered except for one horizontal line. Pai's work was considerably enlarged about +1160 by Khung Chhuan[11], hence it was often afterwards called *Pai Khung Liu Thieh*.[12]

Let us look at a typical entry from the *Chhu Hsüeh Chi*. Opening at random we

---

[a] We mention these two only, but research pursued in the appropriate directions would undoubtedly unearth many more, from this time onwards.

[b] Cf. the masterly work of des Rotours (2).

[1] 編珠   [2] 杜公瞻   [3] 虞世南   [4] 北堂書鈔   [5] 藝文類聚
[6] 歐陽詢   [7] 初學記   [8] 徐堅   [9] 白居易   [10] 六帖事類集
[11] 孔傳   [12] 白孔六帖

fall upon the discussion on the chestnut tree (*li*[1]),[a] *Castanea vulgaris* just as in Europe.[b] The main discussion (*hsü shih*[2]) runs as follows:

In Mao's edition of the *Odes* it says: 'On the hillsides are the lacquer trees,[c] in the low damp grounds grow the chestnuts.'[d] The *Shih I Su*[3] (Explanation of the Ideas in the *Odes*)[e] says: 'All quarters of the country have the chestnut-tree. It is particularly abundant in Chou, Chhin and Wu (States). The elongated chestnuts of Yu-yang and Fan-yang have the best and sweetest flavour, with which no others can compare. Korea and Japan send as tribute splendid chestnuts as large as a hen's egg, but they are shorter and not of good taste. In Kweiyang the chestnut-trees grow thickly, and their fruits in clusters like those of the chu[4] oak.'[f]

The *Chou Li* says: 'The sacrificial food-baskets of the manciples bringing provisions are full of chestnuts.'[g]

The (*Chhien*) *Han Shu* says that Yen Chhin[5] had (the profits of) a thousand chestnut-trees, as good as a marquis of a thousand households, etc.

The *Hsi Ching Tsa Chi* tells us that 'in the Imperial Park (Shang-lin Yuan[6]) there grew the *hou*[7] chestnut, the *kuei*[8] chestnut, the *khuei*[9] chestnut, the *chên*[10] chestnut, and the *I-yang*[11] chestnut'.[h]

Mr Hsin,[12] in his *San Chhin Chi*[13] (Record of the Three Princedoms of Chin) says: 'In the Imperial Orchards of Han Wu Ti there were chestnut-trees with fruit so big that fifteen went to a bushel.'

There follows a 'concordance of contrasts' (*shih tui*[14]), one of which is:

The produce of Nan-an; the sacrifice-offerings of the North (*Nan an chhu. . . Pei shuo chien*.[15])

Wang Pao[16], in his *Thung Yo*[17] (Contract with a Serving-Lad)[i] says: 'In Nan-an you must pick up chestnuts and pluck oranges.' The commentary says that Nan-an (in Szechuan) was famous for fine chestnuts and oranges.

Wang I[18], in his *Li-chih Fu*[19] (Rhapsodic Ode on the Lichi)[j] says: 'The Western guests present the grapes (from beyond) the Khun (-Lun) Mountains, and the sacrifice-offering visitors from Northern Yen bring gifts of huge chestnuts from their boreal river-banks.'

---

[a] Ch. 28, p. 10*a*.   [b] B II 494.

[c] *Rhus verniciflua*, cf. pp. 31, 85, 123.

[d] *Shih Ching*, I, xi, 1, (2), tr. Legge (8), p. 190.

[e] We do not know to what work Hsü Chien was referring here, for all the *Shih I* titles named in the dynastic history bibliographies were Sung or post-Sung.

[f] *Quercus sinensis* = *acutissima* = *serrata*; CC 1639; B II 239, 534; Chhen Jung (*1*), p. 200.

[g] *Chou Li*, ch. 2, p. 9*b*, tr. Biot (1), vol. 1, p. 108.

[h] It would be a good exercise to unravel the reference to these five varieties but here we need perhaps only notice that they were already recognised at this early time.

[i] This is a serio-comic production in verse, of great interest for social history. It was written in −59. One version is in *Chhu Hsüeh Chi*, ch. 19, pp. 18*b* ff.; a complete translation is given by Wilbur (1), pp. 383 ff., cf. pp. 390, 392.

[j] This could be dated approximately + 120. Wang I was one of the first commentators on the *Elegies of Chhu*, and the last contributor to them; among other writings of interest for the history of science and technology was his ode on the loom.

[1] 栗	[2] 叙事	[3] 詩義疏	[4] 杼	[5] 燕秦
[6] 上林苑	[7] 侯栗	[8] 瑰栗	[9] 魁栗	[10] 榛栗
[11] 嶧陽	[12] 辛	[13] 三秦記	[14] 事對	
[15] 南安出 ··· 北朔薦		[16] 王襃	[17] 僮約	[18] 王逸
[19] 荔枝賦				

Thus the historian of today must know how to rummage in these ancient collections for the material of scientific interest which he is looking for. There have been great masters of this art, such as Berthold Laufer. From the flavour of the entry just quoted it will be obvious that Hsü Chien was not trying to write botany or horticulture in any modern sense, he was putting together all that the brightest young administrators and academicians ought to be expected to know, yet in the process he recorded much that is of interest for the history of botany today. Thus on one page we have found several oecological references, intimations of foreign trade and contacts, a statement of five varieties grown in the Imperial quasi-botanic Garden during the Han, and several titbits of economic data. The encyclopaedias of traditional Chinese culture have never yet been brought under systematic contribution by historians of science, and much awaits the investigator willing and able to dig in these quarries.

Here matters rested for four and a half centuries, till towards the end of the Ming period activity began again. The deceptively named *Thang Lei Han*[1] (The Thang Encyclopaedias Conflated), prepared by Yü An-Chhi[2], came out in +1618; not merely a combination of the sources, however, but with additions on politics and administration, seasons and festivals. Natural history was left alone in this, but again found great increase in the enormous compilation of +1701 for which Yü An-Chhi's work had been but the prelude. This was the *Yuan Chien Lei Han*[3] (Mirror of the Infinite) edited by Chang Ying[4] and others, which drew not only from the four Thang encyclopaedias but from 17 others, as well as a variety of other sources down to +1566, embodying of course the work of Yü.[a] So comprehensive is this book that it recalls the *Thu Shu Chi Chhêng*. The natural history chapters occupy 12.6% of the whole 450, the preponderance in this section going to zoology (with 33 chapters) and a special emphasis on birds. In the botanical part (24 chapters) a new category was introduced beside the old ones, that of ornamental plants and garden flowers. If one looks for chestnut-trees,[b] one sees immediately a dozen times more material than there was in the *Chhu Hsüeh Chi* entry sampled above, so for any specific enquiry into a plant the *Yuan Chien Lei Han* would form the best basis, weeding out of course the purely poetic and the merely anecdotal.

The *Yuan Chien Lei Han* was the culmination of this particular line of classical Chinese scholarship, and nothing like it was attempted again. But it was not a far cry to what might be called 'phrase concordances', which would give the origin and the full form of allusions and phrases in common use. The great encyclo-

---

[a] Cf. Mayers (2). Each entry begins with an explanation and a study of the origin and development of the matter, then it goes on to quote factual material from a great variety of sources arranged in chronological order (like *Thai-Phing Yü Lan*); after this it gives instances of parallel phrases (*tui ou*[5]) chosen for their literary merit, selected sentences, and parts of poems and essays (or their texts in full, if short). Yü had given references to the encyclopaedias used but Chang and his colleagues noted only the original sources.

[b] Ch. 403, pp. 1a ff.

[1] 唐類函　　　[2] 俞安期　　　[3] 淵鑑類函　　　[4] 張英　　　[5] 對偶

paedia was seconded therefore by works such as the *Tzu Shih Ching Hua*[1] (Inflorescence of the Philosophers and Historians) produced under the editorship of Yün Lu[2] in +1727. A good example of the sort of thing this was can be found at the beginning of the chapter on fruit-trees—for eight out of the 160 chapters were concerned with natural history. Here we find:

Huai nan chü, Huai pei chih.[3] South of the Huai the orange, north of the Huai the thorny limebush. (From) *Yen Tzu (Chhun Chhiu)*[4]. (The full text is:)

'I, Ying, have heard that when [orange-trees grow south of the Huai] (River) they become good orange-trees, but when they are planted [north of it they turn into thorny limebushes.] The leaves look alike, but fruitlessly, for the taste of the fruits is quite different. How can this be? It is because the water and the soil are not the same.'[a]

To us this extract will be very familiar, for it has been given already, with its full background, on p. 106 above. Each tag is first quoted, and then comes the full text. The work is obviously still useful, but alas, no such concordance was ever made with natural history alone in view.

(v) *Dictionaries of origins, technic and scientific*

Before we turn to the realm of dictionaries proper, about which there will be less to say, there is one special type of encyclopaedia which cannot be left out of this discussion, though the role of natural history in it is somewhat unexpected. This was the genre of lexical works devoted entirely to explaining the origins of things, inventions, customs and affairs—very characteristic of Chinese literature but liable to be puzzling to any Westerners who still cherish the illusion that that civilisation was 'timeless' and 'static'. In fact, it was historical to the core, conscious also of a kind of social evolution from primitive existence, and therefore very much concerned with origins.[b]

The oldest book of this kind that has come down to us is the −2nd-century *Shih Pên*[5] (Book of Origins),[c] one text of which was used by Ssuma Chhien when composing the *Shih Chi* around −100. We have it in a number of versions,[d] some bearing the name of its greatest editor Sung Chung[6], who was working towards the end of Han about +210. Besides the recital of the names of the legendary, semi-legendary, and clearly historical inventors of all kinds of devices, instruments and machines, the book enlarges on the origin of the names of the clans and

---

[a] Ch. 140, p. 20*b*, tr. auct. The parts in square brackets, which repeat the phrase of the entry, are indicated by a vertical caesura sign (the printers' 'upright') for each phrase, so as to show how they fit in to the full text.
[b] See further Sect. 49 in Vol. 7, and meanwhile Needham (55, 56).
[c] We have already discussed this in Vol. 1, pp. 51 ff. under the rubric 'Chinese traditions of inventors'. Fuller information will be found there. For an example of how useful the *Shih Pên* can be, see Vol. 4, pt 2, p. 189.
[d] Eight of them, partly reconstructions from scattered quotations, have been assembled in Anon. (71).

[1] 子史精華   [2] 允祿   [3] 淮南橘淮北枳   [4] 晏子春秋
[5] 世本   [6] 宋衷

gives elaborate genealogies of the ruling houses. Several fresh books of a similar kind were produced during subsequent centuries,[a] and the genre took a new lease of life early in the Sung when between +1068 and +1085 historical perspectives were emphasised more then ever before in the *tshê lun*[1] style of questions in the imperial examinations. Hence at this time two much larger encyclopaedias of origins were set on foot.[b] The *Ku Chin Yuan Liu Chih Lun*[7] (Essays on the Course of Things from Antiquity to the Present Time) was too big to be finished by its first author Lin Kung[8] who started it about +1070, and had to be continued by a much later scholar, Huang Li-Ong[9], who printed it in +1237. All this indeed has nothing to do with natural history, but plants and animals found their way in at last to the second of the two Sung works, the *Shih Wu Chi Yuan*[10] (Records of the Origins of Affairs and Things), produced by Kao Chhêng[11] in +1085, a book which we have had occasion to quote in every volume of the present series. Botany and zoology form a small part of his work, it is true, occupying only the last two out of 55 sections (*pu*[12]), but what he recorded can be very helpful, as is shown by an example on p. 171 above.

By far the largest encyclopaedia of origins, however, was produced in the +18th century by a scholar named Chhen Yuan-Lung[13], and printed between +1717 and +1735. Though entitled *Ko Chih Ching Yuan*[14] (Mirror (or, Perspective Glass)[c] of Scientific and Technological Origins), and faithfully dealing with all trades, industries, arts and sciences, it includes, somewhat surprisingly, chapters forming no less than 42% of its bulk on plants and animals. The wealth of quotations given by Chhen Yuan-Lung, often from rare or now lost books,[d] make it a treasure-house for any specific study in the history of biology. Eighteen chapters[e] are devoted to plants and 24 to animals, with a rather more clear classification than pertained in earlier times;[f] for example, water-plants, epiphytes, vines and lianas are placed under separate sub-headings, while fishes, crustacea, molluscs and aquatic reptiles come in one chapter, and insects, worms, amphibia and land reptiles in another. Each entry begins with a general discus-

---

[a] Notably the *Shih Shih*[2] (Beginnings of all Affairs) by Liu Tshun[3] (probably identical with the mathematician Liu Hsiao-Sun[4]) in the Sui period, c. +610; and the *Hsü Shih Shih*[5], intended as a supplement, by Ma Chien[6] in the Later Shu State in Szechuan c. +960. Ma Chien probably knew Wu Chao-I, the patron of the printers of the first edition of the classics (cf. p. 193 above), and Han Pao-Shêng, the compiler of the *Shu Pên Tshao* pharmaceutical natural history (cf. p. 223 below).

[b] The gnomic style of the *Shih Pên* was perpetuated only in a Ming work of the +15th century, the short *Wu Yuan*[15] (Origins of Things) by Lo Chhi[16].

[c] For facts justifying this translation see Needham & Lu Gwei-Djen (6). One of the several first inventors of the telescope may have been Chinese.

[d] One must beware of his habit of shortening titles.

[e] Including two on perfumes and aromata.

[f] Here the influence of Li Shih-Chen was doubtless important; cf. pp. 310, and 315 below.

[1] 策論	[2] 事始	[3] 劉存	[4] 劉孝孫	[5] 續事始
[6] 馬鑑	[7] 古今源流至論		[8] 林駉	[9] 黃履翁
[10] 事物紀原	[11] 高承	[12] 部	[13] 陳元龍	[14] 格致鏡原
[15] 物原	[16] 羅頎			

sion (*tsung lun*[1]), goes on with more detailed material (*hsiang lei*[2]) and concludes with strange or unusual reports (*chi i*[3]). Larger sections tend to end with miscellaneous species (*chu*[4] *x*) and peculiar events (*i*[5] *x*). This fine work is thus reminiscent of *Thai-Phing Yü Lan*, though a private effort, but the quotations from texts are not placed so systematically in chronological order, and all should be checked against the original if possible, since Chhen and his copyists were none too accurate.

The title of the *Ko Chih Ching Yuan* is an arresting one with a deep philosophical background, and though we have already touched upon this at a much earlier stage,[a] it cries out for more elucidation since the whole basis of the natural sciences in Chinese culture is involved. Here the *Shih Pên* tradition fused with another which derived from the ancient statement *chih chih tsai ko wu*[6]—pregnant words found in a brief text called *Ta Hsüeh*[7] (The Great Learning), traditionally attributed to Confucius as recorded by Tsêng Shen[8] but now considered more probably the work of Yochêng Kho[9], a pupil of Mencius, *c.* −260. In the +12th century Chu Hsi[10], the Aquinas of Neo-Confucianism, extracted it from the *Li Chi*[11], in which it had long been a chapter,[b] and made it one of the great classics in its own right. The proper interpretation of the crucial phrase has for centuries been one of the most disputed questions in all Chinese philosophy, hence it has always caused difficulty for translators. Legge (2), following the great authority of Chu Hsi, gave it as 'the extension of knowledge to the utmost lies in the investigation of things', but it is clear from his explanations that what he would like to have made of it would have been something such as: 'When (self-) knowledge is complete, it is manifested in the right judgment and handling of all things.'[c] Sinologists have always found difficulty in believing that the frank charter for the natural sciences implied by the Neo-Confucian interpretation could have emanated from a milieu so preoccupied with ethics and self-cultivation as that of the early Confucian school. So Wilhelm (6) for example, translated: 'The highest form of understanding is the influencing of reality (i.e. the world of external things).'[d] More moderate, and not unattractively Taoist, is Hughes' 'The extension of knowledge consists in appreciating the nature of things.'[e] One thing is certain, for the past thousand years Chinese scholars concerned with the investigation of Nature have taken the words to apply to scientific studies more or less in our modern sense, and shortening the phrase to *ko wu*[12] or *ko chih*[13] utilised it in the titles of their books so that it became the very watchword of the physical and natural sciences.

---

[a] Vol. 1, p. 48. New information has accrued since then, so the matter is worth going into again. Cf. too pp. 9, 10 above.

[b] Ch. 42; cf. the tr. of Legge (7), vol. 2, p. 412 (as ch. 39).

[c] P. 222, on paras. 4, 5.   [d] Pp. 22, 369.   [e] (2), pp. 146 ff.

[1] 總論   [2] 詳類   [3] 紀異   [4] 諸   [5] 異
[6] 致知在格物   [7] 大學   [8] 曾參   [9] 樂正克   [10] 朱熹
[11] 禮記   [12] 格物   [13] 格致

It is most striking to see how early this began. The oldest book with a title of this kind seems to be the *Ko Wu Tshu Than*[1] (Simple Discourses on the Investigation of Things), a collection of short statements about natural phenomena almost in *Shih Pên* style produced by a learned monk (Lu) Tsan-Ning[2] about +980.[a] There is a good deal of natural history in this, oriented particularly to plant physiology, oecology and properties non-pharmaceutical. 'Let male and female ginkgo trees grow near one another', he says, 'then fruit will form.' 'Tree-peonies flourish if given powdered stalactite.'[b] And there are many things besides about geo-botanical prospecting,[c] insecticidal plants and chemicals,[d] weather forecasting, nutritional science, hygiene, etc. It is certainly a book not to be neglected, and we are fortunate that it has been preserved. The remarkable thing is that in spite of its title Tsan-Ning's book antedates the high Neo-Confucian period; in his time, at the very beginning of the Sung dynasty, there had only been Li Ao[4] (d. +844), regarded as the precursor of the school,[e] and the two founding figures (Shao Yung and Chou Tun-I)[f] did not live until the +11th century. One can only conclude that the phrase in the *Ta Hsüeh* was already attracting the attention of naturalists and philosophers.

All the later titles, however, came during or after the Neo-Confucian synthesis. The oldest of the other stream, the *Ko Chih Yü Lun*[5] (Supplementary Discourse on the Investigation of Things in the Field of Medicine) was written by the great physician Chu Chen-Hêng[6] in +1347 under the Yuan. In his preface he specifically draws attention to the Neo-Confucian significance of the *Ta Hsüeh* statement, and says that his medical studies were needed because it was impossible to go on any longer relying on the medical classics of the Han and the government pharmacy formulae of the Sung. Somewhat earlier there had actually been a lexicographic work with this type of title, the *Ko Wu Lei Pien*[7] (Classified Encyclopaedia of Natural Knowledge), compiled by Phan Ti[8] in the Liao or the Yuan dynasties, but most unfortunately this has not come down to us. However, the watchword was not forgotten, and several +16th and +17th-century works embodied it.[g] In

---

[a] Cf. Vol. 4, pt 1, pp. 77. The book has generally been attributed to the poet Su Tung-Pho[3], but this seems to be due to a confusion with a literary name which the monk adopted. The gnomic style of studiously plain staccato separated statement reminds one also of the *Huai Nan Wan Pi Shu*, on which see Vol. 5, pt 3, pp. 25–6.

[b] Cf. Vol. 3, pp. 605 ff.

[c] Cf. Vol. 3, pp. 675 ff.

[d] Cf. below, p. 471.

[e] See Vol. 2, p. 452. He wrote one of the botanical monographs (see p. 358 below).

[f] Vol. 2, pp. 455, 457 ff.

[g] Notably the *Ko Chih Tshung Shu*[9] collection assembled by Hu Wên-Huan[10] about +1595. This contains 293 books of all periods on classics, history, law, Taoism and Buddhism, divination, astrology, geomancy, longevity techniques, medicine, agriculture, tea technology and the like. In +1620 Hsiung Ming-Yü[11] produced a *Ko Chih Tshao*[12] (Scientific Sketches in Astronomy and Cosmology), part of a larger work in which his son participated. Then in +1670 there was the *Ko Wu Wên Ta*[13] (Questions and Answers about Natural Philosophy) by Mao Hsieh-Shu.[14] Hu's title was used again for a collection on modern science (1901); see Hsü Chien-Yen (*1*).

---

[1] 格物麤談	[2] (錄)贊寧	[3] 蘇東坡	[4] 李翱	[5] 格致餘論
[6] 朱震亨	[7] 格物類編	[8] 潘迪	[9] 格致叢書	[10] 胡文煥
[11] 熊明遇	[12] 格致草	[13] 格物問答	[14] 毛先舒	

this way we find ourselves back again with Chhen Yuan-Lung and his *Ko Chih Ching Yuan*.

It need only be added that the ancient phrase came at last into its own during the nineteenth century, when books on the natural sciences began to stream from the Chinese presses. This occurred in two ways: first in titles for books and tractates on the sciences themselves written either by Chinese or Westerners, or translated from Western languages,[a] and secondly in titles for books dealing really with the history of science. There was a special group of these, written by Chinese scholars who rebelled against the facile assumption that all science was of Western origin—*Hsi-yang-ti Kho-Hsüeh*[1]—and sought to show, not only that China had contributed a vast fund of knowledge and technique in the centuries before the development of modern science, but that a good deal of this had been trans-mitted to Europe and had indeed stimulated the course of discovery there so that its origins had been quite forgotten.[b] The men who joined in this movement were not very distinguished, and as they wrote in a rather unscholarly and sometimes exaggerated style they were not much listened to, but we have more than once had occasion to admire their insight.[c] Both these movements deserve careful studies to themselves, but this is not the place for either. All that has to be noted here is that the encyclopaedias of origins formed a special genre in Chinese literature of much interest for the history of science, though hitherto little drawn upon, that it linked with scientific books which flew the flag of the *Ta Hsüeh*, and that some of these constitute valuable sources for enquiries into the development of botanical and zoological knowledge.

(vi) *Dictionaries based on script, sound or phrase*

Lastly, there is the story of the dictionaries proper. What was the position of Chinese natural history with relation to them? Since the aim of all dictionary-making is brief definition, not explanation, there is much less to say than in the case of the encyclopaedias, and we shall not repeat the information already given

---

[a] As examples we may give *Ko Chih Shih Chhi*[2] (Explanations of Scientific Instruments and Apparatus) translated from Western sources by John Fryer (1839 to 1928), one of the chief translators of the Kiangnan Arsenal; and *Ko Wu Ju Mên*[3] (Introduction to Natural Philosophy), written by W. A. P. Martin a dozen years earlier, in 1868. Between the two came Alexander Williamson's *Ko Wu Than Yuan*[4] (Enquiry into the Principles of Natural Philosophy), printed at Shanghai in 1876. The *Ko Chih Chhi Mêng*[5], mentioned in Vol. 1, p. 49, was an introduction to chemistry by Sir Henry Roscoe, translated at Shanghai by Y. J. Allen in 1885. The Kiangnan translations are an epic in themselves, but there is very little in English on them; see however Bennett (1) and Chhüan Han-Shêng (2).

[b] The *Ko Chih Ku Wei*[6], published by Wang Jen-Chün[7] in 1896, has also been mentioned in Vol. 1, p. 48; its title could be translated 'Scientific Traces in Olden Times'. Since then we have found a better and much less chaotic book, the *Ko Chih Ching Hua Lu*[8] (Record of the Inflorescence of Men of Science in Olden Times), produced by Chiang Piao[9] a year or two later. The only Western-language reference to this group seems to be in Chou Tshê-Tsung (1), but there is a good treatment of it in Chinese by Chhüan Han-Shêng (3).

[c] Cf. Vol. 4, pt 2, p. 525.

¹ 西洋的科學   ² 格致釋器   ³ 格物入門   ⁴ 格物探原   ⁵ 格致啓蒙
⁶ 格致古微   ⁷ 王仁俊   ⁸ 格致精華錄   ⁹ 江標

in connection with the radicals on p. 117 above.[a] The main lines of descent can quickly be sketched out. Canonisation as the father of Chinese lexicographers is always given to Hsü Shen,[1] who in +121 produced his wonderful *Shuo Wên Chieh Tzu*[2] (Explanations of Simple Characters and Analyses of Composite Ones), still to this day in constant use. It was a palaeographic handbook as well as a dictionary, since it explained and analysed the ancient 'small seal' forms of the characters, which head each entry, along with the standard script forms in use from Chhin times onwards. The *Shuo Wên* recognised, as we know, some 540 radicals.[b] Immediately after the Han the dictionary tradition split into two directions, one conserving the graphic arrangement by radicals (*chien tzu shou pu*[3]),[c] the other following a new principle, arrangement by the sound (*yün*[4]) of the characters. Let us look first at the visual system and then at the phonetic one.

The next great graphic dictionary after the *Shuo Wên* was the *Yü Phien*[5] (Jade Pages) of Ku Yeh-Wang[6] produced under the Liang in +543, then extended and edited in the Thang by Sun Chhiang[7] (+674). Each entry gives the *fan-chhieh* 'spelling',[d] a very brief definition, and sometimes a quotation of a few characters from one of the classics such as the *Erh Ya* or the *Tso Chuan*. For example, under duckweed: '*Phing*[13]. *Pu, ting*, therefore pronounced *ping* (now *phing*). A plant without roots which floats on the water.' Or, for a plant not now easily identifiable: '*Yün*[14]. *Ku, chün*, therefore pronounced *kün* (* *gwin*), (now *yün*). A fragrant herb. The *Shuo Wên* says that it looks like alfalfa.' Obviously the historian of biology must use the medieval dictionaries, but they do not exactly press information upon him. After the *Yü Phien* there followed many other dictionaries of essentially the same kind,[e] till a thousand years later, in +1615, Mei Ying-Tso[15] produced his *Tzu Hui*[16] (The Characters Classified)—the first dictionary to reduce the number

---

[a] In considering the next few paragraphs the reader may like to refer again to the notes on language in vol. 1, pp. 27 ff.

[b] Those of botanical interest have already been discussed on pp. 119 ff. above. Those of zoological and pathological interest will be dealt with in Sects. 39 and 44 below. All are grouped by graphic similarity. There is little identification or description of the things referred to in the entries.

[c] There were really three directions which dictionary development could have taken because, besides arranging the characters according to their radical components it would also have been possible to arrange them in accordance with their 'phonetic' components. For example, all words having *kung*[8] as the phonetic could be placed together, in which case *chiang*[9] (river) would come neither under the water radical nor under the rhyme-sound -iang, but with words such as *kung*[10] (tribute) and *hung*[11] (red). This principle did find some use in China, but only within the groupings of dictionaries arranged on the purely phonetic 'rhyme' principle. It was left to Western sinologists to see what could be done with it systematically, as Callery (1) did in 1841, and Chalmers in 1877, writing the *Khang-Hsi Tzu Tien Tsho Yao*[12] (1) entirely in Chinese—a quixotic enterprise. Of course, those sinologists who have carried about with them for decades the pocket dictionary of Soothill (1) find the principle second nature.

[d] Cf. Vol. 1, p. 33.

[e] E.g. the *Lei Phien*[17] (Classified Dictionary) of Ssuma Kuang[18] in +1067, and the *Tzu Thung*[19] (Complete Character Dictionary) of Li Tshung-Chou[20], also in the Sung.

---

[1] 許慎   [2] 說文解字   [3] 檢字首部   [4] 韻   [5] 玉篇
[6] 顧野王   [7] 孫強   [8] 工   [9] 江   [10] 貢
[11] 紅   [12] 康熙字典撮要   [13] 萍   [14] 芸
[15] 梅膺祚   [16] 字彙   [17] 類篇   [18] 司馬光   [19] 字通
[20] 李從周

of the radicals to the present standard figure of 214, and the first to arrange both radicals and characters, as is now so universal, in the order of the number of their strokes.[a] The climax was reached with the *Khang-Hsi Tzu Tien*[1] (Imperial Dictionary of the Khang-Hsi reign-period), still today the best court of appeal in our daily work, commissioned in +1710 and first printed +1716, under the editorship of Chang Yü-Shu[2], Chhen Thing-Ching[3] and others.[b]

Retracing our steps, we find the first of the line of purely auditory or phonetic dictionaries in the *Shêng Lei*[5] (Sounds Classified) prepared by Li Têng[6] in the +3rd century.[c] This move must certainly have had something to do with the work of the great philologist Sun Yen[7], whose preoccupation with Sanskrit, arising out of the many translations of Buddhist texts then being made, led to the invention of the *fan-chhieh* 'spelling' system. Classification of Chinese words by sounds always primarily depended on the 'rhyme' (*yün*[8]) or finial combination of vowels and consonants (cf. Table 3 in Vol. 1, p. 37); and only to a secondary extent on the initial consonant or consonants (*tzu mu*[9]). Like the standard number of radicals, the standard number of recognised rhymes tended to fall as time went on, for while Li Têng and his successors accepted 206, Liu Yuan[10] about +1250 reduced them to 107, and by the early +15th century they had fallen to 76.[d] The successors formed a group unusually continuous in time for they worked on what was virtually the same lexicon from +600 to +1000. First there was the *Chhieh Yün*[11] (Dictionary of Characters arranged according to their Sounds when Split) produced by Lu Fa-Yen[12] in +601. This was later embodied in the *Thang Yün*[13] brought out by Chhangsun No-Yen[14] in +677 and revised by Sun Mien[15] in +751. Finally the product was again revised and enlarged[e] by Chhen Phêng-Nien[16], Chhiu Yung[17] and others, appearing as the *Kuang Yün*[18] in +1011.[f] If we

---

[a] His work was completed in that of Chang Tzu-Lieh[4] a dozen years later; see p. 121 above.

[b] It contains 49,030 characters.

[c] This now exists only in fragmentary form. In later times, the word *shêng*, as in the expression *shêng tiao*[19], was reserved by philologists for the 'tones' (cf. Vol. 1, p. 33). These seem to have been first distinguished by Chou Yung[20] and Shen Yo[21] in the +5th century. It is very relevant here to find that about two hundred years later, in the Thang period, a physician named Hsiao Ping[22] prepared a *Ssu Shêng Pên Tshao*[23], i.e. an encyclopaedia of pharmaceutical natural history in which the entries were arranged following the tone and standard rhyme of the first character of the name of the particular plant or animal.

[d] Just as in the case of the radicals the finally accepted number was an intermediate one, namely 106. Cf. pp. 121, 125 above.

[e] To a total of *c.* 26,000 characters.

[f] Later in the +11th century a different study of the *Chhieh Yün* was made. One defect of the phonetic system was that as pronunciations imperceptibly shifted, older phonetic dictionaries became difficult to use. Someone therefore produced a tabular key entitled *Chhieh Yün Chih Chang Thu*[24], arranging the characters in 'rhyme-tables' (cf. Vol. 1, Fig. 1) according with the Sung pronunciations. This was usually attributed to Ssuma Kuang, but his authorship is doubtful and the work may belong to the following century; cf. Tung Thung-Ho (*1*). Corr. Vol. 1, p. 34. Cf. Vol. 3, p. 107.

---

[1] 康熙字典	[2] 張玉書	[3] 陳廷敬	[4] 張自烈	[5] 聲類
[6] 李登	[7] 孫炎	[8] 韻	[9] 字母	[10] 劉淵
[11] 切韻	[12] 陸法言	[13] 唐韻	[14] 長孫訥言	[15] 孫愐
[16] 陳彭年	[17] 丘雍	[18] 廣韻	[19] 聲調	[20] 切韻指掌圖
[21] 周顒	[22] 沈約	[23] 蕭炳	[24] 四聲本草	

look here for the same two entries as above,[a] we find little difference, though duckweed has been cut down and the fragrant herb somewhat extended.

Phing[1]. The *phing* (plant)[b] that floats on water.

Yün.[2] A fragrant herb. The *Shuo Wên* says that it looks like alfalfa. The *Huai Nan Wang (Wan Pi Shu)*[c] says that it can raise the dead. The *Tsa Li Thu*[3][d] says: 'the *yün* is a kind of *hao*[4] (some species of *Artemisia*).[e] The leaves look like those of the leaning *hao* (*hsieh hao*[5]), but the fragrance is pleasant and the plant is edible.'

The *Kuang Yün* is still in common use today, and the historian of science can consult it with advantage.[f]

Where natural history came into this field much more prominently, however, was in the phonetic dictionaries which set out to define not only single characters but doublet or triplet forms,[g] and even brief common phrases. This began with the *Yün Hai Ching Yuan*[6] (Mirror of the Ocean of Rhymes) produced by Yen Chen-Chhing[7] about +780, and continued with the *Yün Fu Chhün Yü*[8] (Assembly of Jade Tablets; a Word-Store arranged by Rhymes) of Yin Shih-Fu[9] *c.* +1280. A number of other similar works exist, e.g. Ling I-Tung's[10] *Wu Chhê Yün Jui*[11] (Five Cartloads of Rhyme-Inscribed Tablets) of +1592, but all previous efforts were put completely in the shade when the Chhing government organised lexicographic work on an official basis early in the +18th century. Thus the *Phei Wên Yün Fu*[12] (Word-Store arranged by Rhymes, from the Hall of the Admiration of Literature) was commissioned in +1704, completed in +1711 and printed in the following year, its chief editor being Chang Yü-Shu.[13] The system adopted in this was that its 10,257 characters were those which came last in each doublet or phrase.[h] Then the work was repeated so that the material should be attached to those characters which came first in each doublet or phrase. Thus arose the *Phien Tzu Lei Pien*[17] (Classified Collection of Phrases and Literary Allusions),[i] commissioned in +1719, completed in +1726 and printed in +1728, under the chief editorship of Ho Chhuo.[18] If we test the *Phei Wên Yün Fu* for botanical possibilities,

[a] P. 6280 above.

[b] That it was a plant was fairly self-evident in Chinese because of the *tshao-thou* radical.

[c] See Vol. 5, pt 3, pp. 25–6; pt 4, pp. 310–1 and *passim*.

[d] No title approximating to this at such an early date can be found in the bibliographies of the dynastic histories.

[e] The name *hao* is sometimes currently applied to celery, but incorrectly. *Han chhin*[16] is the proper name for *Apium graveolens* (cult. *dulce*); cf. CC 533; Anon. (*109*), vol. 2, p. 1067.

[f] A number of other dictionaries based on this principle also exist, but we omit mention of them here; cf. Bauer (3); Têng & Biggerstaff (1).

[g] Cf. Vol. 1, p. 40.

[h] A supplement, the *Yün Fu Shih I*[14], edited by Wang Yen[15], appeared in +1722. Cf. Mayers (2); Hirth & Edkins (1).

[i] Cf. Mayers (2); Hirth (24).

[1] 萍	[2] 芸	[3] 雜禮圖	[4] 蒿	[5] 邪蒿
[6] 韻海鏡源	[7] 顏眞卿	[8] 韻府羣玉	[9] 陰時夫	[10] 凌以棟
[11] 五車韻瑞	[12] 佩文韻府	[13] 張玉書	[14] 韻府拾遺	[15] 王掞
[16] 旱芹	[17] 駢字類編	[18] 何焯		

we find that it contains a great deal. To take an example at random, one might enquire after the doublet *yün miao*[1], not at all necessarily the same plant as the *yün* of the ancients alone. In fact there is only one entry, after which follows:

The *Shih I Chi*[2] (Memoir on Neglected Matters) says that in Ying-chou there is a herb called *yün miao*, in appearance like the rush (*chhang phu*,[3] *Acorus Calamus*), but if any man eats the leaves he becomes drunk, howbeit if he then eat of the root he will be made sober.'[a]

The fact that Wang Chia's[4] book, datable about +370, contains much legendary and 'superstitious' material, and the strong suspicion that Ying-chou[5] here was the name of one of the fabled isles of the immortals in the Eastern Ocean,[b] are beside the point; for nowadays we are beginning to know much about neura transmitters, hallucinogenic glucosides and the like, to realise, moreover, that the active principles of *Amanita* and *Rauwolfia* gradually being elucidated were anciently within the esoteric knowledge of shamans and doubtless Taoists. Here then we may have the fragment of a chapter on psycho-tropic pharmacology and botany. In any case the point is made that any historian of biology able to use these vast rhyme-based phrase-dictionaries is likely to do so with profit. And it is here that we may say good-bye to lexicographers in general, appropriately drunk with the leaves of detail, but sobered by the root of our continuing plan and purpose.

(2) THE PANDECTS OF NATURAL HISTORY (*Pên Tshao*[1]); A GREAT TRADITION

We are now in a position to look into that notable department of Chinese literature which goes by the name of *pên tshao*.[6] In an epic series of books from the −5th century onwards, the Chinese assembled their knowledge, ever growing, of the natural worlds of mineral substances, plants and animals.[c] There is some room for argument as to how these great treatises should best be designated in English, and in a moment it may be enlightening to consider this, but our preferred name is the 'pandects' of pharmaceutical natural history, not 'herbals', certainly not 'lapidaries' or 'bestiaries', neither *materia medica* handbooks nor always 'pharmacopoeias'. A pandect is only a treatise covering the whole of any subject, as did the digest of Roman law commissioned by Justinian in the +6th century, a hold-all in fact, and that is true of the greater number of the books we have now to describe. Nevertheless there is good reason for placing the main discussion in the

---

[a] P. 782.3, tr. auct.
[b] Cf. Vol. 4, pt 3, pp. 551 ff., Vol. 5, pt 3, pp. 17 ff., Figs. 1343–6.
[c] With respect to the first of these domains we have already brushed against this literature (Vol. 3, pp. 643 ff.) but only in passing. The third, the zoological, receives due consideration in Sect. 39 below.

[1] 芸苗   [2] 拾遺記   [3] 菖蒲   [4] 王嘉   [5] 瀛洲
[6] 本草

section on Botany, for the science of plant forms was always the predominant part in bulk and detail.

At the same time there are many books traditionally listed in the *pên tshao* category, and even with *Pên Tshao* as part of their titles, which we shall here pass over, deferring accounts of them to later more convenient sub-sections. First, there are those, some even of quite early date, which are overwhelmingly pharmacological or pharmaceutical, and most of these will be postponed until the section devoted to this science.[a] Secondly, another strand out of which the *pên tshao* cable was woven were the books which enlarged on the properties of foods and the nature of diet;[b] these will obviously be more suitable for the Section on nutrition.[c] Thirdly, there are works of primarily alchemical and iatro-chemical interest; these have either been dealt with already in the chemical volumes or will come up for consideration in the Sections on medicine and pharmacology.[d] Lastly, one must remember the special treatises dealing with wild plants which could be used for food in times of emergency, but these we shall not postpone so far, only to the sub-section immediately following this one, for they are highly botanical in character.

For many long years those who wished to study the *pên tshao* books in the world literature could have recourse only to the pioneer account of Bretschneider,[e] produced in 1881, but apart from its acceptance of legendary datings this contained so many mistakes that it is now distinctly out of date. No book has replaced that of old Bretschneider, but in recent decades readers have been able to consult a monograph by Liu Ho & Roux (1),[f] as also the more careful list of works in Merrill & Walker,[g] all the better if supplemented by the addenda and corrections offered by Goodrich (18).[h] Now and then a distinguished Chinese pharmacological botanist such as Ching Li-Pin (1, 2) has written a good summary of the story in a Western language,[i] and there have been many descriptions of particular items by Swingle and Hummel[j] as well as by a number of Japanese scholars. However, no Western-language account has approached the excellence of the contributions of Kimura Koichi (1) in Japanese[k] and Tshao Ping-Chang (1) in

---

[a] Sect. 45 below.

[b] Here the key phrase is *shih liao*[1], in accordance with the immortal dictum 'some diseases can be cured by diet alone'; cf. Lu Gwei-Djen & Needham (1).

[c] Sect. 40 below.   [d] Sects. 44 and 45.

[e] (1), vol. 1, pp. 27 ff., 39 ff.; vol. 3, pp. 1 ff.

[f] This marked no advance on Bretschneider in scholarly standard. It is surprising to find that as late as 1927 the *Shen Nung Pên Tshao Ching* could be placed in the −27th century, and Thao Hung-Ching in the +10th. But a wider field was covered and mention made of a number of books not discussed by Bretschneider; the bibliography, too, is useful.

[g] (1), vol. 1, pp. 551 ff.   [h] Cf. Huard & Huang Kuang-Ming (3).   [i] Cf. Kimura Koichi (1).

[j] In the *Annual Reports of the Library of Congress* between 1925 and 1950. Hummel (13) was an important contribution to which we shall refer again.

[k] This includes an excellent chart showing the filiation of the pandect books and their several editions. We have seen a similar survey by Nakao Manzō (2) but it has not been available to us.

[1] 食療

Chinese. The early history of the *pên tshao* literature, especially interesting as we shall see, has been sketched by Shih Tzu-Hsing (*1*) and Nguyen Tran-Huan (*2*). But all pale before the *pièce de résistance* of Lung Po-Chien (*1*), available since 1957, who set out to catalogue all the extant books of the *pên tshao* category. Although they amount to no less than 278 (of which 62 alone have titles beginning with the words *Pên Tshao*), it would be a fair guess that for every book which has come down to us at least three others have perished, so one might not be far wrong in placing the total number at something like 1500 treatises. It may be interesting to see how Lung Po-Chien classified his material:

	Entries
Reconstructions of the *Shen Nung Pên Tshao Ching*[1] and commentaries upon it	24
General pharmaceutical natural histories and pharmacopoeias	140
Monographs on particular drug-plants (cf. Sect. 45 below), *tan yao*[2]	11
Nutritional and dietary treatises, *shih wu*[3]	46
Works on the preparation of *materia medica*, *phao chih*[4]	7
Collections of mnemonic rhymes	32
Miscellaneous	18

Naturally the above remarks are not an exhaustive account of the literature, whether Chinese or Western, on *pên tshao*;[a] nor shall we attempt to give in what follows a full description of all the original works themselves, which would be tedious partly because of much repetitiveness especially in late times. But the landmarks of the literature constitute a grand monument of the Chinese biological sciences, and we must look at them carefully and comparatively.[b]

What was the essential meaning of the phrase *pên tshao*[13]? If the words could be taken as two nouns, the combination, since *pên* has the meaning of 'root', would be wholly botanical—'woody plants with prominent roots, and grassy herbs'.[c] This indeed was the suggestion of an eminent scholar, Yang Ching-Fu, when in 1942 he was sharing with me the hospitality of the National Institute of Pharmacology directed by Ching Li-Pin at Ta-phu-chi in the beautiful hills around Kunming in Yunnan. But the most obvious construction would consider *pên* as adjectival and employ its abstract rather than its concrete meaning, not 'rooted' but 'essential, original or principal', so that one would think of 'basic

---

[a] Recently there appeared a large history of this literature by Unschuld (*1*) in German, but unfortunately too late to be of use to us in our work.

[b] They were, of course, well known to scholars in general, and echoes of them can be found in unlikely places. For instance there could be political nicknames or satire parodying *pên tshao* style. The *Lei Shuo*[5] of +1136 quotes a couple of pieces of this kind from a Thang book, the *Yü Shih Thai Chi*[6] (Memoirs of the Censorate) by Han Yüan[7] or Wei Shu[8]. Chia Chung-Yen[9], it seems, wrote a *Yü Shih Pên Tshao*[10] and Hou Wei-Hsü[11] a *Pai Kuan Pên Tshao*[12]. See *Lei Shuo*, ch. 6, p. 25*a*, *b*.

[c] Cf. p. 127 above on the ancient classification of Theophrastus.

---

[1] 神農本草經   [2] 單藥   [3] 食物   [4] 炮製   [5] 類說
[6] 御史臺記   [7] 韓琬   [8] 韋述   [9] 賈忠言   [10] 御史本草
[11] 侯味虛   [12] 百官本草   [13] 本草

herbs' or, somewhat more sophisticatedly, of 'fundamental simples'.[a] If a grammatical inversion could be permitted, making *tshao* adjectival, the phrase would then mean 'the herbal foundation,[b] or 'the botanical basis (of pharmacy)', or 'the vegetable origin (of the art of healing)', etc. And in spite of philologists, this is just what, in the majority opinion of the competent commentators, it did mean. About +945 Han Pao-Shêng[1], writing the introduction to his *Shu Pên Tshao*[2] (Pharmacopoeia of the State of Szechuan), said:[c]

Among all the *materia medica* there are gems, mineral substances, herbs, products of trees, creeping things and the parts of beasts. Yet we speak of natural history as *pên tshao*[1]. This is because the great majority of the drugs are derived from herbs.

And in the +13th century a Japanese work, the *Pên Tshao Shih*[3] (Explanations of Natural History) has the following words:[d]

Among all the groups of *materia medica* there is none which exceeds the plants and herbs in number. So one goes by the majority, and hence we speak of *pên tshao*.

And in 1843 Mori Tateyuki (*1*), prefacing his reconstruction of the *Shen Nung Pên Tshao Ching*, repeated this venerable view.[e]

Play was also made, of course, with the fact that the character *yao*[7], drug or medicine, has a '*tshao-thou*', i.e. belongs to the plant radical, group no. 140, so fully discussed above (p. 118). The *Shuo Wên* (+121), ignoring the mineral and animal kingdoms, says simply that a *yao* is a herb that can cure disease (*yao, chih ping tshao yeh*[8]).[f] Since the phonetic part of the character is *lo*, happiness (= *yo*, music), someone was sure to assert a semantic significance in the whole, lyrical joy rising out of the relief of pain and suffering. Tamba Motokata[9], in his *Ju I Ching Yao*[10] (Essential Knowledge of the Learned Physician) attributes this (*c.* 1840) to Chao Ching-Chai[11], who also said that the mineral drugs were remedies suitable for immortals (*hsien*[12]) and alchemists,[g] not for ordinary people. Another idea was

---

[a] Even, alternatively, 'what originates from herbs'.

[b] 'Herbal' is, of course, a word with many overtones, and in a moment we shall discuss its applicability to the *pên tshao* literature. We shall give our reasons for not using the term as an appellation for this genre. Nor do we approve of a widespread tendency in recent times to call much of Chinese-traditional therapy 'herbal medicine' (or even 'herbology'). Plant drugs, and drug-plants, we can understand, along with all experimental pharmacologists, but there is no ground for identifying Chinese medicine with the 'herbalism' fad of the West.

[c] Cit. *CLPT, ch. 1, Hsü li shang, p. 1a* (p. 25.1) and *PTKM, ch.* 1A, p. 2b. Cf. Taki Mototane (*1*), p. 105.

[d] Cit. Mori (*1*), pp. 6, 7. His reference seems to mean that it was quoted in a *Chhien Tzu Wên* (presumably an *I Hsüeh Chhien Tzu Wên*[4], or Medical Thousand-Character Primer) written by a distinguished physician in the time of Koreyasu-shinnō, the seventh Kamakura shogun (r. +1266 to +1289). The *Chhien Tzu Wên* itself, in the genre of the *San Tzu Ching* (see Vol. 2, p. 21) but a more elaborate feat, was traditionally attributed to Chou Hsing-Ssu[5] (d. +521) though it must be much later.

[e] *Kai wei yao wu i tshao wei pên*[6], p. 6.

[f] Ch. 1B, (p. 24. 1).

[g] Vol. 2, pp. 139 ff. On Taoist prolongevity techniques, cf. and Vol. 5, pts 2–5; Ho Ping-Yü & Needham (4), p. 245.

[1] 韓保昇　　　[2] 蜀本草　　　　[3] 本草釋　　　[4] 醫學千字文　　[5] 周興嗣
[6] 蓋謂藥物以草爲本　　　　[7] 藥　　　　[8] 藥治病草也　　[9] 丹波元堅
[10] 儒醫精要　　　[11] 趙敬齋　　　[12] 仙

voiced by Ni Chu-Mo[1] in his *Pên Tshao Hui Yen*[2] (A Rearrangement of the Classification in the Pharmaceutical Natural Histories) of +1624; he said that as the great culture-hero Shen Nung (cf. pp. 237–8 below) had tested all the plants, so historically they had precedence over the mineral and animal kingdoms. Hsieh Sung-Mo (*1*) himself, who records all these views, suggests that the reason why even the *Shen Nung Pên Tshao Ching* started with mineral substances was the influence of the Han *fang shih*[3] and alchemists. This may be so, but surely the order minerals-plants-animals constitutes the most primitively obvious *scala naturae*, and if we today think of this in evolutionary terms, Chuang Chou in the −4th century, as we have seen (Vol. 2 pp. 78–9) was not very far from the same conception. In these occasional Chinese assertions of a primacy for the botanical world we might discern a parallel to the Galenical aversion to 'mineral remedies' so characteristic of the Western world, but this would be a mistake, for no Chinese pharmaceutical natural historian ever decried or excluded mineral and animal products. They were indeed in from the very beginning.

In his *Pên Tshao Ko Kua*[4] (*Materia Medica* in Mnemonic Verses) of +1295, Hu Shih-Kho[5] wrote a memorable passage.[a]

The pharmaceutical natural histories [he said], are (to the physician) what the historical books and the (dynastic) histories are to the scholar-official. If the scholars do not read the histories how can they know the qualities, achievements and personalities of the men (who brought about) the prosperity or downfall of countries? If (the physicians) do not read the pharmaceutical natural histories, how can they know the names, virtues, properties and active principles (of the minerals, plants and animals) which bring about health and longevity?

This raises the question, no longer deferrable, of how we are to translate the titles of the books of the *Pên Tshao* literature. We believe that the best way is to use as often as possible the phrase 'Pharmaceutical Natural History'. This can start from Thao Hung-Ching[6] about +500, because he first departed from the *Pên Ching*[b] system of classifying natural objects into the three therapeutic classes, adopting instead a division into minerals, herbs and trees, fruits and vegetables, cereal grains, insects and animals. Very significantly he also was the first, though far from the last, to add a category of *yu ming wei yung*[7], 'things that have a name (and description) but are not used (in medicine)'. As we shall see, the 'natural' classification developed in Sui, Thang and Sung, with such great men as Thang Shen-Wei,[8] and culminated in the work of Li Shih-Chen[9] and his successors from the end of the +16th century onward, but in between there was a return to a pharmacological classification in the work of Chang Yuan-Su[10] and Li Kao[11] (J/Chin and Yuan), though at a much more sophisticated level. The word

---

[a] Quoted by Taki Mototane (*1*), (p. 172).

[b] Here we use the abbreviation which nearly twenty centuries of Chinese biological writing has reserved for the *Shen Nung Pên Tshao Ching*. We shall continue this in what follows.

[1] 倪朱謨   [2] 本草彙言   [3] 方士   [4] 本草謌括   [5] 胡仕可
[6] 陶弘景   [7] 有名未用   [8] 唐愼微   [9] 李時珍   [10] 張元素
[11] 李杲

'Pharmacopoeia' we propose to reserve strictly for those works which were commissioned by imperial authority, as happened from the $+$7th century onwards; with a special exception in the case of the $-$1st-century *Shen Nung Pên Tshao Ching*, for which the translation 'Pharmacopoeia of the Heavenly Husbandman' has become so traditional,[a] and not inappropriately. The expression '*materia medica*' in titles we intend to reserve for books of predominantly pharmaceutical orientation, which will not be considered in the present Section. But what of the appellation 'herbal', which has quite often been applied by sound scholars (such as Swingle and Hummel) to the treatises of the *Pên Tshao* kind?

We have given a good deal of thought to this, and came to feel so uneasy about it that we decided never to use it. A herbal, says Arber (3), 'is a book containing the names and descriptions of herbs, or of plants in general, with their properties and virtues'. Characteristically more trenchant, Singer (14) says flatly 'a collection of descriptions of plants put together for medical purposes'. At first sight these definitions might satisfy, though the hieratic undertones of 'herbal' (cf. missal, processional, manual) are extremely inappropriate for the Chinese naturalists. But then doubts flood in. To begin with, all the Chinese natural histories were 'pandectal' from the beginning, including the mineral and animal kingdoms as well as the vegetable—they were always lapidaries and bestiaries no less than herbals. They also included far more than herbs, for they covered all the known cereal and forest products, seaweeds, algae and other cryptogams. Next comes a very important point, the relatively low nonsense-content of the Chinese treatises.[b] Lapidary, herbal and bestiary are all unsuitable names because they all imply a considerable amount of the fabulous and the magical; the great Renaissance 'herbals' were of course comparatively free from this, but the Chinese natural histories always were, and what nonsense there was, never at any time got into their illustrations. Though Singer said that the writers of the Greek 'herbals', e.g. Dioscorides (*c.* $+$50), believed in the 'direct attack' on disease, with 'no nonsense about theories', he meant here specifically pathological theories, and it was not long before the herbals of the West were swarmed over with all sorts of proto-scientific, quasi-scientific, or even sheerly superstitious, material. One has only to mention the doctrine of signatures,[c] and the close connection of botany with astrology.[d]

---

[a] And also for one or two much older works, now long lost (cf. p. 253 below).

[b] The different judgment of Bretschneider on this (cf. p. 23 above) arose from the fact that he was not well acquainted with the history of medieval botany in the West.

[c] This was the belief that all plants had been stamped by the Creator with some sign which indicated their usefulness to man. The example most commonly given of this is the fact that there are no poisonous Cruciferae, for each one bore in itself the holiest of symbols. But Paracelsus, for example, said that since the flowers of St John's wort go red as they decay, it was clear that they were good for curing wounds. And plants with milky sap were obviously advantageous galactogogues. On the whole subject see Arber (3), ch. 8; Jessen (1), pp. 195 ff.; Thorndike (1), vol. 6, pp. 294, 422 and *passim*; Quecke (1). The origins of the theory are obscure, but it was very powerful in the $+$16th and $+$17th centuries.

[d] Again see Arber (3), ch. 8, and for the earlier Middle Ages E. H. F. Meyer (1). It is strange that like the doctrine of signatures, astrological botany should have been so powerful in men's minds just during the rise of modern science in astronomy and physics. It was, as it were, a counter-revolution of the Middle Ages, doomed to failure, but viable for a while between Leonardo, Galileo and Linnaeus. Cf. Vol. 5, pt 4, p. 122.

> Wonderful tales had our fathers of old,
>   Wonderful tales of the herbs and the stars,
> The Sun was lord of the marigold,
>   Basil and rocket belonged to Mars.
> Pat as a sum in division it goes—
>   Every plant had a star bespoke—
> Who but Venus should govern the rose?
>   Who but Jupiter own the oak?
> Simply and gravely the facts were told,
>   In the wonderful books of our fathers of old.[a]

Then there was the strong current of 'emblems' and religious symbolism, with full many a sermon preached on the ivy, heliotrope or pelican as types and shadows of moral truths;[b] to say nothing of plants and animals purely fabulous like the bausor,[c] the *lignum paradisi* or the *mantichoras*. Sometimes a real plant, like the Solanaceous *Mandragora*, became the focus of a whole corpus of legend.[d] The early medieval herbals, such as the *Herbarium* of Apuleius Platonicus (= Apuleius Barbarus or Pseudo-Apuleius), dating from the +5th or late +4th century, are full of apotropaic magic and charms; there are Anglo-Saxon MSS of this, and it was printed about +1481.[e] One of its ways was to depict, in each crude illustration of a plant, the venomous animal for which it was supposed to be the antidote. We are certainly not saying that there was no nonsense in the contemporary Chinese works, but it was generally the sort of medieval natural history tale which the early Fellows of the Royal Society would have thought worth while investigating.

In a way, we are facing here a situation which we have met with before, the lack of a 'dark ages' in China.[f] It is a striking thought that Thao Hung-Ching was writing his *Pên Tshao Ching Chi Chu* (see below, p. 248) within a few years of +512, the date of preparation of the Anicia Juliana Codex of Dioscorides,[g] and like that *De Materia Medica* it was fundamentally rational and practical. But as we have seen, European plant lore was already then upon its downward way;[h] while Chinese natural history preserved a rational tone, admitting neither heavenly signs nor starry influences, nor holy emblems nor (to any great extent) charms and incantations, but adding many a sceptical 'it is said . . .' Perhaps one could gain an idea of the character of discourse in the *pên tshao* literature by opening the *Pên Tshao Kang Mu* at random and seeing what we find; suppose then that we fell

---

[a] Rudyard Kipling, *Rewards & Fairies*, p. 275.

[b] Cf. the books of Robin (1); Steele (1); Fischer (1).

[c] Arber (3), 1st. ed. pp. 29, 30; 2nd. ed. pp. 31, 32.

[d] Cf. Arber (3), 1st ed. p. 36, 2nd ed. p. 39. It was death to pull up the root (bifurcate like the legs of a human being) without elaborate precautions.

[e] Arber (3), chs. 2 and 3. Modern English version by Cockayne (1). Often the herbs were recommended as talismans, to carry about; cf. Arber, *loc. cit.* 2nd ed., p. 39. This recalls the −6th-century *Shan Hai Ching*; cf. Needham & Lu Gwei-Djen (1), and Sect. 44 below.

[f] Many places could be cited, but we shall refer only to the Section on geography in Vol. 3, p. 587. See too the discussion in Needham (45).

[g] Cf. p. 3 above, and Singer (14).   [h] See p. 5 above.

upon the entry for the centipede (*wu kung*[1]), *Scolopendra* spp. often *morsitans*.[a] What sort of thing were the naturalists affirming? First the *Pên Ching* (*c. −*100) classed its poison as a dangerous drug, able to counteract other poisons.[b] Thao Hung-Ching (+500) said that it was an enemy of snakes, biting their heads and eating their brains, an opinion which he supported from *Chuang Tzu* (*c. −*290) and *Huai Nan Tzu* (−120). He distinguished several species by the colours of the appendages, and recommended salt and mulberry juice in cases of centipede bite. Su Ching (+659) and Su Sung (+1061) agreed with the statements about snakes, and criticised Kuo Pho (*c.* +300) for confusing the centipede with an insect. Han Pao-Shêng (*c.* +940) gave oecological details and identified the best species by colour, as also did Khou Tsung-Shih (+1116), who recommended, however, the excrement of black fowls with garlic as a salve when bitten. Centipedes were, he reported, the natural enemies of slugs, whose paths they feared to cross, hence slugs were an antidote to centipede poisoning. Li Shih-Chen (+1596) added a wealth of further literary quotations, including some rather tall stories about the size of centipedes in the southern regions, and listed the affections for which centipede preparations were given—tetanus, infantile trismus, scrofula, facial paralysis, snake bite and other diseases thought to be caused by snakes, cramps in the extremities, and gangrene of the toes in women. Of course all this is not modern biological and pathological science—but nor is it the wholly unverifiable pseudo-science of the Western herbals.[c] The appearance of scrofula is especially

---

[a] *PTKM*, ch. 42, pp. 12a ff.

[b] There is no doubt about the great toxicity of the venoms of some of the Myriapoda, especially the large tropical and sub-tropical species, but biochemical identification has not yet proceeded far; cf. Phisalix (1); Kaiser & Michl (1). Like the snake venoms, however (see Slotta, 1), they include: (*a*) curare-like neurotoxins with peripheral action, (*b*) circulation poisons lowering blood-pressure catastrophically and giving shock phenomena, (*c*) histolytic, haemolytic and haemorrhagic substances. The old Chinese naturalists were far from imagining things when they made preparations from centipedes, always on the entirely sound Paracelsian principle that the worst poison may be in certain conditions the best of drugs. As is well known, viper venom has found employment in modern dental surgery, for it is haemostatic as well as haemolytic.

Already in +1798 Donovan (1) drew attention to the unwelcome presence of the Scolopendromorpha in China, though perhaps he over-estimated the toxicity of their venom. It is not lethal for man, only intensely painful, because of the presence of the neuro-transmitter serotonin in it; but it has neurotoxic and haemolytic proteins of low molecular weight which paralyse and immobilise the usual prey (Lewis (1), pp. 156 ff.). Venoms are of great interest to pharmacologists today, and much research is going on into their possible use in combating human diseases.

[c] The material challenges us to compare with modern knowledge what the old Chinese naturalists said. First it is indeed the case that centipedes attack and eat slugs occasionally, though they probably avoid their slimy tracks. They also feed upon many insects and their larvae, as well as worms, which they poison first with their venom, but higher animals are not immune from assault, notably small snakes, small gecko lizards, and even small birds; while in captivity they are generally fed with small mice (Cloudsley-Thompson (1), 1st ed., pp. 50–5; Lewis (1), pp. 167 ff., 172 ff., 177 ff., 183 ff.). This carnivorous voracity is not characteristic only of centipedes, but of many related groups such as the Solifugae (false spiders or wind-scorpions) among the Arachnids (Cloudsley-Thompson (1), pp. 87 ff., 90). Centipedes may also feed upon plant tissues and exudates, however. On the other hand, larger reptiles and mammals, but also other arthropods too, including ants and spiders, are among the predators of centipedes (Lewis (1), pp. 153 ff.).

Finally it was very acute to maintain that centipedes were not insects, for indeed they now form a Class, the Chilopoda, quite distinct from the three great Arthropod groups of Crustacea, Arachnida and Insecta (Cloudsley-Thompson (1), pp. 15–6, 40–2).

[1] 蜈蚣

interesting, for it will be remembered that in Europe these tuberculous lymph-glands with suppurating abscesses were 'the King's Evil', curable, men said, by the 'royal touch'.[a] Whatever good Li's centipede venoms did, they could certainly not have done less good than that. Thus the Chinese pharmaceutical natural histories will not be called herbals and bestiaries by us.

The rational basis of the plant and animal drugs in them is really a separate question. 'Most herbal remedies', wrote Singer (14) 'are quite devoid of any rational basis.' But he wrote in 1927, and nowadays, since we know more, we would not be so sure. When we are clear about all the antibiotics, peculiar alkaloids, peptides, polyterpenes, glucosides, trace elements, co-enzymes and vitamins in all plants and animals, it will be time enough to sit in final judgment on the traditional pharmacopoeias.[b] Knowledge daily grows.[c]

At the beginning of her book, Arber (3) made a sharp distinction between the philosophical and the utilitarian attitudes in the study of plants, albeit admitting that they were rarely separated in pre-Renaissance times.[d] From the précis just given, and from our study of technical terms (p. 117 above), it will be seen how natural history as such was always mixed up in China with practical pharma-ceutical needs. But one may say that it became less and less purely practical as time went on. In the *Pên Ching* there is hardly any phytographic description, only the plant name and a recital of the pharmacological properties of the drug. Yet other books of the same period, as we shall shortly see (p. 246), especially those associated with the name of Thung Chün[1], were famous for their botanical descriptions, now alas long lost, though still fully available in Thao Hung-Ching's time.[e] If Singer (14) was right in identifying ten drawings in the Anicia Juliana MS of Dioscorides as belonging to the *Rhizotomikon* (Ριζοτόμικον) of Krateuas (*c.* −70),[f] this very early medical herbal would parallel the *Pên Ching* rather closely,

---

[a] Cf. Castiglioni (1), pp. 385 ff.; Garrison (3), p. 288; and the special monograph by Crawfurd (1). Again the +17th century was a period of great belief in the 'royal touch'. As late as +1712 Dr Samuel Johnson, as a boy of three, was 'touched', unsuccessfully, it seems, by Queen Anne.

[b] See further in Section 45 below. We are now quite convinced that the medieval Chinese pharmacopoeia contained many, if not indeed most, of the powerful drugs classically known in the West, or else alternatives to them, and in addition a number of others not there known; all of course generally unpurified in both civilisations.

Of course Singer's dictum depends on what is meant by 'rational'. All medieval drug-plants had been evaluated perforce empirically, and knowledge must have accumulated by long tradition from master to disciple, some of it certainly unreliable, much of it fully justified. Statistical analysis, which was what they needed, had to await the rise of modern mathematics at the gaming-tables of emergent capitalism. Similarly, modern pharmacological methods of experimental verification had to await the rise of modern chemistry and physiology. So if rational (deductive) is contrasted with empirical (inductive), Singer was right enough; but if it is contrasted with irrational (as most readers would take it to be), he was plainly wrong.

[c] For example, the action of *Schisandra chinensis* (Anon. (*109*), vol. 1, p. 800.2; R 512), one of half-a-dozen well-known anti-viral drugs in the Chinese pharmaceutical natural histories, can now be explained as du to the induction of interferon formation; cf. Yang & Yang (1).

[d] 1st ed., p. 1.

[e] And even as late as the Sui period.

[f] Krateuas was the Royal Physician of Mithridates VI Eupator (−123 to −63) of Pontus, himself a curious enquirer into pharmacology and medicine, and indeed none other than our old friend the owner of the first Western water-mill (cf. Vol. 4, pt 2, p. 366). Gunther (3) gives all the other drawings in the Anicia Juliana MS. as well.

[1] 桐君

for its text had only name and properties. By about +500, however, when Thao was writing, other traditions, such as those of Thung Chün and the *Erh Ya*, had been absorbed into the *pên tshao* complex. Dipping again at random into one of these pandectal books,[a] we may come upon the herb *niu hsi*[1] ('cow's-knee'), *Achryanthes bidentata* (≃*aspera*) of the Amarantaceae.[b] Thao Hung-Ching is quoted as saying:[c]

The plants that grow along the roadsides in Tshaichow are the best. Their leaves are large and glossy, and the stalks have nodes giving the appearance of the kneecaps of cows, hence the name. There are male and female plants, the male having purplish stalks and larger nodes.[d]

Whatever this is, it is no longer mainly pharmaceutics. Then from the middle of the +7th century onwards, with the great Thang compendia, plant description goes further forward, never thereafter looking back.

A still further reason for not calling the Chinese books herbals or bestiaries is that they never had *no* classification. They never descended to the abyss of despair represented by the purely alphabetical order of the ancient, medieval[e] and early Renaissance herbals in the West;[f] here was a case, perhaps, where the alphabet was a snare and a temptation, and the Chinese were lucky not to have one. Hieronymus Tragus (Jerome Bock) was the first to return to the Theophrastean classification (cf. p. 127 above), in +1539, but in China there had always been some framework.[g] As mentioned already, the *Pên Ching*'s framework in the Han was pharmacological, and this was resumed in much more elaborate form for a while in the J/Chin and Yuan periods, but Thao Hung-Ching's before the end of the Southern Chhi was basically naturalistic, in a way 'Theophrastean', and foreshadowing natural families, to which the Chinese books through the centuries slowly approached more closely. Thao's arrangement was at least more logical

---

[a] *CLPT* (+1249), ch. 6, pp. 24*b*, 25*a*.

[b] R 556; CC 1498.

[c] By name. *CLPT* never cites *Pieh Lu* or *Ming I Pieh Lu* (see p. 248 below) so the passage probably comes from the *Pên Tshao Ching Chi Chu* (see also below).

[d] Comment on plant sex reserved for a later point (Sect. 38 (*h*) 2).

[e] Alphabetical order was used in Galen's *De Simplicibus* (properly *Peri Kraseōs kai Dynameōs tōn Haplōn Pharmakōn*, περὶ κράσεως καὶ δυνάμεως τῶν ἁπλῶν φαρμάκων), written just before +180; as also in the *De Virtutibus Pigmentorum vel Herbarum Aromaticarum* written by Theodorus Priscianus, the Royal Physician of the emperor Gratian (+375 to +383). The Apuleius Platonicus MSS are, so far as I know, always alphabetically arranged. In the +4th century even Dioscorides was recast into this form (Singer (14), p. 24).

[f] See Arber (3), 2nd ed., pp. 124, 166. Alphabetical order prevailed in the Latin *Herbarius* of +1484, the German *Herbarius of* +1485, the *Ortus Sanitatis* and all its derivatives from +1491 onwards; even in Leonard Fuchs' *De Historia Stirpium* of +1542, and as late as Wm. Turner's 'Herball' (+1551 to 1568).

[g] The only parallels to European alphabetical order were in the books and tractates written about *materia medica* in mnemonic verses to help students and practitioners by their rhymes. These appeared in several periods. One can mention the *Pên Tshao Yin I*[2] of the naturalist and alchemist Li Han-Kuang[3], dating from about +750, which may have been of this sort; the *Pên Tshao Ko Kua* of Hu Shih-Kho (+1295) just mentioned (p. 6290), and the *Pên Tshao Shih Chien*[4] of Chu Lun[5] produced in +1739. We need not allude to these again. On medical education, see Sect. 44 below, and meanwhile Lu Gwei-Djen & Needham (2).

---

[1] 牛膝   [2] 本草音義   [3] 李含光   [4] 本草詩箋   [5] 朱鑰

than the muddle of Dioscorides.[a] Thus once again there were no 'dark ages' in the structure of the Chinese pharmaceutical natural histories.

Lastly, these very substantial works form landmarks down the centuries more evenly spaced than the herbal literature of the West. Counting all the lost books, Singer (14) began his Greek herbals from Diocles of Karystos (c. −350),[b] going then through the work of Theophrastus (c. −287) so familiar, and adding such little-known perished authors as Mantias (c. −270), Andreas of Karystos (d. −217) and Apollonius Mys (c. −200).[c] Then came Krateuas (c. −70) and Dioscorides (c. +50),[d] after which, broadly speaking, there set in for centuries what might be called the Age of the Copyists. The only important exceptions to this were the work of Galen already mentioned (c. +178) and the herbal of Apuleius Platonicus (first evidenced in the fragmentary Johnson papyrus of c. +400) about which we have also already spoken.[e] All these were then copied and recopied in various recensions traceable in the filiation charts which Singer worked out. Arber (3) on the other hand, began the period of the printed herbals from +1472 or so,[f] and ended it at +1670; this latter date she chose because the opening of modern botanical science, in her view, and very justifiably, began with the discovery of the function of the stamens.[g] After that point, she said, the herbal's line of descent bifurcated into that of the flora on the one hand and the pharmacopoeia on the other. The high period of the herbal, according to Arber, covered less than a single century, from about +1530 to +1614 as outside dates.[h] Here we must not anticipate what the following pages will discuss, but anyone who knows the Chinese literature is aware that there was not a single century between +100 and +1700 that did not see the appearance of at least one new and original work on pharmaceutical natural history, and some which saw many, especially if we include the remarkable monographic literature as well as the *pên tshao* books. In other words, once again, if Chinese botany had no Renaissance and no Linnaeus or Camerarius, it also had no 'dark ages' either (in the West from about +300 to +1500).[i] And the *Pên Tshao* are not 'herbals'.

[a] The sections of the *De Materia Medica* are: I, aromatics, oils, ointments, trees, II, animals and their products, cereals, sharp herbs, III, roots, juices, herbs, IV, herbs, roots, V, vines, wines and mineral substances. Cf. Arber, *loc. cit.* 2nd ed., p. 164.

[b] Approximately contemporary with the *Erh Ya* botanists.

[c] The two poems of Nicander of Colophon (c. −200) on poisons survive. 'In general form', says Singer, 'as well as in irrationality, these are in the main line of herbal tradition' (14), p. 3.

[d] Approximately contemporary, therefore, with the writer or writers of the *Shen Nung Pên Tshao Ching*. Cf. Sarton (1), vol. 1, p. 258.

[e] P. 3 above. About contemporary with the work of Thao Hung-Ching, though in this first form a century older.

[f] This was the date of the printing of the *De Proprietatibus Rerum* of Bartholomaeus Anglicus (cf. Steele, 1).

[g] Suggested by Millington and Grew in +1682 (Sachs (1), p. 382) and proved by R. J. Camerarius in +1691.

[h] The first signalising Otto Brunfels' *Herbarum Vivae Eicones*, illustrated by the woodcuts of Hans Weiditz; the second signalising the copper-plate engraved *Hortus Floridus* of Crispin de Passe.

[i] These dates directly correspond with those in the filiation charts of Singer (14).

(i) *Origins of the name*

The phrase *pên tshao* does not make its earliest appearances in Chinese literature as a component of any extant book title.[a] It occurs rather as the name of an expertise, in the Former Han period, and by implication in the Chhin. The first mentions of this are well worth examining. Though there are none in the *Shih Chi*, we find some in the *Chhien Han Shu*.

Under the emperor Chhêng Ti (-32 to -7) opportunity was taken by certain reforming ministers to urge reductions of the large staff of adepts, priests and magician-technicians at the court. In -31 two memorials were presented by the Prime Minister Khuang Hêng[1], in the course of which he said that there were no less than 683 established posts for these people, and that as many as 475 of them were not in accordance with proper custom; numerous temples and sacrificers had been founded since the time of Kao Tsu, and a lot of them ought to be abolished. Another high official, Chang Than[2], was associated with Khuang Hêng in these remonstrances. What interests us is that he demanded that

the Magician-Technicians in charge of Sacrificing to the Spirits (*hou shen fang shih shih chê*[3]), and the Assistant Experts-in-Attendance (for Observing the Vapours)[b] (*fu tso*[4]), and the Experts-in-Attendance for Pharmaceutical Natural History (*pên tshao tai chao*[5]), should all be sent home (i.e. dismissed from office), to the number of more than seventy people, and should return to their own habitations.[c]

The passage is quite revealing. Yen Shih-Ku added a commentary to the effect that 'the Pên-Tshao Tai-Chao officials were those who understood the botanical basis of pharmacy, and so were maintained in the posts of Experts-in-Attendance[d]'. The post of Adviser or Tai-Chao certainly went back to the Chhin, for Shusun Thung[6], the liturgiologist of the first Han emperor,[e] had been a Tai-Chao Po-Shih[7] or Professor-in-Attendance[f] in the previous dynasty. Now and in

---

[a] Except in so far as we may date much of the content of the *Shen Nung Pên Tshao Ching* as compiled in the Chhin and Chhien Han. We discuss the dating of this book on p. 243 below.

[b] This identification of the function of these experts is due to Yang Shu-Ta (*1*), p. 126. But though we may accept that they knew all about *hou chhi*[8] we cannot be quite sure what branch of proto-scientific activity is here referred to. In Vol. 4, pt 1, pp. 187 ff., we have given a full account of the curious practice of 'observing the *chhi*' or 'watching for the periodic arrival of the *chhi*', and this *hou chhi*[8] may well have been that. One of its founding fathers had only recently died—Ching Fang[9] in -37 (cf. p. 189). And it is a curious coincidence that another of these fathers, the polymathic scholar Tshai Yung[10] (p. 188) was also a botanist, as we shall shortly see (p. 259 below). On the other hand the significance of the phrase could rather be astronomical-meteorological (see Vol. 3, pp. 190, 476, 482), the watch for strange clouds and mists, auroras, sun-spots, etc. which might be important in the State astrology.

[c] *Chhien Han Shu*, ch. 25B, p. 13a, tr. auct. Attention was called to the passage by Nguyen Tran-Huan (2).

[d] These posts are also noted in the *Hsi Han Hui Yao*.

[e] The reader will not have forgotten him, from the colourful story about him in Vol. 1, p. 103.

[f] *Shih Chi*, ch. 99, p. 5b, tr. Watson (1), vol. 1, p. 291, with a slight difference of interpretation from ours. The circumstance is noted also in *Chhin Hui Yao*, p. 219. Shusun Thung was promoted to Po-Shih before the end of the Chhin, so it must have been a higher rank. On Po-Shih, which Dubs translated 'Erudits', we shall have more to say below, p. 268. Meanwhile *Chhin Hui Yao*, p. 135.

[1] 匡衡   [2] 張譚   [3] 候神方士使者   [4] 副佐
[5] 本草待詔   [6] 叔孫通   [7] 待詔博士   [8] 候氣   [9] 京房
[10] 蔡邕

the Later Han such men as Huan Than[1][a] and Ma Yuan[2][b] were at one time or another Tai-Chao.[c] So we have to visualise a whole group of pharmaceutical naturalists at the imperial court of the Han. This is perhaps a surprising fact, indicating as it does the rather high respect which the sprouting sciences of mineral substances, plants and animals received from the Chinese ruling classes of the −1st century.[d]

Whatever exactly happened to the Experts-in-Attendance for Pharmaceutical Natural History in Khuang Hêng's time, the atmosphere changed when power was assumed in the new Hsin dynasty from +9 onwards by Wang Mang, who was a veritable patron of inventors, magician-technicians and proto-scientists of all sorts.[e] It was under his auspices, while still chief minister of the emperor Phing Ti in +5, that what I like to think of as the First Chinese National Science Congress was called together. We have already given the essential quotation in Vol. 1, but we repeat it here, in slightly improved form, because of its great interest for botany in particular as well as the sciences in general.[f]

A convocation (of assembly) at the court was sent out to all persons throughout the empire who were learned in the lost classics and ancient records, in astronomy (*thien wên*[3]), calendrical science and mathematics (*li suan*[4]), and the acoustics of the standard musical tones of bells and drums (*chung lü*[5]); in philology (*hsiao hsüeh*[6]) and historical writings (*shih phien*[7])[g], in magical, medical and technical arts (*fang shu*[8]), and in the botanical basis of pharmacy (*pên tshao*[9]); as also in the (explanation of the) Five Classics[h]

---

[a] Eminent sceptical philosopher (−43 to +28) already encountered in Vol. 2, p. 367 and Vol. 4, pt 2, p. 392. Much new work has been devoted to him by Pokora (2, 3, 4, 8, 9). He almost certainly attended the 'scientific congress' about to be described.

[b] Eminent geographer, hydraulic engineer, and naval commander (*fl.* +20 to +49). Cf. Vol. 4, pt 3, pp. 27, 303, 442 ff.

[c] *Tung Han Hui Yao*, p. 241.

[d] What happened to Khuang Hêng? He lost his job in the following year. Though a majority of the ministers agreed with him in the retrenchment of the clergy and the cosmological rationalisation of the principal temples, the mass of the people feared that fires, tempests and sterility would result from such innovations. So 'reforms' were long held up, and perhaps many pharmaceutical botanists continued to receive their official salaries after all. A detailed account of the events of that time and of Khuang's part in them is given by Loewe (6).

[e] Cf. Vol. 1, p. 109.

[f] *Loc. cit.* p. 110, from *Chhien Han Shu*, ch. 12, p. 9a, tr. auct. adjuv. Dubs (2), vol. 3, p. 84.

[g] Dubs takes these two words to refer to the ancient orthographic word-list about which we have already spoken (p. 197 above) made by Chou[10] the Chronologer-Royal, in the late −9th or early −8th century. The invention of the 'greater seal' (*ta chuan*[11]) script was usually attributed to him. But the title actually given in our main source (*CHS*, ch. 30, p. 13b) is *Shih Chou, shih-wu phien*[12], the *phien* being the number of chapters, fifteen, and not part of the title (cf. further, p. 194 (g) above); and moreover, the other subjects mentioned in the list are certainly not book titles. On the other hand we also know that an important orthographic conference was precisely part of the proceedings of the cultural and scientific congress of +5, its results reaching publication under the editorship of Yang Hsiung, so that Dubs' suggestion has much to commend it. Adoption would mean translating *shih phien*[5] as 'the lexical chapters of the Chronologer-Royal'. We have preferred the more conservative rendering of 'historical writings' as fitting the context better.

[h] The *I Ching* (Book of Changes), *Shih Ching* (Odes), *Shu Ching* (Historical Classic), *Li Chi* (Record of Rites) and the *Chhun Chhiu* (Spring and Autumn Annals).

---

[1] 桓譚	[2] 馬援	[3] 天文	[4] 曆算	[5] 鐘律
[6] 小學	[7] 史篇	[8] 方術	[9] 本草	[10] 籀
[11] 大篆	[12] 史籀十五篇			

and the Confucian Analects (*Lun Yü*), the Filial Piety Classic (*Hsiao Ching*) and the Literary Expositor[a] (*Erh Ya*).[b] (These doctors travelled in) small (one-horse) government chariots bearing single-seal letters of credence. Several thousands of them assembled at the capital.

It is truly a misfortune for the history of science that the proceedings of this congress have not come down to us; they would be as useful as the *Yen Thieh Lun*[c], and far more so than the *Shih Chhü Li Lun*[d] and the *Pai Hu Thung Tê Lun*[e], all conference reports still extant more or less. The latter two deal only with social affairs, rites, ceremonies and precedents, administrative practices, etc., and we would much rather know what stars, what drug-plants, what natural knowledge, was discussed by the doctors of +5. This may be considered the second appearance of the phrase *pên tshao*.

The third occurs, as has been pointed out quite often, in the biography of a famous physician, Lou Hu[1], who was living during the 'science congress', in which, like Huan Than, he almost certainly participated (*fl.* −20 to +10). The *Chhien Han Shu* says:[f]

Lou Hu, whose *tzu* name was Chün-Chhing, was a man of Chhi.[g] Medicine had been the traditional calling in his family, and when he was young he followed his father in his practice at Chhang-an, where they often had occasion to frequent the residences of the relatives of the imperial house. (Lou) Hu studied (hard), chanting over the medical classics (or manuals), (and the traditions and writings on) the botanical basis of pharmacy, as well as the technical practices of adepts and healers (*i ching pên tshao fang shu*[2]) amounting to several tens of thousands of words. His elders were all deeply fond of him and valued him. People said to him: 'You have such natural genius, why not study to be an official?'[h] So after some time he left his father in order to peruse the literary classics, and indeed eventually he became a Metropolitan Hekato-chiliarch (*ching chao li*[3]).

This makes it quite clear that at the time of the congress there were already books on pharmaceutical natural history, some of them perhaps centuries old (cf. p. 253 below). It does not follow that Lou Hu abandoned the practice of medicine when he became an official, for although the post which he attained does not seem a very high one, his renown was so great at the time of the death of his mother in or

---

[a] The botanical significance of this will be apparent from pp. 126 ff. above.

[b] In the Han the Four Books as an entity did not yet exist. Only after +1177 they comprised the *Ta Hsüeh* (Great Learning) and *Chung Yung* (Doctrine of the Mean), both extracted from the *Li Chi*, plus the *Lun Yü* and Mencius (*Mêng Tzu*). *Erh Ya* and *Hsiao Ching* never formed part of either corpus.

[c] 'Discourses on Salt and Iron', the record of a conference that took place in −81. See Vol. 2, p. 251, and Vol. 5 *passim*.

[d] 'Report of the Discussions in the Stone Canal Pavilion', at a conference which was held in −51. See Vol. 1, p. 105, Vol. 2, p. 391.

[e] 'Comprehensive Discussions at the White Tiger Lodge', the proceedings of a conference held in +79. See Vol. 1, p. 105, Vol. 2, p. 391.

[f] Ch. 92, p. 7*b*, tr. auct.

[g] The significance of this will not be lost upon those who have read Vol. 2, pp. 240 ff.

[h] The significance of this too will be obvious to those who have read Vol. 4, pt 2, pp. 39 ff.

[1] 樓護　　　[2] 醫經本草方術　　　[3] 京兆吏

about +4, that her funeral was attended by persons in private carriages to the number of two or three thousand.[a]

While it is true that the bibliography of the *Chhien Han Shu* (though listing many scientific and medical works) contains no book title incorporating the phrase *pên tshao*,[b] this can nevertheless apparently be found in one of the bibliographer's statements about the groups of books in his catalogue. The group concerned is that of Ching Fang[1], which might be translated 'Manuals of Prescriptions'. But from the lilt of the style it seems that we should take *pên* here as a quasi-verb, implying the idea 'to know the fundamental properties of ...' The passage plunges one immediately into the intricacies of Han medical theory, and we shall refer to it again later on.[c] This is how it goes:[d]

To the right are the Ching Fang[1] books, comprising 11 authors with 274 chapters (or scrolls).[e]

Ching Fang means (to know) the fundamental properties, whether algorific or calefacient, of plants and minerals (*pên tshao shih chih han wên*[2]); to be able to estimate the gravity, whether benign or dangerous, of indispositions and illnesses (*liang chi ping chih chhien shen*[3]); to employ the nourishing *wei*[4] (sapidities)[f] of drugs, and take advantage of the resonance[g] of the *chhi*[h]; to distinguish between the Five Yin viscera (lit. acerbities, *khu*[5]) and the Six Yang viscera (lit. acridities, *hsin*[6]); all so as to attain the equalisation of the kidney and heart functions (lit. *shui huo*[7], i.e. Yin and Yang), in order to free what is blocked up and to unravel what is contorted. Prescribing calefacient (drugs) for calescant diseases, and fighting algidolesive illnesses with algorific (drugs), results in failure to achieve this balance, and losing all their advantages; then the internal vital resistance is injured even though one may see no external signs. This is a fundamental mistake. As the proverb says: 'When a disease is not cured, it is often because the physician is mediocre.'

Surely the bibliographer must have been helped by some medical friend to write this rather lapidary text.[i] Though it does not exactly provide a further example of the early use of the phrase, the wording is near enough to show how important the properties of the drug-plants were felt to be, and hence the botanical knowledge which could efficiently distinguish them. Now let us turn to the focal point of this whole literature, the *Shen Nung Pên Tshao Ching* and its dating.

---

[a] Cf. Vol. 4, pt 3, p. 30.   [b] On this see Nakao Manzō (2).

[c] The reader is referred to the appropriate Section, 44, for closer scrutiny of the unusual technical terms which we employ in the translation. Meanwhile, see Needham & Lu Gwei-Djen (9).

[d] *CHS*, ch. 30, p. 51b, tr. auct.

[e] Among the titles are many lost books which sound intriguing. *Thang Yeh Ching Fa*[8] was clearly a treatise on the preparation of decoctions. But what was *Shen Nung Huang Ti Shih Chin*[9], one wonders, a book on possible food substances that were dangerous to eat, or an account of secret immortality drugs?

[f] *Wei* in its simplest acceptation means just 'taste', but for the pharmacists it also meant what we should now call the nature of the active principle, and further it was a Yin quality opposite to the Yang *chhi* of the drug.

[g] On resonance, a basic conception in Chinese naturalist thought, cf. Vol. 2, *sub voce*.

[h] The Yang quality, as just explained, of the drug, but also that of the patient.

[i] In the literary sense, of course.

[1] 經方	[2] 本草石之寒溫	[3] 量疾病之淺深	
[4] 味	[5] 苦	[6] 辛	[7] 水火	[8] 湯液經法
[9] 神農黃帝食禁			

(ii) *The Heavenly Husbandman*

With one single exception,[a] which later we must take a closer book at, no ancient manuscript of the *Shen Nung Pên Tshao Ching*[1] (Classical Pharmacopoeia of the Heavenly Husbandman) has come down to us. Nevertheless, its text remained engraved, as it were, in the minds of all practitioners of medicine through the ages, because it was invariably quoted in full under the separate entries in each succeeding pharmaceutical natural history. From the end of the Ming onwards, therefore, many medical scholars made it their aim to reconstitute the original work by assembling all the quotations. The first to attempt this was Lu Fu[2] in +1616, basing his work on the then recently published *Pên Tshao Kang Mu* alone, but much remained to be done, for the sources of quotations were far wider than that. We need not follow this work in detail here,[b] and it may suffice to say that the best available reconstructions for use today are those of the Japanese Mori Tateyuki[3] done in 1845 (Fig. 36), and the *Shen Nung Ku Pên Tshao Ching*[4] which Liu Fu[5] prepared in 1942. It is surprising, in view of the relatively small size of the work, that no full translation into a Western language has ever been published.[c]

[a] The manuscript described by Kuroda Genji (*1*).

[b] After Lu Fu there was no further straightforward reconstruction of the text of the botanico-pharmaceutical classic until the work of Sun Hsing-Yen[6] & Sun Fêng-I[7] about 1802, based primarily on the *Chêng Lei Pên Tshao* (cf. p. 291). Wider sources were quarried by Ku Kuan-Kuang[8] for another version published in 1844, and Lung Po-Chien (*1*) describes several other works in the same category.

It was quite natural that another genre should have arisen in which the textual reconstructions were mingled with medical and pharmaceutical commentaries reflecting the writer's own opinions or those of his school. Thus in +1625 Miu Hsi-Yung[9] published a *Shen Nung Pên Tshao Ching Su*[10] (Commentary on the Text of the Classical Pharmacopoeia of the Heavenly Husbandman); cf. MW 556, Swingle (11). He probably anticipated the Suns in drawing on the *Chêng Lei Pên Tshao*, but his work has not been highly regarded, chiefly because it was part of a controversy on therapeutic methods with another famous physician Chang Chieh-Pin[11]. Nevertheless it was reprinted in revised form by Wu Shih-Khai[12] as late as 1809. We are fortunate to have in our library a copy of the original Ming edition. Next came the (*Shen Nung*) *Pên* (*Tshao*) *Ching Fêng Yuan*[13] (A reconstruction of the Classical Pharmacopoeia, etc.) with commentary, by Chang Lu[14] in +1695. Like Lu Fu, he depended on the *Pên Tshao Kang Mu*. Later on a selection of freshly culled entries appeared as the *Shen Nung Pên Tshao Ching Pai Chung Lu*[15] (A Hundred Entries from the Classical Pharmacopoeia, etc.), by Hsü Ta-Chhun[16] (Hsü Ling-Thai) in +1736; here the reconstruction was made from a Ming re-issue of the Sung *Ta-Kuan Pên Tshao* (cf. p. 282). Finally there was Tsou Chu's[17] (*Shen Nung*) *Pên* (*Tshao*) *Ching Su Chêng*[18] (Critical Commentary on (a Revised Text of) the Pharmacopoeia, etc.), with its two supplements, all finished by 1840 and printed posthumously nine years later; cf. MW 557, Swingle (6). This was based, it seems, on the ideas of a much earlier book, the *Pên Tshao Shu*[19] (Explanations of Materia Medica), written by Liu Jo-Chin[20] in +1699. An account of it will also be found in Swingle (6).

One can only understand this literature, of which we have named but a few examples, if one realises that the text of the classic was like Holy Writ to the traditional physicians and naturalists. It was an 'inspired document', in a sense not wholly different from that which Europeans thought applicable to their biblical books. No pains could therefore be too great to establish clearly every word of it.

[c] The old translations in du Halde (*1*), vol. 3, pp. 444 ff., and by Williams (*2*) in Bridgman's *Chrestomathy* are not, as sometimes thought, of the whole of the *Pên Ching*, as it was always familiarly called, but of its preface as reproduced by Li Shih-Chen in *PTKM*, ch. 1A (Hsü Li), pp. 43*b* to 55*b* (mod. ed. pp. 29 ff.) under the heading Shen Nung Pên Ching Ming Li. The preface, traditionally printed in white characters on black (cf. p. 250 below)

[1] 神農本草經	[2] 盧復	[3] 森立之	[4] 神農古本草經
[5] 劉復	[6] 孫星衍	[7] 孫馮翼	[8] 顧觀光	[9] 繆希雍
[10] 神農本草經疏	[11] 張介賓	[12] 吳世鎧	[13] 本經逢原
[14] 張璐	[15] 神農本草經百種錄	[16] 徐大椿	[17] 鄒澍
[18] 本經疏證	[19] 本草述	[20] 劉若金	

神農本草經　卷中

五八

不可持物。洗洗酸痛。除大熱煩滿及耳聾。

理石一名立制石。味辛寒。生山谷。治身熱。利胃解煩益

精明目破積聚去三蟲

長石一名方石。味辛寒。生山谷。治身熱。四肢寒厥。利小

便通血脈明目去翳眇去三蟲殺蠱毒久服不飢

膚青味辛平。生川谷。治蠱毒。毒蛇菜肉諸毒惡瘡

鐵落味辛平。生平澤。治風熱惡瘡瘍疽瘡痂疥氣在皮

膚中鐵堅肌耐痛鐵精明目化銅

當歸一名乾歸。味甘溫。生川谷。治欬逆上氣溫瘧寒熱

洗洗在皮膚中。婦人漏下絕子諸惡瘡瘍金創煮飲之。

Fig. 36. A page from Mori Tateyuki's edition of the *Shen Nung Pên Tshao Ching*. Besides several inorganic substances it has the entry for *tang kuei* (*Angelica polymorpha*).

How, one may ask, did the 'Heavenly Husbandman', Shen Nung, come into it? Readers of our book will hardly need to be reminded that this personage was one of the greatest culture-heroes of legendary Chinese antiquity,[a] the second of the 'three primordial sovereigns' (*san huang*[1]),[b] reigning as Yen Ti[2],[c] and the technic deity, arch-inventor and patron saint of all the biological arts—agriculture, tillage, animal husbandry, pharmacy and medicine.[d] This was the soil from which Chinese botany and zoology as sciences emerged. Opening the *Shih Chi*, we find the statement[e] that Shen Nung 'experimented with (lit. tasted, experienced) the hundred herbs, and so began the use of medicaments (*chhang pai tshao shih yu i yao*[3])'. A longer passage occurs in the *Huai Nan Tzu* book (*c.* −120).[f]

Anciently the people lived on plants and drank water, collecting the wild fruits from the trees and eating the flesh of grubs and mussels. They often got ill and were hurt by poisonous things. So Shen Nung began to teach them how to sow (and reap) the five cereal grains, how to assess the different soils and lands, and how to distinguish between the dry and the damp, the rich and the poor, the high and the low. He tested (lit. tasted)[g] the properties (*tzu wei*[4]) of the hundred plants, and the qualities of the water, whether sweet or bitter; and thus he caused the people to know what to avoid and what to accept. At that time in a single day they met with (as many as) seventy (plants with) active principles (lit. poisons).

It is true that in later times another tradition existed which ascribed rather to Huang Ti (the third of the 'three sovereigns'), and his medical counsellor Chhi Po[5],[h] the first systematic examination of medicinal plants,[i] but the fame of Shen Nung was little eclipsed thereby. And so it came about that his name was attached to the book on pharmaceutical natural history which destiny perpetuated

is also found, for example, in *CLPT*, ch. 1 (pp. 30 ff.), and in most of the reconstructed editions of the ancient work. A relatively up-to-date translation not only of the text but of the extensive commentaries also which Li Shih-Chen appended to it, was made by Hagerty (15), and is available, though not so far printed. Its interest is overwhelmingly for pharmacy rather than natural history, so we draw on it below (Sect. 45) and not here.

[a] Cf. Vol. 1, pp. 87, 163; Vol. 2, pp. 51, 120, 327. There is of course much further information in Granet (1, 2); Mayers (1) and similar sources.

[b] The historicisers of later times placed him variously from the −29th to the −27th centuries, but no such dates are acceptable today.

[c] 'The Blazing-Bright Emperor', because he was supposed to have reigned by the virtue of the element Fire. This seems at first sight odd for so biological a character, but the circumstance evokes the words of Sir Thomas Browne: 'Life is a pure flame, and we live by an invisible sun within us.'

[d] It is conventional in Western books on Chinese medicine to illustrate one of the well-known folk-pictures of Shen Nung dressed in a garment of leaves and munching some plant (e.g. Huard & Huang Kuang-Ming (2), opp. p. 48). We can dispense with this here. But see Figs. 38, 39.

[e] Ch. 1, p. 2*b*, tr. auct., adjuv. Chavannes (1), vol. 1, p. 13. This chapter was added by Ssuma Chên about +730, because there had been modifications in Five-Element theory since the time of Ssuma Chhien himself, and these necessitated adjustments in the proto-history of the imperial demigods; see Chavannes, *ibid.* p. ccxiv.

[f] Ch. 19, p. 1*a*, tr. auct., adjuv. Morgan (1), p. 220.

[g] The same word as before is used. Etymological affinity might, I thought, excuse this pun, but in fact test comes from Latin *testa*, a pot (e.g. for assay), and taste is derived from Latin *tangere*, to touch.

[h] His great interlocutor in the *Nei Ching*; cf. Lu Gwei-Djen & Needham (5), pp. 90–1.

[i] For this the *locus classicus* is Huangfu Mi's *Ti Wang Shih Chi* (Stories of the Ancient Sovereigns), written *c.* +270. The passage is quoted in *Yü Hai*, ch. 63, p. 5*a*.

[1] 三皇　　　　[2] 炎帝　　　　[3] 嘗百草始有醫藥　　　　[4] 滋味
[5] 岐伯

Fig. 37. Tshang Chieh as a feathered being, holding a four-stemmed plant, and seeking to attract the attention of the figure on the right, probably Shen Nung. One of the Ying-chhêng-tzu tomb-paintings (Mori Osamu & Naitō Hiroshi (*1*), p. 36 A).

for twenty centuries, ignoring many older and contemporary texts which may well have been equally worthy of preservation, or even more so. We may see him in Fig. 32, taken from a Han Stone cave-tomb relief at I-nan, studying a plant in collaboration with someone we have met before (p. 196), the Orthographical Master, Tshang Chieh, who will know exactly how to write its name.

From legend to fact—what is scientifically interesting is its content and arrangement. As typical examples of the former, almost any entries would do, but we shall choose a couple of them from very different plant families. The first is of the Umbelliferae, a powerful uterine stimulant derived from several species of the genera *Angelica* (*sinensis*, *polymorpha*[a] and *anomala*) and *Ligusticum* (*acutilobum*,[b] *ibukiense* and *japonicum*).[c] The second is from the very peculiar family Gnetaceae

---

[a]   CC 552. The chief source according to R 210.

[b]   Formerly *A. acutiloba*, CC 571. The chief source according to Lu Khuei-Shêng (*1*), p. 211.

[c]   See Chang Chhang-Shao (*1*), p. 100. On the drug, Schmidt, Read & Chhen Kho-Khuei (*1*). The roots contain a wealth of complex polycyclic organic compounds, especially lactones, including butylidene phthalide, but exactly which of them are responsible for the pharmacological effects remains undetermined. An extract used to be marketed in the West under the name Eumenol. See Lin Chhi-Shou (*1*), pp. 222, 227 ff., 230–1, 234 ff., 248; Anon. (*166*), pp. 433, 550–1; Roi (*1*), pp. 245–6.

Fig. 38. A Japanese depiction of Shen Nung, made during the time of the 5th Tokugawa Shōgun, Tsunayoshi (+1680 to +1709). From the Yushima Seidō Confucian temple, in Tokyo.

Fig. 39. Another Japanese representation of Shen Nung, a bronze standing figure holding a medicine gourd (date uncertain; Wellcome Historical Medical Museum).

(or Ephedraceae), intermediate between the gymnosperms and angiosperms, *Ephedra sinica*, which produces the well-known alkaloid ephedrine. It will immediately be seen that as in the case of the earliest Greek 'herbals' no botanical information is given at all. Nor have illustrations of the plants[a] accompanied the text down through the centuries, though it is not unlikely that it originally had them.[b] The text runs:[c]

Tang kuei[1] [also called *kan kuei*[2]];[d] its sapidity is dulcic and calefacient; [it grows in valleys and ravines.] It cures coughing due to ascending adverse *chhi*. Also the chills and fever of

---

[a] See Figs. 40 *a*, *b*, 41 from other sources.

[b] We shall deal shortly (p. 281 below) with the question of the beginnings of plant illustration in China, and return to the subject as a whole in sub-section *g*.

[c] Tr. auct., drawing on five reconstituted versions, pagination not given because different in each. Here we have made use of our standard equivalents of Chinese technical medical terms. On this subject see Needham & Lu Gwei-Djen (9); Needham (64), pp. 83 ff., 305 ff., 403–4, where we consider the question whether this learned language, devised for rendering medieval and traditional medical texts, is applicable without change to ancient ones.

[d] Certain phrases, enclosed in square brackets, may or may not have been in the original text; Mori thought so, most of the other scholars not.

[1] 當歸   [2] 乾歸

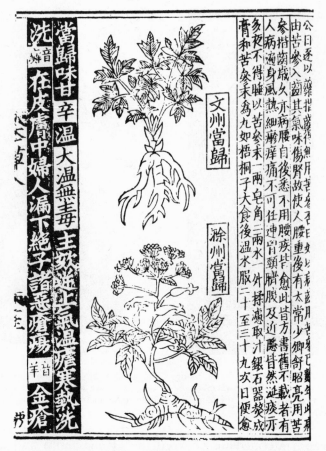

洗_音在皮膚中婦人漏下絕子諸惡瘡瘍_首金瘡

當歸味甘辛溫大溫無毒主欬逆止氣溫瘧寒熱洗

文州當歸

滁州當歸

Fig. 40a. *Tang kuei* (*Angelica polymorpha*) as depicted (from two provenances) in *Chêng Lei Pên Tshao*, ch. 8, p. 13a.

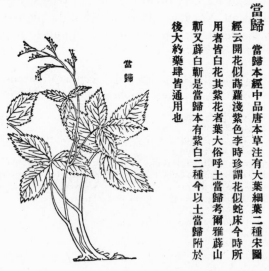

當歸

當歸　當歸本經中品唐本草注有大葉細葉二種宋圖
經云開花似蒔蘿淺紫色李時珍謂花似蛇床今時所
用者皆白花其紫花者葉大俗呼土當歸考爾雅薜山
蘄又薜白蘄是當歸本有紫白二種今以土當歸附於
後大約藥肆皆通用也

Fig. 40b. A late Chinese drawing of *tang kuei*, from *Chih Wu Ming Shih Thu Khao*, p. 583.

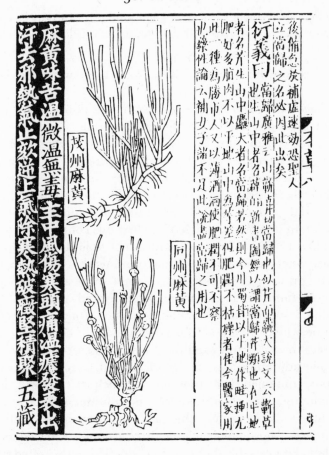

Fig. 41. *Ma huang* (*Ephedra sinensis*) as depicted (from two provenances) in *Chêng Lei Pên Tshao*, ch. 8, p. 14*b*.

the *wên nio*[1] disease[a] (that comes in springtime), with its drops of sweat on the skin. It also cures the vaginal affections of women, with discharges; and cases of infertility. It is further good against all kinds of skin diseases, boils and ulcers, including even wounds. Boil (the roots in water) and drink the decoction.

Ma huang[2] [also called *lung sha*[3]];[b] its sapidity is acerbic and calefacient; [it grows in valleys and ravines.] It cures feverish chills (*chung fêng*[4]),[c] algidolesive fevers (*shang han*[5])[d] and their accompanying headaches, as also the *wên nio*[1] disease. Its effects manifest themselves

[a] A description is given in Anon. (*35*), p. 131. The illness was slight, starting with pyrexia and chills, the fever lasting longer; but often running its whole course within a day and a night.

[b] Cf. p. 172 above. R 783; CC 2115; Lu Khuei-Shêng (*1*), p. 189.

[c] In later times this expression came to refer to all kinds of paralytic and hemiplegic conditions, but it is not used in this sense here. As Anon. (*35*), p. 33, explains, the illness referred to was probably infectious, due to an external malign *fêng* (cf. p. 154), showing a pulse floating and leisurely, with pyrexia, sweating and psychrophobia.

[d] This term has in modern times come to be loosely used for typhoid fever or any similar condition, but that cannot be accepted for the class of pyrexias recognised in traditional Chinese medicine as the algidolesive fevers (see further, Sect. 44 below).

[1] 溫瘧　　　[2] 麻黃　　　[3] 龍沙　　　[4] 中風　　　[5] 傷寒

externally in sweating, it drives away the malign calid *chhi*, it arrests the coughing due to ascending adverse *chhi*, it removes chills and fever, and it disperses obstructions in the bowels.

Next we must see how many entries of this kind there were, and how they were classified.

The *Shen Nung Pên Tshao Ching*, as it has come down to us, contains three chapters, with a supposed total of 365 entries.[a] The significance of the three chapters lies in the classification adopted, which was purely pharmacological, for the items were thus separated into three grades (*phin*[1]). The naming of these was inspired by the bureaucratic order of society, for those in the first chapter (*shang phin*[2]) were known as 'princely' (*chün*[3]), those in the second (*chung phin*[4]) were termed 'ministerial' (*chhen*[5]), while those in the third and lowest (*hsia phin*[6]) were defined as 'adjutant' (*tso shih*[7]).[b] Each of the first two contained 120 items (*chung*[8]) and the third had 125. Now on modern ideas one would expect that all the most powerful drugs, whether botanical, animal or mineral, would have been grouped together in the princely class, but this was not so at all; the mentality of the ancient Chinese naturalists was more sophisticated than that, health- and hygiene-minded, less pharmacodynamic, so to say. For the princely drugs were defined as those which were good for general health, containing no dangerous active principles (*wu tu*[9]), and capable of being taken constantly without untoward effects. The adjutant drugs, on the other hand, were available for therapy in acute infections, contained dangerous active principles (*yu tu*[10]),[c] had to be prescribed in small doses, and should not long be continued.[d] The ministerial drugs occupied an intermediate position. One can appreciate this ancient categorisation well by the aid of a diagram in which the three groups are placed on a graph with the minimal lethal dose as the abscissa, (Fig. 42). This unexpected system gives us two sociological hints which it is hard to ignore. Surely the ancient Chinese classification of drugs in three grades of potency, named as they were to deprecate the element of violence and compulsion or 'armed force' in the human physiological realm, mirrored the subjection of the military to the civil power so deeply characteristic of feudal bureaucratism. And is there not cause to fear that our own instinctively contrary expectation mirrors our own unconscious admiration of

---

[a] One can only speculate whether this number had any cosmological significance. It is in any case not correct as we have the book now, for some 18 duplications bring the total down to 347 items.

[b] One could say 'official' (or 'functionary') in contrast with the two higher ranks, but that would invite confusion with 'officinal', a notion quite out of place here. The phrase has also the sense of 'courier' or 'messenger'; and combines semantically the ideas of effective aid and powerful activity.

[c] Note the idea of 'using poison to drive out poison' (*i tu kung tu*[11]), and the dictum of Paracelsus: 'Alein die Dosis macht das ein Ding kein Gift ist' (in *Sieben Defensiones*, +1537; Sudhoff ed., vol. 11, p. 138; Strebel ed., vol. 1, p. 107.).

[d] Both our examples, *tang kuei* and *ma huang*, belong here.

[1] 品	[2] 上品	[3] 君	[4] 中品	[5] 臣
[6] 下品	[7] 佐使	[8] 種	[9] 無毒	[10] 有毒
[11] 以毒攻毒				

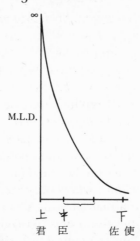

Fig. 42. Diagram to illustrate the principle of drug classification in the *Shen Nung Pên Tshao Ching*. Drugs in the uppermost or princely (*chün*) group were those with the greatest minimal lethal dose (M.L.D.). Drugs in the lowest or adjutant (*tso shih*) category were those with the most powerful active principles, i.e. with the smallest M.L.D. The intermediate or ministerial (*chhen*) class occupied all the intervening positions. Thus the drugs most prized were those which could be taken in any amount without causing harm, and pharmacological force (like all other forms of force) was deprecated.

obvious power, thaumaturgic effectivity and dominance over Nature as such, in the interests of no matter what? It only remains to say that the *Pên Ching* text contains the names of some 170 more or less recognisable diseases, and mentions many terms afterwards in common medical usage.[a] To add more now would be to take us too far away from the botanical sphere.

The dating of this first, or rather oldest extant, book of the *pên tshao* type, must now be brought into focus. Although it is unmistakably a Han work, it is not mentioned by name in any bibliography or other text of that period. It first appears under its present title in the *Chhi Lu*[5] (Bibliography of the Seven Classes of Books)[b] composed by Juan Hsiao-Hsü[6] in +523. Of course this does not prove that it was not circulating long before under another name, possibly current only among the common people outside the literary élite, possibly transmitted in secret by certain groups of physicians as part of their professional arcana.[c] It then appears, indeed in several forms, in most of the bibliographies, as in the dynastic histories,[d] down to the Yuan period. Just at the time when Juan the Taoist librarian was working, there lived one of the most eminent physicians, naturalists and alchemists in all Chinese history, often already mentioned in these pages,[e]

[a] E.g. how to use (*yung fa*[1]), principles of prescribing (*phei wu*[2]), decoction methods (*chih chi*[3]), contra-indications (*chin chi*[4]), etc.
[b] Cf. *Sui Shu*, ch. 33, p. 27*a*.
[c] Cf. the account of Shunyü I below (p. 258) on how he received books from his teacher.
[d] E.g. *Sui Shu*, ch. 34, p. 28*b*.
[e] Vol. 2, *passim*, Vol. 3, pp. 668, 675; Vol. 4, pt 1, pp. 234, 238, pt 2, p. 482; Vol. 5, pts 2–5, *passim*.

[1] 用法   [2] 配伍   [3] 制劑   [4] 禁忌   [5] 七錄
[6] 阮孝緒

Thao Hung-Ching[1]. The Mencius to 'Shen Nung's' Confucius, his work is inextricably bound up with the 'Classical Pharmacopoeia' for he was the first to
comment upon it extensively, and to convert its pharmaceutical grading into a
naturalist's order. His writings present a bibliographical problem of singular
difficulty, upon which we shall have to touch shortly; here it need only be said
(anticipating a quotation worth giving) that he recognised Later Han place-
names in the *Pên Ching* text[a] and suggested therefore that this was the time of its
origin. However, that gave only a *terminus ad quem*, a point after which no further
emendations took place, while the style and manner of the whole, as we can well
appreciate today, suggests rather the Former Han. Besides we know now (p. 255)
of a number of Chou and Chhin works on pharmaceutical natural history, all
long lost, and some perhaps already lost in Thao's own time; this tends to pull
back the *Pên Ching* text to the -2nd or -1st centuries. Such at any rate is the
opinion of the majority of Chinese medical historians.[b]

Thao Hung-Ching's work, the (*Shen Nung*) *Pên Tshao Ching Chi Chu*[2] (Collected
Commentaries, etc.) was therefore one of great importance. But it was lost quite
early on. He was probably finishing it about the year $+492$, before the Liang had
replaced the Southern Chhi dynasty, and it was probably still available in the Sui
time, for it is listed in the *Sui Shu* bibliography.[c] We can believe that it was used by
the Thang naturalists[d] (p. 264 below), but it had completely disappeared before
the time of Thang Shen-Wei (*c.* $+1090$), and the learned men of Ming, like Li
Shih-Chen, certainly never had a sight of it. In our own time, however, a manuscript of the Liu Chhao period (here the $+6$th century)[e] has been found among
the desert collections[f] and printed in facsimile (Fig. 43). Although it is only
fragmentary it is a document of great value, and an instructive passage from its
preface calls for quotation here.[g]

---

[a] There is a special study of the places of origin of the *Pên Ching* drug-plants by Li Ting (*2*).

[b] We may cite in support of this Chhen Pang-Hsien (*1*), p. 42; Chhen Chih (*1*), p. 69; Li Ting (*1*); Chang
Hsin-Chhêng (*1*); Yen Yü (*5*); Huang Kuang-Ming (*1*). A late Han or San Kuo date, however, is still preferred
by Kan To (*1*); Liang Ching-Hui (*1*); Hsieh Sung-Mo (*1*). For further information consult the extensive
collection of arguments and opinions assembled by Chang Hsin-Chhêng (*1*), vol. 2, pp. 964 ff.

[c] Ch. 34, p. 28*b*. Not in large type, however, but under the rubric often found: 'The Liang dynasty possessed
... but it is now lost.' Of course it was in the *Chhi Lu*.

[d] This is almost proved by an inscription dated $+718$ by one Yüchhih Lu-Lin[3] on the last page of the
manuscript. He must have been of Khotanese origin and was perhaps a physician, but though he says 'at the
capital', it is not clear whether he was the owner or one of the copyists.

[e] The dating has been established by the absence of Thang tabu forms of certain characters.

[f] The origin and present location of the manuscript is not clear to us. Lung Po-Chien (*1*), p. 16, says that it
was found at Turfan, presumably by one of the von le Coq expeditions, and is now in Berlin; Wang Chung-Min
(*1*), p. 151, includes it in his *catalogue raisonné* of Tunhuang manuscripts, and implies that it is now in Japan.
The lamentable confusion in current Tunhuang document studies may be appreciated from the account in
Twitchett (*5*). Our facsimile reproduction was edited by Fan Hsing-Chun in 1955. He pays credit to Lo Chen-
Yü for seeing its importance in 1915 and printing it later privately. Cf. Watanabe Kozō (*3*).

[g] Tr. auct. The preface alone was preserved in the *Chhi Lu*, so that Taki Mototane (*1*) was able to quote it
complete (pp. 104, 162, etc. but especially ch. 10, pp. 109 ff.). This text seems rather better than that in the
manuscript, which gives the impression of having been written down rather hastily to dictation. The rest of the
manuscript, of course, has no parallel except occasional short quotations in the pandects. The importance of the
preface was underlined by Hsieh Li-Hêng (*1*).

[1] 陶弘景   [2] 本草經集注   [3] 尉遲盧麟

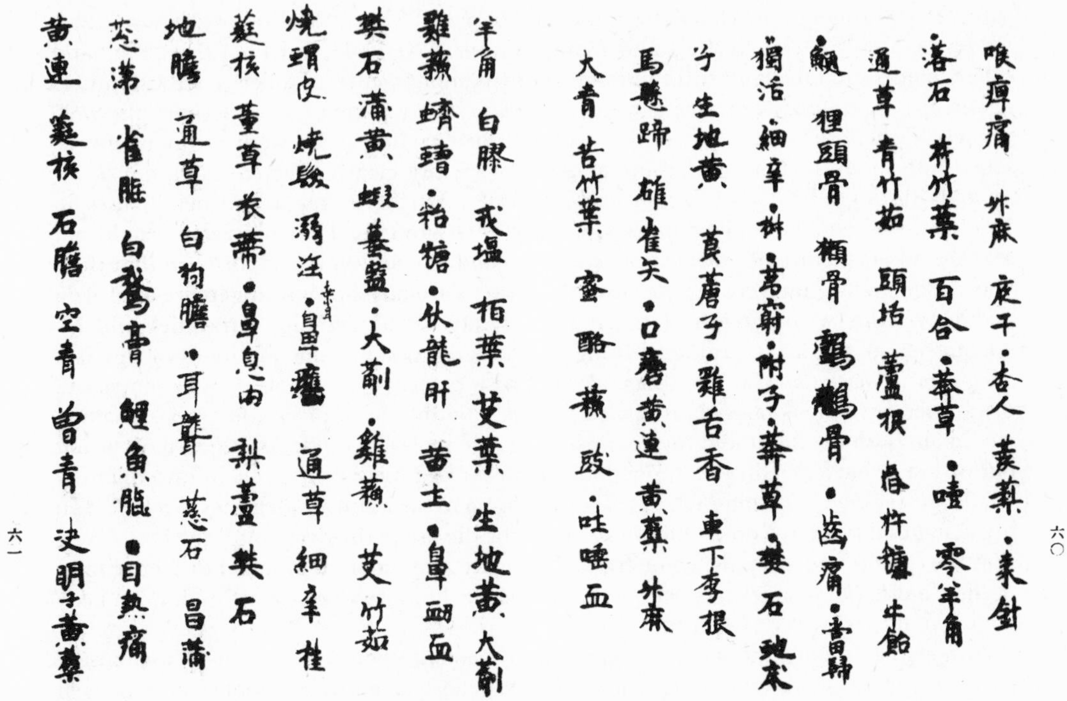

Fig 43. Two pages from the facsimile MS edition of the *Pên Tshao Ching Chi Chu* of Thao Hung-Ching. Note the colour code indicated by the dots above some of the entries.

When Chhin (Shih) Huang (Ti) burnt the books[a] he did not destroy those on medicine and divination, so people could still copy out fully the old material. But afterwards, what with the moving of the capital at the time of Han Hsien Ti,[b] and the disruption under Chin Huai Ti,[c] many literary collections were scattered or lost by fire, so that hardly one book in a thousand remained. Thus (the *Shen Nung Pên Tshao Ching*) as we have it has only four chapters.[d] Now the prefectures and districts mentioned in it as places of origin (of the plants) are mostly names of the Later Han period, so that I suspect it was recorded by (Chang) Chung-Ching[e] and (Hua) Yuan-Hua[f] and other people of that time.

There is also the *Thung Chün Tshai Yao Lu*[1] (Thung Chün's Directions for Gathering Drug-Plants); this gives descriptions of the flowers and leaves with their forms and colours (*shuo chhi hua yeh hsing sê*[2]). Besides this there is the (*Thung Chün*) *Yao Tui*[3] (Thung Chün's Answers to Questions on Drug-Plants); this discusses the 'adjutant' (*tso shih*[4]) drugs (in the

[a] Cf. Vol. 1, p. 101.
[b] R. +189 to +220. It was the period of anarchy which followed the 'Yellow Turbans' rebellion (Vol. 1, p. 112). See *TH*, vol. 1, pp. 798 ff.
[c] R. +307 to +312. At this time the north of China was appropriated by many dynastic houses of nomadic origin, and the Chin dynasty had to retire south to Nanking (Vol. 1. p, 119). See *TH*, vol. 1, pp. 898 ff.
[d] Note the discrepancy with what has come down to us.
[e] The most eminent physician of the Later Han period, (+152 to +219). Cf. p. 248 below. Also named Chang Chi.
[f] Better known as Hua Tho, the outstanding physician of the San Kuo period (+190 to +265). Cf. p. 247 below.

[1] 桐君採藥錄   [2] 說其花葉形色   [3] 桐君藥對   [4] 佐使

three-grade system), and their compatibility (*hsiang hsü*[1]) (with those in the other grades).

During the (San Kuo) Wei, and Chin, periods, Wu Phu[2] and Li Tang-Chih[3] and others made additions and subtractions, some giving 595 entries, others 431, and others again 319. Sometimes the three grades (*san phin*[4]) were mixed up, sometimes the properties of the drug, whether algorific or calefacient (*lêng jê*[5]), were wrongly given, sometimes again the plant and mineral origins were not clearly differentiated, nor lower animals distinguished from higher (*chhung shou*[6]). Furthermore, the principal uses in therapy were sometimes given correctly, sometimes wrongly. Thus physicians could not see (the whole picture of) the *materia medica* in complete and practical form, so that their knowledge was of uneven depth. Now therefore, carefully holding together, as if in a heddle, all the warp threads of the matter, I have examined the verbosities and inadequacies (of the literature), and using the three grades and the 365 items of the *Shen Nung Pên (Tshao) Ching* as my basis (*chu*[7]), I also present 365 supplementary entries of drugs used by famous physicians (*ming i fu phin*[8]), with their appropriate grades, making 730 in all. Both the finest and the coarsest drugs are included, so that nothing may be left out or lost. I have carefully examined all the items and have divided them into different headings (*kho thiao*[9]), giving each its place according to the natural categories (*wu lei*[10]).[a] In my commentary I have noted the (best) times for the use of the drugs, and the best places that the (plants and animals) come from. There are also things mentioned in the writings of the Taoists (*hsien ching*[11]) as necessary for their techniques (*Tao shu*[12])[b]—all have been included, in a total of 7 chapters.[c]

Although I do not claim any very great improvement upon what has been done in the past, the present work represents at least the best efforts of a single school. I shall be well content if after I have departed from this life it can be handed down to those who will understand the music that is contained within it.

There is indeed much to be learnt from this. Apart from the opinion on the date of the 'Classical Pharmacopoeia' in the first paragraph, it is particularly important to be reminded by a near contemporary of the frightful losses of ancient botanical literature that took place in the +2nd and the +4th centuries. Immediately afterwards Thao names the works of Thung Chün,[d] clearly a botanist of real significance, for he 'gave descriptions of the flowers and leaves with their forms and colours'. There is no way of telling his exact date, but it would not be going far wrong to place him about the time of the 'scientific congress' (p. 232 above); in the +1st century perhaps rather than the −1st. It is sad that his work so completely disappeared.[e] Apart from occasional quotations, the same has to be

---

[a] On categories (*lei*) see Vol. 5, pt. 4, pp. 305 ff.

[b] Cf. Vol. 2, pp. 143 ff.　　[c] The manuscript says 3.

[d] On the face of it, the name means Lord Thung or Master Thung, or the Master-under-the-*Paulownia*-Tree. According to some traditions, this personage, a minister of the legendary Yellow Emperor, Huang Ti, experimented (rather than Shen Nung) with the pharmaceutical and other properties of minerals, plants and animals. Teaching his disciples under a *thung* tree, he instituted the three-grade system. Most probably, therefore, the name in these titles is a pseudonym, analogous to Shen Nung in the *Pên Ching*, assumed by the +1st-century botanist, but one cannot quite exclude in this case the real name of a Han person.

[e] A book with a composite title, *Thung Chün Yao Lu*, was still available in the Sui (*Sui Shu*, ch. 34, p. 28*b*).

[1] 相須	[2] 吳普	[3] 李當之	[4] 三品	[5] 冷熱
[6] 虫獸	[7] 主	[8] 名醫副品	[9] 科條	[10] 物類
[11] 仙經	[12] 道術			

said about the books of Wu Phu[a] and Li Tang-Chih;[b] both worked in the Wei State during the San Kuo period, and both were disciples of Hua Tho, Li being the elder of the two. The third paragraph is of great value because it shows how Thao Hung-Ching was striving for a naturalistic classification instead of a pharmacodynamic one. Though we must reserve a closer look at this until a later point (sect. 38 f), it should here be said that his 'natural categories' (*wu lei*) were as follows: gems and minerals, herbs, cereal plants, vegetables, trees, fruit-trees, lower animals and higher animals. The last paragraph contains a personal statement of a kind relatively rare in the early literature of science in China. Thus we see that Thao Hung-Ching opens up the important subject of the lost botanical works of ancient times. To this we must return in a moment, but first it may be interesting to give a couple of later echoes of Thao's pregnant preface.

About +590, Yen Chih-Thui[4], discussing anachronisms in ancient books, pointed out in a discussion the Later Han place-names in the *Pên Ching*, but took a more moderate view about them than Thao Hung-Ching; these he said, do not belong to the original text' (*fei pên wên yeh*[5]).[c] So we still think. Then in the following century came the time when the 'National Pharmacopoeia' of the Thang dynasty (cf. p. 265 below) was being prepared. The *Hsin Thang Shu* enshrines a very interesting discussion which took place in the presence of the emperor about +655, worth reproducing here not only because it refers to Thao's opinions but because of the question it raises of how far medieval Chinese naturalists thought in terms of a long-continuing scientific progress.[d] The text reads:[e]

Before this, Yü Chih-Ning[6] had collaborated with the Minister of Works, Li Chi[7], in (supervising) the amendment and revision of a pharmaceutical natural history, to be in 54 chapters. The emperor (Kao Tsung) said: 'The *Pên Tshao* is something very ancient, yet now you are revising it again; what is the point of making a different compilation?'

(Yü Chih-Ning) replied: 'Formerly Thao Hung-Ching combined the *Shen Nung Ching* with miscellaneous writings of the various schools (*tsa chia pieh lu*[8]), commenting perspicaciously on them; but he was not completely familiar with the regional drugs and prescriptions of the provinces south of the River, so that among the plants and animals there were often mistakes. We have therefore examined and corrected more than 400 of his entries, and we have also added over 100 items of *materia medica* the use of which has been introduced in subsequent generations. Hence the differences (from former works).'

The emperor then remarked: 'Why should the *Pên Tshao* and the miscellaneous writings be considered as two things?'

(Yü Chih-Ning) replied: 'Pan Ku (in the *Chhien Han Shu*) recorded only the *Huang Ti Nei Ching* (The Yellow Emperor's Manual of Corporeal Medicine) and the *Huang Ti Wai*

---

[a] This was the *Wu shih Pên Tshao*[1] (Mr Wu's Pharmaceutical Natural History), *c.* +235.

[b] This was the *Li shih Yao Lu*[2] (Mr Li's Record of Drugs), *c.* +225.

[c] *Yen shih Chia Hsün*[3] (Mr Yen's Advice to his Family), ch. 6, pp. 14*b*, 15*a*.

[d] Cf. the discussion in Needham (56), and Vol. 7, Sect. 49.

[e] *Hsin Thang Shu*, ch. 104, p. 3*a*, tr. auct. The passage is not contained in the corresponding biography of the *Chiu Thang Shu*, ch. 78.

[1] 吳氏本草   [2] 李氏藥錄   [3] 顏之推   [4] 非本文也   [5] 于志寧

[6] 李勣   [7] 雜家別錄   [8] 顏氏家訓

Ching (The Yellow Emperor's Manual of Incorporeal (i.e. Apotropaic) Medicine)[a], but he did not list any (*Shen Nung* or other) *Pên Tshao*. (Books of this kind were) first named in the *Chhi Lu* bibliography of the Southern Chhi dynasty. Tradition has it that Shen Nung experimented with (lit. tasted) the drug-plants to ascertain their properties. Before the time of Huang Ti there were no written records, and knowledge was transmitted orally by personal contact. It was first written down on (bamboo) tablets in the time of Thung (Chün) and Lei (Kung)[b]. As the names of the prefectures and districts mentioned (in the *Shen Nung Pên Tshao Ching*) were those established in (Later) Han times, it was thought that Chang Chung-Ching and Hua Tho had been responsible for the text. The expression *pieh lu* (miscellaneous writings) refers to the works of Wu Phu and Li Tang-Chih, who were active in the (San Kuo) Wei and the Chin periods. They recorded the flowers and leaves, the forms and colours, of plants[c], as well as discussing what is primary and what is supplementary in the compounding of prescriptions, what compatible and what not so. And they included the classical texts in their discussions. Thus (Thao) Hung-Ching combined both sorts of writings (*Pên Ching* and *pieh lu*) and recorded them.'

The emperor said: 'Good!'. And so it came about that the pharmaceutical natural history (of the Thang dynasty) had a wide circulation.[d]

When discussing this a few pages below we must come back to the scientific statesmen Yü Chih-Ning and Li Chi; here what matters is the clear impression given by the dialogue that people in +655 felt themselves to be in possession of a larger amount of scientific knowledge about plants and animals than had been available in +492. This point is well worth making, but the passage leads us back to another of the more bibliographical kind.

(iii)   *The Famous Physicians*

For those who are acquainted with these subjects the words *pieh lu* ring a bell. From the Sui and Thang until our own time the book for which Thao Hung-Ching was best known was not the *Pên Tshao Ching Chi Chu* but the *Ming I Pieh Lu*[1], i.e. 'Informal (or Additional) Records of Famous Physicians (on *Materia Medica*)'. Many have been very puzzled as to just what this was. In the later pharmaceutical natural histories quotations are made sometimes from Thao Hung-Ching by name and sometimes from a *Ming I Pieh Lu*, or more often just *Pieh Lu*.[e] At the same time the *Pên Tshao Ching Chi Chu* is never mentioned.[f] The

---

[a]   On these titles see Needham & Lu Gwei-Djen (8); Needham (64), p. 272.

[b]   See p. 260 below.

[c]   Note the continuation of the botanical Thung Chün tradition.

[d]   A somewhat garbled version of Yü Chih-Ning's speech was reproduced in the *Shih Wu Chi Yuan* (+1085), ch. 7, p. 37*a,b*. The *Hsin Thang Shu* had been finished only twenty years earlier (+1061).

[e]   These latter were collected, about a century ago, so as to reconstitute as much as possible of the book, by Huang Yü (*1*). They formed the basis of the late +15th-century *materia medica* treatise of Wang Lun[2] entitled *Pên Tshao Chi Yao*[3]. Cf. Bretschneider (*1*), vol. 1, p. 53, and Fan Hsing-Chun's postface to the *PTCCC* manuscript.

[f]   Not even, so far as we can see, in Li Chi's *Hsin Hsiu Pên Tshao* (p. 266 below), written in the mid +7th century, when the *PTCCC* was still in circulation. But the *Hsin Hsiu Pên Tshao* is unusual in quoting hardly anybody.

---

[1] 名醫別錄   [2] 王綸   [3] 本草集要

Ming I Pieh Lu (which seems to take the place of it), though not known to the *Chhi Lu* cataloguer, was listed as a separate fully available book in the *Sui Shu*[a], and in the two Thang dynastic bibliographies, but not thereafter. The passages translated above, however, seem to put the key of the mystery in our hands. In his own preface to the *Pên Tshao Ching Chi Chu* Thao Hung-Ching says (p. 246 above) that he added '365 supplementary entries of drugs used by famous physicians, with their appropriate grades (*ming i fu phin*[1])'. Then Yü Chih-Ning, in his address to the emperor, says that *pieh lu* (informal or additional records) meant the work of Li Tang-Chih and Wu Phu in the +3rd century, i.e. exactly the same thing. Moreover, since the two phrases consist of characters easily confused if written cursively, one scents a textual corruption. These identifications explain why the impression got about that there was a *Ming I Pieh Lu* before Thao Hung-Ching, so that he could not have been the writer of it; in fact this was a half-truth, for what the *Ming I Pieh Lu* must have been was a disentanglement, made by other hands in the +6th or +7th century (between +523 and +618 or +656)[b], of the contributions of Li and Wu,[c] and the commentaries of Thao, from the text of the *Pên Ching* itself.[d]

In other words, it was the non-*Pên-Ching* part of the *Pên Tshao Ching Chi Chu*.[e] Alternatively, it was the non-*Pên-Ching*, non-Thao-Hung-Ching part of the *Pên Tshao Ching Chi Chu*. In this case we may suggest that when the Thang and Sung natural histories quoted the *Pieh Lu* or the *Ming I Pieh Lu* they were drawing on Later Han, San Kuo and Chin material, while when they quoted Thao Hung-Ching personally they were using material from his own commentaries in the *Pên Tshao Ching Chi Chu*. The distinction, whatever it meant, continued till the end of the tradition.[f] Since *fu*[4] is so similar a character to *pieh*[5], one thus suspects that a copyist's error may well have been involved somewhere in the naming of the later book.[g] This bit of the history of early scientific literature has itself been awkward

[a] Ch. 34, p. 31b.

[b] The dates of the *Chhi Lu*, the end of the Sui dynasty, and the compilation of the Sui bibliography respectively.

[c] And almost certainly other Hou Han, San Kuo and Chin materials as well.

[d] Already as early as +1236 scholars suspected that the disentanglement had not been perfectly done, for the librarian Chhen Chen-Sun[2], in his *Chih Chai Shu Lu Chieh Thi*[3], said that there was some admixture of *Pên Ching* with later text in the *Ming I Pieh Lu*. Hence indeed the efforts at reconstitution of the *Pen Ching* text in modern times already referred to (p. 235).

[e] Bretschneider (1) was quite acute about this. In vol. 1, p. 42, he accepted the usual view of the *Ming I Pieh Lu* as a separate book authored by Thao Hung-Ching, but in vol. 3, p. 2, he realised, from close reading of the *Pên Tshao Kang Mu* that *pieh lu* of some sort existed before Thao's time, and gave a substantially correct estimate of the contents of the *PTCCC*, though he never himself saw it.

[f] The only difficulty is that if the *Ming I Pieh Lu* contained nothing of Thao's own it is hard to account for the tradition of his authorship of it.

[g] Hsieh Li-Hêng (1), p. 6b, quoting Thao's preface to the *PTCCC*, absent-mindedly wrote *ming i pieh phin*. Hung Kuan-Chih (1), p. 14, corrected this. We ourselves read *fu* as *phin* when studying the *PTCCC* manuscript for the first time. It is interesting that the *Chêng Lei Pên Tshao* of +1249 used both phrases. It quotes (p. 25) Chang Yü-Hsi's preface of the +1060 *Chia-Yu Pu Chu Pên Tshao* (cf. p. 281 below), saying that Thao Hung-Ching added a *Ming I Pieh Lu* of 365 entries; but elsewhere (p. 29) quoting Thao's own preface, it reproduces clearly the words *ming i fu phin*. Hence Taki Mototane (1), p. 110, could get it right.

[1] 名醫副品   [2] 陳振孫   [3] 直齋書錄解題   [4] 副   [5] 別

to disentangle,[a] but if a Chinese emperor of the +7th century had to ask earnestly for explanations, we need not be too ashamed of failing to understand it over a thousand years later.

The *Pên Tshao Ching Chi Chu* is remarkable in yet another way, for Thao Hung-Ching adopted a two-colour technique in composing it. According to tradition, the *Pên Ching* text was written in red (*chu tzu*[1]) while the *pieh lu* additions and Thao's own were written in black (*hei tzu*[2]). This was shown to be true when the Liu Chhao (+6th-century) manuscript material came to light in modern times.[b] But it then appeared that besides this there was a flagging system to indicate the pharmaceutical properties of each drug. At the end of his introduction Thao Hung-Ching says[c] that in order to abbreviate long commentaries he used a red dot (*chu tien*[3]) above the entry to indicate a calefacient medicine (*jê yao*[4]), a black dot (*hei tien*[5]) to indicate an algorific one (*lêng yao*[6]), and if there was no dot at all the medicine was in this respect neutral (*phing yao*[7]). But there is an undertone here of 'peace-giving'—tonic, restorative, nutritive, etc. These dots can be seen in Fig. 43, which illustrates a page of the manuscript facsimile of Fan Hsing-Chun. In addition, most of the text was punctuated with little red dots or circles. It appears that the use of red and black characters, as well as the red markings, was continued in the manuscript treatises and pandects which circulated during the Thang period (cf. p. 264 below), i.e. in the +7th and +8th centuries; but when the Sung came in and printing entered common use, some other method had to be devised, for two-colour printing was at first too difficult, or perhaps not thought of.[d] What was done, therefore, was to reverse the red characters, like Yin seals,[e] so that they came out white on black, thus preserving quite separately the text of the *Pên Ching*. This seems to have been done from the *Khai-Pao Pên Tshao* (+974) onwards (cf. p. 280), and we illustrate here (Fig. 44) a page from the *Chêng Lei Pên Tshao* of +1249 which shows the system very well. Striking in this connection is the fact that when the great historiographer Liu Chih-Chi[8] wrote in +710 his treatise on history-writing (the first in Chinese or any other civilisation)[f] *Shih Thung*[9], he devoted a whole chapter to the help that could be given by the use of coloured inks in the marshalling of material. This was

---

[a] The words *pieh lu* have also proved confusing in another, entirely different, context. From −26 onwards Liu Hsiang and his son Liu Hsin put in order the imperial library, presenting the *Chhi Lüeh* (Seven Summaries) catalogue in −6. As we know, this was the basis of the bibliographical chapter in Pan Ku's *Chhien Han Shu* (*c.* +100). But the Lius had also written a report on each separate book, giving a contents table, a description of the material from which the collators had produced their standard version, a brief account of the writer and his historical background, and finally an opinion on the authenticity, transmission and value of the work. These reports, omitted from the *Chhien Han Shu*, were collected together in a book called *Chhi Lüeh Pieh Lu*, but it lasted only till the Thang. For further information see van der Loon (1).

[b] Cf. Kuroda Genji (*1*), and Fan Hsing-Chun's facsimile edition of *PTCCC*.

[c] Facsimile, p. 51.

[d] On the origins of polychrome printing in China before it attained such fame in Japan see Vol. 5, pt 1.

[e] See Vol. 5, pt 1.

[f] Cf. Pulleyblank (7); Needham (56).

¹ 朱字　　　　² 黑字　　　　³ 朱點　　　　⁴ 熱藥　　　　⁵ 黑點
⁶ 冷藥　　　　⁷ 平藥　　　　⁸ 劉知幾　　　　⁹ 史通

Fig. 44. A page from the *Chêng Lei Pên Tshao* to show the system of indicating quotations from the *Pên Ching* by printing them in white on black. Other examples occur in Figs. 25*a*, 40*a* and 41.

entitled Tien Fan Phien[1] (On the Complications of Dots); and he opened it by recalling how Thao Hung-Ching in former days had kept his data on pharmaceutical natural history straight by using inks of different colours. Thus a technique developed for a science proved helpful as a practical method in humanistic studies.

Natural history books which printed some of the text in Yin or white characters were called *pai tzu pên tshao*.[2] A rather curious passage is given under this head[a] in the *Wei Lüeh*[3] (Compendium of Non-Classical Matters), written by Kao Ssu-Sun[4] towards the end of the +12th century. It runs as follows:[b]

Thêng Yuan-Fa[5] said once that he had known a skilful physician who used only the drugs described in white characters in the *Pên Tshao*; he often tested them and proved their effectiveness.[c]

Su Tzu-Jung[6] (Su Sung[7])[d] said that the black characters indicated what the Han (and later) people had added. How could one not study very carefully such an important work as the *Pên Tshao*?

Chhüan Tê-Yü[8] in one of his poems said:[e] 'The lands of middling altitude and revenues were evaluated in the "Tribute of Yü", And the most beneficial drugs were tested by Thung Chün.'[f]

Li Chhün-Yü[9] in one of his poems said:[g] 'For commenting on drugs there was Thao (Hung-Ching), And for fathoming mountains there was Hsü Yuan-Yu[10]'.[h]

Wang Chi[11] in one of his poems said:[i] 'In travelling (to collect drug plants) one may follow the commentaries of Master Ko (Hung)[12];[j] In testing medicines at home one may follow the agricultural emperor (the Heavenly Husbandman, Shen Nung)'.

Tu Fu[13] says also in a poem[k] (about Chêng Chhien[14]): 'Turning to his pharmacopoeia, he would show you drug-plants from the Farthest West; Pointing to his palm, he would illustrate the ideas of the schools of military art.'

[a] Cf. Okanishi Tameto (2), p. 1224; Taki Mototane (1), p. 106.
[b] Ch. 10, p. 11b, tr. auct.
[c] Thêng was a well-known official and general in the Northern Sung, *fl.* +1080 to +1100. G 1909.
[d] Statesman and scientific man of many parts. Cf. Vols. 3, and 4, pt 2, *passim*. Another mention of these sayings of Thêng and Su occurs in the Sung book *Hou Ching Lu*, ch. 4, p. 5b.
[e] Chhüan was a minister and literary scholar in the Thang, *fl.* +785 to +820. G 507.
[f] Cf. p. 246 above.
[g] Li was an academician of Thang, *fl.* +845 to +860.
[h] This was one of the names of Hsü Mai (*fl.* +340 to +365), an alchemist and Taoist adept of the Chin period. A friend of the great Taoist calligrapher Wang Hsi-Chih, he lived in a mountain hermitage west of Hangchow.
[i] Wang was a scholar and specialist on wine (*fl.* +605 to +645). After holding minor office under the Sui, he retired to the country during the troubles attending the change of dynasty, and made his living by brewing millet wine, and planting and preparing medicinal herbs. Under the Thang he was called to court, and made famous wine under the expert Chiao Ko[15]. Several books and essays on wine and wine-making are due to him.
[j] The outstanding alchemist of Chin times; see Vol. 5, *passim*.
[k] The celebrated +8th-century poet. Chêng Chhien, the friend in whose memory he wrote the verses, was a pharmaceutical naturalist whom we shall meet below (p. 274). The whole of the poem is given in translation in von Zach (7), pp. 475 ff. For the identification of this reference we are much indebted to Professor William Hung.

[1] 點煩篇	[2] 白字本草	[3] 緯略	[4] 高似孫	[5] 滕元發
[6] 蘇子容	[7] 蘇頌	[8] 權德輿	[9] 李羣玉	[10] 許遠游
[11] 王績	[12] 葛洪	[13] 杜甫	[14] 鄭虔	[15] 焦革

Li I[1] says too;[a] 'The herbs and trees are divided into a thousand grades,[b] but the books of (Taoist) adepts speak only of the six conservable grains.'[c]

All these passages refer to this sort of thing.

Perhaps we can follow his drift in the light of what we have already seen in this sub-section.

Before leaving the *Pên Tshao Ching Chi Chu* of Thao Hung-Ching it will be worth while to have a look at one typical plant description of his, and for this purpose we have fallen upon a place where he is talking of thistles.[d] The passage runs:

Fei lien[2] (the welted or plumeless thistle, *Carduus crispus*).[e] This plant occurs nearly everywhere. It looks rather like the *khu yao*[3] (a plumed thistle, *Cirsium nipponicum*)[f] except that along the stems of the *fei lien*, under the leaves, there are fine skinny longitudinal excrescences like (the feathers of) an arrow. The feather-shaped leaves are also different because they have more indentations. The flower is purple in colour.

Ordinary people do not make use of this, but the Taoists take the stems to promote longevity. It is also an ingredient in 'magic pillow' prescriptions (*shen chen fang*[4]).[g]

Besides both these there is the *lou lu*[5]. This is something quite distinct (a globe thistle, *Echinops dahuricus*), and the name is not a synonym.[h]

From this one can see that he was a good enough botanist to distinguish successfully three Compositae that are placed in different genera today, and to describe the prickly stem ridges of *Carduus crispus* (Fig. 45) very clearly. For the end of the +5th century the account is admirably matter-of-fact. Evidently Thao was of the ancient school of Thung Chün, and believed in giving descriptions of 'leaves and flowers, with their forms and colours'. With this, then, we may turn to examine further what can be disinterred of the lost botanical literature.

(iv) *Botanical writings from Chou to Chhen (−6th to +6th century)*

If anyone were to ask about the date of the oldest Chinese written records specifically devoted to the study of natural history, especially that of plants, we should have to name a book and a circle of which we have so far said nothing at all. The title of the book was *Tzu-I Pên Tshao Ching*[6] (The Classical Pharmaco-

---

[a] Li was a poet and imperial tutor in the Thang period (*fl.* +760 to +830). G 1150.

[b] I.e. possessed of a thousand virtues.

[c] I.e. advising against the use of cereals as food.

[d] Cit. *CLPT*, ch. 7, p. 25a (p. 184.2) and *PTKM*, ch. 15, p. 52a. Tr. auct.

[e] CC 45; R 19; Steward (1), p. 412. The illustration is from *WCC/TK*, p. 256. A good modern drawing will be found in Sato (1), p. 12.

[f] CC 68; R 28; Steward (1), p. 413. *C. nipponicum* = *chinense*, formerly *Carduus lanceolatus*.

[g] These words are, it is true, the title of a book (see Okanishi Tameto (2), p. 857) but if the attribution to Sun Ssu-Mo is correct, it could not have existed in Thao Hung-Ching's time. The reference is to the practice of stuffing pillows with herbs considered hygienic, as has been done in China down to this day.

[h] CC 77; R 31; Steward (1), p. 411. *Echinops* has smooth ridges and a pappus of many short scales, unlike the feathery pappus of *Cirsium* and the hairlike bristles of the pappus of *Carduus*.

[1] 李益　　　[2] 飛廉　　　[3] 苦芺　　　[4] 神枕方　　　[5] 漏蘆
[6] 子義本草經

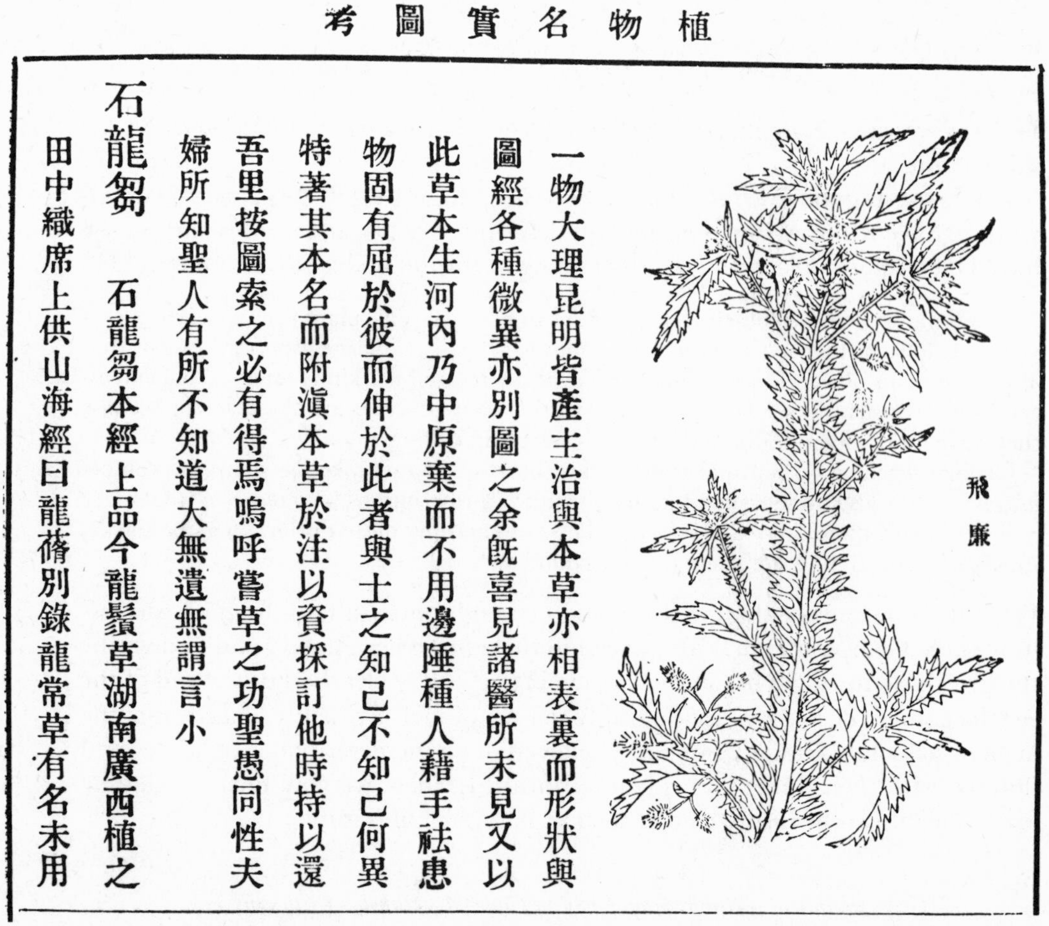

考圖實名物植

石龍芻

田中織席上供山海經曰龍蓨別錄龍常草有名未用

石龍芻本經上品今龍鬚草湖南廣西植之

婦所知聖人有所不知道大無遺無謂言小

吾里按圖索之必有得焉嗚呼菅草之功聖愚同性夫

特著其本名而附滇本草於注以資探訂他時持以還

物固有屈於彼而伸於此者與士之知己不知己何異

此草本生河內乃中原棄而不用邊陲種人藉手祛患

圖經各種微異亦別圖之余既喜見諸醫所未見又以

一物大理昆明皆產主治與本草亦相表裏而形狀與

飛廉

Fig. 45. A late Chinese representation of the thistle *Carduus crispus*, from *Chih Wu Ming Shih Thu Khao*, p. 256. Its Chinese name was always *fei lien*.

poeia of Tzu-I) and the evidence is that it was written during the lifetime of Confucius himself or shortly after.[a] Who was Tzu-I? In the *Chou Li* (which for the present purpose we may take as a −2nd-century text) we find a description of the duties of the Chi I[1] (the Physician-in-Ordinary for Internal Medicine at the imperial court),[b] to which Chêng Hsüan adds the +2nd-century commentary: 'in dispensing he uses the methods of Shen Nung and Tzu-I[2].[c] Then in the +7th century Chia Kung-Yen, most learned of commentators, appends a statement of old tradition that when Pien Chhio[3] attended the prince of Chao he was waited

[a] Attention has been drawn to this ancient work particularly by Tamba Motohiro (*1*), ch. 1, p. 1*b*, and Li Ting (*1*).

[b] Ch. 2, p. 2*b*. Cf. Needham & Lu Gwei-Djen (*1*).

[c] The orthographic difference is not significant.

[1] 疾醫　　　[2] 子儀　　　[3] 扁鵲

on by three chief disciples, Tzu-Ming[1] who knew how to make decoctions, Tzu-I who knew how to interpret the pulse, and Tzu-Shu[2] who understood massage. Pien Chhio's lifetime[a] extended from about -550 to -490, with a firm date for a famous consultation over a prince of Chin in -501,[b] and there is no reason why the three medical students of his should have been apocryphal, though nothing much has come down to us of their subsequent careers.[c] At all events the *Tzu-I Pên Tshao Ching*[1] was recorded in a bibliography called the *Chung Ching Pu*[3] compiled by Hsün Hsü[4] about $+280$, and no doubt it was also listed in the catalogue of his predecessor Chêng Mo[5], the *Chung Ching*[6] of about $+240$.[d] But by the time of the *Chhi Lu* and the subsequent dynastic bibliographies beginning with that of the *Sui Shu* it had quite disappeared. This means that it was available to the *Pên Ching* writer, as also to Li Tang-Chih and Wu Phu, but that it had been lost by about $+500$, so that it is questionable whether Thao Hung-Ching ever saw it. Of course, there is now nothing to prove that the pharmaceutical natural history attributed to Tzu-I was not a writing of the late Warring States time (-4th century), or of the Former Han, or even of the Later Han, fathered upon him by its real writer, just as the Shen Nung *Pên Ching* covered itself with an ancient name. The difference is that Shen Nung was a wholly legendary character, while Tzu-I has at least a very good chance of being a historical one. There is no obvious reason why someone should not have been writing down descriptions of plants on bamboo tablets about -480—the pity is that we do not have them now.

From the -5th century we may pass to the -4th and the -3rd. This is the period of the *Erh Ya* dictionary, already well known to us (pp. 126, 192 above, p. 467 below) for its great botanical importance, but there are one or two other texts of this time which need to be mentioned.[e] There is, for example, the *Shan Hai Ching*[8] (Classic of the Mountains and Rivers)[f], a work which has sometimes been described as the oldest geographical treatise of China. It is a very archaic text, containing some material which must be much older than the -4th century, perhaps even as old as the Shang (before the -1st millennium). The book is ostensibly a geographical account of all the regions of the Chinese culture-area. It

---

[a] Cf. Sect. 44 below. He has appropriately been called the Hippocrates of China, though his name did not become attached to the Chinese Hippocratic Corpus, the *Huang Ti Nei Ching*, on which see Sect. 44.

[b] See Lu & Needham (5), pp. 79 ff. The dating of Pien Chhio is in fact not without difficulties.

[c] An even older character is sometimes thought of (cf. e.g. Li Thao, 9) as having been the first writer on drug-plants, namely Chhang Sang Chün[7]. He is named in Pien Chhio's biography (*Shih Chi* ch. 105) as his teacher—a semi-immortal. But no book is attributed to him by name in any of the bibliographies.

[d] It is interesting that like the *Shen Nung Pên Tshao Ching* it was not included in the *Chhien Han Shu* bibliography.

[e] We omit here any discussion of ancient writings such as the *Shu Ching* (Historical Classic) and *Shih Ching* (Book of Odes), which contain many incidental mentions of plants—the latter especially—but cannot be considered part of the history of botany. As we shall see below (p. 463) the Chinese scholars spent much effort in trying to identify the plants named in them.

[f] Cf. Vol. 3, pp. 504 ff.

[1] 子明   [2] 子術   [3] 中經簿   [4] 荀勗   [5] 鄭默
[6] 中經   [7] 長桑君   [8] 山海經

contains indeed a good deal of mythology about strange beings, gods and local spirits, who were worshipped in different places, yet the tone is surprisingly matter-of-fact and the content includes a good deal of very rational description; for example, the minerals found in different places are recognised, the kinds of trees and animals that abound there, and the difficulties of communications. A count has reckoned (apart from the fabulous) mention of some 49 plants and 64 animals.[a] Among the former some 22 (10 herbs and 12 shrubs) were recommended as medicines, generally, however, not for curing diseases but for preventing their onset.[b] Under their age-old names it is easy to recognise a couple of Umbellifers, *chhiung chhiung*[1] (*Conioselinum vaginatum*)[c] and *mi wu*[2] (*Cnidium monnieri* or a related species)[d], ranunculaceous *shao yao*[3] (*Paeonia lactiflora*)[e]; and *mên tung*[4], which was certainly a lily, either *Asparagus lucidus* or *Liriope spicata*.[f] But the *Shan Hai Ching* is of course in no sense a work on natural history as such.

A very different book is the *Chi Ni Tzu*[5], now extant only in fragments, and also called the *Fan Tzu Chi Jan*[6]. Chi Jan[7] was the name borne in Yüeh State by a philosopher named Hsin Wên-Tzu[8] from Chin. His name is connected with two quite historical figures, Kou Chien[9] the prince or king of Yüeh (r. −496 to c. −467), and his minister Fan Li[10], afterwards a famous merchant; but Chi Jan's existence is more shadowy—he was, according to some traditions, the adviser of Kou Chien and the teacher of Fan Li.[g] Master Chi Ni, whoever he was, was a natural philosopher and (like the writers of the *Kuan Tzu* book far away in the north) interested in economics; at an earlier stage we gave some translations from his text.[h] In spite of the date of the actors, this can hardly have been written down before the time of Tsou Yen in the latter part of the −4th century, and some of it may well be a century later. The interest in all this for us here is that in its third chapter the *Chi Ni Tzu* book gives a list of entries 96 in number, which looks like part of a stock-in-trade inventory of some merchant who was also an apothecary.[i] The best place of origin is always given, and sometimes even the price. Of the list as a whole 64 items are plants, 15 are minerals,[j] 8 are animals or parts of animals, and 9 are manufactured articles or made-up products (Fig. 46). At one point a *Pên Tshao Ching* is quoted,[k] as likely Tzu-I's as that named after Shen Nung. We have not noted any study of this text by historians of botany or pharmacy in China, but without making a census of the items, it springs to the eye that the

---

[a] Chang Tsan-Chhen (2).
[b] Cf. p. 226 (e) above, and Needham & Lu Gwei-Djen (1).
[c] R 216. [d] R 231. [e] R 536. [f] R 676 and R 684 respectively.
[g] Hence the alternative name of the book, which implies that the teachings of the master were recorded by Fan Li.
[h] Cf. Vol. 2, pp. 245, 275 and 544 ff. An improved translation has been given in Needham (50).
[i] In *YHSF*, ch. 69, pp. 34a ff. Possibly also a surviving fragment of State fixed-price market regulations.
[j] Hence the interest of the book for the history of chemical technology and alchemy. On the chemical knowledge in the *Chi Ni Tzu* see Vol. 5, pt 3, pp. 14 ff.
[k] P. 35b.

[1] 芎藭   [2] 蘪蕪   [3] 芍藥   [4] 門冬   [5] 計倪子
[6] 范子計然   [7] 計然   [8] 辛文子   [9] 句踐   [10] 范蠡

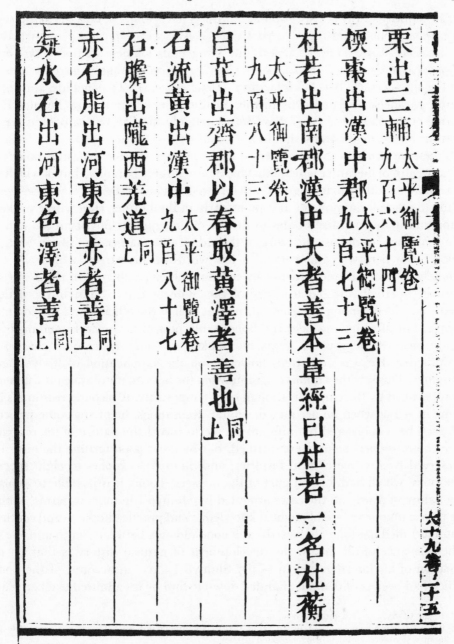

Fig. 46. A page from the *Chi Ni Tzu* book (−4th century), ch. 3, p. 2*b*. Besides sulphur and red bole clay, the writer mentions the chestnut (*Castanea vulgaris*), the date-plum (*juan tsao, Diospyros Lotus*), the *tu jo* (*Pollia japonica* of the Commelinaceae), and *pai chih* (*Angelica anomala*, an Umbellifer). The date-plum is closely allied to the persimmon.

great majority of them are contained in the Shen Nung *Pên Ching*. Taking some at random, one could mention *pa tou*[1] (*Croton Tiglium*)[a] powerful in its action, classical *ma huang*[2] (*Ephedra sinica*), the indigenous Chinese 'pepper', *shu chiao*[3] (*Zanthoxylum piperitum*),[b] the *tang kuei*[4] (*Angelica polymorpha*) just referred to above, and again *shu chhi*[5], 'Szechuan lacquer' (*Orixa japonica*).[c] To conclude, this interesting parallelism seems to us further justification for at least a Former Han dating of the *Shen Nung Pên Tshao Ching*. Certainly the *Chi Ni Tzu* book is a relic that the historian of botany should not neglect.

When we come to the Han period and the later shorter dynasties there is a great wealth of literature which must have been largely botanical but of which we have now only the intriguing titles. Before giving some of these,[d] it will be worth while to pause for a moment at a point early in the Former Han where we have some unusually good fixed dates in the life of the eminent physician Shunyü I.[6] We shall have to discuss the work of this admirable man more fully in Sect. 44 below,[e] but here we cannot avoid a few words on his career, which was fortunately recorded in much detail by Ssuma Chhien in ch. 105 of the *Shih Chi*.[f] Born in −216 in the old territory of the State of Chhi, Shunyü I practised medicine widely among the feudal princes as well as the officials and the common people, but after having held the honourable post of Granary Intendant from −177 onwards he was ten years later taken to court upon some charges of malpractice. He was acquitted, however, on the supplication of his youngest daughter. This was the famous occasion when the laws on mutilative punishments were revoked by the emperor, though only temporarily. A second crisis in his life came in −154 when on account of similar accusations, involving some princely families, he was ordered by imperial decree to reveal the nature of his practice. The commissioners again were satisfied, but by great good fortune the historian preserved twenty-five clinical histories, and Shunyü I's replies to eight specific questions, which had formed part of the dossiers. Today it is possible to explain the nature of nearly all the cases attended by Shunyü I in modern terms, so that we have a unique record of medical knowledge and practice in the −2nd century. Shunyü I died not long afterwards, at a good old age, between −150 and −145. The relevance of all this to the development of natural history is that on the occasion of his interrogation in −167 Shunyü I gave an account of the books which had been confidentially handed down to him by his admired teacher, Yang

---

[a] R 322; CC 857.   [b] R 360; CC 923.

[c] R 353; CC 915; Khung Chhing-Lai et al. (*1*), p. 1232.2. Note among this brief list by chance three products of Szechuan, which were doubtless being marketed in Yüeh.

[d] In the same sort of way we adopted for the literature on astronomy in Vol. 3, pp. 206 ff.

[e] Cf. also Lu Gwei-Djen & Needham (4); Needham & Lu Gwei-Djen (8).

[f] In our own time it has been the subject of what we regard as the most scholarly monograph yet produced by a Westerner in the field of the history of medicine in China—Bridgman (2).

[1] 巴豆   [2] 麻黃   [3] 蜀椒   [4] 當歸   [5] 蜀漆
[6] 淳于意

Chhing[1] (or Kungchhêng Yang-Chhing[2]). The specifically medical treatises, which may have been some early form of parts of the *Huang Ti Nei Ching*, will concern us later on (Sect. 44), but there were two books clearly of pharmaceutical natural history. One of these was entitled *Pien Yao*[3] (Drugs that effect Changes in the Body), evidently based on vegetable (and no doubt also animal) material; the other was called *Lun Shih*[4] (Discussions on (the Use of) Mineral Subtances in Pharmacy).[a] No parallel titles to any of Shunyü I's books can be found in the bibliography of the *Chhien Han Shu*, a fact which adds strength to the suggestion that they were rather the headings of chapters or tractates within a corpus.[b] On the other hand, it is interesting to come across in the *Sui Shu* bibliography a *Shih Lun*[5];[c] it has no author's name or any clue as to date, but it might conceivably be the ancient text which Shunyü I had studied in his time. It is a pity that he was not more of a plant collector, one who knew the fresh plants growing in the field.

From the *Chhi Lu* and *Sui Shu* bibliographies there is generally no way of telling what was the date of composition of the books they mention, but we can group them very provisionally.[d] We may therefore place as Later Han in date (before +220):

Shen Nung Pên Tshao Shu Wu[8] (The Plants, etc. in the 'Pharmacopoeia of the Heavenly Husbandman' arranged in a (Natural) Classification).   Writer unknown   SSL
Shen Nung Ming Thang Thu[9] (Illustrations (of Plants and) of the Parts of the Body (which they control, according to the ('Pharmacopoeia of the) Heavenly Husbandman).[e]
      Writer unknown   SSL
Tshai Yung Pên Tshao[10] (Pharmaceutical Natural History).   Tshai Yung   SSL

The last of these gives us another clear date, for Tshai Yung, whom we have often met before,[f] lived from +133 to +192. Best known as a mathematician, astronomer and acoustics expert, much inclined also to Taoist techniques and alchemy, this title reveals him as given to the study of plants as well. Here then is a botanical treatise which could be dated *c.* +170. The first title is interesting in

---

[a] The interpretation of the titles is rendered particularly difficult by the lack of punctuation, and we have not followed Bridgman (2) closely in this. He made out a *Yao Lun*[6] and a *Shih Shen*[7], the former reasonable, but the latter very bizarre.

[b] See further on p. 397 below.

[c] Ch. 34, p. 29*a*.

[d] The abbreviations SSL indicate that the *Sui Shu* records the book as having been available during the Liang period, and SS that it records it as being available during the Sui. One gets a strong impression in reading the *Sui Shu* list that it is roughly chronological.

[e] This translation is very conjectural; the words Ming Thang (cf. Sect. 44 below, and Vol. 4, pt 3, p. 80) were used later on in titles of anatomical texts. This book may have been concerned only with acupuncture and moxa, without mention of drug-plants.

[f] Cf. Vol. 2, p. 386, Vol. 3, pp. 20, 200, 210, 288, 355, 537, Vol. 4, pt 1, pp. 183, 188, Vol. 4, pt 2, p. 18.

[1] 陽慶   [2] 公乘陽慶   [3] 變藥   [4] 論石   [5] 石論
[6] 藥論   [7] 石神   [8] 神農本草屬物   [9] 神農明堂圖
[10] 蔡邕本草

that it suggests a dissatisfaction with the *Pên Ching's* pharmacodynamic classification system much earlier than Thao Hung-Ching. The second is obviously important for the early history of botanical illustration in China, a subject which we shall here defer (cf. Sect. 38 *g* below).

Possibly San Kuo period in dating come three more treatises on natural history:

Sui Fei Pên Tshao[1] (Pharmaceutical Natural History).   Sui Fei   SSL
Wang Chi-Pho Pên Tshao[2] (Pharmaceutical Natural History).   Wang Chi-Pho   SSL

It is likely that both these men worked about the same time as Li Tang-Chih and Wu Phu, i.e. before the end of the fourth decade of +3rd century. There follows a curious title which one might assign to the Chin (+265 to +420):

Than Tao Shu Pên Tshao Ching Chhao[3] (Manuscript of a Pharmaceutical Natural History for those concerned with Taoist Arts).   Writer unknown   SSL

This would surely have been of interest to Ko Hung and the other great adepts of his time such as Hsü Mai just mentioned (p. 252). Then from the Liu Sung dynasty we have several interesting items, among them the following:

Chhin Chhêng-Tsu Pên Tshao[4] (Pharmaceutical Natural History).
                                                          Chhin Chhêng-Tsu[a]   SSL
Sung Ta Chiang-Chün Tshan-Chün Hsü Shu-Hsiang Pên Tshao Ping Yuan Ho Yao Yao Chhao[5] (Manuscript on the Essentials of Pharmaceutical Natural History, Disease Aetiology and Drug Compounding).
                    Hsü Shu-Hsiang, General of the Sung Dynasty and Chief of Staff[b]   SSL
Hsü Shu-Hsiang têng ssu chia Thi Liao Tsa Ping Pên Tshao Yao Chhao[6] (Manuscript on the Essentials of Curing Various Bodily Ills by Vegetable and other Drugs).
                              Gen. Hsü Shu-Hsiang and three other writers   SSL

It might seem surprising that a high military officer should have been so prominent in the literature of pharmaceutical natural history, but those were days (which continued, after all, till the end of the nineteenth century) when armies were quite at the mercy of epidemic diseases.[c] So a general might well be a bit of a botanist at times. This is the period (+420 to +479) to which belongs another group of books, associated with the name of Lei Hsiao[7], but their interest is so purely medical and pharmaceutical that we shall speak of them elsewhere.[d]

From the +6th century come many more titles. Some that we can provisionally assign to the Liang are as follows:

---

[a] Cf. *PTKM*, ch. 1A, p. 53*b* (Hagerty (1), p. 61) and *CKI*, p. 2290.
[b] Cf. *Nan Shih*, ch. 32, p. 15*a* and *CKI*, pp. 2061, 2063.
[c] Cf. the discussion in Needham (59).   [d] Sect. 44 below.

[1] 隨費本草   [2] 王季璞本草   [3] 談道術本草經鈔
[4] 秦承祖本草   [5] 宋大將軍參軍徐叔嚮本草病源合藥要鈔
[6] 徐叔嚮等四家體療雜病本草要鈔   [7] 雷斆

I Pên Tshao Lu Yao Hsing[1] (The Natures of Plant and other Drugs in the Pharmaceutical
  Natural Histories).   Writer unknown   SS
Ling Hsiu Pên Tshao Thu[2] (Refined and Elegant Pictures of Medicinal Plants and other
  Natural Objects).   Yuan Phing-Chung[3]   SS
Chih Tshao Thu[4] (Illustrations of Mushrooms and Herbs). Writer unknown   SS
Ju Lin Tshai Yao Fa[5] (Methods of Collecting Drug-Plants in the Depths of the
  Forests).   Writer unknown   SS
Thai Chhang Tshai Yao Shih Yüeh[6] (Manual of the Court of Imperial Sacrifices on the Best
  Seasons for Collecting Drug-Plants).   Writer unknown   SS
Ssu Shih Tshai Yao chi Ho (Yao) Mu Lu[7] (Hand-list of the Best Seasons for Collecting and
  for Compounding Vegetable Drugs).   Writer unknown   SS
Chu Yao I Ming[8] (Unusual and Synonymic Names for Vegetable and other Drugs).
                                                         Hsing-Chü[9]   SS[a]
Chung Chih Yao Fa[10] (On the Planting and Cultivation of Drug Plants). Writer unknown
                                                                              SS
Chung Shen Chih[13] (On the Planting and Cultivation of Magic Mushrooms). Writer unknown
                                                                              SS

And then the list goes off into books of prescriptions. But here there is much of
interest. Two more books (the second and third) appear at this early date in the
roll of botanical illustration; the pictures of mushrooms in particular must have
been an extremely early landmark in the history of mycology, which was a rather
late-developing science in the West. The title of the last book of the group shows
that fungi of some kind were being regularly cultivated—hardly as food, with
that special designation, more probably medicinal, conceivably hallucinogenic.[b]
Other titles (the fifth, sixth and eighth) indicate that already at this time there
was systematic cultivation of drug plants in 'physic gardens'.[c] On the other hand,
it was also still necessary to collect plants from their natural habitats in remote
places (as is shown by the fourth item), and this is the kind of source (if only we
still had it) which would yield the greatest amount of really botanical information. Lastly, the work by the monk Hsing-Chü is of much interest in connection
with what we have said above (p. 143) about the efforts that were always being
made to impose order on the luxuriant growth of phytonyms. Although it was lost
after little more than a century, much of what was good in it was probably
incorporated into the conclusions of the naturalists of the Thang.

---

[a] This book was extant in the Sui, but the Thang compilers added a note saying that it had been lost by their
time, the middle of the +7th century.

[b] In the proper place (Vol. 5, pt 2, pp. 121 ff.) we consider the possibility that the 'magic mushrooms'
which took so prominent a part in Taoist religious symbolism from an early date may have owed this to their
hallucinogenic or ecstasy-inducing properties.

[c] There is much information about the organisation of the 'physic gardens' in Thang times (see Lu Gwei-
Djen & Needham, 2). The Court of Imperial Sacrifices comes into it because the Imperial Medical Service was
one of the organisations subordinate to it in the official bureaucratic hierarchy (see des Rotours (1), vol. 1,
pp. 315 ff., 339 ff.); this is also discussed in Lu & Needham (2).

[1] 依本草錄藥性   [2] 靈秀本草圖   [3] 原平仲   [4] 芝草圖   [5] 人林採藥法
[6] 太常採藥時月 [7] 四時採藥及合目錄   [8] 諸藥異名   [9] 行矩
[10] 種植藥法   [11] 種神芝

It seems that we can pin down a few more works to some of the shorter dynasties of the century. So in the Chhen (+557 to +587), if not before, there were written:

Kan Chün-Chih Yung Chü Erh Yen Pên Tshao Yao Chhao[1] (Manuscript on the Essentials of Vegetable and other Medicines for Boils and Ulcers, and in Otology and Ophthalmology).   Kan Chün-Chih   SSL

Pên Tshao Yao Fang[2] (Important Prescriptions based on the Pharmaceutical Natural Histories).   Kan Chün-Chih   SS

Wang Mo Chhao Hsiao Erh Yung Yao Pên Tshao[3] (Practical Materia Medica in Paediatrics; a Manuscript).   Wang Mo   SSL

Chao Tsan Pên Tshao Ching[4] (Manual of Pharmaceutical Natural History). Chao Tsan
        SSL

Here one can see the development of specialisation. From the Western Wei and Northern Chou (+535 to +554 and +557 to +581 respectively) comes the

Chen shih Pên Tshao[5] (Mr Chen's Pharmaceutical Natural History).   SS

and we can assume, as is likely enough, that Chen Luan[6], already known to us as a mathematician and astronomer,[a] was the writer of it. Perhaps we should put about the same time a work by another military officer, the

Yün-Hui Chiang-Chün Hsü Thao Hsin Chi Yao Lu[7] (New Collected Records on Vegetable and other Drugs).   Hsü Thao, Cloud-Signalling General.[b]   SSL

Finally there is no doubt that the *Yao Lu*[8] (Records on Drugs) of Li Mi[9] was written during the Northern Chhi dynasty (+550 to +577).[c]

Here we must end our survey of the lost botanical and quasi-botanical treatises of ancient times,[d] and proceed on our way past the landmarks of the literature which still exist. The foregoing details have not, we hope, proved tedious, even for those whose interests lie on the scientific rather than the historical side; some recognition ought surely to be given to the intensity of the work on drug plants which was done in China so long ago. It seemed a pity to leave all of it to languish entombed in the bibliographies and known only to a handful of humanistic scholars. Besides, we have not been at all exhaustive; there are at least a couple of dozen further books of the same kind listed under these categories at the time, and further research could bring to light much more about them. In its way, the accumulation of knowledge about plants and their properties, growing and en-

---

[a] Cf. Vol. 2, p. 150, Vol. 3, pp. 20, 29 ff., 33, 35, 58 ff., 76, 121, 205, Vol. 4, pt 1, pp. 259 ff.
[b] This title, indicating a general of particularly high rank, was in use from the Liang to the Thang.
[c] Not to be confused with his more famous namesake, the trebuchet artillery officer who founded the Thang dynasty.
[d] Attention has been drawn to them especially by Shih Tzu-Hsing (1) and Hung Kuan-Chih (1) from whose papers we gained much.

[1] 甘濬之癰疽耳眼本草要鈔   [2] 本草要方   [3] 王末鈔小兒用藥本草
[4] 趙贊本草經   [5] 甄氏本草   [6] 甄鸞   [7] 雲麾將軍徐滔新集藥錄
[8] 藥錄   [9] 李密

larging continually through these ages, in spite of all the wars, confusions and social disruptions, constituted a real epic.

One should not underrate the extent to which these botanical-pharmaceutical works were read and studied by scholars of general culture in these centuries. Hsieh Ling-Yün[1], for example (+385 to +433), the son of a Chin army commander, was one of the most prominent men of the Liu Sung dynasty, a wealthy patrician with hundreds of retainers; but he spent much of his life in retirement on his estates in the south, and there it was that he wrote most of his poetry, an ancient Wordsworth among the hills and streams of Chekiang. In his *Shan Chü Fu*[2] (Ode on Dwelling in the Mountains)[a] he wrote:

> What the books of plants and simples (*pên tshao*[3]) record
> Is a wealth of different things from mountain and marsh,
> Set in order by the Venerable Lei[4] and Master Thung[5],[b]
> Judiciously prescribed by the Physicians Ho[6] and Huan[7],[c]
> The three (kinds of) kernels they knew, and the six root (-types),
> The five (sorts of) flowers and the nine (varieties of) seeds.[d]
> The two *tung*[8] have the same names but very divergent natures,[e]
> The three *chien*[9] differ in form but come up at the same place.[f]

[a] Preserved in *Sung Shu*, ch. 67, p. 22a. We translate only a small part in which he mentions some medicinal plants all of which are in the *Pên Ching*. Hsieh Ling-Yün was a remarkable naturalist, interested also in geography, oecology, zoology, etc., and his writings would well repay study from the point of view of the history of science. There is an account of him and his poetry in Burton Watson (3), pp. 79 ff., 85–6, 98.

[b] The first of these is of course Lei Hsiao[10] (Lei Kung[11]) putative author of the *Lei Kung Yao Tui*[12] (Answers of Master Lei to Questions about Drugs) and other works. The reference here is interesting, because they have usually been placed in the neighbourhood of +470 while Hsieh Ling-Yün's poem would have been written about +420. Thus they should be ante-dated by at least half a century. The very obscure character Thung Chün[13] has been discussed above, p. 246.

[c] These are the two most prominent physicians in the *Tso Chuan*. I Huan's[14] *floruit* was about −580, and I Ho's[15] some forty years later. See Sect. 44, and in the meantime Needham (64), pp. 265–6; Lu Gwei-Djen & Needham (4).

[d] The commentary says that the first of these refers to double-kernelled peaches and apricots. From the *Pên Ching* (Mori ed., pp. 86, 107) we know that the third was the plum. The commentary then names the roots of six plants; the flowers of five, including the chrysanthemum; and the seeds or fruits of nine, including the lotus, the *Sophora* tree, the cypress and the dodder.

[e] These are both lilies (cf. Forbes & Hemsley (1), vol. 3, pp. 79, 102). *Thien mên tung*[16] is *Asparagus lucidus* (R 676; CC 1830; B III 176 and Anon. (*182*), p. 537). *Mai mên tung*[17] is *Liriope graminifolia* (= *spicata*), see R 684; CC 1864; Anon. (*182*), p. 535. The properties of their active principles differ considerably.

[f] These are all aconites. The place was Chien-phing[18], prob. mod. Wu-shan[19] district in Szechuan; hence the expression *san chien*. The commentary says that the three are *wu thou*[20], *fu tzu*[21] and *thien hsiung*[22]. *Wu thou* (*PTKM*, ch. 17B, pp. 4b ff.) has usually been identified as *Aconitum fischeri* (= *sinense*, = *japonicum*) as by R 523 and CC 1386, but Anon. (*109*), vol. 1, p. 695 replaces these names by *A. carmichaeli*. Everyone agrees that *fu tzu* were the seeds of this plant (cf. *PTKM*, ch. 17A, pp. 46a ff.). *Thien hsiung* was clearly another species of the genus, generally named *A. hemsleyanum*, as by R 524, but Anon. (*109*), vol. 1, p. 691 makes this *kua yeh wu thou*[23] and drops the name *thien hsiung*. For the classical description see *PTKM*, ch. 17B, pp. 1a ff. There are several other aconites described in the pharmaceutical natural histories, and Anon. (*109*), vol. 1, pp. 685 ff. gives no less than 41 species, so the identifications need some attention. Aconitine is one of the most toxic alkaloids known, and its association with other alkaloids in leaves, roots and seeds of the different species is most probably different. Although it hardly finds any place in modern medicine, one should always be on the look out for evidences of its multifarious pharmacological actions in relevant ancient and medieval Chinese prescriptions.

[1] 謝靈運	[2] 山居賦	[3] 本草	[4] 雷	[5] 桐
[6] 和	[7] 緩	[8] 冬	[9] 建	[10] 雷斆
[11] 雷公	[12] 雷公藥對	[13] 桐君	[14] 醫緩	[15] 醫和
[16] 天門冬	[17] 麥門冬	[18] 建平	[19] 巫山	[20] 烏頭
[21] 附子	[22] 天雄	[23] 瓜葉烏頭		

The *shui hsiang*[1] belongs to autumn's end, and then luxuriates,[a]
The *lin lan*[2] can't wait for the snow to melt before blossoming.[b]
The *chüan po*[3] survives for ages without decay,[c]
And the *fu ling*[4] grows for a thousand years, then gives
Pink petals carried on stems of green
And white buds set on purple branches,[d]
As time goes by its mysterious powers increase,
Powers to dispel all evils and cure the ills of mankind.[e]

(v) *Pharmaceutical natural history in Sui and Thang (+6th to +10th century)*

It is from the Sui and Thang periods onwards that most of our existing literature in this field derives. In the Sui dynasty the most important *pên tshao* specialists were two medical men also named Chen. Chen Luan the mathematician (*fl.* +535 to +577) is not to be confused with two other scholars of the same clan who wrote on pharmaceutical natural history a generation later.[f] Chen Chhüan[10], who died in +640 or thereabouts at the age of 103, and his younger brother Chen Li-Yen[11] (*fl.* +618 to +626) were both famous physicians, the latter being more interested in natural history than the former. Chen Li-Yen wrote a *Pên Tshao Yin I*[12] (Meanings and Pronunciations of Words in Pharmaceutical Natural History) and—perhaps in collaboration with his brother—a *Pên Tshao Yao Hsing*[13]. (The Natures of the Vegetable and other Drugs in the Pharmaceutical Treatises).[g] Both these works are now only accessible through quotations.

[a] The *shui hsiang* is certainly a *Eupatorium*. Khung *et al.* (*1*), p. 218.1; CC 85 and Anon. (*182*), p. 226 followed the traditional name *E. chinense*; but Anon. (*109*) now prefers *E. fortunei* and distinguishes the two species (vol. 4, pp. 410, 411) giving the former the name of *tsê lan*[5]. The last word of the line here, *chhien*[6], also means the dye-plant, madder, *Rubia cordifolia* (*chhien tshao*[7], CC 228, Anon. (*109*), vol. 4, p. 275). This also blooms in autumn, so perhaps the end should be: 'and so does the *chhien*'.

[b] This is clearly *Magnolia denudata* (= *obovata*, = *conspicua*, = *yulan*). See Khung *et al.* (*1*), p. 552.1; CC 1346; Anon. (*109*), vol. 1, p. 786.

[c] Strictly speaking, this is one of the Lycopodiales, *Selaginella involvens*. See R 794; CC 2162; B III 211; Wu Chhi-Chün (*1*), p. 381, (*2*), p. 676. Anon. (*109*), vol. 1, p. 111, prefers *S. tamariscina*, cf. p. 114. In the *Pên Ching* it is in ch. 1 (p. 34), Mori ed. But perhaps Hsieh really meant the *po* tree, or *pien po*[8], the cypress, *Biota* (= *Thuja*) *orientalis* (R 791; CC 2146; Wu Chhi-Chün (*1*), p. 713, (*2*), p. 925; Anon. (*109*), vol. 1, p. 317). In the *Pên Ching* it is in ch. 1 (p. 28), Mori ed., where the seeds alone are specified, and so for centuries thereafter.

[d] The *fu ling* is of course a fungus, *Pachyma Cocos* or *Poria Cocos*, growing on the roots of trees—the tuckahoe or Indian bread. See B III 350; CC 2320. The only explanation for this strange behaviour on the part of a fungus is that there was in China a very ancient association between the parasite *fu ling* and the epiphyte *thu ssu tzu*[9] i.e. the dodder. See on this Vol. 4, pt 1, p. 31. Though there was no visible connection between them, their relations were thought to be those of plant and root, and this was used as an argument for the reality of invisible action at a distance, resonance and sympathy. On the dodder, *Cuscuta chinensis*, see Anon. (*109*), vol. 3, p. 521.

[e] Tr. auct., adjuv. Huang Jen-Yü. All these plants, says the commentary, are things pertaining to the holy immortals and conducing to longevity.

[f] Cf. Taki Mototane (*1*), ch. 12 (pp. 169, 170).

[g] See *TSCCIW*, p. 273, and the biographies in *Chiu Thang Shu*, ch. 191, p. 2b, *Hsin Thang Shu*, ch. 204, p. 1b.

[1] 水香	[2] 林蘭	[3] 卷柏	[4] 茯苓	[5] 澤蘭
[6] 蒨	[7] 茜草	[8] 側柏	[9] 菟絲子	[10] 甄權
[11] 甄立言	[12] 本草音義	[13] 本草藥性		

About fifteen years after the death of Chen Chhüan the botanical-zoological pandects took a new lease of life, emperor Kao Tsung (Li Chih[1]) having come to the throne in +650. In the following year he commissioned Li Chi[2], a famous general,[a] and Yü Chih-Ning[3], a high civil official, to superintend the preparation of a new pharmaceutical natural history, radically revised and improved. We have come across this already in the translation (p. 247 above) of the interesting colloquy that took place between them and the emperor on the matter. But progress must have been unduly slow, for in +657 an expert naturalist, Su Ching[4],[b] memorialised that a new committee should be set up, to be headed by another statesman Chhangsun Wu-Chi[5] and to consist of 22 others including two Academicians, three Imperial Physicians,[c] six Assistant Physicians-in-Waiting, two Directors of the National Medical Administration (Thai I Ling[9])[d], and the Assistant Director. These were joined by the polymath Lü Tshai[10], who at this time was Vice-Minister of the Court of Imperial Sacrifices,[e] and by the Astronomer-Royal (Thai Shih Ling[11]), our old friend Li Shun-Fêng[12] so much in evidence in Sect. 20;[f] with Mr Secretary Su Ching's name bringing up the rear, for all the world like a Committee of the Royal Society.[g] Early in +659 Chhang-sun Wu-Chi fell from power as a result of intrigues connected with the empress Wu Hou, and was obliged to take his own life, but this did not prevent the successful completion of the project later in the same year. It appeared under the title *Hsin Hsiu Pên Tshao*[13] (New (lit. Newly Improved) Pharmacopoeia). It was the first national pharmacopoeia, issued by royal decree, in any civilisation. Nearly another thousand years had yet to pass before a work of the nature of a pharmacopoeia was published under government authority in Europe;[h] this was the *Pharmacorum ... Dispensatorium* of Valerius Cordus, produced by the Municipality of Nuremberg in +1546. But though this was truly officinal, like the *Hsin Hsiu Pên Tshao*, it was not national, and for that one has to wait for the

---

[a] Li Chi's title was Ying Kung or Ying Kuo Kung[6] (Duke of Ying), so that in subsequent times, when the great work was no longer available, it got the name, as in *PTKM*, of *Ying Kung Thang Pên Tshao*[7]. But this was never its correct designation.

[b] For tabu reasons he changed his name to Su Kung[8], a form often found.

[c] Hsü Hsiao-Chhung[14], Hu Tzu-Chia[15] and Chiang Chi-Chang[16].

[d] Cf. Sect. 44 below, as also Lu Gwei-Djen & Needham (2). Their names were Chiang Chi-Yuan[17] and Hsü Hung[18].

[e] Lü Tshai was a cartographer and an expert on the techniques of time-keeping and acoustics. He was a sceptical naturalist in philosophy, and learned in Five-Element theory. See further in Vol. 2, p. 387, Vol. 3, p. 323, 545, and the special study by Hou Wai-Lu & Chao Chi-Pin (1).

[f] Vol. 3, *passim*.

[g] The list of names appears in *TSCCIW*, p. 274 (from *HTS*, ch. 59, p. 21a), with only Yü Chih-Ning's absent.

[h] See Arber (3), 1st ed. p. 66, 2nd. ed. p. 75. The *Dispensatorium* has been published in facsimile by Winckler (1). Cf. Greene (1), p. 271; Tschirch (1). Some consider that it was preceded by the Venetian *Luminare Majus* of +1496 and the Florentine *Recettario* or *Antidotarium* of +1498 (Garrison (3), pp. 229, 817, 819). It was characteristic of European city-state society that the municipalities should have taken the lead in this matter.

---

[1] 李治	[2] 李勣	[3] 于志寧	[4] 蘇敬	[5] 長孫無忌
[6] 英國公	[7] 英公唐本草	[8] 蘇恭	[9] 太醫令	[10] 呂才
[11] 太史令	[12] 李淳風	[13] 許孝崇	[14] 胡子家	[15] 蔣季璋
[16] 蔣季琬	[17] 許弘	[18] 新修本草		

first *London Pharmacopoeia* of +1618, made valid for the whole country by royal proclamation.[a]

But the *Hsin Hsiu Pên Tshao* was a landmark of natural history at least as much as a treatise on *materia medica*.[b] It was, so far as we know, the first of the pandects to be richly illustrated; besides the *chêng ching*[1] or general text, its 54 chapters contained 25 called *yao thu*[2] and 7 more entitled *thu ching*[3], both parts evidently giving pictures (*thu*) of plants, animals and minerals, with their descriptions.[c] Of the entries 361 had already been in the *Pên Ching*, 192 came from the time of Thao Hung-Ching, and 114 were new, while no less than 195 more came under the heading of *yu ming wei yung*[4], i.e. named natural objects not employed in medicine. While Thao had used 6 or 7 natural categories now 9 were differentiated, birds, mammals, fishes and invertebrates all being separately treated; indeed the Thang naturalists much increased the number of animals and their parts described. They knew more, too, about the plants and animals of the southern provinces. Undoubtedly one of the stimuli for Su Ching's work was the larger number of drugs from foreign countries which were becoming known,[d] a notable example being

---

[a] An intermediate position is taken by the *Aqrābādhīn* of the Christian physician of the Jundishāpūr medical school, Sābur ibn Sahl (d. +869), but it is not quite clear how far this was governmentally authorised (Mieli (1), pp. 89 ff.).

[b] Among special studies of it we may mention Ma Chi-Hsing (1); Hung Kuan-Chih (2, 3); Shang Chih-Chün (1); Chhen Thieh-Fan (1). It has not attracted much attention from Western sinologists, but there is a bare mention in Pelliot (51), p. 340.

[c] It is interesting to see that the older books of plant illustrations continued in circulation during the Thang period. The *Ling Hsiu Pên Tshao Thu* and the *Chih Tshao Thu* (p. 261 above) are both listed in the dynastic bibliography (*HTS*, ch. 59, p. 20a, b).
One would like to know something of the names and lives of these early botanical illustrators, but all is dark. Schafer (13), p. 178, has suggested that Wang Ting[5] was one of them, but the facts do not seem to be quite as represented by him. Wang Ting was a famous painter of the +7th century, active between +620 and +650, who did a scroll or a series of scrolls entitled *Pên Tshao Hsün Chieh Thu*[6] (Instructions and Admonitions on Pharmaceutical Natural History). But significantly this work is listed in the dynastic bibliography not with the books on natural history but in the section on 'works of art', where it carries the note 'ordered for the Shang Fang (Imperial Workshops) in the Chên-Kuan reign-period'. Wang Ting was mainly a painter of people and Buddhas, so perhaps the most likely interpretation would be that he was asked to paint a class of medical students receiving instruction in pharmaceutical natural history. Since, as we shall see in a moment, it was just at this time that the Imperial College of Medicine was founded, the circumstances would be appropriate. I fear that Wang Ting was not one of the holy and humble men of heart, the artisans of the brush (*hua kung*[7]), who drew the fresh plants as carefully as they could.

[d] There is a special article on this by Mo Tê-Chhüan (1). See also Schafer (13), pp. 176 ff. At this time numerous plant-collecting expeditions were organised under Chinese government auspices, mostly by Buddhist pilgrims who could combine religious and scientific activities. From Vol. 1, pp. 211 ff. we remember the story of the Chinese monk Hsüan-Chhao[8], who came back from India about +655 by the help of the ambassador Wang Hsüan-Tshê[9], and was sent out again by imperial order in +664 to search for famous physicians in India and to collect medicinal plants. He probably sent many specimens home, but never succeeded in getting back himself, and died in India. A similar character was the Indian or Central Asian monk Nandī (Na-Thi[10] or Fu-Shêng[11]) who came to the Chinese capital in +655 with a large collection of Sanskrit manuscripts, but in the very next year was despatched to the Indies to collect exotic drug-plants. He must have returned successfully for in +663 he was sent out again for the same purpose to Cambodia. All these activities must have been closely connected with the *Hsin Hsiu Pên Tshao* project.
Buddhist temple gardens played a role in the acclimatisation of some of the plants brought back. There are

---

[1] 正經　　　[2] 藥圖　　　[3] 圖經　　　[4] 有名未用　　　[5] 王定
[6] 本草訓誡圖　　[7] 畫工　　　[8] 玄超　　　[9] 王玄策　　　[10] 那提
[11] 福生

the *theriaca* of Byzantium, now discussed in Chinese for the first time.[a] Although existing manuscripts do not all show it, we know that the *Hsin Hsiu Pên Tshao* fully maintained the red and black colour system of Thao Hung-Ching. One feature of its text is particularly noticeable to a reader who is familiar with the style of the later pandects—there is almost no quotation of authorities or differences of opinion; the information is set forth as if it were from the pen of one man. The commission may have been many-headed, but either it knew its own mind to an extent unusual in commissions, or else Su Ching was a strong personality. The result is a singularly fresh approach, as if everything had been written down anew.

The *Hsin Hsiu Pên Tshao* had a rather sad history. Although so great a work, it was produced a couple of centuries before the beginning of printing in China, and must therefore have circulated only in manuscript form on the flimsy medium of paper. So hardly was it treated by the storms of the times, the An Lu-Shan rebellion of the +8th century, the period of the provincial war-lords in the +9th, and the struggles between the Wu Tai States in the +10th, that it had certainly become rare when the *Khai-Pao Pên Tshao* (cf. p. 280) was being compiled about +970. It is just possible that Chang Yü-Hsi (p. 281) was able to use it a century later in +1060, but it is fairly sure that Thang Shen-Wei around +1090 (p. 282) never saw the complete work. Of course a great deal of it found its way down through the trickling rills of quotation in other natural history treatises, whence it could be largely re-composed today. But the actual intact text was not destined to be quite abandoned by fate and time. Two Tunhuang manuscripts,[b] one of them dated +667 and +669, within a decade after the original work appeared, have been preserved until now, though both are fragmentary; we illustrate in Fig. 47 a page from the Stein Collection example. That in the Pelliot Collection preserves the red and black convention. Then in +731 came a Japanese, Tanabe Fubito[3], who copied the whole work and took it home with him, where in course of time part was lost, but the rest treasured up in the library of the Ninnaji[4] temple at Kyoto.[c] Rediscovered and trace-copied about 1848,[d] it was printed in facsimile by Fu Yün-Lung[5] in 1889, so that we can illustrate a page of it in Fig. 48. Putting all the pieces together one can say that we still possess the greater part of 12 out of the 20 text chapters of the *Hsin Hsiu Pên Tshao*; and in addition its contents table

references to the accumulation of exotic drugs in Buddhist monasteries, and the +9th-century poet Phi Jih-Hsiu[1] wrote about an old monk named Yuan-Ta[2], over 80 years of age, who loved to cultivate rare medicinal plants in his garden (*Chhüan Thang Shih*, han 9, tshê 9, ch. 6, p. 13b).

[a] Cf. our fuller treatment of this at a much earlier stage, Vol. 1, p. 205.

[b] S 4534 in London and P 3714 in Paris. The Pelliot example is the dated one, but the Stein manuscript cannot be later than the second half of the +7th century. See the descriptions of Wang Chung-Min (1), pp. 152 ff.

[c] Where I had the pleasure of viewing the original, with Dr Lu Gwei-Djen and Dr Dorothy Needham, in 1964. We owe grateful thanks to Dr Nakayama Shigeru and Prof. Yabuuchi Kiyoshi for arranging this. Full details of the manuscript's history in Japan are given by Ma Chi-Hsing (1). Tanabe Fubito did not bother about the black and red.

[d] Cf. Mori Shikazō (3).

[1] 皮日休   [2] 元達   [3] 田邊史   [4] 仁和寺   [5] 傅雲龍

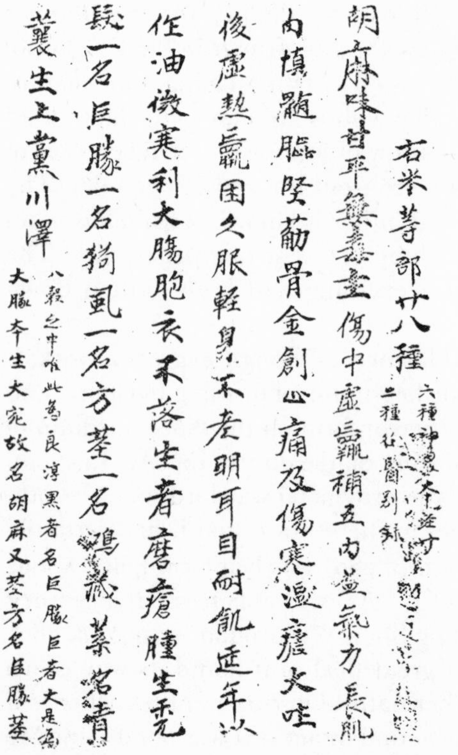

稷米 酢 贊 塩

胡麻味甘平無毒主傷中虛羸補五內益

右米等部合廿八種　六種神農本経
　　　　　　　　　二種名醫別錄

胡麻味甘平無毒主傷中虛羸補五內益
氣力長肌肉填髓腦堅筋骨金創心痛及
傷寒溫虛大吐後虛熱困久服輕身不老
明耳目耐飢延年以作油微寒利大腸胞
衣不落生者摩瘡腫生禿髮一名狗虱一

Fig. 47. The Tunhuang MS. of the *Hsin Hsiu Pên Tshao* dating from the second half of the +7th century (Stein Collection, S 4534 (2). It deals on this page with *hu ma* (*Sesamum indicum*).

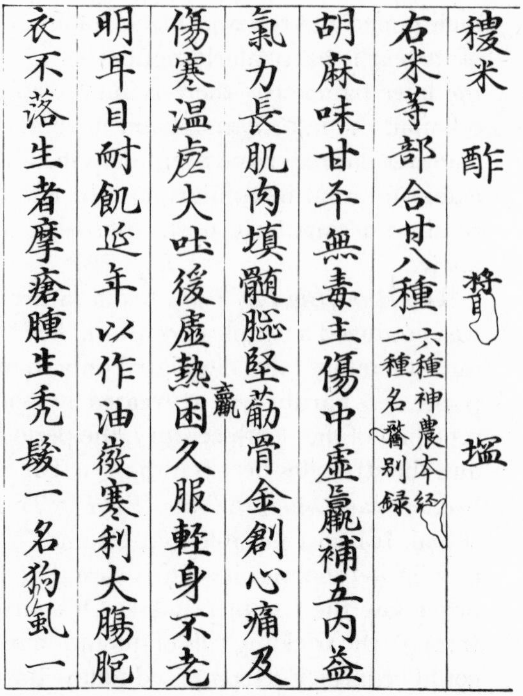

Fig. 48. The same page from the Ninnaji temple MS preserved in Kyoto (ch. 19, p. 1*b*).

has been found preserved in Sun Ssu-Mo's[1] *Chhien Chin I Fang*[2] (Supplement to the Thousand Golden Remedies),[a] written between +660 and +680.

This great project of the +7th century, the *Hsin Hsiu*, has a close connection with two movements not at first sight germane to it, the development of medical and scientific education in China, and the radiation of Chinese iatro-scientific culture to Japan. Without trespassing too much here on a story which we tell properly later in this volume (Sect. 44) it is needful to know that the first decades of full functioning of the Imperial College of Medicine coincided with those in which the *Hsin Hsiu* project was being conceived.[b] The posts of Regius Professor of Medicine (Thai I Po-Shih[3]) and Regius Lecturer in Medicine (Thai I Chu-Chiao[4]) date from +493, when a reorganisation of the bureacracy, including the national medical administration, took place under the Northern Wei; and this

---

[a] Chs. 2 to 4, and rather more accurately than in the copy made by the Japanese. This discovery is due to Hung Kuan-Chih (*3*).
[b] Cf. Lu & Needham (2).

[1] 孫思邈　　　[2] 千金翼方　　　[3] 太醫博士　　　[4] 太醫助教

staff was increased to eight, including two Curators of Physick Gardens (Yao-Yuan Shih[1]), by the Sui emperor in +585. But the whole system of medical education was placed on a new footing as soon as the Thang dynasty came into power, a central Imperial Medical College (Thai I Hsüeh[2])[a] being established a year or two later, and provincial Medical Colleges (I Hsüeh[3]) in every great city in +629. From the rules about qualifying examinations in +758 it is clear that by that time knowledge of *pên tshao* natural history had long been established as one of the great teaching subjects. The extent to which medical education was organised in the early Middle Ages in China has been little appreciated, but it is evident that we have to think of the *Hsin Hsiu* project in just that context.

This too was the period in which Japan was intensely absorbing the medical and biological achievements of the Chinese. There was a continual coming and going of monks, physicians and specialists between the two countries, as can be followed in detail in works such as that of Kimiya Yasuhiko (*1*). As early as +554—we give only an example or two at random—the Korean professor of medicine (I Po-Shih[5]) Wang Yurŭngt'a[6] [b] went to Japan to do something about medical education,[c] taking along with him two Pharmacognostic Masters[d] (Tshai Yao Shih[7]) Pan Kyŏngnye[8] and Chŏng Yut'a[9]. In +562 a Chinese monk took over a library of scientific works,[e] and in +602 a Korean abbot arrived, full of the medicine as well as the astronomy of the Sui.[f] In +608 the Suiko Empress sent the Pharmacist (Yao Shih[10]) Enichi[11], presumably a monk, with Yamato no Aya no fumi no Atahe[12], to study pharmaceutical natural history in China.[g] A wealth of similar examples of contact is known. By a century later, in +702, the Emperor Mommu established an Imperial Medical College in Japan, modelled closely on the Chinese system. In +731, as we have seen, Tanabe Fubito was copying the *Hsin Hsiu Pên Tshao*, and in +757 pharmaceutical natural history was emphasised as a subject of study, while in +787 the request was made to the throne that the

[a] Or more correctly, the I Hsüeh within the Thai I Shu[4], the National Medical Administration under the Thai I Ling Cf. Lu Gwei-Djen & Needham (2).

[b] One does not know how this name should be divided.

[c] Cf. Kim Tujong (*1*), pp. 81, 89, 113; Miki Sakae (*1*), p. 26.

[d] Lit. Masters of Drug Collection, or Production. This must have included botany, drug-plant horticulture, and the assessment of dried products.

[e] This was the learned monk of Wu, Chih-Tshung[13] (Chisō), who travelled in the company of a Japanese general Sadehiko, who had won a victory over the Koreans. Chih-Tshung took with him many books on pharmaceutical natural history (*yao tien*[14]), and anatomy and acupuncture (*ming thang thu*[15]). Cf. Shirai Mitsutarō (*1*); Miki Sakae (*1*), p. 26.

[f] This was Kwŏllŭk[16] (Kwanroku), whom we have already met (Vol. 3, p. 391 d). He introduced to Japan the first of the learned calendar systems, and brought knowledge of astronomical apparatus, but he was also famous for his knowledge of pharmacy and medicine. One of the medical pioneers of Japan, Hinamitachi[17], appears to have been his pupil. See Miki Sakae (*1*), p. 27, with a photograph of a traditional statue of Kwŏllŭk.

[g] Cf. Wang Chi-Min (*1*); Kimiya Yasuhiko (*1*), pp. 70, 71.

[1] 藥園師	[2] 太醫學	[3] 醫學	[4] 太醫署	[5] 醫博士
[6] 王有陵陀	[7] 採藥師	[8] 潘量豐	[9] 丁有陀	[10] 藥師
[11] 惠日	[12] 倭漢直福因	[13] 知聰	[14] 藥典	[15] 明堂圖
[16] 勸勒	[17] 日並立			

Hsin Hsiu should be ordained as the national Japanese pharmacopoeia.[a] This was the period of activity of the famous abbot Kanshin (Chien-Chen[3]), expert philosopher, architect and naturalist, so knowledgeable in the latter field that he was called 'the Shen Nung of Japan'.[b] He was in Japan from +735 to +748 and settled at Nara from +753; at one point he was accompanied by a Persian physician Li Mi-I[4], who must have known much about the drug-plants of East and West.[c] Thus the naturalists of the early Thang may be said to have constituted a great school gathering information from the west and the south, and passing it on to regions further east.

It may be interesting to look at one instance of the new botanical knowledge which came in with the *Hsin Hsiu Pên Tshao*. Let us choose the case of asafoetida. This group of substances, at various times and places of great commercial importance and widespread use in daily life, consists of vegetable products originating from the giant fennels, genera of the Umbelliferae like *Ferula*, *Narthex*, *Scorodosma*, etc., some herbaceous; some shrubby, perennial, and capable of attaining 10 ft in height.[d] The exudates that can be collected from them, notably by wounding or slicing the root-stocks, consist of resins, gums and essential oils in varying proportions; and they often have a very disagreeable smell, hence the name—and the common German expression for asafoetida, *Teufelsdreck*. Although so malodorous, the stuff seems to have the strange property of neutralising other foul or unwanted odours, if used in small amount; and those varieties that did not smell badly were valued as a pleasant spice.[e] Pharmaceutical worth has always been attributed to asafoetida, which was supposed to be carminative, anti-spasmodic and anti-malarial, a vermifuge, a demonifuge and many other things,[f] but pharmacy alone would hardly have accounted for its prominent position in the trade of the ancient Mediterranean region.[g] Of this we have already come upon a trace, when (in Vol. 4, pt 1, p. 24) we referred, in connection with the balance, to the famous dish-painting showing the weighing of silphium before the king of Cyrene, Arcesilaus II, in the −6th century. Silphium seems to have been *Ferula tingitana*,[h] and was exported in immense quantities from North Africa between −600 and −200, after which, for reasons still unknown, it was replaced by *Ferula foetida* from Persia. The sweet was thus supplanted by the foul,[i] but all this remained

---

[a] It was about this time that one of the first pharmaceutical natural histories of Japan was written—c. +790 by the scholar Wake Hiroyo[1] and entitled *Yakkei Taiso*[2]. This was based almost entirely on the *Hsin Hsiu Pên Tshao*, with adaptations for the Japanese flora. Wake Hiroyo is always remembered in Japan as the founder in +800 of the first free College. Cf. Sarton (1), vol. 1, p. 539; Shirai (1).

[b] Cf. Andō Kōsei (1); Chou I-Liang (2)   [c] Takakusu (3); Saeki (1), p. 62.

[d] A good botanical survey of the asafoetida-producing plants is in Holmes (1).

[e] The best account of the Asian side of the story is in Watt (1), pp. 533 ff.; a little in Burkill (1), vol. 1, p. 999; something too in Ainslie (1), vol. 1, pp. 20, 585.

[f] Cf. Stuart (1), p. 173, and many references in Gemmill (1); Schafer (13).

[g] An excellent review of the silphium problem has recently been published by Gemmill (1).

[h] See the interesting account of Theophrastus, VI, ii, 7 ff., iii, i ff.

[i] The Arabic authors, such as al-Rāzī and Ibn Sīnā, like Dioscorides (cf. Gunther ed. p. 328), knew the two kinds of asafoetida, the pleasant and the evil-smelling.

[1] 和氣廣世   [2] 藥經大素   [3] 鑑眞   [4] 李密醫

unknown to the Chinese botanists and pharmacists, who lived in a region where plants of this kind had always been inconspicuous and certainly unexploited.

In Su Ching's time, however, contacts with Persia led to the introduction of a third species to China, *Ferula assafoetida* (= *F. Scorodosma* = *Scorodosma foetida*),[a] and this was what his great work described. It was destined also to be the first representative of these plants to receive a modern-Western description, that of Kaempfer in +1712 (just over a thousand years later), in his *Amoenitates Exoticae* (see Figs. 49*a*, *b*).[b] Naturally the names adopted in Chinese were all transliterations of West or South Asian names—*a-wei*[1,2] from Old Iranian *angkwa*, *a-yü*[3] from Persian *anüzet*, *hsün-chhü*[4] from Sanskrit *hiṅgu*, and *ha-hsi-ni*[5] from Mongol and Persian *kasnī* or *gisnī*.[c] In Thang times asafoetida as a product was imported to China from many Asian countries, partly by land and partly by ship through the South China seas; and it came as a regular tribute article from the Chinese garrison at Beshbalik (Pei-thing[6]) on the north slopes of the Thien-shan in northern Sinkiang.[d] Apparently Thang people added it to tea, as in the poem written by a +9th-century monk Kuan-Hsiu[7]:

> In the quiet room I burn a sandal seal;
> In the deep brazier I heat an iron flask.
> The tea, blended with *a-wei*, is warming,
> The fire, sown with cypress roots, is fragrant.
> Some few single cranes have come flying,
> A good heap of *sūtras* is read through;
> What hinders me from stealing away like Chih-Tun[8]—
> And riding a horse up into the blue darkness?[e]

Apart from its use in medicine and as a condiment, asafoetida was admired in China for its property of absorbing undesirable odours. 'Its most prominent characteristic,' wrote Su Ching, 'is a rank odour, yet it can stop disagreeable smells; it is indeed a strange product (*thi hsing chi chhou, erh nêng chih chhou; i wei chhi wu yeh*[10]).' Excellent examples of this use can be found in the nutritional and food-preservation handbooks, e.g. the *Yin Shan Chêng Yao*[11] (Principles of Correct Diet) by Hu Ssu-Hui[12], written about +1330, where it plays a part in the cooking of mutton dishes and the conserving of the meat.[f] Before leaving this subject there

[a] CC 565; R 220; Khung *et al.* (*1*), p. 534.1. The entry in *PTKM* is ch. 34, pp. 63*a* ff.; it contains a long quotation from the *Hsin Hsiu*.

[b] Pp. 535 ff. and plate. Cf. Woodville (1), vol. 1, p. 22.

[c] This was first demonstrated by Laufer (1), p. 361.

[d] Near the town of Fu-yüan[9] east of Urumchi (Tihua).

[e] *Chhüan Thang Shih*, han 12, tshê 3, ch. 5, p. 5*b*, tr. Schafer (13), p. 188. The 'sandal (-wood) seal' was, of course, an incense-clock; cf. Bedini (5, 6). The 'cypress was *Thuja orientalis* (cf. p. 89). Chih-Tun was a famous monk of the Chin (d. +367) who was particularly fond of horses, and other animals. Kuan-Hsiu himself was a visionary painter as well as a poet.

[f] Ch. 1, pp. 41*a*, 46*a* (p. 34).

[1] 阿魏   [2] 阿虞   [3] 阿虞   [4] 薰渠   [5] 哈昔呢
[6] 北庭   [7] 貫休   [8] 支遁   [9] 孚遠
[10] 體性極臭而能止臭亦爲奇物也   [11] 飲膳正要   [12] 忽思慧

Fasciculus III.   **535**

in principio morbi oportet, ac temperare in progreſſu,
ne ab affluentium copiâ & acredine ulcus aggraveſcat.
Pinguium applicatione, ferventi climate facilè gangræ-
na tibiis inducitur: proinde tutiſſimum eſt, pinguium loco primum tumenti parti imponere cataplaſma, in progreſſu verò locum operiri emplaſtro, eo ſaltem fine, ut
ligula, cui vermis extremitas involuta eſt, retineatur:
nullâ omnino inſertâ turundâ.   Hujus, quod obiter dico,
merus apud noſtrates chirurgos abuſus eſt, vix eâ aliud
agentibus, quàm ut partem offendant, dolores concitent,
humores improbos alliciant, effluentiam puris impediant,
inflammationem cauſent, ſanationem protrahant: magi-
ſtrorum imprudentium ſerviles & ignavi imitatores!

OBSERVATIO V.

حَكَايَت هِبنك دَرْكَاني

Hiſtoria Aſæ fœtidæ Disganenſis.

§. I.

هِبنتِچِكسْتَر *Hingiſeh*

*Umbellifera Leviſtico affinis, foliis inſtar Pæoniæ ra-
moſis; caule pleno maximo; ſemine foliaceo, nudo,
ſolitario, Brancæ urſinæ vel paſtinacæ ſimili; radice aſam fœtidam fundente.*

Hingiſeh *radicem* habet ad plures annos reſtibilem,
magnam, ponderoſam, nudam; exteriori vultu nigram,
in ſolo limoſo lævem, in ſabuloſo ſcabram ac quadantenus
rugoſam; ut plurimum, Paſtinacæ inſtar, ſimplicem, ſæpe
paulo à capite duabus vel pluribus divaricationibus bra-
chiatam, aliis perpendiculariter demiſſis, aliis incondi-
tè & in obliquum extenſis, prout ab objectis flectuntur.
Faſti-

Fig. 49a. The entry for *Ferula assafoetida* in Engelbert Kaempfer's *Amoenitates Exoticae* of +1712.

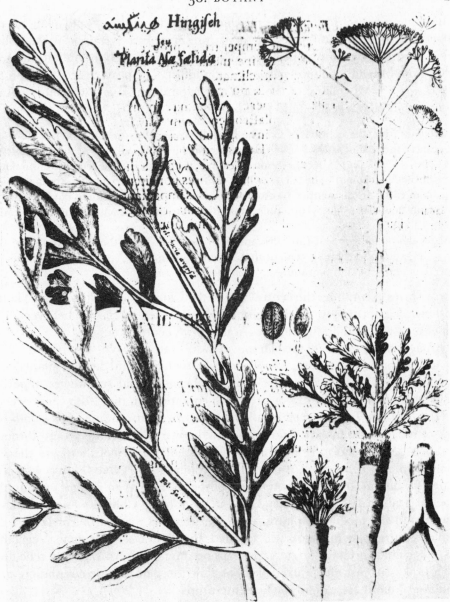

Fig. 49*b*. The drawing of this umbellifer on the plate pertaining thereto.

is one thing more to add, a scene in which we find a Chinese scholar discussing asafoetida with a 'Roman' priest or monk, presumably a Syrian or an Anatolian and of course a Byzantine, perhaps a Nestorian, and also an Indian monk or 'Deva' from Magadha, almost certainly a Buddhist. The passage occurs in the *Yu-Yang Tsa Tsu*[1] (Miscellany of the Yu-Yang Mountain Cave), written by Tuan Chhêng-Shih[2] about +860, and it runs as follows:[a]

A-wei comes from the country of Ga-Shê-Na[3] (Ghazni) which is more or less Northern India; there they call it *hsing-yü*[4] (*hiṅgu*). It also comes from Persia, where the name is *a-yü-chieh*[5] (*anüzet*). The plant (lit. tree) grows to a height of eight or nine feet,[b] and the skin is greenish yellow in colour. The leaves, which come out in the third month, are shaped like the ears of rats.[c] It has neither blossoms nor fruit. If the branches are cut they give a continuous flow of sap like syrup; after a time this hardens and consolidates and is called *a-wei*.

Wan[6], the monk from Fu-Lin[7] (Byzantium)[d], and Thi-Pho[8], the monk from Mo-Ga-Tho[9] (Magadha), were in agreement that a combination of the (dried) gum with rice or bean (flour) crumbled into granules, is what we get as *a-wei*.[e]

How well this exemplifies the central role of the Chinese naturalists of the Thang, and how pleasant it is to catch a glimpse of a learned discussion between three men of such widely differing cultures in the +9th century!

Throughout the rest of the Thang dynasty and the Wu Tai period of independent competing States, knowledge of provincial and foreign plants continued to grow, though of course some of the medicinal substances imported reached China only in the dry or preserved condition. In the +8th century even the *Hsin Hsiu Pên Tshao* proved insufficient, and two outstanding books, both the work of individual scholars, and both now known unfortunately only through quotations, attempted to cope with the flood. The smaller of these was that by Chêng Chhien[10] (d. +764), a man much admired by Hsüan Tsung, who had created for him in +750 a Polymathic College (Kuang Wên Kuan[11]) within the Imperial University,[f] and a friend of Tu Fu. Chêng Chhien was particularly interested in the medicinal plants and other materia medica of Persia and the Arabic countries, so his book was entitled *Hu Pên Tshao*[12] (Iranian Vegetable and Other Drugs). Unfortunately, perhaps because of the An Lu-Shan rebellion, which broke out just afterwards, his book failed to gain a wide circulation, and quotations from it are rare in the later literature.

[a] Ch. 18, p. 8*b*, tr. auct. adjuv. Hirth & Rockhill (1), p. 225; Laufer (1), p. 359.
[b] Most texts say 80 to 90 ft but quotations of the passage in *pên tshao* books, e.g. *CLPT*, ch. 9 (p. 224.1), get it right.
[c] A good description, but rather for the large bracts than the leaves.
[d] Cf. Vol. 1, pp. 186, 205.
[e] Neither Hirth nor Laufer, we think, understood this. What the clergy were talking about was the use of flour as a quasi-adulterative vehicle; cf. Watt (1), p. 535.
[f] Cf. des Rotours (1), vol. 1, p. 451.

[1] 酉陽雜組　　[2] 段成式　　[3] 伽闍那　　[4] 形虞　　[5] 阿虞截
[6] 彎　　[7] 拂林　　[8] 提婆　　[9] 摩伽陁　　[10] 鄭虔
[11] 廣文舘　　[12] 胡本草

Much more successful was the larger work, written rather earlier, about +739, by the naturalist Chhen Tshang-Chhi[1], and entitled *Pên Tshao Shih I*[2] (A Supplement for the Pharmaceutical Natural Histories).[a] A great part of this book could easily be reconstructed today from its quotations in the later pandects. How many new entries Chhen added is not clear, according to Li Shih-Chen it was 368, according to Thang Shen-Wei, who may well have known better, it was nearly five hundred (488). But it is interesting to see a few of the plants which Chhen now described for the first time. There was, for instance, the nutmeg, *jou tou khou*[3] (*Myristica fragrans = moschata = officinalis*),[b] representative of a family with very few genera, which came in first from the East Indies, but was cultivated successfully in Kuangtung by the beginning of the Sung period. Then there was the 'thorn honey' plant from Central Asia, *tzhu mi*[4] (*Alhagi graecorum = manniferum* or *Hedysarum Alhagi ⇋ Alhagi camelorum* or *turcorum*),[c] a leguminous shrub from which a sweet-tasting exudate or gum was collected. Tribute of this had been arriving occasionally since the Liang, but the plant itself was now discussed. It is still not certain whether this form of manna comes directly from the sap or whether (as first suggested by al-Bīrūnī early in the +11th century)[d] it is due to the activities of some aphid like *Gossyparia mannifera*.[e] Thirdly, one might mention another legume, the Indian laburnum, *Cassia fistula*,[f] the black pods of which contain a useful laxative pulp; this was called *a-lê-pho*[5],[g] from Skr. *aragbadha*; or, in comparison with *Gleditsia sinensis*, 'the Brahmin soap-bean tree', *po-lo-mên tsao-chia*.[6] A rather longer and better botanical description of it was given in the following century by Tuan Chhêng-Shih,[h] who so often echoed Chhen's accounts of exotic plants and elaborated upon them.[i] It must not be thought, however, that Chhen concentrated exclusively upon foreign parts; he was the first to describe, for instance, the Chinese plant *Ranunculus acris = japonicus = propinquus*,[j] *mao chien tshao*[7], one of those not without value in medieval times as a mild counter-irritant. He also added much to the Chinese knowledge of poisonous mushrooms and toadstools. So highly was his contribution considered that the *Chêng Lei Pên Tshao* lists the number of items introduced by him in front of the contents-tables at the beginning of each of its chapters, side by side with the

[a] On the dating see Hsieh Thang (*1*).
[b] R 503. See *PTKM*, ch. 14, p. 47*b*.
[c] R 371. See the account in *PTKM*, ch. 33, p. 17*b*.
[d] Cf. Meyerhoff (1).   [e] See Moldenke & Moldenke (1), pp. 31, 278.
[f] R 377. See *PTKM*, ch. 31, p. 26*b*. Descriptions in Watt (1), p. 287; Burkill (1), vol. 1, p. 475. Also called 'purging cassia'.
[g] Li Shih-Chen inadvertently altered the order of the last two syllables, but Laufer (1), pp. 420 ff. restoring the sound from earlier pandects, could establish the equation with Sanskrit.
[h] *Yu-Yang Tsa Tsu*, ch. 18, p. 10*b*, tr. Laufer (1), *loc. cit.*
[i] See Laufer's discussion of this, p. 423, showing that Tuan's information was generally independent of Chhen's, though corroborative.
[j] R 538. See *PTKM*, ch. 17B, p. 52*a*. On the pharmacological activity, Sollmann (1), p. 694; Stuart (1), p. 370.

[1] 陳藏器   [2] 本草拾遺   [3] 肉豆蔻   [4] 刺蜜   [5] 阿勒勃
[6] 婆羅門皂莢   [7] 毛健草

number of those listed in the three greatest of the earlier pandects, and quotes him in detail on his new items at its close.

From the present point of view the great Thang century ended in an anti-climax, for about +775 the physician Yang Sun-Chih[1] (so appropriately named that his *ming tzu* seems almost a sobriquet)[a] wrote his *Shan Fan Pên Tshao*[2] (The Pharmacopoeia Purged). Like the drafters of the Book of Common Prayer, he found the older pandects and their rubrics too complicated and full of matter, so he therefore removed a large part of it, keeping only the drugs most used, and abandoning the *yu ming wu yung* category, i.e. plants with names and notices but no use in medicine—botanically a most reactionary procedure.

Activity was next resumed in the +10th century in the independent Szechuan of the period of contending States between the Thang and Sung unified empires, under two successive local dynasties of Shu. In the Former Shu State, in the capital of Chhêngtu, between the years +919 and +925, one could have met at the court of the reigning house of Wang a remarkable girl named Li Shun-Hsien[3], ornamenting the age by her poetic talent no less than her beauty. Together with her two brothers, the younger Li Hsien[4] and the elder Li Hsün[5], she came of a family of Persian origin which had settled in West China about +880,[b] acquiring wealth and renown as ship-owners and merchants in the spice trade.[c] Li Hsien was a student of perfumes and their distilled attars as well as a merchant,[d] but he also worked on Taoist alchemy and investigated the actions of inorganic medicaments.[e] The one who took up the brush was Li Hsün, for about +923 he produced his *Hai Yao Pên Tshao*[6] (*Materia Medica* of the Countries beyond the Seas),[f] study of 121 plants and animals and their products, nearly all foreign, with at least 15 completely new introductions.[g] His work as a naturalist was highly regarded by subsequent scholars, and often quoted in the later pandects.[h]

Li Hsün was interested in all 'overseas' drugs, whether of the Arabic and Persian culture-areas or of East Indian and Malayo-Indonesian origin. A good example of the kind of thing he talked about is *an-hsi hsiang*[10], the 'Parthian aro-

---

[a] Yang the Subtractor.

[b] The family was fleeing from the rebellion of Huang Chhao in +878, cf. Vol. 1, p. 216. Their grandfather may well have been the Persian incense merchant Li Su-Sha[7], whose dates would be between +820 and +840.

[c] For a long time Li Hsün was ascribed to the +8th century, but Fêng Han-Yung (1) could demonstrate that this was due to confusion with another person of the same name, and clear up the circumstances of the life of the Li family.

[d] This is a focal point in the story of the development of distillation methods in China; see Vol. 5, pt 4, pp. 158 ff.

[e] He was also a notable chess-player.

[f] Alternatively called *Nan Hai Yao Phu*.[8]

[g] See further in Sung Ta-Jen (1); Schafer (2, 13, 16). A monograph on the Li family would be well worth while.

[h] The best biography of Li Hsün and his brother and sister is that by Lo Hsiang-Lin (4, 5). From some of the entries in his book, one can see that Li Hsün, although by origin a Nestorian Christian, acquired a very Taoist belief in medicines which would promote longevity and material immortality. He wrote much poetry in the Northern Sung style. His brother, Li Hsien, was even more Taoist, and had much regard as an adept, engaging in the preparation of *chhiu shih*[9] (steroid sex hormones from urine, cf. Vol. 5, pt 5, pp. 311 ff.).

[1] 楊損之   [2] 删繁本草   [3] 李舜絃   [4] 李玹   [5] 李珣
[6] 海藥本草   [7] 李蘇沙   [8] 南海藥譜   [9] 秋石   [10] 安息香

matic gum', which he thought came both from Persia and from the South Seas.[a] Actually this Chinese term was applied to two different things. In and before the Thang it meant the bdellium myrrh (or gum guggul) of the Bible, product of a tree belonging to the Burseraceae (*Commiphora = Balsamodendron, africana*, and *roxburghii = mukul*, in Arabia and Northwest India respectively), which was doubtless traded over the land routes.[b] But by the end of the Thang this had been replaced by gum benzoin, tapped like rubber from a Malayo-Indonesian member of the Styracaceae (*Styrax Benzoin*, with its more northerly relatives, *benzoides* and *tonkinense*).[c] We met already (Vol. 1, pp. 192 ff., 202) with another 'frankincense', the storax from *S. officinalis* that reached China in earlier times from the Levant; but benzoin now came up from the south. 'The change', as Schafer says, 'signalised the increasing importance of the products of the Indies in the economy of medieval China at the expense of the Syrian and Iranian ones.'[d] How interested Li Hsün and Li Hsien would be if they could know that their benzoin, named by the Portuguese from Mal. *lubān jāwī* (incense of Java) was destined to give its name to the prototype of all cyclic aromatic compounds, benzene, the very foundation-stone of modern organic chemistry, and all that that implies.[e] And to pass from the sublime to the childlike, all those of us who as children were given oil of cloves to ease our aching teeth, will be interested to know that clove 'bark' was recommended by Li Hsün in +923 for just this thing.[f] He described the tree, *Eugenia aromatica*,[g] under its classical Chinese name of *ting hsiang*.[1]

Two other books of this period claim mention. Just at the same time as the Li brothers were studying natural things in Chhêngtu, the learned Japanese physician Fukane no Sukehito[2] was at work on his *Honzō Wamyō*[3] (Synonymic Materia Medica with Japanese Equivalents). Finished in +918, this is a work of real value and importance today.[h] It was based on the *Hsin Hsiu Pên Tshao*, the great influence of which in Japan we have already noted (p. 269). Then a few decades later Szechuanese naturalists were active again. After the Mêng family had superseded the house of Wang and founded the State of Later Shu, Mêng Chhang[4] (r. +934 to +965) commissioned the Academician Han Pao-Shêng[5], with a number of other learned physicians, to revise the *Hsin Hsiu*.[i] The result, finished

[a] See *PTKM*, ch. 34, p. 54*b*. First discussed in *Hsin Hsiu Pên Tshao*.

[b] Cf. Moldenke & Moldenke (1), pp. 81 ff. A good account, for its time, is in *Yu-Yang Tsa Tsu*, ch. 18, p. 7*b*, tr. Laufer (1), p. 466; Hirth & Rockhill (1), p. 202. There is a link here with Central Amerindian culture, for the copal resin so prominent there was from the related tree *Bursera*. See also Yamada (1).

[c] R 185; cf. Laufer (1), pp. 464 ff.; Lin Thien-Wei (*1*), p. 45. A full account of these trees will be found in Burkill (1), vol. 2, pp. 2101 ff.

[d] (13), p. 169. The confusion was already clear to Laufer. Three centuries later there was still uncertainty, for Chao Ju-Kua (*Chu Fan Chih*, ch. 2, p. 4*a*) thought that the South Sea trade purveyed the Persian product, but this is most unlikely; cf. Hirth & Rockhill (1), p. 201.

[e] On all these aromatics, cf. Vol. 5, pt 2, pp. 136 ff.

[f] *PTKM*, ch. 34, p. 32*b*.   [g] R 244.

[h] See the remarks in Vol. 3, p. 645 above. Details in Karow (1).

[i] *ICK*, p. 117; *SIC*, p. 1267.

[1] 丁香   [2] 深根輔仁   [3] 本草和名   [4] 孟昶   [5] 韓保昇

some time between +938 and +950, with a preface by the king himself, is always known as the *Shu Pên Tshao*[1], though the title it actually bore was *Chhung Kuang Ying Kung Pên Tshao*[2]. Later experts considered the task rather indifferently done, but admitted that the illustrations were an improvement on those of the *Hsin Hsiu*; since no edition has come down to us we cannot tell, but it would be reasonable to think that Han and his colleagues added information about the plants of West China which they must have known very much at first hand. The contemporary title certainly indicates an enlargement.

(vi) *Natural history and printing in Sung, Yuan and Ming (+ 10th to + 16th century)*

From the moment when Chao Khuang-Yin[3] was invested with the imperial yellow as the first of the Sung emperors on the thirty-first of January in the year +960 we enter an altogether different world. The transformation of all aspects of life was something like that other change, so hard to define, between what are called the early middle ages in the West and the later middle ages. Poetry and *belles lettres* and religion gave way to prose and to science and technology. 'Whenever one follows up', I wrote, in the introduction to this work, 'any specific piece of scientific or technological history in Chinese literature, it is always at the Sung dynasty that one finds the major focal point';[a] and all our discoveries since then have confirmed this judgment. The Sung was characterised also by two movements closely connected with its science, on the one hand the philosophical naturalism of the Neo-Confucian school,[b] and on the other a great increase of industrial production,[c] containing even, as some believe, the germs of capitalist entrepreneurship, choked as always by the bureaucratic society.[d] Side by side with this went a marked development of maritime commerce and foreign trade,[e] especially after the move of the capital to Hangchow in the south. No less important were the social changes which broadened the basis of the mandarinate, greatly reducing the role of the traditional 'aristocratic' families and bringing in 'new men' through a revivified examination system from far wider origins.[f]

Probably there was no one of these great changes and advances with which the cardinal invention of printing was not connected.[g] Perfected during the late Thang and Wu Tai periods, certainly before the end of the +9th century, it found a vast expansion as soon as the Sung empire had become firmly established.

---

[a] Vol. 1, pp. 134 ff.

[b] Extremely congruent, we found, with the spirit of the natural sciences, and not only in their medieval, but even in their modern Western forms. See Vol. 2, pp. 493 ff. and the preceding parts of Sect. 16.

[c] See, for instance, the interesting, indeed startling, papers of Hartwell (2–4).

[d] Cf. Sect. 48 below.

[e] See, for instance, the monograph of Wheatley (1) and the papers and monographs of Lo Jung-Pang, especially (1).

[f] See the works of Kracke, e.g. (1, 2, 3).

[g] Cf. Sect. 32 in Vol. 5, pt 1, and, as always, Carter (1).

[1] 蜀本草   [2] 重廣英公本草   [3] 趙匡胤

On many a page of what has gone before, the reader will have encountered some book the title of which began with Thai-Phing[1]—this was a reference to the Thai-Phing Hsing-Kuo[2] reign-period (+976 to +983), after which they were named. The presses were indeed now roaring, if such an expression were permissible for the endless laborious carving of pear-wood blocks and the manual impressment on them of the beautiful pages. We quote constantly from the *Thai-Phing Yü Lan*[3] encyclopaedia of excerpts, edited by Li Fang[4] in +981, and the same year saw the appearance of another encyclopaedia of miscellaneous records, including fictional tales and stories, the *Thai-Phing Kuang Chi*[5], also edited by Li. Then there was the *Thai-Phing Huan Yü Chi*[6], or world geography, brought out between +976 and +983; and the *Thai-Phing Shêng Hui Fang*[7], nearer to our present interests, an imperially-commissioned treasury of medical prescriptions (+992). In this same period, between +971 and +983, the *Ta Tsang*[8], the Buddhist Patrology, or *Tripiṭaka*, was first printed, still in the dying glow, as it were, of Thang devotion, and before the chill wind of Neo-Confucian scientific naturalism had arisen. So also in +1019 came the first definitive collection of the *Tao Tsang*[9] or Taoist Patrology, and in +1022 that of its overflow, the *Yün Chi Chhi Chhien*[10] (Seven Bamboo Tablets of the Cloudy Satchel), prepared by its editor Chang Chün-Fang[11]. Printing of these took place between +1111 and +1117. Also in +1111 came the Imperial Medical Encyclopaedia of the Sung, *Shêng Chi Tsung Lu*[12], edited by a committee of 12 physicians of the court.[a]

This then was the different world in which the pharmaceutical naturalists, the botanists, zoologists and mineralogists, as we should now call them according to their predominant interests, found themselves. They seized with both hands the opportunity for woodcut block illustrations of plants and animals, and their writings multiplied bewilderingly.[b] While in former times two or three decades, even half a century, might pass with nothing new to report in this field, now several different editions of books on natural history might come out in one year. This was not on account of any *cacoethes scribendi* among the naturalists, but because standards of criticism were rising higher, and the technique of printing made revision and reproduction much easier. Of course this brings us difficulties as historians. There are pitfalls and contradictions in the original sources; even when these are available, for many have long been lost, and are known only by prefaces and excerpts quoted in later works, or as rare and incomplete manuscripts preserved in other countries such as Japan. Chinese authorities therefore

---

[a] Besides this upsurge of printing, it is good to remember that for the invention of gunpowder the year +1000 is a central point (cf. Sect. 30 in Vol. 5), and for that of the magnetic compass +1044 and +1086 (cf. Sect. 26*i* in Vol. 4, pt 1). Such were the origins of those inventions the world-shaking importance of which Francis Bacon clearly saw, origins still to him 'obscure and inglorious' (cf. Needham, 47).

[b] On the monographic literature see pp. 355 ff. below.

[1] 太平　　　[2] 太平興國　　[3] 太平御覽　　[4] 李昉　　　[5] 太平廣記
[6] 太平寰宇記　[7] 太平聖惠方　[8] 大藏　　　[9] 道藏　　　[10] 雲笈七籤
[11] 張君房　　　[12] 聖濟總錄

sometimes disagree, especially on questions of exact dating.[a] We cannot be sure of the accuracy of all that we have now to unfold, though we have made every effort to attain it.[b]

The period opened with deceptive quiet in the work of an obscure naturalist named Ta Ming[1] (possibly Thien Ta-Ming[2]) who took the name of Jih Hua Tzu[3] (The Sun-rays Master) and in +972 produced a work on plant and animal drugs entitled *Jih Hua Chu Chia Pên Tshao*[4] (Master Jih-Hua's Pharmaceutical Natural History, collected from Many Authorities).[c] He sings in a minor key in the later pandects, never being quoted much at length, so he was probably a practical man and a laconic writer, but as his book, whether printed or not, completely disappeared, we cannot now get to know him well.[d] Much more important in bulk and in influence was the government enterprise set on foot in the following year. In +973 a Chief Physician-in-Attendance at court, Liu Han[5], and a Taoist priest, Ma Chih[6], were commissioned, with a team of seven other naturalists under the administrative direction of a distinguished ambassador and geographer Lu To-Hsün[7], to produce a pharmacopoeia which should supersede all previous pandectal treatises and be multiplied by the typographic art.[e] By the end of the same year they accomplished their work, producing the *Khai-Pao Hsin Hsiang-Ting Pên Tshao*[8], (New and More Detailed Pharmacopoeia of the Khai-Pao reign-period). Some criticism supervening, however, a rapid revision was made in the following year, +974, leading to the issue of a second edition;[f] this was something which would not have been feasible before the days of printing. The work, always afterwards known by its shortened title *Khai-Pao Pên Tshao*, was the earliest pharmaceutical natural history which we can be sure was printed—a long time of course before its parallel in the West, Aemilius Macer's *De Naturis Qualitatibus et Virtutibus . . . Herbarum*, printed at Naples in +1477.[g] We also know that it was the first to introduce the practice of printing texts from the *Pên Ching* in *yin* form,

---

[a] To say nothing of Western writers, with access to fewer old editions, and often dependent on Chinese collaborators equipped with inadequate bibliographical tools.

[b] A special study of the period is given by Li Thao (1). Another, by Okanishi Tameto (3), appeared too late to be of help to us, but the story he tells is closely similar.

[c] Cf. the mention in Thao Tsung-I's *Cho Kêng Lu* (+1366), ch. 24, p. 17b.

[d] Some valuable citations, e.g. on metallurgical matters, have been preserved outside the *pên tshao* tradition, however, e.g. in Kao Ssu-Sun's *Wei Lüeh* (c. +1190), ch. 5, p. 1b.

[e] *Yü Hai*, ch. 63, p. 20a.

[f] Entitled *Khai-Pao Chhung Hsiang-Ting Pên Tshao*[9]. The commission remained the same, with the addition of two academic scholars, and the replacement of Lu To-Hsün in the presidency by Li Fang[10], the polymathic editor just referred to.

[g] Macer's real name is supposed to have been Odo, and he may have been a +10th-century contemporary of Liu Han and Ma Chih; cf. Arber (3), 2nd ed. p. 44. Sarton (1), vol. 1, p. 765, however, identifies him with Odo of Meung (+11th); see p. 3 above. One could select the *De Proprietatibus Rerum* of Bartholomaeus Anglicus for comparison, but it would only take one back to Cologne in +1472. Neither of these authors bears comparison with Liu Han and Ma Chih. There was no original observation in Odo, and all too much magic and symbolism in Bartholomew.

---

[1] 大明   [2] 田大明   [3] 日華子   [4] 日華諸家本草
[5] 劉翰   [6] 馬志   [7] 盧多遜   [8] 開寶新詳定本草
[9] 開寶重詳定本草   [10] 李昉

white on black, while all subsequent additions whatever the date were printed in *yang* form, black on white. Containing 983 entries, of which 123 were new, it is almost certain that the work was well illustrated with the novel wood-blocks, but alas no page of it has come down to posterity.

Nearly a century passed. Further knowledge having accrued, another emperor, Jen Tsung, ordained in +1057 another pandectal treatise, entrusting its preparation to the high official and naturalist Chang Yü-Hsi[1], the famous physician Lin I[2], the outstanding scientific thinker and statesman[a] Su Sung[3] and another naturalist Chang Tung.[4] A new procedure was now adopted, an imperial rescript being sent to all the prefectures and *hsien* cities of the empire instructing governors and magistrates to have drawings made of the most important drug plants of their regions.[b] In due course more than a thousand of these reached the capital. This measure, intended to improve the drawings of the previous century and restore the lost iconography of the Thang, was a striking example of what the bureaucratic organisation of medieval Chinese society was capable of effecting.[c] But the upshot was two treatises rather than one. By +1060 the compilation of Chang Yü-Hsi was ready, appearing as the *Chia-Yu Pu-Chu Shen Nung Pên Tshao*[5] (Supplementary Commentary on the 'Pharmacopoeia of the Heavenly Husbandman' commissioned in the Chia-Yu reign-period); this contained 1082 entries, of which 99 were new.[d] Su Sung in his turn brought out in +1061 the *Pên Tshao Thu Ching*[6] (Illustrated Pharmacopoeia, or Pharmaceutical Natural History),[e] not all pictures but containing much text different from Chang's, as we know from the later pandects, which always quote them separately. Both these works were greatly appreciated—and not in China only.[f] They lasted as late as +1616, so that Li Shih-Chen probably had access to them,[g] but they are not available to us now in their intact form. During the +11th century they doubtless circulated

---

[a] See Vol. 3 and Vol. 4, pt 2, *passim*; also Needham, Wang & Price (1). Chang Yü-Hsi's biography heads *Sung Shih*, ch. 294. He was a geographer as well, and for a time imperial tutor, but in character somewhat eccentric.

[b] *Yü Hai*, ch. 63, p. 20*a, b*.

[c] Other instances have already been given; see for instance Vol. 3, p. 274, Vol. 4, pt 1, p. 45. The point has been emphasised in Needham (45). Cf. also Vol. 7 below, Sect. 48.

[d] We give all such figures subject to reservations, for many different counts have been made, all differing slightly according to the conventions adopted.

[e] This was often afterwards quoted inverted as *Thu Ching Pên Tshao*, even by high authorities, but wrongly, according to the evidence of the Sung bibliographies (see *SSIW*, pp. 179, 529). The term *Thu Ching* applied originally to one of the two illustrated parts (the other being a *Yao Thu*) of the *Hsin Hsiu Pên Tshao* of +659. The retention of the expression emphasises that the work of Su Sung was conceived as a replacement of what had been lost, and of course an improvement on it.

[f] The *Pên Tshao Thu Ching*, especially in its later recension by Chhen Chhêng, seems to have been the inspiration for the works of the Japanese monk Seiken[7] of the Henchi-in[8] temple. About +1200 he wrote three valuable tractates, one on drugs, the *Yakushu-shō*[9], one on cereals, the *Kokurui-shō*,[10] and one on perfumes, the *Kōyaku-shō*.[11] Later Chinese pandects, however, may also have been available to him. See Shirai (1) and Sarton (1), vol. 2, pp. 52 ff., 305, 311, 443 and vol. 3, p. 2151.

[g] *Shih Shan Thang Tshang Shu Mu Lu*, ch. 2, p. 43*a*.

[1] 掌禹錫   [2] 林億   [3] 蘇頌   [4] 張洞
[5] 嘉祐補註神農本草   [6] 本草圖經   [7] 成賢   [8] 遍智院
[9] 藥種抄   [10] 穀類抄   [11] 香藥抄

widely, but not widely enough for the demand, for Lin Hsi[1] tells us significantly that by +1090 or so there were very few naturalists and physicians who could get access to both.[a] This led Chhen Chhêng[2] to produce, as was now so much easier with the printing technique, a kind of conflation of the two works of Chang Yü-Hsi and Su Sung entitled *Chhung-Kuang Pu-Chu Shen Nung Pên Tshao ping Thu Ching*[3] (Enlarged 'Supplementary Commentary on the Pharmacopoeia of the Heavenly Husbandman' with the 'Illustrated Treatise (of Pharmaceutical Natural History)'), out in +1092.[b] This was also influential in Japan.

We have now to retrace our steps a few years. While Chhen Chhêng was doing his scissors-and-paste operations a much more original mind was at work, namely Thang Shen-Wei[5], the writer of the most important and often-modified natural history text of the Sung period. Thang was a physician in Chin-yuan[c], a small town in the irrigated plain of Chhêngtu in Szechuan, so he may well have inherited the learned traditions of Shu in the previous century, of Li Hsün and Han Pao-Shêng, as well as the stimulus of the rich flora and fauna of this Western border of China. Working for just under a decade, he finished his *Ching-Shih Chêng Lei Pei-Chi Pên Tshao*[6] in +1082, and it was first printed in the following year.[d] So important was this 'Classified and Consolidated Armamentarium of Pharmaceutical Natural History' felt to be that a second edition appeared in +1090 edited by Sun Shêng[7], an Academician of the College of All Sages; this was no wonder, since Thang had added over 600 new descriptions, making 1746 in all, consulting 248 previous books on the way.[e] This work had been conceived and carried out as a private enterprise, but it was too good to be allowed to remain so,[f] and in +1108 it was reprinted offiially under the editorship of the Medical Officer Ai Shêng[8] as the *Ta-Kuan Ching-Shih Chêng Lei Pei-Chi Pên Tshao*[9] (The Classified and Consolidated etc. of the Ta-Kuan reign-period).[g] To view this against its proper background one must remember that the court of Hui Tsung,

---

[a] *ICK*, ch. 10 (p. 126). In the preface to Chhen Chhêng's book.

[b] This work was often afterwards given the shorter title *Pên Tshao Pieh Shuo*[4], but incorrectly. The 'additional remarks' were of course those of Chhen Chhêng himself. Chhen was an eminent physician who also edited a compendium of standard formularies as used in the government dispensaries (cf. Sect. 44 below), in +1109. Cf. *SIC*, p. 1287.

[c] Modern Chhung-chhing west of Chhêngtu.

[d] This work and all its many descendants was always abbreviated as *Chêng Lei Pên Tshao*, and it is possible that this shortest form was the original title of Thang Shen-Wei's own MS.

[e] Many sources may be used to get a picture of the central position of Thang Shen-Wei, e.g. Chang Tsan-Chhen (*1, 2*); Ma Chi-Hsing (*3*); Anon. (*65*); Hummel (*13*); Nakao Manzō (*1*); Huard & Huang Kuang-Ming (*3*). He did not get a biography in the *Sung Shih*, unfortunately. Wang Chün-Mo (*2*) has made a special study of the geographical origin of the plants and animals recorded in Thang's book.

[f] Apart from original observations, it had ransacked the classics, the histories, and Taoist and Buddhist books, as well as preserving many prescriptions, some previously secret, for posterity.

[g] It is thought that Ai Shêng now made some additions from the Chhung-Kuang conflation of Chhen Chhêng (p. 281 above).

[1] 林希   [2] 陳承   [3] 重廣補註神農本草並圖經   [4] 本草別說
[5] 唐慎微   [6] 經史證類備急本草   [7] 孫升   [8] 艾晟
[9] 大觀經史證類備急本草

who had come to the throne in +1101, was a veritable entourage of virtuosi, appreciative of all the best that the art, the science and the technology of the time could perform.[a] It was therefore quite in character that Thang Shen-Wei's work should have been laureate with the sign of imperial approval.[b]

Matters did not rest there, however. The scientific interests of the court continuing, two further developments occurred in +1116. First the emperor ordered another Medical Officer, Tshao Hsiao-Chung[3], to revise and reprint Thang's book, which thus appeared again in this year headed by another regnal title as the *Chêng-Ho Hsin-Hsiu Ching-Shih Chêng Lei Pei-Yung Pên Tshao*[4]. This was its final form as such, and when it was reprinted in +1143 by the Chin Tartars up in the north it received a postface by Yüwên Hsü-Chung[9] giving interesting details not recorded elsewhere about the life of Thang Shen-Wei, who had been known to him as a physician when a boy.

The second event was the presentation to the throne of another entirely original treatise on pharmaceutical natural history by a hitherto unknown Medical Officer of the same stature as Thang himself, Khou Tsung-Shih[10]. Entitled *Pên*

---

[a] At earlier stages we have dwelt at some length upon this golden period (cf. Vol. 4, pt 2, pp. 496 ff., 500 ff., 502 ff. and Needham, Wang & Price (1), pp. 118 ff., 123 ff., 125 ff.). In its way Hui Tsung's court may be compared with those of the Caliph al-Ma'mūn (r. +813 to +833), of Alfonso X, king of Castile (r. +1252 to +1284) and Sejong, king of Chosŏn (Korea, r. +1419 to +1450); cf. Vol. 4, pt 2, p. 516. The year +1104 was a high-water mark in the art of iron-casting; between +1094 and +1125 there was great activity in horological engineering, with two books on this subject, by Juan Thai-Fa[1] and (probably) Wang Tzu-Hsi[2], the latter certainly written at this time; the rarities in the Imperial Museum were being catalogued between +1111 and +1125; Taoist metropolitan temples were planned and partly built in +1114, and the *Tao Tsang* (as we have seen) was being printed between +1111 and +1117. What we are now confronting is the biological analogue of these other activities in engineering and proto-chemistry.

[b] The *Ta-Kuan ... Pên Tshao* was many times reprinted. We know of Sung editions in +1185 and +1195; this last was re-issued in 1823 and 1904 (whence the old woodcut illustrations reproduced in Sato, 1) and again in Japan in 1970 (whence Figs. 50a, b). We illustrate *fang fêng*, an umbellifer (*Siler divaricatum*) and the composite, safflower, *hung lan hua* (*Carthamus tinctorius*). Sato Junpei often juxtaposed modern drawings, as in Figs. 51a, b.

Then there was a J/Chin reprint in +1215, and naturally enough this was the basis for the Yuan version of +1302; on which see Wu Kuang-Chhing (1). It led in due course to a Korean one of +1392. After this, there were Ming reprints of +1519, +1577 and +1581; of which last only one chapter (the seventh) is extant, in the Bern Codex (p. 294 below), dated +1249. Subsequently the *Ta-Kuan ... Pên Tshao* was issued twice again before the end of the dynasty (+1600, +1610) and early in the Chhing (+1656). This was one of the more remarkable runs of any pharmaceutical natural history.

According to Nakao Manzō (1), the *Ta-kuan ... Pên Tshao* was published, at least in part, in Japan only half a century after its appearance in China. i.e. in +1156; for two scrolls of very fine pictures and text, dated that year, still exist in Japan. One is entitled *Kōyō-shō Edu*[5] (Illustrations of Odoriferous Plants), the other *Kokurui-shō Edu*[6] (Illustrations of Cereals). Text and captions in both seem to be from the *Ta-Kuan ... Pên Tshao*, but while the illustrations of the perfume scroll agree with those in the contemporary *Shao-Hsing ... Pên Tshao* (see p. 288 below), the illustrations of the cereal one are different, and even finer, therefore possibly later. Then in +1282 Koremune Tomotoshi[7] made an index of the *Ta-Kuan ... Pen Tshao* entitled *Honzō Iroha-sho*[8].

Sarton (1), vol. 2, p. 52, made the general remark that the relation between the Japanese and Chinese works on pharmaceutical natural history was rather like that between the Latin and the Arabic. The Japanese and the Latins were usually a century or so out of date. Thus Wake Hiroyo in the late +8th century depended on the *Hsin Hsiu Pên Tshao* of the middle +7th; Seiken in the early +13th depended upon the *Thu Ching* and the *Chhung-Kuang* of the +11th; and Koremune in the late +13th worked on the *Ta-Kuan* of the early +12th. This comment seems fair.

[1] 阮泰發   [2] 王仔昔   [3] 曹孝忠   [4] 政和新修經史證類備用本草
[5] 香要抄繪圖   [6] 穀類抄繪圖   [7] 惟宗貝俊   [8] 本草以呂波鈔
[9] 宇文虛中   [10] 寇宗奭

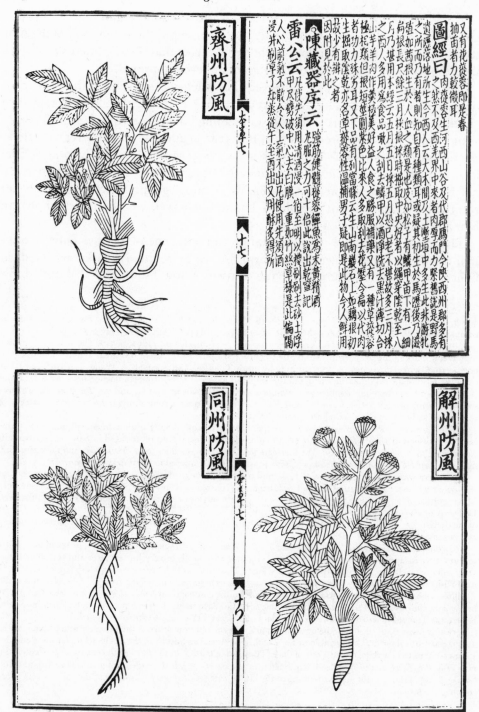

Fig. 50a. Drawings of *fang fêng* (*Siler divaricatum*) from three provenances, in the *Ta-Kuan Pên Tshao* of +1108, ch. 7, pp. 17a, b, 18a, b.

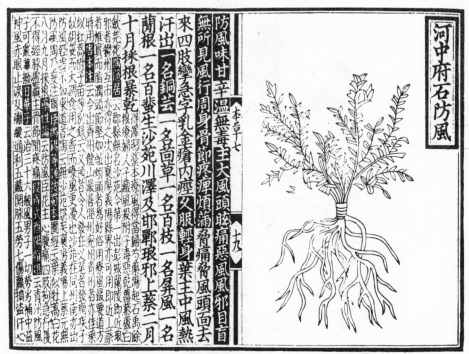

Fig. 50*b*. Drawing and description of another umbellifer, *shih fang fêng* (*Peucedanum terebinthaceum*) from the same work, ch. 7, pp. 19*a*, *b*.

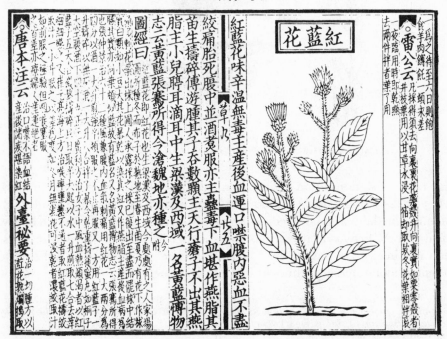

Fig. 51*a*. Drawing and description of the thistle *Carthamus tinctorius* (*hung lan hua*) from the *Ta-Kuan Pên Tshao* of +1108, ch. 9, p. 25*a*, *b*.

Fig. 51*b*. A modern drawing of this *Carthamus species* (from Sato (*1*)), p. 166.

Tshao Yen I[1] (Dilations upon Pharmaceutical Natural History),[a] it was printed in +1119 by the writer's nephew Khou Yo[2], and earned for its author a substantial promotion in the National Medical Service. There is no doubt that Khou was critical of the *Chêng Lei* in its various forms, either because he thought its classification was inferior to that of the Thang *Hsin Hsiu*,[b] or because he adhered to a different school of medical thought from Thang Shen-Wei, or because (with the

---

[a] This bore for a time the title *Pên Tshao Kuang I*[3], perhaps from the moment of its first printing, but the original one was restored in +1195 or a little later, when the third word came into tabu conflict with the *ming tzu* of the reigning emperor. Cf. *SIC*, p. 1278; Pelliot (52). Corr. Sarton (1), vol. 2, p. 248.

[b] So Mori Tateyuki in *SIC, loc. cit.*

[1] 本草衍義　　[2] 寇約　　[3] 本草廣義

usual pendulum-swing of concentration on essentials following super-abundance)
he wanted to discuss only *materia medica* which he felt sure were pharmaceutically
effective. He certainly restricted the number of his entries to 471, and threw out the
yu ming wu yung category, but his work was greatly admired during the Sung, Chin
and Yuan periods.[a] Later on it was customary to criticise him for having confused
lan hua[1] with *lan tshao*[2],[b] and *chüan tan*[3] with *pai ho*[4],[c] as did Li Shih-Chen.[d] But we
ourselves have already had occasion to notice his merits. For instance, he wrote a
passage of cardinal importance on the lodestone and the magnetic needle, recording induction and polarity, with a description of the floating compass, and not
only a statement of the declination but also an attempt to give an explanation of
it, all some sixty years before anything whatever was known about magnetic
polarity in Europe.[e] He also discussed a number of fossils, notably the remains of
gastropods, or cephalopods (ammonites), affirming rightly that these, though
called *shih shê*[13] (stone serpents), had never been, when living, animals of the same
kind as those which are called snakes now.[f] Elsewhere he had interesting things to
say about ore deposit associations.[g] Perhaps he was a better mineralogist and
geologist, according to the lights of his time, than a botanist. Significant in
connection with the Taoist ascendancy of the period is the fact that the *Pên Tshao
Yen I* was incorporated into the *Tao Tsang* as soon as it had been presented to the
emperor; this by Chang Hsü-Pai[14], one of the leading adepts or high priests at the
capital.[h] It is not in the *Tao Tsang* as we have it now, for about a century later a
derivative work was substituted for it there.[i] Before mentioning this, however, we
must see what happened after the catastrophe of +1126.

That was the fateful year when first Hui Tsung, conscious of his incapacity to
organise the defence of the empire, abdicated in favour of his son, and then for
two long periods the capital (Khaifêng) was besieged by the Chin Tartars.[j] In

---

[a] Four works are considered to have been directly inspired by it (*a*) the *Chieh Ku Lao Jen Chen Chu Nang*[5] (Old Master Chieh-Ku's Bag of Pearls) by Chang Yuan-Su[6], *c.* +1200; (*b*) *Yung Yao Fa Hsiang*[7] (Methodology of Pharmaceutic Therapy) by Li Kao[8], *c.* +1220; (*c*) *Thang I Pên Tshao*[9] (*Materia Medica* of Decoctions and Tinctures) by Wang Hao-Ku[10], *c.* +1280; and (*d*) *Pên Tshao Yen I Pu I*[11] (Revision and Amplification of the Dilations ...) by Chu Chen-Hêng[12], *c.* +1330. All these were primarily medical rather than botanical, so we reserve comment on them to the appropriate place (Sect. 44 below). Cf. Anon. (*65*).
[b] I.e. the orchid *Cymbidium virescens* (CC 1725) with the Composite *Eupatorium chinense* (R 33, CC 85). This by a confusion of names; Khou cannot have known the plants. *Lan hua* could also have been *Forsythia suspensa* (R 176; CC 449).
[c] I.e. *Lilium tigrinum* (R 682a, CC 1863) with *Lilium japonicum* = *brownii* (R 682, CC 1852).
[d] Cf. *ICK*, p. 144; *PTKM*, ch. Shou, Su (I piao), p. 34a, ch. 1A, Hsü li, p. 11a.
[e] See Vol. 4, pt 1, pp. 251 ff.   [f] See Vol. 3, p. 618.   [g] Vol. 3, p. 674.
[h] This Taoist, the White-Clouds Master (Pai Yün Tzu[15]) was abbot of the Thai I Kung[16] temple, and kept the manuscript of Khou Tsung-Shih at the Ling Yu Kung.[17] Cf. *SIC*, p. 1280.
[i] See p. 289 below.
[j] Much is said of this siege in Section 30 on military technology.

[1] 蘭花   [2] 蘭草   [3] 卷丹   [4] 百合   [5] 潔古老人珍珠囊
[6] 張元素   [7] 用藥法象   [8] 李杲   [9] 湯液本草   [10] 王好古
[11] 本草衍義補遺   [12] 朱震亨   [13] 石蛇   [14] 張虛白
[15] 白雲子   [16] 太乙宮   [17] 靈佑宮

September it fell at last, and both emperors, the old and the young, were carried away to spend the rest of their lives in captivity in the north. What remained of the court and government fled southwards, moving uncertainly from place to place behind the lines still held by the Sung, till in +1129 Hangchow, the lovely city with its lake and hills on the Chhien-thang estuary beyond the Yangtze,[a] was chosen as the new capital. Government was established there from +1133, but the Sung navy could not make it quite safe from Chin attack until +1139, after which cultural activities and commerce resumed and gradually strengthened, so much so that a century and a half later, in Marco Polo's time, Hangchow was the cultural metropolis of East Asia and surpassed by no city elsewhere in the world. The re-awakening of natural history started from +1157 when a new pandectal treatise was begun; this was the *Shao-Hsing Chiao-Ting Ching-Shih Chêng Lei Pei-Chi Pên Tshao*[1] (Corrected Classified and Consolidated Armamentarium; Pharmacopoeia of the Shao-Hsing reign-period), with a preface dated +1159, the editors being Wang Chi-Hsien[2] and three colleagues.[b] This was essentially a re-do of the Northern Sung *Ta-Kuan* and *Chêng-Ho* pandects,[c] but it differed in being associated from the beginning with the most beautiful and clear woodcut illustrations of any of the treatises we are here describing. Though printed in the year given, the *Shao-Hsing ... Pên Tshao* also circulated in many manuscript copies, some of which had coloured illustrations, so that its bibliography is extremely complicated.[d] Fourteen manuscripts survive in Japan, mostly the illustrations with little text, and one of these, dating from the +12th century, and probably from the Shao-Hsing reign-period, Wang Chi-Hsien's own time, has been reproduced in facsimile by Wada Toshihiko (*1*) and by Karow (*2*).[e] Thence we draw Fig. 52. Another (Fig. 53) was reproduced by Ching Li-Pin (*2*) from a Japanese printed edition. The usual view is that the *Shao-Hsing ... Pên Tshao* did not come into Japan until the general Ukita Hideiye[6] brought back copies and manuscripts of it in +1592, at the time of Hideyoshi's invasion of Korea,[f] but in view of the great influence it is said to have had in Japan this seems very hard to believe.

The next development took place when life had settled down to normal in Hangchow and the great +13th century had begun. One has to think of it

---

[a] Cf. Vol. 4, pt 3, p. 311 and *passim*.

[b] I.e. Chang Hsiao-Chih[3], Chhai Yuan[4] and Kao Shao-Kung[5]. As Wang was a courtier of intriguing character, the real experts in pharmaceutical natural history were probably the other three. For Wang's biography see *Sung Shih*, ch. 470.

[c] Nevertheless the *Shao-Hsing* preface criticised its predecessors for having left many uncertainties about the toxicity of particular drugs. The descriptions of pharmacological properties contained contradictions, it said, and the modes of application, whether internal or external, were not clearly distinguished. Also, worst of all, quotations had been inaccurate, and the *Pên Ching* text sometimes confused with later writings.

[d] On this see the special study of Nakao Manzō (*3*, 1).

[e] This is the copy preserved in the Omori Memorial Library at the Kyoto Botanic Gardens.

[f] As in Sarton (*1*), vol. 2, p. 443. The principal Japanese printed edition was produced in 1836.

---

[1] 紹興校定經史證類備急本草   [2] 王繼先   [3] 張孝直   [4] 柴源

[5] 高紹功   [6] 宇喜多秀家

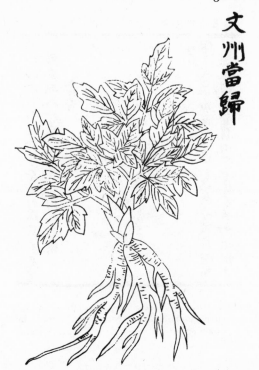

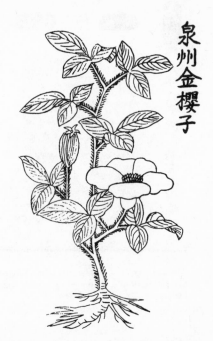

Fig. 52. Drawings from the *Shao-Hsing Pên Tshao* (+1157); *tang kuei*, *Angelica polymorpha*. From Karow (2), p. 39.

Fig. 53. Drawings from the *Shao-Hsing Pên Tshao* (+1157); *chin ying tzu*, *Rosa laevigata* (= *trifoliata*) R 455. From Ching Li-Pin (2), p. 62.

against the background of the sumptuous Imperial Medical College which we describe below.[a] A need was felt to unify the pandects, marrying Thang's *Chêng-Ho ... Yen I*, just as Chhen Chhêng towards the end of the +11th century had combined Chang's *Chia-Yu ...* with Sung's ... *Thu Ching*, making the *Chhung-Kuang ... Pên Tshao* (p. 282 above). This was effected first by Hsü Hung[1] about +1223, but not with great success.[b] His *Thu Ching Yen I Pên Tshao*[2] (Illustrations (and Commentary) for the 'Dilations', etc.) has came down to us in the *Tao Tsang*[c], whence it presumably ousted Khou's own work (p. 287 above), so we can see that he considerably abridged both his sources, omitting, for instance, all the algorific drugs. The illustrations, however, are often original (cf. Figs. 54*a*, *b*, *c*, *d*). A few years later, about +1226, another scholar, Chhen Yen[3], produced a work,

---

[a] Sect. 44; cf. Lu Gwei-Djen & Needham (2).

[b] Hsü Hung was one of the Imperial Physicians, and responsible also for an important repertorium of the standard formularies used in the government dispensaries of his day (see on, Sect. 44).

[c] TT 761. The text contains many mistakes. Cf. Chang Tsan-Chhen (2); Lung Po-Chien (1), nos. 38, 39; Pelliot (52) The original title of Hsü Hung's compilation was apparently different—*Hsin Phien Lei Yao Thu Ching Pên Tshao*[4]. For it he had a co-editor, Liu Hsin-Fu[5].

[1] 許洪   [2] 圖經衍義本草   [3] 陳衍
[4] 新編類要圖經本草   [5] 劉信甫

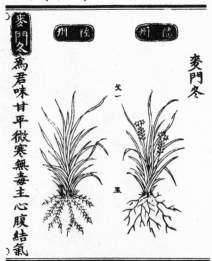

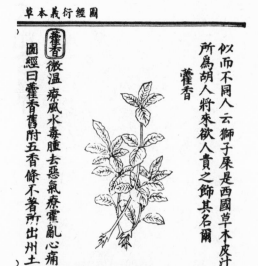

a. *Mai mên tung*, i.e. *Liriope spicata* of the Liliaceae (the black leek, R 684; Anon. (*109*), vol. 5, fig. 7877). Ch. 8, p. 6a.

b. *Huo hsiang*, i.e. *Lophanthus rugosus*, a labiate, allied to betony and bishopswort. Now the source of an important anti-hypertension drug. Ch. 21. p. 30a.

c. *Jou tou khou*, the nutmeg, i.e. *Myristica fragrans*. Ch. 15, p. 13a.

d. *Chih shih*, the thorny limebush, *Poncirus trifoliata*, much discussed above, pp. 103 ff. Ch. 22, p. 33b.

Fig. 54a–d. Pages from the *Thu Ching (Chi-Chu) Yen I Pên Tshao* (TT 761) of +1223.

still extant, different in character from the others we are describing; it was entitled *Pao-Chhing Pên Tshao Chê Chung*[1] (An Evaluation of the Literature on Pharmaceutical Natural History), and constituted a critical account of 21 previous books going back to the *Khai-Pao ... Pên Tshao*.[a] A few years earlier, in + 1220, an excellent painter of flowers, Wang Chieh[2], had written his preface for an album of 205 coloured pictures of plants which he called (from the name of his mountain hall) *Lü Chhan Yen Pên Tshao*[3], and this, in a single Ming MS copy, has survived till the present day.[b]

But curiously enough it was not in the opulent atmosphere of Hangchow that the most lasting work of the period was done. This happened in the less cultivated north, where a scholar-printer, not a naturalist, Chang Tshun-Hui[4], combined the *Chêng-Ho* ... with the ... *Yen I* by the scissors-and-paste method of Chhen Chhêng, yet so honestly and meticulously that it went into more editions through the centuries afterwards than any other of the pandects. Its name was the *Chhung-Hsiu Chêng-Ho Ching-Shih Chêng Lei Pei-Yung Pên Tshao*[5] (New Revision of the Pharmacopoeia of the Chêng-Ho reign-period; the Classified and Consolidated Armamentarium). Chang Tshun-Hui lived and worked at Phing-yang in southern Shansi, first under the Chin Tartars till + 1234 and then under the Mongols of the early Yuan before their conquest of all China. The date of his joint edition, now so familiar in facsimile reproductions, was + 1249; it was well printed[c] and excellently illustrated.[d] An unparalleled number of editions followed down to the + 17th century,[e] the most notable of which perhaps was a Ming reprint of + 1468 in substantial folio size.[f] A page from this is shown in Fig. 56. The *Chhung-Hsiu ...*

[a] It is known only from a rare Yuan edition. One thing it shows is that although there were great losses of books as Sung culture moved south (cf. the bibliographers' tag: 'failed to cross the river'), some parts of scientific literature were not too badly affected.

[b] Lung Po-Chien (*1*), no. 42.

[c] There has been a persistent impression that the work was first done and published in + 1204, but this rests on a mistake. The stele-like inscription on the title-page mentions this year, it is true, but in the following form: 'Finished in the *chi-yu* year of the cycle that began with its *chia-tzu* in the Thai-Ho reign-period of the J/Chin dynasty'. Chang Tshun-Hui had to do this because the Mongols had no reign-periods before the accession of Khubilai in + 1260. Perhaps the mistake started with Sarton (*1*), vol. 2, p. 247.

[d] See Fig. 55. Wu Kuang-Chhing (*1*) avers that a few of the pictures bear the name of their draughtsman, one Chiang I[6] of Phing-Yang. This, if true, is interesting, as we know so few such names, but we have not ourselves been able to confirm it.

[e] There was a Yuan one of + 1306, and at least seven Ming reprints before + 1624 or 1625. See Lung Po-Chien (*1*), no. 33.

Some of the Ming editions, starting with that of + 1557 edited by Wang Ta-Hsien[7] bore the title *Chhung-Khan Ching-Shih Chêng Lei Ta-Chhüan Pên Tshao*.[8] They seem to have been based on the + 1468 edition of the *Chhung-Hsiu* ... and the + 1302 edition of the *Ta-Kuan ...*, which itself was modified by the *Shao-Hsing ...* On this see Ting Chi-Min (*1*).

[f] *Pace* Sarton (*1*), *loc. cit.*, this is the only one which the Library of Congress does not possess, according to Hummel (*13*), p. 156. We had the good fortune, however, to find a copy of it in Peking in 1946, and we have used it here extensively.

[1] 寶慶本草折衷   [2] 王介   [3] 履巉巖本草   [4] 張存惠
[5] 重修政和經史證類備用本草   [6] 姜一
[7] 王大獻   [8] 重刊經史證類大全本草

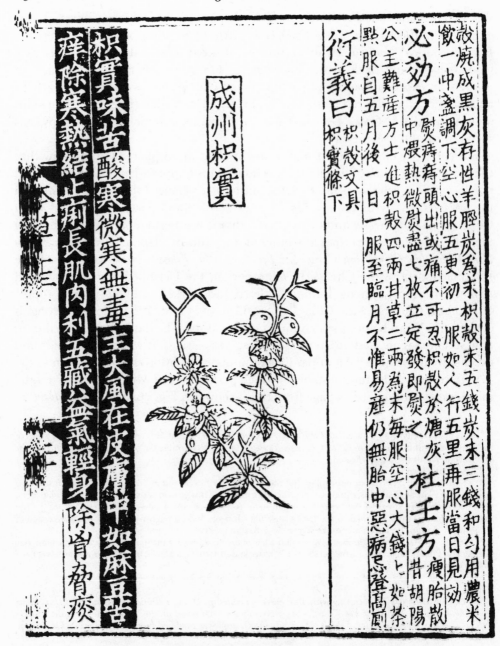

成州枳實

枳實味苦酸寒微寒無毒主大風在皮膚中如麻豆苦痒除寒熱結止痢長肌肉利五藏益氣輕身除肝膈痰

本草經上

衍義曰枳實條下

必効方中漫蒸蒜頭出或痛不可忍枳殼於糖灰內漫熱熨之即熨盡七枚立定發即熨之公主難産方士進枳殼四兩甘草二兩為末每服空心大錢七如茶點服自五月後一日一服至臨月不惟易産仍無胎中惡病忌登高厠

杜壬方昔胡陽瘦胎散

殼燒成黑灰存性羊脛炭為末枳殼末五錢炭末三錢和匀用濃米飲一冲盞調下空心服五更初一服如人行五里再服當日見効

Fig. 55. The thorny limebush, *chih shih* (*Poncirus trifoliata*) R 349, in the *Chêng Lei Pên Tshao* of +1249, ch. 13, p. 20a. See the discussion on pp. 103 ff.

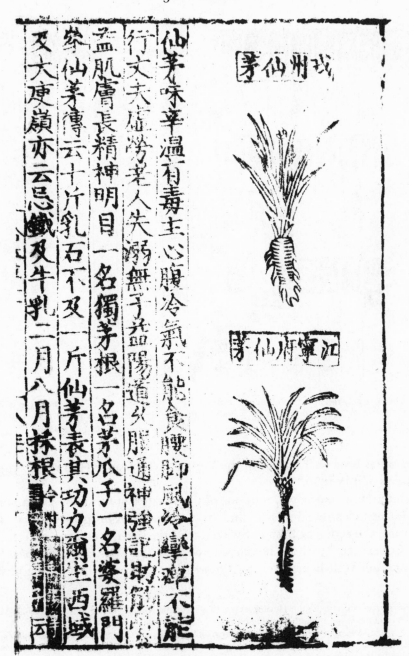

Fig. 56. A page from the rare +1468 edition of the *Chêng Lei Pên Tshao* (ch. 11, p. 38*a*) depicting (from two parts of the country), and describing, star-grass, *hsien mao* (*Curculigo ensifolia*) of the Amaryllidaceae. R 660; CC 1800.

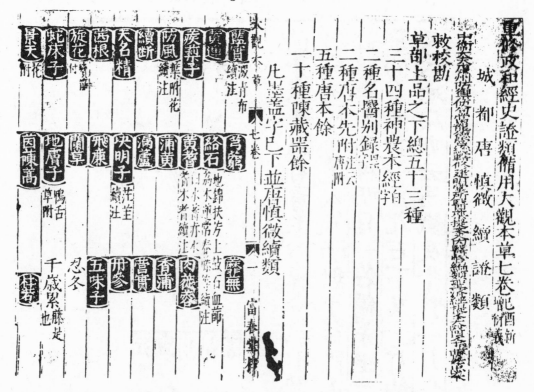

Fig. 57. The extremely rare, perhaps unique, +1581 edition of the *Ta-Kuan Pên Tshao*, preserved in Bern; the contents table of ch. 7. Photo. kindness of Prof. E. Hintzsche.

Pên Tshao has been well known in the Western world since Hummel (13) gave a description of it in 1940.[a]

It is interesting, moreover, that one of the versions of this book was among the first Chinese scientific works to find its way into a European library. In the Bongarsian Collection at Bern in Switzerland[b] the seventh chapter of it[c] is bound up as 'Codex 350'[d] with an incomplete copy of the *Wan Ping Hui Chhun*[1] (The Restoration of Well-Being from a Myriad Diseases), published by the physician

---

[a] Special studies of Chang Tshun-Hui's work are those of Hung Kuan-Chih (4) and Wang Chün-Mo (1).

[b] Now part of the Burgerbibliothek in that city.

[c] There is a mystery about this, since the title bears all three terms, *Chhung-Hsiu*, *Chêng-Ho* and *Ta-Kuan*, a juxtaposition not mentioned by any Chinese bibliographer. Lung Po-Chien (1), no. 33, item 11, however, knows of an edition by the printer of the Bern copy (Fu Chhun Thang). The edition was probably a very late one, perhaps that of +1581, and the Bern copy may be unique. Lung himself has evidently not seen this issue. Cf. Okanishi Tameto (1), p. 1323. It would have been one of the *Ta-Chhüan* family.

[d] The classification of all Chinese printed books as 'manuscripts' is a curious aberration of librarians, dating no doubt from a +17th-century *idée fixe* about the date of the invention of printing, and persisting, I believe, in the British Museum still.

[1] 萬病回春

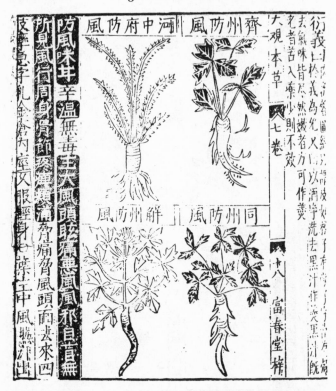

Fig. 58. A page from the same edition of this pharmaceutical natural history, showing *fang fêng* (*Siler divaricatum*) of the Umbelliferae from four different places (ch. 7, p. 18*b*). That from Ho-chung, at the top on the left, must be a different plant. Cf. Fig. 50*a*. Photo. kindness of Prof. E. Hintzsche.

Kung Thing-Hsien[1] in +1615,[a] and interesting, *inter alia*, for its anatomical diagrams.[b] Hintzsche (1) has printed a letter about these two books, written in +1632 by Wilhelm Fabry (G. Fabricius Hildanus), which dates their accession, and quite probably it was his own son Johannes Fabry, also a physician, who had brought them back when he returned from the East Indies early in +1630, obtaining them doubtless in Batavia from Macao. No one in Bern at that time or for long afterwards, could read or identify the books, which were catalogued as 'liber anatomicus et botanicus ex China' in +1634. Some pages of this *Pên Tshao* are reproduced in Figs. 57, 58, 59.

Albrecht von Haller, the great physiologist, who had himself in his younger days been librarian at Bern, referred to Wilhelm Fabry's Chinese anatomical

---

[a] Hintzsche says that the Bern copy has prefaces of +1587, +1596 and +1597, and that the printing was of +1605; this does not agree with the best Chinese bibliographers (e.g. Ting Fu-Pao & Chou Yün-Chhing (3), app. 1, p. 28*a*), on whom we rely, but again the Bern copy may be unique.

[b] See the special study of Hintzsche (2), and further on in Sect. 43.

[1] 龔廷賢

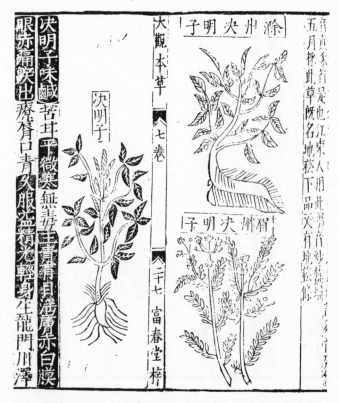

Fig. 59. Another page from the same edition, showing *chüeh ming tzu* (*Cassia tora*), from three provenances. The drawing at the bottom on the right gets the leaflets of this legume best, but the long upright pods of the plant are better shown in the other two. Ch. 7, pp. 27*a*, *b*. Photo. kindness of Prof. E. Hintzsche.

book in his *Bibliotheca Anatomica* (+1774).[a] A short while before, in his *Bibliotheca Botanica* (+1771),[b] he had discussed (without of course knowing title or date) a Chinese book of natural history with 105 plant illustrations similar to those of the Bern *Pên Tshao*, which he had in his own library. Unfortunately it has not so far been possible to identify what book this was.[c] In von Haller's opinion the pictures were generally much better than those of the popular Rembert Dodoens (Dodonaeus)[d], and he felt easily able to determine many of the plants botanically from them, especially the cereals, palms and amaranths; moreover, he admired their workmanship, including the portrayal of the leaves in black, with white veins. Chinese plant depictions in colour, of which he read in du Halde, would be, he thought, immediately recognisable. We reproduce a page of these reactions of von Haller's in Fig. 60.[e]

[a] See vol. 1, pp. 9, 138.   [b] In vol. 1, p. 5.
[c] Chinese characters appear in the work of C. G. von Murr, *Adnotationes ad Bibliotheca Halleriana*, but it does not answer our question.
[d] Herbals, etc. published in +1554, +1566, +1568, Eng. +1578, etc.; cf. Arber (3), 2nd ed. pp. 82 ff., 124 ff.
[e] It was gratifying to us to note the expression used in one of the footnotes (3), for we chanced upon this page long after we had ourselves decided to use the term pandects for the pharmaceutical natural histories of China.

CAP. I. ORIGINES REI HERBARIÆ.

ftrat. Nam horti Malabarici 800 Tabulæ fupra mille ftirpium figuras continent, omnes nominibus Indicis diftinctas, & Coromandeliæ medicis multo majorem in re herbaria diligentiam ineffe, demonftrant, quam vel in Græcis fuerit, vel demum ante CLUSIUM & GESNERUM in Europæis.

Exftat inter mea ficcarum ftirpium volumina non mediocris fafciculus merorum graminum, quæ fuis cum Malabaricis nominibus medici ejus regionis indigenæ Danicis verbi divini Præconibus, hi ILL. olim ARCHIATRO JOHANNI AUGUSTO de HUGO miferunt, ifte pro fuo, quo me profequebatur favore, mihi donavit. Plurima funt, perelegantia, & numero fuo immenfum fuperant, quidquid de graminibus veteres Græci & Latini reliquerunt. Olim STRABO, medicos indicos cibandi ratione, unctione & cataplasmatibus uti, omnino ut Ægyptios. Zeyloniæ incolæ fuis plantis morbos fuos fanant, & decocta aliaque remedia parare norunt (6). CHINENSES plantarum ftudiolissimi (1*), ad HOANGTI (1), tertium gentis a FOHI conditæ Imperatorem, medicæ artis inventionem, & ftirpium nobiliffimarum vires referunt (2), & opera botanica poffident plenitfima. Unicum defcribit J. BAPTISTA DU HALDE, cui titulus *Pent Sao* (3) feculo decimo fexto (4) a medico LITSCHESING compilatum, & anno fere 1597. a fuperftite filio editum (5). Longe fupra mille plantas continet, quarum 265. aromaticæ funt. Eas plantas Chinenfes in fuas claffes dividunt, neque iconum artificium ignorant. Eft etiam inter meos libros Compendium botanicum Chinenfi lingua editum, non pars prioris, quod citavi operis, cum ex omnibus claffibus plantas aliquas contineat, arbores, cerealia, cucurbitaceas, amaranthos, lapatha. In eo opere centum &. quinque icones numeravi, rudes equidem, ejus fere faporis, qui eft in plantis a SCHOIFFERO depictis. Multo tamen meliores funt, quam DIOSCORIDEÆ illæ, quarum aliquas DODONÆUS edidit, & quas nemo adgnofcat. Noftræ Chinenfes ad naturam delineatæ facile ad fuas plantas reducuntur, ut in cerealibus, palmis, amaranthis expertus fum: neque absque fuo funt artificio: · Sunt enim, in quibus nervos albos in nigro folio fculptor accurate expreffit, neque fine difficultate. Catalogum medicameutorum fimplicium, Sinis medicis familiarium, quorum longe major pars eft ex vegetabili regno, CLEYERUS edidit (6). Chinenfium ftirpium defcriptiones, ut ex J. BAPTISTÆ DU HALDE opere video, a coloribus aliisque fignis fponte oculos ferientibus fumuntur, non fine hyperbola: virium longa adeft enumeratio, & fupra verum,

A 3 quod

(6) VALENTYN *Ooftindien* V. p. 43.
(1*) In Cataya (Chinæ parte boreali) omnes fere incolæ plantarum habent cognitionem RUBRUQUIS p. 68.
(1) BOYM *Eph. Nat. Cur.* dec. II. anno 10. *Supplem.*
(2) Ginfeng apud du HALDE T. III.
(3) *Pan San Kan Mau* opus medici LITSCHI SIN, vocat Cl. MALOUIN, qui excerptum legit *Chym. medicin.* T. 1. p. 8. & addit effe *Pandectas medicas.*
(4) Recufum a. 1684. du HALDE.
(5) Du HALDE.
(6) *Medic. Chin.* p. 25. P. II.

Fig. 60. Part of the discussion of Chinese pharmaceutical natural history in Albrecht von Haller's *Bibliotheca Botanica* of +1771, p. 5. The *Pên Tshao Kang Mu* of Li Shih-Chen (+1596) appears under various linguistic disguises, and a footnote recalls that P. J. Malouin in his *Chimie Médicinale ... of* +1755 had dubbed it 'the pandects of medicine'. After complimenting the traditional pharmacists of India and Ceylon, von Haller turns to the Chinese works on botany, recognising that many of their illustrations were at least as good as those of Dodonaeus, and particularly admiring the use of white lines on a black ground for depicting the veins of leaves in wood-cuts (as in the previous two pictures).

We have now completed what had to be said about pandectal natural history in its Thang period of expansion and its Sung period of maturity. In telling this rather complicated story it is hard to avoid the impression of a mere catalogue of editions with unwieldy titles, but it will have been evident that far more than this was involved, for each of the great Sung pharmaceutical naturalists had a very distinct personality, and government patronage exerted at particular moments was an important factor, even to the extent of organising the collection of plant drawings from all over the empire (p. 281). So magisterial was the work of Thang Shen-Wei and Khou Tsung-Shih that the *Chêng Lei Pên Tshao* dominated Chinese drug-plant botany for close on five hundred years, from +1100 to +1600, as we know from the many reprints which were made. Yet from +1300 onwards, during the Yuan (Mongol) and Ming periods, the role of the *pên tshao* literature in Chinese botany as a whole clearly decreases; though maintaining, and indeed increasing, its bulk, it is oriented in rather different directions, to medico-pharmaceutical theory on the one hand, to nutritional theory and dietary prac-tice on the other.[a] Fresh genres now arose to pipe off botanical writing—there was the upgrowth of the monographic style, devoted to particular plants or families, in the Sung and thereafter; there were the relatively new agricultural[b] and then horticultural[c] encyclopaedias; and also the new studies on plants which could be used for food in emergencies.[d] The largest single factor which led in the +16th century to the breaking of the spell of the *Chêng Lei Pên Tshao* was no doubt the steady increase in Chinese knowledge of the plants and plant products not only of Indo-China and the East Indies but of India and all over the Indian Ocean; this of course most intense in and after the period of the great navigations associated with the name of Chêng Ho.[e]

During the Yuan dynasty (*c.* +1280 to +1370) almost nothing was done in the *pên tshao* field of importance from the botanical point of view, though much progress was made in the other directions, pharmacological and nutritional, just mentioned. With the Ming came a certain restoration, and here it is convenient to divide the material, as Li Thao (11) did, into activities before the *Pên Tshao Kang Mu* of Li Shih-Chen (+1596), that second towering mountain of the *pên tshao* landscape, and activities after it. As examples of efforts to keep up with the introduction of new plants from the south and from overseas one might men-tion two books, the *Pên Tshao Fa Hui*[1] (Further Advances in *Materia Medica*) written by Hsü Yung-Chhêng[2] about +1360, and the *Pên Tshao Hui Pién*[3] (The Congregation of the Pharmaceutical Naturalists) due to Wang Chi[4] around

---

[a] Hence we shall deal with it in Sects. 45 and 40 respectively.
[b] Cf. the discussion already given in Vol. 4, pt 2, pp. 166 ff. We say 'relatively' because although the oldest work of the kind dates from +1149, there was only one other of importance (+1274) before the great type-specimen, Wang Chên's *Nung Shu* of +1313. See further, Sect. 41 in Vol. 6, pt 2 below.
[c] Cf. Sect. 38 (*i*), 1 below.   [d] Cf. pp. 328 ff. below.
[e] On this subject see Vol. 4, pt 3, pp. 486 ff.

[1] 本草發揮   [2] 徐用誠   [3] 本草會編   [4] 汪機

Fig. 61. Lan Mao, the author of the *Tien Nan Pên Tshao* (+ 1436), seated against the background of the Kunming Lake; a scroll-painting by Yang Ying-Tieh. From Yü Nai-I & Yü Lan-Fu (*1*).

+1540. Neither of these had illustrations and neither was very influential; indeed Wang Chi earned the criticism of Li Shih-Chen for what he regarded as many mistakes.[a]

Much more important botanically than either of these was a work on the natural history of Yunnan province in the south-west, the *Tien Nan Pên Tshao*[1] written by Lan Mao[2] in +1436, an exploration of the local flora[b] as adventurous in its way as those of far-away Africa superintended by his older contemporary and co-provincial Chêng Ho. The work, which has 448 entries, opens with some 70 illustrations,[c] but the plants are in many cases hard to identify as the names have not been incorporated into the modern Chinese Floras.[d] We give four pages to show the nature of the book (Figs. 62*a, b, c, d*). From p. 6*a* comes an orchid (*lan hua shuang yeh tshao*[5]) *Cypripedium japonicum*,[e] and from p. 6*b* *Viola verecunda* (*ju i tshao*[6]).[f] P. 22*a* gives us a labiate *Caryopteris incana* (*chia su*[7]),[g] and on p. 33*b* there is a rather good drawing of a poisonous member of the Araceae, *Lysichition camtschatense*, the 'water-banana' (*shui pa chiao*[8]).[h] Further research on the work of Lan Mao would be rewarding. His *Tien Nan Pên Tshao* stands with three other books in a class of their own, pandectal in that they dealt with animals as well as plants, but not pandectal because they concentrated attention wholly on a particular geographical area. We have already mentioned the +8th-century *Hu Pên Tshao* and the +10th-century *Hai Yao Pên Tshao* (p. 276 above) but the fourth, inserted here out of order because of its logical relevance, was the *Chih Wên Pên Tshao*[9] (Enquirer's Natural History), published by Wu Chi-Chih[10] about +1765. This was a very original book. Wu was a physician in the Liu-Chhiu Islands, at that time a Chinese colony, and being intrigued by the plants and possible drug-plants of the islands, he sent a collection over each year for the systematic comments of Chinese friends who were scholars, pharmacists and naturalists on the mainland. Eventually he assembled all his notes and the replies into a book, still today appreciated for its quality by modern scientists.[i] This was a striking

---

[a] Nevertheless Wang Chi's work was of some interest from the point of view of classification, for he restored the system of the *Erh Ya* (cf. p. 127 above), combining cereals and vegetables with herbs, and restoring fruit-trees to trees. Here again in a way he was retrograde, since any such extreme 'lumping' tendency militates against the recognition of natural families. Classification by functional pharmacological properties continued in Wang Lun's[3] *Pên Tshao Chi Yao*[4] of +1492.

[b] The reader will remember what has been said of the richness of this region on pp. 38 ff. above.

[c] The best descriptions of it are those of Yü Nai-I & Yü Lan-Fu (*1*) and Tsêng Yü-Lin (*1*). The book seems to have been altered somewhat by its editors of 1887, Kuan Hsüan and Kuan Chün, but an early Chhing edition also exists. Cf. Fig. 61.

[d] The best identifications are doubtless those of Ching Li-Pin, Wu Chhêng-I *et al.* (*1*), but their publication has not been available to us. This I much regret, as I saw the work in progress at Ta-phu-chi near Kunming during the second world war, and often spoke with Dr Ching about it. Cf. Needham (*4*), p. 78, fig. 92; Needham & Needham (*1*), p. 88.

[e] Cf. CC 1729. The drawing of Lan Mao looks more like a *Cymbidium*.

[f] Cf. CC 689; Khung *et al.* (*1*), p. 374. The drawing looks more like *V. japonica* or *patrini*.

[g] CC 342; Khung *et al.* (*1*), p. 867.   [h] CC 1927; Khung *et al.* (*1*), p. 215.

[i] See e.g. Ching Li-Pin (*1, 2*).

[1] 滇南本草   [2] 蘭茂   [3] 王綸   [4] 本草集要   [5] 蘭花雙葉草
[6] 如意草   [7] 假蘇   [8] 水芭蕉   [9] 質問本草   [10] 吳繼志

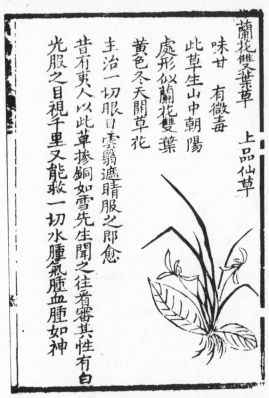

蘭花雙葉草

上品仙草

味甘　有微毒

此草生山中朝陽

處形似蘭花雙葉

黃色冬天開草花

主治一切眼目雲翳遮睛服之即愈

昔有夷人以此草摻銅如雪先生聞之往者審其性有白

光服之目視千里又能救一切水腫氣腫血腫如神

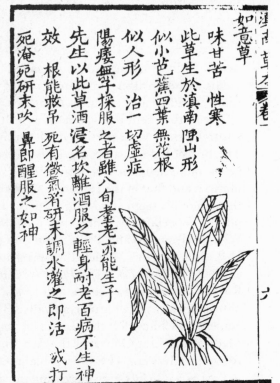

如意草

味甘苦　性寒

此草生於滇南陰山形

似小芭蕉四葉無花根

似人形　治一切虛症

陽痿無孕採服

先生以此草酒浸之名坎離酒服之輕身耐老百病不生神

效　根能救吊死有微氣者研末調水灌之即活　或打

死淹死研末吹

鼻即醒服之如神

之者雖八旬耋老亦能生子

a. An orchid, *Cypripedium japonicum* (*lan hua shuang yeh tshao*), from ch. 1, p. 6a.

b. *Viola verecunda* (*ju i tshao*) from ch. 1, p. 6b.

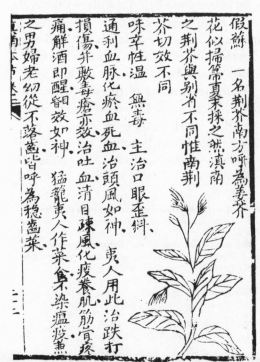

假蘇

一名荊芥南方呼為姜芥

花似掃帚常夏末採之然滇南

之荊芥與別州不同惟南荊

芥切效不同

味辛性溫　無毒　主治口眼歪斜

通利血脉化瘀血死血治頭風如神　夷人用此治跌打

損傷并散毒瘡亦效治吐血清目疏風化疹養肌筋骨疼

痛解酒即醒甚效如神　猛籠夷人作菜食不染瘟疫

之男婦老幼從不落齒皆呼為穩齒菜

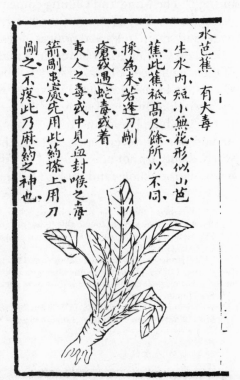

水芭蕉

有大毒

生水內短小無花形似山芭

蕉祇高尺餘所以不同

揉為末若逢刀削

瘡戒遇蛇毒戒著

夷人之毒戒中見血封喉之毒

箭剔夷瘡處先用此藥搽上用刀

剔之不疼此乃麻約之神也

c. The labiate *Caryopteris incana* (*chia su*), but this is now placed in the Verbenaceae. *Nepeta incana* or *japonica* may be the better name (R 133; Anon. (109), vol. 3, fig. 5164)' Ch. 1, p. 22a.

d. The 'water-banana', *shui pa chiao* (*Lysichiton camtschatense*), a poisonous member of the Araceae; ch. 1, p. 33b.

Fig. 62. Pages from the *Tien Nan Pên Tshao*.

case of Chinese naturalists being called upon to name and determine unfamiliar plants entirely within the framework of the traditional botanical nomenclature, Linnaean ideas and practices being not yet known.

The +16th century was notable in that two great works of pharmaceutical natural history were produced within five years of its beginning and its ending. The first of these has been greatly overshadowed by the second because it was never published until our own time, remaining in manuscript in the Palace libraries and that of the Imperial Academy of Medicine (Thai I Yuan[1]). Towards the end of his reign the Ming emperor Hsiao Tsung commissioned a grand new pharmacopoeia, entrusting two Assistant Academicians (Yuan Phan[2]), Liu Wên-Thai[3] and Wang Phan[4], and one of the Imperial Physicians, Kao Thing-Ho[5], with the work, which they presented two years later, in +1505. This was entitled (*Yü Chih*) *Pên Tshao Phin Hui Ching Yao*[6] (Essentials of the Pharmacopoeia Ranked according to Nature and Efficacity); it contained 1815 entries, classified naturalistically not pharmaceutically.[a] Two centuries afterwards, the Khang-Hsi emperor of Chhing felt the need of an amplification, and this was carried out by a Medical Warrant Officer of the Academy (Li Mu[7]), Wang Tao-Shun[8], and an Assistant Physician, (I Shih[9]), Chiang Chao-Yuan[10], who finished their work in +1701 as a ... *Hsü Chi*[11], increasing the size of the whole by a fifth.[b] But again this remained in manuscript and was not printed till it came out with the main work in 1937.[c] The Ming and Chhing emperors were not, it seems, filled with such universal compassion as those of the Sung, who printed medical formularies far and wide, and set up steles and notices of remedies at public cross-roads.[d] A number of Ming and Chhing MSS beautifully illustrated in colour have survived;[e] and two of them have been discussed in detail by Bertuccioli (2, 3). One of these has been since 1877 in the National Library at Rome, having been brought back in 1847 by an Italian bishop, Ludovico de Besi, from the remains of the library of the princely bibliophile Yün-Hsiang[12] (+1686 to +1730).[f] The most magnificent manuscript, however, appeared on the market in Hongkong in 1959 and its whereabouts are not now known.[g] In Fig. 63 we reproduce, however, the first page of the introduction, and in Fig. 64 the opening of the entry for *ai na hsiang*[14],

---

[a] On the scheme adopted, cf. p. 305 below.
[b] By this time they were of course able to use the great *Pên Tshao Kang Mu*, on which see p. 309 ff. below.
[c] Even now the illustrations have never been published.
[d] Cf. Sect. 44 below.
[e] There is one in the National Library at Peking, and another in that of the Tanabe Pharmaceutical Company at Osaka. The latter, by the kindness of Dr Miyashita Saburō, we had the opportunity of examining in 1964. The original paintings were executed by a team of eight led by Wang Shih-Chhang.[13]
[f] See Hummel (2), p. 923.
[g] We are much indebted to Dr S. D. Sturton for providing us with information and photographs at the time. The MS was again offered for sale in London in 1972.

[1] 太醫院	[2] 院判	[3] 劉文泰	[4] 王槃	[5] 高廷和
[6] 御製本草品彙精要	[7] 吏目	[8] 王道純	[9] 醫士	
[10] 江兆元	[11] 本草品彙精要續集	[12] 允祥	[13] 王世昌	
[14] 艾蒳香				

Fig. 63. The preface of the *Pên Tshao Phin Hui Ching Yao* of +1505, an opening page. Photo. kindness of Dr Sturton, from a Hongkong MS. The compilers say that they have examined all the old traditions, rejecting their mistakes, and affirming the health-giving and longevity-promoting truth about the myriad productions of heaven and earth. Illustrations were provided to show both the colours and the forms of the plants and animals discussed. Extant MSS still have coloured illustrations, but the printed versions have none at all.

Blumea balsamifera, an aromatic camphor-yielding Composite growing widely in tropical regions to a height of 10 or 12 feet, and known in Chinese natural history at least since the +10th-century *Khai-Pao Pên Tshao*.[a]

The style of Liu Wên-Thai's pandects is very systematic, rather laconic and businesslike. After recording the presence or absence of powerful active principles, the habit, the oldest ascertainable mention; and the *Pên Ching* text, if any, done in red characters; they divide the material into 24 brief headings under key-words as follows:

1 Synonyms   *ming*[1]
2 Botanical description (here a good deal of lexicographic and other literature   *miao*[2]
  is drawn on as well as the earlier pandects)

[a] The entry occurs in ch. 12 (p. 355); see R 17 and especially Burkill (1), vol. 1, pp. 334 ff.

[1] 名   [2] 苗

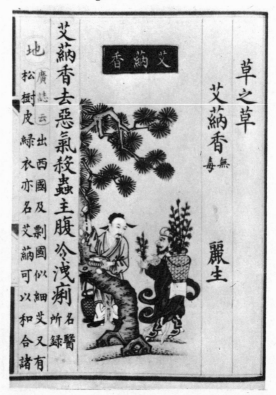

Fig. 64. A page from the same MS depicting the aromatic and camphor-yielding composite *Blumea balsamifera* (*ai na hsiang*). It is also anthelminthic (R 17); cf. Burkill (1), vol. 1, pp. 334 ff. Ch. 12, pp. 5*a*, *b* (p. 355); photo. kindness of Dr Sturton.

3	Places of origin	*ti*[1]
4	Season of sprouting, ripening, gathering	*shih*[2]
5	Methods of preservation	*shou*[3]
6	What parts of plant or animal utilised	*yung*[4]
7	Description of product as *materia medica* (pharmacognostic properties)	*chih*[5]
8	Colour	*sê*[6]
9	Sapidity	*wei*[7]
10	Nature (pharmacological properties)	*hsing*[8]
11	Effectiveness	*chhi*[9]
12	Odour	*chhou*[10]
13	Therapeutic employment (if exhibited as the primary agent)	*chu*[11]
14	Effect on the system of acupuncture tracts (what tract chiefly affected)	*hsing*[12]
15	What drugs adjuvant	*chu*[13]
16	Incompatibility with what other drugs	*fan*[14]

[1] 地	[2] 時	[3] 收	[4] 用	[5] 質
[6] 色	[7] 味	[8] 性	[9] 氣	[10] 臭
[11] 主	[12] 行	[13] 助	[14] 反	

17	Methods of processing	*chih*[1]
18	Other therapeutic virtues	*chih*[2]
19	Effects in combination (synergic effects)	*ho*[3]
20	Cautions	*chin*[4]
21	Imitations and substitutes encountered	*tai*[5]
22	Contra-indications	*chi*[6]
23	Antidotes for excess of the drug	*chieh*[7]
24	How to distinguish genuine from false or fraudulent samples of the drug	*yen*[8]

More interesting, perhaps, from the botanical point of view, are the terms which Liu Wên-Thai and his colleagues systematically used to describe the general nature and habit (*shêng*[9]) of each plant. There were seven of these:

thê[10][a]	unbranched (lit. special, hence unique)
san[11]	straggly, dispersed
chih[12]	upstanding
man[13]	creeping, vine-like
chi[14]	epiphytic
li[15]	climbing creeper
ni[16]	growing in water and mud.

Knowledge of what growth to expect facilitated collection, they said.

So far all is easy going. But we then find that each entry is headed by a mysterious category, such as *mu chih fei*[17], here on the face of it something to do with wood, and something to do with flying. This must evidently be some kind of classification, but exactly what? The answer, which shows the great influence of Neo-Confucian philosophy at the time, is given in the prolegomena (Hsü li[18] and Fan li[19])[b] where Liu Wên-Thai and his colleagues say that they will follow the system of the *Huang Chi Ching Shih Shu*[20], written by Shao Yung[21] about +1060.[c] We have already had occasion to refer in several places[d] to this remarkable work on natural philosophy, the 'Book of the Sublime Principle that Governs All Things within the World'—a text which constitutes one of the clearest and most highly systematised, though of course deeply medieval, Chinese thought-constructions concerned with cosmology, including the place of plants and animals therein. We shall return to it later (see diagram in Fig. 65) because of its influence on fundamental medical thinking; here we have only to explain how it came to be involved

---

[a] It is interesting that this term has persisted into modern botany, but with a quite different meaning; it is applied to plants fully self-supporting in photo-synthesis, as opposed to those which live a wholly or partly parasitic or saprophytic existence.

[b] Pp. (13 ff., 15 ff.).

[c] The work, with its many diagrams, is somewhat abridged in *Hsing Li Ta Chhüan*[22], chs. 7–13, and still more so in *Hsing Li Ching I*[23], ch. 3. On these *Summa* compilations of Neo-Confucian philosophy, see Vol. 2, *sub voce*.

[d] Especially Vol. 2, pp. 455 ff. Cf. Forke (9), pp. 21 ff.

[1] 制	[2] 治	[3] 合	[4] 禁	[5] 代
[6] 忌	[7] 解	[8] 贗, 贋	[9] 生	[10] 特
[11] 散	[12] 植	[13] 蔓	[14] 寄	[15] 麗
[16] 泥	[17] 木之飛	[18] 序例	[19] 凡例	[20] 皇極經世書
[21] 邵雍	[22] 性理大全	[23] 性理精義		

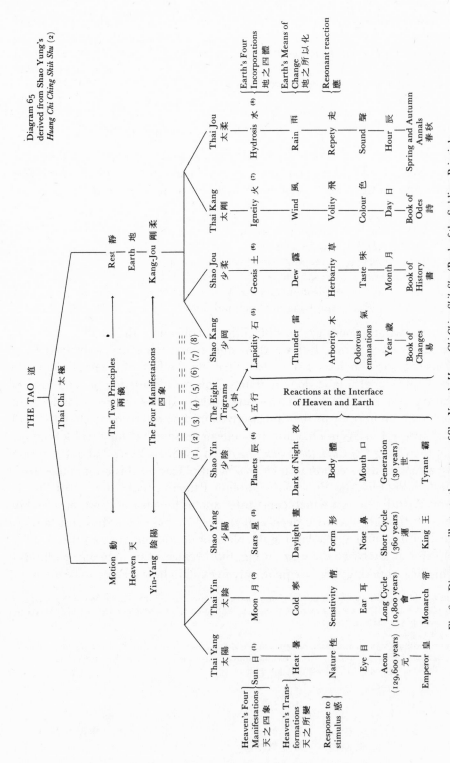

Fig. 65. Diagram to illustrate the system of Sho Yung's *Huang Chi Ching Shih Shu* (Book of the Sublime Principle which governs All Things within the World) of +1060. This was the classification adopted by Liu Wên-Thai and his collaborators in the *Pên Tshao Phin Hui Ching Yao* 450 years later; see text. Chart drawn up by our collaborator Mr Kenneth Robinson.

in a pandectal treatise of natural history. Its section most important for us is entitled Kuan Wu Nei Wai Phien[1], or 'Esoteric and Exoteric Chapters on the Observation of Things'. The doctrine ran as follows.

In the cosmogonic process the Tao generates the two primary manifestations (*liang i*[2]), i.e. motion (*tung*[3]) and rest (*ching*[4]); and these in turn generate the four secondary manifestations (*ssu hsiang*[5]), Yang and Yin in the heavens, and hardness or Durity (*Kang*[6]) and softness or Lenity (*Jou*[7]) on earth.[a] Since each of these exists in two qualities, strong and weak (*thai Yang, shao Yang, thai Yin, shao Yin; thai Kang, shao Kang, thai Jou, shao Jou*[8]),[b] the philosopher ends with eight entities capable of assimilation or correlation with the Eight Trigrams of the *I Ching* (Book of Changes).[c] Into this we will not go here, but only take leave to observe how much his four- and eight-fold system differed from the traditional framework of the Five Elements,[d] as also indeed from the ancient six-fold classification of the medical thinkers (cf. p. 263). Omitting now three levels of generation in the heavenly things, we gain access to those that concern Shao Yung's classification of earthly things by examining two appropriate lower levels. The last four entities in the series bracketed above are represented by four virtues which we may call igneity (*huo*[9]) lapidity (*shih*[10]), hydrosis (*shui*[11]) and geosis (*thu*[12]);[e] these were known as the 'four earthly bodies' (*ti chih ssu thi*[13]) and defined as arising by the exhaustive differentiation of the earthly cosmos (*ti chih thi chin chih i*[14]). Similarly the last four entities were also represented by four other virtues emanating at a still lower level; these we may call flight or volity (*fei*[15]), aspiring dendritic tree-ness or arbority (*mu*[16]), creeping and all terrestrial locomotion or repety (*tsou*[17]), and crouching plant-ness or herbarity (*tshao*[18]). These were defined as the ultimate responses of animal and plant life (*tung chih chih ying chin chih i*[19]). This last was not illogical if

---

[a] I.e. the Yang and Yin incarnated in the qualities, forms and things of the earthly world.

[b] Each term in these two groups of four occupies, of course, a specific place on the cyclical sine-wave curve given in Fig. 277 of Vol. 4, pt 1, and in Fig. 1515 of Vol. 5, pt 4, where we elaborate further on this. It is worth nothing how the 'quality' varying with position on this curve really depends on the 'quantities' of Yin and Yang measured (if they were measurable) along the two parameters at right angles to the time axis. The 'change of quantity into quality' is an idea which would not have been at all foreign to the medieval Neo-Confucian thinkers.

[c] Cf. Vol. 2, pp. 304 ff., 313 ff., Vol. 4, pt 1, p. 296.

[d] Cf. Vol. 2, pp. 242 ff., 253 ff. Five-element theory, it may be noted here, never played any important part in the thought of the pharmaceutical naturalists before J/Chin and Yuan times, when Chu Chen-Hêng was writing (cf. Sect. 44 below).

[e] Though three of these entities were homonyms of three of the five elements they are not to be confused with them. Two of the five (metal and wood) had no correlates in Shao Yung's system, because he considered them, reasonably enough, to be of a derivative or secondary nature.

In coining these neologisms we are in good company, for Peter Roget, when grouping words for his *Thesaurus of English Words and Phrases* in 1852, also experienced a lack of 'substantive terms corresponding to abstract qualities or ideas denoted by certain adjectives'. He points out (in a footnote to p. 15 in the 1930 edition in Everyman's Library, no. 630) that from the adjectives *irrelative, amorphous, sinistral* and *gaseous*, he had felt obliged to frame *irrelation, amorphism, sinistrality* and *gaseity*. We are grateful to Mr Kenneth Robinson for noticing this, and for all the elucidation of Shao Yung's ideas which he has accomplished.

[1] 觀物內外篇　[2] 兩儀　[3] 動　[4] 靜　[5] 四象
[6] 剛　[7] 柔　[8] 太陽少陽太陰少陰太剛少剛太柔少柔
[9] 火　[10] 石　[11] 水　[12] 土　[13] 地之四體
[14] 地之體盡之矣　[15] 飛　[16] 木　[17] 走
[18] 草　[19] 動植之應盡之矣

one regarded the great divide as coming between the aerial and the terrestrial-aquatic environments, analogous in their way to the greater domains of heaven and earth.

How did Liu Wên-Thai and his colleagues incorporate this schematic thought into their natural history? For inorganic things they distinguished between spontaneous and man-made substances, recognising as their categories stone, water, fire and earth, but adding metal (*chin*[1]) to cope with artifacts. But each of these groups was again divided according to the igneitic, lapidic, hydrotic and geotic (plus metalloid) cosmic virtue which inhered in the particular substance under discussion. One could thus have a geotic mineral (*shih chih thu*[2]), talc for example, or a metalloid mineral (*shih chih chin*[3]), e.g. saltpetre. For plants they distinguished between herbs, trees, cereals, vegetables and fruit-trees (*tshao, mu, ku, tshai, kuo*[4]), characterising similarly each item of these by one or other of the four vital virtues already mentioned, so that a herb could be a 'volitic' herb, an 'arboric' herb, a 'repetic' herb or a 'herbaric' herb. And this gives us at last the answer we were seeking, for *mu chih fei*[5] would mean a 'tree with the volitic virtue'.[a]

Exactly how the Ming naturalists decided which of these categories a particular plant or mineral should be put into we do not know, and a good deal of research not yet attempted would be required to find out. It had some relation to the habit and gross morphology, but there was more to it than that. Though the system commanded little future, later writers carefully avoiding it, its appellations were not so bizarre as they might at first sight seem, for they were really highly technical terms expressing sophisticated judgments about the Yin and Yang aspects of the essential nature of the plant or mineral concerned. Although the *Pên Tshao Phin Hui Ching Yao* was never printed in its own time, it was a noble compilation, and it would be a mistake to suppose that it had no influence. One may well wonder whether Li Shih-Chen, who was himself to outdo it, could not have studied it in the library when he was on the staff of the national Academy of Medicine half a century later.

(vii) *The Prince of Pharmacists*

Li Shih-Chen[9] (+1518 to +1593),[b] of whom we have now to speak, was probably the greatest naturalist in Chinese history, and worthy of comparison with the best of the scientific men contemporary with him in Renaissance Europe. His

---

[a] As for the animals, Liu Wên-Thai had less Neo-Confucian philosophy to follow, so he divided them as usual into birds, beasts, insects and fishes (*chhin, shou, chhung, yü*[6]), separating these again into categories based on integument, feathered, hairy, scaly, carapaced and naked (*yü, mao, lin, chia, lo*[7]), and adding as well their method of generation, from wombs, from eggs, by spontaneous generation from moisture, or by metamorphosis (*thai, luan, shih, hua*[8]).

[b] On the dates see Liu Po-Han (*1*).

[1] 金	[2] 石之土	[3] 石之金	[4] 草木穀菜果	[5] 木之飛
[6] 禽獸蟲魚	[7] 羽毛鱗甲臝	[8] 胎卵濕化	[9] 李時珍	

scholarly approach to the wealth of previous literature makes him also the greatest Chinese historian of science before modern times, for his works are an unparalleled source of information on the development of biological and chemical knowledge in East Asia. Such a man deserves, even in so austere a work as this, a few lines of biographical detail.[a] Li Shih-Chen was born in the neighbourhood of Chhichow[1], just north of the Yangtze between the Wuhan lake country and the Poyang Lake at Chiu-chiang; where the traditional calling of his family had been for some generations the practice of medicine. His grandfather, who died when Li Shih-Chen was in the cradle, had been little more than a wandering leech or medical pedlar (*ling i*[2]), but his father Li Yen-Wên[3] had attained the *hsiu-tshai* degree and been for a time a Li Mu (Medical Warrant Officer) at the Imperial Medical Academy in the capital.[b] Seeing his son's fascination with natural objects, he encouraged this bent, and gave him the *Erh Ya* to study at an early age, so that ideas of biological nomenclature and classification were implanted in his mind from boyhood onwards.

Nevertheless Li Shih-Chen was destined by his family for the civil service by way of the imperial examinations, the first stage of which he passed successfully at the age of fourteen. It was only when three later attempts at the higher examinations failed that his father agreed to let him concentrate on the study of medicine.[c] Following his father, therefore, he 'tramped the wards', as one might say, and learnt much from practical experience. Soon he came to feel the need for a better understanding of the principles involved in diagnosis and therapy, but especially for a more precise identification of *materia medica*, greater knowledge of the plants and animals which furnished them, and clearer definitions of their pharmacological properties. Gradually Li Shih-Chen's medical skill gained recognition by his contemporaries, and patients visited him in Chhichow from all parts of the country, often, it is said, receiving treatment without fee.[d] In time his

[a] Just as we felt in the case of Su Sung; cf. Vol. 4, pt 2, pp. 446 ff. There have been many accounts of the life and work of Li Shih-Chen. In Western languages reference may be made to Lu Gwei-Djen (1); Lung Po-Chien, Li Thao & Chang Hui-Chien (1); Chêng Chih-Fan (1) with photographs of Li's tomb and the Hsüan Miao Kuan[4] temple where he did his work; Chang Hui-Chien (1). Huang Kuang-Ming (4). In Chinese, at varying levels of popularisation, one may consult Yen Yü (1), p. 67, (5); Li Thao (8, 9); Chang Hui-Chien (1); Chhen Pang-Hsien (3)

Exhibitions on the life and work of Li Shih-Chen have been held, as at Shanghai in 1954, for which see the catalogue of Wang Chi-Min (2)

It is a sobering comment on sinology that Li's existence was quite ignored by Mayers (1), given only two lines by Wieger (3), and put in as an afterthought by Giles (1), p. 1021. The biographical account by Bretschneider (1), vol. 1, pp. 54 ff. contains many misapprehensions and is now no longer usable. But Goodrich & Fang Chao-Ying (1), p. 859, have a good one.

[b] Cf. p. 302 above.

[c] These failures need be no reflection, as Li Thao (9) has pointed out, upon Li Shih-Chen's intellectual stature, for they may have been sociologically motivated. The Ming inherited from the Yuan a system of special hereditary status groups, of which the physicians were one, and though the prejudice against change of profession weakened as the dynasty went on, it was still strong in Li's time. Between +1526 and +1562 only 7 sons of physicians graduated as *chin-shih* out of some 4000 successful candidates. See Ho Ping-Ti (2), pp. 54 ff., 68. and cf. Chang Tzhu-Kung (2).

[d] This may be a piece of hagiography, but it is pleasant to think of Li Shih-Chen as one of the Anargyroi celebrated in every Orthodox liturgy.

[1] 蘄州   [2] 鈴醫   [3] 李言聞   [4] 玄妙觀

eminence was recognised by the princely family of Chhu, who about +1549 appointed him Superintendent of Sacrifices (Chhu Wang Fu Fêng Tzhu Chêng[1]) with charge of their medical administration.[a] Shortly afterwards he served for a time in the Imperial Medical Academy, holding, according to one source,[b] the fifth-grade rank there (Yuan Phan, Assistant Academician), though perhaps it is more probable that he was a Li Mu like his father. There is reason for thinking that his nominations to such posts were occasioned by striking cures which he had performed in the patrician families. Neither of these official employments lasted very long, but they must have given precious opportunities to read many books— we have just noted a striking example—which were only to be found in princely or imperial Palace or Academy collections, and to examine numerous rare drugs which had come as tribute from overseas. At other times Li Shih-Chen worked mostly at home, a home which his biographers record that he did not leave for a decade, so intent was he upon his studies, allowing no book available to him to be left unread. He himself said that his devotion to the literature of pharmaceutical natural history was like the avidity of some people for sweets.[c] Although his greatness as a naturalist was not recognised in his own time, he was understood to be a man of learning as well as a physician.[d] He could thus number among his friends the famous writer Wang Shih-Chên[5] (+1526 to +1593)[e] who penned the preface ultimately for Li's great book; Ku Wên[6] the Neo-Confucian philosopher and official (*fl.* +1540 to +1575); and Lo Hung-Hsien[7], the geographer (+1504 to +1564) who first printed Chu Ssu-Pên's famous world atlas (*Kuang Yü Thu*).[f] And the Probationer of the Han-Lin Academy, Chhü Chiu-Ssu[8], regarded himself as the pupil of Li as well as of Lo.

Li Shih-Chen wrote a dozen books, but the *Pên Tshao Kang Mu*[10] is by far the greatest of them.[g] About +1547, when he was thirty, he began to be oppressed by

---

[a] This modelled imperial custom; cf. Lu Gwei-Djen & Needham (2), p. 66.

[b] *Chhi-chou Chih*[2] (Ju Lin sect.), ch. 11, pp. 3b ff. We are much indebted to Dr Chhen Tsu-Lung for providing us with a photocopy of the relevant pages of this from Paris. The other important biographical source is the *Pai Mao Thang Chi*[3] by Ku Ching-Hsing[4], who devoted ch. 38 in part to Li Shih-Chen, and ch. 45 in part to his father.

[c] *PTKM*, ch. Shou[9], Su (I piao), p. 33a, and Wang pref., p. 30a.

[d] Yen Yü (5) well emphasises that although Li Shih-Chen died just ten years after Matteo Ricci entered China there is no trace whatever of European influence in his work. Ricci did not reach Nanking until after the death of Li. Here I cannot help mentioning, as in private duty bound, the contemporaneity of Li Shih-Chen and John Caius (+1510 to +1573), two learned Linacres so brilliant in potential acquaintance, yet sundered impenetrably as if living on two different planets of the solar system.

[e] Prolific both in poetry and prose; the putative author of the novel *Chin Phing Mei*. Cf. Hightower (1), p. 95; Balazs (10), p. 13; Chhen Shou-Yi (3), pp. 489 ff.; Nagasawa (1), pp. 283, 305

[f] Cf. Vol. 3, p. 552.

[g] The largest of the others were concerned with sphygmology, e.g. *Phin-Hu Mo Hsüeh*[11] (+1564) and *Chhi Ching Po Mo Khao*[12] (+1572). From the edition of +1603 onwards (with the exception of that of +1606) they were printed as appendixes to the *Pên Tshao Kang Mu*, and survive as useful books today. Other books were lost, e.g. the *Phin-Hu I An*[13], which dealt with clinical histories in Li Shih-Chen's experience; the *Wu Tsang Thu Lun*[14],

[1] 楚王府奉祠正   [2] 蘄州志   [3] 白茅堂集   [4] 顧景星
[5] 王世貞   [6] 顧問   [7] 羅洪先   [8] 瞿九思   [9] 首
[10] 本草綱目   [11] 瀕湖脈學   [12] 奇經八脈考   [13] 瀕湖醫案   [14] 五臟圖論

the confusion which persisted in the pandects of pharmaceutical natural history. Since the days of Thang Shen-Wei and Khou Tsung-Shih, moreover, many new drugs, mostly from overseas, had been introduced by the physicians of Chin Tartar, Yuan and early Ming times, especially since the period of Chinese supremacy in the Indian Ocean in the first half of the +15th century. Some drug materials had been classified under entirely wrong headings, others had been erroneously separated into two or more synonyms, while others again had been insufficiently distinguished and so confused together.[a] Li Shih-Chen therefore decided to devote himself to the colossal task of producing a revised and truly modern encyclopaedia of pharmaceutical natural history. This was an act of real audacity, since in former times works of this magnitude had generally been commissioned by imperial authority and undertaken by whole teams of physicians. Li was well aware of the vast mass of literature which he would have to review, and of the travels which he would have to make to collect pharmacognostic specimens, studying minerals *in situ* as well as the plants and animals in their natural habitats. During the following thirty years, therefore, he read and noted some 800 books, revising his own draft three times.[b] From +1556 onwards he travelled constantly through the main drug-producing provinces, notably Chiangsu, Chiangsi, Anhui, Hopei and Honan (it is not clear whether he was ever in Szechuan), collecting and studying specimens. At the age of seventy, in +1587, his work was completed.[c] But he never lived to see the *magnum opus* in print, though he knew on his death-bed in +1593 that most of the blocks were already cut in Nanking for the printing. Three years later his son,[d] Li Chien-

which must have been a discussion of old illustrated anatomical tractates on the viscera; the *San Chiao Kho Nan Ming-Mên Khao*[1], which must have been a physiological disquisition on the three coctive regions (*san chiao*) and the heimartopyle (*ming-mên*), cf. Sects. 43, 44 below; and finally a *Pai Hua Shê Chuan*[2], which was a monograph on the white spotted snake or embroidered pit-viper, *Agkistrodon halys brevicaudus* (R 114) or *A. acutus* (R 120), or other species (R 488), the flesh of which, in alcoholic extract, was used as a drug in leprosy and other *fêng* diseases. It may be interesting in this connection that the venom of this genus does in fact contain very active proteases and hyaluronidases (Hadidian, 1). The Library of the Chinese Medical Association in Shanghai preserves an old MS under Li Shih-Chen's name, entitled *Thien Khuei Lun*[3]. He may well have written on monsters and teratology, but none of his bibliographies list this title. If genuine then it is unique.

[a] Precise examples of many of these confusions are given in *PTKM*, ch. Shou, Su (I piao), pp. 33*b* ff. This *su*[4] or *i piao*[5] was the letter of presentation to the throne transmitted posthumously by Li's son. Other examples again are to be found in Li's valuable *catalogue raisonnée* of the earlier pandects which forms part of his prolegomena (Hsü li[6]) in ch. 1A, pp. 2*a* ff. More examples have been collected by Tshai Ching-Fêng (1). Cf. Fig. 66*a*.

It is quite interesting to reflect that the Renaissance botanists of the West, such as Brunfels, Fuchs, Cordus and Brassavola, were also stimulated to their work by parallel confusions.

[b] *PTKM*, ch. Shou, Wang pref., p. 30*b*. The bibliography given in ch. 1A, Hsü li, pp. 14*a* ff., embodies 981 titles. We know of no other quite like this in all Chinese scientific literature, the systematic enumeration of the reading of one man.

[c] *PTKM*, ch. Shou, Su (I piao), p. 33*a*, Wang pref. p. 30*b*.

[d] The preliminary pages of the first edition show that all Li's family, sons and grandsons, contributed to the collating of the text, the correction of the proofs and the preparation of the drawings. These last, more than 1100 in number, were primarily due to his son Li Chien-Mu[7], who also acted as general editor. On the illustrations, see Fig. 66*b*, and further pp. 323 below.

[1] 三焦客難命門考   [2] 白花蛇傳   [3] 天傀論   [4] 疏
[5] 遺表   [6] 序例   [7] 李建木

Yuan[1], made the presentation to the imperial throne,[a] and the book became available for the public.[b]

Although by respect for custom we allow 'The Great Pharmacopoeia' as a translation for the title of Li Shih-Chen's work, it is infinitely more than that name would imply. This can be seen only by reading his introduction.[c] All that has been recorded, he said, shall be discussed, whether it has a practical use in medicine or not. The book is thus a pandectal treatise on mineralogy, metallurgy, mycology, botany, zoology, physiology and other sciences in its own right, so far as they could be distinguished in the +16th century.[d] All facts, said Li, shall be presented critically, whether acceptable to particular practitioners or not.[e] This involved him in careful historical accounts of the development of knowledge in the different departments of natural history. In nomenclature he adopted a system of priority, so that the name which had first been given historically to any particular plant or animal was taken as the standard term (*chêng ming*[6]),[f] copious synonymy being provided. Needless to say, this is the same system as that used today by official bodies such as the International Commissions of Botanical and Zoological Nomenclature. It was all part of Li's anxiety to resolve confusions, and he often criticised previous writers for their misapplications of technical names and terms.

[a] The opportunity came because in +1594 the emperor Shen Tsung ordered the collection of materials towards the writing of the history of the dynasty (*chao hsiu kuo shih*[2]; *Ming Shih*, ch. 20, p. 13*a*), and all kinds of books were bought by the government or presented.

[b] According to the brief and inadequate biography of Li Shih-Chen in *Ming Shih*, ch. 299, pp. 19*b* ff. (copied in *TSCC*, I shu tien, ch. 532, p. 31*b*), the emperor admired the work and gave orders that it should be printed to circulate in the world, 'with the result that it became current in the families of scholars and high officials'. In fact the inscription on the MS in the imperial library only said: 'Seen. To be preserved in the Ministry of Rites'. The words of the vermilion pencil are recorded in *PTKM*, ch. Shou, p. 36*a*, and in *Ming Shih Lu* (Shen Tsung Wan-Li), ch. 304. Whether or not the emperor intended to commission an imperial edition, the project was never carried out, because the Wên Yuan Ko[3] Library was burnt down in +1597 and all such undertakings cancelled. So at least says the famous painter Tung Chhi-Chhang[4] (cf. Vol. 4, pt 3, p. 111) in his preface to the rare Hupei edition of +1606. During the +17th century there were five other printings (cf. Ting Chi-Min (*2*); Ting Fu-Pao & Chou Yün-Chhing (*3*), p. 456), but no imperial edition has ever been found. At the present day China conserves two copies of the first edition of +1596, and Japan three, while the Library of Congress at Washington has one (Swingle, 2). The rarest edition is the Hupei one, found only at Shanghai.

The modern Chinese edition (1930, 1954) has a good index in a seventh, supplementary, volume, and there is a Japanese index by Shirai Kōtarō (*1*). Between +1596 and 1954 there were no less than thirteen editions.

[c] Fên li[5], in ch. Shou, pp. 37*a* ff.

[d] Mention of old fragmentary translations, etc. will be made later (Sect. 39), here we need speak only of those studies which one must have constantly at hand. The botanical chapters have never been translated, for Read (*1*), with Liu Ju-Chhiang, indispensable though it is, is but a book of tables. The zoological chapters are translated fairly fully in Read (*2–7*), with Li Yü-Thien & Yu Ching-Mei, but sudden omissions and mis-apprehended nuances make a check with the original text invariably necessary. The mineralogical chapters have been englished only in précis form, by Read & Pak Kyebyŏng (*1*), a handbook which should be supplemented by that of Wang Chia-Yin (*1*).

A bibliographic sketch of the fragmentary translations of, and the foreign works derivative from, the *Pên Tshao Kang Mu*, as well in Japanese as in Western languages (but excluding the Russian and German literature) is due to Huard & Huang Kuang-Ming (*4*).

[e] One senses that Li Shih-Chen, as a private physician with a long and successful practice behind him, was freer to write as he wished than someone with close governmental and Academy connections.

[f] Ch. Shou, Fên li, p. 37*b*.

[1] 李建元   [2] 詔修國史   [3] 文淵閣   [4] 董其昌   [5] 凡例
[6] 正名

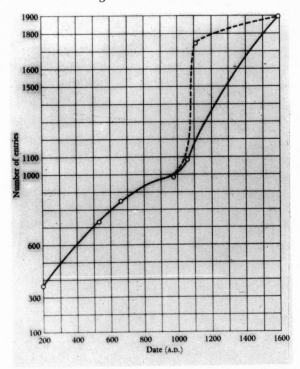

Fig. 66a. Chart showing the increase in the number of entries in the Chinese pharmaceutical natural histories through the centuries (+200 to +1600), from Needham (56), p. 41, after Yen Yü (5). It has been remarked that the unduly sharp rise after +1100 is probably referable to increasing acquaintance with foreign, especially Arabic and Persian, minerals, plants and animals, as also those of South-east Asia, with a consequent synonymic multiplication which afterwards righted itself.

In classification Li Shih-Chen retained both the ancient patterns, the naturalistic and the pharmacodynamic, radically separating them, however, for while the *materia medica* were arranged in the body of the book in a scientific taxonomic order, this was prefaced by another categorisation in which the drugs were placed under the types of diseases in which they ought to be exhibited; the book thus also constituted a general system of medicine, including as it did a wealth of specimen prescriptions (no less than 11,096) and a discussion of the principles of the art of prescribing. What Li's classification was within the natural realm we must look at later more closely (Sect. 38*f*), but here it may be interesting to note that although not comparable in exactness to that of Linnaeus it had a distinctly visible binomial element—in that the genus (of a particular family of trees, the *thung*[1] for example) was recognised, and then a number of different species (or sub-divisions) of it carefully described. Although in modern botany these trees may be classified in as many different families, the fundamental principle was the same, all classification necessarily depending on the definitions and criteria adopted. Among Li's description of the *thung* trees one can also notice his careful delineation of still

[1] 桐

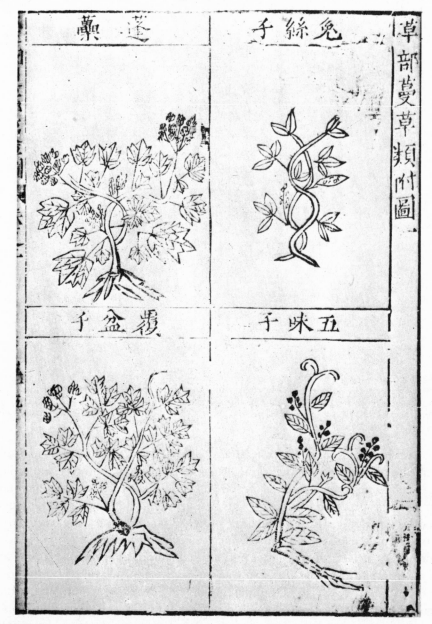

Fig. 66b. A page of illustrations from the first edition of the *Pên Tshao Kang Mu* (+1596); photo. kindness of the Shanghai Library. This edition is extremely rare, and it can be seen that the drawings are rather cruder than they had been in some of the earlier pharmaceutical natural histories, and much less sophisticated than they became in later editions of this one.

Top left: *Phêng lei*, *Rubus hirsutus* (= *thunbergii*), the bramble, R 459, Anon. (*109*), vol. 2, fig. 2279.

Top right: *Thu ssu tzu*, *Cuscuta chinensis*, the dodder, an epiphytic plant, hence no roots (cf. Vol. 4, pt. 1, p. 31), R 156, Anon. (*109*), vol. 3, fig. 4996.

Bottom left: *Fu phên tzu*, *Rubus idaeus* (the *coreanus* of R 457), the wild raspberry, Anon. (*109*), vol. 2, fig. 2297.

Bottom right: *Wu wei tzu*. The northern one is *Schisandra chinensis*, R 512, Anon. (*109*), vol. 1, fig. 1600. The southern one is *Kadsura longipedunculata*, R 507, Anon. (*109*), vol. 1, fig. 1605. Both plants are of the Magnoliaceae.

smaller sub-divisions, the varieties in fact within the species, showing that some common names were true synonyms while others denoted varietal differences. The location of a particular item in Li's natural sphere depended, moreover, not only on its visible morphology (*hsing chih*[1], a Yin feature) but also on its invisible properties or active principles (*chhi*[2], a Yang feature). In general he proceeded very systematically from the smaller to the larger (*tshung wei chih chü*[3])[a] when dealing with plants and animals. For the broader background he certainly had a *scala naturae* very consciously in mind,[b] passing from airs, waters and minerals upwards through plants lower and higher to invertebrates, vertebrates, mammals and finally man—as he himself said 'from the lowliest to the highest' (*tshung chien chih kuei*[4]).[a] Li Shih-Chen's own words are indispensable here.[c]

In the writings of old, gems, minerals, waters and earths were all inextricably confused. Insects were not distinguished from fishes, nor fishes from shellfishes. Indeed some insects were placed in the section on trees, and some trees were placed in that on herbs. But now every group has its own Section. (The sequence is as follows:) at the head come waters and fires, then come earths; for water and fire existed before the myriad (inanimate and animate) things, and earth is the mother of all the myriad things. Next come metals and minerals, arising naturally out of earth, and then in order herbs, cereals, (edible) vege-table plants, fruit-bearing trees, and all the woody trees. These are arranged following their sizes in an ascending order, starting with the smallest and ending with the largest. A Section on objects that can be worn by human beings follows (this is logical since most of them come from the plant world). Then the tale continues with insects, fishes, shellfishes, birds and beasts, with mankind bringing up the rear. Such is (the ladder of beings), from the lowliest to the highest.[d]

In the Hsü li[5] introductory section, Li Shih-Chen describes some of his methods.[e] His title echoed the *Thung Chien Kang Mu* (Short View of the 'Com-prehensive Mirror (of History, for Aid in Government)', classified into Headings and Sub-headings)[f] finished by the great philosopher Chu Hsi and his school in +1189, which Li greatly admired.[g] In his 16 Sections (*pu*[6]), which corresponded to the main headings (*kang*[7]), he worked, as need arose, by sub-division (*fên*[8]), combining together (*ping*[9]), altering the classification (*yi*[10]), supplementing (*tsêng*[11]), etc. Each entry (*chung*[12]) was brought under its proper category (*lei*[13])

---

[a] *PTKM*, ch. Shou, Fên li, p. 37*b*.

[b] On the conception of the *scala naturae* in Chinese thought, cf. Vol. 2, pp. 21 ff.

[c] Ch. Shou, Fên li, p. 37*a*, *b*; partly replicated in slightly different wording in ch. 1A, Hsü li, p. 44*b*.

[d] Though the phrase used here has feudal undertones, they had no influence on Li Shih-Chen, who grouped dragon-like animals with reptiles in ch. 43, and birds of the phoenix sort with birds in ch. 49.

[e] *PTKM*, ch. 1A, pp. 44*b*, 45*a* (tr. Hagerty, 15).

[f] This was the abridgment of the *Tzu Chih Thung Chien* which Ssuma Kuang had completed in +1084; cf. Vol. 1, p. 75, and Sect. 49 in Vol. 7. Meanwhile, see Needham (56), p. 13.

[g] Note here another aspect of the dominance of Neo-Confucian philosophy and historiography in the Ming period; cf. p. 305 above, and Li Thao (9).

[1] 形質	[2] 氣	[3] 從微至巨	[4] 從賤至貴	[5] 序例
[6] 部	[7] 綱	[8] 分	[9] 併	[10] 移
[11] 增	[12] 種	[13] 類		

in a set of 62, these corresponding to the sub-headings (*mu*[1]). The term *kang* was also used in the expression 'lesser *kang*' (*hsiao kang*[2]) to mean the main discourse in each entry, while subsidiary discussions, e.g. on nomenclature, natural history, and history of science, as well as on particular parts of a plant or animal, were considered another kind of *mu*,[1] i.e. 'lesser *mu*' (*hsiao mu*[3]).[a] Each entry is headed by its general name (*tsung ming*[4])[b] after which follows an explanation of the other names (*shih ming*[5]), an assembly of important quotations concerned with the habitat and nature of the thing (*chi chieh*[7]),[c] with sometimes a discussion of doubtful points (*pien i*[8]); and in certain cases information on the processing and preservation of the drug (*hsiu chih*[9]), with details when necessary on the method of extraction (*phao chih*[10]). Next comes a statement of the essential properties (*chhi wei*[11]), the principal uses (*chu chih*[12]), and an account of the origin and development of knowledge of these (*fa ming*[13]).[d] The entry concludes with a collection of prescriptions (*fu fang*[14]). Its rank within the ancient division into three grades (cf. p. 243 above) is mentioned only incidentally as an item of historical interest. The *Pên Tshao Kang Mu*, divided into 52 chapters,[e] contains a total of 1895 entries, of which 275 belong to the mineral kingdom, 446 to the domain of zoology, and 1094 to that of botany.[f] Entries newly added by Li Shih-Chen

---

[a] This would be the logic of it, but *hsiao mu* seems not to be defined exactly. The 'standard name' was also *kang*, or *tá kang*[6], in contrast with its synonyms, which were *hsiao kang*, or *mu*. Cf. ch. Shou, Wang pref., p. 30*b*, Su (I piao), p. 34*b*, Fên li, p. 37*b*; and ch. 1A, Hsü li, p. 45*a*.

[b] And a useful record of the earlier treatise in which it first appeared.

[c] 'Although my arrangement', said Li Shih-Chen, 'seems to have cut up and divided the former documents, yet the tributaries (lit. descent, *chih mo*[15]) (of the development of our knowledge) of each thing have become much more clear' (ch. 1A, Hsü li, p. 45a). And elsewhere he says: 'Every previous author's name has been given as a reference for statements made, so that credit for truth discovered, or blame for mistakes perpetrated, may be laid at the right door' (ch. Shou, Fên li, p. 38*b*; repeated in slightly different words in ch. 1A, Hsü li, p. 45*a*). These are rather striking words, obviously in accord with the traditions of modern science (at least before the current craze for anonymous teams). Is it not said that the free and open publication of scientific research results by responsible individuals was one of the principles on which modern science was built? Yet here is someone who never heard of the Academy of Cesi or the Lynx, nor yet of the Royal Society, stating what historians of science regard as one of their most precious principles. This seems to be another example of that curious spontaneous approximation to the ethos of modern science in the Chinese +16th century which has attracted our attention before (cf. Vol. 4, pt 1, p. 190).

It is interesting that the learned and scientific academies of Europe were springing up just in Li Shih-Chen's lifetime. J. B. della Porta founded the Academia Secretorum Naturae at Naples in +1560, Bernardino Telesio the Accademia Cosentina somewhat later, while the linguistic Accademia della Crusca started in +1583, and the Accademia dei Lincei at Rome in +1603.

The crystallisation of a scientific academy around the figure of Li Shih-Chen would have been a most reasonable thing, but Chinese society being what it was, his only fellow-virtuosos were the members of his own family. Ming China could generate philological Colleges but not scientific Academies—a difference from Europe to which we shall return in Sects. 48 and 49.

[d] For example, under this last head, in his discussion of the grape-vine in ch. 33, (p. 55), Li Shih-Chen cites the opinions of Tshao Phei (emperor Wên of the Wei, r. +220 to +226) and Chu Chen-Hêng (+1281 to +1358) on the medical properties of grapes and the wine made from them. Cf. Vol. 5, pt 4, pp. 137–8.

[e] Not counting two chapters of illustrations.

[f] Thus the biological items amount to 81.3 %.

[1] 目	[2] 小綱	[3] 小目	[4] 總名	[5] 釋名
[6] 大綱	[7] 集解	[8] 辨疑	[9] 脩治	[10] 炮炙
[11] 氣味	[12] 主治	[13] 發明	[14] 附方	[15] 支脈

himself amount to 374, and 39 others were devoted to drugs which had been successfully used by the physicians of the Chin Tartar, Yuan and early Ming dynasties, though not recorded in the pharmacological natural histories before his time.

It is still a little too early for the definitive evaluation of Li Shih-Chen's scientific attainments,[a] but something must be said of them here. His principles of classification, while not identical with those of +18th-century and modern science, were indubitably systematic. Each section (pu^1) opens with a set of definitions, all well worth study today. As Wu Yü-Chhang[2] said in the preface of +1655:[b] 'in everything he sought the taxonomy (*ko tshung chhi lei yeh*[3])'. The motive was twofold, partly the disinterested study of Nature, partly an extremely high standard of responsibility in the use of drugs on human patients.[c] In another preface to the same edition, Wu Thai-Chhung[5] remarked:[d]

One has to know the ways of the starry heavens, the profitable kindly fruits of the earth, the principles of natural things, and all that the mountains and valleys produce in the form of herbs, trees, metals and minerals from gold to iron; likewise (one must penetrate to) the deepest essentials of their manifold changes and transformations, (following their) likenesses and differences, whether on the minutest or the grandest scale—then alone may one venture to employ a single drug.

In Li Shih-Chen's writing we find a firm conviction of the unity and regularity of Nature, great respect for past traditions but no slavish admiration of antiquity, and an unfailing lucidity of expression. Of course he made many mistakes himself, as every man must. Of course he remained a man of his time and culture, faithful in many ways to the traditional and medieval Chinese natural philosophy of the Five Elements,[e] the Microcosm-Macrocosm Analogy, the symbolic correlations, etc.[f]

In the biological field Li Shih-Chen dealt with many phenomena of great interest. He was conscious of adaptation to environment[g] and the influence of the medium on the features of living things;[h] he described stable varieties and linked

---

[a] A serious discussion has been given by Tshai Ching-Fêng (*1*) on the chemical side, and by Mosig & Schramm (*1*) on the biological.

[b] See *PTKM*, ch. Shou, p. 17*b*.

[c] 'Never have I dared', said Li Shih-Chen, 'to go beyond the facts—only to search for the truth (lit. measure), (*fei kan chien yueh shih, pien thao hsün erh*[4])'; ch. 1A, Hsü li, p. 45*a*.

[d] *PTKM*, ch. Shou, p. 22*a, b*.

[e] But he particularly criticised Chu Chen-Hêng of the Yuan for organising his classification according to the Five Elements in his *Pên Tshao Yen I Pu I* (p. 287 above). 'This is to lose the thread (altogether) and to do violence (to the facts), (*i chu yao fên phei wu hsing shih chih chhien chhiang erh*[6])' (ch. 1A, Hsü li, p. 12*a*).

[f] Cf. Vol. 2, pp. 294 ff., 261 ff.

[g] One might mention his remarks on the melanisation of the head louse, *Pediculus capitis*, a case of protective coloration (ch. 40, p. 29*a*).

[h] For example, terrestrial and aquatic life, viviparity and oviparity, etc.; see his successive introductions, for the insects ch. 39, p. 1*a, b*; for the reptiles and fishes, ch. 43, p. 1*a*; for turtles, crustacea and molluscs, ch. 45, p. 1*a*; and for the birds, ch. 47, p. 1*a, b*. We shall return to these matters in Vol. 6, Sect. 39.

¹ 部　　　² 吳毓昌　　　³ 各從其類也　　　⁴ 非敢僭越實便討尋爾
⁵ 吳太沖　　　⁶ 以諸藥分配五行失之牽强耳

genetic characters as in lotuses[a] or domesticated fowls (e.g. black-boned poultry);[b] he recognised hereditary influences, familial traits and the like; and he described artificial selection, whether of cereal grains[c] or domesticated goldfish.[d] Some of his statements, mediated through Jesuit writings, formed, as we shall see, part of Charles Darwin's Chinese sources.[e] From time to time he experimented, or described the results of experiment, as in such questions as spontaneous generation, the diet of ant-eaters,[f] or the synergic action of certain pharmaca.[g] While not denying the existence of all metamorphoses in the animal kingdom, he was very sceptical of the large body of medieval lore which had seen them everywhere, disproving many of the stories by demonstrating the existence of eggs or intra-uterine embryos in the species concerned.[h] This was only one aspect of a sceptical spirit widely found in his writings.[i] Moreover, his enlightened understanding helped his classification, so that we find him, for example, transferring the galls of aphids and the lac accretions of scale-coccids from the section of woods and trees

---

[a] Red, white and pink flower colour, ability to produce double flowers, abundance of edible root, flower perfume, etc.; *PTKM*, ch. 33, p. 19*a*.

[b] His account of the *wu ku chi*[1] is famous (ch. 48, p. 7*b*; cf. R 268*f*). His description of seven other main breeds had much influence later on in the West. One of his interesting discussions of domestication deals with the training of elephants (ch. 51A, p. 13*a*). He generally avoided trespassing, however, on the fields of agriculture and animal husbandry.

[c] Cf. Sect. 39 below.

[d] *PTKM*, ch. 44, p. 25*b*; cf. R 161 and Hervey (1).

[e] Cf. Sect. 39 below.

[f] This was the pangolin (*ling li*[2]) or scaly ant-eater, *Manis delmanni* (R 106). Li Shih-Chen once found a whole pint (*shêng*) of ants being digested in its stomach, doubt about its pabulum having invited a dissection (ch. 43, p. 11*a*, *b*).

[g] The soya-bean, *Glycine Soja* (R 388), *ta tou*[3], was considered an antidote for indigestion and poisoned conditions of the intestinal tract, but Li Shih-Chen found that this never had any effect unless *kan tshao*[4] (*Glycyrrhiza glabra*, R 391) was given with it (ch. 24, p. 4*a*).

[h] We deal fully later on (Sect. 39) with the ideas on spontaneous generation and metamorphosis in ancient and medieval Chinese biology; here it is necessary only to indicate Li Shih-Chen's contribution. There are, to start with, terminological pitfalls, for in his *pu* introductions (see especially ch. 39, p. 1*a*, *b*) he uses the expressions *luan shêng*[5], born from eggs, *thai shêng*[6], born from wombs, *hua shêng*[7], generated by metamorphosis, and *shih shêng*[8]. This last should logically mean generated from humidity, but in fact it does not do so, for it includes the whole of the amphibia; and under 'tadpole', *kho-tou*[9] (ch. 42, p. 10*b*) Li has an excellent account of their generation from eggs and their metamorphosis (R 81). *Shih shêng*[8] must therefore rather be interpreted as 'living aquatically', the ambiguity residing in the second word, which can mean either to be born or to live. As we shall see, Li Shih-Chen did not feel able to rule out all spontaneous generation, but he tended to restrict its role.

Where he gave his scepticism full play was with regard to the more startling legendary metamorphoses. To take only one, botanical, example, the somewhat phallic Balanophoraceous plant *so yang*[10], *Cynomorium coccineum*, a tree-root parasite (R 240) does not, says Li, arise from the seminal emissions of wild horses, but from seeds like any other plant (ch. 12A, p. 46*b*). Zoological parallels will follow. In most of these cases Li ends with a formula of gentle irony: 'therefore they cannot *all* be generated in the way that has been claimed'.

In general Li's position was in advance of his time. Only those who are familiar with the trouble William Harvey had nearly a century later in understanding the mysteries of reproduction (cf. Needham (2) on his *De Generatione Animalium*, +1651, +1653) can estimate it adequately. One may also profitably study the contemporary beliefs of Europeans in vols. 7 and 8 of Thorndike (1).

[i] Here again cf. Vol. 4, pt 1, p. 190 for the spontaneous appearance of modern scientific scepticism in Chinese culture. Cf. also Vol. 5, pt 3, p. 97.

| [1] 烏骨鷄 | [2] 鯪鯉 | [3] 大豆 | [4] 甘草 | [5] 卵生 |
| [6] 胎生 | [7] 化生 | [8] 溼生 | [9] 蝌蚪 | [10] 銷陽 |

to that of the insects.[a] He had a rather modern conception of the nature of parasitic diseases,[b] criticised strongly the old belief that the presence of helminthic worms aided digestion,[c] and noted the perverted appetites sometimes caused by them.[d] To the conditions of storage of *materia medica* he paid particular attention, knowing that there could easily be deterioration and loss of active principles.

In the fields of chemistry and mineralogy Li Shih-Chen's attitude is of special interest. He was strongly opposed to the traditional beliefs of the Taoists concerning alchemical elixirs,[e] so much so as to risk severe unpopularity among their wealthy and influential patrons. But on the other hand he was extremely interested in the techniques of the alchemists and iatro-chemists, such as precipitation, filtration, sublimation, distillation, etc.,[f] and he was fully prepared to use them in accordance with his own judgment. In these domains he recorded for us many things of the highest interest, for example the use of mercury-silver amalgam for the filling of tooth cavities, a practice dating back to the Thang period in China,[g] but not introduced in Europe until the work of Bell in 1819 and Taveau in 1826.[h] Li also recognised the relation between sweet diets and tooth decay. He gave vivid descriptions of industrial diseases such as lead poisoning,[i] anticipating thereby the great work of Bernardino Ramazzini (+1633 to +1714). True to his universalist principles, Li never confined his observations by a narrow conception of pharmaceutical relevance, so that elsewhere we derive from him a wealth of information about the techniques and interpretations current in the iron and steel industry.[j] We also learn much about fermentations and wine-making,[k] with a

[a] *Wu pei tzu*[1] are the galls made by an aphid, *Schlechtendalia chinensis* on *Rhus javanica* (R 316). 'These are the nests of insects', said Li Shih-Chen (ch. Shou, Su (I piao), p. 34a), 'yet they were placed among the trees (*kou chhung kho yeh, erh jen wei mu*[2]). On coccid lac see R 12, and Liu Kan-Chih (1). So also amber (*hu pho*[3]), 'pine resin concreted in the earth', was put in its proper place among tree products (ch. 37, pp. 10b, 11a); cf. Vol. 4, pt 1, p. 238.

[b] Li Shih-Chen warned strongly against the eating of raw or undercooked fish (*yü kuei*[4]), which he said would lead to worm infestations, cf. ch. 44, p. 55a. This may be the oldest reference in any literature to cestode transmission. Cf. Lapage (1), pp. 137 ff.; Neveu-Lemaire (1), p. 395.

[c] The passage is in *PTKM*, ch. 18A, p. 16a.

[d] A boy who ate lamp-wicks (ch. 6, p. 11b), another who chewed raw rice (ch. 3B, p. 43b).

[e] The classical refutation of elixir beliefs is in *PTKM*, ch. 8. p. 6b, partly tr. Ho Ping-Yü & Needham (4); cf. Vol. 5, pt 3, p. 220. A description of mercury poisoning occurs in ch. 9, pp. 12b, 13a (also tr. Ho Ping-Yü & Needham). But Li Shih-Chen had no inhibitions against using mercury, as calomel and cinnabar, in the treatment of the new disease syphilis (see ch. 4B, p. 16a, b; ch. 18B, p. 5b).

[f] On distillation and its history see especially ch. 25, pp. 41b ff. This has been considered in Vol. 5, pt 4, pp. 55 ff. and by Ho Ping-Yü & Needham (3).

[g] Tin, silver and mercury alloy (*yin kao*[5]), ch. 8, pp. 10b, 11a. There is a special paper on this by Chu Hsi-Thao (1).

[h] See Lufkin (1), p. 119.

[i] *PTKM*, ch. 8, p. 21b. He also recognised poisoning by carbon monoxide (ch. 9, p. 64a).

[j] See ch. 8, pp. 35b ff., evaluated already in separate publications (Needham, 31, 32, 72). We also find mention of the hexagonal form of snow crystals (ch. 5, p. 8a; cf. Needham & Lu Gwei-Djen, 5); of the phosphorescence of sea-water (ch. 5, p. 18b; cf. Vol. 4, pt 1, p. 75); and an important account of the nutritionally valuable plant protein (gluten) *mien chin*[6] (ch. 22, p. 24a, b), which has a curious connection with some of the hypotheses of the ironmasters (cf. Needham (32), p. 34, (72), p. 531 and Vol. 5, pt 8). On *mien chin* see further Sect. 40 below.

[k] Ch. 25, pp. 41b ff. Cf. Vol. 5, pt 4 pp. 132 ff.

[1] 五倍子   [2] 構蟲窠也而認爲木   [3] 琥珀   [4] 魚鱠
[5] 銀膏   [6] 麵筋

famous account of the history of the distillation of spirits. He tells us, too, of hygienic practices such as the fumigation of sick-rooms[a] and the steam sterilisation of patients' clothes in epidemics,[b] practices certainly as old as the Sung; and he mentions the use of ice in the treatment of fevers.[c] But perhaps the most outstanding example of Li Shih-Chen's insight and intelligence in iatro-chemical matters is his detailed account of the making of steroid hormone preparations from large quantities of urine, and their use in suitable endocrinological disorders.[d] This was a discovery which had been made in the Sung, and continued still in use by the more expert pharmacists down to modern times. Finally Li Shih-Chen adopted from the medieval alchemical literature certain valuable concepts of affinity, such as the reactivity or non-reactivity of 'things of the same category' (*thung lei*[1]), which he used in his classificatory thinking.[e]

For all these reasons Wang Shih-Chên was justified in the significant statement which he included in his preface:[f]

How could one see [he said] the *Pên Tshao Kang Mu* only as a book about medicines? In actual fact it contains the finest and profoundest understanding of the organic principles of natural things. 'Tis nothing less than the Comprehensive Institutes of the Investigation of Natural Phenomena (*ko wu chih 'Thung Tien*[6])!

The reference here, highly complimentary, was of course to the *Thung Tien* finished by Tu Yu in +812, that great reservoir of source-material on political and social history.[g] And Li Shih-Chen himself, speaking posthumously in his memorial to the emperor (*su*[7] or *i piao*[8]), says that although the title indicates in the eyes of the world a medical treatise, the book is really about the fundamental patterns of natural things (*shih kai wu li*[9]).[h] This he elaborated significantly in his introduction:[i]

Although the natural objects spoken of in this work provide the drugs of varying value used by physicians, yet the researches needed to explain (their identifications and properties) and (to elucidate) the truth about their natures and patterns (*chhi khao shih hsing*

---

[a] Ch. 34, p. 39a, Cf. Sect. 44 below, and meanwhile Needham & Lu Gwei-Djen (1).
[b] Ch. 38, p. 9a. Cf. Needham (87).   [c] Ch. 5, p. 9b.
[d] See Vol. 5, pt 5, pp. 301 ff., and Lu Gwei-Djen & Needham (3).
[e] 'There are things of the same category (*hsiang lei*[2]),' he said (ch. Shou, Fên li, p. 38b), 'some of which have no medicinal value at all and yet must be investigated, while others have considerable medicinal value and yet are very little known. Some of these items are appended to each of the Sections. Some, though hidden from the ancients, are now much used, such as *sha (tshao) kên*[3], which is cognate to *hsiang fu tzu*[4]. Thao Hung-Ching knew nothing about this. So also *pi hui lei*[5] was seldom mentioned by the ancients, yet now it is well known among the repellents. Though it grows in wild, rustic and remote places, it cannot be left out.' The first two plants are indeed varieties of *Cyperus rotundus* (R 724, CC 1964), the same species. The third is a mountain plant which was apparently a repellent for the bamboo snake *Trimeresurus mucrosquamatus*, a venomous species (Tu, Tu *et al.* (1), pp. 1964 ff.; cf. R 121). The plant is described in *PTKM*, ch. 13, p. 63b, but has not yet been clearly identified (cf. R 886). On category theory in general see Vol. 5, pt 4, pp. 305 ff., or Needham (83).
[f] *PTKM*, ch. Shou, p. 31a.
[g] Cf. Vol. 1, p. 127, Vol. 7, Sect. 49, meanwhile Needham (56), p. 12.
[h] Ch. Shou, Su (I piao), p. 35a.   [i] Ch. Shou, Fên li, p. 39a

[1] 同類   [2] 相類   [3] 莎草根   [4] 香附子   [5] 辟虺雷
[6] 格物之通典   [7] 疏   [8] 遺表   [9] 實該物理

li[1]) are (part of) what our (Neo-) Confucian philosophers have called the 'science of the investigation of things' (*shih wu ju ko wu chih hsüeh*[2]); and they can complement the deficiencies of the (ancient books, such as the) *Erh Ya* and the *Shih* (*Ching*) (Book of Odes).

Such were the words which Li Shih-Chen used to describe his own life-work. They are of much interest, for they show that whatever the phrases *ko wu* and *ko chih*[a] may have meant in the +12th century, the time of Chu Hsi, they certainly had for Li Shih-Chen the meaning of the natural sciences as we understand them, an understanding therefore which did not wait for the appearance of the term *kho hsüeh* in the nineteenth century. They also show how deeply rooted in the indigenous tradition was the *Pên Tshao Kang Mu*, one of the finest flowers produced by it before the age of modern science.

(viii) *Last phases of the autochthonous development*

Afterwards nothing was ever quite the same. Nearly all the work of the seventeenth and eighteenth centuries in pharmaceutical natural history derived from that of Li Shih-Chen,[b] and he was deeply influential also in Japan, as we shall see presently when considering the effect of Chinese on world botany (cf. Sect. 38*j*). There was one smaller but rather original tree, however, which grew up in the shadow of his, though apparently without any connection. In the year of Li Shih-Chen's death, a younger naturalist was working further north, at Chhi-hsien near Khaifêng in Honan, Li Chung-Li[5] by name, who printed in +1612 a work called *Pên Tshao Yuan Shih*[6] (Objective Natural History of *Materia Medica*; a True-to-Life Study). The book is very notable for the freshness of the illustrations, all drawn for it by Li Chung-Li himself, and looking like the products of someone with a painter's talent who had studied under masters of the art. One would like to reproduce many of these interesting pictures, so little bookish and evidently done directly from the originals, but a page of four must suffice.[c] In Fig. 67 (*a*) we have one of the Araceae, an *Arisaema*, *thien nan hsing*[7], perhaps *japonica*, more probably *ringens*, because of its conical naked spadix.[d] The roots above the characteristic spherical bulb are well shown. Fig. 67 (*b*) depicts the *chhê chhien*[8], *Plantago major*,[e] and (*c*) *hu ma*[9], *Sesamum indicum* (cf. p. 172 above).[f] For a fruit-tree we give in (*d*) the loquat, *Eriobotrya japonica*,[g] the *phi pa*[10]. Li Chung-Li's drawings, which

---

[a] Cf. Vol. 1, p. 48, Vol. 2, pp. 510, 578. See also pp. 214 ff. above.

[b] How greatly he affected his junior contemporaries may be gauged from the *Pên Tshao Kang Mu Hui Yen*[3], a collection of articles by thirty authors on the principles of pharmacological therapy edited in +1624 by Ni Chu-Mo.[4]

[c] Li Thao (7, 11), who much admired Li Chung-Li, reproduced from his ch. 1, p. 11*a* the lifelike drawing of the root of *chhai hu*[11], the sickle-leaved hare's ear (*Bupleurum falcatum* and *sakhalinense*) which contains a powerful anti-pyretic; cf. R 214; CC 554, 555.

[d] Ch. 3, p. 26*a*. See R 709, CC 1922, 1924; Wu Chhi-Chün (*1*), pp. 562, 563.

[e] Ch. 1, p. 12*b*. R 90; CC 233.   [f] Ch. 5, p. 2*b*. R 97; CC 257.   [g] Ch. 7, p. 15*b*. R 427; CC 1071.

[1] 其考釋性理   [2] 實吾儒格物之學   [3] 本草綱目彙言
[4] 倪朱謨   [5] 李中立   [6] 本草原始   [7] 天南星   [8] 車前
[9] 胡麻   [10] 枇杷   [11] 茈胡

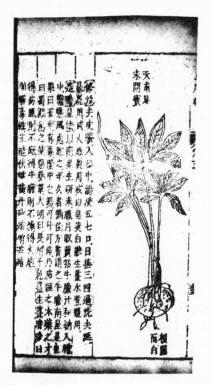

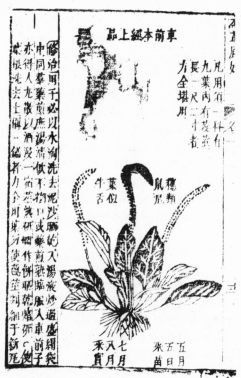

a. *Thien nan hsing, Arisaema thunbergii* (Araceae) allied to jack-in-the-pulpit. Note the tuberous root, said to be round and white. Ch. 3, p. 25*b*.

b. *Chhê chhien, Plantago major* (Plantaginaceae). Ch. 1, p. 12*b*.

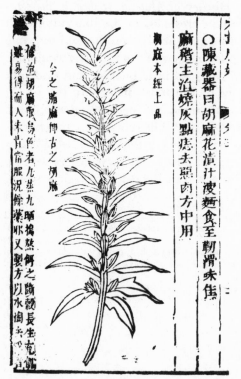

c. *Hu ma, Sesamum indicum* (Pedaliaceae). Ch. 5, p. 2*b*.

d. *Phi pa, Eriobotrya japonica* (Rosaceae), the loquat. Ch. 7, p. 15*b*.

Fig. 67. Illustrations from the *Pên Tshao Yuan Shih* (+1612).

(unlike Li Shih-Chen's) are scattered through the pages of his work, are also remarkable as being perhaps the earliest in Chinese botanical illustration to show particular parts as well as the whole plants; there is a parallelism with European development here which we shall touch on later (p. Sect. 38 g).

During the troubles of the Manchu invasion and the early years of the Chhing dynasty a pause ensued, but the last four decades of the century were prolific in books of the pên tshao type. As they were primarily pharmacological or nutritional, however, we shall reserve mention of them till the appropriate Sections.[a] In Suchow, notable for advances in optics just a little earlier,[b] the naturalists were particularly active, and in Chiangyin a family of painters produced notable plant illustrations for them.[c] At the very end of the century came some new plant descriptions, as for example in the work of Chang Lu[1] whose Pên Ching Fêng Yuan[2] (Additions to Natural History aiming at the original perfection of the 'Classical Pharmacopoeia of the Heavenly Husbandman') was prefaced in +1695 and printed in +1705.[d] The rear was brought up, as it were, by an obscure friar minim, Pedro Piñuela (Shih To-Lu[3]) whose Pên Tshao Pu[4] (A Supplement to the Pharmaceutical Natural Histories) was printed in +1697. This tractate is now an extremely rare book, which few have seen,[e] but in it the Franciscan described four useful plants (not yet identified) which he had introduced from Spain or the Philippines and thought were new to China. He also wrote on five drug-plants previously cultivated there, among which was tobacco. Piñuela's book has some interest as the only contribution to the traditional Pên Tshao literature by a Westerner (cf. Fig. 68), though it is a work of little value.

The middle of the eighteenth century was marked by the appearance of a new compendium, the Pên Tshao Tshung Hsin[5] (New Additions to Pharmaceutical Natural History) by Wu I-Lo[6], published in +1757. Though never illustrated, this was a book of considerable worth; apparently the first of the Chinese systematic treatises to mention quinine.[f] Of course the seeds of the richly-producing Cinchona ledgeriana did not reach East Asia till 1866, when plantations began to be raised in Java. But appropriately enough, the 'Jesuits' Bark' had been introduced

[a] 40 and 45 below.
[b] Cf. Needham & Lu Gwei-Djen (6), and Vol. 5, pt 6.
[c] See below, later in this Section, on the activities of the Chou family.
[d] We have already encountered this book in connection with the reconstruction of the original text of the Pên Ching which it contained, cf. p. 235 above. But it was not confined to this.
[e] There is a copy in the Bibliothèque Nationale at Paris, no. 5332 in the catalogue of Courant (3). No mention in Cordier (8). But the work was quite well known to later naturalists in China, such as Chao Hsüeh-Min (see p. 326 below), as the census of Chang Tzhu-Kung (1) proves. Pedro Piñuela was a Mexican, who came to China by way of the Philippines in +1676 and died there in +1704; but even Cordier (2), vol. 2, col. 1199, who listed his other writings, said nothing of this one. Conversely, Liu Ho & Roux (1), pp. 11, 37, knew of the book, but not the author's name. Merrill & Walker (1), p. 556 alone got both right.
[f] So Ting Fu-Pao (1), but we have not been able to find the passage in the editions available to us. The Chinese names for the outlandish drug were many and various, so we may leave them for a fuller discussion in Sect. 45 below. Meanwhile for the general picture see Duran-Reynals (1); Kreig (1); Burkill (1), vol. 1, p. 538.

[1] 張璐   [2] 本經逢原   [3] 石鐸琭   [4] 本草補   [5] 本草從新
[6] 吳儀洛

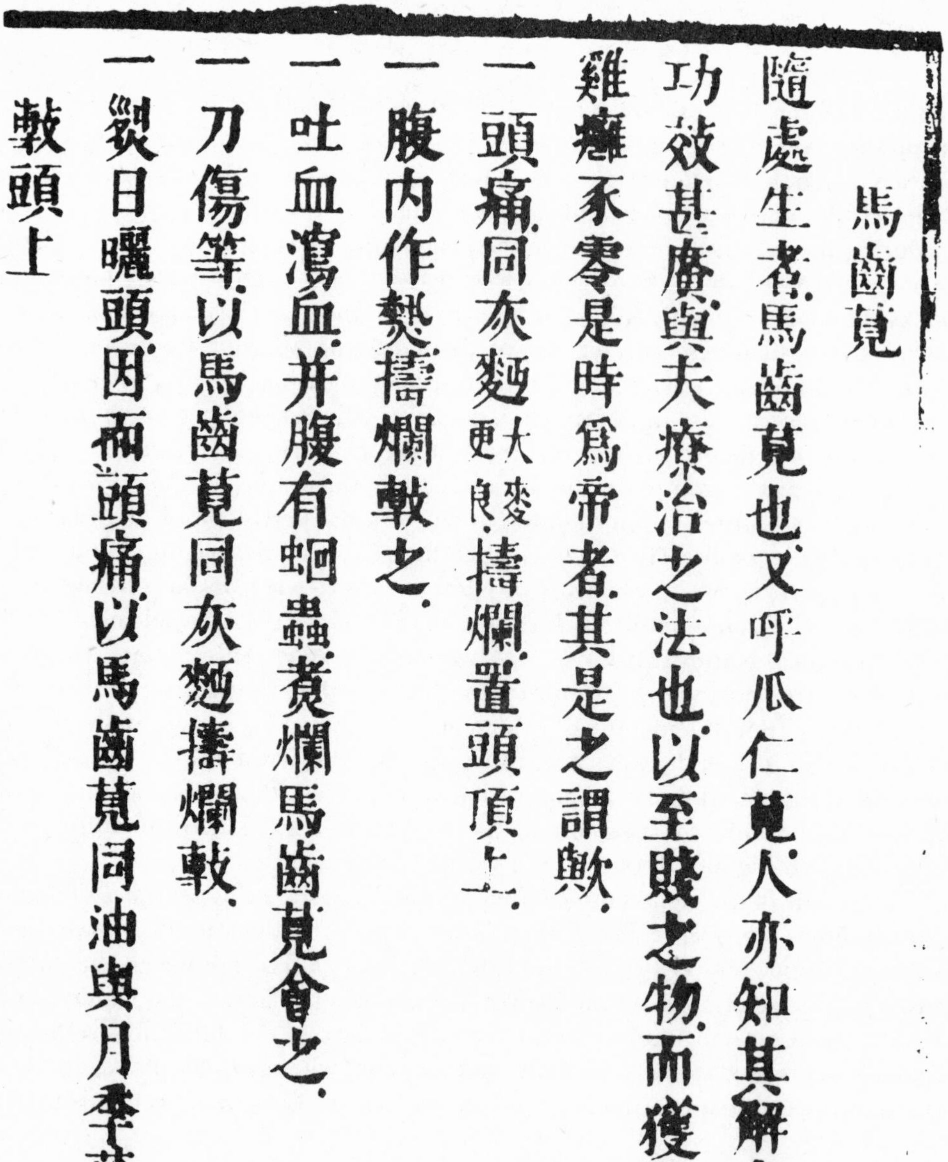

馬齒莧

隨處生者馬齒莧也又呼瓜仁莧人亦知其解毒去熱未審

功效甚廣與夫療治之法也以至賤之物而獲至切之用所

雖癰疢零是時為帝者其是之謂歟

一頭痛同灰麵 大麥更良 擣爛置頭頂上

一腹內作熱擣爛敷之

一吐血瀉血并腹有蛔蟲煮爛馬齒莧食之

一刀傷等以馬齒莧同灰麵擣爛敷

一裂日曬頭因而頭痛以馬齒莧同油與月季華 玫瑰華 擣爛

一敷頭上

Fig. 68. The entry for *ma chhih hsien*, horse-tooth vegetable, i.e. *Portulaca oleracea*, the purslane, from the *Pên Tshao Pu* (Supplement to the Pharmaceutical Natural Histories) written by the Franciscan Shih To-Lu (Pedro Piñuela) and recorded by Liu Ning, in +1697. This contribution by a Westerner was almost the last of the traditional Pên Tshao genre. Although he does not say it on this page, *ma chhih hsien* had been known for centuries in China for two things. First, it is a mercury-accumulator, and metallic mercury had been obtained from it since the +11th. century (cf. Vol. 3, pp. 675 ff., 679; Vol. 5, pt. 3, pp. 147, 207, 231–2). Secondly, it is rich in vitamin B$_1$ and was therefore recognised at least from the +14th. century onwards as a cure for beri-beri (cf. Lu Gwei-Djen & Needham (1), p. 16). Hu Ssu-Hui's depiction of it was quite good, as is shown by modern drawings (Anon. (*109*), vol. 1, fig. 1233). Pedro Piñuela's book is now extremely rare.

to China by the Jesuits themselves nearly a couple of centuries earlier. When the Khang-Hsi emperor was suffering from malaria in +1693 quinine was offered by Claude de Visdelou (Liu Ying[1])[a] and Jean de Fontaney (Hung Jo-Han[2])[b], and the usual dramatic success attended its use.[c] By +1713 a brief account of the tree was given by a Chinese scholar-traveller, Chha Shen-Hsing[3] (+1650 to +1727) in his *Jen Hai Chi*[4] (Notes of Men and Oceans). Thus within sixty or seventy years of the first recorded mention of the drug, by the Augustinian friar Calancha about +1633, it was known to Chinese physicians. One must at the same time remember that they, almost alone of those of the Old World cultures, had been in possession since the −3rd century of a true anti-malarial, *chhang shan*[5], as we shall recount in its proper place.[d] A number of other foreign, especially American, plants besides quinine (such as maize, tobacco, etc.) were also described in Wu I-Lo's book, which he himself regarded as an enlargement and rectification of an earlier work by Wang Ang.[6] This was the *Pên Tshao Pei Yao*[7] (Practical Aspects of *Materia Medica*), published in +1690 (when the author, not himself a medical man, was 80) and again four years later.

Wu I-Lo's time saw also the beginning of the activity of Chao Hsüeh-Min[8], probably the best all-round Chinese naturalist of the +18th century. His book, a really great one, was entitled *Pên Tshao Kang Mu Shih I*[9] (Supplementary Amplifications for the 'Pandects of Natural History'). Born about +1725, he lived till 1803 or soon after, tirelessly investigating for more than fifty years the natures, forms and properties of minerals, plants and animals, often in the botanic garden and laboratory maintained by his own family near Hangchow.[e] There has been confusion about the date of his book, for it contains several dates;[f] it must have been begun about +1760, first prefaced in +1765, and the prolegomena added in +1780, while the latest date mentioned in the text (1803) shows that he went on revising and adding to it until the very end of his life. It was not printed till 1871. What a tribute to Li Shih-Chen it was that Chao should have entitled his *magnum opus* so modestly. As with Li, other books also came from Chao's brush, and we have met with one of them already, the interesting *Huo Hsi Lüeh*[10] or treatise on fireworks, printed in 1813.[g] Another was *Fêng Hsien Phu*[11], a monograph devoted to the genus *Impatiens*, especially *balsamina* and *textori*.[h] A third is of much importance

---

[a] Pfister (1), no. 174.
[b] Pfister (1), no. 170, cf. above Vol. 4, pt 1, p. 310.
[c] The story has been told in detail by Wang Chi-Min (3).
[d] Sect. 45 below.
[e] His biography has been written by Wang Chung-Min (2). Again it is only too characteristic of sinology that there is no biography of him in the excellent compilation of Hummel (2), and needless to say, Giles (1) and Wieger (3) knew him not.
[f] Cf. on this Chang Tzu-Kao (5).
[g] Cf. Vol. 5, pt 6. Tr. Davis & Chao Yün-Tshung (8).
[h] CC 777, 779.

[1] 劉應　　　　[2] 洪若翰　　　[3] 查愼行　　　[4] 人海記　　　[5] 常山
[6] 汪昂　　　　[7] 本草備要　　[8] 趙學敏　　　[9] 本草綱目拾遺
[10] 火戲略　　　[11] 鳳仙譜

for the history of medicine in China. Its title, *Chhuan Ya Nei Phien*[1], is almost untranslatable,[a] but could be freely englished as 'A Pill to Purge the Common Contempt for Traditional Leeches; restoring their Effective Therapies to Proper Recognition' (+1759). Physicians of the socially lowest grade were called, as we have seen (p. 309) 'bell-ringing medicos' (*ling i*[2]) or 'peripatetic doctors' (*tsou fang i*[3]), or 'purging-and-watering leeches' (*chhuan i*[4]). Chao Hsüeh-Min was a scholar who realised what great stores of practical clinical knowledge and skill such men possessed; they had obtained it by discipleship and oral tradition, they could use it with remarkable success, but they could not explain how or why their methods worked. Chao got much of his material from a relative, Chao Po-Yün[5], who was himself a famous *chhuan i*, now about to retire, having travelled the length and breadth of the country for many years. Thus did the peripatetic medicine of the country people find a sympathetic and literate interpreter. Several other works failed to survive, such as his pharmacological tractate *Su Yao Chih*[6] and his notes on irrigated horticulture *Kuan Yuan Tsa Chih*[7]; particularly one regrets the loss of his *Shêng Chiang Pi Yao*[8], a book about chemical operations—sublimation, distillation, precipitation and the like.

Daunting in this connection is the fact established by Chang Tzhu-Kung (*1*) that of all the books quoted by Chao Hsüeh-Min in his major work rather more than half are now lost, though he had access to them only a little over a century and a half ago. People criticised him for wanting to improve on what Li Shih-Chen had done.[b] In his own preface he wrote:[c]

Someone said: 'What is the point of wasting time trying to do something beyond the *Pên Tshao Kang Mu*?'

I replied: 'Of course, you are right. But with the passage of time, species and categories become more numerous (*wu shêng chi chiu, tsê chung lei yü fan*[9]). Even ordinary folks are curious about extraordinary things, so surely (the naturalist should) collect (and describe) them in all their utmost complexity ... [and he gave examples]. These then are new sorts and varieties—if I do not describe them, whoever will get to know them?

We need not suppose that Chao Hsüeh-Min was thinking here of novelties produced by evolutionary change since Li Shih-Chen's time, surely what he had in mind was the production of new varieties in cultivation, the better exploration of

---

[a] The second word harks back, of course, to the encyclopaedic and semantic *Erh Ya* tradition[1], on which we have had much to say already (pp. 126 ff. above). The possible significances of the phrase 'inner chapters' has been discussed by us in Needham (64), pp. 271 ff. *Chhuan* normally means 'strung together' or 'in league with', but its vulgar meaning was 'cathartic', 'laxative' and 'diuretic'.

[b] His principle was to avoid saying anything that Li Shih-Chen had already said. At the beginning of his book there was a special *chêng wu*[10] section setting to rights some mistakes that Li had made. Chao was critical also of some of Li's classifications; thus he felt that Li ought to have separated the *man*[11] (vines and creepers) from the *thêng*[12] (lianas and rattans). Like 'little Dr Harvey', Chao was especially noted for his propensity for conversing with unlettered countrymen, farmers, gardeners, sow-gelders and the like. Cf. Vol. 4, pt 2, p. li.

[c] P. 5a, b.

[1] 串雅內編	[2] 鈴醫	[3] 走方醫	[4] 串醫	[5] 趙柏雲
[6] 疏藥志	[7] 灌園雜志	[8] 升降祕要	[9] 物生既久則種類愈繁	
[10] 正誤	[11] 蔓	[12] 藤		

the wild parts of China's confines, the introduction of strange plants from foreign countries, and in fact the advance of botanical and scientific knowledge in general.

The *Pên Tshao Kang Mu Shih I* is a mine of information, none the less interesting for its relative lateness in the story of indigenous Chinese science. As one would expect from Chao's strong chemical interests, there is a good account (for instance) of nitric acid (*chhiang shui*[1]) and its use in etching with copper plates,[a] while sulphur and other mineral industries are often given fuller treatment than in earlier works. Plants of the New World now come systematically in—tobacco, for example (*yen tshao*[2]),[b] which had been cultivated in Fukien from about +1570 onwards; quinine[c] and the peanut (*lo hua shêng*[3]),[d] which had been introduced in the decades preceding +1538. Then there is an important entry for opium (*ya-phien yen*[4]),[e] and another on cotton.[f] New drugs such as *ya tan tzu*[5], or *Brucea javanica*, one of the Simarubaceae, a valuable anti-dysenteric from the East Indies, are also dealt with.[g] But there were strange natural objects from China too, and Chao Hsüeh-Min was the first to give a full description of the thing called *hsia tshao tung chhung*[6], 'in-summer-a-plant, in winter-an-insect'.[h] This is a caterpillar of various species parasitised by a mould, *Cordyceps sinensis* or *militaris*, which gives it a partly plant-like appearance. The combination, dried for medicinal use, had first been recorded in the *Pên Tshao Tshung Hsin*[i] of +1757. Finally, Chao discussed the edible bird's-nests (*yen wo*[8]) which had long been valued in China for their nutritional and therapeutic properties.[j] These last derived, of course, from the proteins of the adhesive substance, a gelatinous cement regurgitated like 'pigeons' milk' from the crops of certain swifts or swiftlets (*chin ssu yen*),[k] members of the genus *Collocalia*[9] of the Cypselidae (Micropodidae) family, in their nest-

---

[a] Ch. 1, p. 9*b*.

[b] Ch. 2, pp. 14*b* ff. *Nicotiana Tabacum*, CC 304; cf. Burkill (1), vol. 2, pp. 1551 ff. and especially Laufer (41).

[c] Ch. 6, p. 51*a*.

[d] Ch. 7, pp. 84*b* ff. *Arachis hypogaea*, CC 955; cf. Burkill (1), vol. 1, pp. 205 ff. and especially Ho Ping-Ti (1).

[e] Ch. 2, pp. 28*b* ff. *Papaver somniferum* (*ying tzu su*[7]), CC 1311. We shall discuss more fully in Sect. 45 below the history of this in China, which is very different from what is commonly thought.

[f] Ch. 5, pp. 10*a* ff. Cf. Sect. 31.

[g] Ch. 5, pp. 31*b* ff. Apparently = *amarissima*, *sumatrana*; cf. CC 2513; Anon. (*65*), p. 280; Lu Khuei-Shêng (*1*), p. 238; Burkill (1), vol. 1, p. 370.

[h] Ch. 5, pp. 27b ff. Already mentioned in Vol. 2, p. 421. We shall give full details about it in Sect. 45.

[i] Ch. 1, pp. 25*a*, *b*.

[j] Ch. 9B, p. 2*a*. Fuller details will be found in Sect. 40 below. Westerners always regard 'bird's-nest soup' as one of the more bizarre components of Chinese cuisine because they can imagine only the rough twiggy nests of their own hedgerows, not realising that these swift's nests are mainly composed of a white solidified secreted material, which softens and in part dissolves during preparation.

[k] Cf. Needham (12), p. 75 for the comparative background. The exact nature of the process, whether including any contribution of salivary or gastric digestion, is still not settled, but the crop glands must surely take the leading part. As the material contains 30.5 % of carbohydrate as well as 49.8% of protein it must be rich in mucopolysaccharides no doubt partly in protein combination (analysis in Read, Li Wei-Jung & Chhêng Jih-Kuang (1), p. 67, item M 1).

Esp. *esculenta*, *brevirostris*, *francica*, *fuciphaga*, and *innominata*; some names probably synonymic.

---

[1] 鏹水   [2] 煙草   [3] 落花生   [4] 鴉片煙   [5] 鴉膽子

[6] 夏草冬蟲   [7] 罌子粟   [8] 燕窩   [9] 金絲燕

building.[a] When these were first imported into China from sources such as the Niah caves and cliffs in Borneo we do not know exactly, but the trade would probably go back at least to the +15th-century age of navigations,[b] and literary references to the edible bird's-nests begin to become common in the +17th century.[c] As a pendant to this paragraph, and while still with animal life in mind, we may note the mention of mosquito-nets by Chao.[d] All in all, the *Pên Tshao Kang Mu Shih I* constitues a rich source which has so far been insufficiently laid under contribution for the history of science in China.

With it indeed we may end this survey of *pên tshao* literature, necessarily lengthy because so much of the history of indigenous natural history is bound up with it. As the nineteenth century went on, the botany of East and West came closer together at last, and by the sixties Bretschneider would be working away at Peking, though one very great manifestation of traditional Chinese learning was yet to come, the *Chih Wu Ming Shih Thu Khao*[2] of 1848 (Illustrated Treatise on the Names and Natures of Plants), a work which we shall describe in a related sub-section (Sect. 38 *f* below), since it was not in the *pên tshao* genre. That phrase indeed continued alive down to our own time. In 1851 Thu Tao-Ho[3] produced a *Pên Tshao Hui Tsuan*[4] (Classified Compilation of *Materia Medica*), and the most appropriate book with which a biochemist could end the line would be the *Hua-Hsueh Shih-Yen Hsin Pên Tshao*[5] (New Pharmaceutical Natural History based on Chemistry) edited by Ting Fu-Pao[6] in 1934. The modern age had dawned.

(3) STUDIES ON WILD (EMERGENCY) FOOD PLANTS

It is now requisite that we should go back to the beginning of the Ming period, the last years of the +14th century, the time when science in China is often supposed to have been entering a period of 'stagnation'; this to do in order to follow a most remarkable development in Chinese botany, a move in a particular direction, a move which might well be called 'the Esculentist Movement'. Beginning in the second half of the +14th century, it had produced all its chief works by the middle of the +17th, after which little more was done, but while it lasted it gave rise to some veritable monuments and masterpieces of applied botany.[e] The point was, what to do in time of scarcity or famine? As we now so well know,[f] the climate of China has a much more severely fluctuating seasonal rainfall than that

---

[a] See Burkill (1), vol. 1, p. 637; Tu, Tu *et al.* (1), p. 739. The nests are shaped like boats with pointed ends.
[b] See Vol. 4, pt 3, pp. 487 ff.
[c] E.g. *Wu Li Hsiao Shih*[1] (+1636, pr. +1664), ch. 10, p. 4*b*, where interesting physiological discussions are reported.
[d] Ch. 9A, p. 32*b*.
[e] Previous studies of this subject have been those of Needham & Lu Gwei-Djen (10) and of Amano Motonosuke (5).
[f] Cf. Vol. 3, pp. 462 ff., Vol. 4, pt 3, pp. 217 ff.

[1] 物理小識   [2] 植物名實圖考   [3] 屠道和   [4] 本草彙纂
[5] 化學實驗新本草   [6] 丁福保

of Europe (because of its monsoon character), and sustained variations over the years from normal are much more intense, leading only too easily to serious droughts or floods. So in spite of all the great skill of the farmers and the agricultural specialists,[a] and in spite of all that the hydraulic engineers could do with their river conservancy and irrigation canals,[b] China continued to suffer throughout the millennia from periodical food scarcities and often enough devastating famines.[c] In these bad times, when the people were looking for anything with which to fill their stomachs, many wild plants on the outskirts of cultivation and in the remoter woods were used, yet this was to run into other dangers, risking all kinds of poisoning from toxic active principles in roots or berries, or the unbearable irritation of raphides in leaves.[d] And so the botanist 'to the rescue came', extending the frontiers of knowledge to certify what plants were safe and wholesome, what dangerous and evil, for the welfare of the people. Moreover, the line of distinction turned out to be a subtle one, for as in the classical Amerindian case of manioc,[e] many plants containing poisonous substances could prove valuable food resources if you knew exactly which part to eat, or how to prepare the roots or leaves so as to get rid of the toxic alkaloid. Thus it was that in the +14th century the Chinese botanists became involved with iatro-chemical procedures and what we should call biochemistry today; evidently they experimented carefully and widely. Other plants again might be useful because of their mucilaginous components which swell up on wetting,[f] and others might help to fill the

---

[a] Cf. Vol. 4, pt 2, pp. 166 ff. The depredations of insect pests, especially locust plagues (cf. Sects. 41, 42 below) must also be remembered.

[b] Cf. Vol. 4, pt 3, pp. 211 ff.

[c] A valuable study of them from the historical point of view has been written by Chêng Yün-Thê (1). Cf. also Yao Shan-Yu (1, 2, 3). But we have nothing as complete as the register of earthquakes in recorded history compiled by Academia Sinica (Anon. 8). Abundant material is, of course, available in the Thu Shu Chi Chhêng and the dynastic histories.

In spite of appeals in their daily press, Westerners of the 'affluent societies' do not often ponder on what food shortage and famine actually means for individual human beings. A few precise words are therefore necessary. Stored fat lasts but a short time, then there follows wasting of the abdominal and thoracic organs, as well as the muscular system, with decline in blood pressure and pulse rate. Since these organs include the digestive tract, absorptive capacity is weakened as in a vicious circle, leading to failure of the secretion of acid and digestive enzymes, and ending in intractable diarrhoea. Decline in resistance to bacterial attack brings many evils, especially in the susceptibility of children and young people to the characteristic *cancrum oris* or gangrenous stomatitis, a horrible destructive process, caused by haemolytic streptococci and other bacteria, which devours all the tissues round the mouth, destroying the lips and parts of the cheeks (cf. Jefferys & Maxwell (1), 1st ed. p. 215; Snapper (1), pp. 124 ff.; Cecil & Loeb (1), p. 668). Decreased calorie needs and muscular activity may lead to a stabilisation of body weight lasting weeks or even months and hence a prolongation of the agony. All famines injure first and foremost the small children, then the elderly. Of the social disintegration piled on top of all this we need say nothing. See the comments of Mayer (1) and Mayer & Sidel (1), to whom we owe the outline of this note.

[d] Raphides are needle-shaped crystals of oxalic acid. We have had a first-hand account of their potential terrors from Prof. John Corner.

[e] Euphorbiaceous *Manihot Aipi* and *utilissima* is the cassava of the American tropics; a staple food comes from its tuberous starchy root, yet every handful of this useful flour has to be diligently purged of its poisonous juice by extraction and filtration.

[f] This has become the principle of the mucilaginous laxatives much in use today (cf. Laurence (1), p. 463). The seeds of *Linum* spp. and *Plantago* spp. esp. *P. Psyllium*, are cases in point. And indeed *P. major* var. *asiatica* does appear in the Chinese books as one of the valuable famine foods, though mainly for its shoots and leaves (Read (8), 1.11). Another bulk mucilage comes from the nuts of *Firmiana* (= *Sterculia*) *lychnophora* (Burkill (1), vol. 1, p. 1019); this is not Chinese, but the closely related *Firmiana simplex* (CC 724) = *Sterculia platanifolia*, is.

digestive tract although containing (like rice itself) a high proportion of indigestible cellulose fibre, or (like agar and other seaweeds) indigestible polysaccharides.
In a word, everything possible was sought to tide the people over until normal
food became available for them again.

There can be no doubt that the esculentist movement constituted one of the
great Chinese humanitarian contributions. We do not know of any parallel in
European, Arabic or Indian medieval civilisation. For Europe indeed the more
favourable climate may well be part of the explanation, yet there is no reason for
denying to the Chinese scholars a vital motivation in terms of Confucian 'human-
heartedness',[a] all the more effective perhaps because it was so this-worldly, so
distinct from Christian and Buddhist pre-occupations with another life. And like
all actions which seek the kingdom first, the work of the esculentists brought a
great reward. Swingle (1) has written:

The earliest known, and still today the best, of the works of this class was written by a
prince after many years of painstaking research in an effort to alleviate the sufferings and
death that too frequently occurred in China as the result of famine. It is believed by
experts who have studied the food plants of China that these recurring famines have
operated to bring to the attention of the Chinese people every plant that could possibly
serve for human food. As a result of the enforced use of wild plants the best of them came
to be cultivated, and in the hands of the skilful Chinese farmers and gardeners were
rapidly improved until they became standard crop plants. The extreme richness of the
Chinese flora[b] has placed at the disposal of the cultivators vast numbers of plants with
which to experiment, and as a result the Chinese people have today a very large number
of domesticated crop plants, probably ten times as many as those grown in Europe and
twenty times as many as those grown in the United States.[c] It is probable also that the
experiments forced by famine conditions have resulted in the discovery of potent drug
plants that otherwise would have remained unnoticed. To a considerable extent the
Chinese make no sharp distinction between food and drug plants; practically all the food
plants are used in household medicine, or in the prescriptions of physicians for preventing,
curing or alleviating human diseases.

It seems that it was not until the eighteenth century that similar interests began to
prevail in Europe. In +1783, for example, we meet with Charles Bryant's '*Flora
Diaetetica*, or History of Esculent Plants, both Domestic and Foreign ...',[d] where
in the introduction an echoing note is struck. Knowing nothing whatever of the
epic background which we are about to relate, he wrote:

---

[a]   See Vol. 2, p. 11. Also Chhen Jung-Chieh (5, 6, 8); Chou I-Chhing (1); Bodde (20).
[b]   Cf. pp. 33 ff. above.
[c]   Swingle was writing some fifty years ago, but perhaps his estimate needs little change today.
[d]   I owe a knowledge of this book to the fact that there is a copy of it in the library which my predecessor Dr
Martin Davy left to the Lodge at Caius in 1839. Its division, into eleven sections, is mainly morphological, with a
Chinese flavour—roots, shoots, leaves, flowers, berries, stone-fruit, apples, legumens, grains, nuts, funguses.
  Some years before Bryant, A. A. Parmentier had been working in Paris with the same ideas in mind. His
Recherches sur les Végétaux Nourissants ..., a six-hundred page book which came out in +1781, was largely
concerned with the potato, but he did devote some attention to the means of detoxification of plants not
normally usable for food. We have to thank Mr Georges Métailié for our awareness of this. There had also been
the smaller treatise of Zückert, published in Berlin in +1778.

Whether we view Mankind in a natural or civilised state, we shall find that the principal part of his daily food, and also most of the articles necessary to his comfortable enjoyment of Life, are drawn from the vegetable kingdom; every endeavour therefore to point out with precision and accuracy the Species of Plants, immediately adapted to the use of man, must carry with it its own recommendation; for, by furnishing him with the means of distinguishing the different Species of plants clearly, he is thereby enabled to choose such as are most wholesome, and best suited to his palate and constitution, and of rejecting such as are disagreeable and hurtful.

Proceeding then to recommend a knowledge of the Linnaean System to all Gentlemen, whether travelling or 'stationed', he goes on to say:

Having pointed out the principal design, it remains to mention but one circumstance more respecting the succeeding pages; which is, that several plants inserted there were never yet generally introduced into the kitchen, but all of them have been privately tried, and found to be equal, nay even to surpass many, whose uses have long been established. This must prove of public advantage, in particular seasons, as out of such a number, if some should fail, others will be in perfection; and surely no one will object to increasing the Esculent Plants, from an opinion of its tending to promote luxury, especially if he reflect that human health and vigour can never be supported so well, as by a frequent use of vegetable diet, and that by having a great variety to choose from, both the palates and pockets of different people will be the more agreeably accommodated.

The Taoists were guiding Bryant's pen, but how surprised he would have been to learn that just about four hundred years before, at the other end of the Old World, a work after his own heart but much greater than his own had come to fruition, inspired not by the interests of horticultural gentlemen in a polite age, but by the grim necessities of battle against hostile elements in a thickly-populated land of industrious farmers.[a]

Who was the imperial prince mysteriously referred to by Swingle? His name was Chu Hsiao[1], the fifth son of Thai Tsu (the Hung-Wu emperor), and the book which he wrote was called *Chiu Huang Pên Tshao*[2] (Treatise on Wild Food Plants for Use in Emergencies). Born about +1360, he was made Prince of Chou (Chou Wang) in +1378, hence the posthumous title Chou Ting Wang[3] by which he has always been known, and in +1381 'enfeoffed' with the district of Khaifêng in Honan, the old Sung capital, where he was given the former imperial palace as a residence.[b] Apart from a couple of periods of disgrace or banishment,[c] he spent

---

[a] Fifty years or so after Bryant's time the Western world did know about the esculentist botany of China, for in 1846 Stanislas Julien presented a copy of 'the Anti-Famine Herbal' of +1406 to the Academy of Sciences at Paris; and a brief account of this appeared in the *Athenaeum* for 16 Nov. 1846. 'Of this book', it said, 'the Chinese Government annually prints thousands, and distributes them gratuitously in those districts which are most exposed to natural calamities. Such an instance of provident solicitude ... for the suffering classes may be suggestive here at home.' This notice was quoted in the following year by Badham (1) in his *Treatise on the Esculent Funguses of England*.
[b] For his biography see *Ming Shih*, ch. 116, pp. 3a, 9b.
[c] From +1389 to +1391 and +1399 to +1403. This last date was that of the accession of the Yung-Lo emperor, who was a great admirer of Chu Hsiao's.

¹ 朱橚   ² 救荒本草   ³ 周定王

the rest of his life there, gradually organising, as his scientific interests grew, not only a botanical garden but also an experimental one for the acclimatisation and study of as many plants as possible which could be used as food in times of famine.[a] It must have included a laboratory for the biochemical and pharmaceutical tests and preparations. The first edition of the *Chiu Huang Pên Tshao*, which appeared in +1406,[b] was prefaced by a scholar of the prince's household, Pien Thung[1], who wrote:

The Prince of Chou set up private nursery gardens (*phu*[2]) where he experimented with the planting and utilisation of more than four hundred kinds of plants, collected from fields, ditches and wildernesses. He himself followed their growth and development from the beginning to the end of their seasons. Engaging special artists (*hua kung*[3]) to make pictures of each of the plants and trees, he himself set down details of all the edible parts, whether flowers, fruits, roots, stems, bark or leaves, and digested the whole into a book with the title *Chiu Huang Pên Tshao*. He asked me, Thung, to write a preface, which I have done very willingly. Human nature is such that in times when food and clothing is plentiful nobody takes a thought for those who are, or may be, freezing and starving; then when the day comes that they meet with this themselves, they have not the slightest idea what to do and can only wring their hands. Therefore he who would govern himself in order to govern the people should never lose sight of this for a single moment.[c]

There can be little doubt that Pien Thung was among the closest of Chu Hsiao's assistants, but he may have been helped also by one of his sons, Chu Yu-Tun[4], another prince of Chou (Chou Hsien Wang[5]) who survived his father's death in +1425 by fourteen years. We know that the younger man, though greatly troubled by quarrels about the succession, was also interested in botany, for among his writings he left a monograph on tree-peonies entitled *Chhêng-Chai Mu-Tan Ping Pai Yung*[6] (see p. 407 below).[d] It is a pity that more intensive study has not so far been given to the whole circle surrounding Chu Hsiao, for there are intriguing possibilities in his relations with the Jewish community in Khaifêng,[e] and one of his younger brothers, Chu Chhüan[7] (d. +1448) was also an outstanding scientific and literary figure, though interested rather in alchemy, mineralogy, acoustics

[a]   From surviving records it is possible to identify some of the public disasters which happened in Chu Hsiao's earlier years and may have led his mind to the subject of his greatest work. Serious floods occurred in various parts of the country in +1374, +1376 and +1404, droughts in +1370, +1371 (Honan) and +1397, locust plagues in +1373, +1402 and +1403 (Honan itself). See *Ming Hui Yao*, ch. 70 (pp. 1355, 1363, 1367).
[b]   Apparently no copy of this exists today. Swingle (1) reported it from the Library of Congress in 1935 but a re-examination by Wang Chung-Min & Yuan Thung-Li (*1*), p. 488, shows that the book is the +1555 edition.
[c]   The allusion is to the famous sorites in the *Ta Hsüeh*; cf. Legge (2), p. 221.
[d]   In later times, a persistent mistake credited Chu Yu-Tun with the *Chiu Huang Pên Tshao* instead of his father. This can be traced to Lu Chien[8], the editor of the third edition of +1555, who accordingly misled both Li Shih-Chen and Hsü Kuang-Chhi. He may have been led astray by the customary anonymity of books by Ming imperial princes, for Pien Thung's preface of course did not use the posthumous name by which alone one 'Chou Wang' could be distinguished from another. The *Phu Chi Fang* was similarly mis-attributed. One or two medical tractates, such as the *Hsiu Chen Fang*[9], however, which bear Chu Yu-Tun's name in Li Shih-Chen's bibliography, may well have been by him.
[e]   See below, p. 347.

[1] 卜同   [2] 圃   [3] 畫工   [4] 朱有燉   [5] 周憲王
[6] 誠齋牡丹譜並百詠   [7] 朱權   [8] 陸柬   [9] 袖珍方

and geography than in botany or horticulture.[a] This third prince (Ning Hsien Wang[1]) would have been hardly more than a child, it is true, in the years when the *Chiu Huang Pên Tshao* project was in full swing, but later on he may well have come under the influence of the scientific group at Khaifêng. Besides the botanical work which we shall now examine more closely, Chu Hsiao was also the author of a voluminous collection of therapeutic formulae, still in use, entitled *Phu Chi Fang*[2] (Practical Prescriptions for Everyman). We know that he was helped in this by a professor Thêng Shih[3], and the Administrator of the Household Liu Shun[4], who was also the tutor of Chu's sons, so they also may have collaborated in the food-plant project.

The first edition of the *Chiu Huang Pên Tshao* (Treatise on Wild Food Plants for Use in Emergencies)[b] contains descriptions and illustrations of 414 species of plants, 276 of which were entirely new, and only 138 known from earlier books on pharmaceutical natural history. A breakdown of the contents is given in the preliminaries:

[a] Chu Chhüan, the seventeenth son of Thai Tsu, was a distinctly Taoist character, calling himself Chhü Hsien[5] (the Emaciated Immortal) and Han Hsü Tzu[6] (the Full-of-Emptiness Master). His biography (in *Ming Shih*, ch. 117, p. 14a) lists half-a-dozen titles of books written by him, though not the most interesting for us, and hints at many more. The *Kêng Hsin Yü Tshê*[7] (Precious Secrets of the Realm of Kêng and Hsin, i.e. all things connected with metals and minerals, symbolised by these two cyclical signs) on alchemy, metallurgy and pharmacy (cf. Vol. 1, p. 147, Vol. 3, p. 678, Vol. 4, pt 3, p. 531, Vol. 5, pt 3, pp. 210–1) was produced in +1421; we have now given up hope of ever seeing this and believe that it is extant only in the form of quotations. About +1430 came the important geographical encyclopaedia *I Yü Thu Chih*[8] (Illustrated Record of Strange Countries), not certainly attributable to Chu Chhüan but most likely done under his aegis (cf. Vol. 3, p. 513). In later years Chu wrote his *Chhü Hsien Shen Yin Shu*[9] (Book of Daily Occupations for Scholars in Rural Retirement), a regimen of life for country gentlemen desirous of preserving their vitality and increasing their longevity (cf. Wang Yü-Hu (1), p. 128). Much of it concerns horticulture, food preservation, animal husbandry and veterinary science, following a Yuan work of +1314, the *Nung Sang I Shih Tsho Yao*[10] (Selected Essentials of Agriculture, Sericulture, Clothing and Food) by the Uighur Lu Ming-Shan[11], which he greatly admired. He also, as one might expect, wrote a work on tea, the *Chhü Hsien Chha Phu*[12], and another, *Shou Yu Shen Fang*[13] on geriatric prescriptions. Two medical tractates, *Chhien Khun Pi Yün*[14] and *Chhien Khun Shêng I*[15], both dealing with Yin and Yang theory, were available to Li Shih-Chen, though probably lost today. Although his earlier years were spent amidst the vain strife of princes, and once he had to face a charge of witchcraft, Chu Chhüan seems for the most part to have avoided the political troubles that beset Chu Hsiao. Each might merit the title of the Prince Rupert of his age.

[b] The phrase *Chiu Huang* means 'salvation in the midst of desolation', for *huang* is the wilderness where standing crops should be, hence famine with all its dread consequences. But though many books with this phrase in their titles will be found on the shelves of Chinese bookshops, they are not all of direct interest for the history of science. In a wealth of writing the scholar-officials discussed the many measures which it was necessary for the civil service to take when famine threatened or became actual. Economic decisions regarding price-control and transport had to be made, food distributed free from government granaries (*i tshang*[16]), 'ever-normal' granaries (*chhang phing tshang*[17]) set up, arrangements made for the control of emergency population movements, the organisation of police and the suppression of banditry. Large schemes of relief labour on public projects had to be planned and launched. Among the oldest books of this kind are two by Tung Wei[18] in the +12th century, *Chiu Huang Chhüan Shu*[19] and *Chiu Huang Huo Min Shu*[20]. In +1690 Yü Sên[21] collected thirteen of such works on famine relief and administration in his *Huang Chêng Tshung Shu*[22], including a *Chiu Huang Tshê*[23] by Wei Hsi[24], written about +1665. Another book with the same title was produced by the great sceptical philologist Tshui Shu[25] (cf. Vol. 2, p. 393 above), who had personal experiences of great suffering in floods and droughts around his family home in the eastern plains, in +1774 (cf. Hummel (2), p. 770).

[1] 寧獻王	[2] 普濟方	[3] 滕碩	[4] 劉醇	[5] 臞仙
[6] 涵虛子	[7] 庚辛玉冊	[8] 異域圖志	[9] 臞仙神隱書	[10] 農桑衣食撮要
[11] 魯明善	[12] 臞仙茶譜	[13] 壽域神方	[14] 乾坤秘韞	[15] 乾坤生意
[16] 義倉	[17] 常平倉	[18] 董煟	[19] 救荒全書	[20] 救荒活民書
[21] 俞森	[22] 荒政叢書	[23] 救荒策	[24] 魏禧	[25] 崔述

herbaceous plants	245
trees	80
cereals, seeds, legumes	20
fruits	23
vegetables	46
	414

as also an enumeration of the various parts which had been proved to be whole-some and edible. This list is given by combination, plants with edible leaves and seeds being listed separately from those with edible leaves only, or with edible seeds only, and it is instructive to dissect this in order to see the proportions of parts which could be admitted as useful.[a]

roots	51
shoots, stems	8
bark	2
leaves	305
flowers	14
fruits and seeds	114

In a moment we shall give some examples of typical entries, but first the point should be made that the prince's illustrations are of an outstanding excellence; he must have spared no pains to attract the most brilliant 'artisan-limners' and woodcut block-carvers of his time.[b] In +1525 the Governor of Shansi, Pi Mêng-Chai,[1] ordained a second printing,[c] the original having become scarce, and this was prefaced by the physician Li Lien[2].[d]

The climate and soil of the five regions [he said] is not at all the same, so the local plants are also quite different in form and quality. The names are numerous and complicated, and it is difficult to distinguish one from another, the true from the false. If one did not have illustrations and explanations people would confuse *shê-chhuang*[3e] with *mi-wu*[4],[f] or *chhi-ni*[5g] with *jen-shen*[6].[h] Mistakes of this kind can kill people. This was why the *Chiu Huang*

[a] Naturally they add up to a higher figure, 494, because a single plant could be useful in as many as three ways.

[b] We shall return to this in the sub-section on botanical illustration (Sect. 38 *g* below), only remarking that the long priority of Chinese woodcut plant pictures over European counterparts was authoritatively pointed out by Sarton (1), vol. 3, p. 1645, nearly forty years ago.

[c] This edition, extremely close to the original, was reproduced in facsimile in Peking in 1959.

[d] Li Lien was the first to write a history of medicine in China, the *I Shih*[7] of *c.* +1540. Its arrangement was biographical, based on the accounts of famous physicians in the dynastic histories.

[e] Umbelliferous *Selinum japonicum* (R 230; CC 581).

[f] *Selinum* sp. (R 231) = *chhuan-hsiung*[8], *Conioselinum vaginatum* (CC 560; R 216). See Read (8), 1.31. Hemlock parsley.

[g] *Chhi-ni* = *chieh-kêng*[9], *Platycodon grandiflorum* (CC 164), Campanulaceae. See Read (8), 2.1 Broad bluebell. But more properly *chhi-ni* is *Adenophora remotiflora* (CC 156, R 52), a harebell, cf. p. 337 below. This is also in the *Chiu Huang Pên Tshao*, where it is called *ti shen*[10] (earth ginseng) and *shan man-chhing*[11] (mountain rape-turnip); cf. Read (8) 6.22.

[h] *Panax ginseng* (CC 594), Araliaceae.

[1] 畢蒙齋   [2] 李濂   [3] 蛇床   [4] 蘼蕪   [5] 齊苨
[6] 人參   [7] 醫史   [8] 川芎   [9] 桔梗   [10] 地參
[11] 山蔓菁

Pên Tshao was written, with its pictures and descriptions to clarify the form of the plants and to record the methods of their use. In each case the writer first states where the plant grows, then he gives its synonyms and speaks of its Yang properties, whether refrigerant or calefacient, and its Yin sapidities, whether sweet or bitter; finally he says whether one must wash the part to be used (and for how long), soak it, fry it lightly, boil it, steam it, sun-dry it, and so on, with details of whatever method of seasoning is necessary.... If in time of famine people collect (emergency plants) in accordance with the resources of the local flora, there will be no difficulty and many lives will be saved.[a]

A point of nomenclature arises here—how did Chu Hsiao choose the names for the hundreds of plants 'new to science' which he recorded? To answer this in depth would need special research, but we suspect that mostly he adopted or modified the already existing names of the countryfolk; if so, we are brought to realise again that the total fund of botanical knowledge in medieval China was far greater than that contained in the pharmaceutical natural histories alone. Pien Thung, in the same preface quoted above, expressed the hope that the *Chiu Huang* would go down to posterity along with the *Thu Ching Pên Tshao* (see p. 281 above), but in the event it did so much more successfully.[b] Its worth has been well appreciated by modern historians of science,[c] and 87 % of the plants contained in it have been identified (at least provisionally) in the monograph of Read (8). Sarton rightly gave it his highest praise, calling it 'the most remarkable herbal of medieval times.[d]

Looking back at the work of Chu Hsiao one is struck by an impression of great originality. Certainly no previous work of a similar kind has come down to us, though there may have been information in government archives to which we have no access now. For century after century there had been grave concern over famine problems, and sometimes the biological side of it clearly emerges. So for example in +21 there was a terrible scarcity in the north. The *Chhien Han Shu* says:[e]

At Loyang and places to the east grain was fetching two thousand (cash per) picul, (so Wang Mang) sent (one of) the highest ministers and a general to open the various (government) granaries in the eastern provinces and to give and lend to those who were in extremity. He also detailed Commissioners and Expatiators in groups to teach the people

---

[a]   Tr. auct.

[b]   Besides the editions of +1406, +1525 and +1555 already mentioned, there was another in +1586, which reduced the number of entries to 411 from the 440 to which Lu Chien had raised it (cf. Lung Po-Chien (*1*), p. 105); and several further editions followed, including Japanese ones, the first of which was in +1716 (cf. Shirai (*1*), p. 224). Meanwhile, in +1628, Hsü Kuang-Chhi had thought the work so important that he decided to incorporate it in his *Nung Chêng Chhüan Shu*, and when this was printed after his death in +1634 by his editor Chhen Tzu-Lung (+1639) the *Chiu Huang* appeared as chs. 46 to 59 with 413 entries. Many papers of Hsü having recently come to light, Yen Tun-Chieh (*21*) is able to say that Hsü verified experimentally the nutritional conclusions of Chu Hsiao. Comparison of the illustrations shows that some were redrawn for the *Nung Chêng Chhüan Shu* editions, generally being simplified and clarified (cf. Sect. 41 below). Further bibliographical details will be found in Wang Yü-Hu (*1*); Swingle (*1*).

[c]   One may mention Bretschneider (*1*), vol. 1, pp. 49 ff., who identified 43 % of the plants and was the first to recognise the brilliant priority of the woodcut illustrations; then Liu Ho & Roux (*1*); Swingle (*1*); Merrill & Walker (*1*), vol. 1, pp. 553 ff.; Reed (*1*), p. 74; Sarton (*1*), vol. 3, pp. 1170, 1177, 1644, etc.

[d]   (*1*), vol. 3, p. 1170.

[e]   Ch. 24A, p. 20*b*, tr. auct. adjuv. Dubs (*2*), vol. 3, p. 480.

how to cook (parts of) plants and trees so as to make soups or mashes (*lao*[1])[a] of mixed vegetable materials. But these preparations proved to be uneatable, and (the 'missions of instruction') only added to the troubles and disturbances.

Here again we get a glimpse of Wang Mang's characteristic reliance on the scientific knowledge of his day,[b] but unfortunately it was not based on sound experiments such as must have been carried on by Chu Hsiao.[c] Later in the Han, in +154, the cultivation of the rape-turnip[d] as a supplementary food was recommended by imperial decree—only one example of a continuing preoccupation with the diet of the people. As we shall see presently, this took the form in Thang times of special books dealing with food in relation to normal health, e.g. Mêng Shen's[2] *Shih Liao Pên Tshao*[3] of *c.* +670, and the *Shih Hsing Pên Tshao*[4] of Chhen Shih-Liang[5] written about +895, but these did not discuss emergency plants. Thus Chu Hsiao seems to have been a great pioneer, as well as the most weighty writer of his genre. More, he was, we believe, a great humanitarian. Such subjects as famines, floods, droughts, crop failures, epizootics, wars and plagues, must of course be discussed with scientific objectivity but at the same time we are heartless if we fail to visualise something of what they meant in terms of human beings in pain, fear and dispair. 'Ancient chronicles', wrote that great humanist Sarton, 'often use such a sentence as "There was a famine in the land...."; simple words which I cannot read without the appalling vision of starving children, of frantic mothers, of men ghastly or rebellious, without feeling myself surrounded by all the ghosts of human misery in all its forms, physical and mental.'[e] Now that science in evil hands has added to a growing world food inadequacy the perfection of methods for the positive destruction of food crops, his words are even more telling in our ears.

But what of the words of Chu Hsiao himself? Widely though his work has been appreciated, few have been reproduced. Dipping into his book at random, let us give three of his descriptions, plants belonging respectively to the Campanulaceae, Asclepiadaceae and Alismataceae. Fig. 69 shows the drawing of *hsing-yeh sha-shen*[8], the apricot-leaved wilderness-ginseng, i.e. *Adenophora stricta*,[f] newly here described (*hsin tsêng*[9], as he always says), and the statement follows overleaf.[g]

---

[a]   Properly speaking, this word means a milk product of some kind, variously applied to sour milk, fermented milk (kumiss), cream, cheese, dried curd, or even yoghurt. But it could also mean a milk-like drink made, e.g. from almonds, such as one finds today in Spain. Here the term must be purely allusive, hence our translation. On the Chinese words for milk products, and their Hunnish connections, see Pulleyblank (11), pp. 248 ff., and Sect. 40 below.

[b]   Cf. Vol. 1, pp. 109 ff.

[c]   According to Read (8), at least 73 of Chu Hsiao's plants entered Chinese domesticated horticulture, and 16 more were adopted in the diet of Europe or Japan. Watt (1) lists 280 plants used traditionally as famine foods in India, many being identical with Chu Hsiao's.

[d]   *Brassica Rapa-depressa* (R 477; *wu-chhing*[6] = *man-chhing*[7]). The reference is to *Hou Han Shu*, ch. 7, p. 8a.

[e]   (1), vol. 3, p. 281.

[f]   CC 155; Read (8), 8.6.

[g]   Ch. 1B, p. 63*b*.

---

[1] 酪   [2] 孟詵   [3] 食療本草   [4] 食性本草   [5] 陳仕艮

[6] 蕪菁   [7] 蔓菁   [8] 杏葉沙參   [9] 新增

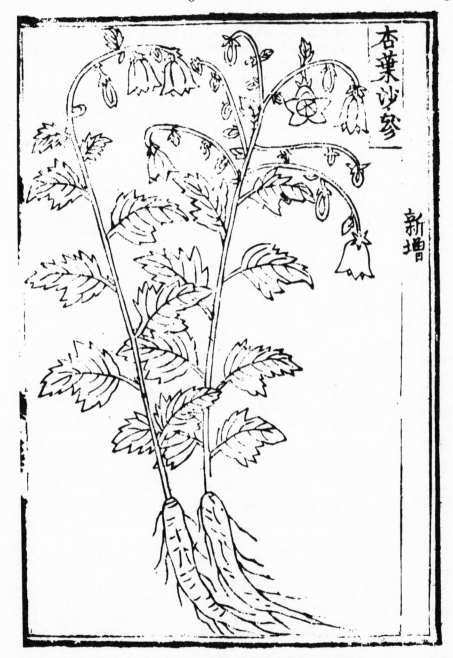

Fig. 69. The *hsing yeh sha shen*, *Adenophora stricta* (= *remotifolia*) of the Campanulaceae, allied to our harebell.
From the *Chiu Huang Pên Tshao* (+1525 edition), ch. 2, p. 63a.

The apricot-leaved wilderness-ginseng is also called *pai-mien-kên*[1] (white flour root). It grows in wild mountain places around Mi-hsien (in Honan, near Khaifêng); the shoots are from 1 to 2 ft in height, the stems are light blue-green, the leaves are like those of apricot trees but smaller, with serrated edges. They also look like the leaves of *shan-hsiao-tshai*[2] (small mountain vegetable)[a] but they are more pointed and have white under-surfaces. The tips of the stems send forth white bowl-shaped flowers with five petals. The roots are shaped like those of the *hu lo-po*[3] (carrot)[b] but are thicker and of an ashy colour, white inside and of a sweet taste. The property (of the root substance) is slightly algorific. The pharmaceutical natural histories (Pên Tshao) have *sha-shen*[4],[c] but the descriptions of the shoots, leaves, roots and stems are all different, so we cannot subsume it under the same entry, and this is (therefore) a separate one. There is another variety of this plant which has caerulean or jade-coloured flowers.

Good against hunger. The shoots and leaves should first be scalded,[d] then soaked in water and washed on a sieve until it runs clear; only after this mix them with oil and salt for eating. As for the roots, they should be well washed and then boiled for food, this is also good.

From this one can see at once several features of Chu Hsiao's methods. Starting with oecological indications, generally concerning places in Honan, he goes on to a very clear description intended to help quite simple people, and naturally to a large extent of the cross-referencing type (cf. p. 140 above). The precision at which he aimed comes out well from this quotation, where we find him differentiating between several species of what we now regard as the *Adenophora* genus. Lastly there is the biochemical technique—something is evidently being got rid of in the extraction with water after destruction of the cell-walls, some bitter or distasteful substance, some compound possibly toxic.

Fig. 70 shows the asclepiadaceous *Cynanchum auriculatum*,[e] *niu phi hsiao*[5], a good example of what one might call the manioc category, in which a dangerous toxin has to be removed. We read:[f]

Ox-skin *hsiao* is a creeper growing wild in hilly places near Mi-hsien; its stems can extend to a length of 4 to 5 ft. The leaves look like those of *ma tou ling*[6] [g] but broader and thinner,

---

[a] *Campanula punctata*, CC 157; Read (8), 5.21. Another emergency plant of Chu Hsiao's.

[b] *Daucus carota*, the garden carrot, R 219. On its introduction from the West via Persia in the Yuan period, hence its name, see Laufer (1), pp. 451 ff.

[c] R 51 gives for this *A. polymorpha*, but probably the *A. verticillata* of CC 153 is a better designation. There are at least 35 species of this genus of 'bluebells', Eurasiatic in distribution. The 'ginseng' references in the names derived undoubtedly from the shape of the root.

[d] The more usual translation of the expression *cha*[7],[8] *shu*[9] used here would be 'thoroughly fried', but the words can also mean 'scalded', and one must assume that in times of scarcity cooking oils were 'in short supply'. Scalding with boiling water denatures the cell-wall proteins and dissolves out pectins and some hemi-celluloses, so that the walls become permeable, turgor 'crispness' is lost, and substances within the cells are free to diffuse out.

[e] CC 416; R 161; Read (8), 8.8 Also new to Chinese botany at the time.

[f] Ch. 1B, pp. 64b, 65a.

[g] *Aristolochia debilis*, birthwort (CC 1559; Read (8), 1.15).

[1] 白麵根	[2] 山小菜	[3] 胡蘿蔔	[4] 沙參	[5] 牛皮消
[6] 馬兜鈴	[7] 煠	[8] 炸	[9] 熟	

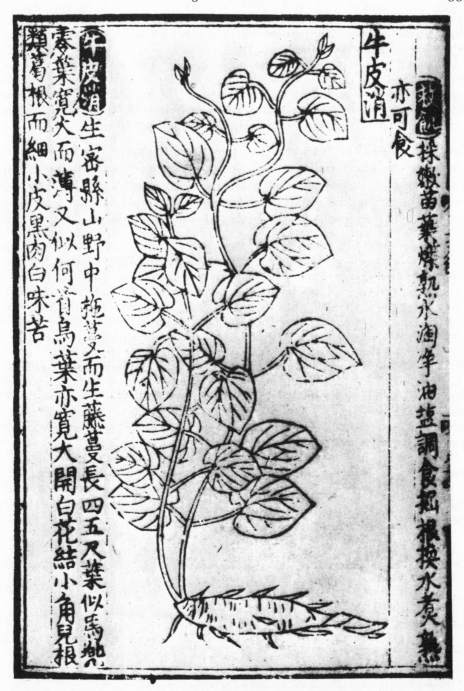

Fig. 70. *Niu phi hsiao*, the ox-hide digester, *Cynanchus auriculatum*, of the Asclepiadaceae; it has a poisonous root, which needs careful preparation before it can be eaten. From the *Chiu Huang Pên Tshao* (+1525 edition), ch. 2, p. 64*b*.

as also like those of *ho shou wu*[1] [a] but broader and larger.[b] It has white flowers which produce small seed pods. The roots are like those of *ko kên*[2] [c] but thinner and smaller with a dark skin and white flesh. Taste bitter.

For relieving hunger collect the leaves and scald them, then soak in water to be rid of the bitter taste, then eat them seasoned with oil and salt. Take off the black skin of the root, cut it into slices, wash them with several changes of water and boil them to remove the bitter taste, wash them again through a sieve till it runs clear, and (finally) boil them with water (for some time) until thoroughly cooked and ready to eat.

Here the repeated aqueous extraction is necessary to remove the active principle cynanchotoxin which would otherwise cause paralysis. A more extreme example is the poke-root, *Phytolacca acinosa*,[d] coarse glabrous perennial of a genus native both to America and East Asia. Phytolaccotoxin is a highly poisonous substance,[e] so that careful preparation before consumption was a necessity.

Chang liu kên[3] [says Chu Hsiao[f]] is the same as *shang lu*[4] ... [and half-a-dozen more synonyms follow, with other names found in the *Erh Ya*, the *Kuang Ya*, etc.]. It grows in river valleys around Hsien-yang, but now pretty well everywhere. The stems are 3 to 4 ft high, rough like those of *chi kuan hua*[5] [g] and angular, somewhat purplish-red in colour. The leaves are green, shaped like those of *niu shê*[6] [h] but slightly broader; the long root is sometimes shaped like a man, and is then particularly potent, there are also red sorts and white sorts. When the flowers are red the roots are red, when the flowers are white the roots are white. The red variety should never be eaten as it is dangerous to man, producing a diarrhoea with loss of blood, which is almost impossible to stop; it is the white variety which is edible. There is yet another variety called *chhih chhang*[7], the shoots and leaves of which look very much the same, but it cannot be used for food so it must be carefully distinguished. The sapidity (Yin) is acrid and oxyntic, though some say acerbic, the nature (Yang) is neutral with a powerful active principle, and some say refrigerant. It is good to take garlic with it.[i]

When there is famine one takes the white-coloured roots. Cut them up into slices, scald, then soak and wash repeatedly (throwing away the extract) until the material is clean; then just eat it with garlic. The best way of preparing the slices is to place them (in a basket) in running (lit. eastward-flowing) water for two days and two nights. Then put them on bean leaves in a steamer[j] and steam from the *wu* double-hour to the *hai* double-

[a] *Polygonum multiflorum*, flowery knotweed (CC 1550; Read (8), 8.16). Cf. p. 358.

[b] It would have been quicker to say 'cordate' as we now should, but that was not in the cross-referencing tradition; cf. p. 140 above.

[c] *Pueraria thunbergiana = hirsuta*, the leguminous vine that yields the textile fibre (cf. p. 86 above). Cf. CC 1038; Read (8), 8.15.

[d] R 555, cf. CC 1494, and Read (8), 6.5, who says that the leaves have long been used as a vegetable in Japan and the Himalayas. Here the starchy root was wanted. Phytolaccaceae. Cf. B III 31; Anon. (*109*), vol. 1, p. 613.

[e] Acting on the medulla and spinal cord, at low doses stimulates, giving spasms, convulsions, nausea, etc; larger doses paralyse (cf. Sollmann (1), pp. 191 ff.).

[f] Ch. 1B, pp. 24*b*, 25*a*.

[g] *Celosia cristata* (R 559), Amarantaceae.

[h] *Plantago major*, the great plantain. Ovate was what he meant.

[i] Down to this point Chu Hsiao was following very closely Su Sung and other earlier naturalists as conflated in *Chêng Lei Pên Tshao*, ch. 11 (p. 263. 1). The plant was not one of his new introductions.

[j] Cf. Vol. 1, p. 82, and Vol. 5, pt 4, pp. 26 ff., 62 ff., 80 ff., as also Ho Ping-Yü & Needham (3).

[1] 何首烏   [2] 葛根   [3] 章柳根   [4] 商陸   [5] 鶏冠花
[6] 牛舌   [7] 赤昌

hour (i.e. from noon till ten o'clock at night). If you don't have such leaves you can use the skin of bean-curd.

Plants with white flowers can (it is said) confer longevity; the immortals collected them to make savouries to take with their wine. For the therapeutic uses, see the pharmaceutical natural histories under 'herbs', the entry for *shang lu*[2].

Here the detail of the instructions for removing the poisonous substance is impressive, including as it does arrangements for continuous extraction. Presumably the leguminous plant material had the property of absorbing the remaining phytolaccotoxin driven out at steam heat.

The Leguminosae, though so useful to man, occasionally conceal serious dangers. Wild and cultivated species of the genus *Lathyrus*, climbing herbs widely distributed through the Old World,[a] contain a paralytic poison in their seeds which generally manifests itself only when men have recourse to this pulse exclusively, in famine times—as e.g. in India, where the 'khessary pea' is *L. sativus*.[b] Impairment or complete inhibition of muscular activity in the lower extremities can occur, and this has become known as lathyrism. The first description in the West was that of Ramazzini arising from an outbreak at Modena in +1690, the standard one was given by Cantani in 1873. Chu Hsiao listed two wild species as useful in time of trouble, *L. palustris* (*shan li tou*[1])[c] and *L. maritimus* (*yeh wan tou*[2]),[d] but said nothing about the potential danger, probably because they were never relied upon exclusively in Chinese conditions. Of the former he wrote:[e]

Shan li tou[1]. Also called *shan wan tou*[2]; grows in wild hilly places near Mi-hsien. The shoots are about a foot high. The stem is grooved on one side, with sharp knifelike longitudinal ridges.[f] The leaves (leaflets) look like bamboo leaves[g] but shorter, and are set opposite one another.[h] Pale purple flowers appear, giving rise to small pods (*chio erh*[3]) with sweet-tasting seeds flat like those of the wild soya-bean (*lao tou*[4]).[i] In famine time, collect the pods and boil them, or else shell the peas and eat them separately.

And of the latter:[j]

Yeh wan tou[5]. Grows in fields and wild places; the shoots come up close to the earth, and the young plant with its small leaves (leaflets) creeps on the ground, later it divides into several stems which can reach more than 2 ft long. The leaves (leaflets) look like those of *hu tou*[6][k] but rather larger, and not unlike alfalfa leaves[l] though also bigger. The flowers are of a pale lavender colour, and the pods resemble those of the cultivated pea[m] only they

---

[a] The familiar sweet pea is *Lathyrus odoratus*, native probably to southern Europe.

[b] Cf. Burkill (1), vol. 2, p. 1322; Garrison (1), p. 320.

[c] CC 1003; Read (8), 7.16.

[d] CC 1001; Read (8), 12.1. *L. japonicus* is now the preferred name.

[e] Ch. 1B, p. 46*b*.   [f] I.e. winged.   [g] I.e. lanceolate.   [h] I.e. paired.

[i] *Glycine ussuriensis*; Read (8), 12.2.   [j] Ch. 2A, p. 53 *a, b*.   [k] *Indigofera decora*; CC 995.

[l] *Medicago sativa*; CC 1018. Cf. p. 164 above.

[m] *Pisum sativum*.

---

[1] 山黧豆   [2] 山豌豆   [3] 角兒   [4] 豺豆   [5] 野豌豆

[6] 胡豆

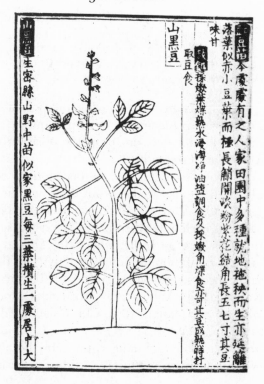

Fig. 71. *Shan li tou, Lathyrus palustris*, one of the barely edible peas. From *Chiu Huang Pên Tshao* (+1525 edition), ch. 4, p. 8*b*.

look like unripe ones and have a bitter taste. In time of famine collect the (young) pods and boil them, or boil the peas themselves to eat, or (dry them and) grind to flour and use like ordinary pea (-flour).

Here there are several things to notice. Chu Hsiao was evidently well able to recognise a leguminous plant when he saw one, and unerringly classified these with the beans. More delicate was the distinction suggested by the two drawings (Figs. 71, 72), namely that in *L. palustris* the peduncles are longer than the leaves while in *L. maritimus* they are shorter. As for the preparation, some significance may lie in the directions to boil, for this is rather unusual in Chinese cooking, and some bitter, if not toxic, substances might be removed by the hot water extraction. Both the species are native to England,[a] and it is interesting that the latter was mentioned by Dr John Caius, the second founder of the College in which all this book has been written, precisely in a context which would have been of particular interest to Chu Hsiao. In his book on unusual animals and plants, *De Rariorum Animalium atque Stirpium Historia*, dedicated to his friend Conrad Gesner and published in +1570, Caius has a story about a maritime pea growing on the

[a] Clapham, Tutin & Warburg (1), p. 361.

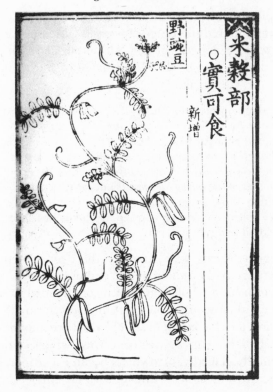

Fig. 72. *Yeh wan tou, Lathyrus maritimus,* another of the same family. From *Chiu Huang Pên Tshao* (+1525 edition), ch. 3, p. 53a.

shingle in great profusion on our Suffolk coasts around Aldeburgh and Orford.[a] On the occasion of a shortage in +1555 the people were saved by using this plant for food, in spite of its nauseous taste, and they thought that it had sprung up there almost miraculously, though doubtless before that time it had not been prominent enough to secure notice.[b] Thus the Chinese prince and the English royal physician were united in their concern for emergency botany even though neither of them could ever have known of the existence of the other.

[a] Under the heading 'De Pisis sponte nascentibus' his words are as follows: 'Pisa in littore nostro Britannico quod orientem solem spectat, certo quodam in loco Suffolciae, inter Alburnum et Ortfordum oppida, saxis incidentia (mirabile dictu) nulla terra circumfusa, autumnali tempore anni 1555 sponte sua nata sunt, adeo magna copia, ut suffecerint vel millibus hominum.' Orig. p. 29b, in *Works*, ed. Roberts (1), p. 63.

[b] See Boswell, Brown, Fitch & Sowerby (1), vol. 3, p. 110. Perhaps the Suffolk men were lucky to get off without lathyrism. If it had occurred, John Caius, a perspicacious physician, would hardly have failed to note it. Perhaps Chu Hsiao could have told them how to get rid of the nauseous taste.

The Orford deliverance was noted by several other English writers, e.g. William Bulleyn, rector of Dunningworth, who wrote a *Book of Simples*. In another work, entitled *A Bulwark of Defence against all Sickness* (+1562) he wrote: 'In the year of our salvation 1555, in a place called Orford in Suffolk, between the haven and the main sea, where never plowe came, nor natural earth was, but stones only, there did Pease grow, whose roots were more than ii fadomes long; and the coddes (pods) did grow upon clusters like the Kaies of ashtrees, bigger than fitches (vetches) and lesse than the field Peason. Very sweet to sate upon, and served many poore people, dwelling there at hande, which els would have perished from hunger; the skarce of bread that yere was so great,

Our last example may illustrate the case of a plant which was first described by Chu Hsiao but later entered the range of normal Chinese crop-plants. This is the arrowhead, *Sagittaria sagittifolia*[a] of the Alismataceae, seen in Fig. 73 as the prince's draughtsmen depicted it.[b] It will bring us back, unexpectedly, to the ingenious Mr Bryant. Chu Hsiao says:[c]

Shui tzhu ku[1]. The common name for this is scissors herb (*chien-tao tshao*[2]) or else *chien ta tshao*[3] (hung-up arrows herb).[d] It grows in water; the stem is grooved on one side (*wa*[4]) and squarish on the other, with stringy fibres. The leaves are three-horned, resembling indeed a pair of scissors. From amidst the leaf-stalks rise up scapes[e] which fork and bear three-petalled white flowers with yellow centres. Each of these gives a blue-green fruit follicle (*ku thu*[5]) like that of the *chhing chhu thao*[6][f] but rather smaller. The (tuberous) root is of the same kind as that of the onion (*tshung*[7])[g] but coarser and larger and of a sweet taste.

In case of hunger collect the young and tender shoots near the (tuberous) roots and scald them, then add oil and salt, and eat.

This is interesting in several ways, first for the degree of precision which the phytographic language had attained by the end of the Yuan period. It is curious that the starchy corms as a whole were not recommended for food, and only the shoots or turions (as they would now be technically termed), though these also would certainly have a rich content of starch; yet there may have been some good reason where the Chinese variety was concerned which could only now be explained experimentally. Thirdly, arrowhead became a crop in China, and Bryant would like to have introduced it in England. In his entry for *Sagittaria sagittifolia* he wrote:[h]

insomuch that the plain poore people did make very much of akornes, and a sickness of strong fever did sore molest them that yere, as none was ever heard of there. Now whether the occasion of these peason, and providence of God came through some shipwrake in mocke misery or els by miracle, I am not able to determine thereof, but sowen by man's hand they were not.' The wild peas can still be found to this day on the peninsula between the river and the sea. As for William Bulleyn, there is a brief account of him in G. E. Evans (1), pp. 136 ff.

    The whole incident invites comparison with the early Mormons of Salt Lake City who believed that they were saved from a plague of locusts by gulls sent by God. Dr Frank Egerton, who told us of this, says that they have a golden statue of the gulls in memory of the event, though certain species of gulls normally inhabit the region.

    [a] CC 2096; R 781; Read (8), 8.25.

    [b] It occurs in several of the other esculentist handbooks, in Japan as well as in China, as witness the *Seikei Zusetsu* (1804) of Sō Han & Shirao Kokujū (*1*); cf. Bartlett & Shohara (1), pp. 150–1.

    [c] Ch. 1B, p. 80b.

    [d] And we too say that the leaves are sagittate.

    [e] More properly, compound dichasia. The whole of this sentence is worth looking at as an example of elegant plant description in the late +14th century; *yeh chung tshuan shêng hêng chha, shao chien khai san pan pai hua huang hsin*[8].

    [f] This is *Broussonetia papyrifera*, the paper mulberry (R 597; Read (8), 11.4). The special technical term used for the fruit follicle is worth noting, but Chu Hsiao was not using it in the narrowly-defined sense which it has come to bear today, the one-carpelled unilocular ovary. *Sagittaria* has numerous carpels.

    [g] *Allium fistulosum* (R 666; CC 1818) the ciboule onion. Here Chu Hsiao did not make the distinction now made between corms (now termed *chhiu hêng*[9], ball stems) and bulbs.

    [h] (1), p. 13.

[1] 水慈菰   [2] 剪刀草   [3] 箭搭草   [4] 窊   [5] 菩荄
[6] 青楮桃   [7] 葱   [8] 葉中攛生莖叉稍間開三瓣白花黃心
[9] 球莖

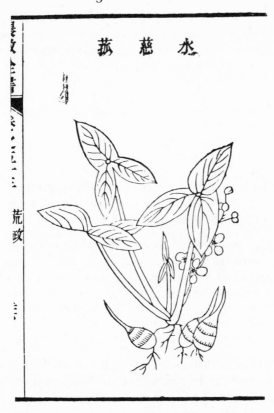

Fig. 73. The *shui tzhu ku* or *Sagittaria sagittifolia*, of the Alismataceae. A drawing from the *Nung Chêng Chhüan Shu* edition of the *Chiu Huang Pên Tshao* (+1630), ch. 53, p. 26a.

This plant grows common in rivulets and water ditches, and often varies much in the size and form of its leaves. Osbeck, in his Voyage to China, says he saw *Sagittaria bulbis oblongis* cultivated in the same field with Rice and *Nymphaea Nelumbo*[a]; it resembled the European *Sagittaria*, but was larger, which might be owing to the culture: the roots of the Chinese sort are the size of a clenched fist, and are oblong, and the Swedish are round, and not much larger than peas. We change the quality of the ground, he remarks, by draining the water, and other arts, till we make it agreeable to our few sorts of corn; but the Chinese make use of so many plants for their subsistence, that they can scarce have any sort of ground, but what will fit some one of them. Thus they do not improve the field for the seed, but chuse the seed for the field.

The *sagittifolia* sends down into the mud many long, slender, brittle fibres, with a bulb[b] suspended at the end of each, which in August is about the size of an Acorn, and of a fine blue colour, streaked with yellow. The inside is white, firm, of a farinaceous taste, but a little muddy. From the crown of this bunch of fibres, shoot many long, spungy stalks, supporting large arrow-shaped leaves, of a fine green colour, and glossy surface. Amidst these rise the flower-stems, higher than the leaves, sustaining at their joints three or four

[a] Cf. p. 135 above.
[b] Bryant was not being more precise here than Chu Hsiao.

white flowers, on long peduncles, each consisting of three roundish petals, which spread open. The uppermost flowers are all male, with many awl-shaped stamina; the lower ones all female, with petals like the male, surrounding many compressed seed-buds, collected in a head, having very short styles, with acute stigmata. These flowers are succeeded by rough heads, containing many small seeds.

I cured some of the bulbs of this plant, in the same manner that Saloop[a] is cured, when they acquired a sort of pellucidness; and on boiling them afterwards they broke into a glutinous meal, and tasted like old peas boiled.

It is indeed instructive to compare this account of $+1783$ with the other of $+1406$ and to note just where the improvement lay, but equally interesting is the admiration for Chinese multiform agriculture arising from Peter Osbeck's observations made in $+1751$. He did use the words attributed to him,[b] and went on to say that after the three crops growing in water

sugarcane and potatoes want a less moist soil. If it is still more dry, it will do for yams. Indigo and cotton grow on the highest mountains. If a mountain should happen to be too dry, it serves for a burying-place. But if a soil be never so wet, the Chinese have a plant that grows in it, and serves as food to men. If we could not imitate the Chinese in our tillage; yet we might manage the pastures in the same manner.... If the husbandman brings such plants upon his meadows as will fit each soil, this would make up what is wanted, and take up the place of such plants as we should like to get rid of.[c]

So far all has been straightforward enough, not of course well known in the West hitherto, but a natural development of the botanical tale we are telling. Yet now must follow a bizarre addendum, the question whether any particle of Chu Hsiao's originality could have derived from contacts with physicians of the house of Israel dwelling in his own princely city? Or conversely could any of his ideas have been transmitted westwards by them? Although the answer, so far as we can gain one, remains obscure and speculative, the juxtaposition is too intriguing to pass without comment.

It is not to anticipate here the sub-section on plant illustration[d] to confess that the time-relations are distinctly disturbing. Chu Hsiao's pictures[e] of $+1406$ preceded the first European woodcut plant illustrations by approximately 64 years, i.e. the $+1470$ edition of Bartholomew of England's encyclopaedia *Liber de Proprietatibus Rerum*.[f] This was just about time enough for the passage of an idea from Khaifêng to Köln, paralleling of course the transmission of printing itself,

---

[a] This was the powdered roots of various orchid species, some indigenous and some imported from the Middle East. They were boiled, skinned, oven-roasted, dried and pounded; the method is given by Bryant, pp. 38 ff. They could also be candied, 'and thus prepared they are very pleasant, and may be eaten with good success against coughs and inward soreness'.

[b] Osbeck, Toreen & Eckeberg (1), vol. 1, p. 334. Osbeck was a pupil of Linnaeus, cf. p. 6.

[c] Cf. the quotation from Swingle above, p. 330.

[d] Sect. 38 *g* below. Meanwhile, see B. Hoppe (2).

[e] Issued, be it remembered, not in some obscure publication but in a noble edition which probably circulated quite widely.

[f] These were hardly more than decorations. See Arber (3), fig. 19.

which started in Europe somewhere around +1440.[a] There followed in +1475 another encyclopaedia, Conrad of Megenberg's *Pûch der Natur*, with plant wood-cuts now for the first time really intended to illustrate the text;[b] and then the sudden uprush of printed herbals from +1484 onwards, in the incunabula period. One could say that the intention of scientific recognisability in Europe dates from +1475, while the first naturalistic and clearly identifiable pictures came with the German *Herbarius* (+1485),[c] and full naturalism with Brunfels' *Herbarum Vivae Eicones* (+1530). Not until then was the Chinese standard attained. Those who have previously discussed this time relationship[d] have not known of the physicians of the lineage of Israel resident in China.

The relations of Jewish with Chinese culture have been touched upon already at many places in the present book.[e] Chiefly relevant here is the fact that from +1163 onwards a community of Jews with their synagogue flourished in Khai-fêng, producing as would be expected several notable physicians.[f] They may have originated from a group of Jewish merchants called Radhanites (al-Rād-hānīyah) who traded regularly between China and Provence as early as the beginning of the +9th century, as we know from reliable Arabic sources.[g] Extant stele inscriptions show[h] that the synagogue at Khaifêng was restored many times, especially in +1279, +1421, +1489 and +1663. The authority to rebuild in +1421 was transmitted to a Jew named An San[1], who had been a soldier in one of the Honan Bodyguard units, by none other than our prince Chu Hsiao; and about the same time An San was given several honours by the emperor including the right to adopt a Chinese family name as Chao Chhêng[2]. Taxts discovered and studied by Fang Chao-Ying (3) show that this was because the Jewish guardsman had lodged accusations against the prince his lord for treasonable activities, and indeed the *Ming Shih* itself confirms[i] that the prince was arraigned before the emperor in +1421 but pardoned on confession of fault, and allowed to return to his estates. It is impossible now to evaluate the seriousness of whatever it was that got Chu Hsiao thus into trouble for the third time;[j] what the evidence does show

[a] This refers to movable-type printing, on which Gutenberg was experimenting between +1436 and +1450. The oldest dated block-print in Europe is of +1423.
[b] Cf. Arber (3), pl. 3.
[c] Anon. (77).
[d] Arber of course ignored Chu Hsiao because she confined herself to the European tradition. All the writers mentioned on p. 330, however, were conscious of the priority.
[e] Cf. Vol. 1, p. 129, Vol. 2, pp. 297 ff., Vol. 3, pp. 89, 252, 257, 311, 575, 681 ff., Vol. 4, pt 2, pp. 231, 236.
[f] Most of what we know of the Khaifêng community is contained in the classical monograph of White & Williams (1), but later researches such as those of Leslie (4, 5) have modified the details of its conclusions.
[g] Radhanite merchants can be seen in some of the frescoes of this period at the Chhien-fo-tung cave-temples near Tunhuang (e.g. cave no. 158).
[h] Translations by Tobar (1); ben Zvi (1); White & Williams, etc.
[i] Ch. 116, p. 10*a, b*.
[j] It may have been merely formal, and seems to have had something to do with the services of certain military bodyguard detachments. For all we know, Chu Hsiao might have roped them in as gardeners or plant-collectors. What speaks for the technicality of his offences is that the emperors kept on increasing his revenues.

[1] 俺三   [2] 趙誠

is that he had connections of some kind with the Jews in his city.[a] The relevant synagogue inscription (+1489) names An San (Chao Chhêng) in the hybrid form of An Chhêng, and dubs him a physician, so perhaps he was some kind of *hakim* but we know nothing of his skill or practice. Although because of the dates there is no reason to think that An San (Chao Chhêng) had played any part in the *Chiu Huang Pên Tshao* project, or in Chu's later medical book (+1418),[b] and although no evidence has so far been found within these writings for any point of scientific or medical contact with Hebrew knowledge; it may still be that other Chinese-Jewish clerks or physicians assisted the prince, and it would certainly be interesting to know what Hebrew, Syriac or Arabic books there were in their libraries.[c] Conversely we know very little of what relations the Jewish community entertained with merchants from afar, occasionally no doubt of their own faith.[d] In all directions, therefore, the verdict must be one of 'not proven', but much indeed remains to stimulate our curiosity, and further research will perhaps throw more light on a channel of communication between East and West that may have had considerable importance. So much for the Jews of Khaifêng.

Once the idea and the concern had been planted in men's minds, there were many during the following couple of centuries who followed the lead of Chu Hsiao, and we can conclude this topic by having a look at what they did. First came Wang Phan.[6] In his *Yeh Tshai Phu*[7] (Treatise on Edible Wild Plants) of +1524, he described and illustrated 60 safe species, many, perhaps most, of them for the first time.[e] His preface tells us how personal experiences had led him to this work.

When the grains do not ripen, (the calamity) is called *chi*[8]. If vegetables do not mature, it is called *chin*[9]. Years when these things happen, even Yao and Thang (the legendary sage-kings) could not avoid. During the Chêng-Tê reign-period (+1506 to +1521) there were floods and droughts in the region between the Yangtze and the Huai Rivers. Crowds of starving people lay exhausted by the roadsides, and although the officials distributed relief they could not succeed in reaching everybody. All the people gathered wild vegetables for food and many were saved thereby, but when mistakes were made they were apt to lose

---

[a] Also in +1421 he presented a gift of incense to the synagogue. Fang Chao-Ying looks upon this as part of his 'humiliation', but the whole story seems too complicated and unclear to warrant the conclusion that his relations with the Jews were always unfriendly.

[b] As suggested by White & Williams, and followed in Vol. 3, p. 682.

[c] In the next century the Jewish medical tradition continued with the Ai family, Ai Ying-Kuei[2], son of the scholar Ai Thien[3], and his grandson Ai Hsien-Shêng[4] both being physicians. It was Ai Thien who paid the celebrated visit to Matteo Ricci in +1605, with its romantic misunderstandings (cf. Trigault, Gallagher tr., pp. 107 ff., or, even more authentically, Ricci himself, in d'Elia (2), vol. 2, pp. 316 ff.). Another physician, Chao Ying-Chhêng[5] (d. *c.* +1657) was so good a Chinese scholar that he, and his younger brother, graduated as *chin shih*, the only two of the community ever to do so. Both had successful careers in the civil service (Leslie, 6).

[d] One must remember, however, that in Chu Hsiao's time there was no longer the xenophilia and cosmopolitanism which had been a feature of China under the Mongols.

[e] Cf. Swingle (1, 12); Lung Po-Chien (*1*), p. 107, no. 188. Hsü Kuang-Chhi thought it worthy of the honour of inclusion in the *Nung Chêng Chhüan Shu*, where it appears as ch. 60 following the *Chiu Huang Pên Tshao*.

[1] 趙俺誠   [2] 艾應奎   [3] 艾田   [4] 艾先升   [5] 趙映乘
[6] 王磐   [7] 野菜譜   [8] 饑   [9] 饉

their lives, since the forms and sorts of the plants seemed to be alike while in fact in wholesomeness or banefulness they were widely different. Therefore it is essential to have manuals on the wild edible plants. Although I have never been of much service to the world, I have always had a mind to save people from afflictions, and this aim I have never forgotten. So while living in the country I occupied myself morning and evening with extensive and detailed investigations, with the result that I found out more than sixty species (of wild plants useful for food). Then I collected their forms in as many drawings so as to make it easy for everyone to recognise them and avoid those mistakes which can lead to fatal injury. Furthermore I made up descriptions of them in rhyming prose based on their names, so that people can learn them and hand them down. It is not on account of my own local countryfolk only that I have done these things, but also for those in every province who are interested in botanising for a useful purpose; such was the main idea, at any rate, of one rustic scholar. If like-minded men far away find it of help in increasing their practical knowledge, he will indeed be happy,[a]

Already with Wang Phan's plants, however, we begin to reach the limits of our knowledge of Chinese plant identifications in modern terms. He himself must have had to face the same nomenclature problem as Chu Hsiao, and presumably solved it in the same way, but modern botanists have not given much attention to identifying the plants of Wang Phan and his successors. Since they are neither ancient nor medicinal they are absent from Bretschneider (1) and Read (1), while Read (8) is strictly confined to the *Chiu Huang* itself. Most of Chu Hsiao's plants are incorporated in Chia & Chia (*1*) and probably Khung *et al.* (*1*), but not those of Wang Phan and the later writers, so there is here an interesting task in the history of botany still to be done. Opening the *Yeh Tshai Phu* at random, we fall upon a plant called *khu ma thai*[1] (Fig. 74) looking vaguely polygonaceous,[b] and the text says, mnemonically rhyming:

(The leaves and stems of) *khu ma thai* have a rather bitter taste, but though to the mouth they are unpleasant, they can help to fill the empty stomach. How we wish we could have good harvests and be able to pay the taxes instead of suffering all the bitternesses of the farmer's life!

To relieve hunger. Pick it in the third month. Pound the leaves and mix with (a little) flour to make bread. The leaves can also be eaten raw.[c]

The next work of this kind was the *Ju Tshao Phien*[2] (Monograph on Uncultivated Vegetables), written by a Taoist naturalist Chou Lü-Ching[3] and prefaced in +1582. It is devoted to all the edible wild plants, not necessarily only those useful in emergencies, which Chou could find in Chekiang.[d] The number freshly described amounts to 105, and two of the four chapters are well illustrated (cf. Figs

---

[a] Tr. auct., adjuv. Hagerty, in Swingle (12).

[b] Out of the 19 species of *Polygonum* listed in Steward (1), p. 96, five were lacking Chinese names. Forbes & Hemsley (1), vol. 3, list no less than 123 Chinese species, while Anon. (*109*), vol. 1, pp. 554 ff. describes and illustrates 29.

[c] Tr. auct. from *NCCS*, ch. 60, p. 18a.

[d] Cf. Swingle (10).

[1] 苦蕒薹　　[2] 菇草編　　[3] 周履靖

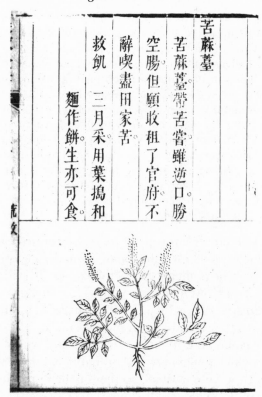

苦蘇薹

苦蘇薹蟞苦薺雖逆口勝

空腸但願收租了官府不

辭喫盡用家苦

救飢

三月采用葉搗和

麵作餅生亦可食。

Fig. 74. An unidentified polygonaceous plant from the *Yeh Tshai Phu* (in *Nung Chêng Chhüan Shu*, ch. 60, p. 18*a*). Although unpalatable, the leaves can be pounded up and mixed with flour to make it go further.

75*a, b*), so the book deserves more study than it has yet received. In +1591 another Taoist naturalist, Kao Lien[1], recorded 64 vegetables and another 100 wild plants of different kinds which were well edible in his *Yin Chuan Fu Shih Chien*[2] (Explanations on Diet and Nutrition),[a] and the new hundred were later re-published under the title *Yeh Su Phin*[3]. The emphasis was now slightly changing, people being advised to search out wild plants not only as a protection against famine times but for varying a health-giving diet, in accordance with the vegetarianism and abstention from excess of cereals which the Taoists had always recommended. Others, moreover, concentrated on the flora of their localities, and about +1600 Chou Lü-Ching's work in Chekiang was extended by a tractate called *Yeh Tshai Chien*[5] by Thu Pên-Chün[6] who described (but like Kao Lien did

---

[a] This was one of the parts of his *Tsun Shêng Pa Chien*[4] (Eight Disquisitions on the Art of Living a Retired Life) which we have had occasion to mention already in Vol. 2, p. 145, and shall encounter again (Vol. 5, pt 5). Cf. Wylie (1), p. 85.

[1] 高濂   [2] 飲饌服食牋   [3] 野蔌品   [4] 遵生八牋   [5] 野菜箋
[6] 屠本畯

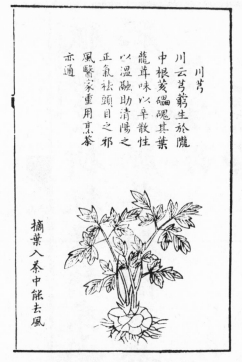

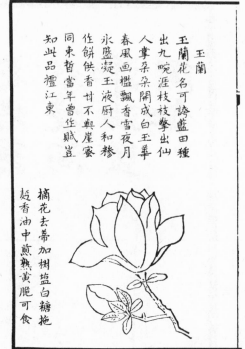

a. Chuan hsiung, i.e. the hemlock parsley, *Conioselinum* (Umbelliferae). The caption says that it comes from Kansu; the leaves, though bitter, can help the *chêng chhi* and drive away malign pneumata of head and eyes if taken in tea (R 216).

b. Yü lan, the white flowers of *Magnolia conspicua* (R 508).

Fig. 75. Illustrations from the *Ju Tshao Phien*.

not illustrate) 22 new plants from that province, carefully excluding any which had previously been recorded by Wang and Chou.[a]

Both these tendencies combined in the important book of Pao Shan[1], finished in +1622 and entitled *Yeh Tshai Po Lu*[2] (Comprehensive Account of Edible Wild Plants), the illustrations in which are considered second only to those of Chu Hsiao himself.[b] One page, that for *Chimonanthus fragrans* (cf. p. 170), the *la mei*[3], is reproduced in Fig. 76. The number of species Pao Shan dealt with was no less than 435, and 43 of these had not been described by any of the previous authors. Although his total number was greater than that of the *Chiu Huang* he included only herbaceous plants (316) and trees (119), but a census of the different plant parts cleared as edible shows a proportion very close to that in the breakdown on p. 334 above. The arrangement of the book is quite similar to Chu Hsiao's. Pao

[a] Cf. Wang Yü-Hu (*1*), 2nd ed., p. 167.
[b] Cf. Swingle (10) and Wang Yü-Hu (*1*), 2nd ed., p. 181.

[1] 鮑山   [2] 野菜博錄   [3] 臘梅

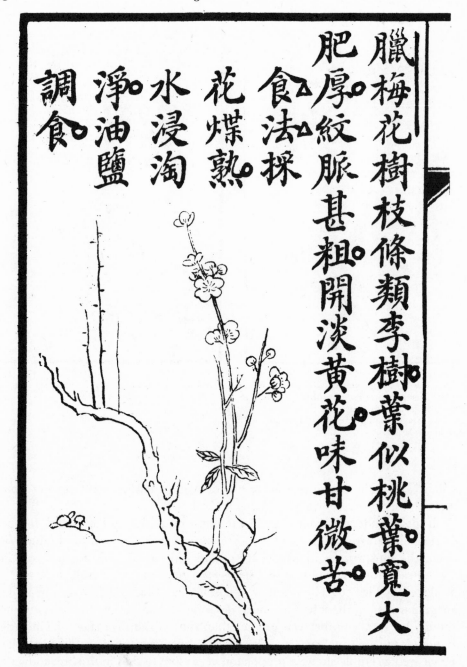

臘梅花樹枝條類李樹葉似桃葉寬大
肥厚紋脈甚粗開淡黃花味甘微苦
食法採
花煠熟
水浸淘
淨油鹽
調食

Fig. 76. Drawing and description of the *la mei* bush, *Chimon anthus fragrans* (= *praecox*) belonging to the Calycanthaceae, sometimes called winter-sweet. From the *Yeh Tshai Po Lu* (+ 1622).

Shan's affiliations were Buddhist rather than Taoist, however; inspired by reading Wang Phan, he built himself a house in the mountains of Huang Shan in southern Anhui and led the life of a hermit, conversing on botany with the vegetarian Buddhist monks who came from all over the country to visit his friend the abbot Phu-Mên[1] in a remote temple. For seven years he spent his time gathering and identifying plants, cultivating them in his own garden plots, experimenting on their nutritional properties and making drawings by which they could be recognised. Eventually he completed the book that he had written 'in order to guard against years of scarcity and make it convenient for people to select plants'. Two postfaces by friends, giving a romantic account of Pao's hermit life, complete the volume.[a] 'Not only', says one of them, 'may these plants be substituted for normal foods in famines, but being innocuous for digestion and life, they will strengthen the vitality, purify the *chhi*, and pave the way to immortality.' Pao Shan knew, he said, of other plants, including fungi, which he believed that adepts could consume 'to promote longevity and make it possible to abstain from cereals', but felt it wiser to omit accounts of these until such effects were more fully verified.[b]

Twenty years later, in +1642, after another devastating famine, another botanical author, Yao Kho-Chhêng[2], brought out a similar book with the title *Chiu Huang Yeh Phu (Pu I)*[3].[c] Selecting 60 useful herbaceous plants from some edition of a *Shih Wu Pên Tshao*[4] then recently published, he added another 45 of his own together with 15 trees. An illustration (not very good) is provided for each plant, and again there are rhyming comments to enable simple people to remember the plants and their forms and properties. Anyone curious to know more about the 'Nutritional Natural History' (*Shih Wu Pên Tshao*) which Yao Kho-Chhêng used will find himself in a bibliographical quagmire.[d] The Ming versions of this had a curious history, starting with an authentic work of +1571 or a little earlier written by Lu Ho[6]. It seems that a few years later this was copied and issued under his own name by Wang Ying[7]. Then it came into the hands of an enthusiastic but unscrupulous naturalist, Chhien Yün-Chih[8], who in his old age

---

[a] In some editions. Partial translations by Hagerty will be found in Swingle (10).

[b] May we not recognise here yet another example of that rise of critical judgment on matters of natural science that occurred in late Ming China paralleling (indeed even sometimes preceeding) the development of scientific scepticism which accompanied the contemporary birth and growth of modern science in Europe? On this see Vol. 4, pt 1, p. 190, and p. 316 above. *Sui yu chhi li erh wei chêng chhi shih*[5], said Pao Shan—'although there is (I believe) this principle, its operations have not yet been demonstrated'.

[c] The title varies, for in some editions Yao's book was added as a supplement to Wang Phan's *Yeh Tshai Phu*, the title of which was changed to *Chiu Huang Yeh Phu*, while in others it stood on its own with this as its substantive title. The question is complex; see Wang Yü-Hu (1), 2nd ed. p. 194; Lung Po-Chien (1), p. 109; Swingle (1), p. 202 (10), p. 189.

[d] The problems are extremely complicated, partly owing to the wide scattering of the various editions in libraries all over the world; see Lung Po-Chien (1), pp. 104 ff.; Swingle (1), pp. 203 ff. (10), p. 190. On the nutritional literature in general see Sect. 40 below. Books with titles such as *Shih Wu Pên Tshao* dealt not only with foods and diet but also with many aspects of household medicine.

---

[1] 普門　　　　[2] 姚可成　　　　[3] 救荒野譜補遺　　　　[4] 食物本草
[5] 雖有其理而未徵其事　　　　[6] 盧和　　　[7] 汪穎　　　　[8] 錢允治

added material from Li Shih-Chen's *Pên Tshao Kang Mu* and brought the whole thing out again in +1620 adding a completely spurious attribution to Li Kao[1], the famous J/Chin physician (+1180 to +1251), and embellishing it with an equally spurious preface purporting to be from the brush of Li Shih-Chen. What Yao the botanist trustingly used was presumably this (though the botany was doubtless more reliable than the philology)—but he may have been in the business himself, for another *Shih Wu Pên Tshao* edition of +1638 attributed to Li Kao as author and Li Shih-Chen as editor is suspected as the compilation of Yao Kho-Chhêng. In any case the historian of botany in Europe will readily admit that such goings-on were not at all unknown there.

Whether or not because the most suitable and relatively common wild food plants had now all been mustered, the movement began to peter out towards the middle of the +17th century.[a] Almost its last representative was the scholar and poet Ku Ching-Hsing[2], who came from Chhichow in Hupei, the home town of Li Shih-Chen, whose biography he wrote. His interest in edible wild plants was somewhat rudely awakened in +1652 when he and his wife returned home in the midst of a vicious scarcity, but with the local people they won through by means of what they could collect, and before the year was out Ku had put 44 of the best plants into a book, together with a song of praise for each, the whole being entitled *Yeh Tshai Tsan*[3] (Eulogies on Edible Wild Plants). This work has survived, but we are not sure that the next (the last that we shall mention) has, yet it is well for it to bring up the rear, for so we can end as we began, with a member of the princely houses of the Ming. About +1630 an obscure descendant of one of these, Chu Yen-Mo[4], wrote a *Yeh Tshai Hsing Wei Khao*[5] (Study of the Natures and Sapidities of Edible Wild Plants).[b] He was an interesting naturalist about whom one would like to learn more, for besides this he wrote a book on fishes, another on ornamental plants and a third on silviculture; we know them mainly by entries in the bibliography of the Hupei provincial gazetteer and it is only too likely that whether or not they saw the printer's blocks they were lost in the upheavals between the Ming and the Chhing.[c] The thought which remains with us, however, is that to anyone who designates the Ming period as one of decline or 'stagnancy' in Chinese scientific endeavour, one can answer that it saw almost the whole of the great and unprecedented effort to extend the realm of botany from plants believed to be of pharmaceutical value to include all those which were useful for the diet of man.

In a way, the esculentists were the vanguard of a great movement still continu-

---

[a] For the Japanese translations and derivative works see the accounts in Bartlett & Shohara (1), pp. 61-2, 118 ff.

[b] He seems to have been a younger brother of a prince of Hsiang-Yin[6], the descendant of the greater Prince of Liao[7].

[c] See Wang Yü-Hu (1), 2nd ed., p. 193.

[1] 李杲   [2] 顧景星   [3] 野菜讚   [4] 朱儼鑷   [5] 野菜性味考

[6] 湘陰王   [7] 遼

ing, immensely enhanced today by the coming of the powerful biochemical technology in which they took the first empirical steps—the movement devoted to extending the possible sources of mankind's food supply. Today this is a question of the first magnitude in world policies and international concerns. Palatable proteins have been prepared from oil and from micro-organisms that can utilise it, from wool, from algae of many different kinds, and from previously discarded cotton or sorghum wastes. Our children may be able to make proteins from the nitrogen of the air. But whatever triumphs science may achieve in the service of the people, men should never forget the systematic enquiries, and the philanthropic compassion, of the economic botanists of Ming China.

(4) BOTANICAL MONOGRAPHS AND TRACTATES

At the very beginning of this volume allusion was twice made (pp. 3, 12) to the almost countless wealth of books and tractates by individual writers on particular plants or groups of plants in Chinese literature throughout the centuries, a phenomenon without parallel in the Western world. Sometimes a whole natural sub-family was discussed, such as the bamboos, sometimes a group of obviously similar genera growing wild (though the criteria were not always the same as those of modern botany), often a great and ever-growing number of varieties of a single species cultivated for its ornamental value or other properties useful to man. This wonderful literature has remained until now very little known internationally, and quite inadequately appreciated. Even the most worthy Western experts on the botany of China have ignored it, E. H. Walker, for example, who prefaced an important paper forty years ago with a brief section entitled 'Pre-scientific study by Chinese'. After mentioning encyclopaedias, local gazetteers and the pharmaceutical natural histories, he remarked:[a]

A few Western scholars of the Chinese language, or Sinologues, have delved into these storehouses of literature and have made translations of scattered portions, but the bulk of this material is still hidden from modern scientists in the intricacies of the Chinese language. It is of relatively little value to us from the purely scientific point of view, but useful in the field of economic botany.

How it could be so dismissed if it was still hidden remains unclear, but more significantly Walker made no mention of the botanical monographs, indispensable though they are, besides these other sources, for any attempt to write the history of botany in an oecumenical way.

There have been very few channels by which the rest of the world could get to know them. Lists and descriptions in Chinese,[b] though precious to those with the golden key of language, do not help for this purpose, nor yet passing mentions by Chinese scientists writing occasionally in English.[c] The great Bretschneider knew them pretty well but never had leisure to devote much time to them. Nevertheless

[a] (1), p. 326.   [b] E.g. Hung Huan-Chhun (1), p. 37.   [c] E.g. Ping Chih & Hu Hsien-Su (1).

his impressive list of Chinese books connected with botany, presented in the first volume of his *Botanicon Sinicum*,[a] contained 1148 entries, and historians of biology during most of the past century have neglected this at their peril. Unfortunately, although it included some hundreds of the monographs, there are also many which it failed to mention, and on the other hand the net was cast perhaps too wide in that it embraced much general literature, philosophy, medicine, travels, encyclopaedias and dictionaries. Today the agricultural bibliography of Wang Yü-Hu[b] is a much finer tool, not only because many errors have been corrected, but also in that each entry has a full description similar to that of the *Ssu Khu Chhüan Shu* bibliographers. Here again, however, many works have been omitted from the 593 entries as being of purely botanical interest or dealing with plant industries, while many books within or bordering upon the fields of agriculture, sericulture, zoology and veterinary medicine are included. We could not dare therefore to take the present sub-section lightly, and it will give us opportunity as we go on to underline the medieval Chinese contributions to the botany of ornamental plants,[c] as well as many striking insights of a taxonomic kind. So far for Western readers the only access to these has been by way of those monographs, less than half a dozen, which have been integrally translated; we shall quote them duly in their proper places.

The historical periods in which the monographic literature was produced are worth looking at again. None of it came as early as the age of Dioscorides, which would correspond to the Later Han, but by the time that Apuleius Platonicus was writing, in the middle of the +5th century (as we suppose) several important Chinese works of this kind had already been produced.[d] Through the time of Isidore of Seville (Sui and early Thang) the movement grew only very slowly, and the first real burst of activity came during the Northern Sung in the +11th century, when so many a botanical scholar took brush in hand to write about some favourite plant group against a background of interest in ornamentals which had evidently been growing during the middle and late Thang periods. Odo (Pseud. Macer Floridus) has to be thought of as contemporary with the court of Hui Tsung (+1100 to +1125)[e] the emperor destined for ruin in the collapse before the Chin Tartars, yet a ruler who himself wrote a work on a botanical subject, as we shall shortly see; the flourishing of the Chinese monograph literature at this time throws into high relief the deficiencies of Odo. After the fall of Khaifêng the increased importance of the southern provinces led to further monographs, notably the one on *Citrus* species and varieties which for me has always been the type-specimen of the genre, so that the whole of the Sung, roughly from +1000 to +1300, was a flourishing period for botanical writing. There followed a temporary decline during the Yuan, when drama and medical and agricultural

---

[a] (1), vol. 1, pp. 138 ff.

[b] (1), 2nd edition, greatly revised and extended.

[c] Here should be mentioned again the unique book of Li Hui-Lin (8), an indispensable companion, though not without a few sinological mistakes.

[d] Cf. pp. 359, 362, 378.   [e] Cf. Vol. 4, pt 2, pp. 497 ff., 501–2.

writing were so strong, but as soon as settled peace came again in the Ming dynasty, the botanical books and tractates poured forth once more. Revolution and war in the middle of the +17th century stopped the flow for some fifty years, but the prosperity of the reigns of the Khang-Hsi and Chhien-Lung emperors opened the sluices again, and individual writings on plants both wild and cultivated continued to appear, not ceasing thereafter before the nineteenth-century period when modern botany and traditional Chinese botany began to combine.

Further by way of introduction it should be remarked that the Chinese botanical monographs did not strictly delimit genera and species as they would be defined today—'bamboos', 'orchids', 'peonies and tree-peonies' sometimes included plants which would be separated from them in modern botany. A particularly good instance of this would be the *thung*[1] trees and the writings on them, but this case we reserve for the sub-section on classification (38 (*f*) below). Yet often medieval Chinese scholars were very conscious that certain plants went together only by custom, so that we find Li Khan[2], for example, devoting the last chapter of his *Chu Phu*[3] of +1299 to two 'borderline' groups—plants that look like bamboos but are not' (*ssu shih erh fei chu phin*[4]), and 'plants that have the name of bamboo but are not' (*yu ming erh fei chu phin*[5]). Here the first group numbered 23, none having the character for bamboo in their names but with deceptively similar leaves and stems, while the second was of 22, all popularly called bamboos but obviously widely divergent in family, as can be seen from Li Khan's drawings—e.g. liliaceous *Polygonatum sibiricum*[a] called 'deer bamboo' (*lu chu*[6]) and looking like our Solomon's Seal. But whatever the exact confines of particular medieval Chinese monographs the extraordinary fact remains that for five centuries before the heyday of the German 'fathers of botany' they were describing literally hundreds of different varieties and many dozens of separate species in a wide range of genera of different families. This taxonomic preoccupation was far in advance of Europe, where down to the middle of the +16th century 'nomina' remained purely generic, unsubsumed in a family and bearing no cognomen. Writing of Otto Brunfels, Greene says:[b]

Most of the genera, with him as with the botanists of antiquity, were monotypic, and the generic name was all that was needed. There was not the shadow of a reason for appending a second name; and he, no more than hundreds of botanical writers before his day, ever thought of such a thing.

One might instance *Adiantum, Althaea, Anethum, Asarum, Asparagus*. At this time (+1530) the genus really meant what we should think of as species or type-species, other obviously related forms being just given descriptive epithets. For example, the buttercup (*Ranunculus acris* to us) was *Pes-corvinus* (un-hyphened

---

[a] R 687; CC 1871.   [b] (1), pp. 187 ff.

[1] 桐   [2] 李衎   [3] 竹譜   [4] 似是而非竹品
[5] 有名而非竹品   [6] 鹿竹

in those days, the Crowfoot), the double garden variety was 'Full-flowered Crowfoot' and *R. bulbosus* was 'Lesser Crowfoot'. With Leonhard Fuchs a dozen years later, the 'genera' are still just type-species and varieties.[a] But already in the +12th century the Chinese had been clear that there was no such thing as *chu*[1] (*Bambusa*) or *chü*[2] (*Citrus*) *tout court*, except perhaps as an abstraction; one had to specify what *chu* or *chü* one was talking about by denoting it by a name of up to four syllables which almost always included the character for the genus. This in a way presaged the Linnaean conviction that every plant must have a specific as well as a generic name.

One aspect of this subject which greatly aroused our interest during the years in which we studied it was the elucidation of the motives of the writers of this literature and the kinds of people that they were. Wandering through the streets of medieval Chhang-an or Yangchow, one could meet professional pharmacists and professional horticulturists, with scholar-officials enamoured of some particular part of the plant world, but never of course professional botanists in the modern sense, since modern science had not yet arisen. Nevertheless many wrote with what we should call today a clearly scientific interest; they discerned what they felt to be a distinct group of plants or trees, and they desired to set down their knowledge of it. Of such a nature was the book probably the first of the whole genre, Tai Khai-Chih's[3] *Chu Phu*[4] (Treatise on Bamboos) in verse and prose, written in the near neighbourhood of +460; as also the notable work on the *Paulownia* and other trees which were than placed in the same family with it, the *Thung Phu*[5], written by Chhen Chu[6] in +1049.[b] Many others followed as time went by, listing and describing in extreme detail the varieties of many plants cultivated in gardens for the beauty of their flowers. Often the titles reflect the urge for the extension of recorded knowledge—*Pu Chha Ching*[7] (A Supplement for the Manuals of Tea)[c] brought out by Chou Chiang[8] in +1008, or *Hsü Chu Phu*[9] (A Supplement to the Treatises on Bamboos) penned by Liu Mei-Chih[10] about the middle of the +14th century. Such men had acquired further information and could not bear to see the available accounts remaining incomplete. Another motive found now and then was pharmacological interest, such as that which led the philosopher[d] Li Ao[11] to write in +840 his tractate on the drug-plant *Polygonum multiflorum*, entitled *Ho Shou Wu Chuan*.[12]

Economic considerations were never very far away from medieval Chinese botanical thinking; Tai Khai-Chih himself had a great deal to say about the practical uses of all the kinds of bamboos he wanted his readers to recognise, and

---

[a] *Ibid.* pp. 207, 216.
[b] This we reserve for the sub-section on classification.
[c] The first *Chha Ching* had been written in the Thang, by Lu Yü[13], about +770.
[d] See Vol. 2, p. 494.

[1] 竹   [2] 橘   [3] 戴凱之   [4] 竹譜   [5] 桐譜
[6] 陳翥   [7] 補茶經   [8] 周絳   [9] 續竹譜   [10] 劉美之
[11] 李翺   [12] 何首烏傳   [13] 陸羽

this remained true of all the monographers of horticultural and orchard botany from lichis to tea and oranges. In very different connections we have already seen how interested provincial governors and magistrates could be in anything which conduced towards the greater prosperity of their bailiwicks,[a] and this certainly accounts for many of the treatises on such economic if not staple plants. Then there was also the regional pride of those who lived or sojourned in particular provinces where certain plants were prominent. Hence the titles of the type of *Yang-Chou Shao-Yao Phu*[1] (Treatise on the Herbaceous Peonies of Yangchow) written by Wang Kuan[2] in +1075, or the *Tien Chung Chha-Hua Chi*[3] (Notes on the Mountain-Tea Plants (*Camellia*) of Yunnan) by Fêng Shih-Kho[4], a work of the Ming. Others again were moved by professional pride, e.g. the officials in charge of the collection of tax tribute. Last but not least there were authors who wrote primarily from a deep love of plants, especially those that they cultivated, often as Taoists, for artistic and spiritual reasons. Here one has to remember the 'fashionable' aspect of ornamental flowering plants; it was the proper thing for successful Confucian officials to take an interest in gardens, a natural love of flowers was widely spread, and habits of keen observation hardly less so. Besides, an intimate knowledge of plants and their accumulated symbolism provided an opportunity of airing possibly dangerous socio-political views within the fabric of apparently innocuous though allusive poetry and literature.

As for the kinds of people who wrote, our survey of motives has already conjured most of them up. One can easily distinguish half a dozen classes. First there were scholar-officials, physicians and others whose duties took them to those places within or on the borders of the empire where special plants flourished; this category trends upon the exotic botany which we discuss in a separate subsection, and the obvious example would be the *Nan Fang Tshao Mu Chuang*[5], the book on the plants of the southern regions written by Hsi Han[6] in +304. Later this writing concentrated on individual plants, as in the treatise on the lichi varieties of Kuangtung, *Tsêng-Chhêng Li-Chih Phu*[7] due to Chang Tsung-Min[8] in +1076, a book which with all others of its kind we must discuss in the horticultural sub-section. Here, however, we shall have much to say about the *Lo-Yang Mu-Tan Chi*[9] (Account of the Tree-Peonies of Loyang) written by the famous scholar Ouyang Hsiu[10] in +1034. Secondly, there were scholar-officials whose duties were specifically concerned with the collection and forwarding of government tribute produce. Later on, in Sect. 42, we shall have to consider several of these when discussing the tea plant and the tea industry; some even have the word *kung*[11] (tax tribute) in their titles, e.g. the *Hsüan-Ho Pei-Yuan Kung Chha Lu*[12] written

---

[a] Cf. Vol. 4, pt 2, pp. 32 ff. on the entourage of engineers and technicians which often surrounded an important official. They needed the protection of a patron, and he in turn was interested in anything they could do to increase the productivity of his province.

[1] 揚州芍藥譜   [2] 王觀   [3] 滇中茶花記   [4] 馮時可   [5] 南方草木狀
[6] 秙含   [7] 增城荔枝譜   [8] 張宗閔   [9] 洛陽牡丹記   [10] 歐陽修
[11] 貢   [12] 宣和北苑貢茶錄

in +1122 by Hsiung Fan[1], a scholar of Chienyang in Fukien whence originated throughout the Sung some of the best of all Chinese tea. An older work of a similar kind was the *Phin Chha Yao Lu*[2] (Essentials of the Grading of Teas) produced in +1078 by Huang Ju[3], not a Fukienese by birth but an official who was stationed there and fell in love with the tea gardens and their crafts. He used the word *phin*[4] (grading) in a practical sense just as we should, but in the branch of botanical literature now to be described we shall often come across it used in an unfamiliar aesthetic-ethical-bureaucratic sense, not at all like the precise and useful significance which it has, for example, in the *Shen Nung Pên Tshao Ching* (see p. 242 above). In perfect accord with the ethos and principles of bureaucratic organisation, the scholar-officials developed a mania for classifying flowers and flower varieties into ranks analogous to those of society—emperor, empress, virtuous minister, local magistrate, postmaster, bandit and so on. This arose in part from the fact that plants and flowers had from early times been emblems of virtues. Here the preface of Liu Mêng's[5] *Chü Phu*[6] (Treatise on Chrysanthemums, +1104) is worth reading. The chrysanthemum is a noble plant, he says, because it flowers beautifully in the autumn, withstanding cold and wind, and avoids like a high-minded recluse the crowded market-place of spring; this is why the great poets of old, Chhü Yuan[a] and Thao Yuan-Ming[b], admired and grew it in their retirement; moreover every part of it can be eaten safely and beneficially, and chrysanthemum wine is good for longevity, while the petals can take the place of tea leaves and be stuffed into fragrant summer pillows; finally even the people of Loyang (sworn devotees of the tree-peony) admit its high eminence. But just how high? The bureaucrats settled down therefore to enjoy the tabulation[c] of botanical eminences; presently we shall see in the +11th-century book of Chhiu Hsüan (p. 406 below) how far these ploys could be carried. There is indeed an analogy to all this in the preacher's symbolisms which flooded the natural history books of the European middle ages, but there were two important differences— the Chinese kept it out of their systematic natural history books (cf. p. 126 above), and the gambit was aesthetic-bureaucratic, not moral-theological.

Other scholar-officials were not particularly concerned with the collection of tribute produce but conceived a great admiration for certain cultivated plants to which the well-being of their provinces was largely due. An example of this kind

---

[a] Cf. Vol. 3, pp. 485–6, Vol. 4, pt 3, pp. 250, 436, Vol. 5, pt 2, p. 98 (−332 to −288). In fact he makes no mention of chrysanthemums recognisable now. The *hsiang tshao*[7] of the *Li Sao* (cf. Hawkes (1), p. 23) might be any 'perfumed herb'. Guesses could include another composite, *Eupatorium chinense* (R 33), the labiate *Ocimum Basilicum* (R 134a), or the leguminous *Melilotus officinalis* (CC 1020), a clover.

[b] Thao Chhien (+365 or +372 to +427). There can be no doubt about his descriptions of chrysanthemums, and he certainly grew a number of varieties of them in his retirement.

[c] Chinese interest in tabulation methods in ancient times recalls us to Vol. 1, pp. 34 ff., and Vol. 3, pp. 106 ff. not forgetting Bodde's study (5). The *Chhien Han Shu* had a chronological table of historical figures arranged in nine ratings of virtue. Assessment of flower qualities was in the same tradition. The V.M.H. of the present day may be surprised to find how far back their competitions go.

[1] 熊蕃　　　[2] 品茶要錄　　[3] 黃儒　　　　[4] 品　　　　　[5] 劉蒙
[6] 菊譜　　　[7] 香草

might be the great Fukienese governor of Fukien, Tshai Hsiang[1], who wrote his monographs on the tea shrub (*Chha Lu*[2]) and the lichi tree (*Li-Chih Phu*[3]) in the close neighbourhood of +1060. Then fourthly there were retired civil servants who devoted their maturer years in retirement to horticultural pursuits. Certainly no less distinguished than Tshai Hsiang was Fan Chhêng-Ta[4] in the following century, the author of one book on chrysanthemums (*Chü Phu*[5]) and another on the Chinese apricot, *Prunus mume* (*Mei Phu*[6]) in +1186, a few years only before his death. Probably belonging to this group was the first of the horticultural encyclopaedists, Chhen Ching-I[7], who in +1256, before the Mongol invasions ruined the peace of the south, produced a splendid work, still extant, entitled *Chhüan Fang Pei Tsu*[8] (Complete Chronicle of Fragrances). As a fifth group in our list come the artists, men who found themselves botanists without knowing it, intending only to tell people how to draw and paint plants but in fact irretrievably involved in fine points of terminology and even classification. Such a one was Li Khan already mentioned (p. 357) and to be mentioned again (p. 387), but one could also instance Sung Po-Jen[9] with his *Mei Hua Hsi Shen Phu*[10] of +1238 (The Spirit of Joy; a Hundred Portraits of the Eight Stages in the Life of the Apricot Flower).[a] Sixthly there arose during the end of the Ming period in the late +16th century a group of scholars, largely of Taoist inspiration, who sought no office and lived in seclusion, cultivating plants and writing about them in order to console the heart in bad times and nourish the spirit (*i chhing yang hsing*[11]).[b] Of these were Kao Lien[12] whose *Tsun Shêng Pa Chien*[16] (Eight Disquisitions on the Art of Living a Retired Life) came out in +1591, and Chang Ying-Wên[17] with his *Chang shih Tshang Shu*[18] of +1596. An important part of the former was the *Hua Chu Wu Phu*[19], five treatises on bamboos, orchids, peonies, tree-peonies and chrysanthemums; while 'Mr Serge-Coat' (Pei-Ho *hsien-sêng*[20]), to give the latter adept his favourite name, expatiated on tea plants and chrysanthemums.

Last but not least of the groups is that wherein we may place members of the imperial families. Here even emperors themselves contributed, such as the ill-fated Hui Tsung[21] of the Sung,[c] Chao Chi[22], who took his 'vermilion brush' in hand

---

[a] Cf. Li Hui-Lin (8), pp. 48 ff., 52 ff. At this first mention of the Chinese *mei hua* it should be emphasised that 'apricot' always needs inverted commas. There are many varieties, some cultivated for their flowers, others for their fruit. The conventional usage is a little confusing because some varieties are not at all like the apricot as cultivated in Western Europe.

[b] It would be superfluous here to enlarge upon the well-known place of flower admiration and arrangement (*ikebana*[13]) in the mystical art of Japan. See Sparnon (1). Like so many other things which the Japanese brought to perfection, it had probably originated in China. See Li Hui-Lin (13), with its complete translation of Chang Chhien-Tê's[14] *Phing Hua Phu*[15] (Record of Vase Flowers), written in +1595.

[c] We have already had a good deal to say about the court of this monarch in Vol. 4, pt 2, pp. 501 ff. It has been compared with that of the emperor Rudolf II at Prague (+1576 to +1611) and that of Frederick II of Sicily (+1194 to +1250). Chao Chi must certainly be numbered among the great patrons of science, learning and culture, with Alfonso X of Castile (r. +1252 to +1282) and our own Charles II.

---

[1] 蔡襄	[2] 茶錄	[3] 荔枝譜	[4] 范成大	[5] 菊譜
[6] 梅譜	[7] 陳景沂	[8] 全芳備祖	[9] 宋伯仁	[10] 梅花喜神譜
[11] 怡情養性	[12] 高濂	[13] 活花	[14] 張謙德	[15] 瓶花譜
[16] 遵生八牋	[17] 張應文	[18] 張氏藏書	[19] 花竹五譜	[20] 被褐先生
[21] 徽宗	[22] 趙佶			

about +1109 to write the 'Discourse on Tea, recorded in the Ta-Kuan reign-period' (*Ta-Kuan Chha Lun*[1]). Though mainly concerned with the grading and classification of the finest teas of the empire, he had a good deal to say also of the harvesting, fermentation and processing of the leaves. After all that has been said about the imperial princes of the Ming, especially Chou Ting Wang, in the preceding subsection under the esculentist rubric, a simple mention may here suffice, but the tradition went back far earlier than the Northern Sung, as an example from the +6th century may show. Thopa Hsin[2], a prince of the Northern Wei (Kuang-Ling Wang[3]) had a passion for horticulture. Not only did his gardens supply all the best fruit of the capital, but we may date to about +540 a book, written by himself as like as not, the *Wei Wang Hua Mu Chih*[4] (Records of the Flowering Plants and Trees in the Gardens of the Prince of the Wei).[a] Thus over the whole range we can see that many men and many motives contributed to swell a literature on plants that remains to this day almost wholly unappreciated by the rest of the world. Let us now look at its several subjects one by one.

Several alternative orders of treatment presented themselves. As most of the plants concerned are ornamentals, it would have been tempting to touch on them in the order in which they occur in the famous ballad of *Mêng Chiang Nü*[5], the folk-song centuries old about the girl who went to search for her husband conscript lost in the building of the Great Wall.[b] Normally this has twelve stanzas one for each month of the year, and each begins with a reference to the flower most characteristic of the time.[c] But our material does not fall easily into this framework, partly because the most monographed species were not the same as those mentioned in the ballad's list, and partly because for historical reasons it is desirable to describe in this place the special literature on bamboos. Moreover, although we shall postpone most of the works on fruit-trees and the like to the sub-section on horticulture (38 (*i*) below)—dates, peaches, lichis, lungans, lotuses, etc.—we are constrained to introduce the citrous fruits at this stage because the very type-specimen of all the botanical monographs is a work of the late +12th century on three genera of the Aurantioideae and their cultivation. We shall thus proceed in the following order: oranges, bamboos, the tree-peony, the herbaceous peony, chrysanthemums, orchids, the Chinese apricot (*Prunus mume*), the crab-apples, two strange and exotic flowering plants cultivated at particular places, then roses and balsams, finally *Camellia* and *Rhododendron*. Thence we shall pass to the general literature on horticulture and traditional botany. In each particular group of monographs we shall take care to mention the earliest one, the best one for plant description or practical value, and the largest one.

---

[a] See further, Sect. 38 (*i*), 2 below.

[b] Cf. Vol. 4, pt 3, p. 53. One of the versions has been translated by Needham & Liao Hung-Ying (1).

[c] The list can be made from the version published by Ku Tzu-Jen (1), no. 16. Starting with the first month, we have the Chinese apricot, then the apricot, the peach, the rose, the pomegranate, the lotus, the balsam (*Impatiens balsamina*), the 'cassia' (*Osmanthus fragrans*), for September the chrysanthemum, for October the hibiscus, snow-crystals in the eleventh month, and last of all the wintersweet *Chimonanthus praecox* = *fragrans*, cf. p. 352.

[1] 大觀茶論   [2] 拓跋欣   [3] 廣陵王   [4] 魏王花木志   [5] 孟姜女

(i) *Citrous fruits*

In order to put the 'orange monograph' in an adequate setting it is really essential to give first a very brief account of the history of oranges and citrous fruits in general—a succulent subject which has attracted the pens of considerable historians.[a] There can be no manner of doubt that the original home and habitat of these trees was on the eastern and southern slopes of the Himalayan massif; a fact which is reflected in the presence of the maximum number of old-established varieties in the Chinêse culture-area, as also in the extreme antiquity of the Chinese literary references. It is also betrayed by the considerable number of single written characters denoting particular species,[b] always a sign of ancientness in the nomenclature. 'Of all ancient records preserved until this day', wrote Tolkowsky, 'in which citrous fruits are mentioned, none go back further than those belonging to the literature of the Chinese people.' And the slow transmission of the fruit-trees westwards is an epic story almost to be compared with that of the transmission of inventions,[c] indeed not at all unworthily so, seeing that the selection, hybridisation and cultivation of sub-species and varieties had been the achievement of the Chinese gardeners and growers working through two thousand years.

Of the early Chinese records of oranges and their congeners something has already been said in the pages devoted above (p. 103) to the oecological story of the *chü* and the *chih*. Without repeating what was said there, some further evidence may be added. Doubtless the oldest words occur in the Yü Kung (Tribute of Yü) chapter of the *Shu Ching*, translated and discussed on p. 89; a text which cannot be later than the early −5th century and may be as old as the early −8th or late −9th. The Domesday provinces of Yang-chou and Ching-chou, it will be remembered, furnished oranges and pommeloes among the tribute sent up to the Chou High King (Fig. 77). Then after the several mentions of the trees in the *Shan Hai Ching*, certainly a Warring States text if not much older, one finds many references to them and their fruit in the works of the −3rd century. The *Han Fei Tzu*[8] book reports a conversation between Yang Hu[9] of Lu and Chao Yang[10] of Chin about −500 in which the opposite characteristics slowly revealed by oranges and thorny limebushes as they grow are taken as a parable of the care needed in choosing young men.[d] The passage in the *Yen Tzu Chhun Chhiu* about Yen Ying's witty speech in the −6th century we have already quoted (p. 106) and it is followed by a second story turning on the etiquette of peeling an orange at a princely court.[e]

---

[a] Gallesio in 1811 was the first modern writer of this kind, but today we refer rather to the absorbing book of Tolkowsky (1) and the clear review of Webber (1). As the revised datings of modern sinology had not reached these authors, the appropriate corrections must be made, but most of their conclusions are unaffected thereby.

[b] Not only *chü*[1] for orange and *yu*[2] for pomelo, but also *kan*[3] for certain kinds of oranges, *chêng*[4, 5] for sweet oranges, *luan*[6] for the sour orange, and *yuan*[7] for the citron. See further below.

[c] Cf. Vol. 4, pt 2, pp. 544, 584.

[d] Ch. 33, p. 5*b*, tr. Liao Wên-Kuei (1), vol. 2, p. 81.

[e] Ch. 6, p. 4*a, b*.

[1] 橘   [2] 柚   [3] 柑   [4] 橙   [5] 棖
[6] 欒   [7] 櫞   [8] 韓非子   [9] 陽虎   [10] 趙鞅

Fig. 77. The tribute of oranges described in the Yü Kung chapter of the *Shu Ching*. A late Chhing representation from the *Shu Ching Thu Shuo*, ch. 6, p. 34*a*.

The Chü Sung[1] (Poetic Eulogy of the Orange) in the *Chhu Tzhu*, generally attributed to Chhü Yuan[2], is almost certainly not his, but it cannot be later than the middle of the −3rd century;[a] and there are several mentions of oranges and pomeloes in the *Lü Shih Chhun Chhiu* of −239, with places where they grow.[b] So also the *Khao Kung Chi* section of the *Chou Li*, a text presumably finalised in the Han, though with much Warring States content, refers not only to the habitat of *chü* and *chih*,[c] but to the use of citrus wood for making bows.[d] Other Han references have already been mentioned (p. 111). From that time onwards a vast number of citations from Chinese literature can be assembled—and indeed they have been in the work edited by Yeh Ching-Yuan (2). Over such a long period it is only to be expected that there should have been much fluctuation of terminology,[e] but when all the evidence is in, the range and expansion of sub-species and varieties will be clearly traceable. There is no other like tradition.

As for systematic commercial production, one of the most striking pieces of evidence is the speech of the statesman Su Chhin[3] (d. −317) when in his youth he was a counsellor of the ruler of the State of Yen (Yen Wên Hou[4], r. −360 to −331). Among other things he said:[f]

The candid and ingenious prince should know how to listen to good advisers. Now the State of Chhi must necessarily gain its wealth from the fisheries of the sea (along the coasts) and the salt which can be extracted from it there. The State of Chhu must necessarily gain wealth from its groves of orange and pomelo trees (*chü yu chih yuan*[5]). So also our State of Yen must necessarily exploit the wealth it has in its felt and furs, its dogs and horses.

Thus it is safe to conclude that citrous fruits were being grown industrially for market in what is now Hupei, Anhui and Hunan for at least half a century before people in Europe encountered the first of the group to become known to them.

'I am in a position to assure you', wrote Athenaeus of Naucratis in +228, 'that Hegesander the Delphian nowhere mentions the citron, for I read through the whole of his "Memorials" with the express purpose of finding out'.[g] Apart from its gentle unintentional satire on the methods of the historian of science and material culture, this passage usefully points to the fact that only slowly over a long period did the peoples of the West become acquainted with the useful and beautiful trees of the orange sub-family. The citron (*Citrus medica*) was the first to

---

[a] *Chhu Tzhu Pu Chu*, ch. 4, pp. 27*b* ff. [b] Ch. 70 (vol. 1, p. 138).
[c] Ch. 11, p. 3*a* (ch. 40), tr. Biot (1), vol. 2, p. 460.
[d] Ch. 12, p. 24*b* (ch. 44), tr. Biot (1), vol. 2, p. 582.
[e] The variation can be confusing. For example Yeh Ching-Yuan (2) criticises Tanaka Tyōzaburō for saying that *huang kan* was superseded by *thien chhêng* as the chief name for *Citrus sinensis* by the Sung; on the contrary it freely occurs down to the end of the Chhing. Cf. p. 376 below.
[f] *Shih Chi*, ch. 69, p. 4*b* (order of sentences inverted), cit. *TPYL*, ch. 966, p. 3*b* (in part), tr. auct.
[g] *Deipnosophistae*, III, 25–9, Yonge tr. vol. 1, pp. 139 ff. I owe the first notice I had of this celebrated remark to my oldest friend, the late Prof. F. P. Chambers. Before we became historians ourselves we thought it wildly funny. Athenaeus puts it into the mouth of Plutarch, who appears as one of his characters.

[1] 橘頌   [2] 屈原   [3] 蘇秦   [4] 燕文侯   [5] 橘柚之園

arrive. Theophrastus has a good description of it about −300, but for him the 'Persian' or 'Median apple' was an exotic plant, not yet acclimatised to Mediterranean shores.[a] Babylonia and Palestine exported it to Europe, and though there are no Biblical references the citron became important in the Jewish Feast of Tabernacles ritual from −136 onwards. By the time of Virgil (c. −30)[b] the trees were growing in Italy, though probably not bearing fruit, while full acclimatisation had taken place when Dioscorides (c. +15)[c] and Pliny (+77)[d] set down their accounts.[e]

Next came the sour orange (*Citrus Aurantium*) and the lemon (*Citrus Limon*), arrivals which coincided rather closely with the opening of the Roman-Indian trade-route directly from the Red Sea towards the end of the −1st century.[f] *Aurantium* was probably a natural development (because of the golden colour of the fruit) from the Indian name *nāranga* which accompanied it; hence also *nerantzion* in Byzantine Greek.[g] The very clear depiction of an orange in full colour on a mosaic at Pompeii[h] attests its cultivation in Italy before +79 though not perhaps its full acclimatisation. A Plinian reference reinforces this,[i] and there is similar iconographic evidence for the lemon. By the beginning of the +4th century a fully indigenous orchard production had been established in Southern Europe.[j]

Yet another route of penetration was constituted by the Maghrib, the northern coast of Africa, after the rise of Islam. The Arabs were very active in the cultivation of citrous fruits, as one can see from numerous references in the literature,[k] and particularly in agricultural writings such as the *Kitāb al-Filāḥa al-Nabaṭīya* (Book of Nabataean Agriculture), written by the Iraqi, Abū Bakr Ibn al-Waḥshīya al-Kaldānī al-Nabaṭī in +904. Half a century earlier (in +851), the traveller Sulaimān al-Tājir had visited China and exclaimed at the abundance of citrous fruits there.[l] Thus the commercial and technological unity of the Arab world brought it about that the pomelo (*Citrus grandis* = *maxima*), transported to the furthest Hesperides from its home on the slopes of the Yangtze Valley from Szechuan to the sea, was flourishing notably in Spain when the greatest Andalusian work on the plant industries was written—that of Abū Zakarīyā al-'Awwām al-Ishbīlī, the *Kitāb al-Filāḥa* (Compendium of Agriculture) about +1180.[m]

---

[a]   *Hist. Plant.* IV, iv, 2–3, Hort tr. vol. I, pp. 310 ff.
[b]   *Georgics*, II, ll. 126–35, Royds tr., p. 97, ll. 150–161; Fairclough tr. vol. I, p. 125.
[c]   *De Mat. Med.* I, 164–6, *cedromeles*; Gunter tr. pp. 84, 85.
[d]   *Hist. Nat.* XII, vii, 15, 16, Rackham tr. vol. 4, pp. 12, 13.
[e]   The citron is little seen in Northern Europe, but flourishes, with its proper French name, *cédrat*, in Corsica, where one may drink an excellent liqueur made from it.
[f]   Cf. Vol. I, p. 178.   [g]   Cf. Yule & Burnell (1), p. 490 as well as Tolkowsky (1).
[h]   Reproduced in Tolkowsky (1), pl. xxxviii.
[i]   *Hist. Nat.* XIII, xxxi, Rackham tr. vol. 4, pp. 160, 161.
[j]   Cf. Tolkowsky (1), pp. 108 ff.
[k]   Hehn (1) and Tolkowsky (1) have much on this.
[l]   See Renaudot (1), p. 17; Sauvaget (2), pp. 11, 48 (text, para. 22). His *turunj* seems to have been collective, covering all citrons, oranges, lemons, etc. Arabic terminology played many changes on this word, which may originally have derived from Skr. *matulunga*, citron; but *naranj* attached only to oranges.
[m]   Cf. Mieli (1), pp. 133, 205 ff. No one who has visited such cities as Granada, Cordoba and Seville can have failed to appreciate the role played by the cultivation of citrus fruits in Muslim Spain.

As for the sweet orange (*Citrus sinensis*), tradition has associated its introduction to the West with the Portuguese trade in the East Indies after Vasco da Gama,[a] but it is now thought to have reached Europe a good deal earlier, *c.* +1470, by way of Genoese contacts through the Levant. After this it spread rapidly through the Mediterranean lands. The lime (*Citrus aurantifolia*) was the next in the series, not known or planted in the West until +17th-century trade brought it from Malaya, though it had been familiar to the Arabs long before. Lastly only in modern times did the loose-skinned tangerine (*Citrus reticulata*) become known.

This sequence may fitly come to an end for us in the London of the Royal Society. About the middle of the +17th century the sweet 'Portugal orange' was displaced in favour by better varieties of 'China orange' brought direct from there, or grown on trees recently brought from there.[b] All the early Portuguese travellers from +1514 onwards, as well as the Jesuits, without exception extolled the citrous fruits of China,[c] and by +1700 the Chinese were exporting sweet oranges individually wrapped in paper all over the East Indies.[d] In +1667 Thomas Sprat, in his book on the Royal Society, had a hortatory heading: 'Mechanics improveable by Transplantations', under which he said: 'The Second Advancement of this Work may be accomplish'd by carrying and transplanting living Creatures and Vegetables from one Climate to another', adding as an example: 'the Orange of China being of late brought into Portugal has drawn a great Revenue every Year from London alone'.[e] So there, i'faith, was Nell Gwynn, standing with her basket of oranges—a remarkably sweet element in Sino-British relations, though not often thought of in that connection. As a character said in a play of +1668:

> The noble peer may to the play repair,
> Court the pert damsel with her China ware—
> Nay marry her if he please—no-one will care ...[f]

Our royal founder could not exactly marry her but he did the next best thing; while the specific gravity and refractive index of China orange oil were duly quizzed in the Society by Mr Fra. Hauksbee, F. R. S.

Utterly unknown to each other, one of al-'Awwām's contemporaries, separated from him by the entire length of the Old World, was working away on the first book in any literature ever to be consecrated to the Aurantioideae. This was Han Yen-Chih[1], Sung governor of Wênchow in Chekiang, who dated the preface of his

---

[a] Mem. Vol. 4, pt 3, pp. 506 ff. The proper interpretation of a famous passage in al-Mas'ūdī seems to indicate that the tree reached the Arabic countries of the Middle East from India via Oman *c.* +912 (cf. Tolkowsky (1), pp. 123, 144).

[b] See Tolkowsky (1), pp. 246 ff., 303.

[c] For example, Gaspar da Cruz (+1569) in Boxer (1), pp. 132 ff., Ricci (+1610) in d'Elia (2), vol. 1, p. 18, Trigault (+1615) in Gallagher (1), p. 11, and Lecomte (+1698), pp. 97 ff.

[d] Rumpf (1), vol. 2, pp. 113 ff.   [e] (1), p. 387.

[f] *A Fool's Preferment* by T. d'Urfey (+1653 to +1723).

[1] 韓彥直

Chü Lu[1] (The Orange Record) in the 10th month of the 5th year of the Shun-Hsi reign-period, i.e. +1178. I shall never forget that sunny day in the spring of 1943 when the botanist Fang Wên-Phei and the citrologist Chang Wên-Tshai from East Malling first introduced me to a knowledge of this book at the Horticultural College of Nanking and Szechuan Universities then evacuated to Chhêngtu in the heart of the orange-growing areas of Szechuan.[a] Ever since that time Han Yen-Chih's book has remained for me the type-specimen of the medieval Chinese monographic literature. More recently we met with his family, for his father was none other than the famous general Han Shih-Chung[2], whose brilliant use of treadmill-operated paddle-wheel warships gained the victory of Huang-thien-tang over the Chin Tartars in +1130.[b] When Han Yen-Chih had been some time at Wênchow he developed an enthusiasm for the citrus industry, and decided to describe it in detail. The *Chü Lu* is thus in three chapters.[c] In the first and second

---

[a] Impressions of that time will be found in Needham (22).

[b] See Vol. 4, pt 2, pp. 418, 432.

[c] By a singular circumstance, my collaborators and I have been enabled to study and cherish the *Chü Lu* in a copy which may be one of the original edition, and consequently the only 'Sung pên', hence the earliest printed book, in England. When Dr Lu Gwei-Djen and I were buying Chinese works connected with the history of science in Shanghai in 1946 one of the booksellers offered us this text quite inexpensively, perhaps inadvertently, to go with the others. When I returned to Cambridge in 1949 and Dr Lu was far away, Dr Wang Ling and I had opportunity to examine this book for the first time, and found to our astonishment that it possessed seals and notes at the front denominating it a unique copy shown at an exhibition of rare books at Suchow in 1937, after having been repaired in 1934. At the back there were two MS postfaces by eminent scholars of the +18th century, affirming that in their view it was a Sung print. Ho Chhuo[3] (+1661 to +1722, cf. Hummel (2), vol. 1, p. 283) who seems to have been at one time the owner, wrote his postface in black ink in +1712, recording Hsü Chhien-Hsüeh[4] (+1631 to +1694, cf. Hummel (2), vol. 1, p. 310) as a still earlier owner; while Huang Phei-Lieh[5] (+1763 to 1825, cf. Hummel (2), vol. 1, p. 340) added a longer critical one in red ink in 1812. Huang in particular was a noted connoisseur of Sung books and their editions.

Friends such as Mr Fu Lo-Huan and the late Dr Chi Chhao-Ting confirmed the authenticity of the postface hand-writings and their seals. The book is printed in a very clear script on thin brown paper which has been carefully repaired, and also strengthened by interleaves in white. It was the only example of printing believed to be Sung which could be shown at the Exhibition of the Arts of the Sung Dynasty organised by the Arts Council of Great Britain in 1960.

Since then we have had an opportunity of examining photocopies (formerly in the possession of Paul Pelliot) of the additional MSS used by Michael Hagerty for his work on Chiang Khang-Hu's draft translation; and one of these, then (1921) in the library of the eminent botanist Shirai Mitsutarō of Tokyo, said to be a trace-copy of a 'Sung pên', has a close similarity in script to ours, the layout being apparently identical. Both have the respectful blank space in front of the words *kuo chhao*[6] 'our present (Sung) dynasty'. The presence in the MS of an unusual tabu form, *chhi*[7], at the left-hand bottom corner of p. 1*a* of the preface, suggested to Pelliot that it might be an avoidance of the name of the emperor Tu Tsung (Chao Chhi[8]) who was on the throne in +1273, the date when the *Chü Lu* was incorporated by Tso Kuei[9] into the first of all tshung-shu, the *Pai Chhuan Hsüeh Hai*[10] (cf. Vol. 1, p. 77). Perhaps this view might be supported by the fact that the Shirai MS continues directly with the *Mu-Tan Jung Ju Chih* (cf. p. 406 below) transcribed in exactly the same style. But our copy does not show the tabu form of *chhi*, nor does it bear any sign of ever having formed part of a collection, so that it may conceivably be what the cover says it is, a copy of the first edition of all, the 'Shun-Hsi pên' (+1178).

On the other hand, some of these same characteristics, taken in conjunction with other evidence, may point rather to the view that our copy is a print of the Ming period. Chhien Tshun-Hsün (3) has compared its typography with that of the Sung tshung-shu edition issued in +1273 (of which only two sets exist, one in China and one in Japan), and with that of the edition of the same collection reprinted by Hua Chhêng[11] in +1501, with the result that our copy seems to be much more like the latter. Chhien also found (*a*) that the *Chü Lu* does not

---

[1] 橘錄	[2] 韓世忠	[3] 何焯	[4] 徐乾學	[5] 黃丕烈
[6] 國朝	[7] 其	[8] 趙禥	[9] 左圭	[10] 百川學海
[11] 華珵				

he gives descriptions under headings of 8 *kan* oranges, 14 *chü* oranges and 4 *chhêng* oranges, besides making passing mentions of the pommelo (*yu*) and the thorny lime (*chih*), the whole incorporating therefore 28 sub-species and varieties, belonging, as we would now say, to three separate genera. In the third he discusses in detail planting, grove management, irrigation, transplanting, grafting (cf. 38 (*i*), 4 below), harvesting, storage, preservation, and medicinal use. A particularly interesting section is devoted to protection against the diseases of the citrus trees. When discussing the sub-species and varieties Han Yen-Chih is strikingly meticulous beneath the cloak of conventional elegant dilettantism. He speaks of the habitat, the tree habit, the shapes of leaves and branches, the size and form of the fruit, its colour, taste, skin thickness, rugosity, essential oil content, peeling properties, number of segments and their separability, number of seeds, and time of ripening; finally he may provide a reason for the name which it bears. As one of the standard descriptions remarks—even nowadays one could hardly find much more to say.[a]

It is clear from all the literature that in quite early times the Chinese understood well that all the indigenous citrus fruits, e.g. *Citrus reticulata* (*chü*[1]), *Citrus junos* (*hsiang chhêng*[2]), *Poncirus trifoliata* (*chih*[3]), *Citrus sinensis* (*thien chhêng*[4]), *Fortunella* spp., the cumquat (*chin kan*[5]), and *Citrus grandis*, the pommelo (*yu*[6]), were of the same group (*lei*[7]).[b] This though the fruit colour could range from green through golden yellow to dark red, and the size, as in the case of the last two mentioned, from that of a marble to that of a football. What is more, after the appearance and acclimatisation of a number of Indian or southern forms, e.g. *Citrus poonensis*

appear in any of the rare book catalogues of Hsü Chhien-Hsüeh, Ho Chhuo or Huang Phei-Lieh, though it could hardly have been omitted if one or other of them had really owned a Sung edition of it, and (*b*) that the texts of the colophons signed by Ho and Huang appear in Sung editions of two other works that they owned, modified by minor adaptations suitable for the *Chü Lu*, the writing and the seals being the same, or cleverly copied. He concluded therefore that it is a Ming print, but made to appear a Sung one by the addition of modified postfaces by two noted collectors, originally written for other books, at some time after 1825. Chhien's arguments appear to be rather convincing, yet if he is right a considerable number of Chinese scholars must have been taken in during the past century and a half. But perhaps the question is still not settled beyond all possible doubt.

[a] Yeh Ching-Yuan (*2*), pp. 5 ff., 26 ff.; Wang Yü-Hu (*1*), 1st ed. p. 77, 2nd ed., p. 93; Reed (*1*), p. 51.

[b] This was the loosely used medieval word. Modern Chinese scientific terminology would say the same *ya kho*[8] (sub-family) consisting of a number of genera (*shu*[9]) and species (*lei*[7]). *Clausena*, for example (cf. 113–14 above), was another genus which Han Yen-Chih would probably have recognised as belonging to his aurantioid or rutaceous *lei*.

It is interesting that the word *shu*[9] was also sometimes anciently used in this very way. In his *Fêng Thu Chi*[10] (Record of Airs and Places), written in the late +3rd century, Chou Chhu[11] said: 'Of the class (*shu*) of oranges (*kan chü*[12]) those with the richest and most exquisite sweet taste mny be divided into the yellow (*huang*[13]) and the red-flushed (*chhen*[14]) sorts; the latter are called 'pot oranges' (*hu kan*[15]).' The passage was preserved in *TPYL*, ch. 966, p. 1*b*.

Originally, *kan*[16] was probably adjectival, 'sweet', and only later became the name of a particular kind or kinds of orange, acquiring the 'wood radical' in the process, *kan*[17].

[1] 橘	[2] 香橙	[3] 枳	[4] 甜橙	[5] 金柑
[6] 柚	[7] 類	[8] 亞科	[9] 屬	[10] 風土記
[11] 周處	[12] 甘橘	[13] 黃	[14] 頳	[15] 壺甘
[16] 甘	[17] 柑			

(*mo kan*[1]),[a] *Citrus Aurantium*, the sour or Seville orange (*chu luan*[2]), *Citrus medica*,[b] the citron (*kou yuan*[3]),[c] and the lime *Citrus aurantifolia*, especially the form of it known as the Canton lemon, *Citrus limonia* (*li mêng*[4]), all were quickly recognised as belonging to the same *lei*[5] as the former.[d] The same was of course true of the common lemon, *Citrus Limon* (*ning mêng*[6]).[e] So far did this recognition go that even fruits of very aberrant character were properly identified. The 'Buddha's fingers' citron (*Citrus medica*, var. *sarcodactylis*), an object which I often met with in Chinese homes as an ornament, is a variety in which the carpels fail to unite to form a rounded fruit, remaining detached as separate fingers, hence the name; it is very fragrant, but inedible unless candied.[f] Indeed this use of it is mentioned in what appears to be the oldest Chinese reference to it, a passage in a book of the Thang period, perhaps +8th century, entitled *Chung Khuei Lu*[7] (Kitchen and Stillroom Records)[g] by an otherwise unknown Mr Wu.[8] But from its first appearance it was unerringly named *fo-shou kan*.[9]

This appreciation of the unity of the Aurantioideae goes back a long way. It appears already in the words of Chang Hua[10], written about +280, where he says:[h]

The number of sorts of trees in the orange-and-pomelo class (*chü yu lei*[11]) is extremely large. There are for example the *kan*[12] and the *chhêng*[13], they all belong to it. The real (i.e. best) ones come from the commandery of Yü-chang (mod. Chiangsi).

And in +300 the first systematic association of *kan*[14] with *chü*[2] appears when Tshui Pao[15], describing a kind of butterfly, says that one finds it 'in the citrous orchards south of the River (*shêng Chiang-nan kan chü yuan chung*[16])', words which indicate

---

[a] Cf. Giles' *Dictionary*, nos. 12764→12753→8128.

[b] From Media, the land of the Medes, of course, not medicine or medium.

[c] One always comes across interesting things while looking for something else. I noticed thus the description of the citron in the late +4th-century *Kuang Chih*, ch. 2, p. 9*a* (*YHSF*, ch. 74, p. 59*a*). In fact this is the oldest literary reference but one to this species in Chinese.

[d] On the lemons and their travels there have been discussions by Laufer (44); Johnson (1) and Glidden (1).

[e] The reservation of the terms *li mêng* and *ning mêng* for the two species mentioned is a quite modern practice; the names (and there are many more) in old Chinese usage were much more fluid. And more than two species may be involved. A quick look would deduce from the double names a probable foreign (Indian or Malayan) origin, but Shao Yao-Nien (1), who discusses the whole question, has brought forward evidence that the lemons may be native to Kuangtung. They still grow wild there. The words may therefore represent an ancient Yüeh tribal name which though southern would not strictly be non-Chinese, rather than a transliteration of a Malayan or Indian one, as usually supposed. Lemons were first discussed in texts of the Sung period, though certainly cultivated in Kuangtung before then, the juice was presented as tribute to the throne in +971, and lemonade sherbet was common under the Mongols by +1299. The modern specific names are associated with Osbeck's voyage of +1751 (cf. Osbeck *et al.* (1), vol. 1, pp. 150, 208, 306, 329).

[f] Cf-Bonavia (1), pls. 139, 140.

[g] Cit. Yeh Ching-Yuan (2), no. 188, pp. 83, 84, from *Shuo Fu*.

[h] *Po Wu Chih*[17], but not in all editions now; we quote following Yeh Ching-Yuan (2), no. 23, p. 19. Tr. auct.

[1] 有柑   [2] 朱欒   [3] 枸櫞   [4] 檸檬   [5] 類
[6] 檸檬   [7] 中饋錄   [8] 吳氏   [9] 佛手柑   [10] 張華
[11] 橘柚類   [12] 甘   [13] 橙   [14] 柑   [15] 崔豹
[16] 生江南柑橘園中   [17] 博物志

again, were it needed, that large-scale production existed in his time.[a] Then about a century later Kuo I-Kung[1] wrote in his *Kuang Chih*[2] (Extensive Records of Remarkable Things):[b]

Of *kan*[3] there are twenty kinds. The (fruits of the) *huang kan* (yellow *kan*) have only one seed (each). The *phing-tai kan*[3] (smooth-peduncle *kan*) of Chhêngtu is as large as a *shêng* (pint measure) with colour verging on the deepest yellow. Nan-an hsien in Chien-wei district[c] is where the *huang kan* come from.

Must this not mean that 'seedless' varieties were already being produced in China before +450? At any rate it is good to cite an ancient reference to the orange growers of Szechuan. This, too, may suffice to show that a wealth of citrus varieties were being cultivated long before Han Yen-Chih took up his brush. But now it is time to hear him in person, and set down some of the words which he wrote.

First let us give his preface:[d]

Very many kinds (*chung*[5]) of oranges (*chü*[6]) are produced in Wên-chün[7] (the commandery of Wênchow in Chekiang). The *kan*[8] is a separate species (*chung*[5]) and separates itself into eight varieties (*chung*[5]). Moreover the *chü*[6] oranges (properly so called) can be divided into fourteen varieties (*chung*[5]). The *chhêng tzu*[9] species (*shu*[10]) belongs to the class (*lei*[11]) of oranges (*chü*[6]) also, and again divides into five varieties (*chung*[5]). In all there are twenty-seven varieties, but the *ju kan*[12] (the orange that tastes sweet like milk) is the best. For this reason the people of Wên (-chou) call it the *chen kan*[13] (true or genuine *kan* orange), the idea being that this is the real *kan* orange and that all others are substitutes.

Chü[6] oranges are also grown in Su-chou (Chiangsu), Thai-chou (Chekiang), in the west at Ching-chou (Hupei),[e] and in the south in several dozen districts of Min (Fukien) and Kuang (Kuangtung and Kuangsi), but they are all *mu chü*[14], 'woody oranges', which cannot dare comparison with the *chü*[6] oranges of Wên (-chou), much less could they dare to compete with the *chen kan*[13] oranges. In Wên (-chou) the *kan*[8] oranges are grown in four districts, but those from Ni-shan[15] are always the best. Ni-shan is a small isolated hill near a place called Phing-yang[16], and looks more or less like a mound of earth with the shape of an upside-down basin. Around the sides of this hill there is a strip of land only two or three *li* wide. Here are no mountain-ranges or valleys of deep shade, and no winds blow in this place, winds which though good in themselves would saturate it with moisture and make it steamingly hot.[f] Only two or three li away the fragrance and flavour of the fruit cannot compare with what grows here.

---

[a] *Ku Chin Chu*[4], ch. 5, tr. auct.
[b] Ch. 2, p. 4a, preserved in *YHSF*, ch. 74, p. 54a, from *TPYL*, ch. 966, pp. 1b, 2a, tr. auct.
[c] We know somebody from there—cf. p. 210 above.
[d] Tr. auct., adjuv. Hagerty (1) with Chiang Khang-Hu.
[e] He was forgetting about the rich orange culture of Szechuan, or perhaps he never knew much of the remote western provinces.
[f] A graphic reference to the monsoon rains of summer.

[1] 郭義恭	[2] 廣志	[3] 平蔕甘	[4] 古今註	[5] 種
[6] 橘	[7] 溫郡	[8] 柑	[9] 橙子	[10] 屬
[11] 類	[12] 乳柑	[13] 眞柑	[14] 木橘	[15] 泥山
[16] 平陽				

How can we investigate the pattern-principles (li^1) of these natural things? Some say that it is because Wên (-chou) is near the sea with a *chhih lu*² soil, salty and alkaline,[a] and that this is good for the *chü* and *kan* oranges. Since Ni-shan is an excellent place with a *chhih lu*² soil especially good, the fruits grow best there, and this is why they are uniquely different from other fruits. I myself do not quite agree with this. Places like Ku-su (Suchow in Chiangsu), Tan-chhiu, the Seven Tribal Regions (of Fukien), and the two Kuang provinces (Kuangtung and Kuangsi) are all near the sea and have *chhih lu*² soil. So why should Wên (-chou) be the only place affected? Furthermore are there not excellent places with *chhih lu*² soil like Ni-shan only two or three *li* away?

During the times of Chhü Yuan³,[b] Ssuma Chhien⁴,[c] Li Hêng⁵,[d] Phan Yo⁶,[e] Wang Hsi-Chih⁷,[f] Hsieh Hui-Lien⁸,[g] and Wei Ying-Wū⁹,[h] authors wrote only of the oranges of Wu (Chiangsu) and Chhu (Hupei), recording nothing of the fruits of Wên (-chou). Wên (-chou) was very late in beginning (the culture of these fruits), but though late in starting, it soon relegated the oranges of all other places to an inferior status. Such transformations of natural things, their rise and fall, prosperity and decline, are vast indeed, almost impossible to be searched out. Although, so far as I can see, during the time of the Chin and Thang dynasties there were no prominent scholars of Wên (-chou) able to compete with the world's learned men, during our present (Sung) dynasty the place began to become rich in culture;[i] and today it may really be considered one which produces literary talents in the greatest abundance. Is it not due to the coming to this land of the *chhi* of Heaven and Earth, brilliant, glorious and imperishable? The overflow of that will affect even ordinary things. Is this not why Ni-shan alone has happened to obtain this beautiful feature in the highest degree?

I am a man from the north, and all my life I have regretted not having seen orange-trees in flower. At various times I bought oranges from market boats but never got any of the fine ones. How could I have had any of the so-called Ni-shan oranges to taste? Last autumn I came here as governor and had the good fortune to see the orange blossoms, and moreover to eat of the fruit. But there is a tradition that the governor may not go far out of the city for pleasure, so I had no excuse to take guests to the fragrant groves of Ni-shan and drink wine at that place. Therefore a friend transported Ni-shan to me and said: 'The delicious quality of the *chü* orange is not less than that of the lichi. Now the lichi has had a monograph written about it, and the *mu-tan* (tree-peony) and *shao-yao* (herbaceous peony) also have their treatises standing side by side—only the orange(-tree) is without one, and yet you are so fond of it. It is as if the orange had been waiting for you, so you really cannot decline.' For this reason, then, I have written this monograph, recklessly following in the footsteps of Ouyang Hsiu¹⁰ and Tshai Hsiang¹¹.[j] Besides, I also wrote it to

---

[a] Here Hagerty translated 'rich in nitre', but this was perhaps unnecessarily medieval. From pp. 80–1 ff. above were are in a position to identify fairly clearly what was meant by *chhih lu* soil—it was the saline-alkaline soil of littoral solonchak type not uncommon in the eastern provinces (see Fig. 21). Chinese agriculturalists in the Sung would have recognised it clearly.

[b] Minister and poet of the 'Odes of Chhu', d. −288.

[c] The great historian, d. −90.   [d] Poet of the Later Han, *fl.* +2nd century.

[e] San Kuo and Chin poet, *fl.* +275.   [f] The great calligrapher, +321 to +379.

[g] Chin and L/Sung poet, +397 to +433.   [h] Thang writer, *c.* +740 to +830.

[i] This reflects the half century of permeation of the south by the literati since the fall of Khaifêng in +1126.

[j] Writers of the tree-peony and lichi monographs; see respectively, pp. 401 ff. below and 361.

---

¹ 理   ² 斥鹵   ³ 屈原   ⁴ 司馬遷   ⁵ 李衡
⁶ 潘岳   ⁷ 王羲之   ⁸ 謝惠連   ⁹ 韋應物   ¹⁰ 歐陽修
¹¹ 蔡襄

show the world that the scholars of Wên (-chou) are worthy to be talked of, and that she is not famous alone for her oranges.

> Given in the 10th month of the 5th year of the Shun-Hsi reign-period (+1178) by Han Yen-Chih of Yenan (Shensi).

Two things here need remark. First, in the opening paragraph, although Han made a loose use of the terms which he chose to express classification, he was at any rate very conscious of the need for ranks of this kind. Secondly, although he preferred a rather vague cosmic theory as the explanation of the excellence of Ni-shan,[a] his account of the materialist reasoning in terms of pedological and edaphic conditions is most interesting because it shows the sort of scientific thought that was going on at the time.

What did he say about the prize 'genuine oranges'?

The true (or perfect) orange (*kan*) is the most valuable among the various sorts and grades of citrous fruits, and is greatly to be admired. The branches, trunk, flowers and fruit are all different from other, ordinary, kinds. The tree is graceful, having elongated leaves and dense foliage, covering the ground with dark shade. In flowering time it is particularly full of poetical purity and elegant remoteness. The fruits are all quite spherical, the texture of the skin being shiny as wax.

'On a morning after early frost when the gardener picks the fruit and presents it (to his master) its beauty impresses everyone. When it is opened, a fragrant mist enchants people.'[b]

Normally northerners never meet with this fruit but as soon as they see it once they recognise that it is the true (or perfect) orange. It is also called the *ju kan* or 'cream orange' because of the resemblance of its (smooth) taste to that of curds.[c] All the four districts of Wên (-chou) grow it, but the produce of Ni-shan is the best. Within one *li* of Ni-shan the oranges measure nearly 7 in. in girth.

'They are thin-rinded and with particularly delicious flavour. The fibres do not adhere to the segment walls, and in eating (the pulp) no refuse is left in the mouth.'[a]

A fruit may contain only one or two seeds, and many have none at all. In recent years the oranges growing on the southern slopes have become more abundant.[d]

And he goes on to quote the lines of some poems written by guests on the occasion of an annual citrus banquet given each autumn by the governor of Wênchow. Here we have a graphic description of a semi-tropical mandarin orange, most

[a] In its embracing of the affairs of Man and Nature as a single unity, this was distinctly Neo-Confucian. Cf. Sect. 16.

[b] The sentences within inverted commas were, as Pelliot recognised, borrowed by Han Yen-Chih from a 'Letter accompanying a Present of Oranges' written by Liu Hsün[1] (+462 to +521). He worked them in, as they were appropriate, just as we might use a few words of Sir Thomas Browne, knowing that all educated readers would respond to the echo. Cf. *CSHK*, Liang sect., ch. 57, p. 1a.

[c] 'Cream' for *ju lao*[2] here, *pace* Hagerty and Tanaka, is not really a good equivalent in the light of all we know of Chinese diet. 'Junket' or 'yoghurt' are nearer, remembering also that curds were habitually made from soya-beans, not necessarily milk at all. Cf. p. 336 above, and Sect. 40 below.

[d] Tr. auct. adjuv. Tanaka (4); Hagerty (1).

[1] 劉峻   [2] 乳酪

varieties of which tend to be almost seedless. Han Yen-Chih's words on the cumquat should also be read:[a]

The *chin kan*[1] is the smallest of the *kan* class. The larger of these fruits are the size of a coin, while the smaller are as little as a *lung mu*[2] fruit.[b] The *chin kan* fruit has a golden colour, a very fine-grained skin, and a spherical form; with its ruddiness it is agreeable and enjoyable to handle. One eats it without peeling off its golden coat. When preserved in honey the flavour is even better. Its pure aroma and delicate taste was described by Ouyang Hsiu in his *Kuei Thien Lu*[3] (On Returning Home).[c] When the fruits are placed in sacrificial dishes or at table they glow and glitter like golden balls (lit. crossbow bullets). They are really precious. Formerly the people of the capital did not value them very highly, but after the Empress Wên Chhêng[d] became very fond of the fruit the price there went up considerably.

This was certainly one of the species of *Fortunella*.

It is hard to avoid the temptation of trying to identify in modern terms the 28 genera, species, sub-species and varieties described by Han Yen-Chih in his book. If, then, we have drawn up the list in Table 12, it is presented with the most extreme reservations, and rather to give an idea of the wealth of varieties that were already being cultivated in his time than with any hope of botanical exactitude.[e] There are indeed some few certainties, as indicated in the left of the three columns of identifications, but in spite of further probabilities much guesswork remains; and one should remember Tanaka's warning (4) that it would be unwarrantable to identify any sort of orange today with a kind grown over 750 years ago. In the intervening centuries there will have been great variations in the cultigens due to selection, climatic and similar changes, while on the other hand not all of Han Yen-Chih's medieval names have persisted with clear meanings until now. Moreover, the taxonomy of the Aurantioideae and the *Citrus* genus in particular is peculiarly difficult, so much so as almost to call in question the applicability of the species concept. All *Citrus* species can hybridise easily and with astonishing variability, while apparently normal seeds will reproduce the parent hybrid exactly because the 'false embryos' readily become invaded by maternal tissue. Hence the field has become a real playing-field, where radical 'splitters' contend against conservative 'lumpers',[f] the former enlarging the number of admitted species, the latter trying to keep them down. Far be it from us to referee.

---

[a] Tr. auct. adjuv. Hagerty (1).
[b] *Euphoria longana* (Sapindaceae), cf. R 302; Chhen Jung (1), p. 683.
[c] A collection of incidents at the imperial court, written in +1067.
[d] Empress of Jen Tsung, *c.* +1025 to +1060.
[e] Besides the general reviews noted below, we have consulted Chia & Chia (1), nos. 896 to 918; Yeh Ching-Yuan (1, 2); Chhen Jung (1), pp. 564 ff.; Burkill (1), vol. 1, pp. 566 ff.; Watt (1), pp. 317 ff.; Bonavia (1).
[f] See Tanaka Tyōzaburō (1) on the history of the disputes. We can only refer the reader to some of the leading accounts of the systematics of *Citrus*, e.g. Swingle (13), and (14) in the great treatise of Webber & Batchelor, who tended to be conservative, and Tanaka himself (1, 2, 3, 2) who tended to be radical. Here we have followed appreciatively the work of Hu Chhang-Chhih (1, 2, 1), who was associated with Tanaka.

[1] 金柑   [2] 龍目   [3] 歸田錄

Table 12. *The Aurantioideae in Han Yen-Chih's Chü Lu*

		Identification (genus and species)	Alternative suggestions	Further possible alternatives (present-day species, sub-species or var.)
I 1 *chen kan* = *ju kan*	眞柑 乳柑	*C. reticulata* = *nobilis*	*C. poonensis* *C. kinokuni* *C. suavissima*	*C. ret.* var. *subcompressa*
2 *shêng chih kan* = *hu kan*	生枝柑 壺柑			*C. succosa* *C. sinensis*, var.
3 *hai hung kan*	海紅柑			*C. poonensis*
4 *tung-thing kan*	洞庭柑			*C. erythrosa*, var.
5 *chu kan*	朱柑			*C. tangerina*, var.
6 *chin kan*	金柑	*Fortunella crassifolia* or *hindsii*	*F. margarita* or *obovata*	*C. microcarpa*
7 *mu kan*	木柑			*C. poonensis*
8 *thien kan*	甜柑	*C. sinensis*	*C. ponki*	
9 *chhêng tzu*	橙子	*C. junos*	*C. Aurantium* var. *junos*	
II (10) *yu*	柚	*C. grandis* = *maxima*		
11 *huang chü*	黃橘		*C. ponki*	*C. nobilis* var. *deliciosa* *C. tangerina*, var.
12 *tha chü*	塌橘			*C. nobilis* var. *deliciosa* *C. tangerina*, var.
13 *pao chü*	包橘	*C. reticulata* = *nobilis*		
14 *mien chü*	綿橘			
15 *sha chü*	沙橘			*C. unshiu* *C. nobilis* var. *unshiu*
16 *li-chih chü*	荔枝橘			
17 *juan thiao chhuan chü*	軟條穿橘			*C. erythrosa* var.
18 *yu chü*	油橘	doubtful whether a *Citrus*, perhaps another of the Aurantioideae		
19 *lu chü*	綠橘			*C. retusa*
20 *ju chü*	乳橘		*C. sunki*	*C. kinokuni*
21 *chin chü*	金橘	*C. mitis*	*Fortunella hindsii*	*C. microcarpa*
22 *tzu-jan chü*	自然橘	wild *Citrus*		*C. ichangensis*?
23 *tsao huang chü*	早黃橘		early ripening variety of *C. sinensis*	
24 *tung chü*	凍橘		*C. sinensis* var. *sekkan*	
25 *chu luan*	朱欒	*C. Aurantium*		
26 *hsiang luan*	香欒		*C. Aurantium*, var.	
27 *hsiang yuan*	香圓	*C. hsiangyuan* = *hybrida* = *wilsonii*		
28 *kou chü* = *chih kho*	枸橘 枳殼	*Poncirus trifoliata*		

NOTE: The number in brackets indicates a mention without a special heading.

The varietal names in Table 12 can be rendered as follows:

1   true (or perfect) *kan* orange (milk-sweet *kan* orange)
2   fresh branch *kan* orange (round pot (-shaped) *kan* orange)
3   littoral red orange
4   variety of *kan* orange grown on Tung-thing Island in Lake Thai-hu (Chiangsu)
5   vermilion *kan* orange
6   golden *kan* orange (the cumquat)
7   woody *kan* orange
8   sweet *kan* orange
9   *chhêng* orange (the sour or Seville orange)
(10) *yu* fruit (the pommelo)
11 yellow *chü* orange
12 flattened *chü* orange
13 clustered *chü* orange
14 soft *chü* orange
15 sandy-soil (or fine) *chü* orange
16 lichi *chü* orange
17 weak-branched hollow-cored *chü* orange (also called maiden's orange)
18 oily *chü* 'orange' (with glossy black skin like a haw; probably of some related rutaceous genus)
19 green *chü* orange
20 milk (-sweet) *chü* orange
21 golden *chü* orange
22 wild (or natural) *chü* orange
23 early yellow *chü* orange
24 ice (winter-fruiting) *chü* orange
25 vermilion *luan* orange
26 fragrant *luan* orange
27 fragrant globe orange
28 spiny *chü* orange (or *chih* rind fruit).

Little comment on the Table is needed except to point out that if we are not misled by a faulty synonymy the existence of the *hsiang yuan* orange (no. 27) demonstrates that hybridisation was being practised (or at least taken full advantage of) long before the time of Han Yen-Chih. For it is known today that *Citrus hsiangyuan* (= *hybrida* = *wilsonii*) is a hybrid of *C. grandis*, the pomelo, with *C. ichangensis*, the most cold-resistant of the group.[a] Resistance to cold, however, was obviously also a feature of other varieties (nos. 23, 24), and every one had some special character such as aroma, oil-abundance, sweetness, use in grafting, etc. which Han Yen-Chih noted.

Now we have only to fit the last bricks into the background niche where stands our image of the good Governor of Wên-chow. In the second half of the + 12th century the natural history scene in Europe was singularly uninspiring; apart

[a] This recalls the much better known hybrid species, *C. paradisi*, the grapefruit. The origin of this is still very obscure. It is not mentioned before the latter half of the + 18th century, when it seems to have arisen in the West Indies. The pomelo was certainly one of its parents.

from a number of treatises on falconry there was only the encyclopaedia of St Hildegard of Bingen (d. +1179) who found she could name and describe about a thousand plants and animals.[a] As usual the Muslim world was more advanced, and we still have a special *Tractate on Lemons* written by the Cairene Jew Khibat-Allāh Ibn Zayn Ibn Jamī (*fl.* +1171 to +1193), the personal physician of Ṣalaḥ al-Dīn (our Saladin),[b] preserved because it was incorporated in the great pharmaceutical natural history of Ibn al-Baythār (son of the veterinary surgeon) al-Mālaqī (+1197 to +1248).[c] From his *Kitāb al-Jāmi' fī al-Adwiya al-Mufrada* (Book of the Whole Gamut of Simple Remedies)[d] Khibat-Allāh's tractate was subsequently extracted, translated into Latin by Andreas Alpagus early in the +16th century, and printed three times by +1758. But it was purely medical-nutritional, however, not botanical-horticultural. Nothing at all like Han Yen-Chih's work was paralleled until about +1500, over three centuries later, when Joh. Jovianus Pontanus published a Latin poem *De Hortis Hesperidum* in nostalgic memory of the orange orchards which he and his wife had looked after in their youth.[e] By +1646 European citrology graduated in the work of a Sienese Jesuit, J. B. Ferrari, whose *Hesperides, sive de Malorum Aureorum Cultura et Usu* contained, besides legend and history, the description of some 1000 varieties. But this was already in the world of nascent modern science.

(ii) *Bamboos*

From the Aurantioideae we pass to an even greater sub-family, the Bambuseae (or Bambusoideae)[f], with its 4000 or more species in some 320 genera (40 % and over 50 % respectively of all the grasses). The group is one of great biological interest as the bamboos are the most primitive of living Gramineae.[g] Nothing could be more deeply characteristic of the Chinese scene than they, or more prominent in Chinese art and technology through the ages.[h] Very natural was it therefore that the oldest of all the botanical monographs was devoted to this group, and that the tradition of botanical and quasi-botanical writing was much

---

[a] See Sarton (1), vol. 2, p. 386, and Singer (3, 16).
[b] Mieli (1), p. 163; Tolkowsky (1), p. 132.
[c] Perhaps the greatest son of Malaga; cf. Mieli (1), p. 212.
[d] Tr. Leclerc (1).
[e] Tolkowsky (1), p. 186. At a previous point (p. 104) we gave evidence of the economic richness of citrous horticulture in China almost two thousand years earlier, in the Chhin and Former Han periods (−3rd century onwards). We mentioned there the term 'wooden slaves' because the groves of orange trees would make and keep a family wealthy with relatively little trouble. The originator of the phrase seems to have been Li Hêng[1], Prefect of Tan-yang about +260, who commended his plantations to his sons with this reflection (*Shui Ching Chu*, ch. 37, p. 18a). Other uses of the term will be found in Morohashi's *Dictionary*, vol. 6, p. 10.3.
[f] The name itself is of course Asian, but its origin is obscure, possibly from Canarese; cf. Yule & Burnell (1), *sub voce*.
[g] As a companion to what follows one cannot do better than study the handbook of McClure (1).
[h] Here worth reading are the articles of Kêng Po-Chieh (1); Li Hui-Lin (5); McClure (2). Cf. Vol. 4, pt 2, pp. 61 ff., 64; pt 3, pp. 102, 191, 328, 391, 597, 664.

[1] 李衡

longer and more continuous than for any other. Seven hundred years before Han Yen-Chih satisfied the expectant oranges of Chekiang, another governor, Tai Khai-Chih[1] in the Liu Sung dynasty, wrote the first book (probably the first in any civilisation) consecrated to the bamboos. This work, which can be dated in the neighbourhood of +460, was entitled *Chu Phu*[2] (Treatise on Bamboos), and a careful look at it in a moment will be well worth while. A bird's-eye view of the main literary landmarks after it would take the following form. Amplifications were probably produced in and before the Thang, though all have been lost, including unfortunately two with the same title as Tai's written in the Sung (+10th and +11th centuries).[a] The *Sun Phu*[3] (Treatise on Bamboo Shoots), due to the scientific monk Tsan-Ning[4] about +970, has survived.[b] Just after the end of the Sung there was the large treatise on bamboos by Li Khan[5] (+1299) again with the same title as Tai's, the writer in principle a painter but (as we have noted already, p. 357) one who still retails for us keen botanical and physiological information, with many interesting technical terms. A little later in the Yuan dynasty came Liu Mei-Chih[6] with his *Hsü Chu Phu*[7], an important enlargement on the subject. After that there were many tractates and publications, among which one need only mention two, the *Hua Chu Wu Phu*[10] (Five Treatises on Flowers and Bamboos) by the Ming recluse Kao Lien[11], and the *Chu Phu* of Chhen Ting[12], a many-sided and much-travelled scholar of the Khang-Hsi reign-period (late +17th century), who wrote on the strange bamboos of the southeastern provinces, Kweichow and Yunnan.

Tai Khai-Chih (c. +420 to c. +485) was a high official and military commander stationed (during the most dramatic part of his life) at Ganhsien in Chiangsi. The Gan River flows northward there from the upper valleys of the 'Five Ranges', the Nan Ling, that divide Kuangtung from the north, valleys richly green with all kinds of bamboos, places very suitable for Tai Khai-Chih's observations.[c] His book was at one time more extensive than it is now, for he is said to have described over 70 species though we count only 47 or so. This was good going for a particular sub-family in the +5th century, especially as he had to deal with plants now placed in a variety of different genera.[d] He wrote his treatise in the form of a poem with four characters to the line rhyming like a rhapsodic ode, and after every few lines inserted a brief prose commentary. This

---

[a] One was by one of the 'high monks' of the early Sung, Hui-Chhung[8], the other by a relatively unknown scholar, Wu Fu[9].

[b] We have encountered him already in Vol. 4, pt 1, p. 77. He was the real writer of the *Wu Lei Hsiang Kan Chih* and the *Ko Wu Tshu Than*, not Su Tung-Pho; corr. Vol. 4, pt 1, pp. 276, 277, 349, 359. Cf. Vol. 5, pt 2, pp. 208, 310, 314–5, pt 4, pp. 149, 199 ff.

[c] He also travelled in Chiao-chou (now Vietnam).

[d] The *Hua Ching* of +1688 only dealt with 39, the *San Tshai Thu Hui* (+1609) with 61. Satow's Japanese source had 51 but half were doubtfully distinct as species.

[1] 戴凱之   [2] 竹譜   [3] 筍譜   [4] 贊寧   [5] 李衎
[6] 劉美之   [7] 續竹譜   [8] 惠崇   [9] 吳輔   [10] 花竹五譜
[11] 高濂   [12] 陳鼎

has often been called the *gāthā* style, under the impression that it originated with the Indian Buddhist scriptures, but in fact Taoist writings which antedate them already used it.[a] The *Chu Phu* opens with an interesting passage on classification, then touches on oecology and physiology, after which it goes on to enumerate bamboo after bamboo, generally giving clear distinguishing characteristics and not forgetting the economic uses of each. Let us select some of its stanzas.[b]

It starts off almost with a riddle.

(1)   Within the vegetable kingdom (*chih lei*[1])
       There is a thing called bamboo (*chu*[2])
       It is neither hard nor soft
       It is neither herb nor tree.

The *Shan Hai Ching* and the *Erh Ya* both speak of the bamboo as a herbaceous plant (*tshao*[3]). Since this parlance was used by the sages and worthies of old it has not been changed. However, calling bamboos herbaceous plants leads to great difficulties. To begin with, the sorts of forms (*hsing lei*[4]) of bamboos are extraordinarily diverse. Then the explanations in the text of the (*Shan Hai*) *Ching* are self-contradictory. It says itself that 'of these plants there are many tribes (*tsu*[5])'. It also says: 'among the bamboos there are many *mei*[6]'.[c] And again: 'On Yün-shan (mountain) there are *kuei chu*[7] (cassia bamboos)'. But if these *chu*[2] are really *tshao*[3] it does not answer to call them *chu*[2]. Actually, since their designation is *chu*[2], that they are not *tshao*[3] may be said to be obvious. *Chu*[2] is a general term for one tribe or class (*tsu*[5]) and a differentiating name for (plants sharing) one particular form. In the vegetable kingdom there are *tshao*[3] (herbaceous plants), *mu*[8] (trees) and *chu*[2] (bamboos) just as in the animal kingdom there are *yü*[9] (fishes), *niao*[10] (birds) and *shou*[11] (mammals). Doubts such as these arise today partly because of the long lapse of time (which separates us from the sages) and partly because of mistakes in the texts that have been handed down. Perhaps it is due not so much to the transgressions of the worthies of old as to the timidity of later scholars who did not dare to discriminate and rectify the ideas in ancient writings. How does this differ from the dread of the very name of Chih Tu[12] which the Huns used to have, and their fear of even the substance of his image?[d]

(2)   That some are hollow and some solid
       Is a difference one may call slight
       That all have culm nodes (*chieh*[13]) and branch buds (*mu*[14])
       Is a resemblance one may call great.

The great majority of bamboos have hollow culms, but now and then about one in ten

---

[a] Cf. Waley (26), p. 159.

[b] Tr. auct. adjuv, Hagerty (2). Cf. Wang Yü-Hu (*1*), 2nd ed., pp. 24 ff. We shall tentatively accept, or suggest, various identifications with modern genera and species as we go along, but the medieval Chinese bamboo literature ought to be elucidated by a professional botanist having at hand (as we have not had) the modern systematic keys and flora of Kêng I-Li (*1, 2*) and Kêng Po-Chieh (*1*), etc.

[c] See p. 385 below.

[d] Chih Tu was an upright but unamiable official and commander (*fl.* −156 to −141), the 'sea-green incorruptible' of the Former Han. The whole story is given in *Shih Chi*, ch. 122, pp. 1*b* ff., tr. Watson (*1*), vol. 2, pp. 420, 422.

[1] 植類   [2] 竹   [3] 草   [4] 形類   [5] 族
[6] 簹   [7] 桂竹   [8] 木   [9] 魚   [10] 鳥
[11] 獸   [12] 邪都   [13] 節   [14] 目

will be found with solid centres.[a] Therefore they are said to be slightly different. But though of course hollowness and solidity are divergent characteristics, there is no bamboo stem which does not have nodes, and that may be said to be a great resemblance.

(3)   Some flourish in sand near the water
      While others thrive on cliffs and uplands.

The *thao-chih*[1] (peach branch bamboo) and the *yün-tang*[2] bamboos are frequently planted on islets. But the *huang*[3] and *hsiao*[4] bamboos have to be grown in high places which are dry.

(5)   The shoots are called *sun*[5] and the sheaths *tho*[6]
      In summer they are numerous and in spring few.
      When the roots and culms are about to rot
      Blossoms and seeds (*fu*[7]) then appear.

When the bamboos produce flowers and seeds, in that year they decay and die.[b] The character for the seeds is pronounced *fu*.

(6)   After sixty years the plants will be dead (*chou*[8])
      And in six more they will have risen again.

Once in sixty years the bamboos 'change their roots' (*i kên*[9]). When they change their roots they suddenly materialise their (flowers and) fruits, then wither and die.[a] When the seeds fall to the ground they grow again and in six years (new plants have) matured and cover a piece of land. The character for a dead bamboo is pronounced *chou*.

(9)   *Kuei*[10] really denotes a tribe (*tsu*[11]) of bamboos
      Having the same name but different origins.

Of the *kuei chu* (cassia bamboos) which grow forty to fifty feet tall, the thick ones have a circumference of two feet, nodes spread wide apart, large leaves shaped like those of the *kan chu*[12] (sweet bamboo) and a red cortex. South of Nan-khang (Ganhsien) they grow in abundance. The *Shan Hai Ching* says that if a person is wounded by a *kuei* bamboo from Ling-yuan he will die.[c] Actually the *kuei*[10] bamboos are of two kinds (*chung*[13]) their names being the same but their substances different. Their forms have not been described in detail.[d]

(11)   The *huang*[14] bamboos are good for poles and flutes
      As the culms are especially hard and round.

[a] Among species exemplifying this one could name *Dendrocalamus strictus*, *Arundinaria prainii* and *Oxytenanthera stocksii*. See Watt (1), p. 105.

[b] This and the next stanza speak of the reproductive phase of the bamboos. Things are of course much more complicated than Tai could say; the duration of the vegetative state and the incidence of flowering and fruiting varies greatly from one species to another. Some bamboos persist almost indefinitely in the former condition, others manifest a continual or annual tendency to flower; but it remains true that the majority of species show a cyclic recurrence which may be anything up to 60 years or even more, at the end of which time all plants of the same generation flower and seed gregariously, and all may also die within a year or so afterwards. The only species which has been observed experimentally under controlled conditions is *Guadua trinii* where the elapsed time from seed to seed was exactly 30 years with common death of the parent plants after two flowerings. See further McClure (1), pp. 82 ff., 275.

[c] This is certainly a reference to the poisonous properties of the appressed hairs carried by the sheaths of many bamboo species, *Bambusa vulgaris* for example, hindering their industrial use in modern times for paper-making. Cf. Burkill (1), vol. 1, pp. 295, 301; Watt (1), p. 109.

[d] This group of bamboos has not so far been identified as the name failed to survive in either learned or common use.

[1] 桃枝	[2] 箮箄	[3] 筐	[4] 篠	[5] 筍
[6] 籜	[7] 蕧	[8] 筹	[9] 易根	[10] 桂
[11] 族	[12] 甘竹	[13] 種	[14] 箽	

Huang (or thicket-) bamboos are hard and have short internodes. Their culms are round and of hard texture, with a cortex as white as frost-powder (purified calomel or white lead). The thick culms are suitable for punt-poles while the slender ones can be used for making flutes.[a] The character for these bamboos is pronounced *huang*.[b]

(12)   The *chi chu*[1] roots intertwine and grow deeply,
        One clump making a whole grove.
        Their rhizomes are like pestles or fat spokes
        Spreading out from the hub of a wheel;
        Their nodes are like a sheaf of needles.[c]
        They are otherwise known as *pa chu*[2]
        And make a sure defence for city walls.
        But if you eat their tender shoots
        The hair on your crown and temples will fall out.[d]

The *chi* (thorny) bamboos grow in all the prefectures of Chiao-chou. Early in its growth a clump will have several dozen clustered culms (*ching*[3]). The largest ones measure two ft in circumference and are rather thick-fleshed with nearly solid centres. The I[4] (tribal) people split and use them to make (cross-) bows. Both branches and nodes have thorns, so they plant them in order to make stockades which soldiers cannot get through. This is what Wan Chen[5] is talking about in his (*Nan Chou*) *I Wu Chih*[6] when he speaks of planted hedges which are more effective than walls several storeys high.[e] Sometimes when these bamboos die and topple over, the exposed rhizomes are as large as 10-*tan* weights.[f] The roots are criss-crossed and intertwined looking like silk-reeling machines (*sao chhê*[7]).[g] An alternative name is *pa chu* (hedge bamboos); see the *San Tshang*[8] dictionary.[h] Eating the shoots causes loss of hair.

---

[a] More correctly, the vertical pipe. See Vol. 4, pt 1, pp. 145, 165.

[b] This bamboo has been identified by Hagerty (2) as *Dendrocalamus* (= *Sinocalamus*?) *latiflorus*, CC 2078; Chhen Jung (*1*), p. 86.

[c] These are the nodal thorns characteristic of such species as *Bambusa stenostachya*, with which this bamboo has been identified by Hagerty. The description given in Chhen Jung (*1*), p. 85 agrees very well.

[d] This may have been just a superstition (though it is echoed elsewhere), but in fact parts of some bamboos other than the sheath hairs have been suspect of pharmacological activities. For example the leaves of *Dendrocalamus sikkimensis* are considered poisonous for domestic animals (Watt (*1*), p. 102) and *Bambusa multiplex* in Malaya seems to have some abortifacient principle in its shoots (Burkill (*1*), vol. 1, p. 299).

[e] *Strange Things of the South*, an important book though extant now only in quotations. *TPYL*, ch. 963, p. 6*a* simply quotes the present passage.

[f] The *tan* or picul was 120 (less often 133) *chin* or catties (equivalent approximately to pounds), so it was rather more than 1 cwt, hence Tai's mass of roots in the clump weighed about half a ton.

[g] For this see Fig. 409 in Vol. 4, pt 2, and the discussions on pp. 107, 382, 404. Tai Khai-Chih probably had in mind the wide-spreading spokes and parallel rods of the main reel on to which the silk is wound. His reference indeed is not without significance for the dating of this important element in textile engineering, cf. Vol. 4, pt 2, p. 269.

[h] This links us with the discussion on the origin of the early dictionaries from the ancient orthographic word-lists given on pp. 194 ff. above. The *San Tshang* was a conflation of the *Tshang Chieh* (*Phien*) of *c.* −220, the *Hsün Tsuan* (*Phien*) of *c.* +6 and the *Phang Hsi* (*Phien*) of *c.* +100 made either by the first editor Chang I[9] *c.* +230 or during the previous century. What we have of it now, reconstructed in *YHSF*, ch. 60, pp. 13*a* ff., lacks all entries for bamboos, unfortunately. The *San Tshang* is a true dictionary, not only an orthographic list, but the definitions and explanations may well be due to the +3rd-century editors.

---

[1] 棘竹	[2] 笆竹	[3] 莖	[4] 夷	[5] 萬震
[6] 南州異物志	[7] 繰車	[8] 三蒼	[9] 張揖	

(14)   The *khu chu*[1] are called very rightly by this name
       And the *kan chu*[2] also is no misnomer.

The *khu* (bitter) bamboos have white sorts and dark red sorts, and they have indeed a bitter taste.[a] The *kan* (sweet) bamboos are similar to the *huang* (thicket-) bamboos and have an abundant foliage.[b] Underneath the nodes there is (a substance of) sweet taste,[c] which people use to put in soups. This kind is to be had everywhere.

(15)   The *kung chu*[4] are like rattans or vines
       Their noded (culms), renouncing straightness,
       Deviously wind about in curves,
       Along the ground they grow profusely
       But rise up aloft when they find a tree.
       They will grow to a length of a hundred fathoms[d]
       Looking as if they would never stop.
       Within their texture there are streaky patterns
       But greasing is needed to bring them out.

The *kung* (curved-like-bows) bamboos grow on all the mountains of the eastern borders. They reach a length of several tens of *chang*[a] curving by abrupt changes of direction at each node. Since they are both long and soft they cannot stand erect by themselves but if they come in contact with a tree they will rely on it (and climb). There are streaks in their substance, but if one wants to bring them out one must rub in some grease and then heat over a fire—thus they will appear. Slats for bamboo sleeping-couches come from this kind.[e]

(17)   There are *yün-tang*[5] bamboos and *shê-thung*[6] bamboos
       The *lin-yü*[7] sort and the *thao-chih*[8] sort;
       They all have long narrow glossy leaves
       And unspotted culms with a thin cortex;
       Hundreds upon thousands grow in confusion
       Yet thickness and thinness is a constant difference.

There are several bamboos which resemble one another in cortex and leaf. The *yün-tang* are the largest, and their thickest pieces are used to make food steamers (*tsêng*[9]);[f] the shoots can also be utilised. The *shê-thung* (quiver, slur-bow or blow-pipe bamboos)[g] are

---

[a] Even in Tai Khai-Chih's time this was probably considered a group rather than a single species, for Li Khan (see below, p. 387) names 22 kinds. Satow (1) identified the name with *Phyllostachys quilioi*, now called *P. bambusoides* (CC 2080), and this may be taken as the type of what Tai was speaking of. It happens to be one of those with the 60-year flowering cycle.

[b] It is thought that the *kan* group of bamboos is the same as the *tan chu*[3] (insipid bamboos), the type of which was identified by Hagerty and Satow as *Phyllostachys henonis*, or *nigra*, var. *henonis*. Another name for this is *P. puberula* (CC 2083; Chhen Jung (1), p. 80; Li Shun-Chhing (1), p. 126).

[c] Many bamboos produce a white exudate on the surface of the internodes ranging from a mere bloom to a fluffy flour-like deposit. It contains carbohydrates as well as waxes and polycyclic compounds related to steroid hormones.

[d] Since a fathom (*hsün*) is 8 ft and a *chang* is 10 ft the first estimate would be 800 ft and the second 2–400 ft. The greatest length observed today is that of *Dinochloa andamanica* which reaches 270 ft (McClure (1), p. 283). Closely related is the *D. scandens* of South-east Asia (Burkill (1), vol. 1, p. 811). We have not found a climbing bamboo listed in the modern Chinese floras, but that is probably only a measure of their inadequacy. See however, Anon. (*109*), vol. 5, p. 788, and esp. Chhen Huan-Yung et al. (*1*), vol. 4, pp. 362–3. Some bamboos twine by the circum-nutation of growing shoots, and others have special morphological adaptations for climbing. Cf. Corner (1), p. 15, and a luminous chapter on the rattans, pp. 201 ff.

[e] The translation of this sentence follows Hagerty (2) but it seems to us to be corrupt.

[f] Cf. Vol. 1, p. 82 and Sect. 40 below.   [g] Cf. Sect. 30.

[1] 苦竹   [2] 甘竹   [3] 淡竹   [4] 弓竹   [5] 籄籝
[6] 射筒   [7] 篍筊   [8] 桃枝   [9] 甑

thin-walled and have the longest internodes. Arrows can be stored and carried within them, hence the name. The *lin-yü* (broad-leaved) bamboos have leaves thin and broad; these are the bamboos that the women of Yüeh used for testing swords.[a] The *thao-chih* (peach-branch) bamboos are the most slender of this class—see the local gazetteers and the rhapsodies that people have written about them. They have a red cortex which is smooth, strong, and may be used in making matting.[b] The bamboos mentioned in the Ku Ming Phien (chapter of the *Shu Ching*)[c] were these.

In its Shih Tshao section, the *Erh Ya* says that 'those with four-inch internodes are *thao-chih* bamboos'. Kuo (Pho) repeats this in his commentary. However, among the *thao-chih* bamboos that I myself have seen, the shorter had internodes of less than one inch while some of the longer had internodes measuring over one foot. As they have these everywhere in Yü-chang (Chiangsi) the proof is not far away. I fear that what was listed in the *Erh Ya*'s plant kingdom section was some other plant called *thao-chih*, not necessarily a bamboo at all. Kuo (Pho) added the word *chu* (bamboo) in his commentary, thereby falling into error.[d] The *Shan Hai Ching* says:

'Among trees there are the *thao-chih* and the *chien-tuan*[3].' Then in its Tshao Mu section the *Kuang Chih* says: 'The *thao-chih* come from Chu-thi Chün (in Szechuan), and these are what Tshao Shuang[4] made use of.'[e] Careful examination of its characteristics shows that it closely resembles a tree, but as we have no details about the plant listed in the *Erh Ya* we cannot tell whether it was the same as this other one. What the *(Shan Hai) Ching* and the *(Erh) Ya* mentioned were two sorts of plants which could certainly not be made into mats. The *Kuang Chih* regarded *tsao*[5] (pond-weeds) as bamboos but this was a mistake, though frequently perpetuated by scholars since then.

(23)   Of the bamboos fit for staffs
    None are better than the *chhiung*[6],
    With protuberances that are quite unusual
    They are fashioned as if by human skill.
    How can they be said to be truly Szechuanese
    When they are produced in other places also?
    In one source they have been called *fu-lao*[7]
    The names being different but the thing the same.

[a] The reference here is a curious one, explained by *Wu Yüeh Chhun Chhiu*, ch. 9 (King Kou Chien, 13th year), pp. 24b, 25a; and Li Shan's commentary on the *Wu Tu Fu* of Tso Ssu, in *Wên Hsüan*, ch. 5, pp. 5b, 6a (the fuller text). In −483 the King of Yüeh was enquiring about swordsmanship, and there was talk of certain women from the southern jungles of the State who were great experts in the art. One of them had fallen in, when travelling, with an old man who introduced himself as Master Yuan (Yuan Kung[1]) and said he would like to test her skill, so he tied back to the ground a clump of *lin-yü* bamboos, and when they were released she cut them through before they had time to stand up—a test involving probably not only speed but an edge that could manage the silica of the skin. When she looked round, however, there was only a white monkey (*pai yuan*[2]) sitting on the branch of a tree. Thanks are due to Dr Michael Loewe for helping to unravel this allusion.

[b] Hagerty could not identify any of the four species named in this stanza. Possibly the *yün-tang* could have been *Bambusa verticillata* (Li Shun-chhing (1), p. 132), and *Arundaria nitida* might agree with the *shê-thung* (Li (1), p. 121). For *thao-chih* one might propose *A. densiflora* (Li (1), p. 124) or *marmorea* (CC 2071), but the whole work deserves much closer attention from specialists in the Bambuseae.

[c] Ch. 42, see Karlgren (12) p. 71; Medhurst (1), p. 298; Legge (1), p. 238.

[d] This is not in the *Erh Ya* now (ch. 13, p. 6b). On the contrary Kuo Pho says the internodes are often a lot longer than four inches.

[e] This is not in what we now have of the *Kuang Chih* (*YHSF*, ch. 74). On Tshao Shuang cf. Vol. 4, pt 2, p. 42.

[1] 袁公   [2] 白猿   [3] 劍端   [4] 曹爽   [5] 藻
[6] 節   [7] 扶老

The *chhiung* bamboos have prominent bulging nodes and solid centres, looking as if they had been carved by man; they make much the best staffs.[a] The *Kuang Chih* says that they come from Chiung-tu[1] in Nan-kuang (in Szechuan), so the name is derived from that of a locality, as in 'kao-liang chin'[2].[b] Now Chang Chhien's[3] biography says that when he was in Ta-Hsia (Bactria) he saw some of these (*chhiung* bamboo staffs) which had come through India (Shen-tu Kuo[4]).[c] The traffic in these it was which finally resulted in opening up (the route to) Yüeh-sui[5], a place which was then in India.[d]

Chang Mêng-Yang[6] says[e] that the *chhiung* bamboos come from Phan-chiang district in Hsing-ku (in Yunnan). The *Shan Hai Ching* calls them *fu*[4] (staff) bamboos and says that they grow in the Hsün-fu Mountains, which are 1120 *li* northwest of the Tung-thing Lake. The (*San Fu*) *Huang Thu* (describing the metropolitan area of Chhang-an in Shensi) says that there are *fu-lao*[4] bamboos in the Hua-Lin Park, indeed three thickets of them. All this shows that they have never been confined to a single region; besides, the rhapsodic odes do not describe them as growing only in Szechuan.

As the *Li Chi* says: 'At fifty a man may lean on his staff at home; at sixty he may do so in and around his village'; this is what is meant by 'the support of the aged' (*fu-lao*[4]). This name arose because the bamboo has a culm solid and stout, but it is simply a question of two names for the same thing.

(24)   The *liao*[14] and *li*[15] bamboos form two classes (*tsu*[16])
   Which are very similar to each other,
   With willow-like leaves they resemble the *khu chu*[17]
   Having short internodes and thin streaky flesh.
   Of a pliable substance they are used for binding
   And almost look like the stalks of male hemp (*hsi*[18]).

 [a] There can be no doubt that Hagerty and Satow were right in identifying this species as *Phyllostachys aurea* (CC 2079); the description is clear.
 [b] Chiung is a town at the edge of the mountains not far west of Chhêngtu, but in Han times there was another (Chiung-tu) south-east of Ning-yuan (Hsi-chhang), i.e. a good deal further south. Hagerty (2) suggests that the *chin* referrd to was some special sort of *Viola* (B II 371) grown at Kao-liang in Kuangsi.
 [c] Cf. Vol. 1, p. 174, where the story is given in translation.
 [d] Here Tai was mistaken. Yüeh-sui was, and is, a town on a trade-route in south-western Szechuan south of Chhêngtu and north of Ning-yuan (Hsi-chhang), i.e. within the huge bend which the Yangtze makes with its tributary the Yalung R. Anciently it exported bamboo objects and special condiments not only south to Burma and India but also east to Kuangtung along the rivers, a fact which when appreciated by strategists was important in opening up more official and regular communications (cf. Vol. 4, pt 3, p. 24). The story of the envoy Thang Mêng[7] encountering a Szechuanese product at the court of the king of Nan Yüeh and finding out how it got there is told in *Shih Chi*, ch. 116. pp. 2b, ff., tr. Watson (1), vol. 2. p. 291. At first sight one might take this *kou chiang*[8] to mean some sauce or marmalade made from the fruit of the thorny lime-bush, *Poncirus trifoliata*, on which we had to expatiate at an earlier stage (pp. 103 ff.). But that would be indeed a pitfall; for *kou chiang* is in fact the proper binome of the betel-vine, *Piper Betle* of the Piperaceae. It has had that name since the Early Han (CC 1709), but by the Thang it had acquired the synonym *chü chiang*[9] (R 628), and others are recorded. These names no doubt derived from the sauce or condiment that was made from its peppery fruits (*chü tzu*[10]; cf. Khung *et al.* (1), p. 1270). The leaves were known as *lou yeh*[11] (Anon. (*109*), vol. 3, p. 343; Chhen Huan-Yung *et al.* (1), vol. 1, p. 331); they were, and are, used to wrap the nut (actually the seed or embryo) of the betel-palm, *Areca catechu* (*pin-lang*[12], Burkill (1), vol. 1, pp. 223 ff., vol. 2, pp. 1736 ff.), all being chewed together with a little lime, and cutch or gambier (see Burkill, vol. 1, p. 15, vol. 2, p. 2198). The habit of betel-chewing, which conduces to sweetness of the breath, originated without doubt in India, but Szechuan produced much betel-leaf, and Hainan island much betel-nut of highest quality (cf. Schafer (18), pp. 37–8, 45–6, 97), so it is not surprising that the chewing habit has been widespread in certain parts of South China for centuries. On this see Imbault-Huart (3).
 [e] Chang Tsai[13], *fl.* +3rd cent.

[1] 邛都	[2] 高梁董	[3] 張騫	[4] 身毒國	[5] 越雟
[6] 張孟陽	[7] 唐蒙	[8] 枸醬	[9] 蒟醬	[10] 蒟子
[11] 蔞葉	[12] 檳榔	[13] 張載	[14] 箹	[15] 籚
[16] 族	[17] 苦竹	[18] 枲		

The *liao* and *li* (bamboos) are two kinds (*chung*[1]) very similar to the 'bitter' bamboos[a] but are slender, bendable and thin-walled. The *liao* shoots are tasteless, but between the Yangtze and the Han Rivers they are called *khu liao*[2] (all the same); see Shen (Jung's[3])[b] (*Lin-Hai Shui Thu I Wu*) *Chih*[4]. The pronunciation is *liao* and *li*, and *chhih*[5] is a technical term meaning a pattern of streaks.

(27)   The *kou chu*[6]   have a furry surface,
        And grow all along the eastern borders—
        What extraordinary things Nature produces,
        Thousands of categories won't suffice for them (*wu lei chung kuei, chhien ho pu chi*[7])!
The 'shaggy dog' bamboos grow among the hills of Lin-hai; their internodes are pubescent.[c] See Shen Jung's book.

(31)   The *mei*[8] bamboos belong to the group (*thu*[9]) called *chhün*[10]
        Their nodes are smooth and their culms short
        They grow in the region between Yangtze and Han
        Where the name they have for them is *khuai*[11] bamboos.
The *Shan Hai Ching* says that the bamboos called *mei* are not grown in one locality but flourish abundantly among all the hills and valleys of Chiang-nan.[d] They belong to the 'arrow-bamboo' class (*chien chu lei*[12]) and have several nodes to the foot. Their leaves are as broad as a sandal and can (be plaited to) make sails. The culms (can be split to) make arrows. Their shoots grow in winter. According to the *Kuang Chih* the governor of Han-chung in Wei times, Wang Thu[13], used to send a present of the shoots to the emperor each winter. In vulgar parlance they are called *khuai-kho*[14]; the character is pronounced *khuai*.

(35)   Then there are the *hai-hsiao*[15] bamboos
        Which grow on the mountainous islands in the sea.
        Their internodes are over one foot long
        While their culms are a full fathom tall.
        In form they are stiff and dry like chopsticks
        But their colour shines out like yellow gold,
        They are the only one of their special kind (*thu*[12a])
        But I do not know to what uses they are put.
These bamboos of the tall sea-islands have solid centres and a cortex so hard that even when forced they will not bend . . . As they grow on high barren land much exposed to sea winds their branches and leaves are sparse. Though they look like chopsticks their texture is different and they are not much good for anything. When people talk of the spreading luxuriant bamboos of Shih-lin Island off Chiao-chou these are what they mean.[e]

   [a] Cf. p. 382.
   [b] He wrote his *Record of the Strange Productions of Lin-hai's Soils and Waters* about +225; it is a part of Chekiang. Most of the book survives.
   [c] If perhaps he originally said the nodes rather than the internodes this might well be a description of *Phyllostachys pubescens*, the nodes of which have just such a velvety texture on them; cf. Li Shun-Chhing (1), p. 126; McClure (1), p. 46. See Hagerty (2), p. 428 for further references in Chinese sources, and Kêng Po-Chieh (2).
   [d] Identification by Hagerty (2): *Sasa tessellata*.
   [e] Perhaps *Bambusa tuldoides* (CC 2077; Li Shun-Chhing (1), p. 132) could be proposed for this littoral island species.

[1] 種	[2] 苦簩	沈瑩	[4] 臨海水土異物志
[5] 齒	[6] 狗竹	[7] 物類衆詭千何不計	[8] 篃
[9] 徒	[10] 箘	[11] 筡	[12] 箭竹類	[13] 王圖
[14] 筡笴	[15] 海篠		

(42)   *Chu* is the generic name for all bamboos
       And the more recondite names specify their sorts,
       Just as we say a cow (*niu*[1]) is not a calf (*tu*[2]);
       The (bulk of the) knowledge that men now have of them
       Is like the ruts and footprints (*kuei chu*[3]) of our fathers before us.
The ruts of a cart are called *kuei*, and a horse's hoofprints are called *chu*.

(43)   As for matters beyond the Chhih Hsien[4] world
       How can we know or record anything of them?
       But if we put aside all arbitrary preconceptions,
       We shall yet extend our insight into encompassing Nature (*i chih pi chih fei mai i chu*[5])

Master Tsou[6] once said: 'The four seas are called Ying Hai[7], and the space within this outer ocean is called Chhih Hsien. Beyond the Ying Hai there are also eight other regions like the Chhih Hsien; therefore these are called the Chiu Chou[8] (the Nine Continents). These are not the same as the Nine Provinces (Chiu Chou[8]) of the Yü Kung.'[a]

Heaven and Earth are boundless, and the Azure Canopy (Nature) brings things into being without limit. If what a man hears and sees be limited to the 'ruts and footprints' of old, how can his knowledge afterwards be worthy of much mention? Other men again use their ears and eyes very little, yet they will readily decide that certain things do not exist; is their knowledge not limited to a very small circle of experience? Master Khung, chief of sages, had no foregone conclusions or arbitrary preconceptions. Mr Chuang, though profoundly enlightened, considered that what men knew was far from equalling what they did not know. Were not these thinkers of infinite perception? Might they not be called model teachers of mankind?

The foregoing extracts may have seemed over-long, but the *Chu Phu* is really a basic document for the assessment of early Chinese natural history. From the extant stanzas we chose some to exemplify Tai Khai-Chih's general outlook and knowledge, and some to follow up specifically the mentions of bamboos which he happened to make in his introductory remarks. Any objective appraiser will be struck by his perspicacity and penetration in observing that part of living Nature which he felt he could distinguish clearly as a whole in itself. One notices his careful and graphic description (e.g. stanzas 2, 11, 12, 14, 15, 17, 35), his recording of peculiar phenomena (e.g. 11, 14, 15, 27), and above all his keen attention to questions of taxonomy (1, 2, 9, 17, 24) and nomenclature (9, 23, 31). Here he was handicapped by the loose use of terms for classes and groups still usual at that time; *phin, lei, tsu, thu* and *chung* all appear, but without any very obvious grading. But Tai's interests went further than this, as we see from his interest in oecology (3, 23, 35), his insight into the physiology of the 'tree-grasses' (5, 6), his notes

---

[a] The reader will be familiar with this ancient cosmology from Vol. 2, pp. 233, 236, and Vol. 3, pp. 565 ff., 568. Tsou Yen's 'Spiritual Continent of the Red Region' (Chhih Hsien), which itself comprised the nine ancient provinces of Yü, was only one of nine great continents separated from each other by an environing ocean which neither man nor beast could cross.

[1] 牛   [2] 犢   [3] 軌躅   [4] 赤縣   [5] 臆之必之匪邁伊矚
[6] 鄒子   [7] 瀛海   [8] 九州

on pharmacological properties (9, 12) and on economic uses, a matter which he rarely forgot (cf. stanzas 11, 12, 17, 23, 24, 31). Strangely modern is his solicitude for quoting even earlier literature (cf. 12, 17, 24, 27), and he would not have been the cultivated scholar that he was if he had not worked in some historical allusions (e.g. 23, 31) when they naturally arose. The whole work ends with a touching statement of his belief in the growth of human natural knowledge—most of what we know comes down to us in the observations and names due to men of old, but there is always far more that is novel, and this when found out by wide experience is not lightly to be dismissed by stay-at-home sceptics (cf. 42, 43). When one remembers that Tai Khai-Chih was probably the first of all the botanical monographers, one realises the extent of his originality, and the standard he set for his descendants.

They may be said to have lived up to it. Omitting the works of intervening centuries,[a] a few words must be said about the monk Tsan-Ning[1] (d. +996) whose *Sun Phu*[2] (Treatise on Bamboo Shoots) appeared almost as soon as the Sung settlement had restored tranquillity after the drums and conquests of the Five Dynasties period. In spite of its name it is concerned with bamboos as a whole. Under the heading of *ming*[3] it discusses names, synonyms and cultivation methods, including details for the harvesting of the edible shoots, while under *chhu*[4] it relates the origins both geographical and literary of no less than 98 bamboo species, with brief entries for each. Monastic vegetarianism was doubtless the background of Tsan-Ning's *shih*[5] section, where the nutritional and medicinal uses of the shoots are described, with elaborate details of preparation, cooking and preservation. Then a historical section (*shih*[6]) gives a wealth of quotations in approximately chronological order from Chou times onward bearing on the bamboos,[b] while some philosophical and other notions bring up the rear in a kind of miscellany (*tsa shuo*[7]).

Li Khan's[8] *Chu Phu Hsiang Lu*[9] (A Detailed Record of Bamboos and the Writings on them), as it was called by later generations, compels us to go a little more slowly. As we know, he began as a painter, so it is natural that his first chapter should be about how one can best draw and paint the tree-grasses. But he was also a marvellous scientific observer, so that his second chapter, *chu thai phu*[10] contains a mass of information on the technical terms used by bamboo specialists—we could almost translate it 'a study of the morphology (lit. the comportment) of bamboos'. Here our attention is at once arrested by the statement that there are two fundamentally different types of bamboo rhizomes,[c] the

[a] But one can never forget the enthusiasm of the Thang poets for the beauty of bamboos; cf. the study of Pai Chü-I by Tsutsumi Tomekichi (*1*). Cf. White (6).

[b] This is how we know of a number of earlier books, now lost, such as a treatise on bamboos by Wang Tzu-Ching[11], probably of the Thang.

[c] Ch. 2, p. 2*b*.

[1] 贊寧　　[2] 筍譜　　[3] 名　　[4] 出　　[5] 食
[6] 事　　[7] 雜說　　[8] 李衎　　[9] 竹譜詳錄　　[10] 竹態譜
[11] 王子敬

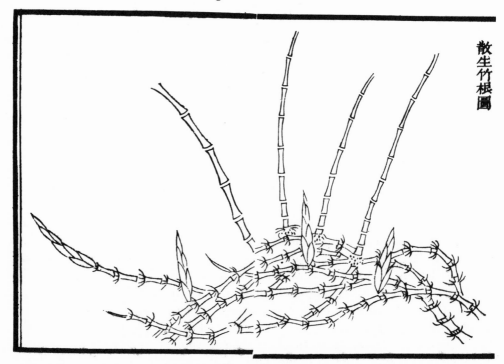

散生竹根圖

Fig. 78. The bamboo type of spreading-root habit, from Li Khan's *Chu Phu* of +1299. Ch. 2, pp. 3a, b.

spreading or dissipating type (*san*[1]) and the clumping type (*tshung*[2])—*chu kên erh chung*[3], he says. Moreover he illustrates this by two type-drawings (Figs. 78, 79). He explains as follows:

> The spreading type (*lei*[4]) gives growth in the first year to diffuse roots running out underground sideways (*hsing kên erh fu shêng*[5]), and only in the second year does it produce shoots and send up culms.
>
> The clumping type, however, does not have to wait to develop sideways-running roots, but in each of several years the shoots develop into culms (*kan*[6]), though only in the following year are branches and leaves fully developed.

And he then gives a list of 22 kinds of bamboos belonging to the former category and 9 belonging to the latter, including several which we have already met with in the discussions of Tai Khai-Chih. If now we turn to a contemporary work on the natural history of bamboos the first thing we find is the statement that 'the rhizome manifests its character in two divergent forms, each with important variations'.[a] The clear distinction between the two basic forms assumed by the

---

[a] McClure (1), pp. 19, 208 ff. The two groups require different procedures in vegetative propagation.

[1] 散   [2] 叢   [3] 竹根二種   [4] 類   [5] 行根而敷生
[6] 竿

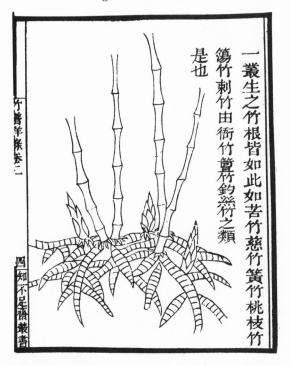

一叢生之竹根皆如此如若竹慈竹簧竹桃枝竹
筱竹刺竹由衙竹簹竹釣絲竹之類
是也

Fig. 79. The bamboo type of clumped-root habit, from Li Khan's *Chu Phu* of +1299. Ch. 2, p. 4*a*.

bamboo rhizome has been accepted by modern bambusiologists for just under a century since Rivière & Rivière found they could divide all bamboos into (*a*) bamboos of spring growth and a generally spreading (*traçant*) diffuse habit, and (*b*) bamboos of autumnal growth and a caespitose or clumping habit, dense like a tuft of grass.[a] Since then other terms have been introduced, and the difference has been recognised as essentially physiological; indeed all kinds of intermediate forms and variations are now known,[b] but the general principle still holds good. The thin wide-ranging rhizomes are now called leptomorph and the thick cigar-shaped clumping ones pachymorph, or monopodial and sympodial respectively, but no-one would question the accuracy with which Li Khan described the two groups at the end of the +13th century. McClure indeed gave all the credit due to the Rivières, but he was a man of understanding who had spent many years in the chair of botany at Lingnan University, so he was careful to say that they were 'the first to publish (in a Western language, at any rate) the clear distinction

---

[a] (1), pp. 312 ff. *Phyllostachys viridis* = *mitis* was taken as the example of the first group, and *Bambusa macroculmis* (really a *Gigantochola*) as that of the second. Photographs showing the striking difference between the grove and clump types can be seen in McClure (1), figs. 16, 17.

[b] Very rarely the two types of rhizome habit can even exist in the same plant; *Chusquea fendleri* is the only case discovered so far.

between the two basic forms'. Perhaps he knew that Li Khan had preceded them by close on six hundred years.[a]

Already in Li's time the terminology differed for the leptomorph and pachymorph forms. Speaking of the former, Li tells us that the part between the culm and the rhizome, i.e. the culm neck, is called *tshan thou*[1], obviously because its segmentation was reminiscent of that of the silkworm caterpillar. The main leptomorph rhizome (*chü*[2] or *li phang*[3]) buds off subsidiary rhizomes (*pien*[4, 5]) just distal to its nodes, and the necks of these are known as secondary or 'false' shoots (*erh sun*[6], *wei sun*[7]). When in active elongation the rhizomes are called *hsing pien*[8], 'on the march', and the many rootlets coming out distal to the nodes in a confusedly repeating manner (*chui*[9]) like beards are the *hsü phang kên*[10]; it is interesting that the primary word for root was here reserved precisely for these.[b] As for the pachymorph rhizomes, which Tai had compared to pestles (p. 381 above), Li differentiated them into those bunching out at the surface of the ground (*chhan tu kên*[11], again the insect segmentation analogy) and those boring down deeply within it (*tsuan ti kên*[12]).[c]

Many other terms were in use. An archaic trait was the use of a whole string of words for the successive stages of development and growth of the shoot,[d] at its first appearing called *mêng*[17], then *hsiai*[18], *jui*[19], *chu thai*[20]; when rather larger *ya*[21], when still larger *chhü*[22] or *thai*[23], finally when with a visible culm *kuan*[24]. Interest was taken in the branch complement, culms having nodes with single branches being called 'male', those with two or more 'female', thus establishing a species characteristic.[e] Every node of every segmented vegetative axis of a bamboo plant bears a sheathing organ, and this 'node-leaf' (*chieh yeh*[25]) had appropriately a special name, *pao tho*[26] (or *chhih*[27]); withered detached ones were called *jo*[28] (cf. Figs. 80a, b). The thick watery material inside the hollow spaces in the culms, says Li, solidifies to a gum or solid, and this is *huang*[29], i.e. tabashir, chiefly silica but long prized in Asian drug-markets.[f] The actual substance of the wall was *yün*[30], the cortex properly

---

[a] A recognition of Li's achievement (and translations of his prefaces and post-face) were given by White (6), discussing a set of +18th-century paintings of bamboos. Satow (1), p. 20 was aware of the passage in the *Chu Phu Hsiang Lu*, but did not know its author's date. Since the Chinese had lived amidst bamboos for millennia, the distinction may well have been appreciated long before Li Khan. For example, Tso Ssu[13], in his *Wu Tu Fu*[14] ode (c. +270) mentions 6 kinds of bamboos growing in 'coppices' (*huang*[15]) and two kinds growing in tufts (*tshung*[16]); see the *Wên Hsüan*, ch. 5, p. 6a. The translation of von Zach (6), vol. 1, p. 60, misses the point.

[b] See McClure (1), figs. 3, 10.

[c] See McClure (1), figs. 2, 12.

[d] Cf. Fig. 82.

[e] Thus *Sasa palmata* always puts forth but one branch, *Phyllostachys elegans* two, and others have complements at characteristic totals. See McClure (1), pp. 51 ff. and Fig. 81.

[f] Details may be found in Burkill (1), vol. 1, p. 296; Watt (1), p. 110.

1 竈頭	2 菊	3 竻竻	4 邊	5 鞭
6 二筍	7 僞筍	8 行邊	9 贅	10 須竻根
11 蟬肚根	12 鑽地根	13 左思	14 五都賦	15 篁
16 叢	17 萌	18 箷	19 蕤	20 竹胎
21 牙	22 笛	23 箈	24 篧	25 節葉
26 筍籜	27 箈	28 箬	29 簧	30 篔

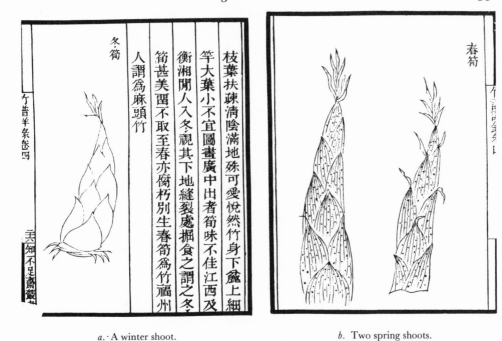

a. · A winter shoot.　　　　　*b.* Two spring shoots.

Fig. 80. Bamboo shoots (*sun*) from Li Khan's *Chu Phu* (+1299), ch. 4, pp. 26*a*, *b*.

known as *min*[1], the green cortex scraped off, *jo*[2]. There was a special name reserved for bamboo leaves, *chha*[3], and the branches should not be called by the ordinary term but rather *thien-chou*[4], nor should the inflorescences have the common name, but *hsüan*[5] instead. Thus one is astonished at the wealth of technical language which the artists and botanists of the Sung had at their disposal for discussing the parts of bamboos.

After all this Li Khan proceeded to his systematic descriptions. He had four grades, first the best all-round species or varieties for economic use (*chhüan tê phin*[6]) comprising 75 species, secondly the bamboos of unusual forms in one way or another (*i hsing phin*[7]) under which he listed 158 different species, and thirdly those of unusual colour (*i sê phin*[8]) which came to as many as 63. Fourth came the bamboos of strange properties (*shen i phin*[9]), 38 in number. And he ended his book with two excellent sections to which we have already alluded (p. 357 above), 'the plants that look like bamboos but are not', and 'the plants that have the name of bamboos but are not'. Since these entries numbered 45, the number of true bamboo species described was 259, an astonishing total for the late +13th century, even allowing for possible duplication, doubtful varieties, and the like. As examples of the quality of Li Khan's illustrations we add Figs. 83, 84.

[1] 筦　　　[2] 筎　　　[3] 箈　　　[4] 天箄　　　[5] 筐
[6] 全德品　　[7] 異形品　　[8] 異色品　　[9] 神異品

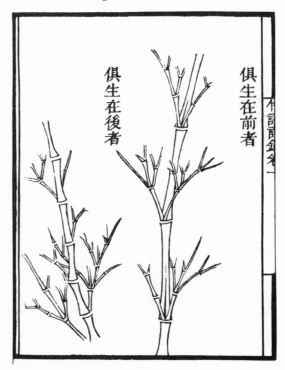

Fig. 81. Bamboo culms having nodes with single branches (left) were called male, those with two or more (right) female. This was afterwards another basis for species differentiation. Li Khan also noticed that there was a tendency for branches coming out at the front of the plant to have their leafy tips behind its culm, and vice versa. *Chu Phu* (+1299), ch. 1, p. 10*b*.

We must now linger no more over the bamboos, yet there are still a few words to be said. Fifty years or so after Li's time another scholar, Liu Mei-Chih[1], produced a continuation entitled *Hsü Chu Phu*[2], in which he added about 20 species or varieties which he felt had not been described before. The *Chu Phu* of the Ming recluse, Kao Lien[3], written just before +1591, was a larger work, however, since it recorded and described many more species, including some especially interesting ones new to science, together with methods of propagation and cultivation. Even so, it was only a part of his *Hua Chu Wu Phu*[4] (Five Treatises on Flowers and Bamboos), and that was but an enlargement of his *Ssu Shih Hua Chi*[5] (Ornamentals of the Four Seasons) which formed a component part of the *Yen Hsien Chhing Shang Chien*[6] (Pleasurable Occupations of a Life of Retirement), itself one of the main sections of the *Tsun Shêng Pa Chien*[7] (Eight Disquisitions on the Art of Living as a Recluse).[a] Lastly needing mention is the *Chu Phu* of Chhen Ting[8] who about +1670 set down details of some of the more extraordinary bamboo

[a] See on Kao Lien, Wang Yu-Hu (*1*), 2nd ed. p. 157, and *SKCS/TMTY*, ch. 123, p. 24*b*.

[1] 劉美之   [2] 續竹譜   [3] 高濂   [4] 花竹五譜   [5] 四時花記
[6] 燕閒清賞牋   [7] 遵生八牋   [8] 陳鼎

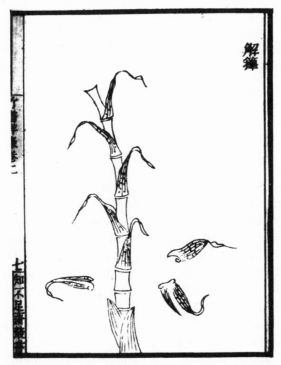

Fig. 82. The sheathing organ or 'node-leaf, *pao tho*, called *jo* when detached and withered.
Chu Phu (+1299), ch. 2, p. 7*a*.

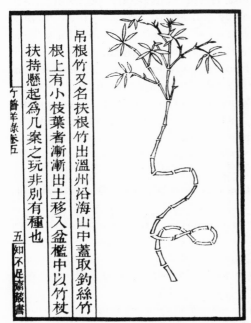

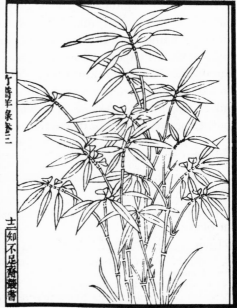

Fig. 83. Li Khan's drawing of one of the rattan creeper species of bamboo; from *Chu Phu*, ch. 5, p. 5*a*.

Fig. 84. Li Khan's drawing of one of the bamboo species characteristic of Hunan and Fukien (Hsiao, Hsiang and Min). From *Chu Phu*, ch. 3, p. 12*a*.

species which he had studied in the wilds of the south-western provinces of Kweichow and Yunnan.[a] Thus by this time the results of exploration were added to the wealth of knowledge about the commoner bamboos of the more populous provinces, inhabited by inquisitive scholars since classical days.[b]

(iii) *Peonies*

The greatest wealth of Chinese botanical monography certainly centered on the plants which were grown for their beauty in gardens.[c] Here it is natural to begin with the genus *Paeonia* of the Ranunculaceae,[d] glorious with their large and showy flowers petalled in wonderful colours from white to deepest murrey, their abundance of brightly-coloured or dark purple anthers and their prominent tripartite bulbous ovaries. It is natural because for traditional Chinese culture this was Hua Wang[3], the king of flowers, known and distinguished since the early Chou and cultivated in hundreds of varieties since the early Thang.

Peonies divide into two great groups, the woody or shrublike kind and the more familiar herbaceous kind.[e] The former has only about half-a-dozen species, including that traditionally cultivated in China, *Paeonia suffruticosa* (formerly *moutan*),[f] eastern and north-western in distribution; but also beautiful wild species domesticated only in much more recent times from the south-west such as *P. delavayi*, *P. lutea* and *P. potaninii*. The Chinese term for them all is *mu-tan*[4], with suitable epithets, hence the old Linnaean specific name, the supersession of which on priority rules has brought regrettable loss. Herbaceous peony species are much more numerous. Some twenty belong to the sub-group *foliolatae*; these include not only the classical Chinese *Paeonia lactiflora* (= *albiflora*) cultivated for so long under the name *shao-yao*[5], but also the fine *P. obovata* wild-growing on the western slopes of the Himalayan massif, *P. emodi* from the southern slopes, *P. daurica* from Yugoslavia, *P. japonica*, and others. The second sub-group, *dissectifoliae*, provides most of the European species, such as *Paeonia officinalis*, known to Theophrastus and Dioscorides as *glykysidē* ($\gamma\lambda\nu\kappa\nu\sigma\acute{\iota}\delta\eta$), or *P. peregrina* from Turkey,[g] but also another beauty of the south-western Chinese forests, *P. veitchii*.[h]

---

[a] Chhen Ting was an interesting man who could be called an ethnological geographer. He took his first wife from the family of a Yunnanese chieftain, and after her death married a distinguished woman mathematician Chhien Chieh[1]. His concubine, Jui-Chu[2], was also a celebrated mathematician and astronomer. Besides bamboos, he wrote also on the lichi.

[b] By this time, too, the Jesuits were beginning to inform Europe about bamboos; cf. Collas (10); Cibot (15).

[c] For all that follows here, it is useful to have at hand the guide of Liu Tzu-Ming (1) as well as Li Hui-Lin's book (8).

[d] Now sometimes separated as Paeoniaceae.

[e] See the standard monograph of Stern (1) and his papers such as (2).

[f] From Sabine (1) in 1826 onwards one may consult Fortune (7); Harding (1); Wister (1); Wister & Wolfe (1) and Smirnow (1).

[g] Grown in England by Parkinson in +1629 (Coats (1), pp. 191 ff.).

[h] We omit here mention of the North American species, which form a separate herbaceous group having petals no longer than their sepals.

[1] 錢潔   [2] 蕊珠   [3] 花王   [4] 牡丹   [5] 芍藥

From the literary records it is possible to trace the way in which the cultivation of tree-peonies moved from one region to another as the centuries passed. First it was notable in Chekiang where by the +4th century people were beginning to transplant the wild species to their gardens and to foster new varieties. Not long before +700 enthusiasm spread to the capital Chhang-an, and this was the great centre during the Thang, but later the art spread both east and west. In the +10th and +11th centuries Loyang was supreme, while in the +11th and +12th Thien-phêng[1] in Szechuan became very important, perhaps because of crossing with the wild species of the west. Then lastly the cultivation spread eastwards again over the North China plain, bringing cities such as Chhen-chow[2], Tshaochow[3] and Pochow[4] into prominence. In its entry for *mu-tan* the *Chhün Fang Phu* says:[a]

Other names for it are *lu-chiu*[5] (deer-scallion), *shu-ku*[6] (rat maiden), *pai liang chin*[7] (worth a hundred ounces of gold) and *mu shao-yao*[8] (woody peony).[b] Nothing is known of it before the time of Chhin and Han. It was Hsieh Khang-Lo[9][c] in the Yung-Chia reign-period (+307 to +312) of the Chin dynasty, who first said (in a poem) that the tree-peony grew abundantly beside the water and among the bamboo groves (in Southern Chekiang). Then in the Northern Chhi dynasty there were paintings of tree-peonies by Yang Tzu-Hua[10] (*fl.* +561 to +565). This shows how old was the origin of the tree-peony. In the Khai-Yuan reign-period (+713 to +741) of the Thang there was great peace on earth (*thien-hsia thai-phing*[11]), so the tree-peony began to (be cultivated and) flourish in Chhang-an. But in the Sung only the flowers of Loyang were really famous. The most distinguished men of the age, like Shao Khang-Chieh[12],[d] Fan Yao-Fu[13][e] and Ouyang Yung-Shu[14][f] were especially fond of them and often mentioned them in their poems and songs.

At Loyang it was always the custom to admire flowers, as one can see in the *Lo-yang Fêng Thu Chi*[15] (Local Character and Manners of Loyang).[g] Thien-phêng in Szechuan was called 'the little Western capital' because the people there had also acquired the love of (tree-peony) flowers, just like those of Loyang itself.

In general the best tree-peony varieties were those of the Yao and Wei families. Before the 'Yao yellow' came out the 'Niu yellow' was the best, before that again the 'Wei flower' was the best, and previously the 'Tso flower' had taken the highest place. Before the 'Tso flower' there were only the 'Su family red', the 'Ho family red', the 'Lin family red' and so on. These three were simple flowers, but then at Loyang only double ones (*chhien yeh*[16])

---

[a] Hua Phu, ch. 2, pp. 1a ff., tr. auct.

[b] This was an ancient name. It occurs in Tshui Pao's *Ku Chin Chu* (+300), and according to Kao Chhêng's *Shih Wu Chi Yuan* (+1085) it was still current in the Thang.

[c] I. e. Hsieh Ling-Yün, actually +385 to +433. Probably a *lapsus* for Yung-Chhu r. p. (L/Sung).

[d] Shao Yung, +1011 to +1077, cf. Vol. 2, pp. 455 ff.

[e] Fan Shun-Jen, +1026 to +1101, cf. p. 401 below.

[f] Ouyang Hsiu, +1007 to +1072, cf. p. 402 below and Vol. 3, pp. 391 ff.

[g] The date and author of this work have not been identifiable. We do not think that it was a book title at all, for the words are the same as those of the heading of Ch. 3 of Ouyang Hsiu's *Lo-Yang Mu-Tan Chi* (cf. p. 403). Wang was just quoting carelessly.

[1] 天彭	[2] 陳州	[3] 曹州	[4] 亳州	[5] 鹿韭
[6] 鼠姑	[7] 百兩金	[8] 木芍藥	[9] 謝康樂	[10] 楊子華
[11] 天下太平	[12] 邵康節	[13] 范堯夫	[14] 歐陽永叔	[15] 洛陽風土記
[16] 千葉				

began to be produced—hence the name 'blossoms of Loyang'. When these began to flourish the glory of the others departed. As time went on people competed to produce new varieties by cultivation and grafting (*phei chieh*[1]), and many exquisite ones appeared, in grades beyond those previously known.

The plants have a nature preferring cold to warmth, and dryness to moisture; when transplanted, the roots flourish. They are happy when set facing the sun, but putting them half in the sun and half shaded is called 'nourishing the flowers' (*yang hua*[2]). Planting them at the best time, and (knowing the methods of) grafting and pruning, is called 'the handicraft of the flower' (*lung hua*[3]). They dread strong winds and weak sunshine, but if they get shade and humidity just right, with transplanting and grafting according to the art, flowers can be produced with seven hundred petals and measuring a foot across. When expert horticulturists select the best sorts for planting, and when every detail is correctly managed, with constant attention, then the flowers will flourish abundantly, and among them there will arise marvellous new grades (of forms and colours) by (spontaneous) transformation. This phenomenon is indeed due to the exertions of man, capturing the powers of Nature (for his purpose).[a]

Surely this is one of the best summaries of the cultivation history of China's favourite flower.

There is rather more to be said about it, however, than what Wang Hsiang-Chin bothered to put into this introduction to the varieties.[b] Notices of the name *shao-yao* (traditionally the herbaceous peony) first occur at a much earlier date than those of the name *mu-tan* (traditionally the tree-peony), but contrary to a common statement, these latter start long before the time of its great development in the Thang. A famous song in the *Shih Ching* (Book of Odes),[c] datable to the −8th or −7th century, begins the story, with a refrain at the end of each verse telling how the young men and women presented each other with *shao-yao*[5]. Since the flowers thus had a significance similar to that of Polynesian hibiscus,[d] it is not surprising that the commentators handed down a tradition that the words meant a magic 'binding herb'. The name *mu-tan* occurs for the first time in a text which we generally attribute to the −4th century, though it must have Han interpolations, the *Chi Ni Tzu* book.[e] What it says is that *mu-tan* grow in Hanchung and Honei (southern Shensi and Shansi), and that the red ones are especially fine

---

[a] *Tzhu tsê i jen li to thien kung chê yeh.*[4] This important statement had appeared first five hundred and more years earlier, in Wang Kuan's *Yang-Chou Shao-Yao Phu* (cf. p. 409). On its far-reaching significance for the philosophy of technology in China, see Vol. 5, pt 5, pp. 293 ff.

[b] In the early nineteenth century interest in this genus was so great that Hoffmann (1) essayed a history of its Chinese background (including even characters, in the horticultural journals of 1848), but neither sinology nor botany were ripe for it at that time.

[c] Mao no. 95, Chên Wei; tr. Legge (8), p. 148 (I, vii, 21); Karlgren (14), p. 61; Waley (1), p. 28.

[d] From references in previous volumes the reader will remember the mating festivals of Chinese antiquity revealed by the classical work of Granet (1, 2), occasions on which the social intercourse of girls and young men was encouraged by custom.

[e] Ch. 3, p. 5*a*, in *YHSF*, ch. 69, p. 38*a*, from *TPYL*, ch. 992, p. 6*b*.

[1] 培接　　　　[2] 養花　　　　[3] 弄花　　　　[4] 此則以人力奪天工者也
[5] 芍藥

(*shan*[1]).[a] Since this is in a context of useful plants and minerals it is likely that we should interpret the approbation as medicinal rather than horticultural. And indeed *mu-tan* appears again soon afterwards in the *Shen Nung Pên Tshao Ching*,[b] where we find the synonyms *lu-chiu* and *shu-ku* already noted in the above quotation, together with the statement that it comes from Pa Commandery, i.e. modern Szechuan, in fact one of the main wild peony habitats. This is evidence for the −2nd century (cf. p. 243). Moreover, it is clear that there were debates in the Early Han about the properties of the plant, for the *Wu shih Pên Tshao* of about +235 says:[c]

The *mu-tan*. Shen Nung and Chhi Po[d] affirm that its sapidity is acrid. Mr Li[e] considers that it is slightly algogenic (cooling). Lei Kung[f] and Thung Chün say that the sapidity is acerbic with no dangerous active principle, but (the) Huang Ti (book) says that though acerbic it does have a dangerous active principle. The leaves are like those of the *phêng*[2] set opposite each other mutually.[g] The colour (of the flower) is yellow. The root is the thickness of a finger, and black, this is where the dangerous active principle resides. The fruits and seeds should be picked between the second and the eighth months, and when dried in the sun can be eaten. They lighten the body and promote longevity.

Here a number of still older texts are quoted, including, it would seem, the *Pên Ching*, and more than one component of what is now the *Huang Ti Nei Ching, Su Wên*, as well as the −1st-century *Thung Chün Tshai Yao Lu* (on which see p. 245). Clearly the interest of scholars was still primarily pharmacological, and *shao-yao* too was listed and described in the *Shen Nung Pên Tshao Ching*.[h]

But the nomenclature was still not sorted out. The *Kuang Ya* of +230 defined *pai shu*[3] as a synonym for *mu-tan*, but the *Ming I Pieh Lu* (presumably +6th century, cf. p. 248) was equally positive that it was a synonym of *shao-yai*.[i] Tshui Pao, in his +4th-century *Ku Chin Chu*, said, revealingly, that 'there are two kinds of *shao-yao*, the herbaceous one and the woody one. The latter has larger flowers more deeply coloured; it is commonly called *mu-tan*, but that is wrong.'[j] One may conclude, therefore, that it was not until the Sui period, the end of the century when Yang Tzu-Hua was making his tree-peony paintings, that the names they ever afterwards bore became firmly attached to the two groups of *Paeonia* species.

[a] On the *Chi Ni Tzu* book see p. 256 above, and Vol. 2, pp. 275, 554, Vol. 3, pp. 218, 402, 643, Vol. 5, pt 3, pp. 14 ff., etc.
[b] Mori ed., ch. 3, p. 93, also in *TPYL*, ch. 992, p. 6b. It belongs to the 'lower' grade. Cf. *PTKM*, ch. 14 (p. 17).
[c] Preserved in *TPYL*, *loc. cit.*; tr. auct. Cf. p. 247 above.
[d] One of the main interlocutors in the present *HTNC/SW*.
[e] The writer of the pharmaceutical natural history just before Wu's, *c.* +225.
[f] Another pharmaceutical writer's pseudonym; cf. Sect. 45.
[g] This probably refers to opposite leaflets of compound leaves. The *phêng* was probably an *Aster* or *Chrysanthemum*, cf. B II 15, 436.
[h] Mori ed., ch. 2, p. 60, placed in the 'middle' grade.
[i] *Kuang Ya Su Chêng*, ch. 10A, p. 18a.
[j] Not in the book now, but preserved by Su Sung in *Pên Tshao Thu Ching* and derivative quotations.

[1] 善　　　[2] 蓬　　　[3] 白茢

If one word more on their etymology may be allowed, it is curious that *tan* may also mean a medicine, like *yao*, and there is no need to seek far for the significance of 'male' (*mu*) when it will be clear from the foregoing argument that the boys and girls of pre-Confucian festivals may have been plighting their troth with tree-peony flowers just as well as with herbaceous ones. Still more curious, though perhaps not our concern, is the fact that in Europe *Paeonia officinalis* was originally *P. foemina*, while another species still to this day retains the name of *P. mascula*.[a]

As soon as the peace and prosperity of the Thang period set in, conditions became propitious for gardeners. Security and quiet permitted nurserymen and private flower-lovers to collect together all the varieties of the *Paeonia* genus that could be found, and to play with them by grafting and all kinds of experimental culture techniques.[b] The time had not yet come to write about them—that happened in the Sung, as we shall see—but there was great enthusiasm for the plant, ranging from the court downwards to the simplest people of town and country. Indeed, the new varieties about which we have already read in the quotation from the *Chhün Fang Phu* began to command extraordinary prices, so that a social situation arose comparable only with the 'tulipomania' of +17th-century Holland.[c] The mere collocation in single gardens of all accessible varieties soon produced more by spontaneous crossing,[d] and this in turn, when chromosome effects were not quite harmonious, led to aberrations in the determination of the organ-rudiments of the flower, stamens particularly being converted into petals.[e] Hence the phenomenon of 'doubling', spectacular when the stamens are as numerous as they are in peonies, an effect sought and perpetuated with delight by the Chinese of the Thang.[f] A mass of literary references attests the important part played by peony horticulture in the social life and technological development of the time.[g] The old paintings of Yang Tzu-Hua were discussed and

[a] Dioscorides, Gunther ed. p. 382. The reason must surely be sought in the shape of the superior lageniform ovaries and styles, so prominent in some species. *P. mascula* = *corallina*; it may be native in England. It still grows wild on Steepholm Is. in the Bristol Channel, and on the edge of the Cotswolds (Coats (1), pp. 191 ff.).

[b] Ouyang Hsiu (see p. 401 below) says that tree-peony culture first flourished in Loyang during the Tsê-Thien period, i. e. the reign of the empress Wu Hou (+684 to +704).

[c] Cf. Jessen (1), pp. 256 ff.; Wright (1), pp. 217, 223 ff., 237; Clifford (1), p. 93.

[d] Selection and sexual or asexual propagation have for their basis (a) chance gene mutations arising in conditions where the plants are under close observation, (b) auto-polyploidy, (c) inter-specific and inter-varietal hybridisation (natural or artificial) with or without chromosome duplication or nuclear aberration, (d) bud-sports, autogenous chimaeras, somatic variations, etc., (e) graft chimaeras, (f) fluctuations caused by environmental conditions (e.g. root-stock, temperature and humidity of culture, etc.). Once particular kinds of plants were assembled, watched, tended and grafted, vast vistas of the possible opened up. Cf. Anderson (1), pp. 59 ff.

[e] Cf. Crane & Lawrence (1), pp. 52, 80, 82, 90b. In many plants doubling depends on specific genes and may be inherited in strict Mendelian fashion; cf. the garden stock, *Matthiola incana*; Saunders (1); Waddington (2). In other cases the effect is very complex, involving many gene interactions; cf. *Dahlia variabilis*. Doubling must occur from time to time in natural plant populations, but in the wild it can hardly ever survive. Cf. Chittenden (1), vol. 2, p. 706.

[f] Doubling in *P. officinalis* seems not to have been noticed in Europe till about +1550—eight hundred years later. See Coats (1), pp. 191 ff.

[g] See especially *Chhün Fang Phu*, Hua Phu, ch. 2, pp. 14a ff. A useful summary of this literature has been made by Ledyard (2).

imitated,[a] the imperial gardens became celebrated for their peonies,[b] and in-dividual horticulturists achieved immortality in the literature. One of these was Sung Shan-Fu[1] (*fl.* +713 to +755). He produced, it is said, a thousand varieties of *mu-tan* tree-peonies, and made a plantation of tens of thousands of them for the emperor at Li-shan.[2] He was called Hua Shen[3] (Flower Genius) or Hua Shih[4] (Master of Flowers).[c] Writing about +860, Li Chao[5] said:

The nobility and gentry of the capital city have been making excursions to admire the peonies for about thirty years past. Every evening in spring-time the carriages and horses take madly to the roads, it being considered shameful not to spend some leisure in enjoying them.[d]

And he goes on to say that the price for a single plant has sometimes amounted to tens of thousands of cash. Since such a cost might be equivalent to a hundred bushels of rice, the possession of a single graft of the most novel or fashionable variety was evidently a luxury of the extremest kind. Only when this is appreci-ated can one get the full force of the famous peony poem of Pai Chü-I, written about +810:[e]

> In the royal city spring is almost over,
> Tinkle, tinkle—the coaches and horsemen pass.
> We tell each other 'This is the peony season',
> And follow with the crowd that goes to the flower market.
> 'Cheap and dear—no uniform price;
> The cost of the plant depends on the number of blossoms.
> For the fine flower,—a hundred pieces of damask;
> For the cheap flower,—five bits of silk.
> Above is spread an awning to protect them,
> Around is woven a wattle-fence to screen them.

---

[a] *Liu Pin-Kho Chia Hua Lu*[6] (Table-Talk of Imperial Tutor Liu) by Wei Hsüan[7], concerning Liu Yü-Hsi[8] (+772 to +842), p. 10*a*, *b*, quoted in *SWCY* (see below). Also *Shang Shu Ku Shih*[9] (Facts about Ancient Records) by Li Chho[10], *c.* +860, p. 18*b*.

[b] Several accounts from books now hard to find are quoted in *TSCC*, Tshao mu tien, ch. 292, pp. 2*a* ff. For example: *Khai-Yuan Thien-Pao I Shih*[11] (reminiscences and Remains from the Khai-Yuan and Thien-Pao reign-periods), i.e. +713 to +755, the reign of Hsüan Tsung (Ming Huang), by Wang Jen-Yü[12] of the Later Chou; *Chü Than Lu*[13] (Records of Interesting Conversations) by Khang Phien[14] of the Thang; and *Chen Chu Chhuan*[15] (The Pearly Boat), material put together by Chhen Chi-Ju[16] of the Ming on lake sports at the Thang court.

[c] Our knowledge of him comes from an *I Jen Lu*[17] (Records of Unusual People) quoted in *Lei Shuo*, ch. 12, p. 28*a*. The same passage is quoted in *TSCC*, Tshao mu tien, ch. 292, p. 1*b* from the *Lung-chhêng Lu*[18] (Records of the Dragon City, Sian) by Liu Tsung-Yuan[19], who would have been almost his contemporary.

[d] *Thang Kuo Shih Pu*[20] (Additional Materials towards a History of the Thang), ch. 2, p. 16*a*, tr. Ledyard (2).

[e] *Pai shih Chhang-Chhing Chi* (Chhien Chi), ch. 2, p. 24, tr. Waley (2), p. 126. Cf. Tatlow (1), pp. 97 ff.; Alley (13) pp. 130–1.

[1] 宋單父　　　[2] 驪山　　　　[3] 花神　　　　[4] 花師　　　　[5] 李肇
[6] 劉賓客嘉話錄　　　　　[7] 韋絢　　　　[8] 劉禹錫　　　[9] 尚書故事
[10] 李綽　　　[11] 開元天寶遺事　　　　　　[12] 王仁裕　　　[13] 劇談錄
[14] 康軿　　　[15] 珍珠船　　　[16] 陳繼儒　　　[17] 異人錄　　　[18] 龍城錄
[19] 柳宗元　　　[20] 唐國史補

If you sprinkle water and cover the roots with mud,
When they are transplanted, they will not lose their beauty.'
Each household thoughtlessly follows the custom,
Man by man, no one realising.
　　There happened to be an old farm labourer
Who came by chance that way.
He bowed his head and sighed a deep sigh;
But this sigh nobody understood.
He was thinking, 'A cluster of deep-red flowers
Would pay the taxes of ten poor houses.'

The best passages on peony cultivation in the Thang occur in the *Yu-Yang Tsa Tsu*[a] of +863 and the *Shih Wu Chi Yuan*[b] of +1085, and it is a pity that we have no room to present them in their entirety. In the latter Kao Chhêng records the tradition that the empress Wu Tsê Thien about +690 banished the tree-peonies from Chhang-an to Loyang because they were late in coming out.[c] In the former Tuan Chhêng-Shih tells a number of stories which show the value then placed upon them. For example:

Towards the end of the Khai-Yuan reign-period (c. +740) Phei Shih-Yen[2] was made a Court Gentleman and given a commission to Yuchow and Chichow. On his way back he arrived at the Chung-Hsiang Ssu (temple) at Fênchow (in Shansi) and obtained one white tree-peony, which he took home and planted in his garden at Chhang-an. During the Thien-Pao reign-period (+742 to +755) it was regarded as rare and praiseworthy by everyone in the capital ...[d]

Or again:

In the Hsing-Thang Ssu (monastery) there is a tree-peony bush which during the Yuan-Ho reign-period (+806 to +820) used to put forth one thousand two hundred blossoms. The colours were in straight shades and intermediate shades—pale reds, pale purples, deep purples, yellows, whites, and some of sandal-wood colour, etc., only the deep reds were lacking. There were also some flowers which developed no centres (gynoecia) at the heart of the petals (*yu hua yeh chung wu mo hsin chê*[3]), and others that had doubled (or repeated) terraces (*chhung thai hua chê*[4]). The diameter of the flowers was as much as seven or eight inches.[e]

This sounds rather fabulous, but some kind of multiple grafting might have been done, or what Tuan took for a single plant may have been many. If this is not the

---

[a] Ch. 19, pp. 3*b* ff.
[b] Ch. 10, p. 31*a*, *b*. Here the starting-point is the emperor Sui Yang Ti. But Tuan had checked in a large book called *Sui Chhao Chung Chih Fa*[1] (Horticultural Methods of the Sui Court) and found no mention of tree-peonies in it.
[c] They were so popular in her time that the tree-peony was commonly called 'emperor of flowers', while the herbaceous peony became the 'dynastic ancestor of flowers'.
[d] Tr. Ledyard (2).
[e] Tr. auct., adjuv. Ledyard (2). Note the mention of sterile flowers without carpels.

[1] 隋朝種植法　　[2] 裴士淹　　　[3] 有花葉中無抹心者　　　[4] 重臺花者

first direct mention of doubling in any literature, another statement follows soon afterwards, in the *Tu-Yang Tsa Pien*[1] of Su Ê[2], written about +890.

> In front of the Audience Hall of Mu Tsung Huang Ti (r. +821 to +824) there were planted thousand-petalled tree-peonies. When the flowers first opened the fragrance of their perfume was perceived by everyone. Each blossom had a thousand petals, large and deeply red. Every time His Majesty gazed upon the sweet-scented luxuriance he would sigh and say 'Surely such a flower has never before existed among men!'. Then each night within the palace grounds tens of thousands of white and yellow butterflies would fly and gather around the flowers, shining and glowing brilliantly in the lights. When the morning came all would vanish away. The palace girls competed to catch them in gauze nets, but none could succeed till the emperor ordered that large nets should be spread in the air, and so several hundreds were captured.[a]

By this time the double flowers would have been cultivated for some considerable period, yet they were still rare and valuable. Some varieties of this kind are listed in the *Chhing I Lu*[3] (Records of the Unworldly and the Strange)[b] written by Thao Ku[4] *c.* +950—e.g. *Pai yeh hsien jen*[5] (Hundred-petalled immortal), pale red, and *Thai-phing lou ko*[6] (Peace-on-earth pavilion), described as thousand-petalled, yellow.

This is the background against which we have to see the monographic literature. During the following eight centuries no less than twenty important works on tree-peonies appeared, to say nothing of the wealth of shorter essays, tractates, memoranda and poems, many of which would enshrine some fact of curious observation or experiment. The first book was by a monk Chung-Hsiu[7], who in +986 brought out his *Yüeh Chung Mu-Tan Hua Phin*[8] (Grades of the Tree-Peonies of Yüeh) i. e. Southern Chekiang, precisely the oldest centre where the wild varieties had been noticed. The monk described 32 of the most beautiful varieties, and one may suppose that he did not fail to transplant them to a paeoniaceous sangha in the abbey gardens. There is some evidence that the next account was that of 'Mr Minister Fan' (Fan Shang Shu[9]), at any rate, there was a treatise by him (*Mu-Tan Phu*[10]) describing 52 varieties, but it has long been lost.[c] Then comes the outstanding book of the whole genre, Ouyang Hsiu's[11] *Lo-Yang Mu-Tan Chi*[12] (Account of the Tree-Peonies of Loyang) written in +1034, which we possess complete and must quote in a moment. After personal and oecological introductions, this great scholar, then a young man, goes on to list 24 varieties of

---

[a] Tr. auct., adjuv. Ledyard (2). The attraction of *Buddleia* for certain butterflies is yearly noticeable in the Lodge garden at Caius. Ch. 2, p. 8*a*.

[b] Ch. 1, p. 33*a*.

[c] It would be convenient if this Fan were the Fan Yao-Fu (Fan Shun-Jen[7]) referred to in the quotation given above from the *Chhün Fang Phu*, but as he was not born till +1026 or +1027 this is not possible. So either the book of Fan came after that of Ouyang Hsiu, Fan Shun-Jen writing it about +1055, or else the author was someone else of the same clan less well known.

[1] 杜陽雜編　　[2] 蘇鶚　　　　[3] 清異錄　　　[4] 陶穀　　　　[5] 百葉仙人
[6] 太平樓閣　　[7] 仲休　　　　[8] 越中牡丹花品　　　　　　[9] 范尚書
[10] 牡丹譜　　　[11] 歐陽修　　　[12] 洛陽牡丹記　[13] 范純仁

P. suffruticosa cultivated in the city in his time. In each case he notes the characteristics of the plant, such as petal structure, colour, leaf-shape and so on, with anything that he knows about the origin of the variety. At the end he records various methods of cultivation, watering, pruning, grafting, sheltering, protection from plant pests, etc., as we shall see. It appears, moreover, that the first editions of his monograph were illustrated.

To begin with Ouyang Hsiu gives his personal experiences.

During my time in Loyang [he wrote] I lived through the spring four times. I first came to Loyang in the third month of the 9th year of the Thien-Shêng reign-period (+1031), but as my arrival was late I saw only the late-flowering varieties. In the following year, together with my friend Mei Shêng-Yü[1][a] (combining pleasure with official business) I travelled about on the Shao-shih ridges of Sung Shan (Mtn.), and on Kou-shih-ling, and we saw the Purple Cloud Grottoes on Shih-thang Shan, but returned to the city too late to see any tree-peonies. Then the year after, I suffered a bereavement in my family and had no leisure to go and look at flowers. Finally in the last year my term of office as Prefectural Judge in the Imperial Custodian's Court came to an end, so that I was released from duty and had to leave (for the capital); thus I could only see the early-flowering varieties. So I have never really seen the flowers at the height of their season. All I can say is that those I did cast my eye upon seemed to me absolutely unsurpassable in loveliness.

When I was living in the prefecture I once paid a visit to Chhien Ssu-Kung[2][b] at the Tower of the Twin Cassias, and there saw a small screen standing behind some chairs. It was all covered over with small-sized characters. Chhien Ssu-Kung pointed to it and said: 'If you want to write a classification or grading of flowers, here are the names of more than ninety varieties of the tree-peony.' Of course I could hardly read and note them all at the time. But those which I have myself seen, and those most valued by the people, amount to about thirty kinds. I do not know where Chhien Ssu-Kung obtained so many varietal names. As for the others, although they are named, some are not well known and some are not particularly good. Therefore those which I record include only the best known ones, which I rank and describe in the following order.[c]

And he goes on to name the twenty-four, from the famous Yao yellow down to the Jade castanet white. Here is his account of what the plants meant to the people.[d]

Almost everyone in Loyang loves these flowers. In springtime the whole population, patricians and commoners alike, decorate themselves and their homes with tree-peonies; even the coolies are no exception. When the flowers bloom, gentry and people vie with each other in making excursions and outings to admire them, often visiting old temples and deserted mansions where stalls are set up beside ponds and terraces. Tents and awnings are raised, while the sounds of singing and the *shêng* organ[e] can everywhere be heard. All this goes on particularly by the Crescent Moon Embankment, at the Chang

---

[a] I. e. Mei Yao-Chhen, +1002 to +1060, Horatian poet who wrote on agricultural techniques, often quoted by Wang Chên in the *Nung Shu*.

[b] I. e. Chhien Wei-Yen, d. +1029, provincial governor and Minister of Works who had known the great architect Yu Hao.

[c] Ch. 1, p. 2*a*, *b*, tr. auct. adjuv. Hagerty (16); Ledyard (2).

[d] Ch. 3, p. 6*a* ff., tr. auct. adjuv. Hagerty (16); Ledyard (2); Li Hui-Lin (8), p. 26.

[e] On the reed month-organ see Vol. 4, pt. 1, pp. 145ff, 211.

[1] 梅聖俞   [2] 錢思公

family garden, in the Crab-apple and Plum Ward, beside Chhang-shou Ssu (temple), in East Lane and near the residence of the Comptroller of the Suburbs. When the flowers fall, all is over.

Between Loyang and the eastern capital (Khaifêng) there are six post-stations. Formerly there was no tribute of these flowers, but after Prime Minister Li Ti[1] of Hsüchow became Resident Imperial Custodian, they began to be offered to the Court. Each year one of the Yamen Collators is despatched (with them), and riding on relays of post-station horses he reaches the capital in one day and one night. Only three or four buds of the Yao yellow or the Wei variety are sent, packed solidly inside little bamboo crates filled with fresh cabbage leaves so that they will not be shaken and jarred on the envoy's horseback journey, all bound up and covered over. Wax is used to seal the stems so that the petals will not fall for several days.[a]

Most Loyang households have tree-peony flowers, but few of the plants grow into big trees, because without grafting the most elegant blooms are not produced. Early in the spring Loyang people go out to the Shou-an Shan (hills) and cut small scions (*tsai tzu*[2]) for sale within the city, cuttings which are called 'mountain combs' (*shan pi tzu*[3]). The townsfolk cultivate the soil around their homes, making little bordered garden plots in which they plant the cuttings, and then when autumn comes they are grafted.[b] The most famous practitioner of this art was a man called Mên the Gardener (Mên Yuan Tzu[4])[c] and among the rich families there were none who failed to employ him. A single graft of the Yao yellow is worth 5000 cash. Contracts are made in the autumn, and when spring comes and the flowers appear, then the full payment is remitted (to the grafter). Loyang people are deeply attached to this variety and loth to allow the spread of its cultivation. If privileged patricians or high officials come looking for this flower they are likely to be given a graft which has been killed by dipping in hot water. When the Wei variety[d] made its first appearance, a single graft bud also had a value of 5000 cash, and even now it is worth 1000.

As for the grafting season, it is essential to use the time between the Shê-jih and Chhung-yang festivals (app. 23 Sept. to 23 Oct.) otherwise it will not succeed. (In making a graft), cut the stock about five to seven inches above the ground. Seal and wrap the joint with mud, and surround the whole with loose soft soil, making a little hood of rush leaves to protect it from the sun and wind, yet leaving an opening to the south for ventilation. When spring comes the covering should be removed. Such is the manner of grafting tree-peonies. [It is also all right to use a tile.][e]

If tree-peonies are to be raised from seed it is necessary to select a suitable place. First remove the old earth completely, and put down fresh fine soil mixed with one catty of the dried powder of the *pai lien*[5] vine.[f] Probably because the roots of the tree-peony are sweet

---

[a] Effective for inhibition of transpiration. Sung people would have said that it prevented loss of *yuan chhi*.[6]

[b] From Fortune (7) and the other standard sources it is clear that the favourite root stocks were always the herbaceous peonies, *shao-yao*.

[c] A comment by the editor, Chou Pi-Ta[7] (+1126 to +1204) points out that his real family name must have been Hsimên or Tungmên. People called Huangfu, for instance, were often known as just Huang.

[d] Flesh-pink, double.

[e] Comment in the text, more likely by Ouyang Hsiu himself.

[f] This is *Ampelopsis serianaefolia* of the Vitaceae (CC 763 = *Vitis s.*, and V. *aconitifolia*, R 287). Its root is employed in medicine (Stuart (1), p. 458), being considered 'anodyne and cooling', but there can be no doubt of its activity as a plant insecticide like *Pyrethrum*. It is interesting that the Thang and Sung horticulturists were so clear about its effects.

[1] 李迪　　　[2] 栽子　　　[3] 山篦子　　　[4] 門園子　　　[5] 白蘞
[6] 元氣　　　[7] 周必大

and pleasant they attract many insects which feed on them, but the *pai lien* powder has the power of killing the grubs and caterpillars. Such is the manner of raising tree-peonies from seed.

For watering, there are appropriate times; either before sunrise, or when the sun is westering. In the ninth month (one should) water once every ten days, in the tenth and eleventh once every two or three days. In the first month water on alternate days, and during the second month every day. Such is the manner of watering tree-peony plants.[a]

If a plant stem develops several buds, select the small ones and remove them, leaving only one or two. This is called 'peeling off' (*ta po*[1]), and is done for fear that the (generative) energy will (over-) divide itself (in too many channels). As soon as a flower has fallen, its pedicel should be snipped off so as to prevent the formation of seeds, for fear that the plant should age too readily. When the hood of rush-leaves is removed in the spring, several branches of the jujube tree (*Zizyphus* spp.) should be placed over the little bush, for since the *chhi* of this plant is warm (*nuan*[2]) it is able to ward off the frost—and the same thing can be done for larger tree-peonies too. Such is the manner of nourishing the flowers.

When the blooms seem to be smaller than they formerly were, there are probably some boring insects (*tu chhung*[3]) injuring the plant. Then it is necessary to search out the cavities they have made and stuff them with sulphur powder. There are also smaller holes like the eyes of needles, in which the insects' (larvae) live; these are called by the gardeners 'vitality vents' (*chhi chhuang*[4]), and taking large needles tipped with sulphur (paste) they introduce it into them—then the insects die. So once again the flowers will flourish. Such are the methods for dealing with diseases of the tree-peonies.[b]

But if cuttle-fish bone is used for puncturing the stalks of the flowers, they will die as soon as the cortex is penetrated. Such is one of the antipathies of tree-peonies.[c]

Thus we have some social description and directions for safe transport, followed by the techniques of grafting, raising from seed, watering, pruning, protection from weather and plant pests, and marketing values. It seems very doubtful whether anything like so sophisticated an account could have been produced by a Saxon England contemporary of Ouyang Hsiu. And one more thing remains. We must listen to him speculating (just as Han Yen-Chih did in another connection, cf. p. 372) about the oecological significance of Loyang as a tree-peony centre.

After saying that the best tree-peonies in the world are cultivated in Loyang, and that the Loyang people are inordinately proud of them, calling them simply 'The Flower', he goes on to say:[d]

[a] This indicates, says Ledyard, the dryness of the soil. The times given have to be advanced about six weeks to make them correspond with the Western calendar.

[b] This method of getting rid of peony borers, says Ledyard, is still used widely in Japan. They are presumably the larvae of pyralid or tortricid moths, or boring beetle larvae. Cf. Dodge & Rickett (1), p. 453; Wardle (1), pp. 256, 266.

[c] The *wu-tsei yü*[5] mentioned is a general term for *Sepia* spp. (R 180). Pharmaceutical tradition, summarised in *PTKM*, ch. 44 (p. 122), recommended the bone (almost wholly calcium carbonate, so far as we know) as a vermifuge, and Ouyang Hsiu was therefore counselling that it was dangerous for the plant as well as the insects. But it would be very hard to see the rationale of this.

[d] Ch. 1, p. 1b ff., tr. auct. adjuv. Hagerty (16); Ledyard (2).

¹ 打剝   ² 暖   ³ 蠹蟲   ⁴ 氣窓   ⁵ 烏賊魚

In discussing these matters many say that of the whole country within the Three Rivers Loyang is pre-eminently the land of antique virtue. Anciently the Duke of Chou investigated here with graduated instruments the waxing and waning of the sun's (gnomon shadows), coming by these measurements to know the (alternation of seasons), the cold and heat, the wind and rain, inhibiting or favouring (men's doings) in this region.[a] And since therefore this region is at the centre of heaven and earth, the flowering plants and trees obtain an abundance of the harmonising power of the *chhi* of the Centre.[b] Naturally therefore they are unique and different from those of all other places. I must say I disagree entirely with this.

It is true that in Chou times Loyang was about the centre of the Nine Provinces, so that tribute coming from all the Four Quarters had approximately the same distance to go to get there. But when one considers the universe around the Khun-Lun Mountains (*Khun-Lun phang po*[1]), Loyang is not necessarily the centre.[c] Besides, even if it were, one would expect that the harmonising *chhi* of heaven and earth would spread all over the four quarters above and below, rather than being limited, as if for private profit, to a central region.

As for 'centrality' and 'harmonious mixing', what (they really mean is this). There is a *chhi* of constancy and normality (*chhang chih chhi*[2]), and when it is manifested in things they have a normal or standard form, being neither particularly beautiful nor particularly ugly. However, when there is a defect in the (inherited) life-force (*yuan chhi chih ping yeh*[3]), beauty and ugliness arise because (the *yuan chhi* are) segregated and cannot attain a harmonious mixture (*pu hsiang ho ju*[4]). So things which possess extreme beauty or extreme ugliness are the results of an imbalance in the vital *chhi* (*chieh tê yü chhi chih phien yeh*[5]). The beauty of a flower, the grotesque ugliness of the twisted bulge of a gnarled tree, are indeed different, yet they are equal in their defectiveness, for each comes from an unbalanced quality in the vital *chhi*.[d]

Now Loyang has a circumference of some tens of *li* if you go round its city-walls. Among the tree-peony varieties of the *hsien* districts round about, none reach the quality of those within the city, while outside their boundaries these cannot be planted (with success). How can it be that the unbalanced *chhi* leading to beauty collects only within the space surrounded by these several tens of *li*? This is indeed one of the great (mysteries) of Nature (lit. heaven and earth) which can hardly be investigated.

Extraordinary things which cause harm to mankind we call calamities (*tsai*[6]), extraordinary but harmless things which only cause wonder and amazement we call marvels or monstrosities (*yao*[7]). The saying has it: 'Heaven contravening the seasons means calamity, Earth conflicting with normal things means a marvel.' The tree-peony is truly a bewitching marvel among the plants, and one of the wonders of the ten thousand things. Unlike the twisted bulge of a gnarled tree (its *chhi* is unbalanced) only on the side of beauty—therefore it finds favour and blessing among men.

[a] This is a reference to the classical Central Observatory of China at Yang-chhêng (now Kao-chhêng) near Têngfêng southeast of Loyang; see Vol. 3, pp. 296 ff.
[b] Cf. p. 86 (*i*) above on the Centre and the Four Quarters.
[c] Here is a Buddhist cosmological echo, the world centered on Mt Meru; cf. Vol. 3, pp. 565 ff., Vol. 4, pt 2, pp. 529 ff.
[d] This important passage on aesthetics was quoted later verbatim in *Chhün Fang Phu*, Hua phu, ch. 2, p. 19*a*.

[1] 崑崙旁礴   [2] 常之氣   [3] 元氣之病也   [4] 不相和入   [5] 皆得於氣之偏也
[6] 災   [7] 妖

Thus Ouyang Hsiu, ignoring the subjectivity of man's appreciations, had re-course to a theory of statistical regularity involving a 'standard mixture' or *krasis* (κρασίς),[a] departures from which in any direction would produce abnormal forms, whether beautiful like the peony or grotesque like an old tree-stump. It was a theory in itself very reasonable, though like all medieval theories hardly capable of quantitative proof or disproof—nevertheless it was a theory, and as in so many other instances disallows the often-held view that +11th-century people, and certainly +11th-century Chinese, did not go in for theories.

At least four other tree-peony books or tractates followed before the century was out. Contemporary or almost so with Ouyang Hsiu's was the *Chi Wang Kung Hua Phin*[1] (Grades of (Tree-Peony) Flowers in the Gardens of the Prince of Chi), i.e. Chao Wei-Chi[2], a grandson of the first Sung emperor. Next came, in +1045, another voice from Chekiang and Chiangsu, Li Ying[3]'s *Wu Chung Hua Phin*[4], followed five years later by a curious work, the *Mu-Tan Jung Ju Chih*[5] (King's Daughters and Humble Handmaids), a classified arrangement of tree-peony varieties by analogy with the ranks of ladies attending upon the emperor.[b] Here Chhiu Hsüan[6] only carried to a fanciful extreme the passion for grading, as if in a modern horticultural show, what would have given more profit by closer botan-ical description.[c] In +1082 the Loyang tree-peonies were monographed again, in a book with the same title as Ouyang's, by Chou Shih-Hou[7].

Immediately the next century began, the activity of other centres manifested itself. Chang Pang-Chi[8] described the tree-peonies of Chhenchow in +1112 or soon after.[d] Some of his words are graphic.[e]

The Niu family nursery developed a special *mu-tan* variety the colour of a newly-hatched goose (pale greenish-yellow). The soft flower corolla (*pa*[9]) had a diameter of 1 ft 3 or 4 in. and formed a mass about a foot high, with 1100 petals piled up layer upon layer. It was developed from the Yao yellow. At the top within the corolla there was a halo of fine thread-like (stamens) covered with golden (pollen) powder.[f] The centre of the flower was purple, but the pistil was also covered over with golden powder.[g]

Grower Niu named it 'Gold-thread yellow' (*Lü chin huang*[10]). He built a marquee of

---

[a] On such conceptions see further in Sect. 44. It is interesting to notice that Ouyang Hsiu's theory regarded the most perfect balance of natural forces (e.g. Yin and Yang) as giving rise to the average between perfect beauty and radical ugliness. The theory he combated, on the other hand, regarded the greatest beauty (and goodness) as resulting from the most perfect balance of natural forces (e.g. Yin and Yang). This latter as perhaps more characteristic of classical Chinese natural philosophy.

[b] In the Thang and Sung the general personnel comprised one empress, three consorts, nine spouses, twenty-seven concubines, and eighty-one assistant concubines. For further information on the system and its cosmolo-gical symbolism see Vol. 4, pt 2, pp. 476 ff.; Needham, Wang & Price (1), pp. 170 ff.

[c] On Chinese categorical thinking, cf. Bodde (5).

[d] This city, now called Huaiyang, is in the east of Honan.

[e] Cit. *TSCC*, Tshao mu tien, ch. 289, p. 4*b*, tr. auct.

[f] *Yü pa ying chih tuan yu chin fên i yün lü chih.*[11]

[g] *Chhi hsin tzu, jui i chin fên lü chih.*[12]

---

¹ 冀王宮花品   ² 趙惟吉   ³ 李英   ⁴ 吳中花品   ⁵ 牡丹榮辱志
⁶ 邱璿   ⁷ 周師厚   ⁸ 張邦基   ⁹ 葩   ¹⁰ 縷金黃
¹¹ 於葩英之端有金粉一暈縷之   ¹² 其心紫蘂亦金粉縷之

bamboo matting around it, with suitable barriers, and spread caerulean silk decorations at the gate, after which guards were set so that only those who paid 1000 cash were allowed in to see it. In ten days the family made hundreds of thousands of cash. I myself was one of those who managed to get in and see it.

Later the governor of the prefecture heard about it, and wanted to send up a cutting in presentation to the imperial court, but all the nurserymen maintained that he must not do so, saying that this was no ordinary flower and might only too easily change. After a while the governor renewed his suggestion, wondering how best to react to the phenomenon, and proposing that a divided root should be presented, but the nurserymen replied politely but firmly in the same way as before. Next year when the flowers opened they had all reverted to the previous (common variety). This was truly a queer marvel (*yao*[1]) of the plant world.

Thus we have a rather striking piece of description, and an account of a reversion to type after an unstable cross, autogenous chimaera or somatic variation, complicated perhaps by root-stock and environmental factors.

The Szechuanese cultivation centres now began to become famous. On the northern edge of the Kuanhsien irrigation system,[a] with the Tibetan foothills to the west, stands the little town of Phêng-hsien, having in its district a hill called Thien-phêng-mên[2]. Here it was that early in the +12th century the air and soil were found to suit tree-peonies capitally, so that before long Thien-phêng became almost as celebrated as Loyang. In +1178 Lu Yu[3] recorded it all in his *Thien-Phêng Mu-Tan Phu*[4], describing 33 varieties from white through yellow to red and deep purple, all entirely different from those previously listed by Ouyang Hsiu at Loyang. Both he and Hu Yuan-Chih,[5] who wrote a further tractate not long afterwards, could trace back the *mu-tan* culture to +915 and document the coming of it to Thien-phêng. Jen Shou[6] also monographed these flowers before the end of the Sung (*Phêng-mên Hua Phu*[7]).

In the Ming period two princes of the imperial house wrote on tree-peony varieties, first Chu Yu-Tun (Chou Hsien Wang) *c.* +1430, as already related (p. 332 above), and then Chu Thung-Chi[8], a descendant of Chu Chhüan[9] (Ning Hsien Wang, see p. 333) in the eighth generation. His *Mu-Tan Chih*[10] (Record of Tree-Peonies), *c.* +1580, is however lost. More important was the work of Hsüeh Fêng-Hsiang[11] about +1610; his *Mu-Tan Pa Shu*[12] (Eight Epistles on the Tree-Peony) was a great amplification of Ouyang's remarks on techniques. In connection with root-stocks he tells us that the art of grafting the special varieties on to roots of the wild forms rather than on to those of the herbaceous peony was mastered only about +1600, and gave much better results. Hsüeh also wrote on the tree-peony varieties of yet another cultivation centre, Pochow,[b] and by this

---

[a] Cf. Vol. 4, pt 3, p. 288.
[b] On the north-western edge of Anhui, now Po-hsien. The titles were: *Po-chou Mu-Tan Shih*[13] and ... *Piao*[14].

[1] 妖 　　　 [2] 天彭門 　　 [3] 陸游 　　 [4] 天彭牡丹譜 　 [5] 胡元質
[6] 任璹 　　　 [7] 彭門花譜 　 [8] 朱統鐥 　 [9] 朱權 　　　 [10] 牡丹志
[11] 薛鳳翔 　　 [12] 牡丹八書 　 [13] 亳州牡丹史 　 [14] 表

time the number of varieties recorded had risen to no less than 266. Several other scholars also wrote on the Pochow tree-peonies, notably Kao Lien[1] that encyclopaedic amateur, in +1591, as part of a *Mu-Tan Phu*[2] which gives a further mass of details on the horticultural techniques in use; and Niu Hsiu[3] in +1683. Niu was an official at Hsiang-chhêng over the border in Honan and never actually went to Po-chow, but he collected all the varieties of the plants and recorded 140 from that place alone.

The last city to become a centre of *mu-tan* culture was Tshaochow[a] in Shantung, very near Honan, and this place, seconded by Pochow, remains the chief headquarters of peony growing and crossing in China to the present day. In +1669 Su Yü-Mei[4] described nurseries with thousands of grafted plants at Tshaochow, at least the equal of the Loyang gardens, and in +1793 Yü Phêng-Nien[5], a painter, noted down the characteristics of 56 Tshaochow varieties; both put the city's name at the head of their titles. In 1809 Chi Nan[6], a great expert, wrote one of the best books on the subject (*1*), describing in detail the cultivation methods for 103 of the best varieties, and thus completed nearly a millennium of close phytotechnical study accomplished for the most part before modern European botany had begun. Tree-peony plants were exported from China to Japan from the +8th century onwards,[b] but they did not reach Europe until much later. They were known there early in the +18th but it was not until Sir Joseph Banks urged Dr Duncan of the Hon. East India Company to bring back some living plants that the first *mu-tan* arrived in England in +1787. When Robert Fortune was sent to China for the Royal Horticultural Society in 1842, *mu-tan* was one of the plants he was particularly to look for, and he did indeed bring back in course of time some forty varieties, now spread all over the world.[c]

The herbaceous peony (*Paeonia lactiflora = albiflora*) also had its devotees through the centuries. As we have already noticed, it was a medicinal plant from antiquity onwards,[d] and in the course of long cultivation for this purpose hundreds of varieties arose. The first to write about it from the purely botanical point of view was a lesser-known scholar-official contemporary with Ouyang Hsiu and like him stationed at Loyang, Chang Hsün[7], whose *Loyang Hua Phu*[8], written about +1045, discussed both *mu-tan* and *shao-yao* at length.[e] However, the centre at which the culture developed best was Yangchow, the fabulous mercantile city on the Grand Canal just north of its crossing with the Yangtze. In the *Chhün Fang Phu* we read:[f]

[a] Now Tshao-hsien.   [b] Cf. Yashiroda (*1*); Miyazawa (*1*).   [c] See Fortune (6, 7); Hoffmann (*1*).
[d] The roots were considered tonic, alterative, astringent, analgesic, diuretic and carminative (Stuart (*1*), p. 300).
[e] It may be of interest that with his brother he had been a pupil of the philosopher Shao Yung (cf. p. 305 above), who had a marked interest in all natural phenomena.
[f] Hua Phu, ch. 4, pp. *1a* ff., tr. auct.

[1] 高濂   [2] 牡丹譜   [3] 鈕琇   [4] 蘇毓眉   [5] 俞鵬年
[6] 計楠   [7] 張峋   [8] 洛陽花譜

The pharmaceutical natural histories all say that the *shao-yao* grows as a plant both elegant and beautiful. Every place has it, but the best grow in Yangchow. It is said thet the climate and soil are just right for it so that it comes there to perfection, just as the tree-peony does at Loyang.... It grows in clusters about 1 to 2 ft high, each stem having three branches and five leaves resembling those of the *mu-tan* but narrower and longer. The flowers bloom in the early summer, red, white, purple, etc. in colour. The yellow ones are generally considered to be the best. There are single (*tan*[1]) varieties, double ones (*chhien yeh*[2]) and piled-up double ones (*lou tzu*[3]). The seeds are like those of the tree-peony but smaller.

Interest seems to have been greater in the Sung than at any other time, for several further monographs were written in the +11th century, generally entitled *Shao-Yao Phu*. Liu Pin's[4] study came in +1073 or just after, for early in that year he was posted to Kuang-Lin (Yangchow), and it may well have been illustrated, for he says that he commissioned artists to make pictures of the varieties.[a] Two other treatises appeared very little later, Wang Kuan's[5] and Khung Wu-Chung's[6], by +1080; the latter almost certainly illustrated.[b] At this time some forty varieties were recorded. Great care was taken in cultivating them—the roots were dug up in the autumn, carefully washed and all dead tissue removed, then divided to stimulate new growth if this had not been done a year or two before, and re-planted in fresh soil which had been heated and sieved. Much attention was given to the admixture of the soil with the right proportion of sand, and with the right kinds of manure. After the Sung there was a long gap but towards the end of the Ming people were writing about the *shao-yao* again, notably Tshao Shou-Chên[7] about +1540 whose interest had been aroused by the work of Liu Pin. His is the most complete monograph with the possible exception of that of Kao Lien[8] (*c.* +1591). *P. lactiflora* was carried to Japan in medieval times, but like the tree-peony did not spread to the West until the +18th century, since when it has been widely grown.[c] Then in very recent times botanical explorations in the remoter regions of west and south-west China have introduced a number of other species of herbaceous peonies to the world (cf. Sect. 38 (*j*), 2).

(iv) *Chrysanthemums*

If the tree-peony was flower-king in China, the chrysanthemum was chief of the scholars and president of the academy. Its great variability made the attainment and preservation of thousands of forms and colours possible, and generated a wealth of botanical-horticultural writings probably greater than for any other

---

[a] Liu Pin was a distinguished scholar and wit, he helped Ssuma Kuang in the *Tzu Chih Thung Chien* project, and made a commentary on the *Chhien Han Shu*.

[b] Wang Kuan's title was *Yang-Chou Shao-Yao Phu*; that of Khung Wu-Chung's was just *Shao-Yao Phu*. The latter is preserved complete in the *Nêng Kai Chai Man Lu*, and almost so in the *Chhüan Fang Pei Tsu*.

[c] The introduction was effected by Pallas in +1784, and to England by Banks in 1805.

[1] 單   [2] 千葉   [3] 樓子   [4] 劉攽   [5] 王觀
[6] 孔武仲   [7] 曹守貞   [8] 高濂

genus. Although the flowering season does not continue for more than about a month and a half it is particularly prominent because it comes in the autumn, when nearly all the other plants of the garden have gone to seed and are beginning to die down. I shall never forget the splendid collection of chrysanthemums which I came upon one day of beginning autumn chill in some pavilions of the Pei Hai park at Peking, rank upon rank in pots assembled, unimaginably rich in their diversity.[a] It is true that there are in China a greater number of fanciful and curious varieties with strikingly beautiful colorations than are in cultivation in the gardens of all the other countries of the world put together.[b] For the garden chrysanthemum, *C. hortorum*, as it is called, is now considered, after much discussion, the product of an inter-specific cross between *Chrysanthemum indicum* (= *japonicum*) and *C. morifolium* (= *sinense*), both of which are wild species native to China.[c] Probably the hardy small-flowered outdoor garden varieties hark back mainly to the former, while the large-flowered greenhouse types contain more traits of the latter. In nature the flower-rays of the former are always yellow, while those of the latter, a stouter plant, are purple, red or white, but not yellow.

The chrysanthemum has always been called *chü*[1, 2, 3], but the way of writing the character varied, and did not settle down to its present form, the third, until the Han.[d] During the Chou period the genus was well recognised and given this single-character name, as we can see from some of the ancient calendar texts. The *Yüeh Ling* (Monthly Ordinances), later incorporated in the *Li Chi*, mentions the yellow flowers of the *chü*[5] as a sign of late autumn,[e] and this takes us back to the −7th century, for on astronomical grounds we have to place the document within a couple of hundred years on each side of −620.[f] Similarly the plant is mentioned at the ninth month in the *Hsia Hsiao Chêng* (Lesser Annuary of the Hsia Dynasty),[g] a book which we situate in the −5th century.[h] The −4th-century *Erh Ya* explains *chü*[6] (now giving it its *tshao-thou*, cf. p. 118) as being the same as *chih-chhiang*[7], and Kuo Pho's commentary of about +300 brings us back to normality by saying that both mean the autumn-flowering *chü*.[8] But *chü hua*[9] had already appeared in the Early Han *Shen Nung Pên Tshao Ching*, taking its place as a safe and beneficial medicament among the *shang phin* group (cf. p. 243).[i] And this was the orthography always subsequently retained.[j]

Why was the chrysanthemum chief of the scholars? The symbolism which

[a] It was a real *chiu hua shan tzu*[4] (cf. *Yen-Ching Sui Shih Chi*, Bodde tr. p. 71).

[b] Li Hui-Lin (8), p. 42.

[c] See Cibot (6); Paxton (1); Hemsley (1); A. Henry (2); Payne (1); E. D. Smith (1); Emsweller (1) and now especially Chhen Fêng-Huai & Wang Chhiu-Phu (1).

[d] See B II130, 404.   [e] *Li Chi*, ch. 6, p. 78a, tr. Legge (7), vol. 1, p. 292.

[f] Cf. Vol. 3, p. 195.

[g] *Hsia Hsiao Chêng Su I*, (p. 47). Also incorporated in *Ta Tai Li Chi*.

[h] Cf. Vol. 3, p. 194. Rickett (1), in the midst of a closely reasoned argument, agrees, (1), pp. 189 ff., but prefers an early −4th-century date for the *Yüeh Ling*.

[i] Mori ed. (p. 31), cf. B III 69.   [j] *Hua*[9] was of course always interchangeable with *hua*.[10]

[1] 鞠	[2] 蘜	[3] 菊	[4] 九花山子	[5] 鞠
[6] 蘜	[7] 治蘠	[8] 菊	[9] 菊華	[10] 花

accreted around this plant deserves some notice. First, autumn was always extremely romantic and melancholy in Chinese feeling ('the falling leaves of a thousand autumns')[a] a season of Yang giving place to Yin and heralding hard winters, a season when sentences of execution were anciently carried out, a time of withdrawal, acceptable to hermits. Then the chrysanthemum symbolised the Confucian scholar, for just as it withstood the cold dews of shortening days, so too he must withstand the disfavour of the emperor and the disapproval of official colleagues and superiors, if necessity arose, when he stood firm in support of some unpopular principle or to rebuke some departure from Confucian ethics. If he failed (and managed to live through it), he might well retire from public life, like a leaf shed from the tree of society, and live in the depths of the country with only Nature and his books. The great Chin poet Thao Yuan-Ming[1] (Thao Chhien[2], +372 to +427) did just this, and it was he more than anyone, perhaps, who gave impetus to the cultivation of chrysanthemums. A phrase derived from his writings: 'Thao Yuan-Ming planted three rows (of chrysanthemums) to give gay colour to his eastern fence' became proverbial,[b] and only in the light of it can some of the titles of later monographs on chrysanthemum culture be understood.[c] The chrysanthemum shunned the company of other flowers by blooming at a different time of year, yet neither gloomily nor enviously, so also the Confucian, become more than half a Taoist, was content to live the life of a recluse. Thirdly there was the theme of longevity and immortality, already present in the description of the plant's virtues in the *Pên Ching*, and permanently associated with the drinking of chrysanthemum wine on the Chhung-Yang[11] festival (the ninth day of the ninth month), when one of the observances was to make an expedition to any local hill-top.[d] Just as the chrysanthemum bloomed in the old age of the year, so might old age be blooming, and even (if Taoist macrobiotic art could succeed) bloom on for ever, overcoming evanescence itself.

Such were some at least of the components of the mental background of those scholars and gardeners who laboured, from Thang times onwards, to see what could be done with *Chrysanthemum hortorum*.[e] The initial hybridisation having been

[a] Cf. Needham (66), p. 185.

[b] *Thao Yuan-Ming chih yü san ching, tshai yü tung li.*[3] 'Eastern Fence' here may rather be a place-name, perhaps that of his garden, which was certainly near Chhai-sang[12], some thirty miles south-west of Hsünyang[13] (mod. Chiu-chiang[14]) in the north of Chiangsi near the Poyang Lake.

[c] E.g. Kao Lien's *San Ching I Hsien Lu*[4] of c. +1585 (cf. p. 361); Thu Chhêng-Khuei's *Tu Hua Chü Tung Li Chi*[5] of c. +1630 (cf. p. 415); and the later books *Tung Li Phin Hui Lu*[6] by Lu Pi[7], and *Tung Li Tsuan Yao*[8] by Shao Chhêng-Hsi.[9]

[d] Cf. Bodde (12), pp. 69 ff.; Bredon & Mitrophanov (1), pp. 425 ff. This was called *têng kao*[10]. Similar excursions also took place on the 7th and 15th days of the first month.

[e] Very little has been written about this epic in Western languages, though one may look at Payne (2); Rivoire (1) and Clément (1). A great quantity of information remains to be disengaged from Chinese original sources.

[1] 陶淵明　　　[2] 陶潛　　　[3] 陶淵明植於三徑采於東籬　　　[4] 三徑怡閒錄
[5] 渡花居東籬集　　　[6] 東籬品彙錄　　[7] 盧璧　　　[8] 東籬纂要
[9] 邵承熙　　　[10] 登高　　[11] 重陽　　　[12] 柴桑　　　[13] 尋陽
[14] 九江

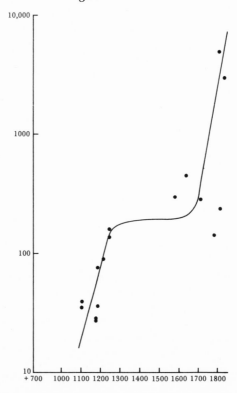

Fig. 85. A semilog plot of the increase in the number of chrysanthemum varieties in China over the centuries.

effected at least as early as Thao Yuan-Ming, there followed century after century of careful selection of the new forms which resulted from continual crossing, mutation, and other sources of novelty. The oldest colours were certainly the yellows, but from the +8th century onwards white-flowered forms were often praised in poems, and somewhat later purple-flowered ones became prominent. The number of varieties cultivated and described increased enormously, and so voluminous is the literature that it is possible to attempt a plot of the number of them at different periods. This is done on semilog paper in Fig. 85, which suggests that after the Sung period there was a lull, renewed activity in breeding resuming towards the end of the Ming. Certainly from about 20 varieties at the beginning of the Sung (*c.* + 1000) the number noted and kept increases to more than 2000 in the early part of the 19th century. Doubling was already common at the beginning of this time, as we shall see in a monent. The plants were raised from seed as well as by asexual propagation, and one interesting technique was used to which we shall later return (Sect. 38 (*i*), 4), namely the shallow planting in the soil of the whole withered flowers themselves, conducive apparently to the production of many fresh forms. Variants were also induced by grafting on to heterologous root-stocks. The genetics of all these was especially complicated because while the initial lines were polyploids (*C. indicum* tetraploid and *C. morifolium* hexaploid), the clones of the infertile propagated stocks are generally aneuploids (with a few chromosomes more or less than the basic polyploids).

As in the case of the peonies, the first monograph came in the Sung. Chrysanthemums were spoken of by Chou Shih-Hou[1] in his *Lo-Yang Hua Mu Chi*[2] (Records of the Ornamental Flowering Plants and Trees of Loyang), written in +1082, but the first book devoted exclusively to them came at the beginning of the following century, when Liu Mêng[3] produced his *Chü Phu* in +1104. At that time there were gardens near the famous Buddhist cave-temples at Lungmên, which he visited and admired so much that he decided to record 35 varieties.[a] The language of this little treatise is very elegant, and we may read part of the section which he entitled 'Resolution of Doubts', because it illustrates the efforts towards botanical precision which Liu Mêng made.

Some say that the *chü*[4] and the *khu-i*[5][b] are two different species. What Thao Hung-Ching and Jih Hua Tzu recorded[c] certainly did not include double forms (*chhien yeh hua*[6]), so certain readers might suspect that this treatise is including some plants not (true) chrysanthemums. But I often read Thao Hung-Ching, who said that when the stems are coloured purplish-caerulean, and when the plant has the odour (*chhi*) of the *hao*[7] and the *ai*[8][d]— that is the *khu-i*.[5] Now what I have put in this book as chrysanthemums (*chü*[4]), though some have caerulean stems they all have a fragrant *chhi* and a sweet sapidity, the leaves and branches being less fibrous. And although some have a bitter sapidity and purple colour, with delicate stems, they still do not have the *chhi* of the *hao*[7] and the *ai*[8]. Furthermore this type has been handed down among the people as *chü*[4] for a very long time; therefore we could not lightly adopt the old saying and leave them out from our treatise.[e]

As for the plants that people have specially selected, if the cultivation and watering are extremely favourable, then the branches, leaves, flowers and seeds are liable to become excessively big. And so in the same way there sometimes occurs an inner collocation of *chhi* within the plant which brings about the happy result of a doubling of the petals, and a duplication of the flowers themselves on their peduncles, not to mention how sometimes a transformation of flowers into the 'thousand-petalled' varieties (*chhien yeh*[6]) occurs.

Jih Hua Tzu says that those plants with large flowers are the 'sweet chrysanthemums', while those with small flowers and bitter taste are the 'wild chrysanthemums'. If the latter

[a] Plus four which he had not actually seen himself, and one of the two wild-growing species.

[b] The *khu-i* is one of the wild parent species, *Chrysanthemum* (= *Pyrethrum*) *indicum* (R 26; CC 53). Also known as *yeh chü*.[9]

[c] The words of the former (*c.* +500) are preserved in *PTKM*, ch. 15, p. 3*b*. The latter's name was Ta Ming (cf. p. 280), writing about +970.

[d] *Artemisia* species, also of the Compositae. Cf. pp. 494–6. The *ai* of Li Shih-Chen (*PTKM*) has long been identified as *Artemisia vulgaris*, var. *indica* (R 9); cf. Anon. (*109*), vol. 4, p. 544; Kimura Koichi (*2*), p. 105 and pl. 53, fig. 1. But in literary writing *ai-hao* was often used vaguely as a double name. Today it denotes one particular species, *A. argyi* (Anon. (*109*). vol, 4, p. 541).

[e] The distinction between the chrysanthemum species was always emphasised in Chinese natural history books. For instance Wu Jui-Chhing[10], in his *Jih Yung Pên Tshao*[11] of +1329 (cf. p. 298) said: 'Those with large fragrant flowers are the sweet chrysanthemums; those with small yellow flowers, the yellow chrysanthemums; those with small flowers and a somewhat disagreeable smell, the wild chrysanthemums.' This looks like *C. morifolium*, *C. hortorum* and *C. indicum* respectively.

[1] 周師厚   [2] 洛陽花木記   [3] 劉蒙   [4] 菊   [5] 苦薏
[6] 千葉花   [7] 蒿   [8] 艾   [9] 野菊   [10] 吳瑞卿
[11] 日用本草

are planted in gardens with very fertile soil, then they can transform into the sweet chrysanthemums. In the same manner it sometimes happens that the single flower (sorts) change into the double or multiple petal (sorts). The tree-peony (*mu-tan*) and the (herbaceous) peony (*shao-yao*) were used (from ancient times) in medicine, but Thao Hung-Ching and others recorded only the red and white forms, never mentioning any double flowers. Nowadays both of these grow wild in the mountains, but always with single flowers simple and small. But when they get into gardens and nurseries where the ground is fertile and they are carefully planted in loose soil and manured, then they all produce double flowers; and as time goes on these large multiple-petalled flower plants produce hundreds of varieties of different forms. Why therefore should one single out chrysanthemums to doubt such changes?

Commentators on the pharmaceutical natural histories give *chü*[1] the alternative name of *jih-ching*[2].[a] According to the *Shuo Wên* it follows *chü*[3] (to nourish) for sound. The *Erh Ya* says that the *chü*[4] is the *chih-chhiang*.[b] The *Yüeh Ling* mentions the *chü*[4] with yellow flowers. I suspect that all such differences (in orthography) are due to the mistakes of copyists.

As for calling the *ma lien*[5][c] the 'purple *chü* (*tzu chü*)[6],[d] and the *chhü mai*[7][e] the 'great *chü* (*ta chü*[8])', and the *wu yuan miao*[9] (crowsbeak)[f] the '*yuan yang chü*[10] (i.e. the mandarin duck[g] chrysanthemum)', and the *hsüan fu hua*[11][h] the 'mugwort *chü* (*ai chü*[12])', etc.—all this is erroneous and unnecessary nomenclature, appropriating the name *chü* (for other plants); and therefore I am not including any of them whatever.[i]

Here perhaps the most interesting features are the use of a biochemical character, the presence of volatile odoriferous substances, to make a specific distinction, and then at the end the acute recognition of the vagaries of vulgar nomenclature in assimilating a number of plants to the *Chrysanthemum* genus which did not belong there. Liu Mêng's treatise was not surpassed by any of those which followed, though five more of importance were written before the end of the Sung, and twelve more before the beginning of the 19th. century.

The next two monographs were written by old acquaintances of the reader. Shih Chêng-Chih[14], Governor of Nanking and Admiral of the Yangtze, was the commander who reported in +1168 the building of an efficient single-wheeled naval paddle-boat, in fact a stern-wheeler.[j] Having kept the Chin Tartars well to the north of the River for many years, however, he retired not long after, and settled at Wu-hsien in Chiangsu to grow chrysanthemums—hence his monograph

[a] This name appeared first in the *Ming I Pieh Lu* (cf. p. 248). It is therefore probably of +3rd-century origin.
[b] See p. 410 just above.
[c] *Iris ensata* (R 655, CC 1779).
[d] Actually this is a synonym of *ma lan*[13], the purple *Aster ageratoides* (R 13, CC 23), which, as another Composite, would be a much more reasonable confusion. But all the texts we have seen write *lien*.
[e] *Dianthus superbus* (Caryophyllaceae), R 547, CC 1461.
[f] The sprouts of *Aconitum uncinatum*, *A. volubile* or *A. kusnetzovii* (R 527, a, b). Hence perhaps a cover-name of morose rural herbalists.
[g] *Aix galericulata*, R 259.
[h] This is *Inula britannica* (R 37, CC 103) or elecampane, another Composite and therefore quite reasonably confused by the vulgar.
[i] Tr. auct.   [j] Cf. Vol. 4, pt 2, p. 422.

[1] 菊	[2] 日精	[3] 鞠	[4] 蘜	[5] 馬蘭
[6] 紫菊	[7] 瞿麥	[8] 大菊	[9] 烏喙苗	[10] 鴛鴦菊
[11] 旋覆花	[12] 艾菊	[13] 馬蘭	[14] 史正志	

of 28 local varieties finished in +1175 with the same title as Liu Mêng's. A dozen years later the eminent scholar Fan Chhêng-Ta[1], one of the Four Great Literati of the Sung, produced another work, always titled the same, but referring to a further group of varieties which he grew near Suchow. In +1191 Hu Jung[2] produced the first illustrated treatise, *Thu Hsing Chü Phu*[3], and in +1213 there was Shen Ching's[4] *Chü Ming Phien*[5]; though both of these were lost, much of their content (unfortunately without the pictures) found its way into the *Pai Chü Chi Phu*[6] (A Treatise on Collections of Chrysanthemum Varieties) of Shih Chu[7], finished in +1242. This was interesting because it set out to catalogue all the varieties from all the culture centres, now reaching up towards a couple of hundred (cf. Fig. 85).

From the time of the Mongol conquest onwards there ensued a long gap both in publications and number of varieties noted, but towards the end of the Ming activity resumed, the first sign being the *I Chü Shu*[8] (Chrysanthemum Cultivation Methods) written by Huang Shêng-Tsêng[9] in +1545. He again is an old friend, for he was the historian of the voyages and explorations of Chêng Ho,[a] as well as the author of several other natural history books,[b] and a worthy pupil of the social philosopher Wang Chhuan-Shan.[c] Before the end of the century a work by an unknown writer described the Shansi varieties,[d] and then there were extensive treatments of the cultivation and grafting techniques, including the control of pests, both incorporated in larger collections, one by Chou Lü-Ching[11] (+1582)[e] and two by Kao Lien[12].[f] Monographs now came thick and fast,[g] but that of Thu Chhêng-Khuei[13] was particularly interesting, for besides giving elaborate technological details he essayed also in his *Tu Hua Chü Tung Li Chi*[14] (Collected Writings of a Lifetime of Chrysanthemums) a general history of the appearance of the varieties, and how one could account for their changes (+1630). These were now approaching some five hundred in number. By this time, of course, monographs on particular plants were beginning to appear in Europe also, and it striking to find that in the following century the Chinese horticultural botanists began to give some attention to chrysanthemums imported from overseas. Thus in +1756 Tsou I-Kuei[15] was commissioned to describe and illustrate 36 varieties

---

[a] Cf. Vol. 4, pt 3. p. 492.

[b] E.g. on the culture of rice varieties and yams, on silkworms and on fishes. Cf. Wang Yü-Hu (*1*), pp. 141 ff.

[c] Cf. Vol. 2, pp. 511 ff.

[d] *Lo Hsiu Yuan Chü Phu*[10] (Chrysanthemum Treatise of the Garden of Happy Retirement).

[e] He was the author of the *Ju Tshao Pien* (cf. p. 349 above), a tea expert, and recorder of much other natural history in his *I Mên Kuang Tu* (cf. Sect. 39).

[f] Cf. p. 361 above, and p. 418 below.

[g] For example, Chhen Chi-Ju's[16] *Chung Chü Fa*[17] of *c.* +1625 was incorporated in Lu Thing-Tshan's[18] *I Chü Chih*[19] of *c.* +1718. Many more books than we have mentioned here will be found in the *bibliographie raisonnée* of Wang Yü-Hu (*1*).

[1] 范成大	[2] 胡融	[3] 圖形菊譜	[4] 沈競	[5] 菊名篇
[6] 百菊集譜	[7] 史鑄	[8] 藝菊書	[9] 黃省曾	[10] 樂休園菊譜
[11] 周履靖	[12] 高濂	[13] 屠承燗	[14] 渡花居東籬集	
[15] 鄒一桂	[16] 陳繼儒	[17] 種菊法	[18] 陸廷燦	[19] 藝菊志

of the plant which had reached the imperial court from abroad.[a] These originated, perhaps, from Japan,[b] Persia or the Philippines rather than from Europe, for the first notice of them there dates only from +1688, when J. Breyn published his account of 'Matricaria japonica maxima' in Prodromus Plantarum Rariorum, saying that six varieties were then being grown in Holland. The definitive introduction was that of Blancard to France in +1789. By 1811 came the Chü Shuo[1] of Chi Nan[2], and though it was far from the last of the Chinese writings on the subject, we may reasonably leave the story at this point, with a vision of the Robert Fortunes and the Augustine Henrys poking about in the Chinese nursery gardens, and carrying off with delight the products of fourteen centuries or more of Chinese plant care and knowledge.

It is sometimes said that the chrysanthemum is mentioned in the Book of Odes (Shih Ching),[c] but this seems to be not exactly so. It is true that two of these −8th-century folk-songs begin with the refrain: 'luxuriantly grow the o[4]',[d] but the problem is to identify the o. Legge translated 'aster-southernwood' and Waley 'tarragon', seeking thereby to give expression to the oldest commentators, who said that the o was a kind of hao (Artemisia),[e] but though this locates us satisfactorily in the Compositae it is not the best bet. Surely what the girls were recognising in the grass was the Chinese aster Callistephus chinensis, now long called lan chü[5] or tshui chü,[6] appropriately, because of its pure blue colour.[f] Under cultivation, which has gone on for a couple of millennia in China, this indigenous annual, the only species in its genus, proved quite variable and developed all kinds of flower colours from purple through lilac and rose to white, though never yellow. Introduced into Europe in +1728, it has been especially useful in modern genetical studies.[g]

This may perhaps be the place to observe how greatly modern genetics has benefited by garden plants of purely Chinese origin. The case of Primula sinensis (tshang pao chhun[7])[h] comes immediately to mind, introduced from Canton to England about 1820 and since then productive of more than thirty mutations.[i]

---

[a] Hence the Yang Chü Phu[3].

[b] Chrysanthemum culture had been transplanted to Japan about +730. The flower became important in Japanese heraldry (cf. McClatchie, 1).

[c] E.g. Li Hui-Lin (8), p. 38.

[d] Mao nos. 176 and 202; tr. Legge (8), pp. 279, 350; Karlgren (14), pp. 119, 152; Waley (1), pp. 104, 316.

[e] Cf. B II 434. The familiar southernwood, so often mentioned in the classics, is A. apiacea, and the mugwort A. vulgaris. Waley doubtless knew that the herb used for tarragon vinegar in France and Spain is A. Dracunculus, wormwood (because of its vermifuge properties), related to A. Absinthium. Cf. Burkill (1), p. 243.

[f] Anon. (109), vol. 4, p. 420; CC 44. Cf. Watson (1). There are other solutions, however. Lu Wên-Yü (1), p. 99, no. 110, believes that the o was the same as pao niang hao[8] or pu niang hao[9], i.e. Descurainea sophia (= Sisymbrium sophia) of the Cruciferae (see Anon. (57), vol. 2, p. 432; Anon. (109), vol. 2, p. 71; CC 1293). This had already been suspected by Bretschneider (loc. cit.). R 104, it is true, has these names as synonyms of lin hao[10], i.e. Pedicularis gloriosa (Scrophulariaceae), but this is probably a mistaken identification. The two possibilities remain open.

[g] Crane & Lawrence (1), p. 80.

[h] CC 502.   [i] Crane & Lawrence (1), pp. 54 ff.

[1] 菊說	[2] 計楠	[3] 洋菊譜	[4] 莪	[5] 藍菊
[6] 翠菊	[7] 藏報春	[8] 抱娘蒿	[9] 柿娘蒿	[10] 蔄蒿

Dianthus chinensis (*shih chu*[1])[a], crossed with the ancient European *D. caryophyllus*, gave the perpetual flowering carnation before +1750.[b] Thus modern science has in a way retraced the steps, but on the far more sophisticated level of a theoretical biology, taken by the Chinese horticultural experimentalists of the middle ages.

(v) *Orchids*

Another family which gave rise to a notable monographic literature in traditional China was that of the orchids.[c] Here we meet with a situation rather different from that just described for peonies and chrysanthemums, in that the number of varieties cultivated was rather fewer but the number of species and even genera much greater. At least a couple of hundred orchid species are native to China, attracting attention from an early time partly because of the delicious and clinging perfumes which many of their peculiar flowers emit, and partly because of the strangeness of their yellowish-green petals often marked with purple blotches or streaks.[d] Though quite a number are epiphytic, those most domesticated by the Chinese, especially the species of *Cymbidium*, are terrestrial, having thick and fleshy roots. Common enough to appear in the standard flora of Chia & Chia (*1*) are 38 species of 23 genera, but only four of these found officinal use.[e] Fifty years ago 15 species of *Cymbidium* along were in cultivation at Canton, and many of these had been growing in the same garden for the previous three hundred years.[f] At the beginning of this century it was customary to find representatives of 50 genera in the nurseries of South China—some indication this of the popularity of orchid culture.[g] But though the botanical net was spread much wider taxonomically, the pattern of monograph production historically was much the same as for the chrysanthemum, namely a rich literature in the Sung (here especially the southern Sung), and then a long gap until the late Ming and late Chhing.

Orchids were prominent among the flowering plants referred to by the poets of Chhu in the Warring States period. In illustration of this one may quote a passage

---

[a] CC 1460.

[b] Crane & Lawrence (1), p. 82.

[c] As companions in this reading Curtis (1); A. D. Hawkes (1) and Kent (1) may be well worth while.

[d] From the Chinese orchid literature we may quote Thang Lan-Chieh (*1*) and Wu Ên-Yuan & Thang Tho (*1*).

[e] *Bletia hyacinthina* (*pai chi*[2]), *Dendrobium monuliforme* (*shih hu*[3]), *Gastrodia elata* (*chhih chien*[4]) cf. p. 139, and *Luisia teres* (*chhai tzu ku*[5]). See respectively R 634, 635, 636, 637, CC 1720, 1731, 1733, 1740.

[f] See Watling (1).

[g] Cf. Westland (1); Herklots (1). Besides genera already mentioned, there were examples of *Arundina*, *Cyrtopera*, *Habenaria*, *Phaius*, *Renanthera*, *Spathoglottis* and *Vanda*. The greatest favourites were however *Cymbidium goeringii* (*chhun lan*[6], *shan lan*[7] or *lan* par excellence), *C. ensifolium* (*chien lan*[8] or Fukienese *lan*), and the *hui*[9], *C. pumilum* (also called *tshao lan*[10] or *hsia lan*[11]). An excellent account of the history and all aspects of orchids in the life and culture of the Chinese people is due to Hu Hsiu-Ying (*12*).

[1] 石竹   [2] 白及   [3] 石斛   [4] 赤箭   [5] 釵子股
[6] 春蘭   [7] 山蘭   [8] 建蘭   [9] 蕙   [10] 草蘭
[11] 夏蘭

written by Huang Thing-Chien[1] about +1090. In some remarks on orchids, he wrote:[a]

The *lan*[2] is like the prince, and the *hui*[3] may be likened to the minister of State.[b] The reason is because in the woods and mountains there are ten *hui* for every *lan*. The *Li Sao* says:

'I had tended the *lan* in nine large gardens
And planted the *hui* in a hundred acres . . .'[c]

Hence we know that the people of Chhu valued the *lan* much more than the *hui*; yet they will both grow thickly together. Planted in sandy soil they come up luxuriantly. If watered with tea they have a beautiful scent. In these characteristics they are the same. When a single stem produces a single flower, and the perfume is penetrating, the orchid is of the *lan* (type), when a single stem produces 5 or 6 flowers and the perfume is not strong, it is of the *hui* (type).

When I lived with the monks at the Pao-an Convent there were windows facing east and west; in the west *hui* used to grow, and in the east *lan*. Visitors used to ask why this was, so I wrote this piece accordingly.

Soon after Huang's time the first of the monographs consecrated entirely to orchids was produced; it was by Chao Shih-Kêng[4], a ninth-generation descendant of the imperial house, who conceived a passion for these plants and wrote in +1233 his *Chin-Chang Lan Phu*[5] (A Treatise on the Orchids of Fukien), giving rather elaborate details on cultivation methods. Orchids from other provinces were dealt with some years later (+1247) by Wang Kuei-Hsüeh[6] in a similar work,[d] generally afterwards considered the best, while in between came a *Lan Phu Ao Fa*[7], on the art and mystery of the culture, still extant, but of uncertain attribution whether to Chao or Wang. Both of these main works, the principal monuments of Sung orchidology, have been translated by Watling (2).[e] After that there was a long lull in the literature until the end of the Ming.

Activity began again with a re-presentation of the Sung material by the indefatigable Kao Lien[13], but soon more original work was done by Chang Ying-Wên[14], who set out in his *Lo-Li Chai Lan Phu*[15] about +1596 to give accounts of many more genera and species than Chao and Wang had known. In +1610 Fêng Ching-Ti[16] carried on the work with his *Lan I*[17], paying particular attention

---

[a] *Lan Shuo*[8], cit. in *Yang Yü Yüeh Ling*[9], Tai Hsi's[10] encyclopaedia of horticulture and animal husbandry of +1633, ch. 27 (p. 251), tr. auct.

[b] *Cymbidium goeriugii* and *pumilum* respectively, but *lan* was a very general word, suitably qualified with epithets when applied to different genera and species.

[c] Cf. D. Hawkes (1), p. 23; Yang & Yang (1), p. 4. Both say 'melilot' for *hui*, probably misled by R 134a, which gives *hui tshao*[11] (*Melilotus indicus*, a legume) as well as the basic name *hsün tshao*.[12] There was perhaps a similarity in the scent.

[d] Also entitled *Lan Phu*.

[e] But with an aberrant Cantonese romanisation which needs revising.

[1] 黃庭堅	[2] 蘭	[3] 蕙	[4] 趙時庚	[5] 金漳蘭譜
[6] 王貴學	[7] 蘭譜奧法	[8] 蘭說	[9] 養餘月令	[10] 戴羲
[11] 蕙草	[12] 薰草	[13] 高濂	[14] 張應文	[15] 羅籬齋蘭譜
[16] 馮京第	[17] 蘭易			

to the habits of the plants, humidity and temperature, edaphic requirements, division and propagation, and the natural orchid pests. Here he advocated isolation methods, separating healthy plants from infected ones, 'just as one isolates human patients attacked by the parasites of tuberculosis (*fang chih ju fang lao chhung yeh*[1])'. Fêng Ching-Ti was a Ming loyalist who died in Japan about +1650 whither he had been sent, vainly of course, to solicit military aid for the collapsing dynasty, and his political activities subsequently gave rise to the idea that his horticultural books had been primarily vehicles of disguised political criticism, but though there may have been an element of this, he is not to be discounted as a genuine horticultural botanist. This was the time when orchids of the south were beginning to be grown in northern greenhouses, a technique which stimulated an interesting paper by the Jesuit Cibot (10) a century or so later.[a] After Fêng there was the *Lan Yen*[2] of Mao Hsiang[3], but then the interest of writers ebbed again until the end of the +18th century, though the gardeners must have gone on producing new varieties and cultivating new wild species as before.

The next great connoisseur was Chu Kho-Jou[4] whose *Ti-i Hsiang Pi Chi*[5] (Notes on the World's Finest Fragrances), an important work of +1796 which, though never printed, still exists in MS.; bound with Chang Kuang-Chao's[6] *Hsing Lan Phu Lüeh*[7], also a MS. Orchids had been cultivated in Chang's family for many generations previously. After this there appeared many monographs of this kind by scholars interested in the Orchidaceae during the decades when the Chinese nurseries were being frequented by Western botanical envoys.[b] Attention was now more and more paid to the provincial specialities; as the title of Chang's book might indicate he was concerned especially with the orchids of Chiangsu, but by this time the Chinese literature begins to include much information on the plants of this family growing in the western provinces such as Yunnan and Kweichow. Wu Chhi-Chün does this, for instance, in his great botanical treatise of 1848 (cf. Sect. 38 *f*), and the quasi-modern Japanese compendia (cf. p. 16) follow suit. So much for the *chün tzu*[10] or *chen jen*[11] among the flowering plants.

(vi) *Rosaceae*

It is quite curious that the Chinese monographers were not necessarily stimulated by ornamental genera in which China was particularly rich, the chief supplier, as time went on, of the gardens of the whole civilised world. Take the case of the Rosaceae. Certain wild roses in their simplicity had been native in all

[a] This was the time of Li Khuei's[8] *Chung Lan Chüeh*[9], though we cannot fix its exact date in the +17th or +18th century.
[b] We may mention only those of Thu Yung-Ning (*1*); Liu Wên-Chhi (*2*); Wu Chhuan-Yün (*1*); Tu Wen-Lan (*1*); Hsü Nai-Ho (*1*).

[1] 防之如防癆蟲也　　[2] 蘭言　　[3] 冒襄　　[4] 朱克柔
[5] 第一香筆記　[6] 張光照　　[7] 興蘭譜略　[8] 李奎　　[9] 種蘭訣
[10] 君子　　[11] 眞人

parts of the European sub-continent from the most ancient times, and some efforts had been made successfully to improve them by cultivation before the +18th century, but the garden roses of modern times are on a quite different level of sophistication, and derive entirely from a wave of inter-specific crossing which occurred after the introduction of roses with many important and valuable qualities from China. Yet although many species of *Rosa* had been assiduously grown in Chinese gardens since the Han and Thang, one scholar only was inspired to write a work devoted to them. Before considering this, however, it will be more logical to look at certain *Prunus* species which excited medieval Chinese writers more.

There is a certain flowering tree, domesticated in China since ancient times, which generation after generation planted around their homes and pavilions, and represented interminably in paintings on silk, paper and porcelain. This is the Chinese 'apricot', *Prunus mume*[a] (*mei hua*[1]), useful for its small hairy fruits which can be preserved for eating, but outstandingly beautiful when it bears its sessile light pink or white and extremely fragrant flowers in spring.[b] The fact that it blooms very early in the year, when snow is still likely to be about, caused it to share in the symbolic virtues of the chrysanthemum (p. 411 above), evoking the adversity-withstanding Confucian hero devoted to ethical principles come what may. And it joined with the evergreen pine and bamboo in proverbial phrase as the 'Three Friends of Winter' (*sui han san yu*[2]).

In the preceding pages we have met several times with the eminent scholar of botanical interests, Fan Chhêng-Ta[6] (+1126 to +1193). If we could visit his gardens and orchards near Suchow about +1186 we should find him growing hundreds of *P. mume* trees in more than 12 different varieties, most of which, he says, he had for the first time brought together. This was indeed the year when he wrote his *Mei Phu*[7], the oldest book in any civilisation devoted entirely to the genus. The zone of origin of *P. mume* in China is not certain. When Li Hui-Lin says that the tree was indigenous to the mountain valleys of Shensi near Hanchung he is relying on good authority, the *Ming I Pieh Lu*,[c] but another centre may have been much further south,[d] the high hills between Chiangsi and Kuangtung, Lo-fu Shan[8] and especially the Ta-yü Ling[9]. Going northeast from Nan-hsiung towards Ganhsien one crosses over an important pass called Mei

---

[a] Often called the 'Japanese apricot', because Europeans first encountered it there, and Siebold gave the Japanese pronunciation as the Linnaean specific name which it still bears. But it was probably not in Japan before the +8th century.

[b] The prevalent translation of *mei* as plum is quite unjustifiable. Plums have hairless fruits and stalked flowers. The Chinese plums were *P. salicina* (*li*[3]) and *P. simonii* (*hung li*[4]), CC 1106, 1107. The edible apricot (*P. Armeniaca*, *hsing*[5]) CC 1098, was native to China, and not in Europe till the +16th century.

[c] Cit. *PTKM*, ch. 29 (p. 41).

[d] See *Hua Ching*, ch. 3 (p. 59); cf. *CFP*, Kuo Phu, ch. 1, p. 8a.

[1] 梅花   [2] 歲寒三友   [3] 李   [4] 紅李   [5] 杏
[6] 范成大   [7] 梅譜   [8] 羅浮山   [9] 大庾嶺

Ling Kuan[1], because of the wealth of *mei hua* trees which grew wild there in old times.[a] The *Chhün Fang Phu* quotes from *Yu-Yang Tsa Tsu* (+863) as follows:[b]

The Ta-yü Ling is one of the Five Ranges (in the Nan Ling separating Kuangtung from the north). When Han Wu Ti was invading Nan Yüeh, Yang Phu[2] [c]ordered his officers to encamp (and make their base) at Yü-shêng[3], therefore the place was called 'Main Granary (Ta-yü). The way over the pass (near here) was steep and hazardous, but since Chang Chiu-Ling[4] (in +716) opened a cutting and engineered a road, it has been suitable for horses and wheeled transport.[d]

As there were many *mei* trees growing there the range is also called Mei Ling[5]. On the wall of a post-station at the foot of the pass there is an inscription written by a woman, saying: 'When I was young I went with my father to Yingchow, where he was military commander, and on the way back we heard that there was the Mei Ling[5] at Ta-yü, but we could not find any *mei*, so we planted 30 beside the road.'

Other books, however, such as the +7th-century *Kua Ti Chih* and the +9th-century *Liu Thieh Shih Lei Chi*, aver that from antiquity there was a forest of *mei* trees on these hills.

By the time of the *Hua Ching* (+1688) there were no less that 21 well-known varieties of *P. mume*, and more than 90 terms (because of synonymic overlap) had grown up for them. A good many of these have since then acquired Latin names because of introduction to the West, and they give some idea of what it was that Fan Chhêng-Ta was cultivating in the +12th century. A form with a green calyx instead of the usual purple one (*lü o mei*[5]) is called *viridicalyx*, one with three carpels looking like the character *phin* (*phin tzu mei*[6]) is known as *pleiocarpa*; the *chao shui mei*[7] variety, with drooping branches and flowers that face downwards so that they can only be seen reflected from water, is naturally *pendula*, but the double dark pink *hung mei*[8] bears the name *alphandii*. One of the rarer varieties has flowers that are so dark as to be almost black; this is the *mo mei*[9] (ink apricot),[e] and Chhen Hao-Tzu says that this was produced by grafting the wild *Prunus mume* on to root-stocks of a quite different family, *Melia Azedarach* (*lien shu*[10]), Persian

[a] I drove twice over this pass in 1944 and recorded it in my log, pausing to visit the nearby open-cast wolframite mines on Hsi-hua Shan. The pass is a famous one; cf. Vol. 4, pt 3, p. 33 and more fully Schafer (13), p. 17; (16), pp. 21, 45.

[b] *Loc. cit.*, but we have not been able to find the passage in the book itself as now printed.

[c] Commander of the Embattled Ships, *fl.* −120 to −100; cf. Vol. 4, pt 3, pp. 441–2.

[d] Chang Chiu-Ling, himself half-Vietnamese, was a poet as well as an engineer. On him see Schafer (16), pp. 45, 273. His own account of the pass road is in *Chhüan Thang Wên*, ch. 291, pp. 1*a* ff.

[e] Cf. Li Hui-Lin (8), p. 53. One ought perhaps to mention that the phrase *mo mei* can also refer to black-and-white paintings of *mei hua* branches. For example, the *Hua-Kuang Mei Phu*[11] is not a botanical or horticultural work, but a book of poems and paintings of *mei* by a Sung abbot. His name was Chung-Jen[12], and he lived at a temple on Hua-kuang Shan, hence the title. According to the story he was accustomed to sleep out of doors under some *mei hua* trees, and waking one moonlit night he was so inspired by the beauty of their shadows on the wall that he took up his pen, and ended by becoming a famous painter (*SKCS/TMTY*, ch. 114, p. 56*b*).

[1] 梅嶺關	[2] 楊僕	[3] 庾勝	[4] 張九齡	[5] 綠萼梅
[6] 品字梅	[7] 照水梅	[8] 紅梅	[9] 墨梅	[10] 棟樹
[11] 華光梅譜	[12] 仲仁			

lilac, which has deep purple flowers.[a] Other varieties have not been latinised—
notably the *huang hsiang mei*[1] the flowers of which are light yellow (*hsiang*[2]) and
highly herfumed; and the *thai ko mei*[3]. This is described as having flowers which
after opening disclose another bud in the centre that opens in its turn, presumably
a double form with outer petals opening widely first.[b]

After Fan Chhêng-Ta's book came another only a few years later, that of
Chang Tzu[4] whose *Mei Phin*[5] of about +1200 again described and graded the
varieties. An unusual work followed in +1238, the *Mei Hua Hsi Shen Phu*[6] by
Sung Po-Jen[7], an artist who drew in minute detail all the stages of the *mei hua*
flowers from the first bud-rudiments to the set fruit, all petals gone. Lastly, and
unusually, a scholar of the Yuan period, Wu Thai-Su[8], wrote a monograph on the
species in +1352.[c]

We call the peach tree *Prunus persica* on account of its presumed home, but there is
every evidence that it really spread from China, as was urged by de Candolle in
1855.[d] Neither the peach nor the apricot (*P. Armeniaca*) were known to Theophrastus, though by the time of Pliny[e] and Columella[f] the former at least had
been introduced to European cultivation.[g] The peach tree (*thao*[12])[h] played an
outstanding role in the most ancient Chinese legend and mythology, its wood
being considered magical when made into images, wands and bows,[i] while its
fruit, and even its gum, from at least the Warring States period onwards, became
symbolical Taoist foods of longevity and immortality (*shou thao*[13]). As for the
deeply pink blossom, it was celebrated by many Chinese poets from the *Shih
Ching*[j] downwards through Thao Yuan-Ming and the laureates of the Thang. As
would be expected if the Chinese sub-continent were the home of the peach,
numerous varieties appeared during the ages of cultivation. Apart from double-
flowered kinds (e.g. *duplex*),[k] and damask kinds striped white and red (e.g. *pi thao*[14],
versicolor), there are also varieties with unusual fruit, especially the *phan thao*[15]
('peen-to', i.e. *platycarpa*), further removed than any other from the natural state
because compressed flat into the shape of a bun,[l] and the *shui mi thao*[16] ('honey-

[a] R 335; CC 886; Anon. (*109*), vol. 2, p. 566. At some time in the Thang period, Hsü Mo[9] wrote a tractate on
a particular variety of pear with deep purple blossoms, the *Tzu Hua Li Chi*.[10] This is still extant.
[b] See on these two *Hua Ching*, ch. 3 (p. 60).
[c] See the study by Shimada Shūjirō (*1*), discussing Wu's *Sung Chai Mei Phu*.[11]
[d] (*1*), pp. 221 ff.
[e] *Nat. Hist.* XV, xi, 39 and xiii, 44, the *persica* or 'Persian apple'.
[f] *De Re Rust.* XI, ii, 11.
[g] *Pace* Laufer (*1*), pp. 539 ff. the evidence for the apricot in Roman times does not seem quite convincing. The
references are Pliny, *Nat. Hist.* XV, xii, 41 and XVI, xlii, 103, mentioning an 'Armenian plum', and Columella, *De
Re Rust.* XI, ii, 96. De Candolle accepted it, however.
[h] CC 1104.   [i] See Granet (*1*), *s.v.*
[j] Cf. Legge (8), pp. 12, 35, 108, 165, 515.
[k] Cf. Guiheneuf & Bean (*1*); L. Henry (*1*). The variety was, it seems, noted by Guy de la Brosse in +1636.
[l] Cf. Evreinov (*1*).

[1] 黃香梅　　　[2] 緗　　　　　[3] 臺閣梅　　　[4] 張鎡　　　　[5] 梅品
[6] 梅花喜神譜　[7] 宋伯仁　　　[8] 吳太素　　　[9] 許默　　　　[10] 紫花梨記
[11] 松齋梅譜　　[12] 桃　　　　　[13] 壽桃　　　　[14] 碧桃　　　　[15] 蟠桃
[16] 水蜜桃

water' Chinese cling), excessively sweet in juice.[a] The nectarine (*yu thao*[1]) was also anciently developed in China.[b] The cultivated peach is now widely considered to have been an interspecific cross between wild *P. persica* and the *shan thao*[2], *P. davidiana*, common in north China and much used as a root-stock.[c] In the light of all this it is singular that no medieval Chinese botanist wrote a monograph consecrated wholly to the peach.[d]

The brilliantly flowering crab-apples interested them a good deal more.[e] Four species of *hai-thang*[3] have been cultivated for centuries in China, three of *Malus* (*spectabilis*, *floribunda* and *halliana*) and one of the quasi-quinces *Chaenomeles* (= *Cydonia*) *lagenaria*. *M. spectabilis*, with its beautiful rose-red flowers, is the *hai-thang* par excellence, *M. floribunda* (probably a hybrid with *M. baccata*, the wild *shan ching tzu*[4]) is known as *hsi fu hai-thang*[5], the 'Western Palace' crab, while *M. halliana* bears the name of 'silk-hangings' crab-apple (*chhui ssu hai-thang*[6]) because of the long pedicels of its flowers. Already in the +3rd century these were firmly placed in what we should call the Rosaceae, for Lu Chi[7] remarked of them, *mei hsing lei yeh*[8].[f] The eminent geographer and cartographer Chia Tan[9] described them in his book *Pai Hua Phu*[10] on ornamental plants about +790 as the 'holy immortals among the flowers' (*hua chung shen hsien*[11]) and the name stuck.[g] Although, as we noted above (p. 276) the prefix *hai* very often denotes some plant introduced from abroad (overseas), there must be some other reason for its appearance here,[h] since all the evidence points to the Chinese crabs being primarily native to Szechuan. As for the *Chaenomeles lagenaria*, with its scarlet and often double almost sessile flowers,[i] one name derived precisely from this characteristic, *thieh kêng hai-thang*,[12] though in even more common usage it was called *mu kua*[13].[j]

The first scholar to write at length exclusively on flowering crab-apples was

---

[a]   One of us (G.D.L.) has vivid memories of a *shui mi thao* orchard belonging to the father of a friend of hers, Dr Catherine Chhen Ying-Mei, at Hangchow, often visited when they were young. Cf. Chhu Hua (*1*).

[b]   The glabrous fruit results from one recessive gene (Crane & Lawrence, 1).

[c]   CC 1105; Carrière (1); L. Henry (2); Sargent (2).

[d]   Though there is much information in the horticultural encyclopaedias.

[e]   The monograph of den Boer (1) and the synopsis of Koidzumi (2) are excellent helps here. Articles by Bean (1); Wyman (1) may also be consulted. See also Anon. (*109*), vol. 2, pp. 234–40 and Csapody & Toth (1).

[f]   'They belong to the same group as the apricots'; in his *Mao Shih Tshao Mu Niao Shou Chhung Yü Su*[14], ch. 1 (p. 34).

[g]   Cit. *PTKM*, ch. 29 (p. 41).

[h]   Perhaps because of a propensity of some species to grow along the sides of lakes or watercourses. The name already puzzled Li Te-Yü[15], another great geographer and like Chia Tan a minister, who remarked on it in his *Phing-Chhüan Shan Chü Tshao Mu Chi*[16] about +820.

[i]   One calls to mind the common 'japonica' of English gardens, *Chaenomeles japonica*, in this case perhaps truly of Japanese origin but long grown in China also.

[j]   Many other crab-apples were known to the old Chinese authors and have been introduced more recently to the West. For example *Malus micromalus* (*shih hai-thang*[17]), *Malus hupehensis* (= *theifera*), the *chha hai-thang*[18], and *Malus prunifolia* (CC 1078, 1080 and 1081) which we shall meet again in a moment.

[1] 油桃	[2] 山桃	[3] 海棠	[4] 山荊子	[5] 西府海棠
[6] 垂絲海棠	[7] 陸璣	[8] 梅杏類也	[9] 賈耽	[10] 百花譜
[11] 花中神仙	[12] 貼梗海棠	[13] 木瓜	[14] 毛詩草木鳥獸蟲魚疏	
[15] 李德裕	[16] 平泉山居草木記		[17] 實海棠	[18] 茶海棠

Shen Li[1] (*fl.* +1023 to +1063) who had been deeply impressed by their beauty and variety when an official in Ichow (southern Szechuan and northern Yunnan). His *Hai-Thang Chi*[2] dates thus from the time of Shen Kua and Su Sung, a flowering-time in its turn of Chinese science and technology, so it is unfortunate that it has not come down to us in full. A second treatise was written, however, after the Chin Tartar wars by Chhen Ssu[3] in +1259, the *Hai-Thang Phu*.[4] A quotation from it will be interesting as showing the efforts made by the scholars of those days to distinguish between the various rosaceous flowering trees and bushes. Chhen Ssu says:[a]

> In the time of the emperor Jen Tsung (*c.* +1030) the scholar Chang Mien[5] wrote a *fu* (rhapsodic ode) on the 'The Hai-Thang Trees of Szechuan'. Shen Li certainly used it in writing his *Hai-Thang Chi*. It says (for example) that 'the wild quince-trees (*mu kua*[6])[b] on the mountains put forth flowers thousands and millions thick; and the wild crab-apples (*lin chhin*[7,8])[c] by the waterside burst out with myriads of blossoms.'
>
> (Shen Li's) commentary says that when the quinces and apples first come into flower they do indeed show that they belong to the same group (*hsiang lei*[9]) as the *hai-thang*,[d] (but they are not the same thing). Possibly what (Chang) Mien had in mind was what the people of Chiangsu rightly call *thang-li hua*[10] (the small coarse wild pear).[e] In fact only the matt coral-red flowered trees are to be called *hai-thang*.
>
> Shen Li's *Record* says that the flowers have five petals, and the buds are so deeply red that they look like dots of rouge or lipstick; when they open they spread out like a halo, and when the petals fall they are pale pink like a woman who has washed her face to go to bed. Analysis shows that this description really applies to the *mu kua*[6] and the *lin chhin*[7]. And if flowers have six petals they are not *hai-thang*.

Thus in the +13th century arguments about the classification of these flowering trees were evidently animated. After that time the subject was discussed in all the books of horticultural botany, but there were no more monographs. We must now look at the situation where but one book was produced in a field where China was pre-eminent, influencing gardens all over the world.

A modern English poet, Walter de la Mare, once wrote:

> Very old are the woods;
>   And the buds that break
> Out of the brier's boughs,
>   When the March winds wake,
> So old with their beauty are—
>   O no man knows,
> Through what wild centuries
>   Roves back the rose.

[a] Tr. auct. from *Shuo Fu*, ch. 70, p. 33*a*.   [b] *Chaenomeles lagenaria*, CC 1067.
[c] *Malus prunifolia* (= *asiatica*), the original *phin kuo*[11], also called *nai*[12] and *hua hung*[13] (CC 1081).
[d] *Malus spectabilis*, CC 1077.   [e] *Pyrus betulaefolia*, also called *tu li*[14] (CC 1124, R 432).

[1] 沈立   [2] 海棠記   [3] 陳思   [4] 海棠譜   [5] 張冕
[6] 木瓜   [7] 林禽   [8] 林檎   [9] 相類   [10] 棠梨花
[11] 蘋果   [12] 柰   [13] 花紅   [14] 杜梨

Perhaps he would have been surprised if he had learnt that tracing the development of this genus would take one back not only to the temples of Cyprus and the mosques of Damascus, but to the gardens of Confucians and Taoists in the Thang and Sung, the genetic wealth of which was celebrated and recorded by a floral bureaucrat of the Chhing some fifty years before it began to re-model radically the roses of the rest of the world.[a] Not long ago in Cambridge a young student of English criticised a seventeenth-century poet, Thomas Carew, for the lines:

> Aske me no more where Jove bestowes
> When June is past, the fading rose . . .

—since everyone ought to know that roses go on blooming all through summer, and far into autumn. But his examiner, a colleague of mine, knew something more, namely that in the +17th century it was only in China that roses did this.

The history of the modern garden roses is really an epic involving many *dramatis personae*, from medieval Chinese scholars and horticulturists through Western plant collectors in nineteenth-century China to the growers, patient breeders and geneticists in many modern countries including China herself. The genus *Rosa* is a highly complex and variable one, so that here we can give only the most naked and unadorned facts.[b] The essential point is that the cultivated roses indigenous to Europe were formerly of indifferent quality, while China was much richer in attractive species; then the introduction of the 'China rose' to Europe towards the end of the +18th century caused a complete revolution in the garden roses of the rest of the world. Hurst has pin-pointed this effect to the transmission of a single Mendelian recessive gene, that for continuous flowering, a heritable constitution which causes the continuous pushing of flowering shoots all through the year. More was involved of course in the introductions than this, for instance the rambling or climbing habit, and various lovely scents and colours, due to other genes; but certainly the ever-flowering gene, originally a mutation peculiar to China, first discovered and utilised by Chinese gardeners a thousand years ago, was destined to effect the greatest improvement in any ornamental plant grown and delighted in by men.

Four main *Rosa* species are wild in Europe, *R. rubra* (= *gallica*),[c] *R. phoenicia*, *R. moschata* (the musk rose) and *R. canina* (the dog-rose or common briar); these between them through cultivation, crossing, polyploidy and muta-

---

[a] There are persistent statements in the horticultural literature that about −500 the library of the High King of the Chou contained many books on rose-cultivation. But since we know nothing of that library, and since no extant MS. on silk or bamboo slips deals with this subject, the story must be considered apocryphal.

[b] To clothe them in appropriate splendour one must read the fundamental papers of Hurst (1); Hurst & Breeze (1) and A. P. Wylie (1). Chittenden (1), pp. 1809 ff. is a useful companion. Pemberton (1) is now rather out of date, and Shepherd (1) a little confusing. The books of Thomas (1, 2, 3) are recommendable. The best classification so far is that of Rehder (1). For illustrations to accompany the story there is the classical Redouté & Thory (1) supplemented by the folio volumes of Willmott (1), but one must beware of pitfalls due to changes in the specific nomenclature.

[c] The oldest representations are Minoan, of the −3rd millennium. This was the species which gave the variety *officinalis* for apothecaries' confections from +1310 onwards. In +1629 Parkinson noted about a dozen varieties.

tion gave rise to all the varieties known in Western gardens down to the end of the + 18th century. Interspecific hybridisation of the first two gave rise to *R. damascena*, the striped and blotched summer damask rose, supposedly an Arabic introduction from Damascus (according to a story of + 1551). Hybridisation of the first and the third gave the autumn damask rose, *R. bifera*, the double blooming-period of which was mentioned by Pliny.[a] But this tetraploid must have been much older than him because it was associated with the cult of Aphrodite on Cyprus, and to this day it is grown commercially because best for rose-water and attar of roses. Hybridisation of *R. damascena* with a white-flowered variety of the last, probably in the Crimea, produced *R. alba*, double flowers of which arose early.[b] Finally, *R. alba* crossing with *R. bifera*, and therefore containing genes of all the basic four species, led to the appearance in + 17th-century Dutch gardens of *R. centifolia*, the Provence or cabbage rose, and this in turn, by a bud mutation in France not later than + 1696, originated all the moss roses (var. *muscosa*).

Enter upon the scene *Rosa chinensis* (= *indica*), the classical *yüeh chi hua*[1] or monthly crop rose,[c] stumbled upon by Peter Osbeck at Canton on 29 Oct. 1751 and first cultivated in England as the 'blush tea' rose by Philip Miller in + 1759. Its ever-flowering character was embodied in the alternative name *yüeh yüeh hung*[2]. This it was which gave the first of the so-called 'four stud roses', Slater's crimson diploid of + 1792; but the other three were crosses with another Chinese species, *R. odorata* (var. *gigantea*), the *hsiang shui yüeh chi*[3]—Parsons' pink (+ 1793),[d] Hume's blush tea (1809) and Parks' yellow tea (1824). Meanwhile, some time in the + 18th century came the Chinese book devoted to these roses, the *Yüeh Chi Hua Phu*[4] by a writer whose name we do not know, for he used an appropriately bureaucratic pseudonym, Phing Hua Kuan Chu[5], the Master of the Flower-Criticism Office.[e] It is a work of value, giving much information on grafting, pruning, climatic and edaphic requirements, and control of pests; and would be well worth translating into some international language. Out of these hybrids came all the bewildering variety of subsequent forms. Parsons', crossed with *R. moschata*, gave all the 'noisettes', with *R. bifera* all the 'Bourbons', and with a Japanese strain of *R. multiflora* all the 'poly-pompoms'. The first hybrid 'perpetual' came in 1816, and when crossed with Middle-Eastern *R. lutea*[f] (the 'Austrian briar') all the Pernet varieties. We must not follow this further, but *R. multiflora*,

---

[a] *Nat. Hist.* XXI, x, 19.

[b] Hence the only ancient European references we have found to doubling. Pliny (*loc. cit.*) speaks of a 'hundred-petalled' rose grown in Campania and Greece though it was not thought attractive either for appearance or scent. Note the use of the term *folia, centifolia*, exactly equivalent to Ch. *pai yeh*.[6] Albertus Magnus later speaks of a white garden rose with 50 or 60 petals.

[c] CC 1138; Anon. (*57*), vol. 3, p. 309 (*109*), vol. 2, p. 252.

[d] An irregular diploid hybrid derived, it must be remembered, after generations of crossings in Chinese gardens.

[e] See Wang Yü-Hu (*1*), 2nd ed., p. 287.

[f] Or better, *R. foetida*, and Mr E. F. Allen tells us.

¹ 月季花   ² 月月紅   ³ 香水月季   ⁴ 月季花譜   ⁵ 評花館主

the classical *yeh chhiang wei*[1] (with its varieties *cathayensis, hua chhiang wei*[2], and *platyphylla, shih tzu mei*[3])[a] was important in another manner, since after its introduction about 1862 it gave rise to all the ramblers and climbing types, some of which are intensely fragrant.

Since then many other Chinese species have spread in horticulture all over the world, and some of these known to the Flower-Criticism Master should be mentioned. *Rosa rugosa* is the ancient *mei kuei*[4], an upright densely-prickly enduring shrub, long used for flavouring a certain kind of Chinese wine.[b] *R. banksiae* (*mu hsiang*[5]), which grows to a great size and has a perfume of violets, has now naturalised itself as an escape in countries as warm as California.[c] *R. rubus* (= *commersonii*), praised by Sung poets, the *thu mi*[6], is the latest to bloom in autumn, clambering around porticoes of temples and dwellings.[d] *R. laevigata* (*chin ying tzu*[7]), introduced so early into America as to have been mistaken for an indigenous plant and dubbed 'Cherokee rose', is an evergreen shrub for which the dismissed name *sinica* would still be more appropriate.[e] *R. xanthina* (*huang tzhu mei*[8]) with its bright yellow flowers, grows wild in North China and Korea.[f] And during the past half-century some twenty further species from China have poured their genetic treasures into the stock of sorts with which the breeders and improvers of the whole world can now work.[g]

(vii) *Other ornamentals*

Next we meet with an entertaining contrast. Having discussed a genus rich in species and of great aesthetic and even economic importance affecting gardens everywhere yet stimulating only one learned work in its own home culture, we may turn to look at two or three plant species or varieties now little known or even positively disappeared which nevertheless gave rise to quite a literature in Chinese traditional botany.

The first of these may be the 'jade stamen flower', *yü-jui hua*.[9] This has been identified by Chia & Chia (*1*) as *Barringtonia racemosa*, a member of the Lecythidaceae. It is a striking shrub or small tree quite common from India through Malaysia to Polynesia and the Philippines, but highly unusual to find in the Yellow River Valley.[h] It reaches a height of some 30 ft with the leaves

---

[a] CC 1144, 1136, 1145.
[b] CC 1147; Anon. (*57*), vol. 3, p. 342 (*109*), vol. 2, p. 247.
[c] CC 1133. [d] CC 1150.
[e] CC 1140; this is mentioned in the *Shu Pên Tshao, c.* +940. Cf. Phei Chien & Chou Thai-Yen (*1*), vol. 2, no. 79; Anon (*57*), vol. 2, p. 214; Fitzherbert (*1*).
[f] Its name records its thorniness and flower-colour.
[g] For further details see Yü Te-Chün (*2*) and Li Hui-Lin (*8*).
[h] For descriptions see Burkill (*1*), vol. 1, pp. 303 ff.; Brown (*1*), vol. 3, pp. 53 ff. and fig. 14; Pham-Hoang Ho & Nguyen-Van-Duong (*1*), pp. 331ff; Anon. (*109*), vol. 2, p. 979. The *yü-jui hua* section in *TSCC* is Tshao mu tien, ch. 297, pp. 1a ff., with a very poor illustration, because the plant was no longer available in the Chhing.

[1] 野薔薇　　　[2] 華薔薇　　　[3] 十姊妹　　　[4] 玫瑰　　　　[5] 木香
[6] 茶蘪　　　　[7] 金櫻子　　　[8] 黃刺玫　　　[9] 玉蕊花

crowded at the ends of the branches; the inflorescences are pendulous, bearing twenty or more flowers, each of white or delicate pink petals, but arresting because of the mass of jade-like stamens radiating in brush-like form twice as long as the petals themselves. The seeds and bark are used as fish-poisons, apparently because of the saponin which they contain. The raw leaves are esteemed as salad, and there is a way of making a wholesome starchy food from the fruit. All the *Barringtonia* species have traditional uses in the South, but in China the interest was in the beauty of the flowers (Fig. 86).

About +1195 Chou Pi-Ta[1], a scholar already often mentioned by us,[a] wrote a study entitled *Yü-Jui Pien Chêng*[2] (On the Identification of the Jade-Stamen Flowers), since the plant was so rare.[b] Originally it was part of his collection, *Phing Yüan Chi*[3] (Writings from the Peaceful Garden), but afterwards it circulated separately, and is still preserved. What is instructive about the passages to be given from the ensuing literature is the care which the pre-Linnaean Chinese botanists took in their descriptions and identifications. About +1630 Wang Hsiang-Chin[4] wrote:[c]

Yü-Jui Hua.[6] A variety of traditions has been handed down about this plant. In the Thang, Li Wei-Kung[7] thought it was the same as *chhiung hua*[8] (the 'nephrite' flower);[d] in the Sung, Lu Tuan-Po[9] thought it must be the same as the *chhang hua*[10];[e] while Huang Shan-Ku[11] identified it with *shan fan*[12], 'mountain alum'[f]—but all were wrong. Chou Pi-Ta of the Sung said that the Thang people greatly prized the *yü-jui hua*; there were some plants in the Thang-chhang Kuan[13] (Taoist temple) in Chhang-an, and also in (the garden of) the Chi Hsien Yüan (College of All Sages),[g] and in (that of) the Han-Lin Yüan (the Academy) as well—no ordinary places. I myself travelled from far away to visit the Chao-yin Ssu[14] (temple), and got one of the these plants.

It spreads out widely (in sprays) like the *thu mi*[15].[h] In the winter (this shrub) dies down and in the spring it flourishes again. Its leaves are like those of the *chê*[16] (the silkworm thorn tree),[i] with purplish stems and petioles (*hêng*[17]). The flower buds (*hua pao*[18]) at the

---

[a] Cf. p. 403 above.

[b] There are references in Thang poetry to a 'rice-bag flower' (*mi nang hua*[5]), and Hung Mai in his *Jung Chai Sui Pi, c. +1200*, opined that this was the same as the *yü-jui hua*. But others have seen in it the first references to the poppy in China (native to Iran and S. W. Asia) because of the large pistil, ovary and fruit. *Papaver somniferum* was first described only in the *Khai-Pao Pên Tshao* of +973. See the interesting paper devoted by Shinoda Osamu (4) to plants in Thang poetry and their identification.

[c] *Chhün Fang Phu*, Hua Phu, ch. 1, p. 12a, tr. auct.

[d] See immediately below, p. 431.

[e] Also, as the name indicates, a flower reminiscent of jade. Li Shih-Chen's identification was (R 352) a synonym of *shan fan*[6], i.e. *Murraya exotica* (Rutaceae), the China box, mountain-alum honey-bush, or orange jasmine.

[f] Wang Hsiang-Chin (*Hua Phu*, ch. shou, p. 3b) took this to be a synonym for *hai-thung*[19] (= *tzhu thung*[20]), i.e. *Erythrina indica* (Leguminosae), the Indian coral tree, CC 984, R 384. Mem. Vol. 1, p. 180.

[g] See Vol. 4, pt 2, pp. 471 ff.

[h] *Rosa rubus*, cf. p. 427 above.

[i] *Cudrania tricuspidata* (Moraceae), CC 1594, R 599; Roi (1), p. 349. The comparison is fair.

[1] 周必大	[2] 玉蕊辨證	[3] 平園集	[4] 王象晉	[5] 米囊花
[6] 玉蘂花	[7] 李衞公	[8] 瓊花	[9] 魯端伯	[10] 瑒花
[11] 黃山谷	[12] 山礬	[13] 唐昌觀	[14] 招隱寺	[15] 荼蘼
[16] 柘	[17] 莖	[18] 花苞	[19] 海桐	[20] 刺桐

Fig. 86. A modern drawing of *Barringtonia racemosa*. From Walker (3), p. 230.

beginning are quite small, but as the months pass they gradually become very large, coming out in the spring with 8 petals[a] and a mass of stamens (*hsü*[1]) like crystal-icy threads bearing at their extremities beads like golden millet-grains (*chin su*[2], the anthers). Besides this, at the centre (*hsin*[3]) of the flower, there is a (structure like a) caerulean tube (*thung*[4]), with the same sort of form as a gall-bladder-shaped vase (*phing*[5], the ovary), and from it there rises up a long floral pre-eminence (*ying*[6], the style) egregious beyond the mass of stamens.[b] Scattered down (the raceme) there are more than ten of these buds (*jui*[7]) looking as if they were carved out of jade—hence the name *yu-jui hua*[8]. Sung Tzu-Ching,[9] Liu Yuan-Fu[10] and Sung Tzhu-Tao[11] all agreed that this was it. I don't know how it ever got confused with the *chhiung hua*.

This really seems an outstanding piece of pre-Linnaean botanical description. Let us go on to read what Chhen Ching-I[12] said in his *Chhüan Fang Pei Tsu*[13] (Complete Chronicle of Fragrances, cf. Sect. 38 (*f*) below) written in +1256, much nearer the investigations of Chou Pi-Ta.

... At the Chao-yin temple ... there is a pavilion called the Yü-jui Thing[14]. There are two *yü-jui*[14] trees standing opposite each other and supported on a trellis like grape-vines—but no vine is to be compared with them. The leaves are ovate-elliptic and acuminate (*yuan chien*[15]), resembling those of the *chê* (silkworm thorn tree), but in thickness like those of the *mei*[16].[c] The flowers are of the same sort as the *mei*, but the very thin petals are shrunk in comparison and smaller, and the centre is slightly yellowish looking like a small translucent bottle (*phing*[17]). It blossoms in late spring and is at its best in early summer. The petals, white like jade, are the last to wither, and the flower has a special perfume. The shrub reaches a height of more than a dozen feet. The local people all say that from the Thang period until now these two trees in this temple are the only ones in the whole world, just as the *chhiung hua* trees are only to be found in Weiyang (Yang-chow). During more than a thousand years there have been several military conflicts and some fires, so that now there are only these two. [But whether these plants are identical with those which grew at Chhang-an at the Pai-yü Ssu[18] and other (Thang) temples, and in front of the residence of the Imperial Censor, is impossible to verify; all we can study now are those in the (Chao-Yin) temple.] I myself would like everyone to realise that this flower is not the same as the (*shan-*) *fan* nor the *chhiung* (*-hua*); it stands out distinctively from all these others and belongs to a special family of its own (*erh tzu chhêng i chia yeh*[19]. Therefore I have carefully recorded its origin and history.[d]

And he signed the piece with his philosophical name Yü I Tzu.[20] Short of naming the family Lecythidaceae and really understanding the physiological significance

---

[a] He was counting the sepals in.
[b] The long projection beyond them is clear in Fig. 86.
[c] *Prunus mume*, cf. p. 420 above.
[d] Ch. 5, pp. 6*b*, 7*a*, cit. *CFP*, Hua Phu, ch. 1, p. 13*a*, tr. auct. Cf. *HCCC*, ch. 552, pp. 39*b*, 40*a*. The sentence within square brackets is found only in the original text and was omitted from the version in *CFP*.

[1] 鬚	[2] 金粟	[3] 心	[4] 筩	[5] 瓶
[6] 英	[7] 蕊	[8] 玉蕋花	[9] 宋子京	[10] 劉原父
[11] 宋次道	[12] 陳景沂	[13] 全芳備祖	[14] 玉蘂亭	[15] 圓尖
[16] 梅	[17] 瓶	[18] 白玉寺	[19] 而自成一家也	
[20] 愚一子				

of the 'translucent bottle' one could hardly expect anyone of the +13th century to go further. While Wang Hsiang-Chin's account seems to hold its own with Bauhin and Parkinson, Chhen Ching-I's far excels the European botanical writing of the time of Albertus Magnus. On the horticultural side our general conclusion would be that some traveller of the early Thang had brought back *Barringtonia* plants from the south, and that some gardener soon afterwards had found himself the happy possessor of a cold-hardy mutant, this perhaps reproduced by asexual propagation and therefore rare in its distribution, lasting down, however, to the very end of the Ming, when Wang obtained a cutting, and afterwards disappearing altogether by the beginning of the +18th century, long before Wu Chhi-Chün could get a sight of it.[a]

What was the *chhiung hua* or 'nephrite flower' which has been mentioned in the preceding paragraphs? The answer is that we do not exactly know. But it gives the opportunity of studying an unusual case, namely the generation of an appreciable literature by a plant which has completely disappeared and at the nature of which we can only guess.[b] The story is somewhat as follows. During the Thang and early Sung the Hou-thu Tzhu[1] temple[c] at Yangchow was famous for one or more shrubs or trees of considerable size which bore an abundance of fragrant white flowers, so much so that it was considered one of the sights of the city. After the fall of Khaifêng in +1126 however the Chin Tartars pressed south and three years later captured and held Yangchow for a short while, long enough alas for them to dig up most of the plants in order to carry them away to the north.[d] In later years fresh shoots producing something very similar grew up, but there was uncertainty whether this hydrangea-like plant was the same as what had been there before. In +1191 therefore a 'man in hempen cloth', i.e. a horticulturist of the people, Tu Yu[6], went to Yangchow to investigate, and learnt from a Taoist, Thang Thai-Ning[7], then eighty years of age, that shoots had come from the old roots about thirty years earlier, and that by a test he had assured himself that this was indeed the same plant.[e] Tu's account, the *Chhiung-Hua Chi*[8], was printed in +1234, but in the meantime another botanical enquirer, Chêng Hsing-I[9], had also gone to the temple (+1210) to look into the

---

[a] There has been some confusion about the Chinese name of *Barringtonia*. Li Hui-Lin (8), p. 227, listed Chou Pi-Ta's book as *Yü Tshan*[1] ... instead of *Yü Jui*[2] ... But the *yü-tshan*, 'jade hairpin', is a quite different plant, the plantain lily *Hosta sieboldiana* (= *glauca*), cf. Li (*loc. cit.*), p. 113, CC 1849; Khung Chhing-Lai *et al.* (*1*), p. 279. True, it also rated a monograph of its own, the *Yü-Tshan Chi*[3] by Kao Lien[4] in +1591. The name of this plant arose, says *CFP*, Hua Phu, ch. 4, pp. 19 ff., from one of Han Wu Ti's consorts née Li, who put them in her hair, after which all the palace ladies followed suit. Cf. Bailey (2).

[b] The following information is collected from *TSCC*, Tshao mu tien, ch. 297, pp. 1a ff.

[c] Votive temple of the tutelary genius of the soil, the Earth Lord, sixth of the six great ministers of the mythical emperor Huang Ti.

[d] The reader familiar with earlier volumes will not miss the parallel here with the splendid hydro-mechanical clock of Su Sung at the capital (cf. Vol. 4, pt 2, p. 497).

[e] It is said that the bark emitted a characteristic pungent smell on burning.

---

[1] 玉簪	[2] 玉蕊	[3] 玉簪記	[4] 高濂	[5] 后土祠
[6] 杜斿	[7] 唐太寧	[8] 瓊花記	[9] 鄭興裔	

matter.[a] His account has the further interest of including an experimental test. We translate it from the *Chhüan Fang Pei Tsu* (+1256).[b]

The *chhiung hua* plant has no equal in the world. Since the northern cavalry (of the Chin Tartars) invaded the city (Yang-chow), (and the Hou-thu temple was burnt) some people say that the bushes growing there now are not the same as the old ones, and suspect that the Taoists planted *chü-pa-hsien*[1] (the assembly-of-the-eight-immortals plant)[c] in their place. It was difficult to know whether this was true. *Chü-pa-hsien* plants might have been brought here from Hofei, and certainly, as one sees them in gardens, they do look at first sight very much like the *chhiung hua* temple flowers (*hsiang lei*[2]) in category. But if one examines them really carefully and handles them understandingly there are three points of difference.

First, the *chhiung hua* flowers are larger, with thicker petals and of a faintly yellow colour, while the *chü-pa-hsien* flowers are smaller, with thinner petals and of a pale caerulean tint; this is the first difference. Then the foliage of the *chhiung hua* is soft and lustrous while that of the *chü-pa-hsien* is rough and hairy; this is the second difference. Then the stamens (*jui*[3]) of the *chhiung hua* are long enough to be level (with the edges of the petals), it produces no seeds, and it gives forth a fragrance, while the *chü-pa-hsien* has stamens shorter (than the edges of the petals), bears fertile seeds and has no scent; that is the third difference.

Even so, I hardly dared to believe in my own observations, so once I made a test of them, mixing the flowers and getting some children to sort them out; and I found that all of them could recognise and distinguish the flowers (without making a single mistake). After that I had no further doubts.

This is surely an excellent piece of botanical work for a date early in the +13th century. Chêng entitled his remarks 'A Tractate on the Identification of the Nephrite Flower' (*Chhiung-Hua Pien*[5]). A very reasonable guess would be that the plant was a sterile hydrangea hybrid, for many of this Saxifragaceous genus are scented, e.g. *Hydrangea anomala*,[d] which climbs up the walls of Cambridge College courts. One can see well how the hydrangeas got the name of *chü-pa-hsien*, the assembly of the immortals, from the large sterile florets round the margins of the flat corymb inflorescence.

Although, so far as we know, no monograph was written in old times on the hydrangeas (or on the rather similar-looking Caprifoliaceous viburnums)[e] the

---

[a] A scholar and relative of the empress, he was magistrate at Yangchow.

[b] Ch. 5.

[c] *Hydrangea opuloides* (= *Hortensia*, = *macrophylla*), CC 1203–5; Khung Chhing-Lai *et al.* (*1*), p. 16. Several varieties grow in China, differing in the degree of differentiation between the outer sterile and inner fertile florets, as well as in other ways. Another common name for the main species is *hsiu-chhiu*[4], 'embroidered ball'. See Bean (2); Wilson (2).

[d] CC 1208; Anon (*109*), vol. 2, pp. 104 ff. On the probability that the *chhiung hua* was a *Hydrangea* species, cf. Li Hui-Lin (8), pp. 103 ff.

[e] China seems to be the developmental centre of these species also; cf. Rehder (*1*), vol. 2, pp. 105 ff.; Osborn (*1*); Bean (3). Chia & Chia (*1*) and Khung Chhing-Lai *et al.* (*1*) do not agree with Li Hui-Lin (8) that *hsiu-chhiu* is a correct name for *Viburnum* spp.; the generic designation *chia-mi*[6] is more proper. The Chinese type species is *V. dilatatum* (CC 205), first described in the *Thang Pên Tshao*.

---

[1] 聚八仙   [2] 相類   [3] 蕊   [4] 繡球   [5] 瓊花辯

[6] 莢蒾

chhiung hua continued for centuries to excite archaeologically minded Chinese botanists. In +1487 a resident of Yangchow, Yang Tuan[1], put together in his *Chhiung-Hua Phu*[2] all the information which he could find on the subject,[a] naturally not excluding a mass of legendary material of Han and earlier date.[b] By this time, however, the exercise was becoming more literary, for the plant itself had definitely disappeared for good, perhaps during the conquests of the Yuan and the Ming.[c] Around +1540 Lang Ying[3], the learned historian of technology, went into the question again, saying that he long thought the *chhiung hua* must have been a *chih tzu*[4] or gardenia,[d] but then he saw a Sung painting of the plant which closely resembled a wild hydrangea, save that the inflorescences had nine sterile florets (immortals) instead of eight.[e] Tshao Hsüan[5] later in the Ming, and Chu Hsien-Tsu[6] in the Chhing, could naturally not add much,[f] but the case remains remarkable for the combination of history and archaeology with botany in the research for the identification of an irrecoverable treasure.

It seems that most of the *Hydrangea* species now in cultivation are of Asian origin.[g] The striking lability of the flower petal colours was early known in China,[h] where horticulturists followed with interest the changes from green to pink and then to blue ending as bluish-green, and actively manipulated the anthocyanins as one can by adding various substances to the soil in which the plants are grown.[i] *H. opuloides* is a very old temple garden plant in China, but *H. paniculata* (*yuan chui hsiu-chhiu*[7])[j] and *H. bretschneideri* (=*pekinensis*, =*heteromalla*), the *thieh-kan hua erh chieh tzu*[8], are even more beautiful and hardier.[k] *H. xanthoneura* (*huang mo hsiu-chhiu*[9])[l] has sterile florets two inches across. So much for the Eight Immortals.

This account may be concluded with a miscellaneous group of ornamental (and useful) plants mostly from parts of China far away from metropolitan cities. First the Balsaminaceae. Of the balsams, *Impatiens textori* and *Noli-tangere* may be native to China,[m] but it is tolerably certain that the best-known one, *Impatiens Balsamina*[n], was an introduction from India in very early times. As the flowers

---

[a] This was the only work on the subject which the Chhien-Lung bibliographers thought worthy of including, see *SKCS/TMTY*, ch. 116, p. 83*b*.
[b] See B II 571 and Mayers (1), no. 317.
[c] So thought Chhen Hao-Tzu in *Hua Ching*, ch. 3, p. 26*a, b* (p. 71).
[d] The Chinese type species is *Gardenia augusta* (= *florida*, = *jasminoides*), CC 221, 222.
[e] The title of his study was *Chhiung-Hua Pien*[10].
[f] In *Chhiung-Hua Chi*[11] and *Chhiung-Hua Chih*[12] respectively.
[g] See McClintock (1); Harworth-Booth (1, 2). Cf. Hemsley (2).
[h] Li Hui-Lin (8), p. 134.
[i] We shall treat briefly of this in the notes on plant physiology, Sect. 38 (*h*) 1 below.
[j] CC 1207.   [k] CC 1199.   [l] CC 1211.   [m] CC 778, 779.
[n] CC 777, Khung Chhing-Lai *et al.* (*1*), p. 1308. The oldest reference comes in the account of what Han Wu Ti tried to transplant to the north after his victory over the Nan Yüeh people in −111; *San Fu Huang Thu* (late +3rd cent.), ch. 3, p. 7*b*. We see no reason for doubting the authenticity of this, as some have done. Next there is *Nan Fang Tshao Mu Chuang*, ch. 2 (p. 8), ed. of Anon (*56*), with ills.; cf. Wu Tê-Lin (*1*). This is datable at +304, cf. pp. 447 ff. below.

[1] 楊端　　[2] 瓊花譜　　[3] 郎瑛　　[4] 梔子　　[5] 曹璿
[6] 朱顯祖　　[7] 圓錐繡球　　[8] 鐵桿花兒結子　　[9] 黃脈繡球
[10] 瓊花辨　　[11] 瓊花集　　[12] 瓊花志

have many colours, from purple through white to yellow,[a] and as the explosive dehiscence of the seed pods affords amusement, the plant was systematically cultivated. The resemblance of the nectiferous spurs of the flowers to the feet of birds led anciently to the standard name of *fêng hsien hua*[1] (Phoenix Immortal), and the bursting open of the pods at a touch gave the synonym *chi-hsing tzu*[2] (hot-tempered). *Impatiens Balsamina* graduated, as it were, about $+1785$, when Chao Hsüeh-Min[3], the great naturalist who wrote the supplementary book to the pharmaceutical botany and zoology of Li Shih-Chen (cf. p. 325 above), penned a *Fêng Hsien Phu*[4]. During the Thang and perhaps earlier the flower pigments of *I. Balsamina* were used for dyeing women's fingernails bright pink or red, a process involving garlic juice and fixation with alum;[b] hence the plant was called also *jan chih chia tshao*.[5] During the Sung however another plant (on which we have remarked earlier in passing, pp. 112, 148), more effective in dyeing nails and hair, was introduced, again from India, none other than henna, *Lawsonia inermis* (*chih chia hua*[6]) of the Lythraceae.[c] This is still widely cultivated in the southern provinces, and its product has indeed great advantages over modern artificial dyes.[d] But it never had any special writing devoted to it.

The *Camellia* genus did, however, produce a literature, though not an early one. Far from being a foreign introduction, many species grow wild in the forested valleys of sub-tropical China, from Yunnan through Kuangsi and Kuangtung to Fukien.[e] The genus gets its name from the Czech Jesuit G. J. Kamel, who had been in the Philippines and whom Linnaeus wished to honour.[f] On any oecumenical scheme of priority it would better have been called *Fêngia*, in remembrance of the man who really was, as we shall see, the first to write a monograph on these plants, some two hundred years earlier. By the intermediation of the Hon. East India Company *Camellia japonica* (so named, though essentially Chinese) reached England in $+1702$ and was described by Petiver.[g] Other species followed,[h] and

---

[a] Li Shih-Chen in *PTKM*, *loc. cit.* below, particularly noticed this, and said that they spontaneously change of themselves.

[b] Here the best references are *Pei Hu Lu* ($+873$), in *Shuo Fu*, ch. 2, p. 39a; and *Kuei Hsin Tsa Chih* (c. $+1280$), *Hsü Chi*, ch. 1, p. 17a, b. Both tr. Laufer (1). One of us (G. D. L.) always used the method in her youth in Nanking.

[c] The entry in *PTKM* is not under *fêng-hsien* (ch. 17B, pp. 37a ff.) but in an appendix to *mo-li* (ch. 14, p. 68b).

[d] On henna in China see Sampson (1); Mayers *et al.* (1); Hirth (1), pp. 268 ff.; and especially Laufer (1), pp. 334 ff. As the $+$rd- and $+$4th-century references given above speak of *chih chia hua*[6], the plants may have been either *Impatiens* or *Lawsonia*.

[e] One sees them as one drives through the passes between Chiangsi and Fukien, as I myself did in 1944 (cf. Needham & Needham (1), pp. 213–4.

[f] So at least on one account; there is some uncertainty. Cf. Bretschneider (10), p. 18.

[g] CC 709; cf. H. H. Hume (1); but especially Leng & Bunyard (1). Engelbert Kaempfer (1), fasc. 5, p. 850 described the Japanese *shan chha* in $+1712$; we reproduce in Figs. 87a, b his page and picture. The camellias of Japan have now been extensively monographed, with colour photographs, by Tuyama (1, *1*); Tuyama & Futakuchi (*1*); and Tuyama et al. (*1*) who enumerate 585 different cultivars.

[h] Cf. Sealy (2, 3, 4); Bean (4); Hanger (1) and the symposium edited by Synge (1).

[1] 鳳仙花   [2] 急性子   [3] 趙學敏   [4] 鳳仙譜   [5] 染指甲草
[6] 指甲花

the flower was at the height of its popularity in the thirties and forties of the nineteenth century.[a]

In China the name for the ornamental camellias is *chha hua*[1] or *shan chha*[2], 'tea-flower' or 'mountain-tea', since for centuries the relationship of these wild species to the cultivated tea plant or bush (*chha*[3]) was clearly recognised.[b] We shall deal faithfully with tea and its industry in the proper place; here it is necessary only to say that although the growing of tea for drinking goes back to the San Kuo period (*c.* +260), the description of the plant did not get into the pharmaceutical natural histories till the Thang (*Hsin Hsiu Pên Tshao*, +659, and *Pên Tshao Shih I*, +725). Similarly the *shan chha* plants did not get in until Li Shih-Chen's work of +1596,[c] though the monographic botanists had been writing about them since the beginning of the Ming. The relationship between camellias and tea caused great difficulties for European botanists, who for long could not decide whether the generic name ought to be *Camellia* or *Thea* (as it had been in Chinese usage). By general consent the name Theaceae is now retained for the family, but tea has become *Camellia sinensis* (= *C. Thea*).[d]

The species which aroused the attention of the Chinese botanists of the +15th century was *Camellia reticulata*, a form native to Yunnan where it grows wild, with rose-red flowers 3 or 4 in. in diameter, but cultivated by many generations of Yunnanese gardeners in double and semi-double varieties with flowers of various red shades nearly 6 in. across.[e] That the leaves were edible (or potable) was recorded in the *Chiu Huang Pên Tshao* of +1406 (cf. p. 331 above).[f] About 170 years later, though the exact date is uncertain, Fêng Shih-Kho[4] produced his *Tien Chung Chha-Hua Chi* (Notes on the Camellias of Yunnan) in which he enumerated 72 varieties, mostly of *C. reticulata*.[g] In a valuable paper Yü Tê-Chün (1), who had studied the camellia horticulture in the province during and after the second world war, was able to give exact modern descriptions of 18 of these.[h] Fêng Shih-Kho was impressed by the size of the trees, which can reach up to 40 ft in Yunnan, by the waxy corollas widely varying in size and colour, by the long-lastingness of the flowers indoors, and by the length of the flowering period. Another study of them, now lost, was written by Chao Pi[6], who took his degree in

---

[a] Mem. *La Dame aux Camélias*, the famous novel of Alexandre Dumas the younger, published in 1848. Tr. Bray (1).

[b] The most authoritative recent flora, Anon. (*109*), vol. 2, pp. 848 ff., distinguishes about 24 species, including the tea-plant itself.

[c] *PTKM*, ch. 36, pp. 69*b* ff. There may have been an entry, however, in the *Pên Tshao Thu Ching* of +1061.

[d] Thus *T. sinensis*, *T. Bohea* and *C. theifera* have all fallen by the wayside.

[e] CC 710; cf. Sealy (1); Wilkie (1).

[f] Ch. 3, p. 15*b*. Read (8), 9.25 misinterpreted it as a *Clethra*.

[g] This is distinguished as *nan shan chha*.[7] Fêng's book seems not now to be extant in full, but there are extensive quotations from it in various places, notably *TSCC*, Tshao mu tien, ch. 296, e.g. i wên, p. 1*a*.

[h] His fullest presentation is Yü Tê-Chün & Fêng Yao-Tsung (1). Colour photographs of the camellias of Yunnan, with careful descriptions of many varieties, have been collected in the monograph by Fêng Kuo-Mei, Hsia Li-Fang & Chu Hsiang-Hung (1).

[1] 茶花　　　[2] 山茶　　　[3] 茶　　　　[4] 馮時可　　　[5] 滇中茶花記
[6] 趙璧　　　[7] 南山茶

850 *Amœnitatum exoticarum Faſciculus V.*

Sátſuki, flore albo duplo, id eſt, lilio gemino, altero alterum complectente.

Satſuma Satſuki, à provinciâ ita dictus, flore valde concineo.

Jedogana Satſuki, flore in candidum purpuraſcente; ab urbe natali nomen adeptus.

茶 ***Sá & Sjùn***, vulgò ***Tſiubakki.*** Frutex flore roſeo, fructu pyriformi tricocco. Notari meretur: Fruticem hunc ſtructurâ ac facie ſuâ exprimere *Theam*; unde Orienti placet, proprio adhuc carenti, characterem mutuari à *Tſiubacki*. Hic alius eſt Sylveſtris, qui ubiq; in dumetis & ſæpimentis fruticat, alius Hortenſis, qui inſitione vel culturâ manſuetior factus, pleno ac pulchriori flore magnificè decoratur. Utriusq; innumeræ dantur varietates, à loco natali, conditione florum, vel figurâ partium denominatæ. Ex primis & maximè obviis eſt:

山茶 ***Sán Sa***, vulgò ***Jamma Tſiubakki***, i. e. *Tſiubakki* montanus ſive ſylveſtris, flore roſeo ſimplici.

Frutex ex brevi caudice ramoſus, arboris æmulatur magnitudinem; *Cortice* veſtitus in glaucum badio, æquali, carnoſo, tenui, à ligno, quod perdurum eſt, difficulter abſcedente. *Pediculis* ſemi uncialibus, ſupino latere compreſſis, *Folia* promiſcuo loco inſiſtunt ſingula, majuſculis foliis Ceraſi hortenſis ad aſſem ſimilia, ſed quodammodo rigidiora, duriora & utrâq; facie ſplendentia. Ex eorum axillis, ſucceſſivè per autumnum una vel gemina prodit *Gemma*, globi ſclopetarii magnitudinis, ex ſquamis herbaceis, concavis, piloſis, plus minus vicenis gradatim imbricata: quâ oſcitante, *petala*

Fig. 87 Engelbert Kaempfer's account of the *Camellia* genus.

a. The beginning of his descriptive text in *Amonitates Exoticae* (+1712), p. 850.

Fig. 87. Engelbert Kaempfer's account of the *Camellia* genus.

b. His plate.

+1472, and then about +1495 came the third book, *Yung-Chhang Erh Fang Chi*[1] (Notes on Two Ornamentals cultivated at Yungchhang in Yunnan) by Chang Chih-Shun[2], which we still have.

Yungchhang is modern Paoshan, a city on the neck of land between the Salween and Mekong Rivers, enjoying the wonderful upland climate of the province combining sub-tropical latitude with high altitude. The two flowers in question were of the genera *Camellia* and *Rhododendron*, and it was the only time that the latter got into disquisitional Chinese literature. Of the first Chang Chih-Shun described 36 *shan-chha* species or varieties, and of the second, which he characterised as sorts of *tu chüan*[3],[a] 20 species or varieties.[b] The name is now reserved for the rose-red azalea, *Rhododendron simsii* (= *indicum*),[c] with synonyms *shan shih-liu*[4] (mountain-pomegranate), *shan chih-chu*[5][d] and *yang-shan-hung*[6]—('the whole mountain is red')—but this, said Chang, was to forget all about the splendid yellows, purples and greens.[e] When one remembers (cf. p. 34 above) that the Sino-Himalayan Node gave rise to 700 rhododendron species, more than two-thirds of all that are known, it is most satisfactory that some of them should have been scrutinised by Chinese botanists of the indigenous tradition before modern times, but this could hardly have been expected earlier than the Ming since cultural development in that part of the world was so late in arising.[f]

In later times the camellias continued to intrigue scholars with an interest in plants, and those of Fukien in the south-east came up for their examination by an official who chose the pseudonym Phu Ching Tzu[7] and finished his *Shan Chha Phu* in +1719. Not only did he ransack the Fukienese forests and nursery gardens, but also obtained plants from Japan, where the culture had become fashionable,[g] describing in all 43 species or varieties and giving much detail about the techniques employed. The last piece that we have is a MS, perhaps never printed, by

---

[a] This name originally meant the hawk-cuckoo, *Hierococcyx sparverioides*. Cf. Hoffmann (*1*), pp. 30, 221, 336; Chêng Tso-Hsin (*3*), pp. 175 ff., (*4*), pp. 281 ff. Probably the association had a seasonal and phenological origin, both bird and bush manifesting themselves at the same time of year.

[b] They could well have included the purple *Rhododendron pulchrum* (which may be a hybrid), the snow azalea *R. mucronatum*, and the scented *R. auriculatum*. Among the best places in England for these beautiful shrubs and trees are the woods of Portmeirion in Gwynnedd, North Wales, so well known to both of us (J. N. and G. D. L.).

[c] CC 530; Anon. (*109*), vol. 3, p. 147. This can be seen in its splendour in the botanic garden on Tresco in the Scilly Isles, as Dr D. M. Needham and I found in the spring of 1976.

[d] Name derived from the related *yang chih-chu*[8], the 'staggers', because containing a substance poisonous for sheep, i.e. *Rhododendron sinense* (= *molle*), CC 523; Anon. (*109*), vol. 3, p. 144. This is a yellow azalea.

[e] Cf. *SKCS/TMTY*, ch. 116, p. 84*b*.

[f] On the other hand, domestication in gardens by lovers of flowers had begun long before. Schafer (*20*) has written of the azaleas cultivated by the Thang minister Li Tê-Yü[9] (+787 to +849) in those famous gardens which he described in his essay *Phing Chhüan Shan Chü Tshao Mu Chi*[10] (Records of the Plants and Trees of the Retreat at Phing-chhüan). There are précis-translations of this by Edwards (*1*), p. 150 and Belpaire (*4*), p. 90; and further information on Li Tê-Yü as a horticulturist can be found in Tuan Chhêng-Shih's *Yu-Yang Tsa Tsu*, ch. 9 (pp. 245 ff.). The poet Pai Chü-I (+772 to +846), Li's contemporary and in this perhaps his inspirer, was also a great enthusiast for azaleas.

[g] As it still is. Today one of the largest studies of the camellias is that of Tuyama Takashi (*1, 1*) already mentioned, who describes 420 species or varieties of the plant.

---

[1] 永昌二芳記   [2] 張志淳   [3] 杜鵑   [4] 山石榴   [5] 山躑躅
[6] 映山紅   [7] 樸靜子   [8] 羊躑躅   [9] 李德裕   [10] 平泉山居草木記

Li Tsu-Wang[1] about 1846. By this time further species were beginning to find their way to the West, such as *Camellia Sasanqua* (*chha mei*[2])[a] with its Japanese specific name, from Japan, and the closely related *C. oleifera*, certainly native to China, where it yields for commerce a cosmetic oil.[b]

Here it is high time that we ended our disquisition on the ornamental flowering plants which received the homage of Chinese scholarly writings, in manuscript and print, during many centuries. But these books and tractates in themselves are by no means a gauge of the total wealth of flowering genera which China contributed to the rest of the world. We have said nothing about the *Magnolia* species, nor of *Wisteria* so beautiful on Cambridge college walls, nor of the trumpet-vine *Campsis*, nor fragrant *Osmanthus*, nor *Forsythia* present today in almost every English garden, nor *Hibiscus* somewhat unfairly associated only with the South Seas.[c] What we had to establish here was the long lead which the pre-Linnaean Chinese botanists had over the rest of the world in their monographic literature, an extraordinary example of what aesthetic and taxonomic motivation could do for biological science.

As we have observed on several occasions it did not follow exactly the lines that might have been expected, for some genera which turned out to be important historically received relatively little attention, while others of seemingly lesser interest were heavily written up. But this must derive partly from a set of aesthetic choices not necessarily the same as those of other cultures, and partly from a corpus of legend and symbolism, together with specific historical circumstances, which were naturally different. If any one factor mattered more than any other in the development of this literature, it was surely the invention and application of the art of printing towards the end of the Thang. The golden age of botanical monography followed upon this, for every scholar could now feel that his interest in certain fascinating plants could be shared with thousands of friends not personally known to him throughout the length and breadth of the empire.

All in all, this literature constitutes an epic, the production of many hundreds of careful botanical scholars, long pre-Linnaean of course, working from the time of Apuleius Platonicus right through to that of Linnaeus himself. There can be no doubt that they produced monographs devoted to particular families, genera and species, as well as a wealth of varieties, centuries before any European dreamed of doing so. Some of it influenced modern universal botany through the interpreters of the nineteenth century, though not very much, and a great deal of historical research remains to be done on it, though it is more and more being studied and used by contemporary botanists. Today it would be inexcusable not to attempt to

[a] CC 711; Kimura Koichi (2), p. 61.
[b] Cf. Sealy (2); Anon. (109), vol. 2, p. 856.
[c] We may return elsewhere to these garden contributions, freely recognised by many scholars, e.g. Schafer (17). The reading of the book of Li Hui-Lin (8) may be a veritable revelation to many garden-lovers in other parts of the world. *Iris* is another genus deserving mention here; cf. the history by Kuribayashi Motojiro (1).

[1] 李祖望　　[2] 茶梅

do justice to this great botanical tradition and achievement, clearly an essential part of any world-wide historical perspective of the natural sciences.

(5) EXOTIC AND HISTORICAL BOTANY

Let us now turn to consider another motivation which through many centuries inspired with enthusiasm some of those scholars of China who interested themselves in the world of plants. One part of this was geographical in origin, the appreciation of the wealth of plant forms which revealed themselves to the astonished eyes of those who were sent out to administer the parts of the Chinese culture-area most distant from the centre. Here especially the southern subtropical and tropical regions (cf. pp. 25 ff. above) were the main stimuli, but the desert north and the mountainous west also played a certain role[a]. The other part was historical, a never quite satisfied urge to identify the plants (and often the animals too) which had been mentioned in the classics, those writings which had come down from as early as the beginning of the −1st millennium. Here the greatest concern was with the *Shih Ching* (Book of Odes),[b] but other pre-Han and even Han texts also claimed attention. In all these studies the aim was clearly to push back the frontiers of knowledge concerning plants, whether in space expanding, or in literature recovering treasure from time past.

Where the history of the natural sciences has been concerned, the Confucians have not had a good press,[c] though the organicism of the medieval Neo-Confucians has been attractive to modern scientific minds.[d] But there is one botanical reference in the *Lun Yü* (Analects), Master Khung talking to his disciples about the Odes.[e]

The Master said: 'Young men, why do none of you study the *Shih* (Book of Odes)? These (ancient) songs elevate the mind, dispose to observation (of man and nature), conduce to sociability (because of the fellow-feeling aroused by an apposite quotation), and sublimate resentment. They help one at home in the service of one's father, and abroad in the service of one's prince. Finally, they greatly increase one's knowledge of the names of birds and beasts, plants and trees'.

Thus the patron saint of all later scholar-officials recorded his approbation of a discerning awareness of the variety of plant and animal forms, the note of nomenclature being accompanied doubtless by familiarity with their practical uses to man as well as with the traditions of their poetical symbolism. However subsidiary these interests may have seemed to the early Confucians obsessed with the problems of human society, the text certainly served later generations of those

---

[a] Of course, a number of exotic plants have come into the argument in foregoing sub-sections (e.g. pp. 159, 417, 427), and we shall avoid repeating discussions of them here.

[b] Cf. Vol. 1, pp. 86, 145, Vol. 2, pp. 105, 264, 391–2.

[c] Cf. Vol. 2, pp. 12 ff.

[d] This has been fully set forth in Vol. 2, pp. 472 ff., 504–5.

[e] XVII, ix, tr. auct., adjuv. Legge (2), p. 187; Waley (5), p. 212; Ware (7), p. 111; Leslie (9), p. 193.

with taxonomic interests as a kind of justification from what came to be taken as holy writ. But when we pass from the −5th century to the +11th, we find that the Neo-Confucian philosophers had a deep appreciation of the significance of biological forms, and a rather clear conception of how they exemplified principles of order and organisation at all levels.

They started from exactly the same point. The text was quoted by Chhêng I-Chhuan[1] (+1033 to +1107) who summarised the whole position of this great philosophical school when he said:[a] '"A wide acquaintance with the names of birds and beasts, plants and trees" is a means of understanding *li* (*so i ming li yeh*[2]).' For the Neo-Confucians, as we saw in Vol. 2 (Sect. 16) the universe was constructed wholly of two entities: *chhi*[3], which today we might well translate as 'matter-energy', and *li*, which designated a principle of pattern or organisation, working at all levels from the smallest, most imperceptible, most trivial, to the largest in the heavens and the earth, and to the highest in the mind and spirit of man. Thus he said that to understand (up to a point) what pattern means, one must look at living organisms. And not just at one, like Wang Yang-Ming sitting meditating for hours in front of a bamboo plant,[b] but at a great variety of the forms of life. In the course of another discussion:

Somebody said: 'In "extending knowledge",[c] what do you say to seeking (the patterns of the world) first in the "Four Beginnings" (*ssu tuan*[4])?'[d]

(The philosopher answered): 'To seek them in our own nature and passions is of course the most direct way; but every plant and every tree has its own pattern-principle (*li*[5]), and it is necessary to investigate them all.'[e]

This was an unmistakable disavowal of the idea that world pattern and organisation could be apprehended by introspective psychology and human sociology alone; the objective world of Nature had at least equal importance. In a way, his words were a charter for the biological sciences, if other influences in Chinese society had made it possible for them to develop. The theme recurs in a passage of the writings of Yang Shih[7] (+1053 to +1135). In a reply to Lü Chü-Jen[8], who had asked about the 'investigation of things', he wrote:[f]

People generally talk in a very approximate way, but the Six Classics have the most meticulous expressions, preserving all that is most abstruse and mysterious in the world for our scrutiny. For instance 'the men of old had a wide acquaintance with the names of

---

[a] *Honan Chhêng shih I Shu*, ch. 25, p. 6*b*, tr. auct. adjuv. Graham (1), p. 79.

[b] See Vol. 2, p. 510.

[c] A quotation from the *Ta Hsüeh* (Great Learning), *c.* −260; 'the extension of knowledge consists in the investigation of things' (*chih chih tsai ko wu*[6]). All through Chinese history the meaning of this was controverted, but it became a watchword of the natural sciences. Cf. Vol. 1, p. 48, Vol. 3, p. 163. See also p. 214 above.

[d] Allusion to *Mêng Tzu*, II (1), vi, 5 (Legge (3), p. 79). Human-heartedness, righteousness, good customs, and knowledge.

[e] *Honan Chhêng shih I Shu*, ch. 18, pp. 8*b*, 9*a*.   [f] *Yang Kuei-Shan Chi*, ch. 3, § 41 (p. 66).

---

[1] 程伊川   [2] 所以明理也   [3]   [4] 四端   [5] 理

[6] 致知在格物   [7] 楊時   [8] 呂居仁

birds and beasts, plants and trees'. How could it be only their names with which they were acquainted? No, they deeply investigated them and earnestly sought out (their natures)—all are included in the 'investigation of things'.

Thus Neo-Confucianism brought philosophical justification for the botanical interests of scholars during the previous millennium and more, as well as for those who followed after.

The men of this school also showed on many occasions a vivid sense of unity with all other living things, while at the same time being able to distinguish well the levels of complexity and organisation. The following dialogue is reported of their greatest figure, Chu Hsi[1] (+1130 to +1200):[a]

Someone said: 'Birds and beasts, as well as men, all have perception and vitality (*chih chio*[2]), though with different degrees of penetration. Is there perception and vitality also in the vegetable kingdom?'

(The philosopher) answered: 'There is indeed. Take the case of a plant in a pot; when watered, its flowers shed forth glory; when pinched, it withers and droops. Can it be said to be without perception and vitality? Chou Tun-I[3][b] refrained from clearing away the grasses from in front of his window, because, he said, 'their vital force is very like my own'.[c] In this he attributed perception and vitality to plants. But the vitality of the animals is not on the same plane as man's vitality, nor is that of plants on the same level as that of animals.'[d]

Thus the evolutionary ascent through successive stages of integration was plainly implicit in Neo-Confucian thought. The same feeling for Nature is recorded of Chhêng Ming-Tao[6], the other great Chhêng brother (+1032 to +1085). Chang Hêng-Phu[7] tells us that

the steps leading up to Ming-Tao's library were thickly covered with grasses. When someone advised him to have it all cut, he said: 'No. I want always to be able to see the vital processes, the formative urges and the shaping forces of the natural world' (*tsao wu shêng i*[8]). He also bought a pond in which he kept small fish, and was always going to look at them. When asked why he kept them, he said: 'I like to watch the myriad things all satisfied with their existences' (*wan wu tzu tê i*[9]). Grasses and fishes of course everybody sees, but only Ming-Tao could (teach us to) see the life-principles in the grasses, and the life-satisfaction in the fish. That was no common seeing, but more like that of the impartial penetrating sun.[e]

And to take one last example, particularly relevant because the flowering parts of plants attracted botanical attention so early, there is a similar story about Chhêng I-Chhuan.[f]

[a] *Chu Tzu Chhüan Shu*, ch. 42, pp. 31*b* ff., tr. Bruce (1), p. 68, mod. auct.
[b] The founder of the Neo-Confucian school, +1017 to +1073.
[c] This report comes originally from *Honan Chhêng shih I Shu*, ch. 3, p. 2*a*, tr. Graham (1), p. 109, mod. auct. A note by the disciple Hsieh Liang-Tso[4] (d. +1121) adds that Chang Tsai[5] (+1020 to +1077) made a similar remark when watching a braying donkey.
[d] A full translation of the whole passage will be found in Vol. 2, p. 569.
[e] *Sung Yuan Hsüeh An*, ch. 14, p. 5*b* (p. 38), partial tr. Graham (1), p. 109, mod. auct.
[f] *I-Chhuan Wên Chi*, Suppl., ch. 1, p. 5*a*, tr. Graham (1), p. 110, mod. auct.

[1] 朱熹　　　　[2] 知覺　　　　[3] 周敦頤　　　[4] 謝良佐　　　[5] 張載
[6] 程明道　　　[7] 張橫浦　　　[8] 造物生意　　[9] 萬物自得意

I-Chhuan came again with Chang Tzu-Chien[1] during the spring. Shao Yung[2][a] invited them to go for a walk with him on the Thien-mên road to look at the flowers. I-Chhuan declined, saying that he had never been in the habit of looking at flowers. Shao Yung replied: 'What harm could there be? All things have their ultimate patterns (li[3]). We look at flowers differently from ordinary people because we are trying to see into the mysteries of Nature's shaping forces.' I-Chhuan replied, 'In that case I shall be glad to accompany you'.

In the light of these glimpses one can see something of the background, explicit or implicit, which made Confucian scholars feel authorised to probe into the strange plants of outlying regions or the problematical plants to which the men of old had alluded. What they wrote about these things constitutes two special genres of botanical literature, and we shall now describe the books or tractates which dealt with the south, those regions which most fascinated the Chinese as they moved outwards to fill the geo-political *oikumene* of their culture. But first a few words must be said about the historical course of this southern penetration. We have had occasion to speak of this several times already during the preceding volumes.[b]

(i) *The exploration of the borderlands*

During the Warring States period the whole region of what is now south China and Vietnam was occupied by tribal peoples known as the Hundred Yüeh (Pai Yüeh[4]). In the east there were the Tung Yüeh[5] of Chekiang and Chiangsu, who had attained the status of a princedom, the Yüeh State, from about −400 onwards. Most of Fukien (Min) was the area of the Min Yüeh[6] people, while to the south in Kuangtung, Kuangsi and Annam there were the many groups of the Nan Yüeh[7]. These had relations especially close with the State of Chhu[8], which acted as an intermediary for their tropical and sub-tropical products, so it was natural enough that after the fall of this State to the first emperor, Chhin Shih Huang Ti, in −223, the conquering organisation of Chhin sought to annex the regions which had been providing such valuables as rhinoceros horns, ivory, jade and pearls, to say nothing of desirable plant products like bananas and lichis. As soon as the country was unified, then, in −221, a great expeditionary force passed southwards between the peaks of the Nan Ling[9] range to subjugate the Yüeh. It took some seven years, and in the course of it a great technical achievement happened, the construction of the most ancient transport contour canal in any civilisation. Started in −219, this was the design of the engineer Shih Lu[10], who united the upper waters of a northward-flowing river with those of a southward-flowing one, with the result that an unending stream of barges with army supplies

---

[a] One of the earlier of the Neo-Confucian school, +1011 to +1077. But he stood rather aside from their main line of development; cf. Vol. 2, pp. 455 ff. We have had occasion to discuss some of his ideas on p. 305 above.

[b] E.g., Vol. 1, p. 101. On the general subject of trade and expansion in Han China there are now books by Yü Ying-Shih (1) and Wiens (3).

[1] 張子堅	[2] 邵雍	[3] 理	[4] 百越	[5] 東越
[6] 閩越	[7] 南越	[8] 楚	[9] 南嶺	[10] 史祿

could reach the neighbourhood of Canton from as far north as the Yellow River.[a] In −214 three regular commanderies were established, including what is now northern and central Vietnam.

But the Chhin dynasty came suddenly to an end, and during the confusion preceding the inauguration of the Han, Southern Yüeh had peace and even independence under the rule of an official who had been appointed from the north, Chao Tho[1]. After −202 the Han emperors desired to reabsorb these southern regions, so Wên Ti sent as an envoy Lu Chia[2] on two occasions, first in −196 and then again around −179, but these visits did not obtain immediate results. It was not until sixty years after his death, and some twenty-five after that of Chao Tho, that the forests and ricefields of Nan Yüeh re-entered the Chinese empire following vigorous action in −111, naval as well as military, by Han Wu Ti.[b] All the south was now divided into nine commanderies, making together Chiao-chou[3], under a Governor of Chiao-chih[4]. During the commotions of the time of Wang Mang at the beginning of the era, and those still greater upheavals which terminated the Han, the southern regions (Ling-Nan[5]) generally enjoyed considerable tranquillity under a number of enlightened officials, some of whom, like Shih Hsieh[6], proclaimed a temporary independence. But in +211 he submitted to Sun Chhüan[7], soon to be the ruler of the State of Wu in the Three Kingdoms from +221 onwards. He used to send him, says the *San Kuo Chih*,[c]

a variety of perfumes, several thousands of pieces of fine *ko*[8] grass-cloth,[d] brilliant pearls, large cowrie-shells, green glass,[e] caerulean jadeite, tortoise-shell, rhinoceros (horns), ivory and other precious things; together with strange products and rare fruits, bananas, coconuts, longans and the like. Every year there came these gifts, and sometimes he presented several hundred horses as well.

Shih Hsieh lived to be ninety, dying in +226. During the course of his long and successful governorship he had welcomed to the peaceful south many scholars from the north who sought new homes and new employments far from the tumults of their own provinces. Such people began to take great interest in the natural products and industries of Nan Yüeh.

Under his successor, Lü Tai[9], Chinese exploratory curiosity pushed out still further towards the Indian culture-area. Between +226 and +230 Lü sent two famous envoys, Chu Ying[10] and Khang Thai[11], to visit the countries of Champa (Lin-I[12]) and Cambodia (Fu-Nan[13]), and so such kingdoms began to pay tribute to the state of Wu.[f] Both of these men wrote books on their experiences, but these

---

[a] See Vol. 4, pt 3, pp. 299 ff.   [b] See Vol. 4, pt 3, p. 441.

[c] Ch. 49, p. 11a, tr. auct., adjuv. Li Hui-Lin (12).

[d] This was the cloth woven from the fibres of the leguminous vine *Pueraria thunbergiana* (R 406); cf. pp. 86, 340 above.

[e] This could well have been something imported from Europe and passed on. Cf. Vol. 1, p. 200.

[f] Cf. *San Kuo Chih*, ch. 60, p. 9a.

[1] 趙佗   [2] 陸賈   [3] 交州   [4] 交趾   [5] 嶺南
[6] 士爕   [7] 孫權   [8] 葛   [9] 呂岱   [10] 朱應
[11] 康泰   [12] 林邑   [13] 扶南

are extant now only in the form of quotations in later works. Chu Ying was the author of the *Fu-Nan I Wu Chih*[1] (Record of the Strange Things in Cambodia); Khang Thai had two titles to his name, the *Fu-Nan Chuan*[2] (Record of Cambodia) and the *Wu Shih Wai Kuo Chuan*[3] (Record of Foreign Countries in the time of the State of Wu), but they may have been the same book. It is certain that both of these writers had something to say of the plants, trees and animals characteristic of these strange countries, but the fragments left have little of this now. In +280 the empire was re-united under the Chin,[a] and the Nan-Yüeh lands down to the middle of Vietnam were once more within the boundaries. A succession of able and benevolent governors, first Thao Huang[4] from +271, and then Wu Yen[5] from +301, gave peace and good administration to the south. We need not pursue further the fluctuating fortunes of Chiao-chou and the two Kuang provinces,[b] because the time of Wu Yen was precisely that of the greatest type-specimen of books on tropical botany, as we shall see in a moment.

We are now to look at the whole succession of writings about the southern regions,[c] a succession which may go back as far as the −2nd century, if the *Nan Yüeh Hsing Chi*[6] (Records of Travels in Southern Yüeh) was genuinely the work of Lu Chia[7] (p. 111 above). Without dispute he was down there, and the book would have been written about −175, but then it may well have become rare by Chin times and lost altogether thereafter. Equally enigmatic is the second account, the celebrated *Lin-I Chi*[8] (Records of the Champa Kingdom); if Tungfang Shuo[9] was really the author of the first version he probably penned it around −100. In any case, it must have been remodelled a good deal later on, not reaching its present form before the latter part of the +5th century. If the following passage on the betel-nut palm, *Areca catechu*, was contained already in the original version, it must be one of the oldest botanical descriptions of southern plants which have come down to us. The *Lin-I Chi* says:[d]

The areca tree (*ping lang shu*[10])[e] is about ten feet in circumference[f] and more than a hundred feet high. Its bark resembles that of the *chhing thung*[11] tree,[g] and it has joints or

---

[a] The first reign-year of the first Chin emperor dated from +265, but the State of Wu was not absorbed until just after +277.

[b] There is a wealth of Chinese secondary sources condensing the history of these times, e.g. Thang Chhang-Ju (*1*); Hsü Tê-Lin (*1*); Li Ssu-Mien (*1, 2*); Wang Chung-Lo (*2*); Li Chien-Nung (*3, 4*).

[c] These have been touched upon before, especially in the geographical Section 22 (Vol. 3, p. 510), where we had some discussion of the literature on southern regions and foreign countries. Among other surveys of this genre, that of Schafer (*16*), pp. 147 ff. is worthy of note, though it accordance with its design it concentrates mainly on writings of the Thang period.

[d] As cited in the +1647 edition of the *Shuo Fu*, ch. 62, tr. Aurousseau (*4*), p. 15, eng. mod. auct. The parallel passage in the *Nan Fang Tshao Mu Chuang* has been translated by Li Hui-Lin (*12*).

[e] We have said something of the betel complex already on p. 384 above.

[f] Aurousseau suspected that the text had originally 'one foot'.

[g] I.e. *Sterculia platanifolia* = *Firmiana simplex* (R 272), the *wu thung*[12], often called the phoenix tree.

[1] 扶南異物志   [2] 扶南傳   [3] 吳時外國傳   [4] 陶璜   [5] 吾彥
[6] 南越行記   [7] 陸賈   [8] 林邑記   [9] 東方朔   [10] 梹榔樹
[11] 青桐   [12] 梧桐

knots like the *pan chu*[1] bamboo.[a] The trunk is more or less cylindrical, not decreasing in diameter towards the upper parts. Everywhere the areca trees form forests of thousands and ten thousands of plants, dense, vigorous, without branches, but sending out at the top leaves in all directions which give shade. Looking up to the top one can hear a raucous noise (like what one would get from) banana leaves fixed to the end of a bamboo stick and waved about. When the wind rises, the leaves seem like feather fans sweeping the heavens. Under the leaves there are attached several spathes (*fang*[2]), each one of which bears clusters of several dozen fruits. Every family possesses several hundred trees, which seem as high as the clouds, with the fruiting branches like cords hanging down.

Even if these are not the words of Tungfang Shuo himself, the document is venerable. Another fragment found its way down through the ages, and can be retrieved from an early +17th-century collection;[b] it describes the wild fruits of *Myrica rubra*,[c] saying:

In Lin-I the wild myricas (*shan yang mei*[3]) have fruits as large as cups and bowls. [When green they are very sour, but when red they taste like wild honey.][d] They are fermented to make a kind of wine called *mei hsiang chhou*[4], but it is kept for offering to patricians and guests of distinction.

So much for the two oldest texts that we have.

But before the end of the Han there was more. About +90 Yang Fu[7] wrote two books which may really have been one, the *Chiao-chou I Wu Chih*[8] and the *Nan I I Wu Chih*[9] (Strange Things in Chiao-chou—or, from the Southern Borders). Perhaps the second title was the older of the two, the first appearing only in the Sui and Thang bibliographies. Extant now in quotations alone, it doubtless contained some botanical material originally. Then during the San Kuo period there were several such books. We noted on p. 444 above the works of Chu Ying and Khang Thai produced about +240, but there was also the book of another traveller, Shen Jung's[10] *Lin-Hai Shui Thu I Wu Chih*[11] (Record of the Strange Productions of Lin-Hai's Soils and Waters),[e] an account of the plants and animals of Chekiang province. We only have quotations from it now. Again, some time between +270 and +310 Wan Chen[12] wrote another 'Strange Things' book, the *Nan Chou I Wu Chih*[13], about the far south in general; and we may remember how valuable were his descriptions of shipping and pearl-diving[f]. Had it survived, his information on plants would have been equally precious.

---

[a] A variety of spotted bamboo frequent in Anhui, Chiangsi and Szechuan. Cf. Stuart (1), p. 63.

[b] The *Tung Hsi Yang Khao*[5] (Studies on the Oceans East and West), written in +1618 by Chang Hsieh[6], ch. 12, p. 10*b*. It also occurs in *PTKM*, ch. 30 (p. 93) with Tungfang Shuo as the source.

[c] R 621, sometimes called the box myrtle.

[d] The sentence in square brackets occurs only in *PTKM* and the *Nan Fang Tshao Mu Chuang* quotation. I can vouch personally for the excellence of this fruit, having bought it with the late Dr Wu Su-Hsüan from Yunnanese countryfolk in the war years.

[e] Idiomatically, this expression, *shui thu*, includes climatic as well as edaphic factors.

[f] See Vol. 4, pt 3, pp. 600, 671.

¹ 斑竹	² 房	³ 山楊梅	⁴ 梅香酎	⁵ 東西洋考
⁶ 張燮	⁷ 楊孚	⁸ 交州異物志	⁹ 南裔異物志	¹⁰ 沈瑩
¹¹ 臨海水土異物志		¹² 萬震	¹³ 南州異物志	

But none of these books was entirely devoted to plants, so that botanically they pale into insignificance by comparison with Hsi Han's [1a] *Nan Fang Tshao Mu Chuang* [2] (A Prospect of the Plants and Trees of the Southern Regions), traditionally dated at +304. We have often had occasion to refer to it at earlier stages, when for example we noted that it contains by far the oldest account of biological pest control, the use of a particular kind of ant for the protection of orange crops;[b] or in connection with the presentation of local bark-cloth ('honey-fragrance paper') by a Roman-Syrian trade mission in +284.[c] Again, we called Hsi Han one of the greatest of all Chinese botanists when we were discussing his friendship with the celebrated alchemist Ko Hung[3].[d] The *Nan Fang Tshao Mu Chuang* is assuredly a text of weight and importance for the history of Chinese botany,[e] but its evaluation is not an entirely simple matter, so it will be worth while to take a closer look at it both biographically and philologically.

Hsi Han was born in +263 and after the unification of China by the Chin dynasty served as a scholar-administrator and poet on the staff of several princes.[f] In +300 he found himself a military commander under the future emperor Huai, but their forces suffered a serious defeat in which Han's uncle, Hsi Shao[4] (son of the famous poet Hsi Khang[5], one of the Seven Sages of the Bamboo Grove)[g] was killed. Hsi Han was then made prefect of Hsiang-chhêng, but when that in turn became untenable he escaped south to Hsiang-yang, where Liu Hung[6], the general of the garrison, recommended him a few years later for the Governorship of Kuangtung. Ko Hung, who then joined his staff, preceded him to that southern province, but he never got there himself for he was assassinated at Hsiangyang in +307 after the death of Liu Hung. If the *Nan Fang Tshao Mu Chuang* is really his, he wrote it on the basis of first-hand accounts from others, rather than from personal observation; and indeed his own preface says that this was exactly what he did. But the text has given rise to considerable philological controversy, and some of the arguments on both sides need examining, to check, as it were, the stability of the foundations of this monument of Chinese botany.

Broadly speaking, the possibilities are as follows: (a) the work is authentic for the date in question; (b) it is authentic on the whole, but with some later interpolations; alternatively (c) it was constructed by the collection of texts from other sources in the Thang period; or (d) as late as the Southern Sung. There are certainly a few odd things about the *Nan Fang Tshao Mu Chuang*. Perhaps Wên

---

[a] In previous volumes we named him Chi Han, but Hsi is the more correct form of his family name.

[b] Vol. 1, p. 118. The subject is fully discussed below, pp. 531 ff.

[c] Vol. 1, p. 198.

[d] Vol. 5, pt 3, p. 80. Other mentions of the *Nan Fang Tshao Mu Chuang* occurred in Vol. 3, p. 710 (in connection with the geographical literature on the southern regions, pp. 510 ff.), and in Vol. 4, pt 3, p. 721.

[e] Our opinion here was only echoing that of the greatest masters, e.g. Bretschneider (1), vol. 1, pp. 38–9; Laufer (1), p. 329; Merrill & Walker (1), p. 553; Goodrich (1), p. 76.

[f] His biography is in *Chin Shu*, ch. 89, p. 3b.   [g] Cf. Vol. 2, pp. 157, 434, 477.

---

¹ 嵇含   ² 南方草木狀   ³ 葛洪   ⁴ 嵇紹   ⁵ 嵇康
⁶ 劉弘

Thing-Shih about 1888 was the first to suspect that something was wrong,[a] because he found a mention of Liu Chüan-Tzu[1] in one passage,[b] though this physician did not live until about +410. Here the family name was probably a mistake, and Wên's other argument was equally weak, namely that Hsi Han never in fact lived in Kuangtung himself; but Wên Thing-Shih in any case accepted the text as a pre-Thang work. Later, Aurousseau[c] cited the quotation from the *Lin-I Chi*[2] (Records of the Champa Kingdom)[d] as an interpolation, since among the extant fragments of that book there are references to late +4th-century kings of Champa, and even an exact date, +413. But it cannot be excluded that Hsi Han knew an earlier version of the same book, and it cannot even be proved that Tungfang Shuo[3] in the −2nd century did not write the first recension of it, Taoist virtuoso that he was.[e] More recently, Ma Thai-Lai[f], failing to find quotations from the *Nan Yüeh Hsing Chi*[4] (Records of Travels to Southern Yüeh) in any other book,[g] dubbed it a forgery of Sung times, along with the *Nan Fang Tshao Mu Chuang* itself. But this is almost captious, for thousands of books were lost throughout Chinese history, and Lu Chia's[5] text may have become rare already in Hsi Han's time. After all, Lu Chia did really go as an envoy to Nan Yüeh from Han Wên Ti, first in −196 and again about −179. Similarly, Pelliot[h] and Laufer[i] felt uneasiness at the names given by Hsi Han for the two kinds of jasmine,[j] especially *yeh-hsi-ming*[10], since a Persian-Arabic term, they thought, would not have been expected so early; but in the light of what is now known about maritime contacts in the Han period, such doubts are rather unnecessary.[k] Neither Aurousseau nor Pelliot nor Laufer, driven though they were to assume interpolations, impugned the authority of the text as a whole.[l]

Nevertheless, the fact remains that the *Nan Fang Tshao Mu Chuang* makes no appearance in Hsi Han's biography, nor is it listed in the bibliographies of the *Sui Shu* or either of the *Thang Shu*. The first trace of it, as a separate work, comes in the

---

[a] (*1*), p. 53 (p. 3755).

[b] The entry on theriaca, no. 19. Liu Chüan-Tzu was the author of the *Kuei I Fang*[6], one of the oldest Chinese books on surgery and external medicine. But as he was said in the text to have achieved longevity by consuming *shu*[7] (*Atractylodes* spp., cf. Vol. 5, pt 3, p. 40), the reference was an obvious scribal error for Chüan Tzu (−2nd cent.) who in the *Lieh Hsien Chuan*, we are told, did just this; no. 11, Kaltenmark tr. (*2*), p. 68.

[c] (*4*), p. 10.

[d] In entry no. 61 on *yang mei*[8] (*Myrica rubra*).

[e] Cf. R. A. Stein (*1*). Admittedly Lin-I was also called Hsiang-Lin[9] in his time (Gerini (*1*), p. 147), but there is always a tendency to up-date place-names.

[f] (*1*) and (*1*), pp. 11, 19, 22.

[g] Here they come in entries no. 2 (jasmine) and 61 (*yang mei*).

[h] (*9*), p. 146.   [i] (*1*), pp. 329 ff.   [j] Cf. p. 111 above.

[k] See Vol. 4, pt 3, pp. 441 ff.

[l] Nor had the editors of the *Ssu Khu Chhüan Shu* collection (*SKCS/TMTY*, ch. 70, p. 64a) who concluded that 'the style is so elegant that no Thang or post-Thang person could have forged it'. Chiang Fan (*1*) about 1825 had no doubts. Nor did Ku Chieh-Kang or Chang Hsin-Chhêng discuss it in their works on texts of dubious authenticity. Nor was Pelliot (*17*) reluctant to accept its genuineness. Finally, Shen Chao-Khuei[11], in his postface of 1916 reprinted in Anon. (*56*), recorded his opinion that the text could in no way be a pastiche of later times.

---

[1] 劉涓子   [2] 林邑記   [3] 東方朔   [4] 南越行記   [5] 陸賈
[6] 鬼遺方   [7] 朮   [8] 楊梅   [9] 象林   [10] 耶悉茗
[11] 沈兆奎

Sui Chhu Thang Shu Mu[1] complied by Yu Mou[2] about +1180,[a] and thereafter the item is repeated in many late Sung and subsequent book-lists. The oldest edition we have now is the one in the *Pai Chhuan Hsüeh Hai*[3] *tshung-shu*,[b] printed by Tso Kuei[4] in +1273. Such a pre-Sung vacuum could not but arouse suspicion, and accordingly Ma Thai-Lai (1) proposed that the work was really a comparatively late collection, artificially put together between +1108 and +1193, year of Yu Mou's death. And in fact Ma was able to establish textual identities or close resemblances between about half of the entries, and passages in books written between the +4th and the +12th centuries.

But this seems a very singular philological method. Why should not the writers of all these books have been quoting from the *Nan Fang Tshao Mu Chuang* without acknowledgement, as authors so often did in all pre-modern cultures? What could have been the motive for such a fabrication? Hardly a +12th-century passion for the history of botany.[c] Moreover, although it is true that the book itself is absent as such from the earlier dynastic bibliographies,[d] the collected works of Hsi Han are there all right (though they were eventually lost);[e] and long ago Yao Chen-Tsung, in his study of the Sui list, supposed that the *Nan Fang Tshao Mu Chuang* had naturally been included in them.[f] Besides, there is independent evidence of Hsi Han's interest in botany, indeed in scientific things in general. He wrote poetical essays on the *Hemerocallis* day-lily,[g] on the fragrant *huai*[5] tree,[h] on *Hibiscus* flowers,[i] millets,[j] evergreen trees,[k] and sweet melons;[l] they are mostly lost but the prefaces survive.[m] He also wrote exordia on the gearing of mills[n] and on the

---

[a] In *Shuo Fu*, ch. 28, p. 18b. [b] Cf. Vol. 1, p. 77.

[c] As we pointed out in Vol. 1, p. 43, no medieval Chinese scholar would ever have dreamed that kudos was to be gained by making a scientific or technological text appear to be much older than it really was. Only humanistic matters had prestige.

[d] It appears for the first time in *Sung Shih* (+1345), ch. 205, p. 22a.

[e] Under the title *Kuang-chou Tzhu-Shih Hsi Han Chi*[6] in *Sui Shu* (+656), ch. 35, p. 7a. Then under the title *Hsi Han Chi*[7] in *Chiu Thang Shu* (+945), ch. 47, p. 14b, and in *Hsin Thang Shu* (+1060), ch. 60, p. 3a. So the collected works must have been lost some time during the +12th or +13th centuries, only the *Nan Fang Tshao Mu Chuang* and a few poems surviving. This would explain why later writers cited it as such, while earlier ones simply lifted passages from the collected works. Perhaps the corpus as a whole failed to 'cross the river' to the south after the fall of the Northern Sung in +1126.

[f] (3), p. 712 (p. 5750).

[g] *I Nan Hua Fu*[8], cited in the *Chhi Min Yao Shu*, preface preserved in *IWLC* and *TPYL*. There was a long tradition in south China that one of the species, *Hemerocallis minor*, with orange flowers, was good for male sex-determination in pregnancy, hence the name.

[h] *Huai Hsiang Fu*[9], preface in *IWLC* and *TPYL*; the tree was *Platycarya strobilacea*, the roots of which are fragrant in the fire.

[i] *Chhao Shêng Mu Lo Shu Fu*[10], preface in *IWLC*.

[j] *Ku Shu Fu*[11], complete in *IWLC*, ch. 85, p. 4b (p. 2166).

[k] *Chhang Shêng Shu Fu*[12], complete in *IWLC*, ch. 98, p. 9b (p. 2292).

[l] *Kan Kua Fu*[13], preface preserved in *TPYL*.

[m] They are conveniently collected together in *CSHK* (Chin sect.), ch. 65, pp. 3b ff., especially pp. 5b to 6b. We find there also the text of an inscription on the chrysanthemum, *Chü Hua Ming*[14] (p. 7b).

[n] *Ibid.* p. 5a, tr. in Vol. 4, pt 2, p. 195.

[1] 遂初堂書目　[2] 尤袤　[3] 百川學海　[4] 左圭　[5] 槐
[6] 廣州刺史稽含集　[7] 稽含集　[8] 宜男花賦　[9] 槐香賦
[10] 朝生暮落樹賦　[11] 孤黍賦　[12] 長生樹賦　[13] 甘瓜賦
[14] 菊花銘

'cooling regimen powder', a mixture of mineral and plant drugs fashionable in his time.[a] Li Hui-Lin (12) the translator of the *Nan Fang Tshao Mu Chuang*, supports a long tradition in noting how elegant is Hsi Han's style in comparison with other contemporary works. His use of archaic and unusual names for certain plants instead of those which later became standard in the Thang and Sung[b] is a clear piece of evidence for the antiquity of the text. It is also relevant that about eight of his plants cannot now be identified at all.[c] He was liable to be confused where later botanists were not; thus he mixed up *Phrynium* with *Zingiber*,[d] and cloves with garroo.[e] At the same time he was aware of distinctions which were later obscured, as between pepper and betel pepper.[f] He also mentions contemporary events, such as the bark-cloth (*mi hsiang chih*[4]) mission of +284 and the presentation of *pao hsiang lü*[5] pattens by the Fu-Nan country in +285.[g] And finally his celebrated description of the entomological control of the insect pests of orange trees was repeated by later authors more than once.[h] The practice has continued down to the present day, and the ant so cultivated is *Oecophylla smaragdina*.[i] All in all, the answer seems to be that basically Hsi Han's text is authentic, though there may have been some later interpolations.[j]

One of the reasons for all this uneasiness has been the existence of a parallel work with an almost identical name, acting like 'noise' in a telecommunication system. This is the *Nan Fang Tshao Wu Chuang*[6] (Prospect of the Plants and

---

[a] *Ibid.* p. 5*a*. Cf. Vol. 5, pt 2, p. 288, pt 3, p. 45. Wagner (1) has given us an exhaustive study of Han Shih San[1] (cooling regimen powder), a tonic and calefacient prescription (supposedly increasing vitality and longevity), which was particularly fashionable between the +3rd and the +6th centuries. It contained four inorganic substances (oxides and carbonates of calcium, magnesium and silicon, with small amounts of manganese and iron), and nine plant powders, including ginger, *Atractylodes ovata* (R/14, an age-old Taoist longevity medicine, cf. Vol. 5, pt 2, p. 150, pt 3, pp. 11, 40, 113, 117), ginseng, and aconite. Some of these certainly contained alkaloids, and all are still used in traditional medicine. Han Shih San needed a cooling regimen because it stimulated heat-production so much; but whether it was addictive, like opium later, remains very doubtful.

[b] E.g. *liu chhiu tzu*[2] for *Quisqualis indica* (entry no. 12) instead of *shih chün tzu*[3] (cf. p. 159 above).

[c] Besides the close study of Li Hui-Lin (12) there is also a valuable paper on the identifications by Wu Tê-Lin (1).

[d] Entry no. 17.   [e] Entry no. 45.   [f] Entries no. 10 and 58.

[g] Entries no. 56 and 57. The 'honey-fragrance paper' is shown by Li Hui-Lin (12), p. 147, to have been a kind of bark-cloth or tapa from the tree *Aquilaria agallocha*, aloes-wood or garroo (gharu), important also as an incense constituent (cf. Vol. 5, pt 2, p. 141). Besides the references he gives, see also the full account in Burkill (1), vol. 1, pp. 197 ff. On the circum-Pacific distribution of the bark-cloth techniques, which cover both China and Meso-america, see Ling Shun-Shêng (7, 6) and Ling Man-Li (1, 1). These ethnologists have also investigated the relation between bark-cloth and the invention of paper.

The 'water-pine pattens', as Li Hui-Lin (12), p. 153, points out, were certainly made from the root protuberances of the water-pine, *Glyptostrobus pensilis*, a tree with soft, light and buoyant wood something like balsa, now used as a substitute for cork. The pattens would have been ancestral to the geta of Japan.

Ma Thai-Lai (2) has attacked both the historical authenticity and the botanical sense of these two accounts, but his arguments are not at all convincing.

[h] E.g. *Yu-Yang Tsa Tsu* (c. +860), ch. 18, p. 3*a*; *Ling Piao Lu I* (c. +915), ch. 1, p. 4*a*, *b*. The wording is not the same as that of Hsi Han. *En passant*, it is worth remarking that even if Ma Thai-Lai's theory about the *Nan Fang Tshao Mu Chuang* were accepted, China would still have a priority of many centuries in the invention of entomological plant protection—the +9th century if not the +3rd. Cf. below, pp. 519 ff.

[i] See Groff & Howard (1).   [j] This is also the opinion of Li Hui-Lin (12).

[1] 寒食散   [2] 留求子   [3] 使君子   [4] 蜜香紙   [5] 抱香履
[6] 南方草物狀

Products of the Southern Regions), written by a much less well-known person, Hsü Chung[1], some time between +280 and +400.[a] Chia Ssu-Hsieh, in his *Chhi Min Yao Shu* of about +540, was fond of quoting from this book, as were others later,[b] often giving *Nan Fang Tshao Mu Chuang* as the title, but since none of the quotations can be found in the work of Hsi Han as it has come down to us, this caused difficulties for those scholars who supposed the two books to have been one and the same. Yü Chia-Hsi (*1*), followed by Wang Yü-Hu,[c] thought that Hsi Han's work had been partly lost, and reconstructed from other sources in the Sung;[d] but Hu Li-chhu,[e] and even more authoritatively Shih Shêng-Han (*3*), urged that the two books were quite distinct. In contrast to Hsi Han's refinement, Hsü Chung's style is plain and rather repetitive. There are five cases of the same plant being described by both authors, and then the entries are generally quite different; but *Nan Fang Tshao Wu Chuang* was never a flora, for it included marine animals and all kinds of natural products.[f] With this we can now return to more botanical realms, and give a couple of quotations from Hsi Han's work.

Divided into three chapters, it treats in turn of 29 herbs, 28 forest trees, 17 fruit trees and 6 bamboos; 80 plants in all. At the outset Hsi Han makes the following statement:

Nan Yüeh and Chiao-chih have the most extraordinary plants of all the four border areas of the empire. No knowledge of them existed before the Chou and Chhin periods. But from the time of Wu Ti onwards, the Han dynasty set up vassal States beyond the frontiers, so that the best products of the land were selected and sent as tribute to the imperial court. People in the central provinces generally have no idea of the form (and aspect) of these plants, so that I have set down herein, with explanations in due order, all that I have heard, for the benefit of future generations.

This seems a very reasonable preface from a man who had never (so far as we can tell from the historical evidence) travelled in the southern regions himself.

His first entry is on the banana. Although it attains a height four or five times that of a man, it is not a tree at all but a rapidly-growing herbaceous perennial belonging to a monocot family, the Musaceae.[g] Linnaeus named it *Musa paradisiaca* in +1753, but later described what he thought was another species, *M. sapientum*, named no doubt for the ancient Indian sages or gymnosophists who according to

---

[a] There is great variation both in the title and the author's given name, but we follow a useful statistical study by Ma Thai-Lai (*1*).

[b] There are about 20 passages in *CMYS* and 5 more in *PTKM*.

[c] (*1*), pp. 23–4.

[d] This was of course much less radical than the view of Ma Thai-Lai (*1, 1, 2*), who found no genuineness in the *Nan Fang Tshao Mu Chuang* at all.

[e] (*1*), p. 89.

[f] Ma Thai-Lai (*1*), while rejecting Hsi Han (as we think, on quite inadequate grounds) accepted the authenticity of the *Nan Fang Tshao Wu Chuang*, or what is left of it. Together with Li Hui-Lin (*12*), we accept both.

[g] See the excellent illustration in Masefield *et al.* (*1*), pp. 108–9.

[1] 徐衷

both Theophrastus[a] and Pliny[b] ate freely of the fruit. Since the plants are all sterile hybrid cultigens, species differentiation is difficult, but *M. sapientum* has generally meant 'dessert bananas' good for eating raw, and *M. paradisiaca* the other varieties which have to be cooked, i.e. the 'plantains'.[c] Already in 1884 de Candolle[d] saw that the plant was native to south and south-east Asia, but cult-ivated back to prehistory, and with no wild forms now. The leaves are large and handsome, but liable to be torn into tatters by the wind. The flowering stem appears at the apex of the plant about a year after growth has begun, and hangs downwards, with the male flowers towards the tip and the female ones more proximally, each developing into seedless fruit without fertilisation.

What Hsi Han himself wrote was this.[e]

Banana plants (*kan-chiao*[4]) look rather like trees. The larger ones have trunks bigger than a man can encircle with his arms. The leaves are ten feet long, or sometimes seven to eight feet, and have a width of more than a foot, sometimes two. The flowers are as big as wine-cups, with the shape and colour of lotuses (*fu-jung*[5]).[f] At the end of the stem there is what is called a spathe (*fang*[6]), which bears more than a hundred fruits all attached together. Their (flesh within the capsule) is sweet and delicious, and can be preserved with honey. The root is like that of the taro (*yü*[7]),[g] the largest as big as the hub of a carriage wheel. Fruiting follows flowering, each flower being a complete whole that gives rise to six fruits developing successively, but the fruits do not all form at the same time nor do the flowers fade and drop off together. It is also called *pa-chiao*[8] or *pa-chü*[9].[h]

Removing the skin of the capsule one finds that the flesh is yellowish-white with a taste like grapes, sweet and firm, satisfying one's hunger. There are three kinds of banana fruits. One is as thick as a thumb, long and pointed, somewhat like a goat's horn in shape, hence the name *yang chio chiao*[10]; this is the best and sweetest. Another kind is as thick as a hen's egg, and because it resembles the udder of a cow it is called *niu ju chiao*[11]; it is inferior to the former. A third kind is as thick as a lotus root, six or seven inches long and tetragonal in cross-section; it is not very sweet and ranks the lowest of the three.

The banana or plantain stem can be retted and separated into silk-like fibres which are rendered supple by treatment with lime, after which they can be woven into a special kind

---

[a] IV, iv, 5, Hort tr., vol. 1, p. 315.

[b] *Nat. Hist.* XII, xii, 24, Rackham tr. vol. 4, pp. 17, 19.

[c] The name *Musa* came almost certainly from the Sanskrit and Pali term *moca*; cf. Reynolds (1). This led Hsüan-Chuang in the *Ta Thang Hsi Yü Chi* to call the banana *mou-chê*[1], neither he in +646, nor perhaps many of his readers, realising that the plant and fruit were the same as that which had been called *chiao*[2] from the Han onwards. There are several occurrences of this name, for example in the entry for the Pan-nu-tsho[3] (Panacha) country in ch. 3, entry no. 7 (Beal (2) tr., vol. 1, pp. 88, 163, vol. 2, p. 66, Calcutta ed., vol. 2, p. 200). Banana itself is probably an African word, and plantain, from *platano*, a Spanish misunderstanding. Reynolds (1) goes into this.

[d] (1), p. 304. There is a map of the distribution from India to New Guinea in Reynolds (1).

[e] Tr. auct., adjuv. Reynolds & Fang (1); Li Hui-Lin (12).

[f] The justness of this can be seen well in any coloured picture of the plant, e.g. Masefield *et al.* (1), *loc. cit.*

[g] *Colocasia antiquorum* (R 710; CC 1926). A fine description is in Burkill (1), vol. 1, pp. 638 ff.

[h] This name has probably nothing to do with Szechuan (Vol. 1, p.97). Its other meaning was 'a hand held wide open', and indeed the half-spirals of fruits are still today called 'hands' in the trade, with the individual fruits as 'fingers'.

---

[1] 茂遮	[2] 蕉	[3] 半筊蹉	[4] 甘蕉	[5] 芙蓉
[6] 房	[7] 芋	[8] 芭蕉	[9] 芭苴	[10] 羊角蕉
[11] 牛乳蕉				

of cloth. This is known as 'banana vine-cloth' (*chiao ko*[1]),[a] and can be either fine or coarse (*chhih chhi*[2]) in quality.[b] Although it is strong and good, yellowish-white in colour, it cannot be compared with the reddish vine-cloth made from the *ko* plant itself.

Bananas grow in both Chiao-chou and Kuangtung. According to the *San Fu Huang Thu*[3] (Illustrated Description of the Three Cities of the Metropolitan Area): 'In the Yuan-Ting reign-period (+111), Han Wu Ti conquered Nan Yüeh and built the Fu-li Kung[4] Palace (in the garden of which) he caused to be planted the rare herbs and strange trees thus obtained. There are still two banana plants in that place.'[c]

This description is not at all bad, though closer observation would have shown that the six fruits in a half-spiral did not come from one ovary. Judging from the account of the three varieties, the first two were of the edible *sapientum* type, while the third would have been a *paradisiaca* that needed cooking. But for us the most surprising thing is the emphasis placed on the banana as a fibre-plant.[d] Actually, the oldest occurrences of the word *chiao*[6] mention no fruit, but only the value of the yarn and the cloth; thus the *Shuo Wên* dictionary (+121) defines it as a 'natural nettle-hemp' (*shêng hsi*[7]),[e] and the *Wu Tu Fu*[8] (c. +270) uses the same expression as Hsi Han, namely *chiao ko*[9].[f] So it looks as if the banana was first of all a textile-producer, and people ate the fruit thinking nothing of it, as a matter of course. This could reasonably explain the origin of the name, for *chiao*[10] means heat, burning or boiling,[g] and that was how the stems had to be treated with lime water to get the fibres. The history of banana cloth has been rather thoroughly written by Chang Tê-Chün (*2*), and the industry revived in South China at the present day for making gunny sacks.[h]

[a] Here the textile made from banana fibres took the name of a variety of the vine-cloth made since ancient times from the leguminous vine *Pueraria thunbergiana* (R 406; CC 1038; B II 390). Burkill (*1*), vol. 2, p. 1837 describes its other uses. The classical description is now Anon. (*109*), vol. 2, p. 502. We recall a fine specimen of the plant in the Geneva Botanic Garden.

[b] For the elucidation of these two technical terms see Li Chhang-Nien (*2*).

[c] Nobody knows the exact date of this work, which gives an account of the three cities of Chhang-an (mod. Sian), Fêng-i and Fu-fêng, nor whether Miao Chhang-Yen[5] was really the writer of it. It may be as early as the Hou Han period (+2nd cent.) or as late as the beginning of Chin (late +3rd). Hsi Han's quotation is all right, though the original says there were twelve trees, but in any case 'because the climates of south and north are so different most of the plants died within the year'. Generally, we are told, 100 specimens of each plant were transported—cinnamon, henna, lungan, lichi, etc.

[d] In this connection, see our discussion of the Yü Kung chapter of the *Shu Ching* on p. 90 above, and also Section 31 on textiles.

[e] Ch. 1B (p. 25.2). 'Nettle-hemp' has nothing to do with hemp as botanically defined (cf. p. 170 above), but comes from an urticaceous plant *Boehmeria nivea* or *chu ma*[11]. See R 592; CC 1576; Anon. (*109*), vol. 1, p. 517. These fibres are the source of what is commonly called in China 'grass-cloth', or 'summer cloth' (*hsia pu*[12]). Rondot (*4*) wrote well on it nearly a century and a half ago.

[f] This is Tso Ssu's[13] ode on the capital of the Wu Kingdom, in the markets of which *chiao ko* was one of the fabrics on sale. *Wên Hsüan*, ch. 5, p. 9*b*, tr. von Zach (*6*), vol. 1, p. 64; Knechtges (*1*), vol. 1, pp. 402–3.

[g] One recalls the great importance of the 'three coctive regions' (*san chiao*[14]) in traditional medical physiology.

[h] It seems that in the Philippines a special variety of banana has been developed for its fibre, *Musa textilis*, which gives abaca or 'Manila hemp', said to be the world's best for cordage (Brown (*1*), vol. 1, p. 422).

[1] 蕉葛   [2] 絺綌   [3] 三輔皇圖   [4] 扶荔宮   [5] 苗昌言
[6] 蕉   [7] 生枲   [8] 吳都賦   [9] 蕉葛   [10] 焦
[11] 苧麻   [12] 夏布   [13] 左思   [14] 三焦

Hsi Han was not the first to write about bananas, because Yang Fu had already done so before the end of the Han (*c.* +90) in his *Nan I I Wu Chih*, and his account also stresses the preparation and use of the fibres. It runs as follows:

The banana (*pa chiao*[1]) has leaves as large as woven bamboo mats, and its stem[a] is like that of the taro. After boiling in cauldrons, the (stem breaks up into) silk-like fibres, and can be used for making a kind of cloth when spun and twisted, reeled and woven. It is women's work to make this material which, whether fine or coarse, is known nowadays as Chiao-chih vine-cloth. The inside heart is shaped like a garlic bulb or the head of a snow-goose, as large as a pint pot. The fruits develop on the spathe, and each one of these has several dozen of them. Their skin is of a fiery red colour, and when it has been peeled off the flesh is dark, edible and as sweet as honey. Four or five fruits are enough for a meal. After eating, the taste lingers on among the teeth. Another name for it is the sweet banana, *kan chiao*[2].[b]

A third account comes early in the +3rd century, a few words by Ku Hui[3] in his *Kuang-chou Chi*[4] (Records of Canton), which indicate that by this time banana culture had spread north to Chiangsu.

The *kan chiao*[1] plant [he says] in Kuang-chou has flowers, fruits, leaves and roots no different from those that grow in the region of Wu. The only distinction is that since the climate of this southern land is warmer, and never experiences any frost or ice, the plant flourishes through all the four seasons. The fruit when ripe is sweet, but bitter and acrid while still green.[c]

We need not follow further here the history of bananas and their culture in Chinese literature, since that subject has been fully treated by Reynolds & Fang Lien-Chê (1).[d] In addition to their numerous references, only the story of the Thang monk Huai-Su[5] may be offered. He was a disciple of the great Hsüan-Chuang about +650, and afterwards lived at a temple in the south, where he grew a large plantation of bananas; upon being asked why, he explained that he was too poor to buy paper, so with the banana cloth that they afforded he could exercise his brush in calligraphy. And indeed he was a great master of *tshao shu*[6] writing.[e]

The above passages from Yang Fu and Ku Hui are, it can be seen, rather different from what Hsi Han said, but there is one more or less parallel passage, briefer but with closely similar wording, which was first quoted by Chia Ssu-

---

[a] This must be a mistake for root.

[b] Pp. 15–6, tr. auct. adjuv. Reynolds & Fang (1).

[c] Quoted in *Chhi Min Yao Shu*, ch. 10 (ch. 92), p. 92, tr. auct. adjuv. Reynolds & Fang (1).

[d] Among other things, they give a complete translation of Li Shih-Chen's entry on the banana in *PTKM*, ch. 15, pp. 81*a* ff. Thao Hung-Ching was the first to incorporate it in the literature of pharmaceutical natural history, *c.* +500.

[e] This story was preserved in the Sung botanical thesaurus *Chhüan Fang Pei Tsu*, (Hou chi), ch. 13, p. 2*a*. We shall discuss this work presently. On 'grass-writing' cf. Vol. 1, p. 219.

[1] 芭蕉   [2] 甘蕉   [3] 顧徽   [4] 廣州記   [5] 懷素
[6] 草書

Hsieh in the *Chhi Min Yao Shu* about +540,[a] and subsequently copied by many writers.[b] Chia attributed this to a *Nan Fang I Wu Chih* with no author,[c] but later books corrected this to *Nan Chou I Wu Chih* and sometimes supplied correspondingly the writer's name, Wan Chen. Since Wan was still living in +290, and therefore a contemporary of Hsi Han, either might have copied from the other;[d] but it is at least equally probable that Chia, like so many old authors, was negligent in his quoting, and relying on memory wrote *Nan Fang I Wu Chih* when he intended to put *Nan Fang Tshao Mu Chuang*.[e] With this we may leave Hsi Han secure in the possession of those banana groves which he himself never actually saw.

Pursuing examples from Hsi Han's book, nothing could be more characteristically tropical and sub-tropical than the *Hibiscus* genus,[f] beloved of writers in the mode of Tahitian exoticism. Many species of this are native to South China, for example *H. Manihot*,[g] with its mucilage used for sizing paper, *H. mutabilis* the cotton rose,[h] *H. Rosa-sinensis* the China rose,[i] and *H. syriacus*, the rose of Sharon.[j] The last three are all medicinal in traditional Chinese pharmacy, but the genus as a whole has been most admired for the beauty of its flowers, which come in all colours from white through yellow to the richest reds and purples. Hsi Han's description of *H. Rosa-sinensis* is particularly worth attention because of its botanical exactness to an extent perhaps remarkable for the early +4th century. He wrote:

The China rose (*chu chin*[7]) has stems and branches resembling the mulberry (*sang*[8]), the leaves on which are glossy and thick. The bushes are only four to five feet tall, but densely branched and leaved. They put forth blossoms from the second month onwards, not ceasing to do so until nearly mid-winter. The flowers are deep red in colour, pentamerous (*wu chhu*[9]), and as large as those of the hollyhock (*shu khuei*[10]).[k] There is a single style (*jui*[11]) exserted beyond the petals (*hua yeh*[12]), and with golden specks or shreds (*chin hsieh*[13])

---

[a] *CMYS*, ch. 10 (ch. 92), p. 92.

[b] For example, *IWLC*, ch. 87, p. 9*b* (p. 2230); *TPYL*, ch. 975, p. 1*a*; *PTKM*, ch. 15, pp. 81*b*, 82*a*; *YCLH*, ch. 404, p. 1*a*.

[c] A book with this exact title, by one Fang Chhien-Li[1], circulated during the Thang and Sung, but it has long been lost and there is no telling what it had in it as a whole. Since not written till +840 Chia cannot have quoted it.

[d] As Reynolds & Fang remarked, (1), p. 174.

[e] There is no other evidence that Wan Chen wrote on bananas.

[f] For more extensive descriptions see Li Hui-Lin (8), p. 137; Burkill (1), vol. 1, p. 1163; Brown (1), vol. 2, p. 414.

[g] Now *Abelmoschus Manihot*, *huang shu khuei*[2], the yellow mallow of Szechuan. R 276; CC 739. A related plant is *A. esculentus*, *chhiu khuei*[3], the pods of which are the vegetable called 'ladies' fingers', see Anon. (*109*), vol. 2, p. 814.

[h] *Mu fu-jung*[4], R 277; CC 740; Anon. (*109*), vol. 2, p. 817.

[i] *Fu-sang*[5], R 278; CC 741; Anon. (*109*), vol. 2, p. 816.

[j] *Mu chin*[6], R 279; CC 742; Anon. (*109*), vol. 2, p. 817. Many species also yield industrially useful fibres.

[k] Mallow of Szechuan, *Althaea rosa*, R 275; CC 735; Anon. (*109*), vol. 2, p. 808.

---

[1] 房千里	[2] 黃蜀葵	[3] 秋葵	[4] 木芙蓉	[5] 扶桑
[6] 木槿	[7] 朱槿	[8] 桑	[9] 五出	[10] 蜀葵
[11] 蘂	[12] 花葉	[13] 金屑		

attached to it. When seen in bright sunlight one would think that the flowers were blazing flames. On a single bush several hundreds of blooms appear each day, opening in the morning and fading towards nightfall. The plant is easily propagated by grafting (*chha chih chi huo*[1]). It grows (especially) in the commandery of Kao-liang.[a] Other names for it are *chhih chin*[2] (the scarlet *chin*) and *jih chi*[3] (lasting but a day).[b]

Here we notice in order, after his broad description of the plant, his clarity about the number of its petals, his recognition of its malvaceous character, and his use even of a technical term to speak of the great prominence in this flower of the gynoecium or pistil. Hsi Han could not have been expected to appreciate, fourteen centuries before Camerarius[c] and Linnaeus, the function of the 'golden specks and shreds' that were the anthers, nor the part that they would one day play (as 'polyandria', 'monadelphia', and attachment to the style) in the sexual system of classification;[d] but he saw them clearly, not knowing what he saw. Lastly there was his practical horticultural note. In later centuries the name *fusang*[4] prevailed over the others, as we have noted, for *Hibiscus Rosa-sinensis*, and thereby hangs a tale, because it was also the name of a fabulous island country far out in the Eastern Ocean, and of the tree growing there on whose boughs the sun rested before rising.[e] But to follow that by-path would lead us too far astray.

How tropical too is the banyan tree, with its vast over-reaching shade and its aerial roots, all redolent of the Indian stories of Kipling, yet native also, as not so many know, to Indo-China and South China. Hsi Han was the first to discuss it in the Chinese literature. Of *Ficus retusa*[f] he said:

Banyan (*jung shu*[5]) trees are frequently planted in Nan-hai and Kuei-lin. The leaves resemble those of hemp, and the fruits are like the berries of the evergreen (*tung chhing*[6]).[g] The trunk is gnarled and twisted, so it cannot be used for carpentry or the making of objects; lower down it has prominent ridges and deep grooves, so it cannot be used as lumber. When burned, the wood gives no flame, so it cannot be used as fuel. But because it is useless it can last a long time without being injured or cut down,[h] shading an area as

---

[a]   Near modern Yangchiang, on the coast of Kuangtung about half-way between the Pearl River estuary and the Leichow peninsula.

[b]   Entry no. 36, tr. Li Hui-Lin (12), mod. auct.

[c]   *De Sexu Plantarum*, +1694. Cf. p. 6.

[d]   On this, embodied in the *Systema Naturae* of +1735, see particularly Stearn (3) and Lee (1).

[e]   See Vol. 3, pp. 567–8, and Vol. 4, pt 3, pp. 540 ff. Conjecturally, the 'trees of the sun and moon', furthest east and furthest west, were imagined as some kind of mulberry, and the name would have been transferred because, as Hsi Han remarks, the *Hibiscus* bush looks at first sight rather like a young mulberry tree.

[f]   CC 1603. According to Anon. (*109*), vol. 1, p. 483 the specific name should be *microcarpa*, but Burkill (1), vol. 1, p. 1004 differentiates another species, *bengalensis*, cf. p. 1014. The genus is a great one, with at least 42 species in south and south-east Asia.

[g]   This is *Ilex pedunculosa*, related to the hollies; R 310; CC 832; Stuart (1), p. 213; Anon (*109*), vol. 2, p. 649. It will be remembered that Hsi Han wrote a poetical piece especially on an evergreen, the *Chhang Shêng Shu Fu*, cf. p. 449 above.

[h]   Here there is a strong echo of the doctrine of the value of uselessness in *Chuang Tzu* and other Taoist books. For example, in ch. 1 Hui Tzu's[7] large useless *Ailanthus*, *shu*[8], tree; in ch. 4, the useless oak trees that attained great age and were studied by Master-Craftsman Shih (Chiang Shih[9]) and the philosopher Nanpo Tzu-Chhi[10]; also

---

[1] 插枝即活   [2] 赤槿   [3] 日及   [4] 扶桑   [5] 榕樹
[6] 冬青   [7] 惠子   [8] 樗   [9] 匠石   [10] 南伯子綦

large as 10 *mou*, and providing a resting-place for people. The branching is dense, and the leaves fine and small. Soft tendrils like vines or rattans hang downwards and gradually reach the ground. When the tips of these sprouts enter the soil, they develop root systems; sometimes a large tree may root thus round about in four or five places. Also lateral branches when reaching neighbouring trees often become joined together with them by akind of natural grafting process. Southerners consider these phenomena as quite normal, and do not regard these trees as particularly auspicious.[a]

Here the chief interest lies in the observations of plant physiology, the exceptional rooting system and the tendency to fuse with branches of other individuals. Banyans are still common as far north as Kuangsi, as Han said, and many famous places are connected with them, for example the Six Banyan Temple (Liu-jung Ssu[1]) in the middle of the city of Canton.[b]

As one last excerpt from this earliest of tropical botanies let us take a look at certain strange fruits of the south characterised by their high tannin content, and therefore useful in many arts and industries. First the black or chebulic myrobalan, a large deciduous combretaceous tree native to India and Burma, *Terminalia chebula*.[c] Hsi Han wrote:[d]

The *ho-li-lê*[3] trees resemble the *mu-huan*[4],[e] and have white flowers. The fruits are shaped like those of the *kan-lan*[5],[f] but have six ridges and a very close attachment of skin and flesh; they can be made into a drink. Also they can change white hair and beards back to black. This plant grows particularly around Chiu-chen[6].[g]

Here the name alone indicates that this was originally an Indian plant (Skr. *haritaki*), but it must have been grown in Indo-china and South China from a very early date. Records collected by Lo Hsiang-Lin (*3*) show the connection of this tree with another Cantonese temple, Kuang-hsiao Ssu[7] in the western part of the city. First Yü Fan[8] of Wu was banished to Kuangchow about +225, and there he planted a whole garden of these trees; later, in +398, it was taken over as a Buddhist temple by the Indian monk Dharmayasa. There are still one or two myrobalan trees there now. As for the botany, we can note once again the typical cross-referencing description.

the classes of animals and men which were judged by the shamans unsuitable for sacrifice, and so lived on; in ch. 17 the determination of Chuang Chou to decline office, and so to live long because useless. See Legge (5), vol. 1, pp. 174–5, 217–8, 219, 220, 390, etc.

[a] Entry no. 33, tr. Li Hui-Lin (12), mod. auct.

[b] Founded in +479 and restored in +1098, on which Su Tung-Pho left an inscription dated +1101. During the Thang the great Chhan patriarch Hui-Nêng[2] was abbot, and a bronze statue of him dated +989 is still preserved there. There is a famous pagoda too, first built in +537. We ourselves recall a memorable visit in 1972.

[c] R 247; CC 624 and Brown (1), vol. 3, pp. 129 ff. prefer the specific name *catappa*, but Burkill (1), vol. 2, pp. 2134 ff. makes them two different species.

[d] Entry no. 47, tr. Li Hui-Lin (12), mod. auct.

[e] This is one of the names for *Sapindus mukorossi*, R 304; CC 787.

[f] The commonest name for the Chinese olive, *Canarium album*, R 337; CC 890.

[g] A place near mod. Vinh, on the coast of Vietnam directly opposite Hainan Island.

[1] 六榕寺   [2] 慧能   [3] 訶梨勒   [4] 木梡   [5] 橄欖
[6] 九眞   [7] 李孝寺   [8] 虞翻

The emblic myrobalan or Indian gooseberry comes from a plant of quite different family, the euphorbs, yet is also very rich in tannin. This is *Emblica officinalis*[a], the Indian origin of which shows clearly in its Chinese polysyllabic name. Hsi Han describes it as follows:

The *an-mo-lê*[1] tree has fine leaves resembling those of the mimosa-tree (*ho-hun hua*[2]).[b] Its fruits resemble the plum and are yellowish-green in colour; the stone is rounded but with six or seven ridges. The flesh when first eaten is bitter and acrid but has a sweet after-taste. Adepts use it to change the colour of white beards and hair, and this has been proved to be effective. It also grows particularly around Chiu-chen.[c]

The name of this tree was certainly derived from Skr. *āmalaka*.[d] The fruit is still used for making pickles and jams, and like the other myrobalan[e] still used as an effective hair-dye. Its curious property of stimulating two tastes successively led to the name by which it has been known more commonly since the Thang, *yü-kan-tzu*[3], the 'sweetness hangover fruit'. All this is far from exhausting the interest of the *Nan Fang Tshao Mu Chuang*, but space does not permit of more, and we must turn to consider the exotic botany of later centuries.

A hundred years later there were more plant descriptions in the *Chiao-chou Chi*[6] of Liu Hsin-Chhi[7] (+410), a scholar-official who had been posted down south on long assignment. He was often quoted by Chia Ssu-Hsieh. More exceptional were the activities of Thopa Hsin[8], a prince of the Northern Wei, Kuang-Ling Wang[9] (*fl.* +480 to +535 or so),[f] one of those interesting patricians who come into our story from time to time with tastes of a scientific character.[g] In this case the imperial prince was a man fond of falconry, hunting, natural history and

---

[a] Formerly *Phyllanthus emblica* but now separated from that genus. See R 330; CC 875; Anon. (*109*), vol. 2, p. 587; Burkill (1), vol. 1, p. 920.

[b] This member of the mimosa group is *Albizzia Julibrissin*, much favoured for the planting of avenues in China today. We recall a beautiful example of this at the great Taoist temple of Chin Tzhu near Thaiyuan in Shansi. See R 370; CC 952; Anon. (*109*), vol. 2, p. 323.

[c] Entry no. 72, tr. Li Hui-Lin (12), mod. auct.

[d] Laufer (1), pp. 378, 551. The name *emblica* itself may come from another Indian form, *ambala*.

[e] In fact there are at least three, the third being *Terminalia belerica*, R 246, *phi-li-lê*[4] from Skr. *vibhītaka*. These together form the well-known laxative and tonic preparation called triphala, the three fruits; cf. Ainslie (1), vol. 1, pp. 236 ff.; Chopra, Badhwar & Ghosh (1). This was known in the Thang as *san kuo chiang*[5], though not perhaps much used—*Hsin Hsiu Pên Tshao*, quoted in *PTKM*, ch. 31 (p. 8).

[f] His dating was established philologically by Hu Li-Chhu (1).

[g] To take only a few examples, there was Liu Chhung[10] (*fl.* +175) with his crossbow sighting devices (see Sect. 30); and Thopa Yen-Ming[11] (*fl.* +520) who could conceivably have known his relative Hsin personally. Yen-Ming was enamoured of mathematics, astronomy and seismology, the patron of a scientific man of real eminence, Hsintu Fang[12] (see Vol. 3, p. 633). Then in the Thang there was Li Kao[13] (*fl.* +785), one of the first pioneers in the construction of paddle-wheel boats (see Vol. 4, pt 2, p. 417). And lastly we need only recall from p. 331 above the name of Chu Hsiao[14] (*fl.* +1380), the patron of the great botanic garden at Khaifêng, as also that of his younger brother Chu Chhüan[15] (Vol. 5, pt 3, p. 210), active in proto-chemistry and chemical medicine.

[1] 菴摩勒	[2] 合昏花	[3] 餘甘子	[4] 毗梨勒	[5] 三果漿
[6] 交州記	[7] 劉欣期	[8] 拓跋欣	[9] 廣陵王	[10] 劉寵
[11] 拓跋延明	[12] 信都芳	[13] 李皋	[14] 朱橚	[15] 朱權

horticulture,[a] who established and maintained a famous garden near the capital (Loyang) about which the *Wei Wang Hua Mu Chih*[1] (The Prince of Wei, his Book of Flowers and Trees) was written. In the *Pei Shih* we read that

he liked organising horticultural production, and cultivated all the arts related to the nurture of trees, so that all the best fruits of the metropolitan area came from his gardens. But those who were around him did not make his name glorious in his generation.[b]

So much for practical botany, in the historian's opinion. Although this book must be the oldest special treatise on ornamentals and fruit-trees in all Chinese literature, and will therefore be referred to again in the next sub-section and under horticulture, it need not have found mention here if Thopa Hsin had not seriously tried to acclimatise the plants of the south in his Loyang gardens. For example there is a note in his book about the bo-tree or pipal, *Ficus religiosa*, under which the Buddha received illumination.[c]

In Han times there was a Taoist from the Western Regions[d] who planted many seeds of the *ssu-wei*[2] tree[e] in the valley below the western peak of Sung Shan.[f] Later they grew very tall and large. There are four trees in all there now, and they blossom three times a year.

The latter part of the Thang was the next period that saw an intensely renewed interest in the botany of the southern regions. Fang Chhien-Li's[4] *Nan Fang I Wu Chih*[5] already mentioned (p. 455) began it about +840, and though this has long been lost, some of the circumstances of it are recoverable from his other book, called *Thou Huang Tsa Lu*[6] (Records of One Cast out in the Wilderness)—since he was banished to the south. There is much of botany (and rather more of zoology) in Tuan Kung-Lu's[7] *Pei Hu Lu*[8] (Records of the Country where the Doors open to the North to catch the Sun, i.e. Jih-Nan and Lin-I);[g] for here in +873 he discusses oranges, pomegranates, orchids, and *yang mei* among other things, with entries rather long, and not averse to an element of the fabulous.[h] A good deal of botanical description is also to be found in Liu Hsün's[9] *Ling Piao Lu I*,[10] written between +895 and +915; and at some time during the +9th century there

---

[a] There is an irresistible echo here of the great Hohenstaufen king of Sicily (r. +1198 to +1250), Frederick II, whose *De Arte Venandi cum Avibus* was such an outstanding biological treatise of the European Middle Ages (see further in Sarton (1), vol. 2, p. 575). A detailed comparison between these two men, Frederick and Thopa Hsin, would make an interesting work, though the former had a much greater scope for his enlightened actions.

[b] Ch. 19, p. 15*a, b*, tr. auct. Cf. *Wei Shu*, ch. 21A, p. 15*b*.

[c] Cit. also in *TSCC*, Tshao mu tien, ch. 5, hui khao, p. 7*a*. Tr. auct.

[d] I.e. a Buddhist.

[e] *Ssu-wei* implies meditation, but the commoner name is *phu-thi shu*[3] or bodhi tree. Cf. Morohashi dict., vol. 4 (p. 995.3); CC 1601; Burkill (1), vol. 1, p. 1013.

[f] One of the Taoist sacred mountains, in Honan. Cf. Mullikin & Hotchkis (1), pp. 28 ff.

[g] *SKCS/TMTY*, ch. 70, p. 64*b*.

[h] Cf. Schafer (16), pp. 148, 204, 211, 232, 244, 249. On the myrica fruits (*yang mei*) see p. 446 above.

[1] 魏王花木志　　[2] 思惟　　　　[3] 菩提樹　　　[4] 房千里　　　[5] 南方異物志
[6] 投荒雜錄　　　[7] 段公路　　　[8] 北戶錄　　　[9] 劉恂　　　　[10] 嶺表錄異

appeared a *Ling-Nan I Wu Chih*[1] from the pen of one Mêng Kuan[2], though it is now almost wholly lost. Schafer[a] devotes a whole chapter to the plants which these men and others of their time discussed, such as the romantic red bean of the south, known in the West as jequirity beads, *Abrus precatorius*, or palms like *Arenga saccharifera*, or *Cocos nucifera*, the coconut characteristic of hot climates. Dealing with Hainan he also tells[b] how the Chinese scholars described the mangosteen, *Garcinia Mangostana*, and the tree from which the ancient anti-leprosy drug, chaul-moogra oil, was obtained, *Hydnocarpus kurzii*. In this connection Read (9) has a special account of the many drugs which came into the Chinese pharmaceutical natural histories from this time onwards as the result of increasing knowledge of the tropical south.

In the Sung the fascination with tropical botany continued. Plants entered only incidentally into the *Nan Pu Hsin Shu*[3], a collection of material about the south taken from stories and memorabilia of the Thang and Wu Tai periods, due to Chhien I[4] about +1015, himself a descendant of the royal house of Wu Yüeh at Hangchow in the previous century.[c] Something similar could be said about an earlier book full of miscellaneous information, the *Chhing I Lu*[5] (Exhilarating Talks on Strange Things), written about +965 by Thao Ku[6], who had held high office under the Later Chou as well as now in the Sung.[d] Partly he excerpted Thang and Wu Tai records, partly he jotted down notes of his own travels in the south.[e] Among his 144 topics on plants, some quite lengthy,[f] he descanted on strange southern names, curious beliefs about certain fruits and vegetables, and lists of ornamentals, drugs, peony varieties, and the like. But the true provincial spirit manifested itself again in another scholar-official, Sung Chhi[7], who in +1057 produced the *I Pu Fang Wu Lüeh Chi*[8], a serious account of the natural curiosities of sub-tropical Szechuan, where he had been posted. With its 52 entries on plants and 13 on animals, this book deserves more study than it has had.[g] He opens by describing the famous timber tree *Machilus nanmu* (*nan*[9]) of the Lauraceae,[h] and goes on to give interesting statements about medicinal rhubarb, green varieties of grapes unlike the purple ones of the north, the special plants of Omei Shan,[i] and the *yü-kan-tzu* myrobalan already mentioned (p. 458). He also

---

[a] (16), pp. 165 ff.   [b] (18), pp. 37 ff.
[c] *SKCS/TMTY*, ch. 140, p. 46*b*; *SKCS/CMML*, ch. 14 (p. 536).
[d] See *SKCS/TMTY*, ch. 142, p. 70*b*.
[e] His book is valuable for many things, for example the earliest reference to the use of cormorants in fishing, computing-rods made of iron, batteries of multiple-bolt arcuballistae, and so on. Cf. Vol. 3, p. 72, Vol. 4, pt 1, pp. 70, 124, 284, Vol. 4, pt 2, p. 468.
[f] To which must be added, 91 on animals.
[g] Each entry consists of a poetic reflection accompanied by an explanation in prose. The work is reproduced almost complete in *TSCC*, Tshao mu tien, ch. 2, i wên 1, pp. 3*a* ff.
[h] R 502; Burkill (1), vol. 2, p. 1385.
[i] How pleased Sung Chhi would have been by the noble flora of Omei produced under such difficult wartime conditions by Fang Wên-Phei (*1*).

[1] 嶺南異物志   [2] 孟琯   [3] 清異錄   [4] 陶穀   [5] 南部新書
[6] 錢易   [7] 宋祁   [8] 益部方物略記   [9] 楠

speaks of the extremely sweet fruit of *Ficus erecta* (*thien hsien kuo*[1], the fig of the celestial immortals).[a]

A similar but more famous study came from the brush of the poet **Fan Chhêng-Ta**[2], who was military governor of Kuangsi in +1172. Later, ordered north, he felt constrained to put on record as much as he could of the things he had seen on his travels, so in his book, the *Kuei Hai Yü Hêng Chih*[3], he included three chapters on flowering plants, fruits, herbaceous plants and trees respectively.[b] He knew how to talk about them as well as anyone else in his time, as the following two excerpts, taken at random, show.[c]

The camellias of the south (*nan shan chha*[6]) have flowers with a corolla (*pa*[7]) and a calyx (*o*[8]) twice as large as those of the central provinces, but the colours are slightly paler. The leaves are thinner and softer, and they have hairs on them. One can see therefore that they are a different species (or variety, *chung*[9]) from those of the centre.

The 'red cardamom' (*hung tou khou*[10]) has flowers close together in groups, and the leaves are narrow like those of reeds (*pi lu*[11]).[d] They come out late in the spring, and the flowers start with a (special) stem growing forth (*chhou*[12]) embraced by a large bract (*tho*[13]), then, this bursting, they blossom on the stem as an inflorescence (*sui*[14]). There are several dozen pistils and stamens (*jui*[15]), pale pink and very beautiful, like those in the flowers of peaches and apricots. As the pistils and stamens get heavier they bend the flowers downwards so that they look like bunches of hanging grapes, (glistening as if) mica (*huo chhi*[16]) with silky tendrils and a beautiful feathery appearance. These flowers do not produce seeds, so it is not the same plant as the 'herbaceous cardamom' (*tshao tou khou*[17]).[e] The style (*jui hsin*[18]) divides into two tips closely connected, so much so that poets liken it to the 'alongside-eyes' (fish, *pi mu*[19]).[f]

What is interesting here is Fan Chhêng-Ta's analytical approach. First he realises that something ought to be done about making a special category for the Szechuanese camellias, specific or varietal as we should say; then he struggles with the cardamoms growing in the province. This was tough, because no less than six genera of these ginger-like plants are now recognised, *Amomum*, *Aframomum*, *Elettaria*, *Alpinia* (now *Languas*), *Riedelia* and *Vanoverberghia*;[g] all producing different

---

[a] R 602. Now *F. beecheyana*, Anon. (*109*), vol. 1, p. 490.

[b] Of the former 15, of the fruits 55, and of the latter 26. The second was reproduced almost complete under the title *Kuei Hai Kuo Chih*[4] in *TSCC*, Tshao mu tien, ch. 15, hui khao 2, pp. 8a ff.; and the third under the title *Kuei Hai Tshao Chih*[5] in *TSSC*, Tshao mu tien, ch. 10, hui khao 1, pp. 14a, b.

[c] P. 21b, tr. auct. On other aspects of Fan's botanical interests, see the translation of Bullett & Tsui Chi (1).

[d] *Phragmites communis* and related species, R 754; Khung *et al.* (*1*), p. 1279.2. The comparison with the leaves of other monocots was very just.

[e] Presumably the fruits were inconspicuous.

[f] The *pu mu yü*[20] is the flounder *Areliscus* (= *Cynoglossus*) *abbreviatus* (R 177). Another identification is the flatfish *Paralicthys* (= *Pseudopleuronectes* = *Pseudorhombus*) *olivaceus*; Tu Ya-Chhüan *et al.* (*1*), p. 182.2. Other poets applied the term to one of the positions in love-making; *Tung Hsüan Tzu*, p. 3a; van Gulik (8), p. 128.

[g] See Burkill (1), vol. 1, pp. 131 ff., 910 ff., vol. 2, pp. 1302 ff. Schafer (16), pp. 193 ff., had trouble with this.

---

[1] 天仙果   [2] 范成大   [3] 桂海虞衡志   [4] 桂海果志   [5] 桂海草志

[6] 南山茶   [7] 葩   [8] 蕚   [9] 種   [10] 紅荳蔻

[11] 碧蘆   [12] 抽   [13] 籜   [14] 穗   [15] 蘂

[16] 火齊   [17] 草荳蔻   [18] 蘂心   [19] 比目   [20] 比目魚

kinds of cardamom and cardamom-substitutes, important spices used in the cooking of many countries. The red-fruited cardamom of Fan is now identified as *Languas galanga*,[a] but it could perhaps have been *L. officinarum* (*kao liang chiang*[1])[b] or even *L. japonica* (*shan chiang*[2]).[c] Fan's herbaceous cardamom was almost surely *L. globosa*.[d] The subject is necessarily confused and confusing, since one has to reckon not only with a still unstabilised botanical nomenclature but also with the uncertainties of usage of names in pre-Thang, Thang and Sung times, as well as later.

Interest in exotic plants seems to have died down during the Yuan period, but the last act of the indigenous story played itself out in the late Ming and Chhing. In +1581 Shen Mou-Kuan[3], about whom otherwise little is known, produced his *Hua I Hua Mu Niao Shou Chen Wan Khao*[4] (Useful Examination of the Flowers, Trees, Birds and Beasts found among the Chinese and the Neighbouring Peoples), now a rare book indeed.[e] It dealt with the flora and fauna of most of the border areas. But here was a turning-point, because it was just at this time that Europeans began to do the same kind of thing, and of the local floras the first, or one of the very first, was Francesco Calzolari's[f] *Viaggio di Monte Baldo*, published in +1566.[g] And Yunnan now began to enter the world of science, for a great scholar, Yang Shen[5], was exiled there for life in +1522; a civilising force, for until he died in +1559 he exerted great cultural influence, made friends with the tribal peoples, founded a college, wrote the history of the region, and described its plants and animals in his collected works, *Shêng-An Ho Chi*[6] and *Shêng-An Wai Chi*,[7] the latter printed in +1616. Later in the +17th century came the *Ling-Nan Tsa Chi*[8] of Wu Chen-Fang[9], and the *Ling-Nan Fêng Wu Chi*[10] of Wu Chhi[11], both treading the now familiar paths among the flora of Kuangtung and points south. This genre continued with the *Nan Yüeh Pi Chi*[12] of +1777 by Li Thiao-Yuan[13], but Yunnan was cultivated again in the *Tien Hai Yü Hêng Chih*[14], with its title modelled on that of Fan Chhêng-Ta, by Than Tshui[15] in +1799.[h] This has extensive and interesting chapters on animals as well as plants. Finally, among relatively modern writings, there was the *Min Chhan Lu I*[16] of Kuo Po-Tshang[17] in 1886, devoted to the characteristic vegetable and animal products of Fukien

[a] Anon. (*109*), vol. 5, p. 594.

[b] R 639a; Anon. (*109*), vol. 5, p. 595.

[c] R 638; Anon (*109*), ibid.; CC 1795. But both these generally have white petals with red stamens, and both have red fruits.

[d] R 643; Burkill, *op. cit.* p. 1303.

[e] We know only the copy in the Library of Congress in Washington.

[f] Commemorated by the genus *Calceolaria*.

[g] Arber (3), 2nd ed., p. 100. We refrain from emphasising that twelve and a half centuries had passed since the *Nan Fang Tshao Mu Chuang*.

[h] But this of course centred on a real lake, the Tien Chhih near Kunming.

[1] 高良薑   [2] 山薑   [3] 慎懋官   [4] 華夷花木鳥獸珍玩考
[5] 楊慎   [6] 升菴合集   [7] 升菴外集   [8] 嶺南雜記   [9] 吳震方
[10] 嶺南風物記   [11] 吳綺   [12] 南越筆記   [13] 李調元   [14] 滇海虞衡志
[15] 檀萃   [16] 閩產錄異   [17] 郭栢蒼

province, an area not much studied specifically by the writers of old times.[a] And in the end the wheel had come full circle with Chiang Fan's[1] *Hsü Nan Fang Tshao Mu Chuang*[2], an explicit enlargement of Hsi Han's work fifteen hundred years after its first appearance.

This procession of exotic and tropical botanists has been surely an impressive one. Neither wars nor tumults, nor the discomforts of exile or assignment in uncomfortable climes, could dim the urge which Chinese scholars felt to write down as well as they could the features of the flora previously unheard-of in their own home-lands. There seems to be no close parallel to this literature in medieval Europe, though of course there were stories of strange things seen in Africa or Central Asia, naturally developed by symbolising clerics for purposes of edification—yet systematic non-fabulous accounts written in a scientific spirit are rather lacking. It is one more piece of evidence for the intrinsic sobriety and curiosity of the Chinese mind. And side by side with it, though on a lesser scale, other scholars were working away in an archaeological sense, to try and identify the plants which had been spoken of in the classics in terms mysterious because long fallen out of use; these works we must now briefly examine.

(ii) *The elucidation of the ancient*

As will have been clear from the earlier volumes of this work, sustained writing in Chinese, going beyond the relatively brief oracle-bone inscriptions, started around the beginning of the −1st millennium.[b] The first texts of that time naturally generated in due course commentaries, but they in their turn were lost, and often only the texts themselves remained, handed down orally in many cases because of their semi-sacred 'Vedic' character. By the San Kuo and Chin periods in the +3rd and +4th centuries, the meaning of many of the allusions, and especially the names of plants and animals, had become quite obscure, so that it was a task of scholarship to clarify them and elucidate the original meanings if possible. The chief beneficiary of this was the *Shih Ching*[3] (Book of Odes), that marvellous collection of ancient folksongs dating from the −11th to the −7th centuries;[c] though doubtless the *Shu Ching*[4] (Historical Classic)[d] would have been investigated in the same way if it had had an equal element of natural history. These studies were extended to certain much later poems, particularly the *Li Sao*[5] (Elegy on Encountering Sorrow),[e] written about −295 by Chhü Yuan[6], the

---

[a] On this work see Swingle (9) and Wang Yü-Hu (1), 2nd ed., p. 282.
[b] Cf. Vol. 1, p. 86.
[c] Cf. Vol. 1, p. 145, Vol. 2, pp. 4, 105, 391–2, Vol. 5, pt 2, pp. 232–3.
[d] The individual documents in this range over a long period, from perhaps the −12th to the −5th and even later.
[e] Cf. Vol. 4, pt 2, p. 573, Vol. 4, pt 3, p. 250, Vol. 5, pt 2, p. 98.

[1] 江藩　　　[2] 續南方草木狀　　　[3] 詩經　　　[4] 書經
[5] 離騷　　　[6] 屈原

famous minister of Chhu State, partly because of its archaising character which soon made the plant names mentioned in it incomprehensible.[a] In our own time similar work has been done on other ancient books, such as the *Shuo Wên Chieh Tzu*[1] (Analytical Dictionary of Characters), due to Hsü Shen[2] in +121;[b] notably by **Hu Hsien-Su** (*3*).

The first salvo in this oppugnaculum was written by Lu Chi[3] about +245 with the title *Mao Shih Tshao Mu Niao Shou Chhung Yü Su*[4] (An Elucidation of the Plants, Trees, Birds, Beasts, Insects and Fishes mentioned in the 'Book of Odes').[c] He called it 'Mao's Odes' because Mao Hêng[5] in the Early Han (*fl.* −220 to −150) had produced a recension of the songs which displaced all others, and his work had been continued by his son Mao Chhang[6], whose commentary is the oldest still surviving.[d] But Lu Chi was more of a naturalist, and in his book he included 137 entries,[e] the 82 discussions of plant names all being in the first of the two chapters.[f] Each one takes a four-character phrase from one of the songs and adds a commentary seeking to identify the plant in question. It is worth while to have a close look at a couple of these. For example:[g]

Shih: In patches grows the *hsing-tshai*,[h]
       To left and to right one must pick it;
       Shy was that good and beautiful girl,
       Day and night he sought her and couldn't get her.

Lu Chi: '*Tshên tzhu hsing-tshai.*'[11]

       The *hsing* is the plant that is also called *chieh-yü*[12]. Its stems are white, and its leaves of a purplish-red colour, quite round and more than an inch in diameter. They float on the water, but the roots go down to the bottom whether the water is deep or shallow. Thick as hair-pin bodkins, they are green above and whitish down below. If one

---

[a] The older translations of d'Hervey St Denys (4); Lin Wên-Chhing (1) and Yang & Yang (1) have now been altogether superseded by that of Hawkes (1).

[b] Cf. Vol. 1, p. 31, Vol. 2, p. 218.

[c] From at least the Ming onwards, Lu Chi (*fl.* +222 to +258) was constantly confused with two other scholars of the same name, all three of the Wu Kingdom in the San Kuo period. The more famous was Lu Chi[7] (+261 to +303), a military commander and outstanding writer and poet, the friend of Chang Hua[8] the naturalist; the other was Lu Chi[9] (*fl.* +220 to +245), astronomer and mutationist. Our author's book on the *Shih Ching* was edited and reprinted by Ting Yen[10] in 1854. On it see Legge (8), vol. 1, pp. 178–9.

[d] One of the best studies of this subject is still that of Legge (8), pp. 10 ff.

[e] More than 90 items of *materia medica* are named in the *Shih Ching*; cf. Chhen Pang-Hsien (1), p. 4. The economic plants which occur, to the number of 61, have been discussed by Kêng Hsüan (1, 1). They divide as follows: cereals, 4, fibre-plants, 3, terrestrial vegetables, 7, wild plants used as vegetables, 11, aquatic vegetables, 1, wild plants used as vegetables, 5, fruit-trees, 7, forest trees 18, and miscellaneous ornamentals and dye-plants, 5.

[f] In modern times there has been an elaborate continuation of Lu Chi's book by a namesake, Lu Wên-Yü (*1*). Quoting him wherever appropriate, he goes much beyond him because he discusses 132 plants named in the *Shih Ching*, only one of which is today still unidentifiable.

[g] Mao, no. 1, tr. auct., adjuv. Legge (8), p. 1; Waley (1), p. 81; Karlgren (14), p. 2. The translations in both these examples are very free and compressed, just to give an idea of the stanzas on which Lu Chi was commenting.

[h] Lit. the *hsing* vegetable.

---

[1] 說文解字   [2] 許慎   [3] 陸璣   [4] 毛詩草木鳥獸蟲魚疏
[5] 毛亨   [6] 毛萇   [7] 陸機   [8] 張華   [9] 陸績
[10] 丁晏   [11] 參差荇菜   [12] 接余

boils the white stems in bitter wine,[a] they give it a fine distinctive flavour, or they can be eaten with wine. [At the end of the stems are the flowers, yellow in colour like those of rushes.[b]][c]

So went Lu Chi's identification in the +3rd century. He must have paid attention to the older commentaries that were available to him, but one suspects that he must also have listened to the old farmers and countryfolk among whom names and properties had been handed down by oral tradition; and clearly he must have botanised himself. The great sinologists of modern times, preoccupied more with the literature than the botany, made rather heavy weather of it—one ventured no further than 'water-plant' (Karlgren), others thought 'duckweed' (Legge, Giles), another 'water-mallow' (Waley), which put it in the wrong family. It is in fact the water-gentian, *Nymphoides peltatum*,[d] still eaten as a vegetable to this day in Hopei province. Most gentians are blue or purple, so this name is not a good one; it looks more like a 'water-buttercup' but it does not belong to the Ranunculaceae. The 'floating-heart' of Kêng Hsüan (1), appropriate because of the peltate leaves, is an American folk-name.[e] So by and large Lu Chi's description was quite good, and his identification not to be doubted.[f]

One more example:[g]

Shih: All among the branches of the *huan-lan* ...
   A boy with a knot-horn[h] at his belt ...
   A boy with an archer's thumb-ring, so free and easy ...

Lu Chi: '*Huan-lan chih chih.*'[1]
   The *huan-lan* is the plant that is also called *lo-mo*[2]. In Yuchow they call it *chhiao-phiao*[3] (sparrow-gourd). It is a creeper with thick, dark green, leaves. When cut (the stems) exude a white juice which is edible if boiled, and has a fine rich taste. The pods are several inches long, and resemble those of gourds (*hu*[4]).[i]

The sinologists did a little better here, though one spoke of 'sedge' (Giles), but they saw that *chhiao-phiao* must be 'sparrow-gourd' (Legge) which made it cucurbitaceous, and, another preferred 'vine-bean' (Waley), which made it a legume, while it really belongs to the Asclepiadaceae. Incredibly unpoetic was the calling of it 'metaplexis' in the translated song (Karlgren), yet this was in fact the correct

[a] Possibly vinegar, but could anything have made that into a drink?
[b] Presumably *Typha* spp.; R 782; Martin (1), pl. 88.
[c] The sentence in square brackets was preserved only in the *Chhi Min Yao Shu* quotation.
[d] Anon. (*109*), vol. 3, p. 414; CC 441; Martin (1), pl. 59.
[e] Strictly, the leaves that float are 'falsely peltate' and more correctly, cordate or sagittate.
[f] Lu Wên-Yü (*1*), p. 1.
[g] Mao no. 60, tr. auct., adjuv. Legge (8), p. 103; Waley (1), p. 55; Karlgren, p. 42.
[h] This was a sign of manhood, perhaps for undoing knots—or, more romantically, the girdles of consenting damsels. The song is, of course, about a young man 'putting on airs', and the girl's complaint.
[i] Here he only meant that the fruits were elongated, like those of the typical plant, the bottle-gourd, *Lagenaria vulgaris* (R 62; CC 178—9).

[1] 芄蘭之支 [2] 蘿摩   [3] 雀瓢   [4] 瓠

identification, for the plant is *Metaplexis japonica*.[a] It is a useful creeper as the stems are good for tying up bundles, the seeds and leaves are medicinal,[b] and the floss in the pod can be substituted for cotton floss (*mien*[1]) when making seal ink pads and pincushions. Of old, people ate parts of it, and later it was listed as a famine plant (cf. pp. 328 ff above);[c] but the leaves when dried and burnt give out a particularly foul smoke.[d] Fascinating though it is to follow out the thoughts and observations of Lu Chi, the two instances we have given must suffice, and it is time to continue with the story of those who followed him as the centuries passed.

His efforts were not forgotten, for at some time in the +5th or +6th century a writer whose name is not now known to us produced a *Mao Shih Tshao Chhung Ching*[3] (Manual of the Plants and Animals mentioned in the 'Book of Odes'). It is extant only in fragmentary form, containing a few notes on mammals.[e] During the same period there appeared the first of the analyses of the *Li Sao* from the botanical point of view; this was the *Li Sao Tshao Mu Su*[4] (On the Plants and Trees mentioned in the 'Elegy on Encountering Sorrow') by Liu Miao[5], but it was lost after the Sui.[f] The work with the same title which did survive was that by Wu Jen-Chieh[6], written in +1197; it contains notices of 51 plants. The kind of difficulty caused by Chhü Yuan's poetic writing has already been noticed in connection with 'melilot' and orchids (p. 418 above), to which Wu devoted some of his longish entries, drawing much, as usual, on the pharmaceutical natural histories available to him. Meanwhile, about the beginning of the Thang, Yang Ssu-Fu[7] produced a *Mao Shih Tshao Mu Chhung Yü Su*[8], but it has long been lost, along with several other books of similar titles which can be found in the dynastic bibliographies. Then came, about +1080, the *Mao Shih Ming Wu Chieh*[9] (Analysis of the Names and Things, including Plants and Animals, in the 'Book of Odes'), written by Tshai Pien[10], which still exists and was helpful to Legge when translating the *Shih Ching*.[g]

After that there was a long gap till in +1617 Hsü Kuang-Chhi[11], the celebrated scholar and friend of the Jesuits, produced his *Mao Shih Liu Thieh Chiang I*[12] (Exposition of the 'Book of Odes' in Six Aspects). One of the aspects was Po Wu[13], the 'investigation of things', which included discussions of plant and animal names

---

[a] Formerly *M. stauntoni* and *M. chinensis*; Anon. (*109*), vol. 3, p. 491; R 165; CC 422.

[b] Stuart (1), p. 264.

[c] *Chiu Huang Pên Tshao*, ch. 5, no. 22 (in *NCCS*, ch. 50, p. 22*a*, *b*), cf. Read (8), p. 28. By Chu Hsiao's time the names were all different, *yang-chio tshai*[2] (goat-horn vegetable), and five others. Cf. Kêng Hsüan (1), p. 402.

[d] Lu Wên-Yü (1), p. 40.

[e] In *YHSF*, ch. 17, pp. 34*a* ff. The greater part was lost during the Sung; perhaps it was one of those books which 'did not cross the river'.

[f] *Sui Shu*, ch. 35, p. 1*a*.

[g] Legge (8), vol. 1, p. 179. Tshai Pien was a politician, and as a pillar of the reform party proposed to destroy in +1094 the great astronomical clock built by Su Sung, who had been inclined to conservatism, cf. Vol. 4, pt 2, p. 497. But at any rate he had the merit of having been earlier a serious antiquarian interested in botany and zoology.

[1] 棉	[2] 羊角菜	[3] 毛詩草蟲經	[4] 離騷草木疏	[5] 劉杳
[6] 吳仁傑	[7] 楊嗣復	[8] 毛詩草木蟲魚疏		[9] 毛詩名物解
[10] 蔡卞	[11] 徐光啓	[12] 毛詩六帖講意		[13] 博物

and properties.[a] This was followed soon afterwards (+1639) by Mao Chin's[1] book, which had exactly the same title as that of Lu Chi, with the addition of the words *Kuang Yao*[2] (An Elaboration of the Essentials in the 'Elucidation . . .'). Here his four-character phrases are in general the same as those chosen for comment by Lu, but there is naturally much further information—for example, in the case of *Nymphoides peltatum* Mao Chin quotes the *Erh Ya*, the *Phi Ya* (cf. pp. 126, 192 above) and several other venerable authorities including the *Yen Shih Chia Hsün*[3] of +590 and the *Ching Tien Shih Wên*[4] of about +600. The activity continued with Hsü Ting's[5] *Mao Shih Ming Wu Thu Shuo*[6], an illustrated work printed in +1769, which assisted Bretschneider as well as Legge,[b] inducing in the former a rhapsodical passage on Chinese scientific scholarship.

It seems [he said], that the Chinese have a predilection for investigating the origins of natural objects. I need only cite the *Ko Chih Ching Yuan*[7] in 100 chapters,[c] published in +1735. In this work the origin and history of every subject is treated of in a long series of quotations from the native literature, ancient and modern; 16 chapters are dedicated to the investigation of the origin of the different plants, and represent therefore a kind of Chinese geographical botany. Another work in this department is the *Mao Shih Ming Wu Thu Shuo*, which contains an enumeration and description of all the plants and animals mentioned in the *Shih Ching*.[d]

Contemporary with Hsü Ting was Chu Huan[8], who six years earlier had published a book in the same genre but not illustrated, the *Mao Shih Ming Wu Lüeh*[9]. He did it in classical *lei shu*[10] style, following the divisions of heaven, earth, men and things; fortunately the things included quite a lot on plants.[e] There the subject rested until modern biologists reopened it.

Now comes an unexpected part of the story, the enthusiasm of Japanese scholars for researches of this kind. It began very early, with one of the Japanese ambassadors to the Thang court, Ono Takamura[11], an eminent writer (+802 to +852). In +836 he resigned from this post which had brought him to China, and was granted the rank of Imperial Vice-Commissary (Fu Shih[12]) at the court, whereupon he painted a series of scrolls, more than a hundred in number, entitled *Mao Shih Tshao Mu Chhung Yü Thu*[13].[f] They were doubtless based on the identifications of Lu Chi and his successors such as Yang Ssu-Fu.

A thousand years later the Japanese were back in the picture, so to say, probably stimulated by the work of Hsü Ting and Chu Huan. For in +1778 Fuchi Zaikan[14] produced his *Mōshi Rikushi Somoku Sozukai*[15] (Illustrated Commentary on Lu Chi's Treatise on the Plants and Trees mentioned in the 'Book of

---

[a] See Legge (8), vol. 1, p. 175.   [b] (8), vol. 1, p. 179.
[c] On this 'Mirror of Scientific and Technological Origins' see Vol. 1, p. 48.
[d] Bretschneider (6), p. 7.   [e] Legge (8), *loc. cit.*   [f] Fu Pao-Shih (1), p. 47.

[1] 毛晉   [2] 廣要   [3] 顏氏家訓   [4] 經典釋文   [5] 徐鼎
[6] 毛詩名物圖說   [7] 格致鏡原   [8] 朱桓   [9] 毛詩名物略
[10] 類書   [11] 小野篁   [12] 副使   [13] 毛詩草木蟲魚圖
[14] 淵在寬   [15] 毛詩陸氏艸木疏圖解

Odes') a work now rather rare.[a] The illustrations were good, and the scope was wider than the title, for it included birds, as well as objects connected with human rites and ceremonies. Oka Kōyoku[1] may well have been working at the same time, for only a few years later, in +1785, there came out at Kyoto his parallel work *Mōshi Himbutsu Zukō*[2]. This book has four chapters on animals as well as the three on plants, each entry based on the usual four-character quotations.[b] The writer was a medical man, known as a distinguished poet in the Chinese style, and greatly interested in the plants, many of which he grew in his own garden. We reproduce here in Fig. 88 his drawing of *Nymphoides peltatum*, illustrating the first of our examples from Lu Chi, and in Fig. 89 *Metaplexis japonica* shown twining round a graminaceous reed-like plant which may be either *Phragmites communis* or *Miscanthus sacchariflorus*.

Next, in the early years of the nineteenth century came work by the eminent scholar Chhêng Yao-Thien[3], who in a number of tractates wrestled with the difficulties of the historic terminology of gourds (*1*), the nine grains (*3*), and various other plants (*5*). And now it was that European botanists awakened to the interest of discerning what really were the plants which had been mentioned in the classics. So in 1822 came Fée's flora of Virgil (*1*),[c] then his study of the botany of Pliny (*2*) eleven years later,[d] and finally his flora of Theocritus (*4*). Later in the century these studies were extended to the Bible by Tristram (*1*), who in a well-known book reviewed the physiography of the Holy Land, and gave 'a description of every plant and animal mentioned in Holy Scripture'. Much further work on these subjects has been done since then, of course,[e] and in earlier European literature there had been tentatives in the same direction, but the beginning of the nineteenth century was a bit late in the day, after all, for elucidating the plants in ancient writings, compared with the work of Lu Chi and his successors.

Last of all, we return to Japan, where in recent times the tradition has been continued by the botanical study of that celebrated poetical anthology, the *Manyōshū*[4], compiled at Nara in +759, not quite a century before Ono Takamura was painting his plants in Chhang-an. Wakahama Sekiko (*1, 2*) has written on the identification of all the plants in those ancient poems, while Koshimizu Takumi (*1*) has brought out the scientific aspects of their nomenclature, as well as publishing a book of colour photographs of the plants in question (*2*). How such a technique would have delighted Lu Chi.

In this sub-section we have united the explorers telling of their encounters with new plant forms, and the literary scholars determined to seek for clarity about

---

[a] A description of it is given by Bartlett & Shohara (*1*), pp. 210 ff.
[b] There are 150 of these for plants; cf. Bartlett & Shohara (*1*), pp. 212–3. The work was subsequently reprinted in China, without the *katakana* soundings.
[c] With 93 entries.   [d] In which he dealt with some 540 species.
[e] As by Moldenke & Moldenke (*1*), following in Tristram's footsteps.

[1] 岡公翼   [2] 毛詩品物圖考   [3] 程瑤由   [4] 萬葉集

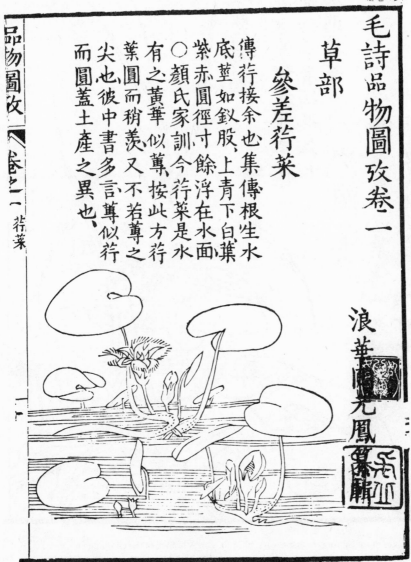

毛詩品物圖攷卷一

草部

參差荇菜

傳荇接余也集傳根生水
底莖如釵股上青下白葉
紫赤圓徑寸餘浮在水面
○顏氏家訓今荇菜是水
有之黃華似蓴按此方荇
葉圓而稍羨又不若蓴之
尖也彼中書多言蓴似荇
而圓蓋土產之異也

品物圖攷　　卷之一　荇菜

Fig. 88. A page from the historical botany of Oka Kōyoku (+1785) *Mōshi Himbutsu Zukō*, ch. 1, p. 1a. He gives a drawing of *Nymphoides peltatum* (Gentianaceae), *shen chha hsing tshai*, CC 441. In the text he criticises the *Yen shih Chia Hsün* pf +590 for confusing it with the *shun* (or *chhun*), i.e. *Brasenia purpurea* of the Nympheaceae (CC 1447), the leaves of which are truly peltate, while those of *Nymphoides* are falsely peltate, i.e. cordate.

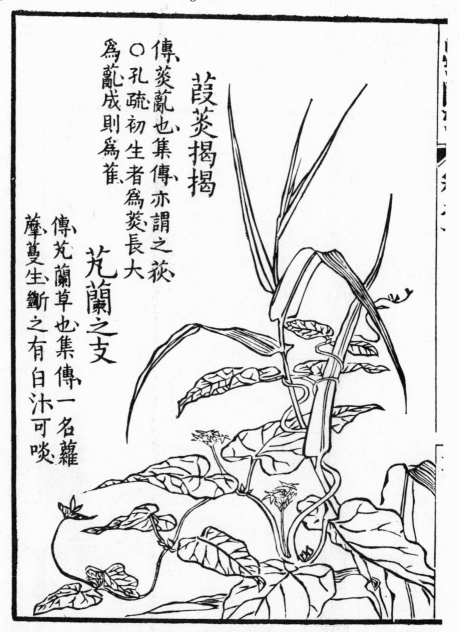

Fig. 89. Another drawing from the same work, ch. 1, p. 11b. A branch of the climbing herb *huan lan* (or *lo mo*) i.e. *Metaplexis japonica* of the Asclepiadaceae, CC 422, is twining round the stem of some water-plant. *Chia than chieh chieh* is a quotation or common expression, meaning 'rushes and sedges growing rank'. Oka quotes the Khung commentary as saying that *chia* is called *than* when young, *luan* when larger, and *huan* when fully grown. *Chia* is usually identified as *Phragmites communis* (CC 2045, R 754), but it could also be *Miscanthus sacchariflorus* (CC 2026), another kind of grass growing in shallow water.

those which the ancients had told of. Both endeavours were epics in their way, and both long anticipated the parallel activities of Europeans.

(e) PLANTS AND INSECTS IN MAN'S SERVICE

(1) NATURAL PLANT PESTICIDES

Since the dawn of civilisation man has had to contend with the assaults of invertebrate pests on his well-being. The victims include his person, his family, his dwelling, his household effects, his food crops and his domestic animals. We shall now examine the development of two of the several technologies the Chinese have evolved to deal with this situation. The first is the use of plant materials to destroy pests or alleviate the damage they cause; and the second is the control of invertebrate pests by manipulating the activity of their natural, biological enemies.

For the purpose of this study we use the term 'invertebrate' to denote the Chinese character *chhung*[1]. Actually, there is no exact English equivalent. An early system of classification in ancient China had divided the animal kingdom into two groups:[a] *ta shou*[2] and *hsiao chhung*[3], which may be translated rather loosely as 'large beasts' and 'little creatures'. The 'large beasts' included mammals, birds and fishes, while the 'little creatures' comprised all invertebrates, amphibians and reptiles. The grouping of amphibians and reptiles with invertebrates was an unfortunate mistake.[b] Even more unfortunately, it was compounded at a time when Chinese writing was still at an early formative stage, and the error became embedded within the structure of the language. The radical for *chhung*[1] became incorporated as part of the word for frog, *wa*[4], and the word for snake, *shê*[5]. Once the idea that *frog* and *snake* were somehow connected with *chhung*[1] and fixed permanently in the written word, it became much more difficult for later naturalists to break through the mental images connected with the characters *wa*[4] and *shê*[5] and recognise thay they were really members of the *shou*[6] family, and should be grouped together with mammals, birds and fishes, rather than with insects, crabs and earthworms. Thus the reader is reminded that in the passages he will encounter later, the word *chhung*[1] may mean not only insects and other invertebrates, but also, occasionally, amphibians and reptiles.[c]

---

[a] *Chou Li*, Tung Kuan, Khao Kung Chi[7].

[b] The Chinese are not the only ancient people who had trouble in classifying amphibians and reptiles. Harpaz (1, pp. 22, 26) points out that 'the Hebrews in Biblical times have terminologically included the insects among a wider group of teeming or creeping creatures embracing reptiles, amphibians, mollusks, arthropods and possibly other invertebrates'. The Hebrew term for these teeming creatures, *sheretz* is reminiscent of the Chinese concept of *hsiao chhung*. Even Aristotle, as pointed out by Balme (1, p. 262), was unsure about the position of snakes and sponges, which he could not fit either in the 'blooded' group (man, viviparous quadrupeds, oviparous quadrupeds, cetaceans, fishes, birds, etc.) or in the 'bloodless' group (molluss, crustaceans, testaceans and insects).

[c] In the preparation of this section we have been aided greatly by the monographs on the history of entomology in China by Chou Yao[8] (1, 2) and Tsou Shu-Wên[9] (1).

---

[1] 蟲	[2] 大獸	[3] 小蟲	[4] 蛙	[5] 蛇
[6] 獸	[7] 冬官考工記	[8] 周堯	[9] 鄒樹文	

(i) *The earliest plant pesticides*

The earliest reference to the application of plant materials to exterminate pests of various types is found in the *Chou Li*[1] (Records of the Rites of Chou), which provides descriptions of all the government posts and their duties during the late Chou dynasty. In Chapter 10, *Chiu Kuan Hsia*[2] (Autumn Official, lower section), we read the following passages:

The *Chu Shih*[3] (Decoction Officer) is in charge of eradicating poisonous pests: he drives them out with conjuratory spells, and suffocates them with *chia*[4] plant smoke.[a] (*Chu shih chang chhu tu ku, i kung shuo khuai chih, chia tshao kung chih.*[5])

The *Kuo Shih*[6] (Frog Controller) is in charge of destroying frogs and other water pests. He burns the male *chü*[7], sprinkles the ashes to kill them, and blows the toxic smoke over the area. All the water pests would be silenced.[b] (*Kuo shih chang chhü wa mêng, fên mu chü, i hui shai chih tsê szu, i chhi yen pei chih, tsê fan shui chhung wu shêng.*[8])

The *Chien Shih*[9] (Exterminator) is in charge of exterminating vermin and pests: he drives them away with prayers, and smokes them out with the *mang tshao*[10] plant.[c] (*Chien shih chang chhu tu wu, i kung jung kung chih, i mang tshao hsün chih.*[11]).

Before we examine the meaning of these passages in detail, a few general observations may be in order. First, we note with interest the existence of special officers responsible for the eradication or control of pests. It testifies to the importance of pest infestation as a problem afflicting ancient Chinese society. The problem evidently declined in importance later, perhaps due to improvements in housing construction, sanitation and public health, since the offices were discontinued in subsequent dynasties. Secondly, in two of three examples cited, assistance from appropriate deities was sought to drive away the pest. This practice may well be associated with the idea that pest infestations are the imposition of divine punishment for the sins of the Emperor or a leading official of the community,[d] a theme to

---

[a] *Chou Li*[1], ch. 10, p. 7. Biot's translation of this passage (1: II, p. 386) reads as follows: 'Le cuiseur (d'herbes) est chargé d'expulser les animaux venimeux. Il les éloigne, par des paroles conjuratoires. Il les attaque par des plantes excellentes.

[b] *Chou Li*, ch. 10, pp. 9, 9*a*. Biot's translation (1: II, p. 9. 390–1) reads: 'Le préposé aux grenouilles est chargé d'éloigner les grenouilles et les crapauds. Il brûle des plantes *khieou* de l'espèce mâle, il les asperge avec les cendres de ces plantes et alors ces animaux meurent.'

[c] *Chou Li*, ch. 10, pp. 8*a*, 9. Biot's translation (1. II, pp. 389–90) reads: 'Le destructeur est chargé d'expulser les teigres. Il les attaque par le sacrifice conjuratoire. Il fait contre eux des fumigations avce la plante *mang*.'

[d] Cf. this volume, p. 549. Conversely, the territories under the administration of virtuous officials will be spared even when the neighbouring areas are plagued by major insect infestations. Two celebrated cases of such events are recorded in the *Hou Han Shu*[12] (History of the Later Han Dynasty), +450, in the biographies of 'Cho Mou'[13] (about +1 to +5) and 'Lu Kung'[14] (in +82). For a discussion of the true meaning of theses accounts, see Tsou Shu-Wên (1) pp. 62–3.

---

[1] 周禮　　　　[2] 秋官下　　　[3] 庶氏　　　　[4] 嘉
[5] 庶氏,掌除毒蟲. 以攻說禬之,嘉草攻之.　　　[6] 蟈氏　　　[7] 鞠
[8] 蟈氏,掌去鼃黽. 焚牡鞠,以灰洒之則死.以其煙被之.則凡水蟲無聲.
[9] 翦氏　　　[10] 莽草
[11] 翦氏,掌除蠹物. 以攻禜攻之,以莽草熏之.　　　[12] 後漢書　　　[13] 卓茂
[14] 魯恭

which we shall have occasion to refer later. Finally, we have in these quotations the earliest record of the use of plant smoke to flush out and destroy invertebrate pests. This concept has been refined through the centuries in China. It is being practised today, in innumerable individual instances, in the form of 'mosquito incense' coils that are so popular throughout much of South China and Southeast Asia.

The English versions of the three passages quoted above are based on the interpretations of the text given by Chêng Hsüan[1] (+127–200), whose commentaries on the *Chou Li*[2] have been accepted by Chinese scholars through subsequent centuries. Yet Chêng Hsüan's interpretations are not always consistent or clear. The prevailing view has been that the *Chou Li*[2] as we know it today was compiled towards the end of the Western Han dynasty.[a] At least part of the sources used, however, was thought to be as old as the time of the Duke of Chou (−1030). Thus the meaning of some of the ancient material might already have become obscure by the time Chêng Hsüan[1] lived to propound his views.

Let us now examine the first passage, on the duty of the *Chu Shih*[3], the Decoction Officer. He is responsible for eradicating the *tu ku*[4], which we have translated as 'poisonous pests'. Actually Chêng Hsüan[1] explained that *tu ku*[4] are invertebrate pests (*chhung wu*[5]) that cause illness in man (*erh ping hai jên che*[6]).[b] According to the *Shuo Wên Chieh Tzu*[7] (Analytical Dictionary of Characters), +121, *tu ku*[4] are *fu chung chhung yeh*[8] (invertebrate pests within the belly), that is, parasitic worms. Later, the term was extended to mean any kind of internal invertebrate pest that causes disease or is harmful to man.

Having clarified the nature of the pest, we can now attempt to understand how the remedy was administered. The text reads: '*i kung shuo khuai chih, chia tshao kung chih*[9]'. The first part deals with the role of the supernatural and is beyond the scope of our present inquiry: we will concern ourselves only with the use of the *chia tshao*[10]. Chêng Hsüan's commentary states, '*chia tshao*[10] is a herbal material, its nature is not known; *kung chih*[11] means to attack with smoke'.[b] The inconsistency of Chêng Hsüan's interpretation was recently pointed out by Hsia Wei-Ying[12].[c] If *tu ku*[4] were an internal parasite, how could we then possibly attack it with the smoke of a dried plant? Chêng Hsüan[1] was probably confused by trying to draw a parallel between the use of *chia tshao*[10] and that of *mang tshao*[13], which is mentioned in the third passage. It would seem likely that the meaning of *chia tshao kung chih*[14] is literally to attack the (pest) by taking *chia tshao*[10] internally. That is to say, the plant

[a] This may no longer be the prevailing view in China today. Hsia Wei-Ying[12] (pp. 3–6) has concluded that the *Chou Li* was compiled during the Warring states period by followers of the Yin-Yang[15] School led by Tsou Yen[16] (cf. Vol 2, pp. 234 ff.) in the State of Chhi[17]. If he is correct, then the *Chou Li* should be dated at about −300, but some of the source material would be a good deal older.

[b] Cf. *Chou Li*[2], ch. 10, p. 7.

[c] Hsia Wei-Ying[12], pp. 30–2.

[1] 鄭玄	[2] 周禮	[3] 庶氏	[4] 毒蠱	[5] 蠱物
[6] 而病害人者	[7] 說文解字	[8] 腹中蟲也	[9] 以攻說禬之,嘉草攻之.	
[10] 嘉草	[11] 攻之	[12] 夏緯瑛	[13] 莽草	[14] 嘉草攻之
[15] 陰陽	[16] 鄒衍	[17] 齊		

was extracted with boiling water and the decoction imbibed in the time-honoured tradition of Chinese plant drugs.

What then is *chia tshao*[1]? Hsia Wei-Ying[2] thought that it was a general term for herbal medicine.[a] Biot translated it as *plantes excellentes*.[b] Chou Yao[3], however, stated that it is *jang ho*[4].[c] In the *Chhi Min Yao Shu*[5] (important Arts for the People's Welfare), +540, there is a quotation from *Sou Shen Chi*[6], which said that *jang ho*[4] was also known as *chia tshao*[1].[d] According to the *Ming I Pieh Lu*[7],[e] there are two types of *jang ho*[4], a red root and a white root. Using these descriptions, Bretschneider[f] and Read[g] have identified the plant as *Zingiber mioga*. The root is officinal, although the leaves also have similar medicinal properties. The *Ming I Pieh Lu*[7] states that *jang ho*[4] is used as a vermifuge. It is also prescribed for malaria, and insect and scorpion bites.[h] From these accounts we can conclude that the *chia tshao*[1] of *Chou Li*[8] is indeed *jang ho*[4], *Zingiber mioga*, Rosc. We may then revise the translation of this passage as follows: 'The Decoction Officer is in charge of eradicating internal parasites: he drives them out with conjuratory spells, and expels them with *chia* herb decoction.'

We shall now consider the next passage, the duty of the *Kuo Shih*[9], the Frog Controller. Chêng Hsüan's commentary indicates that *wa*[10] and *mêng*[11] are different types of frogs, whose loud croaks disrupt the tranquility of the human community. They had therefore to be destroyed. The official burns the *mu chü*[12], which Chêng Hsüan[13] regards as a type of *chü*[14], i.e. chrysanthemum, that does not bear flowers, presumably because the word *mu* means male in reference to a large animal. Biot has translated *mu chü*[12] as male *chü*[14] plants. This view is disputed by Hsia Wei-Ying[2].[i] All accounts in the ancient literature indicate that the *chü*[14] or chrysanthemum as known in early China, flowered in the ninth month of the lunar year and that the flowers were yellow. The ancient chrysanthemum must therefore be what is known today as *yeh chü*[15], or wild chrysanthemum, i.e. *Chrysanthemum indicum*, L. In this context the word *mu*[16] in *mu chü*[10] is simply a connecting word without any special meaning.

Thus, the translation of the second passage as given would be satisfactory if we replaced 'male *chü*[14]' by 'wild chrysanthemum'. After the chrysanthemum plant is burnt, the ashes are sprinkled over the places where frogs or tadpoles congregate. While it is burning the smoke and volatile materials are allowed to fumigate the area and silence all the water pests, *shui chhung*[17], i.e. frogs and presumably some invertebrates, in the vicinity.

---

[a] Hsia Wei-Ying[2], p. 31.   [b] Biot (1) II, p. 386.   [c] Chou Yao[3] (2), p. 81.
[d] *Chhi Min Yao Shu*[5], vol. 3, ch. 28.   [e] Cf. *PTKM*, ch. 15, p. 1006.
[f] B III, item 96.   [g] R (1), item 649.
[h] Cf. *PTKM*, ch. 15, 1006; B III, item 96; and F. Porter Smith & G. A. Stuart (1) pp. 464, 465.
[i] Hsia Wei-Ying[2], pp. 28–30.

[1] 嘉草	[2] 夏緯瑛	[3] 周堯	[4] 蘘荷	[5] 齊民要術
[6] 搜神記	[7] 名醫別錄	[8] 周禮	[9] 蟈氏	[10] 蠹
[11] 黽	[12] 牡蘜	[13] 鄭玄	[14] 蘜	[15] 野菊
[16] 牡	[17] 水蟲			

We now come to the third and probably the most interesting passage. It is about the *Chien Shih*[1], the Exterminator, whose duty it was to exterminate the *tu wu*[2]. According to the *Shuo Wên Chieh Tzu*[3] (Analytical Dictionary of Characters), +121, *tu wu*[2] are insects that bore into wood. *Tu wu*[2] may be regarded as household pests, such as silverfish, moths, boring beetles, termites and the like. In any case the smoke of the *mang tshao*[4] would be an effective way to drive them out.

What then is *mang tshao*[4]? There is an excellent summary of information on it in the *Chih Wu Ming Shih Thu Khao*[5] (Illustrated Investigations of the Names and Natures of Plants), 1848, which we would like to quote in full:[a]

The *mang tshao*[4] is included among the lower class (*hsia phin*[6]) of medicinal substances in the *Shên Nung Pên Tshao Ching*[7]. In Kiangsi and Hunan much is grown. It is commonly called *shui mang tzu*[8] (water *mang* fruit). Its root, which is especially poisonous, grows to a length of a foot or more, and is commonly known as *shui mang thou*[9] (water *mang* bag?). It is also called *huang thêng*[10] (yellow creeper). When soaked in water the extract assumes the colour of *hsiung huang*[11]. Its odour is very foul. The market nurserymen soak them and use the water to kill insects. This practice is very extensive. The leaves are also poisonous. In Southern Kiangsi (Nan-kan[12]) it is called *ta chha yeh*[13] (large tea leaf) and it is not different from the *tuan chhang tshao*[14] (break intestine plant). It is described in detail in the *Mêng Chhi Pi Than*[15].[b] The *Sung Ching*[16,c] says it is without blossom or fruit, but a profound study of it has not been made.

Yü-Lou Nung[17,d] says: 'In all the places I went to in the hills of Kan-chou[18] (Kiangsi), Hêng-chou[19] (Hunan) and Li-chou[20] (Hunan) there was much *mang tshao*[4], but judging from their form and appearance, it is uncertain that they are identical to the plant described in the *Mêng Chhi Pi Than*[15]. The flower is like that of apricot and attractive, but it is quite different from what Li Tê-Yü[21,e] calls the *hung kuei*[22] (red cassia) and what Chin Hsüeh-Yen[23,f] describes as having a red calyx and a white bud. It is because of this that Shen Tshun-Chung[29] (Shen Kua[30]) says the species exists in several different varieties. That produced in Kiang-yu[31] (right of the Yangtze River) have leaves like those of the tea plant and is commonly called *ta chha yeh*[32] (large tea leaves). In Hsiang-chung[33] (i.e. Central Hunan) they use the roots to poison insects. The roots are several feet long and therefore the plant is called *huang thêng*[34], but *shui mang*[35] is the common name. Are there similarities between it and the *shu mang*[36] (rat poison plant)? The poets have often preferred the terms '*wang lu*'[37].

---

[a] Tr. Hagerty, mod., *Chih Wu Ming Shih Thu Khao*[5], ch. 24, pp. 573–4.

[b] *MCPT*, *Pu Pi Than*[25] (Supplement to the *MCPT*), ch. 3, pp. 327–8.

[c] *Sung Ching*[16] refers to Su Sung's *Thu Ching Pên Tshao*[26] (Illustrated Pharmacopoeia) +1061. Cf. *PTKM*, ch. 17, p. 1218, for the quotation from Su Sung[27].

[d] Yü-Lou Nung[17] (Husbandman of Yü-Lou), is the appelation that the author, Wu Chhi-Chün[28], applied to himself.

[e] Li Tê-Yü[21], +787–849, referred to the *hung kuei*[22] in a preface to a book of poems.

[f] Chin Hsüeh-Yen[23] lived in the Ming dynasty. This statement occurs also in the preface on the *mang tshao*, now found in the *TSCC* (Imperial Encyclopaedia) of +1726, *20*, 110.

[1] 鶪氏	[2] 蠹物	[3] 說文解字	[4] 莽草	[5] 植物名實圖考
[6] 下品	[7] 神農本草經	[8] 水莽子	[9] 水莽兜	[10] 黃藤
[11] 雄黃	[12] 南贛	[13] 大茶葉	[14] 斷腸草	[15] 夢溪筆談
[16] 宋經	[17] 雩婁農	[18] 贛州	[19] 衡州	[20] 澧州
[21] 李德裕	[22] 紅桂	[23] 靳學顏	[24] 毒草	[25] 補筆談
[26] 圖經本草	[27] 蘇頌	[28] 吳其濬	[29] 沈存中	[30] 沈括
[31] 江右	[32] 大茶葉	[33] 湘中	[34] 黃藤	[35] 水莽
[36] 鼠莽	[37] 茵露			

Thao Yin-Chü[1,a] considers that *mang*[2] was originally written as *wang*[3]. In the hills many regard plants of the *huang mao*[4] (yellow reed) class as *wang tzu tshao*[5]. Kuo Pho[6], +300, in his commentary (on the *Erh Ya*[7]),[b] under the name *wei chhun tshao*[8], says another name is *mang tshao*[9]. In Sun Yen's[10] commentary (*Erh Ya Chêng I*[11]),[c] he states that the plant is commonly called *wang tshao*[12]. The thorns of the *wang tshao*[12] prick the clothes of people. As it spreads well over depressions and valleys it is referred to in poetry in connection with strolls in the morning and being wet with dew at dawn or late at night.'

Although the *mang tshao*[13] is luxuriant, it is not to be compared with the thorny bramble and hazel. Some say that the *wei*[14] is regarded as the *pai wei*[15] because *wei*[14] and *wei*[16] sound nearly alike, but as for *chhun tshao*[17] being another name for the same plant, it would be difficult to determine. Hsing Ping's[18] commentary[d] (on the *Erh Ya*[7]) says that for the *mang tshao*[13] of the *Pên Tshao*[19] Kuo Pho[6] gave *mang tshao*[9] as an alternative name, but the original plant he saw was different from that in the book. However, since the *Shen Nung Pên Tshao Ching*[20] has been copied by hand so many times and errors are rampant, one cannot but be careful in drawing conclusions. And as for the statement in the *Thu Ching*[21,e] that if boiled into a decoction and while hot held in the mouth for a short time, it will cure toothache and sore throat, how could one so lightly experiment with so poisonous a substance?

According to the *Chou Li*[22], to get rid of boring insects, the official known as *Chien Shih*[23] took *mang tshao*[13] and by burning it smoked them out. The *Fang Yen*[24,f] says that the *hui*[25] is the *mang tshao*[13]. In the region between Tung-yüeh[26] and Yang-chou[27] it is called *hui*[25]. In Nan-chhu[28] (Southern Hupeh) it is called *mang*[2]. According to the *Shuo Wên*[32], *hui*[33] is a general name for *tshao*[34] or herbaceous plants, so it cannot be a name for the poisonous *mang*[35] plant. Today people burn herbs to smoke out insects without having to use the poisonous *mang*[35]. Also the *Shuo Wên*[32] states that just as the dog is skilful in driving out the hare, so does *mang*[35] behave among the other plants. Commenting on the words *Mêng tzu tshao mang chih chhen*[36], Chao Chhi[37,g] says *mang*[2] is also a *tshao*[34] or plant. *Mang*[35], *hui*[33], *tshao*[38] and *mang*[39] have the same meaning. In the *Chhu Tzhu*[40] (Elegies of the Chhu[41] Kingdom)[h] the commentary after *Lang chung chou chih hsiu mang*[43] (grasp China's perennial *mang*[35] plant), states that

[a] Thao Yin-Chü[1] is another name for Thao Hung-Ching[29], the author of *Ming I Pieh Lu*[30] (Informal Records of Famous Physicians), +510.
[b] *Erh Ya*[7] (Literary Expositor or Dictionary), contains Chou material stabilised in Chhin and Han periods. The compiler is unknown. It was enlarged and commented on c. +300 by Kuo Pho[6].
[c] Sun Yen[10] is a contemporary of Kuo Pho's. He also wrote a commentary on the *Erh Ya*, now known as *Erh Ya Cheng I*[11].
[d] Hsing Ping[18], +930–1010, also wrote a commentary on the *Erh Ya*[7].
[e] Cf. *PTKM*, ch. 17, p. 1219.
[f] *Fang Yen*[24] (Dictionary of Regional Dialects), −28, by Yang Hsiung[31].
[g] Chao Chhi[37], +108–201, H/Han, author of the *Mêng Tzu Chang Chü*[44] (Chapters and Sentences on the Book of Mencius).
[h] By Chhü Yuan[42], c. −300.

[1] 陶隱居	[2] 莽	[3] 茵	[4] 黃茅	[5] 茵子草
[6] 郭璞	[7] 爾雅	[8] 弭春草	[9] 芒草	[10] 孫炎
[11] 爾雅正義	[12] 茵草	[13] 莽草	[14] 弭	[15] 白微
[16] 微	[17] 春草	[18] 邢昺	[19] 本草	[20] 神農本草經
[21] 圖經	[22] 周禮	[23] 翦氏	[24] 方言	[25] 莔
[26] 東越	[27] 楊州	[28] 南楚	[29] 陶弘景	[30] 名醫別錄
[31] 楊雄	[32] 說文	[33] 莔	[34] 草	[35] 莽
[36] 孟子草莽之臣	[37] 趙岐	[38] 艸	[39] 蛑	[40] 楚詞
[41] 楚	[42] 屈原	[43] 攬中洲之宿莽		[44] 孟子章句

it is a plant that survives the winter without dying. But this is merely an explantion of the meaning of the word *hsiu*[1].[a]

However, the *Shan Hai Ching*[2] (Classic of the Mountains and Rivers)[b] says: 'The Chhao-Ko[3] Mountains have the *mang tshao*[7] which can be used to poison fish.' Perhaps this is of the water *mang*[15] (*shui mang*[4]) class. But of the *mang*[15] of *Erh Ya*[5] which contains nodes, Kuo Pho's[6] commentary indicates that it is of the bamboo class. Among bamboos there is also a type called *mang*[15]. The term *mang tshao*[7] as used in the histories sometimes denotes the *mang*[8] and sometimes denotes a *mang*[15] of the bamboo class. It is difficult to determine which is which. If we consider the *mang*[15] to be toxic because it poisons fish, how do we then regard *chhiao mai*[9] (buckwheat)? It is also used to stupefy fish, but how can one possibly say that *chhiao mai*[9] is poisonous? I fear people will wrongly regard this *mang tshao*[7] as edible, and therefore emphasise that one must be careful to discriminate between the different kinds of *mang tshao*[7].

Out of these disparate and sometimes contradictory statements the following conclusions seem to emerge. First, a plant poisonous to insects by the name of *mang tshao*[7] has been quoted in the Chinese records from the late Chou until the 19th century. Secondly, it occurs wild and is particularly abundant among the hills south of the Yangtze River. Thirdly, it stupefies fish, and some varieties may also be poisonous to rats. We may, in addition, note that according to Su Sung[10] (+1061) and Khou Tsung-Shih[11] (+1115), the stem and leaves of *mang tshao*[12] resemble those of *shih nan*[13], which has been identified as a species of *Rhododendron*. Su Sung[14] further states that the plant may also grow as a creeper that winds around large rocks.[c]

Bretschneider pointed out that in Japan 'the Chinese character *mang*[15] is applied to *Illicium religiosum*, Sieb., a small tree held sacred by the Japanese and found also in Southern China. Its seeds and leaves possess poisonous properties.'[d] But he added that 'the Chinese poisonous *mang*[15] is quite different'. Although *mang tshao*[12] was included in the *Shen Nung Pen Tshao Ching*[16], it was not an article of commerce. Presumably Bretschneider did not have the opportunity to inspect an authentic specimen and give it a positive identification. On the other hand, Read came to the conclusion that the Chinese *mang*[15] is of the same species as the Japanese *mang*[15], and called it *Illicium religiosum* Sieb. Furthermore, he indicated that its identification as *Illicium anisatum* is wrong.[e] The problem is by no means resolved even today. Chou Yao[17] maintains that it is a type of poisonous *Illicium* related to star anise, *pa chiao*[18] (*Illicium verum* Hook. or *Illicium anisatum* Lour.),[f] while Hsia Wei-Ying[19] states that it is *Illicium anisatum*.[g] A wild star anise (*yeh hui hsiang*[20]) is described in the *Chung-Kuo*

---

[a] Any plant that is planted in one year and is harvested in the following year is considered to be a *hsiu*[1]. Examples are winter wheat and barley. It is also used to denote a thing kept overnight and used the next day.
[b] Author unknown; probably Chou and C/Han.
[c] *PTKM*, ch. 36, p. 2119. cf. B III, item 347; R (1), item 202, on *Rhododendron metternichi* S & Z.
[d] B III, item 158.   [e] R (1), item 505.   [f] Chou Yao (2), p. 81.   [g] Hsia Wei-Ying, p. 21.

[1] 宿	[2] 山海經	[3] 朝歌	[4] 水莽	[5] 爾雅
[6] 郭璞	[7] 莽草	[8] 芒	[9] 蕎麥	[10] 蘇頌
[11] 寇宗奭	[12] 莽草	[13] 石南	[14] 蘇頌	[15] 莽
[16] 神農本草經	[17] 周堯	[18] 八角	[19] 夏緯英	[20] 野茴香

Thu Nung Yao Chih[1] (Record of Indigenous Chinese Agricultural Drugs), 1959, as a plant insecticide.[a] It is identified as *Illicium lanceolatum* L. However, the description also states that there are at least twenty varieties of *Illicium verum* seen south of the Yangtze River, and many of them are poisonous. Further work is needed to classify and identify them, and determine which ones are suitable for extraction of edible oil[b] and which ones for use as insecticides.

For the time being, all we can say is that the historic *mang tshao*[4] of *Chou Li*[2] is a member of the *Illicium religiosum-anisatum-lanceolatum* group. Whether we can ever define its exact identity remains to be seen, but it will always enjoy the distinction of being the earliest example of a plant used by man to fight against insect pests. Thus, in terms of the history of pest control, the three brief passages in the *Chou Li*[2] that we have quoted are of unusual and outstanding interest. The *chia tshao*[3] passage provides the first record of the use of an anthelmintic to get rid of intestinal parasites. The *mang tshao*[4] passage describes the first plant material used as an insecticide. Finally, in the burning of *mu chü*[5] and *mang tshao*[4] we have the earliest example of the use of plant smoke to fumigate and drive away invertebrate and other noxious pests.

(ii) *Legacy of the Heavenly Husbandman*

As we have noted earlier in this volume,[c] the greatest single source of information on the biology of plants in the Chinese literature is the series of *Pên Tshao*[6] compilations (i.e. Pharmacopoeias or Pandects of Natural History), starting with the *Shen Nung Pên Tshao Ching*[7] (Classical Pharmacopoeia of the Heavenly Husbandman), +100, and culminating in the *Pên Tshao Kang Mu*[8] (The Great Pharmacopoeia), +1596. Although plants toxic to insects and other invertebrates were only of peripheral interest to the compilers of the pharmacopoeias, fortunately for us, many of the pesticidal plants known to the ancient Chinese were originally discovered and valued for their medicinal properties and were, therefore, listed and described in them. For this reason, the *Pên Tshao*[6] compilations have also turned out to be a convenient source of information on plants with pesticidal characteristics. But for the same reason, the information they contain is weighted much more on the side of pests that infest man and domestic animals than on pests that damage horticultural and agricultural crops.

The *Shen Nung Pên Tshao Ching*[7] (*Pên Ching*[9]) is a compendium of medicinal plants discovered in antiquity until almost the latter part of the Han dynasty (+100), while the *Pên Tshao Kang Mu*[8] (*Kang Mu*[10]) covers in considerable detail all

---

[a] *Chung Kuo Thu Nung Uao Chih*[11], Anon. (*58*) item 50.
[b] See Vol. 5, pt. 4, pp. 118–19.
[c] See pp. 220ff.

[1] 中國土農藥誌   [2] 周禮   [3] 嘉草   [4] 莽草
[5] 牡蘜   [6] 本草   [7] 神農本草經   [8] 本草綱目   [9] 本經
[10] 綱目

the plants described in the literature from antiquity till the late Ming dynasty
(+ 1600).[a] For developments on pesticidal plants after + 1600, the *Chung Kuo Thu
Nung Yao Chih*[1] (Record of Indigenous Chinese Agricultural Drugs), 1959, has
proved to be a valuable point of reference and source of information. Thus, by
suitable inspection of the three works in parallel, we have been able to sort out the
pesticidal plants discovered in each of the following periods in China: (*a*) antiquity,
− 1000 to + 100; (*b*) the middle period, + 100 to + 1600, and (*c*) + 1600 to the
present day.[b]

Let us consider first the plants discovered in antiquity. In the *Pên Ching*[5], 252
drugs of plant origin are reported. Of these, 36 are indicated in the text or described
in later publications as having some sort of pesticidal activity. They can be divided
in two groups. The first comprises all entries in the upper class (*shang phin*[2]) and
middle class (*chung phin*[3]) of plants, and the second, the lower class (*hsia phin*[4]).
As has been the case with the other herbs in the pharmacopoeia, many of them
have remained as aritcles of commerce through the centuries under their ancient
original names. They were, therefore, included in the medicinal plants studied
by Western scientists such as Henry, Porter Smith, Bretschneider and Read in
the late 19th or early 20th centuries. By examining samples of commercial origin
and inspecting fresh specimens in the field, they were able to assign scientific names
to all such pesticidal plants as well as to the other herbs included in the *Pên Ching*[5]
and the *Kang Mu*[6].

Listed in Table 13 are the Chinese names and the Latin botanical names of the
fifteen pesticidal plants in the first group, together with relevant references to the
Pên Tshao Kang Mu[7], *Botanicum Sinicum*, *Chinese Medicinal Plants from the PTKM*, and
the *Chung Kuo Thu Nung Yao Chih*[1]. For ease of discussion later, each item is given an
identification number. When discrepancies occur in the assignment of a botanical
name, the one listed in the *Chung Kuo Thu Nung Yao Chih*[8] (*Thu Nung Yao Chih*[8]) is
usually, but not always, given preference, since it represents the consensus of the
latest taxomonic studies carried out in China, presumably on authentic specimens
identical to those that were actually being used for controlling pests in the field.[c]

[a] For the purpose of this review we have found it convenient to follow in the text the tradition of abbreviation
adopted in the Chinese pharmacopoeas. Thus, *Pên Ching* is *Shen Nung Pên Tshao Ching*; *Kang Mu* is *Pên Tshao Kang
Mu*; *Pieh Lu* is *Ming I Pieh Lu*; *Thu Nung Yao Chih* is *Chung Kuo Thu Nung Yao Chih*, etc. But in the footnotes the
abbreviations used in other parts of this volume will be followed, e.g. *SNPTC* for *Shen Nung Pên Tshao Ching*,
PTKM for *Pên Tshao Kang Mu*, *TNYC* for *Chung Kuo Thu Nung Yao Chih*, etc. All page references to the *Kang Mu* are
based on the recent (1975) edition, published by the Jen-Min Wei-Shêng Press, Peking.

[b] It is interesting to note that in a recently completed study of the history of plant pesticides in Western Europe
down to 1850, Allan E. Smith & Diane M. Secoy (1) have also found it convenient to divide the development of
the technology into three periods: (a) Classical Mediterranean, antiquity to about + 400; (b) Medieval, + 400 to
about + 1650; (c) + 1650 to 1850.

[c] In the course of the experimental studies leading to the compilation of the *Thu Nung Yao Chih*, 220 field
specimens were identified by the Institute of Botany, Academia Sinica, 404 specimens examined by the Institute of
Entomology, Academia Sinica, 300 specimens by the Institute of Microbiology, Academia Sinica and 323
specimens by the Plant Protection Institute of the Academy of Agricultural Sciences. The system of classification
used is based on *Botany of Pteridophyta* by Chhin Jên-Chhang and *Flowering Plants* by Engler. Cf. *TNYC*, pp. iii–vi.

[1] 中國土農藥誌   [2] 上品   [3] 中品   [4] 下品
[5] 本經   [6] 綱目   [7] 本草綱目   [8] 土農藥誌

Table 13. *Upper and middle-class plants from the* Pen Ching *that show pesticidal activity*

No.	Chinese name	*PTKM* ch.	B III item	R (1) item	*TNYC* item	Botanical name
UPPER CLASS						
1	*Sung chih*	34	301	789	7	*Pinus massoniana*
2	*Tshang chu*	12	12	14	182	*Atractylodes chinensis*
3	*Chü hua*	15	69	27	186	*Chrysanthemum indicum*
4	*Chhang phu*	19	194	704	193	*Acorus calamus*
5	*Kan tshao*	12	1	391	—	*Glycyrrhiza glabra*
6	*Huai shih*	35	322	410	77	*Sophora japonica*
7	*Yün shih*	17	140	374	—	*Caesalpina sepiaria*
MIDDLE CLASS						
8	*Mi wu*	14	48	231	—	*Selinum sp.*
9	*Kao pên*	14	50	224	—	*Ligusticum sinense*
10	*Ko kên*	18	174	406	74	*Pueraria pseudo-hirsuta*
11	*Khu shên*	13	34	409	76	*Sophora flavescens*
12	*Tshin phi*	35	323	177	105	*Celastrus angulata*
13	*Hsi erh*	15	92	50	192	*Xanthium strumarium*
14	*Shui phing*	19	198	702	—	*Lemna minor*
15	*Liao shih*	16	124	573	23	*Persicaria nodosa*

Abbreviations: *PTKM, Pên Tshao Kang Mu*
B III, *Botanicum Sinicum* Vol. 3
R (1), Bernard Read, *Chinese Medicinal Plants from the* PTKM
TNYC, Chung Kuo Thu Nung Yao Chih.

1 松脂	2 蒼朮	3 菊花	4 菖蒲	5 甘草
6 槐實	7 雲實	8 蘪蕪	9 藁本	10 葛根
11 苦參	12 秦皮	13 枲耳	14 水萍	15 蓼實

The names listed in Table 13 follow the order in which they occur in the Mori Tateyuki's recontruction (1845) of the *Shen Nung Pên Tshao Ching*[1]. Of the seven plants from the upper class (*shang phin*[2]), five are included in the *Thu Nung Yao Chih*, and thus may be considered as genuine insecticidal plants. The status of remaining two, *yün shih*[3] (no. 7) and *kan tshao*[4] (no. 5), both well-known medicinal herbs, is equivocal. The *Pên Ching* states that *yün shih*[3], the fruit of the *yün*[5] plant (*Caesalpina sepiara* Roxb.) kills invertebrate internal parasites (*sha chhung ku tu*[6]).[a] It is an effective anthelmintic, but little is known about its potential insecticidal activity. The *Pên Ching* has no comment on the insectical activity of *kan tshao*[4], the root of the Chinese licorice (*Glycyrrhiza glabra* L.), probably the most popular and highly valued drug in the Chinese pharmacopoeia. But there are two references in later horticultural treatises on the use of *kan tshao*[4] for insect control. In the *Chung Shu Shu*[7] (Book of Tree-Planting), +1401, it is stated that sprinkling ground *kan*

[a] *SNPTC*, Mori ed., ch. 1, p. 12.

[1] 神農本草經 [2] 上品 [3] 雲實 [4] 甘草 [5] 雲
[6] 殺蟲蠱毒 [7] 種樹書

tshao[1] on the surface of the root when transplanting a tree will lessen infestation by insects.[a] And the *Hua Ching*[2] (Mirror of Flowers), +1688, advises that placing a piece of *kan tshao*[1] together with a piece of *ta suan*[3] (chive), in the hole before planting will reduce subsequent insect infestation.[b]

Of the five remaining plants, which have all been validated as being insecticidal, the case of chrysanthemum (no. 3) has been treated earlier. Actually, by the time the *Pên Ching*[4] was compiled, two principal types of chrysanthemums were known in China. One is the cultivated *Chrysanthemum sinensis* Sab., and the other the wild variety, *yeh chü*[5], *Chrysanthemum indicum* L. Thao Hung-Ching[13] (+502) says[c] that there are two types of *chü*[6]. One has a purple stem, a fragrant odour and a sweet taste. The leaves can be boiled to give an edible soup. This is the true *chü*. The other has the odour of *ai*[7] (*moxa*) and tastes bitter. It is called *khu i*[8]. In the *Kang Mu*[9], *khu i*[8] is indicated as being synonymous with *yeh chü*[5].[d] It is difficult to say which *chü*[6] was meant by the compilers of the *Pên Ching*[4], but only the *yeh chü*[5] is now included in the *Thu Nung Yao Chih*[10] as a useful insecticidal plant. The flowers of the *yeh chü*[5] contain essential oils, the anthocyanin chrysanthemin and other phenolic compounds. It is not known whether they contain any pyrethrins.[e]

The next plant (no. 2) is *chu*[11] (or *shu*[11]) as listed in the *Pên Ching*[4]. It is also called *tshang chu*[12] in the *Kang Mu*[9]. It has been identified as *Atractylodes chinensis* Koidz. The root is officinal. It is used against unusual parasites in the abdomen. Thao Hung-Ching[13] reported that *chu*[11] can dispel foul air, and people often burn it in case of illness and at year-end to ward off evil spirits.[f] Perhaps there is a good rationale for this practice since the fumes would be toxic to a variety of household pests. The *Thu Nung Yao Chih*[10] recommends the use of *tshang chu*[12] smoke to drive off mosquitoes and to fumigate stored grain. The seeds may also be employed as an anthelmintic. The plant contains essential oil, with atractylol and atractylone as the principal constituents.[g]

The root is also officinal in *chhang phu*[14] (no. 4), the sweet flag, *Acorus calamus* L. It is used in medicine as an antipruritic and an insecticide. Li Shih-Chen[15] says there are five kinds of *chhang phu*[16] grown in different types of environment. It is said to kill pests (*sha chu chhung*[17]) and drive away body lice and fleas (*tuan tsao sê*[18]).[h] A remarkable remedy for ear infestation by fleas or lice is given in the *Shêng Chi Lu*[19],

---

[a] *Chung Shu Shu*; p. 2a.   [b] *Hua Ching*, ch. 2, p. 24.

[c] *PTKM*, ch. 15, pp. 929 ff.   [d] *PTKM*, ch. 15, p. 932.

[e] The pyrethrins are the active components occuring in the flowers of *Chrysanthemum cinerariaefolium* and *Chrysanthemum coccineum*, the two most useful and potent insecticidal plants known to man. Neither plant is native to China. For a brief account of the discovery and utilisation of the two plants in the 19th century, see Shepherd (1), ch. 8, p. 144. They were introduced into China from Japan in 1919, but did not find extensive cultivation until after World War II (cf. Kuan Lung-Chhing and Chhen Tao-Chhuan).

[f] *PTKM*, ch. 12, p. 739, quotation from Thao Yin-Chü.

[g] *TNYC*, p. 156.   [h] *PTKM*, ch. 19, pp. 1357–8.

[1] 甘草	[2] 花鏡	[3] 大蒜	[4] 本經	[5] 野菊
[6] 菊	[7] 艾	[8] 苦薏	[9] 綱目	[10] 土農藥誌
[11] 朮	[12] 蒼朮	[13] 陶弘景	[14] 菖蒲	[15] 李時珍
[16] 菖蒲	[17] 殺諸蟲	[18] 斷蚤虱	[19] 聖濟錄	

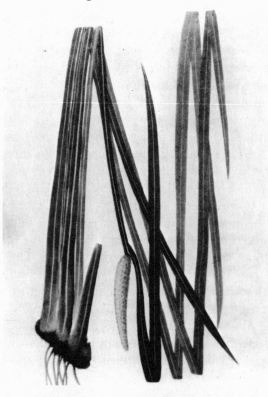

Fig. 90. *Chhang phu, Acorus calamus* L.; *TNYC* no. 11.

'Heat ground *chhang phu*[14] and store the material in a cloth bag. Sleep on it, and you will be cured.'[a] The *Thu Nung Yao Chih*[1] recommends its use as an aqueous infusion on a variety of agricultural pests. The active constituents are various di- and tri-terpenes (Fig. 90).[b]

We are now left with two trees, *sung*[2] (no. 1) and *huai*[3] (no. 6), both of which are important insecticidal plants. Although the *Pên Ching*[4] lists only the resin (*sung chih*[5]) of the pine tree (*Pinus massoniana* Lamb), the *Kang Mu*[6] indicates that in addition to the resin, the twigs (*sung chieh*[7]), leaves (*sung mao*[8]), flowers (*sung hua*[9]), cones (*sung shih*[10]) as well as the bark of the stem and root are all useful as medicine.[c] The *Pên Ching* recognises that the resin cures skin ulcers and itches caused by parasites (*yung chu, chiai sao*[11]), and the *Pieh Lu*[12] states that it also kills insects (*sha chhung*[13]) and other invertebrate pests. According to the *Wu Lei Hsiang Kan Chih*[14] (On the Mutual Responses of Things According to their Categories), +980, pine

---

ᵃ *PTKM*, ch. 19, p. 1360.   ᵇ *TNYC*, p. 166.   ᶜ *PTKM*, ch. 34, pp. 1917 ff.

¹ 土農葯誌   ² 松   ³ 槐   ⁴ 本經   ⁵ 松脂
⁶ 綱目   ⁷ 松節   ⁸ 松毛   ⁹ 松花   ¹⁰ 松實
¹¹ 癰疽, 疥瘡 ¹² 別錄   ¹³ 殺蟲   ¹⁴ 物類相感志

needles can be used to control pests of stored rice.[a] The *Thu Nung Yao Chih* recommends the use of aqueous extracts of pine needles to control field insect pests. That so many different parts of the plant show pesticidal activity is not surprising, since they all contain essential oils, and the terpenes α and β-pinene.[b] Indeed pine oil is even used today as a base or solvent for the formulation of cleaning fluids, deodorants, disinfectants, insecticides and perfumes.

The *Pên Ching* lists only *huai shih*[1] (no. 6), the fruit of the *huai* tree (*Sophora japonica* Linn.), also known as the Japanese pagoda tree. However, the *Kang Mu* states that the flower, the leaves, the stem, the bark of the stem or root, and the resin are all used in medicine. According to the *Kang Mu*, the fruit kills pests (*sha chhung*[18]), i.e. acts as an anthelmintic, and the flowers, which also act as a dye, kill internal parasites (*sha fu tsang chhung*[2]). Boiled extracts of the bark are recommended as an enema to flush out worms from the lower rectum.[c] The *Thu Nung Yao Chih* states that infusions of the leaves and flowers can control various kinds of insect pests. Recent chemical analyses indicate that the leaves and flowers contain a variety of flavonoids, such as rutin, sophoricoside, and sophorobioside.[d]

We now come to the middle class (*chung phin*[3]) of pesticidal activity listed in Table 13. Out of the eight plants, three are not included in the *Thu Nung Yao Chih*, namely, *mi wu*[2], *kao pên*[5] and *shui phing*[6]. The *Pên Ching* notes that the *mi wu*[4] (no. 8), a *Selinium* sp., a member of the Umbelliferae family, eliminates three types of invertebrate pests (*chhu san chhung*[7]) and expels internal parasites (*chhu ku tu*[8]).[e] This statement is in agreement with Porter Smith's observation that the leaves are used as antiseptic and anthelmintic.[f] *Kao pên*[3] (no. 9) is the root of another Umbellifer, *Ligusticum sinense* Oliv. The decoction prepared from it, when used to clean utensils and the dining table, is said to ward off flies.[g] In the *Kang Mu*, it is recommended for use to clean the rashes of children afflicted with mite infestations, and for washing their clothes so as to prevent reinfection.[h] The third item, *shui phing*[4] (no. 14), also known as *fou phing*[9], has been variously identified as *Spirodela polyrhiza* Schleid. or *Lemna minor* L., the duckweed.[i] According to the *Wu Lei Hsiang Kan Chih*[10] 'when dried duckweed is burnt, the smoke will kill mosquitoes'.[j]

[a] *Wu Lei Hsiang Kan Chih*[10], p. 1.   [b] *TNYC*, p. 9.
[c] *PTKM*, ch. 35, pp. 2005 ff., cf. especially pp. 2005, 2007, 2009.
[d] *TNYC*, pp. 67, 68.   [e] *SNPTC*, Mori ed., ch. 2, p. 5.   [f] Porter Smith & Stuart (1), p. 402.
[g] *Wu Lei Hsiang Kan Chih*[10], p. 1. A similar statement is quoted from the *Chhêng Huai Lu*[11] (Reflections of an Innocent Traveller), *c.* +1270, by Chou Mi. See Chou Yao (1), p. 52.
[h] *PTKM*, ch. 14, p. 845.
[i] Presumably different types of duckweed were known in China, Bretschneider, B III, item 198, recognises *Spirodela polyrhiza* and *Lemna minor* as types of *shui phing*[4], while Reed (1), item 702, regards it as *Lemna minor*. Chou Yao (2), p. 81, considers it as a *Spirodela*.
[j] *Wu Lei Hsiang Kan Chih*[10], p. 25. It also lists other plants that when burnt will drive off mosquitoes, viz. *ching yeh*[12] (leaves of *Vitex incisa*, Lam.), *ma yeh*[13] (hemp leaves, probably *Cannabis sativa* L.), *chhen chha*[14] (old tea leaves) residues, and *lou tshung*[15] (a kind of scallion). Indeed, in the account on *mang tshao*[16] quoted from the *Chih Wu Ming Shih Thu Khao*[17], it is said that 'today people burn herbs to smoke out insects without having to use the poisonous *mang*'; cf. p. 476. Thus the use of dried duckweed is by no means unique. Cf. also *Kang Mu*, ch. 19, p. 1366.

[1] 槐實	[2] 殺腹臟蟲	[3] 中品	[4] 蘪蕪	[5] 藁本
[6] 水萍	[7] 去三蟲	[8] 除蟲毒	[9] 浮萍	[10] 物類相感志
[11] 澄懷錄	[12] 荆葉	[13] 麻葉	[14] 陳茶	[15] 樓葱
[16] 莽草	[17] 植物名實圖考		[18] 殺蟲	

The remaining five plants, all included in the *Thu Nung Yao Chih*, show in varying degrees significant activity against invertebrate pests. The first herb is *ko kên*[1] (no. 10), that is, the root of the *ko*[2] vine (*Pueraria hirsuta* or *Pueraria thunbergiana*, order Leguminosae). This plant is the familiar *kudzu* vine of Japan. The leaves, seeds and roots are all used in medicine. The *Kang Mu* states that it is effective in alleviating the injury caused by snake and various invertebrate bites.[a] The *Thu Nung Yao Chih* describes it as *fên ko thêng*[3] (*Pueraria pseudo-hirsuta* Tang et Wang), the leaves and root of which are particularly active as insecticide. Recent work shows that one of the active ingredients is the flavone, kaempferol rhamnoside.[b]

The next item, *khu shên*[4] or bitter root (no. 11) is the root of another *Leguminosae*, *Sophora flavescens* Ait. According to the *Kang Mu*, *khu shên*[4] expels pests that cause skin rashes (*chih chieh sha chhung*[5]), kills intestinal worms (*sha kan chhung*[6]), dispels aches and kills invertebrate pests (*chih fêng sha chhung*[8]).[c] Porter Smith also affirmed its activity as an anthelmintic.[d] The stem and leaves have properties similar to those of the root. The *Thu Nung Yao Chih*[9] describes the use of root, stem and leaves to control a host of agricultural pests, fly maggots and mosquito larvae. The active ingredients consist of cytisine, matrine and other alkaloids (Fig. 91).[e]

The next entry is listed in the *Pên Ching* as *chhin phi*[10] (no. 12), the bark of the *chhin*[11] tree, but the *Kang Mu* says that according to Su Kung[12], *khu shu*[13] is synonymous with *chhin phi*[10]. There is considerable uncertainty about its botanical identity. Porter Smith called it *Fraxinus pubinervus*, while Read preferred *Fraxinus bungeana*.[f] The genus *Fraxinus* would place the plant in the Oleaceae. The *Kang Mu* states that *chhin phi*[10] kills invertebrates (*sha chhung*[14]), and the leaves can be used for washing clothes.[g] Porter Smith confirmed the value of the bark as an astringent, and that it is effective as a wash for snake and insect bites.[h] In the modern literature *khu shu*[13] is the name given to *Celastrus angulata* Maxim., which would make it a member of the Celastraceae.[i] The *Thu Nung Yao Chih* states that both leaves and bark of the stem or root are effective as insecticide. It is thus possible that the *khu shu*[13] (or *chhin phi*[10]) of the *Kang Mu* is not the same as the *khu shu*[13] in the *Thu Nung Yao Chih*.

No such controversy surrounds the identity of the next plant, *hsi erh*[15] (no. 13), also known as *tshang erh*[16]. It is *Xanthium strumarium* L. or cocklebur, a common weed throughout China. The seed, fruit, stem and leaf are all active as medicine or as a

[a] *PTKM*, ch. 18, p. 1277, cited from the *Ta Ming*[7] pharmacopoeia.
[b] *TNYC*, p. 64.   [c] *PTKM*, ch. 13, p. 799.
[d] Porter Smith & Stuart (1), p. 415.   [e] *TNYC*, p. 65.
[f] Porter Smith & Stuart (1), p. 178; R (1), item 177.
[g] *PTKM*, ch. 35, pp. 2011, 2012.
[h] Porter Smith & Stuart (1), p. 178.
[i] *TNYC*, p. 92. Khu shu is one of the more effective plant pesticides known in recent years in China; cf. Chiu Shin-Foon (1).

[1] 葛根	[2] 葛	[3] 粉葛藤	[4] 苦參	[5] 治疥殺蟲
[6] 殺疳蟲	[7] 大明	[8] 治風殺蟲	[9] 土農葯誌	[10] 秦皮
[11] 秦	[12] 蘇恭	[13] 苦樹	[14] 殺蟲	[15] 枲耳
[16] 蒼耳				

Fig. 91. *Khu shên, Sophora flavescens* Ait; *TNYC* no. 76. 1. Stem with flowers, 2. Fruit.

pesticide. The *Kang Mu* reports that the herb will drive out various poisons and venoms, presumably produced by invertebrate pests (*chhu chu tu shih*[1]), kill pests (*sha chhung*[2]) and alleviate the effect of poisonous spiders.[a] In the *Nung Sang I Shih Tsho Yao*[3] (Selected Essentials of Agriculture, Sericulture, Clothing and Food), +1314, we note the following advice on the storage of wheat: 'Sun the wheat for three days. Collect when it is thoroughly dry, and mix in with it *tsang erh*[4] and water pepper, *la liao*[5].'[b] We see a similar passage in the *Chung Shu Shu*[6]: 'When drying wheat grain in the sun, harvest at mid-day, when the grain is hot. Mix in small pieces of *tshang erh*[4] or hemp leaves, then no moths will arise.'[c] Li Shih-Chen also says: 'Before autumn break up some *tshang erh*[4] and sun it together with wheat. The wheat will no longer have insect pests.'[d] The *Thu Nung Yao Chih* states that the stem and leaves, which contain tannins and other bitter substances, are useful for controlling invertebrate pests on agricultural plants.

[a] *PTKM*, ch. 15, pp. 989–94.
[b] *Nung Sang I Shih Tsho Yao*[11], 6th month, p. 96. *La liao*[12] is another name for *ma liao*, the next item to be considered.
[c] *Chung Shu Shu*[6], p. 11.   [d] *PTKM*, ch. 22, p. 1451.

[1] 除諸毒螫   [2] 殺蟲   [3] 農桑衣食撮要   [4] 蒼耳

[5] 辣蓼   [6] 種樹書

The last item in this series is *liao shih*[1] (no. 15), the fruit of the *liao*[2] plant. It is a reed that grows widely in the marshy areas of China. Han Pao-Shêng[3] recorded seven types of *liao*[2]. The two recommended by the *Thu Nung Yao Chih* are the *shui liao*[4], water *liao*[2], (*Persicaria hydropiper* (L.) *Spach.*), and the *ma liao*[5], horse *liao*[2], (*Persicaria nodosa* (Pers.) Opiz.). The whole plant is chopped up and soaked in water. The aqueous infusion is highly effective in controlling agricultural pests, flies and mosquitoes. The *Pên Ching* states that *ma liao*[5] eliminates intestinal worms (*chhu chhang chung chi chhung*[6]). Porter Smith reiterated the observation that the stalk and leaves of *ma liao* were used as a vermicide.[b] The plant contains anthraquinones, flavonols, persicarin, polygonic acid, and other organic acids.[c]

So much then for the pesticidal activity of the upper- and middle-class herbs from the *Pên Ching*[7]. We will now turn our attention to the twenty-one lower-class (*hsia phin*[8]) plants with pesticidal activity. These are listed in Table 14, also in the order that they occur in the *Pên Ching*. We shall comment on each one of them briefly.

Ta huang[9] (no. 16) is the root of the Chinese rhubarb, *Rheum officinale* Baill. It is a well-known purgative. The *Wu Lei Hsiang Kan Chih* (+980) states that the leaves when placed on mats will repel fleas.[d] The *Thu Nung Yao Chih* indicates that the root and stem can be ground and used as an insectide on agricultural crops. The plant contains the well-known laxative emodin and related anthraquinones.[e]

Mang tshao[10] (no. 17) is *Illicium lanceolatum* L. or toxic star anise which we have already discussed earlier. The seed contains shikimitoxin, shikimin, shikimic acid and essential oils. The bark contains quercetin.[f]

Pa tou[11] (no. 18) is the fruit (bean) of *Croton Tiglium* L., a tree of the Euphorbiaceae. The *Pên Ching* says that it kills invertebrates and fish (*sha chhung yü*[12]). The *Kang Mu* affirms that it kills internal parasites (*sha fu tsang chhung*[13]).[g] It is, of course, an anthelmintic and is well known as a drastic purgative. The seed contains crotin and crotonoside, a purine riboside.[h] It is also used externally to control rashes caused by mites and lice. According to the *Thu Nung Yao Chih*, the leaves as well as powdered seed are effective insecticides to control plant pests (Fig. 92).

Yüan hua[14] (no. 19) is the flower of the shrub *Daphne genkwa* Sieb. et Zucc. The *Pên Ching* states that it kills invertebrates and fish (*sha chhung yü*[15]). The *Pieh Lu* calls it 'fish poison' (*tu yü*[16]).[i] The stem as well as flowers are used for the control of agricultural pests. According to the *Nung Sang Chi Yao*[17] (Fundamentals of Agriculture and Sericulture), +1273, if trees are infested with boring insects, these can be controlled by stuffing *yüan hua*[2] into the holes made by them.[j] A similar statement

---

[a] *PTKM*, ch. 16, pp. 1091 ff.   [b] Porter Smith & Stuart (1), p. 344.
[c] *TNYC*, pp. 21–3.   [d] *PTKM*, ch. 17, p. 1122.   [e] *TNYC*, p. 25.
[f] *Anon.* (166), p. 384.   [g] *PTKM*, ch. 35, p. 2053.   [h] *Anon.* (166), p. 378.
[i] *PTKM*, ch. 17, p. 1213.   [j] *Nung Sang Chi Yao*, ch. 5, p. 27.

[1] 蓼實   [2] 蓼   [3] 韓保昇   [4] 水蓼   [5] 馬蓼
[6] 去腸中蛭蟲   [7] 本經   [8] 下品   [9] 大黃   [10] 莽草
[11] 巴豆   [12] 殺蟲魚   [13] 殺腹臟蟲   [14] 芫花   [15] 殺蟲魚
[16] 毒魚   [17] 農桑輯要

Table 14. *Lower-class herbs from the* Pen Ching *that show pesticidal activity.*

No.	Chinese name	PTKM ch.	B III item	R (1) item	TNYC item	Botanical name
16	*Ta huang*	17	130	582	26	*Rheum officinale*
17	*Mang tshao*	17	158	506	50	*Illicium lanceolatum*
18	*Pa tou*	35	331	322	91	*Croton tiglium*
19	*Yüan hua*	17	156	253	119	*Daphne genkwa*
20	*Kou wên*	17	162	174	138	*Gelsemium elegans*
21	*Lang tu*	17	132	526	120	*Stellera chamaejasme*
22	*Pien hsü*	16	127	566	25	*Polygonum aviculare*
23	*Thien hsiung*	17	144	524	—	*Aconitum fischeri*
24	*Wu thou*	17	146	523	36	*Aconitum kusnezoffi*
25	*Fu tzu*	17	143	523a	—	*Aconitum autumnale*
26	*Tsao chia*	35	325	387	68	*Gleditsia sinensis*
27	*Lien shih*	35	321	335	89	*Melia azedarach*
28	*Thung yeh*	35	320	321	90	*Aleurites fordii*
29	*Tzu pai phi*	35	319	98	165	*Catalpa ovata*
30	*Shih nan*	36	347	202	135	*Rhododendron metternichi*
31	*Lang ya*	17	134	440	—	*Potentilla cryptotaenia*
32	*Li lu*	17	142	693	212	*Veratrum nigrum*
33	*Niu pien*	17	161	525	—	*Aconitum lycoctonum*
34	*Pai lien*	18	180	287	112	*Ampelopsis japonica*
35	*Tshao hao*	15	74	4	176	*Artemsia annua*
36	*Chin tshao*	16	128	732	—	*Arthraxon ciliaris*

Abbreviations: *PTKM*, *Pên Tshao Kang Mu*
B III, *Botanicum Sinicum* Vol. 3
R (1), Bernard Read, *Chinese Medicinal Plants from the* PTKM
TNYC, *Chung Kuo Thu Nung Yao Chih*.

16 大黃	17 莽草	18 巴豆	19 芫花	20 鈎吻
21 狼毒	22 萹蓄	23 天雄	24 烏頭	25 附子
26 皂莢	27 棟實	28 桐葉	29 梓白皮	30 石南
31 狼牙	32 藜蘆	33 牛扁	34 白歛	35 草蒿
36 藎草				

also occurs in the *Hua Ching*[1] (+1688).[a] The plant contains genkwanin (5,4′-dihydro-7-methoxy-flavone), genkwanin glycosides and apigenin (5,7,4′-trihydroxyflavone).[b]

Kou wên[2] (no. 20), also known as *hu man thêng*[3] (foreign creeping vine) or *tuan chhang tshao*[4] (break intestine plant), is undoubtedly *Gelsemium elegans* Benth. The root is officinal. The *Pên Ching* states that it is effective in controlling internal parasites and toxins. The *Pieh Lu* recognises its use against skin parasites (lice and mites) and its toxicity towards birds and mammals. The juice should be formulated as an ointment and not taken internally.[c] According to the *Thu Nung Yao Chih* both

[a] *Hua Ching*, ch. 2, p. 23.   [b] *Anon.* (166), p. 426.   [c] *PTKM*, ch. 17, p. 1228.

[1] 花鏡   [2] 鈎吻   [3] 胡蔓藤   [4] 斷腸草

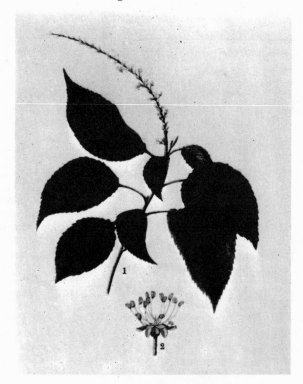

Fig. 92. *Pa tou, Croton Tiglium* L.; *TNYC* no. 91. 1. Stem with flowers, 2. Flower.

the stem and leaves may be soaked in water to give an insecticidal infusion. The active principles are gelsemine, gelsemicine, koumine and related alkaloids.[a] The name *kou wên*[1] has been applied to another vine which may be called the Chinese poision ivy. It is *Rhus toxidodendron* L. In Japan *kou wên*[1] is definitely the name for *R. toxidodendron*, and this may be one cause for confusion in the early identification of the Chinese plant. But there is no doubt that the plant described in the *Kang Mu* is a *Gelsemium*.[b]

Lang tu[2] (no. 21), another poisonous plant, also with the popular name of *tuan chhang tshao*[3], has been identified as *Stellera chamaejasme* L. Again, the root is used in medicine. The *Pên Ching* says that it controls evil spirits and poisonous parasites and kills flying birds and running beasts (*sha fei niao chou shou*[4]). The *Kang Mu* recommends its use in oil-based ointments to cure skin afflictions caused by external parasites and to be taken internally to expel worms. It is effective for persistent

[a] *Anon.* (166), p. 381; *TNYC*, p. 119.
[b] *Gelsemium elegans* is mentioned in the *Nan Fang Tshao Mu Chuang*[5] under the name of *yeh ko*[6], which has often been confused with another *yeh ko*[1], *Rhus toxicodendron* L. For a discussion of the botanical identity of these plants, see Li Hui-Lin (1), p. 72.

[1] 鈎吻   [2] 狼毒   [3] 斷腸草   [4] 殺飛鳥走獸   [5] 南方草木狀
[6] 冶葛   [7] 野葛

rashes caused by fungal infections. The *Thu Nung Yao Chih*[1] has verified the efficacy of aqueous extracts of the root and other parts of the plant to control a variety of insect pests on crop plants. The root contains long-chain alcohols, phenolic substances, amino-acids and organic acids.[a]

Pien hsü[2] (no. 22) is also of the Polygonaceae, *Polygonum aviculare* L. This is the ordinary knot-grass which grows wild over large parts of North and Central China. The whole plant is used in medicine. The *Pên Ching* states that it kills three types of invertebrate pests and cures skin rashes. It is effective as a purgative to expel worms from children.[b] The *Thu Nung Yao Chih* affirms its use as an insecticide. The stem and leaves can be chopped and cooked in water and the decoction sprayed against agricultural pests. It contains tannins, essential oil, wax, anthraquinones and avicularin (quercetin 3-α-arabinoside).[c]

Thien hsiung[3], *wu thou*[4] and *fu tzu*[5] (nos. 23, 24, 25) are three *Aconitum* (monkshood, wolfsbane) plants well known to the ancient Chinese. In each case the root is officinal. All are poisonous. They are constituents of the famous *San-chien-thang*[6] (three-cycle decoction), a useful tonic and depurative.[d] Although different botanical names have been assigned to the three plants, the medieval Chinese herbalists actually considered them to be separate root stages of the same plant. Thao Hung-Ching says that '*fu tzu*[5] and *wu thou*[4] are names applied to roots of the same plant. That taken up in the eighth moon is called *fu tzu*[5], while that dug up in the spring, when the plant begins to sprout, and resembling a crow's head in shape, is called *wu thou*[4]. *Thien hsiung*[3] resembles the *fu tzu*[5] but is more slender.'[e]

The *Kang Mu* states that *thien hsiung*[3] kills birds and poisonous invertebrates (*sha chhin chhung tu*[7]). The *Pên Ching* says that *wu thou*[4] kills birds and mammals.[f] The *Chhi Min Yao Shu*[8] (Important Arts for the People's Welfare), +540, recommends the use of *wu thou*[4] decoction to cure mite sores on cattle.[g] But the most interesting application in agriculture of this group of plants involves *fu tzu*[5] as described in *Fan Shêng-Chih Shu*[9] (The Book of Fan Shêng-Chih), -1st century, and preserved in the *Chhi Min Yao Shu*:

For fields with poor soil which cannot be manured, mix the seeds with silkworm droppings and then plant. The seedlings will be free from insect pests. Or take one picul of chopped horse bones, and heat them with three piculs of water. Boil three times, decant the decoction

---

[a] *PTKM*, ch. 17, pp. 1125, 1126; *Anon.* (*166*), p. 427.
[b] *PTKM*, ch. 16, p. 1101.   [c] *Anon.* (*166*), p. 429.
[d] *PTKM*, ch. 17, p. 1175. See also Porter Smith & Stuart (1), p. 477.
[e] *PTKM*, ch. 17, p. 1158. Even more explicit is the statement from *Kuei-Hsin Tsa Shih*[10] (*c.* +1298), p. 31: 'The three products (i.e. *thien hsiung*, *wu thou* and *fu tzu*) are really different forms of the same thing. When it is one year old, it is called *tse tzu*[11], two years old, *wu hui*[12], three years old, *fu tzu*, four years old, *wu thou*, and five years old, *thien hsiung*.'
[f] *PTKM*, ch. 17, p. 1178, *sha chhin shou*[13].
[g] *Chhi Min Yao Shu*[8], vol. 6, ch. 56.

---

[1] 土農藥誌	[2] 萹蓄	[3] 天雄	[4] 烏頭	[5] 附子
[6] 三建湯	[7] 殺禽蟲毒	[8] 齊民要術	[9] 氾勝之書	[10] 癸辛雜識
[11] 荝子	[12] 烏喙	[13] 殺禽獸		

and steep in it five pieces of *fu tzu*[1]. Three to four days later, remove the *fu tzu*[1] and stir the broth with silkworm excreta or sheep dung until it reaches the consistency of thick congee. Twenty days before sowing, mix the seeds in to give the appearance of cooked wheat meal. Do this on a clear day and dry the mix immediately; or squeeze the material through a piece of thin cloth and then dry. Soak again on the next day but not if the weather is wet. Be sure to dry and store with care. Do not allow the seeds to stay wet. After six to seven soakings sow the seeds together with the decoction that is left. Then the plants will be free from locusts (and other pests).[a]

Of the three plants only *wu thou*[2] is listed in *Thu Nung Yao Chih* as an insecticide for agricultural crops. In theory, the three plants should be equivalent in their action. It could be that only *wu thou*[2] is available on a scale that would permit its use in agriculture. The active ingredients in these plants are aconitine and a host of related alkaloids.[b]

Tsao chia[3] (no. 26) is the fruit of the tree legume *Gleditsia sinensis* Lam. The *Pên Ching* says that it kills noxious pests (*sha kuei ching wu*[4]). According to the *Wu Lei Hsiang Kan Chih*[5], it can be used to ward off ants.[c] The Pharmacopoeia by Ta Ming[6] states that 'it dissolves phlegm and kills invertebrate pests' (*hsiao than sha chhung*[7]).[d] The *Thu Nung Yao Chih* recommends its use to control agricultural pests. The fruit, leaves and bark contain the saponins, gledimin and gledegenin. *Tsao chia*[3] is probably one of the earliest plants recognised for its detergent properties. Its use for washing clothes is recorded in the *Chhi Min Yao Shu*.[e]

Lien shih[8] (no. 27) is the fruit of the *khu lien*[9] (bitter *lien* tree), *Melia azedarach* L. The *Pên Ching* states that it kills three types of pests (*sha san chhung*[10]).[f] According to the *Kang Mu* the bark, leaf and fruit may all be used to expel intestinal worms, and the flowers when placed under a mat will drive off fleas and lice.[g] The *Thu Nung Yao Chih* indicates that the aqueous infusion is highly active against spiders, hoppers, aphids and lepidopterous larvae. They contain azaridine and other alkaloids, and kaempferol and its glycosides. They also contain azadiractin, a limonoid with a powerful feeding-deterrent activity towards insects, which has attracted considerable attention among natural product chemists and entomologists in recent years (Fig. 93).[h]

Thung yeh[11] (no. 28) is the leaf of the well-known *thung*[12] oil tree, *Aleurites fordii* Hemsl. It is listed in the *Kang Mu* as *ying tzu thung*[13]. Although the *Pên Ching* lists only the leaf as a herb, the fruit and oil are also useful as pesticides. The *Pên Ching* says the

---

[a]   Tr. auct. Cf. Shih Shêng-Han (2), p. 11 and *Chhi Min Yao Shu*, vol. 1, ch. 3.   [b]   *Anon.* (166), p. 386.
[c]   *Wu Lei Hsiang Kan Chih*[5], p. 26.   [d]   *PTKM*, ch. 35, p. 2015.
[e]   Shih Shêng-Han (1), p. 89. Cf. *Chhi Min Yao Shu*, vol. 3, ch. 30.
[f]   So we translate, but one must always bear in mind that the expressions "three pests" or "three corpses" mean in Taoist physiology the factors in the human body leading to mortality and ageing (cf. Morohashi dict., i, 136, no. 630; i, 167, no. 1290).
[g]   *PTKM*, ch. 35, p. 2004.
[h]   Cf. Kraus & Bokel (1981); Zhao Shanhuan, Xu Muchheng, Zhang Xing, Chen Sunyuan & Liao Changqing (1982).

[1] 附子	[2] 烏頭	[3] 皂莢	[4] 殺鬼精物	[5] 物類相感志
[6] 大明	[7] 消痰殺蟲	[8] 楝實	[9] 苦楝	[10] 殺三蟲
[11] 桐葉	[12] 桐	[13] 罌子桐		

Fig. 93. *Khu lien, Melia azedarach* L.; *TNYC* no. 89. 1. Stem with flowers, 2. Flower, 3. Stem with fruits.

leaf kills three types of pests (*sha san chhung*[1]). When a decoction is applied externally on skin sores of swine, they (the swine) grow three times as fast.[a] The *Ko Wu Tshu Than*[2] (Simple Discourses on the Investigation of Things), +1080, states that when branches are infested with boring insects, they can be killed by painting *thung*[2] oil around the trunk of the tree.[b] In the same chapter there also occurs what is probably the earliest account of the use of light to trap and destroy insect pests: 'When peach trees are infested by boring insects, hang an old much-used bamboo lamp on the tree and the insects will drop into it.'[b] There is an interesting passage in the *Kuei-Hsin Tsa Shih, Pieh Chi*[3] (Second Addendum to 'Miscellaneous Information from Kuei-Hsin[4] Street'), +13th century, which combines both the use of *thung* oil and the use of the bamboo lamp in one story. We quote: 'Many small insects attack peach trees, spreading all over the branches like black ants. This pest is commonly called *ya chhung*[5] (aphids). Though spraying *thung* oil is effective, not all the pests are eliminated. A special method is to use a bamboo lamp that has been

---

[a] *SNPTC*, Mori ed., ch. 3, p. 6. This is an interesting observation and is reminiscent of the modern practice of adding drugs to the feed of poultry and swine to increase the rate of weight gain.

[b] *Ko Wu Tshu Than*[2], upper ch., pp. 10, 6.

[1] 殺三蟲   [2] 格物麤談   [3] 癸辛雜識別集   [4] 癸辛

[5] 蚜蟲

hung on the wall for many years, and place it on the tree. The insects will fall into the lamp. The principle of this method is not known. Tai Chu-Yü[1] learnt of it from an old gardener.'[a]

According to the *Nung Phu Ssu Shu*[2] (Four Books on Agriculture), +16th century, *thung* oil may be injected into infested holes to control insect pests.[b] The *Hua Ching*[3] states that the oil may be added to manure so as to kill the insect pests before the material is applied on vegetables. Furthermore, *thung* oil-soaked paper may be twisted and used to plug insect-infested holes.[c]

Tzu pai phi[4] (no. 29) is the bark of the *tzu mu*[5] tree, *Catalpa ovata* Don. (also assigned the name *Lindera tzumu*). The *Pên Ching* says it kills three types of pests (*sha san chhung*[6]). When applied on skin sores, it will also triple the rate of growth of swine.[d] According to the *Kang Mu*, it is an anthelmintic and parasiticide.[d] The *Thu Nung Yao Chih* states that the leaves and bark can be extracted with water to give an insectidal spray.

Shih nan[7] (no. 30) is *Rhododendron metternichi* S. & Z. The *Pên Ching* says the fruit is toxic to internal parasites (*sha ku tu*[8]) and the *Kang Mu* notes that the leaves kill invertebrate pests (*sha chhung*[9]).[e] Although *shih nan*[7] itself is not included in the *Thu Nung Yao Chih*, a closely related species, *Rhododendron molle* L., *yang chih-chu*[10] (sheep paralyser) is listed and regarded as one of the more effective plant pesticides known in China today for use against agricultural pests.[f]

Lang ya[11] (no. 31) is the root of the cinquefoil, *Potentilla cryptotaenia* Maxim. It is so named because the root is shaped like the tooth of a wolf. It is a deadly poison. The *Pên Ching* notes that it kills white pests (*chhü pai chhung*[12]), and recommends its use to expel skin parasites and alleviate the sores they cause.[d] It is not included in the *Thu Nung Yao Chih*.

Li lu[13] (no. 32) is identified as *Veratrum nigrum* L. (black hellebore). It was a well-known insecticidal plant in ancient China. The root is used in medicine. The *Pên Ching* states that it kills poisonous invertebrates (*sha chu chhung tu*[14]), and recommends it as a decoction for treating various skin afflictions.[g] The *Chou Hou Pei Chi Fang*[15] (Handbook of Medicines for Emergencies), +340, includes it in a remedy for boils and itches on sheep: 'Take two parts of *li lu*[13] and eight parts of *fu tzu*[16], and use as a salve. The insect pests will move out (of the wound).'[h] In the *Chhi Min Yao Shu* we find this prescription for sheep scabies: 'Chop up the *li lu* root[13], and soak it in rice washing in a full closed bottle. Place it in a warm place next to the stove. After a few days when the bottle smells pleasantly of vinegar, it is ready for

[a] *Kuei Hsin Tsa Shih*, Pieh Chi, pp. 4, 5.   [b] *Nung Phu Ssu Shu*[2], ch. 2, p. 6a.
[c] *Hua Ching*[11], ch. 2, p. 23a.   [d] *PTKM*, ch. 35, pp. 1994, 1995.
[e] *PTKM*, ch. 36, p. 2120.   [f] Cf. Chiu Shin-Foon (1).
[g] *SNPTC*, Mori ed., ch. 3, p. 8a.   [h] *Chou Hou Pei Chi Fang*[15], ch. 5, p. 137.

1 戴祖禹	2 農圃四書	3 花鏡	4 梓白皮	5 梓木
6 殺三蟲	7 石南	8 殺蟲毒	9 殺蟲	10 羊躑躅
11 狼牙	12 去白蟲	13 藜蘆	14 殺諸蟲毒	15 肘後備急方
16 附子				

use. Scrape the scab with brick or tile until it is red. Scabs too hard or thick may be washed with hot water first. When the scabs are cleared off and wiped dry apply the medication twice and the sores will be healed.'[a] The *Pen Tshao Yen I*[1] (The General Ideas of the Pharmacopoeia), +1116, describes ground *li lu*[2] as a remedy for parasitic skin afflictions of horses.[b] According to the *Ko Wu Tshu Than*[3], +1080, *Li lu*[2] can also be used to control flies: 'Take finely ground *li lu*[2] and mix it with wine. Place the paste in a bowl or spread it on willow stems. When flies feed on it, they will become dizzy, pass out, and drop to the ground.'[c] The *Thu Nung Yao Chih* recommends the use of the whole plants, leaves, stems, roots to kill household and agricultural pests. The active principles are jervine, pseudo-jervine and related alkaloids.[d]

Niu pien[4] (no. 33) is the root of another aconite plant, *Aconitum lycoctonum*. It was apparently regarded as sufficiently dissimilar from *thien hsiung*, *wu thou* and *fu tzu* (nos. 23, 24, 25) so as to be placed away from the trio in the *Pên Ching*. The drug was noted for its ability to kill lice and other pests on cattle (*sha niu sê hsiao chhung*[5]).[e] This was presumably its most popular usage. The active ingredients are lycaconitine and other alkaloids.[f] Porter Smith said it was used as a lotion for ulcers or as an insecticide for cattle.[g] It is not included in the *Thu Nung Yao Chih*.

Pai lien[6] (no. 34) is the vine *Ampelopsis japonica* (Thunb.) Mak. The root is officinal. The *Pên Ching* states that it cures rashes and sores, presumably caused by insect parasites.[h] According to the *Ko Wu Tshu Than*[3] (+1080) it is advisable 'when planting peonies, to sprinkle ground *pai lien*[6] near the roots, so that insect infestation will be avoided'.[i] Similar advice is also offered in the *Hua Ching*[7] (+1688).[j] The *Thu Nung Yao Chih* recommends that the entire plant be dried and ground to a powder, to be used as a dust for agricultural crops. Or the chopped plant is soaked in water, and the aqueous infusion used as a spray against insect pests.

Tshao hao[8] (no. 35) as listed in *Pên Ching* is also known as *chhing hao*[9] or *fang kuei*[10]. According to which name is preferred, it could be *Artemisia annua* L. or *Artemisia apiacea* Hance. The whole plant can be used as medicine or as an insecticide. The *Pên Ching* says that *tshao hao*[8] cures itches and sores and kills pests (*sha chhung*[11]).[k] According to the *Thu Nung Yao Chih*, either *Artemisia* is useful as an insecticide for mosquitoes, flies and agricultural pests. *Artemisia annua* contains a host of terpenes, including the recently identified *qinghaosu*[12] (artemisinine), a sesquiterpene lactone with pronounced activity against malarial parasites.[l]

Chin tshao[13] (no. 36) the last entry in Table 14, is *Arthraxon ciliaris*, Beauv. The *Pên*

---

[a] Tr. Shi Shêng-Han (1), mod. *Chhi Min Yao Shu*, vol. 6, ch. 57.
[b] *Pên Tshao Yen I*[3], ch. 11, p. 2. *Ko Wu Tshu Than*[3], upper ch., p. 15.
[d] *TNYC*, p. 180. [e] *SNPTC*, Mori ed., ch. 3, p. 9. [f] R (1), item 525.
[g] Porter Smith & Stuart (1), p. 11. [h] *SNPTC*, Mori ed., ch. 3, p. 9a.
[i] *Ko Wu Tshu Than*[3], upper vol., p. 6. [j] *Hua Ching*[7], ch. 2, p. 24.
[k] *SNPTC*, Mori ed., ch. 3, p. 9a. [l] Hu Shih-Lin *et al.* (1981).

[1] 本草衍義 [2] 藜蘆 [3] 格物麤談 [4] 牛扁 [5] 殺牛虱小蟲
[6] 白蘞 [7] 花鏡 [8] 草蒿 [9] 青蒿 [10] 方潰
[11] 殺蟲 [12] 青蒿素 [13] 藎草

Ching says that it kills small skin pests (*sha phi fu hsiao chhung*[1]), presumably mites, lice and fleas. The whole plant is officinal. It is also used as a yellow dye. The *Thu Nung Yao Chih* does not include it as an insecticidal plant.[a]

We have now examined all thirty-six pesticidal plants from the *Pên Ching* listed in Tables 13 and 14. If we accept the notion that the four *Aconitum* species, i.e. *thien hsiung* (no. 23), *wu thou* (no. 24), *fu tzu* (no. 25) and *niu pien* (no. 33) are virtually equivalent in their action, then all the plants in the lower class (i.e. Table 14), except *ta huang* (no. 16) and *chin tshao* (no. 36), were recognised as pesticidal by the *Pên Ching*. For the 15 upper and middle class herbs (cf. Table 13) only four, *sung chi* (no. 1), *yün shih* (no. 7), *mi wu* (no. 8) and *liao shih* (no. 15) were recognised as pesticidal. This is not surprising since the lower-class herbs are supposed to be more toxic than those in the upper and middle classes. We may also note that all the plants except seven, i.e. *kan tshao* (no. 5), *yün shih* (no. 7), *mi wu* (no. 8), *kao pên* (no. 9), *shui phing* (no. 14), *lang ya* (no. 31) and *chin tshao* (no. 36), have been validated as useful control agents for the protection of crop plants against insects by the *Thu Nung Yao Chih*. Yet references to their use for plant protection in the literature before + 1600 are relatively few. But before we deal with this issue we should move on to the second category of plants, that is, those that were discovered during the medieval period, from about + 100 to + 1600.

(iii) *Gifts of the medieval pharmacists*

In addition to providing much valuable information on the pesticidal activity of plants listed in the *Pên Ching* (cf. Tables 13 and 14), the pharmaceutical natural histories also contributed new entries to our catalogue of pesticidal plants known in China. The names of these plants, together with appropriate references, are presented in Table 15. Although only six of the eleven plants listed (nos. 37, 39, 40, 43, 44 and 47) are included in the *Thu Nung Yao Chih*, they are all of sufficient interest to deserve more than a passing mention in this survey.

The first item, *ai hao*[2] (no. 37) is one of the best known herbs in the Chinese pharmacopeia, since it provides the *moxa* used in the celebrated therapeutic technique of moxibustion.[b] Both Bretschneider and Read identified it as *Artemesia vulgaris* L., but the recent literature from China, such as the *Chung Kuo Thu Nung Yao Chih*[3], 1959, and the *Chung Yao Ta Tzhu Tien*[4] (Encyclopaedia of Chinese Materia Medica), 1978, prefer the name *Artemisia argyi* Levl. et Vant. It is first mentioned in the *Pieh Lu*[5], and Thao Hung-Ching[6] says that it kills intestinal worms (*sha hui chhung*[7]).[c] But for our purpose the most interesting application of *ai hao*[1] as a pesticidal agent is for the preservation of stored grain.[d] In the *Chhi Min Yao Shu*[8]

---

[a] *SNPTC*, Mori ed., ch. 3, p. 10a.   [b] Cf. Lu Gwei-Djen & Needham, J. (5), pp. 170–2.
[c] *PTKM*, ch. 15, p. 936.   [d] Cf. Vol. 6, pt. 2, pp. 378, 475.

[1] 殺皮膚小蟲   [2] 艾蒿   [3] 中國土農藥誌   [4] 中藥大辭典
[5] 別錄   [6] 陶弘景   [7] 殺蚘蟲   [8] 齊民要術

Table 15. *Pesticidal plants listed in the pharmacopoeias between +100 and +1600*

No.	Chinese name	1st ref.	PTKM ch.	B III item	R (1) item	TNYC item	Botanical name
37	Ai hao	PL	15	72	9	178	Artemisia argyi
38	Jang ho	PL	15	96	649	—	Zingiber mioga
39	Pai pu	PL	18	177	694	300	Stemona japonica
40	Ta suan	PL	26	243	672	206	Allium scorodoprasum
41	Ming cha	PL	30	277	425	—	Cydonia vulgaris
42	Chüeh hao	TPT	15	—	100	—	Incarvillea sinensis
43	Wu chiu	TPT	35	—	332	99	Sapium sebiferum
44	Yang mei	KP	30	—	621	12	Myrica rubra
45	Yün hsiang	KM	36	—	352	—	Ruta graveolens
							Murraya exotica
46	Huan hua	KM	19	782a	—	—	Typha latifolia
47	Tsui yü tshao	KM	17	—	172	137	Buddleia lindleyana

Abbreviations: PL, *Ming I Pieh Lu*
TPT, *Thang Pên Tshao*
KP, *Khai-Pao Pên Tshao*
KM, *Kang Mu*
B III, *Botanicum Sinicum* Vol. 3
R (1), Bernard Read, *Chinese Medicinal Plants from the* PTKM
TNYC, *Chung-Kuo Thu Nung Yao Chih*.

37 艾蒿	38 蘘荷	39 百部	40 大蒜	41 檳榔
42 角蒿	43 烏桕	44 楊梅	45 芸香	46 莞花
47 醉魚草				

(Important Arts for the People's Welfare), +540, Chia Ssu-Hsieh[1] says: 'Store them in containers of plaited *moxa*; or keep them in pits or burrows covered with *moxa*. In pitting wheat, sun to dryness and bury while it is still quite warm.'[a] This passage is of particular interest in that it describes the storing of dried grain while it is still hot. The method would be effective in destroying most insect ova and pupae. Shih Shêng-Han[2] (1958) states that the same practice is still popular among wheat farmers in China.[b] The procedure was actually first recommended for the storage of seed corn by Fan Shêng-Chih[3] (−1st century):

How to obtain seed corn for wheat: Watch when wheat ripens. Pick out large and healthy ears, cut them off, tie them up into bundles and stand them on a high and dry place in the (threshing) field. Allow the sun to dry them out to the utmost. Never allow *Lepisma* to remain. Always shake it off when found. Store the dried grains with dry *moxa*. Use one handful of *moxa* (*Artemisia* leaves) for each *shih*[1] (picul) of wheat. Store in earthen or bamboo containers. Later on, sow them in good time. The yield will at least be doubled.[c]

[a] Tr. Shih Shêng-Han, mod. auct. Cf. Shih Shêng-Han (1), pp. 48–9, and *Chhi Min Yao Shu*, vol. 3, ch. 10.
[b] Shih Shêng-Han (1), p. 49.
[c] Tr. Shih Shêng-Han, mod. auct. Cf. Shih Shêng-Han (1), pp. 20–1 and *Chhi Min Yao Shu*, vol. 1, ch. 2.

[1] 賈思勰   [2] 石聲漢   [3] 氾勝之   [4] 石

Fig. 94. *Ai hao, Artemisia argyi* Levl. et Vant.; *TNYC* no. 178. 1. Leaf, 2. Stem with flowers.

According to the *Thu Nung Yao Chih*, aqueous infusions of the leaves are effective in combating aphids, spiders and lepidopterous larvae. The plant contains di and tri-terpenes such as absinthol, thujone, thujyl alcohol, phellendrene ester, cardinene, etc. (Fig. 94).[a]

The next plant (no. 38) is *jang ho*[1], *Zingiber mioga* Rosc., which we had earlier encountered as the *chia tshao*[2] of the *Chou Li*[3]. In spite of its anthelmintic activity, it was described just as a food crop in the *Chhi Min Yao Shu*. However, the *Shih Ching*[4] says that *jang ho*[1] should be pickled in brine-vinegar, a process which may well destroy the anthelmintic agent.[b] *Jang ho*[1] is not included in the *Thu Nung Yao Chih*, but a close relative, ginger i.e. *shêng chiang*[5], *Zingiber officinale* Rosc., is listed as an effective pesticidal plant.

[a] *TNYC*, p. 153.
[b] The *Shih Ching*[4] (Nutrition Manual), *c.* +450, is quoted in the *Chhi Min Yao Shu*, vol. 3, ch. 28. Shih Shêng-Han (1), p. 28, suggests that the author of the *Shih Ching* was the mother of Tshui Hao[6], the Prime Minister to the first emperor of the Later Wei dynasty. It should be noted that according to Thao Hung-Ching, there are two types of *jang ho*, a red type and a white type. The red type is suitable for food, while the white type is superior as medicine. Perhaps only the white type was used as an anthelmintic.

[1] 蘘荷   [2] 嘉草   [3] 周禮   [4] 食經   [5] 生薑
[6] 崔浩

The third item, *pai pu*[1] (no. 39), is one of the best known insecticidal plants in the pharmacopoeia. It has been identified as *Stemona tuberosa* Lour. by Porter Smith and Read, because the root, which is the part of the plant used in medicine, consists of a central mass with ten or more tubers attached.[a] The *Thu Nung Yao Chih*, however, calls it *Stemona japonica* Miq. but lists also two other related species, *Stemona sessifolia* (*chih li pai pu*[2]) and *Stemona tuberosa* (*tui yeh pai pu*[3]), as useful insecticidal plants. Ta Ming's[4] pharmacopoeia (+970) says that it cures intestinal afflictions, and kills roundworms, tapeworms and pinworms.[b] Its ashes quickly destroy wood-boring insects on trees. It also eliminates lice, fleas and midges. Thao Hung-Ching[5] states that *pai pu*[1] decoction is used to wash cattle and dogs to keep them free from lice. For ear infestations one remedy is to mix heat-dried ground *pai pu*[1] with raw oil and stuff the paste into the opening of the ear.[c]

The *Nung Sang Chi Yao*[6] (+1273) and *Chung Shu Shu*[7] (+1401) both advise the use of *pai pu*[1] leaves to plug the opening of tunnels bored by insects on plants as a method of eliminating them.[d] The *Thu Nung Yao Chih* recommends *pai pu*[1] powder for the control of fleas and lice on domestic animals, and the aqueous decoction to kill fly and mosquito larvae as well as insect pests on crops. The active principle has been shown to be a series of alkaloids, stemonine, stemonidine, iso-stemonidine, hodorine, etc.[e]

We now come to *ta suan*[10] (no. 40), *Allium scorodoprasum*, L., rocambole or chive. Also the known as *hu*[11], it is said to have been introduced into China by Chang Chhien[12] in the early Han dynasty. Like its close relative the *hsiao suan*[13], *Allium sativum*, or garlic, it has a pungent taste and a powerful odour. The *Lan Phu Ao Fa*[14] (The Art and Mystery of Orchid Culture), +12th century, says that to get rid of insect larvae 'grind *ta suan*[1] with water; brush the suspension on the leaves until clean and they will be free from insects'.[f] The *Thu Nung Yao Chi* recommends the use of an aqueous extract of ground *ta suan*[10] on agricultural crops to control insect pests. The *Hua Ching*[15] advises placing a piece of *ta suan*[10] together with a piece of *kan tshao*[16] before transplanting a herb, as a measure to reduce insect infestation later on.[g] The plant contains essential oils and allicin, the S-allyl ester of thio-2-propene-1-sulphinic acid, which is responsible for its powerful characteristic odour.[h]

The next two plants of interest are not listed in the *Thu Nung Yao Chih*. No. 41 is *ming cha*[17], the common quince *Cydonia vulgaris* Pers. The fruit has a strong fragrance

---

[a] Porter Smith & Stuart (1), p. 422; Read (1), item 694.
[b] Cf. *PTKM*, ch. 18, p. 1286.
[c] *Shêng Chi Lu*[8], in *PTKM*, ch. 18, p. 1287.
[d] *Nung Sang Chi Yao*[6], ch. 5, p. 27; *Chung Shu Shu*[7], p. 13. See also *Hua Ching*[15], ch. 2, p. 23.
[e] *TNYC*, p. 172.   [f] *Lan Phu Ao Fa*[14], p. 8. See p. 418 above.
[g] *Hua Ching*[15], ch. 2, p. 24.   [h] *TNYC*, p. 176.

[1] 百部   [2] 直立百部   [3] 對葉百部   [4] 大明   [5] 陶弘景
[6] 農桑輯要   [7] 種樹書   [8] 聖濟錄   [9] 花鏡   [10] 大蒜
[11] 葫   [12] 張騫   [13] 小蒜   [14] 蘭譜奧法   [15] 花鏡
[16] 甘草   [17] 榠樝

and may be placed in storage trunks to keep clothes free from insects pests.[a] No. 42 is *chüeh hao*[1], *Incarvillea sinensis*, Lam. It is prescribed for skin ulcers caused by external pests. The *Chhêng Huai Lu*[2] (*c.* +1240) says that when placed among rugs and bedding, or among books, *chüeh hao*[1] will keep the bugs away.[b]

The next item is *wu chiu*[3] (no. 43), *Sapium sebiferum* Roxb., the Chinese tallow tree. Li Shih-Chen says the birds like to eat the fruit, hence the name *wu chiu*[3].[c] The bark of the root is used in medicine. But the unique feature of the plant is the fruit, which consists of a small seed coated with a thick layer of tallow. After removing the tallow, the seed itself can be pressed to give a clear oil, which has been noted to give a bright light when used as fuel for lamps. The oil is also a useful purgative. For children infected with skin parasites, an effective remedy is to paint the oil over old clothing and have the child wear it. After a day all the insects are attracted to the cloth, and can be physically destroyed (Fig. 95).[d]

In the *Nung Chêng Chhüan Shu*[6] (Complete Treatise on Agriculture), +1639, Hsü Kuang-Chhi[7] wrote enthusiastically about the cultivation of *wu chiu*[8] as an economic crop. He noted that the tallow was useful for making candles; and the seed oil was ideal for lighting lamps. Moreover, the oil could also be used to dye hair, dilute lacquer and make oil-paper. The leaves were a source of black dye and the tree was extremely hardy (against insect pests?). He cautioned, however, that the tree should not be placed too close to a fishpond, as the leaves, if they should drop into the water, would turn black and make the fish sick.[e] He made no mention of the effect of the leaves on insects, but one would assume that they must have a large content of tannins and other phenolic substances which, we know, are toxic to insects. Indeed, the *Thu Nung Yao Chih* states that when the ground leaves are soaked in water, the infusion is an effective spray for the control of a variety of agricultural and garden pests.

The eighth plant in Table 15 is *yang mei*[9] (no. 44), which has been identified as *Myrica rubra*, Sieb. & Zucc. It is included in the *Thu Nung Yao Chih*. Its use in medicine was first mentioned in the *Khai-Pao*[10] pharmacopoeia (+976).[f] Ta Ming[11]

---

[a] *PTKM*, ch. 30, *ming cha*, quoted from *Shih Liao Pên Tshao*[4], by Mêng Shên[5] (*c.* +650).
[b] Cf. Chou Yao (*1*), p. 51.
[c] *PTKM*, ch. 35, p. 2050. As Li Shih-Chen said, the birds do love to eat the fruit. In 1772 Benjamin Franklin sent some seeds of the Chinese tallow tree to his friend Noble W. Jones of Wormsloe Plantation near Savannah, Georgia. The tallow trees soon established themselves at the plantation, although the idea of using the tallow from the fruit for making soap did not prove to be commercially feasible. Nevertheless, through the good offices of the birds, the Chinese tallow, *Sapium sebiferum*, spread spontaneously all along the Middle Atlantic coast of the U.S.A. The populations now extend as far north as Richmond County, North Carolina, and as far west as the Texas lowlands along the Gulf of Mexico. Indeed, in the coastal region near Houston, Texas, it has apparently become the dominant tree species. One reason for its successful natural propagation is presumably its hardiness in the face of attacks by insects that normally feed on the native plants of these regions.
[d] *PTKM*, ch. 35, p. 2052.
[e] *Nung Chêng Chhüan Shu*[6], ch. 38, pp. 1065–8.
[f] *PTKM*, ch. 30, p. 1798.

[1] 角蒿   [2] 澄懷錄   [3] 烏桕   [4] 食療本草   [5] 孟詵
[6] 農政全書 [7] 徐光啓   [8] 烏桕   [9] 楊梅   [10] 開寶
[11] 大明

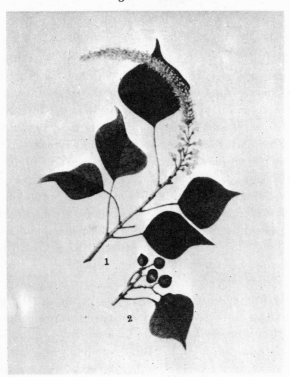

Fig. 95. *Wu chiu*, *Sapium sebiferum* (L.) Roxb.; *TNYC* no. 99. 1. Stem with flowers, 2. Stem with fruits.

(+970) describes a decoction prepared from the bark and root to treat skin rashes caused by exoparasites.[a] Aqueous solutions extracted from the bark or root are recommended by the *Thu Nung Yao Chih* for controlling a variety of agricultural pests.

The next plant is *yün hsiang*[7] (no. 45) or rue. It is not listed in the *Thu Nung Yao Chih*. The earliest record on the insecticidal activity of *yün hsiang*[1] is found in the *Mêng Chhi Pi Than*[2] (Dream Pool Essays), +1086:

The ancients who kept books used *yün*[3] to protect them against boring insects. *Yün*[3] is a fragrant plant which is what we nowadays call *chhi li hsiang*[4] (seven-mile fragrance). Its leaves look like those of *wan tou*[5] (*Pisum sativum*), green pea. It grows in small bunches. The leaves are extremely fragrant. After autumn they turn to a milky white colour. They are most effective in controlling insects that attack books (such as silverfish). People in the South place them under mattresses to get rid of fleas and lice. When I was an official in the Chao Wên Kuan[6], I received several *yün*[9] plants from the house of Lu Kung[7] and planted them at the back of the Pi Ko[8]. But now they are no longer alive. The name *hsiang tshao*[9] (fragrant herb) has been applied to several different kinds of plants. For example, the so-called *lan*

---

[a] *Ibid.* p. 1799.

[1] 芸香   [2] 夢溪筆談   [3] 芸   [4] 七里香   [5] 碗豆
[6] 昭文館   [7] 潞公   [8] 秘閣   [9] 香草

sun[1] or *sun*[2] is today known as *chhang pu*[3] (cf. no. 4); *hui*[4] is present-day *ling ling hsiang*[5] and *chih*[6] is now *pai chih*[7].[a]

According to the *Hsü Po Wu Chih*[8] (Supplement to the Record of the Investigation of Things), mid +12th century, Tshang Chieh[9] in the *Chieh Ku*[10], states that '*yün hao*[11] has the appearance of *hsieh hao*[12] (corrupt artemisia), but it can be eaten'. The *Yü Huan Tien Lüeh*[13] says that '*yün hsiang*[14] protects paper against silverfish and other bookworms. Therefore libraries are now popularly known as *yün thai*[15] (*yün*[16] pavilions).'[b] The *Shuo Wên Chieh Tzu* (Analytical Dictionary of Characters), +121, states that *yün*[16] is similar in appearance to *mu su*[17] (alfalfa).[c] In the *Hsia Hsiao Chêng*[18] (Lesser Annuary of the Hsia Dynasty), −7th to −4th century, we find the statement 'First month, the rue is picked (*yün*[17]). Second month, beautiful is the rue.' The *Li Chi*[19] (Record of Rites), −50, mentions that *yün*[16] begins to grow in the second month of winter: Thus *yün*[16] was evidently well known to the ancient Chinese, and had been used extensively as a herb to ward off bookworms. It appears to be a perennial plant about the size of alfalfa.[c]

If so, we all owe the discoverers of *yün*'s insecticidal activity a debt of gratitude for the preservation of countless volumes of Chinese books through the ages. Yet it is by no means certain that we know today the identity of this important plant. In the *Kang Mu*, Li Shih-Chen lists *yün hsiang*[14] under *shan fan*[20], which Read regarded as the China box, *Murraya exotica*, L. Li Shih-Chen describes *shan fan*[20], which he considers synonymous with *yün hsiang*[14] and *chhi li hsiang*[21], as a tree which may reach a height over ten feet, whereas Shen Kua[22] calls it a '*tshao*[23]' and the older literature considers it a perennial similar in appearance to alfalfa. Wu Chhi-Chün[24] evidently disagrees with Li Shi-Chen's interpretation. In the description of *yün hsiang*[14] in the *Chih Wu Ming Shih Thu Khao*[25], 1848, he completely ignores the reference to *shan fan*[20] in the *Kang Mu*.[d]

Modern botanists in China have assigned the botanical name *Ruta graveolens* L. to *yün hsiang*[14].[e] It is said to have originated in Southern Europe. There is no question that in China today *yün hsiang*[14] is acknowledged to be *Ruta graveolens*. But is it the same plant referred to in the *Li Chi*[19], *Shuo Wên Chieh Tzu*, *Mêng Chhi Pi Than*, etc.? The *Kang Mu* quotes a *Tu Yang Pien*[30] which says that '*Yün hsiang*[14], a *tshao*[23],

[a] *MCPT*, ch. 3, p. 41 (item 53). Lu Kung (+1006–97) was a high official during the Sung dynasty. Pi Ko was the name given to the special library or repository in the palace where rare books were kept. *Hui*[4] or *ling ling hsiang*[5] is *Ocimium basilicum*, R (1), 134a. *Chih*[6] or *pai chih*[7] is *Angelica anomala*, R (1), 207.

[b] *Hsü Po Wu Chih*[8]. ch. 3, p. 7. Cf. also *Chhi Min Yao Shu*, vol. 10, p. 31.

[c] Cf. *PTKM*, ch. 36, pp. 2005, 2006; B II, item 409; R (1), item 352.

[d] *Chih Wu Ming Shih Thu Khao*[25], ch. 25, p. 80.

[e] Phei Chien[26] (1), item 172; Chia Tsu-Shan[27] & Chia Tsu-Chang[28], p. 544. See also Li Chhün[29], p. 8.

[1] 蘭蓀	[2] 蓀	[3] 菖蒲	[4] 蕙	[5] 零陵香
[6] 茝	[7] 白芷	[8] 續博物誌	[9] 倉頡	[10] 解詁
[11] 芸蒿	[12] 邪蒿	[13] 魚豢典略	[14] 芸香	[15] 芸臺
[16] 芸	[17] 苜蓿	[18] 夏小正	[19] 禮記	[20] 山礬
[21] 七里香	[22] 沈括	[23] 草	[24] 吳其濬	[25] 植物名實圖考
[26] 裴鑑	[27] 賈祖珊	[28] 賈祖璋	[29] 李群	[30] 杜陽編

originally came from the Thien[1] kingdom'. Thien[1] is possibly Khotien, a place in Sinkiang, which for thousands of years was a meeting-ground between China and the West. Thus it is possible that both Li Shih-Chen and Shen Kua were correct. They were describing two different plants with the same name, *yun hsiang*[2]. Since both *Murraya exotica* and *Ruta graveolens* belong to the family *Rutaceae*, they may very well be similar in their chemical constitution so that in terms of the function of the leaves in protecting books from insect pests, they may be essentially equivalent.

The next to the last entry, *huan*[3] (no. 46) is probably cat's tail, *Typha latifolia* L. The *Erh Ya*[4] (Literary Expositior), +300, states that it is also called *phu*[5], a rush used for making mats, and is closely related to *chhang phu*[6] (no. 4).[a] It is not listed in the *Thu Nung Yao Chih*, but there is an interesting passage in the *Nung Sang I Shih Tsho Yao*[7] (Outline of Agriculture, Sericulture and Household Arts), +1314, about its insecticidal activity:

To protect leather goods against pests, mix them with finely ground *huan*[3] flower. There will be no insect damage. Or roll *ai*[8] (*Artemisia argyi*, no. 37) in the leather article, place it in an urn, and seal the opening with clay. Or roll *hua shu*[9] (hot pepper) with it and store.

To protect woollen blankets and clothing, mix them with *huan*[3] flower powder, or *huang hao*[10], yellow artemisia (*Artemisia annua* L., no. 35). The latter is collected in the fifth moon. It is dried in the sun, before it is scattered among the woollen material or rolled up in them. Then there will be no insect damage.[b]

The last plant of the series, no. 47, is *tsui yu tshao*[11] (fish-stupefying herb), *Buddleia lindleyana* Fort. The leaves and flowers are used in medicine. Li Shih-Chen warns against the placing of this close to a fish-pond since its leaves are highly toxic to fish.[c] The *Thu Nung Yao Chih* recommends soaking chopped-up leaves and the stem in water, and using the aqueous infusion against a variety of agricultural and household pests.

We have now completed our survey of pesticidal plants known in China from the early days of her recorded history (−1000), until the time of the publication of the Great Pharmacopaeia (*Pen Tshao Kang Mu*), +1596. Before we attempt to evaluate the practical significance of this group of plants as pesticides, we may first try to gain an idea of the way the individual members are distributed botanically. For example: Does any particular group of plants predominate? How wide is the spectrum of genera and families represented?

If we follow Li Shih-Chen's classification, which he used in the organisation of the *Pen Tshao Kang Mu*[12], we will see that the 47 plants are distributed among four divisions (*pu*[13]) and eleven classes (*lei*[14]) in the manner shown in Table 16. The

---

[a] For a useful discussion on the identity of *huan*[3] see B II, item 98, and *PTKM*, ch. 19, pp. 1361 ff.
[b] *Nung Sang I Shih Tsho Yao*[7], 4th month, p. 78.
[c] *PTKM*, ch. 17, p. 1217.

[1] 闐	[2] 芸香	[3] 莞	[4] 爾雅	[5] 蒲
[6] 菖蒲	[7] 農桑衣食撮要		[8] 艾	[9] 花椒
[10] 黃蒿	[11] 醉魚草	[12] 本草綱目	[13] 部	[14] 類

Table 16. *Distribution of pesticidal plants known in ancient and medieval China*
according to Li Shih-Chen's classification

Division Pu[1]	Class	Lei[2]	Species (Chung[3]) plant number
Herbs (Tshao[4])	Mountain plants	Shan tshao[5]	2, 5, 11
	Fragrant plants	Fang tshao[6]	8, 9
	Marsh plants	Shih tshao[7]	3, 13, 15, 22, 35, 36, 37, 38, 42
	Poisonous plants	Tu tshao[8]	7, 16, 17, 19, 20, 21, 23, 24, 25, 31, 32, 33, 47
	Creeping plants	Wan tshao[9]	10, 34, 39
	Water plants	Shui tshao[10]	4, 14, 46
Vegetables (Tshai[11])	Pungent vegetables	Hun hsin lei[12]	40
Fruits (Kuo[13])	Mountain fruits	Shan kuo[14]	41, 44
Trees (Mu[15])	Fragrant trees	Hsiang mu[16]	1
	Stately trees	Chiao mu[17]	6, 12, 18, 26, 27, 28, 29, 43
	Dense trees	Kuan mu[18]	30, 45?

[1] 部   [2] 類   [3] 種   [4] 草   [5] 山草
[6] 芳草   [7] 隰草   [8] 毒草   [9] 蔓草   [10] 水草
[11] 菜部   [12] 葷辛類   [13] 果   [14] 山果   [15] 木
[16] 香木   [17] 喬木   [18] 灌木

Herbs Division (*Tshao Pu*[1]) contains 33 entries. Thirteen are from the *tu tshao*[2] (poisonous plants) class and nine are from the *shih tshao*[3] (marsh plants) class. The next largest class is the *chiao mu*[4] (stately tree) class from the Trees Division (*Mu Pu*[5]), with eight entries. It should be noted that *shih tshao*[3] does not necessarily mean plants that live in the marshes. Presumably, what Li Shih-Chen had in mind are plants that grow well on an adequate supply of water on level ground, since they include such familiar garden plants and weeds as chrysanthemum (no. 3), cocklebur (no. 13), knotgrass (no. 22) and artemisia (nos. 35, 37). The two classes, *tu tshao*[2] and *shih tshao*[3] actually represent small herbaceous plants that grow well on accessible ground and are easy to harvest. Their popularity as herbs or pesticides is, therefore, predictable.

We can also follow modern botanical nomenclature and classify the 47 plants according to their genus and family as shown in Table 17. It turns out that the 47 plants represent 47 species, 42 genera and 25 families. The most numerous family is the Leguminosae with six species from five genera. The next family is the Compositae with five species from four genera, followed by the Ranunculaceae with four species, but all from the genus *Aconitum*. Except for the Polygonaceae and the Euphorbiaceae with three entries each, all the other families are represented by only one or two

[1] 草部   [2] 毒草   [3] 隰草   [4] 喬木   [5] 木部

Table 17. *Distribution of pesticidal plants known in ancient and medieval China according to modern botanical classification*

Angiospermae	Plant Number
Dicotyledonae	
Compositae	2, 3, 13, 35, 37
Bignoniaceae	29, 42
Loganaceae	20, 47
Ericaceae	30
Umbelliferae	8, 9
Thymeleaceae	19, 21
Vitaceae	34
Celastraceae	12
Euphorbiaceae	18, 28, 43
Meliaceae	27
Rutaceae	45
Leguminosae	5, 6, 7, 10, 11, 26
Rosaceae	31, 41
Magnoliaceae	17
Ranunculaceae	23, 24, 25, 33
Polygonaceae	15, 16, 22
Myricaceae	44
Monocotyledonae	
Zingiberaceae	38
Liliaceae	32, 40
Stemonaceae	39
Lemnaceae	14
Araceae	4
Gramineae	36
Typhaceae	46
Gymnospermae	
Pinaceae	1

species. Thus, the 47 plants are distributed quite widely among 25 families, indicating that they were selected at random from a large assortment of plants that constituted the natural botanical background in China from −1000 to about +1600.

Let us now return to the question of the practical value of these plants as agents for the control of invertebrate pests. We can conveniently categorise the target pests that we have reviewed in terms of the kind of hosts that they infest as follows:

(1) Internal parasites of man and domestic animals, i.e. tapeworms, pinworms, roundworms, etc.
(2) External parasites that live on man and domestic animals, i.e. fleas, mites, lice, etc.
(3) Pests that infest man's immediate environment and household possessions, e.g. flies, mosquitoes, ants, silverfish, etc.

Table 18. *Applications of pesticidal plants as reported before +1600*

Host	Pest	Pesticidal plant number
Man and domestic animals	Internal parasites	2, 6, 7, 8, 11, 15, 16, 18, 21, 22, 27, 29, 30, 35, 37, 38, 39
Man and domestic animals	External parasites	1, 4, 9, 10, 11, 12, 13, 19, 20, 21, 22, 23, 24, 25, 27, 28, 29, 30, 31, 32, 33, 34, 35, 36, 39, 42, 43, 45
Home and environment	Mosquitoes, flies, silverfish, etc.	2, 3, 9, 14, 17, 26, 27, 35, 37, 42, 43, 45, 46
Stored grains	Insects	1, 13, 37
Crop plants	Insects, other invertebrates	5, 17, 19, 23, 24, 25, 28, 34, 39, 40

(4) Pests that infest stored grains, e.g. mealworms, flour moths, etc.

(5) Pests that infest crop plants in the field, e.g. mites, aphids, lepidopterous larvae, beetles, etc.

A plant may be toxic to one or more than one category of pests. We have checked each plant against the type of pests it has been reported to control according to the literature before +1600, and the results are summarised in Table 18. They show that 17 plants have been reported as useful for the control of group (1) pests; 28 for group (2) pests; 13 for group (3) pests; 10 for group (5) pests, and 3 for group (4) pests. All the references to pesticidal activity against external and internal parasites of man and domestic animals are found in the pharmacopoeias and cited in the *Kang Mu*. But of the 13 plants listed as useful for the control of pests of home and environment, only five of the relevant references are cited in the *Kang Mu*. The pesticidal activity of the remaining eight plants is described in agricultural treatises and miscellaneous works. None of the references on plants that control pests of field and garden crops are referred to in the *Kang Mu*. All are found in agricultural and horticultural treatises. We are led to the conclusion that while the pharmacists were concerned to a limited extent about the control of pests in the home environment, they evidently had little or no interest at all in the control of pests of crop plants.

Data on the eight plants found useful for the control of horticultural and agricultural pests are recapitulated in Table 19. If we consider the three related *Aconitum* species (nos. 23, 24 and 25) as equivalent in their pesticidal action, then the number goes up to ten. Nevertheless, by the same reckoning we see that out of

Table 19. *Application of plant pesticides known in China before +1600 against horticultural and agricultural pests*

Plant no.	Chinese name	Botanical name	Part used	Mode of usage	Date
5	Kan tshao	*Glycyrrhiza glabra*	Root	Soil treatment	+1406
17	Mang tshao	*Illicium lanceolatum*	Whole plant	Spray aqueous infusion	< +1600
19	Yuan hua	*Daphne genkwa*	Flower, stem	Plug infested holes	+1273
25	Fu tzu	*Aconitum kusnezoffii*	Root	Steep seed	−100
28	Thung yeh	*Aleurites fordii*	Leaf, oil	Paint around trunk, flood infested holes	+1030
34	Pai lien	*Ampelopsis japonica*	Root	Soil treatment	+1030
39	Pai pu	*Stemona tuberosa*	Root, leaf	Plug infested holes	+1273
40	Ta suan	*Allium scorodoprasum*	Whole plant	Spray aqueous infusion	+1200

5 甘草　17 莽草　19 芫花　25 附子　28 桐葉　34 白蘞　39 百部　40 大蒜

the 47 plants that we have picked out in our survey, 35 have been validated by the *Thu Nung Yao Chih* as having at least a modicum of activity for the protection of crop plants against invertebrate pests.

It is disappointing that we have not been able to find more examples in the older literature on the use of plants for the protection of agricultural or horticultural crops than those listed in Table 19. The relatively poor showing in this category stands in sharp contrast to the wealth of examples of plant pesticides that have been used successfully to control invertebrate pests of people, domestic animals and the home environment. All in all, it seems quite clear that the pesticidal plants known to the ancient and medieval Chinese were more effective in controlling pests of people, animals and household goods than they were in controlling pests of crop plants. But perhaps there are good reasons why this should be so.

First of all, there is a problem with the low intrinsic pesticidal potency of plant materials in general. It certainly is the case with the eight plants listed in Table 19. The active principle is often present at a low concentration, so that in order to be effective a relatively large amount of crude material would have to be applied. This does not pose a problem where the area or volume to be treated is limited as, for example, an afflicted part on the skin of an animal, the abdominal cavity of a man, a room in a house, a piece of furniture, articles of clothing, books, or even the contents of a granary. But when it comes to plants in a garden or in a field, it is much more difficult to achieve the concentration of pesticide on all the afflicted plants necessary for efficient control of the pests. Indeed, the examples we have cited are the very ones where it has been possible to target the pesticide on a specific, limited locality, such as the infested hole in a tree, the soil round a root, or the surface of a few leaves. Such methods, while no doubt helpful in their proper context, would be completely ineffective against a major infestation of locusts or caterpillars on an agricultural crop in the field.

Secondly, there is the problem of having sufficient quantities of pesticidal material available when and where needed. Except perhaps for thung oil, the other plant products listed are medicinals usually stocked only by the apothecary in amounts appropriate for their expected usage. One can hardly conceive of such stocks being large enough to meet the needs of a sudden unexpected insect pest infestation. Thus the opportunities to test the efficacy of any one of the plants under proper field conditions must have been few.

A happy solution to both these problems apparently emerged in the use of pesticidal plant materials to control mosquitoes, one of the most important pests that infest man's environment. The extreme seriousness of mosquitoes as a public health meanace in parts of China in the medieval period is illustrated by the following anecdotes from a discourse on mosquitoes in the *Chhi Tung Yeh Yü*[1] (Rustic Talks from Eastern Chhi) by Chou Mi[2], +1290:

Mosquitoes are abundant in the Wu-hsing[3] district. On a summer evening after a bath, if one should be so careless as to loosen and raise one's robe, one would be attacked and bitten

[1] 齊東野語   [2] 周密   [3] 吳興

by incessant swarms of mosquitoes. There would be no respite. The venerable Tung-Pho[1] used to say, 'In the lake country there are lots of mosquitoes and gnats, of which the most vicious is a type called *pao chiao*[2] (leopard's claw)'.[a]

In the summer months, in the vicinity of Hsin-An[3], Tshang[4] and Ching[5] counties, cows and horses are smeared with mud. Otherwise they would be stung to death (by mosquitoes).[b]

Across the Huai[6] River mosquitoes and gnats are particularly plentiful, especially at Lu Chin Miao[7] in the Kao-Yu[8] prefecture. *Sun Kung Than Phu*[9] (The Venerable Mr. Sun's Conversation Garden), +1085, says that in the lowlands west of Thai-Chou[10] mosquitoes abound. Messengers on the road drive them away by using the smoke of moxa (*ai*[11]). A drunken minor official was so badly bitten by mosquitoes that he actually died.[c]

Of special interest to us is the observation in the last quotation that moxa smoke was used to ward off mosquitoes. Moxa is a material that could be easily packaged, distributed, sold and burnt whenever and wherever the need arose. A convenient method to package and apply such botanical pesticides was to incorporate them into an incense product. Although we have no record as to whether *moxa* was its principal constituent, there was a mosquito incense (*wên yên*[12]), according to Chou Mi[2] (+1270), on sale as one of the common articles available in the market in Hangchow.[d] We have no idea of the shape or size of this *wên yên*[12], but we can imagine that it could be similar in form to the mosquito incense that we see today in South China and Southeast Asia. We also have evidence that it was, indeed, used inside the house to ward off mosquitoes, as recorded in a story from the *Hsien Chhuang Kua I Chih*[13] (Strange Things Seen Through a Barred Window), of the Sung Dynasty:

In the Hai-Yen[17] district, a man by the name of Ni Shêng[18] used to incorporate sawdust from miscellaneous timber remnants into an incense called *yin hsiang*[19], which was sold on the local market. One night while he was asleep with mosquito incense burning, some sparks fell on a basket of *yin hsiang*[19] and it immediately started to smoulder ... Soon the whole house caught fire, and Ni Shêng[18], unable to escape, perished with the house.[e]

No doubt it was divine punishment visited upon him for adulterating the incense. Thus we see that the idea of using plant pesticides to control pests in the home

---

[a] *Chhi Tung Yeh Yü*, ch. 10, p. 5. Wu-hsing is located south of Lake Thai[20] in Chekiang Province. The reference must be to the famous poet Su Tung-Pho.

[b] *Chhi Tung Yeh Yü*, ch. 10, p. 5a. Hsin-An[3], Tshang[4] and Ching[5] are counties in Hopei south of Peking. This passage is a repetition of an identical account in the *MCPT*, ch. 23, p. 229, about the deadly action of *wên mêng*[14], which, as pointed out by Li Chhün[15], refers to horseflies, *Tabanus* and *Chrysops*. spp. and not to mosquitoes. We suspect that Chou Mi simply copied the passage but inadvertently failed to distinguish between *wên* and *wên mêng*.

[c] *Chhi Tung Yeh Yü*, ch. 10, pp. 5, 5a. Lu Chin Miao[7], Kao-Yu[8] and Thai-Chou[10] are all in northern Kiangsu, east of the Kao-Yu[8] Lake.

[d] *Wu Lin Chiu Shih*[16] (+1270) ch. 6, pp. 7, 18a.

[e] *Hsien Chhuang Kua I Chih*, p. 23a. Tr. auct.

[1] 東坡	[2] 豹脚	[3] 信安	[4] 滄	[5] 景
[6] 淮	[7] 露筋廟	[8] 高郵	[9] 孫公談圃	[10] 泰州
[11] 艾	[12] 蚊烟	[13] 閑窗括異記	[14] 蚊蝱	[15] 李群
[16] 武林舊事	[17] 海鹽	[18] 倪生	[19] 印香	[20] 太

environment was well accepted. We suspect that the potential of this idea for the control of pests of crop plants was more widely recognised before + 1600 than is apparent from the results of our survey. But before we discuss how this idea was developed further in the next three centuries, we might mention two examples of pesticidal plants which show interesting activity, but whose identity is still uncertain.

In the *Hsü Po Wu Chih*[1] (Supplement to the *Record of the Investigation of Things*), + 12th century, there occurs this passage on a plant called *chien kang*[2]:

Shaped like the *fêng*[3] plant, it has many branches but few leaves. The roots are like silk and the leaves like fans. Even when you do not shake it, the leaves will move by themselves and create a draught. In the kitchen it keeps the air cool and clean, drives away insects and flies and helps to maintain general cleanliness. At the time of the Emperor Yao[4], it was planted by the kitchen to drive away evil influences ...[a]

This would be a useful plant to have, if only we knew what it is. The other plant is *chien*[5] or *lan hsiang tshao*[6], as described in the *Mao Shih Tshao Mu Niao Shou Chhung Yü Su*[7] (An Elucidation of the Plants, Trees, Birds, Beasts, Insects and Fishes described in the Book of Odes), + 3rd century:

The plant *chien*[5] in the Book of Odes is the same as the fragrant plant *lan*[8] This latter is mentioned in the *Tso Chuan*[9]; also in the *Chhu Tzhu*[10] (Elegies of Chhu). Confucius says the *lan*[8] is the fragrancy (fragrant flower) of the king. The stem and leaf are like those of the medicinal plant *tsê lan*[11] (marsh lan[11]). The joints are wide apart, and the stem between them is red, and it rises to a height of four or five feet. In the time of the Han, this plant was cultivated in the Imperial Gardens. It was added to cosmetics, and was used also to preserve clothes and books from insects.[b]

A *lan tshao*[12] is mentioned as an upper class (*shan phin*[13]) herb in the *Pên Ching* and elaborated in the *Kang Mu*. Read has identified this *lan tshao* as *Eupatorium chinense*, the Chinese thoroughwort.[c]

There is considerable doubt whether the *lan tshao*[12] described in the *Kang Mu* is the same as the *lan hsiang tshao*[6] of the *Mao Shih Tshao Mu Niao Shou Chhung Yü Su*[7]. 'Although the flowers of *Eupatorium* exhale an odour which is not unpleasant', Bretschneider did not believe that 'the famous perfume *lan*[8] of the ancient Chinese can be referred to this plant,'[b] He preferred to think that the ancient *lan*[8] was an orchidaceous plant, since *lan*[8] is the common name for orchids in China, valued for the powerful fragrance of their flowers. It is likely that the ancients used two plants for the preservation of books, *yün*[14] and *lan*[8], but by the Sung dynasty in Shen Kua's time, the use of *lan*[8] had alrdady died out.[g]

---

[a] *Hsü Po Wu Chih*[1], ch. 2, p. 5.
[b] Tr. Bretschneider (B II, item 405), mod. *Mao Shih Tshao Mu Niao Shou Chhung Yü Su*[7], ch. 1, p. 1.
[c] R (1), item 33.   [d] B II, p. 228.

[1] 續博物誌	[2] 蓮茴	[3] 蓬	[4] 堯	[5] 蘭
[6] 蘭香草	[7] 毛詩草木鳥獸蟲魚疏	[8] 蘭	[9] 左傳	
[10] 楚詞	[11] 澤蘭	[12] 蘭草	[13] 上品	[14] 芸

(iv) *Meeting of East and West*

Before we discuss the pesticidal plants discovered in China or introduced from abroad after + 1600, we may pause briefly and take stock of the historical significance of the information that we have so far assembled. Two questions are of particular interest. (1) What is the earliest date when a plant against each of the categories of pests (cf. Table 18) was first recognised? And (2) how does the development of the technology of plant pesticides in China compare with that of Western Europe? To answer the first question, we have re-examined our data and arrived at the following propositions:

(1) *Internal parasites of man and domestic animals.* The first anthelmintic used was *jang ho*[1], that is, the *chia tshao*[2] of the *Chou Li*[3] which, if Hsia Wei-Ying[4] is correct, was compiled at about − 300. There is, however, a possibility that some of the anthelmintics mentioned in the *Pên Ching*, such as *khu shên*[5] (no. 11) *liao shih*[6] (no. 15), *pa tou*[7] (no. 18), etc., are equally ancient.

(2) *External parasites of man and domestic animals.* The earliest records of plants toxic to such pests are those described in the *Pên Ching*, e.g. *sung chih*[8] (no. 1), *lang tu*[9] (no. 21) *pien hsü*[10] (no. 22), etc. Their activity was recognised well before + 100, but how much before would be difficult to say.

(3) *Pests of the home environment.* *Chü hua*[11] (no. 3) and *mang tshao*[12] (no. 17) of the *Chou Li*[3] are undoubtedly the earliest plants used to control this category of pests. The date would be before − 300.

(4) *Pests of stored grains.* The earliest record is the use of *ai hao*[13] (no. 37) or moxa, *Artemisia argyi*, in the storage of wheat as described by Fan Shêng-Chih[14], − 1st century.

(5) *Pests of plant crops.* The earliest example is the use of *fu tzu*[15] (no. 25), *Aconitum autumnale*, to prepare an infusion for the steeping of seeds, as recorded by Fan Shêng-Chih, − 1st century.

In summary, we may state that herbs active against invertebrate pests of man and domestic animals and the home environment were recognised early, probably earlier than − 300, but certainly well before + 100. Herbs active against pests of stored grains and crop plants were first recorded at about − 100, but most of the examples of the use of herbs to counter agricultural and horticultural pests did not appear until much later in the medieval period, during the Sung and Yuan dynasties (11th to 13th centuries), as indicated in Table 19. By about + 1600, 47 plants from the pharmacopoeias had been found to possess pesticidal activity. Of these eight were recognised as useful for the protection of plant crops against invertebrate pests.

How does this record compare with the corresponding situation in Europe?

[1] 蘘荷	[2] 嘉草	[3] 周禮	[4] 夏緯瑛	[5] 苦參
[6] 蓼實	[7] 巴豆	[8] 松脂	[9] 狼毒	[10] 萹蓄
[11] 菊花	[12] 莽草	[13] 艾蒿	[14] 氾勝之	[15] 附子

Fortunately for us, Smith & Secoy[a] have recently completed a survey of the literature, the first of its kind ever attempted, to assess the role played by plants for the control of agricultural and horticultural pests in Western civilisation. They found that the written record before 1850 can be divided into three periods:

(1) Classical Greek and Roman, when the approach was entirely empirical (about −300 to +400).
(2) Medieval European, when the principal activity was compilation and translation (+400 to about +1650).
(3) Experimental period, when new substances and techniques began to be tested and used (+1650 to 1850).

Smith & Secoy have essentially accomplished for pesticidal plants in Europe (except for those active against internal parasites) what we have tried to do in this chapter for pesticidal plants in China. Thus it is now relatively easy to compare the status of pesticidal plants in Europe and China at various periods in their history. The results, summarised in Table 20, show that by 1850, there were fifteen plants or plant products known in Western Europe for the protection of crops and the home environment against vertebrate and invertebrate pests. But what about the situation in +1600, the end of the medieval period in China? We see that of the fifteen plants listed in Table 20, four were discovered after +1600, one in the middle ages, and the remaining eleven in the classical period. Thus at +1600 the Western Europeans would have had at their disposal eleven plants that were useful as pesticides. By eliminating from Table 20 the four plant materials discovered after +1600, i.e. the English walnut, strychnine, stavesacre and tobacco, and deleting any information (for example, activity of the European elder and the white hellebore against plant pests) acquired after +1600, we are left with the plants listed in Table 21, which summarises the state of the art as it existed in Europe at about +1600.

We now have a collection of eleven plants, five of which were known to be active against invertebrate plant pests, and seven against rodents. The prominence of rodents in this table suggests that during the classical and medieval periods rodent infestation (rats, mice) must have been a severe problem in the agricultural environment of Europe. Compared to the information presented in Tables 18 and 19, it is clear that by +1600 the Chinese were aware of a much wider array of plant materials that could be used to control pests of people, domestic animals, home environment and crop plants than their contemporaries in Europe.

There are some familiar entries in Table 21. Perhaps the most notable is mugwort, which may signify not only *Artemisia absinthium* but also *A. vulgaris*, *A. abrotanum* and *A. maritima*, since the ancient Greeks and Romans were no more precise than the ancient Chinese in their accounts of the botanical features of

---

[a] Smith & Secoy (1). It should be noted that they did not propose the specific dates associated with the classical and the medieval periods. The dates indicated are suggested by us.

Table 20. *Plants used for pest control in Western Europe before 1850*

Common name	Botanical name	Family	Plant pests	Other pests	Earliest record
Elder	*Sambucus niger* L.	Caprifoliaceae	+	Flies	*Geoponika*,[a] +2nd cent.
English walnut	*Juglans regia* L.	Juglandaceae	+	—	Evelyn, +1664
Colocynth	*Citrullus colocynthis* L.	Cucurbitaceae	+	Rodents	Columella, +50
Strychnine	*Strychnos, nux-vomica* L.	Strychnaceae	–	Rodents	Forsyth, 1802
Black hellebore	*Helleborus niger* L.	Helleboraceae	+	Rodents, birds	*Geoponika*, +1st cent.
Monkshood	*Aconitum napellus* L.	Ranunculaceae	–	Flies	Ibn al-Baythār *c.* +1240
Stavesacre	*Delphinium staphisagria* L.	Ranunculaceae		Exoparasites, rats, ants	Anon. +1788
Houseleek	*Sempervivum tectorum* L.	Crassulaceae	+	—	Pliny, +77
Hemlock	*Conium maculatum* L.	Apiaceae	–	Rodents	*Geoponika*,[a] +2nd cent.
Mugwort	*Artemisia absinthium* L.	Asteraceae	+	—	Pliny, +77
Henbane	*Hyoscyamus niger* L.	Solanaceae	+	Rodents	*Geoponika*,[a] +2nd cent.
Tobacco	*Nicotiana tabacum* L.	Solanaceae	+	—	Evelyn, +1664
Garlic	*Allium sativa* L.	Liliaceae	+	—	Pliny, +77
Squill[b]	*Urginea maritima* (L.)	Liliaceae	+?	Rodents .	Theophrastus, –300
White hellebore	*Veratrum album*	Liliaceae	+	Rodents	*Geoponika*,[a] +1st cent.

[a] A 6th- or 7th-century compilation by Cassianus Bassus of early Greek and Roman agricultural writings. References to specific authors are given in Table 22 and in the text.

[b] Although known to the ancients as an insecticide, it was later found to be a useful rodenticide.

Table 21. *Plants used for pest control in Western Europe in + 1600*

Common name	Botanical name	Activity against	
		Plant pests	Other pests
Elder	*Sambucus niger*	—	Flies, snakes
Colocynth	*Citrullus colocynthis*	Insects, seed steep	Rodents
Black hellebore	*Helleborus niger*	?	Birds, rodents, flies
Monkshood	*Aconitum napellus*	—	Rodents, flies
Houseleek	*Sempervivum tectorum*	Insect (repellent), seed steep	—
Hemlock	*Conium maculatum*	—	Rodents
Mugwort	*Artemisia absinthium*	Insects, grain pests	—
Henbane	*Hyoscyamus niger*	Insects	Birds, rodents
Garlic	*Allium sativa*	Insects	—
Squill	*Urginea maritima*	—	Rodents
White hellebore	*Veratrum album*	—	Rodents

plants. The use of artemisia leaves and shoots in granaries to protect the grain from weevils and other pests was reported by Pliny (+77),[a] which would date this practice in the Roman Empire almost as early as that for *ai hao*[2] (no. 37) reported by Fan Shêng-Chih[1] (−1st century) during the Han dynasty.[b] Another plant with a Chinese counterpart is the white hellebore, *Veratrum album*, as opposed to *li lu*[3] (no. 32), *Veratrum nigrum*, the black hellebore.[c] *V. album* was used by the ancients in Europe to control rodents,[d] while in China *V. nigrum* was recommended mainly for treating mite and flea afflictions of the skin of people and animals.[e]

We also find an *Allium* in Table 21, *Allium sativa*, garlic, as compared with *ta suan*[4] (no. 40), *Allium scorodoprasum*, chive. Pliny thought garlic could intoxicate birds[f] and Palladius considered the fumes of burning garlic to be fatal to caterpillars.[g] Thus, the insecticidal property of garlic was recognised at an earlier date (+380) in Europe than chive in China, the use of which as a pest control agent was not recorded until about +1200. But in the case of the aconite, *Aconitum napellus*, which was known to the Arabs in Spain (*c.* +1200), and mixed with meat in bait to kill

---

[a] Pliny, **18**, ch. 73.   [b] Shih Shêng-Han (2), p. 29.
[c] The 'black hellebore' also serves as the common name for *Helleborus niger* L. Cf. Smith & Secoy (1), p. 14.
[d] *Geoponika*, **13**, ch. 4, 7.   [e] Cf. pp. 492–3.
[f] Pliny, **19**, ch. 34.
[g] Palladius, **1**, chs. 126, 127; *Geoponika*, **5**, chs. 30, 48.

¹ 氾勝之   ² 艾蒿   ³ 黎蘆   ⁴ 大蒜

houseflies,[a] the parallel use of *wu thou*[1] (no. 24) and *fu tzu*[2] (no. 25) in pest control had occured several hundred years earlier in China.[b]

The use of *fu tzu*[2] as a steep for seeds before planting was recorded in the *Fan Shêng-Chih Shu*[3] (− 1st century).[c] There are two examples of plant infusions used to steep seed in Table 21. Columella (+ 50) described the steeping of seed in an infusion from colocynth,[d] and Pliny (+ 77) cited Democritus in advising that cabbage seeds be soaked in houseleek juice before sowing.[e] Both practices are probably about as old as that prescribed by the *Fan Shêng-Chih Shu*.

We have placed the end of the period of antiquity in China at about + 100 and are naturally curious to know how Chinese technology in plant pesticides compared with that existing in Greece and Rome at about the same time. From Smith & Secoy's study and our review, we have culled the relevant information to prepare the parallel lists of plant pesticide usages in China and in the West for the year + 100 shown in Table 22. Plants used against internal pests of man and domestic animals are not a part of this comparison, but we did include plants known in China as toxic to fish and other invertebrates (*sha yü chhung*[4]), since activity against fish is indicative of activity against invertebrate pests that live in water or go through a part of their life cycle in water, such as mosquitoes. One cannot help but notice that more information was available in China in + 100 on the control of pests of man and domestic animals than on the control of pests of crop plants, while the reverse appeared to be true in Europe.[f] On balance, Table 22 suggests to us that in terms of both the breadth and sophistication of the technology, Europe and China were nearly equal in usage of pesticidal plants, with China enjoying perhaps a slight lead in the number of agents available. But we recognise that + 100 was not a particularly appropriate time to take stock of the development of plant pesticides in the West. A more meaningful time would be the end of the classical period in Europe, as defined by Smith & Secoy. Since works on agriculture by Greek and Roman writers continued to appear between the + 1st and + 4th centuries, it is clear that the classical period in Europe did not end until about + 400, after the publication of *On Husbondrie* by Palladius in + 380. If we were to update Table 22 to the year + 400, we would have to add the following entries to the European side of the ledger:

[a] Odish (1), p. 43.
[b] Cf. pp. 489–90.
[c] Shih Shêng-Han (2), p. 11.
[d] Columella, **2**, ch. 9.
[e] Pliny, **18**, ch. 45, and **19**, ch. 58.
[f] This impression may, of course, merely be a reflection of the nature of the sources available to us. For China we are heavily dependent on the *Pên Ching*, a pharmacopoeia, for much of the information. It should be no surprise that the slant is towards pests that cause disease in man and domestic animals. In Europe, however, a good deal of the ancient literature on agriculture has survived, and we are able to consult sources that deal specifically with agricultural matters including the protection of plants against insect pests.

[1] 烏頭　　　　[2] 附子　　　　[3] 氾勝之書　　[4] 殺魚蟲

Target Pests	Plants	Reference
Rats, birds	Hemlock	Apuleius,[a] + 2nd century
	Henbane	Apuleius,[a] + 2nd century
Household pests	Elder	Berytius,[b] + 2nd century
	Black hellebore	Berytius,[b] + 2nd century

We would also have to take note of additional observations by Palladius on colocynth, black hellebore, houseleek, henbane, garlic and squill.[c] But in spite of the new entries and additions to the record, the overall status of the technology of plant pesticides in Europe cannot be said to have moved ahead in any major way between the years + 100 and + 400.

In China, unfortunately, no new information at all on plant pesticides published between + 100 and + 400 has survived to our day.[d] Thus, as far as we can determine, the situation at + 400 must be considered to be about the same as it was in + 100. This leaves us with the conclusion that by + 400 Europe and China were virtually at about the same level of development in the technology and usage of plant pesticides. But in the next twelve hundred years, while the Chinese continued to deepen and expand their knowledge, the West Europeans not only had little to add to the technology of pesticidal plants, but also had to expend much time and effort to recover and re-assimilate their Greek and Roman heritage which had passed into the cultural sphere of the Arabs. At any rate, by + 1600 China was undoubtedly and significantly ahead of Western Europe. After + 1650, however, the start of what Smith & Secoy call the experimental period, Western Europe,[e] no doubt stimulated by her discovery of a vast array of new plant resources from the Americas and East Asia, began to make fresh and steady progress. We can easily gauge the extent of this progress by comparing Table 20 with Table 21. Four new pesticidal plants had been added to the list by 1850, and a wealth of new publications on the use of plant pesticides had made their way into the literature.[f]

There was progress between + 1600 and 1850 in China too. One major pesticidal plant of native origin was discovered, and another major one introduced from abroad. The native plant is *lei kung thêng*[3] (no. 48), thunder-god vine, which has been identified as *Tripterygium wilfordii* Hook (family, Celastraceae). It is also known by the name of *tuan chhang tshao*[4] (break-intestine plant), the third plant with such a name that we have encountered. The earliest reference to this plant occurs in the *Wu Li Hsiao Shih*[5] (Small Encyclopedia of the Principles of Things), + 1664, but

---

[a] *Geoponika*, **13**, ch. 5.   [b] *Geoponika*, **13**, ch. 12.

[c] Palladius, **8**, chs. 122, 126, 131, and 135.

[d] Only fragments of *Fan Shêng-Chih Shu*[1], − 1st century, and *Ssu Min Yüeh Ling*[2] (Monthly Ordinances for the People),[c] + 180, remain. It is possible that material relevant to the use of plant pesticides to protect crop plants was originally present in these agricultural treatises, but was later lost.

[e] Smith & Secoy (1), p. 12.

[f] Smith & Secoy (1) list 70 references relevant to plant pesticides in Europe dated between + 1650 and 1850. Cf. p. 17.

[1] 氾勝之書   [2] 四民月令   [3] 雷公藤   [4] 斷腸草   [5] 物理小識

Table 22. *Comparison of plants used for pest control in China and in Europe in + 100*

Target pests	CHINA		EUROPE	
	Plant no.	Reference[a]	Plant	Reference
Rats, birds	*Lang tu 21*	*SNPTC*	Black hellebore	Paxamus[b]
	Thien hsiung 23	*SNPTC*	White hellebore	Paxamus[c]
	Wu thou 24	*SNPTC*		
Household pests	*Chü hua 3*	*CL*	Mugwort	Pamphilus[d]
	Mang tshao 17	*CL*		
Pests of man and animals	*Thung yeh 28*	*SNPTC*		
	Tzu pai phi 29	*SNPTC*		
	Niu pien 33	*SNPTC*		
	Chin tshao 36	*SNPTC*		
Fish and invertebrates	*Mang tshao 17*	*SNPTC*		
	Pa tou 18	*SNPTC*		
	Yüan hua 19	*SNPTC*		
Pests of stored grain	*Ai hao 37*	*FSCS*	Mugwort	Pliny[e]
Pests of seedlings	*Fu tzu 25*	*FSCS*	Colocynth	Columella[f]
			Houseleek	Pliny[g]
Pests of crop plants	—		Colocynth	Pamphilus[d]
			Mugwort	Pliny[g]
			Garlic	Pliny[h]
			Houseleek	Pliny[g]

[a] *SNPTC, Shen Nung Pen Tshao Ching; CL, Chou Li; FSCS, Fan Shêng-Chih Shu;*
[b] *Geoponika,* **13**, ch. 4 (Paxamus, before + 1st century).
[c] *Geoponika,* **13**, ch. 7 (Paxamus).
[d] *Geoponika,* **13**, ch. 15 (Pamphilus, −2nd century).
[e] Pliny, **18**, ch. 73.
[f] Columella, **2**, ch. 9.
[g] Pliny, **19**, ch. 58; cf. also *Geoponika,* **13**, ch. 1 (Democritus, −300).
[h] Pliny, **19**, ch. 34.

[3] 菊花	[17] 莽草	[18] 巴豆	[19] 芫花	[21] 狼毒
[23] 天雄	[24] 烏頭	[25] 附子	[28] 桐葉	[29] 梓白皮
[33] 牛扁	[36] 蠶草	[37] 艾蒿		

it is described more fully in the *Pên Tshao Kang Mu Shih I*[1] (Supplementary Amplification for the Great Pharmacopoeia), + 1769. 'It is said to be also called *phi li mu*[2] (thundering wood). The most potent product is from Kiangsi. The natives collect it to poison fish, but it also kills clams and snails. It is powerful in action. When its fumes cover silkworm eggs, they would no longer be able to hatch. Silkworm growers, therefore, avoid this plant. The mountain people collect it to smoke out bedbugs.'[a] *Lei kung thêng*[3] is now considered one of major plant

[a] *Pên Tshao Kang Mu Shih I*[1], ch. 7, p. 264.

[1] 本草綱目拾遺　　　[2] 霹靂木　　　[3] 雷公藤

pesticides in China. The *Thu Nung Yao Chih* recommends grinding the bark of the root to a fine powder and using it as a base for dust, or to prepare an aqueous decoction. The active ingredients are a series of alkaloids (tripterygine, willforine, willfordine, etc).

The import is none other than *yen tshao*[1] (no. 49), tobacco, *Nicotiana tabacum* L. (family Solanaceae) one of the less felicitous gifts from the New World to the Old. It was apparently introduced into China about +1620, probably by way of Manila.[a] Its cultivation soon spread widely all over China, which now enjoys the dubious distinction of being the largest producer of tobacco in the world. The use of tobacco for the control of insect pests was first mentioned in the *Liu Yang Hsien Chih*[2] (Chronicle of the Liu-Yang[3] District), which states that in the reign of Tao-Kuang[4] (1821–50), 'when late rice failed regularly to give a good crop (due to caterpillar infestation) tobacco stalks were planted in the soil. The infestation then died out.'[b] A more direct application of tobacco to control mulberry pests is recorded in the *Sang Tshan Thi Yao*[8] (Essentials of Sericulture), c. 1870. After describing the various insects that feed on mulberry leaves, it went on to say:

The leaves may be large and healthy, but after the worms are through with them, they will look like cheesecloth. Even the leaves that come up next year will be less vigorous. To control the pests, beat the leaves with a stick, and they will fall off. Spread a piece of cloth under the tree to catch and destroy them. Another method is to dilute tobacco extract in water or prepare a decoction of *pai pu*[9] (no. 39) and *pa tou*[10](no. 18). Dip a palm leaf broom in the solution and sprinkle heavily on the mulberry leaves. If the tree is too tall use a mechanical sprayer. After the worms feed on the contaminated leaves, they will die. Treatment should be timed early; otherwise the pests would have laid eggs which will cause another infestation next year.[c]

The *Thu Nung Yao Chih* lists *yen tshao*[11], *Nicotiana tabacum* L. as well as *huang hua yen tshao*[12], *Nicotiana rusticum* L. as useful insecticidal plants.[d] The active principle is the alkaloid nicotine, which can be easily extracted from the leaves and stem with hot water. It is recommended for the control of a wide variety of pests in the field.

In Europe the use of decoctions of tobacco as an agricultural pesticide was first noted in the *Kalendarium Hortense* of Evelyn in +1664.[e] It was soon shown to be an extremely effective insecticide, and liquid tobacco extracts, dried tobacco dusts, and tobacco fumigations were used widely in the 18th and 19th centuries. From the

---

[a] Porter Smith & Stuart (1), p. 283. See also Yü Chêng-Hsieh[5], p. 326, which states that '*Yen tshao* originates from Luzon where it is called *tan-pa-ku* (transliteration of tobacco?). During Ming times it came to China by sea by way of Fukien (*Fu-Chien*[6]). So it is still called *chien-yen*[7] (the smoke plant from Fu-Chien).'

[b] *Liu Yang Hsien Chih*[2]. Cf. Chou Yao (1), 53. See also Yü Chêng-Hsieh[5], p. 328, which quotes from other sources that 'if tobacco leaf is rolled in the hollow of a pen, it will be free from insect attack', and 'tobacco can be used to cure head lice', and 'when seedlings in the terraced fields are infested by caterpillars, plant a piece of tobacco stem next to the seedling while transplanting, then the infection will cease'.

[c] *Sang Tshan Thi Yao*. Cf. Tsou Shu-Wên, p. 128.

[d] *TNYC*, pp. 137–40.   [e] Evelyn (1), p. 71.

[1] 煙草   [2] 瀏陽縣誌   [3] 瀏陽   [4] 道光   [5] 俞正燮
[6] 福建   [7] 建煙   [8] 桑蠶提要   [9] 百部   [10] 巴豆
[11] 煙草   [12] 黃花煙草

large number of references in the European literature on the subject during this period,[a] we are led to believe that by 1850 tobacco products were applied as pesticides for plant protection on a more extensive scale and with greater sophistication in western Europe than in China.

By adding *lei kung thêng*[1] (no. 48) and *yen tshao*[2] (no. 49) to Tables 18 and 19, we would update the status of plant pesticide usage in China to the year 1850. The corresponding status in western Europe is already summarised in Table 20. Although China still led in the number of plants recognised as pesticidal, this advantage is probably balanced by the greater usage of tobacco in western Europe. Thus we suspect that in terms of the amount of materials applied and the extent of plant protection achieved, by 1850 the technology of plant pesticides in China and that of western Europe were again about equal. In other words Europe had now caught up.

After 1850 western Europe continued to benefit from imports of plant pesticidal materials from the other parts of the world. The most important products are:

Pyrethrum. Dried flowers or extracts from *Chrysanthemum cinerariaefolium* and *Chrysanthemum coccineum*. Contains pyrethrins or cinerins. Originally *C. cinerariaefolium* was from Dalmatia and *C. coccineum* from northern Iran, but commercial cultivation moved to Japan in the late 19th century, and to Kenya in the 1930's.[b]

Derris. Dried root powder of *Derris elliptica* and *Derris uliginosa* from Malaya, Indonesia or the Philippines. Contains rotenone and other rotenoids.[c]

Cubé. Dried root powder of *Lonchocarpus* spp. from the Amazon valley in South America. Contains rotenone and other rotenoids.[c]

Other plant materials in commercial use included quassia (*Quassia amara*) from Brazil, sabadilla (*Schoenocaulon officinale*) from Venezuela and Mexico, ryania (*Ryania speciosa*) from tropical South America, and hellebore (*Veratrum album*) from Northern Asia.[d] But their use declined in the years before World War II. Indeed, Feinstein & Jacobson stated in 1953 that 'only three dependable plant insecticides have been discovered from the time of the early Romans to the beginning of the twentieth century'.[e] They are tobacco, rotenone and pyrethrum. But in the

---

[a] Smith & Secoy (1), pp. 15–16, list 27 references on the use of tobacco as a pesticide to protect crop plants. One cannot overstate the importance of tobacco in the development of the concept of plants as a source of pesticides. Tobacco was the first botanical pesticide that became a commercial as well as a technological success. We have cited earlier (p. 506) the two problems that impede the use of plant materials for the control of pests of crop plants: (1) the intrinsic low potency of the crude plant product; and (2) the uncertain availability of material when and where it is needed. In the case of tobacco, both problems were solved in a felicitous manner. First, tobacco contains a relatively high concentration of the active ingredient, nicotine, which could be easily extracted from the plant to give a highly potent concentrate. Secondly, tobacco was already cultivated on a large scale to provide the raw material for pipe tobacco, cigars and cigarettes. The dust and residues from the manufacture of smoking products served as an inexpensive raw material for the extraction of nicotine. Thus, large quantities of nicotine concentrates were available to meet the farmer's needs.

[b] Shepard (1), pp. 144 ff.   [c] Fukami & Nakajima (1), pp. 71 ff.
[d] Crosby (1), pp. 177 ff.   [e] Feinstein & Jacobson (1), p. 426.

[1] 雷公藤   [2] 煙草

1950's usage of tobacco and rotenone also declined in the face of the challenge posed by organic chemical insecticides. Today only pyrethrum remains as a plant pesticide of commerce in the West.

After 1850 we saw in China the introduction of both native and foreign rotenone-bearing plants that were not known to the compilers of the pharmacopoeias, namely *Derris* spp. (*marginata, elliptica* and *fordii*), under the general name *yü thêng*[1] (fish creeper), *Millettia pachycarpa* or *chi hsüeh thêng*[2] (chicken blood vine) and the Mexican yam bean *Pachyrhizus erosus* (*tou shu*[3]). Although relatively late-comers to the Chinese scene, they were among the nine insecticidal plants selected for further development in the 1940's by Chiu Shin-Foon.[a] The nine plants are:

Millettia pachycarpa	Leguminosae
Derris fordii	Leguminosae
Pachyhizus erosus	Leguminosae
Tripterygium forrestii	Celastraceae
Celastrus angulata	Celastraceae
Rhododendron molle	Ericaceae
Stellera chamaejasme	Thymelaeaceae
Stemona tuberosa	Stemonaceae
Veratrum nigrum	Liliaceae

It is curious that this group does not include pyrethrum, the most successful and efficacious of all plant pesticides.[b] Apparently seeds of *Chrysanthemum cinerariaefolium* were first brought to Shanghai from Japan in 1919. Small plantings were started in Kiangsu and Chekiang provinces, but they attracted little attention. During World War II, seeds were carried into the interior and cultivation was extended to Kweichow, Kuangsi, Yunnan and Szechuan provinces. Today pyrethrum is grown widely in South China, but most of the cultivation is still in Chekiang.[c]

In 1958, in recognition of the considerable experience and expertise among farmers in the use of indigenous plant pesticides for the control of insect pests on agricultural crops, an interdisciplinary group of scientists in China, entolomogists, botanists, microbiologists, pharmacologists, etc., organised a nation-wide campaign to codify and test a large number of native plants and other materials reputed to have significant pesticidal activity. The result is a compendium entitled *Chung Kuo Thu Nung Yao Chih*[4] (Record of Chinese Indigenous Agricultural Drugs) published in 1959. It lists and describes 522 individual items comprising 19 mineral products, 220 plants of known botanical identity, and 183 plants whose botanical

---

[a] Chiu Shin-Foon (1), p. 276.

[b] Chiu Shin-Foon (private communication, 1984) states that by the 1940's pyrethrum was already considered as a well-established plant insecticide in China. It was, therefore, not included in the group of plants selected for further development.

[c] Kuan Lung-Chhing[5] & Chhen Tao-Chhuan[6] (1), pp. 29 ff.

[1] 魚藤   [2] 鷄血藤   [3] 豆薯   [4] 中國土農藥誌
[5] 關龍慶   [6] 陳道川

identity has not yet been determined. Of the 220 known plants 128 (i.e. 58 per cent) have been described in the *Kang Mu*, and of the 128, 57 (i.e. 45 per cent) had been listed originally in the *Pên Ching*.

The wealth and breadth of the information assembled in this compendium suggest that pesticidal plants are playing and will continue to play a significant role in China in the protection of agricultural and horticultural crops. They also remind us of an issue that we have raised earlier (p. 506) in this chapter. With so many plants described in the pharmacopoeias, and known by the farmers in local communities, presumably through oral dissemination, to have pesticidal activity, why then are there so few references to their use in plant protection in the literature before +1600? One possibility is that the practices were so localised that they never attracted the attention of the *literati* and made their way into the written record. Another possibility is that items of information on pesticidal plants are scattered so widely in minor Chinese literary compilations that they are extremely difficult to identify and collect. What we have assembled so far probably represents a small fraction of the material that is really available.

The *Thu Nung Yao Chih* is also important as a repository of information on potentially useful plants which warrant further study, for example, the isolation and identification of active principles and the determination of their mode of action. The results may provide novel structures which could lead to the synthesis of new pesticides that are effective and oecologically benign. Knowledge of their mode of action may also yield new principles of pest control that are hitherto unknown. Based on life-long experience in the study of pesticidal plants from different parts of the world that have been brought to the United States, Martin Jacobson had recently recommended seven plants for development and commercialisation as insect control agents.[a] Three out of the seven are plants directly listed or closely related to those listed in our compilation: *Acorus calamus* (*chhang phu*[1], no. 4), *Artemisia tridentata* (related to *tshao hao*[2], no. 35, and *ai hao*[3], no. 37), and *Azadirachta indica* (*lien*[4] tree, no. 27). These examples certainly indicate that the potential of many of the lesser known plants we have described deserves to be examined, and if possible, exploited to help tip the scale in favour of man in his struggle against invertebrate pests.

(2) BIOLOGICAL PEST CONTROL

When after two decades of isolation, China resumed cultural and scientific exchanges with the West in the early 1970's, one of the more striking impressions brought home by visiting scientific delegations from America, was the extent to which biological methods were being used in China for the protection of agricul-

---

[a] Jacobson (1), 1983.

[1] 菖蒲　　　[2] 草蒿　　　[3] 艾蒿　　　[4] 楝

tural crops against insect pests.[a] Indeed, biological control, in all its manifestations,[b] is probably being practised in China today on a scale and with an intensity unmatched anywhere else in the world. This seemingly modern development may, in fact, be regarded as the revival of a very ancient tradition, since the concept of biological control has had a remarkably long history in China. We have already, in earlier volumes,[c] referred to the account of the control of plant pests by an insect predator in the *Nan Fang Tshao Mu Chuang*,[1] which probably represents the very first record of the use of biological control in any literature. But before we take up this story in detail, let us attempt to gain an appreciation of the intellectual milieu from which this innovation had sprung by examining what ancient and medieval Chinese writers have written about the interaction between insects and their biological enemies in the natural environment.

(i) *Insect pests and natural enemies*

The *Nan Fang Tshao Mu Chuang*, a discourse on tropical and subtropical botany, is reputed to have been written by *Hsi Han*[2] in +304. But the recognition that certain insects feed on crop plants, and that some insects prey on other insects can be traced back to records from a much earlier age. References to both types of observations can be found in the *Shih Ching*[3] (Book of Odes), a compilation of folk songs and ceremonial odes collected from the −9th to the −5th centuries. For example, in the *Ta Thien*[4], a poem on the cultivation of grain, the farmer is urged to 'drive away the *ming*[5] and *thê*[6], as well as the *mao*[7] and *tsê*[8] (*chü chhi ming tê, chi chhi mao tsei*[9]) from his field, although nowhere does the poem instruct how this was to be done. According to the commentary of Mao Hêng[10] the four classes of insects are[e]:

> *Ming:* insects that eat the heart of seedlings, *shih hsin yüeh ming*.[11]
> *Tê:* insects that eat leaves, *shih yeh yüeh tê*.[18]
> *Mao:* insects that eat roots, *shih kên yüeh mao*.[19]
> *Tsei:* insects that eat nodes, *shih chieh yüeh tsei*.[20]

---

[a] Robert Metcalf & Arthur Kelman, report on 'Plant Protection' in L. Orleans (1), pp. 319–25.

[b] For a comprehensive overview of current biological control activities in China, see Phu Chê-Lung[12] (1).

[c] Cf. Vol. 1, p. 118; Vol. 2, p. 258. Cf. also p. 447 above.

[d] In the preparation of this chapter, we have been aided greatly by the monographs on the history of entomology in China by Chou Yao[13] (1, 2) and Tsou Shu-Wên[14] (1), as well as reviews by Liang Chia-Mien[15] & Phêng Shih-Chiang[16] (1, 2).

[e] The text according to Mao Hêng has been regarded as the authorised version of *Shih Ching* since it appeared in the former Han dynasty (c. −200). Mao's commentaries are commonly referred to as *Mao Chhuan*[17]. Of the four types of insects listed here only *ming*[5] has retained the meaning it originally had in the ancient world. Tsou Shu-Wên[14] (1) p. 19, has concluded that in the *Shih Ching*, *ming*[5] represents boring caterpillars (e.g. *Chilo* spp), *tê*[6] locusts, *mao*[7], crickets and *tsei*[8], armyworms (e.g. *Leucania* or *Pseudolethia* spp.).

[1] 南方草木狀	[2] 嵇含	[3] 詩經	[4] 大田	[5] 螟
[6] 螣	[7] 蟊	[8] 賊	[9] 去其螟螣，及其蟊賊	
[10] 毛亨	[11] 食心曰螟	[12] 蒲蟄龍	[13] 周堯	[14] 鄒樹文
[15] 梁家勉	[16] 彭世獎	[17] 毛傳	[18] 食葉曰螣	[19] 食根曰蟊
[20] 食節曰賊				

This passage is noteworthy in that it attemps to identify insect pests by the parts of the cereal plant that they attack, but of greater interest to us, in the current context, is what presumably is the earliest record of the interaction between an insect predator and its prey. It occurs in the third stanza of the *Hsiao Yüan*[1], a poem in which a minor feudal lord admonishes his followers to maintain their virtue and propriety:

> In the midst of the plains there is pulse,
> And the common people gather it.
> The mulberry insect bears an heir,
> But the sphex carries it away,
> Teach and train your offspring,
> And they will be good as you are yourself.[a]

The meaning of this stanza is clear enough. The first couplet, *Chung yüan yu shu, Shu min tshai chih* and the second couplet *Ming ling*[b] *yu tzu, Kuo lo*[c] *fu chih*, form a complementary pair. The sphex carries away the mulberry worm, just as the common people harvest the pulse (i.e. soybeans, *shu*[8]), to be stored away as food for a rainy day. These are cited as examples of thrift and foresight, qualities which, no doubt, one would like to inculcate in one's children.

And yet what was obviously a reference to a simple predator/prey relationship among insects was given a bizarre intrepretation in the Han [9] dynasty that has befuddled the true meaning of this passage for two thousand years. As the Ju Chia[10], the Confucian school, gained ascendency at the Han court, there was a nation-wide effort to collate and systematise all the literature that had survived the Chhin[11] revolution. During this process, the collection of ancient folk-songs and odes we now call the *Shih Ching* (Book of Odes), was reaffirmed as sacred literature and ranked in honour and prestige with the other books of the Confucian classics. While it was fortunate that this action no doubt helped greatly to preserve the *Shih Ching* faithfully for posterity, at the same time it is unfortunate that as a venerated

---

[a] Tr. Legge (8) p. 334, mod. The original text reads as follows:

> *Chung yüan yu shu*[2]
> *Shu min tshai chi*[3]
> *Ming ling yu tzu*[4]
> *Kuo lo fu chih*[5]
> *Chiao hui êrh tzu*[6]
> *Shih ku ssu chih*[7]

[b] In modern usage *ming ling* is the general term for the larval stage of the Lepidoptera, butterfly and moths. According to Mao Hêng's commentary, the specific insect referred to in this passage is the mulberry caterpillar. Considering the antiquity of sericulture in China, such a pest must have been a common sight to the people at the time when this poem was actually composed.

[c] Chu, H. F.[12] & Kao, C. S.[13] (*1*) have identified the *kuo lo* as members of the family *Trypoxilidae*. Chou Yao (*2*) p. 98, however, after evaluating the evidence in the classical literature, concludes that in the context of this poem, *kuo lo* could also refer to members of the *Eumenidae* and the *Sphecidae*. For convenience we have followed Legge's translation of *kuo lo* as sphex.

[1] 小宛	[2] 中原有菽	[3] 蔗民探之	[4] 螟蛉有子	[5] 蜾蠃負之
[6] 敎誨爾子	[7] 式穀似之	[8] 菽	[9] 漢	[10] 儒家
[11] 秦	[12] 朱弘復	[13] 高金聲		

classic it became subject to the concerted effort of Confucian scholars to thrust upon it a meaning consistent with orthodox views of morality and piety. Often their interpretations were so contrived as to twist the original sense of a poem beyond recognition.

For the passage at hand, the trouble may be said to have originated with Yang Hsiung[1] (+5), who in his 'Adomonitary Sayings' (*Fa Yen*[2]) wrote: 'The offspring of the mulberry worm, in a paralysed state, is picked up by the sphecid wasp, which blesses it with the chant 'Be like me, Be like me'. After a while, the larva assumes the form of the wasp.'[a] This statement undoubtedly derived its inspiration from the prevailing view at that time of the generation of life from other forms of life or inanimate matter, which may be summed up as the concept of *hua shêng*[4] (born of change).[b] A prominent exponent of this theory was Chuang Tzu[5]. In the famous passage from the eighteenth chapter of the Book of Chuang Tzu that we have quoted in Volume 2,[c] several examples of such alleged transformations are given. It is interesting to note that in the parallel passage in *Lieh Tzu*[6] that we have alluded to earlier,[c] we find the additional phrase 'the small *thin-waisted* wasp exists only as males[d] (*chhun hsiung chhi ming chih fêng*[7])'. Elsewhere Chuang Tzu also said 'the narrow-waisted ones undergo transformation[e] (*hsi yao chê hua*[8])' and 'the small wasp cannot transform the large bean caterpillar[f] (*pên fêng pu nêng hua huo shu*[9])', but presumably it can transform the small mulberry worm. These passages gave a measure of authority to Yang Hsiung's comment, ostensibly consistent with the observation that after the young mulberry worm is carried away and sealed in the wasp's nest, a few days later a new wasp would emerge and fly out. The inference naturally was that the mulberry larva had been transformed into a wasp.

A century later a direct rationalisation for this interpretation was attempted by Hsü Shen[10]. In the *Shuo Wên Chieh Tzu*[11] (Analytical Dictionary of Characters), +121, he says:

---

[a] Tr. auct. *Fa Yen*, ch. 2, *Hsüeh Hsing Phien*[12], from '*Ming ling chih tzu, I êrh fêng kuo lo. Chu chih yüeh: lei wo, lei wo. Chiu tsê hsiao chih i*[3].'

[b] Tsou Shu-Wên[13] (*1*), p. 26, noted that the concept of *hua shêng*[4] has been a major impediment to a systematic understanding of the biology of insects by ancient Chinese writers. The expression was adopted by the Confucian school, and used in the Appendices to the *I Ching*[14], commonly known as the *Shih I*[15], the Ten Wings. In the second wing, we find the statement, '*Thien ti kan erh wan wu hua shêng*[16]' which Legge (*9*), p. 238, appendix 1 section 2, translated as: 'Heaven and earth exert their influences, and there ensue the transformation and production of all things.'

In the sixth wing, it is stated: '*Thien ti yin yün, wan wu hua chhun, nan nü kou ching, wan wu hua shêng*[17].' Legge's translation (*9*), p. 393, appendix 3, section 2 reads: 'There is an intermingling of the genial influences of heaven and earth, and transformation in its various forms abundantly proceeds. There is an intercommunication of seed between male and female and transformation in its living types proceeds.' Based on this concept, it is thus perfectly understandable that the mulberry worm would be transformed into a wasp.

[c] Cf. Vol. 2, p. 78. See also footnote 1, p. 79.   [d] Tr. auct. *Lieh Tzu*[6], ch. 1, p. 66

[e] Tr. Shih Chün-Chhao[18] (*1*), *Chuang Tzu*[5], ch. 14.   [f] Tr. auct. *Chuang Tzu*[5], ch. 23.

[1] 楊雄   [2] 法言   [3] 螟蛉之子，殪而逢蜾蠃，祝之曰，類我類我，久則肖之矣

[4] 化生   [5] 莊子   [6] 列子   [7] 純雄其名稌蜂

[8] 細腰者化   [9] 奔蜂不能化藿蠋   [10] 許慎   [11] 說文解字

[12] 學行篇   [13] 鄒樹文   [14] 易經   [15] 十翼   [16] 天地感而萬物化生

[17] 天地絪縕，萬物化醇，男女構精，萬物化生   [18] 史俊超

The *kuo lo*[1], that is, *phu lu*[2] is a mud wasp with a small waist. In accordance with the order of Nature, a thin waist indicates that this wasp exists only in the male form, and is unable to bear children. Therefore, we find in the 'Book of Odes' the statement: The mulberry insect bears an heir and the sphex raises it as its own.[a]

Now in this passage, *fu*[3] is stretched to mean not merely to carry or hold, but to train and transform. This view was adopted by Chêng Hsüan[4] (+200) in his edition of the *Mao Shih Chhuan*[5], known also as the *Mao Shih Chhuan Chêng Chien*[6], which became the standard edition of the *Shih Ching* for the next thousand years. In his commentary on this stanza of the *Hsiao Yüan*[7], he states: 'The *phu lu*[2] takes the offspring of the mulberry worm, protects and comforts it, and raises it to become its own son[b] (*Phu lu chhü sang chhung chih tzu, fu chhih erh chhü. Hsü yü yang chih, i chhêng chih tzu*[8]).' This interpretation was further supported by Lu Chi[9] (+3rd century, early Chin dynasty), who in his 'An Elucidation of the Plants, Trees, Birds, Beasts, Insects and Fish described in the *Mao Shih*', i.e. *Shih Ching*, (*Mao Shih Tshao Mu Niao Shou Chhung Yü Su*[10]), wrote: 'The wasp takes the mulberry larva and places it in a crevice among the wood, the hollow of a writing brush or within a bunch of bamboo strips. In seven days the larva is transformed into a wasp, which makes its presence felt by buzzing the chant *Hsiang Wo, Hsiang Wo*[11].'[c]

During the Han dynasty this interpretation gained such wide acceptance and authority that from then onwards until today, *ming ling tzu*[12] has become the literate, idiomatic expression for 'adopted child' in the Chinese language. Throughout the later centuries we continue to find in scholarly works references to the view that wasps with small waists are exclusively male, and that they capture caterpillars to transform them into their own progeny, as shown in the following tabulation:

Author	Date	Title
Chang Hua[14]	+290	'Records of the Investigation of Things', *Po Wu Chih*[15], ch. 4, p. 4*a*.
Kuo Pho[16]	+324	'Explanations and Commentaries on the *Literary Expositor*', *Êrh Ya Chu Shu*[17], ch. 3, p. 9.
Khung Ying-Ta[18]	+642	'Basic Ideas of Mao's *Book of Odes*', *Mao Shih Chêng I*[19], ch. 12, no. 3.
Li Fang[20]	+982	'Imperial Encyclopaedia of the Thai-phing Reign-period', *Thai Phing Yü Lan*[21], ch. 945, p. 7.

[a] Tr. auct. *Shuo Wên Chieh Tzu*, ch. 13, pt. 2.
[b] Tr. auct. Cf. *Mao Shih Chhuan Chêng Chien*[6], *Hsiao Yüan*[7]
[c] Tr. auct. *Mao Shih Tshao Mu Niao Shou Chhung Yü Su*[10], *Han Wei Tshung Shu*[13] edition. p. 7.

[1] 蜾蠃　　[2] 蒲盧　　[3] 負　　[4] 鄭玄　　[5] 毛詩傳
[6] 毛詩傳鄭箋　[7] 小宛　　[8] 蒲盧取桑蟲之子, 負持而去, 煦嫗養之, 以成之子
[9] 陸璣　　[10] 毛詩草木鳥獸蟲魚疏　　[11] 象我象我　[12] 螟蛉子
[13] 漢魏叢書　[14] 張華　　[15] 博物志　　[16] 郭璞　　[17] 爾雅注疏
[18] 孔穎達　[19] 毛詩正義　[20] 李昉　　[21] 太平御覽

Lu Tien[1]	+1102	'New Additions to the *Literary Expositor*', *Phi Ya*[2] (or 'Classification of Things According to Their Natures', *Wu Hsing Mên Lei*[3]), ch. 11, p. 280.
Chu Hsi[4]	+1177	'The *Book of Odes* With Commentaries' *Shih Chi Chhuan*[5], ch. 12, p. 138.
Chu Kung-Chhien[6]	+1200	'Integrated Commentaries on the *Book of Odes*', *Shih Ching Su I Hui Thung*[7], ch. 12, p. 7.

Although Chu Hsi[4] brought a refreshing open-mindedness to his commentaries on the *Shih Ching*, in his discussion of the *Hsiao Yüan*[8] he embraced without question the traditional view that the wasp was able to transform the mulberry larva into its own offspring. By putting his prestige behind this interpretation, he undoubtedly made it that much more difficult for opposing evidence on the true nature of this phenomenon, which had been gathering steadily during the preceding centuries, to be recognised and accepted. For such evidence has had at least as long a history as Yang Hsiung's mistaken original idea. In the *I Lin*[9] (Forest of Symbols of the *Book of Changes*), which dates about −40, Chiao Kan[10] declared: 'The sphex produces an offspring. With deep eyes and dark features, it closely resembles its mother. Even when right next to it, no one would want to hold it.'[a] This passage is significant in that almost forty years before the publication of Yang Hsiung's *Fa Yen*[12], it was already known that the sphecid wasp does bear an offspring and therefore must exist in the female form. But this passage was either unknown to, or ignored by, Yang Hsiung[13], Hsü Shen[14] and Chêng Hsüan[15].

The first serious challenge to the orthodox interpretation was launched by the Taoist physician and naturalist, Thao Hung-Ching[16], who, in the *Ming I Pieh Lu*[17] (Informal Records of Famous Physicians), +502, thus described his personal observations on this phenomenon.

> The sphecid wasp (*i ong*[18]) occurs in many varieties. Although commonly called the mud wasp (*thu fêng*[19]), it does not necessarily have to burrow in the earth, but may use mud to complete its nest. I have seen one type of wasp, black in colour, with a tiny waist, carry mud to seal crevices in houses or furniture to form a shelter. One favourite site is the hollow of a bamboo tube. It lays an egg the size of a small grain in the cavity, fills it with more than ten captured young grass-spiders, and seals the opening. The spiders will provide food for the growing wasp larva. The type that uses the reed pipes also captures young caterpillars that feed on grass. In the 'Book of Odes' it is stated: 'The mulberry insect bears an heir, And the sphex carries it away.' It has been implied that the thin-waisted wasp does not exist as a female. It takes the mulberry insect, trains and transforms it into its own offspring. Such a view is entirely erroneous.[b]

---

[a] *Kuo lo shêng tzu, shêng mu hei chhou, ssu lei chhi mu, sui huo hsiang chiu, chung mo chhü chih*[11]. *I Lin*, ch. 9, p. 66. tr. auct.
[b] Tr. auct. This passage is quoted by Li Shih-Chen[20] in the entry on *i ong*[18] in the *Pên Tshao Kang Mu*[21] ch. 39.

[1] 陸佃	[2] 埤雅	[3] 物性門類	[4] 朱熹	[5] 詩集傳
[6] 朱公遷	[7] 詩經疏義會通		[8] 小宛	[9] 易林
[10] 焦贛	[11] 蜾蠃生子，深目黑醜，似類其母，雖或相就，衆莫取之			
[12] 法言	[13] 楊雄	[14] 許愼	[15] 鄭玄	[16] 陶弘景
[17] 名醫別錄	[18] 蠮螉	[19] 土蜂	[20] 李時珍	[21] 本草綱目

Thao's observation, however, apparently received little attention in orthodox Confucian circles. It took another three and a half centuries before additional information on this problem was recorded when Tuan Chhêng-Shih [1] (+863) made the following statement on the sphecid wasp (*i ong*) in his *Yu Yang Tsa Tsu* [2] (Miscellany of the Yu-Yang Mountain):

I often encounter such insects in my study. I find them nesting in crevices among the books or in the hollow of a writing brush. They can be detected by the buzzing sound they produce. When I open the nest for examination, I find it packed with little spiders about the size of the common fly eaters (*ying hu* [3]). The nests are sealed with mud. It is only then that I realise that such wasps do not just carry away mulberry worms.[a]

Although Tuan Chhêng-Shih did not refer to Thao Hung-Ching's earlier work, it is obvious his experience favored Thao's explanation over the prevailing orthodox view. Less than a hundred years later Thao Hung-Ching's contribution was recognised by Han Pao-Shêng [4] in a passage on this problem in the *Shu Pên Tshao* [5] of +935 (Pharmacopoeia of the State of ⟨Later⟩ Shu):

According to the commentaries on the 'Book of Odes', 'the *ming ling* is the mulberry worm, and the *kuo lo* is the sphecid wasp' (*ming ling, sang chhung yeh*; *kuo lo, phu lu yeh* [6]), and the sphecid wasp trains and transforms the worm into its own offspring. Other insects may also be carried into the wasp's lair. After a few days the captive becomes a wasp and flies away. I have waited to reopen the nest after it has been sealed. I found an egg about the size of a small grain on the paralysed worm, just as Thao Hung-Ching had previously reported. It would appear that the commentators know only the overall picture, but are ignorant about the details. Such wasps are very common. They may operate singly or in pairs. They make nests everywhere, in crevices in the mud, among stones, and in wood or bamboo cavities.[b]

This point of view was reiterated and reinforced by many writers in the Sung and Ming dynasties as can be seen by the following list of supportive references:

Author	Date	Title
Phêng Chhêng [7]	+1063	'Declamations of a Literary Practioner', *Mo Kho Hui Hsi* [8], ch. 5, p. 1.
Khou Tsung-Shih [9]	+1116	'Dilations upon Pharmaceutical Natural History', *Pên Tshao Yen I* [10], ch. 17, p. 15.
Lo Yüan [11]	+1184	'Wings for the *Literary Expositor*', *Êrh Ya I* [12], ch. 26, p. 3.
Tai Thung [13]	+1275	'The Six Classes of Characters Explained', *Liu Shu Ku* [14], ch. 20, p. 6.

[a] Tr. auct. *Yu Yang Tsa Tsu*, ch. 17, p. 6a. Chou Yao (1), p. 61 has cited this passage in favour of the scientific point of view. Actually, its meaning is ambiguous. It just says that the wasp carries away spiders as well as mulberry worms. Tsou Shu-Wen (1), p. 100 has suggested that traditionally it has been interpreted to mean that the sphex can transform the spider as well as the mulberry worm into its own offspring.

[b] Tr. auct. The *Shu Pen Tshao* is long lost. This passage is from *PTKM*, ch. 39, p. 2232.

[1] 段成式　　[2] 酉陽雜俎　　[3] 蠅虎　　[4] 韓保昇　　[5] 蜀本草
[6] 螟蛉桑蟲也，蜾蠃蒲盧也　　[7] 彭乘　　[8] 墨客揮犀　　[9] 寇宗奭
[10] 本草衍義　　[11] 羅源　　[12] 爾雅翼　　[13] 戴侗　　[14] 六書故

Chhê Jo-Shui[1]	+1300	'Notes by a Beriberi Patient', *Chiao Chhi Chi*[2], ch. 2, p. 9.
Wang Chün-Chhuan[3]	+1538	'Elegant Discourses', *Ya Shu*[4], ch. 55, pp. 27a, 28.
Thien I-Hêng[5]	+1572	'Daily Jottings' *Liu Chhing Jih Cha*[6], ch. 31, pp. 10–11.
Huangfu Phang[7]	+1582	'New Talks on Old Explanations'[a], *Chieh I Hsin Yü*[8] .
Thao Fu[9]	Ming	'Random Notes Written at Dusk', *Sang Yü Man Chi*[10], pp. 15–16.
Li Shih-Chen[11]	+1596	'The Great Pharmacopoeia', *Pên Tshao Kang Mu*[12], ch. 39, Insects.
Mao Chin[13]	+1639	'An Elaboration of the Essentials of the 'Elucidation of the Plants, Trees, Birds, Beasts, Insects and Fishes mentioned in the *Book of Odes*' edited by Mao[14,b], *Mao Shih Tshao Mu Niao Shou Chhung Yü Su Kuang Yao*[15], ch. 2, part 1.

Much useful information providing important details in the main body of evidence against the Confucian mainstream tradition was added to the literature by later authors. For example, Phêng Chhêng[16](+1063) reported that three types of wasps were involved.[c]

Those that build nests in crevices in walls are *kuo lo*[17], those that burrow in the earth are *i ong*[18], and those that inhabit the hollows in books and writing brushes are *phu lu*[19]. The names are different; so are their sizes and life histories. *Kuo lo*[17] and *phu lu*[19] usually capture mulberry caterpillars and tiny spiders (*chih chu*[20]), while *i ong*[18] prefers spiders (*hsiao hsiao*[21]) and crickets. The captives are killed, their legs removed, and they are deposited in the crevice. After the egg is laid, the opening is sealed with mud. A number of days afterwards, an adult wasp emerges to fly away. By then all the captives have been consumed.[d]

Five hundred years later, in the Ming Dynasty, Wang Chün-Chuan[22] (+1538) recorded that the larval wasp turned into a pupa before it was transformed into an adult wasp.[e] Thien I-Hêng[23](+1572) found that the wasp exists in both male and female forms and actually observed the male and female in copulation.[f] Finally, Huang Fu-Phang[24] (+1582) compared the deposition of an egg by the wasp on the paralysed prey to the deposition of the egg by a parasitic fly on the silkworm larva,[g]

---

[a] The relevant passage is quoted in Chou Yao (*1*), p. 61. We have been unsuccessful in obtaining access to the original source.

[b] *Mao Shih*[14] is the commonly accepted name for Mao Hêng's version of the Book of Odes.

[c] *Mo Kho Hui Hsi*, ch. 5, p. 1.

[d] Tr. auct.

[e] *Ya Shu*, ch. 55, p. 28 'after the larva has consumed all the caterpillars and spiders stored in the nest, it turns into a pupa. In a few days, a mature wasp will emerge' (tr. auct.).

[f] *Liu Chhing Jih Cha*, ch. 31, p. 11.

[g] *Chieh I Hsin Yü*, quoted by Chou Yao (*1*), p. 61.

[1] 車若水	[2] 腳氣集	[3] 王浚川	[4] 雅述	[5] 田藝蘅
[6] 留青日扎	[7] 皇甫汸	[8] 解義新語	[9] 陶輔	[10] 桑榆漫記
[11] 李時珍	[12] 本草綱目	[13] 毛晉	[14] 毛詩	
[15] 毛詩草木鳥獸蟲魚疏廣要		[16] 彭乘	[17] 蜾蠃	[18] 蠮螉
[19] 蒲蘆	[20] 蜘蛛	[21] 蠮螉	[22] 王浚川	[23] 田藝蘅
[24] 皇甫汸				

a phenomenon, no doubt, well known to those who had been involved in silkworm cultivation for centuries.

These records were duly considered by Li Shih-Chen[1], who, in his entry on the *i ong* in the *Pên Tshao Kang Mu*[2] (+1596) summed up the controversy as follows:

Different explanations have been offered for the *i ong*. After reviewing all the evidence, examining the egg, and observing pairs of wasps flying hither and thither, I am convinced that they exist both as males and females. The views of Thao[3] and Khou[4] are correct; those of Li[5] and Su[6] are false.[a]

But the diehard orthodox school did not wish to be convinced. In a summary of the literature on the *kuo lo*[14], Fêng Ying-Ching[7] in his 'Commentaries on Names and Things in the *Book of Odes*' (*Liu Chia Shih Ming Wu Su*[8]), +1605, noted that Chêng Chhiao[9] in the Sung dynasty had already supposedly refuted Thao Hung-Ching's[10] observation that the mulberry worm was simply used as a source of food for the newly hatched wasp larva. The fact that by opening the nest one could observe the presence of an egg on the body of the dying worm meant simply to Chêng Chhiao that the transformation was still progressing and therefore incomplete. When one examined the nest after the new wasp had flown, one could see that the empty outer skeleton is in the shape of a snail's shell. It shows clearly that the worm had been transformed. Chêng Chhiao's conclusion was, of course, completely wrong, but his prestige was such that his interpretation was taken seriously for generations to come. It also illustrates the pervasive influence of the concept of *hua shêng*[15], which viewed the transformation of a mulberry worm into a sphecid wasp simply as a manifestation of a general natural phenomenon.

Finally, the great scholar Wang Fu-Chih[16] in the early Chhing dynasty (+1695) again studied previous records, and made his own observations. He concluded in the *Shih Ching Pai Su*[17] (Little Commentaries on the *Book of Odes*), that:

The wasp captures the caterpillar just as the bee collects nectar; both are for feeding their young. When an animal is newly born it has to rely on its mother for food. Mammals suckle their young. Birds bring food to their chicks. The thin-waisted wasp stores food for its child so that it can feed itself. When the food is exhausted, it flies away.[b]

There the matter should have rested once for all in favour of the scientific point of view. Yet the prestige of the orthodox Confucian school was so great that the erroneous interpretation persisted and continued to surface from time to time in publications in the 19th and 20th century. For example, in a different translation of the *Shih Ching*, Legge rendered this stanza from the *Hsiao Yüan*[18] as follows:

---

[a] *Pên Tshao Kang Mu*[2], ch. 39, section on Insects, Entry on *i ong*. tr. auct. In this passage Thao[3] refers to Thao Hung-Ching[10], and Khou[4] to Khou Tsung-Shih[11]. Li[5] is Li Han-Kuang[12], of the Thang dynasty and Su[6] is Su Sung[13] of the Sung dynasty. Both wrote in defence of the orthodox view:
[b] *Shih Ching Pai Su*[17], ch. 2, p. 9, tr. auct.

[1] 李時珍	[2] 木草綱目	[3] 陶	[4] 寇	[5] 李
[6] 蘇	[7] 馮應京	[8] 六家詩名物疏		[9] 鄭樵
[10] 陶弘景	[11] 寇宗奭	[12] 李含光	[13] 蘇頌	[14] 蜾蠃
[15] 化生	[16] 王夫之	[17] 詩經稗疏	[18] 小宛	

> All o'er the plain they gather beans,
> Which they will sow again.
> The grubs hatched on the mulberry tree,
> The sphex bears off to train.[a]

In this case, Legge was obviously following the orthodox view established by Yang Hsiung[1], Chêng Hsüan[2] and Chu Hsi[3]. Other examples can still be found in recent presentations of the *Shih Ching*, in which the original poems are printed together with their equivalent renditions or explanations in modern vernacular Chinese (*pai hua*[4]), such as those edited by Chung Chi-Hua[5], Hung Tzu-Liang[6], and Hsü Hsiao-Thien[7]. It is a situation strongly reminiscent of the printing of an erroneous story on the front page of a newspaper, after which no amount of retractions and corrections on the back pages of subsequent issues can apparently erase it in the mind of many of the paper's readers.

The wasp is by no means the only insect predator recorded in ancient Chinese literature. Another celebrated example is the praying mantis, *thang lang*[8], which was first mentioned as a predator in *Chuang Tzu*[9] (−290):

Chuang Chou[10] was strolling in the Tiao Ling[11] Park when he noticed a strange bird fly in from the south. Its wings had a span of seven feet, and its eyes were an inch in diameter. The bird's wing touched his forehead and it flew on to rest in a chestnut grove. 'What sort of bird is this?' asked Chuang Chou. 'With wings as large as that it does not keep going, and with eyes as big as that it does not see straight.' He lifted his gown, moved towards it, and waited with his sling at the ready. Then he noticed a solitary cicada so intent upon having found a shady spot, that it forgot to protect itself, and so a mantis already poised, seized it. But the mantis too forgot himself, for the strange bird was right behind him. The bird in turn, absorbed by its success, also forgot to protect itself and could easily have been shot. Chuang Chou was so astonished by what he saw that he exclaimed 'Alas, this is how living things control each other, and how loss can follow gain.'[b]

This fable undoubtedly represents the earliest recognition of the interactions between living things as members of a natural food chain. It became a favourite story for quotation by later authors. Thus we find similar versions of the phenomenon, that is, the praying mantis while preying on the cicada is in turn preyed upon by a bird, repeated in the *Shuo Yuan*[12] (+140),[c] the *Han Shih Wai Chuan*[13] (+160)[d] and the *Wu Yüeh Chhun Chhiu*[11] (+100).[e] But in these versions the strange

---

[a] Legge, (12) p. 254. This translation is in rhyme. It is interesting to compare with the literal rendition by Karlgren, (14): 'In the middle of the plain there is pulse, the common people gather it; the mulberry insect has young ones, the solitary wasp carries them on its back; teach and instruct your sons, then in goodness they will be like you.' It would have been instructive to see the translation of this poem by Arthur Waley, but much to our regret it is apparently one of the pieces he did not include in *The Book of Songs*.

[b] *Chuang Tzu*, ch. 20, *Shan Mu Phien*[15], tr. Shih Chün-Chhao (1), p. 244 and Lin Yü-Thang (1), p. 218, mod.

[c] *Shuo Yüan*[12] (Garden of Discourses) by Liu Hsiang[16], ch. 9, p. 4.

[d] *Han Shih Wai Chuan*[13] (Moral Discourses Illustrating the Han Text of the *Book of Odes*) by Han Ying[17], ch. 10, p. 13.

[e] *Wu Yüeh Chhun Chhiu*[14] (Spring and Autumn Annals of the States of Wu and Yueh) by Chao Yeh[18], ch. 15, p. 20.

[1] 楊雄	[2] 鄭玄	[3] 朱熹	[4] 白話	[5] 鍾際華
[6] 洪子良	[7] 許嘯天	[8] 螳螂	[9] 莊子	[10] 莊周
[11] 雕陵	[12] 說苑	[13] 韓詩外傳	[14] 吳越春秋	[15] 山木篇
[16] 劉向	[17] 韓嬰	[18] 趙曄		

bird is replaced by the commen yellow sparrow. In each case the fable is used to warn a feudal prince of impending danger from another direction, while he is deeply absorbed in preparing for a campaign of conquest against some other State.

A third insect predator mentioned in ancient texts is the spider, *chih chu*[1]. In the *Fu Tzu*[2] (+4th century) we identify the following passage:

The prince of Chin[3], Chung Erh[4], fled the kingdom of Chhi[5] with five attending officials and they made their way through a large swamp. He saw a spider weave a net and pull its silk to catch and devour an insect. The prince halted his retinue to take a closer look at this scene. He remarked to his attendant Chiu Fan[6], 'Although the wisdom of the spider is limited, yet it is able to weave a nest, pull a silken thread and capture an insect for its meal. For all of man's wisdom, he is unable to cast a net over heaven, and pull its strings across the earth, so as to control events in a small area. He is not nearly as wise as the spider. How can we call him man?'[a]

The spider's activity is also described in the *Kuang Chih*[7] (+5th century).

The grass spider lives among the grass and is coloured green. The earth spider lives on the ground. In the spring they move among the grass or over the earth. In the autumn they are found in the grass or under implements. Some spin their silk between fences and catch flies that go near them; the long-legged ones weave under the eaves.[b]

Finally, in the *Lun Hêng*[8] of +82 (Discourses Weighed in the Balance), Wang Chhung[9] wrote:

Look at the spiders, how they knit their webs with the view of entrapping flying insects. How are the transactions of men superior to theirs? Using their brains they work out their selfish and deceitful schemes with the object of acquiring the amenities of wealth and long life, paying no heed to the study of the past or the present. They behave just like the spiders.[c]

The fourth insect predator talked about in ancient literature is the dragon-fly, *chhing thing*[10], which was first mentioned in the *Chan Kuo Tshê*[11], 'Records of the Warring States' (pre-Chhin, i.e. before—220). In a meeting with Duke Hsiang[12] of Chhu[13], Chuang Hsin[14] said: 'See, your Grace, the dragon-fly. With six feet and four wings, it soars above between heaven and earth, and captures flies and mosquitoes to eat them.'[d] It is again described in the *Phi Ya*[18](New Additions to the Literary Expositor), +1102, as follows:

The dragon-fly drinks dew, and has six legs and four wings. The wings are as thin and light as those of the cicada. They feed on mosquito larvae. When it rains they congregate on the surface of the water and remain there with their tails in an upright position.[e]

---

[a] Tr. auct. *Fu Tzu*[2] by Fu Lang[15], *YHSF*, ch. 71, pp. 16a, 17.
[b] Tr. auct. *Kuang Chih*[7] by Kuo I-Kung[16], *YHSF*, ch. 74, p. 50. According to Sugimoto Naojiro (*1*), Kuo I-Kung lived some time between +420 and +520.
[c] Tr. Forke, (4), pt. I, p. 105. *Lun Hêng*[8], Pieh Thung Phien[17].
[d] Tr. auct. *Chan Kuo Tshê*: Chapter on Chuang Hsin speaking to Duke Hsiang of the Chhu Kingdom, *Chuang Hsin Wei Chhu Hsiang Wang Chang*[19], p. 153.
[e] Tr. auct. *Phi Ya*[1], ch. 11, p. 282.

[1] 蜘蛛	[2] 符子	[3] 晉	[4] 重耳	[5] 齊
[6] 咎犯	[7] 廣志	[8] 論衡	[9] 王充	[10] 蜻蜓
[11] 戰國策	[12] 襄	[13] 楚	[14] 莊辛	[15] 符朗
[16] 郭義恭	[17] 別通篇	[18] 埤雅	[19] 莊辛謂楚襄王章	

Predators are not the only natural enemies of insects recognized in early Chinese literature. The fact that insects may harbour parasites was first mentioned by Lieh Tzu[1] (−300) who said: 'Among the river banks there lives a tiny insect. It is called *chiao ming*[2]. They swarm and congregate at the eyebrow of a mosquito.'[a] Similar statements have been made by other writers, for example by Yen Ying[3] (−6th century): 'By the Eastern sea there lives an insect which makes its nest at the eyebrow of a mosquito. A swarm may suckle and fly repeatedly without disturbing the host. I do not know its name, but the fishermen call it *chiao ming*[2].'[b] Ko Hung[4] (+320) also wrote: 'The *chiao ming*[3] resides in the eyebrow of the mosquito.'[c] Again Tungfang Shuo[6] (−1st century) said: 'In the south some mosquitoes harbour flying insects under their wings. People with sharp eyes can see them.'[d]

Chou Yao[8], in his *History of Entomology in China*, has suggested[e] that *chiao ming*[3] are tiny mites, and the insects under the mosquito's wings are biting midges. In the light of current knowledge on mites that parasitise mosquitoes, the suggestion appears to be eminently plausible. Some of these mites are a bright orange colour so that even without the aid of a magnifying device they can be seen by the human eye.[f] Indeed, mites are probably one of the major factors that control the population of mosquitoes in their natural environment.

The next example occurs about a thousand years later (+1582) in the *Chieh I Hsin Yu*[14] (New Talks on Old Explanations): 'Today, those who rear silkworms are familiar with fly predators which lay their eggs in the silkworm larvae. After a while the egg develops into an adult fly which emerges by breaking through the cocoon.[g] There is no question that we have here an authentic account of a silkworm parasite. It is interesting to note that the first observation of an insect parasite in the West occurred at about the same time. In +1602 Ulysses Aldrovandi observed cocoons of the internal parasite *Apanteles glomeratus* (L.) in the integument of its host, the cabbage caterpillar *Pieris rapae* (L.), but he thought they were eggs.[h] It remained for Antonio Vallisneri in +1706 to interpret this phenomenon correctly and to extend the interpretation to several other parasites that he had discovered.[h] In this

[a] Tr. auct. *Lieh Tzu*[2], *Thang Wên Phien*[13], ch. 5, p. 98.
[b] Tr. auct. *Ying Tzu Chhun Chhiu*[10], ch. 8, p. 11. Yen Ying died about −500 but the book was probably compiled by later writers during the +1st century.
[c] *Chiao ming thun wên mei chih chung*[5], *Pao Phu Tzu*[11], ch. 3, p. 27, tr. auct.
[d] *Nan fang wên i hsia yu hsiao fei chhung yeh, ming mu chê chien chih*[7], *Shen I Chi*[12], Chou Yao (2), pp. 100–1 tr. auct. Original not consulted.
[e] Chou Yao (2), *Chung Kuo Khun Chhung Hsüeh Shih*[9], p. 100–1. He suggested that the midges are members of the *Ceratopogonidae* but did not attempt to characterise further the mites.
[f] Lanciani (1) has shown that natural populations of the mosquito *Anopheles crucians* are often heavily parasitised by water mites, *Arrenurus* sp. He drew attention (2) to the presence of several mite larvae, easily detected by their bright orange colour, feeding on the neck of a mosquito as it engorges the blood of a human victim, in a dramatic picture published in *Audubon* (July 1979).
[g] Tr. auct., quoted in Chao Yao (1), p. 61.   [h] Cf. Silvestri, (1), tr. Rosenstein, p. 288.

[1] 列子	[2] 蟭螟	[3] 晏嬰	[4] 葛洪
[5] 蟭螟屯蚊眉之中		[6] 東方朔	[7] 南方蚊翼下有小飛蟲焉，明目者見之
[8] 周堯	[9] 中國昆蟲學史	[10] 嬰子春秋	[11] 抱补子
[12] 神異記	[13] 湯問篇	[14] 解義新語	

achievement he might well have been guided by the work of Antonie van Leeuwenhoek who in +1701 had already discussed and illustrated a parasite of a sawfly from willows.[a]

(ii) *The citrus ant story*

Having surveyed the observations in early Chinese records on the interactions between insects and their natural enemies, we may now return to the account by Hsi Han[1] on the use of ants for the control of insect pests on citrus fruits. It occurs in the entry on the mandarin orange (*kan*[2]) in ch. 3 of the *Nan Fang Tshao Mu Chuang*[3] (Records of the Plants and Trees of the Southern Regions), which has recently been given an admirable English translation by Li Hui-Lin[4]. We quote:

> The *kan*[2] is a kind of orange with an exceptionally sweet and delicious taste. There are yellow and red kinds. The red ones are called *hu kan*[5] (jar orange). The people of Chiao-Chih[6] sell in their markets ants in bags of rush matting. The nests are like silk. The bags are all attached to twigs and leaves which, with the ants inside the nests, are for sale. The ants are reddish-yellow in colour, bigger than ordinary ants. In the South, if the *kan* trees do not have this kind of ant, the fruits will all be damaged by many harmful insects, and not a single fruit will be perfect.[b]

This account, dated at +304, has been cited by several Western[c] and Chinese[d] scholars as the earliest reference in any literature to the use of a biological agent for the control of a plant pest. Within the last century, however, questions have been raised on whether the *Nan Fang Tshao Mu Chuang* was indeed written by Hsi Han[1] or compiled later by unknown authors in the Thang to Sung period, thus casting doubt on the date when this practice was first recorded. We have elaborated on the authorship issue earlier[e] and will touch upon it again later on in this chapter. In the meantime let us continue to document the application and expansion of this technology in the centuries that follow Hsi Han's report. We find the next entry of interest to us provided by Tuan Chhêng-Shih[7] in the *Yu Yang Tsa Tsu*[8] (Miscellany of the Yu-Yang Mountains), +863, of the Thang dynasty. He said:

> In Ling-Nan[9] we find ants which are larger than the horse ants (*ma i*[10]) of Shensi. These ants make their nests in the orange trees and move about on the surface of the fruits as they

---

[a] van Leeuwenhoek, A (1), pp. 786–99.

[b] Li Hui-Lin[4] (12), p. 118–19, no. 63. Kan is identified as *Citrus reticulata* Blanco. Formerly a prefecture of Chiao-Chou[11], Chiao-Chih[6] is now the Hanoi region of Vietnam. During the Three Kingdoms period and the Chin dynasty, Chiao-Chou[11] comprised Southern Kuangsi, Southwestern Kuangtung and Vietnam south to Cape Veralla.

[c] Swingle (13), p. 98; Sarton (16), p. 99, Konishi M. & Ito, Y. (1), p. 4; Klemm (1), p. 121; see also Vol. 1, p. 118, Vol. 2, p. 258.

[d] Chou Yao (1), p. 48; Chhen Shou-Chien (1), p. 401; Anon. (257); Phu Chê-Lung (1), p. 1.

[e] Pp. 447 ff. above.

[1] 嵇含	[2] 柑	[3] 南方草木狀	[4] 李惠林	[5] 壺柑
[6] 交趾	[7] 段成式	[8] 酉陽雜組	[9] 嶺南	[10] 馬蟻
[11] 交州				

develop. Therefore, the skin (of the fruit) is thin and smooth. Often the fruits are found actually inside the nests. When they are picked in the depth of winter, their flavour is far superior to that of ordinary oranges.[a]

This report was corroborated by Liu Hsün[1] shortly afterwards in a passage in the *Ling Piao Lu I*[2] (Strange Things Noted in the South), +890, or perhaps early in the +10th century:

In Ling-Nan[3] there are many types of ants. Bags of rush matting containing ant nests are sold in the market. The nests are like sacs of thin silk and are intertwined with twigs and leaves. The ants inside are sold as part of the nest. These are yellow, larger than ordinary ants, and have long legs. It is said that in the South orange trees which are free of ants will have wormy fruits. Therefore, the people race to buy nests for their orange trees.[b]

Later in the 10th century we find this passage by Yüeh Shih[4] in the *Thai-Phing Huan Yü Chi*[5] (General Description of the World during the Thai-Phing Reign-period), +985:

In Tshang-Wu[6] there is a local saying, 'In this prefecture, many *kan*[7] and *chü*[8] oranges are eaten by black ants. The people buy yellow ants and place them in he trees, causing the two

[a] Tr. auct., Pt. 1, ch. 18, Woody Plants (*Chhien Chi, Chüan Shih-Pa, Tshao Mu Phien*[9]). Cited in *TSCC*, ch. 226, *Chi Shih, Kan Pu*[10], p. 36. During the Thang dynasty Lingnan was a major region of the country, a Tao[11] that comprised present day Kuangtung, Kuangsi and Northern Vietnam.

[b] Tr. auct., vol. 3, p. 11. in *Tshung Shu Chi Chhêng*[12] edition. The substance of this passage is identical with that in the account of citrus ants from the *Nan Fang Tshao Mu Chuang*. The close resemblance between them has led Ma Thai-Lai[13](1) to declare *Ling Piao Lu I* as the source of the ant passage in the *Nan Fang Tshao Mu Chuang*, and to cite it as one of the pieces of evidence in support of the view that the latter is a forgery compiled in the late Sung dynasty. Actually, after comparing the two parallel passages, one could just as easily infer that the account in the *Ling Piao Lu I* was copied from the *Nan Fang Tshao Mu Chuang*. Indeed, as pointed out by Phêng Shih-Chiang[14] (2), it would make considerably more sense to do so. As it was a little-known work with limited circulation during the Thang dynasty, Liu Hsün could have copied from *Nan Fang Tshao Mu Chuang* liberally at will without the risk of being discovered. On the other hand, by the late Sung, when printing had greatly enhanced the fidelity and the circulation of all books, and *Ling Piao Lu I* was already a well-known work, extensive copying from it would certainly have aroused suspicion. And yet such major Sung collectors and bibliophiles as Yu Mao[15], Chhen Chên-Sun[16], Tso Kuei[17] and others accepted its authenticity without question, indicating, therefore, that they had independent reason for their confidence.

This passage has been cited by Gaines Liu (1) in 1939 as the 'first written record' of the use of a beneficial insect as a biological control agent (p. 24). Liu has been cited in turn by H. L. Sweetman (2) in 1958 (p. 2) and by Simmonds, Franz & Sailer (1) in 1976 (p. 20). It is clear to us now that even if the *Nan Fang Tshao Mu Chuang* were to be proved unequivocally a work compiled after the Thang dynasty, the reference in *Yu Yang Tsa Tsu* (cf. above) would still take precedence over this passage as the 'first written record' of biological control. While we are discussing this passage, we might also take this opportunity to clear up an ambiguity present in it as quoted by Chou Yao[1] (1), p. 48. The relevant statement is the next to last sentence and reads as follows: '*Yün nan chung kan chü shu, wu i chê shih to chu*'[19]. It is open to two interpretations, depending on the way the sentence is punctuated, viz. Yün-nan chung[20] i.e. in Yunnan (province), or Yün nan-chung[21] i.e. it is said that in the South. Chou Yao subscribed to the former interpretation, while Shih Shêng-Han[22] followed the latter. In his annotated edition of the *Nung Chêng Chhüan Shu* of 1979, Shih Shêng-Han (8), p. 825, has punctuated this sentence to read: '*Yün ⟨Nan chung kan chu shu, wu i chê shih to chu⟩*'[23]. It seems to us Shih Shêng-Han's interpretation is the more reasonable one, since there is no evidence that the cultured citrus ant has been used at all in any part of Yunnan at any time.

[1] 劉恂	[2] 嶺表錄異	[3] 嶺南	[4] 樂史	[5] 太平寰宇記
[6] 蒼梧	[7] 柑	[8] 橘	[9] 前集,卷十八,草木篇	
[10] 紀事,柑部	[11] 道	[12] 叢書集成	[13] 馬泰來	[14] 彭世獎
[15] 尤袤	[16] 陳振孫	[17] 左圭	[18] 周堯	
[19] 雲(云)南中柑橘樹無蟻者實多蛀			[20] 雲南中	[21] 云,南中
[22] 石聲漢	[23] 農政全書	[24] 云⟨南中柑橘樹無蟻者,實多蛀⟩		

types of ants to fight each other. As the black ants are killed, the oranges are then left unmolested to mature.'[a]

Tshang-Wu is modern Wuchow[1] in Kuangsi Province. This is the first time that a specific location where yellow ants are used to control citrus pests was mentioned in print. Another specific location where this control method is practiced is recorded by Chuang Chi-Yü[2] in the *Chi Lei Phien*[3] (Miscellaneous Random Notes), +1130:

In Kuangchow there is a shortage of arable land so people often plant *kan* and *chü* oranges for income; but they suffer considerable losses caused by small insects feeding on the fruits. However, if there are many ants on the trees then the injurious insects cannot survive. Fruit-growing families buy these ants from vendors who make a business of collecting and selling such creatures. They trap them by filling hog's or sheep's bladders with fat and placing them with the cavities open next to the ants' nests. They wait until the ants have migrated into the bladders and then take them away. This is known as *yang kan i*[4], rearing orange ants.[b]

This passage clearly shows that these ants are carnivorous and do not attack vegetation. It is also evident that a significant advance in the technique has been made. Instead of collecting the whole nest, the ants are enticed to entrapment in the bladder from which they could be released to colonise a new tree or orchard.

The next reference is a brief note by Yü Chên-Mu[6] in the *Chung Shu Shu*[7] (Book on Planting Trees), +1401: 'When *kan* and *chü* oranges are eaten by insect pests, place ant nests on the trees; the pests will then be driven away. When one breaks open a large ripe orange a fine mist is released.'[c] The *Chung Shu Shu*[7] was apparently well known during the Ming dynasty among the agricultural literati. The above passage is included in the famous *Nung Chêng Chhüan Shu*[8] (Complete Treatise on Agriculture), +1639 by Hsü Kuang-Chhi[9]. None of the other passages we have quoted are referred to in this well-known treatise, presumably because they were not considered to be in the mainstream of agricultural literature.

An item of considerable interest is reported late in the Ming dynasty by Wu Chen-Fang[10] in the *Ling-Nan Tsa Chi*[11], (Miscellanies from the Southern Regions), +1600:

In Li-chih[12] village, west of Kaochow[13], oranges and pommelos are important secondary crops. Trees planted on several *mou*[14] of land are connected to each other by bamboo strips to

---

[a] Tr. auct. Cited in *TSCC*, ch. 229, *Chi Shih*[5], p. 3*b*. Tshang-Wu was established as a prefecture in the Han dynasty. It was called Wu-Chou in the Thang dynasty. It is located in eastern Kuangsi, bordering on Kuangtung.
[b] Tr. auct. Cited in *TSCC*, ch. 229, *Chi Shih*, p. 5*a*.
[c] Tr. auct., ch. 3, on Fruits, *kuo*[15]. Cited in *TSCC*, ch. 1, p. 21 of *Tshao Mu Tien*[16] Yü Chên-Mu[1] was a supporter of *Hui Ti*[17] in his struggle against his uncle in the early years of the Ming dynasty. *Hui Ti*[17] lost and his uncle ascended the throne in 1404 as Emperor *Chhêng Tsu*[18]. Yü Chên-Mu was beheaded, and declared a traitor. Thus early editions of the *Chung Shu Shu* were attributed to Kuo Tho-Tho[19], a legendary figure of the Thang dynasty. For a discussion on the history of this publication, cf. Shih Shêng-Han[20] (7), p. 63.

[1] 梧州	[2] 莊季裕	[3] 鷄肋篇	[4] 養柑蟻	[5] 紀事
[6] 兪貞木	[7] 種樹書	[8] 農政全書	[9] 徐光啓	[10] 吳震方
[11] 嶺南雜記	[12] 荔枝	[13] 高州	[14] 畝	[15] 果
[16] 草木典	[17] 惠帝	[18] 成祖	[19] 郭橐駝	[20] 石聲漢

facilitate the movement of the large ants which ward off insect pests. The ants build nests among the leaves and branches in the hundreds and thousands. A nest may reach the size of a *tou*[1].[a]

Thus we are now told that the trees are connected by bamboo bridges so that when a nest is established on one tree, the ants can easily spread out to colonise other trees. Soon the whole orchard would be colonised. The last item recorded in the Ming Dynasty is by Fang I-Chih[2] in the *Wu Li Hsiao Shih*[3] (Small Encyclopaedia of the Principles of Things), +1643: 'At Lin-chiang[5] people buy ants and rear them under the orange trees. The ants will control the insect pests.'[b] A publication on oranges in a series entitled 'Selections from Chinese Inherited Agricultural Property' has suggested that Lin-chiang[5] refers to the city of that name in Szechuan.[c] If correct, this is the first indication that the technique was used outside Kuangtung and Kuangsi.

We now come to another brief statement, this time in the *Hua Ching*[7] (Mirror of Flowers), +1688, by Chhen Hao-Tzu[8] in the early Chhing dynasty. In discussing various methods for controlling insect pests, the author indicates that one alternative is: 'To rear ants which eat the insects that prey on oranges.'[d] The next entry was reported at about the same time by Chhü Ta-Chün[9] in the *Kuang-Tung Hsin Yü*[10] (New talks about Kuang-Tung Province), +1700:

In the Kuang region ants proliferate in winter as in summer. There is a large yellow-red ant that lives in the hilly woods, making their nests like those of wild wasps, each of which may be as large as several pints. The local people collect these large ants and feed them. Then the orchard owners buy such nests and place them up on the tree. By connecting the trees with bamboo strips, the ants are able to migrate from tree to tree and ward off injurious insects, which damage the flowers and fruits. This procedure is particularly effective for orange, tangerine and lemon trees, for these are easily infested by caterpillars, which turn into moths and deposit offspring on the trees as young larvae. It is essential to eliminate them so that the trees are not injured. But hand labour is not nearly as efficient as ant power, so horticulturists say that in order to grow flowers one must first rear ants.[e]

This passage suggests that the technique was being used for trees other than citrus, and confirms the practice of connecting the trees with bamboo bridges as described earlier in the *Ling-Nan Tsa Chi*[11]. The last reference of interest in the Chinese

---

[a] Tr. auct., *Lung Wei Pi Shu*[4] edition, ch. 17, pp. 80–1. Kaochow prefecture lies in southwestern Kuangtung. A *mou* equals 6.6 acres and a *tou* is about 1.6 gallons.

[b] Tr. auct., *Tshao Mu Lei*[12], Grasses and Trees, p. 126.

[c] The volume is edited by Yeh Ching-Yüan[13] (1), *Chung Kuo Nung Hsüeh I Chhan Hsüan Chi, Kan Chü, Shang Phien*[6]. It is stated on p. 10 that the above passage describes the use of ants in Lin-chiang[5], Szechuan, to control insects on oranges. As far as we know (private communication, Phu Chê-Lung[14]) there is no indication that this technique is being used in Szechuan today.

[d] Tr. auct., chapter on *Kuo Hua Shih Pa Fa*[15], Eighteen Ways of Training Fruit and Flowers.

[e] Tr. auct., ch. 24, p. 602.

¹ 斗　　　² 方以智　　　³ 物理小識　　　⁴ 龍威祕書　　　⁵ 陷江
⁶ 中國農學遺產選集, 柑橘, 上篇　　　⁷ 花鏡　　　⁸ 陳淏子
⁹ 屈大均　　　¹⁰ 廣東新語　　　¹¹ 嶺南雜記　　　¹² 草木類　　　¹³ 葉靜淵
¹⁴ 蒲蟄龍　　　¹⁵ 果花十八法

literature before the advent of Western scientists on the scene is in the *Nan Yüeh Pi Chi*[1] (Memoirs of the South), +1795: 'There are many kinds of ants in Kuangchow. The nests are sacs of thin silk, interwoven with twigs and leaves, and the people store them in cloth bags and sell them to orange growers to control insect pests.'[a]

The first notice in the Western literature of the use of ants in Kuangtung to control insect damage is a paper by H. C. McCook based on an article in the *North China Herald* of 4 April 1882.[b] Evidently it received very little attention among entomologists and horticulturists at that time. So the first encounter between Western scientists and the citrus ant in the field did not take place until the twentieth century, indirectly as the result of a severe outbreak of citrus canker in the orange groves of Florida in the early 1910s. Walter T. Swingle, a plant physiologist at the Bureau of Plant Industry, U.S. Department of Agriculture in Washington, D.C., was sent to the Far East in 1915 to search for canker-resistant varieties of orange, since most of the varieties grown in the U.S. had originally come from China. In Canton Swingle collaborated with Professor George W. Groff of Lingnan University who, together with his students, made field trips into the countryside at his behest. As Swingle recalled the encounter in 1942, it occurred

... some twenty-five years ago (1918) when a group studying the citrus fruits in the vicinity of Canton for me, under the supervision of Professor George Weidman Groff, found a small village where the inhabitants said their principal business was 'growing ants'. Some of Professor Groff's Chinese student helpers from Lingnan University at Canton made fun of

---

[a] Tr. auct., Li Thiao-Yuan, *Nan Yüeh Pi Chi*, ch. 2, pp. 1–2.

[b] McCook, H.C. p. 263. The quotation is from an article by Dr Magowan of Wênchow on 'Utilization of Ants as Grub-Destroyers in China', *North China Herald*, 4 April 1882.

'Accounts of the depredations of the coccids on the orange-trees of Florida, induce me to publish a brief account of the employment by the Chinese of ants as insecticides. In many parts of the province of Canton, where, says a Chinese writer, cereals cannot be profitably cultivated, the land is devoted to the cultivation of orange-trees, which, being subject to devastation from worms, require to be protected in a peculiar manner, that is, by importing ants from neighboring hills for the destruction of the dreaded parasite. The orangeries themselves supply ants which prey upon the enemy of the orange, but not in sufficient numbers; and resort is had to hill-people, who, throughout the summer and winter find the nests suspended from branches of bamboo and various trees. There are two varieties of ants, red and yellow, whose nests resemble cotton-bags. The orange-ant feeders are provided with pig or goat bladders, which are baited inside with lard. The orifices of these they apply to the entrance of nests, when the ants enter the bags and become a marketable commodity at the orangeries. Orange-trees are colonized by depositing the ants on their upper branches, and to enable them to pass from tree to tree, all the trees of an orchard are connected by bamboo rods.

'Is the orange the only plant thus susceptible of protection from parasitic pests? Are these the only species of ants that are capable of utilization as insecticides? Indubitably not; and certainly entomologists and agricultural-ists would do well to institute experiments with a view to further discovery in this line of research.'

It is a pity that the name of the Chinese writer referred to above was not given. We have no idea how much influence this report might have had among entomologists and horticulturists in the West who were beginning to appreciate the potential of predaceous insects as a means to control noxious insects pests, and were diligently working to apply the concept under practical conditions. The big breakthrough came in 1888, when Albert Koebele introduced the vedalia beetle, *Rodolia cardinalis* (Muls), from Australia into California, to control the cottony-cushion scale, *Icerya purchasi* (Mask), which was threatening the destruction of the infant citrus industry there. The success of this venture was so spectacular that the event is generally regarded today as the milestone that celebrated the birth of the modern Age of Biological Control. For details of this fascinating story, see Richard L. Doutt (1), pp. 21–42.

[1] 南越筆記

this claim, since they saw mulberry trees and silkworms in the village and thought that silk was the true product. The villagers then said: 'True, we have mulberry trees and do grow silkworms, but we feed them before they are full-grown to the ants, which we sell to orange growers for a dollar a nest.'

This ant, *Oecophylla smaragdina*, a well-known tropical or sub-tropical species, constructs silk nests on the trees, into which all the ants retire at night. The orange growers connect the orange trees with one another by means of bamboo poles, over which the ants travel to build nests in all the orange trees. Then the ants no longer eat silkworms but devour the insects which attack the orange trees or fruits. The owner of a lychee tree in his home garden near Canton told me he had purchased a nest of ants (the nests are cut off the tree at night and tied in a tight bag) which had prevented all insect injury to his crop of lychees and were particularly efficient in driving off a large insect, *Tessarotoma papillosa*, a pentatomid bug nearly an inch long.[a]

Later, another collaborator of Swingle's, Kuo Hua-Hsiu[1], observed[b] that the ant was locally known as the *kan i*[2], *ching i*[3] or *ta huang i*[4]. It was used in the orange groves of Hsi-Sha[5] in Ssu-Hui[6] district, the villages of Lo-Kang-Thung[7] and Hsien-Kang[8] in Phan-Yü[9] district, the Yang-Chhun[10] district of Chao-Chhing[11] and the Tien-Pai[12] district of Kaochow[13]. The ants occur naturally in the bamboo groves and Chinese olive trees (*Canarium* sp.).

The original encounter was reported six years later in 1924 in a paper by Groff Howard,[c] who described the life habits of the ants as follows:

Our observations were that these ants are about twelve to fifteen millimeters in length and of a light brown colour. Their small black eyes are very active and they move very rapidly. Their bite is very painful to man and they will attack if teased. Their most interesting habit is that of weaving from the leaves of the trees their most remarkable, silk-structured nests. This habit they share with certain species of *Camponotus* and *Polyrhachs*. The nest is made by drawing together the leaves of the tree in which the ants have chosen to live. Often the nest is formed of a mass of leaves joined into a more or less regular sphere, at least eight to ten inches in diameter. The leaves are drawn together and held by the worker ants while a second set brings up nearly mature larval ants, which they hold in their jaws and move back and forth from one leaf edge to another, until the larva has spun a mass of silk which holds the two leaves firmly together. This process of spinning can be easily observed.[d]

[a] Swingle, (5), pp. 95–6.
[b] Kuo Hua-Hsiu[1], *Kan chü lei tsai phei fa*[14], Citrus Cultural Methods, unpublished monograph written about 1925. A draft translation prepared by Michael J. Hagerty is on file at the Needham Research Institute in Cambridge, England.
[c] Groff, & Howard (1), pp. 108–14.
[d] At one time there was considerable mystery as to how the workers were able to spin silk, since no adult insects were known to produce silk. The mystery was solved by Henry Ridley (1) in Singapore in 1890. He was the first person to show that *Oecophylla* uses its larvae for spinning the silk of its nest. This was no doubt an interesting diversion from his untiring, single-minded effort to introduce the rubber tree as a plantation crop in Malaya, with far-reaching effects on the economy and prosperity of Southeast Asia in the twentieth century.

[1] 郭華秀 [2] 柑蟻 [3] 驚蟻 [4] 大黃蟻 [5] 西沙
[6] 四會 [7] 灘岡同 [8] 逼岡 [9] 番禺 [10] 陽春
[11] 肇慶 [12] 電白 [13] 高州 [14] 柑橘類栽培法

In 1921 specimens of the citrus ant were sent to W. M. Wheeler of Harvard University who identified it as *Oecophylla smaragdina* Fabr.[a] Additional field observations were made a decade later by W. E. Hoffman who published an account of his work in 1936.

Neither Groff & Howard nor Hoffman were convinced of the benefits of establishing yellow ants among the orange trees. Hoffman wrote:

What is the practical effect of this type of ant on the economics of citrus culture? It remains to be determined. While they are undoubtedly able to destroy or drive away caterpillars, stink bugs and other ants which are injurious to citrus, at the same time they tend to protect and service the scale insects which also are highly injurious. Often they transport the scale insects from tree to tree, sometimes to new trees which had not been infested by the scale insect before in order to feed on the nectar they leave behind. Because these insects are protected and cared for by the yellow ant, they can proliferate profusely. They are usually very small, so that the growers can easily overlook them. It should be pointed out that these scale insects occur in very large numbers and they can cause great damage to the citrus trees, even greater than that inflicted by the more visible, larger insects. Furthermore, these ants apparently also offer the same protection and service to another important citrus pest, the aphids, and possibly plant lice.[b]

He went on to say that from his own observations he must conclude that the practice of rearing such ants in orchards probably did more harm than good. Unfortunately, as the orchards where the ants were cultivated were rather far from the city of Canton, Hoffman was unable to carry out extensive investigations of their efficacy.

At any rate, in spite of the reservations voiced by Groff, Howard and Hoffman, the practice evidently survived the dislocations of the Sino-Japanese War, the social upheavals attendant upon the establishment of the People's Communes and the proliferation of highly effective synthetic chemical insecticides in the 1950's. It became better known among both economic entomologists as well as historians of science in China. In 1958 Chhen Shou-Chien[1] initiated a series of field observations in the Kuang-Ssu[2] district to ascertain the efficacy of the yellow citrus ant under practical field conditions. He found that:

When the ant population is adequate, there is excellent control of *Rhychochoris humeralis* Thunberg, with modest control of *Podagricomela nigricollis* Chen, *Hypomeaes squamosus* F. and *Anomala cupripes* Hope. There are 18.3% more healthy leaves on trees with ants than trees without ants. They are without effect on scale insects, *Lowana* sp. and larvae of the longhorn beetle. In fact, they form friendly associations with scale insects, such as *Pseudococcus citriculus* and *Coccus* sp. On the other hand they are harmless to the enemies of the scale insects, *Rodolia rufopilosa* Mulsant and *Rodolia pumila* Weise. In orchards with yellow ants in residence, black ants have all been driven away.[c]

---

[a] Wheeler, (1), p. 544; see also N. Gist Gee (1), pp. 100–7.
[b] Tr. auct. Hoffman, (Ho Fu-Min[3], 1), p. 209–10.
[c] Tr. auct. Chhen Shou-Chien (1), p. 401.

[1] 陳守堅   [2] 廣四   [3] 賀輔民

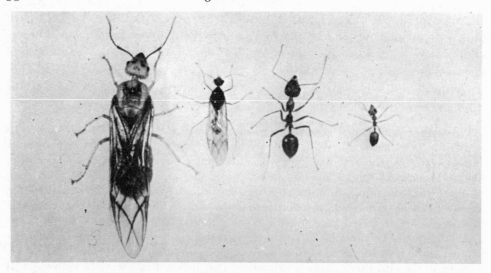

Fig. 96. Yellow citrus ant, *Oecophylla smaragdina* Fabr. (photo. Yang Phei). Left to right: Female, male, mature worker, young worker.

Fig. 97. Ant nest on orange tree, about 54 cm. long, Ssu-hui, Kuangtung 1979 (Photo. Yang Phei).

Fig. 98. Bamboo bridges between trees to facilitate migration of ants; Ssu-hui, Kuangtung 1979 (photo. Yang Phei).

Chhen Shou-Chien[1] noted that the 'yellow citrus ant is not an ideal natural enemy of citrus pests'.[a] He recognised that 'it may be appropriate for use in certain areas where the new chemical insectides might be in short supply and labour scarce'[b] and proposed studies to find ways to improve its performance.

We have now traced the record of the continuing use of the yellow citrus ant in South China from about +300 to the 1960's. Considering the diversity of insect pests that occur in a citrus grove and the changing dynamics of insect populations in an agricultural ecosystem, it is remarkable that the use of one species of ant as the principal control agent in the orchards of Kuangtung has proved to be as effective as it evidently has been over such a long period of time (Figs. 96, 97, 98, 99, 100).[b]

[a] Tr. auct. Chhen Shou-Chien, p. 401, Unfortunately, Chhen was not able to continue his studies. However, since 1978 field observations are again being made in the Ssu-hui District by investigators from the Institute of Entomology, Chung-Shan[2] University, Canton. For a preliminary report see Yang Phei[3] (1), pp. 101–3.

[b] Groff & Howard, Hoffman and Chhen Shou-Chien have all noted that the yellow ant forms friendly associations with scale insects, which are highly destructive to the orange trees. Yet all the growers they talked to were absolutely convinced, on the basis of their own experience, of the efficacy of the use of citrus ants. Yang Phei (1) has found that scale insects on the orange trees on which citrus ants have been established, are heavily infested by parasitic wasps. This observation plus the fact that the ant does not attack other natural enemies of scale insects such as ladybird beetles may account for the relative lack of injury due to scale insects in these orchards.

[1] 陳守堅　　[2] 中山　　[3] 楊沛

Fig. 99. Ant nest under construction on orange tree, Hua-an, Fukien, 1982 (photo. Yang Phei).

Fig. 100. Ants capturing pest larva, Hua-an, Fukien, 1982 (photo. Yang Phei).

If the *Nan Fang Tshao Mu Chuang*[1] is correctly dated at +304, then the ant technology has been in use in South China for almost 1700 years. The book has often been quoted by Chinese authors since the late Sung Dynasty and for centuries its authenticity was never questioned. Chi Yün[2], the editor of the great encyclopaedia *Ssu Khu Chhüan Shu*[3] in his 'Annotated Catalogue', *Tsung Mu Thi Yao*[4] (+1782) stated that, judging from its refined and elegant prose, the *Nan Fang Tshao Mu Chuang*[1] must have been written before the Thang Dynasty.[a] It was not until late in the Chhing Dynasty (1888) that the first doubt about its authorship was expressed by Wên Thing-Shih[5], in the *Pu Chin Shu I Wên Chih*[6] (Additions to the Bibliography[7] of *Chin Shu*[8]).[b] And it was well after the founding of the People's Republic (1962) that Hsin Shu-Chih[9] (*3*) first propounded the idea that the work was not written by Hsi Han, but rather compiled artfully by a forger from miscellaneous extant sources in the late Sung dynasty.[c] Since then the issue has generated considerable confusion and controversy among scholars both in China[d] and in the West.[e] We have ourselves already examined in some detail the major arguments put forth in support of the forgery thesis in an earlier section, but have found them to be neither conclusive nor convincing.[f]

In connection with the citrus ant story, it is interesting to note that the leading authorities on the history of entomology in China have taken somewhat divergent positions on this issue. Chou Yao[17] (1978) has continued to accept Hsi Han's authorship of the *Nan Fang Tshao Mu Chuang*,[g] while Tsou Shu-Wên[18] (1981) has admitted the possibility that the book might have been composed at a later age.[h] Indeed, he went on to quote the ant passage of the *Ling Piao Lu I*[19] as possibly the earliest record of the biological control of insect pests. No such hesitation was shown by Liu Tun-Yüan[20], who stated categorically that *Nan Fang Tshao Mu Chuang* was written by a Sung author.[i]

In the introduction and commentaries that accompany his translation of the *Nan Fang Tshao Mu Chuang*, Li Hui-Lin[16] has examined in exhaustive detail the historical and geographical background to the work, the life and times of the author, and the derivation and botanical identity of each of the items in the text. The results of his painstaking studies led him to say that:

---

[a] Chi Yün[3], in the section on 'Historical Geography', pt. 3, ch. 70, p. 64, considers the *Nan Fang Tshao Mu Chuang* as *Tsui wei wan chêng* and also *tu hsien ê chhüeh*[10], that is, 'a most complete whole' and marred 'by only a few blemishes'.

[b] *Pu Chin Shu I Wên Chih*[6], p. 53.

[c] Hsin Shu-Chih (*3*), pp. 110–8.

[d] Shih Shêng-Han[11] (*3*), p. 745; Chi Tun-Yu[12] (*1*), p. 182; Wu Tê-To[13] (*1*), p. 152; Phêng Shih-Chiang[14] (*1*), p. 75.

[e] Ma Thai-Lai[15] (*1*); Li Hui-Lin[16] (*12*).

[f] See pp. 447 ff., 531 ff, above).   [g] Chou Yao (*2*), p. 77.

[h] Tsou Shu-Wên (*1*), p. 102.   [i] Liu Tun-Yüan (*1*), p. 229.

[1] 南方草木狀	[2] 紀昀	[3] 四庫全書	[4] 總目提要	[5] 文廷式
[6] 補晉書藝文志		[7] 藝文志	[8] 晉書	[9] 辛樹幟
[10] 最爲完整，獨鈔譌闕		[11] 石聲漢	[12] 吉敦諭	[13] 吳德鐸
[14] 彭世獎	[15] 馬泰來	[16] 李惠林	[17] 周堯	[18] 鄒樹文
[19] 嶺表錄異	[20] 劉敦愿			

All these seem to indicate the originality and early derivation of Hsi Han's work. Although we cannot rule out the possibility of interpolations and cannot verify the authenticity of most of the contents we can be reasonably sure that the text, as it has come down to us in its present form since the late Sung period, represents on the whole a historically trustworthy account of the plants treated therein as they appeared in the southern regions around the third and fourth centuries A.D.[a]

We agree completely with Li Hui-Lin's summing-up. It is beyond the scope of the present work to delve any deeper into the question of authenticity of Hsi Han's test. For our purpose, we are content to accept the view that the yellow citrus ant has been an instrument of biological control on citrus in South China for sixteen to seventeen centuries. Thus, it would appear that the claim that it is the earliest example of the use of a biological agent for the control of agricultural pests remains secure.

Or does it? Recently Isaac Harpaz (1973) has cited a remarkable passage in the Talmud which could be construed as the record of an application of biological control:

How are they [the ants] destroyed? Rabban Simeon b. Gamaliel says: 'Soil is fetched from one hole and put into another, and they (the ants of the two nests not knowing each other) strangle each other.' Rabbi Yemmer b. Shelemia said in the name of Abaye: That is (effective) only if (the ants are) situated on two sides of a river; and that if there is no bridge; and if there is not even a crossing-plank; and if there is not even a rope to cross by. How much apart?—up to one parasang (about four miles)'.[b]

Rabban Simeon b. Gamaliel lived in Israel about +140, and Abaye lived in Babylonia about +330. Thus the use of one group of ants to destroy another group of ants was recognised in +2nd century Israel. However, beyond this terse account, there is no record of how extensively or for how long the practice was followed. So the significance of this passage in terms of the history of biological control remains rather uncertain.

Coming down to an age closer to our own, there is a fascinating account by Peter Forskål published in +1755 on the use of a species of red ants for the control of ant pests which infest date-palms. In a compilation of the fauna of Arabia, he wrote as follows:

23. a) FORMICA ANIMOSA, *rubra*. Arab. Kaas.
Minor quam F. 22. Habitat in ligno. Oeconomis grata ob utile odium quo persequitur

---

[a] Li Hui-Lin (12), p. 28. Li's conclusion is really not much different from that of Shih Shêng-Han (3), p. 745, who said (tr. auct.): 'It is not that I doubt Hsi Han is the author of most of the words that comprise the book, but rather that I question whether Hsi Han was actually responsible for arranging the book in its present form.'
[b] Harpaz, (1), p. 35, quoting from the Babylonian Talmud, *Mo'ed Katan*, pp. 6b to 7a.

Formicas DHARR, *Phoenicem dactyliferam* exitiose infestantes. Ad hanc militiam conducitur acervatim Heml (onus Cameli) imperiali pretio.[a]

This observation was later confirmed by Paul-Emile Botta (1841) in the Yemen.

J'appris aussi d'Ezze que le vent est favorable à la végétation du dattier, et que plus les vents sont violents et fréquents, plus les dattes sont abondantes et belles; enfin, j'ai pu vérifier le fait singulier déjà observé par Forskål, que les dattiers sont attaqués dans l'Yemen par une espèce de fourmis qui les ferait périr si chaque année, on n'apportait des montagnes et ne suspendait pas à leur sommet des bûches d'un arbre que je ne connais pas, et qui contiennent les nids d'une autre espèce de fourmis qui détruit celle du dattier.[b]

This practice, it seems, is still alive today, although the nature of the target pest has apparently changed. In the oases in Saudi Arabia, ants are now employed passively to control the caterpillars and pupae of the greater date-moth, *Arenipses sabella* Hemps.[c] Furthermore, in the Tihamma region of the Yemen Arab Republic ants have been used for a long time to control the lesser date-moth, *Betrachedra amydraula*, until about twenty years ago, when the practice was displaced by the introduction of DDT (dichlorodiphenyltrichloroethane). Indications are that this method of biological control may be revived in the near future.[d]

Forskål's account is presumably the earliest record of biological pest control in the Western literature. We have no idea how long the practice had been in existence when Forskål first encountered it in the +18th century. Considering the active maritime communication between China and the Arab countries of the Middle East through the Thang, Sung and Ming dynasties, and that Canton (or Khanfu) was a major port of call for the Arab merchants, it would not be surprising if the concept had travelled from South China to the Yemen.

We cannot close the citrus ant story without noting that ants have been used in China historically for the protection of at least one other crop. In Southern Fukien, a red ant (*hung ma i*[1]), *Tetramorium guineense*, has been used widely to control insect pests in the sugarcane field. This practice is said to have a long history, but its antiquity has yet to be determined.[e] Further studies of the local records in southern

---

[a] Forskål, P. (1), ch. 3, p. 85. A translation in English reads: '*Formica animosa*, red. Arabic *kaas*. Smaller than F. 22. Lives in wood. Welcome to the market gardeners because of the value of the hatred with which it persecutes the ants called *dharr* which excessively infest the date-palms (*Phoenix dactylifera*). Such military service leads to the piling up of *heml* (the bales of dates borne by camels) for the imperial tribute.'

[b] Botta, P. E. (1), p. 130. In English, the passage reads: 'I also learnt from Ezze that the wind is favourable for the date-palms, and that the more frequent and violent it is, the more abundant the harvest of lovely dates; finally, I was able to verify the strange fact, already observed by Forskål, that in the Yemen the date-palms are attacked by a kind of ant which would kill them if it were not for the fact that every year (the gardeners) bring down from the mountains some branches of a tree the name of which I don't know, and hang them from the tops of the palms, for these branches contain the nests of another kind of ant which destroys those that prey upon the date-palms.'

[c] Private communication (1982) from Dr A. S. Talhouk, Riyadh, Saudi Arabia. He has seen such ants in the groves of Bishah, approximately 200 km. northeast of the Red Sea coast, east of the Asir mountains.

[d] El Haidari, H. S. (1), pp. 129–30. The date-growers relate that two types of ants were used, one black and one red. Specimens of the black species were collected and identified as *Crematogaster* sp.

[e] Tsou Shu-Wên (1), p. 9; Phu Chê-Lung (1), pp. 72–7.

[1] 紅螞蟻

Fukien will, we may hope, provide an answer to this problem. One can only speculate whether the red ant application in Fukien was directly inspired by the yellow ant story in the two Kuang provinces. It is safe to assume that the modern development of the large black ant (*ta hei i*[1]), *Polyrhachis* sp. as a biological control agent against forest pests and paddy rice pests has been an extension of traditional citrus ant technology.[a]

Sweetman said in 1936 that 'the use of ants for the protection of orchards from insect pests is a practice of long standing in various Asiatic countries',[b] and Swingle stated in 1942 that ants 'are now employed in Indo-China and New Guinea',[c] although neither gave any specific references to support these claims. However, De Bach related in 1974 that he had personally observed the practice in operation 'in the Shan States of North Burma not long ago'.[d] More recent information on the efficacy of *O. smaragdina* in controlling pests of cocoa and coconuts in the tropics has been documented and summarised by Leston (1973)[e]. In Zanzibar, the only other living relative of *O. smaragdina* in the same genus, *Oecophylla longinoda* (Latr.) was found to be of undoubted value for controlling certain coconut pests, notably *Theraptus* sp. (Coreidae).[f] Moreover, there has been a considerable amount of interest in Europe and in the Soviet Union in the potential of ant species as agents for the control of forest pests.[g] Thus, it is likely that we will be hearing more and more in years to come of the use of selected ant species as agents of biological control for a variety of insect pests.

(iii) *Other insects*

The potential of other insects, besides ants, as biological control agents was also recognised though not extensively exploited. The earliest example of the control of an agricultural pest by a predator was reported by Tho-Tho[2] in the *Liao Shih*[3] (History of the Liao Dynasty), +1343. The event took place in the 9th year of the Hsien-Yung[4] reign-period (+1074), and the record reads as follows: 'Locusts developed in Kuei-i[5] and Chia-shui[6] Districts, and flew in a swarm into Sung[7] territory. Those that remained behind were soon gobbled up by wasps.'[h] The nature of the wasps involved was not described. A similar but more detailed account was given in the *Hsin Hsien Chih*[8] (Chronicles of the Hsin District), 1887:

---

[a] Phu Chê-Lung (*1*), pp. 77–80. The ant has been recently identified as *Polyrhachis dives* F. Smith (private communication by Yang Phei 1982).

[b] Sweetman, (1), p. 1. This observation is corroborated by Clausen (1), p. 308, who stated in 1940 that 'another ant *Dolichoderus bituberculatus* Mayr. is employed in Java for the protection of cacao plantings from injury by *Helopeltis*'.

[c] Swingle, (5), p. 96   [d] De Bach, (1), p. 71.

[e] Leston, (1).   [f] Way, (1, 2).

[g] Sailer, (1); and Adlung, (1).

[h] Tr. auct. *Liao Shih, Pên Chi*[9], no. 23, Tao Tsung[10] 3, p. 2*a*.

[1] 大黑蟻   [2] 脫脫   [3] 遼史   [4] 咸雍   [5] 歸義
[6] 浹水   [7] 宋   [8] 莘縣志   [9] 遼史，本記   [10] 道宗

In the 5th month of the 9th year of the Chia-Chhing[1] reign-period (+1530), massive swarms of locusts were seen drifting north like clouds from the Yen[2] region towards Hsin-hsien, leaving behind them trails of utter crop devastation. Chhen Tung[3], the magistrate, promptly cleansed himself by bathing and fasting, and with his retinue prayed to the Eighth Spirit (*pa cha*[4]). Shortly thereafter swarms of black wasps covered the entire area, attacking and killing the locusts. The thunder and rain that followed shredded the locusts into mud, and the grains in the fields were unharmed.[a]

But the most dramatic and celebrated example of the sudden emergence of a natural predator population to control a major pest infestation is undoubtedly that recorded by Shen Kua[5] in *Mêng Chhi Pi Than*[6] (Dream Pool Essays), +1086:

In the Yuan-Fêng[7] reign-period (+1078 to 1085), in the Chhing-Chou[8] region, an outbreak of *tzu-fang*[9] insects caused serious damage to the crops in the fields in autumn. Suddenly another insect appeared in swarms of thousands and tens of thousands, covering the entire ground area. It was shaped like earth-burrowing *kou-ho*[10] (dog grubs), and its mouth was flanked with pincers. Whenever it met a *tzu-fang*, it would seize it with the pincers and break the poor beast into two bits. Within ten days all the *tzu-fang* had disappeared, so the locality had an abundant harvest. Such kinds of insects have been known since antiquity and the local people call them *phang pu khên*[13] (not allowing other ⟨insects⟩ to be).[b]

The *tzu-fang*[9] is one of the most destructive pests of grain crops known in China.[c] Also called *nien chhung*[14], it has been identified as a member of the family Noctuidae, subfamily Hardeninae, *Leucania separata*.[d] *Phang pu khên*[13] is believed to be a ground beetle.[e] It still flourishes in areas where *tzu-fang* is endemic today. It is also

---

[a] Tr. auct. *Hsin Hsien Chih, Chi I Chih*[11], ch. 4, p. 4. The Eighth Spirit, *Pa Cha*[4], is one of the eight deities receiving prayers and sacrifices offered at the Ta Cha Ceremony (*Ta Cha Li*[12]) to mark the end of each year of agricultural production. For a discussion of the eight deities, see Liu Tun-Yüan (*1*). The Eighth Spirit is the deity of insect pests and weeds.

[b] Tr. auct., ch. 24, p. 21.

[c] Chou Yao (*2*), pp. 211–13, has compiled a table of the major outbreaks of *tzu-fang* infestation recorded in the Chinese literature from +275 to 1825. A total of 44 infestations is listed, indicating a major outbreak every 35 years. For a detailed discussion of the references to and characterisation of *tzu-fang* in the Chinese classical literature, see Tsou Shu-Wen (*1*), pp. 87–93.

[d] Chou Yao (*2*), p. 71, identifies *tzu-fang* as *Leucania separata*. Phu Chê-Lung (*1*), pp. 9, 15, 22, agrees. In volume 2 of the *Khun Chhung Fêng Lei Hsüeh*[15] (1973), 'The Taxonomy of Insects', Tshai Pang-Hua[16] (p. 242), classifies *tzu-fang* as *Pseudolethia separata* (Walker), subfamily Hardeninae, family Noctuidae, and characterises it as an armyworm. The most recent literature from China, e.g. Wu Hsiang-Kuang[17] & Huang Mei-Chên[18] (1982), *Acta Ecologica Sinica*, 2, pp. 39–46 refers to it as *Mythimna separata* Walker, the oriental armyworm.

[e] Chou Yao (*1*), p. 49, (*2*) p. 79, identifies *phang pu khên* as a ground beetle, family Carabidae. Li Chhün[19] in *Mêng Chhi Pi Than Hsüan Tu*[20] (1975), 'Selections From the Dream Pool Essays With Commentaries', pp. 167–8, also lists it as a carabid. The official name is *ti tshan hu*[21] (ground worm tiger), which according to Tshai Pang-Hua[17] would be *Calosoma chinensis* Kirby. A European species, *Calosoma sycophanta* Linn. was introduced into New England early in the twentieth century to help control the gypsy moth (Metcalf, Flint & Metcalf (*1*), p. 64).

---

[1] 嘉靖	[2] 兗	[3] 陳棟	[4] 八蜡	[5] 沈括
[6] 夢溪筆談	[7] 元豐	[8] 慶州	[9] 好蚄	[10] 狗蝎
[11] 莘縣志, 畿異志		[12] 大蜡禮	[13] 旁不肯	[14] 粘蟲
[15] 昆蟲分類學	[16] 蔡邦華	[17] 鄔祥光	[18] 黃美貞	[19] 李群
[20] 夢溪筆談選讀		[21] 地蠶虎		

commonly known as *chhi pu fên*[1] (anger without exasperation), to indicate its undying enmity towards its natural prey.

This story is corroborated by a passage in the *Tung Pho Chih Lin*[2] (Journal and Miscellany of (Su) Tung-Pho[3]) of +1101:

On the tenth day in the fifth month of the eighth year of the Yuan-Yu[4] reign-period (+1093), the magistrate of Yung-chiu[5], Mi Fu[6], stated in a letter that in his district there was an infestation of insects that ate the leaves but not the grain. At that time it happened that an official of the treasury, Chang Yüan-Fang[7], saw the letter and he remarked, 'Pulse and grain crops are seldom infested by insects. When pests are present, it is an unusual happening. But by eating the leaves, they are bound to cause damage to the grain. It is unthinkable that the grain would suffer no injury at all.' Chang Yüan-Fang continued to say that the *tzu-fang*[11] may often cause more injury than the locust. But there is a small beetle, which upon seeing the pest immediately seizes it and breaks its waist. The local people call it *phang pu khên*[9]. I never heard of this phenomenon before, and so I have noted it in my records.[a]

The observation that the *tzu-fang*[11] feeds on leaves but not on grains has facilitated its identification by modern entomologists. Of course, by eating the leaves and weakening the stalk the pest would often indirectly cause the tassel to droop and break.

Considering the numerous outbreaks of *tzu-fang*[11] that have been recorded in the histories between the early Chin dynasty (+275) and the 1800's we may assume that the ground beetle and other natural enemies must have played a major, though unrecognised, role in keeping this pest in check. Sometimes the *tzu-fang* may have been mistaken for locust larvae. There is a passage in the *Pu Huang Yao Chüeh*[8] (Effective Ways to Control Locusts), 1885, which says:

When locust nymphs are seen in abundance, there may suddenly appear small dark red insects flying among the fields. They fly swiftly. As soon as they see the larvae they bite and kill them. The people are jubilant and call them *chhi pu fên*[1] (anger without exasperation). In a few days, none of the nymphs would be left.[b]

Tsou Shu-Wên[10] points out that the so-called locust nymphs described in the *Pu Huang Yao Chüeh*[8], are probably *tzu-fang*[11]. Because the exoskeleton of the locust nymph would be much tougher than that of the *tzu-fang*[11] larvae, they would not have been quite so easily killed by the beetle.

No conscious effort has been made to rear or influence the natural population of the ground beetle. But the release of another insect predator, the *thang-lang*[12],

---

[a] Tr. auct., ch. 5, p. 1
[b] Tr. auct., quoted in Tsou Shu-Wên (*1*) p. 91. The original text was prefaced by Chhien Hsin-Ho[13] (1855), but the author is not identified. It is also known as *Pu Huang Yao Shuo*.[14]

[1] 氣不憤	[2] 東坡志林	[3] 蘇東坡	[4] 元佑	[5] 雍丘
[6] 米芾	[7] 張元方	[8] 捕蝗要決	[9] 旁不肯	[10] 鄒樹文
[11] 好蚄	[12] 螳螂	[13] 錢炘和	[14] 捕蝗要說	

praying mantis, was advocated by Chhêng Tai-An[1], in the *Hsi Wu Chü Lüeh*[2] (Chrysanthemum Album of Western Wu), 1845:

In the fifth month look for egg masts in the *thang-lang*'s nests. Place several in the vicinity of the chrysanthemums. By early autumn, the young mantis will have emerged and will be jumping about among the plants. They do not feed on the leaves, but prey on other insects and drive away the butterflies.[a]

By this time entomologists in the West had also begun to take notice of the beneficial activities of predacious coccinellids such as ladybird beetles (Kirby & Spence, 1815) in England, and the possible use of carabid beetles to control caterpillars of the gypsy moth (Boisgiraud, 1840) in France. Biological control as a modern scientific descipline was about to be born.[b]

(iv) *Vertebrates*

We shall now turn to a uniquely Chinese contribution to the technology of biological control, that is, the exploitation of vertebrates as insect control agents. The first group to be considered is the amphibians, or more strictly, frogs. Although there are no records to show that frogs were cultured or released for use in the fields, their value for the protection of plants against insect pests was well known.[c] Indeed, edicts were promulgated by local officials to prohibit the capture and slaughter of frogs in Chekiang. For example, we find this passage in the *Hsing Ying Tsa Lu*[4] (Random Notes From a Tour of Duty) by Chao Khuei[5], late Sung, about + 1250.

Ma Yü-Chhi[6], prefect of Chhu-Chou[7], decreed that the people were forbidden to capture frogs (for food). One peasant from a village disobeyed. He took winter melons, scooped out their contents, and hid the frogs in the empty cavities. Then at dawn, he carried the melons into the city. At the city gate he was stopped by the sentry and arrested.[d]

In this case it turned out that the peasant's wife, who had been having an affair with another man, had encouraged her husband to break the edict, and informed the sentry on his movements. The prefect was able to get to the bottom of the case and dealt swift justice to the unfaithful wife and her lover.

Another example was recorded by Phêng Chhêng[8] in the *Mo Kho Hui Hsi*[9] (Declamations of a Literary Practioner), of + 1603:

[a] Tr. auct. Edition of *Thiao Shang Chhêng Tai-An Yuan Pên*[3], p. 7. Cf. pp. 409 ff. above.
[b] For accounts of the history of biological control see Doutt, R. L. (1964) and Simmonds, F. J., Franz J. M. & Sailer R. I. (1976). Much useful information on the early recognition of the value of predators and parasites of insects is found in Silvestri, F. (1909). See also Essig (1), pp. 274 ff.
[c] The rearing and release of frogs in rice fields has been practised in Chekiang and Fukien since the 1960's, cf. Phu Chê-Lung (1), pp. 223-8.
[d] Tr. auct. *Hsü Pai Chhuan Hsüeh Hai*[10], p. 1478.

[1] 程岱簝	[2] 西吳菊略	[3] 苕上程岱簝元本	[4] 行營雜錄
[5] 趙葵	[6] 馬裕齊	[7] 處州	[8] 彭乘	[9] 墨客揮犀
[10] 續百川學海			

Fig. 101. Frog avidly eyeing a flying insect; *TSCC* 1062/1.

The people in Chekiang like to eat frogs, but Shen Wên-Thung[1] at Chhien-Thang[2] prohibited their capture. As a result, the population of frogs in the ponds and marshes quickly declined. After Shen Wên-Thung's departure, the people revived the custom of eating frogs, and the frog population sprang back to its former abundance. It would appear that Heaven looks with favour upon the consumption of frogs by human beings, and the practice of eating frogs has become more popular than ever.[a]

We suspect that by prohibiting the capture of the frogs, Shen Wên-Thung may have inadvertently reduced the amount of food available to each individual adult, so that they became more susceptible to disease or less able to resist the attacks of their natural enemies. Hence their population declined (Fig. 101).

The second group of invertebrates used in biological control are birds, and particularly ducks. The earliest reference to birds providing natural control occurs in the *Nan Shih*[3] (History of the Southern Dynasties) by Li Yen-Shou[4], *c.* + 650. In the biography of Prince Hui[5] and his close relatives, of the Liang kingdom, we find that during the reign of the Emperor Liang Wu-Ti[6] (+502–9) there was a locust infestation at the estate of the deputy governor Fan Hung-Wei[7]. Prince Hsiu[8], the

---

[a] Tr. auct., ch. 6, pp. 7–8.

[1] 沈文通　　[2] 錢塘　　[3] 南史　　[4] 李延壽　　[5] 恢
[6] 梁武帝　　[7] 范洪冑　　[8] 修

Governor, was advised to attack and capture the locusts, but he said, 'The visitation is a reflection of my lack of virtue, what good will it do to catch them?' Immediately after he finished speaking: 'Thousands of birds suddenly appeared, shielding the sun. Soon all the locusts were eaten up by the birds, which left the scene shortly afterwards. The birds were not identified.'[a]

Similar stories were recorded in the *Yu Yang Tsa Tsu*[1] (Miscellany of the Yu-Yang Mountain) by Tuan Chhêng-Shih[2] (+863): 'In the second year of the Thien-Pao[3] reign-period (+743) there was an infestation of purple insects feeding on the crop seedlings. From the Northeast, there appeared a swarm of red-headed birds, which ate up the insects'. Again:

In the twenty-third year of the Khai-Yuan[4] reign-period (+735) an infestation of *tzu-fang*[5] occured in Yü-kuan[6] and extended into Phing-chou[7]. Swarms of sparrows arrived to feed on them. Also, during the time of the Khai-Yuan[4] reign-period locusts in Pei-Chou[8] attacked the grain seedlings. Suddenly, thousands of large white birds, and tens of thousands of little white birds, flew down and soon consumed all the insects.[b]

Similar accounts are recorded in the *Chiu Thang Shu Wu Hsing Chih*[9] (Chapter 5 on Strange Events in the Old History of the Thang Dynasty), +945:

In the twenty-fifth year of the Khai-Yüan[4] reign-period (+737) there was an outbreak of locusts at Pei-chou[8]. Thousands and ten thousands of white birds swarmed over them, and in a single night, all the locusts were consumed, so that there was no damage inflicted on the crops. In the twenty-sixth year (+738) there was another *tzu-fang*[5] infestation at Yü-kuan[6]. Swarms of sparrows arrived to feed on the insect pests.[c]

So far, the records show simply what happened. But in the next citation taken from the *Han Shih Lu*[12] (Records of the Posterior Han Dynasty) +948, positive action was taken to hasten the arrival of beneficial birds:

In the early years of the Chhien-Yu[13] reign-period (+948 to 51), in Khai-Fêng[14] prefecture, three districts, Yang-wu[15], Yung-chhiu[16] and Hsiang-i[17] suffered from locust attacks. The Fu I[18] (i.e. the Prefect), Hou I[19], sent representatives to offer wine and sacrificial prayers, so that the locusts in the three districts would be consumed by mynah birds. A decree was then issued to prohibit the shooting or capture of birds since they were deemed invaluable in controlling locusts.[d]

The recognition that birds were an effective instrument to combat insect pests is probably of much greater antiquity than these records suggest. Thus, based on

---

[a] Tr. auct. *Nan Shih, Po Yang Chung Lieh Wang Hui Chuan*[10], biography of Prince Hui, ch. 52, p. 7.
[b] Ch. 16, p. 7a., tr. auct.
[c] Tr. auct., *TSCC*, ch. 176. *Huang Pu Chi Shih*[11], 'On Locusts', p. 4.
[d] Tr. auct., quoted in *TPYL*, ch. 950, p. 5.

[1] 酉陽雜組   [2] 段成式   [3] 天寶   [4] 開元   [5] 好蚄
[6] 榆關   [7] 平州   [8] 貝州   [9] 舊唐書五行志
[10] 南史, 鄱陽忠烈王恢傳   [11] 蝗部紀事   [12] 漢實錄   [13] 乾祐
[14] 開封   [15] 陽武   [16] 雍丘   [17] 襄邑   [18] 府尹
[19] 侯益

previous experience, the official might have had reason to believe that by offering sacrifices and prayers, mynah birds could be summoned to rid the land of this calamity.

Although few records are available, the intervention of birds to control serious outbreaks of insect infestation was also noted in the West. Indeed, Pliny (+23 to 9) reported that when locusts infested the region of Mt Cassius in Syria, Jupiter answered the prayers and supplications of the people by sending the migratory Seleucid birds (rose-coloured starling, *Pastor roseus*) to eradicate the pests.[a]

Natural control achieved by birds has undoubtedly provided the inspiration for the most interesting innovation in biological control to come out of China, namely the use of ducks. Ho Thao[1] (+1540?) described the efficacy of ducks in the control of land crabs. 'In Shun-tê[2], there are crab nymphs that can eat rice seedlings. Only ducks can chew them. So ducks are particularly abundant in Southern Kuang. Here crabs can be used to rear ducks, and ducks used to control crabs. The benefit is in both directions'.[b] The origin of the use of ducks to control insects in the field has been shrouded in uncertainty. It was generally believed to have started some time in the early Chhing Dynasty. Recent research by Min Tsung-Tien[3] has shown that the practice was actually invented by Chhen Ching-Lun[4] in the twenty-fifth year of the Wan-Li[11] reign-period (+1597) in the Ming Dynasty.[c] The circumstances in which this invention took place are related in a relatively unknown work entitled *Chih Huang Chhuan Hsi Lu*[12] (A Legacy of Locust Control) edited by Chhen Shih-Yuan[13], a fifth-generation descendant of Chhen Ching-Lun[14], during the Chhien-Lung[15] reign-period (+1736 to 96).[d]

Chhen Ching-Lun[4] was a native of Fuchow in Fukien province. His father, Chhen Chen-Lung[16], was an agricultural pioneer who advocated the planting of sweet potatoes in China. In the twenty-first year of the Wan-Li[1] reign-period (+1593), Fukien was plagued by a severe drought. Chhen Ching-Lun[4] petitioned the provincial government to promote the cultivation of sweet potatoes, in which effort he was highly successful. Later, based on his observation of the devouring of locusts by natural populations of egrets, he experimented on the use of ducks to

---

[a] Harpaz (1), p. 33. The offering of prayers to natural deities to ward off insect pests is a practice of great antiquity in China. For example, in the *Chou Li*[5], 'Records of the Rites of the Chou Dynasty', ch. 10, *Chhiu Kuan hsia*[6] (Autumn Official, lower section) on the functions of the *Chu Shih*[7] and the *Chien Shih*[8], cited earlier (p. 472).

[b] Tr. auct.; quoted in *Wu Shan Chih Lin, Pien Wu*[9], from *Ho Wên-Min Kung Wên Chi*[10], i.e. 'Collected Essays of the Venerable Ho Wen-Min', ch. 4, p. 1946.

[c] Min Tsung-Tien (1).

[d] According to Min Tsung-Tien, the only extant copy of the *Chih Huang Chhuan Hsi Lu* is preserved in the Fukien Library, Fuchow. For his work, he has used a reproduction available at the Agricultural History Research unit (*Nung Shih Yen Chiu Shih*[17]) at the South China College of Agriculture, *Hua Nan Nung Hsüeh Yuan*[18]. One of us (HTH) had an opportunity to inspect this reproduction at the South China College of Agriculture in October 1982.

[1] 霍韜	[2] 順德	[3] 閔宗殿	[4] 陳經綸	[5] 周禮
[6] 秋官下	[7] 庶氏	[8] 翦氏	[9] 五山志林, 辨物	
[10] 霍文敏公文集		[11] 萬歷	[12] 治蝗傳習錄	[13] 陳世元
[14] 陳經綸	[15] 乾隆	[16] 陳振龍	[17] 農史研究室	[18] 華南農學院

control this scourge. The results were quite encouraging and he described his experience in an article entitled, *Chih Huang Pi Chi*[1] (Pen Notes on Locust Control), which has been preserved as a part of the *Chih Huang Chhuan Hsi Lu*[5]. There he wrote:

The locust has been a scourge of the north and the west for generations, and different dynasties have combated the problem in different ways. When I travelled around the country to promote the cultivation of the sweet potato, I often had to contend with swarms of locusts which decimated the leaves on the vine. Later, I observed groups of flying birds zoom down and quickly consume them all. I noticed that these birds were usually egrets. According to the *Phi Ya*[2] (New Additions to the *Literary Expositor*), locusts are derived from young fish. When there is water, they grow into fish, but if water is scarce, they hover among the reeds along the shore, where being scorched by the heat and humidity, they turn into locusts. Egrets love to eat young fish, but they fly back and forth and it is difficult to tame and domesticate them. It occured to me that ducks live on land yet swim on water, and like the egrets, they are fond of young fish. I have grown several broods and released them where the egrets were wont to gather. Among the reeds along the shore they too gobbled up similar types of prey. In fact, they are even faster than the egrets, since their beaks are broad and flat, and their mouths are large. I have taught the farmers to rear several broods of ducklings, and release them in the spring and summer at various suitable locations, so in those years there were no locust infestations. Since this was the first time this usage had been tried, I have not yet publicised the results. I have just returned home after several tiring journeys, and I wonder whether the efficacy of this procedure will withstand the test of time.[a]

The article is dated in a ting-yu[3] year of the Wan-Li[4] reign-period, i.e. +1597, indicating that the invention was tested some time before that year. In the book, detailed directions are given on how the procedure is to be applied in the field.

After locating an area where locust nymphs congregate, indicate the spot by raising a smoke signal by day, or a bonfire at night. Have several tens of labourers carry baskets of ducklings to the area. Release the ducks so as to surround and destroy the infestation.[b] The best time to release the ducks is about twenty days before the locusts take to the air. It is said 'For twenty days the nymphs test their ability to fly. When the moon is full, they will swarm high in the air.'[c] Once they are airborne, the ducks will have no power over them. One duck can kill more locusts than one man. A single duckling can consume up to a thousand nymphs. Forty ducklings can destroy a forty thousand nymph infestation. One labourer can carry forty ducklings and thus his power is multiplied forty times. Furthermore, rearing ducks is not only an effective way to control locusts, it can also in itself be a profitable undertaking.[d]

[a] Tr. auct. Chhen Ching-Lun[1], *Chih Huang Pi Chi* in Chhen Shih-Yuan, as quoted by min Tsung-Tien (1).
[b] Tr. auct. in Chhen Shih-Yuan. The original reads: '*Chêng huang sha tsai ho fang, Jih tsê chü yen, Yeh tsê fang huo wei hao. Yung fu shu shih jen, Thiao ya shu shih lung, Pa mien huan êrh sha chih.*'[6]
[c] Tr. auct. in Chhen Shih-Yüan. The original reads: '*Liang hsün shih fei, Tsa yüeh kao thêng.*'[7]
[d] Tr. auct., excerpt from Min Tsung-Tien (1), p. 106.

[1] 治蝗筆記　　[2] 埤雅　　　　[3] 丁酉　　　　[4] 萬歷　　　　[5] 治蝗傳習錄
[6] 偵蝗煞在何方，日則舉烟，夜則放火爲號．用夫數十人，挑鴨數十籠，八面環而唼之
[7] 兩旬試飛，匝月高騰

Yet in spite of these apparent advantages, this technology did not catch on for about a hundred and seventy years, until Chhen Chiu-Chen[1], a fifth-generation descendant of Chhen Ching-Lun[2], went to assume an official post in Wu-hu[3]. There he was confronted with a severe locust infestation. We will let him tell what happened in his own words:

I have often thought of the Honourable Ching-Lun[4], my great-great-great grandfather—but have not had a chance to try out his method of locust control. Before I reported to my post at Wu-hu[5], I returned to my ancestral home, and was again briefed about his benevolent acts to promote the planting of sweet potatoes and the control of locusts.—After I assumed the new appointment, I had to take charge of a locust control programme. I tested my ancestor's method and it was found to be highly effective. As a result I was promoted to be the magistrate of Han-shan[5]. I repeated the procedure in several other districts, and the locust ceased to be a problem.[a]

The successful demonstration and dissemination of this invention was celebrated in verse as follows:

> Many are the schemes for the control of locusts,
> But none can equal the employment of ducks.
> They devour the young fish fry with enthusiasm,
> But they like the locust nymphs just as much.
> They seek out other grubs along the shores of ponds and lakes,
> Patrol the fields, and protect the rice-plants.
> New broods are reared throughout the north and the west,
> To free the land from pests that plagued us already before the Thang.[b]

Between +1597, when Chhen Ching-Lun[10] recorded his invention, and +1773 when Chhen Chiu-Chen[11] wrote his introduction to the *Chih Huang Chhuan Hsi Lu*[12], there had been eighty major outbreaks of locusts in China[c]. Why was the innovation not disseminated and utilised sooner? We can see two reasons. First, Fukien is not in a region where locust infestation normally occurs, and so large-scale

[a] Tr. auct. Chhen Chiu-Chen in *Chih Huang Hsü*[6], 'Introduction to the *Chih Huang Chhuan Hsi Lu*[7]' It is not known when Chhen Chiu-Chen went to take up his post in Wu-hu. But it is stated that the introduction was written in the Kuei-Ssu[8] year of the Chhien-Lung reign-period, i.e. +1773. Chiu-Chen was apparently Shih-Yuan's elder brother.

[b] Tr. auct. *Chhu Ya Chih Huang*[13] (Duck Rearing for Locust Control) by Ong Tien-Tui[14], included in *Chih Huang Chhüan Hsi Lu*. The original reads:

> *Hsü to miao tshê shuo pu huang,*
> *Tsung suan wu ju chhu ya chhang.*
> *Yü tzu hsia tshan chen pi chüeh,*
> *Nan chhung sha chin chien liang fang.*
> *Hu ming shui chi hsin mao thêng,*
> *Chieh chen thien chien pao tao liang.*
> *Hsi pei tshun tshun shih yün tzu,*
> *Thun tsai tuan ti lou chhien Thang.*'[9]

[c] Chou Yao (2), pp. 197–204.

[1] 陳九振   [2] 陳經綸   [3] 蕪湖   [4] 經綸   [5] 含山
[6] 治蝗序   [7] 治蝗傳習錄   [8] 癸己
[9] 許多妙策說捕蝗，總算無如畜鴨長．魚子呷殘眞秘訣，蝻蟲唉盡見良方．
    呼名水際尋孟朦，結陣田間保稻粱．西北村村時孕字，吞災端的陋前唐．
[10] 陳經綸   [11] 陳九振   [12] 治蝗傳習錄   [13] 畜鴨治蝗   [14] 翁殿對

field tests could not be easily planned and carried out locally. Secondly, being without power or influence, it was difficult for Chhen Ching-Lun to persuade officials in other provinces to test the technology and put it into practice. But five generations later, both factors converged favourably in the person of Chhen Chiu-Chen. First, he was confronted with a large locust infestation. Secondly, as an official in charge of locust control he was in a position to take a novel approach and to organise and execute a proper field demonstration. Once favourable results were achieved the procedure was quickly adopted and used successfully in later infestations. Its efficacy was soon firmly established.

It is possible that Chhen Ching-Lun's invention had not remained quite as obscure as we are led to believe from the meagre record left by his descendents. In the *Chhu Huang Chi*[1] (Methods for Extermination of Locusts) written by Lu Shih-I[2] in the early Chhing dynasty, we find the following passage on the use of ducks to control locusts: 'Before the nymphs begin to fly, they are easily eaten by ducks. By releasing several hundred ducks into the paddy, the nymphs are soon exterminated. This method is much favoured south of the river.'[a] This account was later quoted in the *Huang Chêng Chi Yao*[3] (Principles of Famine Management), 1806, by Wang Chih-I.[4] Another account is given by Ku Yen[5] (1857) in the *Chih Huang Chhüan Fa*[6] (Complete Methods of Locust Control). He says: 'During the fourth month of the seventh year of the Hsien-Fêng[7] reign-period (1857), there was an infestation of locust nymphs in Chün-Chang[8] hill region in Wu-hsi[9]. Between seven to eight hundred ducks were released, and soon the nymphs were destroyed.'[b] Ducks are particularly suitable for the control of pests on paddy rice. They have now been used continuously in the Pearl River Delta and neighbouring districts in Kuangtung and Kuangsi Provinces for many generations.[c]

The use of vertebrates to control insect pests has also been noted in the West. Sweetman has cited two examples:[d] the importation of the giant toad *Bufo marinus* (L.) from Jamaica to Barbados in 1844, and the introduction of the Indian mynah bird *Acridotheres tristis* (L.) into Mauritius in +1762. The latter has successfully controlled the red locust *Nomadocris septemfasciata* there ever since.

But enough has been said, we believe, in the foregoing discussion to show that before the nineteenth century there was greater recognition and appreciation of biological agents for the control of noxious agricultural pests in China than in the Western world. This tradition has undoubtedly been a source of inspiration for the modern flowering of biological control that has taken root in China in the last thirty years.

[a] Tr. auct. *Chhing, Ching Shih Wên Phien*[10] (Collections of Papers of the Chhing Dynasty), vol. 45, p. 6. Lu Shih-I[11] lived from 1611 to 1672, but the exact date of publication of his book is unknown.
[b] Tr. auct., ch. 1, as quoted by Min Tsung-Tien[12] (1), p. 107.
[c] Phu Chê-Lung (1), p. 233.
[d] Sweetman, (2), p. 2.

[1] 除蝗記	[2] 陸世儀	[3] 荒政輯要	[4] 汪志伊	[5] 顧彥
[6] 治蝗全法	[7] 咸豐	[8] 軍嶂	[9] 無錫	[10] 清, 經世文篇
[11] 陸世儀	[12] 閔宗殿			

BIBLIOGRAPHIES

A   CHINESE AND JAPANESE BOOKS BEFORE + 1800

B   CHINESE AND JAPANESE BOOKS AND JOURNAL ARTICLES SINCE + 1800

C   BOOKS AND JOURNAL ARTICLES IN WESTERN LANGUAGES

In Bibliographies A and B there are two modifications of the Roman alphabetical sequence: transliterated *Chh-* comes after all other entries under *Ch-*, and transliterated *Hs-* comes after all other entries under *H-*. Thus *Chhen* comes after *Chung* and *Hsi* comes after *Huai*. This system applies only to the first words of the titles. Moreover, where *Chh-* and *Hs-* occur in words used in Bibliography C, i.e. in a Western language context, the normal sequence of the Roman alphabet is observed.

When obsolete or unusual romanisations of Chinese words occur in entries in Bibliography C, they are followed, wherever possible, by the romanisations adopted as standard in the present work. If inserted in the title, these are enclosed in square brackets; if they follow it, in round brackets. When Chinese words or phrases occur romanised according to the Wade–Giles system or related systems, they are assimilated to the system here adopted (cf. Vol. 1, p. 26) without indication of any change. Additional notes are added in round brackets. The reference numbers do not necessarily begin with (1), nor are they necessarily consecutive, because only those references required for this volume of the series are given.

Korean and Vietnamese books and papers are included in Bibliographies A and B. As explained in Vol. 1, pp. 21 ff., reference numbers in italics imply that the work is in one or other of the East Asian languages.

ABBREVIATIONS

See also p. xviii

A/AIHS	Archives Internationales d'Histoire des Sciences	BABEL	Babel; Revue Internationale de la Traduction
AAAG	Annals of the Association of American Geographers	BABGPB	Bull. Applied Bot., Genetics and Plant Breeding (USSR)
AAN	American Anthropologist	BAPES	Bull. American Peony Soc.
ABRN	Abr-Nahrain (Annual of Semitic Studies. Universities of Melbourne and Sidney)	BBAE	Bulletin of the Bureau of American Ethnology (Smithsonian Inst.)
ACOM	Acta Comeniana (Prague)	BBPI/DA	Bull. Bureau of Plant Industry, (U.S.) Dept. of Agriculture
ACPP	Annales Cryptogramiae et Phytopathologiae		
ADVS	Advancement of Science (British Assoc., London)	BCGS	Bulletin Chinese Geological Soc.
		BCSA	Bulletin Chrysanthemum Soc. America
AENSAT	Annales de l'Ec. Nat. Sup. Agronomie de Toulouse	BDB/SYS	Bull. Dept. Biol., Coll. of Sci., Sun Yat-Sen University (Canton)
AEST	Annales de l'Est (Fac. des Lettres. Univ. Nancy)	BDBG	Ber. d. deutschen botanischen Gesellschaft
		BDCG	Ber. d. deutsch. chem. Gesellschaft
AGHST	Agricultural History (Washington. D.C.)	BDS	Bulteno de Scienco (Kho-Hsüeh Shih Pao, Peiping)
AGKAW	Abhandlungen a. d. Gebiet d. Klass. Altentumswissenschaft	BE/AMG	Bibliographie d'Etudes (Annales du Musée Guimet)
AGMW	Abhandlungen z. Geschichte d. Math. Wissenschaft	BEFEO	Bulletin de l'Ecole Française de l'Extrême Orient (Hanoi)
AGNT	Archiv f. d. Gesch. d. Naturwiss. u. d. Technik	BFMIB	Bull. Fan Memorial Institute of Biology (Peking)
AGZ	Allgemeine Gartenzeitung	BGCA	Bull. Garden Club of America
AHES/AESC	Annales; Economies, sociétés, civilisations	BIBGEN	Bibliographie Genetica
AHOR	Antiquarian Horology	BIHM	Bulletin of the (Johns Hopkins) Institute of the History of Medicine
AHP	Acta Horti Petropolitani		
AHRA	Agric. History Research Annual (Nung Shih Yen-Chiu Chi-Khan, formerly Nung Yeh I-Chhan Yen-Chiu Chi-Khan)	BJBSP	Bull. du Jardin Bot. de St. Petersburg
		BK	Bunka (Culture), Sendai
		BL	Blumea
		BLSOAS	Bulletin of the London School of Oriental and African Studies
AJCM	Amer. Journ. Chinese Medicine		
AJSC	American Journ. Science and Arts	BMFEA	Bulletin of the Museum of Far Eastern Antiquities (Stockholm)
AK	Arkiv för Kemi		
ALC/DO	See ARLC/DO	BMFJ	Bulletin de la Maison Franco-Japonaise (Tokyo)
AM	Asia Major		
AMSC	American Scientist	BNISI	Bulletin of the National Institute of Sciences of India
ANS	Annals of Science		
ANYAS	Annals of the New York Academy of Sciences	BNYAM	Bull. New York Acad. of Med.
		BOTJ	(Engler's) Bot. Jahrb. (f. Systemat. Pflanzengesich. u. Pflanzengeogr.)
APS	Acta Pedologica Sinica		
APTGB	Acta Phyto-taxonomica et Geo-botanica	BOTM	(Curtis') Botanical Magazine
AQ	Antiquity	BR	Biological Reviews
ARB	Annual Review of Biochemistry	BSA	Bull. de la Société d'Acclimatation (several variations in title)
ARLC/DO	Annual Reports of the Librarian of Congress (Division of Orientalia)	BSAC	Bulletin de la Société d'Acupuncture
ARO	Archiv Orientalni (Prague)	BSBF	Bull. de la Société Chimique de France
ARSI	Annual Reports of the Smithsonian Institution	BSEIC	Bulletin de la Société des Etudes Indochinoises
AS/BIE	Bulletin of the Institute of Ethnology. Academia Sinica (Thaiwan)	BSRCA	Bulletin of the Society for Research in (the History of) Chinese Architecture
AS/BIHP	Bulletin of the Institute of History and Philology. Academia Sinica	BTBC	Bull. of the Torrey Botanical Club
		BZ	Biochemische Zeitschrift
ASEA	Asiatische Studien; Etudes Asiatiques		
AX	Ambix	CAMR	Cambridge Review

CAND	Candollea (Geneva)
CBLSSC	Contributions from the Biological Laboratory of the Science Society of China
CBOT	Chronica Botanica
CCB	Chung-Chi Bulletin (Chhung-Chi Univ. Coll. Hongkong)
CCIT	California Citrograph
CCJ	Chung-Chi Journal (Chhung-Chi Univ. Coll. Hongkong)
CCN	Christian Century (New York)
CEIB	Ceiba (Botany)
CEN	Centaurus
CF	Chinese Forestry
CHIND	Chemistry and Industry (Journ. Soc. Chem. Ind. London)
CHJ/T	Chhing-Hua (Ts'ing-Hua) Journal of Chinese Studies (New Series, publ. Thaiwan)
CHWSLT	Chung-Hua Wên-Shih Lun Tshung (Collected Studies in the History of Chinese Literature)
CJ	China Journal of Science and Arts
CJE	Chinese Journ. Ecology
CJEB	Chinese Journ. Exp. Biol.
CJOP	Chinese Journ. Physiol.
CJST	Chinese Journ. Stomatol.
CKYW	Chung-Kuo Yü Wen (Peking)
CMED	Clio Medica (Internat. Journ. Hist. Med.)
CMJ	Chinese Medical Journal
CMJ/C	Chinese Medical Journal (Chinese edition)
CONBT	Contributions from the Boyce Thompson Res. Inst. (Bot.)
CPICT	China Pictorial
CR	China Review (Hongkong and ·Shanghai)
CREC	China Reconstructs
CRRR	Chinese Repository
CWFLHP	Chih-Wu Fen-Lei Hsüeh-Pao (Acta Phytotaxonomica Sinica)
CWHP	Chih Wu Hsüeh Pao (Acta Botanica Sinica)
CYC	Cycles (Foundation for the Study of Cycles)
D	Discovery
DISS/CKSA	Contributions to the Knowledge of the Soils of Asia (Dokuchaiev Instit. of Soil Science, Moscow)
DWAW/MN	Denkschrifter d. k. Akad. d. Wissenschaften Wien; Math-Naturwiss. kl.
ECB	Economic Botany
EDR	Edinburgh Review
EHOR	Eastern Horizon (Hongkong)
END	Endeavour
EP	Epetéris (Journ. Nat. Research Inst. Leucosia (Nicosia, Cyprus))
EPI	Episteme
FCON	Fortschritte d. Chemie d. organischen Naturstoffe

FEQ	Far Eastern Quarterly (continued as Journal of Asian Studies)
FL	Folklore
FLOSIL	Flora and Silva
FMNHP/AS	Field Museum of Natural History (Chicago) Publications; Anthropological Series
G	Geography
GBSS	Gardens Bulletin of the Straits Settlements (continuation of Agric. Bull. Malay States; later Garden. Bull. Singapore)
GDC	Gardeners' Chronicle
GDN	The Garden
GDNF	Garden and Forest
GESN	Gesnerus
GH	Gentes Herbarum (Ithaca N.Y.)
GJ	Geographical Journal
GR	Geographical Review
HEJ	Health Education Journal
HH	Han Hiue (Han Hsüeh); Bulletin du Centre d'Etudes Sinologiques de Pékin
HJAS	Harvard Journal of Asiatic Studies
HKHS/ON	Hongkong Horticult. Soc. Occasional Notes
HKN	Hongkong Naturalist
HR	Horticultural Register (Paxton's)
HS	Historia Scientiarum (continuation of JSHS)
IEC/AE	Industrial and Engineering Chemistry; Analytical Edition
ISIS	Isis
ISM	Interferon Scientific Memoranda
ISTC	I Shih Tsa Chih (Chinese Journal of the History of Medicine)
JA	Journal Asiatique
JAA	Journal of Asian Art
JAAC	Journ. Agric. Assoc. China
JAHIST	Journ. Asian History (International)
JAN	Janus
JAOS	Journal of the American Oriental Society
JAS	Journal of Asian Studies (continuation of Far Eastern Quarterly, FEQ)
JATBA	Journal d'Agriculture tropicale et de Botanique appliqué
JBSC	Journ. Bot. Soc. China
JCE	Journal of Chemical Education
JCR	Journ. Chem. Research: (S) Synopses; (M) Microfiches
JDN	Jardin
JEH	Journal of Economic History
JESHO	Journal of the Economic and Social History of the Orient
JG	Journ. Genetics
JHMAS	Journal of the History of Medicine and Allied Sciences
JHS	Journal of Hellenic Studies
JLS/B	Journ. Linnean Society, Bot. Sect.
JNMT	Jaarb. d. Nederland. Maatsch. Tuinbouw (Hort. Soc. Holland)

JNYBG	*Journ. New York Bot. Garden*	*MAN*	*Mathematische Annalen*
JOP	*Journal of Physiology*	*MAO*	*Memoires de l'Athenée Oriental* (Paris)
JOSHK	*Journal of Oriental Studies* (Hongkong Univ.)	*MBRF*	*Magazine of Botany and Register of Flowers* (*Paxton's*)
JPISH	*Jih-Pên I Shih-Hsüeh Tsa Chih* (Jap. Journ. Medical History)	*MCHSAMUC*	*Mémoires concernant l'Histoire, les Sciences, les Arts, les Moeurs et les Usages, des Chinois, par les Missionnaires de Pékin* (Paris, 1776)
JRAGS	*Journ. Royal Agricultural Soc.*		
JRAI	*Journal of the Royal Anthropological Institute*	*MDGNVO*	*Mitteilungen d. deutsch. Gesellschaft f. Natur. u. Volskunde Ostasiens*
JRAS	*Journal of the Royal Asiatic Society*	*MED*	*Medicus* (Karachi)
JRAS/M	*Journal of the Malayan Branch of the Royal Asiatic Society*	*MFSA/TIU*	*Memoirs of the Faculty of Science and Agriculture, Taihoku Imp. Univ.*
JRAS/NCB	*Journal (or Transactions) of the North China Branch of the Royal Asiatic Society*	*MGSC*	*Memoirs of the Chinese Geological Survey*
JRHS	*Journ. Royal Horticultural Society* (contd. as *The Garden*)	*MH*	*Medical History*
		MHJ	*Middlesex Hospital Journal*
JSBNH	*Journ. of the Soc. for the Bibliography of Natural History*	*MHP*	*Monspeliensis Hippocrates* (Montpellier)
		MN	*Monumenta Nipponica*
JSFA	*Journal of the Science of Food and Agriculture*	*MNFGB*	*Mitt. d. Naturforschenden Gesellschaft in Bern*
JSHS	*Japanese Studies in the History of Science* (Tokyo)	*MRAB*	*Memorie della R. Accad. Bologna*
JSSI	*Journal of the Shanghai Science Institute*	*MS*	*Monumenta Serica*
JWAS	*Journal of the Washington Academy of Science*	*MSOS*	*Mitteilungen d. Seminars f. orientalische Sprachen* (Berlin)
JWCBRS	*Journal of the West China Border Research Society*	*MSRSL*	*Mem. Soc. Roy. Sci. de Liège*
		MSTRM	*Mainstream* (New York)
JWCI	*Journal of the Warburg and Courtauld Institutes*	*MTC*	*Mem. Tanaka Citrus Exp. Station* (Minomura, Fukuoka-ken)
JWH	*Journal of World History* (UNESCO)	*N*	*Nature*
KBMI	*Kew (Gardens) Bull. of Misc. Information*	*NACJ*	*Nanking Agricultural College Journal*
KHCK	*Kuo Hsüeh Chi Khan* (Chinese Classical Quarterly)	*NAMHN*	*Nouvelles Archives du Museum National d'Histoire Naturelle* (Paris)
		NFLOSIL	*New Flora and Silva*
KHHPN	*Kuo Hsüeh Hui Pien* (Chilu University Journ.)	*NFRB/TB*	*National Forestry Research Bureau, Technical Bulletin*
KHS	*Kho Hsüeh* (Science)		
KHSC	*Kho-Hsüeh Shih Chi-Khan* (Ch. Journ. Hist. of Sci.)	*NHM*	*National Horticultural Magazine* (U.S.A.)
		NMJC	*National Medical Journal of China*
KHTP	*Kho-Hsüeh Thung Pao* (Scientific Correspondent)	*NPA/CIB*	*Contributions from the Institute of Botany, National Peiping Academy*
KKD	*Kiuki Daigaku Sekai Keizai Kenkyūjo Hōkoku* (Reports of the Institute of World Economics at Kiuki Univ.)	*NPA/CIP*	*Contributions from the Institute of Physiology, Nat. Peiping Academy*
		NQCJ	*Notes and Queries on China and Japan*
		NRBGE	*Notes of the Royal Bot. Garden, Edinburgh*
		NRRS	*Notes and Records of the Royal Society*
LAN	*Language*	*NS*	*New Scientist*
LAR	*Lingnan Agricultural Review*	*NSN*	*New Statesman and Nation* (London)
LHP	*Lingnan Hsüeh Pao* (Lingnan University Journal)	*NW*	*Naturwissenschaften*
		NWACJ	*Northwest Agricultural College Journal*
LINN	*Linnaea*		
LP	*La Pensée*		
LSJ	*Lingnan Science Journal*	*OAS*	*Ostasiatische Studien* (Berlin)
LSYC	*Li Shih Yen Chiu* (Journal of Historical Research), Peking	*OAZ*	*Ostasiatische Zeitschrift*
		OE	*Oriens Extremus* (Hamburg)
		OLZ	*Orientalische Literatur-Zeitung*
M	*Mind*	*OP*	*The Optician*
MAAAS	*Memoirs of the American Acadmey of Arts and Sciences*	*ORC*	*Orchid Review*
MAI/NEM	*Mémoires de l'Academie des Inscriptions et Belles-Lettres, Paris* (*Notices et Extraits des MSS.*)	*PA*	*Pacific Affairs*
		PAMAAS	*Proc. Amer. Assoc. for the Advancement of Science*

PANS	Pest Articles and News Summaries	SHIY	Shanghai Chung I Yao Tsa Chih (Shanghai Journ. Traditional Chinese Medicine and Pharmacy)
PANSP	Proc. Acad. Nat. Sci. Philadelphia		
PAPS	Proc. Amer. Philos. Soc.		
PC	People's China		
PCAS	Proc. California Academy of Sciences	SOS	Semitic and Oriental Studies (Univ. of California Publ. in Semitic Philol.)
PEW	Philosophy East and West (Univ. Hawaii)	SPCK	Society for the Promotion of Christian Knowledge
PHY	Physis (Florence)	SPR	Science Progress
PJ	Pharmaceut. Journal (and Trans. Pharmaceut. Soc.)	SS	Science and Society (New York)
		SSIP	Shanghai Science Institute Publications
PL	Philologus, Zeitschrift f. d. Klass. Altertums	STIC	Science and Technology in China
		STK	Sōdai Kenkyu Bunken Teiyō
PLM	The Plantsman	SWAW/PH	Sitzungsberichte d. k. Akad. d. Wissenschaften Wien (Phil.-Hist. Klasse), Vienna
PLS	Proc. Linnean Soc. (London)		
P.NASW	Proc. Nat. Acad. Sci. Washington		
P.NHB	Peking Natural History Bulletin	SYZ	Systematic Zoology
POCH	Pochvovedenie (Soil Science); Moscow (see PVV)		
		TAMS	Transactions of the American Microscopical Society
PP	Past and Present		
PRMS	Proc. Roy. Microscop. Soc	TAPS	Transactions of the American Philosophical Society
PRSB	Proceedings of the Royal Society (Series B)		
PRSM	Proceedings of the Royal Society of Medicine	TAS/J	Transactions of the Asiatic Society of Japan
PS	Palaeontologica Sinica	TAX	Taxon
PSAM	Proc. Symposia in Applied Maths (Amer. Math. Soc.)	TCULT	Technology and Culture
		TFTC	Tung Fang Tsa Chih (Eastern Miscellany)
PSSA	Proc. Soil Science Society of America		
PTRS	Philosophical Transactions of the Royal Society	TG/K	Tōhō Gakuhō, Kyōto (Kyoto Journal of Oriental Studies)
PVV	Pochvovedenie (Soil Science) (Moscow)	TIYT	Trudy Instituta Istorü Yestestvoznania i Tekhniki
QJGS	Quarterly Journal Geol. Soc. (London)	TJTC	Tza-Jan Tsa Chih (Nature Magazine)
QJTM	Quarterly Journ. Thaiwan Museum (Thaipei)	TLS	Transactions of the Linnean Society
		TLTC	Ta Lu Tsa Chih (Continent Magazine), (Thaipei)
RAM	Ramparts (New York)	TNS	Transactions of the Newcomen Society
RBS	Revue Bibliographique de Sinologie	TORR	Torreya
RHORT	Revue Horticole	TP	T'oung Pao (Archives concernant l'Histoire, les Langues, la Géographie, l'Ethnographie et les Arts de l'Asie Orientale), Leiden
RHP	Revue d'Histoire de la Pharmacie		
RHS	Revue d'Histoire des Sciences (Centre Internationale de Synthèse, Paris)		
RHSID	Revue d'Histoire de la Sidérurgie (Nancy)	TRHS	Trans. Royal Horticultural Society
RO	Rocznik Orientalistyczny (Warsaw)	TSGH	Tōkyō Shinagaku-hō (Bull. of Tokyo Sinol. Soc.)
RPBP	Review of Palaeobotany and Palynology		
RPHARM	Repertorium f. d. Pharmacie	TSSC	Transactions of the Science Society of China
RPPCR	Repertorium f. Pharm. u. prakt. Chem. in Russland	TSSV	Travaux de la Societé Scientifique de Varsovie
RSO	Rivista di Studi Orientali	TT	Tools and Tillage
S	Sinologica (Basel)	TYBK	Tōyō Bungaku Kenkyū (Waseda Univ. Tokyo)
SA	Sinica (originally Chinesische Blätter f. Wissenschaft u. Kunst)	TYG	Tōyō Gakuhō (Reports of the Oriental Society of Tokyo)
SACUN	Sacu (Society for Anglo-Chinese Understanding) News (London)		
SAM	Scientific American	VB	Vegetationsbilder (ed. G. Karsten & H. Schenk, Jena)
SAR	Sargentia (Bot.)	VFDM/GNT	Veröffentlichungen des Forschungsinstitut des Deutschen Museums für die Gesch. d. Naturwiss. und der Technik
SBE	Sacred Books of the East (Series)		
SC	Science		
SCI	Scientia		
SCISA	Scientia Sinica (Peking)	W	Weather
SEF	Sefunot (Jerusalem)	WT	Wennti [Wên-thi; The Question] (a bulletin of Chinese Studies, issued
SF	Soils and Fertilisers (Rothamsted)		

by the Hall of Graduate Studies, Yale University, New Haven, Conn.)

YBDA   *Year-book of the (U.S.) Dept. of Agriculture*

YK   *Yü Kung (Chinese Journal of Historical Geography)*

ZANN   *Zoologische Annalen*

ZGEB   *Zeitschrift d. Gesellschaft f. Erdkunde (Berlin)*

A. CHINESE AND JAPANESE BOOKS BEFORE +1800

Each entry gives particulars in the following order:

(a) title, alphabetically arranged, with characters;
(b) alternative title, if any;
(c) translation of title;
(d) cross-reference to closely related book, if any;
(e) dynasty;
(f) date as accurate as possible;
(g) name of author or editor, with characters;
(h) title of other book, if the text of the work now exists only incorporated therein; or, in special cases, references to sinological studies of it;
(i) references to translations, if any, given by the name of the translator in Bibliography C;
(j) notice of any index or concordance to the book if such a work exists;
(k) reference to the number of the book in the *Tao Tsang* catalogue of Wieger (6), if applicable;
(l) reference to the number of the book in the *San Tsang* (Tripiṭaka) catalogues of Nanjio (1) and Takakusu & Watanabe, if applicable.

Words which assist in the translation of titles are added in round brackets.

Alternative titles or explanatory additions to the titles are added in square brackets.

It will be remembered (p. 305 above) that in Chinese indexes words beginning *Chh-* are all listed together after *Ch-*, and *Hs-* after *H-*, but that this applies to initial words of titles only.

Where there are any differences between the entries in these bibliographies and those in Vols. 1–4, the information here given is to be taken as more correct.

An interim list of references to the editions used in the present work, and to the *tshung-shu* collections in which books are available, has been given in Vol. 4, pt. 3, pp. 913 ff., and is available as a separate brochure.

ABBREVIATIONS

C/Han	Former Han.
E/Wei	Eastern Wei.
H/Han	Later Han.
H/Shu	Later Shu (Wu Tai).
H/Thang	Later Thang (Wu Tai).
H/Chin	Later Chin (Wu Tai).
S/Han	Southern Han (Wu Tai).
S/Phing	Southern Phing (Wu Tai).
J/Chin	Jurchen Chin.
L/Sung	Liu Sung.
N/Chou	Northern Chou.
N/Chhi	Northern Chhi.
N/Sung	Northern Sung (before the removal of the capital to Hangchow).
N/Wei	Northern Wei.
S/Chhi	Southern Chhi.
S/Sung	Southern Sung (after the removal of the capital to Hangchow).
W/Wei	Western Wei.

Chan Kuo Tshê 戰國策.
Records of the Warring States [semi-fictional].
Chhin.
Writer unknown.

Chang Shih Tshang Shu 張氏藏書.
Mr Chang's Treasuring.
Ming, +1596.
Chang Ying-Wên 張應文.

Chê Chiang Lu 哲匠錄.
see in Bib. B:
Chu Chhi-Chhien & Liang Chhi-Hsiung (*1* to *6*). Chu Chhi-Chhien, Liang Chhi-Hsiung & Liu Ju-Lin (*1*), and Chu Chhi-Chien & Liu Tun-Chen (*1, 2*).

Chen Chu Chhuan 珍珠船.
The Pearly Boat (sports on the lake at the Court of the Thang).
Ming.
Chhen Chi-Ju 陳繼儒.

Chêng-Ho Hsin-Hsiu Ching-Shih Chêng Lei Pei-Yung Pên Tshao 政和新修經史證類備用本草.
New Revision of the Classified and Consolidated Armamentarium Pharmacopoeia of the Chêng-Ho reign-period.
Sung, +1116; repr. +1143 (J/Chin).
Thang Shen-Wei 唐愼微.
Ed. Tshao Hsiao-Chung 曹孝忠.

Chêng Lei Pên Tshao
Classified Pharmaceutical Natural History.
See Ching-Shih Chêng Lei Pei-Chi Pên Tshao, +1083, repr. +1090. Ta-Kuan Ching-Shih Chêng Lei Pei-Chi Pên Tshao, +1108. Chêng-Ho Hsin-Hsiu Ching-Shih

Chêng Lei Pei-Yung Pên Tshao, +1116, repr. +1143. Shao-Hsing Chiao-Ting Ching-Shih Chêng Lei Pei-Chi Pên Tshao, +1157. Chhung-Hsiu Chêng-Ho Ching-Shih Chêng Lei Pei-Yung Pên Tshao, +1204 & +1249. Ching-Shih Chêng Lei Ta-Chhüan Pên Tshao +1577 & later.
See, e.g. Anon (*65*); Chang Tsan-Chhen (*1, 2*); Hung Kuan-Chih (*4*); Hummel (*13*); Wang Chün-Mo (*1*); Ma Chi-Hsing (*1*); Ting Chi-Min (*1*); Swingle (6).

Chêng Lei Pên Tshao 證類本草.
(*Chhung Hsiu Chêng-Ho Ching Shih Chêng Lei Pei Yung Pên Tshao.*) Reorganised Pharmacopoeia.
N/Sung, +1108, enlarged +1116; re-edited in J/Chin, +1204, and definitively re-published in Yuan, +1249; re-printed many times afterwards, e.g. in Ming, +1468.
Original compiler: Thang Shen-Wei 唐愼微.
Cf. Hummel (*13*); Lung Po-Chien (*1*).

Chêng Lun 政論.
On Government.
H/Han, +155.
Tshui Shih 崔寔.
In *YHSF*, ch. 71, pp. 67*a* ff.

Chêng Tien 政典.
Governmental Institutes [political and social encyclopoedia].
Thang, +732.
Liu Chih 劉秩.
Now known only as embodied in the *Thung Tien* of Tu Yu (q.v.).

Chêng Tzu Thung 正字通.
Complete character Orthography [dictionary].
Ming, +1627.
Chang Tzu-Lieh 張自烈.

Chi Chiu (Phien) 急就(篇).
Handy Primer [orthographic word-lists intended
for verbal exposition, connected with a continuous thread of text, and having some
rhyme arrangements].
C/Han, between −48 and −33.
Shih Yu 史游. With +7th cent. commentary
by Yen Shih-Ku 顏師古 and +13th cent.
commentary by Wang Ying-Lin 王應麟.

Chi Ni Tzu 計倪子.
[*Fan Tzu Chi Jan* 范子計然].
The Book of Master Chi Ni.
Chou (Yüeh), −4th century.
Attrib. Fan Li 范蠡, recording the philosophy
of his master Chi Jan 計然.

Chi Than Lu 劇談錄.
Records of Entertaining Conversations.
Thang, *c.* +885.
Khang Phien 康駢 or 軿.

Chi Wang Kung Hua Phin 冀王宮花品.
Grades of (Tree-Peony) Flowers in the Gardens
of the Prince of Chi.
Sung, *c.* +1035.
Chao Wei-Chi 趙惟吉.
Now extant only in quotations.

Chia-Yu Pu-Chu Shen Nung Pên Tshao 嘉祐補註
神農本草.
Supplementary Commentary on the
Pharmacopoeia of the Heavenly Husbandman,
Commissioned in the Chia-Yu reign-period.
Sung, commissioned, +1057, finished +1060.
Chang Yü-Hsi 掌禹錫.
Lin I 林億.
& Chang Tung 張洞.

Chiao-Chou Chi 交州記.
Records of Chiao-Chow (District), [mod.
Annam].
Chin. *c.* +410.
Liu Hsin-Chhi 劉欣期.

Chiao-Chou I Wu Chih 交州異物志.
Strange Things (incl. plants and animals) in
Chiao-Chow (District), [mod. Annam].
H/Han, *c.* +90.
Yang Fu 楊孚.
Perhaps the later title of his *Nan I I Wu Chih*.
Extant only in quotations.

Chieh-Ku Lao Jen Chen Chu Nang 潔古老人珍
珠囊.
Old Master Chieh-Ku's Bag of (Pharma-
ceutical) Pearls.
J/Chin *c.* +1200.
Chang Yuan-Su [Chieh-Ku] 張元素.
In *Chi Shêng Pa Sui*, no. 5.

Chih Chai Shu Lu Chieh Thi 直齋書錄解題.
Analytical Catalogue of (3070) Books Preserved
in the Library of (Mr Chhen) Chih-Chai.
Sung, *c.* +1236.
Chhen Chen-Sun 陳振孫.

See Têng & Biggerstaff (1), p. 21.

Chih Huang Chhüan Fa 治蝗全法.
Complete Handbook of Locust Control.
See Ku Yen (1) in Bib. B.

Chih Wên Pên Tshao 質問本草.
The Candid Enquiror's Pharmaceutical Natural
History (of the Liu-Chhiu Islands).
Chhing, *c.* +1765.
Wu Chi-Chih 吳繼志.

Chih Wu Ming Shih Thu Khao (and *Chhang Phien*).
See Wu Chhi-Chün (1, 2).

Chin-Chang Lan Phu 金漳蘭譜.
A Treatise on the Orchids of Fukien.
Sung, +1233.
Chao Shih-Kêng 趙時庚.
Tr. Wathing (2).

Chin Shu 晉書.
History of the Chin Dynasty [+265 to +419].
Thang, +635.
Fang Hsüan-Ling 房玄齡.
A few chs. tr. Pfizmaier (54–57); the astrono-
mical chs. tr. Ho Ping-Yü (1). For transla-
tions of passages see the index of Frankel (1).

Chin Ssu Lu 近思錄.
Summary of Systematic Thought; or,
Reflections on Things at Hand.
Sung, +1175.
Chu Hsi & Lü Tsu-Chhien 朱熹, 呂祖謙.
Trs. Graf (1); Chhen Jung-Chieh (11).

Chin Wên Shang Shu 今文尚書.
The 'New Text' version of the *Historical Classic*
(collected fragments, readings and commen-
taries).
Chou, Han & Chin.
Ed. Ma Kuo-Han 馬國翰.
YHSF, ch. 9, p. 2a.

Ching-Chai Ku Chin Chu 敬齋古今黈.
The Commentary of (Li) Ching-Chai (Li Yeh)
on Things Old and New.
Sung, +13th century.
Li Yeh 李冶.

Ching-Shih Chêng Lei Pei-Chi Pên Tshao 經史證類
備急本草.
The Classified and Consolidated Armamen-
tarium of Pharmaceutical Natural History.
Sung, +1083, repr. +1090.
Thang Shen-Wei 唐慎微.

Ching-Shih Chêng Lei Ta-Kuan Pên Tshao 經史證類
大觀本草.
See *Ta-Kuan Ching-Shih Chêng Lei Pei-Chi Pên
Tshao*, +1108.

Ching-Tê-Chen Thao Lu 景德鎮陶錄.
See Lan Phu (1).

Ching Tien Shih Wên 經典釋文.
Textual Criticism of the Classics.
Sui, *c.* +600.
Lu Tê-Ming 陸德明.

Chiu Chang Suan Shu 九章算術.
Nine Chapters on the Mathematical Art.
H/Han, +1st century (containing much
material from C/Han and perhaps Chhin).
Writer unknown.

Chiu Huang Chhüan Shu 救荒全書.
The Whole Art of Famine Relief.
Sung, +12th century.
Tung Wei 董煟.
In *Huang Chêng Tshung Shu*, q.v.

Chiu Huang Huo Min Shu 救荒活民書.
The Rescue of the People; a Treatise on Famine
Prevention and Relief.
Sung, +12th century.
Tung Wei 董煟.

Chiu Huang Pên Tshao 救荒本草.
Treatise on Wild Food Plants for Use in
Emergencies.
Ming +1406, repr. +1525, +1555, etc.
Chu Hsiao 朱橚 (prince of the Ming) Chou
Ting Wang 周定王.
Incl. as chs. 46 to 59 of *Nung Chêng Chhüan Shu*,
(q.v.).

Chiu Huang Tshê 救荒策.
Plans for Famine Relief (and Administration).
Chhing, c. +1665.
Wei Hsi 魏禧.

Chiu Huang Tshê 救荒策.
Plans for Famine Relief (and Administration).
Chhing, +1774.
Tshui Shu 崔述.

Chiu Huang Yeh Phu 救荒野譜.
Alternative title later given to the *Yeh Tshai Phu*
of Wang Phan (q.v.).

Chiu Huang Yeh Phu (Pu I) 救荒野譜(補遺).
(Supplement to) A Treatise on Edible Wild
Plants for Emergency Use [partly based on
the *Shih Wu Pên Tshao* of Lu Ho, q.v.].
Ming or Chhing, c. +1642.
Yao Kho-Chhêng 姚可成.
Cf. Wang Yü-Hu (*1*), p. 194; Swingle (*1*) pp.
202 ff. (*10*) pp. 189 ff.; Lung Po-Chien (*1*),
pp. 104 ff.

Chiu Thang Shu 舊唐書.
Old History of the Thang Dynasty [+618 to
+906].
Wu Tai, +945.
Liu Hsü 劉昫.
Cf. des Rotours (*2*), p. 64.
For translations of passages see the index of
Frankel (*1*).

Cho Kêng Lu 輟耕錄.
[sometimes *Nan Tshun Cho Kêng Lu*].
Talks (at South Village) while the Plough is
Resting.
Yuan, +1366.
Thao Tsung-I 陶宗儀.

Chou Li 周禮.
Record of the Institutions (lit. Rites) of (the)
Chou (Dynasty) [descriptions of all govern-
ment official posts and their duties].
C/Han, perhaps containing some material from
late chou.
Compilers unknown.
Tr. E. Biot (*1*).

Chou Li 周禮.
Record of the Rites of (the) Chou (Dynasty)
[descriptions of all government official posts
and their duties].
C/Han, perhaps containing some material from
late Chou, esp. Khao Kung Chi which may
be from archives of Chhi State.
Compilers unknown.
Tr. E. Biot (*1*).

Chu Fan Chih 諸蕃志.
Records of Foreign Peoples (and their Trade).
Sung, c. +1225. (This is Pelliot's dating; Hirth
& Rockhill favoured between +1242 and
+1258.).
Chao Ju-Kua 趙汝适.
Tr. Hirth & Rockhill (*1*).

Chü Hua Phu 菊花譜.
A Treatise on Chrysanthemum Flower
(Varieties) [part of *Hua Chu Wu Phu*, q.v.].
Ming, +1591.
Kao Lien 高濂.

Chü Lu 橘錄.
The Orange Record [monograph on citrus
horticulture].
Sung, +1178.
Han Yeh-Chih 韓彥直.
Tr. Hagerty (*1*) with Chiang Khang-Hu.

Chu Phu 竹譜.
A Treatise on Bamboos (and their Economic
Uses; in verse and prose) [probably the first
monograph on a specific class of plants].
L/Sung, c. +460.
Tai Khai-Chih 戴凱之.
Tr. Hagerty (*2*).

Chü Phu 菊譜.
(= *Fan Tshun Chü Phu, Shih-Hu Chü Phu*).
Treatise on the Chrysanthemum.
Sung, +1186.
Fan Chhêng-Ta 范成大.

Chü Phu 菊譜.
Treatise on the Chrysanthemum.
Sung, +1104.
Liu Mêng 劉蒙.
In *Kuang Chhün Fang Phu, Thu Shu Chi Chhêng*,
abridged in *Chih Wu Ming Shih Thu Khao*.

Chü Phu 菊譜.
(= *Shih shih Chü Phu, Shih Lao Phu Chü Phu*).
Treatise on the Chrysanthemum.
Sung, +1175.
Shih Chêng-Chih 史正志.
In *Kuang Chhün Fang Phu, Thu Shu Chi Chhêng,
Chih Wu Ming Shih Thu Khao*.

Chü Phu 菊譜.
A Treatise on Chrysanthemums [part of *I Mên
Kuang Tu*, q.v.].
Ming, +1582 (+1598).
Chou Lü-Ching 周履靖.

Chu Phu 竹譜.
A Treatise on the Bamboos [of Kweichow and
Yunnan].
Chhing, c. +1670.
Chhen Ting 陳鼎.

Chu Phu 竹譜.
Treatise on Bamboos.

Chu Phu (*cont.*)
  Yuan, +1299.
  Li Khan 李衎.
Chu Phu Hsiang Lu 竹譜詳錄.
  Detailed Records for a Treatise on Bamboos
    (original and later title of the *Chu Phu* of Li
    Khan, q.v.).
Chu Ping Yuan Hou Lun 諸病源候論.
  Treatise on Diseases and their Aetiology.
  Sui, *c.* +610.
  Chhao Yuan-Fang 巢元方.
    = *Chhao Shih Chu Ping Yuan Hou* (*Tsung*) *Lun.*
Chü Shu 菊書.
  A Writing on the Chrysanthemum.
  Ming, *c.* +1596.
  Chang Ying-Wên 張應文.
Chü Shuo 菊說.
  A Discourse on the Chrysanthemum.
  See Chi Nan (2).
Chü Than Lu
  See *Chi Than Lu.*
Chu Tzu Chhüan Shu 朱子全書.
  Collected Works of Master Chu (Hsi).
  Sung (ed. Ming) *editio princeps*, +1713.
  Chu Hsi 朱熹.
  Ed. Li Kuang-Ti 李光地 (Chhing).
  Partial trs. Bruce (1); le Gall (1).
Chuang Tzu 莊子.
  (= *Nan Hua Chen Ching*.)
  The Book of Master Chuang.
  Chou, *c.* −290.
  Chuang Chou 莊周.
  Tr. Legge (5); Fêng Yu-Lan (5); Lin Yü-Thang
    (1).
  Yin-Tê Index no. (suppl.) 20.
Chün Phu 菌譜.
  A Treatise on Fungi.
  Sung, +1245.
  Chhen Jen-Yü 陳仁玉.
Chung Chih Shu 種植書.
  Book of Sowing and Planting.
  Han.
  Fan Shêng-Chih 氾勝之.
  (or Fan Shêng)
  (Only preserved in quotations, esp. in *Chhi Min
    Yao Shu*).
Chung Ching Pu 中經薄.
  Notes on Bibliography [a catalogue raisonné of
    books].
  Chin, *c.* +280.
  Hsün Hsü 荀勗.
  Ed. Wang Jen-Chün (Chhing) 王仁俊.
Chung Chü Fa 種菊法.
  Methods of Cultivating Chrysanthemums.
  Ming, *c.* +1625.
  Chhen Chi-Ju 陳繼儒.
Chung Hua Ku Chin Chu 中華古今注.
  Commentary on Things Old and New in China.
  Wu Tai (H/Thang), +923 to +926.
  Ma Kao 馬縞.
  See des Rotours (1), p. xcix.
Chung Khuei Lu 中饋錄.

  Kitchen and Stillroom Records.
  Thang, *c.* +8th.
  Mr Wu 吳氏.
Chung Lan Chüeh 種蘭訣.
  Directions for the Cultivation of Orchids.
  Chhing.
  Li Khuei 李奎.
Chung Shu Kuo Tho-Tho Chuan 種樹郭橐駝傳.
  The Story of Camel-Back Kuo the Fruit-grower.
  Thang, *c.* +800.
  Liu Tsung-Yuan 柳宗元.
Chung Shu Shu 種樹書.
  Book of Forestry (also contains material on
    agriculture).
  Thang, prob. +8th century.
  Kuo Tho-Tho 郭橐駝.
  (Preserved only in quotations).
  Cf. Wang Yü-Hu (1), p. 99; *I Mên Kuang Tu*,
    pp. 39–40.
Chung Shu Shu 種樹書.
  Book of Tree Planting.
  Yuan or Ming.
  Yü Tsung-Pên 俞宗本.
Chung Yü Fa 種芋法.
  On the Cultivation of Yams.
  Ming, *c.* +1538.
  Huang Shêng-Tsêng 黃省曾.
Chha Ching 茶經.
  The Manual of Tea (*Camellia* (*Thea*) *sinensis*).
  Thang, *c.* +770.
  Lu Yü 陸羽.
Chha Lu 茶錄.
  A Record of Tea (*Camellia* (*Thea*) *sinensis*).
  Sung, *c.* +1060.
  Tshai Hsiang 蔡襄.
Chhang Wu Chih 長物志.
  Records of Precious Things.
  Ming.
  Wên Chen-Hêng 文震亨.
Chhao Shih Chu Ping Yuan Hou (*Tsung*) *Lun* 巢氏諸
    病源候總論.
  (= *Chu Ping Yuan Hou Lun*).
  Mr Chhao's Systematic Treatise on Diseases and
    Their Aetiology.
  Sui, *c.* +610.
  Chhao Yuan-Fang 巢元方.
Chhao Yeh Chhien Tsai 朝野僉載.
  Stories of Court Life and Rustic Life [or,
    Anecdotes from Court and Countryside].
  Thang, +8th century, but much remodelled in
    Sung.
  Chang Tso 張鷟.
Chhen-Chou Mu-Tan Chi 陳州牡丹記.
  Essay on the Tree-Peonies of Chhenchow
    (Honan).
  Sung, soon after +1112.
  Chang Pang-Chi 張邦基.
  In *TSCC*, Tshao mu tien, ch. 289.
Chhêng Chai Mu-Tan Phu ping Pai Yung 誠齋牡丹
    譜並百詠.
  A Treatise on Tree-Peonies from the Sincerity
    Studio, with a Hundred Poems for Chanting.

Chhêng Chai Mu-Tan Phu ping Pai Yung (cont.)
Ming, *c.* +1430.
Chu Yu-Tun (Chou Hsien Wang) 朱有燉 (周 憲王).
Cf. Wang Yü-Hu (*1*), p. 128.

Chhi-Chou Chih 蘄州志.
Local History and Topography of Chhichow.
Chhing, +1664.
Lu Hung 盧紘.
Enlarged and ed. 1884.
Fêng Wei-Jêng 封蔚礽.

Chhi Lu 七錄.
Bibliography of the Seven Classes of Books.
Liang, +523.
Juan Hsiao-Hsü 阮孝緒.

Chhi Lüeh 七略.
The Seven Summaries [bibliography].
C/Han, −6.
Liu Hsiang 劉向, completed by his son Liu Hsin 劉歆.
Now extant only as incorporated into the *Chhien Han Shu* bibliography (*I wên chih*).

Chhi Min Yao Shu 齊民要術.
Important Arts for the People's Welfare [lit. Equality].
N/Wei (and E/Wei or W/Wei), between +533 and +544.
Chia Ssu-Hsieh 賈思勰.
See des Rotours (*1*), p.c.; Shih Shêng-Han (*1*).

Chhieh Yün 切韻.
Dictionary of Characters arranged according to their Sounds when Split [rhyming phonetic dictionary; the title refers to the *fan-chhieh* method of 'spelling' Chinese Characters—see Vol. 1, p. 33].
Sui, +601.
Lu Fan-Yen 陸法言.
Now extant only within the *Kuang Yün* (q.v.).
Têng & Biggerstaff (*1*), p. 203.

Chhieh Yün Chih Chang Thu 切韻指掌圖.
Tabular Key (lit. Finger-Reckoning) for the *Dictionary of Characters arranged according to their Sounds when Split* [the *Chhieh Yün*, q.v.].
Sung, *c.* late +11th (with addendum of explanations in late +14th).
Attrib. Ssuma Kuang 司馬光.
Addendum by Shao Kuang-Tsu 邵光祖.
Cf. Têng & Biggerstaff, p. 204.

Chhien Chin Fang.
A Thousand Golden Remedies.
See *Chhien Chin Yao Fang.*

Chhien Chin I Fang 千金翼方.
Supplement to the *Thousand Golden Remedies* [i.e. Revised Prescriptions for Saving Lives, worth a Thousand Ounces of Gold].
Thang, +660
Sun Ssu-Mo 孫思邈.

Chhien Chin Yao Fang 千金要方.
A Thousand Golden Remedies [i.e. Essential Prescriptions for Saving Lives, worth a Thousand Ounces of Gold].
Thang, between +650 and +659.

Sun Ssu-Mo 孫思邈.

Chhien Han Shu 前漢書.
History of the Former Han Dynasty, −206 to +24.
H/Han, *c.* +100.
Pan Ku 班固, and after his death in +92 his sister Pan Chao 班昭.
Partial trs. Dubs (*2*), Pfizmaier (*32–34, 37–51*), Wylie (*2, 3, 10*), Swam (*1*), etc.
Yin-Tê Index, no. 36.

Chhien Khun Pi Yün 乾坤秘韞.
The Hidden Casket of Chhien and Khun (Kua, i.e. Yang and Yin) Open'd.
Ming, *c.* +1430.
Chu Chhüan 朱權.
(Ning Hsien Wang, prince of the Ming) 寧獻王.

Chhien Khun Shêng I 乾坤生意.
Principles of the Coming into Being of Chhien and Khun (Kua, i.e. Yang and Yin).
Ming, *c.* +1430.
Chu Chhüan 朱權.
(Ning Hsien Wang, prince of the Ming) 寧獻王.

Chhien Tzu Wên 千字文.
The Thousand-Character Primer.
Ascr. S/Chhi, +520.
Attrib. Chou Hsing-Ssu 周興嗣.
Tr. St Julien (*10*).

Chhin Hui Yao 秦會要.
History of the Administrative Statutes of the Chhin Dynasty.
See Sun Chhiai (*1*).

Chhin-Ting Ku Chin Thu Shu Chi Chhêng.
See *Thu Shu Chi Chhêng.*

Chhing Chhao Thung Chih 清朝通志.
The *Historical Collections* (continued) for the Chhing Dynasty (see *Thung Chih*).
Chhing, commissioned +1767, not completed till after +1785.
Ed. Hsi Huang 嵇璜 *et al.*

Chhing I Lu 清異錄.
Exhilarating Talks on Strange Things.
Wu Tai & Sung, *c.* +965.
Thao Ku 陶穀.

Chhing-Li Hua Phu 慶曆花譜.
Treatise on Flowers of the Chhing-Li reign-period (+1041–8).
See *Lo-Yang Hua Phu.*

Chhiu Hua Phu 秋花譜.
(= *Hsü Yuan Chhiu Hua Phu*).
Treatise on the Flowers of Autumn.
Chhing, +1682.
Wu I-I 吳儀一.

Chhiu Yuan Tsa Phei 秋園雜佩.
Miscellaneous Notes on the Autumn Garden.
Chhing.
Chhen Chên-Hui 陳貞慧.

Chhiung-Hua Chi 瓊花記.
A Record of the 'Nephrite' Flower.
Sung, +1191, pr. +1234.
Tu Yu 杜斿.

Chhiung-Hua Chi (cont.)
Contained in Yang Tuan's *Chhiung-Hua Phu*, q.v. and in *Thu Shu Chi Chhêng*.

Chhiung-Hua Chi 瓊花集.
Accounts of the 'Nephrite' Flower [a plant for which Yangchow was famous; probably a sterile *Hydrangea* hybrid].
Ming.
Tshao Hsüan 曹璿.

Chhiung-Hua Chih 瓊花志.
A Study of the 'Nephrite' Flower [a plant for which Yangchow was famous; probably a sterile *Hydrangea* hybrid].
Chhing.
Chu Hsien-Tsu 朱顯祖.

Chhiung-Hua Khao 瓊花考.
An Investigation of the 'Nephrite' Flower.
Alt. title of *Chhiung-Hua Phu*, q.v.

Chhiung-Hua Phu 瓊花譜.
A Study of the 'Nephrite' Flower [a plant for which Yangchow was famous; probably a sterile *Hydrangea* hybrid].
Ming, +1487.
Yang Tuan 楊端.

Chhiung-Hua Pien 瓊花辯.
A Tractate on the Identification of the 'Nephrite' Flower.
Sung, *c.* +1210.
Chêng Hsing-I 鄭興裔.
In *Thu Shu Chi Chhêng*.

Chhiung-Hua Pien 瓊花辨.
A Discriminatory Essay on the 'Nephrite' Flower.
Ming, *c.* +1540.
Lang Ying 郎瑛.
In *Thu Shu Chi Chhêng*.

Chhou Jen Chuan 疇人傳.
Biographies of Mathematicians and Astronomers.
Chhing, +1799.
Juan Yuan 阮元.
With continuations by Lo Shih-Lin 羅士琳, Chu Kho-Pao 諸可寶 and Huang Chung-Chün 黃鍾駿.
In *HCCC*, chs. 159 ff.

Chhü Hsien Chha Phu 臞仙茶譜.
A Treatise on Tea, by the Emaciated Immortal.
Ming, *c.* +1430.
Chu Chhüan 朱權.
(Ning Hsien Wang, prince of the Ming 寧獻王).

Chhü Hsien Shen Yin Shu 臞仙神隱書.
Book of Daily Occupations for Scholars in Rural Retirement, by the Emaciated Immortal.
Ming, *c.* +1440.
Chu Chhüan (Ning Hsien Wang) 朱權 (寧獻王).
Cf. Wang Yü-Hu (*1*), p. 128.

Chhu Hsüeh Chi 初學記.
Entry into Learning [encyclopaedia].
Thang, +700.
Hsü Chien 徐堅.

Chhu Tzhu 楚辭.
Elegies of Chhu (State) [or, Songs of the South].
Chou, *c.* −300, (with Han additions).
Chhü Yuan 屈原 (& Chia I 賈誼, Yen Chi 嚴忌, Sung Yü 宋玉, Huainan Hsiao-Shan 淮南小山 *et al.*).
Partial tr. Waley (23); tr. Hawkes (*1*).

Chhu Tzhu Pu Chu 楚辭補註.
Supplementary Annotations to the *Elegies of Chhu*.
Sung, *c.* +1140.
Ed. Hung Hsing-Tsu 洪興祖.

Chhüan Fang Pei Tsu 全芳備祖.
Complete Chronicle of Fragrances [thesaurus of botany].
Sung, +1256.
Chhen Ching-I 陳景沂.

Chhüan Nan Tsa Chih 泉南雜志.
Misallany of the Garden South of the Springs.
Ming.
Chhen Mou-Jen 陳懋仁.

Chhuan Phu Shih Chih Chi 孱圃蒔植記.
Notes on the Cultivation of Plants from the Garden of (Refreshed) Fatigue.
See *Shih Chih Chi*.

Chhüan Thang Wên.
See Tung Kao (*1*).

Chhuan Ya Nei Pien 串雅內編.
Combined Collections of Prescriptions of Folk Medicine.
Ching, preface of +1759.
Chao Hsüeh-Min 趙學敏.
Peking, 1956.

Chhun Chhiu 春秋.
Spring and Autumn Annals [i.e. Records of Springs and Autumns].
Chou; a chronicle of the State of Lu kept between −722 and −481.
Writers unknown.
Cf. *Tso Chuan*; *Kungyang Chuan*; *Kuliang Chuan*.
See Wu Khang (*1*); Wu Shih-Chhang (*1*); van der Loon (*1*).
Tr. Couvreur (*1*); Legge (*11*).

Chhun Chhiu Ta Chuan 春秋大傳.
The Great Tradition (or Commentary) on the Spring and Autumn Annals.
Chou.
Writer unknown.
YHSF, ch. 31, p. 41*a*.

Chhun Chhiu (Wei) Ming Li Hsü 春秋(緯)命歷序.
(Apocryphal Treatise on the) *Spring and Autumn Annals*; Preface to the Ordained Calendar.
C/Han −1st.
Writer unknown.

Chhün Fang Phu 羣芳譜.
The Assembly of Perfumes [thesaurus of botany].
Ming, +1630.
Wang Hsiang-Chin 王象晉.

Chhün Shu Khao So 羣書考索.
A Critical Guide through the Multitude of Books.
Sung, *c.* +1220.
Chang Ju-Yü 章如愚.

Chhung-Hsiu Chêng-Ho Ching-Shih Chêng Lei Pei-Yung Pên Tshao 重修政和經史證類備用本草.

New Revision of the Pharmacopoeia of the Chêng-Ho reign-period; the Classified and Consolidated Armamentarium. (A Combination of the *Chêng-Ho ... Chêng Lei ... Pên Tshao* with the *Pên Tshao Yen I*).

Yuan +1249; reprinted many times afterwards, esp. in the Ming, +1468, and at least seven Ming editions, the last in +1624 or +1625.

Thang Shen-Wei 唐慎微.

Khou Tsung-Shih 寇宗奭.

pr. (or ed.) Chang Tshun-Hui 張存惠.

Chhung-Khan Ching-Shih Chêng Lei Ta-Chhüan Pên Tshao 重刊經史證類大全本草.

The Complete and Newly Printed Classified and Consolidated Pharmacopoeia.

Alternative title for some of the Ming editions (+1577, +1600, +1610) of the *Ching-Shih Chêng Lei Ta-Kuan Pên Tshao* (= *Ta-Kuan Ching-Shih Chêng Lei Pei-Chi Pên Tshao*), q.v., combined with the *Chhung-Hsiu Chêng-Ho Ching-Shih Chêng Lei Pei-Yung Pên Tshao* of +1468.

This was first done by Wang Ta-Hsien 王大獻, +1577.

See Ting Chi-Min (*1*); Lung Po-Chien (*1*), no. 34.

Chhung-Kuang Pu-Chu Shen Nung Pên Tshao Ping Thu Ching 重廣補註神農本草并圖經.

Enlarged *Supplementary Commentary on the Pharmacopoeia of the Heavenly Husbandman*, with the *Illustrated Treatise (of Pharmaceutical Natural History)*.

Sung, +1092.

Chang Yü-Hsi 掌禹錫.

Su Sung 蘇頌.

Enlarged and ed. Chhen Chhêng 陳承.

Chhung-Kuang Ying Kung Pên Tshao 重廣英公本草.

Revision and Enlargement of the Pharmaeopoeia of the Duke of Ying (i.e. the *Hsin Hsin Pên Tshao*, q.v.).

Original title of the *Shu Pên Tshao* (q.v.).

Erh Ya 爾雅.

Literary Expositor [dictionary].

Chou material, stabilised in Chhin or C/Han.

Compiler unknown.

Enlarged and commented on *c.* +300 by Kuo Pho 郭璞.

Yin-Tê Index no. (suppl.) 18.

Erh Ya Chu Su 爾雅注疏.

Explanations of the Commentaries on the *Literary Expositor*.

Sung, *c.* +1000.

Hsing Ping 邢昺.

Erh Ya Hsin I 爾雅新義.

Fresh Interpretations of the *Literary Expositor*.

Sung, +1099.

Lu Tien 陸佃.

Erh Ya I 爾雅翼.

Wings for the *Literary Expositor*.

Sung, +1174.

Lo Yuan 羅願.

Fan Chiang (Phien) 凡將篇.

Most Important Phrases [orthographic word-list].

C/Han, *c.* −140.

Ssuma Hsiang-Jo 司馬相如.

Fragmentary reconstruction in *YHSF* ch. 60, pp. 3*a* ff.

Fan Shêng-Chih Shu 汜勝之書.

The Book of Fan Shêng-Chih (on Agriculture).

C/Han, −1st century.

Fan Shêng-Chih 汜勝之.

Comm. Wan Kuo-Ting (*1*).

Tr. Shih Shêng-Han (*2*).

Fan Tshun Chü Phu 范村菊譜.

The Fan (Family) Garden Chrysanthemum Treatise.

See the *Chü Phu* of Fan Chhêng-Ta.

Fan Tshun Mei Phu 范村梅譜.

The Fan (Family) Garden Apricot Treatise.

See the *Mei Phu* of Fan Chhêng-Ta.

Fan Tzu Chi Jan 范子計然.

See *Chi Ni Tzu*.

Fang Yen 方言.

Dictionary of Local Expressions.

C/Han, *c.* −15 (but much interpolated later).

Yang Hsiung 揚雄.

Fang Yen Su Chêng 方言疏證.

Correct Text of the *Dictionary of Local Expressions*, with Annotations and Amplifications.

Chhing, +1777.

Tai Chen 戴震.

Fêng Hsien Phu 鳳仙譜.

Treatise on the Flower of the Phoenix-Immortal [balsam; *Impatiens Balsamina*].

Chhing, *c.* +1785.

Chao Hsüeh-Min 趙學敏.

Fêng Thu Chi 風土記.

Record of Airs and Places [local customs].

Chin, +3rd century.

Chou Chhu 周處.

Fu Hon Ku Chin Chu 伏侯古今注.

Commentary of the Lord Fu on Things New and Old.

H/Han, *c.* +140.

Fu Wu-Chi 伏無忌.

(Only fragments, as in *YHSF*, ch. 73.)

Fu Huang Khao 捕蝗考.

A Treatise on Catching Locusts.

Chhing, +18th century.

Chhen Fang-Sheng 陳芳生.

Fu-Nan Chi 扶南記.

Record of Cambodia.

= *Fu-Nan Chuan* (Khang Thai), q.v.

Fu-Nan Chuan 扶南傳.

A Record of Cambodia.

San Kuo (Wu), *c.* +240.

Khang Thai 康泰.

Extant only in quotations. Perhaps the same as *Wu Shih Wai Kuo Chuan*, q.v.

Fu-Nan I Wu Chih 扶南異物志.
Record of the Strange Things in Cambodia (plants, animals, etc.).
San Kuo (Wu), *c.* +240.
Chu Ying 朱應.
Extant only in quotations.

Hai-Thang Chi 海棠記.
Memoir on Flowering Crab-Apple Trees.
Sung, *c.* +1050.
Shen Li 沈立.
Partially preserved in *Thu Shu Chi Chhêng* and *Lei Shuo*.

Hai-Thang Phu 海棠譜.
Treatise on Flowering Crab-Apple Trees.
Sung, +1259.
Chhen Ssu 陳思.
In *Pai Chhüan Hsüeh Hai* and *Shuo Fu*.

Hai Yao Pên Tshao 海藥本草.
Drugs of the Southern Countries beyond the Seas [or Pharmaceutical Codex of Marine Products].
Thang, *c.* +775 (or early +10th century).
Li Hsün (acc. to Li Shih-Chen) 李珣.
Li Hsien (acc. to Huang Hsiu-Fu) 李玹.
Preserved in *Pên Tshao Kang Mu*, etc.

Han Fei Tzu 韓非子.
The Book of Master Han Fei.
Chou, early −3rd century.
Han Fei 韓非.
Tr. Liao Wên-Kuei (1).

Ho Ming Pên Tshao.
See *Honzō Wamyō*.

Ho Ping Tzu Hsüeh Chi Phien 合幷字學集篇.
Collected Papers on Unified Graphology [lexical; 200 radicals].
Ming, +15th.
Hsü Hsiao 徐孝.

Ho Shou Wu Chuan 何首烏傳.
Treatise on the Ho-shou-wu Plant (*Polygonum multiflorum*, R/576).
Thang, *c.* +840.
Li Ao 李翱.

Honan Chhêng shih I Shu 河南程氏遺書.
Remaining Records of Discourses of the Chhêng brothers of Honan [Chhêng I and Chhêng Hao, +11th. century Neo-Confucian philosophers].
Sung, +1168; pr. *c.* +1250.
Chu Hsi (ed.) 朱熹.
In *Erh Chhêng Chhüan Shu*, q.v.
Cf. Graham (1), p. 141.

Honan Chhêng Shih Wên Chi 河南程氏文集.
Literary Remains of the Chhêng brothers of Honan [Chhêng I and Chhêng Hao, +11th-century Neo-Confucian philosophers].
Sung, +1107; pr. *c.* +1150.
Coll. Hu An-Kuo 胡安國.
In *Erh Chhêng Chhüan Shu*, q.v.
Cf. Graham (1), p. 143.

Honzō-Wamyō 本草和名.
Synonymic Materia Medica with Japanese Equivalents.

Japan, +918.
Fukane no Sukehito 深根輔仁.
Cf. Karow (1).

Hou Han Shu 後漢書.
History of the Later Han Dynasty [+25 to +220].
L/Sung, +450.
Fan Yeh 范曄. The monograph chapters by Ssuma Piao 司馬彪 (d. +305), with commentary by Liu Chao 劉昭 (*c.* +570), who first incorporated them into the work.
A few ch. trs. Charannes (6, 16); Pfizmaier (52, 53). Yin-Te Index, no. 41.

Hu Pên Tshao 胡本草.
Materia Medica of the Western Countries.
Thang, between +740 and +760.
Chêng Chhien 鄭虔.
Now extant in rare quotations only.

Hua Ching 花經.
The Flower Manual.
Sung.
Chang I 張翊.

Hua Ching (= Pi Chhuan Hua Ching) 花鏡.
The Mirror of Flowers [horticultural, botanical and zoo-technic manual].
Chhing, +1688.
Chhen Hao-Tzu 陳淏子.
Tr. Halphen (1).

Hua Chu Wu Phu 花竹五譜.
Five Treatises on Flowers (tree-peonies, peonies, chrysanthemums and orchids, etc.) and Bamboos [see *Tsun Shêng Pa Chien*].
Ming, +1591.
Kao Lien 高濂.

Hua Hsiao Ming 花小名.
Popular Names of Flowers.
Chhing, +17th century.
Chhen Yü-Wên 陳羽文.

Hua I Hua Mu Niao Shou Chen Wan Khao 華夷花木鳥獸珍玩考.
Investigations on Flowers, Trees, Birds, Beasts, Gems and Curios, both Chinese and Foreign.
Ming, +1581.
Shen Mou-Kuan 慎懋官.

Hua-Kuang Mei Phu 華光梅譜.
Studies on the *mei hua (Prunus mume)* from the Hua-kuang Temple [manual of black and white paintings of this tree].
Sung, *c.* +1110.
Chung-Jen 仲仁.

Hua Li 花曆.
A Calendar of Ornamental Flowering Plants.
Chhing, +17th century.
Chhen Yü-Wên 陳羽文.

Hua Mu Chi 花木記.
A Record of Flowering Plants and Trees.
Yuan or pre-Yuan.
Li Tsan-Huang 李贊皇.

Hua Mu Niao Shou Chi Lei 花木鳥獸集類.
Accounts of Flowers, Trees, Birds and Beasts.
Chhing.
Wu Pao-chih 吳寶芝.

Hua Phu 花譜.
A Treatise on Ornamental Flowering
Plants.
Chhing, late +18th century.
Than Tshui 檀萃.

Hua Shih 花史.
A History of Flowers.
Ming.
Wu Yen-Khuang 吳彥匡.

Hua Shih Tso Phien 花史左編.
Supplement to the *History of Flowers*.
Ming, +1617.
Wang Lu 王路.

Hua Su 花疏.
(Miscellaneous) Notes on Ornamental
Flowering Plants.
A Section of *Hsüeh Phu Tsa Su*, q.v.

Hua Yang Kuo Chih 華陽國志.
Records of the Country South of Mount Hua
[historical geography of Szechuan down to
+138].
Chin, +347.
Chhang Chhü 常璩.

Huai Nan Tzu 淮南子.
[= *Huai Nan Hung Lieh Chieh*.]
The Book of (the Prince of) Huai-Nan [com-
pendium of natural philosophy].
C/Han, c. −120.
Written by the group of scholars gathered by
Lui An (prince of Huai-Nan) 劉安.
Partial trs. Morgan (1); Erkes (1); Hughes (1);
Chatley (1); Wieger (2).
Chung-Fa Index, no. 51.
TT/1170.

Huai-Nan (Wang) Wan Pi Shu 淮南(王)萬畢術.
[Prob. = *Chen-Chung Hung-Pao Yuan-Pi Shu* and
variants.]
The Ten Thousand Infallible Arts of (the Prince
of) Huai-nan [Taoist magical and technical
recipes].
C/Han, −2nd century.
No longer a separate book but fragments con-
tained in *TPYL*, ch. 736 and elsewhere.
Reconstituted texts by Yeh Tê-Hui in *Kuan Ku
Thang So Chu Shu*; Sun Fêng-I in *Wên Ching
Thang Tshung-Shu*; and Mao Phan-tin in *Lung
Chhi Ching Shih Tshung-Shu*.
Attrib. Liu An 劉安.
See Kaltenmark (2), p. 32.
It is probable that the terms *Chen-Chung* 枕中
Confidential Pillow-Book; *Hung-Pao* 鴻寶
Infinite Treasure; *Wan-Pi* 萬畢 Ten
Thousand Infaltible; and *Yuan-Pi* 苑祕
Garden of Secrets; were originally titles of
parts of a *Huai-Nan Wang Shu* 淮南王書
(Writings of the Prince of Huai-Nan) forming
the Chung Phien 中篇 (and perhaps also
the Wai Shu 外書) of which the present
Huai Nan Tzu Book (q.v.) was the nei
Shu 內書.

Huang Chhing Ching Chieh 皇清經解.
Collection of (more than 180) Monographs on

Classical Subjects written during the Chhing
Dynasty.
See Yen Chieh (1) (ed.).

Huang Chi Ching Shih Shu 皇極經世書.
Book of the Sublime Principle which governs All
Things within the World.
Sung, c. +1060.
Shao Yung 邵雍.
TT/1028. Abridged in *Hsing Li Ta Chhüan* and
Hsing Li Ching I.

Huang Lan 皇覽.
Imperial Speculum.
San Kuo & E/Chin.
Edited c. +220 by Miu Pu 繆卜 enlarged by
Ho Chhêng-Thien 何承天 in early +5th
century.
No longer extant, except in quotations.

Huang Ti Nei Ching, Ling Shu 黃帝內經靈樞.
The Yellow Emperor's Manual of Corporeal
(Medicine); the Vital Axis [medical phys-
iology and anatomy].
Probably C/Han, c. −1st century.
Writers unknown.
Edited Thang, +762, by Wang Ping 王冰.
Analysis by Huang Wên (1).
Tr. Chamfrault & Ung Kang-Sam (1).
Commentaries by Ma Shih 馬蒔 (Ming) and
Chang Chih-Tshung 張志聰 (Chhing) in
TSCC, I shu tien, chs. 67 to 88.

Huang Ti Nei Ching, Su Wên 黃帝內經素問.
The Yellow Emperor's Manual of Corporeal
(Medicine); Questions (and Answers) about
Living Matter [clinical medicine]. (Cf. *Pu Chu
Huang Ti Nei Ching, Su Wên*.)
Chou, remodelled in Chhin and Han, reaching
final form c. −2nd century.
Writers unknown.
Ed. & comm., Thong (+762), Wang Ping
王冰; Sung (c. +1050), Liu I 林億.
Partial trs. Hübotter (1), chs. 4, 5, 10, 11, 21;
Veith (1); complete, Chamfrault & Ung
Kang-Sam (1).
See Wang & Wu (1), pp. 28 ff.; Huang Wên (1).

Huang Ti Nei Ching, Su Wên, Chi Chu 黃帝內經
素問集註.
The *Yellow Emperor's Manual of Corporeal
(Medicine); Questions (and Answers) about Living
Matter*; with Commentaries.
Chhing, +1672 (See Anon (*83*), p. 83).
Chang Chih-Tshung 張志聰.
(Comm. repr. with other commentaries in
TSCC, I Shu tien, chs. 21–66.)

Huang Ti Nei Ching, Su Wên, Chu Chêng Fa Wei
黃帝內經素問註證發微.
An Elucidation of the Manifold Subtleties of
the *Yellow Emperor's Manual of Corporeal
(Medicine); Questions (and Answers) about Living
Matter*.
Ming, +1586 (see Anon. (*83*). p. 83).
Ma Shih 馬蒔.
ISK/39. Comm. repr. with other commentaries
in *TSCC, I Shu tien*, chs. 21–66.

Huang Ti Nei Ching Su Wên I Phien 黃帝內經素問遺篇.

The Missing Chapters from the *Questions and Answers* of the *Yellow Emperor's Manual of Corporeal (Medicine)*.

Asor. pre-Han.

Sung, preface, +1099.

Ed. (perhaps written by) Liu Wên-Shu 劉溫舒.

Often appended to his *Su Wên Ju Shih Yün Chhi Ao Lun* (q.v.) 素問入式運氣奧論.

Huang Ti Nei Ching, Su Wên, Ling Shu, Chi Chu 黃帝內經素問靈樞集注.

Collected commentaries on the *Yellow Emperor's Manual of Corporeal (Medicine); Questions (and Answers) about Living Matter, and the Vital Axis*.

Chhing, +1672 (See Anon (*83*), p. 83).

Chang Chih-Tshung 張志聰.

Huang Ti Nei Ching, Thai Su 黃帝內經太素.

The Yellow Emperor's Manual of Corporeal Medicine; the Great Innocence [i.e. the YE'S M of C (M), Q (QA) about LM, and the VA, arranged in their original form].

Chou, Chhin & Han, essentially in present form by −1st century, commented upon in Sui, +605 to +618.

Ed. and comm. Yang Shang-Shan 楊上善.

Identifications of chapters and component passages with the corresponding texts in the *Questions (and Answers)* about Living Matter and the Vital Axis in Wang Ping's recension, and in the *Chen Chiu Chia I Ching* (Treatise on Acupuncture and Moxibustion) of Huangfu Mi; by Hsiao Yen-Phing 蕭延平 (1924).

Hui Chhen Lu 揮塵錄.

Flicking Away the Dust (of the World's Affairs).

Sung Chhien Lu, +1166, Hon Lu, +1194, San Lu, +1195, Yü Hua, +1197.

Wang Ming-Chhing 王明清.

Hung Ho Phu 項荷譜.

A Treatise on the Cultivation of the Lotus in Huge Pots,

See Yang Chung-Pao (*1*).

Huo Hsi Lüeh 火戲略.

A Treatise on Fireworks.

Chhing, +1753; pr. 1833.

Chao Hsüeh-Min 趙學敏.

Tr. Davis & Chao Yün-Tshung (*9*).

Hsi Ching Tsa Chi 西京雜記.

Miscellaneous Records of the Western Capital.

Liang or Chhen, mid +6th century.

Attrib. Liu Hsin 劉歆 (C/Han) or Ko Hung 葛洪 (Chin), but probably Wu Chün 吳均 (Liang).

Hsi Han Hui Yao 西漢會要.

History of the Administrative Statutes of the Former (Western) Han Dynasty.

Sung, +1211.

Ed. Hsü Thien-Lin 徐天麟.

Cf. Têng & Biggerstaff (*1*), p. 158.

Hsi Hsüeh Fan 西學凡.

A Sketch of European Science and Learning [written to give an idea of the contents of the 7000 books which Nicholas Trigault had brought back for the Pei-Thang Library].

Ming, +1623.

Ai Ju-Lüeh (Giulio Aleni) 艾儒略.

Hsia Hsiao Chêng 夏小正.

Lesser Annuary of the Hsia Dynasty.

Chou, between −7th and −4th century.

Writers unknown.

Incorporated in *Ta Tai Li Chi*, q.v.

Tr. Grynpas (*1*); R. Wilhelm (*6*); Soothill (*5*).

Hsia Hsiao Cheng Su I 夏小正疏義.

Commentary on the *Lesser Annuary of the Hsia Dynasty*.

Chhing.

Hung Chen-Hsuan 洪震煊.

Hsiao Ching 孝經.

Filial Piety Classic.

Chhin and C/Han.

Attrib. Tsêng Shen (pupil or Confucius) 曾參.

Tr. de Rosny (*2*); Legge (*1*).

Hsiao Erh Ya 小爾雅.

The *Literary Expositor* Abridged.

Ascr. Chhin (or Han); but perhaps Sung, *c.* +1060.

Attrib. Khung Fu 孔鮒.

Perhaps composed, doubtless in part from ancient fragments, by the commentator.

Sung Hsien 宋咸.

Hsiao Hsüeh Kan Chu 小學紺珠.

Useful Treasury of Elementary Knowledge [lit. the Purple Beads..., an allusion to the good memory of chang Yüeh (q.v.) of the Thang, whose powers were assisted by such beads worn by him].

Sung, *c.* +1270, but not pr. till +1299.

Wang Ying-Lin 王應麟.

Hsiao Wei Yuan Shen Chhi 孝緯援神契.

Apocryphal Treatise on the *Filial Piety Classic*; Documents adducing the Evidence of the Spirits.

Han.

Writer unknown.

In *YHSF*, ch. 58.

Hsin Hsiu Pên Tshao 新修本草.

The New (lit. Newly Improved) Pharmacopoeia.

Thang, +659.

(Ed.) Su Ching (= Su Kung) 蘇敬(蘇恭) and a commission of 22 collaborators under the direction first of Li Chi & Yü Chih-Ning 李勣,于志寧 then of Chhangsun Wu-Chi 長孫無忌.

This work was afterwards commonly but incorrectly known as *Thang Pên Tshao*. It was lost in China, apart from MS fragments at Tunhuang, but copied by a Japanese in +731 and preserved in Japan though incompletely. The name of the Japanese physician or medical student was Tanabe Fubito 田邊史 and eleven chapters of the whole are preserved in

Hsin Hsiu Pên Tshao (cont.)

the copy of the MS at Ninnaji 仁和寺 in Kyoto.

Other MS copies have been in circulation and one, bought by the arehoeologist Lo Chen-Yü 羅振玉 in Japan in 1901, was published at Shanghai in 1982.

Hsin Phien Lei Yao Thu Ching Pên Tshao 新編類要 圖經本草.

Newly Prepared Classified and Illustrated Manual of the Essentials of Pharmaceutical Natural History.

The original title, in all probability, of the first (+1220) version of Hsü Hung's *Thu Ching Yen I Pên Tshao* (q.v.). For this he had a co-editor, Liu Hsin-Fu 劉信甫, but he did the second alone.

Hsin Thang Shu 新唐書.

New History of the Thang Dynasty [+618 to +906].

Sung, +1061.

Ouyang Hsiu 歐陽修 & Sung Chhi 宋祁.

Cf. des Rotours (2), p. 56.

Partial trs. des Rotours (1, 2); Pfizmaier (66–74). For translations of passages see the index of Frankel (1).

Yin-Tê Index, no. 16.

Hsing Lan Phu Lüeh 興蘭譜略.

Brief Treatise on the Orchids of (I-) Hsing (in Chiangsu).

See Chang Kuang-Chao (1).

Hsing Li Ching I 性理精義.

Essential Ideas of the Hsing-Li (Neo-Confucian) School of Philosophers [a condensation of the *Hsing Li Ta Chhüan*, q.v.].

Chhing, +1715.

Li Kuang-Ti 李光地.

Hsing Li Ta Chhüan (Shu) 性理大全.

Collected Works of (120) Philosophers of the Hsing-Li (Neo-Confucian) School [*Hsing* = human nature; *Li* = the principle of organisation in all Nature].

Ming, +1415.

Ed. Hu Kuang *et al.* 胡廣.

Hsing Yuan Hua Phu 醒園花譜.

A Record of Flowers in the Arousal Garden.

Chhing, c. +1780.

Li Thiao-Yuan 李調元.

Cf. Wang Yü-Hu (1), 1st. ed. p. 172, 2nd. ed. p. 234.

Hsiu Chen Fang 袖珍方.

Precious Prescriptions to Keep under One's Hat.

Ming, c. +1435.

Chu Yu-Tun (Chou Hsien Wang) 朱有燉 (周憲王).

Hsiu Wên Tien Yü Lan 修文殿御覽.

Imperial Speculum of the Hall of the Cultivation of Literature, revised.

N/Wei, c. +5th century.

Tsu Hsiao-Chêng 祖孝徵.

No longer extant, except in quotations.

Hsü Chu Phu 續竹譜.

A Continuation of the *Treatise on Bamboos*.

Yuan, +14th century.

Liu Mei-Chih 劉美之.

Hsü Nan Fang Tshao Mu Chuang 續南方草木狀.

See Chiang Fan (1).

Hsü Shih Shih 續事始.

Supplement to the *Beginnings of All Affairs* (Cf. *Shih Shih*).

H/Shu, c. +960.

Ma Chien 馬鑑.

Hsü Thung Chih 續通志.

The *Historical Collections* Continued (see *Thung Chih*) [to the end of the Ming Dynasty].

Chhing, commissioned +1767, pr. c. +1770.

Ed. Hsi Huang 秙璜 *et al.*

Hsü Yuan Chhiu Hua Phu 徐園秋花譜.

Treatise on the Autumn Flowers cultivated in Mr Hsü's Garden [Hsü Shih-Shu] 徐時叔.

Chhing, +1682.

Wu I-I 吳儀一.

Hsüan-Ho Pei-Yuan Kung Chha Lu 宣和北苑 貢茶錄.

The Pei-Yuan Record of Tribute Tea (from Chienyang in Fukien, from the Beginning of the Sung) to the Hsüan-Ho reign-period.

Sung, +1122.

Hsiung Fan 熊蕃.

Hsüeh Phu Tsa Su 學圃雜疏.

Miscellaneous Notes on Horticultural Science.

Ming, +1587.

Wang Shih-Mou 王世懋.

Hsün Tsuan (Phien) 訓纂篇.

Instruction on Selected Words [orthographic word-list prepared as a result of the philological conference of +5].

C/Han c. +6.

Yang Hsiung 揚雄.

Fragmentary reconstruction in *YHSF*, ch. 60, pp. 6a ff.

Hsün Tzu 荀子.

The Book of Master Hsün.

Chou, c. −240.

Hsün Chhing 荀卿.

Tr. Dubs (7).

I-Chhuan Wên Chi.

See *Honan Chhêng Shih Wên Chi*.

I Ching 易經.

The Classic of Changes [Book of Changes].

Chou with C/Han additions.

Compilers unknown.

See Li Ching-chhih (1, 2); Wu Shih-Chhang (1).

Tr. R. Wilhelm (2), Legge (9), de Harlez (1).

Yin-Tê Index, no. (suppl.) 10.

I Chou Shu 逸周書.

[= *Chi Chung Chou Shu.*]

Lost Records of the Chou (Dynasty).

Chou, −245 before, such parts as are genuine. (Found in the tomb of An Li Wang, a prince of the Wei State, r. −276 to −245; in +281).

Writers unknown.

I Chü Chih 藝菊志.
An Account of the Cultivation of the Chrysanthemum.
Chhing, *c.* +1718.
Lu Thing-Tshan 陸廷燦.

I Chü Shu (or *Phu*) 藝菊書 (譜).
On the Cultivation of the Chrysanthemum [part of *Nung Phu Ssu Shu*, q.v.].
Ming, *c.* +1545.
Huang Shêng-Tsêng 黃省會.

I Hua Phu 藝花譜.
A Treatise on the Culture of Garden Flowering Plants.
Alternative title for *ssu Shih Hua Chi* q.v.

I Mên Kuang Tu 夷門廣牘.
Archives of the Hermit's Home [a collection].
Ming, +1598.
Chou Lü-Ching 周履靖.
SKCS/TMTY, ch. 134. p. 97a.

I Pu Fang Wu Lüeh Chi 益部方物略記.
Classified Notes on the Creatures (Plants and Animals) Characterstic of I-pu (Szechuan and the South-West).
Sung, +1057.
Sung Chhi 宋祁.

I Shêng 醫膡.
See Tamba Motohiro (1).

I Shih 醫史.
A History of Medicine.
Ming, *c.* +1540.
Li Lien 李濂.
Now extant only in a rare Japanese edition.

I Wên Lei Chü 藝文類聚.
Art and Literature Collected and Classified [encyclopaedia].
Thang, *c.* +640.
Ouyang Hsün 歐陽詢.

I Wu Chih 異物志.
Record of Strange Things.
See *Chiao-chou IWC* (*Yang Fu*).
Fu-nan IWC (Chu Ying).
Lin-hai Shui Thu IWC (Shen Jung).
Ling-nan IWC (Mêng Kuan).
Nan Chou IWC (Wan Chen).
Nan Fang IWC (Wan Chen), (Fang Chhien-Li).
Nan I IWC (Yang Fu).

I Wu Chih 異物志.
Memoris of Marvellous Things.
San Kuo.
Hsüeh Yü 薛珝.

I Yü Thu Chih 異域圖志.
Illustrated Record of Strange Countries.
Ming, +1392 to +1430 (*c.* +1420). Pr. +1489.
Compiler unknown, perhaps Chu Chhüan.

Jen Hai Chi 人海記.
Notes on Men and Oceans.
Chhing, *c.* +1713.
Chha Shen-Hsing 查慎行.

Jih Hua (Chu Chia) Pên Tshao 日華 (諸家) 本草.
Master Jih-Hua's Pharmacopoeia (of All the Schools).
Sung, *c.* +970.
Ta Ming (Jih-Hua Tzu) 大明 (日華子).

Jih Yung Pên Tshao 日用本草.
The Pharmaceutical Natural History of Food Substances in Daily Use.
Yuan, +1329.
Wu Jui-Chhing 吳瑞卿.

Ju Tshao Phien 茹草編.
A Monograph on Uncultivated Vegetables.
Ming, +1582.
Chou Lü-Ching 周履靖.

Ju-Nan Phu Shih 汝南圃史.
An Account of the Gardens of Ju-nan (in Honan).
Ming, *c.* +1620.
Chou Wên-Hua 周文華.

Jung Chai Sui Pi 容齋隨筆.
Miscellanies of Mr [Hung] Jung-Chai [collection of extracts from literature, with editorial commentaries].
Sung, 1st part pr. *c.* +1185, 2nd part +1192, 3rd part +1196, 4th part after +1202.
Hung Mai 洪邁.

Ka-i 華彙.
Classified Selection of Flowering Plants.
Japan, +1759, +1765.
Shimada Mitsufusa 島田充房 & Ono Ranzan 小野蘭山.
See Bartlett & Shohara (1). pp. 60, 63, 133; Merrill & Walker (1). p. 561.
Tr. Savatier (1), for whom it was 'Kwa-wi'.

Kêng Hsin Yü Tshê 庚辛玉冊.
Precious Secrets of the Realm of Kêng and Hsin (i.e. all things connected with metals and minerals, symbolised by these two cyclical characters) [on alchemy and pharmaceutics Kêng-Hsin is also an alchemical synonym for gold].
Ming, +1421.
Chu Chhüan 朱權.
Ning Hsien Wang (prince of the Ming) 寧獻王.
Extant only in quotations.

Khai-Pao Chhung Hsin Hsiang-Ting Pên Tshao 開寶重新詳定本草.
Revised More Detailed Pharmacopoeia of the Khai-Pao reign-period.
Sung, +974.
Liu Han 劉翰, Ma Chih 馬志, and 8 or 9 other scholars and naturalists, under the direction of Li Fang 李昉.

Khai-Pao Hsin Hsiang-Tung Pên Tshao 開寶新詳定本草.
New and More Detailed Pharmacopoeia of the Khai-Pao reign-period.
Sung, +973.
Liu Han 劉翰, Ma Chih 馬志, and 7 other naturalists, under the direction of Lu To-Hsün 盧多遜.

Khai-Yuan Thien-Pao I Shih 開元天寶遺事.
Reminiscences and Remains of the Khai-Yuan
and Thien-Pao Reign-Periods (of the Thang
dynasty; +713–55).
Wu Tai (H/Chou), betw. +950 and +960.
Wang Jen-Yü 王仁裕.

Khang-Hsi Tzu Tien 康熙字典.
Imperial Dictionary of the Khang-Hsi reign-
period.
Chhing, +1716.
Ed. Chang Yü-Shu 張玉書.

Khao Kung Chi Thu 考工記圖.
Illustrations for the *Artificers' Record* (of the *Chou
Li*) (with a critical archaeological analysis).
Chhing, +1746.
Tai Chen 戴震.
In *HCCC*, chs. 563, 564; reprinted Shanghai
1955.
See Kondō (1).

Khung Tzu Chia Yü 孔子家語.
Table Talk of Confucius. [or, School Sayings of
Confucius].
San Kuo, c. +240. (but compiled from much
earlier sources).
Ed. Wang Su 王肅.
Partial trs. Kramers (1); A.B Hutchinson (1); de
Harlez (2).

Kimmōzui 訓蒙圖彙.
Illustrated Compendium for the Relief of
Ignorance [an encyclopaedia for the young].
Japan, +1666. Repr. +1789 as *Kimmōzui Taisei*.
Nakamura Tekisai 中村暢齋.
Cf. Kimura Yōjiro (1).

Ko Chih Chhi Mêng.
See Lo Ssu-Ku (1).

Ko Chih Ching Yuan 格致鏡原.
Mirror (or Perspective Glass) of Scientific and
Technological Origins.
Chhing, +1735.
Chhen Yuan-Lung 陳元龍.

Ko Chih Shih Chhi.
See Fu Lan-Ya (1).

Ko Chih Tshao 格致草.
Scientific Sketches [astronomy and cosmology;
part of *Han Yü Thung* q.v.].
Ming, +1620 pr. +1648.
Hsiung Ming-Yü 熊明遇.

Ko Chih Tshung-Shu 格致叢書.
The 'Investigation of Things' Collection [293
books of all periods on classics, history, law,
Taoism, Buddhism, divination, astrology,
geomancy, longevity techniques, medicine,
agriculture, tea technology, etc.].
Ming, c. +1595.
Ed. Hu Wên-Huan 胡文煥.

Ko Chih Yü Lun 格致餘論.
Supplementary Discourse on the Investigation of
Things (in the field of Medicine).
Yuan, +1347.
Chu Chen-Hêng 朱震亨.

Ko Wu Ju Mên.
See Ting Wei-Liang (1).

Ko Wu Lei Pien 格物類編.
Classified Encyclopaedia of Natural Knowledge.
Liao or Yuan, +11th to +14th century.
Phan Ti 潘廸.
Now extant only as occasional quotations in
encyclopaedias.

Ko Wu Than Yuan.
See Wei Lien-Chhen (1).

Ko Wu Tshu Than 格物麤談.
Simple Discourses on the Investigation of
Things.
Sung, c. +980.
Attrib. wrongly to Su Tung-Pho 蘇東坡.
Actual writer (Lu) Tsan-Ning (錄)贊寧
(Tung-Pho hsien-sêng).
With later additions, some concerning Su Tung-
Pho.

Ko Wu Wên Ta 格物問答.
Questions and Answers about Natural
Philosophy.
Chhing, c. +1670.
Mao Hsien-Shu 毛先舒.

Ku Chin Chu 古今註.
Commentary on Things Old and New.
Chin, c. +300.
Tshui Pao 崔豹.
See des Rotours (1), p. xcviii.

Ku Chin Thu Shu Chi Chhêng
See *Thu Shu Chi Chhêng.*

Ku Chin Wei Shu Khao 古今偽書考.
Investigations into the Authenticity of Ancient
and Recent Works.
Chhing, c. +1695.
Yao Chi-Hêng 姚際恆.

Ku Chin Yuan Liu Chih Lun 古今源流至論.
Essays on the Course (of Things and Affairs)
from Autiquity to the Present Time.
Sung, begun c. +1070, completed +1233, pr.
+1237.
Begun by Lin Kung 林駉.
Completed by Huang Li-Ong 黃履翁.

Kua Ti Chih 括地志.
Comprehensive Geography.
Thang, +7th century.
Wei Wang-Thai 魏王泰.
(Fragments reconstituted by Sun Hsing-yen in
+1797.)

Kuan Tzu 管子.
The Book of Master Kuan.
Chou and C/Han. Perhaps mainly compiled in
the Chi-Hsia Academy (late −4th century)
in part from older materials.
Attrib. Kuan Chung 管仲.
Partial trs. Haloun (2, 5); Than Po-Fu *et al.* (1).

Kuang Chhün Fang Phu 廣羣芳譜.
The *Assembly of Perfumes* Enlarged [thesaurus of
botany].
Chhing, +1708.
Wang Hao (ed.) 王灝.

Kuang Chih 廣志.
Extensive Records of Remarkable Things.
Chin, late +4th century.

Kuang Chih (cont.)
    Kuo I-Kung　郭義恭.
    YHSF, ch. 74.
Kuang-chou Chi　廣州記.
    Records of Canton (and Kuangtung Province).
    San Kuo, *c.* +220.
    Ku Hui　顧徽.
Kuang-chou Chi　廣州記.
    Records of Canton (and Kuangtung Province).
    Chin, +4th century.
    Phei Yuan　裴淵.
Kuang Chün Phu　廣菌譜.
    Extensive Monograph on the Fungi.
    Ming, *c.* +1550.
    Phan Chih-Hêng　潘之恆.
    Reprinted in Wu Chhi-Chün (*1*), ch. 3.
Kuang Ya　廣雅.
    Enlargement of the *Erh Ya*; *Literary Expositor*
        [dictioNary].
    San Kuo (Wei), +230.
    Chang I　張揖.
Kuang Ya Su Chêng　廣雅疏證.
    Correct Text of the *Enlargement of the Erh Ya*,
        with Annotations and Amplifications.
    Chhing, +1796.
    Wang Nien-Sun　王念孫.
Kuang Yü Thu　廣輿圖.
    Enlarged Terrestrial Atlas.
    Yuan, +1320.
    Chu Ssu-Pên　朱思本.
    First printed, and the word *Kuang* added, by Lo
        Hung-Hsien　羅洪先, Ming *c.* +1555.
Kuang Yün　廣韻.
    Revision and Enlargement of the *Dictionary of
        Characters arranged according to Their Sounds when
        Split* [rhyming phonetic dictionary, based on,
        and including, the *Chhieh Yün* and the *Thang
        Yün*, q.v.].
    Sung, +1011.
    Chhen Pheng-Nien,　陳彭年.
    Chhiu Yung, *et al.*　丘雍.
    T & B, p. 203.
Kuangtung Hsin Yü　廣東新語.
    New Talks about Kuangtung Province.
    Chhing　*c.* +1690.
    Chhü Ta-Chün　屈大均.
Kuei Hai Yü Hêng Chih　桂海虞衡志.
    An Account of the Notable Things of the
        Southern Provinces (especially Kuangsi, with
        its flowers and fruits), [lit. Perpending the
        Curiosities of the Cinnamon Shores, i.e. the
        South].
    Sung, +1175.
    Fan Chhêng-Ta,　范成大.
    Cf. Wang Yü-Hu (*1*), 1st. ed., p. 76, 2nd. ed.
        p. 91.
Kuei-Hsin Tsa Chih　癸辛雜識.
    Miscellaneous Information from Kuei-Hsin
        Street (in Hangchow).
    Sung, late +13th century, perhaps not finished
        before +1308.
    Chou Mi　周密.

See des Rotours (*1*), p. cxii; H. Franke (*14*).
Kuei-Hsin Tsa Chih Hsü Chi　癸辛雜識續集.
    Miscellaneous Information from Kuei-Hsin
        Street (in Hangchow), First Addendum.
    Sung or Yuan, *c.* +1298.
    Chou Mi　周密.
    See des Rotours (*1*), p. cxii.
Kuei-Hsin Tsa Chih Pieh Chi　癸辛雜識別集.
    Miscellaneous Information from Kuei-Hsin
        Street (in Hangchow), Final Addendum.
    Sung or Yuan, *c.* +1298.
    Chou Mi　周密.
    See des Rotours (*1*), p. cxii.
Kuei I Fang　鬼遺方.
    Procedures handed down by Spirits [the oldest
        Chinese book on external medicine, i.e.
        surgery].
    Probably early Sung, +10th century.
    Authorship purports to be by Liu Chüan
        Tzu　劉涓子 (+5th century).
    Preface purports to be of +483 by Kung
        Chhing-Hsüan　龔慶宣.
Kuei Thien Lu　歸田錄.
    On Returning Home.
    Sung, +1067.
    Ouyang Hsiu　歐陽修.

Lan Hui Ching　蘭蕙鏡.
    Mirror of *Lan* and *Hui* Orchids.
    See Thu Yung-Ning (*1*).
Lan I　蘭易.
    The Orchid *Book of Changes*.
    Ming, *c..* +1610.
    Attrib. Lu Thing-Ong (Sung)　鹿亭翁.
    Actual writer Fêng Ching-Ti　馮京第. (Tien
        Chhi Tzu)　簟溪子.
Lan I Shih-erh I　蘭易十二翼.
    Part of *Lan I*, q.v.
Lan Phu　蘭譜.
    Treatise on Orchids.
    Sung, +1247.
    Wang Kuei-Hsüch　王貴學.
    Repr. in *Pai Chhuan Hsüeh Hai*, *Shuo Fu*, etc.
    Tr. Watling (*2*).
Lan Phu　蘭譜.
    Treatise on Orchids [part of *Hua Chu Wu Phu*,
        q.v.].
    Ming, +1591.
    Kao Lien　高濂.
Lan Phu Ao Fa　蘭譜奧法.
    The Art and Mystery of Orchid Culture; a
        supplement to the Treatise.
    Sung, *c.* +1240.
    Chao Shih-Kêng　趙時庚.
    or Wang Kuei-Hsüeh　王貴學.
    In *I Mên Kuang Tu*, etc.
Lan Shih　蘭史.
    Part of *Lan I*, q.v.
Lan Yen　蘭言.
    A Brochure on Orchids.
    Chhing, late +17th century.
    Mao Hsiang　冒襄.

Lei Kung Yao Tui 雷公藥對.
  Answers of the Venerable Master Lei (to Questions) concerning Drugs.
  Perhaps L/Sung, at any rate before N/Chhi.
  Attrib. Lei Hsiao 雷斅.
  Later attrib. a legendary minister of Huang Ti.
  Comm. by Hsü Chih-Tshai (N/Chhi) 徐之才 (+565).
  (Now extant only in quotations).

Lei Phien 類篇.
  Classified Dictionary.
  Sung, +1067.
  Ssuma Kuang 司馬光.

Lei Shuo 類說.
  A Classified Commouplace-Book [a great florilegium of excerpts from Sung and pre-Sung books, many of which are otherwise lost].
  Sung, +1136.
  Ed. Tsêng Tshao 曾慥.

Lei Tsuan Ku Wên Tzu Khao 類纂古文字考.
  Study of the Classification of the Ancient Literary Characters [dictionary; 314 radicals].
  Ming, c. +1590.
  Tu Yü 都兪.

Lei Yuan Hsiang Chu 類苑詳註.
  Garden of Classified Facts, with Commentary (material from Thang and Sung encyclopaedias, with Ming commentary).
  Ming, +1575.
  Wang Shih-Chên 王世貞.

Li Chi 禮記.
  [= *Hsiao Tai Li Chi*]
  Record of Rites [compiled by Tai the Younger] (cf. *Ta Tai Li Chi*).
  Ascr. C/Han, c. −70/−50, but really H/Han, between +80 and +105, though the earliest pieces included may date from the time of the *Analects* (c. −465/−450).
  Attrib. ed. Tai Shêng 戴聖.
  Actual ed. Tshao Pao 曹褒.
  Trs. Legge (7); Couvreur (3); R. Wilhelm (6).
  Yin-Tê Index, no. 27.

Li-Chih Phu 荔枝譜.
  A Treatise on the Lichi (*Nephelium litchi*).
  Sung, +1059.
  Tshai Hsiang 蔡襄.

Li Hai Chi 蠡海集.
  The Beetle and the Sea [title taken from the proverb that the beetle's eye view cannot encompass the wide sea—a biological book].
  Ming, late +14th century.
  Wang Khuei 王逵.

Li Sao 離騷.
  Elegy on Encountering Sorrow [ode].
  Chou (Chhu), c. −295, perhaps just before −300. Some scholars place it as late as −269.
  Chhü Yuan 屈原.
  Tr. Hawkes (1).

Li Sao Tshao Mu Su 離騷草木疏.
  On the Trees and Plants mentioned in the *Elegy on Encountering Sorrow*.
  Sung, +1197.

Wu Jen-Chieh 吳仁傑.

Li shih Yao Lu 李氏藥錄.
  Mr Li's Record of Drugs.
  San Kuo (Wei), c. +225.
  Li Tang-chih 李當之.
  Extant now only in quotations

Li Tang-chih Pên Tshao Ching 李當之本草經.
  Li Tang-Chih's Manual of Pharmaceutical Natural History Alternative title of *Li shih Yao Lu* q.v.

Lieh Hsien Chuan 列仙傳.
  Collection of the Biographies of the Immorials.
  Chin, +3rd or +4th century.
  Attr. Liu Hsiang 劉向.

Lin-Hai I Wu Chih 臨海異物志.
  Strange Things in Lin-Hai (mod. Chekiang province, incl. plants and animals)
  = *Lin-Hai Shui Thu I Wu Chih* (Shen Jung), q.v.

Lin-Hai Shui Thu I Wu Chih 臨海水土異物志.
  Record of the Strange Productions of Liu-hai's Soils and Waters [natural history of part of Chekiang].
  San Kuo (Wu), c. +270.
  Shen Jung 沈瑩.
  Extant only in quotations.

Lin-I Chi 林邑記.
  Records of Champa (Kingdom).
  Chin and L/Sung completed late +5th century (present version).
  Attrib. Tungfang Shuo 東方朔 (−2nd century).
  Subsequent writers unknown.
  Extant only in quotations, e.g. in *Shui Ching Chu* Tr. Auroussean (4).

Ling-Nan Fêng Wu Chi 嶺南風物記.
  An Account of the Customs and Products of Kuangtung (lit. the Land South of the Ranges).
  Chhing, late +17th century.
  Wu Chhi 吳綺.

Ling-Nan I Wu Chih 嶺南異物志.
  Strange Things of Kuangtung (lit. South of the Ranges).
  Thang, +9th century.
  Mêng Kuan 孟琯.
  Extant now only in quotations, as in *Thai-Phing Kuang Chi*.

Ling Nan Tsa Chi 嶺南雜記.
  Miscellaneous Notes on Kuangtung Things (lit. in the Land South of the Ranges).
  Chhing, late +17th century.
  Wu Chen-Fang 吳震方.

Ling Piao Lu I 嶺表錄異.
  Strange Southern Ways of Men and Things [on the special characteristics and natural history of Kuangtung].
  Thang & Wu Tai between c. +895 and +915.
  Liu Hsün 劉恂.

Ling Wai Tai Ta 嶺外代答.
  Information on What is Beyond the Passes (lit. a book in lieu of individual replies to questions from friends).

Ling Wai Tai Ta (*cont.*)
Sung, +1178.
Chou Chhü-Fei 周去非.

Liu Pin-Kho Chia Hua Lu 劉賓客嘉話錄.
Table-Talk of the Imperial Tutor Liu (Liu Yü-Hsi).
Thang, *c.* +845.
Wei Hsüan 韋絢.

Liu Pu Chhêng Yü Chu Chieh 六部成語註解.
The terminology of the Six Boards, with Explanatory Notes.
Chhing, text +1742, notes *c.* 1875.
Winter unknown.
Commentator unknown.
Tr. Sun Jen I-Tu (1).

Liu Shu Ku 六書故.
A History of the Six Graphs [History of the Six Principles of Formation of the Chinese Characters].
Sung, +1275, printed +1320.
Tai Thung 戴侗.
Partial tr. Hopkins (36).

Liu Shu Pên I 六書本義.
Basic Principles of the Six Graphs [the Six Principles of Formation of the Chinese Characters].
Yuan, *c.* +1380.
Chao Wei-Chhien 趙撝謙.

Liu Thieh 六帖.
See Liu Thieh Shih Lei Chi.

Liu Thieh Shih Lei Chi 六帖事類集.
The "Six Slips" Collection of Classified Quotations (The reference was to the six slips of paper on which the candidates in the Thang imperial examinations had to complete whole sentences or passages chosen by the examiner, the text of the classic being covered except for one horizontal line.)
Thang, +802 or between +840 and +845, Enlarged *c.* +1160.
Ed. Pai Chü-I 白居易.
Enlarged by Khung Chhuan 孔傳.

Lo Hsiu Yuan Chü Phu 樂休園菊譜.
Chrysanthemum Treatise of the Garden of Happy Retirement.
Ming, *c.* +1575.
Writer unknown.
In *Chih Wu Ming Shih Thu Khao* and *Thu Shu Chi Chhêng*.

Lo-Li Chai Lan Phu 羅籬齋蘭譜.
Treatise on Orchids from the Lo-Li Studio [part of *Chang Shih Tshang Shu*, q.v.].
Ming, *c.* +1596.
Chang Ying-Wên 張應文.

Lo-Yang Hua Mu Chi 洛陽花木記.
Records of the Ornamental Flowering Plants and Trees of Loyang.
Sung, +1082.
Chou Shih-Hou 周師厚.

Lo-Yang Hua Phu 洛陽花譜.
Treatise on the (Tree-Peony and Peony) Flowers of Loyang.

Sung, *c.* +1045.
Chang Hsün 張峋.

Lo-Yang Ming Yuan Chi 洛陽名園記.
Notes on the Famous Gardens of Loyang.
Sung.
Li Ko-Fei 李格非.

Lo-Yang Mu-Tan Chi 洛陽牡丹記.
Account of of the Tree-Peonies of Loyang.
Sung, +1034.
Ouyang Hsiu 歐陽修.
In *Thu Shu Chi Chhêng* and *Chih Wu Ming Shih Thu Khao*, etc.

Lo-Yang Mu-Tan Chi 洛陽牡丹記.
Record of the Tree-Peonies of Loyang.
Sung, +1082,
Chou Shih-Hou 周師厚.
In *Thu Shu Chi Chhêng* and *Chih Wu Ming Shih Thu Khao*.

Lü Chhan Yen Pên Tshao 履巉巖本草.
(Pharmaceutical) Natural History of the Lü Chhan Yen (Mountain Hall).
[an album of 205 coloured pictures of plants, surviving as a Ming MS copy].
Sung, +1220.
Wang Chieh 王介.
Cf Lung Po-Chien (1), no. 42.

Lu Shih 路史.
The Grand History (of Antiquity) [a collection of fabulous and legendary material put together in the style of the dynastic histories, but containing much curious information on techniques].
Sung.
Lo Pi 羅泌.

Lü Shih Chhun Chhiu 呂氏春秋.
Master Lü's Spring and Autumn Annals [compenduim of natural philosophy].
Chou (Chhin), −239.
Written by the group of scholars gathered by Lü Pu-Wei 呂不韋.
Tr. R. Wilhelm (3).
Chung-Fa Index, no. 2.

Lun Yü 論語.
Conversations and Dis Discourses (of Confucius) [perhaps Discussed Sayings, Normative Sayings, or Selected Sayings]; Analects.
Chou (Lu), *c.* −465 to −450.
Compiled by disciples of Confucius (chs. 16, 17, 18 and 20 are later interpolations).
Tr. Legge (2); Lyall (2); Waley (5); Ku Hung-Ming (1).
Yin-Tê Index no. (suppl.) 16.

Lung Chhêng Lu 龍城錄.
Records of the Dragon City (Sian).
Thang, *c.* +800.
Liu Tsung-Yuan 柳宗元.

Lung-Khan Shou Chien 龍龕手鑑.
Handbook of the Dragon Niche [a dictionary in which the number of the radicals was reduced to 240].
Liao, +997.
Hsing-Chün 行均.

Mao Shih Liu Thieh Chiang I 毛詩六帖講意.
An Exposition of the 'Book of Odes' in Six Slips
(i.e. six aspects) [cf. the *Liu Thieh* dictionary].
Ming, +1617.
Hsü Kuang-Chhi 徐光啓.

Mao Shih Ming Wu Chieh 毛詩名物解.
Analysis of the Names and Things (including
Plants and Animals) in the 'Book of Odes'.
Sung, *c.* +1080.
Tshai Pien 蔡卞.

Mao Shih Ming Wu Lüeh 毛詩名物略.
Classified Explanations of Names and Things
(including Plants and Animals) in the 'Book
of Odes'.
Chhing, +1763.
Chu Huan 朱桓.

Mao Shih Ming Wu Thu Shuo 毛詩名物圖說.
Illustrated Explanation of the Names and
Things (including Plants and Animals) in the
'Book of Odes'.
Chhing, +1769.
Hsü Tiñg 徐鼎.

Mao Shih Tshao Chhung Ching 毛詩草蟲經.
Manual of the Plants and Animals mentioned in
the 'Book of Odes'.
L/Sung or Liang, +5th or +6th century.
Writer unknown.
Extant only as a fragment in *YHSF*, ch. 17 pp.
34*a* ff. containing a few notes on mammals.

Mao Shih Tshao Mu Niao Shou Chhung Yü Su 毛詩草
木鳥獸蟲魚疏.
An Elucidation of the Plants, Trees, Birds,
Beasts, Insects and Fishes mentioned in the
Book of Odes edited by Mao (Hêng and Mao
Chhang).
San Kuo (Wu), +3rd century (*c.* +245).
Lu Chi 陸璣.

*Mao Shih Tshao Mu Niao Shou Chhung Yü Su Kuang
Yao* 毛詩草木鳥獸蟲魚疏廣要.
An Elaboration of the Essentials in the
'Elucidation of the Plants, Trees, Birds,
Beasts, Insects and Fishes mentioned in the
Book of Odes edited by Mao (Hêng and Mao
Chhang)'.
Ming, +1639.
Mao Chin 毛晉.

Mei Hua Hsi Shen Phu 梅花喜神譜.
The Spirit of Joy; A Hundred Portraits of the
(Eight Stages in the Life of the) Apricot
Flower (*Prunus mume*).
Sung, +1238.
Sung Po-Jen 宋伯仁.

Mei Phin 梅品.
On the Grading of Varieties of the Chinese
Apricot (*Prunus mume*).
Sung, *c.* +1200.
Chang Tzu 張鎡.

Mei Phu (= *Fan Tshun Mei Phu*) 梅譜.
A Treatise on the Chinese Apricot (*Prunus
mume*).
Sung, *c.* +1186.
Fan Chhêng-Ta 范成大.

Mên Shih Hsin Hua 捫蝨新話.
More Conversations in Odd Moments (lit. while
cracking lice between one's thumb-nails).
Sung, *c.* +1150.
Chhen Shan 陳善.

Meng Chhi Pi Than 夢溪筆談.
Dream Pool Essays.
Sung, +1086; last supplement dated +1091.
Shen Kua 沈括.
Ed. Hu Tao-Ching (*1*); cf. Holzman (*1*).

Mêng Tzu 孟子.
The Book of Master Mêng (Mencius).
Chou, *c.* −290.
Mêng Kho 孟軻.
Tr. Legge (3); Lyall (1).
Yin-Tê Index, no. (suppl.) 17.

Min Chhan Lu I 閩產錄異.
Records of the Strange Products of Fukien.
See Kuo Po-Tshang (*1*).

Min Pu Shu 閩部疏.
Flowers of Fukien.
Ming.
Wang Shih-Mou 王世懋.

Ming Hui Yao 明會要.
History of the Administrative Statutes of the
Ming Dynasty.
See Lung Wên-Pin (*1*).

Ming I Pieh Lu 名醫別錄.
Informal (or Additional) Records of Famous
Physicians (on Materia Medica).
Ascr. Liang, *c.* +510.
Attrib. Thao Hung-Ching 陶弘景.
Now extant only in quotations in the pharmaceutical natural histories, and a reconstitution by Huang Yü (*1*).
This work was a disentanglement, made by
other hands between +523 and +618 or
+656, of the contributions of Li Tang-Chih
(*c.* +225) and Wu Phu (*c.* +235) and the
commentaries of Thao Hung-Ching (+492)
from the text of the *Shen Nung Pên Tshao Ching*
itself. In other words it was the non-*Pên-Ching*
part of the *Pên Tshao Ching Chi Chu* (q.v.). It
may or may not have included some of all of
Thao Hung-Ching's commentaries.

Ming Shih 明史.
History of the Ming Dynasty [+1368 to
+1643].
Chhing, begun +1646, completed +1736 first
pr. +1739.
Chang Thing-Yü 張廷玉 *et al.*

Ming Shih Lu 明實錄.
Veritable Records of the Ming Dynasty.
Ming, collected early +17th century.
Official compilation.

Mo Kho Hui Hsi 墨客揮犀.
Fly-Whisk Conversations of a Literary Person.
Sung, *c.* +1080.
Phêng Chhêng 彭乘.

Mo Tzu (incl. *Mo Ching*) 墨子.
The Book of Master Mo.
Chou, −4th century.

Mo Tzu (incl. *Mo Ching*) (*cont.*)
Mo Ti (and disciples) 墨翟.
Tr. Mei Yi-Pao (1); Forke (3).
Yin-Tê Index, no. (suppl.) 21.
TT/1162.

Mōshī Himbutsu Zukō 毛詩品物圖考.
Illustrated Study of the Creatures (Plants and
Animals) in the 'Book of Odes'.
Japan, +1785 (Kyoto).
Subsequently reprinted in China without the
Kotakona.
Oka Kōyoku 岡公翼.

Mōshī Rikushi Sōmoku Sozukai 毛詩陸氏艸木
疏圖解.
Illustrated Commentary on Lu (Chī's) Treatise
on the Plants and Trees mentioned in the
'Book of Odes'.
Japan, +1778; (Miyako) +1779.
Fuchi Zaikan 淵在寬.

Mu-Tan Jung Ju Chih 牡丹榮辱志.
King's Daughters and Humble Handmaids;
A Classified Arrangement of (the Varieties of)
the Tree-Peony (by analogy with the ranks of
ladies attending upon the emperor).
Sung, c. +1050.
Chhiu Hsüan 邱璿.

Mu-Tan Pa Shu 牡丹八書.
Eight Epistles on the Tree-Peony [*Paeonia
montan*].
Ming, c. +1610.
Hsüeh Fêng-Hsiang 薛鳳翔.
In *Thu Shu Chi Chhêng*.

Mu-Tan Phu 牡丹譜.
A Tractate on the Tree-Peonies [of Thien-
phêng, Szechuan].
Sung, +12th or +13th century.
Hu Yuan-Chih 胡元質.
In *Thu Shu Chi Chhêng* and *Chih Wu Ming Shih
Tha Khao*.

Mu-Tan Phu 牡丹譜.
A Treatise on Tree-Peonies [part of *Hua Chu Wu
Phu*, q.v.].
Ming, +1591.
Kao Lien 高濂.

Mu-Tan Phu 牡丹譜.
A Treatise on Tree-Peonies.
See Chi Nan (1).

Nan Chou Chi 南州記.
Record of the Southern Provinces = *Nan Fang
Tshao Wu Chuang* (Hsü Chung), q.v.

Nan Chou I Wu Chih 南州異物志.
Strange Things of the Southern Provinces.
San Kuo or Chin between +270 and +310.
Wan Chen 萬震.
Extant only in quotations.

Nan Chou Tshao Mu Chuang 南州草木狀.
A Prospect of the Plants and Trees of the
Southern Province = *Nan Fang Tshao Wu
Chuang* (Hsü Chung), q.v.

Nan Chung Tsou 南中奏.
Memorials from the South = *Nan Fang Tshao
Wu Chuang* (Hsü Chung), q.v.

Nan Fang Chi 南方記.
A Record of the Southern Regions = *Nan Fang
Tshao Wu Chuang* (Hsü Chung), q.v.

Nan Fang I Wu Chih 南方異物志.
= *Nan Chou I Wu Chih* (Wan Chen), q.v.
The latter is the correct title. In the Thang and
Sung periods there circulated a work with the
present title by Fang Chhien-Li 房千里
but it does not exist in full now.

Nan Fang I Wu Chih 南方異物志.
Record of the Strange Things of the Southern
Regions.
Thang, c. +840.
Fang Chhien-Li 房千里.
Extant now only in quotations, as in *Thai-Phing
Kuang Chi*.

Nan Fang Tshao Mu Chuang 南方草木狀.
A Prospect of the Plants and Trees of the
Southern Regions.
Chin, +304.
Hsi Han 秸含.

Nan Fang Tshao Wu Chih (or *Chi*) 南方草物
志 (記).
Record of the Plants and Products of the
Southern Regions = *Nan Fang Tshao Wu
Chuang* (Hsü Chung), q.v.

Nan Fang Tshao Wu Chuang 南方草物狀.
A Prospect of the Plants and Products of the
Southern Regions.
Chin, +3rd or +4th century.
Hsü Chung 徐衷.
Extant only in quotations, esp. in *Chhi Min Yao
Shu* and *TPY(L)*.

Nan I I Wu Chih 南裔異物志.
Strange Things from the Southern Borders.
H/Han, c. +90.
Yang Fu 楊孚.
Perhaps the earlier title of his *Chiao-chou I Wu
Chih*, q.v.
Extant only in quotations.

Nan Pu Hsin Shu 南部新書.
A New Collection of Southern Matters (from the
stories and memorabilia of Thang and Wu
Tai times).
Sung, c. +1015.
Chhien I 錢易.

Nan Shih 南史.
History of the Southern Dynasties [Nan Pei
Chhao period, +420 to +589].
Thang, c. +670.
Li Yen-Shou 李延壽.
For translations of passages see the index of
Frankel (1).

Nan Yüeh Hsing Chi 南越行記.
Records of Travels to Southern Yüeh
(Kuangtung and Kuangsi).
C/Han, c. −175.
Attrib. Lu Chia 陸賈.
Known only by quotations in *Nan Fang Tshao
Mu Chuang*.

Nan Yüeh Pi Chi 南越筆記.
Notes on Kuangtung (and its Plant Products).
Chhing, +1777.

Nan Yüeh Pi Chi (*cont.*)
Li Thiao-Yuan 李調元.
Nei Ching 內經.
See *Huang Ti Nei Ching, Su Wên* and *Huang Ti Nei Ching, Ling Shu.*
Nêng Kai Chai Man Lu 能改齋漫錄.
Miscellaneous Records of the Ability-to-Improve-Oneself Studio.
Sung, mid +12 century.
Wu Tshêng 吳曾.
Nung Chêng Chhüan Shu 農政全書.
Complete Treatise on Agriculture.
Ming. Composed +1625 to +1628; printed +1639.
Hsü Kuang-Chhi 徐光啓.
Ed. Chhen Tzu-Lung 陳子龍.
Nung Phu Ssu Shu 農圃四書.
Four Books on Agriculture and Horticulture.
Ming, *c.* +1545.
Huang Shêng-Tsêng 黃省曾.
Nung Sang Chi Yao 農桑輯要.
Fundamentals of Agriculture and Sericulture.
Yuan, +1273. Preface by Wang Phan(b) 王磐.
Imperially commissioned, and produced by the Agricultural Extension Bureau (Ssu Nung Ssu) 司農司.
Probable editor, Mêng Chhi 孟祺.
Probable later editors, Chhang Shih-Wên 暢師文 (*c.* +1286). Miao Hao-Chhien 苗好謙 (*c.* +1318).
Cf. Lui Yü-Chhüan (*1*).
Nung Sang I Shih Tsho Yao 農桑衣食撮要.
Selected Essentials of Agriculture, Sericulture, Clothing and Food.
Yuan, +1314 (again pr. +1330).
Lu Mung-Shan (Uighur) 魯明善.
Nung Shu 農書.
Agricultural Treatise.
Sung, +1149, pr. +1154.
Chhen Fu 陳旉.
Nung Shu 農書.
Agricultural Treatise.
Yuan, +1313.
Wang Chen 王禎.
Textual refs. are to the 22 *chüan* Palace ed. of 1783, prefaced 1774.
Nung Shu 農書.
Agricultural Treatise.
Late Ming. *c.* +1620.
Master Shen 沈氏.
Tr. into modern Chinese in Chhen Heng-Li & Wang Ta. Tshan (*1*), 210–50.

Pai Chhuan Hsüeh Hai 百川學海.
The Hundred Rivers Sea of Learning [a collection of separate books; the first *tshung-shu*].
Sung, +1273.
Compiled and edited by Tso Kuei 左圭.
Pai Chü Chi Phu 百菊集譜.
A Treatise on Collections of Chrysanthemum (Varieties).
Sung, +1242.

Shih Chu 史鑄.
Pai Hu Thung Tê Lun 白虎通德論.
Comprehensive Discussions at the White Tiger Lodge.
H/Han, *c.* +80.
Pan Ku 班固.
Tr. Tsêng Chu-Sên (*1*).
Pai Hua Phu 百花譜.
A Treatise on the Hundred Flowers.
Thang, *c.* +790.
Chia Tan 賈耽.
Pai Khung Liu Thieh 白孔六帖.
See *Liu Thieh Shih Lei Chi.*
Pai Mao Thang Chi 白茅堂集.
A Collection of Notes from the Hall of Fragrant Grasses.
Chhing, *c.* +1655.
Ku Ching-Hsing 顧景星.
Pai Shih Liu Thieh Shih Lei Chi 白氏六帖事類集.
See *Liu Thieh Shih Lei Chi.*
Pao-Chhing Pên Tshao Chê Chung 寶慶本草折衷.
An Evaluation of the Literature on Pharmaceutical Natural History, done in the Pao-Chhing reign-period.
Sung, *c.* +1226.
Chhen Yen 陳衍.
Pao Phu Tzu 抱樸子 (朴).
Book of the Preservation-of-Solidarity Master.
Chin, early +4th century., prob. *c.* +320.
Ko Hung 葛洪.
Fr. Ware (5), Nei Phien chs. only.
Partial trs. Feifel (*1, 2*); Wu & Davis (*2*); etc.
Pei Chi Chhien Chin Yao Fang 備急千金要方.
The Thousand Golden Remedies for Use in Emergencies.
See *Chhien Chin Yao Fang*, of which it is the full title.
Pei Hu Lu 北戶錄.
Records of (the Country where) the Doors (open to) the North (to catch the Sun).
[i.e. Jih-Nan and Lin-I].
Thang, *c.* +873.
Tuan Kung-Lu 段公路.
Pei Shih 北史.
History of the Northern Dynasties [Nan Pei Chhao period, +386 to +581].
Thang, *c.* +670.
Li Yen-Shou 李延壽.
For Translations of passages see the index of Frankel (*1*).
Pei Shu Pao Wêng Lu 北墅抱甕錄.
An Account of the Flowering Plants Treasured and Cultivated in the Pei Shu (Northern Lodge) Garden (at Hangchow).
Chhing, +1690.
Kao Shih-Chhi 高士奇.
Pei Thang Shu Chhao 北堂書鈔.
Book Records of the Northern Hall [encyclopaedia].
Thang, *c.* +630.
Yü Shih-Nan 虞世南.
Pên Ching.
See *Shen Nung Pên Tshao Ching.*

Pên Ching Fêng Yuan　本經逢原.
　(Additions to Natural History).
　Aiming at the Original Perfection of the *Classical
　　Pharmacopoeia (of the Heavenly Husbandman)*.
　Chhing, +1695, pr. +1705.
　Chang Lu　張璐.
　LPC, no. 93.
Pên Ching Su Chêng (with supplements)　本經疏證.
　Critical Commentary on (a Revised Text of) the
　　Classical Pharmacopoeia of the Heavenly Husbandman.
　See Tsou Chu (*1*).
Pên Tshao Chhiu Chen　本草求眞.
　Truth Searched out in Pharmaceutical Natural
　　History.
　Chhing, +1773.
　Huang Kung-Hsiu　黃宮繡.
Pên Tshao Chi Yao　本草集要.
　Summary of the Most Important Facts in
　　Materia Medica [drugs arranged according to
　　their pharmacological properties, with special
　　emphasis on the *Ming I Pieh Lu* (q.v.)].
　Ming, +1492.
　Wang Lun　王綸.
Pên Tshao Ching Chi Chu　本草經集注.
　Collected Commentaries on the *Classical
　　Pharmacopoeia (of the Heavenly Husbandman)*.
　S/Chhi, +492.
　Thao Hung-Ching　陶弘景.
　Now extant only in fragmentary form as a
　　Tunhuang or Turfan MS, apart from the
　　many quotations in the pharmaceutical
　　natural histories, under Thao Hung-Ching's
　　name.
Pên Tshao Ching Chieh Yao　本草經解要.
　An Analysis of the Most Important Features of
　　the *Pharmacopoeia (of the Heavenly Husbandman)*.
　Chhing, +1724.
　Attrib. Yeh Kuei　葉桂.
　Actual author Yao Chhiu　姚球.
　Cf. *LPC* no. 8; Swingle (6).
Pên Tshao Fa Hui　本草發揮.
　Further Advances in Materia Medica.
　Yuan, *c.* +1360.
　Hsü Yung-Chhêng　徐用誠.
　Cf. Swingle (6).
Pên Tshao Ho Ming.
　See *Honzō-Wamyō*.
Pên Tshao Hsing (Shih) Lei　本草性(事)類.
　The Natures, Effects (and Contra-Indications)
　　of Drugs.
　Thang.
　Tu Shan-Fang　杜善方.
Pên Tshao Hui　本草滙.
　Needles from the Haystack; Selected Essentials
　　of Materia Medica.
　Chhing, +1666, pr. +1668.
　Kuo Phei-Lan　郭佩蘭.
　LPC, no. 84.
　Cf. Swingle (4).
Pên Tshao Hui Chien　本草彙箋.
　Classified Notes on Pharmaceutical Natural
　　History.

　Chhing, begun +1660, pr. +1666.
　Ku Yuan-chiao　顧元交.
　LPC, no. 83; cf. Swingle (8).
Pên Tshao Hui Pien　本草會編.
　The Congregation of the Pharmaceutical
　　Naturalists; Correlated Notes on Materia
　　Medica.
　Ming, *c.* +1540.
　Wang Chi　汪機.
Pên Tshao Hui Tsuan　本草彙纂.
　A Classified Materia Medica compiled (from the
　　literature of pharmaceutical natural history).
　See Thu Tao-Ho (*1*).
Pen Tshao Huī Yen　本草彙言.
　A Rearrangement of the Classification in the
　　Pharmaceutical Natural Histories.
　Ming, +1624.
　Ni Chu-Mo　倪朱謨.
Pên Tshao Kang Mu　本草綱目.
　The Great Pharmacopoeia; or, The Pandects of
　　Natural History.
　Ming, +1596.
　Li Shih-Chen　李時珍.
　Paraphrased and abridge tr. Read & Collabora-
　　tors (*1–7*) and Read & Pak (*1*) with indexes.
Pên Tshao Kang Mu Hui Yen　本草綱目彙言.
　A Collection of Articles on the *Pandects of
　　Pharmaceutical Natural History* (thirty authors
　　on the principles of pharmacological therapy).
　Ming, +1624.
　Ni Chu-Mo　倪朱謨.
　LPC, no. 68; cf. Swingle (5); Li Thao (7, 11).
Pên Tshao Kang Mu Shih I　本草綱目拾遺.
　Supplementary Amplifications for the *Pandects of
　　Natural History* (of Li Shih-Chen).
　Chhing begun *c.* +1760, first prefaced +1765,
　　proligomena added +1780, last date in text
　　1803. First pr. 1871.
　Chao Hsüeh-Min　趙學敏.
　LPC, no. 101.
　Cf. Single (11).
Pên Tshao Ko Kua　本草謌括.
　Materia Medica in Mnemonic Verses.
　Yuan, +1295.
　Hu Shih-Kho　胡仕可.
Pên Tshao Kuang I　本草廣義.
　Alternative name of *Pên Tshao Yen I* (q.v.),
　　perhaps used at its first printing in +1119,
　　but definitively abandoned after +1195
　　because of a tabu conflict with the imperial
　　personal name.
Pên Tshao Mêng Chhüan　本草蒙筌.
　Enlightenment on Pharmaceutical Natural
　　History.
　Ming, +1565.
　Chhen Chia-Mo　陳嘉謨.
Pên Tshao Pei Yao　本草備要.
　Practical Aspects of Materia Medica.
　Chhing, *c.* +1690; second ed. +1694.
　Wang Ang　汪昂.
　LPC, no. 90; *ICK*, pp. 215 ff.
　Cf. Swingle (4).

Pên Tshao Phin Hiu Ching Yao 本草品彙精要.
Essentials of the Pharmacopoeia Ranked according to Nature and Efficacity (Imperially Commissioned).
Ming, +1505.
Liu Wên-Thai, 劉文泰, Wang Phan 王槃 & Kao Thing-Ho 高廷和.

Pên Tshao Phin Hiu Ching Yao Hsü Chi 本草品彙 精要續集.
Continuation of the *Essentials of the Pharmacopoeia Ranked according to Nature and Efficacity.*
Chhing, +1701.
Wang Tao-Shun 王道純, & Chiang Chao-Yuan 江兆元.

Pên Tshao Pieh Shuo 本草別說.
Additional Remarks on Pharmaceutical Natural History.
See *Chhung-Kuang Pu-Chu Shen Nung Pên Tshao ping Thu Ching,* +1092, for which title it was erroneously used, probably as an abbreviation of a description, by Li Shih-Chen, *PTKM,* ch. 1 (p. 6).

Pên Tshao Pu 本草補.
A Supplement to the Pharmaceutical Natural Histories.
Chhing, +1697.
Shih To-Lu 石鐸琭. (Fr. Pedro Piñuela, OFM) Recorded by Liu Ning 劉凝.

Pên Tshao Shih Chien 本草詩箋.
Materia Medica in Tasteful Verse.
Chhing, +1739.
Chu Lun 朱鑰.
Cf. Swingle (12).

Pên Tshao Shih I 本草拾遺.
A Supplement for the Pharmaceutical Natural Histories.
Thang, *c.* +725.
Chhen Tshang-Chhi 陳藏器.
Now extant only in numerous quotations.

Pên Tshao Shu 本草述.
Explanations of Materia Medica.
Chhing, before +1665, first pr. +1700.
Liu Jo-Chin 劉若金.
LPC, no. 79; cf. Swingle (6).

Pên Tshao Shu Kou Yuan 本草述鉤元.
Essentials Extracted from the *Explanations of Materia Medica.*
See *Yang Shih-Thai* (1).

Pên Tshao Thu Ching 本草圖經.
Illustrated Pharmacopoeia; or, Illustrated Treatise of Pharmaceutical Natural History.
Sung, +1061.
Su Sung *et al.* 蘇頌.
Now preserved only in numerous quotations in the later pandects of pharmaceutical natural history.

Pên Tshao Thung Hsüan 本草通玄.
The Mysteries of Materia Medica Unveiled.
Chhing begun before +1655, pr. just before +1667.
Li Chung-Tzu 李中梓.

LPC, no. 75.
Cf. Swingle (4).

Pên Tshao Tshung Hsin 本草從新.
New Additions to Pharmaceutical Natural History.
Chhing, +1757.
Wu I-Lo 吳儀洛.
LPC, no. 99.

Pên Tshao Yao Hsing 本草藥性.
The Natures of the Vegetable and Other Drugs in the Pharmaceutical Treatises.
Thang, *c.* +620.
Chen Li-Yen 甄立言 & (perhaps) Chen Chhüan 甄權.
Now extant only in quotations.

Pên Tshao Yen I 本草衍義.
Dilations upon Pharmaceutical Natural History.
Sung, pref. +1116, pr. +1119, repr. +1185, +1195.
Khou Tsung-Shih 寇宗奭.
See also *Thu Ching Yen I Pên Tshao* (*TT*/761).

Pên Tshao Yen I Pu I 本草衍義補遺.
Revision and Amplification of the *Dilations upon Pharamaceutical Natural History.*
Yuan, *c.* +1330.
Chu Chen-Hêng 朱震亨.
LPC, no. 47; cf. Swingle (12).

Pên Tshao Yin I 本草音義.
Meanings and Pronunciations of Words in Pharmaceutical Natural History.
Sui, *c.* +600.
Chen Li-Yen 甄立言.
Now extant only in quotations.

Pên Tshao Yin I 本草音義.
Materia Medica Classified according to Rhyme.
Thang, *c.* +750.
Li Han-Kuang 李含光.
Now extant only in quotations.

Pên Tshao Yuan Shih 本草原始.
Objective Natural History of Materia Medica; a True-to-Life Study.
Chhing, begun +1578, pr. +1612.
Li Chung-Li 李中立.
LPC, no. 60.

Phang Hsi (Phien) 滂喜篇.
The Copious Enjoyment Primer [orthographic word-list or dictionary, one of the three incorporated later into the *San Tshang* dictionary].
H/Han, *c.* +100.
Chia Fang 賈魴.
Now extant only within the reconstructed *San Tshang,* q.v.

Phei Wên Yün Fu 佩文韻府.
Word-store arranged by Rhymes, from the Hall of the Admiration of Literature [phrase—dictionary based on the last character of each phrase].
Chhing, commissioned, +1704, completed +1711, pr. +1712.
Ed. Chang Yü-Shu 張玉書 *et al.*

Phêng-mên Hua Phu 彭門花譜.
A Treatise on the (Tree-Peony) Flowers of Phêng-hsien.
Sung, *c.* +1260.
Jen Shou 任璹.
Now extant only in quotations.

Phi Ya 埤雅.
New Edifications on (i.e. Additions to) the *Literary Expositor*.
Sung, +1096.
Lu Tien 陸佃.

Phien Tzu Lei Phien 駢字類編.
Classified Collection of Phrases and Literary. Allusions [phrase dictionary based on the first character of each phrase].
Chhing, commissioned +1719, completed +1726, pr. +1728.
Ed. Ho Chhuo 何焯 *et al.*

Phin Chha Yao Lu 品茶要錄.
Essentials of the Grading of Teas.
Sung, +1078.
Huang Ju 黃儒.

Phin-Hu Mo Hsüeh 瀕湖脈學.
(Dr. Li) Phin-Hu's Treatise on sphygmology.
Ming, +1564.
Li Shih-Chen (Phin-Hu) 李時珍.
Usually appended to the *Pên Tshao Kang Mu*.
Abridged tr. Hübutter (1), p. 179.

Phing-Chhüan Shan Chü Tshao Mu Chi 平泉山居草木記.
Record of the (Notable) Plants and Trees growing (in the Gardens of) the Country Residence of Phing-Chhüan [about 10 miles from Loyang].
Thang, *c.* +820.
Li Tê-Yü 李德裕.

Phing Hua Phu 瓶花譜.
A Treatise on Flowers suitable for Vases.
Ming, +1595.
Chang Chhien-Tê 張謙德.
Tr. Li Hui-Liu (13).

Phing Shih 瓶史.
The History of the Vase; Studies in Flower Arrangement.
Ming, +16th or early +17th century.
Yuan Hung-Tao 袁宏道.

Phing Yuan Chi 平園集.
Collection of Writings from the Peaceful Garden.
Sung, late +12th century.
Chou Pi-Ta 周必大.

Phu Chi Fang 普濟方.
Practical Prescriptions for Everyman.
Ming, *c.* +1418.
Chu Hsiao (Chou Ting Wang) 朱橚 (周定王).
ISK, p. 914.

Pi Chhuan Hua Ching 秘傳花鏡.
The Mirror of Flowers; Family Records of the Art and Mystery of Horticulture.
See *Hua Ching*.

Pien Chu 編珠.
Strung Pearls (of Literature) [the second oldest private encyclopaedia of brief quotations].
Sui, *c.* +605.
(Not all of what is now preserved is considered original.)
Tu Kung-Chan 杜公瞻.

Pien Min Thu Tsuan 便民圖纂.
Everyman's Handy Illustrated Compendium; or, the Farmstead Manual.
Ming, +1502; repr. +1552, +1593.
Ed. Kuang Fan 鄺璠.

Po-Chou Mu-Tan Chi 亳州牡丹記.
An Account of the Tree-Peonies of Pochow (in Anhui).
Chhing, +1683.
Niu Hsiu 鈕琇.

Po-Chou Mu-Tan Piao 亳州牡丹表.
A List of the Varieties of Pochow Tree-Peonies according to their Grades.
Ming, *c.* +1610.
Hsüeh Fêng-Hsiang 薛鳳翔.
In *Thu Shu Chi Chhêng*.

Po-Chou Mu-Tan Shih 亳州牡丹史.
The History of the Tree-Peonies of Pochow (in Anhui).
Ming, *c.* +1610.
Hsüeh Fêng-Hsiang 薛鳳翔.
In *Thu Shu Chi Chhêng* and *Chih Wu Ming Shih Thu Khao*.

Po Hsüeh (Phien) 博學篇.
Extensive Knowledge of Words [orthographic primer].
Chhin, *c.* −215.
Huwu Ching 胡毋敬.
Incorporated by Han times in the *Tshang Chieh (Phien)*, q.v.

Po Wu Chi 博物記.
Notes on the Investigation of Things.
H/Han, *c.* +190.
Thang Mêng (b) 唐蒙.

Po Wu Chih 博物志.
Records of the Investigation of Things (cf. *Hsü Po Wu Chih*).
Chin, *c.* +290 (begun about +270).
Chang Hua 張華.

Po Ya 博雅.
Alternative name for *Kuang Ya* (q.v.) used from the Sui onwards.

Pu Chha Ching 補茶經.
A Supplement to the Manual of Tea.
Sung, +1008.
Chou Chiang 周絳.

Pu Nung Shu 補農書.
Supplement to the Treatise on Agriculture [of Mr Shen].
Ming, *c.* +1620.
Chang Li-Hsiang 張履祥.

San Chhin Chi 三秦記.
Record of the Three Princedoms of Chhin [into which that State was divided after the Chhin and before the Han].

San Chhin Chi (*cont.*)
Chin.
Sometimes attributed to a Mr Hsin 辛氏.
Writer unknown.

San Ching I Hsien Lu 三徑怡閒錄.
Records of Leisurely Chrysanthemum Culture
[the title taken from the 'three rows of chrysanthemums planted along the eastern fence'
by Thao Yuan-Ming].
Ming, *c.* +1585.
Kao Lien 高濂.

San Fu Chüeh Lu 三輔決錄.
A Considered Account of the Three Cities of the
Metropolitan Area (Chhang-an, Fêng-i and
Fu-fêng) [the earliest book of the gazetteer
genre].
H/Han, +153.
Chao Chhi 趙岐.

San Fu Huang Thu 三輔黃圖.
Illustrated Description of the Three Cities of the
Metropolitan Area (Chhang-an (mod. Sian),
Fêng-i and Fu-fêng).
Chin, original text late +3rd century, or per-
haps H/Han; present version stabilised, in-
cluding much older material, between +757
and +907.
Attrib. Miao Chhang-Yen 苗昌言.
Cf. des Rotours (1), p. lxxxvi.

San Kuo Chih 三國志.
History of the Three Kingdoms [+220 to
+280].
Chin, *c.* +290.
Chhen Shou 陳壽.
Yin-Tê Index, no. 33.
For translations of passages see the index of
Frankel (1).

San Nung Chi 三農紀.
Records of the Three Departments of
Agriculture.
Chhing, +1760.
Chang Tsung-Fa 張宗法.

San Tshai Thu Hui 三才圖會.
Universal Encyclopaedia.
Ming, +1609.
Wang Chhi 王圻.

San Tshang 三蒼.
The Three Tshang [orthographic primer and
dictionary; a conflation of the *Tshang Chieh
(Phien)* of *c.* −220, the *Hsün Tsuan (Phien)* of
c. +6, and the *Phang Hsi (Phien)* of *c.* +100;
with added explanations].
Chhin to San Kuo.
Conflation either by Chang I (Wei) 張揖 the
first editor, *c.* +230, or during the previous
century.
Second editor Kuo Pho (Chin) 郭璞 *c.* +300.
Reconstruction in *YHSF*, ch. 60, pp. 13*a* ff.

San Tzu Ching 三字經.
Trimetrical Primer.
Sung, *c.* +1270.
Wang Ying-Lin 王應麟.

Shan Chha Phu 山茶譜.
A Treatise on *Camellia* (Species and Varieties)
[of Fukien and Japan].
Chhing, +1719.
Phu Ching Tzu (ps.) 樸靜子.

Shan Chü Fu 山居賦.
Ode on Dwelling in the Mountains.
L/Sung, *c.* +420.
Hsieh Ling-Yün 謝靈運.

Shan Fan Pên Tshao 刪繁本草.
The Pharmacopoeia Purged.
Thang, *c.* +775.
Yang Sun-Chih 楊損之.

Shan Hai Ching 山海經.
Classic of the Mountains and Rivers.
Chou and C/Han.
Writers unknown.
Partial tr. de Rosny (1).
Chung-Fa Index no. 9.

Shan Thang Khao So 山堂考索.
See *Chhün Shu Khao So*.

Shang Shu Ku Shih 尚書故實.
Facts and Corrections about Ancient
Records.
Thang, *c.* +860.
Li Chho 李綽.

Shang Shu Wang Shih Chu 尚書王氏注.
Mr Wang's Commentary on the *Historical
Classic*.
San Kuo, *c.* +245.
Wang Su 王肅.
YHSF, ch. 11, p. 3*a*.

*Shao-Hsing Chiao-Ting Ching-Shih Chêng-Lei Pei-Chi Pên
Tshao* 紹興校定經史證類備急本草.
The Corrected Classified and Consolidated
Armamentarium; Pharmacopoeia of the Shao-
Hsing Reign-Period.
S/Sung, pres. +1157, pr. +1159, often copied
and repr. especially in Japan.
Thang Shen-Wei 唐慎微.
Ed. Wang Chi-Hsien 王繼先 *et al.*
Cf. Nakao Manzō (*1, 1*); Swingle (11).
Illustrations reproduced in facsimile by Wada
(*1*); Karow (2).

Shao-Yao Phu 芍藥譜.
A Treatise on the Herbaceous Peony (and its
Varieties).
Sung, +1073.
Liu Pin 劉攽.

Shao-Yao Phu 芍藥譜.
A Treatise on the Herbaceous Peony (and its
Varieties).
Sung, *c.* +1080.
Khung Wu-Chung 孔武仲.

Shao-Yao Phu 芍藥譜.
A Treatise on the Herbaceous Peony (and its
Varieties) [part of *Hua Chu Wu Phu*, q.v.].
Ming, +1591.
Kao Lien 高濂.

Shen Nung Ku Pên Tshao Ching 神農古本草經.
The Ancient Text of the *Classical Pharmacopoeia of
the Heavenly Husbandman*.
See Liu Fu (5).

Shen Nung Pên Tshao Ching 神農本草經.
Classical Pharmacopoeia of the Heavenly Husbandman.
C/Han, based on Chou and Chhin material, but not reaching final form before the +2nd century.
Writers unknown.
Lost as a separate work, but the basis of all subsequent compendia of pharmaceutical natural history, in which it is constantly quoted.
Reconstituted and annotated by many scholars; see Lung Po-chien (*1*), pp. 2 ff., 12 ff.
Best reconstructions by Mori Tateyuki (1845) 森立之, Liu Fu (1942) 劉復.

Shen Nung Pên Tshao Ching Fêng Yuan.
See *Pên Ching Fêng Yuan*.

Shen Nung Pên Tshao Ching Khao I 神農本草經考異.
A Reconstruction of the Text of the *Classical Pharmacopoeia of the Heavenly Husbandman*, with an Analysis of Textual Variations.
See Mori Tateyuki (*1*).

Shen Nung Pên Tshao Ching Pai Chung Lu 神農本草經百種錄.
A Hundred Entries (reconstructed from) the *Classical Pharmacopoeia of the Heavenly Husbandman*.
Chhing, +1736.
Hsü Ta-Chhun (= Hsü Ling-Thai) 徐大椿.

Shen Nung Pên Tshao Ching Su 神農本草經疏.
Commentary on the Text of the *Classical Pharmacopoeia of the Heavenly Husbandman*.
Ming, +1625.
Miu Hsi-Yung 繆希雍.
LPC, no. 62; cf. Swingle (11).

Shen Nung Pên Tshao Ching Su Chêng.
See *Pên Ching Su Chêng* under Tsou Chu (*1*).

Shen Nung Pên Tshao Ching Su Chi Yao 神農本草經疏輯要.
Essentials of the *Commentary* [of Miu Hsi-Yung] on the Text of the Classical Pharmacopoeia of the Heavenly Husbandman.
See Wu Shih-Khai (*1*).

Shen Nung Pên Tshao Pu-Chu 神農本草補註.
Alternative title for Chang Yu-Hsi's *Chia-Yu Pu-Chu Shen Nung Pên Tshao*, +1060 (*Shih Shan Thang Tshang Shu Mu Lu*, ch. 2, p. 43*a*).

Shen Nung Pên Tshao Thu Ching 神農本草圖經.
Alternative title for Su Sung's *Pên Tshao Thu Ching*, +1061 (*Shih Shan Thang Tshang Shu Mu Lu*, ch. 2, p. 43*a*).

Shen Yin Shu 神隱書.
See *Chhü Hsien Shen Yin Shu*.

Shêng-An Ho Chi 升菴合集.
Collected Writings of (Yang) Shêng-An.
Ming, +1541; coll. & repr. *c*. 1890.
Yang Shen 楊慎.
Ed. Chêng Pao-Chhen 鄭寶琛 & Wang Wên-Lin 王文林.

Shêng-An Wai Chi 升菴外集.
Additional Collection of the Writings of (Yang) Shêng-An (astronomy, botany and zoology).
Ming, +1530–59; pr. +1616; repr. 1844.
Yang Shen 楊慎.
Ed. Chiao Hung 焦竑.

Shêng Chi Tsung Lu 聖濟總錄.
Imperial Medical Encyclopaedia (lit. General Treatise (on Medical Care) Commissioned by the Majestic Benevolence) [issued by authority].
Sung, *c*. +1111–18. Repr. Yuan, +1300.
Ed. by twelve physicians, headed by Shen Fu 申甫.
SIC, pp. 1002 ff.

Shêng Lei 聲類.
The Sounds Classified; a Character Dictionary [the oldest with a phonetic arrangement, according to final syllables or 'rhymes'].
San Kuo, +3rd century.
Li Têng 李登.
Now preserved only in fragments.

Shêng Sung Chha Lun 聖宋茶論.
The Imperial Discourse on Tea, recorded in the Sung Dynasty.
See *Ta-Kuan Chha Lun*.

Shih Chhü Li Lun 石渠禮論.
Report of the Discussions in the Stone Canal Pavilion.
C/Han, −51
Attrib. Tai Shêng 戴聖; tr. Tsêng Chu-Sên (*1*), pp. 128 ff.
(*YHSF* ch. 28 p. 31*a*).

Shih Chi 史記.
Historical Records [or perhaps better: Memoirs of the Historiographer (-Royal); down to −99].
C/Han, *c*. −90 [first pr. *c*. +1000].
Ssuma Chhien 司馬遷, and his father Ssuma Than 司馬談.
Partial trs. Chavannes (1); Pfizmaier (13–36); Hirth (2); Wu Khang (1); Swann (1), etc.
Yin-Tê Index, no. 40.

Shih Chien Pên Tshao 食鑑本草.
The Dietary Mirror; a Pharmaceutical Natural History of Nutritional Substances.
Ming, *c*. +1540.
Ning Yuan 寧源.

Shih Chih Chi 蒔植記.
(= *Chhuan Phu Shih Chih Chi*).
Notes on the Cultivation of Plants (at the Chhuan Phu Villa).
Chhing, +1684.
Tshao Jung 曹溶.

Shih Ching 詩經.
Book of Odes [ancient folksongs].
Chou, −11th to −7th century (Dobson's dating).
Writers and compilers unknown.
Tr. Legge (8); Waley (1); Karlgren (14).

Shih Chou (Phien) 史籀篇.
Chou the Chronologer-Royal, his book [orthographic word-list or glossary].
Chou, *c*. −800.

Shih Chou (Phien) (cont.)
(Shih) Chou 史籀.
Partial reconstruction in *YHSF*, ch. 59, pp. 3*a* ff.

Shih Chuan Chu Su 詩傳注疏.
Commentaries on the Traditions concerning the
Book of Odes.
Sung, *c.* +1270.
Hsieh Fang-Tê 謝枋得.

Shih Hsing Pên Tshao 食性本草.
The Natural Properties of Foods; a Pharmaceu-
tical History.
Thang, *c.* +895.
Chhen Shih-Liang 陳仕良.

Shih-Hu Chü Phu 石湖菊譜.
Mr (Fan) Shih-Hu's Chrysanthemum Treatise.
See the *Chü Phu* of Fang Chhêng-Ta.

Shih Lao Phu Chü Phu 史老圃菊譜.
Old Gardener Shih's Treatise on
Chrysanthemums.
See the *Chü Phu* of Shih Chêng-Chih.

Shih Liao Pên Tshao 食療本草.
Nutritional Therapy; a Pharmaceutical Natural
History.
Thang, *c.* +670.
Meng Shen 孟詵.

Shih Ming 釋名.
Expositor of Names.
Early +2nd century.
Liu Hsi 劉熙.

Shih Ming Su Chêng Pu 釋名疏證補.
See Wang Hsien-Chhien (*3*).

Shih Pên 世本.
Book of Origins [imperial genealogies, family
names, and legendary inventors].
C/Han (incorporating Chou material). −2nd
century.
Ed. Sung Chung (H/Han) 宋衷.

Shih Shan thang Tshang Shu Mu Lu 世善堂藏
書目錄.
Catalogue of the Library of Shih Shan Thang.
Ming, +1616.
Chhen Ti 陳第.

Shih Shih 事始.
The Beginnings of all Affairs.
Sui, +605–16.
Lin Tshun 劉存, or Liu Hsiao-Sun 劉孝孫.

Shih Shih Chü Phu 史氏菊譜.
See the *Chü Phu* of Shih Chêng-Chih.

Shĭ Thung 史通.
Summa Historiae [the first treatise on his-
toriography in Chinese or any other
civilisation].
Thang, +710.
Liu Chih-Chi 劉知幾.

Shih Tshao Hsiao Chi 釋草小記.
Brief Notes on the Names of Herbaceous Plants.
Chhing, late +18th.
Chhêng Yao-Thien 程瑤田.

Shih Wu Chi Yuan 事物紀原.
Records of the Origins of Affairs and Things.
Sung, *c.* +1085.
Kao Chhêng 高承.

Shih Wu Pên Tshao 食物本草.
Nutritional Natural History.
Ming, +1571 (repr. from a slightly earlier
edition).
Attrib. Li Kao (J/Chin) 李杲 or Wang Ying
(Ming) 汪穎 in various editions; actual
writer Lu Ho 盧和.
The bibliography of this work in its several
different forms, together with the questions of
authorship and editorship, are complex.
See Lung Po-Chien (*1*) pp. 104, 105, 106; Wang
Yü-Hu (*1*) 2nd ed. p. 194; Swingle (*1*, *10*).

Shih Wu Pên Tshao Hui Tsuan 食物本草會纂.
Newly Compiled *Pharmaceutical Natural History of
Foods*.
Chhing, +1691.
Shenli Lung 沈李龍.

Shih Yao Erh Ya 石藥爾雅.
The Literary Expositor of Chemical Physic;
or, Synonymic Dictionary of Minerals and
Drugs.
Thang, +806.
Mei Piao 梅彪.
TT/894.

Shih Yuan 事原.
On the Origins of Things.
Sung.
Chu Hui 朱繪.

Shou Shih Thung Khao 授時通考.
Compendium of Works & Days.
Chhing, +1742.
Compiled by imperial order under the Direction
of O-Erh-thai 鄂爾泰.
Textual refs. all to the 1847 repr. of the original
1742 Palace ed.

Shou Yü Shen Fang 壽域神方.
Magical Prescriptions of the Land of the Old.
Ming, *c.* +1430.
Chu Chhüan 朱權 (Ning Hsien Wang, prince
of the Ming) 寧獻王.

Shu Ching 書經.
Historical Classic [or, Book of Documents].
The 29 'Chin Wên' chapters mainly Chou
(a few pieces possibly Shang); the 21 'Ku
Wên' chapters a 'forgery' by Mei Tsê 梅賾,
c. +320, using fragments of genuine antiquity. Of the former, 13 are considered to go
back to the −10th century, 10 to the −8th,
and 6 not before the −5th. Some scholars
accept only 16 or 17 as pre-Confucian.
Writers unknown.
See Wu Shih-Chhang (*1*); Creel (*4*).
Tr. Medhurst (*1*); Legge (*1*, *10*); Karlgren (*12*).

Shu Hsü Chih Nan 書叙指南.
The Literary South-Pointer [guide to style in
letter-writing, and technical terms].
Sung, +1126.
Jen Kuang 任廣.

Shu I Shu 樹藝書.
A Dissertation on Methods of Planting.
Ming.
Chou Chih-Yü 周之瑀.

Shu Pên Tshao 蜀本草.
(= *Chhung Kuang Ying Kung Pên Tshao*).
Pharmacopoeia of the State of (Later) Shu [Szechuan].
Wu Tai (H/Shu), betw. +938 and +950.
Ed. Han Pao-Shêng 韓保昇.

Shui Ching Chu 水經注.
Commentary on the *Waterways Classic* [geographical account greatly extended].
N/Wei, late +5th or early +6th century.
Li Tao-Yuan 酈道元.

Shui Mi Thao Phu 水蜜桃譜.
A Treatise on the 'Honeydew' Peach.
See Chhu Hua (*1*).

Shuo Fu 說郛.
Florilegium of (Unofficial) Literature.
Yuan, *c.* +1368.
Ed. Thao Tsung-I 陶宗儀.
See Ching Phei-Yuan (*1*); des Rotours (*4*), p. 43.

Shuo Wên.
See *Shuo Wên Chieh Tzu*.

Shuo Wên Chieh Tzu 說文解字.
Analytical Dictionary of Characters (lit. Explanations of Simple Characters and Analyses of Composite Ones).
H/Han, +121.
Hsü Shen 許慎.

Shuo Wên Thung Hsün Ting Shêng.
See Chu Chün-Shêng (*1*).

Shuo Yuan 說苑.
Garden of Discourses.
Han, *c.* −20.
Liu Hsiang 劉向.

Ssu Khu Chhüan Shu Tsung Mu Thi Yao 四庫全書總目提要.
Analytical Catalogue of the *Complete Library of the Four Categories* (made by imperial order).
Chhing, +1782.
Ed. Chi Yün 紀昀.
Indexes by Yang Chia-Lo; Yü & Gillis. Yin-Tê Index, no. 7.

Ssu Min Yüeh Ling 四民月令.
Monthly Ordinances for the Four Sorts of People (Scholars, Farmers, Artisans and Merchants).
H/Han, *c.* +160.
Tshui Shih 崔寔.

Ssu Shêng Pên Tshao 四聲本草.
Materia Medica Classified according to the Four Tones (and the Standard Rhymes), [the entries arranged in the order of the pronunciation of the first character of their names].
Thang, *c.* +775.
Hsiao Ping 蕭炳.

Ssu Shih Hua Chi 四時花記.
On the Flowers of the Four Seasons [part of *Yen Hsien Chhing Shang Chien*, q.v.].
Ming, +1591.
Kao Lien 高濂.

Ssu Shih Tsuan Yao 四時纂要.
Important Rules for the Four Seasons (agricul-

ture) (preserved only in quotations).
Thang.
Han É 韓諤.

Sui Chhao Chung Chih Fa 隋朝種植法.
Horticultural Methods of the Sui Court.
Thang, +8th century.
Writer unknown.
Now extant only in quotations.

Sui Chhu Thang Shu Mu 遂初堂書目.
Bibliography of the Sui Chhu Library.
Sung, *c.* +1180.
Yu Mou 尤袤.

Sui Shu 隋書.
History of the Sui dynasty [+581 to +617].
Thang, +636 (annals and biographies); +656 (monographs and bibliography).
Wei Chêng 魏徵 *et al.*
Partial trs. Pfizmaier (61–65); Balazs (7, 8); Ware (*1*).
For translations of passages see the index of Frankel (*1*).

Sun Phu 筍譜.
Treatise on Bamboo Shoots.
Sung, *c.* +970.
Tsan-Ning (monk) 贊寧.

Sung Chai Mei Phu 松齋梅譜.
The Pinetree Studio Treatise on the Chinese Apricot (*Prunus mume*).
Yuan, *c.* +1352.
Wu Thai-Su 吳太素.
Cf. Shimada Shūjirō (*1*).

Sung I-chhien I Chi Khao 宋以前醫籍考.
Comprehensive Annotated Bibliography of Chinese Medical Literature in and before the Sung Period.
See Okanishi Tameto (*2*).

Sung Yuan Hsüeh An 宋元學案.
Schools of Philosophers in the Sung and Yuan Dynasties.
Chhing, *c.* +1750.
Huang Tsung-Hsi 黃宗羲; & Chhüan Tsu-Wang 全祖望.

Ta Hsüeh 大學.
The Great Learning or, [the Learning of Greatness].
Chou, *c.* −260.
Traditionally attributed to Tsêng Shen 曾參. but probably written by Yochêng Kho 樂正克.
Tr. Legge (2); Hughes (2); Wilhelm (6).

Ta-Kuan Chha Lun 大觀茶論.
The Discourse on Tea, recorded in the Ta-Kuan reign-period (+1107–110).
Sung, *c.* +1109.
Chao Chi (emperor of the Sung; Sung Hui Tsung) 趙佶.

Ta-Kuan Ching-Shih Chêng Lei Pei-Chi Pên Tshao 大觀經史證類備急本草.
The Classified and Consolidated Armamentarium; Pharmacopoeia of the Ta-Kuan reign-period.

Ta-Kuan Ching-Shih Chêng Lei Pei-Chi Pên Tshao (cont.)
　Sung, +1108; repr. +1121, +1214 (J/Chin),
　　+1302 (Yuan).
　Thang Shen-Wei 唐愼微.
　Ed. Ai Shêng 艾晟.

Ta Kuar Pên Tshao.
　See *Ta-Kuar Ching-Shih Chêng Lei Pei-Chi Pên
　　Tshao.*

Ta Sung Chhung Hsiu Kuang Yün 大宋重修廣韻.
　See *Kuang Yün.*

Ta Tai Li Chi 大戴禮記.
　Record of Rites [compiled by Tai the Elder] (cf.
　　Hsiao Tai Li Chi; Li Chi).
　Ascr. C/Han, c. -70 to -50, but really H/Han,
　　between +80 and +105.
　Attrib. ed. Tai Tê 戴德 in fact probably ed.
　　Tshao Pao 曹褒.
　See Legge (7).
　Trs. Douglas (1); R. Wilhelm (6).

Ta Thang Hsi Yü Chi 大唐西域記.
　Record of (a Pilgrimage to) the Western
　　Countries in the time of the Thang.
　Thang, +646.
　Hsüan-Chuang 玄奘.
　Text by Pien-Chi 辯機.
　Tr. Julien (1); Beal (2).

Ta Tsang 大藏.
　The Buddhist Patrology (*Tripitaka*).
　All dates from the +2nd century onwards, or
　　earlier, if translations.
　Writers numerous.

Tao Tsang 道藏.
　The Taoist Patrology [containing 1464 Taoist
　　works].
　All periods, but first collected in the Thang
　　about +730, then again about +870 and
　　definitively in +1019. First printed in the
　　Sung (+1111 to +1117). Also printed in
　　J/Chin (+1168 to +1191), Yuan (+1244),
　　and Ming (+1445, +1598 and +1607).
　Writers numerous.
　Indexes by Wieger (6), on which see Pelliot's
　　review (58); and Ong Tu-chien (Yin-Tê
　　Index, no. 25).

Thai-Phing Huan Yü Chi 太平寰宇記.
　Thai-Phing reign-period General Description of
　　the World [geographical record].
　Sung, +976–83.
　Yüeh Shih 樂史.

Thai-Phing Kuang Chi 太平廣記.
　Copious Records collected in the Thai-Phing
　　reign-period [anecdotes, stories, mirabilia and
　　memorabilia].
　Sung, +978.
　Ed. Li Fang 李昉.

Thai-Phing Shêng Hui Fang 太平聖惠方.
　Prescriptions collected by Imperial Solicitude in
　　the Thai-Phing reign-period.
　Sung, commissioned, +982; completed +992.
　Ed. Wang Huai-Yin et al. 王懷隱.
　SIC, p. 921.

Thai-Phing Yü Lan 太平御覽.

Thai-Phing reign-period Imperial Encyclo-
　paedia (lit. the Imperial Speculum of the
　Thai-Phing reign-period i.e. the Emperor's
　Daily Readings).
　Sung, +983.
　Ed. Li Fang 李昉.
　Some chs. tr. Pfizmaier (84–106).
　Yin-Tê Index no. 23.

Thang Chhang Yü Jui Pien Chêng 唐昌玉蕊辨證.
　On the Identification of the Jade-Stamen
　　Flower (Trees) at the Thang-Chhang (Taoist
　　temple, at Chhang-an).
　See *Yü Jui Pien Chêng.*
　Cf. *SKCS/TMTY*, ch. 116, p. 83b.

Thang I Pên Tshao 湯液本草.
　The Materia Medica of Decoctions and
　　Tinctures.
　Sung or Yuan c. +1280.
　Wang Hao-Ku 王好古.

Thang Kuo Shih Pu 唐國史補.
　Additional Materials towards a History of the
　　Thang.
　Thang, c. +860.
　Li Chao 李肇.

Thang Lei Han 唐類函.
　Classified Treasure-chest of the Thang; or, The
　　Thang Encyclopaedias Conflated [i.e. the *Pei
　　Thang Shu Chhao*, the *I Wên Lei Chü*, the *Chhu
　　Hsüeh Chi* and the *Liu Thieh Shih Lei Chi*].
　Ming, +1618.
　Yü An-Chhi 兪安期.

Thang Pên Tshao 唐本草.
　See *Hsin Hsiu Pên Tshao.*

Thang Sung Pai Khung Liu Thieh 唐宋白孔六帖.
　See *Liu Thieh Shih Lei Chi.*

Thang Yün 唐韻.
　Thang Dictionary of Characters arranged
　　according to their Sounds [rhyming phonetic
　　dictionary based on, and including, the *Chhieh
　　Yün*, q.v.].
　Thang, +677, revised and republished +751.
　Chhangsun No-Yen (+7th) 長孫訥言, and
　　Sun Mien (+8th) 孫愐.
　Now extant only within the *Kuang Yün*, q.v.

Thang Yün Chêng 唐韻正.
　Thang Dynasty Rhyme Sounds (compared with
　　those of antiquity).
　Chhing, +1667 (in *Yin Hsüeh Wu Shu*).
　Ku Yen-Wu 顧炎武.

Thao Chih Thu Shuo 陶治圖說.
　[= *Thao Yeh Thu* and *Thao Yeh Thu Shuo*].
　Illustrations of the Pottery Industry, with
　　Explanations.
　Chhing, +1743.
　Thang Ying 唐英.
　Tr. Julien (7), pp. 115 ff.; Bushell (4), pp. 7 ff.;
　　Sayer (1), pp. 4 ff.

Thien Chia Wu Hsing 田家五行.
　The Farmer's Guide to Nature (the Five
　　Elements).
　Sung.
　Lou Yuan-Li 婁元禮.

Thien Khuei Lun 天傀論.
A Study of Naturally-Occurring Monsters and Abnormalities [teratological].
Ascr. Ming, *c.* +1580.
Attrib. Li Shih-Chen 李時珍.
Unique MS in the Library of the Chinese Medical Association at Shanghai. Not among the titles of Li's known works.

Thien Kung Khai Wu 天工開物.
The Exploitation of the Works of Nature.
Ming, +1637.
Sung Ying-Hsing 宋應星.

Thien-Phêng Mu-Tan Phu 天彭牡丹譜.
A Treatise on the Tree-Peonies of Thien-phêng (near modern Phêng-hsien in Szechuan).
Sung, +1178.
Lu Yu 陸游.

Thou Huang Tsa Lu 投荒雜錄.
Miscellaneous Jottings far from Home (lit. Records of One Cast out in the Wilderness).
Thang, *c.* +835.
Fang Chhien-Li 房千里.
Cf. Schaper (16), p. 149.

Thu Ching Chi-Chu Yen I Pên Tshao 圖經集注衍義本草.
Illustrations and Collected Commentaries for the *Dilations upon Pharmaceutical Natural History.*
TT/761, (ong index, no. 767).
See also *Thu Ching Yen I Pên Tshao.*
The *Tao Tsang* contains two separately catalogued books, but the *Thu Ching Chi-Chu Yen I Pên Tshao* is in fact the introductory 5 chapters, and the *Thu Ching Yen I Pên Tshao* the remaining 42 chapters of a single work.

Thu Ching (Pên Tshao) 圖經(本草).
Illustrated Treatise (of Pharmaceutical Natural History). See *Pên Tshao Thu Ching.*
The term *Thu Ching* applied originally to one of the two illustrated parts (the other being a *Yao Thu*) of the *Hsin Hsiu Pên Tshao* of +659 (q.v.); cf. *Hsin Thang Shu*, ch. 59, p. 21*a* or *TSCCIW*, p. 273. By the middle of the +11th century these had become lost, so Su Sung's *Pên Tshao Thu Ching* was prepared as a replacement. The name *Thu Ching Pên Tshao* was often afterwards applied to Su Sung's work, but (according to the evidence of the *Sung Shih* bibliographies, *SSIW*, pp. 179, 529) wrongly.

Thu Ching Yen I Pên Tshao 圖經衍義本草.
Illustrations (and commentary) for the *Dilations upon Pharmaceutical Natural History.* (An abridged conflation of the *Chêng-Ho* ... *Chêng Lei* ... *Pên Tshao* with the *Pên Tshao Yen I.*)
Sung, *c.* +1223.
Thang Shen-Wei 唐慎微, Khou Tsung-Shin 寇宗奭, ed. Hsü Hung 許洪.
TT/761; see also *Thu Ching Chi-Chu Yen I Pên Tshao.*
Cf. Chang Tsan-Chhen (2); Lung Po-Chien (1), nos. 38, 39.

Thu Shu Chi Chhêng 圖書集成.
Imperial Encyclopaedia [or: Imperially Commissioned Compendium of Literature and Illustrations, Ancient and Modern].
Chhing, +1726.
Ed. Chhen Meng-Lei 陳夢雷.
Index by L. Giles (2).

Thung Chien Kang Mu 通鑑綱目.
Short View of the *Comprehensive Mirror (of History, for Aid in Government)* [the *Tzu Chih Thung Chun* condensed, with headings and sub-headings].
Sung (begun +1172), +1189.
Chu Hsi 朱熹 (and his school).
With later continuations, *Thung Chien Kang Mu Hsü Pien* and *Thung Chien Kang Mu San Pien.*
Partial definitive edition, with all commentaries, etc. *c.* +1630.
Ed. Chhen Jen-Hsi 陳仁錫. Tr. Wieger (1).

Thung Chih 通志.
Historical Collections.
Sung, *c.* +1150.
Chêng Chhiao 鄭樵.
Cf. des Rotours (2), p. 85.

Thung Chih Lüeh 通志略.
Compendium of Information [part of *Thung Chih*, q.v.].

Thung Chün Tshai Yao Lu 桐君採藥錄.
Thung Chün's (or Master Thung's) Directions for Gathering Drug-Plants.
C/Han or H/Han.
Thung Chün 桐君.
(The name, like Shen Nung in *Shen Nung Pên Tshao Ching*, is probably a pseudonym, for Thung Chün was one of the legendary ministers of Huang Ti, the Yellow Emperor).
Extant now only in quotations.

Thung Chün Yao Tui 桐君藥對.
Thung Chün's (or Master Thung's) Answers to Questions about Drug-Plants.
C/Han or H/Han.
Thung Chün 桐君.
(The name, like Shen Nung in *Shen Nung Pên Tshao Ching*, is probably a pseudonym, for Thung Chün was one of the legendary ministers of Huang Ti, the Yellow Emperor).
Extant now only in quotations.

Thung Phu 桐譜.
A Treatise on Thung Trees (*Paulownia* and certain others).
Sung, +1049.
Chhen Chu 陳翥.
Repr. in *Chih Wu Ming Shih Thu Khao* (Chhang Phien), ch. 20.

Thung Tien 通典.
Comprehensive Institutes [reservoir of source material on political and social history].
Thang, *c.* +812 (completed by +801).
Tu Yu 杜佑.

Ti-i Hsiang Pi Chi 第一香筆記.
Notes on the World's Finest Fragrances [orchids].
Chhing, +1796.

Ti-i Hsiang Pi Chi (cont.)
Chu Kho-Jou 朱克柔.
MS, in the Peking National Library.

Ti Wang Shih Chi 帝王世紀.
Stories of the Ancient Sovereigns.
San Kuo or Chin, c. +270.
Huangfu Mi 皇甫謐.

Tien Chung Chha-Hua Chi 滇中茶花記.
Notes on the Mountain-Tea Plants (*Camellias*) of
Yunnan.
Ming.
Fêng Shih-Kho 馮時可.
Cit. in *Thu Shu Chi Chhêng*.

Tien Hai Yü Hêng Chih 滇海虞衡志.
An Account of the Geography (and Products) of
Yunnan.
Chhing, +1799.
Than Tshui 檀萃.

Tien Nan Pên Tshao 滇南本草.
Pharmaceutical Natural History of Southern
Yunnan.
Ming, +1436.
Lan Mao 蘭茂.
Ed. (and perhaps modified) by Kuan Hsüan &
Kuan Chün, 1887 管暄, 管濬.
LPC/49, 135; cf. Yü Nai-I & Yü Lan-Fu (*1*);
Tsêng Yü-Lin (*1*).

Tsêng-chhêng Li-chih Phu 增城荔枝譜.
A Treatise on the Lichis of Tsêng-chhêng (in
Kuangtung).
Sung, +1076.
Chang Tsung-Min 張宗閔.

Tshai Yao Lu 採藥錄.
See *Thung Chün Tshai Yao Lu*.

Tshan Thung Chhi Wu Hsiang Lei Pi Yao 參同
契五相類祕要.
Arcane Essentials of the Similarities and
Categories of the Five (Substances) in the
Kinship of the Three (sulphur, realgar, orpi-
ment, mercury and lead).
Liu Chhao, possibly Thang; prob. between
+3rd and +7th cents., must be before the
beginning of the +9th. cent., though ascr.
+2nd.
Writer unknown (attrib. Wei Po-Yang).
Comm. by Lu Thien-Chi 盧天驥.
Sung +1111 to +1117, probably +1114.
TT/898.
Tr. Ho Ping-Yü & Needham (*2*).

Tshang Chieh Hsün Ku 蒼頡訓詁.
Instructions and Explanations for the *Book of
Tshang Chieh* [especially on the pronunciations
of the rarer words and names in that ortho-
graphic primer].
H/Han, c. +46.
Tu Lin 杜林.
Fragmentary reconstruction in *YHSF*, ch. 60,
pp. 9a ff.

Tshang Chieh (Phien) 蒼頡篇.
Book of Tshang Chieh [legendary inventor of
writing; an orthographic primer].
Chhin, c. −220.

Li Ssu 李斯.
Edited by Chang I (San Kuo, Wei) 張揖 and
Kuo Pho (Chin) 郭璞.
Reconstruction in *YHSF*, ch. 59, pp. 18a ff.

Tshao-Chou Mu-Tan Phu 曹州牡丹譜.
A Treatise on the Tree-Peonies of Tshaochow
(in Shantung).
Chhing, +1793.
Yü Phêng-Nien 俞鵬年.

Tshao Hua Phu 草花譜.
A Treatise on Herbaceous Garden Flowering
Plants.
Alternative title for *Ssu Shih Hua Chi*, q.v.

Tshao Nan Mu-Tan Phu 曹南牡丹譜.
A Treatise on the Tree-Peonies South of
Tshaochow (in Shantung).
Chhing, +1669.
Su Yü-Mei 蘇毓眉.

Tshê Fu Yuan Kuei 册府元龜.
Collection of Material on the lives of Emperors
and Ministers.
[lit. (Lessons of) the Archives, (the True)
Scapulimancy] [a governmental ethical and
political encyclopaedia.] commissioned
+1005.
Sung, pr. +1013.
Ed. Wang Chhin-Jo 王欽若 & Yang I
楊億.
Cf. des Rotours (*2*), p. 91.

Tso Chuan 左傳.
Master Tsochhiu's Tradition (or Enlargement)
of the *Chhun Chhiu* (*Spring and Autumn Annals*)
[dealing with the period −722 to −453].
Late Chou, compiled from ancient written and
oral traditions of several states between −430
and −250, but with additions and changes by
Confucian scholars of the Chhin and Han,
especially Liu Hsin. Greatest of the three
commentaries on the *Chhun Chhiu*, the others
being the *Kungyang Chuan* and the *Kuliang
Chuan*, but unlike them, probably originally
itself an independent book of history.
Attrib. Tsochhiu Ming 左邱明.
See Karlgren (*8*); Maspero (*1*); Chhi Ssu-Ho
(*1*); Wu Khang (*1*); Wu Shih-Chhang (*1*)
vander Loon (*1*); Eberhard, Müller &
Henseling (*1*).
Tr. Couvreur (*1*); Legge (*11*); Pfizmaier (*1–12*).
Index by Fraser & Lockhart (*1*).

Tsu Hsiang Hsiao Phu 祖香小譜.
A Little Treatise on the Ancestor of all
Fragrances.
Original title of *Ti-i Hsiang Pi Chi*, q.v.

Tsun Shêng Pa Chien 遵生八牋.
Eight Disquisitions on Putting Oneself in Accord
with the Life-Force [a collection of works].
Ming, +1591.
Kao Lien 高濂.
For the separate parts see:
1. Chhing Hsiu Miao Lun Chien (chs. 1, 2).
2. Ssu Shih Thiao Shê Chien (chs. 3–6).
3. Chhi Chü An Lo Chien (chs. 7, 8).

Tsun Shêng Pa Chien (cont.)
    4. Yen Nien Chhio Ping Chien (chs. 9, 10).
    5. Yin Chuan Fu Shih Chien (chs. 11–13).
    6. Yen Hsien Chhing Shang Chien (chs. 14, 15).
    7. Ling Pi Tan Yao Chien (chs. 16–18).
    8. Lu Wai Hsia Chü Chien (ch. 19).
Tu Hua Chü Tung Li Chi 渡花居東籬集.
  Collected Writings of a Lifetime of
    Chrysanthemums [the title taken from the
    "three rows of chrysanthemums planted along
    the eastern fence" by Thao Yuan-Ming].
  Ming, *c.* +1630.
  Thu Chhêng-Khuei 屠承熉.
Tu Tuan 獨斷.
  Imperial Decisions and Definitions [on the rites
    and customs of the Later Han court].
  H/Han, *c.* +190.
  Tshai Yung 蔡邕.
Tu-Yang Tsa Pien 杜陽雜編.
  The Tu-yang Miscellany.
  Thang, end +9th century.
  Su Ê 蘇鶚.
Tung Han Hui Yao 東漢會要.
  History of the Administrative Statutes of the
    Later (Eastern) Han Dynasty.
  Sung, +1226.
  Ed. Hsü Thien-Lin 徐天麟.
  Cf. Têng & Biggerstaff (1), p. 159.
Tung Hsi Yang Khao 東西洋考.
  Studies on the Oceans East and West.
  Ming, +1618.
  Chang Hsieh 張燮.
Tung Hsüan Tzu 洞玄子.
  Book of the Mystery-Penetrating Master.
  Pre-Thang, perhaps +5th century.
  Writer unknown.
  In *Shuang Mei Ching An Tshung Shu*.
  Tr. van Gulik (3).
Tung Li Phin Hui Lu 東籬品彙錄.
  On the Grading of Chrysanthemum Varieties
    [title taken from the 'three rows of
    chrysanthemums planted along the eastern
    fence' by Thao Yuan-Ming].
  Chhing, *c.* +1798.
  Lu Pi 盧璧.
Tung Li Tsuan Yao 東籬纂要.
  See Shao Chhêng-Hsi (1).
Tzu Chih Thung Chien Pien Wu 資治通鑑辯誤.
  Correction of Errors in the *Comprehensive Mirror*
    (of History), for Aid in Government.
  Sung & Yuan, *c.* +1275.
  Hu San-Hsing 胡三省.
Tzu Hua Li Chi 紫花梨記.
  On a Variety of Pear with Purple Flowers.
  Thang.
  Hsü Mo 許默.
Tzu Hui 字彙.
  The Characters Classified [the first dictionary to
    reduce the number of radicals to the present
    standard number of 214, and the first to
    arrange the radicals and characters in the
    order of the number of their strokes].

  Ming, +1615.
  Mei Ying-Tsu 梅膺祚.
Tzu Shih Ching Hua 子史精華.
  Essence of the philosophers and Historians
    [dictionary of quotations].
  Chhing, +1727.
  Yün Lu 允祿 *et al.*
Tzu Thung 字通.
  Complete Character Dictionary [89 radicals, the
    most drastic of all the reductions].
  Sung.
  Li Tshung-Chou 李從周.

Wai Kuo Chuan 外國傳.
  Records of Foreign Countries.
  See *Wu Shih Wai Kuo Chuan* (Khang Thai).
Wakan Sanzai Zue 和漢三才圖會.
  The Chinese and Japanese Universal Encyclo-
    paedia (based on the *San Tshai Thu Hui*)
  Japan, +1712.
  Terashima Ryōan 寺島良安.
Wamyō Ruijūshō 和名類聚抄 (or 倭).
  General Encyclopaedic Dictionary.
  Japan (Heian), +934.
  Minamoto no Shitagau 源順.
Wamyōshō 和名抄.
  See *Wamyō Ruijūshō*.
Wan Ping Hui Chhun 萬病回春.
  The Restoration of Well-Being from a Myriad
    Diseases.
  Ming, +1587 pr. +1615.
  Kung Thing-Hsien 龔廷賢.
Wang Hsi-Lou Yeh Tshai Phu 王西樓野菜譜.
  See *Yeh Tshai Phu*.
Wang shih Lan Phu 王氏蘭譜.
  See the *Lan Phu* of Wang Kuei-Hsüeh.
Wei Lüeh 緯略.
  Compendium of Non-Classical Matters.
  Sung, +12th century (end) *c.* +1190.
  Kao Ssu-Sun 高似孫.
Wei Shu 魏書.
  See *San Kuo Chih*.
Wei Wang Hua Mu Chih 魏王花木志.
  A Book of Flowers and Trees by Prince (Hsin, of
    Kuangling, of the Northern) Wei (Dynasty).
  N/Wei, between +480 and +535.
  Thopa Hsin 拓跋欣, (Prince of Kuangling)
    廣陵王, or one of his secretaries.
Wei-Yang Shao-Yao Phu 維揚芍藥譜.
  Treatise on the Herbaceous Peonies of
    Yangchow.
  Alternative name for Liu Pin's *Shao-Yao Phu*,
    q.v.
Wei-Yang Shao-Yao Phu ho Tsuan 維揚芍藥
  譜合纂.
  A Conflation of Treatises on the Herbaceous
    Peonies of Yangchow.
  Ming, *c.* +1550.
  Tshao Shou-Chên 曹守貞.
Wên Hsien Thung Khao 文獻通考.
  Comprehensive Study of (the History of)
    Civilisation (lit: Complete Study of the

Wên Hsien Thung Khao (cont.)
    Documentary Evidence of Cultural
        Achievements (in Chinese Civilisation)).
    Sung & Yuan, begun perhaps as early as +1270
        and finished before +1317, printed +1322.
    Ma Tuan-Lin 馬端臨.
    Cf. des Rotours (2), p. 87.
    A few chs. tr. Julien (2); d'Hervey St Denys (1).
Wên Hsüan 文選.
    General Anthology of Prose and Verse.
    Liang, +530.
    Ed. Hsiao Thung (prince of the Liang) 蕭統.
    Comm. Li Shan 李善 *c.*+670.
    Tr. von Zach (6).
Wu Chhê Yün Jui 五車韻瑞.
    Five Cartloads of Rhyme-Inscribed Tablets
        [phrase-dictionary phonetically arranged].
    Ming, +1576; pr. +1592.
    Ling I-Tung 凌以棟.
Wu Ching Wên Tzu 五經文字.
    Characters of the Five Classics [dictionary; 160
        radicals].
    Thang, *c.* +770.
    Chang Shen 張參.
Wu Chung Hua Phin 吳中花品.
    Grades of (Tree-Peony) Flowers in the Region
        of Wu
    Sung, +1045.
    Li Ying 李英.
    Now extant only in quotations.
Wu Lei Hsiang Kan Chih 物類相感志.
    On the Mutual Responses of Things according
        to their Categories.
    Sung, *c.* +980.
    Attrib. wrongly to Su Tung-Pho 蘇東坡.
    Actual writer (Lu) Tsan-Ning (monk)
        錄贊寧.
Wu Li Hsiao Shih 物理小識.
    Small Encyclopaedia of the Principles of Things.
    Ming and Chhing, finished by +1643, sent to
        his son Fang Chung-Thung in +1650, finally
        pr. +1664.
    Fang I-Chih 方以智.
    Cf. Hou Wai-Lu (*3, 4*).
Wu shih Pên Tshao 吳氏本草.
    Mr Wu's Pharmaceutical Natural History.
    San Kuo (Wei), *c.* +235.
    Wu Phu 吳普.
    Extant only in quotations in later literature.
Wu Shih Wai Kuo Chuan 吳時外國傳.
    Records of the Foreign Countries in the Time of
        the State of Wu.
    San Kuo (Wu), *c.* +240
    Khang Thai 康泰.
    Only in fragments in *TPYL* and other sources.
Wu Tu Fu 吳都賦.
    Rhapsodic Ode on the Capital of Wu
        (Kingdom).
    San Kuo, *c.* +260.
    Tso Ssu 左思.
Wu Yin Lei Chü Ssu Shêng Phien Hai 五音類聚
        四聲篇海.

Ocean of Characters arranged according to
    the Five Rhymes and the Four Tones
    [dictionary].
J/Chin, +1208.
Han Tao-Chao 韓道昭.
Wu Yuan 物原.
    The Origins of Things.
    Ming, +15th century.
    Lo Chhi 羅頎.
Wu Yüeh Chhun Chhiu 吳越春秋.
    Spring and Autumn Annals of the States of Wu
        and Yüeh.
    H/Han.
    Chao Yeh 趙曄.

Yamato Honzō 大和本草.
    The Medicinal Natural History of Japan.
    Japan, +1708, +1715.
    Repr. 1932, 1936, ed. Shirai Mitsutarō.
    Kaibara Ekiken 貝原益軒.
    See Bartlett & Shohara (1), pp. 58, 63, 114–15;
        Merrill & Walker (1), p. 560.
Yang-Chou Chhiung-Hua Chi 揚州瓊花集.
    A Collection on the 'Nephrite' Flower of
        Yangchow. Alt title of *Chhiung-Hua Phu*, q.v.
Yang-Chou Shao-Yao Phu 揚州芍藥譜.
    A Treatise on the Herbaceous Peonies of
        Yangchow.
    Sung, +1075.
    Wang Kuan 王觀.
Yang Chü Phu 洋菊譜.
    A Little Treatise on Foreign Chrysanthemums
        (illustrated by Paintings).
    Chhing, +1756.
    Tson I-Kuei 鄒一桂.
Yang Kuei-Shan Chi 楊龜山集.
    Collected Writings of Yang Shih (Yang Kuei-
        Shan).
    Sung, *c.* +1130.
    Yang Shih 楊時.
Yang Yü Yueh Ling 養餘月令.
    Monthly Ordinances for Superabundance
        [encyclopaedia of horticulture and animal
        husbandry].
    Ming, +1633.
    Tai Hsi 戴羲.
Yeh Su Phin 野蔌品.
    (A Hundred) Wild Vegetables (for Healthy
        Diet) according to their Grades [part of *Yin
        Chuan Fu Shih Chien*, q.v.].
    Ming, +1591.
    Kao Lien 高濂.
Yeh Tshai Chien 野菜箋.
    Papers on Edible Wild Plants.
    Ming, *c.* +1600.
    Thu Pên-Chün 屠本畯.
Yeh Tshai Hsing Wei Khao 野菜性味考.
    A Study of the Natures and Sapidities of Edible
        Wild Plants.
    Ming, *c.* +1630.
Chu Yen-Mo 朱儼鑣.
    It is uncertain whether this book still exists.

Yeh Tshai Phu 野菜譜.
  A Treatise on Edible wild Plants.
  Ming, +1524.
  Wang Phan 王磐.
  Incl. as ch. 60 in *Nung Chêng Chhüan Shu*, (q.v.).
Yeh Tshai Po Lu 野菜博錄.
  Comprehensive Account of Edible wild Plants.
  Ming, +1622.
  Bao Shan 鮑山.
Yeh Tshai Tsan 野菜讚.
  Eulogies on Edible Wild Plants.
  Chhing, +1652.
  Ku Ching-Hsing 顧景星.
Yen-Ching Sui Shih Chi. 燕京歲時記.
  See Tun Li-Chhen (1)
Yen Chhi Yu Shih 巖棲幽事.
  Peaceful Occupations of a Mountain Hermitage.
  Ming.
  Chhen Chi-Ju 陳繼儒.
Yen Hsien Chhing Shang Chien 燕閒清賞牋.
  Pleasurable Occupations of a Life of Retirement
  [the sixth part of *Tsun Shêng Pa Chien*, q.v.].
  Ming, +1591.
  Kao Lien 高濂.
Yen shih Chia Hsün 顏氏家訓.
  Mr Yen's Advice to his Family.
  Sui, *c.* +590.
  Yen Chih-Thui 顏之推.
Yen Thieh Lun 鹽鐵論.
  Discourses on Salt and Iron [record of the
  debate of −81 on state control of commerce
  and industry].
  C/Han, *c.* −80 to −60.
  Huan Khuan 桓寬.
  Partial tr. Gale (1); Gale, Boodberg & Lin (1).
Yen Tzu Chhun Chhin 晏子春秋.
  Master Yen's Spring and Autumn Annals.
  Chou, Chhin or C/Han, proceeding from oral
  tradition but not stabilised before the −4th
  century.
  Attrib. Yen Ying (−6th) 晏嬰.
  But in fact a collection of stories about him.
Yin Chuan Fu Shih Chien 飲饌服食牋.
  Explanations on Diet, Nutrition and Clothing
  [the fifth part (chs. 11–13) of *Tsun Shêng Pa
  Chien*, q.v.].
  Ming, +1591.
  Kao Lien 高濂.
Yin Shan Chêng Yao 飲膳正要.
  Principles of Correct Diet [on deficiency dis-
  eases, with the aphorism 'many diseases can
  be cured by diet alone'].
  Yuan, +1330, re-issued by imperial order in
  +1456.
  Hu Ssu-Hui 忽思慧.
  See Lu and Needham (1).
Ying (Kuo) Kung Thang Pên Tshao 英(國)公唐
  本草.
  The Pharmacopoeia of Duke Ying of the Thang
  Dynasty = *Hsin Hsin Pên Tshao* q.v.
Yü Chih Pên Tshao Phin Hui Ching Yao.
  See *Pên Tshao Phin Hui Ching Yao.*

Yü Hai 玉海.
  Ocean of Jade [encyclopaedia of quotations].
  Sung +1267, but not pr. till +1337/+1340, or
  perhaps +1351.
  Wang Ying-Lin 王應麟.
  Cf. des Rotours (2), p. 96. Têng & Biggerstaff
  (1), p. 122.
Yu Hsüeh Ku Shih Chhiung Lin 幼學故事瓊林.
  The Red Jade Forest of Historical and Mytho-
  logical Allusions used among Cultured
  Persons.
  See *Yu Hsüeh Ku Shih Hsün Yuan Hsiang Chieh*. 幼學
Yu Hsüeh Ku Shih Hsün Yuan Hsiang Chieh
  故事尋源詳解.
  (= *Yu Hsüeh Ku Shih Chhiung Lin = Chhêng Yü
  Khao*).
  Studies in the Historical Elements of Basic
  Culture [a handbook of historical and
  mythological allusions].
  Ming, *c.* +1480.
  Chhiu Chün 邱濬.
  Comm. by Yang Ying-Hsiang 楊應象.
Yü-Jui Pien Chêng 玉蕊辨證.
  On the Identification of the Jade-Stamen
  Flowers (*Barringtonia racemosa*, Lecythidaoeae),
  [originally part of *Phing Yuan Chi*, q.v.].
  Sung, late +12th century.
  Chou Pi-Ta 周必大.
Yü Kung 禹貢.
  The Tribute of Yü.
  (A chapter of the *Shu Ching*, q.v.).
Yü Kung Chui Chih 禹貢錐指.
  A Few Points in the Vast Subject of the *Tribute
  of Yü* [the geographical chapter in the *Shu
  Ching*] (lit. 'Pointing at the Earth with an
  Awl') (including the set of maps, *Yü Kung Thu*).
  Chhing, +1697 and +1705.
  Hu Wei 胡渭.
Yü Kung Shuo Tuan 禹貢說斷.
  Discussions and Conclusions regarding the
  Geography of the *Tribute of Yü*.
  Sung, *c.* +1160.
  Fu Yin 傅寅.
Yü Lin 語林.
  Forest of Anecdotes.
  Chin, +4th century.
  Phei Chi 裴啟.
Yü Phien 玉篇.
  Jade Page Dictionary.
  Liang, +543.
  Ku Yeh-Wang 顧野王.
  Extended and edited in the Thang (+674) by
  Sun Chhiang 孫強.
Yu-Yang Tsa Tsu 酉陽雜俎.
  Miscellany of the Yu-yang Mountain (cave) [in
  S.E. Szechuan].
  Thang, *c.* +860.
  Tuan Chhêng-shih 段成式.
  See des Rotours (1), p. civ.
Yü-Tshan Chi 玉簪記.
  On the Identification of the Jade-Hairpin
  Flowers (Plantain Lilies, *Hosta* spp.).

Yü Tshan Chi (cont.)
    Ming, +1591.
    Kas Lien　高濂.
Yuan Chien Lei Han　淵鑑類函.
    Mirror of the Infinite; a Classified Treasure-
        chest [great encyclopaedia; the conflation of 4
        Thang and 17 other encyclopoedias].
    Chhing, presented, +1701, pr. +1710.
    Ed. Chang Ying　張英 *et al.*
Yuan Lin Tshao Mu Su　園林草木疏.
    Studies on the Plants and Trees of Garden and
        Grove.
    Thang, *c.* +690.
    Wang Fang-Chhing　王方慶.
    Preserved in incomplete form.
Yuan Ma (Phien)　爰麼篇.
    Explanation of Difficult Words [orthographic
        primer].
    Chhin, *c.* −215.
    Chao Kao　趙高.
    Incorporated by Han times in the *Tshang Chieh
        (Phien)*, q.v.
Yuan Shang Phien　元尙篇.
    Ancient Traditional Terms.
    C/Han, Late −1st century.
    Li Chhang　李長.
    Extant only in quotations.
Yuan Thing Tshao Mu Su　園庭草木疏.
    Prob. orig. title of *Yuan Lin Tshao Mu Su*, q.v.
        (see Wang Yü-Hu (*1*), 2nd, ed. p. 38).
Yüeh Chi Hua Phu　月季花譜.
    A Treatise on the *Yüeh-chi-hua* (*Rosa cheninsis*, the
        'monthly rose').
    Chhing.
    Phing Hua Kuan Chu (ps.)　評花館主(the
        Master of the Flower-Criticism office).
Yüeh Chung Mu-Tan Hua Phin　越中牡丹花品.
    Grades of the Tree-Peonies of Yüeh (Shao hsing,
        Chekiang).
    Sung, +986.
    Chung-Hsiu (monk)　仲休.
    Preserved in *Yung-Lo Ta Tien*.
Yüeh Ling　月令.
    Monthly Ordinances (of the Chou Dynasty).
    Chou, between −7th and −3rd centuries.
    Writers unknown.

Incorporated in the *Hsiao Tai Li Chi* and the *Lü
    Shih Chhun Chhin*.
Tr. Legge (7), R. Wilhelm (3) q.v.
Yu Huan Chi Wên　游宦紀聞.
    Things seen and Heard on my official Travels.
    Sung, +1233.
    Chang Shih-Nan　張世南.
Yün Chi Chhi Chhien　雲笈七籤.
    The Seven Bamboo Tablets of the Cloudy
        Satchel [an important collection of Taoist
        material made by the editor of the first
        definitive form of the *Tao Tsang* (+1019), and
        including much material which is not in the
        Patrology as we now have it].
    Sung, *c.* +1022.
    Chang Chün-Fang　張君房.
    TT/1020.
Yün Fu Chhün Yü　韻府羣玉.
    The Assembly of Jade (Tablets); a Word-Store
        arranged by Rhymes [rhyming phrase-
        dictionary using only 107 sounds instead of
        the previously current 206].
    Sung or Yuan, *c.* +1280.
    Yin Shih-Fu　陰時夫.
Yün Fu Shih I　韻府拾遺.
    A Supplement for the *Word-Store Arranged by
        Rhymes*.
    Chhing, commissioned +1716, completed
        +1720, pr. +1722.
    Ed. Wang Yen　王捵 *et al.*
Yün Hai Ching Yuan　韻海鏡源.
    Mirror of the Ocean of Rhymes [phrase-
        dictionary phonetically arranged].
    Thang, *c.* +780.
    Yen Chen-Chhing　顏眞卿.
Yung-Chhang Erh Fang Chi　永昌二芳記.
    Notes on Two Flowering Plants (*Camellia* and
        Rhododendron) Cultivated at Yungchhang (in
        Yunnan).
    Ming, *c.* +1495.
    Chang Chih-Shun　張志淳.
Yung-Chia Chü Lu　永嘉橘錄.
    Record of the Oranges of Yung-chia
        (Wênchow).
    Alternative title for Han Yen-Chih's *Chü Lu*, q.v.

B. CHINESE AND JAPANESE BOOKS AND JOURNAL ARTICLES SINCE +1800

Akiyasu Yasuji (*1*) (ed.). 秋保安治.
Yedo Jidai no Kagaku 江戸時代の科學.
Science in the Yedo Period (+1603–1867), [the
  Tokugawa Shogunate].
Tokyo Science Museum, Tokyo, 1934.

Amano Motonosuke (*4*). 天野元之助.
Chūgoku Nōgyōshi Kenkyū 中國農業史研究.
Researches into Chinese Agricultural History.
Tokyo, 1962; 2nd, expanded ed. 1979.

Amano Motonosuke (*5*). 天野元之助.
Mindai ni okeru Kyūkō Sakubutsu Chojutsukō
  明代における救荒作物著述考.
A Study of the Works on Plants for Famine
  Relief written in the Ming Period.
TYG, 1964, **47** (no. 1), 32.

Andō Kōsei (*1*) 安藤更生.
Kanshin 鑑真.
Life of Chien-Chen (+688–763), [outstanding
  Buddhist missionary to Japan, skilled also in
  medicine and architecture].
Bijutsu Shuppansha, Tokyo 1958, repr. 1963.
Abstr. *RBS*, 1964, **4** no. 889.

Anon. (*1*).
*Chung Kuo Ku Tai Kho Hsüeh Chi Shu Chu Yao
  Chhêng Chiu Piao* 中國古代科學技術主
  要成就表.
Chart of the Principal Scientific and Tech-
  nological Achievements of Ancient China.
Peking University 1976 北京大學物理系
  理論小組編.

Anon. (*8*) (ed.).
Chung-Kuo Ti-Chen Tzu-Liao Nien Piao 中國
  地震資料年表.
Register of Earthquakes in Chinese Recorded
  History (−1189–+1955).
2 vols., Kho-Hsüeh, Peking, 1956.

Anon. (*35*).
Chung I Chhang Yung Ming Tzhu Chien Shih
  中醫常用名詞簡釋.
Glossary of Traditional Chinese Medicine.
Szechuan Jen-min, Chhêngtu, 1959.

Anon. (*56*).
Nan Fang Tshao Mu Chuang 南方草木狀.
[Hsi Han's] 'Records of the Plants and Trees of
  the Southern Regions' [illustrated with a set
  of plant paintings of uncertain date preserved
  in the Shanghai Municipal Library, and
  formerly in the possession of Wu Yün 吳雲
  (1811–83)].
With postface by Shen Chao-Khuei 沈兆奎
  (1916).
Com. Press, Shanghai, 1955.

Anon. (*57*).
Chung Yao Chih 中藥志.
Repertorium of Chinese Materia Medica (Drug
  Plants and their Parts, Animals and
  Minerals).
4 vols.
Jen-min Wei-shêng, Peking, 1961.

Anon. (*58*).
Chung-Kuo Thu Nung Yao Chih 中國土
  農藥誌.
Repertorium of Plants used in Chinese
  Agricultural Chemistry.
Kho-Hsüeh, Peking, 1959.

Anon. (*59*).
Ma-Lai Yeh-Sêng Shih-Yung Chih-Wu Thu Shuo
  馬來野生食用植物圖說.
Illustrated Guide to the Edible Wild Plants of
  Malaya [50 entries].
Japanese Army Publishing Bureau, [Botanic
  Gardens], Singapore, 1944.

Anon. (*60*).
Shih-Yung Yeh-Sêng Tung Chih Wu 食用野生
  動植物.
Illustrated Guide to the Edible Wild Animals
  and Plants [of Malaya].
Japanese Army Publishing Bureau [Botanic
  Gardens], Singapore, 1944.

Anon. (*61*).
Chung-Kuo Chu-Yao Chih-Wu Thu Shuo (Tou Kho)
  中國主要植物圖說(豆科).
Atlas of the Principal Plants of China
  (Leguminosae).
For Academia Sinica, Kho-Hsüeh, Peking,
  1955.

Anon. (*64*).
Chhüan Kuo Chung Yao Chhêng-Yao Chhu Fang Chi
  全國中藥成藥處方集.
National Pharmacopoeia (Standard Formularies
  of Chinese Drug Prescriptions).
Ed. Pharmacological Division of Shenyang
  Pharmaceutical College and Pharmacological
  Institute of the Nat. Acad. Chinese Trad.
  Medicine, Peking.
Jen-Min, Peking, 1964.

Anon. (*65*).
Chung Yao Hsüeh 中藥學.
Pharmacology of Chinese Drugs.
Ed. Nanking College of Chinese Medicine, and
  Chiangsu Provincial Chinese Medicine
  Research Institute.
Jen-Min, Peking, 1959.

Anon. (*71*) (ed.).
Shih Pên Pa Chung 世本八種.
Eight Versions of the Text of the *Book of Origins*
  (−2nd century.).
Com. Press, Shanghai, 1957.

Anon. (*109*).
　Chung-Kuo Kao Têng Chih-Wu Thu Chien
　中國高等植物圖鑑.
　Iconographia Cormophytorum Sinicorum (Flora of
　　Chinese Higher Plants).
　Kho-Hsüeh, Peking, 1972– (for Nat. Inst. of
　　Botany). Vols. 1 and 2, 1972; Vol. 3, 1974;
　　Vol. 4, 1975; Vol. 5, 1976.
Anon. (*110*).
　Chhang Yung Chung Tshao Yao Thu Phu 常用
　中草藥圖譜.
　Illustrated Handbook of the Most Commonly
　　Used Chinese Plant Materia Medica (pre-
　　pared by the Chinese Academy of medicine
　　and the Chekiang Provincial College of
　　Traditional Medicine).
　With Index of Latin binomials as well as
　　Chinese names.
　Jen-min wei-shêng, Peking, 1970.
Anon. (*166*).
　Chung Tshao Yao Yu Hsiao Chhêng-Fên ti Yen-Chiu
　中草藥有效成分的研究.
　A Study of the Chemical Constituents of
　　Chinese Drug-Plants (Vol. 1).
　Jen-Min, Peking 1972.
Anon. (*176*).
　Fuchien Chung Tshao yao 福建中草藥.
　[Illustrated Handbook of] Drug-Plants in
　　Fukien Province.
　Index of Chinese names only.
　I-Yao Yen-Chiu-So, Fuchow, 1970.
Anon. (*177*).
　Ninghsia Chung Tshao Yao Shou-Tshê 寧夏
　中草藥手冊.
　[Illustrated] Handbook of Drug-plants in
　　Ninghsia.
　With index of Latin binomials as well as Chinese
　　names.
　Jen-min, Ninghsia, 1971.
Anon. (*178*).
　Pei-Fang Chhang Yung Chung Tshao Yao Shou-Tshê
　北方常用中草藥手冊.
　[Illustrated] Handbook of Drug-plants in
　　Common use in Northern China.
　Index of Chinese names only.
　Jen-min Wei-shêng, Peking, 1971.
Anon. (*179*).
　Kansu Chung Tshao Yao Shou-Tshê 甘肅中草
　藥手冊.
　[Illustrated] Handbook of Drug-plants in Kansu
　　Province, 2 vols.
　With index of Latin binomials as well as Chinese
　　names.
　Jen-min, Lanchow, 1971.
Anon. (*180*).
　Hunan Nung Tshun Chhang Yung Chung Yao Shou-
　Tshê 湖南農村常用中草藥手冊.
　Illustrated] handbook of Drug-plants in
　　Common Use in Rural Hunan.
　Index of Chinese names only.
　Hunan Jen-min, Chhangsha, 1970.
Anon. (*181*).

Tung-pei Chhang Yung Chung Tshao Yao Shou-Tshê
　東北常用中草藥手冊.
　[Illustrated] Handbook of Drug-plants in
　　Common use in North-east China.
　Index of Chinese names only.
　Hsin-Hua Shu-Tien, Shenyang, 1970.
Anon. (*182*).
　Hopei Chung Yao Shou-Tshê 河北中藥手冊.
　[Illustrated] Handbook of Drug-plants in Hopei
　　Province.
　Index of Latin binomials as well as Chinese
　　names.
　Kho-Hsüeh, Peking, 1970.
Anon. (*183*).
　Shensi Chung Tshao Yao 陝西中草藥.
　[Illustrated Handbook of] Drug-plants in
　　Shensi Province.
　Index of Latin binomials as well as Chinese
　　names.
　Kho Hsüeh, Peking, 1971.
Anon. (*184*).
　Chekiang Min Chien Chhang Yung Tshao Yao
　浙江民間常用草藥.
　[Illustrated Handbook of] Drug-plants com-
　　monly used among the People of Chekiang
　　Province, 2 vols.
　Chekiang Jen Min, Hangchow, 1970.
Anon. (*185*).
　Kueichow Tshao Yao 貴州草藥.
　[Illustrated Handbook of] the Drug-plants of
　　Kueichow Province, 2 vols.
　Indexes of Chinese names only.
　Kweichow Jen-min, Kueiyang, 1970.
Anon. (*186*).
　Yünnan Chung Tshao Yao 雲南中草藥.
　[Illustrated Handbook of] Drug-plants in
　　Yünnan.
　Index of Latin binomials as well as Chinese
　　names.
　Yunnan Jen-min, Kunming, 1971.
Anon. (*187*).
　Nei Mêngku Chung Tshao Yao 內蒙古中
　草藥.
　[Illustrated Manual of] the Drug-plants of
　　Inner Mongolia.
　With index of Latin binomials as well as Chinese
　　names.
　Nei Mêng-ku Tzu-chih-chü Jen-min, Huhehot,
　　1972.
Anon. (*188*).
　Shansi Chung Tshao Yao 山西中草藥.
　[Illustrated Manual of] the Drug-plants of
　　Shansi Province.
　Index of Latin binomials as well as Chinese
　　names.
　Shansi Jen-min, Thaiyuan, 1972.
Anon. (*189*).
　Hunan Yao Wu Chih 湖南藥物志.
　Flora of Hunanese Drug-plants, vol. 1.
　Index of Latin binomials as well as Chinese
　　names.
　Hunan Jen-min, Chhangsha, 1970.

Anon. (*190*).
　　Chhang Yung Chung Tshao Yao Shou-Tshê 常用中
　　草藥手冊.
　　[Illustrated] Handbook of the Most Commonly
　　Used Chinese Plant-drugs.
　　Index of Chinese names only.
　　Com. Press, Hongkong, 1970.
Anon. (*191*).
　　*Chhang-Yung Chung Tshao Yao Tshai-Sê Thu-
　　Phu* 常用中草藥彩色圖譜.
　　Manual of the Commonly Used Chinese Drug-
　　plants, with Coloured Illustrations.
　　Vol. 1, Jen Min Chhu Pan, Canton, 1970.
Anon. (*193*).
　　Chiangsi Tshao yao 江西草藥.
　　Plant Drugs of Chiangsi.
　　Hsin Hua Shu Tien, Chiangsi, 1970.

Chan Yo-Han (= John Chalmers) & Wang Yang-
　　An (*1*) 湛約翰, 王揚安.
　　Khang-Hsi Tzu Tien Tsho Yao 康熙字
　　典撮要.
　　The Essentials of the *Imperial Dictionary of the
　　Khang-Hsi reign-period* [a rearrangement ac-
　　cording to the phonetic component, not the
　　radical component or the sound (rhyme) of
　　each character].
　　London Missionary Society, Canton, 1878.
Chang Chhang-Shao (*1*) 張昌紹.
　　Hsien-tai-ti Chung Yao yen-Chiu 現代的
　　中藥研究.
　　Modern Researches on Chinese Drugs.
　　Kho-Hsüeh Chi-Shu, Shanghai, 1956.
Chang Chhi-Yün (*1*) (ed.) 張其昀 with 10 col-
　　laborators.
　　Chung Hua Min Kuo Ti Thu Chi 中華民國
　　地圖集.
　　Atlas of the Republic of China (title on some
　　volumes: National Atlas of China).
　　5 vols.
　　Nat. Defence Res. Inst. (Nat. War College) &
　　Chinese Geographical Res. Institute.
　　Yang-ming Shan, Thaipei, 1959, 2nd. ed. 1963.
　　Vol. 1 Thaiwan; 2, Tibet, Sinkiang &
　　Mongolia; 3, North China; 4, South China; 5
　　General.
Chang Chia-Chü (*1*) 張家駒.
　　Shen Kua 沈括.
　　Biography of Shen Kua (scientist and high
　　official, +1031–95).
　　Shanghai Jen-min, Shanghai, 1962.
　　Abstr. in *RBS*.
Chang Han-Chieh (*1*) 張漢潔.
　　*Wo Kuo Ku-Tai tui 'Thu-Jang Ti-Li'-ti Yen-Chiu
　　ho Kung-Hsien* 我國古代對「土壤地理」
　　的研究和貢獻.
　　Investigations and Achievements in Ancient
　　China on the Geographical Distribution of
　　Soil Types.
　　APS, 1959 **7** (no. 1–2), 23.
Chang Hsin-Chhêng (*1*) 張心澂.
　　Wei Shu Thung Khao 僞書通考.

A General Study of Books of Uncertain Date
　　and Authorship.
　　2 vols., Shanghhai, 1939, revised ed. 1955.
Chang Hui-Chien (*1*) 張慧劍.
　　Li Shih-Chen 李時珍.
　　Biography of Li Shih-Chen (+1518–93; the
　　great pharmaceutical naturalist).
　　Jen Min Pub. House, Shanghai, 1954; repr.
　　1955.
Chang Hung-Chao (*1*) 章鴻釗.
　　Shih Ya 石雅.
　　Lapidarium Sinicum; a Study of the Rocks,
　　Fossils and Minerals as known in Chinese
　　Literature.
　　Chinese Geol. Survey, Peking: 1st ed. 1921, 2nd
　　ed. 1927.
　　MGSC (ser. B), no. 2, 1–432 (with Engl.
　　summary.) Crit. P. Demiéville, *BEFEO*, 1924,
　　24, 276.
Chang Kuang-Chao (*1*) 張光照.
　　Hsing Lan Phu Lüeh 興蘭譜略.
　　Brief Treatise on the Orchids of (I–) Hsing (in
　　Chiangsu) 1816.
　　Ms in the Peking National Library.
Chang Tê-Chün (*2*) 張德鈞.
　　*Liang Chhien Nien Lai Wo Kuo Shih Yung Hsiang-
　　Chiao Hêng Hsien Wei Chih Pu Khao Shu* 兩千
　　年來我國使用香蕉莖纖維織布考述.
　　An Investigation of the Fine Banana Stem Fibre
　　Cloth made in Ancient and Mediaeval China,
　　and the Techniques used for it.
　　CWHP, 1596, **5** (no. 1), 103.
Chang Tsan-Chhen (*1*) 張贊臣.
　　Chung-Kuo Li-Tai I-Hsüeh Shih Lüeh 中國歷
　　代醫學史略.
　　A Brief History of Chinese Medicine.
　　1st ed, 1933; 2nd ed. Chhing Chhien Thang Shu
　　Chü, Shanghai, 1954.
Chang Tsan-Chhen (*2*) 張贊臣.
　　Wo Kuo Li-Tai Pên Tshao ti Pien Chi 我國歷
　　代本草的編輯.
　　A History of the Chinese Pharmaceutical
　　Natural Histories.
　　ISTC, 1955, **7** (no. 1), 3.
Chang Tzhn-Kung (*1*) 章次公.
　　'*Pên Tshao Kang Mu Shih I' Yin Shu Phien Mu*
　　「本草綱目拾遺」引書編目.
　　On the Books Quoted in the *Supplementary
　　Amplifications for the Pandects of Natural History*
　　(+1765–1803).
　　ISTC, 1948, **2** (nos. 3/4), 20.
Chang Tzhu-Kung (*2*) 章次公.
　　*Ming-Tai Kua Ming I Chi chih Chin-Shih Thi Ming
　　Lu* 明代掛名醫籍之進士題名錄.
　　A Register of the Chin-Shih Graduates of Ming
　　Times who practised Medicine.
　　ISTC, 1948, **2** (no. 1/2), 5.
Chang Tzhu-Kung (*3*) 章次公.
　　Man-Kung Shih Chi Khao 曼公事跡考.
　　A Study of (Tai) Man-Kung (Late Ming physi-
　　cian in Japan).
　　ISTC, 1951, **3** (no. 1), 35.

Chang Tzu-Kao (5)　張子高.
Chao Hsüeh-Min 'Pen Tshao Kang Mu Shih I' Chu
　Shu Nien-Tai, Chien-Lun Wo-Kuo Shou-Tzhu
　Yung Chhiang-Shui Kho Thung Pan Shih
　趙學敏「本草綱目拾遺」著述年代
　兼論我國首次用强水刻銅版事.
On the Date of Publication of Chao Hsüeh-
　Min's *Supplement to the Great Pharmacopoeia*, and
　the Earliest Use of Acids for Etching Copper
　Plates in China.
KHSC, 1962, **1** (no. 4), 106.

Chang Tzu-Kung (1)　張資珙.
Lüch Lun Chung-Kuo ti Nieh Chih Pai-Thung ho tha
　tsai Li-Shih shang yü Ou-Ya Ko Kuo ti Kuan-Hsi
　略論中國的鎳質白銅和他在歷史
　上與歐亞各國的關係.
On Chinese Nickel and Paktong, and on their
　Role in the Historical Relations between Asia
　and Europe.
KHS, 1957, **33** (no. 2), 91.

Chang Yung-Yen (1)　張永言.
Lun Hao I-Hsing ti 'Erh Ya I Su'　論郝懿行
　的「爾雅義疏」.
A Discussion of Hao I-Hsing's 'Commentary on
　the *Erh Ya*' (1882).
CKYW, 1962, 495, 502.
Abstr. *RBS*, 1969, **8**, no. 505.

Chao Yü-Huang (1)　趙燏黃.
Chung-Kuo Hsin Pên Tshao Thu Chih　中國
　新本艸圖誌.
Neuer Pharmakognostischer Atlas der chinesis-
　chen Drogen; auf der Grundlage des alten
　chinesischen Arzneibuches Pên Tshao mit
　modernen Methode bearbeitet.
Vol. 1, pts. 1, 2 (no more published). Pt. 1 deals
　only with *Glycyrrhiza* and *Astragalus*. Pt. 2 only
　with *Panax*.
National Research Institute of Chemistry,
　Academia Sinica, Shanghai, 1931, 1932.
　(Memoirs of the Nat. Res. Inst. of Chem.
　No. 3).

Chao Yü-Huang (2) = (1)　趙燏黃.
Chêng Li Pên Tshao Yen-Chiu Kuo Yao chih Fang An
　chi chhi Shih-Li; 1, Chhichow Yao chih Yen-chiu;
　A, shu yü Chü Kho chi chhuan Hsü Tuan Kho chih
　Yao Tshai　整理本草研究國藥之方案
　及其實例；1, 祁州藥之研究；A, 屬於
　菊科及川續斷科之藥材.
A Programme, together with Concrete Research
　Examples, to make Intensive Study of Chinese
　Materia Medica; 1, The Study of Chhichow
　Drugs [birthplace of Li Shih-Chen, and a
　famous drug market]; A. Compositae and
　Dipsacaceae Dept. of Pharmacognosy,
　Institute of Chinese Drugs, College of
　Medicine, National Peking University,
　Peiping, 1941.

Chêng Tso-Hsin (1)　鄭作新.
Phu-Thung Tung-Chih-Wu-Hsüeh Ming Tzhu
　普通動植物學名辭.
Preliminary Glossary of Technical Terms in
　Zoology and Botany [English-Chinese].

Fukien United University Biological Labora-
　tories (Chienyang), Fukien, 1942.

Chêng Tso-Hsin (3)　鄭作新.
Chung-Kuo Niao Lei Fên Pu Mu-Lu　中國
　鳥類分布目錄.
Taxonomic Index of the Groups of Chinese
　Birds.
Kho-Hsüeh, Peking, 1955.

Chêng Tso-Hsin (4)　鄭作新.
Chung-Kuo Niao Lei Fên Pu Ming Lu　中國
　鳥類分布名錄.
The Nomenclature and Classification of the
　Birds of China.
Kho-Hsüeh, Peking, 1976.

Chêng Yün-Thê (1)　鄭云特.
Chung-Kuo Chiu Huang Shih　中國救荒史.
A History of Famines and Famine Relief in
　China.
San Lien, Peking, 1958.

Chi Nan (1)　計楠.
Mu-Tan Phu　牡丹譜.
A Treatise on Tree-Peonies.
1809.

Chi Nan (2)　計楠.
Chü Shuo　菊說.
A Discourse on the (Varieties of the)
　Chrysanthemum.
1803, with postface of 1811.

Chia Liang-Chih & Kêng I-Li (1)　賈艮智,
　耿以禮.
Hua-Nan Ching-Chi Ho Tshao Chih-Wu
　華南經濟禾草植物.
Economic Grasses of Southern China.
Kho-Hsüeh, Peking, 1955 (S. China Bot. Res.
　Bur. Acad. Sin. monographs (3rd ser.), no. 2).

Chia Tsu-Chang & Chia Tsu-Shan (1)　賈祖璋,
　賈祖珊.
Chung-Kuo Chih-Wu Thu Chien　中國植物
　圖鑑.
Illustrated Dictionary of Chinese Flora [ar-
　ranged on the Engler system; 2602 entries].
Chung-hua, Peking, 1936; repr. 1955, 1958.

Chiang Chien-Min & Tshang Tung-Chhing (1)
　蔣劍敏, 倉東卿.
Wa Chien ti Chien Tu yü Pan Chieh　瓦礆的
　礆度與板結.
The Alkalinity and Crust of Tile alkali Soil.
APS, 1964, **12**, 320.

Chiang Fan (1)　江藩.
Hsü Nan Fang Tshao Mu Chuang　續南方
　草木狀.
A Continuation of the 'Prospect of thePlants and
　Trees of the Southern Regions'.
c. 1825.

Chiang Piao (1)　江標.
Ko chih Ching Hua Lu　格致精華錄.
Record of the Inflorescence of Men of Science
　(in Olden Times).
Shanghai, *c.* 1897.

Chiang Ying (1)　蔣英.
Tui 'Tung-Ya Chih-wu-hsüeh wen-hsien' Fu-lu-chung
　'Chung-Kuo Ku-tai Wen-hsien' Pu-fen ti Ting-

Chiang Ying (*1*) (*cont.*)
Cheng 對'東亞植物學文獻'附錄中
'中國古代文獻'部分的訂正.
A commentary on the Part of Appendix (Older
Chinese Works) in Merrill and Walker *A
Bibliography of Eastern Ariatic Botany*.
APTS, 1977, **15**, (no. 4), 95.

Ching Li-Pin, Wu Chhêng-I, Khuang Kho-Jen &
Tshai Tê-Hui (*1*) 經利彬, 吳徵鎰,
匡可任, 蔡德惠.
'*Tien Nan Pên Tshao*' *Thu Phu* 滇南本
草圖譜.
Illustrations (and Identifications, of Plants) for
the *Pharmaceutical Natural History of Yunnan*
(+1436).
Nat. Inst. Pharmacol. Kunming, 1945.

Chou I-Liang (*2*) 周一良.
*Chien-Chen ti Tung Tu yü Chung Jih Wên-Hua
Chiao-Liu* 鑑眞的東渡與中日文化
交流.
The Mission of Chien-Chen (Kanshin) to Japan
(+735-48) and Cultural Exchanges between
China and Japan.
WWTK, 1963 (no. 9), 1.

Chu Chhi-Chhien 朱啟鈐 & Liang Chhi-Hsiung
梁啟雄.
Chê Chiang Lu [parts 1 to 6] 哲匠錄.
Biographies of [Chinese] Engineers, Architects,
Technologists and Master-Craftsmen.
BSRCA, 1932, **3** (no. 2), 125; 1932, **3** (no. 1)
123; 1932, **3** (no. 3), 91; 1933, **4** (no. 1), 82;
1933, **4** (no. 2), 60,; 1934, **4** (nos 3 and 4),
219.

Chu Chhi-Chhien 朱啟鈐, Liang Chhi-Hsiung
梁啟雄 and Liu Ju-Lin (*1*) 劉儒林.
Chê Chiang Lu [part 7] 哲匠錄.
Biographies of [Chinese] Engineers, Architects,
Technologists and Master-Craftsmen
(continued).
BSRCA, 1934, **5** (no. 2), 74.

Chu Chhi-Chhien 朱啟鈐 & Liu Tun-Chên *1*, *2*
劉敦楨.
Chê Chiang Lu [parts 8 and 9] 哲匠錄.
Biographies of [Chinese] Engineers, Architects,
Technologists and Master-Craftsmen
(continued).
BSRCA, 1935, **6** (no. 2), 114; 1936, **6** (no. 3), 148.

Chu Chi-Hai (*1*) 朱季海.
'*Chhu Tzhu*' *Chieh Ku Shih I* 「楚辭」解故
識遺.
An Analysis of Certain Verses in the 'Odes of
Chhu' (with a detailed study of some bot-
anical terms and names in them).
CHWSLT, 1962, no. 2, 77.

Chu Chün-Shêng (*1*) 朱駿聲.
Shuo Wên Thung Hsün Ting Shêng 說文通訓
定聲.
Investigation into the Sounds and Meanings of
Characters in the *Shuo Wên* (dictionary).
c. 1850, pres. 1851, pr. 1870.

Chu Hsi-Thao (*1*) 朱希濤.
Wo Kuo Shou-Hsien Ying-Yung Hung ho Chin Chhung

Thien Ya-Chhih-ti Kuang-Jung Shih 我國首先
應用汞合金充塡牙齒的光榮史.
The Glorious History of the Chinese Invention
and Practical Application of Mercury
Amalgams for the Filling of Tooth Cavities.
CJST, 1955, (no. 1).
Abstr. in *ISTC*, 1955, **7** (no. 2), 116. *CMJ*,
1955, **73** (no. 3).

Chung Chao-Hsiang, Kuan Phei-Shêng & Chiang
Jun-Hsiang (*1*) 莊兆祥, 關培生, 江潤祥.
Pên Tshao Yen-Chiu Ju Mên 本草研究入門.
An Introduction to the Study of the [Chinese
Pharmaceutical Natural Histories; the] *Pên
Tshao* [Genre].
University Press, Chinese University of Hong-
kong, Shatin, 1983.

Chhen Chih (*1*) 陳直.
*Hsi Yin Mu Chien chung Fa-Hsien ti Ku-Tai I-Hsüeh
Shih-Liao* 璽印木簡中發現的古代醫
學史料.
Ancient Chinese Medicine as recorded in Seals
and on Wooden Tablets.
KHSC, 1958, **1**, 68; *ISTC*, 1958, no. 2, 139.

Chhen Chhung-Ming (*1*) 陳重明.
Wu Chhi-Hsün to Chih-Wu Ming Shih Thu Khao
吳其濬和植物名實圖考.
Wu Chhi-Hsün and *Explications and Illustrations of
Plants*.
CMJ/C, 1980, no. (2), 65-70.

Chhen Chhung-Ming (*2*) 陳重明.
Tui '*Chih-Wu Ming Shih Thu Khao*' *San-Shih-Liu
Chung Chih-Wu ti Ting-Cheng* 對《植物名實
圖考》三十六種植物的訂正.
Revision of [the names] of 36 Botanical Species
in *Explications and Illustration of Plants*.
CWFLHP, 1981, 19(1), 136-9.

Chhen Fêng-Huai & Wang Chhiu-Phu (*1*)
陳封懷, 王秋圃.
Chü Hua Than Yuan 菊花探源.
A Study of the Origin of the Chrysanthemum.
TJTC, 1979, **2** (no. 10), 652.

Chhen Huan-Yung *et al.* (*1*) (ed.) 陳煥鏞.
Hainan Chih Wu Chih 海南植物志.
Flora Hainanica (Hainan Island, Kuangtung).
Vol. 1, 1964; vol. 2, 1965; vol. 3, 1974 (Anon.
ed.).
Kho-Hsüeh, Peking, 1964- .

Chhen Jen-Shan (*1*) 陳仁山.
Yao Wu Shêng-Chhan Pan 藥物生產辨.
Pharmagnosy of Plant Drugs.
1930.

Chhen Jung (*1*) 陳嶸.
Chung-Kuo Shu Mu Fên Lei Hsüeh 中國樹木
分類學.
Illustrated Manual of the Systematic Botany of
Chinese Trees and Shrubs.
Agricultural Association of China Series.
Nanking, 1937.

Chhen Pang-Hsien (*1*) 陳邦賢.
Chung-Kuo I-Hsüeh Shih 中國醫學史.
History of Chinese Medicine.
Com. Press, Shanghai, 1937, 1957.

Chhen Pang-Hsien (*3*)　陳邦賢.
　　Li Shih-Chen 李時珍.
　　Biography of Li Shih-Chen.
　　Art. in Li Nien (*27*), 1st ed. p. 161, 2nd ed. p. 171.
Chhen Thieh-Fan (*1*)　陳鐵凡.
　　'Thang Pên Tshao' Khao　「唐本草」考.
　　A Study of the *Thang Pên Tshao* (the *Hsin Hsiu
　　　Pên Tshao*).
　　TLTC, 1970, **40** (no. 10), 1.
Chhen Tshun-Jen (*1*) *et al.*　陳存仁.
　　Chung-Kuo Yao-Hsüeh Ta Tzhu Tien 中國藥學
　　　大辭典.
　　Encyclopaedia of Chinese Materia Medica. 2
　　　vols. with a supplementary volume illus-
　　　trations, of Shih-Chieh, Shanghai, 1935.
Chhen Tsu-Kuei (*1*) (ed.)　陳祖槼.
　　Chung-Kuo Nung-Hsüeh I-Chhan Hsüan Chi; Mien
　　　中國農學遺產選集；棉.
　　Anthology of Quotations [illustrating the
　　　History of] Chinese Agricultural Science and
　　　Production; Cotton.
　　Vol. 1.
　　Chhung-hua, Peking, 1957.
　　Joint Committee on Agricultural History
　　　(National Academy of Agriculture and
　　　Nanking College of Agriculture), ser. A, no. 5.
Chhêng Yao-Thien (*1*)　程瑤田.
　　Kuo-Lo Chuan Yü Chi　果贏轉語記.
　　On the Kuo-Lo (gourds) and similar [bio-
　　　logical] Doublet Words.
　　Peking, *c.* 1810.
Chhêng Yao-Thien (*3*)　程瑤田.
　　Chiu Ku Khao　九穀考.
　　A Study of the Nine Grains (monograph on the
　　　history of cereal agriculture).
　　Peking, *c.* 1805.
　　In *HCCC*, ch. 548.
Chhêng Yao-Thien (*5*)　程瑤田.
　　Shih Tshao Hsiao Chi　釋草小記.
　　A Minor Treatise on Certain Plants.
　　Peking, *c.* 1810.
　　In *HCCC*, ch. 552.
Chhien Chhung-Shu & Chhen Huan-Yung (*1*)
　　　錢崇澍, 陳煥鏞 (ed.).
　　Chung-Kuo Chih-Wu Chih　中國植物誌.
　　Flora Reipublicae Populris Sinicae (delectis
　　　Florae Reip. Pop. Sin. Agendae Academiae
　　　Sinicae Edita, Red. Princ. Chhien Chhung-
　　　Shu & Chhen Huan-Yung).
　　80 vols. planned.
　　2 published 1965 vol. 2 Pteridophyta ed. Chhin
　　　Jen-Chhang 秦仁昌 vol. 11 Cyperaceae ed.
　　　Thang Chin 唐進 & Wang Fa-Tsuan
　　　王發纘.
Chhu Hua(*1*)　褚華.
　　Shui Mi Thao Phu　水蜜桃譜.
　　A Treatise on the 'Honeydew' Peach (Varieties
　　　of *Prunus persica*, grown at Hangchow, Tientsin
　　　and elsewhere).
　　1813.
Chhü Wan-Li (*1*)　屈萬里
　　'Shang Shu', 'Kao Yao Mo' Phien Chu-Chhêng ti Shih-
Tai　尚書皐陶謨篇著成的時代.
　　On the Dating of the 'Counsels of Kao Yao'
　　　chapter [and other chapters] of the *Historical
　　　Classic*.
　　AS/BIHP, 1957, **28**, 381.
　　Abstr. *RBS*, 1962, **3**, no. 121.
Chhüan Han-Shêng (*3*)　全漢昇.
　　*Chhing Mo ti 'Hsi-Hsüeh Yuan Chhu Chung-Kuo'
　　　Shuo* 清末的「西學原出中國」說.
　　A Research on the 'Theory of the Chinese
　　　Origin of Western Science' at the End of the
　　　Chhing Dynasty.
　　LHP, 1935, **4** (no. 2), 57-102.

Fang Wên-Phei (*1*) = (*2*)　方文培.
　　Omei Chih-Wu Thu Chih 峨眉植物圖誌.
　　Icones Plantarum Omeiensium (Flora of Mount
　　　Omei, Szechuan).
　　Nat. Szechuan Univ, Chhêngtu 1942-5. Vol. 1
　　　(nos 1 & 2), Vol. 2 (no. 1) all pub.
　　Text in English and Chinese.
Fang Wên-Phei (*2*)　方文培.
　　Wo-mên so chih-tao-ti Omei Shan Chih-Wu
　　　我們所知道峨眉山植物.
　　What we know about the plants of Omei Shan
　　　(in Szechuan).
　　KHS, 1957, **33** (no. 2), 115.
Fêng Chhêng-Chün (*1*)　馮承鈞.
　　Chung-Kuo Nan Yang Chiao Thung Shih 中國南
　　　洋交通史.
　　History of the contacts of China with the South
　　　Sea Regions.
　　Com. Press, Shanghai, 1937, repr. Thaiphing,
　　　Hongkong, 1963.
Fêng Han-Yung (*1*)　馮漢鏞.
　　'Hai Yao Pên Tshao' Tso-Chê Li Hsün Khao
　　　「海藥本草」作者李珣考.
　　A Study of Li Hsün, the writer of the *Materia
　　　Medica of the (Southern) Countries Beyond the Seas*
　　　(*c.* +766).
　　ISTC, 1957, **8** (no. 2), 122.
Fêng Kuo-Mei, Hsia Li-Fang & Chu Hsiang-Hung
　　　(*1*)　馮國楣, 夏麗芳, 朱象鴻.
　　Yunnan Shan Chha Hua (Japanese translation by
　　　Tuyama Takashi, with foreword).　雲南山
　　　茶花 (雲南のツバキ).
　　The *Camellias* of Yunnan.
　　Yunnan Jen-min, Kunming, 1981.
　　Nihon Hoso Suppan Kyokai, Tokyo, 1981
　　　(Japan Broadcast Pub. Co.).
Fu Lan-Ya (*1*) (tr.) (= J. Fryer).　傅蘭雅.
　　Ko Chih Shih Chhi　格致釋器.
　　Explanations of Scientific Instruments and
　　　Apparatus.
　　Shanghai, at various times. Chs. 2, 8, 9, a tr. of
　　　J. J. Griffin's *Chemical Handicraft* were out by
　　　1864. Ch. 1 (Meteorol.), 1880.
Fu Lan-Ya (*1*)　傅蘭雅.
　　Ko Chih Shih Chhi　格致釋器.
　　Explanations of Scientific Instruments and
　　　Apparatus.
　　Kiangnan Arsenal, Shanghai, 1880.

Fu Pao-Shih (*1*) 傅抱石.
  Chung-Kuo Mei Shu Nien Piao 中國美術年表.
  A Chronological Dictionary of the Fine Arts in
    China.
  Thaiphing, Hongkong, 1963.
Fu Shu-Hsia (*1*) 傅書遐.
  *Chung-Kuo Chu-Yao Chih-Wu Thu Shuo (Chüeh Lei
    Chih-Wu Mên)* 中國主要植物圖說
    (蕨類植物門).
  Descriptive Atlas of the Most Important Chinese
    Flora (Ferns; Pteridophyta).
  Kho-Hsüeh, Peking, 1957.
Fu-Chha Tun-Chhung (*1*) 富察敦崇.
  Yenching Sui Shih Chi 燕京歲時記.
  Annual Customs and Festivals of Peking.
  Peking, 1900.
  Tr. Bodde (12).

Hao I-Hsing (*1*) 郝懿行.
  Erh Ya I Su 爾雅義疏.
  Commentary on the *Literary Expositor* [with
    special reference to plant and animal
    names].
  Peking, 1822.
  Cf. Chang Yung-Yen (*1*).
Ho Ping-Ti (*1*) 何炳棣.
  Huang Thu yü Chung-Kuo Nung Yeh ti Chhi-Yuan
    黃土與中國農業的起源.
  The Loess lands and the Origins of Chinese
    Agriculture.
  Chinese Univ. Press, Shatin, Hongkong, 1969.
Hou Hsüeh-Yü, Chhen Chhang-Tu & Wang Hsien-
    Phu (*1*) 侯學煜, 陳昌篤, 王獻溥.
  *Chung-Kuo Chih Pei yü Chu-Yao Thu Lei ti Kuan-
    Hsi* 中國植被與主要土類的關係.
  The Vegetation of China with Special Reference
    to the Main Soil Types.
  APS, 1957, **5** (no. 1), 19.
  Engl. précis in Hou, Chhen & Wang (*1*).
Hou Khuan-Chao (*1*) & Sixteen collaborators
    侯寬昭.
  Kuang-chou Chih-Wu Chien So Piao 廣州植物
    檢索表.
  Index of Cantonese Plants.
  Kho-Hsüeh, Peking, 1957.
Hou Khuan-Chao (*2*) & Sixteen collaborators
    侯寬昭.
  Kuang-chou Chih-Wu Chih 廣州植物誌.
  Flora of Canton.
  Kho-Hsüeh, Peking, 1956.
Hou Khuan-Chao & Hsü Hsiang-Hao (*1*) 侯寬昭,
    徐祥浩.
  *Hainan Tao ti Chih-Wu ho Chih-Pei yü Kuangtung
    Ta Lu Chih-Pei Kai-Khuang* 海南島的植物
    和植被與廣東大陸植被概況.
  Wild and Cultivated Flora of Hainan Island
    and Kuangtung Province.
  Kho-Hsüeh, Peking, 1955.
Hon Wai-Lu & Chao Chi-Pin (*1*) 侯外廬,
    趙紀彬.
  Lü Tshai ti Wei-Wu-Chu-I Ssu-Hsiang 呂才的
    唯物主義思想.

On the Materialist Philosophy of Lü Tshai (in
    the Thang Dynasty).
  LSYC, 1959, (no. 9), 1.
Hu Chhang-Chhih (*1*) 胡昌熾.
  *Chung-Kuo Kan Chü Tsai-Phei chih Li-Shih yü Fên-
    Pu* 中國柑橘栽培之歷史與分佈.
  The History and Distribution of Citrus Fruits in
    China.
  JAAC, 1934 (nos. 126/127), 1–79.
Hu Chhang-Chhih (*2*) 胡昌熾.
  Kuan-yü Chung-Kuo Kan Chü Lei chih Thiao-Chha
    關於中國柑橘類之調查.
  On the Principal Cultivated Varieties of Citrus
    in China.
  JAAC, 1930 (nos. 75/76), 1.
Hu Hsi-Wên (*1*) 胡錫文.
  *Chung-Kuo Hsiao Mai Tsai-Phei Chi-Shu Chien-
    Shih* 中國小麥栽培技術簡史.
  Brief History of the Cultivation of Wheat in
    China.
  AHRA, 1958, **1**, 51.
Hu Hsien-Su (*1*) 胡先驌.
  Ching-Chi Chih-Wu Shou Tshê 經濟植物手冊.
  Handbook of Economic Plants. (No index.).
  Kho-Hsüeh, Peking, 1955.
Hu Hsien-Su (*2*) 胡先驌.
  Chih-Wu Hsüeh Hsiao Shih 植物學小史.
  A Short History of Botany (based on Harvey-
    Gibson (1), *Outlines of the History of Botany*).
  Com. Press, Shanghai, 1930.
Hu Hsien-Su (*3*) 胡先驌.
  'Shuo Wên' Chih-Wu Ku Ming Chin Chhêng 「說
    文」植物古名今證.
  The Scientific Names of Some Plants mentioned
    in the *Shuo Wên*.
  KHS, 1916, **2**, 311.
Hu Hsien-Su & Chhen Huan-Yung (*1*) 胡先驌,
    陳煥鏞.
  Chung-Kuo chih-Wu Thu Phu 中國植物圖譜.
  Icones Plantarum Sinicarum.
  Com. Press, Shanghai, 1927–37, in 5 parts
    containing 250 entries with large drawings
    and modern plant descriptions; text both in
    English and Chinese.
Hu Hsien-Su & Chhen Huan-Yung (*2*) 胡先驌,
    陳煥鏞.
  Chung-Kuo Sên Lin Shu Mu Thu Chih 中國森林
    樹木圖誌.
  The Silva of China; a Description of the Trees
    which Grow Naturally in China.
  2 vols. Fan Mem. Inst. Biol. Peiping, and Nat.
    For. Res. Bur. (Min. Ag. & For.), Peiping,
    1946–8.
Hu Hsien-Su & Chhin Jen-Chhang (*1*) = (*1*)
    胡先驌, 秦仁昌.
  Chung-Kuo Chüeh Lei Chih-Wu Thu Shuo 中國蕨類
    植物圖譜.
  Icones Filicum Sinicarum [in English and
    Chinese with large illustrations; 250 entries].
  Acad. Sin. Nanking and Fan Mem. Inst.
    Peiping. Pt. 1 1930; Pt. 2 (Chhin only) 1934;
    Pt. 3 (Chhin only) 1935.

Hu Li-Chhu (*1*) 胡立初.
'*Chhi Min Yao Shu' Ying Yung Shu Mu Khao Chêng*
「齊民要術」引用書目考證.
A Critical Bibliography of the Books quotes in
the *Important Arts for the People's Welfare*.
KHHPN, 1934, **2**, 52–111.

Hu Shih (*3*) 胡適.
Hsien Chhin Chu Tzu Chin Hua Lun 先秦諸子
進化論.
Theories of Evolution in the Philosophers before
the Chhin Period).
KHS, 1917, **3**, 19.

Huang Yü (*1*) 黃鈺.
Ming I Pieh Lu 名醫別錄.
(A Reconstitution of the) *Informal (or Additional)
Records of Famous Physicians (on Materia Medica)*
(from the quotations in the later pharmaceu-
tical natural histories).
1869.
In *Chhen Hsiu-Yüan I Shu*.

Hung Huan-Chhun (*1*) 洪煥椿.
*Shih chih Shih-San Shih-Chi Chung-Kuo Kho-Hsüeh-ti
Chu-Yao Chhêng-Chiu* 十至十三世紀中
國科學的主要成就.
The Principal Scientific (and Technological)
Achievements in China from the +10th to
the +13th centuries (inclusive) [the Sung
period].
LSYC, 1959, **5** (no. 3), 27.

Hung Kuan-Chih (*1*) 洪貫之.
Chung-Kuo Ku-Tai Pên Tshao Chu Shu Shih Lüeh
中國古代本草著述史略.
A Short History of the Ancient Works on
Pharmaceutical Natural History.
ISTC, 1948, **2** (nos 1 & 2), 13.

Hung Kuan-Chih (*2*) 洪貫之.
*Ti-i-Pu Yao Tien; 'Hsin Hsiu Pên Tshao' Chien
Chieh* 第一部藥典「新修本草」簡介.
The First Official Pharmacopoeia; the *Newly
Improved Pharmacopoeia* (+659).
SHIY, 1957 (no. 10), 468.

Hung Kuan-Chih (*3*) 洪貫之.
*Thang Hsien-Chhing 'Hsin Hsiu Pên Tshao' Yao
Phin Tshun-Mu-ti Khao-Chha* 唐顯慶「新修
本草」藥品存目的考察.
A Study of the Preservation of the Index of the
(Lost) [+659] *Newly Improved Pharmacopoeia*
(in the *Supplement to the Thousand Golden
Remedies* [+660–80]).
ISTC, 1954, **6** (no. 4), 239.

Hung Kuan-Chih (*4*) 洪貫之.
'*Chêng Lei Pên Tshao' yü 'Pên Tshao Yen I' ti chi-ko
Wên-Thi* 「證類本草」與「本草衍義」的
幾個問題．
On the Question of the Relations between the
Classified Pharmaceutical Natural History (+1094)
and the *Dilations upon Pharmaceutical Natural
History* (+1116).
ISTC, 1954, **6** (no. 2), 100.

Hsia Wei-Ying (*1*) 夏緯英,
'*Erh Ya' chung so Piao-Hsien-ti Chih-Wu Fên-Lei*
「爾雅」中所表現的植物分類.

On the Classification of Plants in the *Erh Ya*
[ancient dictionary].
KHSC, 1962, (no. 4), 41.

Hsia Wei-Ying (*2*) 夏緯英.
'*Kuan Tzu' Ti Yuan Phien Chiao Shih* 「管子」
地員篇校釋.
The Chapter in the *Kuan Tzu* Book on the
'Variety of Earth's Products' Emended and
Explained.
Chung-Hua, Peking & Shanghai, 1958.

Hsia Wei-Ying (*3*) 夏緯英.
'*Lü Shih Chhun Chhiu' Shang Nung Têng Ssu Phien
Chias Shih* 「呂氏春秋」上農等四篇
校釋.
An Analytic Study of the 'Exaltation of
Agriculture' and Three Similar Chapters in
Master Lü's Spring and Autumn Annal (−239).
Chung-Hua, Peking, 1956.

Hsiao han (*1*) 曉菡.
*Chhangsha Ma-Wang-tui Han Mu Po Shu Kai
Shu* 長沙馬王堆漢墓帛書概述.
Brief Notes on the Silk Manuscripts of Ancient
Books found in the Han Tomb (no. 3) at Ma-
wang-tui near Chhangsha (−168).
WWTK, 1974, (no. 9), no. 220; 40.

Hsieh Khung (*1*) 謝堃.
Hua Mu Hsiao Chi 花木小記.
A Short Account of Some Flowering Plants and
Trees.
1820–50.

Hsieh Li-Hêng (*1*) 謝利恆.
Chung-Kuo I-Hsüeh Yüan Liu Lun 中國醫學
源流論.
A Discourse on the Historical Development of
Chinese Medicine.
Chhêng Chai I Shih, Shanghai, 1935.

Hsieh Sung-Mo (*1*) 謝誦穆.
*Chung-Kuo Li-Tai I-Hsüeh Wei Shu Khao; I, Pên
Tshao* 中國歷代醫學偽書考；上，本草.
An Investigation of the Authenticities of Ancient
and Mediaeval Chinese Medical Books; I,
The Pharmaceutical Natural Histories.
ISTC, 1947, **1** (no. 1), 57.

Hsieh Thang (*1*) 核堂.
I Shih Chih Yen 醫史卮言.
A Note on Beri-beri and Other Points in
Medical History.
ISTC, 1948 **1** (no. 3/4) 37.

Hsin Shu-Chhih (*1*) 辛樹幟.
'*Yü Kung' Chih Tso Shih-Tai ti Thui Tshe (Chhu
Kao)* 禹貢制作時代的推測(初稿).
A Preliminary Study on the Date of the *Tribute
of Yü* [chapter of the *Historical Classic*].
NWACJ, 1957, **3**, 1.

Hsü Chien-Yen (*1*) 徐建寅.
Ko Chih Tshung-Shu 格致叢書.
Compendium of General Science. [A collection
of many short introductory works on science
and engineering, including translations by
the staff of the Kiangnan Arsenal, J. Fryer
et al.].
Shanghai, 1901.

Hsü Hsün-Chan 許訓湛 & Chhen Ta-Chang (1)
陳大章.
Têng-Hsien Tshai-Sê Hua Hsiang Chuan Mu
鄧縣彩色畫象磚墓/.
The Tomb of the Painted Brick Reliefs at Têng-
hsien (Honan), [+5th century, N/Wei].
Wên-Wu, Peking, 1958.

Hsü Nai-Ho (1) 許鼐穌.
Lan Hui Thung Hsin Lu 蘭蕙同心錄.
On Understanding *Lan* and *Hui* Orchids
[illustrated].
1865, pr. 1890.

Hsü Tê-Lin (1) 徐德嶙.
San Kuo Shih Chiang Hua 三國史講話.
Lectures and Discussions on the History of the
Three Kingdoms Period.
Shanghai, 1955.

Iinuma Yokusai (1) 飯沼慾齋.
Shintei Sōmoku Zusetsu 新訂草木圖說.
Iconography of [Japanese] Plants, Indigenous
and Introduced [arranged according to
natural families, with hand-coloured
illustrations].
Tokyo 1832, repr. 1856, 2nd ed. (Shintei) 1874
(with index by Tanaka Yoshio & Ono
Motoyoshi), 3rd ed. 1907.
See Bartlett & Shohara (1), p. 145; Merrill &
Walker (1), p. 204.
Bretschneider's 'Somoku', see (1), vol. 1.
p. 101; vol. 2. p. 17.
The 'Soo- bokf' of Miquel and Maximowicz.

Itō Keisuke (1) 伊藤圭介.
Nihon Sambutsushi 日本產物志.
Record of the (Plant) Products of Japan.
Tokyo, 1872.

Iwasaki Tsunemasa (1) 岩崎常正, with illus-
trations by Okada Seifuku *et al.*
Honzō Zufu 本草圖譜.
Illustrated Manual of Medicinal Plants.
Yedo, 1828–56. Repr. in 93 vols. Tokyo 1920–2
with index.
See Bartlett & Shohara (1), pp. 63 ff., 131 ff.,
135; Merrill & Walker (1), pp. 214, 560.
Bretschneider's 'Phonzo', see (1), vol. 1, p. 100;
vol. 2. p. 17.
Cf. also Rudolph (6, 9).

Kan To (1) 干鐸.
*Chung-Kuo Lin-Yeh Chi-Shu Shih-Liao Chhu-Pu Yen-
Chiu* 中國林業技術史料初步研究.
Preliminary Researches on the History of
Forestry in Chinese Culture.
Nung-yeh, Peking & Shanghai, 1964.

Kanchira Ryōzō (1) 金平亮三.
Formosan Trees 臺灣樹木誌.
Japanese Government, Thaipei, 1917; 2nd. ed.
1936.

Kao Jun-Shêng (1) 高潤生.
'Erh Ya' Ku Ming Khao 爾雅穀名考.
A Study of Cereal and Crop Plant Names in the
Literary Expositor.
1915.

Kariyone Tatsuo (1) 刈米達夫
Yakuyō Shokubutsu Zufu 藥用植物圖譜.
Illustrations and Descriptions of Medicinal
Plants.
Kanehara Suppan, Tokyo, 1961.

Katayama Naoto (1) 片山直夫.
Nihon Chiku-fu 日本竹譜.
A Manual of Japanese Bamboos [with many
Chinese names].
1884.
Tr. Satow (1).

Kawabara Keiga (1) 川原慶賀.
Sōmoku Ka Kajitsu Shashin Zufu 草木花果實寫
眞圖譜.
A Collection of Illustrations of Plants and Trees,
Flowers and Fruits.
Maekawa Zembei, Osaka, 1842.
A Hand-coloured edition of this work is in the
Nagasaki University Library.

Kêng Hsüan (1).
'*Shih Ching*' *chung-ti Ching-Chi Chih-Wu* 詩經中
的經齊植物.
The Economic Plants mentioned in the *Book of
Odes*.
Com. Press, Thaipei, 1974 (Everyman's Library,
nos. 2099–100).

Kêng I-Li (1) 耿以禮.
*Chung-Kuo Chu-Yao Ho Pên Chih-Wu Shu Chung
Chien-So-Piao; fu Hsi Thung Ming Lu* 中國主
要禾本植物屬種檢索表; 附系統名錄.
Claves Generum et Specierum Graminearum
Primarum Sinicarum; Appendice Nomen-
clatione Systematica.
Biol. Dept. Univ. Nanking, 1957.
Bot. Inst. Academia Scinica, Peking, 1957.

Kêng I-Li (2) 耿以禮.
*Chung-Kuo Chu-Yao Chih-Wu Thu Shuo, Ho Pên
Kho* 中國主要植物圖說, 禾本科.
Flora Illustralis Plantarum Prinarum Sini-
carum, Gramineae.
Acad. Sin., Peking, 1959.

Kêng Po-Chieh (1) 耿伯介.
Chung-Kuo Chu Lei Chih-Wu Chih Lüeh 中國竹
類植物誌略.
A Preliminary Study of the Bamboos of China.
NFRB/TB, 1948 (no. 8).

Kêng Po-Chieh (2) 耿伯介.
Mao Chu ti Chih-Wu-Hsüeh Hsing-Chih 毛竹的
植物學性質.
The Botanical Characteristics of the Pubescent
Bamboo (*Phyllostachys edulis*).
CF, 1954, **4**, 14.

Khung Chhing Lai *et al.* (1) 孔慶萊 (13
collaborators).
Chih-Wu-Hsüeh Ta Tzhu Tien 植物學大辭典.
General Dictionary of Chinese Flora.
Com. Press, Shanghai and Hongkong, 1918;
repr. 1933 and often subsequently.

Kim Tujong (1) 金斗鍾.
Hanguk Ŭihak Sa 韓國醫學史 (上中世編)
A History of Medicine in Ancient and
Mediaeval Korea.

Kim Tujong (*1*) (*cont.*)
   Chŏngsŭm Sa, Seoul, 1955.
   With mimeographed summary in English.
Kimiya Yasuhiko (*1*)　木宮泰彦.
   Nikka Bunka Kōryūshi　日華文化交流史.
   A History of Cultural Relations between Japan
    and China.
   Fwzambō, Tokyo, 1955.
   RBS, **2**, no. 37.
Kimura, Koichi (*1*)　木村康一.
   Honzō(Pên Tshao);　本草.
   [a historical bibliography of Chinese Botany &
    Zoology] in Shinakagaku Keizaishi (*A History*
    of Chinese Science and Economics)　支那科學
    經濟史.
   Vol. 8 of *Shina Chirirekishin-taike* (Chinese History
    & Geography)　支那地理歷史大系.
   Tokyo, 1942.
Kimura Koichi (*2*)　木村康一.
   Sōtennen-shoku Nihon no Yakuyō Shokubutsu
   總天然色日本の藥用植物.
   Japanese Medicinal Plants [Album of Colour
    Photographs with text].
   2 vols.
   Hirokawa, Shoto, Tokyo, 1st ed. 1960; 2nd ed.
    1962.
Kimura Koichi (*3*)　木村康一. 植物の漢名に
   ついて.
   On Chinese Plant Nomenclature.
   Tokyo.
Kimura Koichi & Kimura Takeatsu (*1*)　木村康一,
   木村孟淳.
   Genshoku Nihon Yakuyō Shokubutsu Zukan　原色
   日本藥用植物圖鑑.
   Medicinal Plants of Japan, with Colour
    [Photograph] Illustration.
   Hoikusha, Osaka, 1964 (Hoikusha no Genshoku
    Zukan, no. 39).
Kitamura Shirō, Murata Gen, Hori Masaru &
   Hirosuke Ishizu (*1*)　北村四郎, 村田源,
   堀勝.
   Genshoku Nihon Shokubutsu Zukan; Somokuhen I.
    Gobenkarui　原色日本植物圖鑑草木編
    vol. 1　合弁花類.
   Herbaceous Plants of Japan, with Colour
    [Paintings]; Vol. 1 Sympetalae [Families].
   Hoikusha, Osaka, 1958.
Kitamura Shirō & Okamoto Syōgo (*1*)　北村四郎,
   岡本省吾.
   Genshoku Nihon Jiumoku Zukan　原色日本樹
   木圖鑑.
   Trees and Shrubs of Japan, with Colour
    [Photograph] Illustrations.
   Hoikusha, Osaka, 1958.
Kiyohara Shigeomi (*1*)　清原重巨.
   Sōmoku Seifu　草木性譜.
   Treatise on the Natures of Plants.
   Tokyo, 1823.
   See Merrill & Walker (1), p. 560.
Kōno Reizo (*1*)　河野齡藏.
   Nihon Shokubutsu Zusetsu　日本高山植物
   圖說.

Alpine Flora of Japan.
   Hobundo, Tokyo, 1931.
Koshimizu Takuji (*1*)　小清水卓二.
   'Manyōshu' Shokubutsu to Kodaijin no Kagaku Sei
   万葉集植物と古代人の科学性.
   The Plants mentioned in the *Anthology of a*
    Myriad Leaves; an example of Ancient
    Botanical Science.
   Osaka, 1950.
Koshimizu Takuni (*2*)　小清水卓二.
   'Manyōshu' Shokubutsu Shashin to Kaisetsu　万葉
   集植物写真と解説.
   (Colour) Photographs, with Explanations, of the
    Plants mentioned in the *Anthology of a Myriad*
    Leaves.
   Sanseido, Osaka, 1941.
Ku Yen (*1*)　顧彥.
   Chih Huang Chhüan Fa　治蝗全法.
   Complete Handbook of Locust Control.
   Huan-chhêng 1857.
Kuo Po-Tshang (*2*)　郭栢蒼.
   Min Chhan Lu I　閩彥錄異.
   Records of the Strange Products of Fukien
    (especially plants and animals).
   1886.
   Cf. Wang Yü-hu (1), 2nd ed; p. 282; Swingle
    (9).
Kuribayashi Motojirō (*1*).　栗林元次郎.
   Hana-Shobu Dai-Zukan.　花昌蒲大圖鑑.
   A History of the *Iris* genus.
   Asahi Shimbunsha, Tokyo, 1971.
   With Engl. abstr. of 28 pp.

Lan Phu (*1*)　藍浦.
   Chung-Tê-Chen Thao Lu　景德鎮陶錄.
   The Ching-Tê-Chen Record of the Potteries of
    China [Lan Phu (Pin-Nan) resided at this
    capital of the pottery and porcelain industry].
   1815 (but written *c.* +1795).
   Tr. Julien (7); Sayer (1).
Lêng Fu-Thien & Chao Shou-Jen (*1*)　冷福田,
   趙守仁.
   Chiangsu Shêng Ling Hai Ti-Chü Yen Chi Thu Fa
    Sêng Kuo Chhêng Chi Yen Chi Thê-Hsing ti Chuan
    Hua　江蘇省沿海地區鹽漬土發生過
    程及鹽漬特性的轉化.
   The Genesis and Properties of the Saline Soils
    along the Eastern Coasts of Chiangsu Province.
   APS, 1957, **5**, 195.
Li Chien-Nung (*3*)　李劍農.
   Hsien Chhin Liang Han Ching-Chi Shih Kao
   先秦兩漢經濟史稿.
   Sketch for an Economic History of the Chhin
    and Han Periods.
   Sanlien, Peking, 1957.
Li Chien-Nung (*4*)　李劍農.
   Wei, Chin, Nan Pei Chhao, Sui, Thang, Ching-Chi
    Shih Kao　魏晉南北朝隋唐經濟史稿.
   Sketch for an Economic History of the Wei,
    Chin, Northern and Southern Dynasties, and
    Sui and Thang Periods.
   Chunghua, Peking, 1963.

Li Chhang-Nien (*1*) *et al.* 李長年.
 Chung-Kuo Nung Hsüeh I-Chhan Hsüan Chi; Tou Lei
 中國農學遺產選集; 豆類.
 Anthology of Quotations [illustrating the
 History of] Chinese Agricultural Science and
 Production; Beans and other Legumes.
 Vol. 4, Nung-Yeh, Peking, 1959.
Li Chhang-Nien (*2*) *et al.* 李長年.
 *Chung-Kuo Nung-Hsüeh I-Chhan Hsüan Chi; Ma Lei
 Tso Wu* 中國農學遺產選集; 麻類作物.
 Authology of Quotations [illustrating the
 History of] Chinese Agricultural Science and
 Production; Fibre-crop and Oilseed Plants
 [*Cannabis, Boehmeria, Abutilon, Linum, Corchorus,
 Pueraria*].
 Vol. 1, Nung-Yeh, Peking 1962.
 Joint Committee on Agricultural History
 (National Academy of Agricuture and
 Nanking College of Agriculture), ser. A, no. 8.
Li Chhing-Khuei & Chang Hsiao-Nien (*1*)
 李慶逵, 張效年.
 Chung-Kuo Hung Jang ti Hua-Hsüeh Hsing-chih
 中國紅壤的化學性質.
 Chemical Characteristics of Krasnozems in
 China.
 APS, 1957, **5**, 78.
 Engl. tr. Soil Sci. Soc. China Reports to the VI
 International Congress of Soil Science,
 Peking, 1956.
Li Kuang-Pi 李光璧 & Chhien Chün-Yeh (*1*)
 (ed.) 錢君曄.
 *Chung-Kuo Kho-Hsüeh Chi-Shu Fa-Ming ho Kho-
 Hsüeh Chi-Shu Jen Wu Lun Chi* 中國科學
 技術發明和科學技術人物論集.
 Essays on Chinese Discoveries and Inventions in
 Science and Technology, and on the Men
 who made them.
 San-lien Shu-tien, Peking, 1955.
Li Lai-Jung (*1*) 李來榮 等.
 Nan Fang-ti Kuo Shu Shang Shan 南方的果
 樹上山.
 Mountain Cultivation of the Fruits of the South.
 Kho-Hsüeh, Peking, 1956.
Li Lai-Jung (*2*) 李來榮 等.
 Kuan Yü Li-Chih Lung-Yen-ti Yen-Chiu 關於荔
 枝龍眼的研究.
 Researches on the Lichih and the Lungyen
 Fruits.
 Kho Hsüeh, Peking, 1956.
Li Lien-Chieh, Yeh Ho-Tshai, Hou Kuang-Chhiung,
 Huang Jui-Tshai & Tsu Khang-Chhi (*1*) 李
 連捷, 葉和才, 侯光炯, 黃瑞采, 祖康
 祺.
 Thu Jang 土壤.
 An Introduction to Soil Science and the Soils of
 China.
 Nung-Yeh, Peking 1963, Shanghai 1964
 (Agricultural Production Fundamental
 Knowledge Series).
Li Ssu-Kuang (*1*) 李四光.
 'Tshang Sang Pien Hua' ti Chieh-Shih 「滄桑變
 化」的解釋.

An Elucidation of the Phrase 'Changing from
 Blue Sea to Mulberry Groves'.
 Tzhu-Hsing Local Government, Tzhu-Hsing.
 1942.
Li Ssu-Mien (*1*).
 Chhin Han Shih 秦漢史.
 A History of the Chhin and Han Dynasties.
 Shanghai, 1947.
Li Ssu-Mien (*2*).
 Wei, Chin, Nan Pei Chhao Shih 魏晉南北朝史.
 A History of the Wei, Chin, and Northern and
 Southern Dynasties.
 Shanghai, 1948.
Li Thao (*7*) (= 11) 李濤.
 Ming Tai Pên-Tshao-ti Chhêng-Chiu 明代本草
 的成就.
 Achievements of Pharmaceutical Natural
 History in the Ming Period.
 ISTC, 1955, **7** (no. 1), 9.
Li Thao (*8*) 李濤.
 Wei-Ta ti Yao-Hsüeh Chia Li Shih-Chen 偉大的藥
 學家; 李時珍.
 The Great Pharmacologist Li Shih-Chen
 (+1518–93) (biography), 20 pp.
 Peking, 1955.
Li Thao (*9*) 李濤.
 Li Shih-Chen ho 'Pên Tshao Kang Mu' 李時珍
 和「本草綱目」.
 Li Shih-Chen and the *Pandects of Pharmaceutical
 Natural History*.
 ISTC, 1954, **6** (no. 3), 168.
Li Ting (*1*) 李鼎.
 *Khao-Chha Pên-Tshao-ti Chu-Shu Hsiu-Ting Ho
 Kai-I* 考察本草的著述修訂和改移.
 An Analysis of Editions and Changes in the
 Botanical-Pharmaceutical Literature.
 ISTC, 1955, **7** (no. 2), 90.
Li Ting (*2*) 李鼎.
 'Pên Tshao Ching' Yao-Wu Chhan-Ti Piao Shih
 「本草經」藥物產地表釋.
 Explanation and Tabulation of the Places
 of Origin of the Materia Medica of the
 Pharmacopoeia (of the Heavenly Husbandman).
 ISTC, 1952, **4** (no. 4), 167.
Liang Chhi-Chhao (*6*) 梁啓超.
 Ku Shu Chen Wei chi chhi Nien-Tai 古書眞僞
 及其年代.
 On the Authenticity of Ancient Books and their
 Probable Datings.
 Lectures recorded by Chou Chuan-Ju
 周傳儒, Yao Ming-Ta 姚名達 & Wu
 Chhi-Chhang 吳其昌.
 Chung-Hua, Peking, 1955, repr. 1957.
Liang Ching-Hui (*1*) 梁景暉.
 'Shen Nung Pên Tshao Ching' Nien-Tai-ti Than-Thao
 「神農本草經」年代的探討.
 A Discussion on the Date of the *Classical
 Pharmacopoeia of the Heavenly Husbandman*.
 ISTC, 1957, **8** (no. 2), 114.
Lin Chhi-Shou (*1*) 林啟壽.
 Chung Tshai Yao Chhêng Fên Hua Hsüeh 中草藥
 成分化學.

Lin Chhi-Shou (*1*) (*cont.*)
Analytical Organic Chemistry of Chinese Drug-Plants.
Kho-Hsüeh Chhu Pan Shih, Peking, 1977.
Lin Thien-Wei (*1*) 林天蔚.
Sung-Tai Hsiang Yao Mou-I Shih Kao 宋代香藥貿易史稿.
A History of the Perfume (and Drug) Trade of the Sung Dynasty.
Chung-Kuo Hsüeh-Shê, Hongkong, 1960.
Ling Man-Li (Mary) (*1*) 凌曼立.
Thaiwan yü Huan Thai-Phing Yang ti Shu Phi Pu Wên-Hua 臺灣與環太平洋的樹皮布文化.
Bark-Cloth in Thaiwan and the Circum-Pacific Area of Cultures.
In Ling Shun-Shêng (*7*), p. 211.
Inst. of Ethnol., Acad. Sin., Nankang, Thaiwan, 1963, (Monograph Series, no. 3).
First published in *AS/BIE*, 1960, no. 9, 313.
Ling Shun-Shêng (*7*) = (*6*) 凌純聲.
Shu Phi Pu, Yin Wên Thao, yü Tsao Chih Yin Shua Shu Fa Ming 樹皮布印文陶與造紙印刷術發明.
Bark-Cloth, Impressed Pottery, and the Invention of Paper and Printing.
Inst. of Ethnol., Acad. Sin., Nankang, Thaiwan, 1963 (Monograph Series, no. 3).
Papers first published in *AS/BIE* 1961, no. 11, 1; 1962, no. 13, 213, no. 14, 193; 1963, no. 15, 1.
With Three others and an introduction, concluding with a contribution by Ling Man-Li (*1*).
Liu Fu (*1*) 劉復.
Hsi Han Shih-Tai ti Jih Kuei 西漢時代的日晷.
Sun-Dials of the Western Han Period.
KHCK, 1932, **3**, 573.
Lui Fu (*5*) 劉復.
Shen Nung Ku Pên Tshao Ching 神農古本草經.
Reconstruction of the Ancient Text of the *Classical Pharmacopoeia of the Heavenly Husbandman*.
Chung-Kuo Ku I-Hsüeh Hui, 1942.
Liu Pao-Nan (*1*) 劉寶楠.
Shih Ku 釋穀.
On the Cereal Grains [historical and philological].
Peking, 1855.
In *HCCC* (*HP*), chs. 1075–8.
Lui Po-Han (*1*) 劉伯涵.
Kuan-yü Li Shih-Chen Sheng-Tsu-ti Than-So 關於李時珍生卒的探索.
On the Dates of the Birth and Death of Li Shih-Chen.
ISTC, 1955, **7** (no. 1), 1.
Liu Shen-O (*1*) (ed.) et al. = (*1*) 劉慎諤.
Chung-Kuo Pei Pu Chih-Wu Thu Chih 中國北部植物圖誌.
Flore Illustreé du Nord de la Chine (in French and Chinese).
Nab. Peiping Acad. Peiping.

1931 Pt. 1 Convolvulaceae, by Liu Shen-O & Lin Jung 劉慎諤, 林鎔.
1933 Pt. 2 Gentianaceae, by Lin Jung 林鎔.
1934 Pt. 3 Caprifoliaceae, by Hao Ching-Shêng 郝景盛.
1935 Pt. 4 Chenopodiaceae, by Khung Hsien-Wu 孔憲武.
1937 Pt. 5 Polygonaceae, by Khung Hsien-Wu 孔憲武.
All. pub.
Liu Shen-O (*2*) = (*2*) 劉慎諤.
Essai sur la Géographie Botanique du Nord et de l'Ouest de la Chine.
NPA/CIB, 1934, **2**, 423.
Liu Thang-Jui (*1*) 劉棠瑞.
With illustrations by Chhen Chien-Chu 陳建鑄.
Thaiwan Mu Pên Chih-Wu Thu Chih 臺灣木本植物圖誌.
Illustrations of Native and Introduced Ligneous Plants of Thaiwan (Formosa).
Thaiwan Univ. Agric. Series no. 8 (Forestry series no. 1), Thaipei, 1960.
Liu Tzu-Ming (*1*) 柳子明.
Chung-Kuo Chu-Ming-ti chi Chung Hua Hui 中國著名的幾種花卉.
Some of the Famous Flowering Plants of China and their Cultivation.
Hunan Jen-Min, Chhangsha, 1959.
Liu Wên-Chhi (*2*) 劉文淇.
I Lan Chi 藝蘭記.
On the Cultivation of Orchids.
c. 1840.
Liu Yü-Chhüan (*1*) 劉毓瑮.
'*Nung Sang Chi Yao*' *ti Tso-chê, Pan-pên ho Nei-jung* 農桑輯要的作者版本和內容.
On the Author of the *Fundamentals of Agriculture and Sericulture*, its Editions and Content.
AHRA, 1958, **1**, 215.
Lo Fu-I (*2*) 羅福頤.
Hsi Chhui Ku Fang-Chi Shu Tshan Chüan Hui Phien 西陲古方技書殘卷彙編.
Fragments of Medical and Technical Texts (from Tunhuang). Preserved in Western Libraries.
ISTC, 1953, **5** (no. 1), 27.
Lo Hsiang-Lin (*3*) 羅香林.
Thang Tai Kuang-chou Kuang-Hsiao Ssu yü Chung-Yin Chiao-Thung chih Kuan-Hsi 唐代廣州光孝寺與中印交通之關係.
The Kuang-Hsiao Temple at Canton during the Thang period, with reference to Sino-Indian Relations.
Chung-Kuo Hsüeh Shê, Hongkong, 1960.
Lo Hsiang-Lin (*4*) 羅香林.
Hsi Chhu Pho-ssu chih Li Hsün chi chhi '*Hai Yao Pên Tshao*' 系出波斯之李珣及其「海藥本草」.
Li Hsün of Persia and his 'Exotic Pharmacopoeia'.
Art. in *Symposuim of Chinese Studies commemorating the Golden Jubilee of the University of Hongkong*.
Repr. in Lo Hsiang-Lin (*5*), p. 97.

Lo Hsiang-Lin (5) 羅香林.
Thang Yuan Erh Tai chih Ching Chiao 唐元二
代之景教.
Nestorianism in the Thang and Yuan Dynasties.
Inst. of Chinese Culture, Hongkong, 1966.

Lo Ssu-Ku (1) (= Sir Henry Roscoe) 羅斯古.
Ko Chih Chhi Mêng 格致啓蒙.
Introduction to (Chemical) Science (Science
Primer, no. 2).
Shanghai, 1885.
Tr. Lin Lo-Chih (Y. J. Allen) 林樂知.

Lu Khuei-Shêng (1) (ed.) 陸奎生.
Chung Yao Kho-Hsüeh Ta Tzhu-Tien 中藥科
學大辭典.
Dictionary of Scientific Studies of Chinese
Drugs.
Shanghai Pub. Co., Hongkong, 1957.

Lu Wên-Yü (1) 陸文郁.
Shih Tshao Mu Chin Shih 詩草木今釋.
A Modern Elucidation of the Plants and Trees
mentioned in the Book of Odes.
Jen-Min, Tientsin, 1957.
RBS, 1964, 4, no. 956.

Lung Po-Chien (1) 龍伯堅.
Hsien Tshun Pên Tshao Shu Lu 現存本草
書錄.
Bibliographical Study of Extant Pharma-
copoeias (from all periods).
Jen-min Wei-Sêng, Peking, 1957.

Ma Chi-Hsing (1) 馬繼興.
Tsai Wo-Kuo Li-Shih-shang Tshui-Tsao-ti I-Pu
Yao-Tien-Hsüeh Chu-Tso; Thang 'Hsin Hsiu Pên
Tshao' 在我國歷史上最早的一部藥典
學著作;唐「新修本草」.
The Oldest official Pharmacopoeial Writings in
our Culture; the Newly Improved Pharmacopoeia
of the Thang period (+659).
ISTC, 1955, 7 (no. 2), 83.

Ma Chi-Hsing (3)
Kuan-Yü 'Chêng Lei Pên Tshao' ti i-hsieh Wên Thi ti
Shang-Chüeh 關於「證類本草」的一些問
題的商榷.
A Critical Discussion of Some Questions con-
cerning the Classified Pharmaceutical Natural
History (+1083 and +1096).
ISTC, 1955, 7 (no. 3), 182.

Ma Thai-Lai (1) 馬泰來.
Lu Chia 'Nan Yüeh Hsing Chi'; Tungfang Shuo 'Lin
I Chi'—Chhuan Pên 'Nan Fang Tshao Mu
Chueng' Pien-Wei Chü Yü 陸賈「南越行
記」,東方朔「林邑記」—傳本「南方
草木狀」辨僞舉隅.
On the 'Records of Travels to Southern Yüeh'
attributed to Lu Chia, and on the 'Records
of the Champa Kingdom' attributed to
Tungfang Shuo—Doubts on the Authenticity
of the received text of the 'Prospect of the
Plants and Trees of the Southern Regions'.
TLTC, 1969, 38 (no. 1). 20.

Ma Thai-Lai (2) 馬泰來.
Mi-hsiang Chih, Pao-hsiang Li—Chhuan Pên 'Nan

Fang Tshao Mu Chuang' Pien-Wei Chü Yü
蜜香紙, 抱香履—傳本「南方草木狀」
辨僞舉隅.
On the 'honey fragrance paper' and the 'water
pine pattens' (discussed by Hsi Han in the
Nan Fang Tshao Mu Chuang)—Doubts on
the Authenticity of the received text of the
'Prospect of the Plants and Trees of the
Southern Regions'.
TLTC, 1969, 38 (no. 6), 25.

Ma Yung-Chih (1) 馬溶之.
Kuan-yü Wo Kuo Thu-Jang Fên-Lei Wên-Thi di
Sheng-Chhüch 關於我國土壤分類問題
的商榷.
A Discussion on the Question of Soil Classifica-
tion in China.
APS, 1959, 7 (no. 3/4), 115.

Makino Tomitaro (1) 牧野富太郎.
Nihon shokubutsu zukan 日本植物圖鑑.
An Illustrated Flora of Japan, with the
Cultivated and Naturalised Plants.
Hokuryokan, Tokyo, n.d. (1956).

Makino Tomitaro (2) 牧野富太郎.
Futsu Shokubutsu Zufu 普通植物圖譜.
Illustrated Treatise on Common [Japanese]
Plants.
5 vols.
Univ. Tokyo, Tokyo 1912–13.

Mao Chhun-Hsiang (1) 毛春翔.
Ku Shu Pan Pên Chhang-Than 古書版本常談.
A Brief Discussion on the Printing Styles of Old
Books.
Chung-hua, Shanghai, 1962.

Matsumura Jinzō (1) 松村任三.
Shokubutsu Mei-i 植物名彙, (orig. Nihon
Shokubutsu Mei-i).
Index Plantarum Japoricarum; Enumeration of
Selected Scientific Names of both Native and
Foreign Plants, with romanised Japanese
Names and in many cases Chinese Characters
[Main listing in Latin binomials, romanised
Japanese index, no Chinese name index,
characters not always given].
1st. ed. Maruzen, Tokyo, 1884, 2nd. ed. greatly
revised 3 vols. Tokyo 1895. Repr. 1897,
revised ed. Maruya, Tokyo 1966; 9th ed.
enlarged Tokyo 1915–16.
Index by Byrd & Wead (1) for Chinese plant
names.

Mirjoshi Manabu & Makino Tomitaro (1)
三好學, 牧野富太郎.
Nihon Kōzan Shokubutsu Zufu 日本高山植物
圖譜.
Pocket Atlas of the Alpine Plants of Japan.
2 vols.
2nd. ed. Seibido, Tokyo, 1907.

Miki Sakae (1) 三木榮.
Chōsen Igakushi oyobi Shippeishi 朝鮮醫學史及
疾病史.
A History of Korean Medicine and of Diseases
in Korea.
Sakai, Osaka, 1962.

Miki Shigeru (*1*) 三木茂. メタセ
  コイア（生ける化石植物）.
  On *Metasequoia*, Fossil and Living.
  Nihon Kōbutsu Shumi-no-Kai 日本礦物趣
    味の會.
  Kyoto, 1953.
Mizukami Shizuo (*1*) 水上静夫.
  *Ashi to Chūgoku Nōgyō; Awasete sono Shinkō Kigen-ni
    Oyobu* 葦と中國農業；併せてその信
    仰起源に及ぶ.
  The Worship of Reeds in Ancient China.
  TSGH, 1957, **3**, 51.
  Abstr. *RBS*, 1962, **3**, no. 790.
Mizuno Tadaaki (*1*) 水野忠暁.
  Sōmoku Kinyōshū 草木錦葉集.
  Collection of Trees with Ornamental Foliage.
  Tokyo, 1829.
  See Bartlette Shohara (*1*), pp. 173–4.
Mo Tê-Chhüan (*1*) 穆德全.
  *Sung-Tai I-Chhien-ti Wai Lai Yao-Wu Chi Chhi tsai
    Fang-Chi-chung-ti Ying-Yung* 宋代以前的外
    來藥物及其方劑中的應用.
  On the Drugs from Foreign Countries which
    came into China Before the Sung, and on
    their Use in Prescribing.
  SHIY, 1957, (no. 9), 388.
Mori Osamu (*1*) & Naitō Hiroshi 森修，内藤寛.
  *Ying Chhêng Tzu; Chhien-Mu-Chhêng-I fukin no
    Kandai hekiga sembo* 營城子；前牧城驛附
    近の漢代壁畫甎墓.
  Ying Chhêng Tzu; (Two) Han Brick Tombs
    with Fresco Paintings near Chhien-mu-
    chhêng-i (in South Manchuria).
  With preface by Kusaka Tatsuta and appen-
    dices by Hamada Kōsaku & Mizuno Seiichi.
  Tōa Kōkogaku Kwai, Tokyo and Kyoto, 1934,
    (Archaeologia Orientalis, no. 4).
Mori Shikazō (*3*) 森鹿三.
  Shinshū Honzō to Kojima Hōso
    新修本草と小島實素.
  'The New Pharmacopoeia' (+659) and
    Kojima Hōso (+1797–1848).
  TG/K, 1940, **3**, 66.
Mori Tateyuki (*1*) (ed.) 森立之.
  Shen Nung Pên Tshao Ching Khao I 神農本草
    經考異.
  A Reconstruction of the Text of the *Classical
    Pharmacopoeia of the Heavenly Husbandman*, with
    an Analysis of Textual Variations.
  1845.
  Chün-Lien, Shanghai, 1955 (with postface by
    Fan Hsing-Chun)
Murakoshi Michio (*1*) 村越三千男.
  Naigai Shokubutsu Genshoku Daizukan 内外植物
    原色大圖鑑.
  General Description of Japanese and Foreign
    [other East Asian] Plants, with Illustrations in
    Colour [4339 entries].
  Tokyo, 1928, under title *Dai Shokubutsu Zukan*;
    2nd ed. 1932; 3rd ed. 1935; 4th ed. 1944.
  See Merrill & Walker (*1*), p. 339.
Murakoshi Michio (*2*) 村越三千男.

Genshoku Zusetsu Shokobutsu Daijiten 原色圖説
  植物大辭典.
General Dictionary of [Japanese] Plants with
  Illustrations in Colour (Latin, Chinese and
  Japanese name indexes).
Tokyo, 1938.

Nagano Yoshio (*1*) 永野芳夫.
  Tōyō Ranpu 東洋蘭譜.
  Oriental Miniature Orchids.
  Kashima Shoten, Tokyo, 1959.
Nakai Takenoshin *et al.* (*1*) 中井猛之進.
  Tōa Shokubutsu Zusetsu 東亜植物圖説.
  Iconographia Plantarum Asiae Orientalis [with
    descriptions in Latin as well as Japanese].
  5 vols.
  Shunyodo Shoten, Tokyo, 1935.
Nakao Manzō (*1*) 中尾万三.
  Shokuryō-honsō no Kōsatsu 食療本草の考察.
  A Study of the [Tunhuang MS. of the] *Shih Liao
    Pên Tshao* (Nutritional Therapy; a Pharma-
    ceutical Natural History), [by Mêng Shen,
    c. +670].
  SSIP, 1930, **1** (no. 3), 1–222.
Nakao Manzō (*2*) 中尾万三.
  *Kansho Geibunshi yori Honzō Engi ni itaru Honzō
    Shomoku no Kōsatsu* 漢書藝文志より本
    草縁起に至る本草書目の考察.
  A Study of the Bibliography of the Pên Tshao
    Literature (on pharmaceutical natural his-
    tory) and its Origins in the Light of the
    Bibliographical Chapter of the History of the
    (Former) Han Dynasty.
  Kyoto, 1928.
  Abstr. *STK*, no. 5252.
Nakao Manzō (*3*) = (*1*) 中尾万三.
  Shōkō Kōtei Keishi Shōrui Bikyū Honzō no Kōsatsu
    「紹興校定經史證類備急本草」の考察.
  An Investigation of the *Shao-Hsing Chiao-Ting Ching-
    Shih Chêng-Lei Pei-Chi Pên Tshao* [of the period
    +1131 to +1162; 4 copies preserved in Japan].
  With Chinese résumé.
  SSIP, 1933, **2** (no. 2), 1–52.
Nakao Manzō & Kimura Koichi (*1*) 中尾万
  三，木村康一.
  Kanyaku Shashin Shūsei 漢藥寫真集成.
  Photographic catalogue of Chinese Drugs, with
    Sketches and Explanations.
  SSIP 1929 **1** (no. 2), 1–103, with indexes, entries
    1–69. 1930 **1** (no. 5), 1–109, with indexes,
    entries 70–99.
Nalan Yung-Shou (*1*) 納蘭永壽.
  Shih Wu Chi Yuan Pu 事物紀原補.
  A Supplement for the *Records of the Origins of
    Affairs and Things*.
  1806.
Noda Mitsuzō (*1*) 野田光藏.
  Chūgoku Tohoku-Ku (Manshu) no Shokubutsu-shi
    中國東北區（滿洲）の植物誌.
  A Flora of the North-Eastern Region of China
    (Manchuria).
  Tokyo, 1971.

Ogawa Tamaki (*1*) 小川環樹.
Sō Ryō Kin Jidai no Jisho 宋遼金時代の字書.
On the Dictionaries of the Sung, Liao and
J/Chin Dynasties.
In Tohō Gakkai 東方學會創立十五周年
記念 (Fifteenth Anniversary Volume of the Society
for East Asian Studies).
Tokyo, 1962.
Okanishi Tameto (*1*) 岡西為人.
Hsü Chung-Kuo I-Hsüeh Shu Mu 續中國醫學
書目.
Continuation of the Classified Bibliography of
Chinese Medical Books.
Res. Inst. for East Asian Med.
Tokyo, 1942.
Okanishi Tameto (*2*) 岡西為人.
Sung I-chhien I Chi Khao 宋以前醫籍考.
Comprehensive Annotated Bibliography of
Chinese Medical Literature in and before the
Sung Period.
Jên-min Wei-shêng, Peking, 1958.
Okanishi Tameto (*3*) 岡西為人.
Chūgoku Honzō no Dentō to Kin Gen no Honzō
中國本草の傳統と金元の本草.
On the Transmission of Pharmaceutical Natural
History in China especially in the J/Chin and
Yuan Periods.
Art. in Yabunchi Kiyoshi (*26*), p. 171.
Okanishi Tameto (*4*) 岡西為人.
Tan Fang chih Yen-Chiu 丹方之研究.
Index to the 'Tan' Prescriptions in Chinese
Medical Works.
In Huang Han I-Hsüeh Tshung-Shu, 1936 vol. 11.
Okanishi Tameto (*5*) 岡西為人.
Chhung Chi 'Hsin Hsiu Pên Tshao' 重輯「新脩
本草」.
Newly Reconstituted Version of the New and
Improved Pharmacopoeia (of +659).
National Pharmaceutical Research Institute,
Thaipei, 1964.
Okanishi Tameto (*6*) 岡西為人.
Chūgoku Isho Honzō-Kō 中國醫書本草考.
A Study of the Medical and Pharmaceutical
Natural Histories in Chinese Literature.
Tokyo, 1974.

Phei Chien & Chou Thai-Yen (*1*) 裴鑑, 周太炎.
Chung-Kuo Yao Yung Chih Wu Chih 中國藥用植
物誌.
Illustrated Repertorium of Chinese Drug-Plants.
4 parts.
Kho-Hsüeh, Peking, 1951–6.
Pheng Shih-Chiang (*1*) 彭世獎.
Wo Kuo Ku Tai Nung Yeh Chi Shu Ti Yu Liang
Chhuan Thung Chih I—Shêng Wu Fang Chih
我國古代農業技術的優良傳統之一
一生物防治.
Biological Control—An Excellent Tradition
in the Agricultural Technology of Ancient
China Chung Kuo Nung Yeh Kho Hsüeh 中國
農業科學 1983 No. 1. pp. 92–96.
Pu Yü-Sên (*1*) 步毓森.

Ying Yung Tou-kho Chih-Wu Kai Lun 應用豆
科植物概論.
General Treatise on the Applied Biology (and
Biochemistry of Leguminous Plants.
Commercial Press, Shanghai 1934, 1935.

Sato Junpei (*1*) 佐藤潤平.
Kanyaku no Genshokubutsu 漢藥の原植物.
On the Chinese Medical Plants [of the North].
Jap. Assoc. for the Adv. of Sci. Tokyo, 1959.
Shang Chih-Chün (*1*) 尚志鈞.
Wo Kuo Tshui-Tsao-ti Yao-Tien, 'Thang Pên
Tshao' 我國最早的藥典「唐本草」.
Our Earliest official Pharmacopoeia, issued in
the Thang period (+659) [the Newly Improved
Pharmacopoeia].
ISTC, 1957, **8** (no. 4), 275.
Shao Chhêng-Hsi (*1*) 邵承熙.
Tung Li Tsuan Yao 東籬纂要.
Essentials of Chrysanthemum Culture [title
taken from the 'three rows of chrysan-
themums planted along the eastern fence' by
Thao Yuan-Ming].
1889.
Shen Yuan (*1*) 沈元.
'Chi Chiu Phien' Yen-Chiu 「急就篇」研究.
A Study of the Handy Primer (Han orthographic
word-list).
LSYC, 1962, **9** (no. 3), 61.
Abstr. RBS, 1969, **8**, no. 90.
Shih Shêng-Han (*2*) 石聲漢.
Ssu Min Yüeh Ling Chiao Chu 四民月令校注.
An Analytical Commentary on the Monthly
Ordinances for the Four Sorts of People (c. +160).
Chung-Hua, Peking, 1965.
Shih Sheng-Han (*3*) 石聲漢.
'Chhi Min Yao Shu' Chin Shih 齊民要術今釋.
A Modern Translation of Chhi Miu Yao Shu.
4 vols, Science Press, Peking, 1957.
Shih Shêng-Han (*4*) 石聲漢.
Tshung 'Chhi Min Yao Shu' Khan Chung-Kuo Ku-
Tai-ti Nung-Yeh Kho-Hsüeh Chih-Shih 從「齊民
要術」看中國古代的農業科學知識.
The Important Arts for the People's Welfare
[c. +540] and the Light it throws on the
Development of Chinese Agricultural
Industry and Knowledge.
Kho-Hsüeh, Peking, 1957.
Shih Tzu-Hsing (*1*) 石子興.
Chung-Kuo Pên Tshao Hsin Lun 中國本草
新論.
A Fresh Discussion of the Pharmaceutical
Natural History Literature of China.
BDS, 1946, **11** (no. 1), 23.
Shih Ya-Fêng (*1*) 施雅風.
Chung-Kuo Ku-Tai chih Thu Jang Ti Li 中國古
代之土壤地理.
Soil Science and Geography in Ancient China.
TFTC, 1944, **41** (no. 9), 33.
Shimada Shūjirō (*1*) 島田脩二郎.
Shōsai Baifu Teiyō 「松齋梅譜」提要.
A Study of the Pinetree Studio Treatise on the

Shimada Shūjirō (*1*) (*cont.*)
  Chinese Apricot (by Wu Thai-Su, *c.* +1352).
  BK, 1956, **20**, 211.
  Abstr. *RBS*, 1959, **2**, no. 343.
Shinoda Osamu (*4*) 篠田統.
  Tōshi Shokubutsu-Shaku 唐詩植物釋.
  Botany in Thang Poetry.
  Art. in Yabuuchi Kiyoshi (*25*), p. 341.
Shirai Mitsutarō (*1*) (tr.) 白井光太郎.
  Kokoyaku Honzō Komoku 國譯本草綱目.
  The *Pandects of Pharmaceutical Natural History* in
    Japanese (with elaborate indexes according to
    Japanese pronunciation).
  15 vols.
  Shunyōdō, Tokyo, 1929.
Shirai Mitsutarō (*1*) 白井光太郎.
  Shokubutsu Yōiko 植物妖異考.
  Malformations and Curiosities in Plants.
  Tokyo, 1914.
Sō Senshun & Shirao Kokujū (*1*) 曾槃,
    白尾國柱.
  Seikei Zusetsu 成形圖說.
  Illustrated Treatise on Plant Forms [a treatise
    on agricultural botany and wild food plants,
    commissioned in 1703 by the Lord of Satsuma
    and planned in 100 pên, but the blocks being
    twice destroyed by fire only 30 pên were
    actually produced].
  Yedo (Edo, Tokyo), 1804. Repr. in 4 vols. 1974.
  Cf. Bartlett & Shohara (*1*), pp. 68–9, 149,
    150–1.
Sugimoto Naojiro (*1*) 杉本直治郎.
  Kaku Gikyō no Kōshi 郭義恭の廣記 Kuo I-
    Kung and *Kuang Chih*.
  Tōyōshi Kenkyū, 東洋史研究, 1964, 23:3,
    88–107.
Sun Chia-Shan (*1*) 孫家山.
  Pên Tshao Hsüeh ti Chhi-Yuan chi chhi Fa-Chan
    本草學的起源及其發展.
  The Origins and Development of Pharmaceu-
    tical Natural History and Botany [in China].
  AHRA, 1959, **1**, 101.
Sun Tai-Yang & Liu Fang-Hsün (*1*) 孫岱陽,
    劉昉勳.
  Thien-chien-ti Tsa Tshao 田間的雜草.
  Miscellaneous Weeds of Arable Land.
  Kho-Hsüeh, Peking, 1955.
Sun Yün-Yü (*1*) 孫雲蔚.
  Hsi-Pei-ti Kuo Shu 西北的果樹.
  Fruit-Trees of the North-West.
  Kho-Hsüeh, Peking, 1962.
Sung Ta-Jen (*1*) 宋大仁.
  Chung-Kuo ho A-La-Po ti I Yao Chiao-Liu
    中國和阿拉伯的醫藥交流.
  Medical and Pharmaceutical Intercultural
    Contacts between China and the Arabs.
  LSYC, 1959, **5** (no. 1), 78.
Sung Ta-Jen (*6*) 宋大仁.
  Chung-Kuo Pên-Tshao Hsüeh Fa-Chan Shih Lüeh
    中國本草學發展史略.
  Towards a Historical Classification of
    Pharmaceutical Natural History and its

Development in China.
  MS in East Hsian History of Science Library,
    recd. 1978.
Suzuki Shūji (*1*) 鈴木修次.
  Nihon Kango to Chūgoku 日本漢語と中國.
  Japanese Expressions that entered Chinese.
  Tokyo, 1979.

Takeda Hisayoshi (*1*) 武田久吉.
  Genshoku Nihon Kōzan Shokubutsu Zukan 原色日
    本高山植物圖鑑.
  Alpine Flora of Japan, with Colour
    [Photograph] Illustrations.
  Hoikusha, Osaka, 1959.
Taki Mototane (*1*) 多紀元胤.
  I Chi Khao (*Iseki-Kō*) 醫籍考.
  Comprehensive Annotated Bibliography of
    Chinese Medical Literature (Lost or Still
    Existing).
  c. 1825, pr. 1831, repr. Tokyo, 1933, and
    Chinese-Western Medical Research Society,
    Shanghai, 1936 with introd. by Wang Chi-
    Min.
Takizawa Toshizuke (*1*) 瀧澤俊亮.
  '*Thu Shu Chi Chhêng*' *Fên Lei So Yin* 「圖書集
    成」分類索引.
  Classified Index to the *Imperial Encyclopaedia and
    Florilegium.*
  Shih Yu Wên Ko, Dairen, 1933.
Tamba Motokata (*1*) 丹波元堅.
  Ju I Ching Yao 儒醫精要.
  Essential Knowledge of the Learned Physician.
  Tokyo, *c.* 1840.
Tamba Motoyasu (*1*) 丹波元簡.
  I Shêng 醫賸.
  Medical Miscellany (lit. Medical
    Supererogations).
  Tokyo, 1809.
Tanaka Chyōzaburō (*1*) 田中長三郎.
  Kankitsu Bunrui Ronsō shi 柑橘分類論爭史.
  A History of the Disputes in *Citrus* Classification.
  SC, 1935, **7** (no. 1), 1.
Tanaka Tyōzaburō (*2*) 田中長三郎.
  Kankitsu Zenrui no Kenkyū ni tsuite 柑橘全類の
    研究に就こ.
  Results of Researches in the Classification of all
    the *Citrus* Species.
  SC, 1934, **6** (no. 2), 149.
Tanaka Yoshio & Ono Motoyoshi (*1*) 田中芳男,
    小野職愨; with illustrations by Hattori
    Setsusai 服部雪齋.
  Yūyō Shokubutsu Zusetsu 有用植物圖説.
  Illustrations and Descriptions of Useful Plants
    [arranged according to natural families, with
    characters and romanised forms of Japanese
    names, and Latin binomicals].
  Dai Nihon Nōkai (Agric. Soc. of Japan), Tokyo,
    1891.
  Companion Volume, all in English, *Useful Plants
    of Japan Described and Illustrated*, Tokyo, 1895.
  Index entitled *Yūyō Shokubutsu Zusetsu Mokuroku
    Oyobi Sakuin*, Tokyo, 1891; repr. 1902.

Tanaka Yoshio & Ono Motoyoshi (*1*) (*cont.*)
    See Merrill & Walker (1), p. 490; Bartlett &
      Shohara (1), pp. 143–4.
Than Ping-Chieh (*1*) 譚.
    Important Medical Plants of Szechuan (with
      Latin names as well as Chinese).
    JAAC, 1937, **155**, 106.
Thang Chhang-Ju (*1*) 唐長孺.
    Wei, Chin, Nan Pei Chhao, Shih Lun Tshung
    魏晉南北朝史論叢.
    A Compendious History of the Wei, Chin and
      Northern and Southern Dynasties Periods.
    Peking, 1957.
Thang Chhang-Ju (*2*) 唐長孺.
    *Wei, Chin, Nan Pei Chhao, Shih Lun Tshung, Hsü
    Pien* 魏晉南北朝史論叢續編.
    Continuation of the 'Compendious History of
      the Wei, Chin and Northern and Southern
      Dynasties Periods'.
    Peking, 1959.
Thang Lan-Chieh (*1*) 湯蘭階.
    Lan Lin Pai Chung 蘭林百種.
    The Forest of the Orchids [illustrated].
    1922; pr. 1939.
Thu Tao-Ho (*1*) 屠道和.
    Pên Tshao Hui Tsuan 本草彙纂.
    A Classified Materia Medica compiled (from the
      literature of pharmaceutical natural history).
    Pref. 1851; pr. 1863.
    LPC, no. 114; cf. Swingle (6).
Thu Yung-Ning (*1*) 屠用寧.
    Lan Hui Ching 蘭蕙鏡.
    Mirror of *Lan* and *Hui* Orchids.
    1811.
Ting Chi-Min (*1*) 丁濟民.
    *Yuan Chhien 'Ching-Shih Chêng-Lei Ta-Kuan Pên
    Tshao' Pa* 元槧「經史證類大觀本草」
    跋.
    A Postface for the Yuan edition of the *Classified
    and Consolidated Pharmacopoeia of the Ta-Kuan
    reign-period*.
    ISTC, 1948, **2** (nos. 1/2), 26.
Ting Chi-Min (*2*) 丁濟民.
    Pa Ming Chin-ling Khan Pên 'Pên Tshao Kang Mu'
    跋明金陵刊本「本草綱目」.
    A Postface for the (first,) Chin-ling (i.e.
      Nanking), edition of the *Pandects of Pharma-
    ceutical Natural History* (+1596).
    ISTC, 1948, **2** (nos. 3–4), 39.
Ting Fu-Pao (*1*) 丁福保.
    Hua-Hsüeh Shih-Yen Hsin Pên Tshao 化學實
    驗新本草.
    A New Pharmaceutical Natural History based
      on Chemistry and Biochemistry.
    I-Hsüeh Shu Chü, Shanghai, 1934.
Ting Fu-Pao & Chou Yün-Chhing (*3*) 丁福保,
    周雲青.
    Ssu Pu Tsung Lu I Yao Pien 四部總錄醫藥篇.
    Bibliography of Medical and Pharmaceutical
      Books to supplement the *Ssu Khu Chhüan Shu*
    encyclopaedia.
    3 vols.

Com. Press, Shanghai, 1955.
Ting Kuang-Chhi & Hou Khuan-Chao (*1*)
    丁廣奇, 侯寬昭.
    Chih-Wu Chung Ming Shih 植物種名釋.
    Chinese Equivalents of Latin Botanical
      Nomenclature [based on Zimmer (1) and
      Bailey (1)].
    Kho-Hsüeh, Peking, 1957.
Ting Wei-Liang (*1*) (= W. A. P. Martin) 丁韙良.
    Ko Wu Ju Mên 格物入門.
    Introduction to Natural Philosophy.
    Peking, 1868.
Tsêng Chao-Yü, Chiang Pao-Kêng & Li Chung-I
    (*1*) 曾昭燏, 蔣寶庚, 黎忠義.
    I-nan Ku Hua Hsiang Shih Mu Fa-Chüeh Pao-Kao
    沂南古畫像石墓發掘報告.
    Report on the Excavation of an Ancient [Han]
      Tomb with Sculptured Reliefs at I-nan [in
      Shantung] (*c.* +193).
    Nanking Museum, Shantung Provincial Dept. of
      Antiquities, and Ministry of Culture,
      Shanghai, 1956.
Tsêng Yü-Lin (*1*) 曾育麟.
    *Tui 'Tien-Nan Pên Tshao' ti Khao-Chêng yü Chhu-
    Pu Phing-Chia ti Liang Tien Shang-Chüeh* 對
    「滇南本草」的考證與初步評價的兩點
    商榷.
    A Critique of (Yü Nai-I & Yü Lan-Fu's)
      'Preliminary Evaluation of the *Pharmaceutical
    Natural History of Yunnan*' on Two Points of
    Difference.
    ISTC, 1958, **9** (no. 1), 59.
    Reply by Yü & Yü, *ISTC*, 1958, **9** (no. 12), 136.
Tsêng Yün-Chhien (*1*) 曾運乾.
    Shang Shu Chêng Tu 尚書正讀.
    Commentary on an Emended Text of the *Shang
    Shu* (the *Shu Ching*, Historical Classic).
    Chung-Hua, Peking, 1964.
Tshai Ching-Fêng (*1*) 蔡景峯.
    *Shih Lun Li Shih-Chen chi chhi tsai Kho-Hsüeh-shang-
    ti Chhêng-Chiu* 試論李時珍及其在科學
    上的成就.
    A Study of the Contributions to Science made
      by Li Shih-Chen in his *Pên Tshao Kang Mu*).
    KHSC, 1964 (no. 7), 63.
Tshao Ping-Chang (*1*) 曹炳章.
    Tshao Pa 曹跋.
    A Postface (to the *Encyclopaedia of Chinese Materia
    Medica*). [see Chhen Tshun-Jen (1). The
      postface is a brief account of the history of the
      Chinese literature of pharmaceutical natural
      history.]
Tshao Wan-Ju (*1*) 曹婉如.
    *Wu Tsang Shan Ching ho 'Yü Kung' chung ti Ti-Li
    Chih-Shih* 五藏山經和禹貢中的地理
    知識.
    On the Geographical Knowledge found in
      the [first five chapters of the] *Classic of the
    Mountains and Rivers* and in the *Tribute of Yü*.
    KHSC, 1958, **1** (no. 1), 77.
Tshen Chung-Mien (*2*) 岑仲勉.
    Huang-Ho Pien Chhien Shih 黃河變遷史.

Tshen Chung-Mien (2) (cont.)
History of the Changes of Course of the Yellow
River.
Jen-Min, Peking, 1957.
Tshui Yu-Wên (1) 崔友文.
Hua-Pei Ching-Chi Chih-Wu Chih Yao 華北經
濟植物誌要.
Important Economic Plants of North China.
Kho-Hsüeh, Peking, 1953.
Tsou Chu (1) 鄒澍.
Pên Ching Su Chêng 本經疏證.
Critical Commentary on (a Revised Text of)
the Classical Pharmacopoeia of the Heavenly
Husbandman.
1837 pr. 1849.
With two supplements:
Pên Ching Hsü Su 本經續疏.
1839, pr. 1849.
And Pên Ching Hsü Su Yao 本經續疏要.
1840 pr. 1849.
LPC, no. 11; cf. Swingle (6).
Tsutsumi Tomekichi (1) 堤留吉.
Haku Rokuten to Take 白樂天と竹.
Pai Chü-I and Bamboos.
TYBK, 1960, 8, 21.
Abstr. RBS, 1968, 7, no. 513.
Tu Wên-Lan (1) 杜文瀾.
I Lan Ssu Shuo 藝蘭四說.
Four Explanations on the Cultivation of
Orchids.
c. 1860.
Tu Ya-Chhüan, Tu Chiu-Thien et al. (1) 杜亞泉,
杜就田.
Tung Wu Hsüeh Ta Tzu Tien 動物學大辭典.
A Zoological Dictionary.
Com. Press, Shanghai, 1932; repr. 1933.
Tun Li-Chhen 敦禮臣
See Fu-Chha Tun-Chhung (Fuchha was a
Manchu).
Tung Kao (1) (ed.) 董誥 et al.
Chhüan Thang Wên 全唐文.
Collected Literature of the Thang Dynasty.
1814.
Cf. des Rotours (2), p. 97.
Tung Thung-Ho (1) 董同龢.
Chhieh Yün Chih Chang Thu chung Chi-Ko wên-thi
「切韻指掌圖」中幾個問題.
The Problem of the Authorship and Dating of
the Tabular Key to the Dictionary of Characters
arranged according to their Sounds when Split.
AS/BIHP, 1948, 17, 193.
Tsuyama Takashi (1) = (1) 津山尚.
Nihon no Tsubaki 日本の椿.
Camellias of Japan.
Takedo Science Foundation & Hirokawa,
Osaka, 1968.
Tsuyama Takashi & Futakuchi Yoshio (1) 津山尚,
二口善雄.
Nihon Tsubaki-shū 日本椿集.
The Camellia Cultivars of Japan.
Heibon-sha, Tokyo, 1966.
Tuyama Takashi et al. (1) 津山尚.

Gendoi Tsubaki-shū 現代椿集.
Encyclopaedia of Camellias in Colour.
Camellia Society of Japan, Tokyo, 1972.

Wada Toshihiko (1) (ed.) 和田利彦.
Shao-Hsing Chiao-Ting Ching-Shih Chêng Lei Pei-Chi
Pên Tshao 紹興校定經史證類備急
本草.
Facsimile Edition of the Corrected, Classified and
Consolidated Armamentarium; Pharmacopoeia of the
Shao-Hsing reign-period, from a +12th-century
Manuscript (perhaps +1159) preserved in
the Omori Memorial Library of the Kyoto
Botanic Gardens.
Shunyōdō, Tokyo, 1933.
Cf. Karow (2).
Wakahama Sekiko (1) 若浜汐子.
'Manyōshu' Shokubutsu Gaisetsu 萬葉集植物
概説.
Explanations and Identifications of the Plants
mentioned in the Anthology of a Myriad Leaves.
Tokyo, 1959.
Wakahama Sekiki (2) 若浜汐子.
'Manyōshu' Shokubutsu Zenkai 萬葉集植物
全解.
A Complete Study of the Plants mentioned in
the Anthology of a Myriad Leaves.
Tokyo, 1959.
Wan Kuo-Ting (1) 萬國鼎.
Fan Shêng-Chih Shu Chi Shih 氾勝之書輯釋.
Explanations of the Most Important Matters in
the (Early Han) Book of Fan Shêng-Chih (on
Agriculture).
Chung-hua, Peking, 1957.
Wan Kuo-Ting (2) 萬國鼎.
Chung-Kuo Ku-Tai tui-yü Thu-Jang Chung-Lei chi
Chhi Fên-Pu ti Chih-Shih 中國古代對於土
壤種類及其分布的知識.
On the Knowledge of Soil Types and their (Geo-
graphical) Distribution in Ancient China.
NACJ, 1956, 1, 101.
Wang Chen-Ju (1) 汪振儒.
Chha-Erh Lin-Nei (+1707–78) Shih Lüeh 卡爾
林內 (1707—1778) 事略.
A Biobibliography of Carolus Linnaeus.
KHSC, 1958, 1, 11.
Wang Chi-min (1) 王吉民.
Tsu Kuo II Yao Wên-Hua Liu Chhuan Hai-Wai
Khao 祖國醫藥文化流傳海外考.
On the Transmission of Chinese Medical
Culture beyond the Seas.
ISTC, 1957, 8 (no. 1), 8
Wang Chi-min (2) 王吉民.
Li Shih-Chen Wên-Hsien Chan-Lan Hui Thê-Khan
李時珍文獻展覽會特刊.
Catalogue of the Exhibition on the Contri-
butions of Li Shih-Chen.
Museum of the History of Medicine, Shanghai,
1954.
Wang Chi-Min (3) 王吉民.
Kuan-Yü Chin-Chi-Na Chhuan Ju Wo Kuo Ti Chi
Tsai 關於金雞納傳入我國的紀載.

Wang Chi-Min (3) (cont.)
A Memoir on the Introduction of Chin-chi-na (Cinchona, Quinine) to China.
ISTC, 1954, **6** (no. 1), 28.

Wang Chia-Yin (1) 王嘉蔭.
'*Pên Tshao Kang Mu*' *ti Kung Wu Shih Liao* 「本草綱目」的礦物史料.
Historical Materials on the Mineralogy of the *Great Pharmacopoeia* [by Li Shih-Chen, +1596].
Ko-Hsüeh, Peking, 1957.

Wang Chün-Mo (1) 王筠默.
'*Chêng Lei Pên Tshao*' *yü* '*Pên Tshao Yen I*' *ti chi-ko Wên-Thi* 「證類本草」與「本草衍義」的幾個問題.
On the Question of the Relations between the *Classified Pharmaceutical Natural History* (+1094) and the *Dilations upon Pharmaceutical Natural History* (+1116).
ISTC, 1954, **6** (no. 4), 242.

Wang Chün-Mo (2) 王筠默.
Tshung '*Chêng Lei Pên Tshao*' *Khan Sung-Tai Yao-Wu Chhan-Ti ti Fên-Pu* 從「證類本草」看宋代藥物產地的分佈.
On the Distribution of the Places of Origin of Sung Materia Medica as seen from the *Classified . . . Pharmaceutical Natural History* (with map).
ISTC, 1958, **9** (no. 2), 114.

Wang Chung-Lo (2) 王仲犖.
Wei, Chin, Nan Pei Chhao, Sui, Chhu Thang Shih 魏晉南北朝隋初唐史.
A History of the Wei, Chin, Northern and Southern Dynasties, Sui and Early Thang Periods.
Jenmin, Shanghai, 1961.

Wang Chung-Min (1) 王重民.
Tun-huang Ku Chi Hsü Lu 敦煌古籍叙錄.
Descriptive Catalogue of the Old Manuscripts found at (the) Tunhuang (Cave-Temples).
Com. Press, Peking and Shanghai, 1958.

Wang Chung-Min (2) 王重民.
Chao Hsüeh-Min Chuan 趙學敏傳.
A Biography of Chao Hsüeh-Min [pharmaceutical naturalist and chemist, c. +1725– c. 1804].
ISTC, 1951, **3** (no. 3), 43.

Wang Chung-Min (3) 王重民.
Pên Tshao Ching Yen Lu 本草經眼錄.
Notes on Some Pharmaceutical Natural Histories.
ISTC, 1952, **4** (no. 1) 31; **4** (no. 3) 157.

Wang Chung-Min & Yuan Thung-Li (1) 王重民, 袁同禮.
Mei-Kuo Kuo-Hui Thu-Shu-Kuan Tshang Chung-Kuo Shan Pên Shu Mu 美國國會圖書館中國善本書目.
A Descriptive Catalogue of Rare Chinese Books in The Library of Congress (at Washington, D.C.).
2 vols. Library of Congress, Washington, D.C., 1957.

Wang Hsün-Ling (1) 王勛陵.

Shih Lun Chung-Kuo Ku-Tai-ti Shêng Thai Ti Chih-Wu Hsüeh 試論中國古代的生態地植物學.
An Exploration of the Decology and Geobotany in Ancient China.
CJE, 1982, **3**, 38.

Wang Kuang-Wei (1) 王光瑋.
'*Yü Kung*' *Thu-Jang-ti Than Thao* 「禹貢」土壤的探討.
A Critical Discussion of the Soil Types mentioned in the 'Tribute of Yu' [chapter of the *Historical Classic*].
YK, 1935, **2** (no. 5), 14.

Wang Kuo-Wei (5) 王國維.
Wu Tai Chien Pên Khao 五代監本考.
A Study of the (First) Printing (of the Classics) by the (Imperial) University in the Wu Tai Period.
KHCK, 1923, (no. 1), 139.

Wang Kuo-Wei (6) 王國維.
'*Tshang Chieh Phien*' *Tshan Chien Khao Shih* 「蒼頡篇」殘簡考釋.
A Study of the Fragmentary Bamboo Slips bearing parts of the Text of the *Word-List of Tshang Chieh* [by Lissu].
In Wang shih Kao, p. 159 王氏稿.

Wang Ta (1) 王達.
Kuan Tzu Ti Yuan Phien ti Chhü Hsing Than Thao 管子地員篇的區性探討.
A Critical Discussion of the Natures of the Regions mentioned in the 'Variety of Earth's Products' Chapter in the *Kuan Tzu* book (arising from the book of Hsia Wei-Ying, 2).
AHRA, 1960, **2**, 207.

Wang Yü-Hsin (1) 王與新.
Wo Kuo Nung-Yeh Kai-Khuang 我國農業概況.
A Brief Survey of Chinese Agriculture.
Nung-Yeh, Peking, 1964.

Wang Yü-Hu (1) 王毓瑚.
Chung-Kuo Nung Hsüeh Shu Lu 中國農學書錄.
Bibliography of Agricultural Books.
Chung-Hua, Peking, 1957; 2nd revised and enlarged ed.
Nung-Yeh, Peking, 1964, 1979.

Watanabe Kiyohiko (1) 渡邊清彥.
Nan-Fang Chhüan Yu-Yung Chih-Wu Thu Shuo; 1, Yao Yung Chih-Wu 南方圈有用植物圖說; 1, 藥用植物.
Illustrated Manual of the Useful Flora of the Southern Regions; Vol. 1, Medicinal Plants [194 entries].
Japanese Army Publishing Bureau.
[Botanic Gardens], Singapore, 1944.
Index in Latin and Japanese, with Sino-Japanese Medical Index.

Watanabe Kiyohiko (2) 渡邊清彥.
Nan-Fang Chhüan Yu-Yung Chih-Wu Thu Shuo; 2, Shih Yung Chih-Wu 南方圈有用植物圖說; 2, 食用植物.
Illustrated Manual of the Useful Flora of the

Watanabe Kiyohiko (2) (cont.)
Southern Regions; Vol. 2, Edible Plants [700
entries].
Japanese Army Publishing Bureau.
[Botanic Gardens], Singapore, 1945.
Indexes in Latin and Japanese.

Watanabe Kōzō (3)　渡邊幸三.
*Lo Chen-Yü Tunhuang Pên 'Pên Tshao Chi Chu' Hsü
Lu Pa ti Shang Chüeh*　羅振玉敦皇「本草集
注」序錄跋的商榷.
A Critique of the Exactness of Lo Chen-Yü's
Postface to the Publication of the Tunhuang
copy of (Thao Huang-Ching's) *Collected
Commentaries on the Pharmacopoeia (of the
Heavenly Husbandman)*.
ISTC, 1957, **8** (no. 4), 310.
Tr. from the Japanese by Wang Yu-Shêng
王有生.

Wei Lien-Chhen (1)　韋廉臣.
(= Alexander Williamson)
Ko Wu Than Yuan　格物探原.
An Enquiry into the Principles of Natural
Philosophy.
Shanghai, 1876, 1880.

Wên Huan-Jan & Lin Ching-Liang (1)　文煥然,
林景亮.
*Chou Chhin Liang Han Shih-Tai Hua-Pei Phing-yuan
yü Wei-Ho Phing-yuan Yen Chien Thu ti Fên-Pu
chi Li-Yung Kai-Liang*　周秦兩漢時代華
北平原與渭河平原鹽碱土的分佈及
利用改良.
The Distribution and Reclamation of the Saline-
Alkaline Soils of the North China Plain and
the Wei River Plain during the Chou, Chhin
and two Han Dynasties.
APS, 1964, **12** (no. 1), 1.

Wen Thing-Shih　文廷式.
Pu Chin Shu I Wen Chih　補晉書藝文志.
Additions to the I Wen Chih of Chin Shu
c. 1888.
Erh Shih Wu Shih Pu Pien　二十五史補編.
Additions to the Twenty-Five Histories.
Chung Hua, Peking, 1956.

Wên Thing-Shih (1)　文廷式.
Pu Chin Shu I Wên Chih　補晉書藝文志.
A Bibliography to supplement the 'History of
the Chin Dynasty'.
c. 1888.
In *Erh-shih-wu Shih Pu Pien*, vol. 3 (p. 3703).

Wên Yu (1)　聞宥.
Szechuan Han-Tai Hua Hsiang Hsüan Chi
四川漢代畫象選集.
A Collection of the Han Reliefs of Szechuan
(album).
Chhün-Lien, Shanghai, 1955.

Wu Chêng-I (1)　吳徵鎰.
*Chung-Kuo Chih-Wu Li-Shih Fa-Chan-ti Kuo-Chhêng
ho Hsien-Khuang*　中國植物歷史發展的
過程和現況.
Stages in the Development of Botany in China
and its Present State.
KHTP, 1953, **2**, 12.

Wu Chhi-Chün (2)　吳其濬.
Chih Wu Ming Shih Thu Khao Chhang Phien
植物名實圖考長編.
Comprehensive Treatise on the Names and
Natures of Plants.
Peking, 1848.
Repr. Com. Press. Shanghai, 1919 (with index).

Wu Chhuan-Yün (1)　吳傳澐.
I Lan Yao Chüeh　藝蘭要訣.
The Chief Points about the Cultivation of
Orchids [illustrated].
c. 1860.

Wu Ên-Yuan & Thang Tho (1)　吳恩元,唐駝.
Lan Hui Hsiao Shih　蘭蕙小史.
Brief Natural History of *Lan* and *Hui* Orchids
[illustrated].
Chung-Hua, 1923.

Wu Hsiang-Kuang and Huang Mei-Chên　鄔祥光,
黃美貞.
Study on the Spatial Distribution of Population
of Oriental Armyworm. 粘蟲種羣空間
結構的探討.
Acta Ecologica Sinica　生態學報, 1982, **2**,
39–46.

Wu Khuan-Chao & Chhien Chhung-Shu (1)
吳寬昭, 錢崇澍.
Chung-Kuo Tsai-Phei-ti An Shu　中國栽培的
桉樹.
Cultivated Eucalypts of China.
Hsin-hua, Peking, 1954 (S. China Bot. Res. Bur.
Acad. Sin. monographs (3rd ser.), no. 1).

Wu Shih-Khai (1) (rev. & ed.)　吳世鎧.
Shen Nung Pên Tshao Ching Su Chi Yao　神農本
草經疏輯要.
Essentials of the *Commentary* [of Miu Hsi-Yung]
on the *Text of the Classical Pharmacopoeia of the
Heavenly Husbandman*.
1809.

Wu Tê-Lin (1)　吳德鄰.
*Chhüan Shih Wo Kuo Tsui-tsao-ti Chih-Wu Chih—
Nan Fang Tshao Mu Chuang*　詮釋我國最早
的植物誌—南方草木狀.
A Commentary on Hsi Han's 'Account of the
Plants and Trees of the Southern Regions; our
oldest Botanical Monograph.
CWHP, 1958, **7**, 27.

Yabuuchi Kiyoshi (26) (ed.)　藪內清.
Sō Gen Jidai no Kagaku Gijutsushi　宋元時代の
科學技術史.
(Essays) On the History of Science and
Technology in the Sung and Yuan Periods.
Jimbun Kagaku Kenkyusō, Kyoto, 1967; repr.
1970.

Yang Chung-Pao (1)　楊鍾寶.
Hung Ho Phu　珥荷譜.
A Treatise on the Cultivation of the Lotus in
Huge Pots.
1808.

Yang Min (1)　楊旻.
Ku Chin Shih Wu Kho-Hsüeh Tsa Than　古今事
物科學雜談.

Yang Min (*1*) (*cont.*)
Miscellaneous Talks on Scientific Matters, Old and New.
Com. Press. Peking, 1959 repr. 1960.

Yang Phei (*1*)　楊沛.
Huang kan i shen wu hsüeh thê hsing chi chhi yung yü fang chih kan chü hai chhung tê chhu pu yen chiu
黃柑蟻生物學特性及其用于防治柑橘害虫的初步研究.
A Preliminary Study of the Biological Characteristics of the Yellow Citrus Ant and its use in Controlling Pests of Oranges.

Yang Shu-Ta (*1*)　楊樹達.
Han Shu Khuei Kuan　漢書窺管.
A Microscope for the Text of the *History of the (Former) Han Dynasty* (commentary and textual criticism).
Kho-Hsüeh, Peking, 1955.

Yao Chen-Tsung (*1*)　姚振宗.
'*Hou Han (Shu)*' *I Wên Chih*　「後漢書」藝文志.
Bibliography of Books mentioned in the *History of the Later Han Dynasty*.
1895.
In *Erh-shih-wu Pien*, vol. 2 (p. 2305).

Yeh Ching-Yuan (*2*)　葉靜淵.
Chung-Kuo Nung-Hsüeh I-Chhan Hsüan Chi; 中國農學遺產選集; *Kan Chü*　柑橘.
Anthology of Quotations [illustrating the History of] Chinese Agricultural Science and Production; Citrus Fruits Vol. 1.
Chung-Hua, Peking, 1958.
Joint Committee on Agricultural History (National Academy of Agriculture and Nanking College of Agriculture), ser. A, no. 14.

Yao Chen-Tsung (*3*)　姚振宗.
'*Sui Shu*' *Ching Chi Chih Khao Chhêng*　「隋書」經籍志考證.
Researches on the Bibliography of the *History of the Sui Dynasty*.
1897.
In *Erh-shih-wu Pu Pien*, vol. 4 (p. 5039).

Yeh Ching-Yuan (*1*)　葉靜淵.
Chung-Kuo Wên Hsien Shang-ti Kan Chü Tsai Phei　中國文獻上的柑橘栽培.
On the Cultivation of Citrus Fruits according to the Evidence of (Ancient and Medieval) Chinese Literature.
AHRA, 1958, **1**, 109.

Yeh Ching-Yuan (*2*)　葉靜淵.
Chung-Kuo Nung-Hsüeh I-Chhan Hsüan Chi; 中國農學遺產選集; *Kan Chü*　柑橘.
Anthology of Quotations [illustrating the History of] Chinese Agricultural Science and Production; Citrus Fruits Vol. 1.
Chung-Hua, Peking, 1958.
Joint Committee on Agricultural History (National Academy of Agriculture and Nanking College of Agriculture), ser. A, no. 14.

Yen Chieh (*1*) (ed.)　嚴杰.
Huang Chhing Ching Chieh　皇清經解.

Collection of [more than 180] Monographs on Classical Subjects written during the Chhing Dynasty.
1829; 2nd ed. Kêng Shen Pu Khan, 1860.
Cf. Wang Hsien Chien (*1*).

Yen Tun-Chieh (*21*)　嚴敦傑.
Hsü Kuang-Chhi　徐光啓.
A Biography of Hsü Kuang-Chhi.
Art. in *Chung-Kuo Ku-Tai Kho-Hsüeh Chia*, ed. Li Nien (*27*), 2nd. ed. p. 181.

Yen Yü (*1*)　燕羽.
Chung-Kuo Li-Shih shang ti Kho [Hsüeh] Chi [-Shu] Jen Wu　中國歷史上的科技人物.
Lives of [twenty-two] Scientist and Technologists eminent in Chinese History [including e.g. Chang Hêng, Tsu Chhung-Chih, Yü Yün-Wên and Yüwên Khai].
Chün-Lien, Shanghai, 1951

Yen Yü (*5*)　燕羽.
Shih-liu Shih-Chi ti Wei Ta Kho-Hsüeh Chia; Li Shih-Chen　十六世紀的偉大科學家; 李時珍.
A Great Scientist of the +16th Century; Li Shih-Chen (pharmaceutical naturalist).
Essay in Li Kuang-Pi & Chhien Chün-Yeh (*1*), p. 314.
Peking, 1955.

Yü Chia-Hsi (*1*)　余嘉錫.
Ssu Khu Thi Yao Pien Chêng　四庫提要辨證.
A Critical Study of the Annotations in the 'Analytical Catalogue of the *Complete Library of the Four Categories* (of Literature)'.
Peking, 1937; repr. 1958.

Yü Nai-I & Yü Lan-Fu (*1*)　于乃義, 于蘭馥.
'*Tien Nan Pên Tshao*' *ti Khao-Chêng Yü Chhu-Pu Phing-Chia*　「滇南本草」的考證與初步評價.
A Study of the *Pharmaceutical Natural History of Yunnan* (+1436) and a Preliminary Analysis of its Value.
ISTC, 1957, **8** (no. 1), 24.
With critique by Tsêng Yü-Lin　曾育麟.
ISTC, 1958, **9** (no. 1), 59.
and reply by Yü Lan-Fu　于蘭馥.
ISTC, 1958, **9** (no. 2), 136.

Yü Tê-Chün (*2*)　喻德浚.
Chung-Kuo chih Chhing Wei　中國之薔薇.
The Roses of China.
JBSC, 1935, **2**, 501.

Yü Tê-Chün *et al.* (*1*)　俞德浚.
Hua-Pei Hsi-Chien Kuan Shang Chih-Wu　華北習見觀賞植物.
Familiar Plants of North China.
3 vols.
Kho-Hsüeh, Peking, 1958.

Yü Tê-Chün & Fêng Yao-Tsung (*1*)　俞德浚, 馮耀宗.
Yunnan Shan Chha Hua Thu Chih　雲南山茶花圖誌.
Illustrated Handbook of the *Camellia* (Species and Varieties) of Yunnan.
Kho-Hsüeh, Peking, 1958.

Yu Yü (*1*) 友于.
'*Kuan Tzu*' *Tu Ti Phien Than Wei* 管子度
地篇探微.
A Minute Investigation of the [57th Chapter
of the] *Kuan Tzu* Book: 'Consideration of
Topography and Hydrology for Land Use'.
AHRA, 1959, **1**, 1.
Cf. Rickett (1), p. 72.
Yu Yü (*2*) 友于.
'*Kuan Tzu*' *Ti Yuan Phien Yen-Chiu* 管子地
員篇研究.

Researches on the [58th Chapter of the] *Kuan
Tzu* Book; 'On the Variety of what Earth
produces'.
AHRA, 1959, **1**, 17.
Yü Yün-Hsiu (*1*) 余雲岫.
Ku-Tai Chi Chi Ping Ming Hou Su I 古代疾
病名候疏義.
Explanations of the Nomenclature of Diseases in
Ancient Times.
Jen-Min Wei-Shêng, Shanghai, 1953; Rev.
Nguyen Tran-Huan, *RHS*, 1956, **9**, 275.

C. BOOKS AND JOURNAL ARTICLES IN WESTERN LANGUAGES

ADANSON, MICHEL (1). *Familles des Plantes.* 2 vols. Vincent, Paris, 1763. 2nd. ed. of vol. 1 only, posthumous; with many additions *Familles Naturelles; Première Partie, comprenant l'Histoire de la Botanique*, ed. A. Adanson & J. Payer, Paris, 1847 (1864).

ADLUNG, KARL G. (1). A Critical Evaluation of the European Research on Use of Red Wood Ants for the Protection of Forests against Harmful Insects. *Zeitschrift für Angewandte Entomologie* 1966, **57**, pp. 167–183.

AINSLIE, W. (1). *Materia Indica; or, some Account of those Articles which are employed by the Hindoos and other Eastern Nations in their Medicine, Arts and Agriculture; comprising also Formulae, with Practical Observations, Names of Diseases in various Eastern Languages, and a copious List of Oriental Books immediately connected with General Science, etc. etc.* 2 vols. Longman, Rees, Orme, Brown & Green, London, 1826.

ALLAN, MEA (1). *Plants that changed our Gardens.* David & Charles, Newton Abbot, 1974.

ALLEN, B. SPRAGUE (1). *Tides in English Taste (+1619–1800); a Background for the Study of Literature.* 2 vols. Harvard Univ. Press, Cambridge, Mass. 1937.

ALLEN. R. C. (1). 'The Influence of Aluminium on the Flower Colour of *Hydrangea macrophylla.*' *CONBT*, 1943, **13**, 201.

ALLEY, REWI (9). 'A Visit to Hsishuangbana and the Thai Folk of Yunnan.' *EHOR*, 1966, **5** (no. 5), 6.

ALLEY, REWI (13), (tr.). *Bai Juyi [Pai Chü-I]: Two Hundred Selected Poems, translated by R. A.* New World, Peking, 1983.

ALPAGUS, ANDREAS (1) (tr.).
De Limonibus Tractatus Embitar [Ibn al-Bayṭār] *Arabis per Andream Bellunensem Latinitate Donatus.* Paris, 1602.
In *Ebenbitar Tractatum de Malis Limoniis Commentaria Pauli Valcarenghi.* Cremona, 1758.

ANDERSON, E. B., FINNIS, V., FISH, M., BALFOUR, A. P. & WALLIS, M. (1). *The Oxford Book of Garden Flowers*, with illustrations by B. Nicholson, Oxford, 1970.

ANDERSON, EDGAR (1). *Plants, Man and Life.* Melrose, London, 1954.

ANON (76); a Botanical Society at Lichfield. [Erasmus Darwin, Sir Brooke Boothby & John Jackson, with advice from Dr Samuel Johnson.] *A System of Vegetables, according to their Classes, Orders, Genera and Species, with their Characters and Differences....* translated from the 13th ed. of the *Systema Vegetabilium* of the late Professor Linneus, and from the *Supplementum Plantarum* of the present Professor Linneus. 2 vols. Jackson, Leigh & Sotheby. Lichfield, 1783. The principal editor and translator was Erasmus Darwin himself.

ANON. (77). [Anonymus Traveller & Johann von Cube.]. *The German Herbarius*, beginning "Offt und vil hab ich" [= Herbarius zu Tentsch, German Ortus Sanitatis, smaller Ortus, Johann von Cube's Herbal, etc.] [Schöffer], Mainz, 1485. Another ed. Augsburg, 1485.

ANON. (79). *General Alphabetical List of Chinese Medicines* (1884–5). Inspectorate-General of Customs, Shanghai, 1889.

ANON. (102). *Historical Note on the Opium Poppy in China.* Statistical Dept. of the Inspectorate-General of Customs, 1889.

ANON. (195). 'The Chinese Predatory Stink-bug' [used in biological plant protection]. *CPICT*, 1983 (no. 5), 16.

ANON. & SPINNING JENNY (ps.) (1). 'The China Grass Plant.' *NQCJ*, 1868, **2**, 24; 1870, **4**, 123.

APPLEBY, J. H. (1). 'Ginseng and the Royal Society.' *NRRS*, 1983, **37** (no. 2), 121. (No sinological background material.)

ARBER, AGNES (1). 'Analogy in the History of Science.' In *Studies and Essays in the History of Science and Learning.* Sarton Presentation Volume, Schuman, New York, 1944.

ARBER, AGNES (2). *The Natural Philosophy of Plant Form.* Cambridge, 1950.

ARBER, AGNES (3). *Herbals, their Origin and Evolution; a Chapter in the History of Botany, 1470 to 1670.* Cambridge, 1912. 2nd edn. greatly revised and enlarged, 1938, repr. 1953.

ARCHER, M. (1). *Natural History Drawings in the India Office Library.* H.M.S.O. London, 1962. Rev. J. Théodoridès, *A/AIHS*, 1965, **18**, 122.

ARMSTRONG M. & THORNBER, J. J. (1). *Field-book of Western [North American] Wild Flowers.* Putnam, New York and London, 1915. Often repr.

ASH, H. B. (1) (tr.). *Lucius Junius Moderatus Columella 'On Agriculture', with a Recension of the Text and an English Translation.* 3 vols. Heinemann, London, 1948. (Loeb Classics Edition.)

ATTIRET, J. D. (1). *A Particular Account of the Emperor of China's Gardens near Pekin: in a Letter from F. Attiret, a French Missionary, now employ'd by that Emperor to paint the Apartments in those Gardens, to his Friend at Paris.* Translated from the French [Letters Édif. et Cur. vol. 3, pp. 786 ff.], by Sir Harry Beaumont pseudonym for J. Spence. Dodsley & Cooper, London, 1752.

Aurousseau, L. (4). Review of G. Maspero's *Le Royaume de Champa* (Brill, 1914, repr. from *TP*, 1910–13), including many textual passages and translations concerning Lin-I. *BEFEO*, 1914, **14** (no. 9), 8.

Avery, A. G., Satina, Sophie & Rietsema, Jacob (1). *Blakeslee: The Genus 'Datura'*. Ronald, New York, 1959.

de Bach, Paul. *Biological Control by Natural Enemies*. Cambridge University Press, 1974.

Badham, C. D. (1). *A Treatise on the Esculent Funguses of England, containing an Account of their Classical History, Uses, Characters, Development, Structure, Nutritious Properties, Modes of Cooking and Preserving, etc.* Reeve, London, 1847.

Bailey, Sir Harold (4). '*Madu*; a Contribution to the History of Wine.' Art. in Silver Jubilee Volume of the Jimbun Kagaku Kenkyūsō, Kyoto University. Kyoto, 1954, p. 1.

Bailey, Liberty H. (1).
How Plants get their Names
Macmillan, New York, 1933.
Photolitho reproduction, Dover, New York, 1963.

Bailey, Liberty H. (2). '*Hosta*, the Plantain Lilies'. *GH*, 1930, **2**, 119.

Bailey, Liberty H. (3). *Standard Cyclopaedia of Horticulture*. 6 vols. New York, 1914–17; 2nd ed. 1928.

Balazs, E. (=S.) (9). 'Historical Compilations as Guides to Bureaucratic Practice—the Monographs [in the Dynastic Histories], Encyclopaedias, and Collections of Statutes' (in French). Contribution to the Far East Seminar in the Conference on Asian History, London School of Oriental Studies, July 1956. Published as: 'L'Histoire comme Guide de la Pratique Bureaucratique (les Monographies, les Encyclopédies, les Recueils des Statuts', in *Historians of China and Japan* ed. W. G. Beasley & E. G. Pulleyblank. Oxford Univ. Press, 1961, p. 78.

Balazs, E. (=S.) (10). *Political Theory and Administrative Reality in Traditional China*. Luzac, for School of Oriental Studies, Univ. of London, London, 1965. (Three lectures given in 1963.)

Balfour, F. H. (1) (tr.). *Taoist Texts, ethical, political, and speculative* (incl. *Tao Tê Ching, Yin Fu Ching, Thai Hsi Ching, Hsin Yin Ching, Ta Thung Ching, Chih Wên Tung, Chhing Ching Ching, Huai Nan Tzu* ch. 1, *Su Shu* and *Kan Ying Phien*). Kelly and Walsh, Shanghai, n.d. but prob. 1884.

Balss, H. (2). *Albertus Magnus als Zoologe*. Münchner Drucke, München, 1928. (*Münchener Beiträge Z. Gesch, u. Lit. a. Naturwiss. u. Med.* no. 11/12.)

Balss, H. (3). *Albertus Magnus als Biologe; Werk und Ursprung*. Wissenschaftliche Verlagsgesellschaft, Stuttgart, 1947.

Bartholomaeus Anglicus (1). (Also sometimes known as de Glanville.) *Liber de Proprietatibus Rerum*. Cologne, 1472.

Bartlett, H. H. & Shohara Hide (1). *Japanese Botany during the Period of Wood-Block Printing*. Pt. 1, An Essay on the Development of Natural History, especially Botany, in Japan; on the Influence of Early Chinese and Western Contacts; on Japanese Books and Wood-Block Illustrations; Pt. 2, An Exhibition of Japanese Books and Manuscripts, mostly Botanical, held at the Clements Library of the University of Michigan in Commemoration of the Hundredth Anniversary (1954) of the First Treaty between the United States and Japan. Dawson, Los Angeles, Calif. 1961.

Bauer, W. (3). 'The Encyclopaedia in China.' *JWH*, 1966, **9**, 665.

Bauhin, Kaspar (1). *ΠΙΝΑΞ Theatri Botanici, sive Index in Theophrasti, Dioscoridis, Plinii et botanicorum qui a saeculo scripserunt, opera: Plantarum circiter sex millium ab ipsis exhibitarum nomina cum earundem Synonymiis et Differentiis methodice secundum earum et genera et species proponens. Opus XL annorum hactenus non editum summopere expetitum et ad auctores intelligendos plurimum faciens*. Regis, Basel, 1623.

Bawden, F. C. (1). *Plant Viruses and Virus Diseases*. Ronald, New York, 1964.

Beal, S. (2) (tr.). '*Si Yu Ki [Hsi Yü Chi]*', *Buddhist Record of the Western World, translated from the Chinese of Hiuen Tsiang [Hsüan-Chuang]*. 2 vols. Trübner, London, 1881, 1884; 2nd ed. 1906. Repr. in 4 vols. with new title. *Chinese Accounts of India . . . .* Susil Gupta, Calcutta, 1957–8.

Bean, W. J. (1). 'The *Cydonias*.' *GDC*, 1903 (3rd ser.), **34**, 434.

Bean, W. J. (2). 'Hydrangeas (with a coloured plate of *Hydrangea Hortensia*, var. *japonica rosea*).' *GDN*, 1896, **50**, 122.

Bean, W. J. (3). 'Viburnums.' *GDC*, 1901 (3rd ser.), **30**, 320.

Bean, W. J. (4). 'The Camellias'. *NFLOSIL*, 1930, **2**, 75.

Beck, H. (1). *Alexander von Humboldt*. 2 vols. Wiesbaden, 1959–61.

Bedini, S. A. (5). 'The Scent of Time; a Study of the Use of Fire and Incense for Time Measurement in Oriental Countries.' *TAPS*, 1963 (W.S.) **53**, pt. 5, 1–51. Rev. G. J. Whitrow, *A/AIHS*, 1964, **17**, 184.

Bedini, S. A. (6). 'Holy Smoke; Oriental Fire Clocks.' *NS*, 1964, **21** (no. 380), 537.

Belpaire, B. (4). '*T'ang Kien We Tse*'; *Florilège de Littérature des T'ang*. Paris, 1957.

Bentham, G. (1). '*Flora Hongkongensis*'; *a Description of the Flowering Plants and Ferns of the Island of Hongkong*. Reeve, London, 1861. (No illustrations, no Chinese names or characters.) Supplement by H. F. Hance (2).

Bentham, G. & Hooker, J. D. (1). *Handbook of the British Flora; a Description of the Flowering Plants and Ferns indigenous to, or naturalised in, the British Isles*. 6th ed. 2 vols. (1 vol. text, 1 vol. drawings). Reeve, London, 1892. Repr. 1920.

BERENDES, J. (1). *Die Pharmacie bei den alten Culturvölkern; historisch-kritische Studien.* 2 vols. Tausch & Grosse, Halle, 1891.

BERG, L. S. (1)
'Les Kak Produkt Vyvetrivaniya i Pochvoobrazovaniya' in *Klimat i Zhizn (Climate and Life)*, vol. 3. Acad. Sci. Moscow, 1960.
Eng. tr. A. Gourevitch:
'Loess as a Product of Weathering and Soil Formation.' Israel Programme for Scientific Translations, Jerusalem, Oldbourne, London, 1964.

BERNAL, J. D. (1). *Science in History.* Watts & Co., London, 1954.

BERTUCCIOLI, G. (2). 'A Note on Two Ming Manuscripts of the *Pên Tshao Phin Hui Ching Yao*'. *JOSHK*, 1956, **3**, 63. Abstr. *RBS*, 1959, **2**, no. 228.

BERTUCCIOLI, G. (3). 'Nota sul *Pên Tshao Phin Hui Ching Yao*.' *RSO*, 1954, **29**, 1.

BIOT, E. (1) (tr.). *Le Tcheou-Li ou Rites des Tcheou.* 3 vols. Imp. Nat., Paris, 1851. (Photographically reproduced, Wêntienko, Peking, 1930.)

BLUNT, W. & STEARN, W. T. (1). *The Art of Botanical Illustration.* Collins, London, 1950.

BOCK, JEROME, (HIERONYMUS TRAGUS) (1).
New Kreutterbuch von Underscheydt, Würckung und Namen der Kreutter ... Rihel, Strasburg, 1539. Repr. 1546.
Latin ed. *De Stirpium, maxime earum, quae in Germania nostra nascuntur ...* Rihel, Strasburg, 1552.

BODDE, D. (1). *China's First Unifier, a study of the Ch'in Dynasty as seen in the Life of Li Ssu (−280/−208).* Brill, Leiden, 1938. (Sinica Leidensia, no. 3.)

BODDE, D. (5). Types of Chinese Categorical Thinking. *JAOS*, 1939, **59**, 200.

BODDE, D. (12). *Annual Customs and Festivals in Peking, as recorded in the 'Yenching Sui Shih Chi.'* [by Tun Li-Chhen]. Vetch, Peiping: 1936. (Revs. J. J. L. Duyvendak, *TP*, 1937, **33**, 102; A. Waley, *FL*, 1936, **47**, 402).

BODDE, D. (16). 'Early References to Tea Drinking in China.' *JAOS*, 1942, **62**, 74.

BODDE, D. (20). 'On Translating Chinese Philosophical Terms.' *FEQ*, 1955, **14**, 231.

DEN BOER, A. F. (1). *Ornamental Crab-Apples.* Amer. Assoc. Nurserymen, [New York], 1959.

BOISGIRAUD, *Calosoma sycophante* kill gypsy moth on poplars, 1840. *Revue Zoologique*, 1843. Societa Carvieriana, Paris.

BOLENS, L. (1). 'De l'Idéologie Aristotélicienne à l'Empirisme Médiéval; Les Sols dans l'Agronomie Hispano-Arabe'. *AHES/AESC*, 1975, **30** (no. 5), 1062.

BONAVIA, E. (1). *The Cultivated Oranges and Lemons of India and Ceylon; with Researches into their Origin and the Derivation of their Names, and other Useful Information with an Atlas of Illustrations.* 2 vols, Allen, London, 1888, 1890. 1 vol. text, 1 vol. plates.

BOSWELL, J. T., BROWN, N. E., FITCH, W. H. & SOWERBY, J. E. (1). *English Botany; or, Coloured Figures of British Plants.* 13 vols. (incl. index vol.) 3rd ed. Bell, London, 1887. (Boswell is also known as Syme or Boswell-Syme.)

BOWRA, E. C. (1). *Index Plantarum; Sinice et Latine* [a list of Latin binomial equivalents following Chinese romanised plant names with characters]. In Doolittle (1), pp. 419 ff.

BOXER, C. R. (1). (ed.). *South China in the Sixteenth Century; being the Narratives of Galeote Pereira, Fr. Gaspar da Cruz, O.P., and Fr. Martin de Rada, O.E.S.A. (1550–1575).* Hakluyt Society, London, 1953. (Hakluyt Society Pubs. 2nd series, no. 106.)

BOYKO, H. (1) (ed.). *Saline Irrigation for Agriculture and Forestry.* Junk, The Hague, 1968. (Proc. Internat. Symposium on Plant-growing with Highly Saline Water or Sea-water, with or without Desalination, Rome, 1965. World Acad. of Art and Science Pubs. no. 4.)

B[OYM], M[ICHAEL] (1). *Briefve Relation de la notable conversion des personnes royales et de l'estat de la Religion Chrestienne en Chine, faicte par le très R.P. M[ichel] B[oym], de la Compagnie de Jésus, envoyé par la Cour de ce Royaume-là, en qualité d'Ambassadeur au Saint Siège Apostolique et récitée par luy-mesme dans l'Eglise de Smyrne, le 29 Septembre de l'an 1652.* Cramoisy, Paris, 1654. Repr. in M. Thévenot's *Voyages* vol. 2, Langlois, Paris, 1730, German tr., Friessen, Cologne, 1653; Straub, München, 1653, abridged and modified in *Der Neue Weltbott*, Augsburg & Graz, 1726, vol. 1, no. 13. Polish tr. Warsaw, 1756. Italian tr. Rome, 1652; Parma, 1657.

BOYM, MICHAEL (2). *'Flora Sinensis,' Fructus Floresque humillime porrigens serenissimo ac potentissimo Principi, ac Domino Leopoldo Ignatio, Hungariae Regi florentissimo, etc. Fructus saeculo promittenti Augustissimos, emissa in publicum a R. P. Michaele Boym Societatis Iesu Sacerdote, et a domo professa ejusdem Societatis Viennae Majestati Suae una cum faelicissimi Anni apprecatione oblata Anno Salutis MDCLVI.* Richter, Vienna, 1656. Repr. in M. Thévenot's *Voyages*, vol. 2. Langlois, Paris, 1730.

BOYNTON, G. (1). 'Translation of certain sections of Chhen Hao-Tzu's *Mirror of Flowers*' (+1688). *BGCA* (6th ser.), **6**, 9.

BOYSEN-JENSEN, P. (1). *Growth Hormones in Plants.* McGraw Hill, New York and London, 1936. Eng. tr. of *Die Wuchsstofftheone* by G. S. Avery, P. R. Burkholder, H. B. Creighton & B. A. Scheer.

BRAUN, R. & LYE, W. J. (1). *List of Medicines exported from Hankow and other Yangtze Ports; and, Tariff of Approximate Values of Medicines etc.* [Miscellaneous Goods and Furs] *exported from Hankow.* (3rd Issue), Inspectorate General of Customs, Shanghai, 1917. (China Maritime Customs Pubs., II Special Ser., No. 8.)

BRAY, BARBARA (1), (tr.). *La Dame aux Camélias* [The Lady of the Camellias] translated from the French of

Alexandre Dumas the Younger by B. B. Folio Society, London, 1975.

BRAY, F. (1). 'Swords into Ploughshares, a study of agricultural technology and society in early China.' *TCULT* (1978), 19, **1**, 1–31.

BRETSCHNEIDER, E. (1). *Botanicon Sinicum; Notes on Chinese Botany from Native and Western Sources.* 3 vols.
   Vol. I (Pt. I, no special sub-title) contains:
   ch. 1 Contribution towards a History of the Development of Botanical Knowledge among Eastern Asiatic Nations.
   ch. 2 On the Scientific Determination of the Plants Mentioned in Chinese Books.
   ch. 3 Alphabetical List of Chinese Works, with Index of Chinese Authors.
   app. Celebrated Mountains of China (list).
   Trübner, London, 1882 (printed in Japan).
   Vol. II, Pt. II, *The Botany of the Chinese Classics*, with Annotations, Appendixes and Indexes by E. Faber contains Corrigenda and Addenda to Pt. I
   ch. 1 Plants mentioned in the *Erh Ya*.
   ch. 2 Plants mentioned in the *Shih Ching*, the *Shu Ching*, the *Li Chi*, the *Chou Li* and other Chinese classical works. Kelly & Walsh, Shanghai etc., 1892. Also pub. *JRAS/NCB*, 1893 (N.S.), **25**, 1–468.
   Vol. III, Pt. III, *Botanical Investigations into the Materia Medica of the Ancient Chinese* contains
   (ch. 1) Medicinal Plants of the *Shen Nung Pên Tshao Ching* and the [*Ming I*] *Pieh Lu* with indexes of geographical names, Chinese plant names and Latin generic names. Kelly & Walsh, Shanghai etc., 1895. Also pub. *JRAS/NCB*, 1895 (N.S.), **29**, 1–623. Also pub. *JRAS/NCB*, 1881 (N.S.), **16**, 18–230 (in smaller format).

BRETSCHNEIDER, E. (6). 'On the Study and Value of Chinese Botanical Works; with Notes on the History of Plants and Geographical Botany from Chinese Sources.' Rozario & Marcal, Fuchow, 1871. First published in *CRR*, 1870, **3**, 157, 172, 218, 241, 264, 281, 290, 296. Chinese tr. by Shih Shêng-Han (down to p. 24, omitting the discussion on Palmae, but with the addition of critical notes) *Chung-Kuo Chih-Wu-Hsüeh Wên-Hsien Phing-Lun*. Nat. Compilation & Transl. Bureau, Shanghai, 1935. Repr. Com. Press Shanghai, 1957.

BRETSCHNEIDER, E. (9). 'Early European Researches into the Flora of China.' *JRAS/NCB*, 1880 (N.S.) **15**, 1–194. Sep. pub. Amer. Presbyterian Mission Press, Shanghai, 1881; Trübner, London, 1881.
   ch. 1 Botanical Information with respect to China supplied by the Jesuits.
   ch. 2 James Cunningham (+1702).
   ch. 3 Swedish Collectors of Plants in South China (+1751, +1766).
   ch. 4 Early Researches into the Flora of Peking.
   ch. 5 Sonnerat.
   ch. 6 Loureiro.
   Reproduced, in much abridged form, and larger format, at the beginning of Bretschneider (10).

BRETSCHNEIDER, E. (10). *History of European Botanical Discoveries in China.* 2 vols.
   Sampson Low & Marston, London, 1898.
   Photolitho reproduction, Koehler, Leipzig, 1935.
   This work, though much larger, does not supersede Bretschneider (9).

BRETSCHNEIDER, E. (12). 'Jasmine in China', *CRR*, 1871, **3**, 225.

BRETSCHNEIDER, E. (13). 'Les Palmiers de la Chine.' *NQCJ*, 1869, **3**, 139, 150.

BRETZL, HUGO (1). *Botanische Forschungen des Alexanderzuges.* Teubner, Leipzig, 1903.

BREYN, J. P. (1). *Dissertatio Botanico-Medica de Radice Gin-sen, seu nisi, et Chrysanthemo Bidente Zeylanico Acmella dicto.* Inaug. Diss., Danzig. Schreiber, Gedani (Gdansk), 1789. First pub. 1700. Repr. in *Prodromi Fasciculi Rariorum Plantarum* . . . . 1739. Cf. Merrill & Walker (1), vol. 1, p. 53.

BRIDGMAN, E. C. (1). *A Chinese Chrestomathy, in the Canton Dialect.* S. Wells Williams, Macao, 1841.

BRIDGMAN, E. C. & WILLIAMS, S. WELLS (1). 'Mineralogy, Botany, Zoology and Medicine' [sections of a Chinese Chrestomathy], in Bridgman (1), pp. 429, 436, 460 and 497.

BRIDGMAN, R. F. (2). 'La Médicine dans la Chine antique' (Extrait des *Mélanges Chinois et Bouddhiques* publiés par l'Institut Belge des Études Chinoises, vol. X.) Bruges, 1955.

BRINK, C. O. (1). Art. 'Peripatos [the Peripatetic School in Greece]' in Pauly-Wissowa, *Realenzyklopädie d. klass. Altertumswissenschaft.* Suppl. Vol. 7, cols. 899 ff.

BRONGNIART, A. T. (1). *Sur la Classification et la Distribution des Végétaux fossiles, en général, et sur ceux des Terrains de Sédiment supérieur en particulier.* Paris, 1822. Orig. pub. Mém. du Muséum d'Hist. Nat. vol. 8.

BROWN, W. H. (1). *Useful Plants of the Philippines.* 3 vols. completed 1935. Sands & McDougall, Melbourne, 1950. (Commonwealth of the Philippines, Dept. of Agric. & Commerce, Manila, Techn. Bull. no. 10.)

BRUNFELS, OTTO (1). *Herbarium Vivae Eicones* . . . Scholt, Strasburg, 1530. Repr. 1531, 1536. German edns. *Contrafayt Kreüterbůch* . . . Scholt, Strasburg, 1532. Repr. 1537.

BRYANT, CHARLES (1). *Flora Diaetetica*; or History of Esculent Plants, both Domestic and Foreign, in which they are accurately described, and reduced to their Linnaean Generic and Specific Names, with their English names annexed, and ranged under Eleven General Heads, Viz: Esculent (1) Roots, (2) Shoots, Stalks, etc. (3) Leaves, (4) Flowers, (5) Berries, (6) Stone-fruit, (7) Apples, (8) Legumens, (9) Grain, (10) Nuts, (11) Funguses; And, a particular Account of the Manner of Using them; their native places of Growth; their several Varieties, and Physical Properties; together with whatever is otherwise curious, or very remarkable in each Species: the Whole so methodised as to form a Short Introduction to the Science of Botany. White, London, 1783.

BUCHANAN, F. (1). *A Journey from Madras through the Countries of Mysore, Canara and Malabar, etc.* [1800–1]. London, 1807.

BUCHANAN, K. (1). 'Reshaping the Chinese Earth; Agricultural Change in the Loess and Laterite Lands of China.' *EHOR*, 1966, **5** (no. 11), 28.

BUCHANAN, K., FITZGERALD, C. P. & RONAN, C. A. (1). *China; the Land and the People; the History, the Art and the Science.* Crown, New York, 1981.

BUCKMAN, T. R. (1) (ed.). *Bibliography and Natural History; Essays presented at a Conference convened in 1964 by T.R.B.* Univ. of Kansas Libraries, Lawrence, Kansas, 1966.

BULLETT, GERALD & TSUI CHI (1). *The Golden Year of Fan Chhêng-Ta.* CUP, Cambridge, 1946.

BUNGE, A. (1). *Plantarum Mongholico-Chinensium Decas Prima.* Kazan, 1835.

BUREAU, E. & FRANCHET, A. (1). *Plantes Nouvelles du Thibet et de la Chine Occidentale receuillies pendant le Voyage de Mons. Bonvalot et du Prince Henri d'Orléans en 1890.* Mersch, Paris, 1891. (No Chinese names, no characters.)

BURGES, A. (1). *Micro-Organisms in the Soil.* Hutchinson, London, 1958.

BURKILL, I. H. (1). *A Dictionary of the Economic Products of the Malay Peninsula.* 2 vols., published for the Malay Govt. by Crown Agents, London, 1935.

BURKILL, I. H. (3). 'A List of Oriental Vernacular Names of the Genus *Dioscorea* [yams].' *GBSS*, 1924, **3** (nos. 4–6), 121–244. (Chinese names but no characters.)

BURKILL, I. H. (4). '*Chapters in the History of Botany in India.*' Bot. Survey of India, Calcutta and Delhi, 1965. Rev. N. L. Bor, *N*, 1966, **212**, 1297.

BUTLER, A. R., GLIDEWELL, C. & NEEDHAM, JOSEPH (1). 'The Solubilisation of Cinnabar; Explanation of a Sixth-Century Chinese Alchemical Recipe.' *JCR*, 1980, no. 2, 47.

BYRD, C. R. & WEAD, K. H. (1). *Radical and Subradical Index to Chinese Plant Names listed in J. Matsumura's ... 'Shokubutsu Mei-I' ....* 1920.

CAIN, S. A. (1). *Foundations of Plant Geography.* New York, 1944.

CAIUS, JOHANNES (1). *De Rariorum Animalium atque Stirpium Historia.* Seres, London, 1570. Repr. in E. S. Roberts (1).

CALLERY, J. A. (1). *Systema Phoneticum Scripturae Sinicae* (a dictionary which arranged the characters according to their phonetic components, not their radical components or their sounds). Macao, 1841.

CALZOLARI, FRANCISCO (1). *Viaggio di Monte Baldo.* 1566. Repr. in P. Mattioli's *Compendium de Plantis Omnibus ...* Venice, 1571, and in his *De Plantis Epitome Utilissima ...* Frankfurt, 1586.

CAMERARIUS, R. J. (1). *De Sexu Plantarum Epistola* (to Valentin). Tübingen, 1694. Frankfurt, 1700, 1749. Also in Valentinus' *Declamationum Panegyricarum* 1701. Repr. by J. G. Koelreuter in *R. J. Camerarii Opuscula Botanici Argumenti.* Prague 1797.

CAMP, W. H. *et al.* (1), (ed.). *The International Rules of Botanical Nomenclature.* Chronica Botanica, Waltham, Mass., 1952.

DE CANDOLLE, ALPHONSE (1). *The Origin of Cultivated Plants.* Kegan Paul, London, 1884 (International Scientific Series, no. 49.) Translated from the French edition, Geneva, 1883. Engl. 2nd ed. London, 1886, reproduced photolithographically, Hafner, New York, 1959.

DE CANDOLLE, ALPHONSE (2). *Géographie Botanique Raisonnée; ou Exposition des Faits Principaux et des Lois concernant la Distribution Géographique des Plantes de l'Époque Actuelle.* 2 vols. Geneva 1855; Masson, Paris, 1885.

DE CANDOLLE, AUGUSTIN P. (1). 'Essai Élémentaire de Géographie Botanique.' Art. in *Dict. des Sci. Nat.* vol. 18. Lévrault, Paris, 1820.

CAPELLE, W. (1). 'Zur Geschichte d. griechischen Botanik.' *PL*, 1910, **69**, 264.

CARRIÈRE, E. A. (1). 'Armeniaca [Prunus] Davidiana.' *RHORT*, 1879, 236.

CARTER, T. F. (1). *The Invention of Printing in China and its Spread Westward.* Columbia University Press, New York, 1931.

CASTIGLIONI, ARTURO (1). *A History of Medicine.* Tr. & ed. E. B. Krumbhaar. 2nd ed. revised and enlarged, Knopf, New York, 1947.

CECIL, EVELYN (1). *A History of Gardening in England.* Murray, London, 1910.

CECIL, R. L. & LOEB, R. F. (1). *Textbook of Medicine.* W. B. Saunders Company, Philadelphia and London, 1951 (8th ed.).

CESALPINO, ANDREA (1). *De Plantis Libri* XVI Marescotti, Florence, 1583.

CHAMBERS, SIR WM. (2). *A Dissertation on Oriental Gardening ... ; To which is annexed. An Explanatory Discourse by Tan Chet-Qua, of Quang-chew-fu, Gent.* 2nd ed., with additions, Griffin, Davies, Dodsley, Wilson, Nicoll, Walter & Emsley, London, 1773.

CHAMFRAULT, A. & UNG KANG-SAM (1), with illustrations by M. Rouhier. *Traité de Médecine Chinoise; d'après les Textes Chinois Anciens et Modernes.* Coquemard, Angoulême, 1954–.
  Vol. 1. Traité, Acupuncture, Moxas, Massages, Saignées. 1954.
  Vol. 2. (tr.). Les Livres Sacrés de Médecine Chinoise (*Nei Ching, Su Wên* and *Nei Ching, Ling Shu*). 1957.

Vol. 3. Pharmacopée [372 entries from the *Pên Tshao Kang Mu*]. 1959.

Vol. 4. Formules Magistrales. 1961.

Vol. 5. De l'Astronomie à la Médecine Chinoise; Le Ciel, La Terre, l'Homme. 1963.

CHANG CHHANG-SHAO (1). 'The Present Status of Studies on Chinese Anti-Malarial Drugs.' *CMJ*, 1945, **63A**, 126.

CHANG CHHANG-SHAO, FU FÊNG-YUNG, HUANG K. C. & WANG C. Y. (1). 'Pharmacology of *chhang shan* (*Dichroa febrifuga*), a Chinese Antimalarial Herb.' *N*, 1948, **161**, 400. With comment by T. S. Work.

CHANG CHI-YÜN (1), (ed.). *National Atlas of China* 5 vols. National War College (Taiwan), Chinese Geographical Institute, Taipei, Taiwan, 1960–3.

CHANG HSIN-TSHANG (3). 'Hsü Wei: Seven Stanzas on the Lotus.' Art. in *Essays offered to G. H. Luce by his Colleagues and Friends in honour of his 75 th Birthday*. Artibus Asiae, Ancona, 1966, p. 102.

CHANG HUI-CHIEN (1). *Li Shih-Chen; Great Pharmacologist of Ancient China*. Foreign Languages Press, Peking, 1960.

CHANG JEN-HU (1). 'The Climate of China according to the new Thornthwaite Classification. *AAAG*, 1955, **45**, 393.

CHAO YÜ-HUANG (1) = (2). *A Programme, together with Concrete Research Examples, to make Intensive Study of Chinese Materia Medica*; 1. The Study of Chhichow Drugs [the birthplace of Li Shih-Chen, and a famous drug market]; A, Compositae and Dipsacaceae. Dept. of Pharmacognosy, Institute of Chinese Drugs, College of Medicine, National Peking University, Peiping, 1941.

CHAO YUAN-JEN (4). 'Popular Chinese Plant Words; a Descriptive Lexico-Grammatical Study.' *LAN*, 1953, **29**, 379.

CHAO YUAN-JEN (5). 'Graphic and Phonetic Aspects of Linguistic and Mathematical Symbols.' *PSAM*, 1961, **12**, 69.

CHAVANNES, E. (1). *Les Mémoires Historiques de Se-Ma Ts'ien* [Ssuma Chhien]. 5 vols. Leroux, Paris, 1895–1905. (Photographically reproduced, in China, without imprint and undated.)

CHAVANNES, E. (5). 'Le T'ai Chan [Thai Shan]; Essai de Monographie d'un Culte Chinois.' *BE/AMG*, 1910, no. **21**, 1–591. (With appendix: 'Le Dieu du Sol dans la Chine Antique'.)

CHÊNG CHIH-FAN (1). 'Li Shih-Chen and his *Materia Medica*.' *CREC*, 1963, **12** (no. 2), 29.

CHÊNG TÊ-KHUN (9). *Archaeology in China*.

Vol. 1, *Prehistoric China*. Heffer, Cambridge, 1959.

Vol. 2, *Shang China*. Heffer, Cambridge, 1960.

Vol. 3, *Chou China*. Heffer, Cambridge, and Univ. Press, Toronto, 1963.

Vol. 4, *Han China* (in the press).

CHÊNG TSUNG-HAI (1). 'A Historical Study on the Use of Illustrations in Chinese Educational Books.' *ACOM*, 1961, **20**, 104.

CHEVALIER, A. J. B. (1). *Michel Adanson; Voyageur, Naturaliste et Philosophe*. Larose, Paris, 1934.

CHHEN JUNG-CHIEH (5). Contributions to *A Dictionary of Philosophy*, ed. D. D. Runes. Philos. Lib. New York, 1942. Notably *Chhi* (pneuma, matter-energy), p. 50; *Jen* (human heartedness) p. 153; and *Li* (Neo-Confucian organic pattern), p. 168 [our definitions, not his]. Also *PEW*, 1952, **2**, 166.

CHHEN JUNG-CHIEH (6). 'The Evolution of the Confucian Concept of *Jen*.' *PEW*, 1955, **4**, 295.

CHHEN JUNG-CHIEH (8). 'The Concept of Man in Chinese Thought.' Art. in *The Concept of Man; a Study in Comparative Philosophy* ed. S. Radhakrishnan & P. T. Raju. Allen & Unwin, London, 1960, p. 158.

CHHEN JUNG-CHIEH (11) (tr.). '*Reflections on Things at Hand*' [*Chin Ssu Lu*]; the Neo-Confucian Anthology compiled by Chu Hsi and Lü Tsu-Chhien translated with notes .... Columbia Univ. Press, New York and London, 1967.

CHEN, K. K. See Chhen Kho-Khuei.

CHEUNG, S. C., KWAN, P. S. & KONG, Y. C. See Chuang Chao-Hsiang, Kuan Phei-Shêng & Chiang Jun-Hsiang (1) = (1).

CHHEN KHO-KHUEI, MUKERJI, B. & VOLICER, L. (1) (ed.). *The Pharmacology of Oriental Plants*. Pergamon Press, London, 1965. Czechoslovak Med. Press, Prague, 1965. (Proc. 2nd International Pharmacological Meeting, Prague, 1963, vol. 7.)

CHHEN SHOU-YI (3). *Chinese Literature; a Historical Introduction*. Ronald, New York, 1961.

CHHIEN CHHUNG-SHU & FANG WÊN-PHEI (1). 'The Geographical Distribution of Chinese *Acer* [maples].' *Proc. 5th Pacific Science Congress*, 1934, vol. 4, p. 3305.

CHHIEN TSHUN-HSUN (3). 'On Dating the Edition of the *Chü Lu* at Cambridge University.' *CHJ/T*, 1973, n.s. **10** (no. 1), 106.

CHHIN JEN-CHHANG (1). 'The Present Status of our Knowledge of Chinese Ferns. *PNHB*, 1933, **7**, 253.

CHI YÜEH-FÊNG & READ, BERNARD E. (1). 'The Vitamin C Content of Chinese Foods and Drugs.' *CJOP*, 1935, **9**, 47.

CHIEN S. S. (Sung-Shu). See Chhien Chhung-Shu.

CHING LI-PIN (1). 'Les *Pên Tshao*; la Pharmacopée Chinoise.' *CJEB*, 1940, **1**, 435.

CHING LI-PIN (2). 'Notes pour servir à l'Étude des Matières Médicales en Chine.' *NPA/CIP*, 1936, **4**, 53.

CHING, R. C. (Ren-Chang). See Chhin Jen-Chhang.

CHITTENDEN, F. J. (1) (ed.). *The Royal Horticultural Society Dictionary of Gardening; a Practical and Scientific Encyclopaedia of Horticulture*. Oxford 1951. With subsequent supplements.

CHMIELEWSKI, JANUSZ (1). 'The Problem of Early [i.e. pre-Buddhist] Loan-Words in Chinese, as illustrated by the word *phu-thao* [grape-vine].' *RO*, 1958, **22** (pt. 2), 7. Abstr. *RBS*, 1964, **4**, no. 563.

CHMIELEWSKI, JANUSZ (2). 'Two Early Loan-Words in Chinese' [*mu-su*, alfalfa, lucerne, and *shan-hu*, coral]. *RO*, 1960, **24** (pt. 2), 65. Abstr. *RBS*, 1967, **6**, no. 413.

CHOATE, HELEN A. (1). 'The Earliest Glossary of Botanical Terms; Fuchs, 1542.' *TORR*, 1917, **17**, 186.

CHOPRA, R. N., BADHWAR, R. L. & GHOSH, S. (1). *Poisonous Plants of India*. 2 vols. Delhi, 1949.

CHOU HAN-FAN (1). *The Familiar Trees of Hopei*. Peking Nat. Hist. Bull. Peiping, 1934 (PNHB Handbook, no. 4). Many illustrations, and Chinese characters generally given, but no index.

CHOU I-CHHING (1). *La Philosophie Morale dans le Neo-Confucianisme* (*Tcheou Touen-Yi*) [*Chou Tun-I*]. Presses Univ. de France, Paris, 1954. (Includes tr. of *Thai Chi Thu Shuo, Thai Chi Thu Shuo Chieh* and of *I Thung Shu*.)

CHOU TSHÊ-TSUNG (1). 'The Anti-Confucian Movement in Early Republican China.' Art. in *The Confucian Persuasion*, ed. Wright (8), p. 288.

CHOW YIH-CHING. See Chou I-Chhing.

CHOWDHURY, K. A., GHOSH, A. K. & SEN, S. N. (1). '[The History of] Botany [in India].' Art. in *A Concise History of Science in India*, ed. Bose, D. M., Sen, S. N. & Subbarayappa, B. V. (1). New Delhi, 1971. p. 371.

CHU, COCHING. See Chu Kho-Chen.

CHU KHANG-KHUNG & YANG JEN-CHHANG (1). 'On the Rainfall and Meteorology of China.' In Kovda (1), pp. 43 ff.

CHU KHO-CHEN (3). 'Climatic Pulsations in China.' *GR*, 1926, **16**, 274.

CHU KHO-CHEN (4). 'Climatic Changes during Historic Time in China.' *TSSC*, 1932, **7**, 127; *JRAS/NCB*, 1931, **62**, 32.

CHU KHO-CHEN (5). *The Climatic Provinces of China*. Memoir No. 1, Academia Sinica Nat, Inst. of Meteorology, Nanking, 1930.

CHU KHO-CHEN (9). 'A Preliminary Study on the Climatic Fluctuations during the last 5000 Years in China.' Peking, 1966, pp. 1–26. Enlarged and updated versions: *SCISA*, 1973, **16**, 226; *CYC*, 1974, **25**, 243.

CHUANG CHAO-HSIANG, KUAN PHEI-SHÊNG & CHIANG JUN-HSIANG (1). *An Introduction to the Study of the* [Chinese Pharmaceutical Natural Histories; the] *Pên Tshao* [Genre]. Univ. Press, Chinese Univ. of Hongkong, Shatin, 1983.

CHUNG HSIN-HSÜAN (1). *A Catalogue of the Trees and Shrubs of China*. Science Society of China, Shanghai, 1924. Repr. Chhêng-Wên, Thaipei, 1971. (Memoirs of the Science Society of China, no. 1.)

[CIBOT, P. M.] (6).
'Notices de quelques Plantes, Arbrisseaux, etc., de la Chine;
    (1) Nénuphar de Chine
    (2) Yu-lan [magnolia]
    (3) Ts'ieou-hai-tang [begonia]
    (4) Mo-li-hoa [jasmine]
    (5) Châtaigne d'eau
    (6) Lien-chien ou Ki-teou
    (7) Kiu-hoa ou Matricaire de Chine [chrysanthemum]
    (8) Mou-tan ou Pivoine [peonies, a long account]
    (9) Yê-hiang-hoa
    (10) Pé-gé-hong (pai jih hung)
    (11) Jujubier (*Zizyphus*)
    (12) Chêne
    (13) Châtaigner
    (14) Oranges-Coings; usage de la Greffe'
MCHSAMUC, 1778, **3**, 437–99. Cf. Payne (1) on item (7).

[CIBOT, P. M.] (10). 'Les Serres Chinoises.' *MCHSAMUC*, 1778, **3**, 423.

CIBOT, P. M. (15). 'Sur le Bambou.' *MCHSAMUC*, 1777, **2**, 623.

CLAUSEN, CURTIS, P. *Entomophagous Insects*. McGraw Hill Book Co. New York & London, 1940.

CLAPHAM, A. R., TUTIN, T. G. & WARBURG, E. F. (1). *Flora of the British Isles*. 2nd ed. Cambridge, 1962.

CLÉMENT, G. (1). 'Historique des Cultures du Chrysanthème.' *RHORT*, 1936, **108**, 283.

CLIFFORD, H. T. & STEPHENSON, W. (1). *An Introduction to Numerical Classification*. Academic Press, New York, 1977.

CLIFFORD, DEREK (1). *A History of Garden Design*. Faber & Faber, London, 1962.

CLOUDSLEY-THOMPSON, J. L. (1). *Spiders, Scorpions, Centipedes and Mites; the Oecology and Natural History of Woodlice, 'Myriapods' and Arachnids*. Pergamon, London, 1958. 2nd, revised, ed. 1968.

COATS, A. M. (1). *Flowers and their Histories*. Hulton, London, 1956.

COCKAYNE, T. O. (1). 'Leechdoms, Wortcunning and Starcraft of Early England.' In *Chronicles and Memorials of Great Britain and Ireland during the Middle Ages*. Rolls series, no. 1. 3 vols. London, 1864–66.

[COLLAS, J. P. L.] (2). 'Observations sur les Plantes, les Fleurs et les Arbres de la Chine, quil est possible et utile de se procurer en France.' *MCHSAMUC*, 1786, **11**, 183–297.

COLLAS, J. P. L. (10) (posthumous). 'Notice sur le Bambou.' *MCHSAMUC*, 1786, **11**, 353.

COLLETT, SIR H. (1). *Flora Simlensis; a Handbook of the Flowering Plants of Simla and the Neighbourhood.* Thacker & Spink, Calcutta and Simla, 1902.

COLLINS, V. D. (1). 'Sorgo or Northern Sugar-Cane.' *JRAS/NCB*, 1865, (n.s.) **2**, 85.

COLLISON, R. L. W. (1). *Encyclopaedias; their History throughout the Ages—a Bibliographical Guide with extensive historical notes to the General Encyclopaedias issued throughout the world from* −350 *to the present day. Hafner, New York and London, 1964. 2nd ed. 1966.*

COLONNA, FABIO (1). *ΦΥΤΟΒΑΣΑΝΟC sive Plantarum aliquot Historia.* Carlino & Pace, Naples, 1592.

COMENIUS (KOMENSKY), JAN AMOS (1). *A Reformation of schooles, designed in two excellent Treatises; the first whereof summarily sheweth, the great necessity of a generall reformation of common learning; what grounds of hope there are for such a reformation, and how it may be brought to pass; followed by a Dilucidation answering certaine objections made against the Endeavours and Means of Reformation in Common Learning, expressed in the foregoing discourse.* (Tr. Sam. Hartlib.) London, 1642.

CONDIT, IRA J. (1). *The Fig.* Chronica Botanica, Waltham, Mass. 1947. (New Series of Plant Science Books, no. 19.)

CORDIER, H. (2). *Bibliotheca Sinica; Dictionnaire bibliographique des Ouvrages relatifs à l'Empire Chinois.* 3 vols. Ec. des Langues Orientales Vivantes, Paris, 1878–95. 2nd ed. 5 vols. Pr. Vienna, 1904–24.

CORDUS, VALERIUS (1). *In Hoc Volumine continentur Valerii Cordii ... Annotations in Pedacii Dioscoridis ... de Materia Medica ... eiusdem Val. Cordi Historiae Stirpium Libri. IV ... omnia ... Conrad Gesneri ... collecta, et praefationibus illustrata.* Rihel, Strasburg, 1561.

CORNER, E. J. H. (1). *The Natural History of Palms.* Weidenfeld & Nicolson, London, 1966.

COURANT, M. (3). *Catalogue des Livres Chinois, Coréens, Japonais, etc. dans le Bibliothèque Nationale, Département des Manuscrits.* Leroux, Paris, 1900–12.

COURTNEY-PRATT, J. S. (1). 'Symbols in Scientific Typescripts.' *D*, 1958, 104.

COWAN, J. McQUEEN (1) (ed.). *The Journeys and Plant Introductions of George Forrest, V. M. H.* Roy. Hort. Soc. & Oxford Univ. Press, London, 1952.

COX, E. H. M. (1). *Plant-Hunting in China; a History of Botanical Exploration in China and the Tibetan Marches.* Collins, London, 1945. Photolitho edition, Scientific Book Guild, London, 1945.

COX, E. H. M. (2). *The Plant Introductions of Reginald Farrer.* New Flora & Silva, London, 1930.

CRANE, M. B. & LAWRENCE, W. J. C. (1). *The Genetics of Garden Plants.* 4th ed. Macmillan, London, 1952.

CRANMER-BYNG, J. L. (2) (ed.). *An Embassy to China; being the Journal kept by Lord Macartney during his Embassy to the Emperor Chhien-Lung,* +*1793 and* +*1794.* Longmans, London, 1962. Macartney (1, 2); Gillan (1).

CRAWFURD, R. (1). *The King's Evil.* Oxford, 1911.

CREEL, H. G. (4). *Confucius; the Man and the Myth.* Day, New York, 1949; Kegan Paul, London, 1951. Review D. Bodde, *JAOS*, 1950, **70**, 199.

CRESSEY, G. B. (1). *China's Geographic Foundations: A Survey of the Land and its People.* McGraw Hill, New York, 1934.

CRESSEY, G. B. (3). *Land of the Five Hundred Million; a Geography of China.* McGraw Hill, New York, 1957.

CROIZAT, L. (1). 'History and Nomenclature of the Higher Units of Classification.' *BTBC*, 1945, **72**, 52.

CROSLAND, M. P. (1). *Historical Studies in the Language of Chemistry.* Heinemann, London, 1962.

CSAPODY, V. & TOTH, I. (1). *A Colour Atlas of Flowering Trees and Shrubs.* Akadémiai Kiadó, Budapest, 1982.

VON CUBE, JOHANN. See Anon. (77).

CUNDALL, J. (1). *A Brief History of Wood-Engraving, from its Invention.* Sampson Low & Marston, London, 1895.

CURREY, F. & HANBURY, DANIEL (1). 'Remarks on *Sclerotium stipitatum* Berk. et Curr., *Pachyma cocos* Fries. and some Similar Productions'. *TLS*, 1860, **23**, 93. Repr. in Hanbury (1), pp. 200 ff.

CURTIS, C. H. (1). *Orchids; their Description and Cultivation.* Putnam, London, 1950.

CURWEN, E. C. & HATT, G. (1). *Plough and Pasture; the Early History of Farming:* Pt. I, *Prehistoric Farming of Europe and the Near East;* Pt. II, *Farming of Non-European Peoples.* Schuman, New York, 1953. (Life of Science Library, no. 27.)

CUSHING, H. (1). *A Bio-bibliography of Andreas Vesalius.* Schuman, New York, 1943.

DALZIEL, J. M. (1). 'The Useful Plants of West Tropical Africa' (an appendix to the *Flora of West Tropical Africa* by J. Hutchinson & J. M. Dalziel). Crown Agents for the Colonies, London, 1937.

DANNENFELDT, K. H. (1). *Leonhard Rauwolf, Sixteenth-Century Physician, Botanist and Traveller.* Harvard Univ. Press, Cambridge, Mass. 1968.

DARAPSKY, L. (1). *Zur Geschichte der Zelléntheorie.* Inaug. Diss. Würzburg, 1880.

DARLINGTON, C. D. (1). *Chromosome Betany and the Origins of Cultivated Plants.* 2nd ed. Allen & Unwin, London, 1963. Rev. C. G. G. S. van Steeris, *MAN*, 1965, nos. 172–7, 164.

DARWIN, CHARLES (1). *The Variation of Animals and Plants under Domestication.* 2 vols. Murray, London.

DAUBENMIRE, R. F. (1). *Plants and Environment; a Textbook of Plant Autecology.* 2nd ed. Wiley, New York, 1959; Chapman & Hall, London, 1959.

DAVIS, TENNEY L. & CHAO YÜN-TSHUNG (8). 'Chang Po-Tuan, Chinese Alchemist of the +11th Century.' *JCE*, 1939, **16**, 53.

DEAM, C. C. (1). *Flora of Indiana.* Pr. pr. Indianapolis, 1940.

DEBEAUX, J. O. (1). *Essai sur la Pharmacie et la Matière Médicale des Chinois.* Baillière & Challamel, Paris, 1865.

DELPINO, G. G. F. (1). 'Studi di Geografia Botanica secondo un Nuovo Indirizzo,' *MRAB*, 1898 (Sc ser), **7**.

DENGLER, R. E. (1) (tr.). *Theophrastus 'De Causis Plantarum'*, Bk. 1; Text, Critical Apparatus, Translation and Commentary. Pr. pr. (Westerbrook), Philadelphia, 1927.

DHÉRÉ, C. (1). 'Michel Tswett [1872/1920], le Créateur de l'Analyse chromatographique par Adsorption; sa Vie, ses Travaux sur les Pigments Chlorophylliens.' *CAND*, 1943, **10**, 23.

DIELS, H. (1). 'Über die Pflanzengeographie von Innern China nach den Ergebnissen neuerer Sammlungen.' *ZGEB*, 1905, 708.

DIELS, L. (1). 'Plantae Chinenses Forrestianae.' *NRBGE*, 1912, **25**, 161–308. (New and Imperfectly Known Species); 191, **31**, 1–411. (Catalogue of all Plants Collected.)

DIMBLEBY, G. (1). *Plants and Archaeology.* London, 1975.

DODGE, B. O. & RICKETT, H. W. (1). *Diseases and Pests of Ornamental Plants.* Cattell, Lancaster, Pa. 1943.

DONOVAN, EDWARD (1). *Natural History of the Insects of China.* London, 1798. Germ. tr. ed. J. G. Gruber, Leipzig, 1802; 2nd ed., ed. J. O. Westwood, Bohn. London, 1842.

DOOLITTLE, J. (3). 'Flowers and Fruits according to their Time of Blossoming' [a list compiled from the works of Morrison, Medhurst and Williams, applicable to Southern and Central China; Latin Binomials and some English common names followed by Chinese characters and romanised plant names]. In Doolittle (1), pp. 657 ff.

DRURY, H. (1). *The Useful Plants of India; with Notices of their Chief Value in Commerce, Medicine and the Arts.* 2nd ed. Allen, London, 1873.

DUBS, H. H. (2). (tr., with the assistance of Phan Lo-Chi and Jen Thai). *'History of the Former Han Dynasty', by Pan Ku, a Critical Translation with Annotations.'* 3 vols. Waverly, Baltimore, 1938–.

DUBS, H. H. (8) (tr.) *The Works of Hsün Tzu.* Probsthain, London, 1928.

DUBS, H. H. (9). 'The Political Career of Confucius.' *JAOS*, 1946, **66**, 273.

DUBS, H. H. (27). 'On the Supposed Monosyllabic Myth' (i.e. Chinese as a monosyllabic language). *JAOS*, 1952, **72**, 82.

DUCHAUFOUR, P. (1). *Précis de Pédologie.* Masson, Paris, 1965.

DUGGAR, B. M. & SINGLETON, V. L. (1). 'New Pharmacological Discoveries.' *ARB*, 1953, **22**, 478.

DUMAS, ALEXANDRE (the younger), (1). *La Dame aux Camélias.* Paris, 1848. Eng. tr. by Barbara Bray, London, 1980.

DUNN, S. T. (1). *'A Supplementary List of Chinese Flowering Plants, 1904–1910'* [to Forbes & Hemsley, (1)]. *JLS/B*, 1911, **39**, 411–506.

DUNN, S. T. & TUTCHER, W. J. (1). *Flora of Kuangtung and Hongkong (China); being an Account of the Flowering Plants, Ferns and Fern Allies together with Keys for their Determination, preceded by a Map and Introduction.* HMSO, London, 1912, (Royal Bot. Gdns. Kew Bull. Misc. Inf. Add. Ser. no. 10). No Chinese names or characters, no illustrations.

DURAN-REYNALS, M. L. (1). *The Fever-Bark Tree.* Allen, London, 1947.

EBERT, FELIX (1). *Beiträge z. Kenntnis d. chinesischen Arzneischatzes; Früchte und Samen.* Inaug. Diss., Zürich, 1907.

ECKEBERG, C. G. See Osbeck.

EDWARDS, E. D. (1). *Chinese Prose Literature of the Thang Period.* 2 vols. Probsthain, London. 1937.

EGERTON, F. N. (1). Notes on both parts of E. L. Greene's 'Landmarks of Botanical History....' q.v. Unpub.

D'ELIA, PASQUALE (2) (ed.). *Fonti Riccianè; Storia dell'Introduzione del Cristianesimo in Cina.* 3 vols. Libreria dello Stato, Rome, 1942–1949. Cf. Trigault (1); Ricci (1).

D'ELIA, PASQUALE (9). 'Le "Generalità sulle Scienze Occidentali"; Hsi Hsüeh Fan di Giulio Aleni.' *RSO*, 1950, **25**, 58.

D'ELIA, PASQUALE 'Recent Discoveries and New Studies (1938–1960) on the World Map in Chinese of Father Matteo Ricci, S. J.' *MS*, 1961, **20**, 82.

EMSWELLER, S. L. (1). 'The Chrysanthemum; its Story through the Ages.' *JNYBG*, 1947, **48**, 26.

ENGLER, A. (1). *Versuch einer Entwicklungsgeschichte der Pflanzenwelt, insbesondere der Florengebiete, seit der Tertiärperiode.* 2 vols. Leipzig, 1878–82.

ENGLER, A. (2). 'Geographische Verbreitung d. Coniferae.' Art. in *Natürliche Pflanzenfamilien*, ed. A. Engler & ■. Brandtl. 2nd. ed. 1926, vol. 13, p. 166.

EVANS, G. E. (1). *Ask the Fellows who Cut the Hay* (English rural life and customs), Faber & Faber, London, 1956.

EVREINOV, V. A. (1). 'Les Pêches "Peen-too".' *RHORT*, 1934, **106**, 11.

EYRE, S. R. (1). *Vegetation and Soils; a World Picture.* Arnold, London, 1963.

FABER, E. (posthumous) & McGREGOR, D. (1). 'Contributions to the Nomenclature of Chinese Plants'. *JRAS/NCB*, 1907, **37**, 97–164. Latin binomials according to Families, Latin names alphabetically with Chinese characters (no Chinese index), List of Chinese plant names for which the English equivalent was not known.

FAIRCLOUGH, H. RUSHTON (1) (tr.). *Virgil, with an English Translation....* 2 vols. Heinemann, London, 1960. (Loeb Classics edition).

FANG CHAO-YING (3). 'Notes on the Chinese Jews of Khaifêng.' *JAOS*, 1965, **85**, 126. On Chao An-Chhêng (An San) and his obscure relations with the prince Chou Ting Wang (Chu Hsiao).

FANG CHIH-THUNG (Achilles), (2), (tr.). 'A Bookman's Decalogue' [Yeh Tê-Hui's *Tshang Shu Shih Yo*]. *HJAS*, 1950, **13**, 133. The best woods to use for boards and boxes for conserving Chinese books.

FANG WÊN-PHEI (1). *A Monograph of Chinese Aceraceae*. Sci. Soc. of China, Nanking, 1939 (Contributions from the Biol. Lab. Sci. Soc. Ch., Bot. Ser. no. 11). No Chinese names or characters, no illustrations.

FANG WÊN-PHEI (2) = (1). *Icones Plantarum Omeiensium*. Nat. Szechuan Univ., Chhêngtu, 1942–5. 3 parts only pub. Text in Chinese and English.

FARRADANE, J. (1). 'On the History of Chromatography.' *N*, 1951, **167**, 120.

FARRER, R. (1). *On the Eaves of the World*. 2 vols. Arnold, London, 1917.

FARRER, R. (2). *The Rainbow Bridge*. Arnold, London, 1926.

FAUVEL, A. A. (1). *Promenades d'un Naturaliste dans l'Archipel de Chusan* [Chou-shan] *et sur les Côtes de Chekiang*. Cherbourg, 1880.

FÉE, A. L. A. (1). *Flore de Virgile, ou Nomenclature méthodique et critique des Plantes, Fruits et Produits Végétaux mentionnés dans les Ouvrages du Prince des Poëtes Latins*. Didot, Paris, 1822.

FÉE, A. L. A. (2). *Commentaires sur la Botanique et la Matière Médicale de Pline*. 3 vols. Paris, 1833.

FÉE, A. L. A. (3). *Les Jussieu et la Méthode Naturelle*. Silbermann, Strasbourg, 1837. Orig. pub RDA 1837. (Discours d'Ouverture du Cours de Botanique de la Faculté de Médecine, 3 May 1837).

FÉE, A. L. A. (4). *Flore de Théocrite et des autres Bucoliques Grecs*. Paris, 1832.

FEINSTEIN, L. & JACOBSON, M. (1). 'Insecticides occurring in the Higher Plants.' *FCON*, 1953, **10**, 423.

FERRARI, J. B. (1). *Hesperides, sive De Malorum Aureorum Cultura et Usu libri IV*. Rome, 1646.

FINAN, J. J. (1). *Maize in the Great Herbals*. Chronica Botanica, Waltham, Mass., 1950.

FINET, A. & GAGNEPAIN, F. (1). 'Contributions à la Flore de l'Asie Orientale.' *BSBF*, 1903–4. Sep. pub. Libr. Impr. Reúnies, Paris, 1905.

FISCHER, HERMANN (1). *Mittelalterliche Pflanzenkunde*. Münchner Drucke, Munich, 1929. Repr., with foreword by J. Stendel, 1966.

FITZHERBERT, S. W. (1). 'The "Cherokee rose", *Rosa laevigata*, and its Forms.' *FLOSIL*, 1903, **1**, 294.

FLORKIN, M. (1). *Naissance et Déviation de la Théorie Cellulaire dans l'Oeuvre de Théodore Schwann*. Hermann, Paris, 1960; Vaillant-Carmanne, Liège, 1960. (Actualités Sci. & Industr. no. 1282.).

FLORKIN, M. (2) (ed.). 'Lettres de Theódore Schwann.' *MSRSL*, 1961 (5e sér.), **2** (no. 3), 1–274.

FORBES, F. B. & HEMSLEY, W. B. (1) with more than twelve collaborators. *Index Florae Sinensis*; Enumeration of all the Plants known from China Proper, Formosa, Hainan, Korea, the Liu-Chu Archipelago and the island of Hongkong, together with their Distribution and Synonyms. *JLS/B* 1886 **23** 1–489 (521); 1889 **26** 1–592; 1905 **36** 1–449 (686). Also sep. pub. 3 vols. London, 1906; Photolitho reprint, Peking, 1938 (with original pagination and addition of the Latin title). Historical preface to vol. 3 also pub. in *KBMI*, 1905, 64. First supplement by M. Smith (1). Second supplement by S. T. Dunn (1). See Merrill & Walker (1), p. 122.

FORBES, R. J. (12). *Studies in Ancient Technology*. Vol. 3, *Cosmetics and Perfumes in Antiquity; Food, Alcoholic Beverages, Vinegar; Food in Classical Antiquity; Fermented Beverages*, −500 to +1500; *Crushing; Salts, Preservation Processes, Mummification; Paints, Pigments, Inks and Varnishes*. Brill, Leiden, 1955.—(Crit. Lynn White, *ISIS*, 1957, **48**, 77.).

FORBES, R. J. (14). *Studies in Ancient Technology*. Vol. 5, *Leather in Antiquity; Sugar and its Substitutes in Antiquity; Glass*. Brill, Leiden, 1957.

FORBES, R. J. (18). 'Food and Drink [from the Renaissance to the Industrial Revolution].' Art. in *A History of Technology*, ed. C. Singer *et al.*, vol. 3, p. 1. Oxford, 1957.

[FORD, C.] (1). *Index of Chinese Plants in the 'Journal of Botany', vols. 1–18*. Noronha, Hongkong, 1883.

FORKE, A. (4) (tr.). '*Lun Hêng*', *Philosophical Essays of Wang Chhung*.
Vol. 1, 1907 Kelly & Walsh, Shanghai; Luzac, London; Harrassowitz, Leipzig.
Vol. 2, 1911 (with the addition of Reimer, Berlin) (*MSOS*, Beibände **10** and **14**).
Photolitho repr., Paragon, New York, 1962.

FORKE, A. (9). *Geschichte d. neueren chinesischen Philosophie* (i.e. from beg. of Sung to modern times). de Gruyter, Hamburg, 1938. (Hansische Univ. Abhdl. a.d. Geb. d. Auslandskunde, no. 46 (Ser. B, no. 25).)

FORKE, A. (20). *Yen Ying, Staatsmann und Philosoph, und das 'Yen-Tsê Tsch'un-Tch'iu'* [Yen Tzu Chhun Chhiu]. Hirth Anniversary Volume (*Asia Major* Introductory Volume). N. d. (1923), pp. 101–44. Abridged version in Forke (13), p. 82.

FORTUNE, R. (1). *Two Visits to the Tea Countries of China, and the British Tea Plantations in the Himalayas, with a Narrative of Adventures, and a Full Description of the Culture of the Tea Plant, the Agriculture, Horticulture and Botany of China*. 2 vols. Murray, London, 1853.

FORTUNE, R. (2). *Three Years' Wanderings in the Northern Provinces of China, including a Visit to the Tea, Silk and Cotton Countries; with an Account of the Agriculture and Horticulture of the Chinese, New Plants, etc*. Murray, London, 1847. Abridged as vol. 1 of Fortune (1).

FORTUNE, R. (3). *Journey to the Tea Countries*. Murray, London, 1852. Abridged as vol. 2 of Fortune (1).

FORTUNE, R. (4). *A Residence among the Chinese; Inland, on the Coast and at Sea; being a Narrative of Scenes and Adventures during a Third Visit to China from 1853 to 1856, including Notices of many Natural Productions and Works of Art, the Culture of Silk etc., with Suggestions on the Present War.* Murray, London, 1857.

FORTUNE, R. (5). *Yedo and Peking; a Narrative of a Journey to the Capitals of Japan and China, with Notices of the Natural Productions, Agriculture, Horticulture and Trade of those Countries, and other things met with by the Way.* Murray, London, 1863.

FORTUNE, R. (6). 'Sketch of a Visit to China, in search of New Plants.' *JRHS*, 1846, **1**, 208.

FORTUNE, R. (7). 'The Chinese Tree-Peony.' *GDC*, 1880 (n.s.), **13**, 179.

FOURNIER, P. (1). *Voyages et Découvertes Scientifiques des Missionnaires Naturalistes Français à travers le Monde pendant cinq siècles* (15ᵉ au 20ᵉ). Lechevalier, Paris, 1932. 2 vols.

    Vol. 1 'Les Voyageurs Naturalistes du Clergé Français avant la Révolution.'

    Vol. 2 'La Contribution des Missionnaires Français au progrès des Sciences Naturelles aux 19ᵉ et 20ᵉ siècles.'

FOX, HELEN M. (1) (tr. & ed.). *Abbé David's Diary; being an Account of the French Naturalist's Journeys and Observations in China in the Years 1866 to 1869.* Harvard Univ. Press, Cambridge, Mass, 1949.

FRANCHET, A. (1). *Plantae Davidianae ex Sinarum Imperio.* 2 vols.

    Vol. 1 'Plantes de Mongolie du Nord, et du Centre de la Chine.'

    Vol. 2 'Plantes du Thibet Oriental.' Masson, Paris, 1884.

FRANCHET, A. (2). *Plantae Delavayanae sive Enumeratio Plantarum quas in Provincia Chinensi Yunnan collegit.* J. M. Delavaye. Masson, Paris, 1884.

FRANCHET, A. & SAVATIER, L. (1). *Enumeratio Plantarum in Japonia sponte crescentium, hujusque cognitarum adjectis descriptionibus specierum pro regione novarum, quibus accedit determinatio herbarum in libris japonicis Somoku Zoussets* [Sōmoku Zusetsu] *Zylographice delineatum.* 2 vols. Savy, Paris, 1875–9. The *Sōmoku Zusetsu* (Iconography of Japanese Plants) in 50 *pên* had been compiled by Iinuma Yokusai, 1832, 1856; cf. Bartlett & Shohara (1), p. 145.

FRANKE, H. (19) (ed.). *Sung Biographies.* 3 vols, Steiner, Wiesbaden, 1976. (Münchener Ostasiatische Studien, no. 16 pts 1, 2, and 3.).

FRANKE, O. (9). 'Zwei wichtige literarische Erwerbungen des Seminars für Sprache und Kultur Chinas zu Hamburg' (on the *Yung Lo Ta Tien* and the *Thu Shu Chi Chhêng*). *Jahrb. d. Hamburgischen wissenschaftl. Anstalten*, 1914, **32**. (Reprinted in Franke (8), p. 91.).

FREAR, D. E. H. (1). *A Catalogue of Insecticides and Fungicides.* 2 vols. Chronica Botanica, Waltham, Mass., 1950. Orig. appeared, in *ACPP*, 1947, **7** and 1948, **8**.

FREAR, D. E. H. (2). *The Chemistry of Insecticides and Fungicides.*

FRIES, T. M. (1). *Linné, Lefnadsteckning.* 2 vols. Stockholm, 1903. Abbreviated English version by B. Daydon Johnson (1).

FU FÊNG-YUNG & CHANG CHHANG-SHAO (1). 'Chemotherapeutic Studies on *chhang shan* (*Dichroa febrifuga*): III. Potent Antimalarial Alkaloids from *chhang shan*.' *STIC*, 1948, **1** (no. 3), 56.

FUCHS, LEONHARD (1). *De Historia Stirpium ...* Isingrin, Basel, 1542. Repr. 1545. German ed. *New Kreüterbuch.* Isingrin, Basel, 1543.

FULDER, S. (1). 'Ginseng; useless Root or Subtle Medicine?' *NS*, 1977, **73**, 138.

GALLAGHER, L. J. (1) (tr.). *China in the 16th Century; the Journals of Matthew Ricci, 1583–1610.* Random House, New York, 1953. [A complete translation, preceded by inadequate bibliographical details, of Nicholas Trigault's *De Christiana Expeditione apud Sinas* (1615).] Based on an earlier publication: *The China that Was; China as discovered by the Jesuits at the close of the 16th Century: from the Latin of Nicholas Tregault.* Milwaukee, 1942. [Identifications of Chinese names in Yang Lien-Shêng (4).].

GALLESIO, G. (1). *Traité du Citrus.* Paris, 1811. Eng. tr. *Orange Culture; a Treatise on the Citrus Family.* Jacksonville, Florida, 1876.

GARDENER, W. B. (1). 'The Development of Comprehensive Records of the Chinese Flora.' In the press for *JSBNH.*

GARDENER, W. B. (2). 'A Summary of the Chinese Dogwoods.' *PLM*, 1979, **1** (no. 2), 85.

GARDENER, W. B. (3). 'A Note on *Stellera chemaejasme* L. (Thymelaceae.) *JRHS*, 1979, **104**, 464.

GARDNER, C. S. (3). *Chinese Traditional Historiography.* Harvard Univ. Press, Cambridge, Mass., 1938; repr. 1961. (Harvard Historical Monographs no. 11.)

GARNIER, P. (1). *Essoi de Phytonymie Populaire comparée (noms populaires des plantes) du Lituanien au Zoulou et du Navaho au Chinois; Études et Recherches de Corrélations en une Cinquantaine de Langues.* Inaug. Diss., Lyon, 1983.

GARRISON, FIELDING H. (1). 'History of Drainage, Irrigation, Sewage-Disposal, and Water-Supply.' *BNYAM*, 1929, **5**, 887.

GARRISON, FIELDING H. (3). *An Introduction to the History of Medicine; with Medical Chronology, Suggestions for Study, and Bibliographic Data*, 4th ed. Saunders, Philadelphia and London, 1929.

GARVEN, H. S. D. (1). *Wild Flowers of North China and South Manchuria*, Peking Nat. Hist. Bull. French Bookstore, Peiping, 1937, (PNHB Handbooks, no. 5.).

GAUGER, GUSTAV (1). 'Chinesische Roharzneiwaren.' *RPPCR*, 1848, **7** (no. 12), 565. (Descriptions and drawings of 54 Chinese drugplants based on the collection formed by P. Y. Kirilov of the Russian Ecclesiastical Mission in

Peking.) Abstr. in *RPHARM*, 1848, **100**, 662, Cf. Merrill & Walker (1), vol. 1, p. 135.

GEMMILL, C. L. (1). 'Silphium'. *BIHM*, 1966, **40**, 295.

GERASIMOV, I. P. (1). 'The Chief Genetic Soil Types of China and their Geographical Distribution' (in Russian). *POCH*, 1958 (no. 1), 3.

GERASIMOV, I. P. & GLAZOVSKAIA, M. A. (1).
Osnovi Pochvovedeniya i Geografia Pochv. Gosudarstvennoe Izdatelstvo Geograficheskoi Literatury Moscow, 1960. Eng. tr.:
Fundamentals of Soil Science and Soil Geography. Israel Programme for Scientific Translations, Jerusalem, Israel, 1965.

GERINI, G. E. (1). *Researches on Ptolemys Geography of Eastern Asia (Further India and Indo-Malay Peninsula)*. Royal Asiatic Society and Royal Geographical Society, London, 1909. (Asiatic Society Monographs, no. 1.)

GIBERT, L. (1). *Dictionnaire Historique et Géographique de la Mandchourie*. Société des Missions Etrangères, Hong Kong, 1934.

GIBSON, H. E. (3). 'Agriculture in the Shang Pictographs.' Appendix B in Sowerby (1), p. 180.

GILES, H. A. (1). *A Chinese Biographical Dictionary*. 2 vols. Kelly & Walsh, Shanghai, 1898; Quaritch, London, 1898. Supplementary Index by J. V. Gillis & Yü Ping-Yüeh, Peiping, 1936. Account must be taken of the numerous emendations published by von Zach (4) and Pelliot (34), but many mistakes remain. Cf. Pelliot (35).

GILES, H. A. (2). *Chinese-English Dictionary*. Quaritch, London 1892, 2nd ed. 1912.

GILES, L. (2). *An Alphabetical Index to the Chinese Encyclopaedia (Chhin Ting Ku Chin Thu Shu Chi Chhêng)*. British Museum, London, 1911.

GILMOUR, J. S. L. (1). 'Gardening Books of the Eighteenth Century.' Art. in *Catalogue of Botanical Books in the Collection of R. McM. M. Hunt*. Hunt Foundation, Pittsburgh, Pa, 1961, vol. II, p. 1.

GLIDDEN, H. W. (1). 'The Lemon in Asia and Europe' *JAOS*, 1937, **57**, 381.

GODWIN, H. (1). 'The Ancient Cultivation of Hemp.' *AQ*, 1967, **41**, 42. (The last three pages were inadvertently omitted but printed in the following number; the offprints are complete.)

GODWIN, H. (2). 'Pollen-Analytic Evidence for the Cultivation of Cannabis in England.' *RPBP*, 1967, **4**, 71.

GODWIN, H., WALKER, D. & WILLIS, E. H. (1). 'Radiocarbon Dating and Post-Glacial Vegetational History; Scaleby Moss.' *PRSB*, 1957, **147**, 352.

GODWIN, H. & WILLIS, E. H. (1).
'Cambridge University Natural Radiocarbon Measurements, 1.' *AJSC* (Radiocarbon Supplement), 1959, **1**, 63.
'Radiocarbon Dating of the Late-Glacial Period in Britain.' *PRSB*, 1959, **150**, 199.

GOERKE, H. (1). *Carl von Linné–Artz, Naturforscher, Systematiker, +1707 bis +1778*. Liebing, Würzburg, 1966 (Grosse Naturforscher Series, no. 31). Rev. D. Guinot *A/AIHS*, 1966, **19**, 400.

GOOD, R. (1). *The Geography of the Flowering Plants*. Longmans, London, 1947; 3rd ed. 1964.

GOODRICH, L. CARRINGTON (1). *Short History of the Chinese People*. Harper, New York, 1943.

GOODRICH, L. CARRINGTON (3). 'Cotton in China.' *ISIS*, 1943, **34**, 408.

GOODRICH, L. CARRINGTON (18). 'Some Bibliographical Notes on Eastern Asiatic Botany.' *JAOS*, 1940, **60**, 258. A sequel to his review of Merrill & Walker (1) in *JAOS*, 1939, **59**, 138; and mainly on Hu Ssu-Hui's *Yin Shan Chêng Yao*.

GOODRICH, L. CARRINGTON (20). 'Early Notices of the Peanut.' *MS*, **2**, 405.

GOODRICH, L. CARRINGTON & FANG CHAO-YING (1) (ed.). *Dictionary of Ming Biography, +1368 to +1644*. 2 vols. Columbia Univ. Press, New York, 1976.

GOODRICH, L. CARRINGTON & WILBUR, C. M. (1). 'Additional Notes on Tea.' *JAOS*, 1942, **62**, 195.

GORDEEV, T. P. & JERNAKOV, V. N. (1). 'Material Related to the Study of the Soils and Plant Associations of Northeastern China and the Autonomous Region of Inner Mongolia, collected in 1950.' *APS*, 1954, **2**, 270.

GOTO, S. (1). 'Le Goût Scientifique de Khang-Hsi, Empereur de Chine.' *BMFJ*, 1933, **4**, 117.

GOULD, S. W. (1). 'Permanent Numbers to supplement the Binomial System of Nomenclature.' *AMSC*, 1954, **42**, 269.

GOURLIE, NORAH (1). *The Prince of Botanists* [Linnaeus]. London, 1953.

GOUROU, P. (1a). *La Terre et l'Homme en Extrême-Orient*. Colin, Paris, 1947.

GOUROU, P. (1b). 'Notes on China's Unused Uplands.' *PA*, 1948, **21**, 227.

GOW, A. S. F. & SCHOLFIELD, A. F. (1) (ed. & tr.). *Nicander; the Poems and Poetical Fragments*. Cambridge, 1953.

GRAHAM, A. C. (1). *The Philosophy of Chhêng I-Chhuan (+1033 to +1107) and Chhêng Ming-Tao (+1032 to +1085)*. Inaug. Diss. London, 1953. *Two Chinese Philosophers* Lund Humphries, London, 1958, see (15). Rev. Chhen Jung-Chieh, *JAOS*, 1959, **79**, 150.

GRANEL, F. (1). 'Les Étapes Scientifiques d'Auguste [P.M.A.] Broussonet.' *MHP*, 1967, **10** (no. 37), 25.

GRANET, M. (1). *Danses et Légendes de la Chine Ancienne*. 2 vols. Alcan, Paris, 1926.

GRANET, M. (2). *Fêtes et Chansons Anciennes de la Chine*. Alcan, Paris, 1926; 2nd ed. Leroux, Paris, 1929. Eng. tr. by E. D. Edwards, Routledge, London, 1932.

GRANET, M. (4). *La Religion des Chinois*. Gauthier-Villars, Paris, 1922.

GRANET, M. (5). *La Pensée Chinoise*. Albin Michel, Paris, 1934. (Evol. de l'Hum. series, no. 25 *bis*.)

GRANT, C. J. (1). *The Soils and Agriculture of Hongkong*. Govt. Press, Hongkong, 1960.

GRASSMANN, H. (1). 'Der Campherbaum.' *MDGNVO*, 1895, **6**, 277.

GRAY, ASA (1). 'Analogy between the Flora of Japan and that of the United States.' *AJSC*, 1846 (2nd ser.), **2**, 135.

GRAY, ASA (2). 'Diagnostic Characters of New Species of Phanerogamous Plants collected in Japan by Charles Wright, Botanist of the U.S. North Pacific Exploring Expedition, with Observations upon the Relations of the Japanese Flora to that of North America and of other parts of the North Temperate Zone.' *MAAAS*, 1859 (n.s.), **6**, 377. Address also in *PAMAAS*, 1872, **31**.

GRAY, ASA (3). 'Forest Geography and Archaeology.' *AJSC*, 1878 (3rd ser.), **16**, 85, 183.

GREENE, E. L. (1). *Landmarks of Botanical History; a Study of Certain Epochs in the Development of the Science of Botany*. Pt. 1, Prior to +1562 (all pub.) Smithsonian Institution, Washington, 1909 (Smithsonian Miscellaneous Collections **54**, no. 1870.) Pt. 2, with extensive notes on both parts by F. N. Egerton, in the press.

GREW, NEHEMIAH (1). *The Anatomy of Vegetables Begun*. London, 1672.

GREW, NEHEMIAH (2). *The Anatomy of Plants*. London, 1682.

GRISEBACH, A. (1). *Die Vegetation der Erde*. Göttingen, 1872.

GROFF, G. W. & HOWARD, C. W. (1). 'The Cultured Citrus Ants of South China.' *LAR*, 1924, **2** (no. 2), 108.

GUIHENEUF, D. & BEAN, W. J. (1). 'Double-flowered Peaches.' *GDN*, 1899, **56**, 516.

VAN GULIK, R. H. (8). *Sexual Life in Ancient China; a preliminary Survey of Chinese Sex and Society from c. −1500 to +1644*. Brill, Leiden, 1961. Rev. R. A. Stein, *JA*, 1962, **250**, 640.

GUNTHER, R. T. (3) (ed.). *The Greek Herbal of Dioscorides, illustrated by a Byzantine in +512, englished by John Goodyer in +1655, edited and first printed, 1933*. Pr. pr. Oxford, 1934, photolitho repr. Hafner, New York, 1959.

HAAS, H. (1). *Spiegel der Arznei; Ursprung, Geschichte und Idee der Heilmittelkunde*. Springer, Berlin, Göttingen and Heidelberg, 1956.

HAAS, P. & HILL, T. G. (1). *Introduction to the Chemistry of Plant Products*. 2 vols. Longmans Green, London, 1928.

HADIDIAN, Z. (1). 'Proteolytic Activity and Physiological and Pharmacological Actions of *Agkistrodon piscivorus* [water-moccasin] Venom.' Art. in *Venoms*, Papers Presented to the 1st. Internat. Conference on Venoms, Berkeley, Calif. 1954, ed. E. E. Buckley & N. Porges, (Amer. Ass. Adv. Sci. Pub. no. 44.) Berkeley, 1956, p. 205.

HAGBERG, K. (1). *Carl Linnaeus*. Stockholm, 1939. Eng. tr. by A. Blair, London, 1952.

HAGERTY, M. J. (1) (tr.). (with Chiang Khang-Hu). 'Han Yen-Chih's *Chü Lu* (Monograph on the Oranges of Wên-Chou, Chekiang),' with introduction by P. Pelliot. *TP*, 1923, **22**, 63.

HAGERTY, M. J. (2) (tr. and annot.). 'Tai Khai-Chih's *Chu Phu*; a Fifth-Century Monograph on Bamboos written in Rhyme with a Commentary.' *HJAS*, 1948, **11**, 372.

HAGERTY, M. J. (3) (with Wu Mien) (tr.). 'The *Mu Mien Phu* (Treatise on Cotton), by Chhu Hua (*c.* +1785); a draft Translation.' Unpublished *MS*, 1927.

HAGERTY, M. J. (4) (tr.). 'Translation of the Description of *mu mien*, cotton (*Gossypium herbaceum*) as found in the *Thu Shu Chi Chhêng*, Tshao mu tien, ch. 303. Unpublished MS, n.d.

HAGERTY, M. J. (4a) (tr.). 'Supplementary Translation of the Account of Cotton in Li Shih-Chen's *Pên Tshao Kang Mu*, ch. 36.' Unpublished MS, n.d.

HAGERTY, M. J. (4b) (tr.). 'Supplementary Translation of the Explanation of the Characters in the Names *chi pei* and *mu mien* given in Yü Chêng-Hsieh's collection of Essays entitled *Khuei Ssu Lei Kao*, ch. 7.' Unpublished MS, n.d.

HAGERTY, M. J. (4c) (tr.). 'Translation of an Article entitled "Mu Mien Khao" or "Researches concerning Cotton" in ch. 14 of the *Khuei Ssu Lei Kao*, a Collection of Writings made in the Khuei-Ssu year (1833) by Yü Chêng-Hsieh (T. Li-Chhu), a man of the Manchu Period and a native of Huichow in Anhui.' Unpublished MS, n.d.

HAGERTY, M. J. (6) (with Wu Hsien) (tr.). 'Draft Translation of the "Ya Pien Yen Shih Shu" or "Historical Account of Opium in China", in the *Khuei Ssu Lei Kao* by Yü Chêng-Hsieh (1833).' Unpublished MS, 1928.

HAGERTY, M. J. (7) (with Wu Mien) (tr.). 'Yang mei (*Myrica rubra* = *sapida*) the Chinese strawberry; the complete Account as given in the *Thu Shu Chi Chhêng*, Tshao mu tien, ch. 278.' Unpublished MS, 1926.

HAGERTY, M. J. (8) (tr.). 'Mang tshao (*Illicium religiosum*); the Description in the *Chih Wu Ming Shih Thu Khao* (Illustrated section), ch. 24.' Unpublished MS, n.d.

HAGERTY, M. J. (9) (tr.). 'Pai mao (*Imperata arundinacea*); the Description in Li Shih-Chen's *Pên Tshao Kang Mu*, ch. 13.' Unpublished MS, n.d.

HAGERTY, M. J. (10) (tr.). 'Ta fêng tzu (seeds of *Hydnocarpus*); the Description in Li Shih-Chen's *Pên Tshao Kang Mu*, ch. 35 B.' Unpublished MS, n.d.

HAGERTY, M. J. (11) (tr.). 'Translation of Material concerning Citrous Fruits from the *Nung Chêng Chhüan Shu*, ch. 30.' Unpublished MS, 1917.

HAGERTY, M. J. (12). 'A Description of the *Chhi Min Yao Shu* [(Important Arts for the People's Welfare) by Chia Ssu-Hsieh, *c.* +540], with a Translation of the Titles of its 92 section-headings.' Unpublished MS, n.d.

HAGERTY, M. J. (12a) (tr.). 'Draft Translation of the Preface by Chia Ssu-Hsieh for his *Chhi Min Yao Shu*.' Unpublished MS, 1938.

HAGERTY, M. J. (12b) (tr.). 'The *Chhi Min Yao Shu* on the *shih* or persimmon (*Diospyros kaki*), ch. 40.' Unpublished MS, 1938.

HAGERTY, M. J. (13). 'Draft Contributions to a Dictionary or Manual of Chinese Economic Plants.' Unpublished MS, n.d. (Entries for genera *Aleurites, Calophyllum, Coix lachryma-Jobi, Colocasia, Crataegus, Eleocharis, Erythrina, Eucommia, Euryale, Firmiana, Paulownia, Phellodendron, Populus, Sterculia, Tribulus.*)

HAGERTY, M. J. (14). '*Tu chung*, or *Eucommia ulmoides*; a translation of the Description given in Li Shih-Chen's *Pên Tshao Kang Mu*, ch. 35A.' Unpublished MS, n.d.

HAGERTY, M. J. (15). '(The Preface of the) *Shen Nung Pên Tshao Ching* [Classical Pharmacopoeia of the Heavenly Husbandman]; (according to the) account given in Li Shih-Chen's *Pên Tshao Kang Mu* (under the heading) Shen Nung Pên Ching Ming Li (Terminology and Arrangement of the Materia Medica of Shen Nung), in ch. 1A, pp. 43*b* to 55*b*—a tentative draft translation [of the commentaries as well as the main text].' Unpublished MS.

HAGERTY, M. J. (16) (tr.). 'The *Lo-yang Mu-Tan Chi*, or Treatise on the Tree-Peonies of Loyang, Honan, by Ouyang Hsiu of the Sung period, with textual notes by Chou Pi-Ta, also of the Sung period.' Unpublished MS, n.d.

HAHN, E. (1). 'Das Auftreten des Hopfens bei der Bierbereitung und seine Verbreitung in der Frühgeschichte der Völker.' In Huber, E. (3), *Bier und Bierbereitung bei den Völkern d. Urzeit*, vol. 2, p. 9.

EL HAIDARI, H. S. (1). 'The Use of Predator Ants for the Control of Date Palm Insect Pests in the Yemen Arab Republic.' *Date Palm J.*, 1981, **1** (1), 129–130.

DU HALDE, J. B. (1). *Description Geographique, Historique, Chronologique Politique et Physique de l'Empire de la Chine et de la Tartarie Chinoise.* 4 vols. Paris, 1735, 1739; The Hague 1736. English tr., R. Brookes, London, 1736, 1741. German tr. Rostock, 1748.

HALES, STEPHEN (1). *Vegetable Staticks; or, an Account of some Statical Experiments on the Sap in Vegetables, being an Essay towards a Natural History of Vegetation; also, a Specimen of an Attempt to Analyse the Air by a great variety of Chymio-Statical Experiments; which were read at several Meetings before the Royal Society.* Innys & Woodward, London, 1727. Repr. with foreword by M. A. Hoskin, Oldbourne, London, 1961.

VON HALLER, ALBRECHT (1). *Bibliotheca Botanica, qua Scripta ad Rem Herbariam facientia a Rerum initiis recensentur Autore A. H. cum additionibus Abraham Kall et indice emendato a. J. Christiano Bay perfecto.* Orell, Gessner & Fuessli, Tiguri (Zürich), 1771–2. Photolitho repr. Forni, Bologna, 1965.

HANBURY, DANIEL (1). *Science Papers, chiefly Pharmacological and Botanical.* Macmillan, London, 1876.

HANBURY, DANIEL (2). 'Notes on Chinese Materia Medica.' *PJ*, 1861, **2**, 15, 109, 553; 1862, **3**, 6, 204, 260, 315, 420. German tr. by W. C. Martins, (without Chinese characters), *Beiträge z. Materia Medica Chinas.* Kranzbühler, Speyer, 1863. Revised version, with additional notes, references and map, in Hanbury (1), pp. 211 ff.

HANBURY, DANIEL (4). 'Note upon a Green Dye from China.' *PJ*, 1856, **16**, 213. Repr. in Hanbury (1), pp. 125 ff.

HANBURY, DANIEL (5). 'Some Rare Kinds of Cardamom.' *PJ*, 1855, **14**, 352, 416. Repr. in Hanbury (1), pp. 93 ff.

HANBURY, DANIEL (10). 'Illustrated MS. List of Drugs, with Chinese Characters, Transliterations and Notes, prepared for the exhibition of H. W. Carey.' P 273 MS [1–80], Library of the Pharmaceutical Society, London.

HANCE, H. F. (2). '*Flora Hongkongensis* προσθήκή; a Compendious Supplement to Mr. Bentham's Description of the Plants of the Island of Hongkong.' *JLS/B*, 1873, **13**, 95 and sep. (No illustrations, no Chinese names or characters.)

HANDEL-MAZZETTI, H. (1). 'Das Pflanzengeographische Gliederung und Stellung Chinas.' *BOTJ*, 1931, **64**, 309.

HANDEL-MAZZETTI, H. (2). 'The Phytogeographic Structure and Affinities of China.' (Contribution to Symp. on the *Flora of China*.) Proc. 5th Internat. Botanical Congr., Cambridge, 1930. (Cambridge, 1931), p. 513. English abridgment of (1).

HANDEL-MAZZETTI, H. (3). 'Mittel China,' *VB*, 1922, **14** (no. 2/3), pls. 7–18. Illustrated account of his floristic region ④.

HANDEL-MAZZETTI, H. (4). 'Hochland und Hochgebirge von Yünnan und Südwest-Szechuan; I, Die subtropische und warmtemperierte Stufe.' *VB*, 1930, **20** (no. 7), pls. 37–42. Illustrated account of his floristic region ⑥.

HANDEL-MAZZETTI, H. (5). 'Hochland und Hochgebirge von Yünnan und Südwest-Szechuan; II, Die temperierte Stufe.' *VB*, 1932, **22** (no. 8), pls. 43–8. Illustrated account of his floristic region ⑥.

HANDEL-MAZZETTI, H. (6). 'Hochland und Hochgebirge von Yünnan und Südwest-Szechuan; III, Die Kalttemperierte u. Hochgebirgs Stufe.' *VB*, 1937, **25** (no. 2), pls. 7–12. Illustrated account of his floristic region ⑥.

HANDEL-MAZZETTI, H. (7). 'Das Nordost-Birmanisch-West-Yünnanesische Hochbirgsgebiet.' *VB*, 1927, **17** (no. 7/8), pls. 37A–48. Illustrated account of his floristic region ⑧. With map of his floristic regions as a whole.

HANDEL-MAZZETTI, H. (8). *Naturbilder aus Südwest-China.* Vienna & Leipzig, 1927.

HANDEL-MAZZETTI, H. (9) (ed.). *Symbolae Sinicae; botanische Ergebnisse d. Expedition d. Akad. d. Wissenschaft in Wien nach Südwest China, 1914–18.* 7 vols. in 3 pts. Vienna, 1929–37.

HANGER, F. (1). *Camellias and their Culture. JRHS*, 1947, **72**, 59.

HAO CHIN-SHEN (1). *Synopsis of Chinese 'Salix.'* Berlin, 1936. (Repertorium Specierum Novarum Regni Vegetabilis, Beiheft no. 93.)

HAO KIN-SHEN. See Hao Chin-Shen.

HARDING, ALICE (MRS. E.) (1). *The Book of the Peony*. Lippincott, Philadelphia and London, 1917.

HARTWELL, ROBERT M. (2). 'A Revolution in the Chinese Iron and Coal Industries during the Northern Sung (+960 to +1126).' *JAS*, 1962, **21**, 153.

HARTWELL, ROBERT M. (3). 'Markets, Technology and the Structure of Enterprise in the Development of the +11th-Century Chinese Iron and Steel Industry.' *JEH*, 1966, **26**, 29.

HARTWELL, ROBERT M. (4). 'A Cycle of Economic Change in Imperial China; Coal and Iron in North-east China, +750 to +1350.' *JESHO*, 1967, **10** (pt. 1), 102.

HARVEY-GIBSON, R. J. (1). *Outlines of the History of Botany*.

HARWORTH-BOOTH, M. (1). *The Hydrangeas*. Constable, London, 1950.

HARWORTH-BOOTH, M. (2). 'Further Notes on Hydrangeas.' *JRHS*, 1948, **73**, 112.

HAUKSBEE, FRANCIS (1). 'A Description of the Apparatus for making Experiments on the Refractions of Fluids; with a Table of the Specifick Gravities, Angles of Observations, and Ratio of Refractions of Several Fluids.' *PTRS*, 1710, **27**, 204.

HAWKES, A. D. (1). *Encyclopaedia of Cultivated Orchids*. Faber & Faber, London, 1965.

HAWKES, D. (1) (tr.). *'Chhu Tzhu'; the Songs of the South—an Ancient Chinese Anthology*. Oxford, 1959. (Rev. J. Needham, *NSN*, 18 Jul. 1959.)

HAZARD B. H., HOYT, J., KIM HATHAE, SMITH, W. W. & MARCUS, R. (1). *Korean Studies Guide*. Univ. Calif. Press, Berkeley and Los Angeles, 1954.

HEGI, G. (1). *Illustrierte Flora von Mittel-Europa*. 2nd ed. 2 vols.

HEHN, V. (1). *Kulturpflanzen und Hausthiere in ihren Übergang aus Asien nach Griechenland und Italien so wie in das übrige Europa*. 4th ed., Berlin, 1883; 8th ed., Berlin, 1911.

VON HEINE-GELDERN, R. & EKHOLM, G. F. (1). 173. 'Significant Parallels in the Symbolic Arts of Southern Asia and Middle America.' *Proc. XXVIIth Internat. Congr. Americanists*, vol. 1, p. 299. New York, 1949 (1951).

HELLER, J. L. & STEARN, W. T. (1). 'An Appendix to the [Facsimile Edition of the] *Species Plantarum* of Carl Linnaeus.' Postfaced to the Ray Society Facsimile Edition of the 1st edition, Stockholm, 1753. 2 vols. Quaritch, London, 1957, 1959. Vol. 2, at end (Ray Soc. Pubs. no. 142).

HEMSLEY, W. B. (1). 'The Wild Progenitors of the Chrysanthemum.' *JRHS*, 1890, **12**, 111.

HEMSLEY, W. B. (2). 'The Hydrangeas (with a coloured plate of *Hydrangea paniculata* var. *grandiflora*).' *GDN*, 1876, **10**, 264.

HENREY, BLANCHE (1), (ed.). *British Botanical and Horticultural Literature before 1800*. Oxford, 1974.

HENRY, AUGUSTINE (1). 'Chinese Names of Plants' [in Colloquial Use at Ichang.] *JRAS/NCB*, 1887 (n.s.), **22**, 233–83.

HENRY, AUGUSTINE (2). 'The Wild Forms of the Chrysanthemum.' *GDC*, 1902 (3rd ser.), **31**, 301.

HENRY, L. (1). 'Les Pêchers de Chine à Fleurs Doubles.' *JDN*, 1892, **6**, 93.

HENRY, L. (2). 'L'Amandier de David.' *RHORT*, 1902, 290.

HERBARIUS ZU TENTSCH. See Anon. (77).

HERDAN, G. (1). *The Calculus of Linguistic Observations*. Monton, 's-Gravenhage, 1962 (Janua Linguarum, Series Major no. 9).

HERDAN, G. (2). *The Structuralistic Approach to Chinese Grammar and Vocabulary*. Mouton, The Hague, 1964 (Janua Linguarum, Series Practica no. 6).

HERKLOTS, G. A. C. (1). 'The Cultivated Orchids in Hongkong.' *HKHS/ON*, 1933, **2**, 1–23.

HERKLOTS, G. A. C. (2). *Vegetable Cultivation in Hongkong*. South China Morning Post, Hongkong, 1947.

HERKLOTS, G. A. C. (3). *Vegetables in South-east Asia*. South China Morning Post, Hongkong, for Allen & Unwin, London, 1972.

HERRMANN, A. (1). *Historical and Commercial Atlas of China*. Harvard-Yenching Institute, Cambridge Mass., 1935. Crit. P. Pelliot, TP 1936, 363 2nd ed.: *An Historical Atlas of China*, ed. N. Ginsburg, with preface by P. Wheatley, Edinburgh Univ. Press, Edinburgh 1966; Aldine, Chicago, 1966; Rev. J. Needham *SACUN*, 1968 (Jun/Jul).

HERRMANN, A. (10). 'Die älteste Reichsgeographie Chinas und ihre Kulturgeschichtliche Bedentung.' *SA*, 1930, **5**, 232.

HERS, J. (1). 'Liste des Essences Ligneuses observées dans le Honan Septentrional' [List of Trees and Woody Shrubs recorded along the Lunghai Railway route in N. Honan]. n.p. (Chêngchow), 1922. First appeared in *JRAS/NCB*, **53**.

HERVEY, GEORGE (1). *The Goldfish of China in the Eighteenth Century*, with foreword by A. C. Moule. China Society, London, 1950. Based on de Sauvigny (1) and other +18th-century Memoirs.

D'HERVEY ST. DENYS, M. J. L. (2). *Recherches sur l'Agriculture et l'Horticulture des Chinois; et sur les Végétaux, les Animaux et les Procédés agricoles que l'on pourrait introduire avec avantage dans l'Europe occidentale et le Nord de l'Afrique, suivies d'une Analyse de la grande Encyclopédie 'Shou Shih Thung Khao.'* Allouard & Kaeppelin, Paris, 1850.

D'HERVEY ST. DENYS, M. J. L. (4). *Le 'Li Sao', Poème du 3e Siècle avant notre Ère, traduit et publié avec le Texte Original.* Paris, 1870.

HEYWOOD, V. H. (1). *Plant Taxonomy.* Arnold, London, 1970 (Institute of Biology Studies in Biology, no. 5).

HIGHTOWER, J. R. (1). *Topics in Chinese Literature; Outlines and Bibliographies.* Harvard Univ. Press, 1950 (Harvard-Yenching Institute Studies, no. 3).

HINTZSCHE, E. (1). 'Analyse des Berner Codex 350; ein bibliographischer Beitrag zur chinesischen Medizin und zu deren Kenntnis bei [Gulielmus] Fabricius Hildanus und [Albrecht von] Haller.' *GESN*, 1960, **17**, 99.

HINTZSCHE, E. (2). 'Über anatomische Tradition in der chinesischen Medizin' (on the *Wan Ping Hui Chhun* of Kung Thing-Hsien (+1615) deposited in a Swiss library by +1632). *MNFGB*, 1957, **14**, 81.

HIROE MINOSUKE (1). *Umbelliferae of Asia* (excluding Japan). Eikodo, Kyoto, 1958 (Synonymy, but no Chinese names or characters.).

HIRTH, F. (1). *China and the Roman Orient: Researches into their Ancient and Medieval Relations as represented in Old Chinese Records.* Kelly and Walsh, Shanghai, 1885. Photo-reprinted in China, 1939.

HIRTH, F. (2) (tr.). 'The Story of Chang Chhien, China's Pioneer in West Asia. *JAOS*, 1917, **37**, 89. (Translation of ch. 123 of the *Shih Chi*, containing Chang Chhien's Report; from §18–52 inclusive and 101 to 103. §98 runs on to §104, 99 and 100 being a separate interpolation. Also tr. of ch. 111 containing the biogr. of Chang Chhien.).

HIRTH, F. (24). [Notes on] Chinese Books' (*Phien Tzu Lei Phien*, etc.). *JRAS/NCB*, 1887, **22**, 109.

H[IRTH], F. & EDKINS, J. (1). '[Notes on] Chinese Books' (*Phei Wên Yün Fu, Thu Shu Chi Chhêng, Pên Tshao Mêng Chhüan*, etc.). *JRAS/NCB*, 1886, **21**, 321.

HIRTH, F. & ROCKHILL, W. W. (1) (tr.). *Chau Ju-Kua; His work on the Chinese and Arab Trade in the 12th and 13th centuries, entitled 'Chu-Fan-Chi'.* Imp. Acad. Sci., St Petersburg, 1911. (Crit. G. Vacca, *RSO*, 1913, **6**, 209; P. Pelliot, *TP* 1912, **13**, 446; E. Schaer, *AGNT* 1913, **6**, 329; O. Franke, *OAZ* 1913, **2**, 98; A. Vissière, *JA* 1914 sér.), **3**, 196.).

HITTI, P. K. (1). *History of the Arabs.* 4th ed. Macmillan, London, 1949; 6th ed. 1956.

HO, PING-TI (1). The introduction of American food plants into China, *AAN*, 1955, 57, **2**, pp. 191–201.

HO PING-TI (2). *The Ladder of Success in Imperial China; Aspects of Social Mobility,* +1368 to 1911. Columbia Univ. Press, New York, 1962.

HO PING-YÜ & NEEDHAM, JOSEPH (1). 'Ancient Chinese Observations of Solar Haloes and Parhelia,' *W*, 1959, **14**, 124.

HO PING-YÜ & NEEDHAM, JOSEPH (2). 'Theories of Categories in Early Mediaeval Chinese Alchemy' (with transl. of the *Tshan Thung Chhi Wu Hsiang Lei Pi Yao, c.* +6th to +8th. cent.). *JWCI*, 1959, **22**, 173.

HO PING-YÜ & NEEDHAM, JOSEPH (3). 'The Laboratory Equipment of the Early Mediaeval Chinese Alchemists.' *AX*, 1959, **7**, 57.

HO PING-YÜ & NEEDHAM, JOSEPH (4). 'Elixir Poisoning in Mediaeval China.' *JAN*, 1959, **48**, 221.

HOEPPLI, R. & CHHIANG I-HUNG (1). 'The Louse, Crab-louse and Bed-bug in Old Chinese Medical Literature, with special considerations on Phthiriasis.' *CMJ*, 1940, **58**, 338.

HOFFMANN, J. (1). 'Notes relating to the History, Distribution and Cultivation of the Peony in China and Japan.' *MBRF*, 1849, **16**, 85, 109. Tr. by Polman Mooy from 'Bijdragen tot de Geschiedenis, Verspreiding en Kultuur der Pioenen in China en Japan.' *JNMT*, 1848, 19.

VON HOFSTEN, NILS G. E. (1). *Zur älteren Geschichte des Diskontinuitätsproblems in der Biogeographie.* Kabitzsch, Würzburg, 1916. (From *ZANN*, **8**.).

HOLMES E. M. (1). 'The Asafoetida Plants.' *PJ*, 1889 (3rd ser.), **19**, 21, 41, 365.

HONEY, W. B. (2). *The Ceramic Art of China; and other Countries of the Far East.* Faber & Faber, London, 1945.

HOOK, B. (1), (ed.). *The Cambridge Encyclopaedia of China.* Cambridge, 1982.

HOOPER, D. (1). 'On Chinese Medicine; Drugs of Chinese Pharmacies in Malaya.' *GBSS*, 1929, **6** (no. 1), 1–163. (Chinese characters and romanisations of plant names given as well as Latin binomials.)

HOOPER, W. D. & ASH, H. B. (1). *Marcus Porcius Cato 'On Agriculture'; Marcus Terentius Varro 'On Agriculture'; with an English Translation . . .* Heinemann, London, 1954. (Loeb Classics ed.).

HOPKINS, L. C. (3). *The Development of Chinese Writing.* China Society, London, n.d.

HOPKINS, L. C. (4). 'L'Écriture dans l'Ancienne Chine.' *SCI*, 1920, **27**, 19.

HOPKINS, L. C. (10). 'Pictographic Reconnaissances, VI.' *JRAS*, 1924, 407.

HOPKINS, L. C. (11). 'Pictographic Reconnaissances, VII.' *JRAS*, 1926, 461

HOPKINS, L. C. (25). 'Metamorphic Stylisation and the Sabotage of Significance; a Study in Ancient and Modern Chinese Writing,' *JRAS*, 1925, 451.

HOPKINS, L. C. (27). 'Archaic Sons and Grandsons; a Study of a Chinese Complication Complex.' *JRAS*, 1934, 57.

HOPKINS, L. C. (31). 'The Cas-Chrom v. the Lei-Ssu; A Study of the Primitive Forms of Plough in Scotland and Ancient China.' I, *JRAS*, 1935, 707. II, *JRAS*, 1936, 45.

HOPKINS, L. C. (36) (tr.). *The Six Scripts or the Principles of Chinese Writing*, by Tai Thung. Amoy, 1881. Reprinted by photolitho, with a memoir of the translator by W. Perceval Yetts, Cambridge, 1954. Rev. J. Needham, *CAMR*, 1954.

HOPPE, BRIGITTE (1). *Das Kräuterbuch des Hieronymus Bock als Quelle der Botanik—und Pharmakologie—Geschichte*

(Inaug. Diss.) Goethe University, Frankfurt a/Main, 1964. Introduction repr. in *VFDM/GNT*, Reihe A, no. 25, 1967.

HORT, SIR A. (1) (tr.). *Theophrastus' 'Enquiry into Plants' and Minor Works on Odours and Weather Signs.* 2 vols. Heinemann, London, 1916; repr. 1949. (Loeb Classics ed.).

HOU HSÜEH-YÜ (1). *The Soil Communities of Acid and Calcium Soils in Southern Kweichow.* Nat. Geol. Survey of China (Special Soils Bulletin, no. 5.).

HOU HSÜEH-YÜ, CHHEN CHHANG-TU & WANG HSIEN-PHU (1). *The Vegetation of China with Special Reference to the Main Soil Types.* Soil Sci. Soc. China, Reports to the VIth International Congress of Soil Sci., Peking, 1956. Abstract in Trans. VIth. Int-Congr. Soil Sci. 1956, vol. I, p. 255.

HOU KUANG-CHHIUNG (1). *English-Chinese Vocabulary of Soil Terms.* Geol. Survey of China, Nanking, 1935 (special Soils Publication, no. 2).

HOWES, F. N. (1). *Vegetable Gums and Resins.* Chronica Botanica, Waltham, Mass., 1949. (New Series of Plant Science Books, no. 20.).

HSIUNG, Y. & JACKSON, M. L. (1). 'Mineral Composition of the Clay Fraction; III, Some Main Soil Groups of China.' *PSSA*, 1952, **16**, 294.

HU HSEN-HSÜ. See Hu Hsien-Su.

HU HSIEN-SU & CHHEN HUAN-YUNG (1) = (1). *Icones Plantarum Sinicarum.* Com. Press, Shanghai, 1927–37 (in 5 parts containing 250 entries with large drawings and modern plant descriptions; text both in Chinese and English).

HU HSIU-YING (6). 'The Genus *Ilex* in China.' *JAA*, 1949, **30**, 233; 1950, **31**, 39, 214.

HU HSIU-YING (9). *The Problem of the Preparation of a Flora of China.* Continental Development Foundation, New York, 1953. *The* [Harvard] *Flora of China Project at the Age of Two.* Mimeographed report, Cambridge, Mass., 1955.

HU HSIU-YING (13). *The Genera of Orchidaceae in Hongkong.* Chinese University Press, Shatin, N.T. Hongkong, 1977.

HU CHHANG-CHHIH (1). 'Citrus Culture in China.' *CCIT*, 1931, **16** (no. 11), 502.

HU HSIEN-SU (1). 'A Preliminary Survey of the Forest Flora of Southeast China.' *CBLSSC*, 1926, **2**, 1.

HU HSIEN-SU (2). 'Further Observations on the Florest Flora of Southeast China.' *BFMIB*, 1929, **1**, 51.

HU HSIEN-SU (3). 'The Nature of the Forest Flora of Southeast China.' *PNHB*, 1929, **4** (no. 1), 47.

HU HSIEN-SU & CHANEY, R. W. (1). 'A Miocene Flora from Shantung Province.' *PS*, 1940 (n.s.), **1**, 1. Also sep. in Carnegie Inst. Washington Pubs., 1938, no. 507.

HU HSIU-YING (1). *Malvaceae.* Arnold Arboretum, Harvard Univ. Cambridge, Mass., 1955. (Family 153 in the Harvard Flora of China Project.) Chinese plant names given throughout, but Latin index only.

HU HSIU-YING (2). 'A Monograph of the Genus *Philadelphus*' [pre-Linn. *Syringa*]. *JAA*, 1954, **35**, 275; 1956, **36**, 52; 1956, **37**, 15. Also issued separately in one volume.

HU HSIU-YING (3). 'Statistics of Compositae in Relation to the Flora of China.' *JAA*, 1958, **39**, 347, 380.

HU HSIU-YING (4). 'A Monograph of the Genus *Paulownia*.' *QJTM*, 1959, **12**, 1–54.

HU HSIU-YING (5). 'Chinese Hollies.' *CHJ/T*, 1959, Special Number (Nat. Sci.) 1, 150.

HU HSIU-YING (7). 'A Revision of the Genus *Clethra* in China.' *JAA*, 1960, **41**, 164.

HU HSIU-YING (8). *An Enumeration of the Food Plants of China* (mimeographed list). Harvard Flora of China Project, Arnold Arboretum, Harvard Univ. Cambridge, Mass., 1957.

HU HSIU-YING (10). 'Some Interesting and Useful Plants of Hongkong.' *CCB*, 1968, **44**, 10.

HU HSIU-YING (11). 'Floristic Studies in Hongkong.' *CCJ*, 1972, **2** (no. 1), 1.

HU HSIU-YING (12). 'Orchids in the Life and Culture of the Chinese People.' *CCJ*, 1971, **10** (nos. 1, 2), 1.

HUANG HSING-TSUNG (1). 'Peregrinations with Joseph Needham in China, 1943–4.' Art. in *Explorations in the History of Science and Technology in China*, ed. Li Kuo-Hao *et al.* (1), p. 39.

HUANG KUANG-MING (WONG MING) (1). 'La Première Materia Medica Chinoise.' *CMED*, 1967, **2**, 335.

HUANG KUANG-MING (WONG MING) (2). 'Li Che-Tchen [Li Shih-Chen] et l'Apogée de la Médécine tradition-nelle Chinoise.' *EPI*, 1970 (no. 2), 168.

HUANG KUANG-MING (WONG MING) (3). 'Contribution à l'Histoire de la Matière Médicale Végétale Chinoise.' *JATBA*, 1969, 1970.

HUANG KUANG-MING (WONG MING) (4). 'La Chine et les Sciences de la Vie an 16ᵉ Siècle.' *CMED*, 1969, **4**, 173.

HUANG KUANG-MING (WONG MING) (5). *La Médecine Chinoise par les Plantes.* Tchou, Paris, 1976.

HUANG WÊN-HSI & CHIANG PHÊNG-NIEN (1). 'Research on Characteristics of Materials of Dams Constructed by Dumping Soils into Ponded Water.' *SCISA*, 1963, **12**, 1213.

HUARD, P. (3). 'Introduction à l'Étude de la Médecine Chinoise.' *BSAC*, 1960 (no. 35), 19.

HUARD, P. & HUANG KUANG-MING (M. WONG) (2). *La Médecine Chinoise au Cours des Siècles.* Dacosta, Paris, 1959.

HUARD, P. & HUANG KUANG-MING (M. WONG) (3). 'Évolution de la Matière Médicale Chinoise.' *JAN*, 1958, **47**. Sep. pub. Brill, Leiden, 1958.

HUARD, P. & HUANG KUANG-MING (M. WONG) (4). 'L'Oeuvre d'un grand Pharmacologue Chinois, Li Che-Tchen [Li Shih-Chen], (+1518 à +1593).' A bibliography of translations of, and foreign works derivative from, the *Pên Tshao Kang Mu. RHP*, 1956, (no. 150), 390. The offprint bears erroneously only the name of the writer of a brief foreword, P. Bedel, on its cover.

HUARD, P. & HUANG KUANG-MING (M. WONG) (9). 'Bio-bibliographie de la Médecine Chinoise.' *BSEIC*, 1956, **31** (no. 3), 181.

HUBOTTER, F. (6). *Chinesisch-Tibetische Pharmakologie und Rezeptur* (with no Chinese characters or words in Tibetan script, and no index). Haug, Ulm, 1957 (Panopticon Medicum ser. No. 6). Second ed. of *Beiträge z. Kenntnis d. chinesischen sowie der tibetisch-mongolischen Pharmakologie*, mimeographed in author's own German script with Ch. characters and Tibetan words inserted; with good index. Urban & Schwarzenborg, Berlin & Vienna, 1913.

HUGHES, E. R. (2) (tr.). *The Great Learning and the Mean-in-Action.* Dent, London, 1942.

HULME, F. E. (1). *Familiar Wild Flowers.* In 5 Parts (series) with separate pagination. Cassell, London, n.d. (1st ed., 1883, 2nd ed., 1897).

VON HUMBOLDT, ALEXANDER (2). *Asie Centrale, Recherches sur les Chaînes de Montagnes et la Climatologie Comparée.* 3 vols. Gide, Paris, 1843.

VON HUMBOLDT, ALEXANDER (3). *Examen Critique de l'Histoire de la Géographie du Nouveau Continent, et des Progrès de l'Astronomie Nautique au 15ᵉ et 16ᵉ Siècles.* 5 vols. Gide, Paris, 1836–1839.

VON HUMBOLDT, ALEXANDER (5). 'Essai sur la Géographie des Plantes.' In von Humboldt A. & Bonpland, A. *Voyages,* pt. 5 'Physique Générale et Géologie.' Shoell, Paris, 1807. Germ. tr. Tübingen.

HUME, H. H. (1). 'Forms of *Camellia japonica*.' Art. in *Camellias and Magnolias*, Roy. Hort. Soc. Conf. Rep., ed. P. M. Synge, 1950, p. 27.

HUMMEL, A. W. (2) (ed.). *Eminent Chinese of the Ch'ing Period.* **2** vols. Library of Congress, Washington, 1944.

HUMMEL, A. W. (13). 'The Printed Herbal of +1249'. *ISIS*, 1941, **33**, 439; *ARLC/DO*, 1940, 155.

HUNGER, F. W. T. (1) (ed.). *The Herbal of Pseudo-Apuleius, from the +9th-Century Manuscript in the Abbey of Monte Cassino* (Codex Casinensis no. 97), *together with the first printed edition of Joh. Phil. de Lignamine* (editio princeps, Rome, +1481), *in Facsimile, described and illustrated . . .* Brill, Leiden, 1935.

HURST, C. C. (1). 'Notes on the Origin and Evolution of our Garden Rose.' *JRHS*, 1941, **66**, 73, 242, 282. Repr., with additional illustrations and charts, in G. S. Thomas (1), pp. 59–97.

HURST, C. C. & BREEZE, M. S. G. (1). 'Notes on the Origin of the Moss Rose,' *JRHS*, 1922, **47**, 26. Repr., with additional material, in G. S. Thomas (1), pp. 98–123.

HUTCHINSON, J. (1). *The Families of Flowering Plants (Angiospermae).* Vol. 1, 'Dicotyledons', London, 1926, 1934. 2nd ed. Oxford, 1959, 1964. Rev. A. C. Smith, *SC*, 1965, **147**, 1561.

HUTCHINSON, J. & MELVILLE, R. (1). *The Story of Plants and their Uses to Man.* London, 1948.

IMBAULT-HUART, C. (3). 'Le Bétel [et la Noix d'Arec; et les Coutumes et Usages se rapportant au Masticatoire composé de Noix d'Arec, Feuille de Bétel et Chaux].' *TP*, 1894, **5**, 311.

IRMSCHER, E. (1). *Die Begoniaceae Chinas, und ihre Bedeutung f.d. Frage der Formbildung in polymorphen Sippen.* n.p. n.d. (No Chinese names or characters, few illustrations.).

JACKS, G. V. & WHYTE, R. O. (1). *The Rape of the Earth; a World Survey of Soil Erosion.* Faber & Faber, London, 1939.

JACKSON, B. D. (1). *A Glossary of Botanic Terms, with their Derivation and Accent.* 4th ed. Duckworth, London, 1928.

JACOBSEN, T. & ADAMS, R. M. (1). 'Salt and Silt in Ancient Mesopotamian Agriculture; Progressive Changes in Soil Salinity and Sedimentation that contributed to the Break-up of Past Civilisations,' *S*, 1958, **128**, 1251.

JAEGER, E. C. (1). *A Source-Book of Biological Names and Terms.* Thomas, Springfield, Ill. 1st ed. 1944; 2nd ed. 1950.

JANG, C. S. See Chang Chhang-Shao.

JARRETT, V. H. C. (1). *Familiar Wild Flowers of Hongkong.* S. China Morning Post, Hongkong, n.d. (1937). (Good photographic illustrations and a few Chinese names, but no Chinese characters.).

JEFFERYS, W. H. & MAXWELL, J. L. (1). *The Diseases of China, including Formosa and Korea.* Bale & Danielsson, London, 1910. 2nd ed., re-written by Maxwell alone, ABC Press, Shanghai, 1929.

JESSEN, KARL F. W. (1). *Botanik der Gegenwart und Vorzeit, in Culturhistorischer Entwicklung; ein Beitrag zur Geschichte der abendländischen Völker.* Brockhaus, Leipzig, 1864. Photolitho repr. with Gothic type unchanged, Chronica Botanica, Waltham, Mass., 1948 (Pallas ser. no. 1). Rev. F. E. Fritsch, *N*, 1949, **163**, 115.

JOFFE, J. S. (1). *Pedology.* New Brunswick, N.J., 1949.

JOHNSON, A. T. & SMITH, H. A. (1). *Plant Names Simplified.* Collingridge, London, 1931; repr. 1937. Revised and enlarged ed. 1947.

JOHNSON, B. DAYDON (1). *Linnaeus (afterwards Carl von Linné); the Story of his Life, adapted from the Swedish of Theodor Magnus Fries.* London, 1923.

JOHNSON, H. M. (1). 'The Lemon in India.' *JAOS*, 1936, **56**, 47.

JOHNSTONE, G. H. (1). *Asiatic Magnolias in Cultivation.* Royal Hort. Soc., London, 1955.

JONES, W. H. S. (1). tr. *Pliny Natural History, with an English Translation.* Vols. VI and VII. Loeb Classical Library, Heinemann, London, 1952.

JULIEN, STANISLAS (tr.) (7). *Histoire et fabrication de la Porcelaine Chinoise: tr. par S. Julien: notes et additions par A. Salvétat: avec une Mémoire Sur la Porcelaine du Japon tr. par J. Hoffman.* Mallet-Bachelier, Paris, 1856.

JULIEN, STANISLAS & CHAMPION, P. (1). *Industries Anciennes et Modernes de l'Empire Chinois, d'après des Notices traduites du Chinois.* . . . (paraphrased précis accounts based largely on *Thien Kung Khai Wu*; and eye-witness descriptions from a visit in 1867). Lacroix, Paris, 1869.

JUNG, JOACHIM (1). *Isagoge Phytoscopica* (posthumous), ed. Johann Vagetius, Hamburg, 1678.

634 BIBLIOGRAPHY C

JUNG, JOACHIM (2). *Doxoscopiae Physicae Minores* (posthumous), ed. Martin Fogel, Hamburg, 1662.
JUNG, JOACHIM (3). *Opuscula Botanico-Physica* (posthumous). Coburg, 1747.
DE JUSSIEU, A. L. (1). *Genera Plantarum secundum Ordines Naturales Disposita, juxta Methodum in Horto Regio Parisiensi exaratam anno MDCCLXXIV.* Herissant & Barrois, Paris, 1789.
DE JUSSIEU, A. L. (2). 'Principes de la Méthode Naturelle des Végétaux.' Art. in *Dict. des Sci. Nat.* vol. 30, and sep. Lévrault, Paris, 1824.

KAEMPFER, ENGELBERT (1). *Amoenitatum Exoticarum Fasciculi V; quibus Continentur Variae Relationes, Observationes et Descriptiones Rerum Persicarum et Ulterioris Asiae, multa attentione, in peregrinationibus per universum Orientem, collectae ...* Meyer, Lemgoviae, 1712.
KAEMPFER, ENGELBERT (2). *Geschichte und Beschreibung von Japan.* Edited from the original MS by C. W. Dohm (+1777–9), with an introduction by H. Beck for the photolitho reprint of Dohm's edition. 2 vols. Brockhaus, Stuttgart, 1964. Rev. P. Huard, *A/AIHS*, 1965, **18**, 100.
KAISER, E. & MICHL, H. (1). *Die Biochemie der tierischen Gifte.* Deuticke, Vienna, 1958.
KALE, F. S. (1). *The Soya Bean.* Chronica Botanica, Waltham, Mass., 1937.
KALTENMARK, M. (2) (tr.). *Le 'Lie Sien Tchouan'* [Lieh Hsien Chuan]; *Biographies Légendaires des Immortels Taoistes de l'Antiquité.* Centre d'Etudes Sinologiques Franco-Chinois (Univ. Paris). Peking, 1953. (Crit. P. Demiéville, *TP*, 1954, **43**, 104.).
KAPLAN, F. M., SOBIN, J. M. & ANDORS, S. (1). *Encyclopaedia of China Today.* Harper & Row, New York, 1980.
KARABACEK, JOSEPH (1). *Codex Aniciae Julianae Picturis Illustratus, nunc Vindobonensis Med. Gr. I. Phototypice Editus moderante Josepho de Karabacek.* Leiden, 1906.
KARLGREN, B. (1). 'Grammata Serica; Script and Phonetics in Chinese and Sino-Japanese' (Chung Jih Han Tzu Hsing Shenglun). *BMFEA*, 1940, **12**, 1. (Photographically reproduced as separate volume, Shanghai (?) 1941.) (Cf. Kao Pên-Han (*1*).)
KARLGREN, B. (2). 'Legends and Cults in Ancient China.' *BMFEA*, 1948, **18**, 199.
KARLGREN, B. (4). *Sound and Symbol in Chinese.* Oxford, 1923; repr. 1946. (Eng. tr. of *Ordet och Pennan i Mittens Rike.* Stockholm, 1918.) Repr. Hongkong Univ. Press, Hongkong, 1962 (Chinese Companion Series, no. 1). Crit. P. Pelliot, *TP*, 1923, 315.
KARLGREN, B. (5). *Philology and Ancient China.* Aschehong (Nygaard), Oslo, 1926. (Institutet för Sammenlignende Kulturforskning; A, Forelesninger, no. 8.).
KARLGREN, B. (12) (tr.). 'The Book of Documents' (*Shu Ching*). *BMFEA*, 1950, **22**, 1.
KARLGREN, B. (14) (tr.). *The Book of Odes; Chinese Text, Transcription and Translation.* Museum of Far Eastern Antiquities, Stockholm, 1950. (A reprint of the text and translation only from his papers in *BMFEA*, **16** and **17**; the glosses will be found in **14**, **16** and **18**.)
KAROW, O. (1). 'Der Wörterbücher der Heian-zeit und ihre Bedeutung für das japanische Sprach-geschichte; I, Das *Wamyōruijushō* des Minamoto no Shitagau' (Contains, p. 185, particulars of the *Wamyō-honzō* (Synonymic Materia Medica with Japanese Equivalents) by Fukane no Sukehito, +918). *MN*, 1951, **7**, 156.
KAROW, O. (2) (ed.). *Die Illustrationen des Arzneibuches der Periode Shao-Hsing (Shao-Hsing Pên Tshao Hua Thu) vom Jahre +1159, ausgewählt und eingeleitet.* Farbenfabriken Bayer Aktiengesellschaft (Pharmazeutisch-Wissenschaftliche Abteilung), Leverkusen, 1956. Album selected from the *Shao-Hsing Chiao-Ting Pên Tshao Chieh-Thi* published by Wada Toshihiko, Tokyo, 1933.
KÊNG Hsüan (1). 'Economic Plants of Ancient North China as mentioned in the *Shih Ching* (Book of Poetry).' *ECB*, 1974, **28** (no. 4), 391. See Kêng Hsüan (*1*) for the full publication.
KEES, KEYS, OR KEYES, JOHN. See Caius, Johannes.
KELLNER, L. (2). *Alexander von Humboldt.* Oxford, 1963.
KÊNG PO-CHIEH (1). 'Bamboo; China's Most Useful Plant.' *CREC*, 1956, **5** (no. 5), 14.
KENNEDY, G. A. (1). 'The Monosyllabic Myth' (regarding the Chinese language). *JAOS*, 1951, **71**, 161.
[KENNEDY, G. A.] (2). 'The Butterfly Case' (Part I). *WT*, 1955 (no. 8), 1–48, with supplementary note 1956 (no. 9),69. (Part II was never printed.).
KENNEDY, PETER (1). *An Essay on External Remedies.* London, 1715.
[KENT, A. H.] (1). *A Manual of Orchidaceous Plants.* 2 vols. Veitch, London, 1887–94.
KEYNES, SIR GEOFFREY (1). *John Ray; a Bibliography.* Faber, London, 1951.
KEYS, J. D. (1). *Chinese Herbs; their Botany, Chemistry, and Pharmacodynamics.* Tuttle, Rutland (Vermont) and Tokyo, 1976.
KIMURA KOICHI (1). 'Important Works in the Study of Chinese Medicine' [primarily on the *Pên Tshao* literature]. *CJ*, 1935, **23**, 109.
KIMURA YOJIRO (1). 'Les Illustrations Botaniques du 17e Siècle publieés an Japon' (on the *Kimonōzui* of Nakamura Tokisai). Communication to the XIIth International Congress of the History of Science, Paris, 1968. *Résumés des Communications*, p. 118.
KING, F. H. (3). *Farmers of Forty Centuries; or, Permanent Agriculture in China, Korea and Japan.* Cape, London, 1927.
KING LI-PIN. See Ching Li-Pin.
KLEMM, MICHAEL (1). 'Entomologie und Pflanzenschutz in China.' *Nachrichtenblatt des Deutschen Pflanzen-schutzdienstes*, 1959, **11**, 121–124.

KNECHTGES, DAVID R. (1), (tr. and ed.). 'Wên Hsüan', or, Selections of Refined Literature [assembled by] Hsiao Thung [+ 501–31], translated, with Annotations and an Introduction, by D. R. K. Vol. 1 'Rhapsodies on Metropolises and Capitals. Princeton Univ. Press, Princeton, N.J., 1982 (Princeton Library of Asian Translations, no. 3).

K[NOWLTON], M. J. (1). 'Grapes in China' (Vitis, spp.). NQCJ, 1869, 3, 50.

KNOWLTON, M. J. & HANCE, H. F. (1). 'The Kin-keo [chin kou] Plum' (Hovenia dulcis, Rhamnaceae, R/289). NQCJ, 1868, 2, 107, 124.

KOBUSKI, C. E. (1). 'Synopsis of the Chinese species of Jasminum. JAA, 1932, 13, 145.

KOIDZUMI GENICHI (1). Florae Symbolae Orientali-Asiaticae; Contributions to the Knowledge of the Flora of Eastern Asia[from European herbarium material]. (Imp. Univ.), Kyoto, 1930.

KOIDZUMI, GENICHI (2). 'A Synopsis of the Genus Malus.' APTGB, 1934, 3, 179.

KOMAROV, V. L. (1). 'Prolegomena ad Floras Chinae necnon Mongoliae; Introduction to the Floras of China and Mongolia' (in Russian). AHP, 1908, 29, 1–176. Part II, Generis Caraganae Monographia, 177–388.

KOMENSKÝ, J. A. See Comenius, J. A.

KOO, T. Z. See Ku Tzu-Jen.

KOVDA, V. A. (1). Ocherki Prirody i Pochv Kitaya. Acad. Sci. Moscow, 1959. Soils and the Natural Environment of China. Eng. tr. (2 vols, xero-typescript) by U.S. Joint Publications Research Service; Washington (JPRS, no. 5967), issued by Photoduplication Service, Library of Congress, Washington (no half-tone illustrations, no map, no index, and sometimes no keys for the photo-copied diagrams), 1960.

KOVDA, V. A. & KONDORSKAIA, N. I. (1). 'Novaia Pochvennaia Karta Kitai (The New Soils Map of China).' in Russian (no English, French or German summary) PVV, 1957 (no. 12), 45, with unbound map, on scale of 1 : 10,000,000. Abstr. SF, 1957, 21, 624. Map by Ma Yün-chih, Sung Ta-Chhêng, Li Chhang-Khuei, Hsiung I, Hou Kuang-Chhiung, Hou Hsüeh-Yu, Li Lien-Chieh, Wên Chen-Wang & Wang An-Chiu, with V. A. Kovda & W. I. Kondorskaia.

KRACKE, E. A. (1) Civil Service in Early Sung China (+960–1067), with particular emphasis on the development of controlled sponsorship to foster administrative responsibility. Harvard Univ. Press, Cambridge, Mass., 1953. (Harvard-Yenching Institute Monograph Series, no. 13.) (revs. L. Petech, RSO, 1954, 29, 278; J. Průsek, OLZ, 1955, 50, 158).

KRACKE, E. A. (2). 'Sung Society; Change within Tradition.' FEQ, 1954, 14, 479.

KRACKE, E. A. (3). 'Family versus Merit in Chinese Civil Service Examinations under the Empire (analysis of the lists of successful candidates in +1148 and +1256). HJAS, 1947, 10, 103.

KRAMERS, R. P. (1) (tr.). Khung Tzu Chia Yü'; the School Sayings of Confucius (chs. 1–10). Brill, Leiden, 1950. (Sinica Leidensia, no. 7.).

KREIG, M. B. (1). 'Green Medicine; the Search for Plants that Heal'. Harrap, London, etc. 1965.

KU TZU-JEN (1) = (1) (ed.). 'Songs of Cathay; an Anthology of those current in various Parts of China among her People.' 5th ed. Kuang Hsüeh, Kelly & Walsh, & Assoc. Press Shanghai, 1931.

KUDO YUSHUN (1). 'Labiatarum Sino-japonicarum Prodromus; Kritische Besprechung des Labiaten Ostasiens.' MFSA/TIV, n.d. (1929), 2 (no. 2), 1.

KÜHN, K. G. (1) (tr.). Galen 'opera'. 20 vols. Leipzig, 1821/33. (Medicorum Graecorum opera quae exstant, nos. 1–20.).

KUHN, R. & LEDERER, E. (1). 'Zerlegung des Carotins in seine Komponenten.' BDCG, 1931, 64, 1349.

LANGKAVEL, B. (1). Botanik der späteren Griechen, vom Dritten bis Dreizehnten Jahrhunderte. Berlin, 1866. Repr. (photooffset) Hakkert, Amsterdam, 1964.

LANJOUW, J. (1) (ed.). 'Botanical Nomenclature and Taxonomy; with a Supplement to the International Rules of Botanical Nomenclature, embodying the alterations made at the 6th International Botanical Congress, Amsterdam, 1935, compiled by T. A. Sprague.' CBOT, 1950, 12 (no. 1/2), 1–87. (Papers of an IUBS Symposium, Utrecht, 1948.)

LAUFER, B. (1). Sino-Iranica; Chinese Contributions to the History of Civilisation in Ancient Iran. FMNHP/AS, 1919, 15, no. 3 (Pub. no. 201) (rev. & crit. Chang Hung-Chao, MGSC, 1925 (ser. B), no. 5).

LAUFER, B. (24). 'The Early History of Felt.' AAN, 1930, 32, 1.

LAUFER, B. (27). 'Malabathron.' JA, 1918. (11ᵉ sér), 12, 5.

LAUFER, B. (36). The Introduction of Maize into Eastern Asia. Proc. XVth Internat. Congr. Americanists. Quebec, 1906 (1907), vol. 2, p. 223.

LAUFER B. (37). The Introduction of the Ground-Nut into China. Proc. XVth Internat. Congr. Americanists. Quebec, 1906 (1907), vol. 2. p. 259.

LAUFER, B. (41). 'Die Sage von der goldgrabenden Ameisen.' TP, 1908, 9, 429.

LAUFER, B. (42). Tobacco and its use in Asia. Field Mus. Nat. Hist., Chicago, 1924. (Anthropology Leaflet, no. 18.)

LAUFER, B. (44). 'The Lemon in China and Elsewhere.' JAOS, 1934, 54, 143.

LAURENCE, D. R. (1). Clinical Pharmacology. Churchill, London, 1966 (3rd ed.).

LAWRENCE, G. H. M. (1). Taxonomy of Vascular Plants. Macmillan, New York, 1951.

LAWRENCE, G. H. M. (2), with illustrations by Sheehan, M. R. An Introduction to Plant Taxonomy. Macmillan, New York, 1955.

LÊ THÀNH-KHÔI (1). Le Viet-Nam; Histoire et Civilisation. Editions de Minuit, Paris, 1955.

LECLERC, L. (1) (tr.). 'Traité des Simples par Ibn al-Beithar [al-Bayṭār]'. *MAI/NEM*, 1877, **23**; 1883, **25**.

LECOMTE, H. (1). 'Lauracées de Chine et d'Indochine.' *NAMHN*, 1914 (5ᵉ sér.), **5**, 43–120. (No Chinese names or characters.)

LECOMTE, H., GAGNEPAIN F. & HUMBERT, H. (1) (ed.) *et al. Flore Générale de l'Indochine.* 7 vols. Paris 1907–38.

LECOMTE, LOUIS (1). *Nouveaux Mémoires sur l'État présent de la Chine.* Anisson, Paris, 1696. (Eng. tr. *Memoirs and Observations Topographical, Physical, Mathematical, Mechanical, Natural, Civil and Ecclesiastical, made in a late journey through the Empire of China, and published in several letters, particularly upon the Chinese Pottery and Varnishing, the Silk and other Manufactures, the Pearl Fishing, the History of Plants and Animals, etc. translated from the Paris edition, etc.* 2nd ed. London, 1698. Germ. tr. Frankfurt, 1699–1700. Dutch tr. 's Graavenhage, 1698.)

LEDERBERG, J. (1). The Topological Mapping of Organic Molecules. *PNASW*, 1965, **53**, 134.

LEDERBERG, J. (2). *Dendral-64; a System for Computer Construction, Enumeration and Notation of Organic Molecules as Tree Structures.* (U.S.) Nat. Aeronautics & Space Administration, Washington, 1964. CR Report no. 57029.

LEDERER, E. & LEDERER, M. (1). *Chromatography; a Review of Principles and Applications.* Elsevier, Amsterdam and London, 1957.

LEDYARD, G. (2) (tr. and ed.). '"Notice on the Tree-Peonies of Loyang" (*Loyang Mu-Tan Chi*) by Ouyang Hsiu (+1007–72), translated, with introduction and annotations....' Unpublished MS. 1961.

LEE, H. (1). *The Vegetable Lamb of Tartary; a curious Fable of the Cotton Plant.* London, 1887.

LEE, J. S. See Li Ssu-Kuang.

LEE, JAMES (1) *An Introduction to Botany, containing an Explanation of the Theory of that Science, extracted from the Works of Doctor Linnaeus ...* Rivington, Davis, White, Crowder, Dilly, Robinson, Cadell & Baldwin, London, 1788.

LEGEZA, I. L. (1). *A Guide to Transliterated Chinese in the Modern Peking Dialect*; I, Conversion Tables of the Currently-Used International and European Systems, with comparative Tables of Initials and Finals; II, Conversion Tables of the Outdated International and European Individual Systems, with comparative Tables of Initials and Finals. 2 vols. Brill, Leiden, 1968–9.

LEGGE, J. (1). *The Texts of Confucianism, translated. Pt I, The Shu King, the Religious portions of the Shih King, the Hsiao King.* Oxford, 1879. (*SBE*, vol. 3; reprinted in various eds.; Com. Press, Shanghai.)

LEGGE, J. (2). *The Chinese Classics etc.: Vol. 1. Confucian Analects, The Great Learning, and the Doctrine of the Mean.* Legge, Hongkong, 1861; Trübner, London, 1861.

LEGGE, J. (5) (tr.). *The Texts of Taoism.* (Contains (a) *Tao Tê Ching*, (b) *Chuang Tzu*, (c) *Thai Shang Kan Ying Phien*, (d) *Chhing Ching Ching*, (e) *Yin Fu Ching*, (f) *Jih Yung Ching.*) 2 vols. Oxford, 1891; photolitho reprint, 1927. (*SBE*, 39 and 40.)

LEGGE, J. (7) (tr.). *The Texts of Confucianism*: Pt III. *The 'Li Chi'.* 2 vols. Oxford, 1885; reprint, 1926. (*SBE*, nos. 27 and 28.)

LEGGE, J. (8) (tr.). *The Chinese Classics etc.*: Vol. 4, Pts 1 and 2. '*Shih Ching*'; *The Book of Poetry.* Lane Crawford, Hongkong and Trübner, London, 1871. Com. Press. Shanghai, n.d. Photolitho re-issue, Hongkong Univ. Press, Hongkong, 1960, with supplementary volume of concordance tables, etc.

LEMEE, A. (1). *Dictionnaire descriptif et synonymique des Genres de Plantes phanérogames.* Brest, 1929.

LEMMON, K. (1). *The Golden Age of the Plant Hunters.* Phoenix, London, 1968.

LENG B. & BUNYARD, E. A. (1). 'The *Camellia* in Europe; its Introduction and Development.' *NFLOSIL*, 1933, **5**, 123.

LENZ, H. O. (2). *Botanik der alten Griechen und Römer, deutsch in Auszügen aus deren Schriften, nebst Anmerkungen.* Thienemann, Gotha, 1859.

LESLIE, D. (1). *Man and Nature; Sources on Early Chinese Biological Ideas* (especially the *Lun Hêng*). Inaug. Diss. Cambridge, 1954.

LESLIE, D. (3). 'Contribution to a New Translation of the *Lun Hêng*.' *TP*, 1956, **44**, 100.

LESLIE, D. (4). 'The Chinese-Hebrew Memorial Book of the Jewish Community of Khaifêng.' *ABRN*, 1964, **4**, 19; 1965, **5**, 2; 1966, **6**, 2.

LESLIE, D. (5). 'Some Notes on the Jewish Inscriptions of Khaifêng.' *JAOS*, 1962, **82**, 346.

LESLIE, D. (6). 'The Khaifêng Jew, Chao Ying-Chhêng, and his Family.' *TP*, 1967, **53**, 147.

LESLIE, D. (8). 'The Survival of the Chinese Jews; the Jewish Community of Khaifêng.' Brill, Leiden, 1972. (T'oung Pao Monographs, no. 10.)

LESLIE, D. (9). *Confucius* (with a translation of the *Lun Yü* by D. Leslie & Z. Mayani). Seghers, Paris, 1962. (Philosophes de tous les Temps, no. 3.)

LESLIE, D. (10). 'The Judaeo-Persian Colophons to the Pentateuch of the Khaifêng Jews.' *ABRN*, 1969, **8**, 1.

LEVEILLE, H. (1). '*Catalogue des Plantes du Yunnan....*' pr. pr., Le Mans, 1915–17. (Illustrations but no Chinese names or characters.)

LEVEILLE, H. (2). '*Flore du Koui-Tcheou*' [Kueichow]. Pr. litho pr. le Mans, 1914–15.

LEWIS, JOHN G. E. (1). *The Biology of Centipedes.* Cambridge, 1981.

LEWIS, WALTER H. & ELVIN-LEWIS, MEMORY P. F. (1). *Medical Botany; Plants affecting Man's Health.* Wiley, New York, 1977.

LI HUI-LIN (1). 'A case for pre-Columbian transatlantic travel, by Arab ships.' *HJAS*, 1961, **23**, pp. 114–26.

LI HUI-LIN (2). *Woody Flora of Thaiwan.* Livingston, Narberth, Pa. 1963. (Illustrations but neither Chinese names nor characters.)

Li Hui-Lin (3). 'The Phytogeographic Divisions of China, with special reference to the Araliaceae.' *PANSP*, 1944, **96**, 249.

Li Hui-Lin (4). 'The Araliaceae of China.' *SAR*, 1942, **2**, 1–134.

Li Hui-Lin (5). 'Bamboos and Chinese Civilisation.' *JNYBG*, 1942, **43**, 213.

Li Hui-Lin (6). 'An Archaeological and Historical Account of *Cannabis* in China.' *ECB*, 1974, **28**, 437.

Li Hui-Lin (7). 'The Origin and Use of *Cannabis* in Eastern Asia; Linguistic-Cultural Implications.' *ECB*, 1974, **28**, 293.

Li Hui-Lin (8). [correct from (1) in Vol. 5, pts 2, 3 and 4] '*The Garden Flowers of China*'. Ronald, New York, 1959. (Chronica Botanica series, no. 19.)

Li Hui-Lin (9). 'Floristic Relations between Eastern Asia and Eastern North America.' *TAPS*, 1952, **42**, 371.

Li Hui-Lin (10). 'Eastern Asia/Eastern North America Species-Pairs in Wide-ranging Genera.' Art. in *Floristics and Palaeofloristics of Asia and Eastern North America*, ed. A. Graham, Elsevier, Amsterdam, 1972, p. 65 (ch. 5).

Li Hui-Lin (11). 'Plant Taxonomy and the Origin of Cultivated Plants.' *TAX*, 1974, **23**, 715.

Li Hui-Lin (12). 'The *Nan Fang Tshao Mu Chuang*', a Fourth-century Flora of South-east Asia; Introduction, Translation and Commentaries. Chinese University Press, Hongkong. 1979.

Li Hui-Lin (14). *Contributions to Botany; Studies in Plant Geography, Phylogeny and Evolution, Ethnobotany and Dendrological and Horticultural Botany*. Epoch, Taipei (Taiwan), 1982.

Li Kuo-Hao, Chang Mêng-Wên, Tshao Thien-Chhin & Hu Tao-Ching (1) (ed.). *Explorations in the History of Science and Technology in China; a Special Number of the 'Collections of Essays on Chinese Literature and History'* (compiled in honour of the eightieth birthday of Joseph Needham). Chinese Classics Publishing House, Shanghai, 1982.

Li Shun-Chhing (Lee Shun-Ching) (1). *Forest Botany of China*. Com. Press, Shanghai, 1935.

Li Shun-Chhing (2). 'The Oecological Distribution of Chinese Trees.' *PNHB*, 1934, **9**, 1.

Li Ssu-Kuang (J. S. Lee) (1). *The Geology of China*. Murby, London, 1939.

Li Thao (1). 'Achievements of Chinese Medicine in the Northern Sung Dynasty (+960 to +1127).' *CMJ*, 1954, **72**, 65.

Li Thao (7). 'Achievements of Chinese Medicine in the Sui (+589–617) and Thang (+618 to +907) Dynasties.' *CMJ*, 1953, **71**, 301.

Li Thao (9). 'Achievements of Chinese Medicine in the Chhin (−221 to −207) and Han (−206 to +219) Dynasties.' *CMJ*, 1953, **71**, 380.

Li Thao (11). 'Achievements in Materia Medica during the Ming Dynasty (+1368 to +1644).' *CMJ*, 1956, **74**, 177.

Liao Wên-Kuei (1) (tr.). *The Complete Works of Han Fei Tzu; a Classic of Chinese Legalism*. 2 vols. Probsthain, London, 1939, 1959.

Liljestrand, S. H. (1). 'Observations on the Medical Botany of the Szechuan-Tibetan Border, with notes on general flora.' *JWCBRS*, 1922–3, 37.

Lin Wên-Chhing (1) (tr.). *The 'Li Sao'; an Elegy on Encountering Sorrows, by Chhü Yüan of the State of Chhu (c. 338 to 288 b.c.)*.... Com. Press, Shanghai, 1935.

Lindley, J. (1). *An Introduction to Botany*. 1st ed. London; 1832. 3rd ed. 1839 much changed, the sections on Taxonomy, Geography and Morphology being omitted; and those on Organography, Physiology, Glossology and Phytography enlarged.

Lindley, J. (2). '*Flora Medica*'; a Botanical Account of all the More Important Plants used in Medicine in different parts of the World. Longman, Orme, Brown, Green & Longmans, London, 1838.

Ling Man-Li (Mary) (1) = (1). 'Bark-Cloth in Thaiwan and the Circum-Pacific Area of Cultures.' Art. in Ling Shun-Shêng (6), p. 253. First pub. in *AS/BIE*, 1960, **9**, 355.

Ling Shun-Shêng (6) = (7). *Bark-Cloth, Impressed Pottery, and the Inventions of Paper and Printing*. Inst. of Ethnol., Academia Sinica, Nankang, Thaiwan, 1963. (Monograph Series, no. 3.) Papers first published in *AS/BIE*, 1961, **11**, 29; 1962, **13**, 195, **14**, 215; 1963, **15**, 48, with three others and an introduction, concluding with a contribution by Ling Man-Li (1).

Linnaeus. See von Linné, Carl.

von Linné, Carl (the elder; Linnaeus) (1). *Systema Naturae*. Leiden, 1735, 10th ed. Stockholm, 1758–9. This is the edition internationally accepted as the starting-point for modern zoological nomenclature. 12th ed. Stockholm, 1767–8. This was the edition used by J. A. Murray for the preparation of the *Systema Vegetabilium* (see 1a). It supplements the 6th ed. of *Genera Plantarum* (1764) and the 2nd ed. of *Species Plantarum* (1762–3). Another ed. 4 vols. Vienna, 1767–70.

von Linné, Carl (the elder; Linnaeus) (1a). *Systema Vegetabilium*. Göttingen and Gotha, 1774. This, termed the 13th edition, was the revised form of the botanical part of the 12th edition of the *Systema Naturae*, prepared from Linnaeus' own annotated copy by J. A. Murray. 2nd ed. Göttingen, 1784. For the English translation see Anon. (76).

von Linné, Carl (the elder; Linnaeus) (2). *Bibliotheca Botanica*. Amsterdam, 1736.

von Linné, Carl (the elder; Linnaeus) (3). *Fundamenta Botanica*. Amsterdam, 1736.

von Linné, Carl (the elder; Linnaeus) (4). *Hortus Cliffortianus*. Amsterdam, 1737 (actually 1738).

von Linné, Carl (the elder; Linnaeus) (5). *Methodus Sexualis*. Leiden, 1737.

VON LINNÉ, CARL (the elder; LINNAEUS) (6). *Genera Plantarum*. 1st ed. Leiden, 1737. 5th ed. Stockholm, 1754. This is the edition internationally accepted, together with the 1st ed. of the *Species Plantarum* (1753–4), as the starting-point for modern botanical nomenclature. Facsimile Edition, with notes in Japanese by Nakai Takenoshi, Tokyo, 1939. 6th ed. Stockholm, 1764. Associated with 2nd ed. of *Species Plantarum* (1762–3) to which it contains 'emendanda'. Repr. Vienna, 1767. This 6th ed. also contains a later attempt at a natural classification under the title *Ordines Naturales* (with rules).

VON LINNÉ, CARL (the elder; LINNAEUS) (7). *Critica Botanica*. Leiden, 1737. Facsimile Edition issued by the Ray Society, with translation by Sir A. Hort, London, 1938. (Ray Soc. Pubs. no. 124.) The work devoted to the statement of Linnaeus' taxonomic principles and practice.

VON LINNÉ, CARL (the elder; LINNAEUS) (8). *Classes Plantarum*. Leiden, 1738. This was the work which contained the important sketch for a possible natural classification *Fragmenta Methodi Naturalis*.

VON LINNÉ, CARL (the elder; LINNAEUS) (9). *Öländska och Gothländska Resa*. Stockholm and Upsala, 1745. This was the work which, together with (10) and (11) first introduced the binomial system of nomenclature. See especially its index, reproduced in facsimile by Stearn (3), betw. pp. 50 and 51.

VON LINNÉ, CARL (the elder; LINNAEUS) (10). Gemmae Arborum (On the Buds of Woody Plants). Upsala, 1749; as Dissertation no. 24 in *Amoenitates Academicae* (see 15) defended by P. Löfling, who in this case drafted the text himself. See note on von Linné (9).

VON LINNÉ, CARL (the elder; LINNAEUS) (11) *Pan Suecius* (a work on cattle fodder plants). Upsala, 1749; as Dissertation no. 26 in *Amoenitates Academicae* (see 15) defended by N. L. Hesselgren though drafted by Linnaeus. See note on von Linné (9).

VON LINNÉ, CARL (the elder; LINNAEUS) (12). *Philosophia Botanica*. Stockholm, 1751. This was the work in which the binomial nomenclature was first enunciated systematically. It also contains a guide to botanical Latin.

VON LINNÉ, CARL (the elder; LINNAEUS) (13).

Species Plantarum 1st ed. Stockholm, 1753. This is the edition internationally accepted, together with the 5th ed. of the *Genera Plantarum* (1754), as the starting-point for modern botanical nomenclature.

Facsimile Edition issued by the Ray Society, with Introduction by W. T. Stearn (see (3)), and a Key to Linnaeus' abbreviations by J. L. Heller, 2 vols. Quaritch, London, 1957. (Ray Soc. Pubs. no. 140.)

Earlier Facsimile Editions, Junk, Berlin, 1907; Shokubutsu Bunken Kankokwa, Tokyo, 1934.

Vol. 1 exists in two states, some pages having been cancelled and replaced while going through press.

2nd ed., polished and improved, with additional material, Stockholm, 1762–3. 3rd ed., a re-issue with errata, Vienna, 1764. 4th ed. 6 vols. Berlin 1797–1830, ed. C. L. Willdenow, F. Schwägrich & H. F. Link greatly enlarged.

VON LINNÉ, CARL (the elder; LINNAEUS) (14). *Praelectiones in Ordines Naturales Plantarum* (posthumous), ed. Giseke, 1792. Lectures given in 1764 and 1771 which contained Linnaeus' mature views on natural classification.

VON LINNÉ, CARL (the elder; LINNAEUS) (15) (ed.). *Amoenitates Academicae* 1st ed. P. Camper, Haak, Leiden, 1749. 4th ed. J. C. D. von Schreber, Erlangen 1790. 186 Upsala doctoral dissertations from 1743 to 1776 relating to natural history and medicine, by the students of Linnaeus, i.e. the 'respondents', who defended theses actually originated, or even wholly written, by their professor. They were thus collaborative papers rather than doctoral dissertations in the modern sense.

VON LINNÉ, CARL (the elder; LINNAEUS) (16). *Mantissa Plantarum*. Stockholm, 1767. A supplement to the second volume of the 12th ed. of *Systema Naturae* (see 1), to vol. 3 of which (1768) was appended *Mantissa Plantarum Altera*, Stockholm, 1771 (a further supplement). *In re* see also von Linné Carl (the younger) (1).

VON LINNÉ, CARL (the elder; LINNAEUS) (17). *Sponsalia Plantarum*, Stockholm, 1746. Inaugural Dissertation defended by J. G. Wahlbom *Amoenitates Academicae*, no. 12. On the sexuality of plants, anemophilous pollination etc. (see Stearn (3). Introd. p. 168). A development of the MS *Praeludia Sponsaliarum Plantarum* which Linnaeus had presented to his patron O. Celsius in 1729 or 1730.

VON LINNÉ, CARL (the elder; LINNAEUS) (18). *Demonstrationes Plantarum in Horto Upsaliensi, 1753*. Upsala 1753. Inaugural dissertation defended by J. C. Höjer (*Amoenitates Academicae*, no. 49.). This was the first publication after the *Species Plantarum* of 1753 to employ systematically the binomial nomenclature. Deals with 1434 species belonging to 541 genera.

VON LINNÉ, CARL (the elder; LINNAEUS) (19). *Fundamentum Fructificationis*. Upsala, 1762. Inaugural dissertation defended by J. M. Gråberg (*Amoenitates Academicae*, no. 123). This was the publication in which Linnaeus first suggested that God had originally created only one species for each natural order (or family) the present genera and species having arisen subsequently by hybridisation.

VON LINNÉ, CARL (the elder; LINNAEUS) (20). *Caroli Linnaei Sueci Methodus*. Sylvius, Leiden, 1736 (a broadside, later inserted in many copies of the 1st ed. of the *Systema Naturae*, 1735, and reprinted in the 2nd to 9th editions). Facsimile published by the Swedish Royal Academy, Stockholm, 1907. Eng. tr. K. P. Schmidt (1); Heller & Stearn (1).

VON LINNÉ, CARL (the elder; LINNAEUS) (21). *Plantae Camschatcenses Rariores*. Upsala, 1750. Inaugural Dissertation defended by J. P. Halen (*Amoenitates Academicae*). Contains the prophetic recognition of similarity between plants of North America and Siberia.

VON LINNÉ, CARL (the younger) (1). *Supplementum Plantarum*. 1781. Supplements the 13th ed. of *systema Vegetabilium*

(1774) the 6th ed. of the *Genera Plantarum* (1764) and the 2nd ed. of the *Species Plantarum* (1762–3).

LIOU HO. See Liu Ho.

LIOU T. N. (Tchen-Ngo). See Liu Shen-O.

LIU HO (1). *Lauracées de Chine et d'Indochine; Contribution à l'Étude Systématique et Phytogéographique.* Hermann, Paris, 1934. (No illustrations, no Chinese characters or names, only a preface in Chinese.)

LIU HO & RONX, C. (1). *Aperçu Bibliographique sur les anciens Traités Chinois de Botanique, d'Agriculture, de Sériculture et de Fungiculture.* Bosc & Riou, Lyon, 1927.

LIU I-JAN (1). *Systematic Botany of the Flowering Families in North China; 124 Illustrations of Common Hopei Plants.* Vetch, Peiping, 1931. (Chinese names and characters, but no illustrations.)

LIU, J. C. See Liu I-Jan; Liu Ju-Chhiang.

LIU JU-CHHIANG (1). 'The Cowdry Collection of Chihli Flora.' *PNHB*, 1927, **2** (no. 3). 47.

LIU SHEN-O *et al.* (1) = (1). *Flore Illustrée du Nord de la Chine; Hopei et ses Provinces Voisines.* 5 vols. Nat. Peiping Acad. Peiping, 1931–7. (Text in Chinese and French.)

LIU SHEN-O (2) = (2). 'Essai sur la Géographie Botanique du Nord et de l'Ouest de la Chine.' *NPA/CIB*, 1934, **2**, 423.

LO JUNG-PANG (1). 'The Emergence of China as a Sea-Power during the late Sung and early Yuan Periods.' *FEQ*, 1955, **14**, 489. Abstr. *RBS*, 1955, **1**, 66.

LO KAI-FU (1). 'The Basic Geography of China.' *CREC*, 1956, **5** (no. 12), 18.

LOBOVA, E. V. & KOVDA, V. A. (1). *A Soils Map of Asia.* Trans. 7th Internat. Congress of Soil Science, 1960. Abstr. *SF*, 1962, **25**, 976.

LOEWE, M. (4). *Records of Han Administration.* 2 vols. Cambridge, 1967. (Univ. Cambridge Oriental Pubs. no. 11.)

LOEWE, M. (6). 'Khuang Hêng and the Reform of Religious Practices (−31).' *AM*, 1971, **17**, 1.

LONES, T. E. (1). *Aristotle's Researches in Natural Science.* West & Newman, London, 1912.

VAN DER LOON, P. (1). 'On the Transmission of the *Kuan Tzu* Book.' *TP*, 19, **41**, 357.

DE LOUREIRO, JUAN (1). *Flora Cochinchinensis; sistens Plantas in Regno Cochinchina nascentes; Quibus accedunt aliae Observatae in Sinensi Imperio, Africa Orientali, Indiaeque Locis Variis; Omnes dispositae secundum Systema Sexuale Linneanum.* Acad. Sci. Lisbon, 1790. See Merrill (2).

LOWDERMILK, W. G. & WICKES, D. R. (3). 'History of Soil Use in the Wu-Thai Shan Area.' *JRAS/NCB*, Special Monograph, 1938.

LU GWEI-DJEN (1). 'China's Greatest Naturalist; a Brief Biography of Li Shih-Chen.' *PHY*, 1966, **8**, 383. Abridgement in Proc. XI Internat. Congress of the History of Science, Warsaw, 1965, p. 50.

LU GWEI-DJEN & NEEDHAM, JOSEPH (1). 'A Contribution to the History of Chinese Dietetics.' *ISIS* 1951, **42**, 13 (submitted 1939, lost by enemy action; again submitted 1942 and 1948).

LU GWEI-DJEN & NEEDHAM, JOSEPH (2). 'China and the Origin of (Qualifying) Examinations in Medicine.' *PRSM*, 1963, **56**, 63.

LU GWEI-DJEN & NEEDHAM, JOSEPH (3). 'Mediaeval Preparations of Urinary Steroid Hormones.' *MH*, 1964, **8**, 101. Prelim. pub. *N*, 1963, **200**, 1047. Abridged account, *END*, 1968, **27** (no. 102), 130.

LU GWEI-DJEN & NEEDHAM, JOSEPH (4). 'Records of Diseases in Ancient China.' Art. in *Diseases in Antiquity*, ed. D. Brothwell & A. T. Sandison, Thomas, Springfield, Illinois, 1967, p. 222. Repr. *AJCM*, 1976, **4** (no. 1), 3.

LU GWEI-DJEN & NEEDHAM, JOSEPH (5). *Celestial Lancets; a History and Rationale of Acupuncture and Moxa.* Cambridge, 1980.

LUFKIN, A. W. (1). *A History of Dentistry.* 2nd ed. Lea & Febiger, Philadelphia, 1948.

LUNG PO-CHIEN, LI THAO & CHANG HUI-CHIEN (1). 'Li Shih-Chen—Ancient [i.e. + 16th-century] China's Great Pharmacologist.' *PC*, 1955 (no. 1), 31.

MA, TAILOI. See Ma Thai-Lai (1).

MA THAI-LAI (1). 'The Authenticity of the *Nan Fang Tshao Mu Chuang*.' *TP*, 1978, **64**, 218.

MA YUNG-CHIH (1). *General Principles of the Geographical Distribution of Chinese Soils.* Soil Sci. Soc. China, Reports to the VIth International Congress of Soil Sci., Peking, 1956. Abstr. in *Trans. VIth Int. Congr. Soil Sci.*, 1956, vol. 1, p. 257 (V–139).

McCLINTOCK, E. (1). 'A Monograph of the Genus *Hydrangea*.' *PCAS*, 1957 (4th ser.), **29**, 147–256.

McCLURE, F. A. (1). *The Bamboos; a Fresh Perspective.* Harvard Univ. Press, Cambridge, Mass., 1966.

McCLURE, F. A. (2). 'Bamboo in the Economy of Oriental Peoples.' *ECB*, 1956, **10**, 335.

MACMILLAN, H. F. (1. *Tropical Planting and Gardening, with special reference to Ceylon.* 5th ed. Macmillan, London, 1962.

MAEYAMA, Y. (1). 'The Oldest Star Catalogue of China, Shih Shen's *Hsing Ching*.' Art. in *Prismata; Naturwissenschaftsgeschichtliche Studien*, ed. Y. Maeyama & W. G. Saltzer, Wiesbaden, 1977, p. 211.

MAEYAMA, Y. (2). 'On the Astronomical Data of Ancient China (*c.* −100 to +200); a Numerical Analysis.' *A/AIHS*, 1975, **25** (no. 97); 1976, **26** (no. 98).

MAGNOL, P. (1). *Prodromus Historiae Generalis Plantarum in quo Familiae Plantarum per Tabulas Disponuntur.* Paris, 1689.

MAJUMDAR, G. P. (1). 'The History of Botany and Allied Sciences (Agriculture, Medicine and Arbori-horticulture) in Ancient India (−2000 to +100).' *A/AIHS*, 1951, **4**, 100.

MALOUIN, P. J. (1). *Chimie Médicinale, contenant la Manière de préparer les Remèdes les plus usités, et la Méthode de les employer pour la Guérison des Maladies*. 2nd ed. 2 vols. d'Houry, Paris, 1755; 1st ed. Cavelier, Paris, 1734.

MAQSOOD ALI, S. ASAD & MAHDIHASSAN, S. (1). 'Bazaar Medicines of Karachi; [I], Fresh Herbs.' *MED*, 1961, **21** (no. 6), 264.

MAQSOOD ALI, S. ASAD & MALIDIHASSAN, S. (3). 'Bazaar Medicines of Karachi; [V], Vegetable Drugs in Stock at Herbalists.' *MED*, 1962, **23**, 243.

MAQSOOD ALI, S. ASAD, TASNIF, MOHAMMED, ZAFARUL HASSAN, S. & MAHDIHASSAN, S. (1). 'Bazaar Medicines of Karachi; [II], Drugs of Pavement Herbalists.' *MED*, 1961, **23**, 24.

MARBUT, C. F. (1). *Soils of the United States* (Atlas of American Agriculture, pt III). U.S. Dept. of Agric., Bur. of Chem. and Soils, Washington, 1935.

MARGULIS, H. (1). 'Aux Sources de la Pédologie.' *AENSAT*, 1954, **11** (Supplement).

MARTIN, W. KEBLE (1). *The Concise British Flora in Colour*, with nomenclature edited by D. H. Kent. Ebury & Joseph, London, 1965.

MASEFIELD, G. B., WALLIS, M., HARRISON, S. G. & NICHOLSON, B. E. (1). *The Oxford Book of Food Plants*, Oxford 1969. Several times reprinted.

MASPERO, H. (8). Légendes Mythologiques dans le Chou King.' *JA*, 1924, **204**, 1.

MATTIOLI, PIERANDREA (1). *Di Pedacio Dioscoride Anazarbeo Libri Cinque della Historia et Materia Medicinale tradotta in Lingua Volgare Italiana ...* Bascarini, Venice, 1544. Latin ed. Valgrisi, Venice, 1564. Revised ed. with larger and better illustrations, Valgrisi, Venice, 1565.

MAXIMOWICZ, C. J. (1). *Flora Tangutica, sive Enumeratio Plantarum Regionis Tangut (Amdo) Provincial Kansu, necnon Tibetiae praesertim Orientali-borealis atque Tsaidam, ex Collectionibus N.M. Przewalski atque G. N. Pontanin.* Imp. Acad. Sci. St Petersburg, 1889.

MAXIMOWICZ, C. J. (2). *Enumeratio Plantarum hucusque in Mongolia necnon adjacente Parte Turkestaniae Sinensis Lectarum.* Imp. Acad. Sci. St Petersburg, 1889.

MAYER, JEAN (1). 'The Rape of the Crops.' *RAM*, 1967, **10**, 50.

MAYER, JEAN & SIDEL, V. W. (1). 'Crop Destruction in South Vietnam.' *CCN*, 1966 (29 June).

MAYERS, W. F. (1). *Chinese Reader's Manual*. Presbyterian Press, Shanghai, 1874; reprinted, 1924.

MAYERS, W. F. (2). 'Bibliography of the Chinese Imperial Collections of Literature' (i.e. *Yung-Lo Ta Tien; Thu Shu Chi Chhêng; Yuan Chien Lei Han; Phei Wên Yuan Fu; Phien Tzu Lei Pien; Ssu Khu Chhüan Shu*). *CR*, 1878, **6**, 213, 285.

MAYERS, W. F. (7). *The Chinese Government; a Manual of Chinese Titles categorically arranged and explained, with an Appendix*. Kelly & Walsh, Shanghai, etc. 1877. 2nd ed. with additions by G. M. H. Playfair, Kelly & Walsh, Shanghai, etc. 1886.

MAYERS, W. F. *et al.* (1). 'Henna in China.' *NQCJ*, 1867, **1**, 40; 1868, **2**, 11, 29, 33, 41, 46, 78, 180; 1869, **3**, 30.

MAYERS, W. F. & BUSHELL, S. W. (1). 'Maize in China.' *NQCJ*, 1867, **1**, 89; 1870, **4**, 87.

MEDHURST, W. H. (1) (tr.). *The 'Shoo King' [Shu Ching], or Historical Classic, being the most ancient authentic record of the Annals of the Chinese Empire, illustrated by later commentators*. Mission Press, Shanghai, 1846. (Word by word translation with inserted Chinese characters.)

VON MEGENBERG, CONRAD (+1309–74) (1). Begins 'Hye nach volget das Pŭch der Natur ...' Bamler, Augsburg, 1475.

MERRILL, E. D. (1). *An Interpretation of Rumphius' 'Herbarium Amboinense'*. Bureau of Science, Manila, 1917 (Pub. no. 9). Repr. in abridged form as: 'Amboina Floristic Problems in Relation to the Early Work of Rumphius' in Merrill (5), p. 181.

MERRILL, E. D. (2). 'A Commentary on Loureiro's *Flora Cochinchinensis*.' *TAPS*, 1935, **24**, 1. Repr. in abridged form as: 'On Loureiro's *Flora Cochinchinensis*' in Merrill (5), p. 243.

MERRILL, E. D. (3). 'On the Significance of certain Oriental Plant Names in Relation to Introduced Species.' *PAPS*, 1937, **78**, 112. Repr. in Merrill (5), p. 295.

MERRILL, E. D. (4). 'Some Economic Aspects of Taxonomy.' *TORR*, 1943, **43**, 50. Repr. in Merrill (5), p. 346.

MERRILL, E. D. (5). 'Merrilleana; a Selection from the General Writings of Elmer Drew Merrill ...' *CBOT*, 1946, **10** (no. 3/4), 127–394.

MERRILL, E. D. (6). *Plant Life of the Pacific World*. Macmillan, New York, 1945.

MERRILL, E. D. (7). 'An Enumeration of Hainan Plants.' *LSJ*, 1927, **5** (nos. 1/2), 1–186. (Good local Chinese nomenclature with characters, but no illustrations.)

MERRILL, E. D. (8). 'Observations on Cultivated Plants with reference to certain American Problems.' *CEIB*, 1950, **1**, 3.

MERRILL, E. D. (9). 'The Botany of Cook's Voyages and its unexpected Significance in Relation to Anthropology, Biogeography and History.' *CBOT*, 1954, **14** (no. 5/6), 1–384.

MERRILL, E. D. & WALKER, E. H. (1). *A Bibliography of Eastern Asiatic Botany*. Arnold Arboretum of Harvard Univ., Cambridge, Mass., 1938. Supplementary Volume, Amer. Inst. Biol. Sciences, Washington D.C., 1960.

MÉTAILIÉ, G. (1). *La Terminologie Botanique en Chinois Moderne*. Inaug. Diss., Paris, 1973.

MÉTAILIÉ, G. (2). 'À propos des Noms de Plantes d'Origine étrangère introduites en Chine.' Contrib. to *Langues et Techniques; Nature et Societé*, vol. 1, 'Approche Linguistique', ed. J. M. C. Thomas & L. Bernot. Klincksieck, Paris, 1970, p. 321.

MÉTAILIÉ G. (3). 'Cuisine et Santé dans la tradition chinoise,' *Communications*, 1979, 31, 119.

METCALF, F. P. (1). 'Travellers and Explorers in Fukien before +1700.' *HKN*, 1934, 5, 252.

METCALF, F. P. (2). *Flora of Fukien and Floristic Notes on Southeastern China*. Lingnan University, Hsien-jen-miao, Kuangtung (but pr. in U.S.A.), 1942. 1st fascicle all published. (No illustrations, no Chinese names or characters.)

MEYER, ADOLF (4) (ed.). *Joachim Jungius; Zwei Disputationen ü.d. Prinzipien (Teile) der Naturkörper (+1642), in der Übersetzung von Emil Wohlwill herausgegeben und mit einer Einleitung versehen . . .*, with two facsimile title-pages. Christensen & Hartung, Hamburg, 1928. (Festgabe d.90 Versammlung Deutscher Naturforscher und Ärzte.)

MEYER, ADOLF (5). 'Joachim Jungius' geistesgeschichtliche Gestalt.' Art. in *Naturforschung und Naturlehre im alten Hamburg*. Hamburgischen Staats- u. Universitats-Bibliothek, Hamburg, 1928, p. 3.

MEYER, ERNST H. F. (1). *Geschichte der Botanik* 4 vols. Bornträger, Königsberg, 1854–7. Offset reprint, Asher, Amsterdam, 1965, with introduction by F. Verdoorn (sep. pub. as Communicationes Biohistoricae Ultrajectinae, no. 4).

MEYER, ERNST H. F. (2).
'Albertus Magnus; ein Beitrag z. Gesch. d. Botanik in dreizehnter Jahrh.' *LINN*, 1836, 10, 641–741.
'Albertus Magnus; zweiter Beitrag z. erneuerten Kenntniss seiner botanischer Leistungen.' *LINN*, 1837, 11, 545–95.
Rrpr. in part, in Meyer (1), vol. 4, pp. 9–84.

MEYER, F. N. (1). 'China a Fruitful Field for Plant Exploration.' *YBDA*, 1915, 205 (211), with map.

MEYER, F. N. (2). 'Agricultural Explorations in the Fruit and Nut Orchards of China.' *BBPI/DA*, 1911, no. 204.

MEYER, F. N. (3). *Chinese Plant Names*. Electrotyped for the Office of Foreign Seed and Plant Introduction, Bureau of Plant Industry, U.S. Department of Agriculture, Washington, D.C. by the Chinese and Japanese Pub. Co. New York, 1911. (468 plants with Latin binomials and Chinese character names collected by F.N.M. during his travels in North China 1905–8; romanisations according to Wade-Giles.)

MEYERHOFF, M. (1). 'The Earliest Mention of a Manniparous Insect' [al-Biruni]. *ISIS*, 1947, 37, 31.

MIALL, L. C. (1). *The Early Naturalists; their Lives and Work* (+1530–1789). Macmillan, London, 1912.

MIYAZAWA, B. (1). 'The History and Present State of the Chinese Peony in Japan.' *BAPES*, 1932, 31, 3.

MÖBIUS, M. A. J. (1). *Geschichte der Botanik von der ersten Anfängen bis zur Gegenwart*. Fischer, Jena, 1937.

MOLDENKE, H. N. & MOLDENKE A. L. (1). *Plants of the Bible*. Chronica Botanica, Waltham, Mass., 1952 (New Series of Plant Science Books, no. 28).

MORGAN, E. (1) (tr.). *Tao the Great Luminant; Essays from 'Huai Nan Tzu', with introductory articles, notes and analyses*. Kelly & Walsh, Shanghai, n.d. (1933?).

MORI, TAMEZO (1). *An Enumeration of Plants hitherto known from Korea*. Govt. of Chosen, Seoul, 1922. (Latin binomials in Families; Korean, Japanese and Chinese character and name indexes.)

MORISON, ROBT. (1). *Plantarum Umbelliferarum Distributio Nova*. Oxford, 1672.

MORISON, ROBT. (2). *Historia Plantarum Universalis Oxoniensis*. 3 vols. Oxford, 1680–99.

MORTON, A. G. (1). *History of Botanical Science; an Account of the Development of Botany from Ancient Times to the Present Day*. Academic Press, New York, London, etc., 1981.

MOSIG, A. & SCHRAMM, G. (1). *Der Arzneipflanzen—und Drogen-Schatz Chinas; und die Bedentung des 'Pên Tshao Kang Mu' als Standardwerk der chinesischen Materia Medica*. Volk und Gesundheit, Berlin, 1955 (Beihefte der *Pharmazie*, no. 4).

MOULE, A. C. (17). 'Gingko biloba or *yin hsing*.' *AM*, 1949, 1, 16.

MOYER, R. T. (1). 'Introduction to a Study of the Soils of Shansi Province.' *DISS/CKSA*, 1932, 2, 9–15.

MÜLLER, P. (1). *Studien über die natürlichen Humus-formen*. Berlin, 1887.

MULLINS, L. J. & NICKERSON, W. J. (1). 'A Proposal for Serial Number Identification of Biological Species' [Zoological]. *CBOT*, 1951, 12 (no. 4/6), 211.

VON MURR, C. G. (1). *Adnotationes ad Bibliotheca Halleriana*. Erlangen, 1805.

NAGASAWA, K. (1). *Geschichte der Chinesischen Literatur, und ihrer gedanklichen Grundlage*. Transl. from the Japanese by E. Feifel. Fu-jen Univ. Press, Peiping, 1945.

NAHAS, G. (1). *Haschich, Cannabis et Marijuana; le Chanvre Trompeur*. Presses Univ. de France, Paris, 1976.

NAKAO MANZŌ (1). 'Notes on the *Shao-Hsing Chiao-Ting Ching-Shih Chêng-Lei Pên Tshao* [The Classified and Consolidated Armamentarium; Pharmacopoeia of the Shao-Hsing Reign-Period]—the Ancient Chinese Materia Medica revised in the Sung Dynasty (+1131–62).' *JSSI*, 1933, Sect. III, 1, 1. (English version of the introduction to Nakao, 2.)

NASR, SEYYED HOSSEIN. See Said Husain Nasr.

NEEDHAM, JOSEPH (1). *Chemical Embryology*. 3 vols. Cambridge, 1931.

NEEDHAM, JOSEPH (2). *A History of Embryology*. Cambridge, 1934. 2nd ed., revised with the assistance of A. Hughes. Cambridge, 1959; Abelard-Schuman, New York, 1959.

NEEDHAM, JOSEPH (4). *Chinese Science*. Pilot Press, London, 1945.

NEEDHAM, JOSEPH (12). Cf. Porkert (1); Needham & Lu (9). *Biochemistry and Morphogenesis*. Cambridge, 1942;

Repr. 1950, repr. 1966, with historical survey as foreword.

NEEDHAM, JOSEPH (17). *Science and Society in Ancient China.* Watts, London, 1947. (Conway Memorial Lecture, South Place Ethical Society.) Revised ed. *MSTRM,* 1960, **13** (no. 7), 7.

NEEDHAM, JOSEPH (22). 'Science in Western Szechuan; II, Biological and Social Sciences.' *N,* 1943, **152**, 372. Reprinted in Needham & Needham (1).

NEEDHAM, JOSEPH (30). 'Prospection Géobotanique en Chine Médiévale.' *JATBA,* 1954, **1**, 143.

NEEDHAM, JOSEPH (31). 'Remarks on the History of Iron and Steel Technology in China' (with French translation; 'Remarques relatives à l'Histoire de la Sidérurgie Chinoise'). In *Actes du Colloque International 'Le Fer à travers les Ages',* pp. 93, 103. Nancy, Oct. 1955. (*AEST,* 1956, Mémoire no. 16.)

NEEDHAM, JOSEPH (32). *The Development of Iron and Steel Technology in China* (Dickinson Lecture, 1956). Newcomen Society, London, 1958, repr. Heffer, Cambridge, 1964. Précis in *TNS,* 1960, **30**, 141. Rev. L. C. Goodrich, *ISIS,* 1960, **51**, 108. French tr. (unrevised, with some illustrations omitted and others added by the editors), *RHSID,* 1961, **2**, 187, 235; 1962, **3**, 1, 62.

NEEDHAM, JOSEPH (34). 'The Translation of Old Chinese Scientific and Technical Texts.' Art. in *Aspects of Translation* ed. A. H. Smith. Secker & Warburg, London, 1958, p. 65. (Studies in Communication, no. 2.) (And *BABEL,* 1958, **4** (no. 1), 8.)

NEEDHAM, JOSEPH (35). '*Chinese Astronomy and the Jesuit Mission; an Encounter of Cultures.*' China Society, London, 1958.

NEEDHAM, JOSEPH (45). 'Poverties and Triumphs of the Chinese Scientific Tradition.' Art. in *Scientific Change; Historical Studies in the Intellectual, Social and Technical Conditions for Scientific Discovery and Technical Invention from Antiquity to the Present,* ed. A. C. Crombie, p. 117. Heinemann, London, 1963. With discussion by W. Hartner, P. Huard, Huang Kuang-Ming, B. L. van der Waerden & S. E. Toulmin (Symposium on the History of Science, Oxford, 1961). Also, in modified form: 'Glories and Defects ...' in '*Neue Beiträge z. Geschichte d. Alten Welt,*' vol. 1, *Alter Orient und Griechenland,* ed. E. C. Welskopf, Akad, Verl. Berlin, 1964. French tr. (of paper only) by M. Charlot 'Grandeurs et Faiblesses de la Tradition Scientifique Chinoise', *LP,* 1963, no. 111. Abridged version, 'Science and Society in China and the West', *SPR,* 1964, **52**, 50.

NEEDHAM, JOSEPH (47). 'Science and China's Influence on the West.' Art. in *The Legacy of China,* ed. R. N. Dawson. Oxford, 1964, p. 234. Paperback ed. 1971; Dutch tr. 1973.

NEEDHAM, JOSEPH (48). 'The Prenatal History of the Steam-Engine.' (Newcomen Centenary Lecture.) *TNS,* 1963, **35**, 3–58.

NEEDHAM, JOSEPH (50). 'Human Law and the Laws of Nature.' Art. in *Technology, Science and Art; Common Ground.* Hatfield Coll. of Technol., Hatfield, 1961, p. 3. A lecture based upon (36) and (37), revised from vol. 2. pp. 518 ff. Repr. in *Social and Economic Change* (Essays in Honour of Prof. D. P. Mukerji), ed. B. Singh & V. B. Singh. Allied Pubs. Bombay, Delhi etc., 1967, p. 1.

NEEDHAM, JOSEPH (53). 'Science and Society in East and West.' Art. in J. D. Bernal Presentation Volume *The Science of Science,* ed. M. Goldsmith & A. McKay, Souvenir, London, 1964. Also in *SS,* 1964, **28**, 385, and *CEN,* 1964, **10**, 174.

NEEDHAM, JOSEPH (55). 'Time and Knowledge in China and the West.' Art. in *The Voices of Time; a Cooperative Survey of Man's Views of Time as expressed by the Sciences and the Humanities,* ed. J. T. Fraser. Braziller, New York, 1966, p. 92.

NEEDHAM, JOSEPH (56). *Time and Eastern Man.* (Henry Myers Lecture, Royal Anthropological Institute) 1964. Royal Anthropological Institute, London, 1965.

NEEDHAM, JOSEPH (59). 'The Roles of Europe and China in the Evolution of Oecumenical Science.' *JAHIST,* 1966, **1**, 1. As Presidential Address to Section X, British Association, Leeds, 1967, in *ADVS,* 1967, **24**, 83.

NEEDHAM, JOSEPH (ed.) (63). *The Teacher of Nations; Addresses and Essays in Commemoration of the Visit to England of the great Czech Educationalist, Jan Amos Komensky, Comenius, 1641.* Cambridge, 1942. (With a chronological table showing the events in the life of Comenius, by R. Fitzgibbon Young, and a select bibliography of his works, by A. Heyberger.

NEEDHAM, JOSEPH (64). *Clerks and Craftsmen in China and the West* (Collected Lectures and Addresses). Cambridge, 1970. Based largely on collaborative work with Wang Ling, Lu Gwei-Djen & Ho Ping-Yü. Cf. Porkert (1); Needham & Lu (9).

NEEDHAM, JOSEPH (66). *Within the Four Seas; The Dialogue of East and West* (Collected Addresses). Allen & Unwin, London, 1969.

NEEDHAM, JOSEPH (69). 'The Development of Botanical Taxonomy in Chinese Culture.' In *Actes du XIIe Congrès International d'Histoire des Sciences.* Paris, 1968, vol. 8, p. 127.

NEEDHAM, JOSEPH (72). 'The Evolution of Iron and Steel Technology in East and South-east Asia.' Contribution to the Cyril Stanley Smith Presentation Volume *The coming of the Age of Iron,* ed. T. A. Wertime & J. D. Muhly (1). 1980, p. 507.

NEEDHAM, JOSEPH (83). 'Category Theories in Chinese and Western Alchemy; a Contribution to the History of the Idea of Chemical Affinity.' *EP,* 1979, **9**, 21.

NEEDHAM, JOSEPH (87). *The Guns of Kaifêng-Fu; China's Development of Man's First Chemical Explosive.* Creighton

Lecture. Univ. of London, London, 1979. Repr. *TLS*, 1980, no. 4007, 39; *HS*, 1980, **19**, 11.

NEEDHAM, JOSEPH & LESLIE, D. (1). 'Ancient and Mediaeval Chinese Thought on Evolution.' *BNISI*, 1952, **7** (Symposium on Organic Evolution). Reprinted as art. in *Theories and Philosophies of Medicine; with particular reference to Graeco-Arabic Medicine, Ayurveda and Traditional Chinese Medicine*, ed. Abdul Hamid Hamdard Institute of the History of Medicine and Medical Research, Delhi, 1962, p. 362.

NEEDHAM, JOSEPH & LIAO HUNG-YING (1) (tr.). 'The Ballad of Mêng Chiang Nü weeping at the Great Wall.' *S*, 1948, **1**, 194.

NEEHAM, JOSEPH & LU GWEI-DJEN (1). 'Hygiene and Preventive Medicine in Ancient China,' *JHMAS*, 1962, **17**, 429. Abridged in *HEJ*, 1959, **17**, 170.

NEEDHAM, JOSEPH & LU GWEI-DJEN (5). 'The Earliest Snow Crystal Observations.' *W*, 1961, **16**, 319.

NEEDHAM, JOSEPH & LU GWEI-DJEN (6).
'The Optick Artists of Chiangsu.' *PRMS* (Oxford Symposium Volume), 1967, **2**, 113. Abstr. *PRMS*, 1966, **1** (pt 2), 59. Cf. H. Solomons, *OP*, 1966, 352.
Also in *Studies in the Social History of China and South-east Asia* (Victor Purcell Memorial Volume), ed. J. Chhen & N. Tarling Cambridge, 1970, p. 197.
The first version is the more complete of the two.

NEEDHAM, JOSEPH & LU GWEI-DJEN (8). 'Medicine and Culture in China.' Art. in *Medicine and Culture*. Symposium of the Wellcome Historical Medical Museum and Library and the Wenner-Gren Foundation, London, 1966. Repr. in Needham (64), p. 263.

NEEDHAM, JOSEPH & LU GWEI-DJEN (9). 'Manfred Porkert's Interpretations of Terms in Mediaeval Chinese Natural and Medical Philosophy.' *ANS*, 1975, **32**, 491.

NEEDHAM, JOSEPH & LU GWEI-DJEN (10). 'The Esculentist Movement in Mediaeval Chinese Botany; Studies on Wild (Emergency) Food Plants. *A/AIHS*, 1968, **21** (no. 84–5), 225.

NEEDHAM, JOSEPH & NEEDHAM, DOROTHY (1) (ed.). *Science Outpost*. Pilot Press, London, 1948.

NEEDHAM, JOSEPH, WANG LING & PRICE, DEREK J. DE S. (1). *Heavenly Clockwork; the Great Astronomical Clocks of Medieval China*. Cambridge, 1960. (Antiquarian Horological Society Monographs, no. 1.) Prelim. pub. *AHOR*, 1956, **1**, 153.

NELSON, A. (1). *Introductory Botany*. Chronica Botanica, Waltham, Mass., 1949; 2nd ed. 1962.

NEVEU-LEMAIRE, M. (1). *Traité d'Helminthologie Médicale et Vétérinaire*. Vigot, Paris, 1936.

NGUYÊN TRÂN-HUÂN (1). 'Esquisse d'une Histoire de la Biologie Chinoise; des Origines jusqu'au +4ᵉ Siècle.' *RHS*, 1957, **10**, 1.

NGUYÊN TRÂN-HUÂN (2). 'Notes sur l'Origine des Pên-Tshao en Extrême Orient.' *A/AIHS*, 1961, **14**, 98.

NIELSON-JONES, W. (1). *Plant Chimaeras and Graft Hybrids*. Methuen, London, 1934.

NISSEN, CLAUS (1). *Kräuterbücher aus fünf Jahrhunderten; medizinhistorischer und bibliographische Beitrag*. Wölfle & Weiss-Hesse, Zürich, München and Olten, 1956. Eng. tr. by W. Bodenheimer & A. Rosenthal, *Herbals of Five Centuries . . .*

NISSEN, CLAUS (2). *Die Botanische Buchillustration; ihre Geschichte und Bibliographie*. 2 vols in one. 2nd ed. Hiersemann, Stuttgart, 1951–2. Vol. 3 Suppl. Hiersemann, Stuttgart, 1966. Revs. M. Rooseboom, *A/AIHS* 1952, **5**, 408; 1967, **20**, 219.

NISSEN, CLAUS (3). *Die Zoologische Buchillustration; ihre Bibliographie und Geschichte*. 2 vols (appearing in successive Lieferungen). Stuttgart, 1966–. (No illustrations in Vol. 1. Vol. 2 contains (pp. 413 ff.) 'Zoologische Illustration in China und Japan' by H. Walravens.)

NISSEN, CLAUS (4). *Die illustrierte Vogelbücher; ihre Geschichte und Bibliographie*. Stuttgart, 1953.

OHWI, JISABURŌ (1). *Flora of Japan (in English); a combined much revised and extended Translation by the Author of his 'Nihon Shokubutsuji' (Flora of Japan, 1953) and 'Nihon Shokubutsuji' (Flora of Japan, Pteridophyta, 1957)*. Ed. F. G. Meyer & E. H. Walker, Smithsonian Institution, Washington, D. C. 1965.

OLSCHKI, L. (7). *The Myth of Felt*. Univ. of California Press, Los Angeles, Calif., 1949.

OLSON, L. (1). 'Columella and the Beginning of Soil Science.' *AGHST*, 1943, **17**, 65.

OLSON, L. (2). 'Cato's Views on the Farmer's Obligation to the Land.' *AGHST*, 1945, **19**, 129.

DA ORTA, GARCIA (1). *Colloquies on the Simples and Drugs of India* (with the annotations of the Conde de Ficalho). Sotheran, London, 1913. Eng. tr. by Sir Clements Markham. *Coloquios dos Simples e Drogas he Cousas Mediçinais da India, compostos pello Doutor G. da O. . . .* de Endem, Goa, 1563. Latin epitome by Charles de l'Escluze, Plantin, Antwerp, 1567, repr. 1574, and later standard edition, ed. Conde de Ficalho, Lisbon 1895.

OSBECK, P., TOREEN, O. & ECKEBERG, C. G. (1). *A Voyage to China and the East Indies* [1751]*; together with a Voyage to Suratte by* [Rev.] *Olof Toreen; and an account of the Chinese Husbandry by Capt. C. G. Eckeberg* [all of the Swedish East India Company]*; To which are added a Faunula and 'Flora Sinensis'*. Tr. from Germ. by J. R. Forster. 2 vols. White, London, 1771. First published in Swedish, 1757. (Osbeck's account ends with a congratulatory letter to him from his teacher Linnaeus.).

OSBORN, A. (1). 'Viburnums; the Asiatic Species.' *GDN*, 1924, **88**, 221.

OSTENFELD, C. H. & PAULSEN, O. (1). 'A List of Flowering Plants from Inner Asia collected by Sven Hedin.' In

Southern Tibet by S. Hedin, Swedish Army Lithogr. Inst., Stockholm, 1922.

OTSUKA YASUO (1). 'Kurse Geschichte von einem chinesischen Heilkraut *thu fu ling* (China wurzel).' *JPISH*, 1968, **13** (no. 3), 1.

PAGEL, W. (5). 'The Vindication of "Rubbish".' *MHJ*, 1945. Cf. Neugebauer (4).

PALIBIN, J. W. (1). 'Quelques Mots sur le Nénuphar de la Chine (*Nelumbo nucifera* Gaertn.) et sa Porteé Économique.' *BJBSP*, 1904, **4**, 60.

PAMPANINI, R. (1). *Le Piante Vascolari raccolte dal Rev. P. C. Silvestri nell' Hupei durante gli Anni 1904–07ᶜ negli Anni 1909–10.* Pellas (Chiti), Florence, 1911.

PARMENTIER, A. A. (1). *Recherches sur les Végétaux nourissants, qui, dans les temps de Disettes, peuvent remplacer les Alimens ordinaires.* Impr. Royale, Paris, 1781. Eng. tr. *Observations on such Nutritive Vegetables as may be substituted in the place of ordinary Food in Times of Scarcity,* London, 1783.

PARTINGTON, J. R. (6). 'The Origins of the Planetary Symbols for Metals.' *AX*, 1937, **1**, 61.

PAXTON, J. (1). 'History and Culture of the Chinese Chrysanthemum.' *HR*, 1834, **3**, 469. Germ. tr. *AGZ*, 1835, **3**, 53.

PAYNE, C. H. (1). 'The Chrysanthemum in China.' *GDC*, 1918 (3rd ser.), **64**, 233. Based on item (7) of Cibot (6).

PAYNE, C. H. (2). 'A Brief History of the Chrysanthemum.' *JRHS*, 1890, **12**, 115.

PEI TÊ-AN (1). 'Making the Red Soil Fertile' (the lateritic clays of Chiangsi). *CREC*, 1963, **12** (no. 4). 8.

PELLIOT, P. (9) (tr.). 'Memoire sur les Coutumes de Cambodge' (a translation of Chou Ta-Kuan's *Chen-La Fêng Thu Chi*). *BEFEO*, 1902, **2**, 123. Revised version: Paris, 1951, see Pelliot (33).

PELLIOT, P. (17). 'Deux Itinéraires de Chine à l'Inde à la Fin du 8ᵉ Siècle.' *BEFEO*, 1904, **4**, 131.

PELLIOT, P. (51). 'Notes de Bibliographie Chinoise, I.' *BEFEO*, 1902, **2**, 315.

PELLIOT, P. (52). 'Notes de Bibliographie Chinoise, III.' *BEFEO*, 1909, **9**, 211, 424.

PEMBERTON, J. H. (1). *Roses, their History, Development and Cultivation.* Longmans Green, London, 1908; 2nd ed. 1920.

PENMAN, H. L. (1). *Vegetation and Hydrology.* Commonwealth Bureau of Soils, Technical Communication no. 53. Harpenden, 1963.

PERCIVAL, JOHN (1). *Agricultural Botany, Theoretical and Practical.* 8th ed., Duckworth, London, 1945.

PERNY, P. (1). *Dictionnaire Français-Latin-Chinois de la Langue Mandarine Parlée.* 2 vols. Didot, Paris, 1869–72.
The second volume is entitled 'Appendice' and contains; among other notes:
1. Une Notice sur l'Académie Impériale de Pékin.
3. Une Notice sur la Botanique des Chinois.
4. Une Description Gérérale de la Chine.
7. La Liste des Empereurs de la Chine avec la Date et les Divers Noms des Années de Régne.
9. Le Tableau des Principales Constellations.
12. La Hiérarchie Complète des Mandarins Civils et Militaires.
18. La Nomenclature des Villes de la Chine avec leur Latitude.
15. Le Livre dit des *Cents Familles* avec leurs Origines.
14, 16. Une Notice sur la Musique Chinoise et sur le système Monétaire.
19. La Synonymie la plus Complète qui ait été donnée jusqu'içi sur toutes les Branches de l'Histoire Naturelle de la Chine, etc.
See Bretschneider (1), p. 130, (10), p. 546.

PERRIN, R. M. S. (1). 'The Formation and Composition of Soils.' Art. in *Teaching Symposium* No. 1, British Academy of Forensic Sciences, London, 1963.

PERRY, LILY M. & METZGER, J. (1). *Medicinal Plants of East and South-east Asia; Attributed Properties and Uses.* M. I. T. Press, Cambridge, Mass., 1980.

PERRY, LYNN, R. (1). *Bonsai; Trees and Shrubs—a Guide to the Methods of Murata Kyuzo.* Ronald, New York, 1964.

PFISTER, L. (1). *Notices Biographiques et Bibliographiques sur les Jésuites de l'Ancienne Mission de Chine (+1552–1773).* 2 vols. Mission Press, Shanghai, 1932 (*VS*, no. 59).

PFIZMAIER, A. (77) (tr.). 'Die Toxicologie des Chinesische Nahrungsmitteln. *SWAW/PH*, 1865, **51**, 257. (Tr. ch. 24. *I Tsung Chin Chien.*)

PFIZMAIER, A. (104). (tr.). 'Denkwürdigkeiten von dem Baümen China's.' *SWAW/PH*, 1875, **80**, 191, 198, 205, 213, 220, 234, 240, 251, 264. (Tr. chs. 952, 953, 954, 955, 956, 957, 958, 959, 960 (in part), *Thai Phing Yü Lan.*)

PFIZMAIER, A. (105) (tr.). 'Ergänzungen zu d. Abhandlung von dem Baümen Chinas.' *SWAW/PH*, 1875, **81**, 143, 160, 167, 177, 188, 189, 192, 196. (Tr. chs. 960 (in part), 961, 962, 963, 969 (in part), 972 (in part), 973 (in part), 974 (in part), *Thai Phing Yü Lan.*)

PFIZMAIER, A. (106) (tr.). 'Denkwürdigkeiten v. den Früchten Chinas.' *SWAW/PH*, 1874, **78**, 195, 202, 214, 222, 230, 238, 244, 249, 260, 267, 274, 280. (Tr. chs. 964, 965, 966, 967, 968, 969 (in part), 970, 971, 972 (in part), 973 (in part), 974 (in part), 975, *Thai Phing Yü Lan.*)

PHẠM-HOÀNG-HỘ & NGUYÊN-VĂN-DƯƠNG (1). *Cây-cỏ* Miền Nam Việt-nam [Flore Générale de Vietnam]. Bộ Quóc-gia Giáo-dục Xuát-bản, Saigon, 1960.

PHISALIX, M. (1). *Animaux Venimeux et Venins.* 2 vols. Masson, Paris, 1922.

PING CHIH & HU HSIEN-SU (1). 'The Recent Progress of Biological Science' [in Chinese Culture]. Art. in

Symposium on Chinese Culture, ed. Sophia H. Chen Zen (Chhen Hêng-Chê, Mrs. Jen Hung-Chün), China Institute of Pacific Relations, Shanghai, 1931.

PULTENEY, R. (1). *Historical and Biographical Sketches of the Progress of Botany in England.* 2 vols. London, 1790.

PULTENEY, R. (2). *A General View of the Writings of Linnaeus.* 2 vols. London, 1781. 2nd. ed. London, 1805, with a memoir of the author.

PIRONE, P. P., DODGE, B. O & RICKETT, H. W. (1). *Diseases and Pests of Ornamental Plants.* 3rd ed. Ronald New York, 1960.

PLAYFAIR, G. M. H. (3). 'Ginger in China.' *JRAS/NCB*, 1885 (n.s), **20**, 91.

POKORA, T. (2). 'The Dates of Huan Than.' *ARO*, 1959, **27**, 670; 1961, **29** (no. 4).

POKORA, T. (3). 'Huan Than's *fu* on Looking for the Immortals' [*Wang Hsien Fu*, −14]. *ARO*, 1960, **28**, 353.

POKORA, T. (4). 'An Important Crossroad of Chinese Thought' (Huan Than, the first coming of Buddhism, and Yogistic trends in ancient Taoism). *ARO*, 1961, **29**, 64.

POKORA, T. (8). 'Once more the Dates of Huan Than.' *ARO*, 1961, **29**, 652.

POKORA, T. (9). 'The Life of Huan Than.' *ARO*, 1963, **31**, 1.

POKORA, T. (12). 'Komenský and Wang Kuo-Wei; a Note on the Influence of the educational opinions of [Jan Amos] Komenský (Comenius) upon Educational Reforms in China before the Revolution of 1911.' *ARO*, 1958, **26**, 626.

POLUNIN, O. & HUXLEY, A. (1). *Flowers of the Mediterranean.* Chatto & Windus, London, 1965. French tr. (adapted) by G. E. Aymonin, Nathan, Paris, 1967.

PONG, S. M. See Wu Y. C., Huang K. K. & Phêng, S. M. (1).

PONTANUS, JOH. JOVIANUS (1). *De Hortis Hesperidum, sive De Cultu Citrorum libri II.* Florence, 1514.

POPOV, V. V. (ed.) (1). *Lessy Severnogo Kitaya*, by A. K. Ivanov, V. A. Obruchev, V. I. Pavlinov, A. S. Kes', Chang Tsung-Hu, Yang Chieh, Yang Chhung-Chien & Sun Mêng-Lin (a collective work of separate contributions). Acad Sci. Moscow, 1959 (Trudy Komissü po Izucheniyu Chetvertichnogo Perioda, no. 14). Eng. tr. A. Gourevitch, adv. E. Grause: *The Loess of Northern China.* Israel Programme for Scientific Translations, Jerusalem; Oldbourne, London, 1961.

PORTERFIELD, W. M. (1). *Wayside Plants and Weeds of Shanghai.* Kelly & Walsh, Shanghai, 1933. (All 115 plants illustrated, and Chinese characters given in nearly all cases.)

POTTER, S. (1). *Language in the Modern World.* Penguin, London, 1960; 2nd rev. ed, 1961.

PRITZEL, G. A. (1). '*Iconum Botanicarum Index Locupletissimus*'; *Verzeichniss der Abbildungen sichtbar blühender Pflanzen und Farnkräuter aus der botanischen und Gartenliteratur des 18 u. 19 Jahrhunderts in alphabetischer Folge zusammengestellt.* Nicolai, Berlin, 1855; 2nd ed. 1866.

PRITZEL, G. A. (2). *Thesaurus Literaturae Botanicae Omnium Gentium.*... Berlin, 1871. Photolitho reproduction, Görlich, Milan, 1950.

PRSCHEWALSKI, N. M. See von Przywalski.

PRZEWALSKY, N. M. See von Przywalski.

PRZHEVALSKI, N. M. See von Przywalski.

VON PRZYWALSKI, N. M. (1). *Reisen in die Mongolei, im Gebiet der Tanguten und den Wüsten Nordtibets, in den Jahren 1870 bis 1873.* Germ. tr. by A. Kohn, with additional notes of *Mongolia and the Country of the Tanguts* (in Russian), St Petersburg, 2 vols. 1875. Costenoble, Jena 1877. Eng. tr. by E. D. Morgan with intrd. and notes by H. Yule, 1876.

PULLEYBLANK, E. G. (7). 'Chinese Historical Criticism; Liu Chih-Chi and Ssuma Kuang.' Art. in *Historians of China and Japan*, ed. W. G. Beasley & E. G. Pulleyblank, p. 135. Oxford Univ. Press, London, 1961.

PULLEYBLANK, E. G. (11). 'The Consonantal System of Old Chinese.' *II, AM*, 1964, **9**, 206.

QUECKE, K. (1). *Die Signaturenlehre im Schrifttum des Paracelsus.* Volk und Gesundheit, Berlin, 1955. (Beihefte der *Pharmazie*, no. 2; Beiträge z. Gesch. d. Pharmazie und ihrer Nachbargebiete, no. 1.)

QUIN, J. J. (1). 'The Lacquer Industry of Japan.' *TAS/J*, 1881, **9**, 1.

RACKHAM, H. (2) (tr.). *Pliny 'Natural History', with an English Translation.* 10 vols. Heinemann, London, 1938, revised ed. 1949.

RANDS, R. L. (1). 'The Water-Lily in Mayan Art; a Complex of alleged Asiatic Origin.' *BBAE*, 1953, **151**, 75.

RAUH, W. (1). *Alpenpflanzen.* 4 vols. Winter Universitätsverlag, Heidelberg, 1951–3. Repr. 1958. (Sammlung naturwissenschaftlicher Taschenbücher, nos. 15, 16, 21, 22.)

RAVEN, C. E. (2). *John Ray, Naturalist; his Life and Works.* Cambridge, 1942; 2nd ed. 1950.

RAVEN, C. E. (3). *The English Naturalists from Neckam to Ray.* Cambridge, 1947.

RAY, JOHN (1). *Miscellaneous Discourses concerning the Dissolution and Changes of the World, wherein the Primitive Chaos and Creation, the General Deluge, Fountains, Formed Stones, Sea-shells found in the Earth, Subterranean Trees, Mountains, Earthquakes, Volcanoes*.... *are largely examined.* Smith, London, 1692.

RAY, JOHN (2). *Methodus Plantarum Nova, Brevitatis et Perspicuitatis causa Synoptice in Tabulis exhibita; cum Notis Generum tum summorum tum subalternorum Characteristicis, Observationibus nonnullis de Seminibus Plantarum, et Indice Copioso.* Faithorne & Kersey, London 1682.

RAY, JOHN (3). *Historia Plantarum* 3 vols. Clark & Faithorne, then Smith & Walford, London, 1686–1704.

RAY, JOHN (4). *Catalogus Plantarum circa Cantabrigiam Nascentium.* Cambridge, 1660.

RAY, JOHN (5). *Catalogus Plantarum Angliae et Insularum Adjacentium tum Indigenas tum in Agris passim Cultas Complectens.* Martyn, London, 1670. *Fasciculus Stirpium Britannicarum* ... (an appendix). Faithorne, London, 1688.

RAY, JOHN. See Keynes, Sir Geoffrey (1), Raven C. E. (2).

READ, BERNARD E. (1). (with Liu Ju-Chhiang). *Chinese Medicinal Plants from the 'Pen Tshao Kang Mu' A. D. 1596... a Botanical Chemical and Pharmacological Reference List* (Publication of the Peking Nat. Hist. Bull.), 1931. Sold by French Bookstore, Peiping, 1936. (Chs. 12–37 of *Pên Tshao Kang Mu*) (re W. T. Swingle, *ARLC/DO*, 1937, 191.)

READ, BERNARD E. (2) (with Li Yü-Thien). 'Chinese Materia Medica; I–V, Animal Drugs.' *PNHB*, 1931, **5** (no. 4), 37–80; **6** (no. 1), 1–102. Sep. pub. French Bookstore, Peiping, 1931. (Chs. 50 to 52 of *PTKM*; domestic animals, wild animals, rodentia and man.)

READ, BERNARD E. (3) (with Li Yü-Thien). 'Chinese Materia Medica; VI, Avian Drugs.' *PNHB*, 1932, **6** (no. 4), 1–101. Sep. pub. French Bookstore, Peiping, 1932. (Chs. 47 to 49 of *PTKM*; birds.)

READ, BERNARD E. (4) (with Li Yü-Thien). 'Chinese Materia Medica; VII, Dragon and Snake Drugs.' *PNHB*, 1934, **8** (no. 4), 297–357. Sep. pub. French Bookstore, Peiping, 1934. (Ch. 43 of *PTKM*; reptilia.)

READ, BERNARD E. (5) (with Yu Ching-Mei). 'Chinese Materia Medica; VIII, Turtle and Shellfish Drugs.' *PNHB*, 1939 (Suppl.), 1–136. Sep. pub. French Bookstore, Peiping, 1937. (Chs. 45 and 46 of *PTKM*; reptilia and invertebrata.)

READ, BERNARD E. (6) (with Yu Ching-Mei). 'Chinese Materia Medica; IX Fish Drugs.' *PNHB*, 1939 (Suppl.). Sep. pub. French Bookstore, Peiping, n. d. prob. 1939. (Ch. 44 of *PTKM*; fishes.)

READ, BERNARD. E. (7) (with Yu Ching-Mei). 'Chinese Materia Medica; X, Insect Drugs.' *PNHB*, 1941 Suppl.). Sep. pub. Lynn, Peiping, 1941. (Chs. 39 to 42 *PTKM*; insects, including arachnidae.)

READ, BERNARD E. (8). *Famine Foods listed in the 'Chiu Huang Pên Tshao'* Lester Institute, Shanghai, 1946.

READ, BERNARD E. (9). 'Influences des Régions Méridionales sur les Médecines Chinoises.' *BUA*, 1943 (3e sér), **4**, 475.

READ, BERNARD E. (10). 'Contributions to Natural History from the Cultural Contacts of East and West' *PNHB*, 1929, **4** (no. 1), 57.

READ, BERNARD E. (13). 'A Review of the Scientific Work done on Chinese Materia Medica.' *NMJC*, 1928, **14** (no. 5), 312.

READ, BERNARD E. (14). *Botanical, Chemical and Pharmacological Reference List to Chinese Materia Medica.* Bureau of Engraving and Printing, for the Department of Pharmacology and Physiological Chemistry, Peking Union Medical College, Peking, 1923. (The first version of Read (15); later Read, with Liu Ju-Chhiang (1) q.v..)

READ, BERNARD E. (15). *Bibliography of Chinese Medicinal Plants from the 'Pên Tshao Kang Mu'* (1596). Flora Sinensis, Ser. A, vol. 1, 'Plantae Medicinalis Sinensis', 2nd ed. Dept. of Pharmacology, Peking Union Medical College, and Peking Laboratory of Natural History. Peiping, 1927. Cf. Read (14) and Read, with Liu Ju-Chhiang (1).

READ, BERNARD E. (16). *Indigenous [Chinese] Drugs.* Chinese Medical Association, Shanghai, 1940. (CMA Special Report series, no. 13.)

READ, BERNARD E. (17). *Chinese Medicinal Plants, 'Ephedra', 'ma huang'.* 2 parts. Flora Sinensis, Ser. B, vol. 24. Dept. of Pharmacology, Peking Union Medical College and Peking Laboratory of Natural History, Peiping, 1930. No. 1, pt. 2 'The Botany of *ma huang*, abstracted from the work of Liu Ju-Chhiang, with additional notes from Stapf, Meyers, Groff and others.'

READ, BERNARD E. (18). *English Outline of the Chinese Manual of Toxicology published by the Chinese Medical Association.* CMA, Shanghai, 1932.

READ, BERNARD E. See Wang Chi-Min (2), biography no. 49.

READ, BERNARD E., LI WEI-JUNG & CHHÊNG JIH-KUANG (1). *Shanghai Foods* [; Analyses]. 4th ed. China Nutritional Aid Council, Shanghai, 1948.

READ, BERNARD E. & LIU JU-CHHIANG (1). 'Chinese Materia Medica; the Importance of Botanical Identity.' Contrib to *Trans. 6th Congress of the Far East Assoc. Tropical Medicine*, Tokyo, 1925, p. 987.

READ, BERNARD E. & PAK, C (PAK KYEBYŎNG) (1). 'A Compendium of Minerals and Stones used in Chinese Medicine, from the *Pên Tshao Kang Mu* by Li Shih-Chen (+1596),' *PNHB*, 1928, **3** (no. 2) 1–vii, 1–120. Revised and enlarged ed., French Bookstore, Peiping, 1936. Serial nos. 1–135; corresp. with chs. 8–11 of *PTKM*.

REDI, FRANCESCO (1). *Esperienze intorno à diverse Cose Naturali & particolamente à quelle che son portate dell'Indie.* Florence, 1671; repr. 1686. Abstr. in *PTRS*, 1673, **8**, 6001. Latin edns. Amsterdam 1675, 1685.

REDOUTÉ, P. J. & THORY, C. A. (1). *Les Roses.* Dufart, Paris, 1828–9 (3 vols.); 2nd ed. 1835 (4 vols.). Based on the collection of the Empress Josephine at La Malmaison.

REED, H. S. (1). *A Short History of the Plant Sciences.* Chronica Botanica, Waltham, Mass., 1942. (New Series of Plant Science Books, No. 7.) Repr. Ronald, New York, 1965.

REGENBOGEN, O. (1). 'Theophrastos' [of Eresos]. Art. in Pauly-Wissowa *Realenzyklopädie d. klass. Altertumswissenschaft*, Suppl. Vol. 7, cols. 1354 ff. (botany, cols. 1435 ff.).

REHDER, A. (1). *Manual of Cultivated Trees and Shrubs Hardy in North America.* 2nd. ed., New York, 1940.

R EID C & R EID E. M. (1). *The Pliocene Floras of the Dutch-Prussian Border*. Med. Rijksopsporing van Delfstoffen, 1915 (no. 6).

R EID, E. M. (1). 'A Comparative Review of Pliocene Flora, based on the study of Fossil Seeds.' *QJGS*, 1920, **76**, 145.

R ÉMUSAT, J. P. A. (13). *Mélanges Posthumes d'Histoire et de Littérature Orientales*. Imp. Roy., Paris, 1843.

R ENAUDOT, E. (1). *Ancient Account of China & India*. London, 1733.

R ENKEMA, H. W. (1). 'Oorspong, Beteeknis en Toepassing van de in de Botanie gebruikelije Teekens ter Aanduiting van het Geslacht en den Levensduur. Art. in *Gedenkb. J. Valckenier-Suringar*, p. 96. Nederl. Dendrolog. Vereeniging Wageningen, 1942.

R ENOU, L. & F ILLIOZAT, J. (1). *L'Inde Classique; Manuel des Études Indiennes*.
Vo. 1, with the collaboration of P. Meile, A. M. Esnoul & L. Silburn. Payot, Paris, 1947.
Vol. 2, with the collaboration of P. Demiéville, O. Lacombe & P. Meile, École Française d'Extrême Orient, Hanoi 1953; Impr. Nationale, Paris, 1953.

R EYNOLDS, P. K. (1). 'The Earliest Evidence of Banana Culture.' *JAOS*, 1951 (Suppl. no. 12), 1–27.

R EYNOLDS, P. K. & F ANG L IEN-C HÊ (C. Y.) (1). 'The Banana in Chinese Literature.' *HJAS*, 1940, **5**, 165.

R ICCI, M ATTEO (1). *I Commentarj della Cina*, 1610. MS unpub. till 1911 when it was edited by Ventun (1); since then it has been edited and commented on more fully by d'Elia (2).

R ICHARDSON, H. L. (1). 'Szechuan during the War' (World War II). *GJ*, 1945, **106**, 1.

R ICHARDSON, H. L. (2). *Soils and Agriculture of Szechuan*. Nat. Agric. Research Bureau, Ministry of Agriculture and Forestry, Chungking 1942. Special Pub. no. 27. (In English with Chinese summary.)

R ICHARDSON, H. L. (3). 'The Ice Age in West China.' *JWCBRS*, 1943, **14** B, 1.

R ICHTER, A. A. & K RASNOSSELSKAIA, T. A. (1) (ed.). *Chromatographic Adsorption Analysis*. Selected papers by M. S. Tswett (Mikhail Semionovitch Cvett). Academy of Sciences, Moscow, 1946.

R ICKETT, W. A. (1), (tr.). '*Kuan-tzu*'; *a Repository of Early Chinese Thought*. Hong Kong Univ. Press, 1965. Rev. T. Pokora, *ARO*, 1967, **35**, 169.

R IDLEY, H. N. (1). *The Flora of the Malay Peninsula*. 5 vols. 1922–5.

R IVIÈRE, A. & R IVIÈRE, C. (1). 'Les Bambous,' *BSA*, 1879 (3ᵉ sér.), **5**, 221, 290, 392, 460, 501, 597, 666, 758. (Preprinted in the previous year with continuous pagination as a separate publication, unamended.)

R IVOIRE, P. (1). 'Le Chrysanthème en Chine au dix-septième Siècle'. *RHORT*, 1928, 211.

R OBERTS, E. S. (1) (ed.). *The Works of John Caius, M. D., Second Founder of Gonville and Caius College, and Master of the College, 1559 to 1573; with a Memoir of his Life by John Venn*. Cambridge, 1912. (400th Birthday Anniversary Volume.)

R OBIN, P. A. (1). *Animal Lore in English Literature*. Murray, London, 1932.

R OBINSON, G. W. (1). *Soils; their Origin, Constitution and Classification; an Introduction to Pedology*. 2nd ed. Murby, London, 1936.

R OHDE, E. S. (1). *The Old English Herbals*. Longmans Green, London, 1922.

R OI, J. (1). *Traité des Plantes Médicinales Chinoises*. Lechevalier, Paris, 1955. (Encyclopédie Biologique ser. no. 47.) No Chinese characters, but a photocopy of those required is obtainable from Dr Claude Michon, 8 bis, Rue Desilles, Nancy, Meurthe & Moselle, France.

R ONDOT, N. (4). 'An Account of the Cultivation of Hemp and the Manufacture of Grass-Cloth.' *CRRR*, 1849, **18** (no. 4), 210.

R OSENTHAL, E. (1). *Pottery and Ceramics; from Common Brick to Fine China*. Penguin, London, 1949; 2nd ed. 1954.

DE R OSNY, L. (4). 'Botanique du Nippon; Aperçu de Quelques Ouvrages Japonais relatifs à l'Étude des Plantes, accompagné de Notices Traduites pour la première fois sur les Textes Originaux.' *MAO*, 1872, 123.

DES R OTOURS, R. (1). *Traité des Fonctionnaires et Traité de l'Armée, traduits de la Nouvelle Histoire des T'ang* (ch. 46–50). 2 vols. Brill, Leiden, 1948 (Bibl. de l'Inst. des Hautes Etudes Chinoises, vol. 6). (Rev. P. Demiéville, *JA*, 1950, **238**, 395.)

DES R OTOURS, R. (2) (tr.). *Traité des Examens* (translation of chs. 44 and 45 of the *Hsin Thang Shu*). Leroux, Paris, 1932. (Bibl. de l'Inst. des Hautes Etudes Chinoises, no. 2.)

R OXBY, P. M. (2). 'The Major Regions of China' *G*, 1938, **23**, 9.

R OXBY, P. M. (5). *China*, Oxford, 1942. (Oxford Pamphlets on World Affairs, no. 54.)

R OXBY, P. M. & O'D RISCOLL, P. (1) (ed.).
China Proper
Vol. 1: Physical Geography, History and Peoples, by P. M. Roxby, T. W. Freeman *et al.* (1944).
Vol. 2: Modern History and Administration, by P. M. Roxby *et al.* (1945).
Vol. 3: Economic Geography, Ports and Communications, by B. M. Husain, P. O'Driscoll *et al.* (1945).
Naval Intelligence Division, London, 1944–5. (Geographical Handbook Series, B. R. 530 B, 'Restricted'.) (At one time not generally available, but now in free circulation.)

R OYDS, T. F. (1) (tr.). *The 'Eclogues' and 'Georgics' of Virgil, translated into English Verse*. . . . Dent. London, n.d. (1907). (Everyman Edition.)

R OZANOV, A. N. (1). 'Old Ploughed Soils of the Loess Province in the Yellow River Basin' (in Russian). *POCH*, 1959 (no. 5), 1.

ROZANOV, A. N. (2). 'The Kheilutu Soils of the Loess Province in the Yellow River Basin" (in Russian).' *POCH*, 1959 (no. 10), 59.

RUDOLPH, R. C. (9). 'Illustrated Botanical Works in China and Japan. Art. in Buckman (1), p. 103.

RUMPF, G. E. (RUMPHIUS) (1). *Het Amboinsche Kruid-Boek*, or *Herbarium Amboinense; plurimas complectens arbores, frutices, herbas, plantas terrestres et aquatices, quae in Amboina et adjacentibus reperiuntur insulis ... c. 1700.* Pr. Amsterdam, 1741 to 1750.

SABINE, J. (1). 'On the *Paeonia moutan* and its Varieties.' *TRHS*, 1826, **6**, 465.

VON SACHS, JULIUS (1). *History of Botany* (+1530–1860). Tr. from the Germ. ed. of 1875 by H. E. F. Garnsey; revised by I. B. Balfour. Oxford, 1889. Repr. 1906.

SAEKI, P. Y. (1). *The Nestorian Monument in China.* SPCK, London, 1916.

SAFFORD, W. E. (1). *The Useful Plants of the Island of Guam; with an Introductory Account of the Physical Features and Natural History of the Island, of the Character and History of its People, and of their Agriculture.* Govt. Printing Office, Washington D.C. 1905. Contribs. from the U.S. National Herbarium, no. 9 (Smithsonian Institution.)

SAID HUSAIN NASR (1). *Science and Civilisation in Islam* (with a preface by Giorgio di Santillana). Harvard University Press, Cambridge, Mass. 1968.

SAILER, R. I. (1). 'Invertebrate Predators'. *Proceedings of the Third Annual Northeastern Forest Insect Work Conference.* U.S.D.A. Forest Service Research Paper NE-194, 1971.

SAMPSON, T. (1). 'Cotton in China.' *NQCJ*, 1868, **2**, 74.

SAMPSON, T. (2). 'The China Pine' (*Pinus sinensis*). *NQCJ*, 1868, **2**, 52.

SAMPSON, T. (3). 'Chinese Figs' (*Ficus*, spp.). *NQCJ*, 1869, **3**, 18.

SAMPSON, T. (4). 'The Banyan or *Yung* [*Jung*] Tree (*Ficus retusa*). *NQCJ*, 1869, **3**, 72.

SAMPSON, T. (5). The *Phu-Thi* Tree' [*Borassus flabellifera*]. (The ola-leaf palm, cf. Burkill (1), vol. 1, p. 347.) *NQCJ*, 1869, **3**, 100. In China the name 'Bodhi-tree' is applied mainly to *Ficus religiosa* (CC/1601), but also to a lime *Tilia paucicostata* = *Miqueliana* (CC/756). The name 'Bodhi-beads plant', yielding seeds so called, also applies to *Coix Lachryma-Jobi* (R/737; CC/2003), and especially to *Sapindus Mukorossi* (R/304; CC/787).

SAMPSON, T. (6). 'Palm Trees.' *NQCJ*, 1869, **3**, 115, 129, 147, 170.

SAMPSON, T. (7). 'Tea.' *NQCJ*, 1869, **3**, 110.

[SAMPSON, T.] PS. CANTONIENSIS, DEKA (PS.) & K. (INIT). (1). 'Grafting (*po shu*).' *NQCJ*, 1867, **1**, 157; 1870, **4**, 6.

SAMPSON, T. & HANCE, H. F. (1). 'The Fung [*Fêng*] Tree' (*Liquidambar formosana*). *NQCJ*, 1869, **3**, 4, 31.

SAMPSON, T., McCARTEE, D. B., KNOWLTON, M. J., HANCE, H. F., X. (INIT.) & L. (INIT.) (1). 'The Tallow Tree (*Chhiu Shu*), *NQCJ*, 1868, **2**, 43, 76, 112; 1870, **4**, 5, 27, 64.

[SAMPSON, T.] PS. CANTONIENSIS, M[AYERS], W. F., HANCE, H. F., TAINTOR, E. C. & SCHLEGEL, G. (1). 'Henna in China (*Lawsonia inermis*).' *NQCJ*, 1867, **1**, 40; 1868, **2**, 11, 29, 33, 41, 46, 78, 180; 1869, **3**, 30.

SARGENT, C. S. (2). '*Prunus Davidiana*' *GDNF*, 1897, **10**, 503.

SARGENT, C. S. (1). *Plantae Wilsonianae; an Enumeration of the Woody Plants collected in Western China for the Arnold Arboretum of Harvard University during the years 1907, 1908 and 1910 by E. H. Wilson.* 3 vols. Cambridge, Mass. 1911–17. Pubs. of the Arnold Arboretum, no. 4. (No Chinese names, no illustrations.)

SARTON, GEORGE (1). *Introduction to the History of Science.* Vol. 1, 1927; Vol. 2, 1931 (2 parts); Vol. 3, 1947 (2 parts). Williams & Wilkins, Baltimore. (Carnegie Institution Pub. no. 376.)

SARTON, GEORGE (6). 'Arabic Scientific Literature.' In *Ignace Goldzieher Memorial Volume.* Budapest, 1948, pt I, p. 55.

SARTON, G. (9). *The Appreciation of Ancient and Medieval Science during the Renaissance (1450–1600).* University of Pennsylvania Press, Philadelphia, 1955.

SARTON, GEORGE (12). '*Horus; a Guide to the History of Science; a First Guide for the Study of the History of Science; with Introductory Essays on Science and Tradition.*' Chronica Botanica, Waltham, Mass. 1952.

SAUNDERS, E. R. (1). '*Matthiola*' [the garden stock]. *BIBGEN*, 1928, **4**, 141.

SAUVAGET, J. (2) (tr.). *Relation de la Chine et de l'Inde, redigée en +851 (Akhbār al-Sīn wa'l-Hind).* Belles Lettres, Paris, 1948. (Budé Association; Arab Series.)

SAVAGE, G. (1). *Porcelain through the Ages; a Survey of the Main Porcelain Factories of Europe and Asia....* Penguin, London, 1954.

SAVAGE, SPENCER (1). 'Studies in Linnaean Synonymy; I, Caspar Bauhin's *Pinax* and Burser's Herbarium.' *PLS*, 1935, **148**, 16.

SAVATIER, L. (1). *Botanique Japonaise; Livres 'Kwa-wi [Ka-i]' traduit du Japonais avec l'aide de M. Saba par le Dr L. S....* Paris, 1875. The *Ka-i* (Classification of Flowering Plants) had been compiled by Shimada Mitsufusa & Ono Ranzan by +1759 (another ed. + 1765); cf. Bartlett & Shohara (1), p. 133.

SAVELIEV, D. N. (1). 'O Dinamika Soderzhaniya Karotina v Kormovyh Rasteniya' (Changes in the Carotene Content of Fodder Plants with Time of Day).' *ZN*, 1968, no. 12. 47. Eng. abstr. *NAR*, 1970, **40**, 42.

SAYER, G. R. (1). '*Ching Tê-Chên Tao Lu': or the Potteries of China; being a translation with Notes and an Introduction.* Routledge & Kegan Paul, London, 1951.

SCHAFER, E. H. (1). 'Ritual Exposure [Nudity, etc.] in 'Ancient China.' *HJAS*, 1951, **14**, 130.

SCHAFER, E. H. (2). 'Iranian Merchants in Thang Dynasty Tales.' *SOS*, 1951, **11**, 403.

SCHAFER, E. H. (4). 'The History of the Empire of Southern Han according to chapter 65 of the *Wu Tai Shih* of Ouyang Hsiu.' Art. in *Silver Jubilee Volume of the Zinbun Kagaku Kenkyusō*. Kyoto University, Kyoto 1954, p. 339. (*TG/K*, 1954, **25**, pt. 1.)

SCHAFER, E. H. (11) (tr. & comm.). *Tu Wan's 'Stone Catalogue of Cloudy Forest [Yün Lin Shih Phu].'* Univ. Calif. Press, Berkeley & Los Angeles, 1961, Rev. M. Loehr, *JAOS*, 1962, **82**, 262.

SCHAFER, E. H. (12). *The Conservation of Natural Resources in Mediaeval China*. Contrib. to Xth Internat. Congr. of the History of Science, Ithaca, 1962. Abstract vol. p. 67.

SCHAFER, E. H. (13). *The Golden Peaches of Samarkand; A Study of T'ang Exotics*. Univ. California Press, Berkeley, 1963.

SCHAFER, E. H. (14). 'The Last Years of Chhang-an.' *OE*, 1963, **10**, 133–79.

SCHAFER, E. H. (15). 'Notes on a Chinese word for Jasmine [*su hsing*].' *JAOS*, 1948, **68**, 60.

SCHAFER, E. H. (16). *The Vermilion Bird; Thang Images of the South*. Univ. of Calif. Press, Berkeley and Los Angeles, 1967. Rev. D. Holzman, *TP*, 1969, **55**, 157.

SCHAFER, E. H. (17). 'The Idea of Created Nature in Thang Literature' (on the phrases *tsao wu chê* and *tsao hua chê*). *PEW*, 1965, **15**, 153.

SCHAFER, E. H. (18). *Shore of Pearls* (a study of Hainan Island and its history). Univ. of California Press, Berkeley and Los Angeles, 1970.

SCHAFER, E. H. (20). 'Li Tê-Yü and the Azalea.' *ASEA*, 1965, **19**, 105.

SCHAFER, E. H. (21). 'Notes on Thang Culture.' *MS*, 1962, **21**, 194–321.
  Includes (1) miniature gardens;
           (2) coloured glass windows;
           (3) alligators and crocodiles;
           (4) the God of the South Seas, with tr. of an eulogy by Han Yü engraved on a stele (+820).

SCHAFER, E. H. (22). 'The Conservation of Nature under the Thang Dynasty.' *JESHO*, 1962, **5**, 279. Abstr. *RBS*, 1969, **8**, no. 153.

SCHEUCHZER, J. J. (1). *Herbarium Diluvianum, collectum a J. J. S. Gessner, Tiguri (Zürich)*, 1709.

SCHINDLER, B. (6). 'The Development of Chinese Writing from its Elements.' *OAZ*, **3**, 453.

SCHINDLER, B. (7). 'Prinzipien d. chinesischen Schriftbildung.' *OAZ*, **4**, 297.

SCHLEGEL, G. (5). *Uranographie Chinoise*, etc. 2 vols. with star-maps in separate folder. Brill, Leyden, 1875.

VON SCHLOTHEIM, E. F. (1). *Die Petrefaktenkunde auf ihrem jetzigen Standpunkte durch die Beschreibung seiner Sammlung versteinerter und fossile Ueberreste des Thier- und Pflanzen-reichs der Vorwelt erläutert*. Gotha, 1820.

SCHMID, A. (1). *Über alte Kräuterbücher*. Bern, 1939. (Schweizer Beiträge zur Buchkunde)

SCHMIDT, C. F., READ B. E. & CHHEN KHO-KHUEI (1). 'Experiments with Chinese Drugs; I, Tang-kuei.' *CMJ*, 1924, **38**, 362.

SCHMIDT, K. P. (1). 'The *Methodus* [Broadside] of Linnaeus, +1736.' *JSBNH*, 1952, **2**, 369.

SCHMITZ, R. & TAN, F. TEK-TIONG (1). 'Die *Radix Chinae* in der *Epistola de Radicis Chinae . . .* des Andreas Vesalius (+1546).' *AGMW*, 1967, **51**, 217. Cf. Tan, F. Tek-Tiong (1).

SCHOUW, J. F. (1). *Grundtraek til en almindelig Plantegeographie* (in Danish). Gyldendal, Copenhagen, 1822. Germ. tr. *Grundzüge einer allgemeine Pflanzengeographie*. Berlin, 1823.

SCHULTES, J. A. (1). *Grundriss einer Geschichte und Literatur der Botanik von Theophrastos Eresios bis auf die neuesten Zeiten, nebst einer Geochichte der botanischer Gärten* (second title-page). Schaumburg, Vienna, 1817.
  This work has two title-pages, the first as follows:
  Anleitung zum gründlichen Studium der Botanik zum Gebrauche bey Vorlesungen und zum Selbstunterrichte.
  Index sep. pub. in slightly smaller format:
  Vollständiges Register by J. Schultes, with foreword by L. Radlkofer, Ackermann, München, 1871.

SCHULTES, R. A. (3) (ed.). 'Recent Advances in American Ethnobotany.' *CBOT*, 1953, **15** (no. 1/6).

SCHULTES, R. E. (4). 'The Future of Plants as Sources of New Biodynamic Compounds.' Art. in *Plants in the Development of Modern Medicine*, ed. T. Swain.

SCHUSTER, JULIUS (1). 'Jungius' Botanik als Verdienst und Schicksal.' Art. in *Beiträge Z. Jungius-Forschung*, ed. Adolf Meyer (6), p. 27.

SCOTT, J. CAMERON (1). *Health and Agriculture in Asia; a Fundamental Approach to some of the Problems of World Hunger*. Faber & Faber, London, 1952.

SEALY, J. R. (1). 'Camellia reticulata.' *BOTM*, 1935, **158**, pl. 9397.

SEALY, J. R. (2). 'Camellia oleifera.' *BOTM*, 1954, **170**, pl. 221.

SEALY, J. R. (3). 'Camellia Species.' Art. in *Camellias and Magnolias*, p. 27. Roy. Hort. Soc. Conf. Rep., ed. P. M. Synge, 1950.

SEALY, J. R. (4). 'Species of Camellia in Cultivation.' *JRHS*, 1937, **62**, 352.

SELIGMAN C. G. (6). 'Note on the Preparation and Use of the Kenyah [Sarawak] Dart-Poison, *ipoh*,' *JRAI*, 1902, **32**, 239. 'On the Physiological Action of the Kenyah Dart-Poison *ipoh*, and its active principle Antiarin.' *JOP*, 1903, **29**, 39.

SENN, G. (1). *Die Entwicklung der biologischen Forschungsmethode in der Antike und ihre grundsätzliche Förderung durch Theophrast von Eresos.* Sauerländer, Aaran, 1933. (Veröffentlichungen d. Schw. Gesellsch. f. Gesch. d. Med. u.d. Naturwissenschaften, no. 8.)

SENN, G. (2). 'Hat Aristoteles eine selbstständige Schrift über Pflanzen verfasst?' *PL*, 1930, **85** (n.f. *39*) 113.

SHAO YAO-NIEN (1). 'The Lemons of Kuangtung, with a Discussion concerning Origins.' *LSJ*, 1933, **12** (Suppl.), 271. Tr. from a Chinese paper in the *Lingnan Ta-Hsüeh Nung-Hsüeh Hui Chi-Khan* by Li Lai-Jung & G. W. Groff.

SHAW, C. F. (1). *The Soils of China; a Preliminary Map of Soil Regions in China.* Soil Bulletin no. 1. Geological Survey of China, Nanking and Peiping, 1930–1, 1–30. Chinese tr. by Shao Tê-Hsing. Special Report of the Institute of Geology, no. 1 National Peiping Academy, 1931, 1–50.

SHAW, C. F. (2). 'A Preliminary Field Study of the Soils of China.' *DISS/CKSA*, 1932, **2**, 17–47.

SHEN TSUNG-HAN (1). *Agricultural Resources of China.* Cornell Univ. Press, Ithaca, N.Y. 1951.

SHEPHERD, R. E. (1). *History of the Rose.* Macmillan, New York, 1954.

SHERBORN, C. D. (1) (ed.). *Index Animalium.* Brit. Mus. Nat. Hist., London.

SHIH SHÊNG-HAN (1). *A Preliminary Survey of the book 'Chhi Min Yao Shu', an Agricultural Encyclopaedia of the +6th century.* Science Press, Peking, 1958.

SHIH SHÊNG-HAN (2). *On the 'Fan Shêng-Chih Shu', an Agricultural Book written by Fan Shêng-Chih in −1st Century China.* Science Press, Peking, 1959.

SHIRAI, MITSUTARŌ (1). 'A Brief History of Botany in Old Japan.' Art. in *Scientific Japan, Past and Present*, ed. Shinjo Shinzo. Kyoto, 1926. (Commemoration Volume of the 3rd Pan-Pacific Science Congress.)

SHIU IU-NIN. See Shao Yao-Nien.

SIMMONDS, P. L. (1). *The Commercial Products of the Vegetable Kingdom considered in their various Uses to Man and in their relation to the Arts and Manufactures; forming a practical Treatise and Handbook of Reference for the Colonist, Manufacturer, Merchant and Consumer on the Cultivation, Preparation for Shipment, and Commercial Value, of the various substances obtained from Trees and Plants entering into the Husbandry of the Tropical and Sub-tropical Regions.* Day, London, 1854.

SINGER, C. (1). *A Short History of Biology.* Oxford, 1931.

SINGER, C. (3). The Scientific Views and Visions of St. Hildegard. Art. in Singer (13), vol. 1, p. 1. Cf. Singer (16), a parallel account.

SINGER, C. (6). 'Galen as a Modern.' *PRSM*, 1949, **42**, 563.

SINGER, C. (14). 'The Herbal in Antiquity.' *JHS*, 1927, **47**, 1.

SINGER, C. (15). 'Greek Biology and its Relation to the Rise of Modern Biology.' Art. in Singer (13), vol. 2, p. 1.

SINGER, C. (17). 'Early Herbals.' Art. in Singer (4), p. 168. Orig. Pr. in *EDR*, 1923.

SION, J. (1). *Asie des Moussons Pt. I.: Généralités Chine-Japon* ('Géographie Universelle, Vol. IX). Armand Colin, Paris, 1928.

SLOTTA, K. (1). 'Chemistry and Biochemistry of Snake Venoms.' *FCON*, 1955, **12**, 407.

SMIRNOW, L. (1). 'Tree-Peonies.' *JRHS*, 1953, **78**, 214.

SMITH, E. D. (1). 'Ancient History of the Chrysanthemum.' *BCSA*, 1935, **3**, 6.

SMITH, F. PORTER. See Stuart, G. A. (1) and Wang Chi-Min (2), biography no. 12.

SMITH, G. M. (1). 'The Development of Botanical Micro-Technique.' *TAMS*, 1915, **34**, 71–129.

SMITH, G. M. (2) (ed.). *Manual of Phycology; an Introduction to the Algae and their Biology.* Chronica Botanica, Waltham, Mass., 1951.

SMITH, JOHN (botanist of Kew) (1). *A Dictionary of Popular Names of the Plants which furnish the Natural and Acquired Wants of Man in all Matters of Domestic and General Economy.* Macmillan, London, 1882.

SMITH, KENNETH M. (1). *Textbook of Agricultural Entomology.* Cambridge, 1931.

SMITH, M. (1). 'List of the Genera and Species discovered in China since the Publication of the various parts of the 'Enumeration' [Forbes & Hemsley, (1)] from 1886 to 1904, alphabetically arranged.' *JLS/B*, 1905, **36**, 451–530.

SNAPPER, I. (1). *Chinese Lessons to Western Medicine; a Contribution to Geographical Medicine from the Clinics of Peiping Union Medical College.* Interscience, New York, 1941.

SOLLMANN, T. (1). *A Textbook of Pharmacology and some Allied Sciences.* Saunders, 1st ed. Philadelphia and London, 1901; 8th ed. extensively revised and enlarged, Saunders, Philadelphia and London, 1957.

SOOTHILL, W. E. (1). *The Students Four Thousand Character and General Pocket Dictionary.* Presbyterian Mission Press, Shanghai, 1899; often reprinted.

SOUBEIRAN, J. L. & DE THIERSANT, P. DABRY (1). *La Matière Médicale chez les Chinois; précédé d'un Rapport à l'Académie de Médecine de Paris par Prof. A. Gubler.* Masson, Paris, 1874.

SOWERBY, A. DE C. (1). *Nature in Chinese Art* (with two appendices on the Shang pictographs by H. E. Gibson). Day, New York, 1940.

SOWERBY, J. E. of *Sowerby's Botany*. See Boswell, Brown, Fitch & Sowerby (1).

SPRAGUE, T. A. (1). 'The Herbal of Otto Brunfels.' *JLS/B*, 1928, **48**, 79–124.

SPRAGUE, T. A. & NELMES, E. (1). 'The Herbal of Leonhart Fuchs.' *JLS(B)*, 1928, **48**, 545–642.

SPRAGUE, T. A. & SPRAGUE, M. S. (1). 'The Herbal of Valerius Cordus' (+1515–44). *JLS/B*, 1939, **52**, 1–113.

SPRENGEL, KURT (1). *Historia Rei Herbariae.* 2 vols. Amsterdam, 1808. Germ. tr. *Geschichte der Botanik.* 2 vols. Brockhaus, Altenburg & Leipzig, 1818.

SPURR, S. (1). *Forest Ecology.* Ronald, New York, 1964.

STADLER, H. (1). 'Theophrast und Dioskorides.' *AGKAW* (W. von Christ Festschrift), 1891, 176.

STAFLEU, F. A. (1). 'Adanson and the "Familles des Plantes".' Art. in *Adanson; the Bicentennial of Michel Adanson's 'Familles des Plantes',* ed. G. H. M. Lawrence, 2 vols. Hunt Library, Pittsburg. 1963.

STAKMAN, E. C. & HARRAR, J. G. (1). *Principles of Plant Pathology.* Ronald, New York, 1957.

STANNARD, J. (2). 'Bartholomeus Anglicus and the Influences shaping +13th-Century Botanical Nomenclature.' Communication to the XIIth International Congress of the History of Science, Paris, 1968. *Résumés des Communications,* p. 219.

STEARN, W. T. (1). 'Botanical Gardens and Botanical Literature in the Eighteenth Century.' Art. in *Catalogue of Botanical Books in the Collection of R. McM. M. Hunt,* vol. II, pp. xliii–cxl. Hunt Foundation, Pittsburgh, Pa. 1961.

STEARN, W. T. (2). 'The Origin of the Male and Female Symbols in Biology.' *TAX,* 1962, **11**, 109.

STEARN, W. T. (3). *An Introduction to the 'Species Plantarum' and Cognate Botanical Works of Carl Linnaeus,* prefaced to the Ray Society Facsimile Edition of the *Species Plantarum,* 1st ed. Stockholm, 1753. 2 vols. Quaritch, London, 1957, 1959. vol. I, pp. 1–176 (Ray Soc. Pubs. no. 140.)

STEARN, W. T. (4). 'Linnaeus' *Species Plantarum* and the Language of Botany.' *PLS,* 1953, **165**, 158.

STEARN, W. T. (5). *Botanical Latin; History, Grammar, Syntax, Terminology and Vocabulary.* Nelson, London, 1966.

STEARN, W. T. (6). '*Epimedium* and *Vancouveria* (Berberidaceae); a Monograph.' *JLS/B,* 1938, **51**, 409.

STEARN, W. T. (7). 'The Use of Bibliography in Natural History.' Art. in Buckman (1), p. 1.

STEARN, W. T. (8). 'The Background of Linnaeus' Contribution to the Methods and Nomenclature of Systematic Botany.' *SYZ,* 1959, **8**, 4.

STEELE, R. (1) (ed.). *Mediaeval Lore; an Epitome of the Science, Geography, Animal and Plant Folklore and Myth of the Middle Age; being Classified Gleanings from the Encyclopaedia of Bartholomew Anglicus on the Properties of Things,* with a preface by William Morris. Elliot Stock, London, 1893.

VAN STEEMIS-KRUSEMAN, M. J. (1). *Malaysian Plant Collectors and Collections; a Cyclopaedia of Botanical Exploration in Malaysia.* Ryksherbarium, Leiden, 1950 (Flora Malesiana, Ser. I. no. 1.)

STEIN, SIR AUREL (11). 'On the Ephedra, the *Hūm* Plant and the Soma.' *BLSOAS,* 1932, **6**, 501.

STEIN, W. H. & MOORE, S. (1). 'Chromatography.' *SAM,* 1951, **184** (no. 3), 35. Comment by H. Weil & T. I. Williams, 1951, **184** (no. 6), 2; and reply by the authors.

STEIN, R. A. (1). 'Le Lin-Yi; sa localisation, sa contribution à la formation du Champa, et ses liens avec la Chine.' *HH,* 1947, **2** (nos. 1–3), 1–300.

STEIN, R. A. (2). 'Jardins en Miniature d'Extrême-Orient; le Monde en Petit.' *BEFEO,* 1943, **42**, 1–104.

STEINMETZ, E. F. (1). *Vocabularium Botanicum.* Pr. pr. Amsterdam, 1949. 2nd ed. 1953. Polyglot dictionary of botanical technical terms, Latin, Greek, Dutch, German, English, French.

STEINMETZ, E. F. (2). '*Codex Vegetabilis*'; *Botanical Drugs and Spices.* Pr. pr. Amsterdam, 1948. Dictionary giving for each drug name the Family, Latin binomial, and common name in Dutch, German, English, French.

STEP, E. (1). *Wayside and Woodland Trees; a Pocket Guide to the British Sylva.* Warne, London, 1905.

STERN, F. C. (1). *A Study of the Genus 'Paeonia.'* Royal Horticultural Soc., London, 1946.

STERN, F. C. (2). 'Peony Species.' *JRHS,* 1931, **56**, 71.

VON STERNBERG, K. M. (1). *Versuch einer geognostisch-botanischen Darstellung der Flora der Vorwelt.* Prague, 1820–32; 2nd ed. 1838.

STEWARD, A. N. (1). *The Polygonaceae of Eastern Asia.* Cambridge, Mass, 1930. Contributions from the Gray Herbarium, Harvard University, no. 88. (No Chinese names or characters.)

STEWARD, A. N. (2). *Manual of Vascular Plants of the Lower Yangtze Valley, China.* Oregon State College, Corvallis, Ore. 1958. (Excellent coverage of Chinese names with characters, and glossary of Chinese terms, but no index of Chinese names.)

STRASBURGER, E. (1) with the collaboration of F. Noll, H. Schenck & G. Karsten. *A Textbook of Botany* 3rd Engl. ed. revised with the 8th Germ. ed. by W. H. Lang. Macmillan, London, 1908.

STRÖMBERG, R. (1). *Theophrastea; Studien zur botanischen Begriffsbildung.* Elander, Göteborg, 1937.

STUART, G. A. (1). *Chinese Materia Medica, Vegetable Kingdom.* American Presbyterian Mission Press, Shanghai, 1911.

SULLIVAN, MICHAEL (9). *The Birth of Landscape Painting in China.* Routledge & Kegan Paul, London, 1962. With appendix attempting to identify botanically plants and trees in Han art. Abstr. *RBS,* 1969, **8**, no. 408.

SUN, E-TU ZEN. See Sun, Jen I-Tu.

SUN, I. (1). 'Regional Division of the Saline Soils of China" (in Russian). *POCH,* 1956 (no. 11), 6.

SUN, JEN I-TU (2). 'Wu Chhi-Chün [+1789–1847]; Profile of a Chinese Scholar-Technologist. *TCULT,* 1965, **6**, 394.

SWAIN, T (1) (ed.). *Plants in the Development of Modern Medicine.* Symposium, 1968. Harvard Univ. Press, Cambridge, Mass., 1972.

SWANN, NANCY L. (1) (tr.). *Food and Money in Ancient China; the Earliest Economic History of China to + 25* (with tr. of

[*Chhien*] *Han Shu*, ch. 24 and related texts [*Chhien*] *Han Shu*, ch. 91 and *Shih Chi*, ch. 129). Princeton Univ. Press, Princeton, N.J., 1950 (rev. J.J. L. Duyvendak. *TP*, 1951, **40**, 210; C. M. Wilbur, *FEQ*, 1951, **10**, 320; Yang Lien-Shêng, *HJAS*, 1950, **13**, 524.)

SWINGLE, W. T. (1). 'Noteworthy Chinese Works on Wild and Cultivated Food Plants.'' [The 'Famine Herbals'] *ARLC/DO*, 1935, 193.

SWINGLE, W. T. (2). 'New and Old Chinese Treatises on Materia Medica.' *ARLC/DO*, 1937, 189. (On the work of Yang Hua-Thing and Chou Liu-Thing.)

SWINGLE, W. T. (3). 'Four Medicinal Formularies of the Thang Dynasty.' *ARLC/DO*, 1937, 194.

SWINGLE, W. T. (4). 'Chinese and other East Asiatic Books added to the Library of Congress 1925–1926.' *ARLC/DO*, 1925/1926, 313. (On the *Pên Tshao Thung Hsüan* (*c.* +1666), the *Pên Tshao Hui* (+1666), and the *Pên Tshao Pei Yao* (+1694), together with other works on botany, and on smallpox.)

SWINGLE, W. T. (5). 'Notes on Chinese Accessions to the Library of Congress.' *ARLC/DO* 1927/1928, 287. (On editions of the *Pên Tshao Kang Mu*; with descriptions of the *Pên Tshao Hui Yen* and the *Shih Wu Pên Tshao Hui Tsuan*. Also on the *Thien Kung Khai Wu* of Sung Ying-Hsing and the *Yünnan Kung-Chhang Kung Chhi Thu Lüeh*, Wu Chi-Chün's book on mining.)

SWINGLE, W. T. (6). 'Notes on Chinese Accessions on Medicine and Materia Medica, and on Nashi Pictographic MSS.' *ARLC/DO* 1929/1930, 368. (On the *Pên Tshao Fa Hui*, the *Pên Tshao Shu*, the *Pên Tshao Ching Chieh Yao*, the *Pên Ching Su Chêng* and the *Chêng Lei Pên Tshao*, etc.)

SWINGLE, W. T. (7). 'Notes on Chinese, Korean and Japanese Accessions on Materia Medica, Medicine and Agriculture. *ARLC/DO*, 1930/1931, 290. (On Korean and Japanese botanical books.)

SWINGLE, W. T. (8). 'Notes on Chinese Herbals and other Works on Materia Medica.' *ARLC/DO*, 1931/1932, 199. (On the first edition of the *Pên Tshao Kang Mu* in relation to the question of the origin of maize in China; also on the *Pên Tshao Hui Chien*,)

SWINGLE, W. T. (9). 'Notes on Early Chinese Records of Maize, on Natural Products, and on Medicine.' *ARLC/DO*, 1932/1933, 8. (On maize, the *Shang Han Lun*, marine products, the sweet potato, tobacco, etc.)

SWINGLE, W. T. (10). 'Chinese Famine Herbals, and Nashi Pictographic MSS.' *ARLC/DO*, 1936, 184. [With some translations by M. J. Hagerty.]

SWINGLE, W. T. (11). 'Chinese and other East Asiatic Books added to the Library of Congress, 1926–27.' *ARLC/DO*, 1926/1927, 245. (On editions of the *Pên Tshao Kang Mu*, the *Pên Tshao Kang Mu Shih I*, the *Shen Nung Pên Tshao Ching Su*, and on the *Shao-Hsing Pên Tshao*.)

SWINGLE, W. T. (12). 'Notes on Chinese Accessions; chiefly Medicine, Materia Medica and Horticulture.' *ARLC/DO*, 1928/1929, 311. (On the *Pên Tshao Yen I Pu I*, the *Yeh Tshai Phu*, etc.; including translations by M. J. Hagerty.)

SWINGLE, W. T. (13). 'A New Taxonomic Arrangement of the Orange sub-Family, Aurantioideae.' *JWAS*, 1938, **28**, 530.

SWINGLE, W. T. (14). 'The Botany of *Citrus* and its Wild Relatives of the Orange sub-Family.' Ch. in *The citrus Industry*, ed. Webber H. J., & Batchelor, L. D. Univ. Calif. Press, 1943, vol. 1, pp. 129–474. Also sep. pub.

SYNGE, P. M. (1) (ed.). *Camellias and Magnolias* (Report of a Royal Horticultural Society Conference). Royal Hort. Soc., London, 1950.

SYRENIUSZ, SZYMON (1). *Zielnik Herbarzem z Języka Łacinskiego zowią to jest Opisanie Wtasne Imion, Kszattu, Przyrodzenia, Skutkow, y Mocy Ziöt Wszelakich. . . . Pilnie Zebrane a Porządnie Zpisane przez D. S. S.* Kraków (Cracow), 1613.

VON TAKÁCS, Z (1). *Early Chinese Writing as a Source of Chinese Landscape Painting.* Hirth Anniversary Volume (AM Introductory Volume), 1923, 400; with remarks by B. Schindler, p. 640.

TAKAKUSU, J. (3). 'Le Voyage de Kanshin [Chien-Chen] en Orient.' *BEFEO*, 1928, **28**, 1, 441; 1929, **29**, 47.

TAKEDA, HISAYOSHI (1). *Alpine Flowers of Japan; Descriptions* [and Photographs] *of 100 Select Species together with Culture Methods.* Sanseido, Tokyo, 1938.

TAN, F. TEK-TIONG (1). *Vesals 'Epistola de Radicis Chinae Usu' in ihrer Bedeutung f.d. pharm. Verwendung von 'Smilax China.'* Inaug. Diss., Marburg, 1966. Cf. R. Schmitz & F. Tee-Tiong Tan (1).

TANAKA, T. (1). 'The Taxonomy and Nomenclature of Rutaceae, Aurantioideae.' *BL*, 1931, **2**, 101.

TANAKA, T. (2). 'The Citrus Culture of the Pacific Region.' *MTC*, 1927, **1** (no. 1), 1.

TANAKA, T. (3). 'Botanical Discoveries on the Citrus Flora of China.' *MTC*, 1932, **1** (no. 2), 12.

TANAKA, T. (4). 'A Monograph on the Satsuma Orange, with Special Reference to the Occurrence of New Varieties through Bud Variation.' *MFSA/TIU*, 1932, **4**, 1–626.

TANSLEY, A. G. (1) (ed.). *Types of British Vegetation.* Cambridge, 1911.

TATARINOV, A. (1). 'Die chinesische Medizin.' Art. in *Arbeiten d. k. Russischen Gesandschaft in Peking über China, sein Volk, seine Religion, seine Institutionen, socialen Verhältnisse, etc.*, ed. C. Abel & F. A. Mecklenburg, vol. 2, p. 423. Heinicke, Berlin, 1858.

TATLOW, ANTONY (1). *Brechts Chinesische Gedichte.* Suhrkamp, Frankfurt a/Main, 1973. Repr. 1983.

TEMPLE, SIR WILLIAM (1). 'Upon the Gardens of Epicurus, or Of Gardening' (1685). In *Essays*, vol. 2, pt. 2, p. 58 (1690).

TEMPLE, SIR WILLIAM (2). *Miscellanea.* Simpson, London, 1705, (Contains the essays 'On Ancient and Modern Learning' (1690) and 'Of Heroick Virtue', both of which deal with Chinese questions.)

TENG SSU-YÜ & BIGGERSTAFF, K. (1). *An Annotated Bibliography of Selected Chinese Reference Works.* Harvard-Yenching Inst. Peiping, 1936. (Yenching Journ. Chin. Studies, monograph no. 12.)

THAN PO-FU, WÊN KUNG-WÊN, HSIAO KUNG-CHÜAN & MAVERICK, L. A. (1), (tr.). *Economic Dialogues in Ancient China; Selections from the 'Kuan Tzu' (Book)* ... Pr. pr. Carbondale, Illinois and Yale Univ. Hall of Graduate Studies, New Haven, Conn. 1954. (Rev. A. W. Burks, *JAOS*, 1956, **76**, 198.)

THANG, T. Y. (1). 'The Present Development of Soil Study in China.' *Trans. 3rd. Internat. Congr. Soil Science*, 1936, vol. 3, p. 136.

THÉVENOT, MELCHISEDECH (1). *Relations de divers Voyages curieux qiu n'ont point été publiés et qu'on a traduits ou tirés des Originaux des Voyageurs* ... 4 pts. in 2 vols., articles separately paginated. Paris, 1663–72.

THIMANN, K. V. (1) (ed.). 'The Action of Hormones in Plants and Invertebrates' (Chs 2–5 of *The Hormones*, vol. 1). Academic Press, New York, 1952.

THOMPSON, D'ARCY W. (1). 'Excess and Defect; or the Little More and the Little Less.' *M*, 1929, **38**, 43.

THOMAS, G. S. (1). *The Old Shrub Roses.* Phoenix, London, 1955. Includes reprints of Hurst (1) and Hurst & Breeze (1).

THOMAS, G. S. (2). *Shrub Roses of Today.* Phoenix, London, 1962.

THOMAS, G. S. (3). *Climbing Roses Old and New.* Phoenix, London, 1965.

THOMPSON, D'ARCY W. (3). *On Aristotle as a Biologist; with a Proemion on Herbert Spencer.* Clarendon, Oxford, 1913 (Herbert Spencer Lecture).

THOMPSON, H. S. (1). 'On the Absorbent Power of Soils.' *JRAGS*, 1850, **11**, 68.

THOMPSON, R. CAMPBELL (2). *A Dictionary of Assyrian Botany.* British Academy, London, 1949.

THOMSON, M. H. (1) (tr.). *Textes Grecs Inédits relatifs aux Plantes.* Belles lettres, Paris, 1955. (Assoc. Guillaume Budé, Nouv. Coll. de Textes et Documents.)

THORNDIKE, LYNN (10). *The Herbal of Rufinus* [+ 13th century]. 1946.

THORNDIKE, LYNN (11). 'Rufinus, a forgotten Botanist of the Thirteenth Century.' *ISIS*, 1932, **18**, 63.

THORNTHWAITE, G. W. (1). 'An Approach to a Rational Classification of Climate.' *GR*, 1948, **38**, 85.

THORNTHWAITE, C. W. (2). 'A Re-examination of the Concept and Measurement of Potential Transpiration.' Art. in *The Measurement of Potential Evapo-transpiration.* Publications in Climatology, no. 7, ed. J. R. Mather. Seabrook, New Jersey, 1954.

THORP, J. (1). *Geography of the Soils of China* (map in pocket). Geol. Survey of China; Inst. Geol. Nat. Acad. Peiping; China Foundation for the Promotion of Education and Culture, Nanking, 1936.

THORP, J. (2). 'Geographic Distribution of the Important Soils of China.' *BCGS*, 1935, **14**, 119–46. With tentative folding map and better-quality plates than Thorp (1), 'A Provisional Soil Map of China, with Notes on Chinese Soils', *Trans. 3rd. Internat. Cogr. Soil Science*, 1935, vol. 1, p. 275. (An abstract—and no map.)

THORP, J. (3). 'Soil Profile Studies as an Aid to understanding Geology.' *BCGS*, 1935, **14**, 359.

THORP, J. & CHAO, T. Y. (1). *Notes on Shantung Soils; a Reconnaissance Soil Survey of Shantung.* Soil Bulletin no. 14, 1–130. Peiping, 1936.

THROWER, S. L. (1). *Plants of Hongkong.* Longman, Hongkong, 1971. (Includes colour photographs, and some Chinese characters.)

TOBAR, J. (1). *Inscriptions Juives de K'ai-fong-fou* [*Khaifêng*]. Shanghai, 1900; repr. 1912. (VS. no. 17.)

TOREEN, OLOF, See Osbeck.

TONKIN, I. M. & WORK, T. S. (1). 'A New Antimalarial Drug.' *N*, 1945, **156**, 630.

TOLKOWSKY, S. (1). *Hesperides; a History of the Culture and Use of Citrus Fruits.* Bale & Curnow, London, 1938.

DALLA TORRE, C. G. & HARMS, H. (1). *Genera Siphonogamarum.* Berlin, 1900–7.

DE TOURNEFORT, J. P. (1). *Institutiones Rei Herbariae.* Royal Typ. Paris, 1700. 3rd ed. 1719.

TRAGUS, HIERONYMUS. See Bock, Jerome.

TREGEAR, T. R. (1). *A Geography of China.* Univ. London Press, London, 1965.

TRIGAULT, NICHOLAS (1). *De Christiana Expeditione apud Sinas.* Vienna, 1615; Augsburg, 1615. Fr. tr.: *Histoire de l'Expédition Chrétienne au Royaume de la Chine, entrepris par les. PP. de la Compagnie de Jésus, comprise en cinq livres* ... *tirée des Commentaires du P. Matthieu Riccius, etc.* Lyon, 1616; Lille, 1617; Paris, 1618. Eng. tr. (partial): *A Discourse of the Kingdome of China, taken out of Ricius and Trigautius.* In *Purchas his Pilgrimes.* London, 1625, vol. 3, p. 380. Eng. tr. (full): see Gallagher (1). Trigault's book was based on Ricci's *I Commentarj della Cina* which it follows very closely, even verbally, by chapter and paragraph, introducing some changes and amplifications, however. Ricci's book remained unprinted until 1911, when it was edited by Venturi (1) with Ricci's letters; it has since been more elaborately and sumptuously edited alone by d'Elia (2).

TRISTRAM, H. B. (1). *The Natural History of the Bible; being a Review of the Physical Geography, Geology and Meteorology of the Holy Land; with a Description of every Animal and Plant mentioned in the Holy Scripture.* 1st ed, London, 1867; 3rd ed. 1873; 6th ed. 1880; 7th ed. 1883.

TRISTRAM, H. B. (2). *The Survey of Western Palestine, its Fauna and Flora.* London, 1884.

TSCHIRCH, (1). 'Die Pharmacopoë, ein Spiegel ihrer Zeit.' *JAN*, 1905, **10**, 281, 337, 393, 449, 505.

Tshao Thien-Chhin, Ho Ping Yü & Needham, Joseph (1). 'An Early Mediaeval Chinese Alchemical Text on Aqueous Solutions' (the *San-shih-liu Shui Fa*, early +6th century). *AX*, 1959, **7**, 122. Chinese tr. by Wang Khuei-Kho (*1*), *KHSC*, 1963, no. 5, 67.

Tsien Tsuen-Hsuin. See Chhien Tshun-Hsün.

Tswett, M. S. (1). 'Über eine neue Kategorie von Adsorptions—erscheinungen und ihre Anwendung in der biochemischen Analyse.' *TSSV*, 1903, **14**, 1.

Tswett, M. S. (2). 'Physikalisch-chemische Studien ü das Chlorophyll; die Adsorptionen.' *BDBG*, 1906, **24**, 316. 'Adsorptionsanalyse und chromatographische Methode; Anwendung auf die Chemie des Chlorophylls.' *BDBG*, 1906, **24**, 384.

Tswett, M. S. (3). 'Zur Chemie der Chlorophylls ...' *BZ*, 1907, **5**, 6.

Tswett, M. S. (4). *Chromofilli w Rastitelnom i Schivotnom Mirje* (*The Chromophylls in the Plant and Animal World*), in Russian. Warshawskago Utsch. Okr., Warsaw, 1910.

Tswett, M. S. See Richter & Krasnosselskaia (1).

Twitchett, D. C. (5). 'Chinese Social History from the +7th to the +10th Centuries; the Tunhuang Documents and their Implications.' *PP*, 1966 (no. 35), 28.

Unger, F. (1). *Versuch einer Geschichte der Pflanzenwelt*. Braumüller, for K. K. Akad. Wiss., Vienna, 1852.

Unger, F. (2). *Chloris Protogaea; Beiträge z. Flora der Vorwelt*. Engelmann, Leipzig, 1841–7.

Unger, F. (3). *Sylloge Plantarum Fossilium; Sammlung fossiler Pflanzen besonders aus der Tertiär-formation*. Vienna, 1859–65. Orig. pub. *DWAW/MN*, **19, 22**, and **25**.

Unger, F. (4). *Synopsis Plantarum Fossilium*. Voss, Leipzig, 1845.

Unschuld, P. U. (1). '*Pên Tshao*'; *Zwei Tausend Jahre traditionelle pharmazeutische Literatur Chinas*. Moos, München, 1973.

Vavilov, N. I. (1). 'The Problem of the Origin of the World's Agriculture in the Light of the Latest Investigations., In *Science at the Cross-Roads*. Papers read to the 2nd International Congress of the History of Science and Technology. Kniga, London, 1931.

Vavilov, N. I. (2). *The Origin, Variation, Immunity and Breeding of Cultivated Plants; Selected Writings*. Tr. from Russian by K. Starr Chester, Chronica Botanica, Waltham, Mass., 1950; and as *CBOT*, 1951, **13** (no. 1/6), 1–364. (Chronica Botanica International Collection, no. 13.) Repr. Ronald, New York, 1965.

Vavilov, N. I. (3). 'The Role of Central Asia in the Origin of Cultivated Plants.' *BABGPB*, 1931, **26** (no. 3), 3. (In Russian and English.)

Verdoorn, F. (1), (ed.). 'Plant Genera, their Nature and Definition.' (A symposium by G. H. M. Lawrence, I. W. Bailey *et al.*) *CBOT*, 1953, **14** (no. 3), 1–160.

Verhaeren, H. (1). *L'Ancienne Bibliothèque du Pé-T'ang* [*Pei Thang*]. Lazaristes Press, Peiping, 1940.

Vogelstein, H. (1). *Die Landwirtschaft in Palästina z. Zeit der Mishnah: I, Getreidebau*. Inaug. Diss. Berlin, 1894.

Vyazmensky, I. S. (1). 'Iz Istorü Drevnei Kitaiskoi Biologü i Medizinü (On the History of Biology and Medicine in Mediaeval China).' *TIYT*, 1955, **4**, 1–68.

Waddington, C. H. (2). 'Pollen Germination in Stocks, and the possibility of applying a Lethal Factor Hypothesis to the Interpretation of their Breeding.' *JG*, 1929, **21**, 193.

Wagner, R. G. (1). 'Lebens-stil und Drogen im chinesischen Mittelalter.' *TP*, 1973, **59**, 79–178. (An exhaustive study of the tonic Han Shih San, a powder of four inorganic substances (Ca, Mg, Si) and nine plant substances containing alkaloids.)

Wagner, W. (1). *Die chinesische Landwirtschaft*. Parey, Berlin, 1926.

Waley, A. (1) (tr.). *The Book of Songs*. Allen and Unwin, London, 1937.

Waley, A. (2) (tr.). *One Hundred and Seventy Chinese Poems*. Constable, London, 1918; often reprinted.

Waley, A. (5) (tr.). *The Analects of Confucius*. Allen and Unwin, London, 1938.

Waley, A. (26). *The Opium War through Chinese Eyes*. Allen and Unwin, London, 1958.

Walker, E. H. (1). 'The Plants of China and their Usefulness to Man. *ARSI*, 1943 (1944), 325.

Walker, E. H. (2). *Fifty-one Common Ornamental Trees of the Lingnan University Campus*. Lingnan Univ. Canton, 1930. Lingnan Sci. Bulls. no. 1. (Excellent Chinese nomenclature, with illustrations throughout.)

Walker, E. H. (3). *Important Trees of the Ryukyu* [*Liu-Chhiu*] Islands. Forestry Bureau, Ryuku Govt. Naha, Okinawa; Pacific Science Board, Nat. Acad. Sci. Washington, D.C.; Smithsonian Institution, U. S. Nat. Museum, Washington, D.C.; U. S. Civil Administration, Ryukyu Islands, 1954 Special Bull. No. 3.

Wang Chun-Hêng (1). *A Simple Geography of China*. Foreign Languages Press, Peking, 1958.

Wang Ling (3). 'The Development of Decimal Fractions in China. *Proc. VIIIth Internat. Congress of the History of Science*, Florence, 1956, p. 13.

Ward, F. Kingdon (1). 'Tibet as a Grazing Land.' *GJ*, 1947, **110**, 60.

Ward, F. Kingdon (2). *From China to Hkamti Long*. Arnold, London, 1924.

Ward, F. Kingdon (3). *The Land of the Blue Poppy; Travels of a Naturalist in Eastern Tibet*. Cambridge, 1913.

Ward, F. Kingdon (5). *In Farthest Burma*. Seeley Service, London, 1921.

WARD, F. KINGDON (6). *Plant Hunting in the Wilds*. Figurehead, London, n.d. (1931).

WARD, F. KINGDON (7). *Burma's Icy Mountains*. Cape, London, 1949.

WARD, F. KINGDON (8). *The Riddle of the Tsangpi Gorges*, Arnold, London, 1926.

WARD, F. KINGDON (9). *Assam Adventure*. Cape, London, 1941.

WARD, F. KINGDON (10). *Plant Hunter in Manipur*. Cape, London, 1952.

WARD, F. KINGDON (11). *Plant Hunting on the Edge of the World*. Gollancz, London, 1930.

WARD, F. KINGDON (12). *Return to the Irrawaddy*. Melrose, London, 1956.

WARD, F. KINGDON (13). *Plant Hunter's Paradise*. Cape, London, 1937.

WARD, F. KINGDON (14). *The Romance of Plant Hunting*. Arnold, London, 1924.

WARD, F. KINGDON (15). *Pilgrimage for Plants*. With biographical introduction and bibliography by W. T. Stearn. Harrap, London, 1960.

WARD, F. KINGDON (16). *The Mystery Rivers of Tibet* ... Seeley Service, London, 1923.

WARD, F. KINGDON (17). 'The Sino-Himalayan Node' (Contribution to Symp. on the Flora of China). *Proc. 5th. Internat. Botanical Congress*, Cambridge, 1930 (Cambridge, 1931), p. 520.

WARD, F. KINGDON (18). *Plant Hunter in Tibet*. Cape, London, 1934.

WARD, J. KINGDON (1). *My Hill so Strong*. Cape, London, 1952.

WARDLE, R. A. (1). *The Problems of Applied Entomology*. Manchester Univ. Press, Manchester, 1929.

WARE, J. R. (7) (tr.). *The Sayings of Confucius [Lun Yü]; a New Translation*. Mentor (New American Library), New York, 1955. Often repr.

WASSON, R. G. (3). *Soma, Divine Mushroom of Immortality*. Harcourt Brace & World, New York, 1968; Mouton, the Hague, 1968. Ethno-Mycological Studies, no. 1. (With extensive contributions by W. D. O'Flaherty.)

WATLING, H. (1). 'Orchid Cultivation in China. *ORC*, 1928, **36**, 234, 293.

WATLING, H. (2). 'Researches into Chinese Orchid History.' *ORC*, 1928, **36**, 235, 295.

WATSON, BURTON (1) (tr.). *Records of the Grand Historian of China, translated from the 'Shih Chi' of Ssuma Chhien*. 2 vols. Columbia Univ. Press, New York, 1961.

WATSON, BURTON (2). *Ssuma Chhien, Grand Historian of China*. Columbia Univ. Press, New York, 1958.

WATSON, BURTON (3). *Chinese Lyricism; 'shih' poetry from the + 2nd to the + 12th Century, with Translations* ... Columbia Univ. Press, New York, 1971.

WATSON, E. (1). *The Principal Articles of Chinese Commerce (Import and Export), with a Description of the Origin, Appearance, Characteristics and General Properties of each Commodity; an Account of the Methods of Preparation or Manufacture; together with various Tests etc., by means of which the different products may be readily identified*. 2nd ed. Shanghai, 1930, (Inspectorate-General of Chinese Maritime Customs, Publications II, Special Series, no. 38). (Includes many products of interest for economic botany.)

W[ATSON], W. (1). '*Callistephus hortensis* (the Chinese aster), with a coloured plate.' *GDN*, 1898, **53**, 258.

WATT, SIR G. (1). *Dictionary of the Economic Products of India*. 6 vols. in 9 parts, plus index volume. Govt. Printing Office, Calcutta, 1889–96. (With index of plant-names in Indian languages.) Abridged to one-volume as *The Commercial Products of India*. Murray, London, 1908.

WATT SIR G. (2). *The Wild and Cultivated Cotton Plants of the World; a Revision of the Genus 'Gossypium' framed primarily with the object of aiding planters and investigators who may contemplate the systematic improvement of the Cotton Staple*. 1907. (Includes the East Asian species.) Partial Chinese tr. by Fêng Tsê-Fang, see Merrill & Walker (1), vol. 1, p. 117.

WAY, J. T. (1). 'On the Power of Soils to Absorb Manure.' *JRAGS*, 1850, **11**, 313; 1852, **13**, 123.

WEBBER, H. J. (1). 'The History and Development of the Citrus Industry.' Ch. in *The citrus Industry*, ed. Webber, H. J. & Batchelor, L. D., Vol. 1. pp. 1–40. Univ. Calif. Press, 1943.

WEBBER, H. J. & BATCHELOR, L. D. (1) (ed.). *The Citrus Industry*. 2 vols. Univ. Calif. Press, Berkeley and Los Angeles, 1943, 1948.

WEDEMEYER, A. (1). 'Wie heisst der Ginko-Baum in China und Japan, und was bedeutet sein Name?' *OAS* (Ramming Festschrift), 1959, 216. Abstr. *RBS*, 1965, **5**, no. 849.

WEIL, H. & WILLIAMS, T. I. (1). 'On the History of Chromatography.' *N*, 1950, **166**, 1000; 1951, **167**, 906.

WENT, F. W. & THIMANN, K. V. (1). *Phytohormones*. Macmillan, New York, 1937. Experimental Biology Monographs, no. 5.

WESTLAND, A. B. (1). 'Chinese Orchids.' *GDNF*, 1894, **7**, 76.

WHEATLEY, P. (1). 'Geographical Notes on some Commodities involved in Sung Maritime Trade.' *JRAS/M*, 1959, **32** (pt. 2), 1–140.

WHEWELL, WILLIAM (1). *History of the Inductive Sciences*. Parker, London, 1837. 3 vols. 2nd ed., revised 1847, 3rd ed. 1857 crit. G. Sarton, *A/AIHS*, 1950, **3**, 11 and in (12) pp. 49 ff., 121.

WHITE, W. C. (6). *An Album of Chinese Bamboos; a Study of a Set of Ink Bamboo Drawings* [by Chhen Liu, c. + 1785]. Univ. Toronto Press, Toronto, 1939.

WHITE, W. C. & WILLIAMS, R. J. (1). *Chinese Jews; a Compilation of Matters relating to Khaifêng* Univ. Press, Toronto, 1942. 3 vols. Vol. 1 Historical. Vol. 2 Inscriptional. Vol. 3 [with R. J. Williams], Genealogical. (Royal Ontario Museum Monograph Series, no. 1)

WHITEHEAD, P. J. P. & EDWARDS, P. I. (1). *Chinese Natural History Drawings* [or Paintings] *selected from the Reeves*

Collection in the British Museum (Natural History). Alden & Mowbray, for the Trustees of the British Museum (Natural History), London, 1974.

WHITTAKER, R. H. (1). 'Gradient Analysis of Vegetation.' *BR*, 1967, **49**, 207–64.

WICKES, D. R. (1). *Flowers of Peitaho*. Peking Leader Press, Peiping, 1926. (Well illustrated and with Chinese characters but no index of Chinese names.)

WIEGER, L. (3). *La Chine à travers les Ages; Précis, Index Biographique et Index Bibliographique*. Mission Press, Hsienhsien, 1924. Eng. tr. E. T. C. Werner.

WIENS, H. J. (3). *China's March toward the Tropics; a Discussion of the Southward Penetration of China's Culture, Peoples and Political Control in relation to non-Han Chinese Peoples of South China, and in the Perspective of Historical and Cultural Geography*. Shoestring, Hamden, Conn. 1954.

WILBUR, C. M. (1). 'Slavery in China during the Former Han Dynasty (−206 to +25). *FMNHP/AS*, 1943, **34**, 1–490 (Pub. no. 525).

WILHELM, HELLMUT (12). 'Shih Chhung and his Chin Ku Yuan.' *MS*, 1959, **18**, 314.

WILHELM, RICHARD (3) (tr.). *Frühling u. Herbst d. Lü Bu-We* (the *Lü Shih Chun Chhiu*). Diederichs, Jena, 1928.

WILHELM, RICHARD (4) (tr.). *'Liä Dsi' [Lieh Tzu]; Das Wahre Buch vom Quellenden Urgrund, 'Tschung Hü Dschen Ging'; Die Lehren der Philosophen Liä Yü-Kou and Yang Dschu*. Diederichs, Jena, 1921.

WILHELM, RICHARD (6) (tr.). *Li Gi, das Buch der Sitte des älteren und jungeren Dai* [i.e. both *Li Chi* and *Ta Tai Li Chi*]. Diederichs, Jena, 1930.

WILKIE, D. (1). '*Camellia reticulata*.' *GDC* 1930 (3rd ser), **87**, 284.

WILLDENOW, C. L. (1). *Grundriss der Kräuterkunde zu Vorlesungen entworfen*. Berlin, 1792; 2nd ed. Vienna, 1798. Eng. trs. 1805 and 1811. Sect. 7 is entitled 'Geschichte der Pflanzen'.

WILLIAMS, S. WELLS (1). *The Middle Kingdom; A Survey of the Geography, Government, Education, Social Life, Arts, Religion, etc. of the Chinese Empire and its Inhabitants*. 2 vols. Wiley, New York, 1848; 4th ed. 1861, London, 1883, 1900.

WILLIAMS, S. WELLS. (2). Translation of Material relating to the Preface of the *Shen Nung Pên Tshao Ching* (Classical Pharmacopoeia of the Heavenly Husbandman).' In Bridgman's *Chrestomathy*, pp. 508 ff. See Bridgman (1).

WILLIAMS, T. I. & WEIL, H. (1). 'The [Historical] Phases of Chromatography.' *AK*, 1953, **5**, 283.

WILLIS, J. C. (1). *A Dictionary of the Flowering Plants and Ferns*. 5th ed. Cambridge, 1925 (first ed. 1897).

WILMOTT, ELLEN (1). *The Genus 'Rosa'*. 2 vols. in folio, Murray, London, 1910, 1914.

WILSON, E. H. (1). *The Lilies of Eastern Asia*. Dulau, London, 1925. (No Chinese characters and but few Chinese names.)

WILSON, E. H. (2). 'The Hortensias *Hydrangea macrophylla* and *Hydrangea serrata*. *JAA*, 1923, **4**, 233.

WINCKLER, L. (1). *Das 'Dispensatorium' des Valerius Cordus; Faksimile des im Jahre 1546. . . . ersten Druckes durch Joh. Petreium in Nürnberg*. Gesellsch. f. Gesch. d. Pharmazie, Mittenwald, 1934.

WISTER, J. C. (1). 'The Moutan Tree-Peony.' In *Peonies; Manual of the American Peony Society*, ed. J. Boyd, New York, 1928.

WISTER, J. C. & WOLFE, H. E. (1). 'The Tree-Peonies.' *NHM*, 1955, **34**, 1–61.

DE WIT, H. C. D. (1). 'In Memory of G. E. Rumphius [on the occasion of the 250th Anniversary of his death] 1702 to 1952. *TAX*, 1952, **1**, 101.

WITTFOGEL, K. A., FÊNG CHIA-SHÊNG et al. (1). 'History of Chinese Society (Liao), +907 to +1125. *TAPS*, 1948, **36**, 1–650. (Revs. P. Demiéville, *TP*, 1950, **39**, 347; E. Balazs, *PA*, 1950, **23**, 318.)

WOENIG, F. (1). *Die Pflanzen im alten Aegypten; ihre Heimat, Geschichte, Kultur, und ihre mannigfache Verwendung im sozialen Leben, in Kultus, Sitten, Gebräuchen, Medizin, Kunst*. Friedrich, Leipzig, 1886.

WOLF, R. (1). *Handbuch d. Astronomie, ihrer Geschichte und Litteratur*. 2 vols. Schulthess, Zürich, 1890.

WOLFF, W. (1). 'Die Boden von China in Beziehung auf China, Vegetation und Landwirtschaft.' *NW*, 1939, **27**, 217, 233.

WONG, K. K. See Wu Y. C., Huang K. K. & Phêng S. M. (1).

WONG, M. See Huard & Huang Kuang-Ming.

WOODVILLE, W. (1). *Medical Botany, containing Systematic and General Descriptions, with Plates of all the Medicinal Plants, Indigenous and Exotic, comprehended in the Catalogues of the Materia Medica as published by the Royal Colleges of Physicians of London and Edinburgh; accompanied with a circumstantial Detail of their Medicinal Effects, and of the Diseases in which they have been most successfully Employed*. 3 vols. Phillips, London, 1790–3. With supplementary volume of the *Principal Medicinal Plants not included in the . . . Collegiate Pharmacopoeias*. Phillips, London, 1794.

WOOTTON, A. C. (1). *Chronicles of Pharmacy*. 2 vols. Macmillan, London, 1910.

W[ORLIDGE], J[OHN], GENT. (1). *Systema Agriculturae; the Mystery of Husbandry Discovered*. London, 1669, 1675.

WRIGHT, RICHARDSON (1). *The Story of Gardening; from the Hanging Gardens of Babylon to the Hanging Gardens of New York*. Dodd & Mead, New York, 1934. Repr. Dover, New York, 1963.

WU KUANG-CHHING (1). 'Chinese Printing under Four Alien Dynasties.' *HJAS*, 1950, **13**, 447.

WU Y-C, HUANG, K-K, & PHÊNG S-M (1). 'Polypodiaceae Yaoshanensis, Kuangsi,' *BDB/SYS*, 1932, no. 3, 1–372.

WULFF, E. V. (1). *An Introduction to Historical Plant Geography*. Tr. from the Russian by E. Brissenden, foreword by E. D. Merrill. Chronica Botanica, Waltham, Mass. 1943. (New Series of Plant Science Books, no. 10).

WYLIE, ANN P. (1). 'The History of Garden Roses.' *JRHS*, 1954, **79**, 555; 1955, **80**, 8, 77; *END*, 1955, **14**, 181.

WYLIE, ARTHUR (1). *Notes on Chinese Literature*. First ed. Shanghai, 1867. Ed. here used Vetch, Peiping, 1939 (photographed from the Shanghai 1922 ed.).

WYMAN, D. (1). 'Oriental Flowering Crab-Apples.' *NHM*, 1940, **19**, 149.

YABUUCHI KIYOSHI (6). 'Astronomical Tables in China [the 'Calendars'] from the Wu Tai Period to the Chhing Dynasty' *JSHS*, 1963, **2**, 94.

YABUUCHI, KIYOSHI (9). 'Astronomical Tables (Calendars) in China from the Han to the Thang Dynasty.' Eng. art. in Yabuuchi Kiyoshi (*25*) (ed.), *Chūgoku Chūsei Kagaku Gijutsushi no Kenkyū* (Studies in the History of Science and Technology in Mediaeval China). Jimbun Kagaku Kenkyusō, Tokyo, 1963.

YABUUCHI KIYOSHI (10) = (*28*). 'The Observational Date of the *Shih Shih Hsing Ching*.' Art. in *Explorations in the History of Science and Technology in China*, ed. Li Kuo-Hao, Chang Mêng-Wên, Tshao Thien-Chhin & Hu Tao-Ching, p. 140. Shanghai, 1982.

YAGODA, H. (1). 'Applications of Confined Spot Tests in Analytical Chemistry.' *IEC/AE*, 1937, **9**, 79.

YAMADA, KENTARO (1). *A Short History of Ambergris [and its Trading] by the Arabs and the Chinese in the Indian Ocean*. Kinki University, 1955, 1956. (Reports of the Institute of World Economics, KKD, nos. 8 and 11.)

YAMADA, KENTARO (2). 'A Study of the Introduction of *An-hsi hsiang* to China and of Gum Benzoin to Europe.' *KKD*, 1954 (no. 5); 1955 (no. 7).

YANG CHIN-HOU & KOVDA, V. A. (1). 'The Systematics of Chinese Soils (in Russian). *POCH*, 1956 (no. 1), 89.

YANG CHUNG-TAI & YANG, R. (1). 'The Relationship between some traditional Chinese Medicinal Plants and the Interferon System.' *ISM*, 1982 (Dec), 1, (Memo I-A1233/1).

YANG HSIEN-YI & YANG, GLADYS (1) (tr.). *The 'Li Sao' and other Poems of Chu* [Chhü] *Yuan*. Foreign Languages Press, Peking 1953.

YANG LIEN-SHÊNG (3). *Money and Credit in China; A Short History*. Harvard Univ. Press, Cambridge, Mass. 1952. (Harvard-Yenching Institute Monograph Series, no. 12.)(Rev. R. S. Lopez, *JAOS*, 1953, **73**, 177; L. Petech, *RSO*, 1954, **29**, 277.)

YANG LIEN-SHÊNG (4). *Topics in Chinese History*. Harvard Univ. Press, Cambridge, Mass. 1950. (Harvard-Yenching Institute Studies, no. 4.) Additions and corrections in *HJAS*, 1950, **13**, 585.

YAO SHAN-YU (1). 'The Chronological and Seasonal Distribution of Floods and Droughts in Chinese History (−206 to +1911)' *HJAS*, 1942, **6**, 273.

YAO SHAN-YU (2). 'The Geographical Distribution of Floods and Droughts in Chinese History (−206 to +1911).' *FEQ*, 1943, **2**, 357.

YAO SHAN-YU (3). 'Flood and Drought Data in the *Thu Shu Chi Chhêng* and the *Chhing Shih Kao*.' *HJAS*, 1944, **8**, 214.

YASHIRODA, K. (1). 'Tree-Peonies in Japan.' *GDC*, 1929 (3rd ser.), **86**, 131.

YONGE, C. D. (1) (tr.). *The 'Deipnosophists', or Banquet of the Learned, of Athenaeus* [+228] *... with an Appendix of Poetical Fragments, rendered into English Verse by various Authors....* 3 vols. Bohn, London, 1854.

YÜ T. T. See Yü Tê-Chün.

YÜ TÊ-CHÜN (1). 'The Varieties of *Camellia reticulata* in Yunnan.' Art. in *Camellias and Magnolias*. Roy. Hort. Soc. Conf. Rep., ed. P. M. Synge, p. 13, 1950.

YÜ YING-SHIH (1). *Trade And Expansion in Han China*. Univ. of California, 1967.

YULE, H. & BURNELL, A. C. (1). *Hobson-Jobson: being a Glossary of Anglo-Indian Colloquial Words and Phrases....* Murray, London, 1886.

VON ZACH, E. (6). *Die Chinesische Anthologie; Übersetzungen aus dem 'Wên Hsüan'*. 2 vols. Ed. I. M. Fang. Harvard Univ. Press, Cambridge, Mass., 1958. (Harvard-Yenching Studies, no. 18.)

VON ZACH, E. (7). *Tu Fu's Gedichte* (translations collected from numerous periodical sources). Harvard Univ. Press, Cambridge, Mass. 1952. (Harvard-Yenching Institute Studies, no. 8.)

ZECHMEISTER, L. (1). 'History, Scope and Methods of Chromatography.' *ANYAS*, 1948, **49**, 145, 220.

ZECHMEISTER, L. (2). 'On the History of Chromatography.' *N*, 1951, **167**, 405.

ZECHMEISTER, L. (3). 'Michael Tswett, the Inventor of Chromatography.' *ISIS*, 1946, **36**, 108.

ZIMMER, G. F. (1). *A Popular Dictionary of Botanical names and Terms*. 1912. Repr. London, 1949.

ZIRKLE, C. (1). 'Species before Darwin.' *PAPS*, 1959, **103**, 636.

VON ZITTEL, K. A. (1). *Geschichte d. Geologie u. Paläontologie bis Ende des 19 Jahrhunderts*. München & Leipzig, 1899. (Gesch. d. Wissenschaft in Deutschland, no. 23.) Eng. tr. M. M. Ogilvie-Gordon. *History of Geology and Palaeontology to the End of the 19th Century*. London, 1901. Repr. Cramer, Weinheim, 1962 (Historiae Naturalis Classica, no. 22).

ZÜCKERT, J. F. (1). *Von den Speisen aus dem Pflanzenreiche*. Berlin, 1778.

BEN-ZVI, IZHAK (1). *The Stone Tablets of the Old Synagogue in Khaifêng* (in Hebrew with English summary). ben-Zvi Institute, Hebrew Univ., Jerusalem, 1961. Also in *SEF*, 1961, **5**, 29.

ADDENDA TO BIBLIOGRAPHIES

These addenda consist of supplementary entries required for the two sections by Dr Huang: *Natural Plant Pesticides* and *Biological Pest Control*. References not in these addenda will be found in the main bibliography.

BIBLIOGRAPHY A

Chi Lei Phien 鷄肋篇.
  Miscellaneous Random Notes.
  Sung, +1130.
  Chuang Chi-Yu 莊季裕.

Chiao Chhi Chi 脚氣集.
  Notes by a Beriberi Patient.
  Yuan, +1300.
  Chhê Jo-Shui 車若水.

Chieh I Hsin Yü 解義新語.
  New Talks on Old Explanations.
  Ming, +1582.
  Huangfa Fang 皇甫汸.

Chih Huang Chhuan Hsi Lu 治蝗傳習錄.
  Legacy of a Technique for Locust Control.
  Chhing, +1776.
  Chhen Shih-Yüan 陳世元.
  Copy extant in Fukien Library.
  Reproduction in South China Agricultural
    College.

Chou Hou Pei Chi Fang 肘後備急方.
  Handbook of Medicines for Emergencies.
  Chin, *c.* +340.
  Ko Hung 葛洪.

Chung Shu Shu 種樹書.
  Book of Tree Planting.
  Ming, +1401.
  Yü Chên Mu 俞貞木.

Chhêng Huai Lu 澄懷錄.
  Reflections of an Innocent Traveller.
  Sung, *c.* +1280.
  Chou Mi 周密.

Chhi Tung Yeh Yü 齊東野語.
  Rustic Talks in Eastern Chhi.
  S/Sung, *c.* +1290.
  Chou Mi 周密.

Chhu Huang Chi 除蝗記.
  Methods of Locust Extermination.
  Chhing, *c.* 1650.
  Lu Shih-I 陸世儀.
  Chhing, Chin Shih Wên Phien 清經世文篇.

Fa Yen 法言.
  Admonitory sayings [in admiration, and
    imitation, of the *Lun Yü*].
  Hsin, +5.
  Yang Hsiung 楊雄.

Fu Tzu 符子.
  The Book of Master Fu.
  Eastern Chin, *c.* +360.
  Fu Lang, 符朗, related to Fu Chien 符堅 of
    Former Chhin 前秦 +357–69).

Han Shih Wai Chüan 韓詩外傳.
  More Discourses Illustrating the Han Text of
    the 'Book of Odes'.
  C/Han, *c.* −135.
  Han Ying 韓嬰.

Hsien Chhuang Kua I Chih 閑窗括異志.
  Strange Things Seen through a Barred
    Window.
  Sung.
  Lu Ying-Lung 魯應龍.

Hsing Ying Tsa Lu 行營雜錄.
  Random Notes on a Tour of Duty.
  Sung, *c.* +1250.
  Chao Khuei 趙葵.
  Hsü Pai Chhuan Hsüeh Hai 續百川學海.

Hsü Po Wu Chih 續博物志.
  Supplement to the *Record of the Investigation of
    Things* (cf. *Po Wu Chih*).
  Sung, mid +12th century.
  Li Shih 李石.

I Lin 易林.
  Forest of Symbols of the (*Book of*) *Changes*
    [for divination].
  C/Han, *c.* −40.
  Chiao Kan 焦贛.

Liao Shih 遼史.
  History of the Liao (Chhi-tan) Dynasty
    [+916 to +1125].
  Yuan, *c.* +1350.
  Tho-Tho (Toktaga) 脫脫 & Ouyang
    Hsüan 歐陽玄.
  Partial tr. Wittfogel, Fêng Chia-Shêng
    et al.
  Yin-Tê Index no. 35.

Lieh Tzu [= *Chhung Hsü Chen Ching*] 列子.
  The Book of Master Lieh.
  Chou and C/Han −5th to −1st century.
    (Ancient fragments of miscellaneous
    origin finally cemented together with
    much new material about +380.)
  Attrib. Lieh Yü-Khou 列禦寇.
  Tr. R. Wilhelm (4); L. Giles (4); Wieger (7).
  TT/663.

Liu Chia Shih Ming Wu Shu 六家詩名物疏.
  Six Commentaries on Names and Things in the
    'Book of Odes'.
  Ming, +1605.
  Fêng Ying-Ching 馮應京.

Liu Chhing Jih Cha 留青日扎.
  Daily Jottings.
  Ming +1572.
  Thien I-Hêng 田藝蘅.

Liu Shu Ku 六書故.
  The Six Classes of Characters Explained.
  S/Sung, +1184.
  Tai Thung 戴侗.

Lun Hêng 論衡.
    Discourses Weighed in the Balance.
    H/Han, +82 or +83.
    Wang Chhung 王充.
    Tr. Forke (4).
    Chung-Fa Index no. 1.
Mao Shih Chêng I 毛詩正義.
    Basic Concepts of Mao's 'Book of Odes'.
    Thang, +642.
    Kung Ying-Ta 孔穎達.
Sang Yü Man Chi 桑楡漫記.
    Random Notes Written at Dusk.
    Ming.
    Thao Fu 陶輔.
Shen I Ching 神異經 (or *Chi* 記).
    Book of the Spiritual and the Strange.
    Ascr. Han, but prob. +4th or +5th century.
    Attrib. Tungfang Shuo 東方朔.
Shih Chi Chuan 詩集傳.
    The Book of Odes with Commentaries.
    Sung, +1177.
    Chu Hsi 朱熹.
Shih Ching Pai Shu 詩經稗疏.
    Little Commentaries on the 'Book of Odes'.
    Chhing, +1695.
    Wang Fu-Chih 王夫之.
Shih Ching Shu I Hiu Thung 詩經疏義會通.
    Integrated Commentaries on the 'Book of Odes'.
    Sung +1200.
    Chu Kung-Chhien 朱公遷.
Sou Shen Chi 搜神記.
    Reports on Spiritual Manifestations.
    Chin, *c.* +348.
    Kan Pao 干寶.
    Partial tr. Bodde (9).
Sun Kung Than Phu 孫公談圃.
    The Venerable Mr Sung's Conversation
    Garden.
    Sung, *c.* +1085.
    Sun Shêng 孫升.
Tung Pho Chih Lin 東坡志林.
    Journal and Miscellany of (Su) Tung-Pho.
    Sung, +1101.
    Su Tung-Pho 蘇東坡.
Wu Lin Chiu Shih 武林舊事.
    Institutions and Customs of the Old Capital
    Hangchow.
    Sung, *c.* +1270 but referring to events from
    about +1165 onwards.
    Chou Mi 周密.
Wu Shan Chih Lin, Pien Wu 五山志林, 辨物.
    Notes among the Five Hills; Differentiation of
    Things.
    Ming *c.* +1540.
    Huo Thao 霍韜.
    (in *Huo Wên-Min Kung Wên Chih* 霍文敏公文志,
    Collected Essays of the Venerable Huo Wên-
    Min.)
Ya Shu 雅述.
    Elegant Discourses.
    Ming, +1538.
    Wang Chün-Chhuan 王浚川.

BIBLIOGRAPHY B

Anon. (*258*)
    Chung Yao Ta Tzhu Tien 中藥大辭典.
    Cyclopedia of Chinese traditional Drugs
    (compiled by the New Kiangsu Medical
    College).
    Shanghai, 1978.
Chao Shan-Huan, Hsü Mu-Chhêng, Chang Hsing,
    Chen Hsün-Yuan & Liao Chhang-Chhing (*1*)
    趙善歡, 許木成, 張興, 陳循淵, 廖長青.
    *Ying Yung Lien Kho Chih Wu Fang Chih Kan Chü
    Hai Chhung Shih Yen* 應用楝科植物防治
    柑桔害蟲試驗.
    Experiments on the Application of the Seeds of
    Meliaceae for the Control of Citrus Insects.
    Acta Phytophylacica Sinica, 1982, **9**, 271–9.
Chêng Tai-An (*1*) 程岱葊.
    Hsi Wu Chü Lu 西吳菊略.
    Chrysanthemum Album for Western Wu
    1845.
    Thiao Shang Chhêng Tai-An Yuan Pên 苕上程岱
    葊元本.
Chi Tun-Yü (*1*) 吉敦諭.
    Thang Pien 糖辨.
    History of Sugar.
    Shê Hui Kho Hsüeh Chan Hsien 社會科學戰綫,
    1980, 181–6.
Chou Yao (*1*) 周堯.
    Chung Kuo Tsao Chhi Khun Chhung Hsüeh Yen Chiu

    Shih, Chhu Kao 中國早期昆蟲學研究史,
    初稿.
    Preliminary Draft on the early History of
    Entomology in China.
    Science Press, Peking, 1957.
Chou Yao (*2*) 周堯.
    Chung Kuo Khun Chhung Hsüeh Shih 中國昆蟲
    學史.
    History of Entomology in China.
    Entomotaxonomia, Wugong, Shensi, 1980.
Chu Hung-Fu & Kao Chin-Shêng (*1*) 朱弘復,
    高金聲.
    *Pên Tshao Kang Mu' Khun Chhung Ming Chhing
    Chu* 〈本草綱目〉昆蟲名稱註.
    A Commentary on the Insects mentioned in the
    'Pên Tshao Kang Mu'.
    Acta Entomol. Sinica 1950, *1*, 234.
Chung Chi-Hua (*1*) 鍾際華.
    Shih Ching Pai Hua Chieh 詩經白話解.
    The 'Book of Odes' in the Vernacular.
    Wên-Hua. Thaipei, 1968.
Chhen Shou-Chien (*1*) 陳守堅.
    '*Shih Chieh shang tsui ku lao ti Sheng Wu Fang
    Chih—Huang Kan I tsai kan chü yüan chung ti fang
    ssu chi chhi li yung chia-chih*' 世界上最古老的
    生物防治—黃柑蟻在柑橘園中的放飼
    及其利用價値.
    The Earliest Example of Biological Control:

Chhen Shou-Chien (*1*) (*cont.*)
    Yellow citrus ants in orange groves, their
    cultivation, utilization and efficacy.
    Acta Entomol. Sinica, 1962, **11**, 401.
Chhien Hsin-Huo (*1*) (Preface by) 錢炘和(序).
    Pu Huang Yao Chüeh 捕蝗要決.
    Effective Ways to Control Locusts.
    1855.
    Author unknown.
Chhin Jen-Chhang (*1*) 秦仁昌.
    Chueh Lei Chih Wu 蕨類植物.
    Biology of Pteridophyta.
Fang Ta-Shih (*1*) 方大湜.
    Sang Tshan Thi Yao 桑蚕提要.
    Essentials of Sericulture.
    Chhin, *c.* 1862–74.
Ho Fu-Min (W. E. Hoffman) (*1*) 賀輔民.
    Kan Chü Shu Khun Chhung Chih 柑桔樹昆蟲
    誌.
    Tr. Chou Yu-Wen 周郁文.
    Lingnan Agric. J. 1936, **2**, 165.
Hu Shih-Lin, Hsü Chhi-Chhu, Liu Chü-Fu & Ku
    Yün-Hsia (*1*) 胡世林, 徐起初, 劉菊福,
    古雲霞.
    Chhing Hao Su Ti Chih Wu Tzu Yüan Yen Chiu
    青蒿素的植物資源研究.
    Researches on Plant Resources of Qinghaosu.
    Bulletin of Chinese Materia Medica, 1981, **6**, 13.
Hung Tzu-Liang (*1*) 洪子良.
    Shih Ching Pai Hua Hsin Chieh 詩經白話新
    解.
    'The Book of Odes': A new vernacular
    rendition.
    Northwest Press 西北出版社 Thainan, 1979.
Hsia Wei-Ying (*1*) 夏緯英.
    *Chou Li Shu Chung Yu Kuan Nung Yeh Thiao Wen
    Ti Chiai Shih* 周禮書中有關農業條文
    的解釋.
    Explanations of Passages related to Agriculture
    from the 'Chou Li'.
    Agricultural Press
    Peking, 1979.
Hsin Shu-Chhih (*3*) 辛樹幟.
    Wo Kuo Kuo Shu Li Shih Ti Yen Chiu
    我國果樹歷史的研究.
    Studies on the History of Fruit Trees in China.
    Peking, 1962.
Hsü Hsiao-Thien (*1*) 許嘯天.
    Shih Ching Yuan Wen Tui Chao 詩經言文對照.
    'The Book of Odes': Original text and verna-
    cular rendition.
    Tatung Press 大東書局. Thainan, 1969.
Khung Kuang-Hai (*1*) (ed) 孔廣海.
    Hsin Hsien Chih 莘縣志.
    Chronicles of the *Hsin* District.
    1887.
Kuan Lung-Chhing 關龍慶 & Chhen Tao-
    Chhuan (*1*) 陳道川.
    Chhu Chhung Chü 除蟲菊.
    Pyrethrum.
    Chih Wu Tsa Chih 植物雜志.
    Botanical News, 1981, no. 6, 29.

Kuo Hua-Hsiu (*1*) 郭華秀.
    Kan Chü Lei Tsai Phei Fa 柑橘類栽培法.
    Citrus Cultural Methods.
    c. 1925 (Unpublished MS on file at the
    Needham Research Institute).
Li Chhün (*1*) 李羣.
    Mêng Chhi Pi Tan Hsüan Tu 夢溪筆談選讀.
    Selections from the Dream Pool Essays.
    Science Press, Peking 1975.
Liang Chia-Mien & Phêng Shih-Chiang (*1*)
    梁家勉, 彭世獎.
    *Wo Kuo Ku Tai Fang Chih Nung Yeh Hai Chhung
    Ti Chih Shih* 我國古代防治農業害蟲的
    知識.
    Knowledge of Agricultural Pest Control in
    ancient China.
    in *Chung Kuo Ku Tai Nung Yeh Kho Chi* 中國
    古代農業科技.
    Agricultural Science and Technology in Ancient
    China.
    Agricultural Press, Peking, 1980.
Liu Tun-Yuan (*1*) 劉敦愿.
    *Chung Kuo Ku Tai Tui Yü Tung Wu Thien Ti
    Kuan Hsi ti Jen Shih Huo Li Yung* 中國古代
    對於動物天敵關係的認識和利用.
    The Recognition and Application of Predator—
    Prey Relationships in Animals in Ancient
    China.
    Chung Kuo Ku Tai Nung Yeh Kho Chih 中國古代
    農業科技.
    Agricultural Press, Peking, 1980.
Ming Tsung-Tien (*1*) 閔宗殿.
    *Yang Ya Chih Chhung yu 'Chih Chhung Chhuang Hsi
    Lu'* 養鴨治蟲與〈治蝗傳習錄〉.
    Ducks for Pest Control and the 'Legacy of a
    Technique for Locust Control'.
    Nung Yeh Khao Ku, 1981, **1**, 106 （農業考古）.
Phei Chien & Chou Thai-Yen (*1*) 裴鑑周太炎.
    Chung-Kuo Yao Yung Chih Wu Chih 中國藥用
    植物誌.
    Illustrated Repertorium of Chinese Drug-Plants.
    4 parts Kho-Hsüeh, Peking, 1951–56.
Phêng Shih-Chiang (*1*) 彭世獎.
    *'Nan Fang Tshao Mu Chuang' Chuan Ché, Chuan
    Chhi ti Jo Kan Wen Thi*, 南方草木狀,
    撰者撰期的若干問題.
    Certain problems associated with the Author-
    ship of the 'Nan Fang Tshao Mu Chuang'.
    Nung Shih Yen Chiu 農史研究, 1980, **1**,
    75–80.
Phu Chih-Lung (*1*) (ed.) 蒲蟄龍.
    Hai Chhung Shêng Wu Fang Chih Ti Yüan Li
    Ho Fang Fa 害蟲生物防治的原理和方
    法.
    Principles and Practice of the Biological Control
    of Insect pests.
    Science Press, Peking, 1978.
Shih Shêng-Han (*7*) 石聲漢.
    Chung-Kuo Ku Tai Nung Shu Phing Chieh 中國
    古代農書評介.
    Introduction to Ancient Chinese Agriculture;
    A Critical Treatise.

Shih Shêng-Han (7) (cont.)
　　Agricultural Press, Peking, 1980. 農業
　　出版社.
Shih Shêng-Han (8) ed. 石聲漢.
　　Nung Cheng Chhüan Shu, Chiao Chu　農政全書
　　校注.
　　Complete Treatise on Agriculture with
　　Commentaries.
　　Classics Publishing House, Shanghai, 1979.
　　上海古籍出版社 Ku Chi Chhu Pan Shê.
Tshai Pang-Hua (1)　蔡邦華.
　　Khun Chhung Fêng Lei Hsüeh　昆蟲分類學.
　　Vol. 2. The Taxonomy of Insects.
　　Science Press, Peking, 1973.
Tsou Shu-Wen (1)　鄒樹文.
　　Chung Kuo Khun Chhung Hsüeh Shih　中國昆蟲
　　學史.
　　History of Entomology in China.
　　Science Press, Peking, 1981.
Wang Chih-I (1)　汪志伊.
　　Huang Chêng Chi Yao　荒政輯要.
　　Principles of Famine Management.
　　1806.
Wu Chhi-Chün (1)　吳其濬.

Chih Wu Ming Shih Thu Khao　植物名實圖考.
　　Illustrated Investigations of the Names and
　　Natures of Plants.
　　1848.
Wu Tê-To (1)　吳德鐸.
　　Ta Thang Pien　答糖辨.
　　Response to 'History of Sugar'.
　　Shê Hui Kho Hsüeh Chan Hsien　社會科學戰綫,
　　1981, 150-4.
Yang Phei (1)　楊沛.
　　Huang kan i shên wu hsüeh thê hsing chi chhi yung yü
　　fang chih kan chü hai chhung t chhu pu yen chiu
　　黃柑蟻生物學特性及其用于防治柑橘
　　寶蟲的初步研究.
　　A Preliminary Study of the Biological Char-
　　acteristics of the Yellow Citrus Ant and its use
　　in Controlling Pests of Oranges.
　　Acta-Scientiarum Naturalium Universitatis Sunyatseni
　　中山大學學報, 1982, no. 3, 102-5.
Yü Chêng-Hsieh (1)　俞正燮.
　　Kuei-Ssu Tshun Kao　癸巳存稿.
　　Remnant Drafts from the Kuei-Ssu Year.
　　1849.

BIBLIOGRAPHY C

DE BACH, PAUL (1). *Biological Control by Natural Enemies.* Cambridge University Press, 1974.

BALME, D. M. (1). Aristotle, Natural History and Zoology.' *Dictionary of Scientific Bibliography*, Vol. 1, ed. C. C. Gillespie, New York, 1970, pp. 258–66.

BOTTA, P. E. (1) *Relation d'un voyage dans l'Yemen.* Duprat, Paris, 1841.

CASSIANUS BASSUS (1) ed. *Geoponika: Agricultural pursuits*, translated by T. Owen. London, 1805.

SHIH-CHÜN CHAO 超史俊 (1) (tr.). *The Sayings of Chuang Tzu* 英譯莊子, Chih Wên Publishers 志文出版社 Hong Kong, 1973.

CHIU, SHIN-FOON (CHOU SHAN-HUAN) (1). 'Effectiveness of Chinese Insecticidal Plants with Reference to the Comparative Toxicity of Botanical and Synthetic Insecticides.' *JSFA* Sept. 1950, 276–86.

COLUMELLA, LUCIUS JUNIUS MODERATUS (1). *De re rustica & De arboribus.* Translated by H. Rackham, H. B. Ash, E. S. Forster & E. Heffner. Loeb Classical Library, Heinemann, London, 1968.

CROSBY, D. G. (1). 'Minor Insecticides of Plant Origin.' *Naturally Occuring Insecticides*, ed. by Jacobson, Martin and Crosby, D. G. Marcel Dekker Inc. New York, 1971, p. 177.

DOUTT, R. L. (1). 'The Historical Development in Biological Control.' *Biological Control of Insects and Weeds*, ed. DeBach, Paul and Schlinger, E. I. Reinhold, New York, 1964, pp. 21–42.

EVELYN, J. (1), 'Kalendarium hortense', 1st edition, London 1664.

FEINSTEIN, L. & JACOBSON M. (1). 'Insecticides Occuring in Higher Plants.' *Fortschritte der Chemie organischer Naturstoffe*, 1053, **10**, 423.

FORSKÅL, P. (1). *Descriptiones animalium avium, amphibium, piscum, insectorum, vermium; quae in itinere orientali, observavit P. Forskal*; post mortem autoris edidit, Carsten Niebuhr. Hauniae, Moeller, 1775.

GEE, N. GIST (1). 'A Preliminary List of Ants recorded from China.' *Lingnan Agric. Review*, 1924, **2**, 100–7.

GROFF, G. W. & HOWARD, C. W. (1). 'The Cultured Citrus Ants of South China'. *Lingnan Agric. Review*, 1924, **2**, 108–14.

EL HAIDARI, H. S. (1). 'The Use of Predator Ants for the Control of Date Palm Insect Pests in the Yemen Arab Republic.' *Date Palm J.*, 1981, **1** (1), 129–30.

HARPAZ, ISSAC (1). 'Early Entomology in the Middle East'. *History of Entomology*, ed. Ray F. Smith, Thomas E. Mittler & Carroll N. Smith. Annual Reviews Inc., Palo Alto CA., 1973.

JACOBSON, MARTIN (1). 'Insecticides, Insect Repellants and Attractants from Arid/Semiarid-Land Plants.' *Workshop Proceedings: Plants, the Potential for Extracting Protein, Medicines, and Other Useful Chemicals. 1983*, p. 38–46. U.S. Government Printing Office, Washington D.C.

KARLGREN, B. (14) (tr.). *The Book of Odes; Chinese Text, Transcription and Translation.* Museum of Far Eastern Antiquities, Stockholm, 1950. (A reprint of the translation only from his papers in *BMFEA*, **16** and **17**.)

KONISHI, M. & Y. ITO (1). 'Early Entomology in East Asia' 1–7 in *History of Entomology*, ed. Smith, R. F., Mittler, T. E. & Smith C. N. Annual Reviews, Inc. Palo Alto, 1973.

KIRBY, W. & SPENCE, W. (1). *An Introduction to Entomology*, Longman, Brown, Green and Longmans, 1815.

KRAUS, W. & BOKEL, M. (1). 'Neue tetranortriterpenoids aus *Melia azedarach* Linn (Meliaceae)'. *Chem. Ber.* 1981, **114**, 267–75.

LANCIANI, C. A. (1). 'Water mite-induced Mortality in a Natural Population of the Mosquito *Anopheles crucians*'. *J. Med. Entomol.* 1979, **15**, 529–32.

LANCIANI, C. A. (2). 'Biting a Mosquito.' *Audubon*, 1979, 54, 160.

VAN LEEUWENHOEK, A. (1). 'Part of a letter concerning excrescenses growing on willow leaves, etc.' *Phil. Trans. Roy. Soc. London.* 1701, **22**, 786–99.

LEGGE, J. (8) (tr.). *The Chinese Classics, etc.*: Vol. 4, Pts. 1 and 2. *The Book of Poetry*. Lane Crawford, Hongkong, 1871; Trübner, London, 1871.

LEGGE, J. (12) (tr.). *The Book of Poetry*, Chinese text with English translation, Comm. Press, Shanghai, n.d.

LESTON, D. (1). 'The ant mosaic: tropical tree crops and the limiting of pests and disease.' *PANS* 1973, **19**, 311–41.

LIU, GAINES (1). 'Some Extracts from the History of Entomology in China.' *Psyche*, A Journal of Entomology, 1939, **46**, 23–8.

LU GWEI-DJEN & NEEDHAM, JOSEPH (5). *Celestial Lancets, a History and Rationale of Acupuncture and Moxa*. Cambridge Univ. Press. 1980.

McCOOK, H. C. (1). 'Ants as Beneficial Insecticides.' *Proceedings of the Academy of Natural Sciences of Philadelphia*, 1882, **34**, 263–71.

METCALF, ROBERT L. & KELMAN, ARTHUR (1). 'Plant Protection.' *Science in Contemporary China*, edited by Orleans, Leo A. Stanford University Press, 1980.

ODISH, G. (1). *The Constant Pest.* New York, Ch. Scribner & Sons, 1976.

PALLADIUS (1). *On husbondrie*, ed. B. Lodge. Trubner & Co., London, 1873.

PLINY, THE ELDER (Caius Plinius Secundus) (1). *Natural History*, books 17–19, translated by H. Rackham. Loeb Classical Library, London, Heinemann, 1971.

RIDLEY, H. (1). 'On Oecophylla.' *JRAS/Str. 1980*, 345.

SAILER, R. I. (1). 'Invertebrate Predators.' *Proceedings of the Third Annual Northeastern Forest Insect Work Conference*. U. S. D. A. Forest Service Research Paper, NE-194, 1971.

SARTON, GEORGE (16). 'Query No. 139—The earliest example of biological control of insects, recorded by Chi Han?' *ISIS*, 1953, **44**, 99.

SHEPARD, HAROLD H. (1). *The Chemistry and Action of Insecticides*, McGraw Hill, New York, 1951.

SILVESTRI, F. (1). 'A Survey of the Actual State of Agricultural Entomology in the United States of North America'. *Hawaiian Forester and Agriculturalist*, 1909, **6**, 287. (Tr. from the Italian by J. Rosenstein.)

SIMMONDS, F. J., J. M. FRANZ & R. I. SAILER (1). 'A History of Biological Control.' *Theory and Practice of Biological Control*, ed. C. B. Huffaker & P. S. Messenger, Academic Press, New York, 1976.

SMITH, A. E. & SECOY, D. M. (1). 'Plants used for agricultural pest control in western Europe before 1850.' *CHIND*, 1981, 12.

SMITH F. PORTER & G. A. STUART (1). *Chinese Medicinal Herbs*. Compiled by Li Shih-Chen, Georgetown Press, San Francisco, 1973.

SWEETMAN, H. L. (1). *The Biological Control of Insects*, Comstock Publishing Co. Inc., 1936,.

SWEETMAN, H. L. (2). *The Principles of Biological Control*, 'Interrelation of Hosts and Pests and Utilization.' *Regulation of Animal and Plant Populations*. Wm. C. Brown Co. Dubuque, Iowa, 1958.

SWINGLE, W. T. (13). 'Our Agricultural Debt to Asia.' *The Asian Legacy and American life*, edited by Arthur E. Christy, pp. 84–114. John Day Co. Inc., 1942; Greenwood Press, NY, 1968.

WHEELER, W. M. (1). 'Chinese Ants.' *Bull Museum Comparative Zoology*, 1921, **64**, 529–47.

GENERAL INDEX

by Stephen Jones

Notes

(1) Articles (such as 'the', 'al-', etc.) occurring at the beginning of an entry, and prefixes (such as 'de', 'van', etc.) are ignored in the alphabetical sequences. Saints appear among all letters of the alphabet according to their proper names. Styles such as Mr, Dr, if occurring in book titles or phrases, are ignored; if with proper names, printed following them.

(2) The various parts of hyphenated words are treated as separate words in the alphabetical sequence. It should be remembered that, in accordance with the conventions adopted, some Chinese proper names are written as separate syllables while others are written as one word.

(3) In the arrangement of Chinese words, Chh- and Hs- follow normal alphabetical sequence, and *ü* is treated as equivalent to *u*.

(4) References to footnotes are not given except for certain special subjects with which the text does not deal. They are indicated by brackets containing the superscript letter of the footnote.

(5) Explanatory words in brackets indicating fields of work are added for Chinese scientific and technological persons (and occasionally for some of other cultures), but not for political or military figures (except kings and princes).

夏	HSIA kingdom (legendary?)		c. −2000 to c. −1520
商	SHANG (YIN) kingdom		c. −1520 to c. −1030
周	CHOU dynasty (Feudal Age)	Early Chou period	c. −1030 to −722
		Chhun Chhiu period 春秋	−722 to −480
		Warring States (Chan Kuo) period 戰國	−480 to −221

First Unification 秦 CHHIN dynasty — −221 to −207

漢	HAN dynasty	Chhien Han (Earlier or Western)	−202 to +9
		Hsin interregnum	+9 to +23
		Hou Han (Later or Eastern)	+25 to +220

三國 SAN KUO (Three Kingdoms period) +221 to +265

First Partition			
	蜀	SHU (HAN)	+221 to +264
	魏	WEI	+220 to +265
	吳	WU	+222 to +280

Second Unification	晉	CHIN dynasty: Western	+265 to +317
		Eastern	+317 to +420

劉宋 (Liu) SUNG dynasty +420 to +479

Second Partition — Northern and Southern Dynasties (Nan Pei chhao)

	齊	CHHI dynasty	+479 to +502
	梁	LIANG dynasty	+502 to +557
	陳	CHHEN dynasty	+557 to +589
魏		Northern (Thopa) WEI dynasty	+386 to +535
		Western (Thopa) WEI dynasty	+535 to +556
		Eastern (Thopa) WEI dynasty	+534 to +550
北齊		Northern CHHI dynasty	+550 to +577
北周		Northern CHOU (Hsienpi) dynasty	+557 to +581

Third Unification	隋	SUI dynasty	+581 to +618
	唐	THANG dynasty	+618 to +906

Third Partition 五代 WU TAI (Five Dynasty period) (Later Liang, Later Thang (Turkic), Later Chin (Turkic), Later Han (Turkic) and Later Chou) +907 to +960

遼	LIAO (Chhitan Tartar) dynasty		+907 to +1124
	West LIAO dynasty (Qarā-Khiṭāi)		+1124 to +1211
西夏	Hsi Hsia (Tangut Tibetan) state		+986 to +1227
宋	Northern SUNG dynasty (Fourth Unification)		+960 to +1126
宋	Southern SUNG dynasty (Fourth Unification)		+1127 to +1279
金	CHIN (Jurchen Tartar) dynasty		+1115 to +1234
元	YUAN (Mongol) dynasty		+1260 to +1368
明	MING dynasty		+1368 to +1644
清	CHHING (Manchu) dynasty		+1644 to +1911
民國	Republic		+1912

N.B. When no modifying term in brackets is given, the dynasty was purely Chinese. Where the overlapping of dynasties and independent states becomes particularly confused, the tables of Wieger (1) will be found useful. For such periods, especially the Second and Third Partitions, the best guide is Eberhard (9). During the Eastern Chin period there were no less than eighteen independent States (Hunnish, Tibetan, Hsienpi, Turkic, etc.) in the north. The term 'Liu chhao' (Six Dynasties) is often used by historians of literature. It refers to the south and covers the period from the beginning of the +3rd to the end of the +6th centuries, including (San Kuo) Wu, Chin, (Liu) Sung, Chhi, Liang and Chhen. For all details of reigns and rulers see Moule & Yetts (1).

ROMANISATION CONVERSION TABLES

BY ROBIN BRILLIANT

PINYIN/MODIFIED WADE–GILES

Pinyin	Modified Wade–Giles	Pinyin	Modified Wade–Giles
a	a	chou	chhou
ai	ai	chu	chhu
an	an	chuai	chhuai
ang	ang	chuan	chhuan
ao	ao	chuang	chhuang
ba	pa	chui	chhui
bai	pai	chun	chhun
ban	pan	chuo	chho
bang	pang	ci	tzhu
bao	pao	cong	tshung
bei	pei	cou	tshou
ben	pên	cu	tshu
beng	pêng	cuan	tshuan
bi	pi	cui	tshui
bian	pien	cun	tshun
biao	piao	cuo	tsho
bie	pieh	da	ta
bin	pin	dai	tai
bing	ping	dan	tan
bo	po	dang	tang
bu	pu	dao	tao
ca	tsha	de	tê
cai	tshai	dei	tei
can	tshan	den	tên
cang	tshang	deng	têng
cao	tshao	di	ti
ce	tshê	dian	tien
cen	tshên	diao	tiao
ceng	tshêng	die	dieh
cha	chha	ding	ting
chai	chhai	diu	tiu
chan	chhan	dong	tung
chang	chhang	dou	tou
chao	chhao	du	tu
che	chhê	duan	tuan
chen	chhên	dui	tui
cheng	chhêng	dun	tun
chi	chhih	duo	to
chong	chhung	e	ê, o

Pinyin	Modified Wade–Giles	Pinyin	Modified Wade–Giles
en	ên	jia	chia
eng	êng	jian	chien
er	êrh	jiang	chiang
fa	fa	jiao	chiao
fan	fan	jie	chieh
fang	fang	jin	chin
fei	fei	jing	ching
fen	fên	jiong	chiung
feng	fêng	jiu	chiu
fo	fo	ju	chü
fou	fou	juan	chüan
fu	fu	jue	chüeh, chio
ga	ka	jun	chün
gai	kai	ka	kha
gan	kan	kai	khai
gang	kang	kan	khan
gao	kao	kang	khang
ge	ko	kao	khao
gei	kei	ke	kho
gen	kên	kei	khei
geng	kêng	ken	khên
gong	kung	keng	khêng
gou	kou	kong	khung
gu	ku	kou	khou
gua	kua	ku	khu
guai	kuai	kua	khua
guan	kuan	kuai	khuai
guang	kuang	kuan	khuan
gui	kuei	kuang	khuang
gun	kun	kui	khuei
guo	kuo	kun	khun
ha	ha	kuo	khuo
hai	hai	la	la
han	han	lai	lai
hang	hang	lan	lan
hao	hao	lang	lang
he	ho	lao	lao
hei	hei	le	lê
hen	hên	lei	lei
heng	hêng	leng	lêng
hong	hung	li	li
hou	hou	lia	lia
hu	hu	lian	lien
hua	hua	liang	liang
huai	huai	liao	liao
huan	huan	lie	lieh
huang	huang	lin	lin
hui	hui	ling	ling
hun	hun	liu	liu
huo	huo	lo	lo
ji	chi	long	lung

Pinyin	Modified Wade–Giles	Pinyin	Modified Wade–Giles
lou	lou	pa	pha
lu	lu	pai	phai
lü	lü	pan	phan
luan	luan	pang	phang
lüe	lüeh	pao	phao
lun	lun	pei	phei
luo	lo	pen	phên
ma	ma	peng	phêng
mai	mai	pi	phi
man	man	pian	phien
mang	mang	piao	phiao
mao	mao	pie	phieh
mei	mei	pin	phin
men	mên	ping	phing
meng	mêng	po	pho
mi	mi	pou	phou
mian	mien	pu	phu
miao	miao	qi	chhi
mie	mieh	qia	chhia
min	min	qian	chhien
ming	ming	qiang	chhiang
miu	miu	qiao	chhiao
mo	mo	qie	chhieh
mou	mou	qin	chhin
mu	mu	qing	chhing
na	na	qiong	chhiung
nai	nai	qiu	chhiu
nan	nan	qu	chhü
nang	nang	quan	chhüan
nao	nao	que	chhüeh, chhio
nei	nei	qun	chhün
nen	nên	ran	jan
neng	nêng	rang	jang
ng	ng	rao	jao
ni	ni	re	jê
nian	nien	ren	jên
niang	niang	reng	jêng
niao	niao	ri	jih
nie	nieh	rong	jung
nin	nin	rou	jou
ning	ning	ru	ju
niu	niu	rua	jua
nong	nung	ruan	juan
nou	nou	rui	jui
nu	nu	run	jun
nü	nü	ruo	jo
nuan	nuan	sa	sa
nüe	nio	sai	sai
nuo	no	san	san
o	o, ê	sang	sang
ou	ou	sao	sao

Pinyin	Modified Wade–Giles	Pinyin	Modified Wade–Giles
se	sê	wan	wan
sen	sên	wang	wang
seng	sêng	wei	wei
sha	sha	wen	wên
shai	shai	weng	ong
shan	shan	wo	wo
shang	shang	wu	wu
shao	shao	xi	hsi
she	shê	xia	hsia
shei	shei	xian	hsien
shen	shen	xiang	hsiang
sheng	shêng, sêng	xiao	hsiao
shi	shih	xie	hsieh
shou	shou	xin	hsin
shu	shu	xing	hsing
shua	shua	xiong	hsiung
shuai	shuai	xiu	hsiu
shuan	shuan	xu	hsü
shuang	shuang	xuan	hsüan
shui	shui	xue	hsüeh, hsio
shun	shun	xun	hsün
shuo	shuo	ya	ya
si	ssu	yan	yen
song	sung	yang	yang
sou	sou	yao	yao
su	su	ye	yeh
suan	suan	yi	i
sui	sui	yin	yin
sun	sun	ying	ying
suo	so	yo	yo
ta	tha	yong	yung
tai	thai	you	yu
tan	than	yu	yü
tang	thang	yuan	yüan
tao	thao	yue	yüeh, yo
te	thê	yun	yün
teng	thêng	za	tsa
ti	thi	zai	tsai
tian	thien	zan	tsan
tiao	thiao	zang	tsang
tie	thieh	zao	tsao
ting	thing	ze	tsê
tong	thung	zei	tsei
tou	thou	zen	tsên
tu	thu	zeng	tsêng
tuan	thuan	zha	cha
tui	thui	zhai	chai
tun	thun	zhan	chan
tuo	tho	zhang	chang
wa	wa	zhao	chao
wai	wai	zhe	chê

Pinyin	Modified Wade–Giles	Pinyin	Modified Wade–Giles
zhei	chei	zhui	chui
zhen	chên	zhun	chun
zheng	chêng	zhuo	cho
zhi	chih	zi	tzu
zhong	chung	zong	tsung
zhou	chou	zou	tsou
zhu	chu	zu	tsu
zhua	chua	zuan	tsuan
zhuai	chuai	zui	tsui
zhuan	chuan	zun	tsun
zhuang	chuang	zuo	tso

MODIFIED WADE–GILES/PINYIN

Modified Wade–Giles	Pinyin	Modified Wade–Giles	Pinyin
a	a	chhio	que
ai	ai	chhiu	qiu
an	an	chhiung	qiong
ang	ang	chho	chuo
ao	ao	chhou	chou
cha	zha	chhu	chu
chai	chai	chhuai	chuai
chan	zhan	chhuan	chuan
chang	zhang	chhuang	chuang
chao	zhao	chhui	chui
chê	zhe	chhun	chun
chei	zhei	chhung	chong
chên	zhen	chhü	qu
chêng	zheng	chhüan	quan
chha	cha	chhüeh	que
chhai	chai	chhün	qun
chhan	chan	chi	ji
chhang	chang	chia	jia
chhao	chao	chiang	jiang
chhê	che	chiao	jiao
chhên	chen	chieh	jie
chhêng	cheng	chien	jian
chhi	qi	chih	zhi
chhia	qia	chin	jin
chhiang	qiang	ching	jing
chhiao	qiao	chio	jue
chhieh	qie	chiu	jiu
chhien	qian	chiung	jiong
chhih	chi	cho	zhuo
chhin	qin	chou	zhou
chhing	qing	chu	zhu

Modified Wade–Giles	Pinyin	Modified Wade–Giles	Pinyin
chua	zhua	huan	huan
chuai	zhuai	huang	huang
chuan	zhuan	hui	hui
chuang	zhuang	hun	hun
chui	zhui	hung	hong
chun	zhun	huo	huo
chung	zhong	i	yi
chü	ju	jan	ran
chüan	juan	jang	rang
chüeh	jue	jao	rao
chün	jun	jê	re
ê	e, o	jên	ren
ên	en	jêng	reng
êng	eng	jih	ri
êrh	er	jo	ruo
fa	fa	jou	rou
fan	fan	ju	ru
fang	fang	jua	rua
fei	fei	juan	ruan
fên	fen	jui	rui
fêng	feng	jun	run
fo	fo	jung	rong
fou	fou	ka	ga
fu	fu	kai	gai
ha	ha	kan	gan
hai	hai	kang	gang
han	han	kao	gao
hang	hang	kei	gei
hao	hao	kên	gen
hên	hen	kêng	geng
hêng	heng	kha	ka
ho	he	khai	kai
hou	hou	khan	kan
hsi	xi	khang	kang
hsia	xia	khao	kao
hsiang	xiang	khei	kei
hsiao	xiao	khên	ken
hsieh	xie	khêng	keng
hsien	xian	kho	ke
hsin	xin	khou	kou
hsing	xing	khu	ku
hsio	xue	khua	kua
hsiu	xiu	khuai	kuai
hsiung	xiong	khuan	kuan
hsü	xu	khuang	kuang
hsüan	xuan	khuei	kui
hsüeh	xue	khun	kun
hsün	xun	khung	kong
hu	hu	khuo	kuo
hua	hua	ko	ge
huai	huai	kou	gou

Modified Wade–Giles	Pinyin	Modified Wade–Giles	Pinyin
ku	gu	mu	mu
kua	gua	na	na
kuai	guai	nai	nai
kuan	guan	nan	nan
kuang	guang	nang	nang
kuei	gui	nao	nao
kun	gun	nei	nei
kung	gong	nên	nen
kuo	guo	nêng	neng
la	la	ni	ni
lai	lai	niang	niang
lan	lan	niao	niao
lang	lang	nieh	nie
lao	lao	nien	nian
lê	le	nin	nin
lei	lei	ning	ning
lêng	leng	niu	nüe
li	li	niu	niu
lia	lia	no	nuo
liang	liang	nou	nou
liao	liao	nu	nu
lieh	lie	nuan	nuan
lien	lian	nung	nong
lin	lin	nü	nü
ling	ling	o	e, o
liu	liu	ong	weng
lo	luo, lo	ou	ou
lou	lou	pa	ba
lu	lu	pai	bai
luan	luan	pan	ban
lun	lun	pang	bang
lung	long	pao	bao
lü	lü	pei	bei
lüeh	lüe	pên	ben
ma	ma	pêng	beng
mai	mai	pha	pa
man	man	phai	pai
mang	mang	phan	pan
mao	mao	phang	pang
mei	mei	phao	pao
mên	men	phei	pei
mêng	meng	phên	pen
mi	mi	phêng	peng
miao	miao	phi	pi
mieh	mie	phiao	piao
mien	mian	phieh	pie
min	min	phien	pian
ming	ming	phin	pin
miu	miu	phing	ping
mo	mo	pho	po
mou	mou	phou	pou

Modified Wade–Giles	Pinyin	Modified Wade–Giles	Pinyin
phu	pu	tên	den
pi	bi	têng	deng
piao	biao	tha	ta
pieh	bie	thai	tai
pien	bian	than	tan
pin	bin	thang	tang
ping	bing	thao	tao
po	bo	thê	te
pu	bu	thêng	teng
sa	sa	thi	ti
sai	sai	thiao	tiao
san	san	thieh	tie
sang	sang	thien	tian
sao	sao	thing	ting
sê	se	tho	tuo
sên	sen	thou	tou
sêng	seng, sheng	thu	tu
sha	sha	thuan	tuan
shai	shai	thui	tui
shan	shan	thun	tun
shang	shang	thung	tong
shao	shao	ti	di
shê	she	tiao	diao
shei	shei	tieh	die
shên	shen	tien	dian
shêng	sheng	ting	ding
shih	shi	tiu	diu
shou	shou	to	duo
shu	shu	tou	dou
shua	shua	tsa	za
shuai	shuai	tsai	zai
shuan	shuan	tsan	zan
shuang	shuang	tsang	zang
shui	shui	tsao	zao
shun	shun	tsê	ze
shuo	shuo	tsei	zei
so	suo	tsên	zen
sou	sou	tsêng	zeng
ssu	si	tsha	ca
su	su	tshai	cai
suan	suan	tshan	can
sui	sui	tshang	cang
sun	sun	tshao	cao
sung	song	tshê	ce
ta	da	tshên	cen
tai	dai	tshêng	ceng
tan	dan	tsho	cuo
tang	dang	tshou	cou
tao	dao	tshu	cu
tê	de	tshuan	cuan
tei	dei	tshui	cui

Modified Wade–Giles	Pinyin	Modified Wade–Giles	Pinyin
tshun	cun	wang	wang
tshung	cong	wei	wei
tso	zuo	wên	wen
tsou	zou	wo	wo
tsu	zu	wu	wu
tsuan	zuan	ya	ya
tsui	zui	yang	yang
tsun	zun	yao	yao
tsung	zong	yeh	ye
tu	du	yen	yan
tuan	duan	yin	yin
tui	dui	ying	ying
tun	dun	yo	yue, yo
tung	dong	yu	you
tzhu	ci	yung	yong
tzu	zi	yü	yu
wa	wa	yüan	yuan
wai	wai	yüeh	yue
wan	wan	yün	yun